A TEXTBOOK OF BOTANY

Volume I

A TEXTBOOK OF BOTANY

FOR DEGREE STUDENTS

Volume I

Gopinath Hait MSc, PhD

Principal, Gobardanga Hindu College, North 24 Pgs, WB

Formerly Reader, Dept of Botany, City College, Kolkata, WB

Kashinath Bhattacharya MSc, PhD, FIAS, FICAAI, FAScT

Professor, Department of Botany

Visva-Bharati University, Santiniketan, WB

Ashim Kumar Ghosh MSc, PhD

Reader and Head, Department of Botany

City College, Kolkata, WB

New Central Book Agency (P) Ltd

Co-Published by

Bijay Publications

E-mail : bijaypublications@gmail.com

Exclusively Distributed by

Bijay Pustak Bhandar (P) Ltd

90/6A M. G. Road, Bhawani Dutta Lane, Kolkata-700007

✆ : (033) 4605-7856, 8777898576, 8274916104

E-mail : bijaypustakbhandar@gmail.com

NCBA

REGD OFFICE
8/1 Chintamoni Das Lane, Kolkata 700 009, India
email: ncbakolkata@gmail.com

OVERSEAS
NCBA (UK) Ltd, 149 Park Avenue North, Northampton, NN3 2HY, UK
email: ncbauk@yahoo.com

CORPORATE OFFICE
501 A / 1 Right Side [Upper Ground Floor]
Govind Puri, Kalkaji, New Delhi 110 019
email: delhincba@gmail.com

BIJAY PUBLICATIONS
90/6A M. G. Road, Bhawani Dutta Lane, Kolkata-7
E-mail : bijaypublications@gmail.com

BIJAY PUSTAK BHANDAR (P) LTD
90/6A M. G. Road, Bhawani Dutta Lane, Kolkata-7
E-mail : bijaypustakbhandar@gmail.com

A TEXTBOOK OF BOTANY [Volume I] ● Gopinath Hait, Kashinath Bhattacharya, Ashim Kumar Ghosh

First Published: 2007
Reprinted: 2008, 2009, 2010, 2011, 2012, 2013, 2014,
2015, 2016, 2017, 2018, 2019, 2020, 2021, 2022, 2023, 2024
Thoroughly Revised Edition: 2025

PUBLISHER • TYPESETTER • PRINTER
New Central Book Agency (P) Ltd
8/1 Chintamoni Das Lane, Kolkata 700 009

CO-PUBLISHED BY
Bijay Publications
90/6A M G Road, Bhawani Dutta Lane, Kolkata 700007

TECHNICAL EDITOR
Dr Mita Sen

PROJECT TEAM
Prabhat Jas, Ashit Ghosh

ISBN: 978 81 7381 547 8

Price ₹ 935.00

Cover Photographs:

Front Cover:
1. Pogonatum
2. Oedogonium
3. Pteris vittata
4. Ganoderma

Back Cover:
1. TMV
2. Helicobacter

Photograph courtesy:
DR M R Mazumder, Dr Samit Ray,
Dr Janojit Roy, Mr Monojit Debnath

Dedicated
to the memory of
Our revered teachers

Dr. Narendra Mohan Dutta
&
Dr. Prasad Kumar Mukherjee

Contents

Preface

A Textbook of Botany (Volume 1) Second Edition has been thoroughly revised and significantly updated in line with the latest UGC syllabus revisions. Designed to provide a clear and comprehensive overview, the book focuses on both the fundamental and applied aspects of microorganisms, algae, fungi, and related groups. Unlike many existing textbooks in the Indian market, which are either outdated or only partially cover the required syllabus, this volume addresses the full scope of the curriculum, helping students navigate the subject with ease. The book is organized into five well-structured chapters, beginning with an introduction to basic concepts in botany, followed by detailed sections on bacteria and viruses, algae, fungi and allied organisms, and plant pathology. With its updated content and illustrative approach, this book aims to be a valuable resource for undergraduate students, helping them grasp core topics and stay informed with the latest developments in the field.

Our earnest thanks are due to all our friends and colleagues who have extended their cooperation and help during the period the book was in preparation. We are obliged to mention about the help rendered by Prof. B. P. Mishra and Prof. O. P. Tripathi of Mizoram University; Prof. B. K. Datta of Tripura University; Prof. P. P. Baruah and Prof. Neelakshi Devi of Gauhati University; Prof. N. C. Mandal, Prof. Jnanendra Rath,and Dr. Bomba Dam of Visva-Bharati; Prof. Debdulal Banerjee of Vidyasagar University; Prof. Zahed Hossain and Dr. Neera Sen Sarkar of Kalyani University; Prof.. Subrata Raha of Sidho Kanho Birsha University; Prof. Sikha Datta of Burdwan University; Dr. Dola Boral of Murshidabad University; Dr. Madhumanjari Mandal (Principal) and Dr. Amitava Roy of Scottish Church College; Dr. Nandini Chakrabarty and Dr. Partha Karak of City College; Dr. Debnath Palit (Principal) of Durgapur Govt. College; Dr. Saswati Laha of Bethune College; Dr. Sucheta Sinha of Lady Brabourne College; Dr. Santanu Saha,. Dr. Sukumar Sarkar and Dr. Souvik Das of Bidhannagar Govt. College; Dr. Debisri Datta of Vidyasagar College; Dr. Brij Kumar Tiwari of Chandernagar Govt, College; Dr. Pampa Chakraborty of Narasinha Dutt College; Dr. Mouli Saha of A.P.C. College; Dr. Nisith Ranjan Sarkar of Burdwan Raj College; Dr. Dilip Kumar Mandal of Bolpur College; Dr. Manojit Debnath of Panskura Banamali College; Dr. Tanmoy Basak of Medinipur College; Dr. Sikha Mandal and Dr. Amarjit Kaur of Habra Sri Chaitanya College; Dr. Pritam Chattopadhyay and Dr. Swarnendu Mandal of Burdwan Women's College; Dr. Sunrit Basu Sarbadhikari of Asansol B. B. College, and others.

We are greatly thankful to Mr. Amitabha Sen, and Dr. (Mrs.) Mita Sen, Directors of New Central Book Agency (P) Ltd, for undertaking the publication of the book.

We are thankful to Mr. Sunil Pal, Mr. Prasenjit Haldar and Mr. Narayan Sur for illustrations. Thanks are due to Mr. Ashit Kr. Ghosh of New Central Book Agency (P) Ltd for his continuous help in composing, typesetting, drawing and overall supervision throughout the progress of the book. Thanks are also due to Mr. Jaydeep Ghosh and Mr. Soumen Paul of New Central for their help in various ways.

We welcome comments, criticisms and suggestions for future improvement of the book in the subsequent editions.

Kolkata
27th July 2025
Rathajatra

Gopinath Hait
Kashinath Bhattacharya
Ashim Kumar Ghosh

1 Introduction

The term **Biology** (*bios,* life; *logos,* discourse) was first used by Lamarck and Treviranus (1801). **Biology** (*the study of life*) is a branch of science which deals with living organisms, including both plants and animals. But **Aristotle** (384-322 B.C.E.) was the first person to study animals in sufficient detail and so he is known as **"Father of Biology"**. The branch of biological science which deals with plants is called **Botany** (*botane* – herb) and that which deals with animals is called Zoology (*zoion,* an animal; *logos,* discourse). The term 'botany' derives its origin from the Greek word *'botane'* which means herbs i.e., plants. The Greek naturalist **Theophrastus** (372-286 B.C.E.), a student of Aristotle, prepared a list of 500 plants with their names and description, presented in *"Historia Plantarum"*. Based on habit, he classified plants into four groups: Herb, Undershrub, Shrub and Tree. It is the oldest botanical work in existence and so Theophrastus is called the **"Father of Botany"**.

Humans showed interest in plant life for its various uses. Some are:

1. *Foods:* Our principal foods, like cereals, pulses, fruits and others are derived from the plants.
2. *Fibres:* Plant fibres like cotton, jute, flax and hemp are used in clothing and other purposes.
3. *Wood:* Stem of different plants are used as wood in the preparation of furniture, boat, bridge etc.
4. *Other useful things:* Vegetable oil, resin, rubber, medicines, dyes etc., are directly or indirectly collected from the plants.
5. *Fossil fuels:* Fossil fuels, such as coal, petroleum and natural gases are the products of plants of past geological period.
6. *Soil erosion:* Trees and grasses that grow in different places, help to retain water and thus prevent soil erosion.
7. *Shelter:* Plants in the forest provide good shelter and food for the birds and wild animals.
8. *Supply of nitrogen:* Some plants, like algae and bacteria can fix atmospheric nitrogen and increase soil fertility.
9. *Aesthetic pleasure:* Plants give us aesthetic pleasure, which cannot be measured in terms of money.
10. *Ecological balance:* In addition to the above importance, plants are with us as a friend to sustain our environment by absorbing excess CO_2 from the nature and releasing O_2 to the atmosphere and thereby they help us to survive in our immediate environment. Thus the human beings are now realising the importance of the plants.

The role of plants in maintaining ecological balance can be discussed in brief: Plants are the silent workers of our planet, they want to live and also help others (animals including human beings) to live. They have well-developed metabolic system, which includes both anabolism (synthesis) and catabolism (breakdown). On the other hand, animals have only catabolic system. Anabolic system, like photosynthesis, is present in most of the plants and catabolic system, like respiration, is present in all plants. Photosynthesis is dependent on light, but the respiration occurs both in day and night (throughout their life period). Photosynthesis requires CO_2, water, chlorophyll and light. Plants can absorb CO_2 from the atmosphere and also utilize the CO_2 produced from its own respiration. Water is being absorbed by plants from the soil. Chlorophyll absorbs light mainly from the sun (like an antenna of a T.V.). The resultant product is the carbohydrate (mainly glucose) and O_2 is released in the atmosphere as by-product. This O_2 is utilised by both plants and animals. The above fact indicates that the rate of O_2 production in plant is much more than that of respiration. Thus, the produced O_2 becomes available to animals in sufficient amount. The product of photosynthesis – the carbohydrates is being utilised by the plants during respiration and the excess food is being stored in different organs. *We snatch the food and other products from their storehouse*. On the other hand, during respiration the O_2 is being used up and CO_2 released.

The CO_2 level in the atmosphere is being increased day by day, mainly due to three reasons:

1. Rapid destruction of forest by natural fire or artificial destruction of forests to get more agricultural land and also for urbanisation,
2. Combustion on a large scale of wood and fossil fuels, such as coal, natural gas and mineral oil etc., and
3. From respiration by the overgrowing human population (more than 700 crores) of the world.

(The Greenhouse Effect is the rise in temperature that the earth experiences because of certain gases in the atmosphere such as water vapour, CO_2, nitrous oxide, methane, tropospheric ozone, chlorofluorocarbons and halons. *Without these gases, heat would escape back into space and Earth's average temperature would be about –18°C*. Thus, these gases are referred to as *greenhouse gases*.)

If the number of population of plant gradually reduce (by natural hazards or artificial destruction by human beings), the amount of CO_2 will also gradually increase (CO_2 plays the most vital role in trapping the heat radiated from the earth) to cause global warming due to greenhouse effect. The CO_2 contributes about 60% of the greenhouse gases. (*Besides CO_2, the principal greenhouse gases are water vapour, methane, nitrous oxide, tropospheric ozone, chlorofluorocarbons and halons*). Due to greenhouse effect, the temperature of the earth is gradually increasing which eventually will cause more melting of ice-caps and glaciers of the polar regions of the earth (Greenland, Antarctica), the floating ice on the water of the seas may also melt partially. These will cause expansion of the volume of sea water, as a result the sea water level may rise. The rise of sea water level may cause the reduction of our agricultural and living area due to flood.

So plants are very essential for the sustainability of animals, including human beings, to get O_2 for respiration, for reducing the amount of CO_2 from the atmosphere, for food, medicine, social functions etc. So, throughout history, men showed their keen interest to study the plant as a whole.

Characteristics of plants

The important characteristics of the plants are:

1. **Life cycle.** Each plant has a definite life cycle and within that cycle each plant carries out different activities like growth, reproduction etc. in a definite sequence from birth to death.
2. **Cell structure.** Plant body is made up of one or more structural unit, the cell. Protoplasm of each cell is enclosed in a membrane, called cell membrane. The cell membrane is further enclosed by a hard covering, the cell wall.
3. **Protoplasm** (*Protos – first; plasma* – form). It is the actual living matter of a cell, which performs all the vital activities. Cell contains chloroplasts which perform photosynthesis.
4. **Metabolism.** Sum-total of biochemical changes that takes place in the protoplasm of a living cell is called metabolism. Metabolism includes both anabolism and catabolism. Anabolism is a process by which the simple food particles are produced and stored in the protoplasm, e.g., photosynthesis. On the other hand, catabolism is the

process by which the stored potential energy (food) in the protoplasm is converted into kinetic energy, e.g., respiration. In this process the stored food materials are utilised.

5. **Nutrition**. Plants require nutrients for the synthesis of their protoplasm and also for their growth. Plants can synthesise their foods through photosynthesis.
6. **Respiration**. Respiration is a continuous process in plants, in which they usually take up O_2 from the atmosphere for oxidation of food materials (carbohydrate, protein etc.) and, as a result, release CO_2 and energy. This energy is required to perform all the vital activities of plants.
7. **Growth**. It is the permanent increase in form as well as volume of plant, usually accompanied by increase in dry weight. The growth is the final product of successful metabolism, where the rate of anabolism is much higher than catabolism. In many plants, growth continues indefinitely.
8. **Reproduction**. All plants possess the capacity to produce offspring of their own type by the following three means: vegetative, asexual and sexual.
9. **Movement**. Most of the plants are not able to move from one place to other and generally they perform the movement of curvature. Only some lower plants (e.g., *Chlamydomonas, Volvox* etc.) can move from one place to other.
10. **Adaptability**. They have the power to adjust with the environment.
11. **Sensitivity**. The plants are very much sensitive to change in the environment.
12. **Excretion**. They can produce nitrogenous (Quinine, Nicotine, Atropine, Caffeine etc.) and non-nitrogenous (Resin, Gum, Latex, Tannin etc.) substances as by-products of metabolism, which are either stored or excreted from their body.

BRANCHES OF BOTANY

Botany is a subject which deals with the study of plants. This study includes their external and internal structures, physiological process, cell division, heredity, adaptation to environment, geographical distribution, economic importance, etc.

Based on the above line of studies, this subject has been divided into the following branches:

1. **Morphology** [Greek (= Gr), *morphe*, form; *logos*, study]. The study of external form and structure of the plants is called morphology.
2. **Taxonomy** (Gr. *taxis*, arrangement; *nomos*, principles). This branch deals with the principles of classifications of plants including identification, nomenclature and classification.
3. **Histology** (Gr. *histos*, tissue; *logos*, study). This deals with the study of tissue and cells with the help of microscope.
4. **Anatomy** (Gr. *ana*, above; *temnis*, dissection). The study of internal structure of the plants, mainly with the help of section cutting is known as anatomy.
5. **Physiology** (Gr. *physis*, nature; *logos*, study). This branch deals with the various vital activities like photosynthesis, respiration, etc.
6. **Cytology** (Gr. *kytos*, hollow; *logos*, study). This branch deals with the study of cells, including the structure and function of different cell organelles, nucleus and cell inclusions. It also includes different types of cell division, etc.
7. **Genetics** (Gr. *genesis*, origin). This branch includes the study of heredity and variation of plants.
8. **Plant breeding**. This branch deals with the improvement of plants by different genetic manipulations, like crossing, etc.
9. **Ecology** (Gr. *oikos*, house; *logos*, study). This branch deals with the study of relationship between plants and their environment. It also includes the study of soil erosion, soil conservation and pollution, etc.
10. **Plant geography**. This branch includes the study of plants on different regions on the earth and its reasons thereof.
11. **Palaeobotany** (Gr. *palaios*, ancient). This branch deals with the principles of forma-

tion of fossils in respect of time scale, evaluation of fossils, etc.

12. **Economic botany**. This branch deals with the plants with reference to their products like medicine, oil, fibres, wood etc., which are of great economic value.
13. **Mycology** (Gr. *mykes*, mushroom or fungus; *logos*, study). This branch deals with the study of fungi (Sing. fungus).
14. **Plant pathology** (Gr. *pathos*, suffering; *logos*, study). This branch deals with the diseases of plants.
15. **Embryology** (Gr. *em*, inside; *bryo*, to become swollen, *logos*, study). This branch deals with the study of formation and development of embryo.

Some related branches of botany based on modern technology are:

16. **Biophysics.** This branch deals with the plant activities based on the principles of physics.
17. **Biochemistry** (phytochemistry). This branch deals with the chemistry of plants and their various products.
18. **Microbiology**. The branch which deals with the study of microorganisms including virus, bacteria, microfungi, microalgae and protozoa.
19. **Molecular biology**. This branch of biology deals with the study of bio-macromolecules, especially their structure, sub-cellular localisation, biosynthesis, metabolic roles and regulations.
20. **Biometrics**. It deals with the statistical analysis of different results of biological experiments.
21. **Palynology**. This branch deals with the study of pollen grains and spores in relation to taxonomy, evolution etc.
22. **Genetic Engineering**. This branch deals with adding, removing or repairing parts of genetic material, thereby the changed phenotype of the plant becomes available as desired character.
23. **Biotechnology**. This branch of biology deals with the exploitation of biological molecules for industrial and other processes, especially genetic manipulation of microorganisms (for the production of antibiotic, hormone etc.).
24. **Environmental Botany**. The branch deals with the study of inter-relationships of plants with their physical and biotic environments.

Some other branches allied to Botany are:

25. **Agronomy**. This branch of science (especially agriculture) deals with soil management and crop production.
26. **Horticulture** (L. *hortus*, garden). This branch of science deals with the art of garden cultivation.
27. **Pharmacognosy**. This branch of science deals with drugs in their crude state of medicinal plants.
28. **Forestry**. The branch of science deals with the management of forests.

PLANT GROUPS

At present, about 400,000 different species of plants are recorded throughout the world in different habitat. So, it is very difficult to study the plants without following a definite system. It is mentioned in *Upanishads* that Indians had studied the plants from medical point of view. But, the scientific attempt on the classification of plants began with the classical work of the Greek philosopher, Theophrastus (372-286 B.C.E.) in "*Historia Plantarum*". Later, in 16th century, many herbalists like Brunfels, Fuchs made some valuable contributions towards the study of Systematic Botany. After a long gap, in 18th century, remarkable contribution was made by a Swedish botanist, Carolus Linnaeus or Carl Linne (1707-1778). His classification was first appeared in *Hortus Uplandicus* (1732), with more detail in *Genera Plantarum* (1737), and later he completed his work and presented it in *Species Plantarum* (1753). Thus, he is considered as the "**Father of Modern Botany**".

Later on, many workers like de Candolle, Bentham and Hooker, Eichler, Engler and Prantl, Hutchinson, etc., and recently Takhtajan, Cronquist, Dahlgren and Thorne have proposed their classifications. The classification of Bentham and Hooker appeared in *Genera Plantarum*

(1862-1883) – it is being followed in Indian Herbaria and also found to be easy to understand by the beginners. Thus, the brief outline of classification of Bentham and Hooker is mentioned:

The plant kingdom has been classified into two **sub-kingdoms:** 1. ***Cryptogamia***, the non-flowering plants and 2. ***Phanerogamia***, the flowering plants.

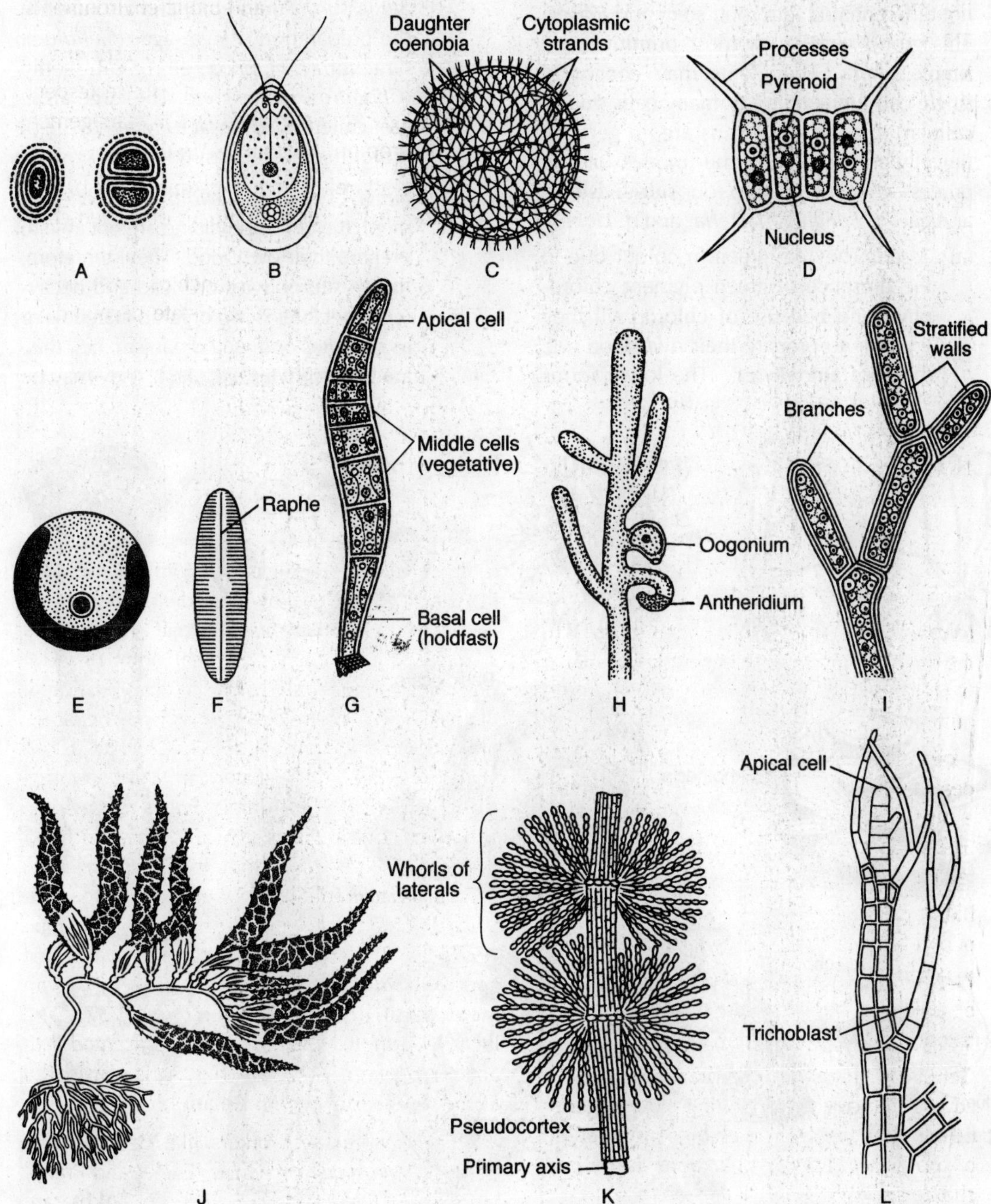

Fig. 1.1: Different types of algae: A. *Gloeocapsa*, B. *Chlamydomonas*, C. *Volvox*, D. *Scenedesmus*, E. *Chlorella*, F. Diatom (*Achnanthes*), G. *Ulothrix*, H. *Vaucheria*, I. *Cladophora*, J. *Macrocystis*, K. *Batrachospermum*, and L. *Polysiphonia*

The cryptogamia has been further divided into three **divisions:** A. *Thallophyta*, B. *Bryophyta*, and C. *Pteridophyta*:

A. *Thallophyta*. Plant body is thallus-like i.e., not differentiated into root, stem and leaves. This group consists of most primitive and simple plants. The plants may consist of single cell (unicellular) or many cells (multicellular) and the sex organs are, in general, unicellular. Based on the modes of life process, it is divided into four **sub-divisions:** a. *Algae*, b. *Fungi*, c. *Bacteria*, and d. *Lichen*.

(a) *Algae*: They are green in colour due to the presence of green pigment chlorophyll and because of chlorophyll they are able to prepare their own food i.e., they are **autophytes**. The lower forms are unicellular, but the plants of higher forms are multicellular (Fig. 1.1). The algae are used as food (*Laminaria, Alaria, Durvillea* etc.), fodder (*Laminaria, Sargassum, Fucus* etc.), and also used in the preparation of agar-agar (*Gelidium, Gracilaria, Hypnea* etc.), carragenin (*Chondrus crispus*), etc. The large algae are called **Kelp** (*Macrocystis pyrifera*, more than 60 metres in length).

(b) *Fungi*: They are devoid of green pigment, the chlorophyll and are not able to prepare their own food. They, therefore, live as *saprophytes* on dead and decaying vegetables or on organic materials or as *parasites* on living organism, i.e., they are **heterophytes** (Fig. 1.2). They may be

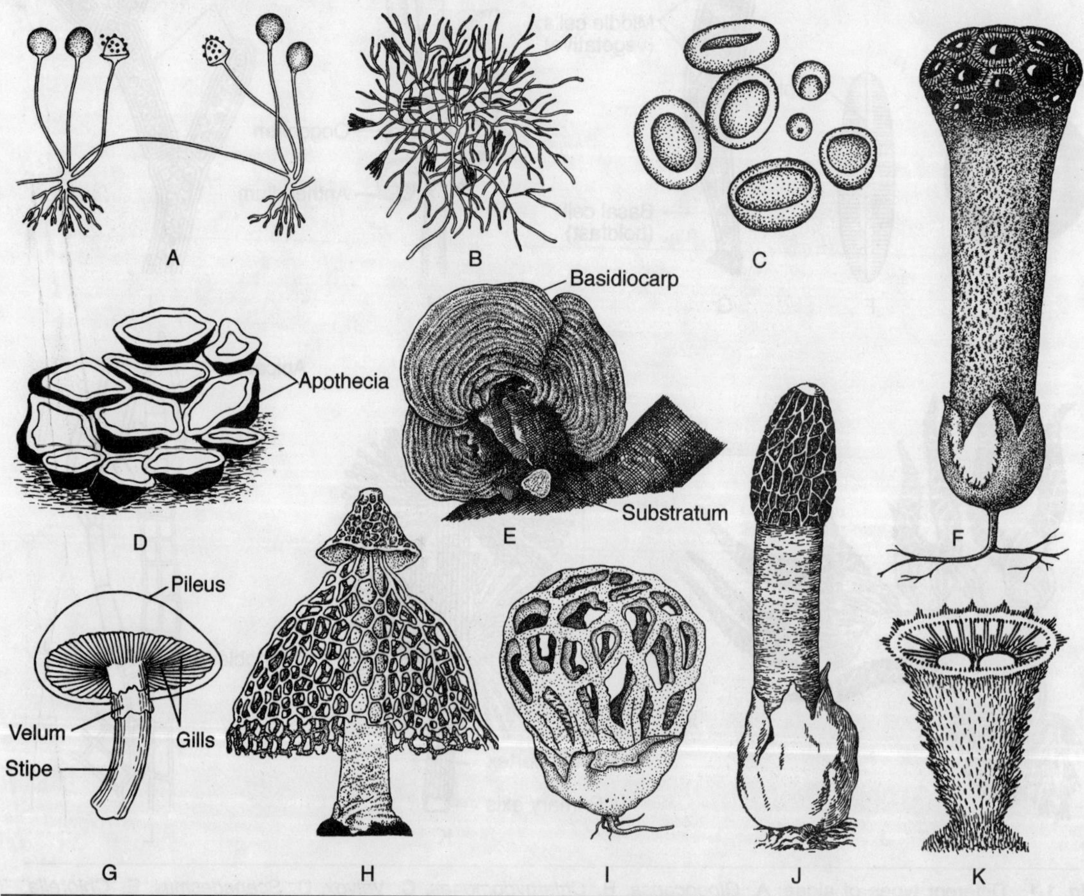

Fig. 1.2: Different types of fungi : A. *Rhizopus*, B. *Penicillium*, C. *Ascobolus*, D. *Peziza*, E. *Polyporus*, F. *Simblum*, G. *Agaricus*, H. *Dictyophora*, I. *Clathrus*, J. *Phallus* and K. *Cyathus*

unicellular (*Saccharomyces*) or multicellular (*Agaricus, Polyporus*) and the **largest known organism** in this world is the fungus *Armellaria solidipes* (mycelial body) reported from America, which covered an area of more than 3.7 sq. miles (9.6 sq. km.). Some fungi are destructive in nature, cause diseases of plants [brown spot of rice, late blight of potato caused **famine in Bengal (1943)** and **Ireland (1845)** respectively]. Some of them are very useful, as they yield valuable drugs (*Penicillin* from *Penicillium notatum* and *P. chrysogenum*) and some others are used as food, the mushrooms (*Agaricus brunnescens, A. campestris, Pleurotus sajor-caju* etc.)

(c) *Bacteria:* These are microscopic, unicellular microorganisms with a saprophytic mode of nutrition (Fig. 1.3). They may be pathogenic (*Xanthomonas campestris* pv. *oryzae*, caused Bacterial leaf blight

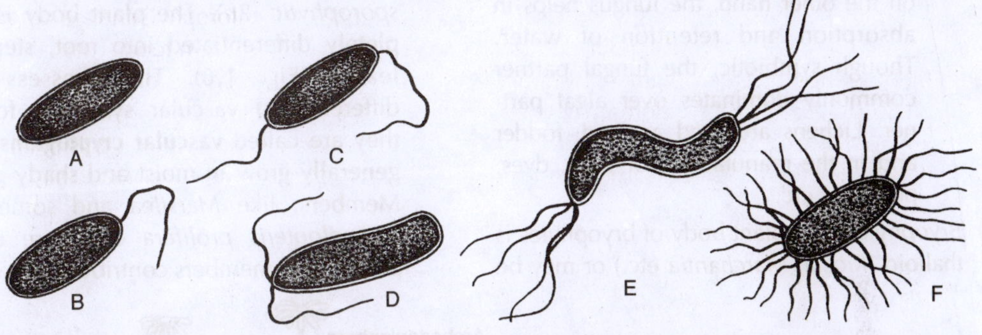

Fig. 1.3: Different types of bacteria

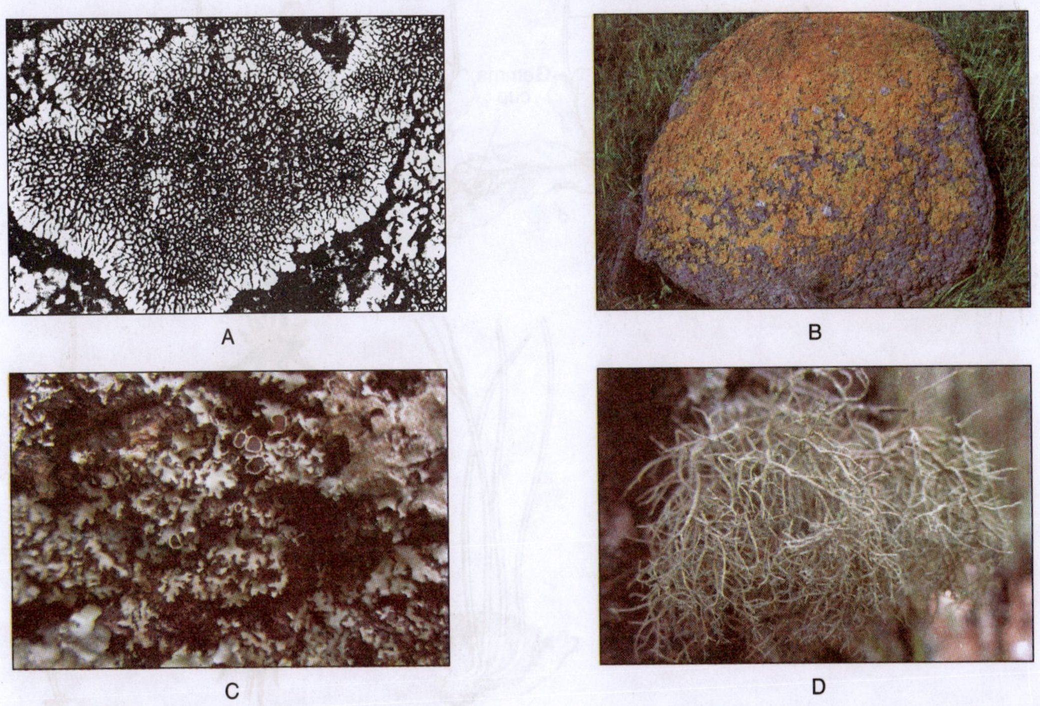

Fig. 1.4: Different types of Lichens: A-B. Crustose type (A. *Graphis* sp. and B. *Caloplaca* sp.) and C. Foliose type (*Parmelia* sp.) and D. Fruticose type (*Usnea* sp.)

of rice, BLB) and at times cause alarming situation in the plants and animals including human beings. Some are useful to increase the soil fertility by adding nitrogen through fixation of atmospheric nitrogen.

(d) *Lichens:* These are specialised organisms, composed of algae and fungi (Fig. 1.4). They live together in a close symbiotic association. The algal partner is capable of manufacturing food and, on the other hand, the fungus helps in absorption and retention of water. Though symbiotic, the fungal partner commonly dominates over algal partner. Lichens are used as food, fodder and in the manufacture of soap, dyes, perfumes etc.

B. *Bryophytes:* The plant body of bryophytes is thalloid (*Riccia, Marchantia* etc.) or may be differentiated into stem and leaf-like organs (*Funaria, Polytrichum* etc.), but does not possess any true root (Fig. 1.5). They are usually found to grow in moist and shady places and are of little economic importance. *Starting from Algae up to Bryophytes, the dominating plant body is gametophyte i.e., haploid* (*n*).

C. *Pteridophytes:* It is the highest group of cryptogams. *Starting from Pteridophyte onwards, the dominating plant body is sporophytic* (2*n*). The plant body is completely differentiated into root, stem and leaves (Fig. 1.6). They possess well-differentiated vascular system – for that they are called **vascular cryptogams**. They generally grow in moist and shady places. Members, like *Marsilea* and some ferns (*Ampelopteris prolifera* etc.), are edible. Some fossil members contribute the world's

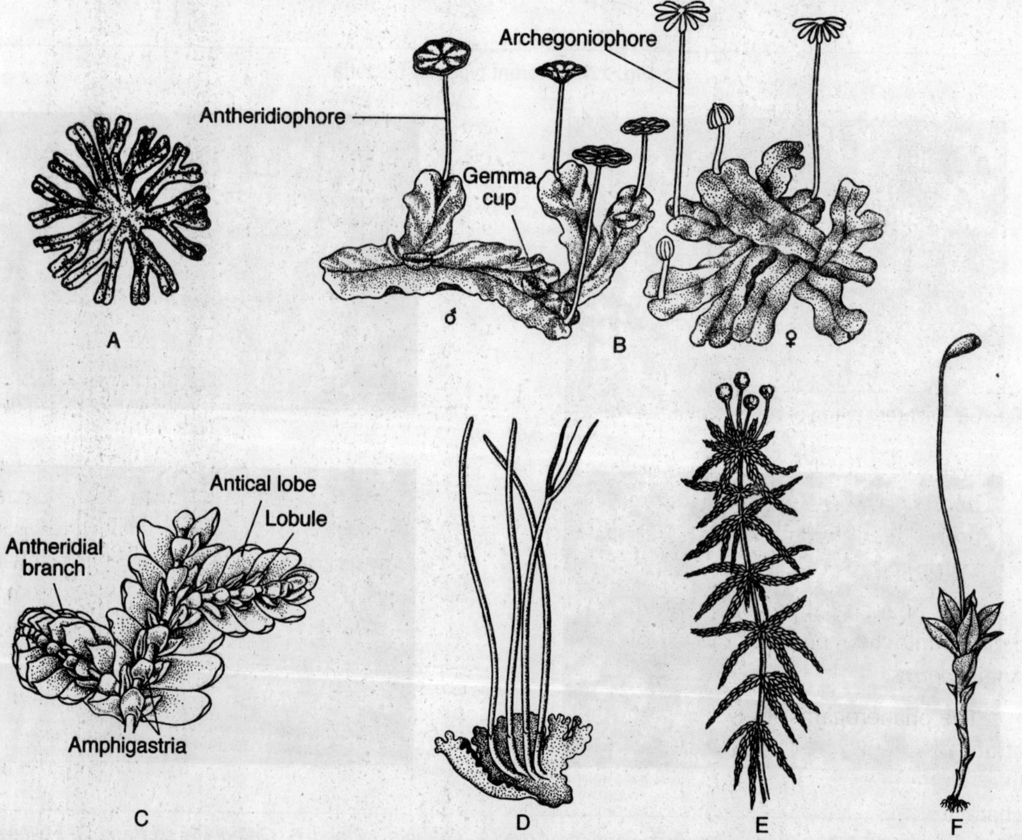

Fig. 1.5: Different types of bryophytes: A. *Riccia*, B. *Marchantia*, C. *Porella*, D. *Anthoceros*, E. *Sphagnum*, and F. *Funaria*

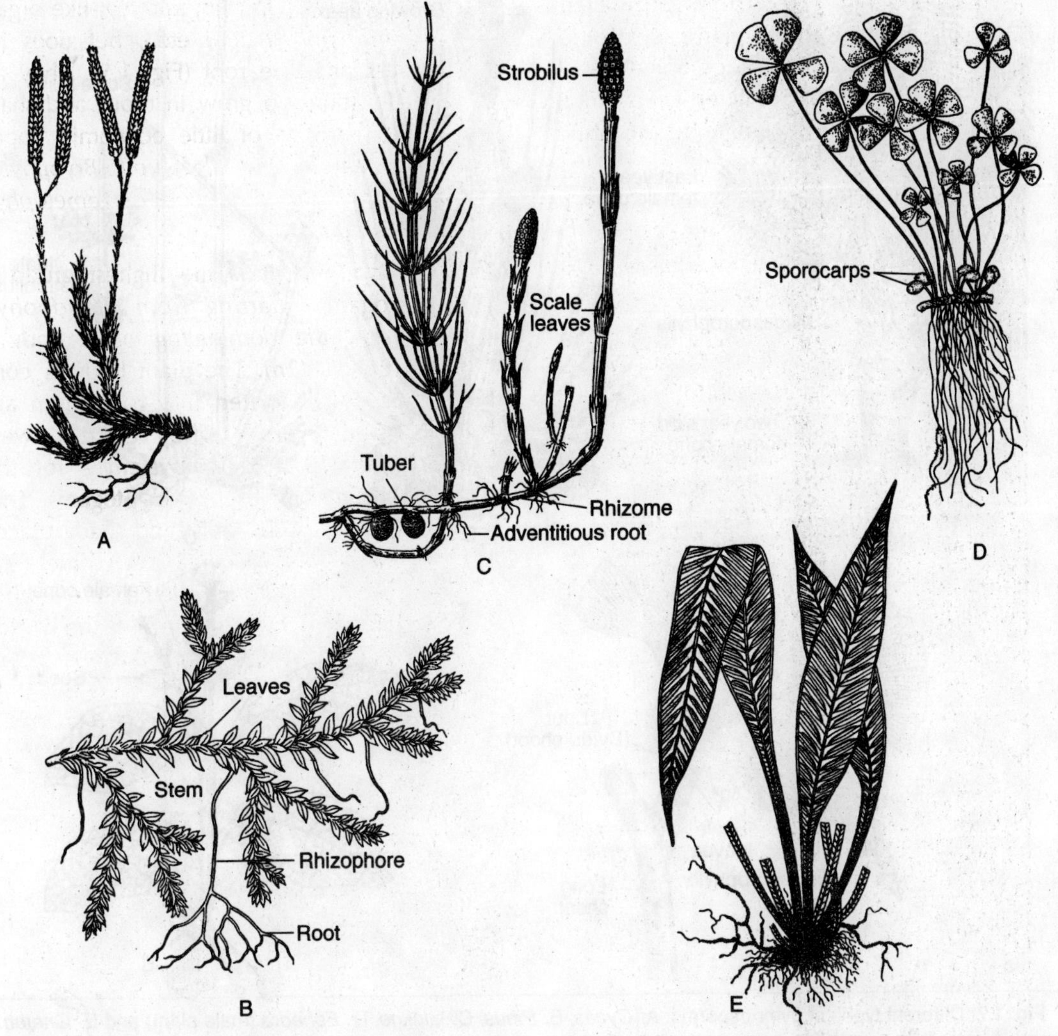

Fig. 1.6: Different types of pteridophyta : A. *Lycopodium*, B. *Selaginella*, C. *Equisetum*, D. *Marsilea* and E. *Asplenium*

major fuel, the coal. The cone of *Lycopodium* is used in the production of an useful homeopathic medicine, named as Lycopodium.

The phanerogams sub-kingdom is further divided into two divisions: Gymnosperms and Angiosperms.

The phanerogams show complete differentiation of plants into root, stem and leaves like Pteridophytes. In addition to the above characteristics, they also produce seeds. The dominating plant body of phanerogams is sporophytic (2n).

A. *Gymnosperms*: These are naked-seeded plants, i.e., the ovules are not enclosed in the ovary and thereby the seeds remain exposed or naked. Most of the plants are big trees (Fig. 1.7). The group is economically important:

1. As good source of timber.
2. Some yield tar, charcoal, methyl alcohol, tannin, etc.
3. Some are used in the manufacture of paper, ink, etc.
4. The valuable medicine ephedrine is prepared from *Ephedra*.

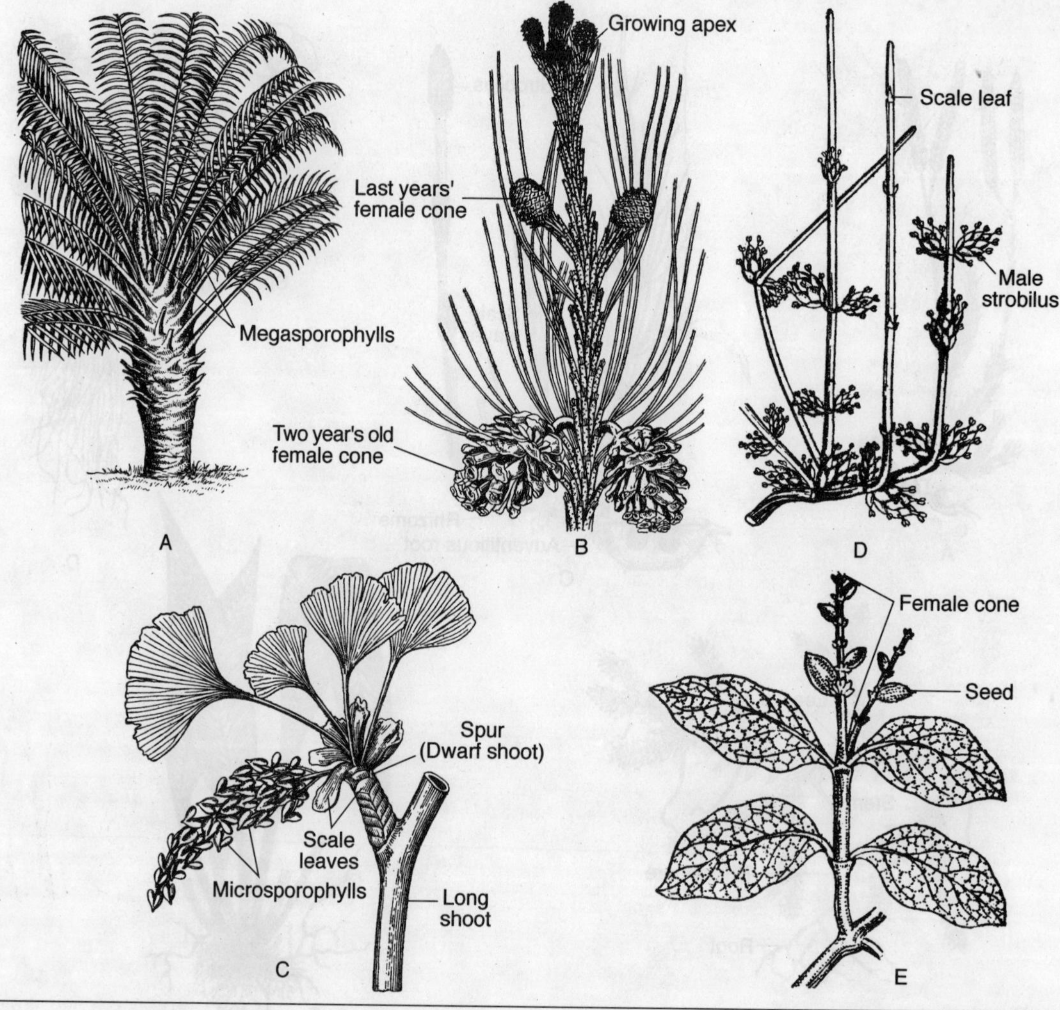

Fig. 1.7: Different types of gymnosperms: A. *Cycas*, B. *Pinus*, C. *Ginkgo*, D. *Ephedra* (male plant) and E. *Gnetum*

5. Like Pteridophytes, its fossil members contribute fossil fuels like coal, lignite, etc.

Common members are *Cycas, Pinus, Cedrus, Ephedra, Ginkgo,* etc.

B. *Angiosperms*: These are closed-seeded plants, i.e., the seeds (ovules) are enclosed in the ovary of the flower and thereby they remain enclosed within the fruits (Fig. 1.8). This group is further divided into two classes: *Dicotyledons* and *Monocotyledons*. The class *Dicotyledons* comprises of the plants having two cotyledons or seed-leaves, e.g., Pea, Gram, Castor, Mango etc., and the plants of *Monocotyledons* have one cotyledon or seed-leaf, e.g., Rice, Wheat, Maize, Coconut etc. The plants under the *Angiosperms* show high economic value.

Mode of nutrition in plants

Based on the *mode of nutrition,* the plants are divided into *two* groups:

A. **Autophytes** or **Autotrophic plants**. These plants are able to produce their own food. These are mainly of two types: (a) Phototrophic plants, and (b) Chemotrophic plants.

(a) Phototrophic plants. These plants can produce their own food through photo-

Fig. 1.8: Different types of angiosperms. A–D. Monocotycedons. A. Banana, B. Maize, C. Bamboo, D. Coconut, E–G. Dicotyledons. E. Pea, F. Banyan and G. Mango

synthesis by utilising light energy – mainly from the sun trapped by chlorophyll, water from soil and CO_2 from air. The common green plants and epiphytes belong to this category. The epiphytes are the plants which remain attached to other plants mainly for mechanical support and not for nourishment. They are entirely cut off from soil contact and adopt various contrivances to absorb and retain atmospheric moisture and rain water for future use.

1. Some plants, like orchids (*Vanda roxburghii, Dendrobium nobile* etc.) develop a special type of water-absorbing tissue (*velamen*) on the outer surface of the aerial roots, which can absorb and retain

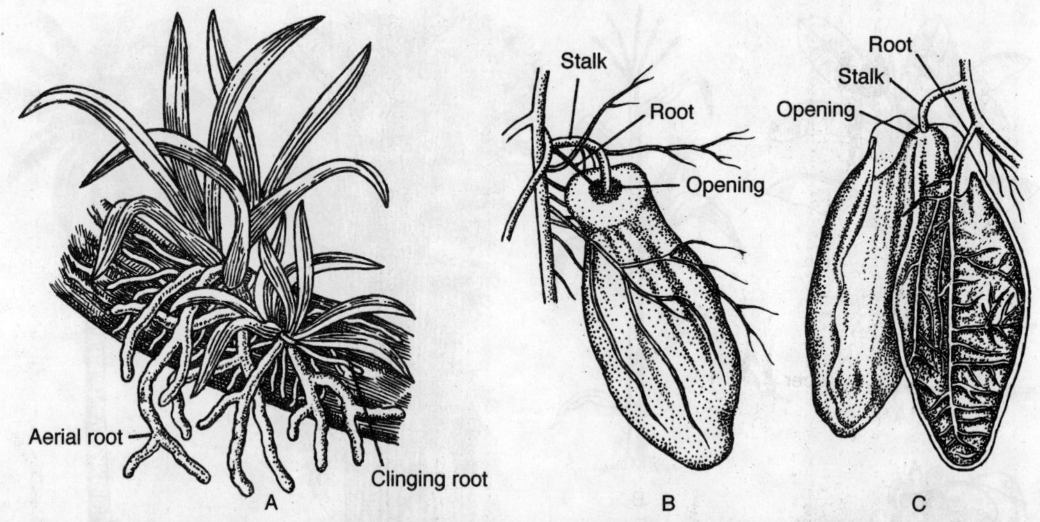

Fig. 1.9: A–C. Phototrophic epiphytic plants. A. Orchid with clinging and aerial roots, B. *Dischidia rafflesiana* with pitcher-like leaves, and C. Pitcher of *Dischidia*, cut open to show roots inside the cavity

water for future use. Another type of roots (clinging roots) are present in Orchids to get attached firmly with the host plant (Fig. 1.9A).

2. In some other orchids, ferns like *Asplenium nidus,* etc., some receptacles are formed by the leaves or aerial roots which collect humus and other material and get spongy. This receptacle with spongy substratum acts as reservoir to retain water and the absorbing roots derive water with nutrients from it.

3. In *Dischidia rafflesiana* (Fig. 1.9B, C) of Asclepiadaceae, the storing and constant supply of water is being performed by the pitcher-like leaves developed at the nodes. Rain water is collected inside the pitcher. This water is being absorbed by the much-branched adventitious roots.

(b) Chemotrophic plants. Bacteria are the plants that can produce their own food by utilising chemical energy. These include iron bacteria, sulphur bacteria, nitrifying bacteria, etc.

B. **Heterophytes** or **Heterotrophic plants**

These plants are not able to produce their own food and depend on others for food. These are mainly of two types: (a) parasites, and (b) saprophytes.

(a) *Parasites*. Some plants, which derive nourishment from other plants or animals are the parasites. The plant on which parasite grows is called **host**. When the parasites are usually confined in the roots of the host, they are called **root parasites,** and, on stem, called **stem parasites**. A close relationship exists between the host and the parasite. Depending on the degree of parasitism, it is of two types:

1. Total parasites or holoparasites. Sometimes the parasites may be wholly dependent on the hosts, these are **total parasite** or **holo-parasite**. The parasites draw the nutrition from the host by root-suckers (**haustoria**) or by parasitic roots, those establish connections with their food conducting channels i.e., the phloem. It is of two types: Root parasites and stem parasites.

Root-parasites:

(i) The example of a leaf-less root parasite without any chlorophyll is *Balanophora dioica* (Fig. 1.10A) of Balanophoraceae.

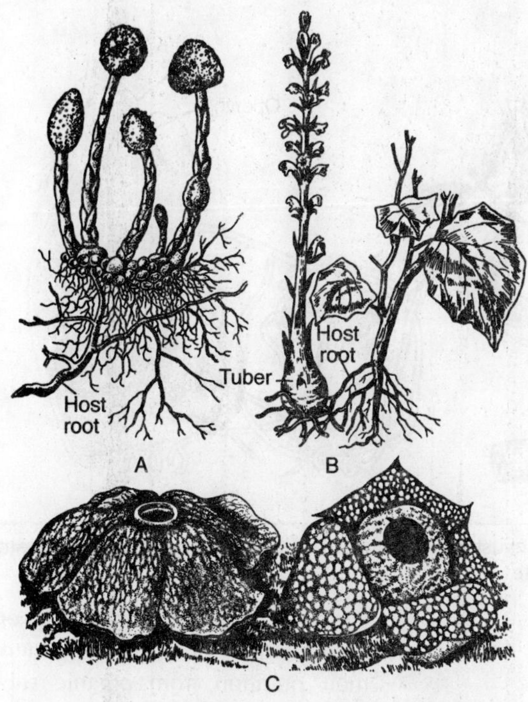

Fig. 1.10: Total root-parasites: A. *Balanophora dioica* developing on host root, B. *Orobanche* plant with flowers growing on roots of Brinjal, and C. Flowers of *Rafflesia arnoldi* (15 kg in weight)

(ii) A root-parasite herb may contain little or no chlorophyll, e.g., *Orobanche caryophyllacea* (Fig. 1.10B) of Orobanchaceae.

(iii) A root-parasite at times becomes stem-parasite with a flower of about 1 metre (3 feet) in diameter, e.g., *Rafflesia arnoldi* (Fig. 1.10C) of Rafflesiaceae found in Sumatra and Borneo.

Stem-parasites:

(i) A stem-parasite may be a thread-like twiners, e.g., Dodder, *Cuscuta reflexa* (Fig. 1.11) of Convolvulaceae.

(ii) A stem-parasite may be a thread-like herb e.g., *Cassytha filiformis* of Lauraceae.

(iii) A stem-parasite, is a **smallest angiosperm** (dicotyledonous plant). e.g., *Arceuthobium minutissimum* of Loranthaceae.

2. Partial parasites or semi-parasites or hemi-parasites. Sometimes the parasites may be partially dependent on the hosts as they are capable of producing part of their own food, called **partial parasites** or **semi-parasites** or **hemi-parasites.** They are of two types:

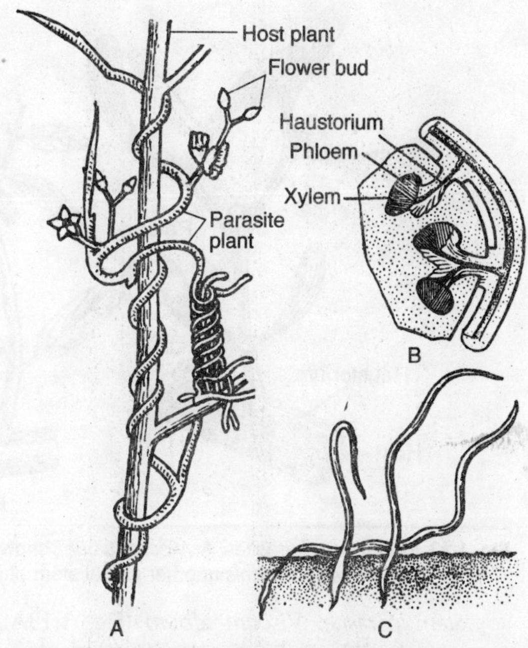

Fig. 1.11: Total stem-parasites: A. *Cuscuta reflexa*, twining on the host, B. T.S. of stem showing an intimate contact of haustoria with vascular bundle, and C. Seedling on soil

Root parasite. Sandalwood tree, *Santalum album*, and *Thesium alpinum* (Fig. 1.12) of Santalaceae are examples of root parasites.

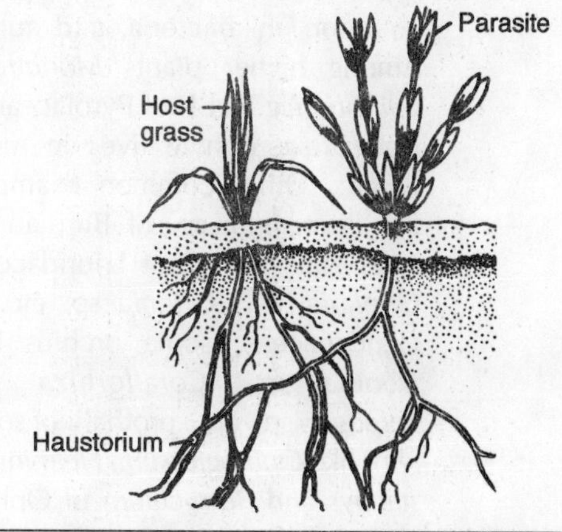

Fig. 1.12: Partial root parasites: *Thesium alpinum* with haustorial suckers attached with grass roots

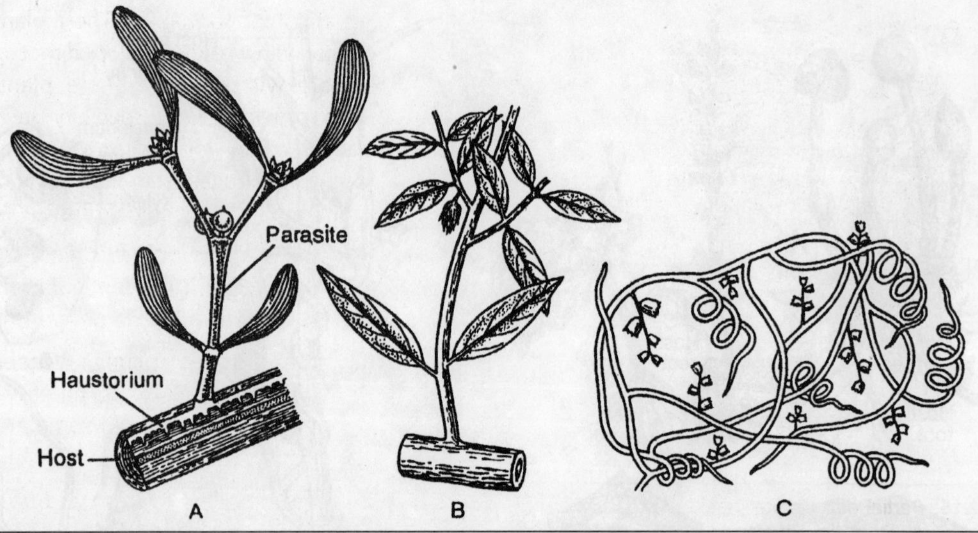

Fig. 1.13: Partial stem parasites. A. *Viscum album*, showing haustoria inside host, B. *Loranthus longiflorus* on host, and C. *Cassytha filiformis* another partial stem parasite

Stem parasite. *Viscum album* (Fig. 1.13A), *Loranthus longiflorus* (Fig. 1.13B) of Loranthaceae; *Cassytha filiformis* (Fig. 1.13C) of Lauraceae are a few common stem parasites.

(b) *Saprophytes*. These plants obtain their food (organic) entirely from dead plants or animals and decaying organic matters. They draw only the organic food which makes sharp difference with the autophytes and heterophytes. They are divided into two groups: 1. **Total saprophytes** or **holosaprophytes**, and 2. **Partial saprophytes**.

1. Total saprophytes. This type is common in bacteria and fungi. Among higher plants *Monotropa uniflora* (Fig. 1.14) of Pyrolaceae, a leafless saprophyte lives in moist humus. Other common examples are *Burmannia* sp. of Burmanniaceae, *Andruris* sp. of Triuridaceae, *Obolaria* sp., *Bartonia* sp. etc. of Gentianaceae; some orchids like *Neottia* sp., *Corallorhiza* sp., *Epipogon* sp. etc.; prothalli of some ferns like *Ophioglossum, Helminthostachys* and *Botrychium* of Ophioglossaceae and many other members among higher plants.

2. Partial saprophytes. These are green plants which obtain a portion of their nutrition from organic substances, but are not found to utilise it directly. This is done through some fungi living in the roots of higher plants. This type of asso-

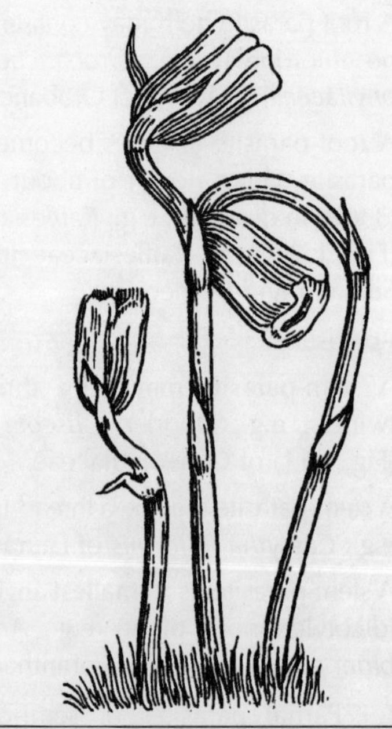

Fig. 1.14: Total saprophyte. *Monotropa uniflora*, a leafless saprophyte with flowers

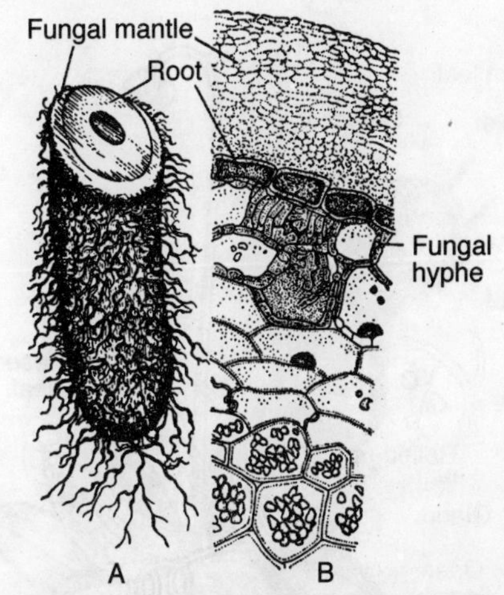

Fig. 1.15: Partial saprophyte (mycorrhizal association). A. Root tip enveloped by fungal mycelium, B. T.S. of Mycorrhizal root of *Pinus* showing fungal mantle and fungal mycelium inside

ciation is called mycorrhizal association (Fig. 1.15).

(c) *Insectivorous* or *carnivorous*. These are the plants which grow in localities where the soil is deficient in nitrogen content – essential for normal life. These plants are either with weakly developed root system or are without roots. These plants are also provided with specially modified leaves meant for capturing insects and securing nitrogen from their body. Some common Indian members are:

(i) the bladderwort, *Utricularia stellaris* (Fig. 1.16A, B) of Lentibulariaceae – commonly grows in old ponds,

(ii) the pitcher plants, *Nepenthes khasiana* (Fig. 1.16C, D) of Nepenthaceae, common in some areas of Assam with high humidity,

(iii) the sundew, *Drosera burmanii*, *D. rotundifolia* (Fig. 1.17A) and *D. indica* are found in Birbhum and Bankura districts of West Bengal, Panchmari hills of M.P. etc.

(iv) the venus flytrap, *Aldrovanda vesiculosa* (Fig. 1.17B-D) of Droseraceae, a submerged plant in salty water, and

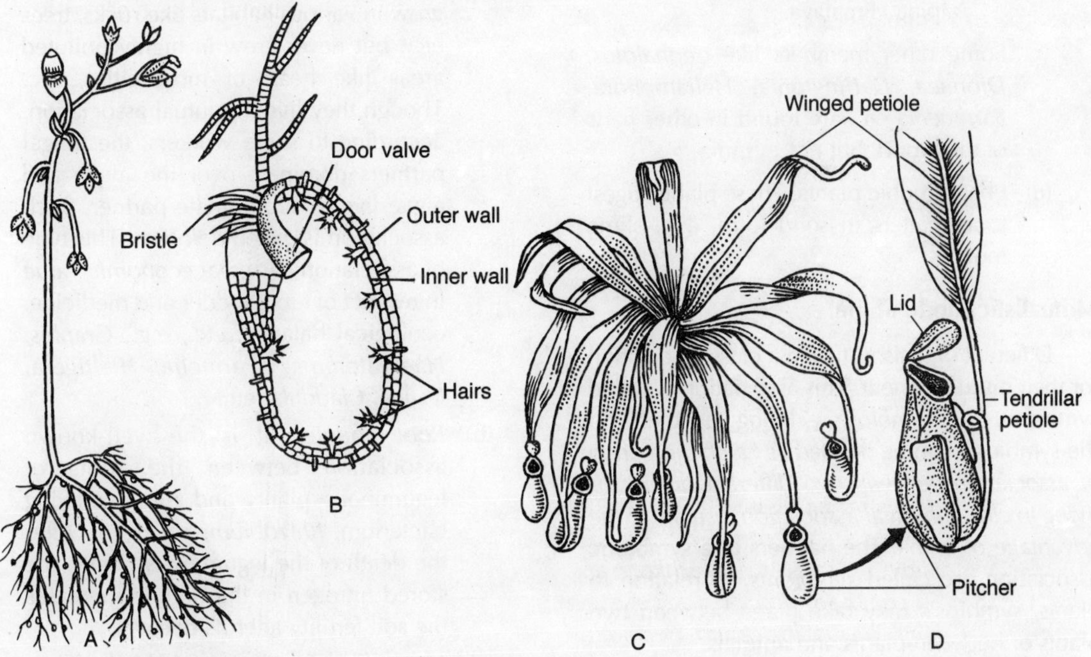

Fig. 1.16: Insectivorous plants: A–B. *Utricularia stellaria*, the bladderwort. A. Entire plant with flowers and bladders, B. L.S. of bladder, C. *Nepenthes khasiana* with pitcher at the leaf apex and D. An enlarged pitcher with a part of the wall removed

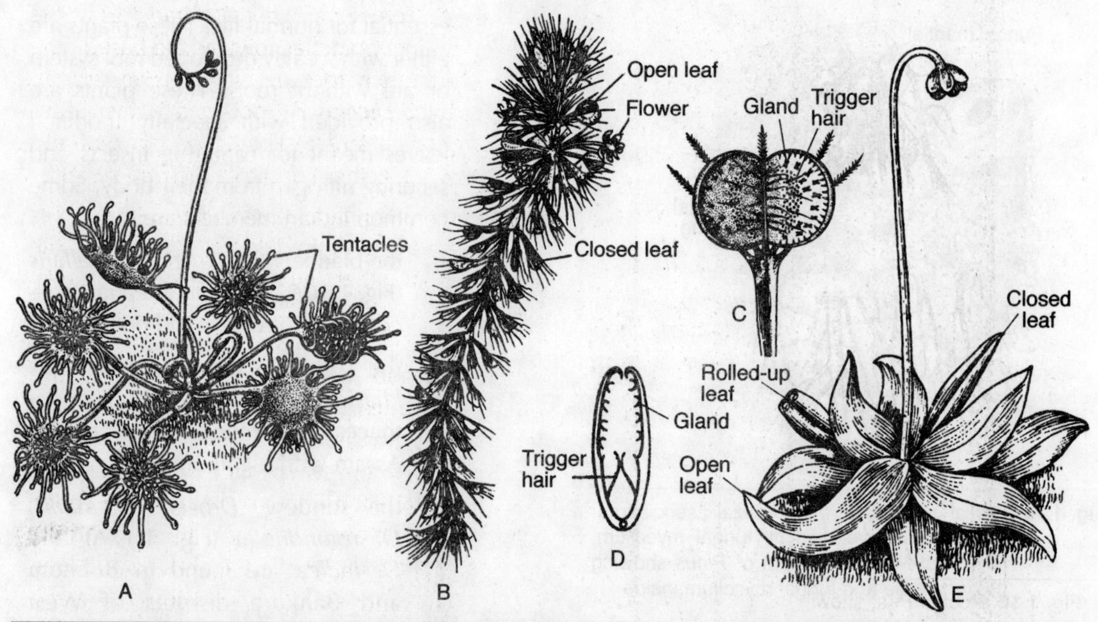

Fig. 1.17: Insectivorous plants: A. *Drosera rotundifolia* (Sundew), an entire plant with inflorescence and lower spoon-like leaves with tentacles (some tentacles curve inwardly to catch insect), B-D. *Aldrovanda vesiculosa*: B. A portion of floating plant, C. An open leaf, D. T.S. of a closed leaf, E. *Pinguicula* catch insect by rolling leaves

(v) only one member of *Pinguicula, P. alpina* (Fig. 1.17E), reported from alpine Himalaya.

Some other members like *Cephalotus, Dionaea, Darlingtonia, Heliamphora, Sarracenia* etc. are found in other parts of the world, but not in India.

(d) Phagotrophic plants. These plants ingest food matters in solid form, e.g., slime moulds.

Mutualistic mode of life

Different organisms live in a close association for their mutual benefit. This association is called symbiosis (Gr. *sumbiosis* – living together). So the symbiosis can be defined as "*a phenomenon of association between two different organisms living in close physical association, usually to the advantage of both*". The partners of a symbiotic association are called *symbionts*. In relation to plants, symbiosis may take place between two plants or between plants and animals.

1. Association between **two plants**:

(a) *Lichen*: The well-known association between algae and fungi is the *lichen*. The lichens are of various forms and grow in various habitats like rocks, trees etc., but never grow in highly polluted areas like heart of most cities, etc. Though they live in mutual association, according to some workers, the fungal partners dominate over the algae and algae live as subordinate partner. Such association is called *helotism*. This type of association has *great economic value* in respect of food, fodder and medicine, ecological balance, etc., e.g., *Graphis, Haematomma, Parmelia, Peltigera, Usnea, Cladonia,* etc.

(b) *Root nodules*: It is the well-known association between the roots of leguminous plants and nitrogen-fixing bacterium, *Rhizobium* (Fig. 1.18). After the death of the leguminous plants, the stored nitrogen in the nodule increases the soil fertility after its decay.

(c) *Mycorrhiza*: It is the association between fungi and the root of higher plants. Various types of mycorrhizal asso-

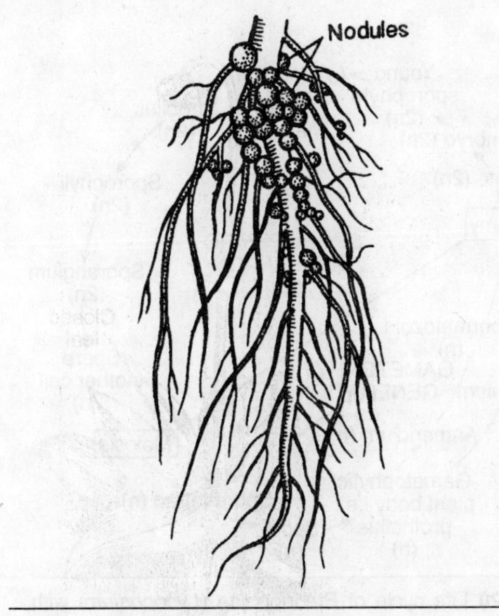

Fig. 1.18: Root of Pea, showing attached nodules

ciation are now recognised. Two common types are *Ectomycorrhiza* (when the fungus lives mostly outside the root tissue) and *Endomycorrhiza* (when the fungus grows mostly inside the root cortex). The fungus helps the plant to get more nutrients in exchange of small amount of food required for its living. The fungus never takes the entry inside the protoplasm, but remains in-between cell wall and cell membrane.

2. Association between **plants** and **animals**:

Fungus and white ant: The fungus *Termitomyces* sp. is an edible mushroom, that associates with the white ants. They remain deep in the soil and fungus develops fruit body once in a year. Fungi do not grow in artificial mushroom bed. The white ant can degrade the soil, thereby the fungus can grow well and, on the other hand, the young white ants grow very well by feeding the apical region of the growing mycelium.

PARTS OF ANGIOSPERMIC PLANTS

An angiospermic plant consists of two main parts – the **root system** i.e., the underground part, and the **shoot system** i.e., the aerial part.

The root system consists of a main axis, the **tap root**, from which **lateral roots** are developed in **acropetal order** i.e., younger branches at the apex and older branches towards the base of the root. The branch root developed from the tap root is called **secondary roots** and further branches are called **tertiary roots**. The tap root and lateral roots together constitute the **root system**.

On the other hand, the shoot system consists of an erect aerial main axis, the **stem**. From the stem, similar other lateral members are developed, called **branches**. From the main axis and also from branches, dissimilar lateral members are developed called **leaves**. Both lateral branches and leaves are developed in **acropetal order**. The regions of the stem from where leaves are developed are called **node** and the regions between two consecutive nodes are called **internodes**. The angle between the leaf and the stem is called **axil**. Beside the above structures, each leaf is subtended by one condensed shoot, i.e., bud, in their axil, and is called **axillary bud**. The stem and its branches are terminated in a bud, called **terminal bud**. With maturity, the plants bear flowers, fruits and seeds. Different organs perform different functions. Based on function, the plant organs are divided into two groups, **vegetative organs**, i.e., stem, branches, leaves etc., necessary for vegetative growth and **reproductive organs** i.e., the flowers, fruits and seeds, to keep their representative in this world.

Alternation of generations

Plants belonging to the different groups normally have life cycle. Each cycle consists of two generations, **sporophytic** i.e., diploid (2n) and **gametophytic** i.e., haploid (n). Plant belonging to Thallophyta and Bryophyta have a life cycle where the dominant stage is gametophytic and the sporophytic stage is completely or partially dependent on it (Fig. 1.19). On the other hand, from Pteridophyte onwards the dominant stage is sporophytic and the gametophytic stage is completely or partially dependent or independent (Fig. 1.20). The two generations regularly alternate with one another in a cyclic pattern and this phenomenon is called **alternation of generations**. The schematic representation of the life cycles of Bryophytes and Pteridophytes are given in next page.

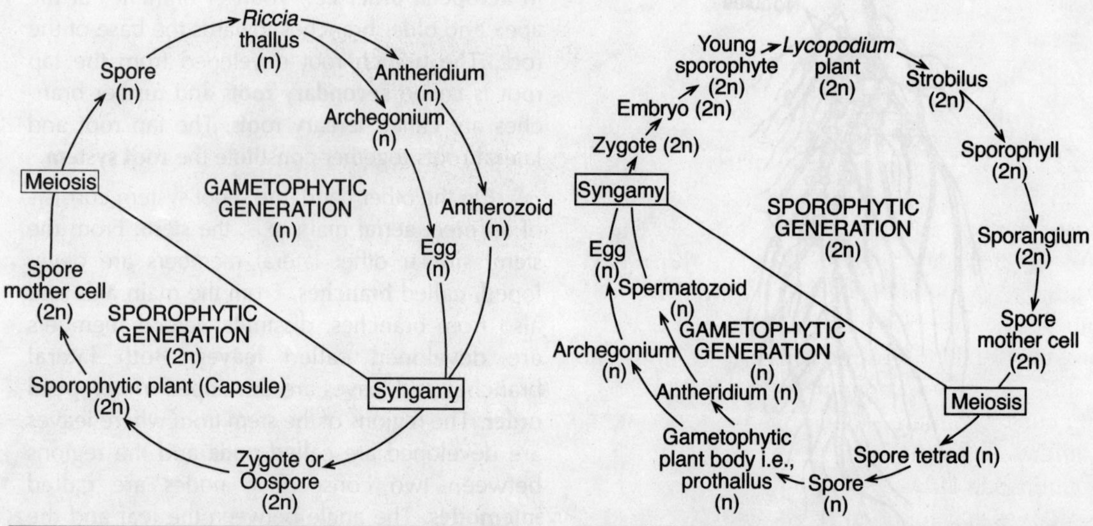

Fig. 1.19: Life cycle of a Bryophyte (*Riccia*) with dominant gametophyte (n)

Fig. 1.20: Life cycle of Pteridophyte (*Lycopodium*) with dominant sporophyte (2n)

MODERN CONCEPT ON CLASSIFICATION OF LIVING WORLD

The hierarchy of biological classification consists of eight major taxonomic ranks (Fig. 1.21). It was Carolus Linnaeus — The Father of Modern Taxonomy — who introduced the rank-based system of nomenclature in 1735, where the highest rank was given the name "**Kingdom**" (Latin: *regnum*, plural *regna*) followed by four other principal ranks namely, **class, order, genus** and **species.** Linnaeus recognised 3 kingdoms of nature namely, *Regnum Vegetable* (plants), **Regnum Animale** (animals) and **Regnum Lapideum** (minerals) which has long been abundoned. Linnaeus' lower level taxonomic categories was found to be acceptable for more than 250 years. Even today, the starting point of Botanical Nomenclature is Linnaeus' 'Species Plantarum' in 1753 and of Zoological Nomenclature is the 10th edition of the 'Systema Naturae' in 1758. Since Linnaeus, the increase in number of described species needs an expansion of the number of hierarchial ranks, hence categories of phyla or division and family were introduced in the early 19th century and many intermediate categories have been added.

In 1987, two other principal ranks were introduced by Thomas Cavallier-Smith, making the sequence kingdom, phylum or division, class, order, family, genus, and species. In 2004, the rank of **domain** has further introduced by Thomas Cavallier-Smith above the kingdom. In addition, prefixes can be added to make sub-kingdom (*subregnum*) and infra-kingdom (also called *infraregnum*) as the two ranks immediately below kingdom. Superkingdom may often be considered as an equivalent of **Domain** or **Empire**, or as an independent rank between Domain and Kingdom.

In 2005, Adl *et al*. presented a nested eukaryote — only **cladification** (classification with regard to cladistic approach), where 6 **superorders**, such as **Amoebozoa, Opisthokonta, Rhizaria, Excavata, Chromalveolata**, and **Archaeplastida** (plants) were suggested as the highest ranked eukaryote groups. The same was again updated by Adl *et al*. in 2012, where Rhizaria and Chromalveolata were replaced by SAR with 4 small *Hacrobian groups*. Recently, an additional rank **Branch** (Latin: *ramus*) has been put between subkingdom and Infrakingdom by Ruggiero *et al*. (2015). They presented a unified and coherent hierarchial ranked system called **'Catalogue of Life (COL)'** where superkingdom rather than Empire or Domain was used above the kingdom. For taxonomic purpose there are three separate codes that govern nomenclature, each with different terminology and requirements. These include (i) **ICN** (International Code of Nomenclature for algae, fungi and plants), in

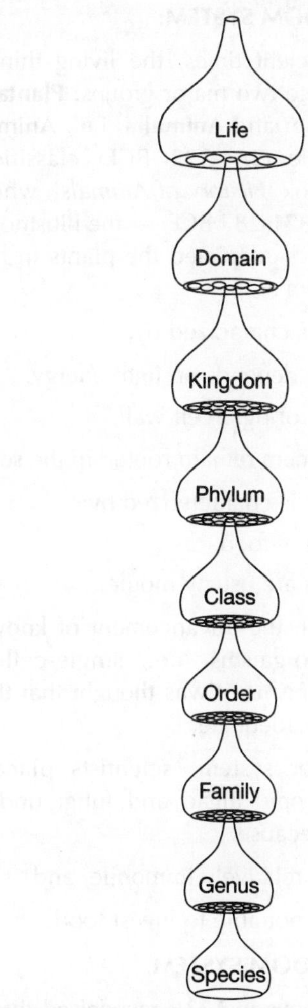

Fig. 1.21: The hierarchy of biological classification

which principle of priority does not apply above the rank of family; (ii) **ICZN** (International Code of Zoological Nomenclature), in which priority does not apply above the family-group names and (iii) **ICNB** (International Code of Nomenclature of Bacteria), where only the categories ranked as class and below are covered by the code.

In some recent classifications based on cladistic approach, the 'kingdom' terminology has been abandoned as the traditional kingdoms are not **monophyletic** (do not consist of all the descendants of a common ancestor).

The gradual development of the concept regarding the grouping of living organisms is briefly described:

TWO-KINGDOM SYSTEM:

From the ancient times, the living things were classified into two major groups: **Plantae**, i.e., Plant Kingdom and **Animalia**, i.e., Animal Kingdom. Arstotle (384-322 BCE) classified animals in his book *'History of Animals'*, while Theophrastus (C. 371-287 BCE) — the illustrious pupil of Aristotle — depicted the plants in his *'Historia Plantarum'*.

The Plantae is characized by:

1. Nutrition depends on light energy,
2. Presence of rigid cell wall,
3. Most of them remain rooted in the soil.

The Animalia is characterized by:

1. Ingestion of food,
2. Members are usualy motile.

Later on, with the advancement of knowledge on microorgaisms, i.e., single-celled microscopic organisms, it was thought that the above system is inadequate.

In the earlier system, scientists placed bacteria, microscopic algae and fungi under plant kingdom, because:

1. They are relatively immobile, and
2. They are not able to ingest food.

THREE-KINGDOM SYSTEM:

Carolus Linnaeus in 1735 recognised three kingdoms of nature. He recognised 2 kingdoms of living things namely, *Regnum vegetabile* (Vegetable kingdom for plants), *Regnum Animale* (animal kingdom). He included minerals by placing them in the third kingdom called *'Regnum Lapideum'*.

Difficulty arose due to the discovery of certain microorganisms having characteristics shared with both plants and animals. With the advancement of more knowledge on microorganisms, many scientists proposed to place them in a separate kingdom. Based on the above concept, **Earnest Haeckel** (1834-1919), a German zoologist proposed (1866) the name of 3rd kingdom, the **Protista** (Gr. *Protistos*, first or primitive), includes all microscopic, unicellular, motile and non-motile, photosynthetic or non-photosynthetic, free and autonomously living organisms (neither typical plant nor animal). Thus all unicellular microscopic organisms of algae, fungi, protozoa and also the bacteria are included under the Kingdom **Protista**.

The viruses were known since the end of 19th century (D. Iwanowski, 1892; a Russian scientist). Due to their non-cellular nature, they were not included in the classification of living organisms.

Based on the more detailed knowledge on cell structure during late 1940s with the aid of powerful magnification provided by electron microscopy, it was found that living organisms are comprised of two types of cells:

1. **Prokaryotic** (Gr. *Pro*, primitive; *karyon*, nucleus), characterised by:
   i) Cell has a non-membrane bound DNA,
   ii) Semi-rigid cell wall composed of murine complex.

This group includes Bacteria and Blue green algae.

2. **Eukaryotic** (Gr. *Eu*, true; *karyon*, nucleus), characterized by:
   i) Cell has a well-organised nucleus,
   ii) Well-developed mitochondria, endoplasmic reticulum.
   iii) Flagella consist of 9 + 2 fibrillar structure.

This group includes green, brown and red algae, protozoa, fungi, higher plants and animals.

The prokaryotes are considered more primitive than the eukaryotes. Based on the above idea, the Kingdom Protista was divided into two divisions:

1. **Lower Protists** i.e., Prokaryotic Protista, and
2. **Higher Protists** i.e., Eukaryotic Protista.

FOUR-KINGDOM SYSTEM:

With the event of above heterogeneity of protista, **Herbert F. Copeland** (1938) proposed the four-kingdom system. In his system, lower protists were placed under a separate kingdom, **Monera**, while higher protists were retained in the kingdom **Protista**. Thus, the four kingdoms are **Monera, Protista, Metaphyta** (green plants) and **Metazoa** (animals).

The four, kingdom system, though highly developed, suffers with several deficiencies. Important shortcomings are:

1. Lack of fundamental difference between fungi and other eukaryotic protists.
2. Based on various characteristics, such as histone distribution and chromosomal organisation, ecological and nutritional role, cytochrome c, amino acid sequence, lack of phagocytosis and pinocytosis, etc., it is difficult to place the fungi (at least the higher ones) in any one of the above kingdoms.

Later, the four-kingdom system was also followed by Barkley (1968).

FIVE-KINGDOM SYSTEM:

Above knowledge on Prokaryotes and Eukaryotes shows many differences and it was thought that different life forms (eukaryotes) have evolved polyphyletically from a variety of primitive ancestral forms (prokaryotes). From the above knowledge, many scientists decided to give the status of Kingdom to the different life types. Based on the above idea, **Robert Whittaker** (1969), an American ecologist, proposed a **Five-Kingdom System** (Fig. 1.22) of classification based on at least two major criteria:

(i) 3 levels of cellular structure: Prokaryotes, unicellular eukaryotes and multicellular eukaryotes, and

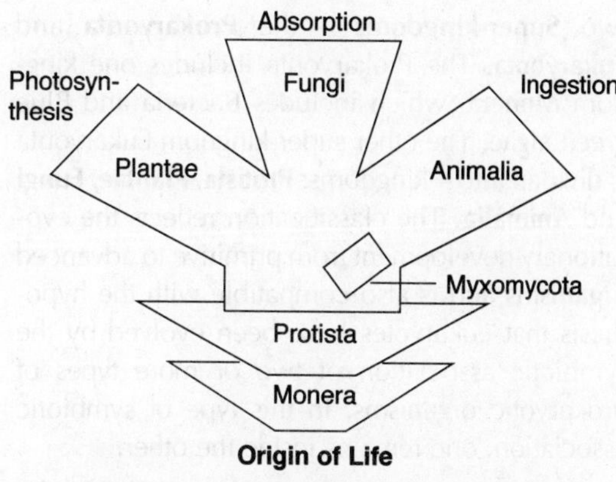

Fig. 1.22: A five-kingdom system for the classification of living organisms (Whittaker, 1969)

(ii) 3 principal modes of nutrition: Photosynthesis (in plants), absorption (in fungi) and ingestion (in animals).

The kingdoms are:

1. **Monera**: Includes all prokaryotes and their hypothetical ancestors.
2. **Protista**: Includes unicellular eukaryotes, such as algae, protozoa and slime molds.
3. **Plantae** (plants): Includes multicellular eukaryotes, i.e., higher algae and green plants.
4. **Fungi**: Includes only fungi.
5. **Animalia**: Includes multicellular animals.

Knowledge about the cells became enriched during the 1970s, with the advancement of different technologies such as electron microscope etc. Later on, Woese and Fox (1977) observed that there are two types of prokaryotes. These are: 1. **Archaebacteria** and 2. **Eubacteria**. Archaebacteria have the following unique characteristics:

(i) Unique wall composition,

(ii) Unique DNA and RNA sequence,

(iii) Plasma membrane is unlike any other living organisms.

But it becomes difficult to place some of the organisms in either of the above two groups.

Later, Edwards (1976), and Whittaker and Margulis (1978) divided the living organisms into two **Super-kingdoms** – the **Prokaryonta** and **Eukaryonta**. The Prokaryonta includes one kingdom **Monera**, which includes Bacteria and Blue green algae. The other super-kingdom Eukaryonta is divided into 4 kingdoms: **Protista, Plantae, Fungi** and **Animalia**. The classification reflects the evolutionary development from primitive to advanced organisms and is also compatible with the hypothesis that eukaryotes have been evolved by the symbiotic association of two or more types of prokaryotic organisms. In this type of symbiotic association, one remains inside the other.

Due to many distinctive characteristics, the fungi are placed in a separate Kingdom.

Due to the presence of differentiation of different algal groups in respect of cellular organisation, physiology and biochemistry; the **unicellular eukaryotic algae** are placed under the kingdom Protista, along with Protozoa and, on the other hand, the higher algae (eukaryotic algae: members of Chlorophyceae, Phaeophyceae, Rhodophyceae etc.) have been included under Kingdom **Plantae**.

Based on biochemical and electron microscopic studies, Stanier *et al*. (1971) and Stanier and Chohen-Brazire (1977) stated that blue green algae are more similar to bacteria, therefore they should be considered as blue green bacteria (**Cyanobacteria**).

Though virus is not a cellular organism, it has been included under the super-kingdom Prokaryonta for easy handling.

Later on, 5-kingdom system was not accepted by many biologists for the following reasons:

(i) Lack of differentiation between Archaea and Bacteria, and

(ii) The boundaries among the Kingdoms – such as Protista, Plantae and Fungi – are not well-demarcated.

SIX-KINGDOM SYSTEM:

For the above problems with five-kingdom system of R. H. Whittekar, alternative suggestion has been made and later six-kingdom system was proposed by Carl Woese et al. in 1977. In his six-kingdom system, Kingdom **Monera** or **Prokaryotae** has been divided into two kingdoms — the **Eubacteria** and **Archaebacteria**. The other kingdoms remain same as suggested by Whittekar (1969). So the six-kingdoms are Eubacteria, Archaebacteria, Protista, Plantae, Fungi and Animalia. In 1990, Woese et al. divided the entire living world into three **Domains** such as **Bacteria, Archaea** and **Eukarya** (including all eukaryotic organisms like Protista, Algae, Fungi, Plants and Animals).

Thomas Cavalier-Smith in 1998 (revised in subsequent papers) divided all living organisms into six kingdoms such as **Bacteria, Protozoa, Chromista, Plantae, Fungi** and **Animalia**. The Bacteria is the sole kingdom of Empire Prokaryota, while the rest five kingdoms are included under the Empire Eukaryota. He created two sub-kingdoms (Unibacteria and Negibaceria) under the kingdom Bacteria based on their membrane topologies (bimembranous-unimembranous transition). He treated Archaebacteria as a phylum of the sub-kingdom Unibacteria. Cavalier-Smith did not accept the requirement for considering a valid taxa to be monophyletic (in his terminology it is 'holophyletic'). He defined Prokaryota, Bacteria, Negibacteria Unibacteria and Posibacteria as valid paraphyta taxa.

EIGHT-KINGDOM SYSTEM:

Attempts have been made by different scientists to divide **Protists** into several well-defined kingdoms. In this regard, the eight-kingdom system of Thomas Cavalier-Smith is an ideal one.

According to him, the cell structure and genetic organisation are very much important to study phylogeny. He considered the ultra-structural characteristics as well as r-RNA sequence and also other molecular characteristics in his classification. He divided the living organisms into 2 **Empires** (Bacteria and Eukaryota) and **8 Kingdoms**.

The first **Empire Bacteria** is divided into *two* kingdoms: (i) **Eubacteria,** and (ii) **Archaebacteria**. Eubacteria has further been divided into two sub-kingdoms: Negibacteria (Gram-negative bacteria) and Posibacteria (Gram-positive bacteria).

The second **Empire Eukaryota** (the eukaryotic organisms) is divided into *six* kingdoms:

Table 1.1: The modification of Linnaean system of classification of living world over time

Linnaeus 1735 (excluding 3rd Kingdom Minerals)	Haeckel 1866	Copeland 1938	Whittaker 1969	Woese *et al.* 1977	Woese *et al.* 1990	Cavalier-Smith 1993	Cavalier-Smith 1998	Cavalier-Smith and his team (Ruggiero *et al.* 2015)
2-kingdoms	**3-kingdoms**	**4-kingdoms**	**5-kingdoms**	**6-kingdoms**	**3-Domains**	**8-kingdoms**	**6-kingdoms**	**7-kingdoms**
		Monera	Monera	Eubacteria	Bacteria	Eubacteria	Bacteria	Bacteria
				Archaebacteria	Archaea	Archaebacteria		Archaea
(Not treated)	Protista	Protista	Protista	Protista		Archaeozoa		
						Protozoa	Protozoa	Protozoa
					Eukarya	Chromista	Chromista	Chromista
Vegetabilia	Plantae	Plantae	Plantae	Plantae		Plantae	Plantae	Plantae
			Fungi	Fungi		Fungi	Fungi	Fungi
Animalia	Animalia	Animalia	Animalia	Animalia		Animalia	Animalia	Animalia

***Maintain the level of arrangement as marked by colour shed

(i) **Archaezoa** (Unicellular organisms having 70S ribosomes and devoid of golgi apparatus, chloroplast, mitochondrion and peroxisomes), e.g., *Gigardia* etc.,

(ii) **Chromista** (Photosynthetic organs having chloroplast within the lumen of the rough endoplasmic reticulum, containing Chlorophyll C), e.g., Diatoms, Cryptomonads, Brown algae and Oomycetes,

(iii) **Protozoa**,

(iv) **Plantae**,

(v) **Fungi,** and

(vi) **Animalia**.

SEVEN-KINGDOM SYSTEM: (Recent Classification)

Thomas Cavalier-Smith, with his collaborators (Ruggiero et al. 2015), raised his 7-kingdom classification of living world in 2015. In this revised scheme, they divided the prokaryotes into two kingdoms: **Bacteria** (Eubacteria) and **Archaea** (= Archaebacteria). The eukaryotes are divided into five kingdoms such as **Protozoa, Chromista, Plantae, Fungi** and **Animalia**. The classification is based on the consensus in the **Taxonomic Outline of Bacteria and Archaea (TOBA)** and the **Catalogue of Life (COL).**

This is a unified and coherent, hierarchial ranked system of higher level classification of all living organisms based on the available data of **1.6 million species**, so far identified and described by more than **3000 taxonomic experts** through consultation of **more than 200 sources**. The information on the species is based on a variety of characters such as morphological, molecular (genes, metagenome, and metabolome), and ethno-ecological. About 1.9 million species (c. 82% of the global species) are provided at the rank of class or below, and about 1.0 million species (c. 63% of the global species) are provided at the rank of order or below. The authors of this classification have only named such groups that are generally considered to have had a **monophyletic origin**, though some of them may be **paraphyletic**, and others (e.g. Euglenozoa, Rhizaria, Cercozoa, etc.) have evolved either by the **symbiogenetic merger** of two fundamentally different lineages or have had **infusions of genes** from elsewhere. Authors also agreed on splitting of paraphyletic taxa into holophyletic groups (e.g. classical bryophytes). They treated Proteobacteria and Cyanobacteria as **holophyletic phyla** though these two phyla exclude their mitochondrial and chloroplast descendants and neither of which is now a bacterium, but an evolutionarily chimaeric cell organella. Several group such as Archaea have been retained conservatively even though the evidence for paraphyly or holophyly is contradictory.

In this classification two-*Superkingdoms*. **Prokaryota** and **Eukaryota**, and seven-kingdoms have been recognised, which are found to be a practical extension of Cavalier-Smith's six-kingdom scheme of 1998. A total of seven kingdoms have been recognised, of which **Bacteria (Eubacteria) and Archaea (Archaeobacteria)** are included under superkingdom **Prokaryota** and **Protozoa, Chromista, Plantae, Fungi** and **Animalia** are under superkingdom **Eukaryota**. Fourteen (14) ranks from superkingdom to order have been retained (Table 1.2).

Table 1.2: List of ranks used in the hierarchy with the number of taxa per rank (after Ruggiero, 2015)

Rank	Number of Taxa
Superkingdom	02
Kingdom	**07**
Subkingdom	11
Infrakingdom	08
Superphylum	06
Phylum	**96**
Subphyllum	60
Infraphyllum	04
Superclass	12
Class	**351**
Subclass	145
Infraclass	23
Superorder	52
Order	**1,467**

* Main ranks are in bold type.

The basic schematic diagram of the classification is mentioned below:

Superkingdom Prokaryota

Kingdom Bacteria (= Eubacteria)

Subkingdom Negibacteria

Subkingdom Posibacteria

Kingdom Archaea (= Archaebacteria)

Superkingdom Eukaryota

Kingdom Protozoa

Subkingdom Eozoa

Infrakingdom Euglenozoa

Infrakingdom Excavata

Subkingdom Sarcomastigota

Kingdom Chromista

Subkingdom Hacrobia

Subkingdom Harosa (= "Supergroup SAR")

Infrakingdom Halvaria

Infrakingdom Rhizaria

Kingdom Plantae

Subkingdom Biliphyta

Subkingdom Viridiplantae

Infrakingdom Chlorophyta

Infrakingdom Streptophyta

Superphylum Charophyta

Superphylum Embryophyta

Kingdom Fungi

Subkingdom Eomycota

Subkingdom Dikarya (= Neomycota)

Kingdom Animalia

Subkingdom N. N.

Subkingdom Bilateria

Infrakingdom Protostomia

Infrakingdom Deuterostomia

The superkingdoms and kingdoms along with their key taxonomic issues are discussed below:

Prokaryota: A prokaryote organism lacks a membrane-bound nucleus, mitochondria or any other membrane-bound organelle, having 70s ribosomes. The higher classification of prokaryotes is still an unsolved problem as no prokaryote names above the rank of class are covered by the International Code of Nomenclature of Bacteria (I CNB). In the present classification (Ruggiero et al. 2015), Bacteria (Eubacteria) and Archaea (Archaebacteria) have been recognised as two separate kingdoms under the superkingdom Prokaryota. In the earlier classifications, Woese and Fox (1977) also treated Eubacteria and Archaebacteria as separate kingdoms of Domain Prokaryote. Margulis and Schwartz (2001) recognised Prokarya as superkingdom comprising of one kingdom Bacteria in which Archaea is treated as a subkingdom. Cavalier Smith (1998) recognised Eubacteria and Archaebacteria as two separate subkingdoms under kingdom Bacteria. It is suggested that more use of multigene analysis than the over reliance on rRNA gene tree alone may eventually provide the higher level classification of prokaryotes.

Protozoa and Chromista: **Protists** (Protozoan) are unicellular eukaryotes comprising a polyphyletic group, which do not undergo tissue formation through embryogenesis. Protists include ancestrally unicellular eukaryotes and are direct descendant of bacteria. The vegetatively wall-less typical phagotrophic members like Microsporidia and Rozellids are some examples of protists. They are the first eukaryotic diverged clades which showed remarkable similarily both morphologically and biologically and all other eukaryotic life forms got evolved from it. Protozoa comprises seven phyla, of which four are clades (early members) and three are paraphyletic.

It is difficult to mark a distinct boundary between Protozoa and **Chromista**. The phylum chromista includes all chromophytic algae (those with chlorophyll c, not b) which have evolved by symbiogenetic enslavement of another eukaryote (a red alga) as well as all heterophytic protists descended from them by loss of photosynthesis or entire plastids. Chromista includes all Harosa as well as other former Protozoa that are related to haptophytes or cryophytes. Sometimes chromista and Plantae form a clade. This is due to the consequence of the eukaryote-eukaryote chimaeric history of chromists that acquired some genes from red algae or of the very rapid basal radiation of the robust corticate clade (i.e. Plantae pus Chromista).

Plantae: The kingdom plantae includes green, red and glaucophyte algae, and land plants (embryophytes), characterized by having cellulosic cell wall, reserved food as starch and photosynthetic pigments like chlorophyll a and b (often a and d). **The structure of plastid genomes and the derived chloroplast protein-import machinery support a single origin of phylum plantae**. The early land plant (especially ancestral embryophyte) is believed to have originated from relatives of charophyta (charales or coleochaetales). Jeffery (1971) first treated charophytes and embryophytes as a clade **streptophyta** which was validated as a superphylum by Cavalier-Smith in 1993, while it was reduced to phylum by Bremer in 1985. The present classification recognised four embryophyte phyla—three of bryophytes (liverworth, hornworth and mosses) and a single phylum Tracheophyta characterized by a diploid phase having xylem and phloem. The angiospermae (Flowering seed plants, seeds are enclosed within fruits) has been treated as a superclass under subphylum spermatophytae of Phylum Tracheophyta.

Fungi: Fungi are heterotrophic eukaryotes, characterised by having **chitin in their cell wall** and **ergosterol in their cell membrane**. Thus Oomycetes, formerly treated as Fungi, is no more placed in Fungi, belong instead in Phylum **Pseudofungi** of the heterokont chromista. The kingdom Fungi has been divided into two sub-kingdom: **Eomycota** (including Chytrides, Zygomycetes and Glomeromyceyes) and **Neomycota** or **Dikarya**. All Ascomycetes and Basiodiomycetes are included under Dikarya as major parts of their life cycle showed dikaryotic nuclei.

Animalia: In the present classification 34 animal phyla have been proposed. It is interesting to note that the conventional first animal phylum protozoa is no more an animal phyla, rather it is placed as a separate kingdom comprise of unicelllar, vegetatively wall-less, phagotrophic eukaryotes (aleady discussed).

2 Microbial World

The Microbial World includes countless of micoorganisms ubiquitous in their distribution i.e. found every where. Microbes live in water you drink, the food you eat, and the air you breathe. Billions of microbes are swimming in your belly and mouth, and crawling on your skin. But don't worry, as more than 95% are good for you and rest may cause harm.

The term microbes in short form of micro-organisms, which mean small organisms. Based on them the subject Microbiology develops.

Microbiology (Gr. *Micros*, small; *bios*, life and *logos*, study or discourse) is the study of microorganisms, which are unicellular or cell-cluster microscopic organisms. This includes eukaryotes such as **fungi** and **protists** and prokaryotes such as **bacteria** and **certain algae (cyanobacteria)**. **Viruses** are also included in the subject. They are **widespread in nature** and are beneficial to life, but some can cause serious harmful effect in plants and animals including human beings. They are found to grow in liquid water, including soil, hot springs, on the ocean floor, high in the atmosphere, and deep inside rocks within the Earth's crust. Some unicellular protists and **bacteria, like *Thiomargarita namibiensis*, are macroscopic** and visible to the naked eye. Microbiology subdivided into divisions including bacteriology, virology, mycology and others. A scientist who specializes in the area of microbiology is called a **Microbiologist**.

According to the present concept "the microbiology is the study of small life". The microbes can be divided into two categories: the **cellular microbes** (or organisms) and the **acellular microbes** (or agents). The cellular group includes bacteria, the archaea, the fungi, and the protists (composed of algae, protozoa, slime molds, and water molds). Cellular microbes can be either **unicellular**, where one cell is the entire organism, or **multicellular**, where hundreds, thousands or even billions of cells can make up the entire organism. The acellular group includes viruses and other infectious agents, such as prions, viroids etc.

It is to be mentioned that Prions, the "infectious proteins", have been investigated by virologists, but **never considered as microorganisms**, however, as the clinical effects traced to them were originally presumed due to chronic viral infections and virologists took search in discovering them.

The existence of microorganisms was predicted many centuries ago, by the **Jains** in India (*based on Mahavira's teachings as early as 6th century BCE i.e. Years: c. 500 BCE - c. 400 BCE*) and **Marcus Terentius Varro** in ancient Rome, before they were first observed. The first record through microscopic observation was the fruiting bodies of **moulds**, by **Robert Hooke** in 1666, but the Jesuit priest **Athanasius Kircher** was likely the first to see microbes, which he mentioned observing in milk and putrid (rotten) material in 1658. **Antonie van Leeuwenhoek** (1676) of Holland (Fig. 2.1), an amateur lens grinder, was the person to make glass lens powerful enough and observed microorganisms. He was the first who gave an illustrated description of microorganisms and communicated his findings to the Royal Society of London. Leeuwenhoek observed a raindrop under simple microscope of his own design and found microscopic creatures, named them as **animalcules**. He is known as **Father of Microbiology**. Scientific microbiology was developed in the 19th century through the work of **Louis Pasteur** and in medical microbiology by **Robert Koch.**

Fig. 2.1: Antonie van Leeuwenhoek (1632-1723)

Sub-disciplines of Microbiology

Microbiology can be divided into several sub-disciplines. These are:

1. **Microbial physiology**: The study of how the microbial cell functions biochemically. It also includes the study of microbial growth, metabolism and cell structure.
2. **Molecular microbiology**: The study of the molecular biology and genomics of microorganisms.
3. **Medical microbiology**: The study of the role of microbes in human diseases. It includes the study of microbial pathogenesis, epidemiology etc.
4. **Veterinary microbiology**: The study of the role of microbes in veterinary medicine or animal diseases.
5. **Environmental microbiology**: The study of the function and diversity of microbes in their natural environments. It includes the study of microbial ecology, microbe-mediated nutrient cycling, geomicrobiology, microbial diversity, and bioremediation and also the habitat such as the rhizosphere and phyllosphere.
6. **Evolutionary microbiology**: The study of the evolution of microbes. It also includes the study of bacterial systematics and taxonomy.
7. **Industrial microbiology**: The exploitation of microbes for use in industrial processes such as industrial fermentation and wastewater treatment. It also includes brewing and baking industry.
8. **Aero-microbiology**: The study of airborne microorganisms.
9. **Food Microbiology**: The study of microorganisms causing food spoilage.
10. **Pharmaceutical microbiology**: It is the study of microorganisms causing pharmaceutical contamination and spoilage.

Importance of Microorganisms

Microorganisms have many importances in the environment. These are:

1. The microorganisms maintain the ecological balance on earth.
2. Some microorganisms live in human beings and other animals in symbiotic mode.
3. Some are used for the production of food and chemicals.
4. Some microorganisms cause disease in both plants and human beings, i.e., pathogenic.

Importance of microbes on Pathogenic ecology

Although many microorganisms are beneficial to human beings, but many are pathogenic. Many of them are pathogenic bacteria, cause diseases such as plague, tuberculosis, anthrax etc. It is very difficult to destroy them and are considered responsible for diseases like cystic fibrosis, Legionnaires' disease etc. They produce dental plaque, colonize catheters, transcutaneous, and orthopedic devices; and also infect open wounds and burned tissue of animals including human beings.

They also produce foodborne diseases as they colonize the surfaces of food and food-processing equipment. They are threatening continuously as because they are resistant to most of the methods used to control microbial growth. Moreover, the excessive and unscientific use of antibiotics has resulted a major global problem since resistant forms of bacteria have been selected over time. A very dangerous strain, methicillin-resistant *Staphylococcus aureus*, in recent year caused severe hazards in human life.

In addition, protozoans are also known to cause diseases such as malaria, sleeping sickness, and toxoplasmosis, while fungi can cause diseases like **ringworm** (*Trichophyton rubrum*, itchy, red rash has a ring-like appearance), **candidiasis** (*Candida*, caused infection in vaginal area), **histoplasmosis** (*Histoplasma capsulatum*) etc. Likewise virus caused severe diseases such as influenza, yellow fever, and AIDS are alarming in destruction of human life.

Food-borne diseases result from the consumption of contaminated food, pathogenic bacteria, viruses, or parasites that contaminate food. "Hygiene" is the avoidance of infection or food spoiling by eliminating microorganisms from the surroundings. Different microorganisms especially bacteria are found virtually everywhere, the levels of harmful microorganisms can be reduced to acceptable levels with proper hygiene techniques.

Soil microbiology (Soil Bacteria, Soil Actinomycetes, Soil Fungi, Soil Algae and Soil Protozoa):

Soil microbiology is a branch of Microbiology deals with the study of microorganisms in soil, their activities including how they affect soil properties. Soil microorganisms are important as they affect soil structure and fertility. Both plants and animals constitute an important component of soil.

Early microbiologists observed that the soil was not an inert static material but a medium containing with life. The soil is now considered as a dynamic or rather a living system, containing a dynamic population of organisms including microorganisms. Cultivated soil has comparatively more population of microorganisms than the fallow land, and the soils rich in organic matter having much more population than sandy soils. Microbes in the soil are important to us in maintaining soil fertility, cycling of nutrient elements in the biosphere and sources of industrial products like enzymes, antibiotics, vitamins, hormones, organic acids etc. At the same time certain soil microbes cause diseases of both human and plants.

Soil microorganisms are of different types such as Bacteria, Actinomycetes, Fungi, Algae and Protozoa. These groups of organisms have their own characteristic and functions in respect of soil. In the rhizosphere region huge number of microorganisms is generally present for food and shelter, and in exchange some of them also help by supplying nutrients.

Types of organisms present in Soil

The Soil's organisms are broadly classified into two groups are the Soil flora and Soil fauna. The detailed classification of which is given below:

A. Soil Flora

(a) Microflora

1. **Bacteria.** The bacteria are further classified into two groups:
   (i) ***Heterotrophic***: symbiotic and non-symbiotic N_2 fixers, Ammonifier, Cellulose decomposers, Denitrifier
   (ii) ***Autrotrophic***: *Nitrosomonas, Nitrobacter*, Sulphur oxidizers, etc.
2. **Fungi.** Like Molds, Yeast and Mushrooms.
3. **Actinomycetes.**
4. **Algae.** BGA, Yellow Green Algae, Golden Brown Algae.

(b) Macroflora: Roots of higher plants.

B. Soil Fauna

(a) Microfauna: Protozoa and Nematodes

(b) Macrofauna: Earthworms, moles, ants and others.

As soil harbours several diverse groups of organisms, but the most important amongst them are the microorganisms like bacteria, actinomycetes, fungi, some algae and protozoa. The name, characteristics and their functions or role in the soil are described below:

1. Soil Bacteria

Amongst the different types of microorganisms inhabiting in the soil, bacteria are the most abundant and predominant organisms. These are primitive, microscopic, prokaryotic and unicellular microorganisms devoid of chlorophyll.

Morphologically, soil bacteria are divided into three groups:

(i) *Cocci* (round or spherical), *Bacillus* (rod-shaped)

(ii) *Spirilla* or *Spirullum* (cells with long wavy chains).

(iii) Bacilli are most numerous followed by Cocci and *Spirilla* in soil.

The bacterial genera involved in diffrerent processes or reactions are given below (Table 2.1):

Winogradsky (1925), on the basis of ecological characteristics classified soil microorganisms in general and bacteria in particular into two broad categories, i.e., **Autochthonotus** (Indigenous species) and **Zymogenous** (fermentative) type.

(i) **Autochthonous**: Bacterial population is uniform and constant in soil, since their nutrition is derived from native soil organic matter, e.g., *Arthrobacter* and *Nocardia* etc. and

(ii) **Zymogenous**: Bacterial population in soil is low, as they require an external source of energy, e.g., *Pseudomonas* and *Bacillus*. The population of Zymogenous bacteria increases gradually when a specific substrate is added to the soil. This category includes cellulose decomposers, nitrogen utilizing bacteria and ammonifiers.

Role of Bacteria

Bacteria bring about a number of changes and biochemical transformations in soil (Table 2.2) and thus directly or indirectly help in the nutrition of higher plants growing in the soil. The important transformations and processes in which soil bacteria play vital role are:

(i) Decomposition of cellulose and other carbohydrates,

(ii) Ammonification (proteins-ammonia),

(iii) Nitrification (ammonia-nitrites-nitrates),

(iv) Denitrification (release of free elemental nitrogen),

Table 2.1: Bacterial genera involved in different processes or reactions

Process/reaction	Bacterial genera
Cellulose decomposition (celluloytic bacteria): Most cellulose decomposers are mesophilic	(a) ***Aerobic***: *Bacillus, Cellulomonas, Angiococcus, Cytophaga, Polyangium, Sporocytophyga, Achromobacter* (b) ***Anaerobic***: *Methanosarcina, Methanococcus, Clostridium*
Ammonification (Ammonifiers)	*Bacillus, Pseudomonas*
Nitrification (Nitrifying bacteria)	*Nitrosomonas, Nitrosococcus, Nitrobacter*
Denitrification (Denitrifies)	*Achromobacter, Pseudomonas, Micrococcus*
Nitrogen fixing bacteria	(a) ***Symbiotic***: *Rhizobium, Bradyrrhizobium* (b) ***Non-symbiotic***: (i) **Aerobic**—*Azotobacter, Beijerinckia* (in acidic soils), (ii) **Anaerobic**—*Clostridium*

Table 2.2: Different soil bacteria capable of degrading various plant residues present in soil

Cellulose	Hemicelluloses	Lignin	Pectin	Proteins
Pseudomonas	*Bacillus*	*Pseudomonas*	*Erwinia*	*Clostridium*
Cytophaya	*Vibrio*	*Micrococcus*		*Proteus*
Spirillum	*Pseudomonas*	*Flavobacterium*		*Pseudomonas*
Actinomycetes	*Erwinia*	*Xanthomonas*		*Bacillus*
Cellulomonas		*Streptomyces*		

(v) Biological fixation of atmospheric nitrogen (symbiotic and non-symbiotic) oxidation and

(vi) Reduction of sulphur and iron compounds.

All these processes play a significant role in plant nutrition.

2. Soil Actinomycetes

These are the organisms having characters common with both bacteria and fungi but possess some distinctive features to keep them into a different category. Based on taxonomic approach, Actinomycetes are clubbed with bacteria in the same class Schizomycetes and confined under the order Actinomycetales.

They are unicellular like bacteria, but produce a mycelium which is non-septate (coenocytic) and more slender, like true bacteria. They do not have distinct cell-wall and their cell wall is without chitin and cellulose (commonly found in fungi). On culture media unlike slimy distinct colonies of true bacteria which grow quickly, Actinomycetes colonies grow slowly, show powdery consistency and stick firmly on agar surface. They produce hyphae and conidia or sporangia like fungi. Certain Actinomycetes whose hyphae undergo segmentation resemble with bacteria, both morphologically and physiologically.

Actinomycetes are numerous and widely distributed in soil and are next to bacteria in abundance. They are widely distributed in the soil, compost etc. They are sensitive to acidity (optimum pH range, 6.5 to 8.0) and waterlogged soil conditions. The population of Actinomycetes increases with depth of soil even upto horizon 'C' of a soil profile. They are heterotrophic, aerobic and mesophilic (25-30°C) organisms and some species are commonly present in compost and manures are thermophilic growing at 55-65°C temperature (e.g., *Thermoatinomycetes, Streptomyces*).

Actinomycetes belonging to the order of Actinomycetales are grouped under four families viz., Mycobacteriaceae, Actinomycetaceae, Streptomycetaceae and Actinoplanaceae. Some genera of the families like Actinomycetaceae and Streptomycetaceae are important both agriculturally and industrially.

In order of abundance in soils, the common genera of Actinomycetes are *Streptomyces* (nearly 70%), *Nocardia* and *Micromonospora* although others are also generally found.

Role of Actinomycetes:

The functions of Actinomycetes are as follows:

1. Actinomycetes are able to decompose all sorts of organic substances like cellulose, polysaccharides, proteins, fats, organic acids etc.
2. The organic residues or substances added to soil are first attacked by bacteria and fungi and later by Actinomycetes, because they are slow in activity and growth than bacteria and fungi.
3. They decompose or degrade the more resistant and indecomposable organic substances and produce a number of dark black to brown pigments, thereby soil humus becomes dark coloured.
4. They are also responsible for decomposition of humus (resistant material) in soil.

5. They are responsible for earthy smell of freshly ploughed soils.
6. Many genera, species and strains (e.g., Streptomyces of Actinomycetes produce a number of antibiotics like Streptomycin, Terramycin, and Aureomycin etc.
7. One of the species of Actinomycetes, *Streptomyces scabies* caused Potato scab disease.

3. Soil Fungi

Fungi in soil are present as mycelial form, rhizomorph or as spores. Their number varies from a few thousand to a few–million per gram of soil. The mycelium composed of individual hyphae. The hyphae may be aseptate, i.e., coenocytic (Mastigomycotina and Zygomycotina) or septate (Ascomycotina, Basidiomycotina and Deuteromycotina).

The most commonly available genera of fungi in soil are: *Alternaria, Aspergillus, Botrytis, Cladosporium, Cephalosporium Chaetomium, Fusarium, Gliocladium, Mucor, Monilia, Penicillium, Pythium, Rhizopus, Trichoderma, Verticillium* etc. (Jackson, 1975). The fungal genera belong to Deuteromycotina (Fungi imperfecti) which lacks sexual mode of reproduction.

As the soil fungi are aerobic and heterotrophic, they require abundant supply of oxygen and organic matter in soil. Fungi are dominant in acid soils, because acidic environment is not suitable for the bacteria and Actinomycetes. The optimum pH range for fungi lies between 4.5 to 6.5. They may grow in neutral and alkaline soils and some can even tolerate above pH 9.0.

Role of Fungi

1. Fungi plays significant role in soils and also in plant nutrition.
2. They plays important role in the decomposition of cellulose, hemicellulose, starch, pectin, lignin in the organic matter added to the soil.
3. Lignin is mainly decomposed by fungi.
4. They also serve as food for bacteria.
5. Certain fungi belonging to Zygomycotina and Deuteromycotina are predaceous in nature and attack protozoa and nematodes present in soil and thus, maintain biological equilibrium in soil.
6. They also plays important role in soil aggregation and also in humus formation.
7. Some soil fungi are parasitic and cause number of plant diseases such as wilts, root rots, damping-off and seedling blights e.g., *Pythium, Phytophthora, Fusarium, Verticillium* etc.
8. Number of soil fungi forms mycorrhizal association with the roots of higher plants and helps in mobilization of soil phosphorus and nitrogen e.g., *Amanita, Boletus, Entoloma, Lactarius* (Ectomycorrhiza) and *Glomus, Gigaspora, Aculospora,* (Endomycorrhiza).

4. Soil Algae

Algae are present in almost all types of soils, with the availability of both moisture and sunlight. About 100-10,000 algae are usually present per gram of soil. They are commonly photoautotrophic, aerobic organisms and obtain CO_2 from air and energy from sunlight, thereby synthesize their own food. They are unicellular, filamentous or colonial in form. The soils algae are divided into 4 main groups are as follows:

1. Cyanophyta (Blue-green algae),
2. Chlorophyta (Grass-green algae),
3. Xanthophyta (Yellow-green algae) and
4. Bacillariophyta (diatoms or golden-brown algae)

Out of these 4 groups (classes), blue-green algae and green algae are more common in soil. The green algae and diatoms are dominant in the soils of temperate zone while blue-green algae are in tropical soils. Green-algae prefer acid soils while blue green algae are commonly found in neutral and alkaline soils. The most common genera of green algae found in soil are: *Chlorella, Chlamydomonas, Chlorococcum, Protosiphon* etc. and that of diatoms are *Navicula, Pinnularia, Synedra, Frangilaria,* etc.

Blue green algae (BGA) are unicellular, photo autotrophic prokaryotes containing phycocyanin pigment in addition to chlorophyll. They do not possess flagella and do not reproduce sexually. They are common in neutral to alkaline soils. The dominant genera of BGA in soil are: *Chroococcus, Phormidium, Anabaena, Aphanocapra, Oscillatoria,* etc. Some BGA possess specialized cells known as "**Heterocyst**" which is the sites of nitrogen fixation. BGA fixes nitrogen (non-symbiotically) in puddle paddy or water logged paddy fields (20-30 kg/ha/season). There are certain BGA which possess the character of symbiotic nitrogen fixation in association with other organisms like fungi, mosses, liverworts and aquatic ferns, *Azolla*; e.g., *Anabaena-Azolla* association fix nitrogen symbiotically in rice fields.

Role of algae or BGA

1. Algae play important role in the maintenance of soil fertility especially in tropical soils.
2. They add organic matter to soil when die and thus increase the amount of organic carbon in soil.
3. Most of soil algae (especially BGA) act as cementing agent in binding soil particles and thereby reduce or prevent soil erosion.
4. Mucilage secreted by the BGA is hygroscopic in nature and thus helps in increasing water retention capacity of soil for longer time.
5. Soil algae through the process of photosynthesis, liberate large quantity of oxygen in the soil environment and thus facilitate the aeration in submerged soils or oxygenate the soil environment.
6. They help in checking the loss of nitrates through leaching and drainage especially in fallow soils.
7. They help in weathering of rocks and building up of soil structure.

5. Soil Protozoa

These are unicellular, eukaryotic, colourless, and animal like organisms. They are larger than bacteria and size varying from few microns to a few centimeters. Their population in arable (fit for tillage) soil ranges from 10,000 to 1,00,000 per gram of soil and are abundant in surface soil. They can withstand adverse soil conditions as they are characterized by "**cyst stage**" in their life cycle. Except few genera which reproduce sexually by fusion of cells, rest of them reproduces asexually by fission or binary fission. Most of the soil protozoa are motile by flagella or cilia or pseudopodia as locomotor organs. Depending upon the type of appendages provided for locomotion, protozoa are:

1. Rhizopoda (Sarcondia)
2. Mastigophora
3. Ciliophora (Ciliata)
4. Sporophora (not common Inhabitants of soil)

Class-Rhizopoda: It consists protozoa without appendages usually have naked protoplasm without cell-wall, pseudopodia as temporary locomotory organs are present some times. Important genera are *Amoeba, Biomyxa, Euglypha*, etc.

Class Mastigophora: It belongs to flagellated protozoa, which are predominant in soil. Important genera are: *Allention, Bodo, Cercobodo, Cercomonas, Entosiphon, Spiromonas, Spongomions* and *Testramitus*. Many members are saprophytic and some possess chlorophyll and are autotrophic in nature. In this respect, they resemble unicellular algae and hence are known as "Phytoflagellates".

Class Ciliophora: The soil protozoa belonging to the class ciliophora are characterized by the presence of cilia (short hair-like appendages) around their body, which helps in locomotion. The important soil inhabitants of this class are *Colpidium, Colpoda, Balantiophorus, Gastrostyla, Halteria, Uroleptus, Vortiicella, Pleurotricha* etc.

Protozoa are abundant in the upper layer (within 15 cm depth) of soil. Soil moisture, aeration, temperature and pH are the important factors affecting soil protozoa.

Role of Protozoa

1. Most of protozoans derive their nutrition by feeding or ingesting soil bacteria belonging

to the genera *Enterobacter, Agrobacterium, Bacillus, Escherichia, Micrococcus,* and *Pseudomonas* and thus, they play important role in maintaining microbial or bacterial equilibrium in the soil.

2. Some protozoa have been recently used as biological control agents against phytopathogens.
3. Species of the bacterial genera such as *Enterobacter* and *Aerobacter* are commonly used as the food base for isolation of soil protozoans.
4. Several soil protozoa cause diseases in human beings which are carried through water and other vectors, e.g. amoebic dysentery (C.O. *Entomobea histolytica*).

Microbes in air: Aeroallergens

Microbes remain suspended in the air. The study of living microbes suspended in the air is the Aeromicrobiology. These microbes are called as bioaerosols (Brandl *et. al.,* 2008). Although they are present in less number than soil and in water, but they are still huge in number that affect the atmosphere in various way (Amato, 2012). The air borne microbes may travel long distances with the help of wind and precipitation. Thereby these aerosols are ecologically important and many of them may cause diseases of plants and animals including human beings. Typically microbes will remain suspended in clouds, where they are able to perform processes that alter the chemical composition of the cloud, and may even induce precipitation (Amato, 2012). In the indoor air, the most common four bacteria are *Micrococcus, Staphylococcus, Bacillus* and *Pseudomonas*.

Microbes in the air may come from various sources. These are aerosols, wind-blown, man's actions, water droplets, and human or animals. The different sources in detail are given below (Fig. 2.2).

Allergy

Anton Van Pirquet (1906) first introduced the term allergy to denote any altered capacity of body to react with a foreign substance. Allergy is a disorder caused by an abnormal reaction to a harmless substance. It occurs when the body's immune system sees a substance as harmful and over reacts to it. The substances that cause allergic reactions are allergens. Allergies are one of the most common chronic diseases. When someone has allergies, their immune system makes an antibody called immunoglobulin E (IgE). These antibodies respond to allergens and triggers the release inflammatory substance like histamine.

Aeroallergens

Allergy is caused by certain living and non-living agents present in the atmosphere. The

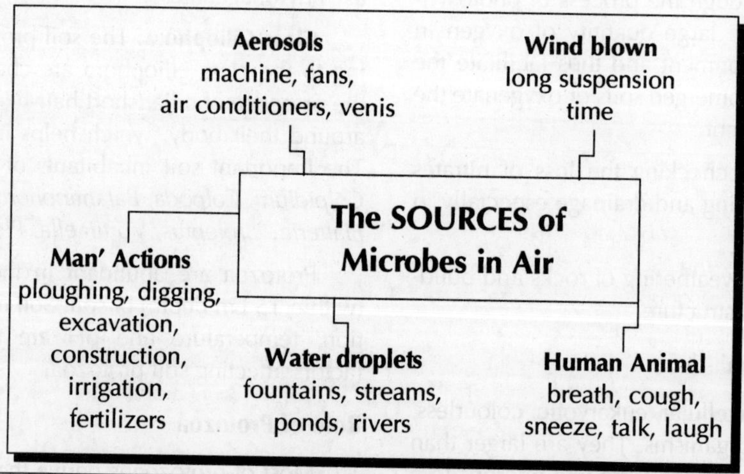

Fig 2.2: Different sources of microbes in the air

allergy causing agents present in air are called aeroallergens, and the allergy caused by them is called aero-allergy. Some of the prominent aeroallergens are house dusts, pollen grains, cosmetics, microbial spores or cells, etc.

Microbes as aeroallergens

Different microorganisms can be present in aerosol formed in the atmosphere. These are mainly bacteria, fungi, viruses and also yeasts and protozoans. These microbes are able to adapt in some harsh climatic condition for their survival, such as temperature, gases and humidity. Sometimes they form endospores, by which they can withstand extreme climatic conditions.

List of some common microbial aeroallergens are given below (Table.2.3):

Table 2.3: Some microbes as aeroallergens

Microbes	Name of some microbes
1. **Cyanobacteria** (i.e. Algae)	*Lyngbya, Nostoc, Oscillatoria, Gloeotrichia* etc.
2. **Fungi**	*Alternaria, Aspergillus, Candida, Chaetomium, Curvularia, Fusarium, Monilia, Penicillium, Phoma, Trichoderma* etc.
3. **Bacteria**	*Bacillus anthracis, Corynebacterium diphtheriae, Legionella* spp., *Mycobacterium tuberculosis, Klebsiella pneumoniae, Bordetella pertussis*;
4. **Viruses**	Picorna virus, Herpes virus, Flavi virus, Hanta virus, Hepatitis virus, Influenza virus etc.

Many of these microorganisms are commonly associated with specific and some commonly known diseases of both plants and animals including human being. Some of the aeroallergens along with the associated diseases are given below:

1. Bacteria

Bacillus anthracis is one of the important airborne human pathogens (Table 2.4). The bacterium infects human beings, cause severe respiratory disease, the **anthrax**. The bacterium can resist environmental stresses. They develop spores with extremely thick cell walls which can remain inactive for many years. It can survive in extreme temperatures, chemical contaminations, and low nutrient environments (Gatchalian 2010).

Table 2.4: Some airborne human pathogens

Human diseases	Pathogenic fungi
Bacterial diseases	
1. Diptheria	*Corynebacterium diphtheriae*
2. Legionellosis	*Legionella* spp.
3. Pulmonary tuberculosis	*Mycobacterium tuberculosis*
4. Pneumonia	*Klebsiella pneumoniae*
5. Pulmonary anthrax	*Bacillus anthracis*
6. Whoopping cough	*Bordetella pertussis*
Fungal diseases	
1. Aspergillosis	*Aspergillus fumigatus*
2. Coccidioidomycoses	*Coccidioides immitis*
Viral diseases	
1. Common cold	Picorna virus
2. Chicken pox	Herpes virus
3. Dengue fever	Flavi virus
4. Hantavirus pulmonary syndrome	Hanta virus
5. Hepatitis	Hepatitis virus
6. Influenza	Influenza virus

2. Fungi

Aspergillus fumigatus is one of the important airborne human pathogens. The fungus is the major airborne plant pathogen (Table 2.5). This pathogen is capable of causing many human diseases when conidia are inhaled into the lungs. On the other hand, *A. fumigatus* lacks virulence traits and is very adaptable to changing in environmental conditions and still capable of mass infection. (McCormick 2010).

Table 2.5: Some airborne plant pathogenic fungi

Plant diseases	Pathogenic fungi
1. Downy mildew	*Pseudoperonospora humili*
2. Dutch elm disease	*Ceratocystis ulmi*
3. Late blight of potato	*Phytophthora infestans*
4. Leaf rust	*Puccinia recondita*
5. Loose smut of wheat	*Ustilago segetum* var. *tritici*
6. Powdery mildew of barley	*Blumeria graminis* f. sp *hordei*
7. Rust of maize	*Puccinia sorghi*
8. Southern corn leaf blight	*Helminthosporium maydis*

3. Virus

The Avian Influenza Virus, is a airborne viral pathogen. It is a single stranded RNA virus that can infect a broad range of animals as well as human beings and cause the Avian Influenza disease.

What is an allergic reaction?

Normally, the immune system functions as the body's defense against invading germs like bacteria and viruses. When an allergic person first comes in contact with an allergen, the immune system treats the allergen as an invader and gets ready to attack. In most allergic reactions, however, the immune system is responding to a false alarm.

The immune system does this by producing large amount of one type of antibody called **immunoglobulin E** (i.e. **IgE**). Each IgE antibody is specific for one particular substance.

The IgE molecules are special because IgE is the only type of antibody that attaches tightly to the body's mast cells (tissue cells), and to basophils (blood cells). When the allergen next encounters its specific IgE, it attaches to the antibody like a lock and key system. This action, signals the cell to which the IgE is attached to release or produce powerful chemicals like **histamine**, which cause inflammation. These chemicals act on tissues in various parts of the body, such as the respiratory system, and cause the symptoms of allergy.

Symptoms

The signs and symptoms of airborne allergies are:

(i) Sneezing, often with a runny or clogged nose.
(ii) Coughing and post-nasal drip.
(iii) Itching eyes, nose, and throat.
(iv) Watering eyes.
(v) Conjunctivitis etc.

In case of non-allergic person, the mucus in the nasal passages simply moves foreign particles to the throat, where they are swallowed or coughed out.

Microbes of water: Water blooms, Biological indicators

Water blooms

The dense aquatic population of microscopic photosynthetic organisms grows in abundance with nutrient salts in surface water, along with adequate sunlight for photosynthesis is called **water bloom.**

The microorganisms or their released toxic substances may discolour the water, deplete its oxygen content, poison aquatic animals and waterfowl, and irritate the skin and respiratory tract of human beings. Single species of algae (cyanophyceae), diatoms, or dinoflagellates like *Gymnodinium*, reproducing every few hours, may dominate a blooming population. The number of individuals per litre of water may range from 1,000 to 60 million.

Algal blooms

Algal bloom is the rapid increase in the population of algae in freshwater or marine water systems, and is recognized by the discolouration of water from their pigments. Blooms which can destroy animals and the ecosystem are called Harmful Algal Blooms (HAB), and can lead to death of fishes, cutting off water to resident areas etc.

Since 'algae' is a broad term including organisms of widely varying sizes, growth rates and nutrient requirements, there is no demarcated threshold level to define a bloom. In some algae, blooming is considered at concentrations reaching millions of cells/ml., while others form blooms of tens of thousands of cells/lt. The colour of the bloom depends on the colour of the photosynthetic pigments of the respective algae. It's colour varies from greenish, yellow, brown, red etc., depending on the species and the type of pigment they produce.

Fig. 2.3: Freshwater algal bloom in river

A. Fresh water algal blooms

Freshwater algal blooms (Fig. 2.3) are caused by an excess accumulation of nutrients, particularly some **phosphates**. The excess of nutrients may be added through the application of fertilizers in agricultural field or from household cleaning products containing **phosphorus**. These nutrients can be added to ponds or lakes through water runoff. Excess **carbon** and **nitrogen** have also been suspected as the causes of blooming. The presence of residual **sodium carbonate** acts as catalyst for the algae to bloom by providing dissolved CO_2, which enhanced photosynthesis rate in presence of nutrients.

When phosphates are introduced into water systems, its concentration increases, caused increase in the growth of algae and plants. Algae tend to grow very quickly under high nutrient availability. As the alga are short-lived, results in accumulation of high concentration of dead organic matter which starts to decay. The decaying process consumes dissolved O_2 in the water, results the depletion of O_2 level in water. Without sufficient dissolved O_2 in the water, animals and plants may die off in large numbers.

Blooms may also be observed in freshwater aquariums when fishes are overfed and excess nutrients are not absorbed by plants. These are generally harmful for fishes, and the situation can be corrected by changing the water in the tank and then reducing the amount of food supply.

B. Harmful Algal Bloom (HAB)

It is an algal bloom that causes negative impacts with associated other organisms via production of natural toxins, mechanical damage or by other means. HABs are often associated with large-scale mortality of marine animals and also associated with shellfish poisonings. In human being it has been significantly related with the risk of non-alcoholic liver disease and caused death.

Harmful Phytoplankton

Phytoplankton or microalgae are the single-celled, microscopic organisms naturally occur in the wet-surface of any object in marine environment. This type of algal populations form the base of the food web upon which nearly all other marine organisms depend. Out of more than 5000 species of marine phytoplankton, about 2% are known to be **toxic** to other animals. These algae have significant impact on marine ecosystems at different levels based on the species, growing environment and mechanism to exert negative impact.

These harmful algal bloom cause adverse effect to different aquatic organisms, specially noted in marine mammals, sea turtles, sea birds and fin-fish. It affects their developmental, immunological, neurological, or reproductive capacities. Some of the remarkable incidences are given below:

(i) Contaminated menhaden (a forage fish): In 2004 a mass mortality of 107 "bottle nose dolphins" was observed in Florida panhandle in the spring due to ingestion of contaminated menhaden (a forage fish) with high levels of brevetoxin.

(ii) Dinoflagellate (*Alexandrium fundyense*) and copepod: During summertime, the highly endangered North Atlantic Right Whale, have been exposed to neurotoxins by preying on highly contaminated zooplankton. Due to seasonal blooms of the toxic dinoflagellate *Alexandrium fundyense*, and subsequent copepod grazing, foraging right whales will ingest large concentrations of these contaminated copepods. Ingestion of such contaminated prey can affect respiratory capabilities, feeding behavior, and ultimately the reproductive condition of the population.

(iii) Contaminated prey: **Brevetoxin** exposure, via inhalation of aerosolized toxins and ingestion of contaminated prey, caused increased lethargy and muscle weakness in critically endangered "loggerhead sea turtles" causing these animals to wash ashore in a decreased metabolic state.

HABs are found many regions of the world and have negative economic and also health impacts. Some of the important incidents are given below:

(i) The Gulf of Maine (North America) frequently experiences blooms of the dinoflagellate *Alexandrium fundyense*, an organism that produces **saxitoxin**, the neurotoxin responsible for paralytic shellfish poisoning.

(ii) In the Gulf of Mexico, the well-known "Florida red tide" is a HAB caused by another dinoflagellate (*Karenia brevis*), which produces **brevetoxin**, the neurotoxin responsible for neurotoxic shellfish poisoning.

The common harmful effects of HABs are:

(i) Neurotoxins cause mass mortalities in fish, seabirds, sea turtles, and marine mammals.

(ii) Human illness or death via consumption of seafood contaminated by toxic algae.

(iii) Mechanical damage such as disruption of epithelial gill tissues in fish, resulting in asphyxiation.

(iv) Oxygen depletion in the water (hypoxia or anoxia) from cellular respiration and also by bacterial degradation.

Biological indicators

Biological indicators i.e. Bio-indicators are living organisms such as plants, planktons, animals, and microbes, utilized to screen the health of ecosystem of the environment.

These are used for assessing environmental health and biogeographic changes that takes place in the environment. Each organic entity inside a biological system provides an indication regarding the health of its surroundings areas such as plankton responding rapidly to changes that takes place in the surrounding environment and serving as an important biomarker for assessing the quality of water as well as an indicator of water pollution. In addition, the health of aquatic flora is best reflected by plankton, which gives an early warning signal.

The advantages associated with using Bio-indicators are as follows:

(a) Biological impacts can be determined.

(b) To monitor synergistic and antagonistic impacts of various pollutants on a creature.

(c) Early stage diagnosis as well as harmful effects of toxins to plants, as well as human beings, can be monitored.

(d) Can be easily counted, due to their prevalence.

Economically viable alternative when compared with other specialized measuring systems.

Microbial indicators

Microorganisms are often used as health indicators of aquatic ecosystems. Due to their abundance, they are readily available and easy to test. Some microorganisms when exposed to **benzene** and **cadmium** contaminants develop new proteins known as **stress proteins**, which can be used as early warning sign (Khatri & Tyagi, 2015). In ocean the microorganisms are the important part of marine biomass and are responsible for the major productivity and nutrient cycle in the marine ecosystem. They have a rapid rate of growth, and react to even low levels of contaminants and other physicochemical and biological changes. They give an important signs of environmental changes (Underwood and Shapiro 1999).

In addition to different microbial indicators, the **bioluminescent bacteria** can be used to detect environmental pollutants in water. The presence of toxins in waters can be easily monitored either by changes in the digestion system of microbes which is hindered or disturbed by the presence of toxins which may result in changes in the amount of light emitted by the bacteria (Arora 1966). In comparison to other available traditional tests, these tests are very quick to monitor; however, their limitation is that, they can only indicate the changes in the organisms in presence of toxins (Malik and Bharti, 2012).

The **bacterium**, *Vogesella indigofera* reacts to heavy metals quantitatively. In absence of metal pollution, this bacterium produces blue pigmentation, an important marker of morphological change can be observed visually. Alternatively, in presence of hexavalent **chromium**, the production of pigment is blocked. This pigment production can be attributed due to the relationship between concentration of chromium and the generation of blue pigmentation by the bacterium (Arora 1966).

Other anaerobic bacteria *Bifidobacterium* and *Bacteroides* are potential indicators. *Bifidobacterium* are primarily associated with human contamination.

Some genetically engineered microbes are designed to remove arsenic and other heavy metal from contaminated water.

The presence or absence of algae can provide important clues about environmental health, or they can be accumulators of metal or their metabolism product, for example, the total algal biomass in aquatic system serves as an important metrics for organic pollution and nutrient loading such as **nitrogen** and **phosphorous**. Algae are increasingly being used as highly effective and sensitive tools for recognizing and predicting environmental stresses. The problem of water pollution has become acute now-a-days due to industrialization and urbanization.

Aquatic algae provide useful information on the status of aquatic environment. Some examples of algae as pollution indicators are as follows:

(i) *Synechococcus leopoliensis* (blue-green alga) and *Dunaliella* showed heavy metal pollution by accumulation, and are good water pollution indicator.

(ii) Cellular malformation, chlorosis and significant increase in heterocyst frequency have been seen in *Anabaena cylindrica* under **cadmium stress.**

(iii) Toxicity tests can be performed on *Chlorella vulgaris, Scapricornatum, Senedesmus subspicatus* by observing culture conditions, culture volume and light intensity.

(iv) A change in species diversity of phytoplankton is indicator of polluted aquatic ecosystem, for example, *Euglena clastica, Phacous tortus and Trachelon anas.*

(v) Charophytes are affected by moderate eutrophication. So, they are often used as bioindicators for good water quality.

(vi) *Eloderdensa* can be used for mercury pollution by observing chloroplast structure and their membranes.

How various bio-indicators tell about the environmental health?

1. If toxins are present, certain plants may not be able to grow in the area affected.
2. Monitoring population of animals may indicate damage to the ecosystem in which they live.

3. Algae blooms are often used to indicate large increases of nitrates and phosphates in lakes and rivers.
4. If pollution causes the reduction of an important food source, the animals dependent on it for food may also decrease. Animals may also change their behaviour or physiology if a toxin is present.
5. The levels of certain liver enzymes in fish increase if they are exposed to pollutants in the water.
6. Changes in the functioning of the nervous systems of worms are used to measure levels of soil pollution.
7. The increase in the number of mutated frogs found in the USA is used as an indicator of toxins in their environment.
8. Microorganisms can also be used as indicators of toxins in an ecosystem. Some microorganisms will produce **stress proteins** if exposed to certain pollutants. By measuring the levels of stress proteins, we can get an idea of the level of pollution present in the environment.

Microbial Nutrition, growth and metabolism

A. Microbial Nutrition

Living organisms collect nutrients from the surrounding environment. These nutrients are utilized for the synthesis of cellular material and for the production of required energy. Microbes are diverse in their specific physiological properties and differ in their nutritional requirements and also in adaptation to the physical environment of their living areas.

Presence of sufficient supply of nutrients from the surrounding environment for the microbes is required for their proper growth. In the microbial cell, water is about 80-90% of the total weight. Thus water is the major essential requirement in terms of quantitative requirement. The water acts as a solvent for nutrients and carries them across the membrane. In addition, because of the polar nature and also the capacity to bind with other polar molecules, it also influences the shape and architecture of many biological polymers like protein etc.

In microbes the nutritional requirements are of two types. These are: 1. Macro-elements i.e. Macro-nutrients and 2. Micro-element i.e. Micro-nutrients.

1. Macro-elements i.e. Macro-nutrients

The macro-elements or macro-nutrients are required by the microbes in comparatively large amount. These are: Carbon, Hydrogen, Oxygen, Nitrogen, Phosphorus, Sulphur, Potassium, Calcium, Magnesium and Iron.

(i) **Carbon**—Based on the requirement of carbon, the microbes are of two types: Autotrophs and Heterotrophs.

(a) ***Autotrophs***: All photosynthetic i.e. autotrophic microbes i.e. Cyanobacteria (Cyanophyceae), obtain their required energy from sun through reduction of inorganic compounds like CO_2 to carbohydrates as principal carbon source.

(b) ***Heterotrophs***: On the other hand, huge number of heterotrophic microbes are not able to use CO_2 and instead of that they depend on the organic compounds available in nature for carbon.

(ii) **Hydrogen**—Hydrogen has a number of important roles in microbial life. These are:

(a) It is a structural atom in organic molecules.

(b) It is a participant in the energy generation process.

(c) In plasma membrane of many microbes, protons are involved in the production of ATP through ATP synthase system.

(iii) **Oxygen**—All microbes require the elemental oxygen to build up their different biochemical components. Based on the requirement of oxygen, the microbes are of three types: Obligate aerobes, Obligate anaerobes and Facultative anaerobes.

(a) ***Obligate aerobes***: They depends on aerobic respiration for their metabolic energy.

(b) ***Obligate anaerobes***: Molecular oxygen is toxic to them as it either inhibits their growth or kills them.

(c) ***Facultative anaerobes***: They are able to grow either in presence or absence of oxygen.

(iv) **Nitrogen**—Microbes require nitrogen for the synthesis of different cellular components such as amino acids, purines, pyrimidines, co-factors and other substances. Some archebacteria and eubacteria are able to fix atmospheric nitrogen. Some others feed on some inorganic compounds like nitrates, nitrites or ammonium. Some microbes derive nitrogen from organic compounds like amino acids.

(v) **Phosphorus**—The phosphorus is supplied to microbes in the form of phosphate. It is an essential component of various substances like nucleic acids, phospholipids, ATP, several co-factors, proteins and also other cell components. Most of the microbes use free Pi containing organic compounds as a source and incorporated directly in the cell.

(vi) **Sulphur**—The sulphur is needed for the synthesis of certain amino acids like cysteine, methionine etc., carbohydrates and B-vitamins like biotin and thiamine. Some others require organic sulphur compounds, some are able to utilize inorganic sulphur and some others use elemental sulphur.

(vii) **Potassium**—The potassium maintains the turgour pressure of the cell. It is also utilized in many enzymatic reactions as co-factor and also in protein synthesis.

(viii) **Calcium**—The calcium contributes in heat resistance in endospore of bacteria and also have few others functions. In some cases spore contains calcium and also dipicolinic acid.

(ix) **Magnesium**—The magnesium serves as a co-factor for many enzymes. It also stabilizes ribosomes and cell membrane of microbes.

(x) **Iron**—The iron is a part of cytochromes and co-factor of different enzymes. It is also essential in electrone carrying proteins.

Out of the above, some elements like Ca, Mg, K and Fe are required in small quantities and thus these are called minor elements.

2. Micro-elements i.e. Micro-nutrients

Microbial cells require some elements in very small amount are called **micro-elements** or **trace elements**. Normally they are the part of the enzyme and co-factors and help in the catalysis of reactions. These are Zinc, Manganese, Molybdenum and Cobalt.

(i) **Zinc**—The zinc is present in the active sites of some enzymes.

(ii) **Manganese**—The manganese aids many enzymes in catalyzing the transfer of phosphate group.

(iii) **Molybdenum**—The molybdenum is required in the process of nitrogen fixation.

(iv) **Cobalt**—Cobalt is a component of vitamin B12 and related co-enzymes.

In addition to the above requirements, some microbes may have some specific mineral requirements. Bacteria those grow in saline lakes and oceans require large amount of sodium. Most microbes are able to synthesize all the required structure from the macro- and micro-elements. But if they can't do it, must be absorbed from the surrounding environment or the medium where it grows. Such type of building blocks is called **growth factors**.

Growth factors

The growth factors are of three types: Amino acids, Purines and pyrimidines, and Vitamins.

(i) **Amino acids**—Amino acids are required in the medium for the growth, where the microbes can't synthesize more than one or few of the 20 amino acids that are required for the synthesis of protein. It can be said that the lactic acid bacteria require the whole complement. On the other hand, most of the strains of *Staphylococcus epidermidis* (inhabitant of human skin) require six different amino acids are proline, arginine, tryptophan, valine, histidine and leucine.

(ii) **Purines and pyrimidines**—Both are required for the synthesis of nucleic acid. The free living soil microbes rarely needed these, but most often required in lactic acid bacteria.

(iii) **Vitamins**—The vitamins are the small organic molecules, functions as co-enzymes or are the building blocks of co-enzymes. Very small amount of vitamins are required for the growth of microbes, such as biotin, thiamin and nicotinic acid.

Nutritional Types of Microorganisms

Microbiologists commonly use the term growth to indicate an increase in a population of microbes rather than increase in size. Growth of microbes depends on the metabolism of nutrients, and results the formation of a discrete colony, an aggregation of cells arising from a single parent cell. Nutrients indicate any chemical, required for growth of microbial population. The most important of these are the compounds containing carbon, oxygen, nitrogen, and/or hydrogen. All microbes require source of energy and hydrogen atoms or electrons for their growth.

There are different sources of energy available to microbes, and based on this they are divided into two groups. These are:

(i) **Phototrophs**: In this type, energy for growth is derived from sunlight.

(ii) **Chemotrophs**: In this type, energy for growth is derived from oxidation of organic or inorganic compounds.

Microbes also have two sources for hydrogen atoms and electrons, and based on this they are divided into two groups. These are:

(i) **Lithotrophs**: In this type, source of electrons is reduced inorganic compounds.

(ii) **Organotrophs**: In this type, source of electrons and hydrogen is the organic compounds.

Although the microbes show great metabolic diversity, yet based on energy, electrons or hydrogen and carbon sources, they are divided into different nutritional types. These are:

(i) **Photo-lithotropic autotrophs**: Microbes of this group is also called photo-autotrophs, as they utilize light as the source of energy and CO_2 is utilize as source of carbon. These are cyanobacteria (cyanophyceae), purple sulphur bacteria and green sulphur bacteria. Here the cyanobacteria use water as the electron donor and O_2 is liberated. On the other hand, purple sulphur bacteria and green sulphur bacteria, can't oxidize water but use the electrons from inorganic donors like hydrogen and H_2S.

(ii) **Photo-organotropic heterotrophs**: Microbes of this group, such as green non-sulphur and purple non-sulphur bacteria, use the organic compounds as electron donors and carbon sources. The bacterium, *Rhodospirillum rubrum* used succinate as an electron donor.

(iii) **Chemo-organotrophic heterotrophs**: Microbes of this group are also called chemoheterotrophs. They utilize organic compounds like amino acids and sugars, as source of energy, carbon and hydrogen. Majority of the pathogenic microbes are the chemoheterotrophs.

(iv) **Chemo-lithotropic autrophs**: Microbes of this group oxidise inorganic substance like sulphur, iron or nitrogen to drive energy and electron for different biosynthetic processes. They utilize CO_2 as the carbon source. In addition to that, a few chemolithotrophs use organic substances as carbon source are called **chemolithotrophic heterotrophs**. Some microbes like bacteria that obtain energy by using inorganic electron donors but obtain carbon from organic compounds are the **mixotrophs**.

The above facts can be summarized in Table 2.6 given below:

Table 2.6: Brief representation of the different nutritional groups of microbes along with the source of energy, hydrogen and carbon.

Nutritional types	Organism involve	Source of energy	Source of hydrogen/electron	Source of carbon
(i) Photo-lithotropic autotrophs	Cyanobacteria, purple and green sulphur bacteria	Light substance	Inorganic	CO_2
(ii) Photo-organotropic heterotrophs	Purple and green sulphur bacteria	Light substance	Organic (CO_2 also be used)	Organic carbon
(iii) Chemo-organotrophic heterotrophs	Most of the non-photo synthetic bacteria	Chemical (organic)	Organic substance	Organic carbon
(iv) Chemo-lithotropic autrophs	Sulphur oxidizing bacteria, hydrogen bacteria, nitrifying bacteria and iron bacteria.	Chemical (inorganic)	Inorganic substance	CO_2

B. Microbial growth

The growth of microbes commonly takes place by cell division through binary fission. It is very common in bacteria. To study the microbial growth, instead of measuring the length/weight it is more practical to count the increase in the number of cells.

Population growth curve (Fig. 2.22) consists of four distinct growth phases. These are lag phase, exponential phase, stationary phase and death or decline phase. Each growth phase shows variation based on changes in environment and also metabolism of the cells.

Cell growth can be measured commonly using counting chambers, viable counts, MPN (most probable number) and OD (optical density) measurements. Different methods of measurement of microbial growth, its use and limitations can be summarized and represented in tabular form shown in the Table 2.7.

C. Microbial Metabolism

Microbial metabolism is the process by which microbes obtains the energy and nutrients (e.g. carbon) it requires for all the live processes. Microbes use different types of metabolic strategies and based on metabolic characteristics species can often be differentiated from one another. In other way, it can be referred as the sum total of all chemical reactions takes place within a living organism. The chemical reactions either require or release energy. Thus the metabolism can be referred as an energy-balancing process.

Table 2.7: Different methods of measurement of microbial growth, its use and limitations

Methods of measurement	Use	Limitations
1. Direct count through microscope	It is the rapid laboratory enumeration of cell suspension.	It requires large number ($\geq 10^6$ cells/ml) of cells for accuracy.
2. Viable count	Here enumeration of viable cells is done in milk, water and many other products.	It is time consuming, requires proper medium for growth.
3. Membrane filtration	Here enumeration of bacterial cells is done in milk, water and many other products, when numbers are low.	It is also time consuming, requires proper medium for growth.

Contd.

Contd. Table 2.7

Methods of measurement	Use	Limitations
4. Most probable number (MPN)	Here enumeration of bacterial cells is done in milk, water and other products.	It is time consuming and requires proper medium for growth, provides indirect estimate of numbers.
5. Turbid metric measurement	Used for the rapid estimation of cell density in a suspension.	It does not differentiate between viable and non-viable cells, provides only estimate of cell density, requires > 10^6 cells/ml.
6. Dry weight determination	Determination of cell mass for industrial and laboratory applications.	It is time consuming, does not differentiate between viable and non-viable cells.

Metabolism → Anabolism (Require energy)
→ Catabolism (Release energy)

Anabolism:

It is the enzyme-regulated energy requiring reactions, where complex organic molecules are formed from simpler ones. These reactions are also called **biosynthetic** or anabolic and they are generally ***dehydration synthesis reactions*** (i.e. reactions that release water), and are ***endergonic*** (i.e. consume more energy than they produce) (Fig. 2.4). These are formation of proteins from amino acids, polysaccharides from simple sugars etc.

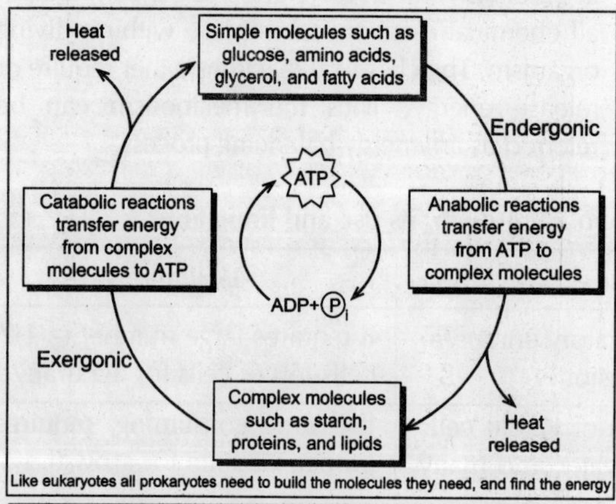

Fig. 2.4: Cyclic representation of exergonic and endergonic reaction in microbes

Catabolism:

It is the enzyme-regulated chemical reactions that release energy, where complex organic compounds are broken down into simpler ones. These reactions are called **degradative** or catabolic and they are generally ***hydrolytic reactions*** (i.e. reactions that use water and chemical bonds are broken), and are **exergonic** (i.e. produce more energy than they consume) (Fig. 2.4). These reactions break down sugars into CO_2 and H_2O.

The growth of a microorganism requires energy. This coupling of energy requiring and energy-releasing reactions is performing through the molecule, the **adenosine triphosphate** (ATP). ATP stores energy derived from catabolic reactions and releases it later to drive anabolic reactions and perform different cellular works.

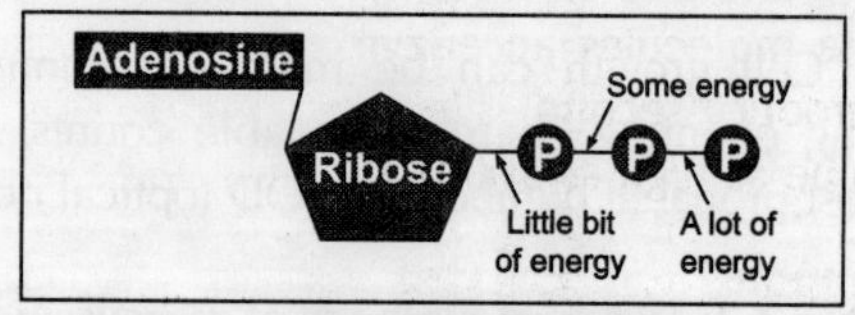

Fig. 2.5: Single ATP molecule

ATP is composed of an adenine, a ribose sugar and 3 phosphate groups (Fig. 2.5). When the terminal phosphate group (Pi) is released ADP is formed and some energy is being released to drive the anabolic reactions.

Thus, ATP → ADP + Pi + energy (released).

Later, when energy is being released from catabolic reactions is used to combine ADP and Pi.

Thus, ADP + Pi + energy (accumulated) → ATP is resynthesized.

Thus, an **anabolic reaction** – coupled to ATP breakdown and **catabolic reactions** – ATP synthesis.

Enzymes:

To speed up a chemical reaction a **catalyst** is required. The catalysts are the substances that can speed up a chemical reaction without being permanently altered themselves. In all living cells, enzymes serve as biological catalysts. Enzymes are specific and acts on specific substances are called **substrates** (sing. substrate) and each enzyme catalyzes only one reaction. Such as, the enzyme sucrase catalyzes the hydrolysis of sucrose (substrate) into glucose and fructose.

Specificity and efficiency of enzyme:

The enzymes are large globular proteins that range in Molecular Weight (MW) of about 10,000 to several millions. Each enzyme has a characteristic three-dimensional shape with a specific surface configuration as a result of its primary, secondary and tertiary structures. This enables them to find out the correct substrate from among the huge number of diverse molecules present in the cell. These are extremely efficient. Under optimum condition it can catalyze reactions at a rate 10^8 to 10^{10} times higher than those of comparable reactions without enzymes. The **Turnover number** (maximum number of substrate molecules an enzyme molecule converts to product/ second) is between 1 and 10,000 and can be as high as 500,000.

Components of enzyme

Some enzymes are solely made up of protein. But most of them consist of two parts, the protein part is called **apoenzyme** and non-protein part is called **co-factor.** The co-factors are commonly the ions of zinc, magnesium, iron and calcium, but if it is an organic molecule, called a **co-enzyme**. The apoenzyme and co-factor together called a **holoenzyme**. In this case, if the co-factor is being removed, the apoenzyme becomes non-functional.

Co-enzymes assist the enzyme by accepting atoms removed from substrate or by donating atoms required by the substrate. Some of them act as electron carriers, removes electrons from the substrate and donating them to other molecule in the subsequent reaction.

Many coenzymes are derived from vitamins. The most important coenzymes in cellular metabolism are NAD^+ and $NADP^+$. NAD^+ involved in catabolic (energy-yielding reactions) and $NADP^+$ involved in anabolic (energy-requiring reactions).

The other flavin coenzyme, such as flavin mononucleotide (FMN) and flavin adenine dinucleotide (FAD), contains derivatives of the B vitamin riboflavin and are also electron carriers.

Coenzyme A (CoA) contains derivatives of pantothenic acid, another B vitamin, which plays an important role in the synthesis and breakdown of fats and also in a series of oxidizing reactions called the Krebs cycle.

Some metal ions of Fe, Cu, Mg, Mn, Zn, Ca and Co acts as co-factors. They form a bridge between the enzyme and a substrate. Such as Mg^{2+} is required by many phosphorylating enzymes, which transfer a phosphate group from ATP to another substrate.

Production of Energy

There are two general process of energy production, i.e. A. the oxidation-reduction reaction and B. the mechanism of ATP generation.

A. Oxidation – reduction reactions:

Oxidation is the reaction of removal of electron from a molecule or an atom, a reaction that often produces energy. On the other hand, reduction is the addition of one or more electrons to a molecule or an atom.

Oxidation and reductions reactions are always being coupled. The pairing of these reactions is called oxidation-reduction or the redox reactions. Most cellular oxidation reactions involve the loss of hydrogen atoms i.e. the dehydrogenation. Hydrogen atom contains both electrons and protons and in the cellular oxidations, electrons and protons are being removed at the same time. An organic molecule is oxidized by the loss of two hydrogen atoms, and a molecule of NAD^+ is reduced by accepting two electrons and one proton. Thus one proton is left over and is released into the surrounding medium. The reduced coenzyme contains more energy than NAD^+. This energy can be used to generate ATP in the following reactions.

Living cells use oxidation-reduction reactions in catabolism to release energy from the nutrient molecules. Such as cells oxidizes a molecule of glucose to CO_2 and H_2O. The energy in the glucose molecule is removed in a stepwise manner and ultimately it is trapped by ATP, serve as an energy source for energy requiring reactions.

B. ATP generation:

Much of the energy being released during oxidation – reduction reactions and is trapped within the cell by the production of ATP. A phosphate group (Pi) is added to ADP with the input of energy to form ATP. Addition of a phosphate to a chemical compound is called **phosphorylation**. Living cells use three mechanisms of phosphorylation to generate ATP from ADP. These are:

(a) **Substrate level phosphorylation**: In this process ATP is generated when a high energy phosphate is directly transferred from a phosphorylated compound (a substrate) to ADP. Such as 1,3-bisphophoglycerate to 3-phosphoglycerate (in glycolysis).

(b) **Oxidative phosphorylation**: Electrons are transferred from organic compounds to one group of electron carriers (usually to NAD^+ and FAD). Then, the electrons are passed through a series of different electron carriers to molecules of O_2 (in electron transport chain) or other oxidized inorganic and organic molecules. This process occurs in the plasma membrane of prokaryotes and in the inner mitochondrial membrane of eukaryotes. The transfer of electrons from one electron carrier to the next releases energy, some of which is used to generate ATP from ADP through a process called the **chemiosmosis**.

(c) **Photophosphorylation**: It occurs only in photosynthetic cells (cells containing chlorophylls), where organic molecules, especially sugars, are synthesized with the energy of light with CO_2 and H_2O. Photophosphorylation starts this process by converting light energy to the chemical energy of ATP and NADPH, which in turn, are used to synthesize organic molecules.

Grouping of microbial metabolism

All microbial metabolisms can be grouped based on three principles. These are:

1. **How the organism obtains *carbon* for synthesizing cell mass.**
   (i) Autotrophic: Where carbon is obtained from carbon dioxide (CO_2),
   (ii) Heterotrophic: Where carbon is obtained from organic compounds,
   (iii) Mixotrophic: Where carbon is obtained from both organic compounds and by fixing carbon dioxide (CO_2).
2. **How the organism obtains *reducing equivalents* used either in energy conservation or in biosynthetic reactions:**
   (i) Lithotrophic: Where reducing equivalents are obtained from inorganic compounds.
   (ii) Organotrophic: Where reducing equivalents are obtained from organic compounds.
3. **How the organism obtains *energy* for living and growing:**
   (i) Chemotrophic: Where energy is obtained from external chemical compounds.
   (ii) Phototrophic: Where energy is obtained from light.

In practice, the above terms are almost freely combined. These are as follows:

1. **Chemolithoautotrophs**: They obtain energy from the oxidation of inorganic compounds and carbon from the fixation of CO_2. Such as Nitrifying bacteria, Iron-oxidizing bacteria Sulfur-oxidizing bacteria etc.
2. **Photolithoautotrophs**: They obtain energy from light and carbon from the fixation of CO_2,

using reducing equivalents from inorganic compounds. Such as Cyanobacteria (H_2O as reducing equivalent donor), Chromatiaceae (H_2S as reducing equivalent donor), *Chloroflexus* (H_2 as reducing equivalent donor) etc.

3. **Chemolithoheterotrophs**: They obtain energy from the oxidation of inorganic compounds, but cannot fix CO_2. Such as some *Thiobacilus*, some *Beggiatoa*, some *Nitrobacter* spp., *Wolinella* (with H_2 as reducing equivalent donor), some sulfate-reducing bacteria etc.

4. **Chemoorganoheterotrophs**: They obtain energy, carbon, and reducing equivalents for biosynthetic reactions from organic compounds. Such as *E. coli, Bacillus* spp., *Actinobacteria* etc.

5. **Photoorganoheterotrophs**: They obtain energy from light, carbon and reducing equivalents for biosynthetic reactions from organic compounds. Some species are strictly heterotrophic; many others can also fix carbon dioxide and are mixotrophic. Such as *Rhodobacter, Rhodospirillum* etc.

Diversity of electron acceptors in respiration

It takes place in two ways: Aerobic and Anaerobic respiration.

In this process most of the microorganisms oxidize the carbohydrates as the main source of cellular energy. They use two main general processes: 1. Aerobic respiration and 2. Fermentation. They also use anaerobic pathway to oxidize glucose.

1. Aerobic respiration

In aerobic respiration, the ultimate electron e-acceptor is O_2 and the reduced form is H_2O. There are **four stages** in aerobic respiration:

Oxygen extracts chemical energy from glucose; the glucose molecule must be split up into two molecules of pyruvate. This process also generates 2 molecules of ATP as an immediate energy yield and 2 molecules of NADH.

$C_6H_{12}O_6 + 2ADP + 2Pi + 2NAD^+ \rightarrow$
$2CH_3COCOO^- + 2ATP + 2NADH + 2H_2O + 2H^+$

The citric acid cycle (Kreb's cycle) begins with the transfer of a two-carbon acetyl group from acetyl-CoA to the four-carbon acceptor compound, the oxaloacetate; to form a six-carbon compound, the citric acid.

The citrate then goes through a series of chemical transformations, losing 2 carboxyl groups as CO_2. The carbons lost as CO_2 originate from what was oxaloacetate, not directly from acetyl-CoA. The carbons donated by acetyl-CoA become a part of the oxaloacetate carbon backbone after the first turn of the citric acid cycle (TCA cycle). Loss of the acetyl-CoA-donated carbons as CO_2 requires several turns of the citric acid cycle. However, because of the role of citric acid cycle in anabolism, they may not be lost, since many TCA cycle intermediates are also used as precursors for the biosynthesis of other molecules.

Most of the energy made available by the oxidative steps of the TCA cycle is transferred as energy-rich electrons to NAD^+, forming NADH. For each acetyl group that enters the citric acid cycle, 3 molecules of NADH are produced.

Electrons are also transferred to the electron acceptor Q, forming QH_2.

At the end of each cycle, the four-carbon oxaloacetate has been regenerated, and the cycle continues as usual.

2. Anaerobic respiration

The microbes are capable of using all sorts of other terminal electron accepters besides oxygen. A few examples of anaerobic respiration, where the final electron acceptor is an inorganic substance other than O_2.

(i) Some bacteria such as *Bacillus* and also *Pseudomonas* can use nitrate ion (NO_{-3}), in the presence of an enzyme called nitrate reductase, as a final electron acceptor, the nitrate ion is reduced to nitrite ion (NO_{2-}).

Nitrite ion can be converted to nitrous oxide (N_2O), or nitrogen gas (N_2) (denitrification process) which helps in recycling of nitrogen.

(ii) Other bacteria like *Desulfovibrio* use sulfate (SO_4^{2-}) as the final electron acceptor and forms hydrogen sulfide (H_2S).

(iii) Still other bacteria use carbonate (CO_3^{2-}) to form methane (CH_4).

(iv) Anaerobic respiration by bacteria using nitrate and sulfate as final electron acceptors is essential for the nitrogen and sulfur cycles that occur in nature.

Amount of ATP generated varies with the organisms and the pathway. Because only a part of the Krebs cycle operates and since not all the carriers in the electron transport chain participate, ATP yield is less and thus anaerobes tend to grow more slowly than aerobes.

Microbial fermentation

The fermentation is a specific type of heterotrophic metabolism that uses organic carbon instead of oxygen as a terminal electron acceptor. Thus these organisms do not use an electron transport chain to oxidize NADH to NAD^+ and therefore they must have an alternative method of using this reducing power and maintain a supply of NAD^+ for the proper functioning of normal metabolic pathways (e.g. glycolysis). As oxygen is not required, fermentative organisms are anaerobic type.

Many organisms can use **fermentation** under anaerobic conditions and **aerobic respiration** when oxygen is present. These organisms are **facultative anaerobes**. To avoid the over production of NADH, obligately fermentative organisms usually do not have a complete citric acid cycle. Instead of using an ATP synthase as in respiration, ATP in fermentative organisms is produced by **substrate-level phosphorylation** where a phosphate group is transferred from a high-energy organic compound to ADP and form ATP. As a result of the need to produce high energy phosphate-containing organic compounds (generally in the form of CoA-esters) fermentative organisms use NADH and other co-factors to produce many different reduced metabolic by-products, often including hydrogen gas (H_2). These reduced organic compounds are generally small organic acids and alcohols derived from pyruvate, the end product of glycolysis, such as ethanol, acetate, lactate and butyrate. Fermentative organisms are very important industrially and are used to make many different types of food products. The different metabolic end products produced by each specific bacterial species are responsible for the different tastes and properties of each food.

The two main types of fermentation are: alcoholic fermentation and lactic acid fermentation (Fig. 2.6).

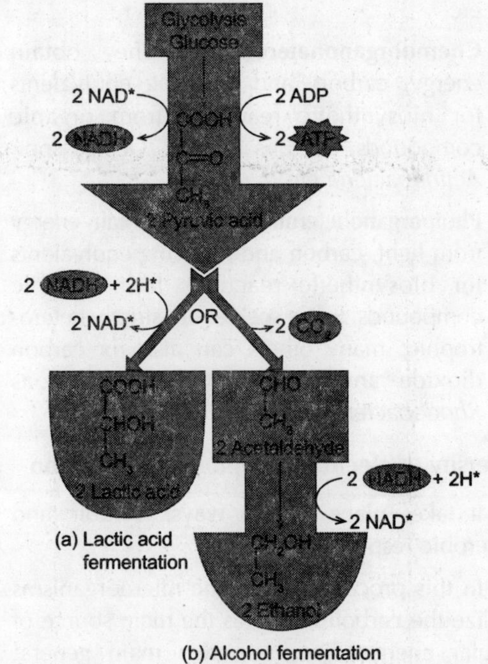

Fig. 2.6: Alcoholic (ethanol) and Lactic acid fermentations

Both types have the same reactants: Pyruvic acid and NADH, both of which are products of glycolysis. In alcoholic fermentation, the major products are ethanol and CO_2. But in lactic acid fermentation, the major product is lactic acid.

For both types of fermentation, there is a side product: NAD^+ which is recycled back to glycolysis so that small amount of ATP can continue to be produced in absence of oxygen.

The chemical equations below summarize the fermentation of sucrose ($C_{12}H_{22}O_{11}$). One mole of sucrose is converted into 4 molecules of ethanol (C_2H_5OH) and 4 moles of CO_2.

$$C_{12}H_{22}O_{11} + H_2O + \text{Invertase} \rightarrow 2C_6H_{12}O_6$$
Sucrose Glucose

$$C_6H_{12}O_6 + \text{Zymase} \rightarrow 2C_2H_5OH + 2CO_2$$
Glucose Ethanol

The process of lactic acid fermentation using glucose is 2 types: (i) Homolactic and (ii) Heterolactic fermentation. These are given below.

(i) In **Homolactic fermentation**, one molecule of glucose is converted to two molecules of lactic acid:

$$C_6H_{12}O_6 \rightarrow 2CH_3CHOHCOOH$$
Glucose Lactic acid

(ii) In **Heterolactic fermentation**, the reaction proceeds as follows, with one molecule of glucose converted to one molecule of lactic acid, one molecule of ethanol, and one molecule of carbon dioxide:

$$C_6H_{12}O_6 \rightarrow CH_3CHOHCOOH + C_2H_5OH + CO_2$$
Glucose Lactic acid Ethanol

BACTERIA

INTRODUCTION

The world of bacteria present around us influmen our life either directly or indirectly. As the bacteria are not visible with naked eye, the bacterial world was unknown until the invension of microscope. Jensen (1590) was the first person who magnified objects with hand lens. Later, Antonie van Leeuwenhoek (1676) worked on microscopic organisms, using simple microscope of his own design. Later, Robert Hooke (1635-1703), developed compound microscope in 1665 and had done his historical study of cork cells under that microscope. After the development of compound microscope these organisms were studied in detail. **Ehrenberg** (1829) studied and called these microbes as **Bacteria. Louis Pasteur** (Fig. 2.7) worked (1860-1890) on fermentation and reported that it took place by bacteria. Later, he proved the rotting of substances and some diseases of plants and animals including humam beings are caused by bacteria. He was the first to culture bacteria. Lister (1867) gave the concept of antiseptic surgery, the process of washing wounded region with spirit, carbolic acid etc. to remove or keep away the pathogenic microbes. Later, **Robert Koch** in 1881 (Fig. 2.8) isolated and cultured bacteria that cause anthrax of cattle and tuberculosis of man. He demanded that a specific bacterium be held as the cause of a disease. For such important contribution, he was awarded the **Nobel Prize** in 1905. For the important contributions made in the identification of micro-organisms, **Louis Pasteur** is known as **Father of Bacteriology** and **Robert Koch** as **Father of Medical Microbiology**. After detailed study of the above microorganisms i.e., the bacteria, it became evident that they did not belong to either plant or animal kingdom. They shared the characters of both the kingdoms, plant and animal. In 1894, a German biologist, **Ernst Haeckel** proposed a new kingdom **Protista** comprising of microorganisms. This kingdom has been further divided into two large groups: Lower Protists and Higher Protists. The lower protists having prokaryotic cellular organisation, include bacteria and cyanobacteria (blue green algae) and higher protists comprising of eukaryotic cell organisation include, algae, fungi and protozoa. Murray (1968) included all lower prokaryotes devoid of nuclear membrane and basic protein in the nucleoplasm under the group **Prokaryotae.** Later Bergey (1974)

Fig. 2.7: Louis Pasteur (1822-1895)

Fig. 2.8: Robert Koch (1843-1910)

placed all bacteria in the Kingdom Prokaryotae in his book *"Manual of Determinative Bacteriology"*.

CLASSIFICATION

In the early stage of bacteriology, classification of bacteria was done based on **morphology, flagellation, nutrition** and **spore formed**. Brief outline of the classifications are discussed along with the general characteristics of bacteria.

Bergey's Manual
(For classification and identification)

In the early part of 20th century David Hendricks Bergey (1860-1937) published a manual, known as **Bergey's Manual** which is a handy and standard book for the classification and identification of bacteria. The full name of the book is *Bergey's Manual of Determinative Bacteriology*. The first edition of the book was published in 1923, where the classification of bacteria was done based mainly on morphological characteristics. This was revised time to time as and when more informations came from researches at biochemical and molecular level. The 8th edition of the Manual was published in 1974, with thorough revision. Based on the prokaryotic structure, the bacteria were placed in a separate kingdom, Prokaryotae (also spelled as Procaryotae). Based on the cellular characteristics, the Phylum was further divided into two Divisions—Cyanobacteria and the Bacteria. The Division–Bacteria was further sub-divided into nineteen (19) groups.

An outline of the classification as adopted in 1974 edition of Bergey's Manual is given below.

Kingdom 1. Prokaryotae

Division 1. Cyanobacteria

These are single, simple or branched chain of prokaryotic cells having chlorophyll and phycobiliproteins and the electron donor is water in their photosynthetic process like higher plants, e.g., *Gloeocapsa, Dermocarpa, Oscillatoria*, etc.

Division 2. Bacteria

They are unicellular, characterised by a rigid cell wall of peptidoglycans. The photosynthetic pigment is bacteriochlorophyll, and the electron donor in the photosynthetic process is a substance other than water.

The division is further divided into nineteen (19) groups based on gram stain, photosynthetic efficiency, type of movement, etc. The groups are as follows : 1. Phototropic Bacteria, 2. Gliding Bacteria, 3. Sheathed Bacteria, 4. Budding Bacteria, 5. Spirochaetes, 6. Spiral and Curved Bacteria, 7. Gram-negative Aerobic Rods and Cocci, 8. Gram-negative Facultative Anaerobic Rods, 9. Gram-negative Anaerobic Bacteria, 10. Gram-negative Cocci and Cocco-bacilli, 11. Gram-negative Anaerobic Cocci, 12. Gram-negative Chemolithotropic Bacteria, 13. Methane-producing (Methanogenic) Bacteria, 14. Gram-positive Cocci, 15. Endospore forming Rods and Cocci, 16. Gram-positive Asporogenous Rod-shaped Bacteria, 17. Actinomycetes and Related Organisms, 18. Rickettsias, and 19. Mycoplasmas.

Later on, with the progress of knowledge, the book was retitled **"Bergey's Manual of Systematic Bacteriology"** and the 1st edition was brought out in four volumes in 1984, 1986, 1989 and 1991, respectively. Later on, the progress in prokaryotic taxonomy took place by studying the sequencing of DNA, rRNA and protein. The phylogenetic studies have also been done. Consequently, the 2nd edition of **"Bergey's Manual of Systematic Bacteriology"** was published. Thereby the 2nd edition of classification became largely phylogenetic rather than phenetic and it is quite different from the 1st edition. The 1st volume was published in 2001 and the others will be published sequentially in five (5) volumes.

Contents in different volumes are as follows:

Volume 1. The Archaea and the deeply branching and phototrophic bacteria.

Volume 2. The Proteobacteria.

Volume 3. The Low G + C Gram+ve Bacteria.

Volume 4. The High G + C Gram+ve Bacteria.

Volume 5. The Planctomycetes, Spirochaetes, Fibrobacteres, Bacteroidetes and Fusobacteria.

In this book the **Prokaryotes** are classified into 2 Domains — **Archaea** and **Bacteria**. The Domain Archaea is divided into 2 Phyla, 8 Classes and, on the other hand, Domain Bacteria is divided into 23 Phyla and 34 Classes. The classes are then divided into orders, families, genera and, finally, into species.

An outline of the 2nd edition of "**Bergey's Manual of Systematic Bacteriology**" (2001) is given below:

Domain *Archaea*

Phylum I. *Crenarchaeota*

Class I. *Thermoprotei*

Order I. *Thermoproteales*

Family I. *Thermoproteaceae*

Genus I. *Thermoproteus*

Genus II. *Caldivirga* etc.

Family II. *Thermofilaceae*

Genus I. *Thermofilum*

Order II. *Desulfurococcales*

Family I. *Desulfurococcaceae*

Genus I. *Desulfurococcus*

Genus II. *Acidolobus* etc.

Family II. *Pyrodictiaceae*

Genus I. *Pyrodictium*

Genus II. *Hyperthermus* etc.

Order III. *Sulfolobales*

Family I. *Sulfolobaceae*

Genus I. *Sulfolobus*

Genus II. *Acidianus* etc.

Phylum II. *Euryarchaeota*

Class I. *Methanobacteria*

Order I. *Methanobacteriales*

Family I. *Methanobacteriaceae*

Genus I. *Methanobacterium*

Genus II. *Methanobrevibacter* etc.

Family II. *Methanothermaceae*

Genus I. *Methanothermus*

Class II. *Methanococci*

Order I. *Methanococcales*

Family I. *Methanococcaceae*

Genus I. *Methanococcus*

Genus II. *Methanothermococcus*

Family II. *Methanocaldococcaceae*

Genus I. *Methanocaldococcus*

Genus II. *Methanotorris*

Order II. *Methanomicrobiales*

Family I. *Methanomicrobiaceae*

Genus I. *Methanomicrobium*

Genus II. *Methanoculleus* etc.

Family II. *Methanocorpusculaceae*

Genus I. *Methanocorpusculum*

Family III. *Methanospirillaceae*

Genus I. *Methanospirillum*

Genera incertae sedis

Genus I. *Methanocalculus*

Order III. *Methanosarcinales*

Family I. *Methanosarcinaceae*

Genus I. *Methanosarcina*

Genus II. *Methanococcoides* etc.

Family II. *Methanosaetaceae*

Genus I. *Methanosaeta*

Class III. *Halobacteria*

Order I. *Halobacteriales*

Family I. *Halobacteriaceae*

Genus I. *Halobacterium*

Genus II. *Haloarcula* etc.

Class IV. *Thermoplasmata*

Order I. *Thermoplasmatales*

Family I. *Thermoplasmataceae*

Genus I. *Thermoplasma*

Family II. *Picrophilaceae*

Genus I. *Picrophilus*

Family III. *Ferroplasmataceae*

Genus I. *Ferroplasma*

Class V. *Thermococci*

Order I. *Thermococcales*

Family I. *Thermococcaceae*

Genus I. *Thermococcus*

Genus II. *Palaeococcus* etc.

Class VI. *Archaeoglobi*

Order I. *Archaeoglobales*

Family I. *Archaeoglobaceae*

Genus I. *Archaeoglobus*
Genus II. *Ferroglobus* etc.

Class VII. *Methanopyri*
Order I. *Methanopyrales*
Family I. *Methanopyraceae*
Genus I. *Methanopyrus*

Domain *Bacteria*

Phylum I. *Aquificae*
Class I. *Aquificae*
Order I. *Aquificales*
Family I. *Aquificaceae*
Genus I. *Aquifex*
Genus II. *Calderobacterium* etc.
Genera incertae sedis
Genus I. *Desulfurobacterium*

Phylum II. *Thermotogae*
Class I. *Thermotogae*
Order I. *Thermotogales*
Family I. *Thermotogaceae*
Genus I. *Thermotoga*
Genus II. *Fervidobacterium* etc.

Phylum III. *Thermodesulfobacteria*
Class I. *Thermodesulfobacteria*
Order I. *Thermodesulfobacteriales*
Family I. *Thermodesulfobacteriaceae*
Genus I. *Thermodesulfobacterium*

Phylum IV. *"Deinococcus-Thermus"*
Class I. *Deinococci*
Order I. *Deinococcales*
Family I. *Deinococcaceae*
Genus I. *Deinococcus*
Order II. *Thermales*
Family I. *Thermaceae*
Genus I. *Thermus*
Genus II. *Meiothermus*

Phylum V. *Chrysiogenetes*
Class I. *Chrysiogenetes*
Order I. *Chrysiogenales*
Family I. *Chrysiogenaceae*
Genus I. *Chrysiogenes*

Phylum VI. *Chloroflexi*
Class I. *"Chloroflexi"*
Order I. "Chloroflexales"
Family I. "Chloroflexaceae"
Genus I. *Chloroflexus*
Genus II. *Chloronema* etc.
Family II. Oscillochloridaceae
Genus I. *Oscillochloris*
Order II. "Herpetosiphonales"
Family I. "Herpetosiphonaceae"
Genus I. *Herpetosiphon*

Phylum VII. *Thermomicrobia*
Class I. *Thermomicrobia*
Order I. *Thermomicrobiales*
Family I. *Thermomicrobiaceae*
Genus I. *Thermomicrobium*

Phylum VIII. *Nitrospira*
Class I. "Nitrospira"
Order I. "Nitrospirales"
Family I. "Nitrospiraceae"
Genus I. *Nitrospira*
Genus II. *Leptospirillum* etc.

Phylum IX. *Deferribacteres*
Class I. *Deferribacteres*
Order I. *Deferribacterales*
Family I. *Deferribacteraceae*
Genus I. *Deferribacter*
Genus II. *Denitrovibrio* etc.
Genera *incertae sedis*
Genus I. *Synergistes*

Phylum X. *Cyanobacteria*
Class I. *"Cyanobacteria"*
Subsection I.
Family I.
Form genus I. *Chamaesiphon*
Form genus II. *Chroococcus* etc.
Subsection II.
Family I.
Form genus I. *Cyanocystis*
Form genus II. *Dermocarpella* etc.

Family II.

Form genus I. *Chroococcidiopsis*

Form genus II. *Myxosarcina* etc.

Subsection III.

Family I.

Form genus I. *Arhrospira*

Form genus II. *Borzia* etc.

Subsection IV.

Family I.

Form genus I. *Anabaena*

Form genus II. *Anabaenopsis* etc.

Family II.

Form genus I. *Calothrix*

Form genus II. *Rivularia* etc.

Subsection V.

Family I.

Form genus I. *Chlorogloeopsis*

Form genus II. *Fischerella* etc.

Phylum XI. *Chlorobi*

Class I. *"Chlorobia"*

Order I. *Chlorobiales*

Family I. *Chlorobiaceae*

Genus I. *Chlorobium*

Genus II. *Ancalochloris* etc.

Phylum XII. *Proteobacteria*

Class I. "Alphaproteobacteria"

Order I. *Rhodospirillales*

Family I. *Rhodospirillaceae*

Genus I. *Rhodospirillum*

Genus II. *Azospirillum* etc.

Family II. *Acetobacteraceae*

Genus I. *Acetobacter*

Genus II. *Acidiplilium* etc.

Order II. *Rickettsiales*

Family I. *Rickettsiaceae*

Genus I. *Rickettsia*

Genus II. *Orientia* etc.

Family II. *Anaplasmataceae*

Genus I. *Anaplasma*

Genus II. *Aegyptianella* etc.

Family III. "Holosporaceae"

Genus I. *Holospora*

Genus II. *Caedibacter* etc.

Order III. "Rhodobacterales"

Family I. "Rhodobacteraceae"

Genus I. *Rhodobacter*

Genus II. *Ahrensia* etc.

Order IV. "Sphingomonadales"

Family I. *Sphingomonadaceae*

Genus I. *Sphingomonas*

Genus II. *Blastomonas* etc.

Order V. *Caulobacterales*

Family I. *Caulobacteraceae*

Genus I. *Caulobacter*

Genus II. *Asticcacaulis* etc.

Order VI. "Rhizobiales"

Family I. *Rhizobiaceae*

Genus I. *Rhizobium*

Genus II. *Agrobacterium* etc.

Family II. *Bartonellaceae*

Genus I. *Bartonella*

Family III. *Brucellaceae*

Genus I. *Brucella*

Genus II. *Mycoplana* etc.

Family IV. "Phyllobacteriaceae"

Genus I. *Phyllobacterium*

Genus II. *Allorhizobium* etc.

Family V. "Methylocystaceae"

Genus I. *Methylocystis*

Genus II. *Albibacter* etc.

Family VI. "Beijerinckiaceae"

Genus I. *Beijerinckia*

Genus II. *Chelatococcus* etc.

Family VII. "Bradyrhizobiaceae"

Genus I. *Bradyrhizobium*

Genus II. *Afipia* etc.

Family VIII. *Hyphomicrobiaceae*

Genus I. *Hyphomicrobium*

Genus II. *Ancalomicrobium* etc.

Family IX. "Methylobacteriaceae"

Genus I. *Methylobacterium*

Genus II. *Protomonas* etc.

Family X. "Rhodobiaceae"

Genus I. *Rhodobium*

Genus II. *Roseospirillum*

Class II. "Betaproteobacteria"

Order I. "Burkholderiales"

Family I. "Burkholderiaceae"

Genus I. *Burkholderia*

Genus II. *Cupriavidus* etc.

Family II. "Oxalobaceraceae"

Genus I. *Oxalobacter*

Genus II. *Duganella* etc.

Family III. *Alcaligenaceae*

Genus I. *Alcaligenes*

Genus II. *Achromobacter* etc.

Family IV. *Comamonadaceae*

Genus I. *Comamonas*

Genus II. *Acidovorax* etc.

Genera *incertae sedis*

Genus I. *Aquabacterium*

Genus II. *Ideonella* etc.

Order II. "Hydrogenophilales"

Family I. "Hydrogenophilaceae"

Genus I. *Hydrogenophilus*

Genus II. *Thiobacillus*

Order III. "Methylophilales"

Family I. "Methylophilaceae"

Genus I. *Methylophilus*

Genus II. *Methylobacillus* etc.

Order IV. "Neisseriales"

Family I. *Neisseriaceae*

Genus I. *Neisseria*

Genus II. *Alysiella* etc.

Order V. "Nitrosomonadales"

Family I. "Nitrosomonadaceae"

Genus I. *Nitrosomonas*

Genus II. *Nitrosolobus* etc.

Family II. *Spirillaceae*

Genus I. *Spirillum*

Family III. *Gallionellaceae*

Genus I. *Gallionella*

Order VI. "Rhodocyclales"

Family I. *"Rhodocyclaceae"*

Genus I. *Rhodocyclus*

Genus II. *Azoarcus* etc.

Order VII. *"Procabacteriales"*

Family I. *"Procabacteriaceae"*

Genus I. *"Procabacter"*

Class III. "Gammaproteobacteria"

Order I. "Chromatiales"

Family I. *Chromatiaceae*

Genus I. *Chromatium*

Genus II. *Allochromatium* etc.

Family II. *Ectothiorhodospiraceae*

Genus I. *Ectothiorhodospira*

Genus II. *Alcalilimnicola* etc.

Order II. *Acidithiobacillales*

Family I. *Acidithiobacillaceae*

Genus I. *Acidithiobacillus*

Family II. *Thermithiobacillaceae*

Genus I. *Thermithiobacillus*

Order III. "Xanthomonadales"

Family I. "Xanthomonadaceae"

Genus I. *Xanthomonas*

Genus II. *Frateuria* etc.

Order IV. "Cardiobacteriales"

Family I. *Cardiobacteriaceae*

Genus I. *Cardiobacterium*

Genus II. *Dichelobacter* etc.

Order V. "Thiotrichales"

Family I. "Thiotrichaceae"

Genus I. *Thiothrix*

Genus II. *Achromatium* etc.

Family II. "Piscirickettsiaceae"

Genus I. *Piscirickettsia*

Genus II. *Cycloclasticus* etc.

Family III. "Francisellacae"

Genus I. *Francisella*

Order VI. "Legionellales"

Family I. *Legionellaceae*

Genus I. *Legionella*

Family II. "Coxiellaceae"

Genus I. *Coxiella*

Genus II. *Rickettsiella*

Order VII. "Methylococcales"

Family I. *Methylococcaceae*

Genus I. *Methylococcus*

Genus II. *Methylobacter* etc.

Order VIII. "Oceanospirillales"

Family I. "Oceanospirillaceae"

Genus I. *Oceanospirillum*

Genus II. *Balneatrix* etc.

Family II. *Alcanivoraxaceae*

Genus I. *Alcanivorax*

Family III. "*Hahellaceae*"

Genus I. *Hahella*

Family IV. *Halomonadaceae*

Genus I. *Halomonas*

Genus II. *Carnimonas* etc.

Family V. *Oleiphilaceae*

Genus I. *Oleiphilus*

Order IX. *Pseudomonadales*

Family I. *Pseudomonadaceae*

Genus I. *Pseudomonas*

Genus II. *Azomonas* etc.

Family II. *Moraxellaceae*

Genus I. *Moraxella*

Genus II. *Acinetobacter* etc.

Order X. "Alteromonadales"

Family I. *Alteromonadaceae*

Genus I. *Alteromonas*

Genus II. *Alishewanella* etc.

Order XI. "Vibrionales"

Family I. *Vibrionaceae*

Genus I. *Vibrio*

Genus II. *Allomonas* etc.

Order XII. "Aeromonadales"

Family I. *Aeromonadaceae*

Genus I. *Aeromonas*

Genus II. *Oceanimonas* etc.

Family II. *Succinivibrionaceae*

Genus I. *Succinivibrio*

Genus II. *Anaerobiospirillum* etc.

Order XIII. "Enterobacteriales"

Family I. *Enterobacteriaceae*

Genus I. *Escherichia*

Genus II. *Alterococcus* etc.

Order XIV. "Pasteurellales"

Family I. *Pasteurellaceae*

Genus I. *Pasteurella*

Genus II. *Actinobacillus* etc.

Class IV. "Deltaproteobacteria"

Order I. "Desulfurellales"

Family I. "Desulfurellaceae"

Genus I. *Desulfurella*

Genus II. *Hippea*

Order II. "Desulfovibrionales"

Family I. "Desulfovibrionaceae"

Genus I. *Desulfovibrio*

Genus II. *Bilophila* etc.

Family II. "Desulfomicrobiaceae"

Genus I. "*Desulfomicrobium*"

Family III. "Desulfohalobiaceae"

Genus I. *Desulfohalobium*

Genus II. *Desulfomonas* etc.

Family IV. "*Desulfonatronumaceae*"

Genus I. *Desulfonatronum*

Order III. "Desulfobacterales"

Family I. "Desulfobacteraceae"

Genus I. *Desulfobacter*

Genus II. *Desulfobacterium* etc.

Family II. "*Desulfobulbaceae*"

Genus I. *Desulfobulbus*

Genus II. *Desulfocapsa* etc.

Family III. "*Nitrospiraceae*"

Genus I. *Nitrosrina*

Order IV. "Desulfuromonadales"

Family I. "Desulfuromonadaceae"

Genus I. *Desulfuromonas*

Genus II. *Desulfuromusa* etc.

Family II. "Geobacteraceae"

Genus I. *Geobacter*

Genus II. *Trichlorobacter*

Order V. "Syntrophobacterales"

Family I. "Syntrophobacteraceae"

Genus I. *Syntrophobacter*

Genus II. *Desulfacinum* etc.

Family II. "Syntrophaceae"

Genus I. *Syntrophus*

Genus II. *Desulfobacca* etc.

Order VI. "Bdellovibrionales"

Family I. "Bdellovibrionaceae"

Genus I. *Bdellovibrio*

Genus II. *Bacteriovorax* etc.

Order VII. *Myxococcales*

Suborder I. *Cystobacterinea*

Family I. *Cystobacteraceae*

Genus I. *Cystobacter*

Genus II. *Archangium* etc.

Family II. *Myxococcaceae*
Genus I. *Myxococcus*
Genus II. *Corallococcus* etc.
Suborder II. *Soranginea*
Family I. *Polyangiaceae*
Genus I. *Polyangium*
Genus II. *Byssophaga* etc.
Suborder III. *Nannocystinea*
Family I. *Nannocystacceae*
Genus I. *Nannocystis*
Family II. *Kofleriaceae*
Genus I. *Kofleria*

Class V. "Epsilonproteobacteria"

Order I. "Campylobacterales"
Family I. *Campylobacteraceae*
Genus I. *Campylobacter*
Genus II. *Arcobacter* etc.
Family II. "Helicobacteraceae"
Genus I. *Helicobacter*
Genus II. *Thiovulum* etc.
Family III. *"Nautilaceae"*
Genus I. *Nautilia*
Genus II. *Caminibacter*

Phylum XIII. *Firmicutes*

Class I. "Clostridia"

Order I. *Clostridiales*
Family I. *Clostridiaceae*
Genus I. *Clostridium*
Genus II. *Acetivibrio* etc.
Family II. "Lachnospiraceae"
Genus I. *Lachnospira*
Genus II. *Acetitomaculum* etc.
Family III. "Peptostreptococcaceae"
Genus I. *Peptostreptococcus*
Genus II. *Anaerococcus* etc.
Family IV. "Eubacteriaceae"
Genus I. *Eubacterium*
Genus II. *Acetobacterium* etc.
Family V. *Peptococcaceae*
Genus I. *Peptococcus*
Genus II. *Anaeroarcus* etc.
Family VI. "Heliobacteriaceae"
Genus I. *Heliobacterium*
Genus II. *Heliobacillus* etc.
Family VII. "Acidaminococcaceae"
Genus I. *Acidaminococcus*
Genus II. *Acetonema*
Family VIII. *Syntrophomonadaceae*
Genus I. *Syntrophomonas*
Genus II. *Acetogenium* etc.
Order II. "Thermoanaerobacteriales"
Family I. "Thermoanaerobacteriaceae"
Genus I. *Thermonaerobacterium*
Genus II. *Ammonifex* etc.
Order III. *Haloanaerobiales*
Family I. *Haloanaerobiaceae*
Genus I. *Haloanaerobium*
Genus II. *Halocella* etc.
Family II. *Halobacteroidaceae*
Genus I. *Halobacteroides*
Genus II. *Acetohalobium* etc.

Class II. *Mollicutes*

Order I. *Mycoplasmatales*
Family I. *Mycoplasmataceae*
Genus I. *Mycoplasma*
Genus II. *Eperythrozoom* etc.
Order II. *Entomoplasmatales*
Family I. *Entomoplasmataceae*
Genus I. *Entomoplasma*
Genus II. *Mesoplasma* etc.
Family II. *Spiroplasmataceae*
Genus I. *Spiroplasma*
Order III. *Acholeplasmatales*
Family I. *Acholeplasmataceae*
Genus I. *Acholeplasma*
Order IV. *Anaeroplasmatales*
Family I. *Anaeroplasmataceae*
Genus I. *Anaeroplasma*
Genus II. *Asteroleplasma*
Order V. *Incertae sedis*
Family I. *"Erysipelotrichaeceae"*
Genus I. *Erysipelothrix*
Genus II. *Bulleidia* etc.

Class III. "Bacilli"

Order I. *Bacillales*
Family I. *Bacillaceae*
Genus I. *Bacillus*
Genus II. *Amphibacillus* etc.

Family II. *"Alicyclobacillaceae"*
Genus I. *Alicyclobacillus*
Genus II. *Pasteuria* etc.
Family III. *Caryophanaceae*
Genus I. *Caryophanon*
Family IV. "Listeriaceae"
Genus I. *Listeria*
Genus II. *Brochothrix*
Family V. *"Paenibacillaceae"*
Genus I. *Paenibacillus*
Genus II. *Ammoniphilus* etc.
Family VI. *Planococcaceae*
Genus I. *Planococcus*
Genus II. *Filibacter* etc.
Family VII. *"Sporolactobacillaceae"*
Genus I. *Sporolactobacillus*
Genus II. *Marinococcus*
Family VIII. *"Staphylococcaceae"*
Genus I. *Staphylococcus*
Genus II. *Gemella* etc.
Family IX. *"Thermoactinomycetaceae"*
Genus I. *Thermoactinomyces*
Family X. *"Turicibacteraceae"*
Genus I. *Turicibacter*
Order II. *"Lactobacillales"*
Family I. *Lactobacillaceae*
Genus I. *Lactobacillus*
Genus II. *Paralactobacillus* etc.
Family II. *"Aerococcaceae"*
Genus I. *Aerococcus*
Genus II. *Abiotrophia* etc.
Family III. *"Carnobacteriaceae"*
Genus I. *Carnobacterium*
Genus II. *Agitococcus* etc.
Family IV. *"Enterococcaceae"*
Genus I. *Enterococcus*
Genus II. *Atopobacter* etc.
Family V. *"Leuconostocaceae"*
Genus I. *Leuconostoc*
Genus II. *Oenococcus* etc.
Family VI. *Streptococcaceae*
Genus I. *Streptococcus*
Genus II. *Lactococcus*
Family VII. *Incertae sedis*
Genus I. *Acetoanaerobium*
Genus II. *Oscillospira* etc.

Phylum XIV. *Actinobacteria*

Class I. *Actinobacteria*

Subclass I. *Acidimicrobidae*
Order I. *Acidimicrobiales*
Suborder IV. *"Acidimicrobineae"*
Family I. *Acidimicrobiaceae*
Genus I. *Acidimicrobium*
Subclass II. *Rubrobacteridae*
Order I. *Rubrobacterales*
Suborder V. *"Rubrobacterineae"*
Family I. *Rubrobacteraceae*
Genus I. *Rubrobacter*
Subclass III. *Coriobacteridae*
Order I. *Coriobacteriales*
Suborder VI. *"Coriobacterineae"*
Family I. *Coriobacteriaceae*
Genus I. *Coriobacterium*
Genus II. *Atopobium* etc.
Subclass IV. *Sphaerobacteridae*
Order I. *Sphaerobacterales*
Suborder VII. *"Sphaerobacterineae"*
Family I. *Sphaerobacteraceae*
Genus I. *Sphaerobacter*
Subclass V. *Actinobacteridae*
Order I. *Actinomycetales*
Suborder VIII. *Actinomycineae*
Family I. *Actinomycetaceae*
Genus I. *Actinomyces*
Genus II. *Actinobaculum* etc.
Suborder IX. *Micrococcineae*
Family I. *Micrococcaceae*
Genus I. *Micrococcus*
Genus II. *Arthrobacter* etc.
Family II. *Bogoriellaceae*
Genus I. *Bogoriella*
Family III. *Rarobacteraceae*
Genus I. *Rarobacter*
Family IV. *Sanguibacteraceae*
Genus I. *Sanguibacter*
Family V. *Brevibacteriaceae*
Genus I. *Brevibacterium*
Family VI. *Cellulomonadaceae*
Genus I. *Cellulomonas*
Genus II. *Oerskovia* etc.

Family VII. *Dermabacteraceae*
Genus I. *Dermabacter*
Genus II. *Brachybacterium*
Family VIII. *Dermatophilaceae*
Genus I. *Dermatophilus*
Family IX. *Dermacoccaceae*
Genus I. *Dermacoccus*
Genus II. *Demetria* etc.
Family X. *Intrasporangiaceae*
Genus I. *Intrasporangium*
Genus II. *Janibacter* etc.
Family XI. *Jonesiaceae*
Genus I. *Jonesia*
Family XII. *Microbacteriaceae*
Genus I. *Microbacterium*
Genius II. *Agreia* etc.
Family XIII. *"Beutenbergiaceae"*
Genus I. *Beutenbergia*
Genus II. *Georgenia* etc.
Family XIV. *Promicromonosporaceae*
Genus I. *Promicromonospora*
Genus II. *Cellulosimicrobium*
Suborder X. *Corynebacterineae*
Family I. *Corynebacteriaceae*
Genus I. *Corynebacterium*
Family II. *Dietziaceae*
Genus I. *Dietzia*
Family III. *Gordoniaceae*
Genus I. *Gordonia*
Genus II. *Skermania*
Family IV. *Mycobacteriaceae*
Genus I. *Mycobacterium*
Family V. *Nocardiaceae*
Genus I. *Nocardia*
Genus II. *Rhodococcus*
Family VI. *Tsukamurellaceae*
Genus I. *Tsukamurella*
Family VII. *"Williamsiaceae"*
Genus I. *Williamsia*
Suborder XI. *Micromonosporineae*
Family I. *Micromonosporaceae*
Genus I. *Micromonospora*
Genus II. *Actinoplanes* etc.
Suborder XII. *Propionibacterineae*
Family I. *Propionibacteriaceae*
Genus I. *Propionibacterium*
Genus II. *Luteococcus* etc.
Family II. *Nocardioidaceae*
Genus I. *Nocardioides*
Genus II. *Aeromicrobium* etc.
Suborder XIII. *Pseudonocardineae*
Family I. *Pseudonocardiaceae*
Genus I. *Pseudonocardia*
Genus II. *Actinoalloteichus* etc.
Family II. *Actinosynnemataceae*
Genus I. *Actinosynnema*
Genus II. *Actinokineospora* etc.
Suborder XIV. *Streptomycineae*
Family I. *Streptomycetaceae*
Genus I. *Streptomyces*
Genus II. *Kitasatospora* etc.
Suborder XV. *Streptosporangineae*
Family I. *Streptosporangiaceae*
Genus I. *Streptosporangium*
Genus II. *Acrocarpospora* etc.
Family II. *Nocardiopsaceae*
Genus I. *Nocardiopsis*
Genus II. *Streptomonospora* etc.
Family III. *Thermomonosporaceae*
Genus I. *Thermomonospora*
Genus II. *Actinomadura* etc.
Suborder XVI. *Frankineae*
Family I. *Frankiaceae*
Genus I. *Frankia*
Family II. *Geodermatophilaceae*
Genus I. *Geodermatophilus*
Genus II. *Blastococcus* etc.
Family III. *Microsphaeraceae*
Genus I. *Microsphaera*
Family IV. *Sporichthyaceae*
Genus I. *Sporichthya*
Family V. *Acidothermaceae*
Genus I. *Acidothermus*
Family VI. *"Kineosporiaceae"*
Genus I. *Kineosporia*
Genus II. *Cryptosporangium* etc.
Suborder XVII. *Glycomycineae*
Family I. *Glycomycetaceae*
Genus I. *Glycomyces*

Order II. *Bifidobacteriales*
Family I. *Bifidobacteriaceae*
Genus I. *Bifidobacterium*
Genus II. *Falcivibrio* etc.
Family II. Unknown Affiliation
Genus I. *Actinobispora*
Genus II. *Actinocorallia* etc.

Phylum XV. *Planctomycetes*
Class I. *"Planctomycetacia"*
Order I. *Planctomycetales*
Family I. *Planctomycetaceae*
Genus I. *Planctomyces*
Genus II. *Gemmata* etc.

Phylum XVI. *Chlamydiae*
Class I. *"Chlamydiae"*
Order I. *Chlamydiales*
Family I. *Chlamydiaceae*
Genus I. *Chlamydia*
Genus II. *Chlamydophila*
Family II. *Parachlamydiaceae*
Genus I. *Parachlamydia*
Genus II. *Neochlamydia*
Family III. *Simkaniaceae*
Genus I. *Simkania*
Family IV. *Waddliaceae*
Genus I. *Waddlia*

Phylum XVII. *Spirochaetes*
Class I. *"Spirochaetes"*
Order I. *Spirochaetales*
Family I. *Spirochaetaceae*
Genus I. *Spirochaeta*
Genus II. *Borrelia* etc.

Phylum XVIII. *Fibrobacteres*
Class I. *"Fibrobacteres"*
Order I. *"Fibrobacterales"*
Family I. *"Fibrobacteraceae"*
Genus I. *Fibrobacter*

Phylum XIX. *Acidobacteria*
Class I. *"Acidobacteria"*
Order I. *"Acidobacteriales"*
Family I. *"Acidobacteriaceae"*
Genus I. *Acidobacterium*
Genus II. *Geothrix* etc.

Phylum XX. *Bacteroidetes*
Class I. *"Bacteroides"*
Order I. *"Bacteroidales"*
Family I. *Bacteroidaceae*
Genus I. *Bacteroides*
Genus II. *Acetofilamentum* etc.
Family II. *"Rikenellaceae"*
Genus I. *Rikenella*
Genus II. *Marinilabilia* etc.
Family III. *"Porphyromonadaceae"*
Genus I. *Porphyromonas*
Genus II. *Dysgonomonas* etc.
Family IV. *"Prevotellaceae"*
Genus I. *Prevotella*

Class II. *"Flavobacteria"*
Order I. *"Flavobacteriales"*
Family I. *Flavobacteriaceae*
Genus I. *Flavobacterium*
Genus II. *Arenibacter* etc.
Family II. *"Myroidaceae"*
Genus I. *Myroides*
Genus II. *Psychromonas*
Family III. *"Blattabacteriaceae"*
Genus I. *Blattabacterium*

Class III. *"Sphingobacteria"*
Order I. *"Sphingobacteriales"*
Family I. *Sphingobacteriaceae*
Genus I. *Sphingobacterium*
Genus II. *Pedobacter*
Family II. *"Saprospiraceae"*
Genus I. *Saprospira*
Genus II. *Haliscomenobacter* etc.
Family III. *"Flexibacteraceae"*
Genus I. *Flexibacter*
Genus II. *Cyclobacterium* etc.
Family IV. *"Flammeovirgaceae"*
Genus I. *Flammeovirga*
Genus II. *Flexithrix* etc.
Family V. *Crenotrichaceae*
Genus I. *Crenthrix*
Genus II. *Chitinophaga* etc.

Phylum XXI. *Fusobacteria*

Class I. *"Fusobacteria"*

Order I. *"Fusobacteriales"*

Family I. *"Fusobacteriaceae"*

Genus I. *Fusobacterium*

Genus II. *Ilyobacter* etc.

Family II. Incertae sedis

Genus I. *Cetobacterium*

Phylum XXII. *Verrucomicrobia*

Class I. *Verrucomicrobiae*

Order I. *Verrucomicrobiales*

Family I. *Verrucomicrobiaceae*

Genus I. *Verrucomicrobium*

Genus II. *Prosthecobacter*

Family II. *"Opitutaceae"*

Genus I. *Opitutus*

Family III. *"Xiphinematobacteriaceae"*

Genus I. *Xiphinematobacter*

Phylum XXIII. *Dictyoglomus*

Class I. *"Dictyoglomi"*

Order I. *"Dictyglomales"*

Family I. *"Dictyoglomaceae"*

Genus I. *Dictyoglomus*

Table 2.8 : Brief outline on the organisation of Bergey's Manual of Systematic Bacteriology with (2nd edition, 2001) representative genera

Taxonomic Rank	Representative Genera
Volume 1. *The Archaea and the Deeply Branching and Phototrophic Bacteria*	
Domain Archaea	
Phylum *Crenarchaeota*	*Thermoproteus, Pyrodictium, Sulfolobus*
Phylum *Euryarchaeota*	
Class I. *Methanobacteria*	*Methanobacterium*
Class II. *Methanococci*	*Methanococcus*
Class III. *Halobacteria*	*Halobacterium, Halococcus*
Class IV. *Thermoplasmata*	*Thermoplasma, Ferroplasma, Picrophilus,*
Class V. *Thermococci*	*Thermococcus, Pyrococcus*
Class VI. *Archaeoglobi*	*Archaeoglobus*
Class VII. *Methanopyri*	*Methanopyrus*
Domain *Bacteria*	
Phylum *Aquificae*	*Aquifex, Hydrogenobacter*
Phylum *Thermotogae*	*Thermotoga*
Phylum *Thermodesulfobacteria*	*Thermodesulfobcterium*
Phylum "Deinococcus-Thermus"	*Deinococcus, Thermus*
Phylum *Chrysiogenetes*	*Chrysogenes*
Phylum *Chloroflexi*	*Chloroflexus, Herpetosiphon*
Phylum *Thermomicrobia*	*Thermomicrobium*
Phylum *Nitrospira*	*Nitrospira*
Phylum *Deferribacteres*	*Geovibrio*
Phylum *Cyanobacteria*	*Prochloron, Oscillatoria, Synechococcus, Pleurocapsa, Anabaena, Stigonema, Nostoc*
Phylum Chlorobi	*Chlorobium, Pelodictyon*

Contd.

Contd. Table 2.8

Taxonomic Rank	Representative Genera
Volume 2. *The Proteobacteria*	
Phylum *Proteobacteria*	
Class I. Alphaproteobacteria	*Rhodospirillum, Caulobacter, Rickettsia, Rhizobium, Nitrobacter, Brucella, Beijerinckia, Methylobacterium, Hyphomicrobium*
Class II. Betaproteobacteria	*Neisseria, Alcaligenes, Burkholderia, Comamonas, Nitrosomonas, Methylophilus, Thiobacillus*
Class III. Gammaproteobacteria	*Chromatium, Legionella, Leucothrix, Pseudomonas, Azotobacter, Vibrio, Escherichia, Proteus, Klebsiella, Salmonella, Yersinia, Shigella, Haemophilus*
Class IV. Deltaproteobacteria	*Desulfovibrio, Myxococcus, Bdellovibrio, Polyangium*
Class V. Epsilonproteobacteria	*Helicobacter, Campylobacter*
Volume 3. *The Low G + C Gram-positive Bacteria*	
Phylum *Firmicutes*	
Class I. Clostridia	*Clostridium, Eubacterium, Peptostreptococcus, Desulfotomaculum, Heliobacterium, Veillonella*
Class II. Mollicutes	*Mycoplasma, Ureaplasma, Acholeplasma, Spiroplasma*
Class III. Bacilli	*Bacillus, Paenibacillus, Caryophanon, Thermoactinomyces, Lactobacillus, Streptococcus, Enterococcus, Listeria, Staphylococcus, Leuconostoc*
Volume 4. *The High G + C Gram-positive Bacteria*	
Phylum *Actinobacteria*	
Class *Actinobacteria*	*Actinomyces, Arthrobacter, Micrococcus, Corynebacterium, Mycobacterium, Nocardia, Actinoplanes, Streptomyces, Propionibacterium, Thermomonospora, Frankia, Actinomadura, Bifidobacterium*
Volume 5. *The Planctomycetes, Spirochaetes, Fibrobacteres, Bacteriodetes and Fusobacteria*	
Phylum *Planctomycetes*	*Planctomyces, Gemmata*
Phylum *Chlamydiae*	*Chlamydia*
Phylum *Spirochaetes*	*Spirochaeta, Leptospira, Treponema*
Phylum *Fibrobacteres*	*Fibrobacter*
Phylum *Acidobacteria*	*Acidobacterium*
Phylum *Bacteroidetes*	*Bacteroides, Prevotella, Porphyromonas, Flavobacterium, Flexibacter, Sphingobacterium, Cystophaga*
Phylum *Fusobacteria*	*Fusobacterium, Streptobcillus*
Phylum *Verrucomicrobia*	*Verrucomicrobium*
Phylum *Dictyoglomus*	*Dictyoglomus*

CHARACTERS OF THE DIFFERENT GROUPS OF PROKARYOTES

Domain: Archaea

The important characters of **Domain Archaea** (Gr. *archaios*, ancient) are:

Important characters

1. Members of the Domain Archaea i.e., **Archebacteria** can live in different habitats like
   (i) Extreme acidic and hot (above 100°C) environment,
   (ii) Extreme saline condition, and
   (iii) In anaerobic condition.
2. They may be spherical, rod-shaped, spiral, lobed, plate-shaped, and irregularly shaped or pleomorphic.
3. Some members are unicellular, but others form filaments or aggregations.
4. Cells are 0.1 μm to more than 1.5 μm in diameter and some filaments are about 200 μm in length.
5. Cell wall composed of pseudomurine, polysaccharide or glycoproteins and other proteins (*none have muramic acid and D-amino acids, characteristic of bacterial peptidiglycan*).
6. They are either Gram-positive or Gram-negative in nature.
7. The membrane lipids have branched chain hydrocarbons connected to glycerol by ether links.
8. They have a covalently linked closed circular genome and its size is generally smaller than other prokaryotes. The size of genome ranges between 0.8–1.1 × 10^9 daltons.
9. G+C content of DNA varies between 21-68 moles %.
10. Ribosomes are of 70S type.
11. Reproduction takes place by binary fission, budding, fragmentation, etc.
12. Nutritionally, they range from chemolitho-autotrophs to organotrophs.

They share the characters of both Prokaryotes and Eukaryotes:

A. **The Prokaryotic characters are:**
   (i) **About 1 μm in diameter.**
   (ii) **Lack of membrane–bound organelles.**
   (iii) **Nucleic acid is not bounded by nuclear membrane.**
   (iv) **Ribosome 70S type.**

B. **The Eukaryotic characters:**
   (i) **Presence of cell wall but devoid of peptidoglycan.**
   (ii) **Mechanism of protein synthesis and structure of protein.**
   (iii) **Genes have introns (present occasionally).**

C. **Unique character i.e., character of their own: Members contain branched chain lipids with ether, these enable them to tolerate extreme pH and heat.**

The Domain Archaea is divided into two Phyla — Crenarchaeota and Euryarchaeota, based on rRNA sequences:

1. *Crenarchaeota*. The group includes thermophilic and hyperthermophilic, generally sulphur metabolising organisms.

2. *Euryarchaeota*. The group includes primarily methanogenic, halophilic, thermophilic and also sulphur reducing organisms.

This group of bacteria lives in such conditions that existed as far back as Pre-Cambrian Era i.e., before 3-4 million years. Thus, they are called ancient group.

Domain: Bacteria

All prokaryotes, except the members of Archaea, are included under the Domain Bacteria. Bacteria of this group are also called **eubacteria**. The group has been classified on the basis of 16S rRNA homology. The important characters of this Domain are:

Important Characters

1. They are available in all possible habitat.
2. Mode of nutrition is mostly heterotrophic, where some may be parasitic, saprophytic or

symbiotic; and others are autotrophic due to the presence of photosynthetic pigment, the bacteriochlorophyll.

3. They are unicellular, having cell wall composed of three main constituents (i) N-acetyl glucosamine, (ii) N-acetyl muramic acid, and (iii) a peptide chain.
4. Inside the wall, a semipermeable cytoplasmic membrane is present, composed of double layer of phospholipid molecules.
5. The mesosome, which forms by the localised infolding of cytoplasmic membrane, functions as mitochondria of higher plant.
6. Bacteriochlorophyll — if present — is located in the involuted cytoplasmic membrane.
7. The DNA reticulum is not surrounded by membrane.
8. Profuse number of ribosomes are present with low rate of sedimentation than eukaryotes.
9. They also contain the extrachromosomal hereditary material, the plasmid.
10. The absence of endoplasmic reticulum and golgi apparatus.
11. The common method of multiplication is binary fission.
12. In the true sense, sexual reproduction is absent, but genetic recombination takes place by conjugation, transformation and transduction.
13. Some bacteria possess one or more flagella at one or more points, composed of several chains of flagellin (a protein molecule).
14. Gram-negative bacteria bear very small cytoplasmic appendages known as Pilli, composed of pilin, a protein.

Table 2.9: Differences between Archaea and Bacteria

Characters	Archaea	Bacteria
A. Growth		
(a) At 100°C or above	+	—
(b) In hypersaline solution (1.5 M NaCl or above)	+	—
B. Cell wall		
(a) Muramic acid	—	+
(b) Cross-linked by	L-amino acids	D-amino acids
C. Membrane lipids	Alcohols bound to glycerol by ether linkage	Fatty acids bound to glycerol by ester linkage
D. Protein synthesis		
(a) Chain initiating amino acid	Methionine	Formylmethionine
(b) Inhibitors		
(i) Kanamycin	—	+
(ii) Chloramphenicol	—	+
(iii) Anisomycin	+	—
(iv) Diphtheria toxin (inhibition function of EF-2, the chain elongation factor)	+	—
E. DNA-dependent RNA polymerase		
(a) Number of enzymes	One	Several
(b) Structure	Consists of 8–12 subunits	Consists of 4 subunits
F. Metabolic activity		
(a) Methanogenesis	+	—
(b) Similar ATPase	+	—
(c) Photosynthetic ability	—	+

Phylum. Proteobacteria (Enterobacteria)

Important characters

1. They are the largest and diverse group of bacteria.
2. Members of this group are sometimes called purple bacteria, because of the inclusion of purple photosynthetic bacteria in several of its sub-groups.
3. Members show variation in morphology, metabolism and reproduction.
4. They range from cocci, rod to others with prosthecae, buds and even fruit bodies.
5. Bacteria are of Gram-negative type.
6. Physiologically they may be photoautotrophs, chemolithotrophs, or chemoheterotrophs, etc.

Based on 16S rRNA sequence, the phylum is divided into the following 5 classes:

(a) Alphaproteobacteria (contains 6 orders and 18 families), e.g., *Rhizobium, Nitrococcus,* etc.

(b) Betaproteobacteria (contains 6 orders and 12 families), e.g., *Nitrosomonas, Thiobacillus,* etc.

(c) Gammaproteobacteria (contains 12 orders and 20 families), e.g., *Escherichia, Salmonella,* etc.

(d) Deltaproteobacteria (contains 7 orders and 17 families), e.g., *Desulphovibrio, Desulfuromonas,* etc.

(e) Epsilonproteobacteria (contains 1 orders and 2 families), e.g., *Campylobacter, Helicobacter,* etc.

Low G+C Gram-positive Bacteria

Phylum. Firmicutes

This group includes the Gram-positive bacteria with low G+C content in their DNA. Most of the members are heterotrophic and Gram-positive. Because of low G+C content, the mycoplasmas are also included in this group though they lack cell wall and are Gram-negative in nature. Many are endospore forming. They show variation in morphology such as cocci, rods and mycoplasmas are pleomorphic.

Class 1. Clostridia

The class contains three orders and eleven families. The members of this group are anaerobic in nature. Important characters of this group are given below:

Important characters

1. The members show much variation in the morphology and size.
2. Members are anaerobic in nature.
3. Some genera produce endospore (*Clostridium, Sporohalobacter* etc.), but many others are non-sporing type.

Clostridium is one of the largest bacterial genera.

Class 2. Mollicutes

The members of the class are commonly known as **mycoplasmas**. The important characters of the class are given below:

Important characters

1. Shape of the members of this group varies from cocci, rods, ring-shaped to filamentous or without any particular shape.
2. Diameter of the cells varies from 0.15 to 0.3 μm, thereby considered as smallest free-living bacteria.
3. Cells are surrounded by a triple-layered double membrane containing sterols.
4. Bacteria of this group cannot synthesize peptidoglycan precursors, the main component of bacterial cell wall and thus lack cell wall.
5. Due to absence of cell wall, they are highly plastic and pleomorphic.
6. They are Gram-positive in nature.
7. Members have genome size ranging from 400 to 1,000 megadaltons. 400 megadaltons is the smallest genome found in *Mycoplasma genitalium,* a human pathogen.
8. G+C content of DNA varies between 23 and 40 moles %.
9. Cells multiply by binary fission like other bacteria, but genomic replication is not

synchronised with cytoplasmic division which lags behind.

10. Cells are devoid of flagella and they are generally non-motile. Some show gliding movement (*Mycoplasma gallisepticum, M. pneumoniae*) or flexous motility (*Spiroplasma*).

e.g., *Mycoplasma, Spiroplasma, Anaeroplasma*, etc.

Class 3. Bacilli

It is a large class consisting of a wide variety of Gram-positive aerobes and facultative anaerobes. Important characters are given below:

Important characters

1. Members are Gram-positive in nature.
2. G+C content is 50 moles % or less.
3. The class includes cocci, endospore forming rod and non-sporing rods, and also other forms.

The class has been divided into orders, Bacillales (includes 9 families) and Lactobacillales (includes 6 families).

Order: Bacillales

Important characters

1. Members include different shapes like cocci, rods, mycelial form etc.
2. Both endospore forming and non-sporing rods are present.
3. They may be non-motile or motile. When motile they are with peritrichous flagella.
4. They are catalase positive.
5. They are aerobic to facultative anaerobic.
6. They are commonly saprophytic, but a few are pathogenic. e.g., *Bacillus, Staphylococcus*, etc.

Order: Lactobacillales

Important characters

1. Members include cocci and non-sporing rods.
2. Cells are non-motile.
3. They are catalase negative.
4. Members are anaerobes and facultative anaerobes, e.g., *Lactobacillus, Lactococcus, Streptococcus, Leuconostoc* etc.

High G+C Gram-positive Bacteria

Phylum. Actinobacteria

This group includes the members with diverse morphological assemblage having pathogenic and economically important members. The important characters of the phylum are given below:

Important characters

1. Morphologically they are of various types ranging from cocci, rods, rods with rudimentary branching to elaborate mycelial forms.
2. They are of Gram-positive type.
3. G+C content of DNA is 50 moles % or more, designated as high G+C group.
4. The Phylum contains one class (also called by the same name as Actinobacteria) and six orders.

The phylum includes both useful and harmful bacteria.

Useful members:

(i) Many species of *Streptomyces* are useful in the commercial production of antibiotics.

(ii) *Propiniobacterium* and *Brevibacterium* are useful in dairy industry.

(iii) *Frankia* spp. are able to infect many *non-leguminous plants* like *Casuarina, Myrica, Alnus, Coriaria* etc., and *produce nitrogen-fixing root nodules.*

Harmful members:

(i) *Mycobacterium leprae* causes leprosy.

(ii) *Corynebacterium diphtheriae* causes diphtheria in humans.

(iii) *Actinomyces israelii* causes abscesses of connective tissues in man.

Gram-negative Bacteria

Phylum. Planctomycetes

The Phylum Planctomycetes includes only one class, one order and four genera.

Important characters

1. They are spherical or oval, budding Gram-negative bacteria.
2. Cell walls have distinctive crateriform structure or pits, devoid of peptidoglycan.
3. Nuclear body is membrane bounded (*Gemmata obscuriglobus* and *Pirullela*).
4. The genus *Planctomyces* is attached to surface through stalk and hold fast, which is absent in other genera, e.g., *Planctomyces, Gemmata, Pirullela* etc.

Phylum. Chlamydiae

Earlier the Chlamydiae and rickettsias were placed together based on the fact that both are intercellular obligate parasite and Gram-negative in nature. Based on 16S rRNA data, rickettsias have been separated out and placed in the Phylum α-Proteobacteria. The important characters of the phylum Chlamydiae are given below:

Important characters

1. Cells of Chlamydiae are very small, coccoid in shape (0.2 to 1.5μ), non-motile and Gram-negative in nature.
2. Cell wall is devoid of muramic acid and peptidoglycan.
3. The size of genome is 4 to 6 x 10^8 daltons, which is very small and thus they have limited metabolic capacity.
4. G+C content of cells ranges between 41 to 44 moles %.
5. They are unable to synthesise ATP from carbohydrate or other substances and thus they depend absolutely on host cell for energy, called **energy-parasite**.
6. They have an unique membrane-located translocase to receive ATP from the host cell.
7. They are obligate parasite and not able to grow in culture medium.
8. Infection to host cell takes place by an elementary body (EB), a small spherical structure of about 0.3 to 0.4 μm in size.
9. They are obligatory intracellular parasites causing diseases of mammals including human beings, birds, spiders etc. e.g., *Chlamydia*.

Three species of *Chlamydia* are important pathogens of human beings and other animals. These are:

(a) *C. psittaci* causes psittacosis in humans. It also infects many other animals like parrot, cats, cattle, sheep etc. and invades respiratory, intestinal, genital tracts and also other parts.

(b) *C. trachomatis* causes trachoma, non-gonococcal urethritis and other diseases of humans.

(c) *C. pneumoniae* causes pneumonia of humans.

Phylum. Spirochaetes

The Spirochaetes (Gr. *spira* — a coil and *chaete* — hair) are distinguished from others by their flexible spiral cells and characteristic motility. The important characters of this phylum are given below:

Important characters

1. Cells of Spirochaetes are spiral, long and flexible, they give the characteristic flexuous movement. They also exhibit creeping or crawling movement in contact with solid surface.
2. Cells are thin and measure 0.1 to 3.0 μm × 5.0 to 250 μm.
3. The characteristic movement of the cells is due to the presence of an unique type of flagellar complex, the axial filament, consists of 2-more than 100 prokaryotic flagella extending from the poles and lying inside the outer membrane.
4. They are Gram-negative and chemoheterotrophic in nature.
5. G+C content of cells ranges between 25–65 moles %.
6. Most of the members grow in aquatic (both fresh water and marine) habitat (*Spirochaeta, Leptospira* etc.) and a few cause disease in human beings (*Treponema pallidum* causes syphilis, *Borrelia burgdorferi* causes lyme disease). e.g., *Treponema, Leptospira, Borrelia,* etc.

Phylum. Bacteroidetes

The Phylum is divided into three classes: *Bacteroides, Flavobacteria* and *Sphingobacteria*. Some important and well-known genera are *Flexibacter* (gliding bacteria), *Flavobacterium, Bacteroides* and *Cytophaga*.

ECOLOGY AND DISTRIBUTION

Bacteria (sing., Bacterium) were among the first life forms to appear on Earth, and are present almost everywhere in air, water, soil, food and in the bodies of plants and animals including human beings. Bacteria inhabit soil, water (fresh and saline water), acidic hot springs, radioactive waste, and the deep portions of Earth's crust. Bacteria also live in symbiotic and parasitic relationships with both plants and animals. Still, most bacteria have not been characterized, and only about half of the bacterial phyla have species that can be grown in the laboratory.

Almost 40 million bacterial cells are present in a gram of soil and a million bacterial cells in a milliliter of fresh water. There are approximately 5×10^{30} bacteria on Earth, forming a biomass which is more than that of all plants and animals.

GENERAL CHARACTERISTICS

Size: The bacterial cells vary greatly in their size. The average diameter ranges from 0.5 μm to 2.0 μm. The size of bacteria also varies according to shape. The cocci measures 0.5 μm to 1.25 μm in diameter. The bacillus or rod shaped bacterium measures 0.5–1.0 μm × 2–3 μm. The helical or spiral bacteria are larger in size, about 1.5 μm in diameter and up to 15 μm in length. Recently some bacteria have been identified which are much larger than the common ones. *Epulopiscium fishelsohnii* (200 μm × 80 μm) and *Thiomargarita namibiensis* (750 μm diameter) are so large that they can be visible to the naked eye.

Morphology (shape): The bacterial cells vary greatly in their shape (Fig. 2.9). All the individuals of a species have almost the same shape. True bacteria are unicellular. In some forms, a number of cells are grouped together and covered by mucilage layer. Based on morphology, the true bacteria are **classified** or divided into the following three main groups:

1. **Spherical** or **Coccus bacteria**: The ellipsoidal or spherical bacteria are called cocci (singular coccus). The cocci measure 0.5–1.25 μm in diameter. The cocci are without flagella (atrichous), thereby non-motile. They may occur singly or in groups in different orientation. Based on the number of cells and its arrangement in an aggregation, the cocci are of different types (Fig. 2.9A–F):

   (a) Micrococci. When the coccus occurs singly, it is called micrococcus; e.g., *Micrococcus nigra, M. lutens, M. cerolíticus.*

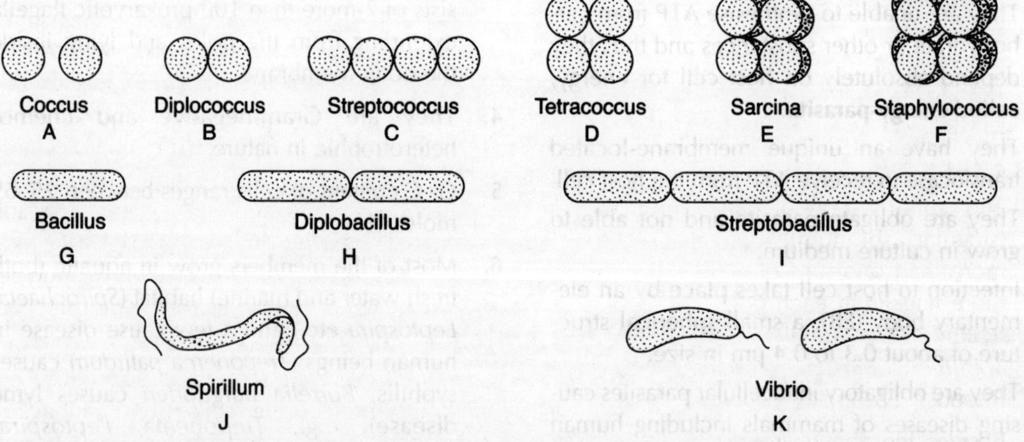

Fig. 2.9: Different shapes of bacteria. A-F: Different cocci, G-I: Different bacilli, J-K: Helical bacteria

(b) Diplococci. When cocci occur in pairs, they are called diplococci, e.g., *Diplococcus pneumoniae*.

(c) Streptococci. When cocci occur in long chain, they are called *streptococci*, e.g., *Streptococcus lactis, S. pyogenes*.

(d) Tetracocci. When cocci form a group of four cells, they are called tetracocci, e.g., *Tetracoccus cechil, Gaffkya tetragena, Pedicoccus cerevisiae*.

(e) Sarcinae. When spherical bacteria are arranged like a cube, made up of eight or more cells, they are called sarcinae, e.g., *Sarcina verticuli, S. lutea*.

(f) Staphylococci. When spherical bacteria form an irregular group, they are called staphylococci, e.g., *Staphylococcus aureus, S. albus*.

2. **Rod-shaped** or **Bacilli bacteria**: The bacterial cells are rod-shaped, cylindrical or rod-like, called bacilli (singular, bacillus). They may be motile or non-motile. The rods may be very short or long, narrow and have blunt or round ends. They may occur either singly or in groups (Fig. 2.9G–I):

   (a) Bacillus. When rod-shaped bacteria occur singly, they are called bacilli, e.g., *Bacillus polymyxa, B. anthracis, Lactobacillus*.

   (b) Diplobacillus. When rod-shaped bacteria occur in pairs, they are called diplobacilli, e.g., *Corynebacterium diphtheriae*.

   (c) Streptobacillus. When bacilli occur in chain, they are called streptobacilli, e.g., *Bacillus cereus, B. tuberculosis*.

3. **Spiral bacteria**: The cells are slightly larger and spirally coiled rods, called spirilli (singular, spirillum). Each bacterium has more than one turn of a helix and one or more flagella at each pole (Fig. 2.9J). They occur either singly or in chain, e.g., *Spirillum minus, S. volutans, S. undulum, Rhodospirillum, Microspora*.

Besides the above groups, following other shapes of bacteria are also present:

A. **Vibrios**. The cylindrical cell is curved and looks like the sign of comma (,) with a single flagellum at its tip, called "comma bacterium" (Fig. 2.9K). It measures about 10 µm × 1.5–1.7 µm, e.g., *Vibrio cholerae, V. coli*.

B. **Filamentous**. Some bacteria are filament-like, e.g., *Beggiatoa, Thiothrix*.

C. **Pleomorphic**. Some are able to change their shape and size in response to variation in the surrounding environment. *Acetobactor* may occur as single rod (bacillus) or chain of small rods (*Streptobacillus*), in response to environmental variation.

Cell Structure

It became possible to know the detailed structure of bacterial cell only after the discovery of electron microscope and various staining techniques. A bacterial cell (Fig. 2.10) shows a typical prokaryotic structure. The cytoplasm is enclosed by three layers, the outermost **slime** or **capsule**, the middle **cell wall** and **inner cell membrane.** The major cytoplasmic contents are nucleoid, plasmid, ribosome, mesosome etc., and the cell is devoid of endoplasmic reticulum, mitochondria, centrosome and golgi bodies.

A. Slime layer, Capsule and Glycocalyx

An amorphous viscid secretion of bacterial cell is present as a loose undemarcated region outside the cell, called *slime layer* (e.g., *Leuconostoc*). But, when it originates as a sharply defined structure outside the cell wall, it is called *capsule* (e.g., *Pneumococcus*). The capsule is about 0.2 µm in width. The capsules those are much narrower than true capsule, called *microcapsule* (e.g., *Haemophilus influenzae*). Both the layers usually consist of polysaccharide and occasionally polypeptide. The capsule contains 2% solid and 98% water. The solid portion is comprised of complex polysaccharide (e.g., *Pneumococcus, Enterobacter*) or polypeptide (e.g., anthrax bacillus) or hyaluronic acid (e.g., *Streptococcus*).

Both the layers serve as a protective covering and protect the cells from antibacterial substances like bacteriophage, enzymes etc. They enhance the virulence of bacteria.

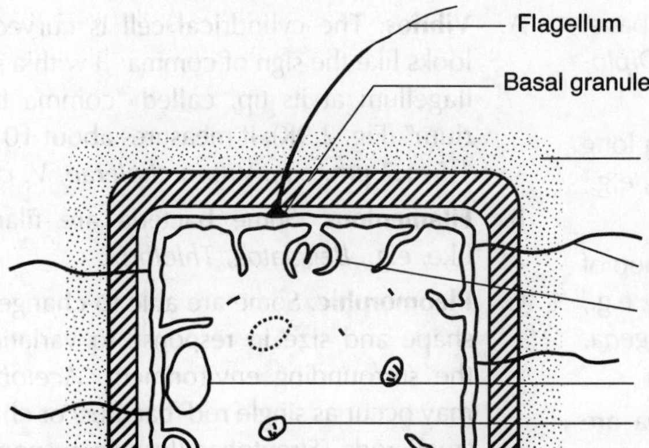

Flagellum
Basal granule
Slime layer (Capsule)
Pilus
Mesosome
Granule of poly β-hydroxy butyric acid
Chromatophore
Nucleoid
Ribosome
Cell membrane
Inclusion body (lipid, starch, sulphur etc.)
Cell wall
Periplasmic space
Cytoplasm

Fig. 2.10: Bacterial cell (a composite diagram-showing various structures)

The glycocalyx (Fig. 2.11) is an extracellular, hygroscopic network of mostly polysaccharides and occasionally of polypeptide or of both. The chemical components of glycocalx are synthe-

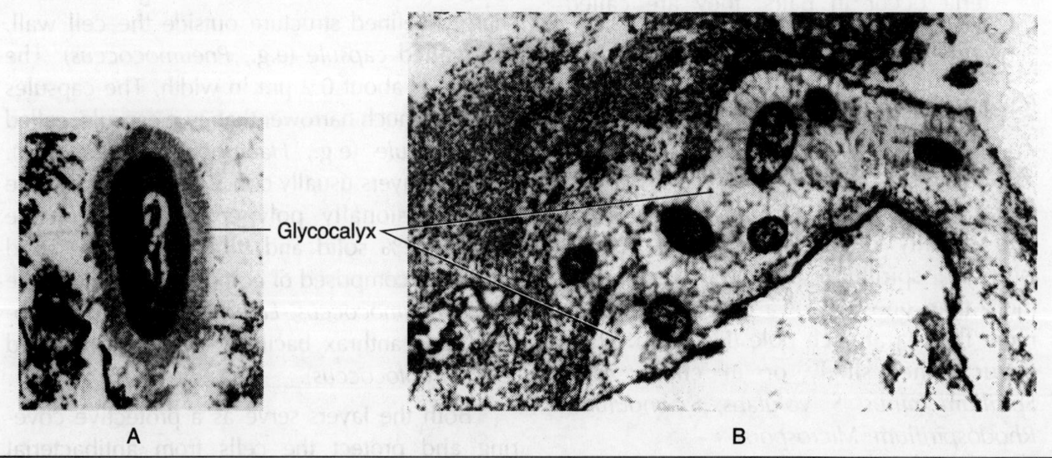

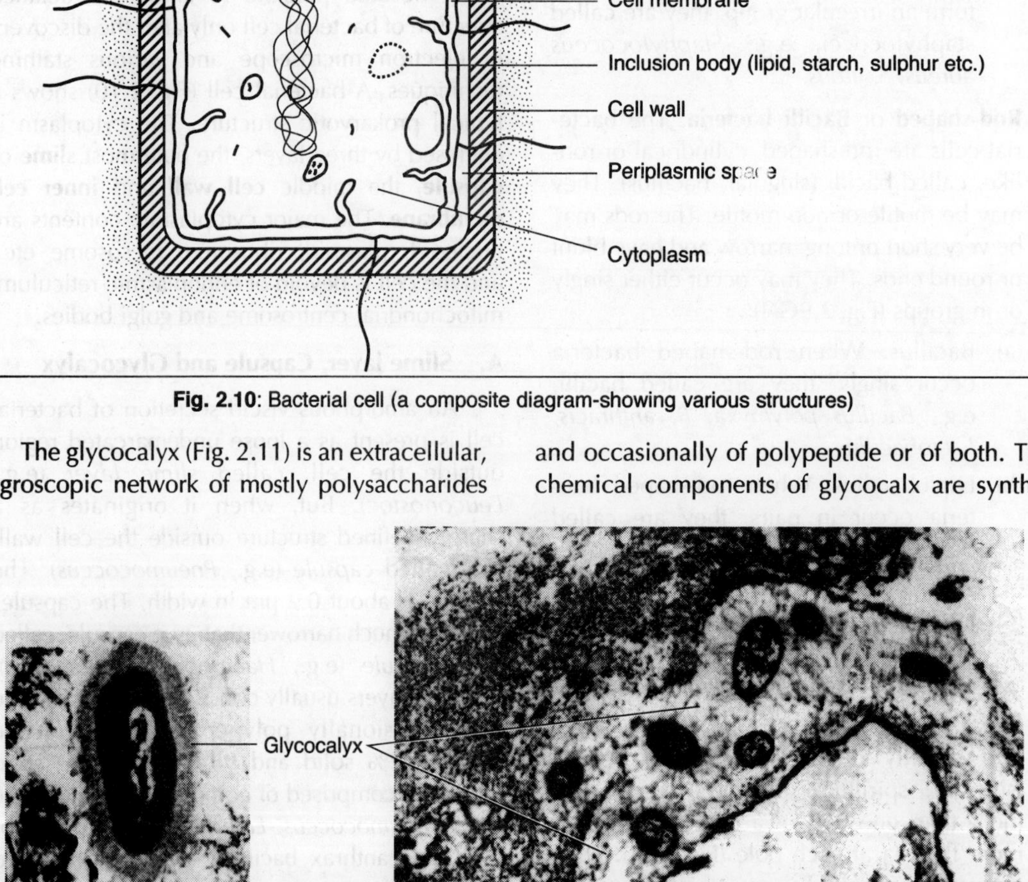

Fig. 2.11: Glycocalyx. A. Glycocalyx on bacteroid, and B. Bacteria are connected with each other and with intestinal wall by glycocalyx.

sized by the cell and transported through the cell membrane and cell wall, and finally deposited outside the cell to form extracellular covering.

The polysaccharides of glycocalyx contain many different sugars such as glucose, galactose, rhamnose and sugar acids like glucuronic acid etc.

Function:

1. It gives *protection* to the cell from desiccation under natural condition.
2. It helps the bacteria to *attach* the surface of solid objects, in aqueous medium or to tissue surface of animals and plants.

Streptococcus mutans, a bacterium associated with the **caries of teeth**, is attached to the dental surface with the help of glycocalyx, consists of water-insoluble polymer, glucan.

B. Cell wall

The bacterial cell wall is tough and rigid due to the presence of strong fibres composed of heteropolymers called mucopeptides, peptidoglycans, mucocomplex, murein etc. The peptidoglycan is composed of alternate units of N-acetyl muramic acid and N-acetyl glucosamine residues, cross-linked with tetra-peptide subunits (Fig. 2.12A). These fibres form a three-dimensional, tough meshwork rather than a solid one (Fig. 2.12B), thereby inward movement of minerals, amino acids, water etc. and outward movement of wastes take place normally. The peptide linked with muramic acid varies in their compo-

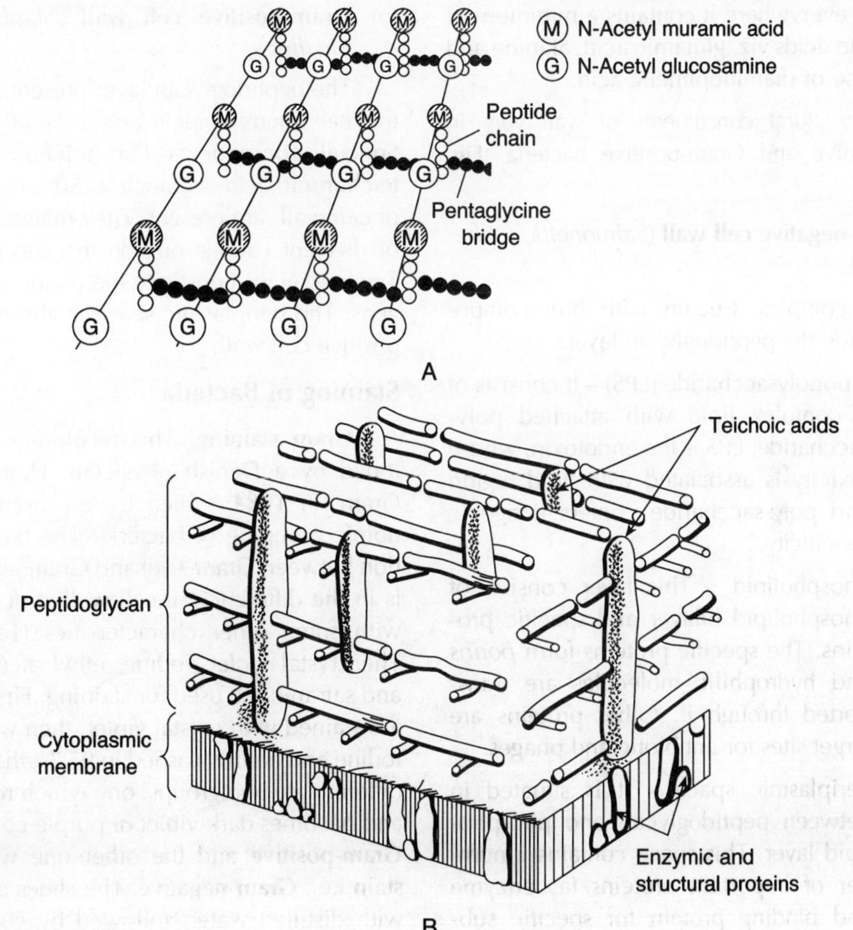

Fig. 2.12: A. Chemical structure of cell wall of Gram-positive bacteria, B. Three-dimensional representation of Gram-positive bacterial cell wall.

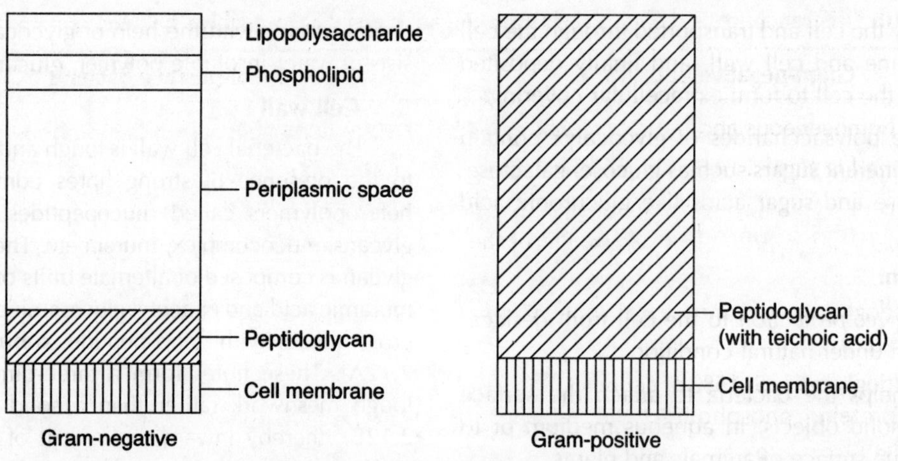

Fig. 2.13: Comparison of cell walls of bacteria

sition, but everywhere it contains a minimum of three amino acids viz. glutamic acid, alanine and either lysine or diaminopimelic acid.

The structural constituents of wall vary in Gram-negative and Gram-positive bacteria (Fig. 2.13).

(a) **Gram-negative cell wall** (*Salmonella, Escherichia*)

It is a complex structure with three components outside the peptidoglycan layer:

(i) Lipopolysaccharide (LPS) – It consists of a complex lipid with attached polysaccharide. LPS is the endotoxin, whose toxicity is associated with lipid region and poly-saccharide confers antigenic specificity.

(ii) Phospholipid – This layer consists of phospholipid bilayer and specific proteins. The specific proteins form *porins* and hydrophilic molecules are transported through it. Other proteins are target sites for antibiotic and phages.

(iii) Periplasmic space – It is situated in between peptidoglycan and phospholipid layer. This space contains a number of important proteins (as enzyme and binding protein for specific substrate) and oligosaccharides (help in osmo-regulation).

(b) **Gram-positive cell wall** (*Staphylococcus, Bacillus*)

The peptidoglycan layer present just outside the cell membrane is about 16-80 nm thick. Special components like teichoic acid and teichuronic acid (as much as 50% of dry weight of cell wall) are present. They maintain the level of divalent cations outside the cell membrane. Teichoic acids constitute the major surface antigens. The periplasmic space is absent in Gram-positive cell wall.

Staining of Bacteria

Gram staining: This technique was developed by a Danish physician, Hans Christian Gram, in 1884; which is very useful in taxonomic grouping of bacteria. The basic distinction between Gram(+ve) and Gram(–ve) bacteria is in the difference of cell wall structure, along with some other characteristics (Table 2.10). The crystal violet, iodine, ethyl alcohol (95%) and safranin are used for staining. First, the cells are stained with crystal violet, then with Gram's iodine and finally washed in 95% ethanol which differentiate two groups: one which retains stain and becomes dark-violet or purple coloured i.e., **Gram-positive** and the other one which loses stain i.e., **Gram-negative**. The slides are washed with distilled water, followed by counter-stain with safranin for 30 sec. and finally the slides are again washed with distilled water. Gram-

Table 2.10: Difference in the cell wall of Gram-negative and Gram-positive bacteria

Gram-negative bacteria	Gram-positive bacteria
1. Less homogeneous and more complex.	1. More homogeneous and amorphous.
2. Cell wall is elastic.	2. Cell wall is rigid.
3. Cell wall is multilayered, thinner, 10–15 nm thick, 10–20% of the total dry weight of the cell.	3. Cell wall is single layered, thicker, 25–30 nm thick, 20–40% of the total dry weight of the cell.
4. Peptidoglycan is 2 nm thick, one to few layered.	4. Peptidoglycan is 16–80 nm thick, many layered (up to 40).
5. Peptidoglycan is 5–15% of cell wall, and rest are protein, phospholipid and lipopolysaccharide.	5. Peptidoglycan is 20–80% of cell wall and rest are protein, polysaccharide and teichoic acid.
6. Principal surface antigen is lipopolysaccharide.	6. Principal surface antigen is teichoic acid.

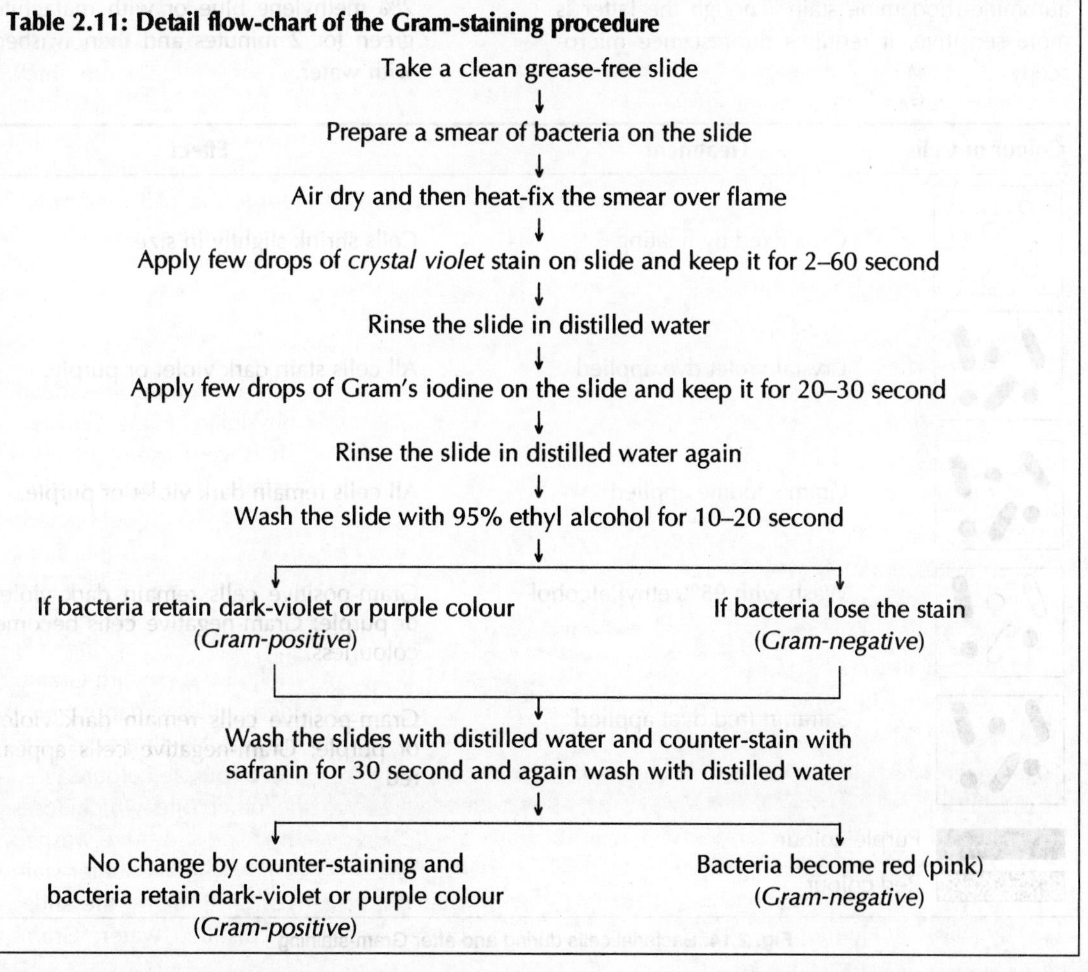

Table 2.11: Detail flow-chart of the Gram-staining procedure

Take a clean grease-free slide
↓
Prepare a smear of bacteria on the slide
↓
Air dry and then heat-fix the smear over flame
↓
Apply few drops of *crystal violet* stain on slide and keep it for 2–60 second
↓
Rinse the slide in distilled water
↓
Apply few drops of Gram's iodine on the slide and keep it for 20–30 second
↓
Rinse the slide in distilled water again
↓
Wash the slide with 95% ethyl alcohol for 10–20 second
↓

If bacteria retain dark-violet or purple colour (*Gram-positive*) | If bacteria lose the stain (*Gram-negative*)

↓
Wash the slides with distilled water and counter-stain with safranin for 30 second and again wash with distilled water
↓

No change by counter-staining and bacteria retain dark-violet or purple colour (*Gram-positive*) | Bacteria become red (pink) (*Gram-negative*)

positive bacteria remain dark-violet or purple coloured, but Gram-negative bacteria become red (pink) (Fig. 2.14). This grouping does not depend on the shape of bacteria, e.g., the bacillus-shaped bacteria like *Bacillus subtilis* is Gram-positive, while the other rod-shaped bacteria like *Coxiella burnetii* is Gram-negative. The coccus-shaped bacteria such as *Escherichia coli* is Gram-negative, but *Sporosarcina* is Gram-positive.

The detailed flow-chart of the Gram-staining procedure is given in Table 2.11.

Acid-fast staining: This technique was developed by Ehrlich and later modified by Ziehl and Neelsen.

The *Mycobacteria* can be stained by acid-fast stain (Ziehl-Neelsen technique) or by auromine-rhodamine stain. Though the latter is more sensitive, it requires fluorescence microscopy.

Ziehl-Neelsen staining:

Procedure:

1. Prepare the smear on a clean grease-free slide.
2. Heat-fix the smear and stain by hot carbol fuchsin for about 10 minutes and wash in tap water.
3. The smear is then decolourised by 20% H_2SO_4 (sulphuric acid) or 1% HCl (hydrochloric acid) in 70% alcohol for 1 minute.
4. The decolourisation is continued with 70% neutral alcohol for 2–3 minutes, until the smear becomes pinkish in colour. (Decolourisation of *M. leprae* should be done with 5% H_2SO_4 only).
5. The smear is then counterstained with 2% methylene blue or with malachite green for 2 minutes and then washed with water.

Colour of Cells	Treatment	Effect
	Cells fixed by heating	Cells shrink slightly in size.
	Crystal violet dye applied	All cells stain dark violet or purple.
	Gram's iodine applied	All cells remain dark violet or purple.
	Wash with 95% ethyl alcohol	Gram-positive cells remain dark violet or purple; Gram-negative cells become colourless.
	Safranin (red dye) applied	Gram-positive cells remain dark violet or purple; Gram-negative cells appear red.

Purple colour
Red colour

Fig. 2.14: Bacterial cells during and after Gram-staining

Observation:

The Bacilli after staining appear **red** in blue background of epithelial cells or pus cells.

1. *Mycobacteria* like *M. leprae* (c.o. of Leprosy), *M. tuberculosis* (c.o. of Tuberculosis) can be stained by this method (***M. tuberculosis* is a acid fast as well as Gram-positive bacterium**).
2. Others like *Staphylococcus aureus*, Ascospores of some yeast, bacterial spores etc., can be stained following the above procedure.

Stains used for Gram-staining:

1. **Crystal violet**

Solution A

(i) Crystal violet (90% dry content)... 2 gm
(ii) Ethyl alcohol (95%) ... 20 ml

Solution B

(i) Ammonium oxalate ... 0.8 gm
(ii) Distilled water ... 80 ml

Mix solution A and B.

2. **Iodine solution**

(i) Iodine ... 1 gm
(ii) Potassium iodite ... 2 gm
(iii) Distilled water ... 300 ml

3. **Safranin**

(i) Safranin (2.5% soln. in 95% ethanol) ... 10 ml
(ii) Distilled water ... 100 ml

Stains used for acid fast staining:

1. Ziehl's carbol fuchsin

Solution A

(i) Basic fuchsin (90% dry content) ... 0.3 gm
(ii) Ethyl alcohol (95%) ... 10 ml

Solution B

(i) Phenol ... 5 gm
(ii) Distilled water ... 95 ml

Mix solution A and B.

2. Loeffler's methylene blue.

Solution A

(i) Methylene blue (90% dry content) ... 0.3 gm
(ii) Ethyl alcohol (95%) ... 30 ml

Solution B

(i) Dilute KOH (0.01 by wt.) ... 100 ml

Mix solution A and B.

C. Cytoplasmic membrane

The cytoplasmic layer is the boundary layer of the protoplast, situated beneath the cell wall (Fig. 2.10). It is thin (5–10 nm), elastic and semi-permeable layer (Fig. 2.15). In section, it appears as a triple-layered structure consisting of a bilayer region of phospholipid molecules, with polar heads on the surface and fatty-acyl chains towards the inner side. The proteins are found embedded in the lipid bilayer.

Functions

(i) *Transport:*

(a) *Active*: Being the site of many enzymes like oxidase, polymerase etc., it is involved in the active transport of selective nutrients. It is impermeable to ionised substances and macromolecules.

(b) *Passive*: The passive transport of fat soluble micromolecular solutes take place by diffusion.

(ii) *Energy production:* It is the site of electron flow in both respiration and photosynthesis leading to phosphorylation and, therefore, the membrane is the site of carriers and enzymes in these reactions.

(iii) *Polymer production:* Cell membrane is the site of polymerising enzymes necessary for synthesis of cell wall.

D. Cytoplasm

The cytoplasm is a colloidal system containing both organic and inorganic substances. *It lacks mitochondria, endoplasmic reticulum, centrosome and golgi bodies*. It contains many ribosomes, few mesosomes, some inclusions and vacuoles (Fig. 2.10).

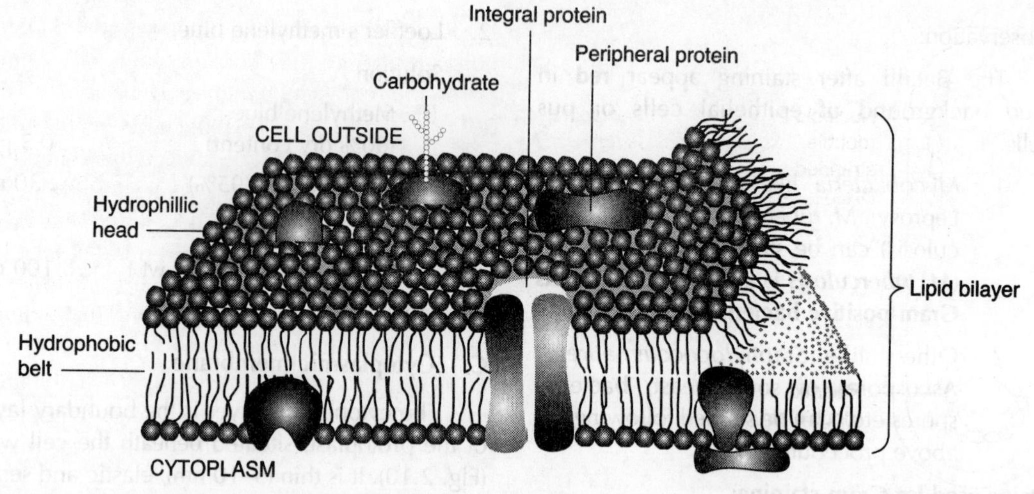

Fig. 2.15: Cytoplasmic membrane (fluid mosaic model)

Ribosome: It is a complex substance of 10–20 nm size and of 70S (S = sedimentation co-efficient) type having two subunits, 50S and 30S. The larger one comprises of molecules of RNA and 35 amino acids and the smaller one has one molecule of RNA and 20 amino acids. They are the sites of protein synthesis. The ribosomes are held together on m-RNA (messenger RNA) strands, known as polysomes or polyribosomes.

Mesosomes (Chondroids): These are convoluted multi-laminated localised infoldings of the cytoplasmic membrane into the cytoplasm (Fig. 2.16). Their number is usually 2–4, but often found to be more in cells with high respiratory activity, e.g., *Nitrosomonas*. Perhaps it serves to accommodate more spaces for respiration. In photosynthetic bacteria (*Rhodopseudomonas*), they are the site of photosynthetic pigments. The mesosomes are of two types — septal mesosome and lateral mesosome. The septal mesosomes are involved in DNA segregation and in the formation of transverse septum during cell division.

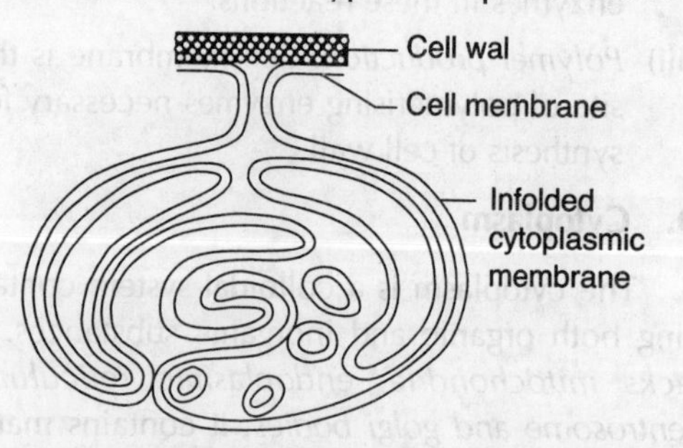

Fig. 2.16: Single mesosome

Chromatophores: These are the pigment-bearing structures, found in photosynthetic bacteria. They are found in all members of Chromatiaceae and Rhodospirillaceae in different forms such as membranes, vesicles, tubes, bundle tubes etc. or as thylakoids, as found in Cyanobacteria.

Cytoplasmic inclusion: These are the sources of stored energy, characteristic of different species of bacteria such as lipid (poly β-hydroxybutyrate), volutin (polymetaphosphate), starch or glycogen (polysaccharide) and granules of sulphur.

E. Genetic material

The genetic material is present both in nucleoid and plasmid (Fig. 2.17).

Nuclear material: Under light microscope nuclear body cannot be differentiated in the cytoplasm, but it is differentiated only under electron microscope as a central area of lower electron dense region than rest of the cytoplasm.

The bacterial nuclear body is devoid of nuclear membrane, nucleolus and nuclear sap and is known as **genophore** or **nucleoid.** The genophore is composed of a single or double stranded circular DNA. When straightened the

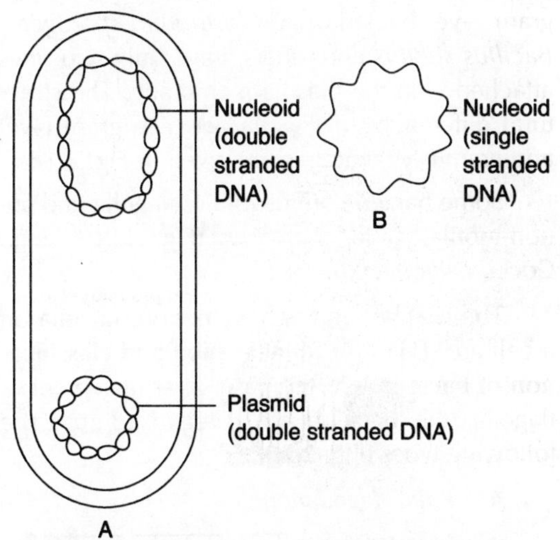

Fig. 2.17: A. Bacterial cell having nucleoid (double stranded DNA) and plasmid (extrachromosomal carrier of gene). B. Nucleoid (single stranded DNA)

DNA measures 1000 μm. It has approximately 5×10^6 base pairs and a molecular weight of about 3×10^9. It is devoid of any basic protein. The DNA is haploid — it undergoes semiconservative replication by simple fission and maintains genetic characteristics.

Plasmid: Bacterial cytoplasm may contain some genetic material excepting the genophore, called plasmid or episomes. Lederberg (1952) termed as plasmid those extragenophoral genetic materials.

Plasmids are ring-like double stranded DNA molecules which may contain about 100 genes having the molecular weight range from 5×10^7 to 7×10^7 or less. The replication of plasmid seems self-controlled. They contain different non-essential characters. Based on host properties, the plasmids are classified into different types. These are (i) "F-factor" for fertility (ii) "Col-factor" for colicinogeny (iii) "R-factor" for resistance (iv) Tumor inducing plasmid (e.g., *Agrobacterium tumifaciens*), (v) Degradative plasmid (e.g., *Pseudomonas*), (vi) Pathogenecity to mammals, (vii) Penicillase plasmid (e.g., *Staphylococcus*), (viii) Mercury resistance, and (ix) Cryptic plasmids.

F. Flagella

Most of motile bacteria (e.g., *Spirochaetes*) possess long (5–20 μm), thin (12–30 nm), helical appendages, called *flagella*. Electron microscopy shows that the flagellum consists of three distinct regions — *filament, hook* and *basal body* (Fig. 2.18). Filament is attached at one end through the cell wall to the cell membrane by the hook, which, in turn, is attached to the basal body. The rings of basal body remain attached to the cell membrane and cell wall. The filament lies external to the cell.

Filament is made up of identical subunits (3 or more), arranged helically along the axis to give a hollow tube (Fig. 2.19). These subunits are

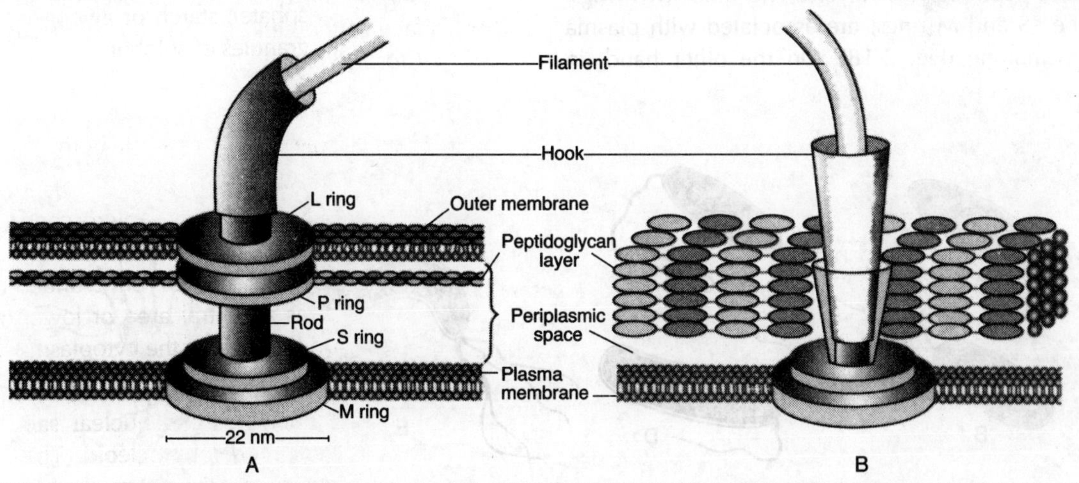

Fig. 2.18: The ultrastructure of bacterial flagella showing flagellar basal bodies and hooks. A. Gram-negative bacteria and B. Gram-positive bacteria.

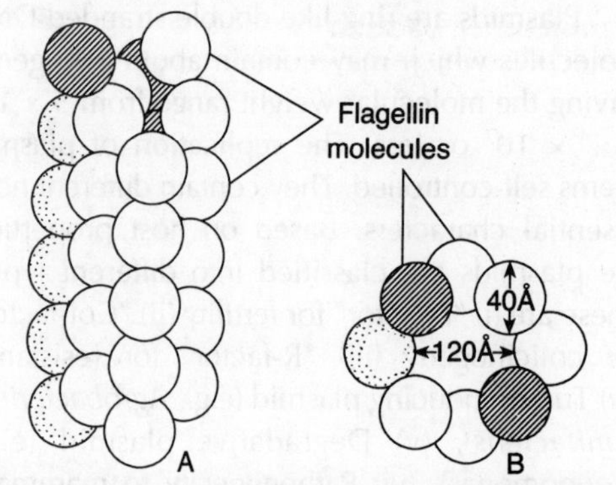

Fig. 2.19: Structure of filament. A. Arrangement of flagellin molecules showing hollow cylinder and B. T.S. of a filament.

made up of protein molecule, called *flagellin* (Fig. 2.19). Each subunit is about 4.5 nm thick.

The filament ends with a capping protein. Some bacteria have sheath surrounding the flagella (e.g., *Vibrio cholerae* has a sheath of lipopolysaccharide).

Structurally, the hook and basal body are quite different from the filament (Fig. 2.18). The hook is slightly wider than the filament and made up of protein subunits.

The structure of basal body is quite different in gram (–ve) and gram (+ve) bacteria. Gram (–ve) bacteria like *E. coli* and others have 4 rings connected to a central rod. The outer L and P-rings are associated with the lipopolysaccharide and peptidoglycan layers. The inner two rings, i.e., S and M rings, are associated with plasma membrane (Fig. 2.18). On the other hand, in gram (+ve) bacteria like *Clostridium sporogens, Bacillus subtilis* and others, have only two rings attached with the plasma membrane. The structural differences of flagella between gram (–ve) and gram (+ve) bacteria are given in Fig. 2.18.

Some bacteria are devoid of flagella and are non-motile, called *atrichous* (Fig. 2.20A) e.g., Cocci, *Lactobacillus,* etc.

The number and arrangement of flagella on a cell are useful for **identification and classification of bacteria**. On the basis of arrangement of flagella, the bacteria are categorised into the following types (Fig. 2.20):

A. *Polar flagellation:*

(a) *Monotrichous* (Fig. 2.20B). Single flagellum at one pole of the cell, e.g., *Vibrio cholerae.*

(b) *Amphitrichous* (Fig. 2.20C). Each single flagellum is attached at both ends, e.g., *Alkaligenes faecalis, Nitrosomonas.*

(c) *Cephalotrichous* (Fig. 2.20D). Two or more flagella at one end only, e.g., *Pseudomonas fluorescens.*

(d) *Lophotrichous* (Fig. 2.20E). A tuft of flagella at both ends. e.g., *Spirillum volutans.*

B. *Non-polar flagellation:*

(a) *Peritrichous* (Fig. 2.20F). Numerous flagella are distributed all over the surface of the cell e.g., *Bacillus typhosus, Clostridium.*

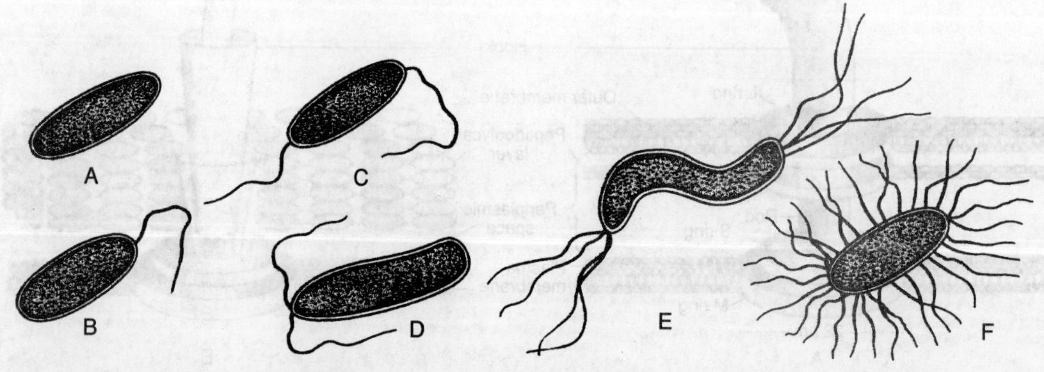

Fig. 2.20: Different types of flagellation: A. Atrichous B. Monotrichous, C. Amphitrichous, D. Cephalotrichous, E. Lophotrichous, and F. Peritrichous

Table 2.12: Differences between Gram-negative and Gram-positive bacteria

Property	Gram-negative	Gram-positive
1. Cell shape	Oval, spherical, straight or slightly curved rods, helical or filamentous; some of the members have capsule or sheath.	Spherical, rod-shaped or filamentous; true branching is present in some members.
2. Motility	Motile or non-motile. Motile members have flagellation of different types: polar, lophotrichous or peritrichous.	Commonly non-motile, motile members have peritrichous flagella.
3. Appendages	Develop various types of appendages like pili, fimbriae etc.	Usually lack appendages.
4. Cell wall	Walls have inner 2–7 nm thick peptidoglycan layer and 7–8 nm thick outer membrane of lipid, protein and polysaccharide. Sometimes third outermost layer of protein may be present.	Walls have a homogeneous, 20–80 nm thick wall, mainly composed of peptidoglycan. Other polysaccharides and teichoic acids may also be present.
5. Metabolism	Various types: Phototrophic, Chemolithoautotrophic or Chemoorganotroheterotrophic.	Usually chemoorganotroheterotrophic.
6. Reproduction	Commonly binary fission, sometimes by budding.	By binary fission.
7. Endospore formation	Endospore does not form.	Formed in some groups of bacteria.

G. Fimbriae or Pili

The term *fimbriae* (sing. *fimbria*) was introduced by Duguid *et al.* (1955) and *pili* (sing. *pilus*) by Brinton (1959). Fimbriae (Fig. 2.21A) are observed mostly in Gram-negative rods (*Salmonella typhi*, typhoid fever; *Shigella dysenteriae*, bacillary dysentery) and also in cocci (*Neisseria gonorrhoeae*, gonorrhoea). Gram-positive bacilli like *Corynebacterium renale* also have fimbriae or pili. These are extremely thin and short, filamentous, non-flagellar appendages projecting peritrichously from cell surface. Their number is 100–500 per cell and measure 0.5–20 μm in length and 3–25 nm in diameter. They are

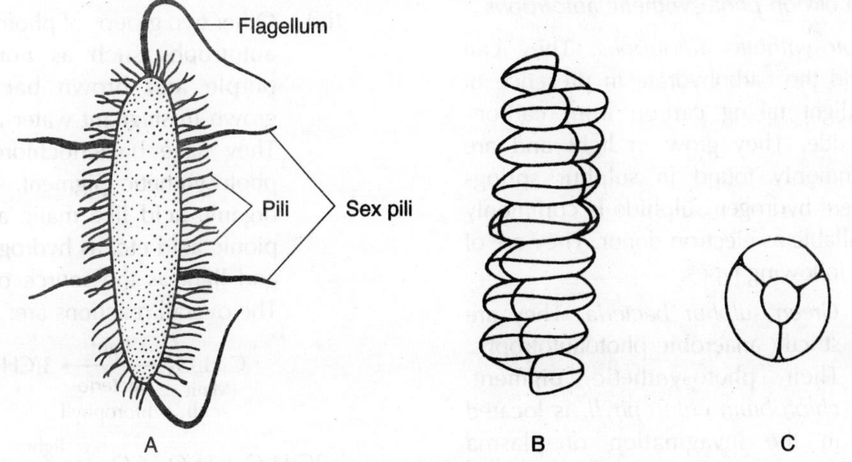

Fig. 2.21: Pili of Bacteria: A. Single bacterium showing polar flagella and two types of pilli, B. Arrangement of pilin molecules and C. T.S. of pilus.

made up of subunits of protein, the *pilin,* arranged helically and form hollow filament (Fig. 2.21B, C). They are antigenic. Pili are of three types:

Common pili, sex pili or F (fertility) pili and col I (colicin) pili (Fig. 2.21A). The common pili are responsible for adhesion; the sex pili (1–5 per cell) help in the transfer of genetic material during conjugation and the col I for colicin or hemolysin production.

H. Spinae

Spinae are tubular, pericellular rigid appendages. They are made up of single protein moiety, the *spinin.* They possibly help to acclimatize the cells in different environmental conditions such as salinity, temperature etc. Spinae have been reported in some Gram-positive bacteria.

Nutrition

Bacteria are devoid of chlorophyll and they perform *heterotrophic* mode of nutrition. However, there is a small group of bacteria which are *autotrophic.* Based on nutrition, bacteria are classified or divided into the following groups:

1. *Autotrophic bacteria.* They can synthesise carbohydrate food from CO_2 like that of chlorophyll containing plants. They are of two types: *Photosynthetic bacteria* or *photosynthetic autotrophs* and *chemosynthetic bacteria* or *non-photosynthetic autotrophs.*

(a) *Photosynthetic autotrophs.* They can build the carbohydrate in presence of sunlight taking carbon from carbon-dioxide. They grow in light and are commonly found in sulphur springs where hydrogen sulphide is commonly available as electron donor. They are of the following types:

(i) *Green sulphur bacteria.* They are strictly anaerobic photoautotrophs. Their photosynthetic pigment, *chlorobium chlorophyll,* is located in the invagination of plasma membrane towards the cytoplasm. These bacteria use hydrogen sulphide (H_2S) or other reduced inorganic sulphur compounds as the donor of hydrogen. Light splits H_2S, and forms hydrogen which combines with CO_2 to form CH_2O. Sulphur is excreted out and deposited extracellularly e.g. *Chlorobium Pelodictyon* etc. The overall reaction is:

$$2H_2S + CO_2 \xrightarrow[\text{Chlorobium chlorophyll}]{\text{light}} CH_2O + 2S + H_2O$$

(ii) *Purple sulphur bacteria.* They are anaerobes and the photosynthetic pigment *bacteriochlorophyll* is located in the invagination of plasma membrane. The H_2S is oxidised anaerobically via elemental sulphur to sulphate. The by-product sulphur is stored in the cell as globules, except *Ectothiorhodospira,* where sulphur is excreted out like green sulphur bacteria. e.g., *Thiocystis, Thiocapsa, Thiospirillum, Rhodospirillum* etc. The overall reaction is:

$$2CO_2 + H_2S + 2H_2O \xrightarrow[\text{Bacterio-chlorophyll}]{\text{light}} 2CH_2O + H_2SO_4$$

In addition to above, some sulphur bacteria can use other sulphur compounds like sulphite, thiosulphate etc., instead of H_2S.

(iii) Other two groups of photosynthetic autotrophs such as non-sulphur purple and brown bacteria are grown in stagnant water and mud. They have bacteriochlorophyll as photosynthetic pigment. They use organic acid like malic acid, propionic acid etc. as hydrogen donor and light as the source of energy. The overall reactions are:

$$\underset{\text{(Malic acid)}}{C_4H_6O_5} \xrightarrow[\text{Bacterio-chlorophyll}]{\text{light}} 3(CH_2O) + CO_2$$

$$\underset{\text{(Propionic acid)}}{2C_3H_6O_2} + H_2O + CO_2 \xrightarrow[\text{Bacterio-chlorophyll}]{\text{light}} 7(CH_2O)$$

e.g., *Rhodospirillum, Rhodomicrobium, Rhodopseudomonas* etc.

(b) *Chemosynthetic autotrophs:* Chemosynthetic bacteria are more abundant in nature than that of photosynthetic bacteria. They are non-photosynthetic and for synthesis of food they get energy by the oxidation of certain inorganic substances such as nitrates, nitrites, ferrous compound, hydrogen sulphide and many other metallic or non-metallic substances. The bacteria absorb inorganic molecules in their body where energy is released by breaking down of chemical bonds of the molecules. This energy is used to combine CO_2 and water into food molecules. This process is called *chemosynthesis*. It is an exothermic reaction. This group of bacteria are independent of light and organic material.

Depending on the substrate specificity, they are of the following types:

(i) *Sulphur bacteria*. These bacteria oxidise sulphur compounds, thus energy being released and is utilised for food synthesis. The sulphur remains as granular residue in the bacterial cytoplasm, e.g., *Beggiatoa*

$$H_2S + \frac{1}{2}O_2 \longrightarrow S + H_2O + \text{Energy}$$

Another type which utilises free sulphur and thereby produces sulphuric acid is *Thiobacillus thiooxidans*.

$$S + H_2O + 1\frac{1}{2}O_2 \longrightarrow H_2SO_4 + \text{Energy}$$

(ii) *Iron bacteria*. These bacteria oxidise ferrous compounds into ferric form and liberate energy. The energy is utilised in the synthesis of organic compounds, e.g., *Gallionella, Leptothrix* etc.

$$Fe^{2+} \longrightarrow Fe^{3+} + e^-$$

The ferric iron being deposited as insoluble ferric hydroxide is mentioned below:

$$2Fe(HCO_3)_2 + O + H_2O \longrightarrow 2Fe(OH)_3 + 4CO_2 + \text{Energy}$$

(iii) *Nitrifying bacteria*. This group of bacteria helps in increasing the nitrogen source in nature. They oxidise ammonia to nitrate in two steps: ammonia to nitrous acid and nitrous acid to nitric acid, in which acids react with metal ions to produce the corresponding salts, i.e., nitrite and nitrate. The steps are bacteria specific. The **first step** involves the oxidation of ammonia to nitrous acid by nitrosifying bacteria like *Nitrosomonas, Nitrococcus* and *Nitrospira* is called **nitrosification**. The process is catalysing by the enzyme, ammonia dehydrogenase. The **second step** involves oxidation of nitrous acid to nitric acid by *Nitrobacter, Nitrococcus* and *Nitrospira* is called **nitrification.** The reacion is carrying out by nitrous acid dehydrogenase enzyme.

Step–I: $2NH_4^+ + 3O_2 \longrightarrow 2NO_2^- + 4H^+ + 2H_2O$

Step–II: $2NO_2^- + O_2 \longrightarrow 2NO_3^-$

(iv) *Hydrogen bacteria*. This group of bacteria oxidises molecular hydrogen and produces water and energy, e.g., *Pseudomonas facilis, Nocardia opaca, Alcaligenes eutrophs*.

$$H_2 + \frac{1}{2}O_2 \longrightarrow H_2O$$

Differences between Photosynthesis and Chemosynthesis

Photosynthesis	Chemosynthesis
1. It takes place in presence of light	No light is required.
2. Reactions are endothermic	Reactions are exothermic.
3. The energy available in Photosynthesis is more than that of Chemosynthesis.	The energy available in Chemosynthesis is much less than Photosynthesis.

2. *Heterotrophic bacteria*. These bacteria obtain their food from any organic source. They are of two types: *Saprophytic* and *Parasitic*.

(a) *Saprophytic (Saprobic) bacteria*. These bacteria grow on decaying organic matters and live by digesting and absorbing them. By secreting enzymes they break complex organic compound into simpler forms. These simpler forms are rendered into soluble form and are absorbed by the bacteria as food. The breakdown of compounds may be of two types: *Fermentation* and *Putrefaction*.

(i) *Fermentation*. The breakdown of carbohydrate is known as *fermentation*. Some bacteria (*Escherichia coli*) are able to ferment glucose and galactose, thereby CO_2 is released.

On the other hand, some bacteria like *Lactobacillus* ferment milk and produce an organic acid, the lactic acid which causes the souring and curding of milk. Carbon dioxide is not evolved in this process.

(ii) *Putrefaction*. The breakdown of protein material is known as *putrefaction*. The earlier stage takes place in absence of oxygen by some anaerobic bacteria, where peptone, peptide, polypeptide and amino acids are produced. Some of the substances have unpleasant odour. The later stages of breakdown require oxygen where substances like amino acids are further decomposed into sulphur substance such as methane, nitrogen, hydrogen, ammonia etc. A few species are able to degrade fats into fatty acids and glycerine.

(b) *Parasitic bacteria*. They grow on or within living organisms like plants and animals. They draw the organic food from their host. These are of two types: *Symbiotic* and *Pathogenic*:

(i) *Symbiotic bacteria*. These bacteria are useful to the host on one hand and receive food and shelter in return. Different types of coli bacteria are inhabiting the intestine of man and other organisms, which help in the digestion of cellulose by various enzymes secreted by them. These bacteria also deposit vitamins which can be used by the host. In return they take shelter and food from the host.

In leguminous plants, the bacterium *Rhizobium* develops root nodules. The *Rhizobium* possesses nitrogenase enzyme and thus can fix atmospheric nitrogen useful to the plants and, in return, it also takes shelter and food from its host. *Frankia* spp. of Actinomyceteous fungi develop root nodules in about 178 species of non-leguminous plants like *Casuarina, Alnus, Myrica, Elaeagnus, Coriaria, Ceanothus* etc. They also fix atmospheric N_2 with the help of nitrogenase enzyme.

(ii) *Pathogenic bacteria*. These bacteria cause diseases of plants and animals including man. They cause diseases on the host either by direct attack or by liberation of toxic substances which affect directly or indirectly. Some common human diseases caused by pathogenic bacteria are Cholera (*Vibrio cholerae*), Diphtheria (*Corynebacterium diphtheriae*), Tuberculosis (*Mycobacterium tuberculosis*), Typhoid (*Salmonella typhi*), Pneumonia (*Mycoplasma pneumoniae*) etc.

The plant diseases caused by bacteria are *citrus* canker (*Xanthomonas axonopodis* pv. *citri*), Bacterial blight of rice (*Xanthomonas campestris* pv. *oryzae*), Tundu disease of wheat (*Clavibacter tritici*), Bacterial blight of cotton (*Xanthomonas malvacearum*), etc.

Growth of Bacteria

The growth of bacteria requires proper nutrition and proper environment:

1. **Requirement of nutrition**. The nutrition includes essential elements, mineral sources and organic growth factors.

   (a) *Essential elements*. The essential elements required for the synthesis of structural compounds such as nucleic acid, protein, fat and carbohydrate are carbon, nitrogen, hydrogen and oxygen. In addition to the above elements, sulphur and phosphorus are also required for bacterial growth. These elements have to be supplied in the medium for proper growth.

   (b) *Mineral sources*. The minerals such as calcium, magnesium, potassium, iron, cobalt, copper, manganese, zinc and molybdenum are required as trace elements for the function of different enzymes.

   (c) *Organic growth factors*. Some bacteria require organic compounds for their growth and these are called growth factors or vitamins. These are required at micro-level as *essential* (required for their growth) factor or accessory (required for enhancing growth) factor.

2. **Environmental factors**. The environmental factors like water, oxygen, carbon dioxide, light, temperature, pH and osmotic pressure are essential for the growth of bacteria.

(a) *Water*. Water is an essential requirement for growth, because approximately 80% of the cell volume consists of water.

(b) *Oxygen*. The bacteria differ in the requirement of oxygen. They are of two types: *aerobes* and *anaerobes*. The aerobes require oxygen for growth. The aerobes again can be divided into obligate (e.g., *Pseudomonas*) or facultative anaerobes i.e., they are normally aerobes, but can grow without oxygen (e.g., *Vibrio cholerae*). The anaerobes do not require oxygen for their growth and they may die in presence of oxygen (e.g., *Clostridium*).

(c) CO_2. Very small amount of CO_2 is required for all bacteria which may be available endogenously or may be collected from atmosphere.

(d) *Temperature*. Temperature is essential for growth, but optimum temperature for growth varies with different bacterial species. Bacteria may be grouped as *Psychrophile* (grows below 20°C), *Mesophile* (grows between 25–40°C) and *Thermophile* (grows between 55–88°C).

(e) *pH*. Most of the bacteria associated with human diseases require 7.2–7.4 pH, but *Vibrio cholerae* requires 10.5 pH. Bacteria like *Lactobacillus* and *Thiobacillus thioxidans* require 3.0 pH for their growth.

(f) *Light*. Bacteria generally prefer to grow in dark, but some phototropic species grow in light.

(g) *Osmotic pressure*. Bacteria are able to grow in a wide range of osmotic pressure, because of rigid and tough cell wall.

Cell Division and Growth of Bacteria

Bacteria generally divide by binary fission into two equal cells. The daughter cells in some species may remain attached even after cell division.

Generation time. The interval of time between two cell divisions is known as *generation time*. The generation time varies with bacterial species. Generally it ranges between 20–30 minutes. The *E. coli* takes 20 minutes, *Mycobacterium tuberculosis* 20 hours, whereas lepra bacillus (*Mycobacterium leprae*) takes 24 days. In culture, the cell division ceases after some time and bacteria start to die. The death takes place due to exhaustion of nutrient and also for the accumulation of toxic products.

Bacterial count. The total count is the total number of bacteria in the medium which include both living and dead cells. The viable count measures the number of living bacteria in the sample.

Bacterial growth cycle

When a sterile liquid nutrient medium is inoculated by a particular bacterium, the bacterial cell undergoes multiplication and increases in number which follows a definite pattern. When periodical determination of bacterial count from such culture was made and plotted

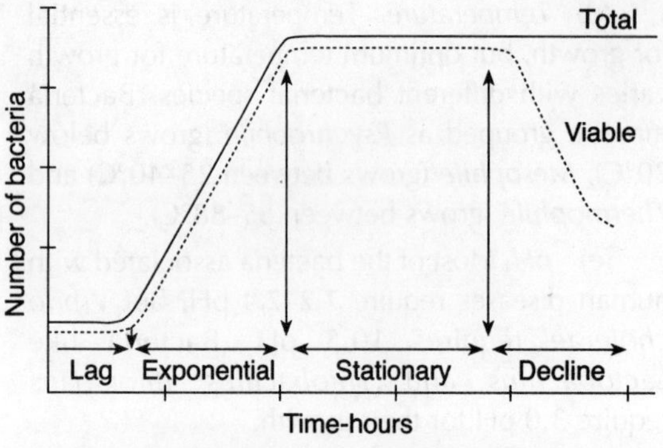

Fig. 2.22: Bacterial growth curve

against time, a curve is usually obtained, known as **growth curve** (Fig. 2.22). The growth curve has *four* main phases:

A. *Lag phase*. After inoculation into the sterile nutrient medium, the bacterium first undergoes a period of acclimatisation. At that time, necessary enzymes and intermediate metabolites are synthesised, thereby bacterium reaches a critical stage before multiplication. No multiplication takes place at this stage.

The duration of lag phase depends on the type of bacteria, quality of culture medium, size of inoculum and several environmental factors such as CO_2, temperature, pH, etc. The average time of lag phase is 2 hours, although it varies from species to species (1–4 hours).

B. *Log phase* or *Exponential phase*. In this phase, the bacteria undergo cell division and their population (number) increase exponentially at a logarithmic rate. The number of viable count, when plotted against time, gives a straight line of inclined fashion. The average time of log phase is 8 hours, though it varies in different species.

C. *Stationary phase*. In this phase, the growth i.e., cell division, almost ceases due to exhaustion of nutrients and also the accumulation of toxic products. At this stage the cell death starts at a slow rate and is compensated by the formation of new cell through cell division. The total cell number increases at a slow rate, but the viable count remains almost constant. The duration of this phase is variable which ranges from few days to few hours. Secondary metabolites like antibiotics, toxins etc. are produced in this phase.

D. *Decline phase*. In the phase of decline, the total number of cells remain constant, but the number of viable cells gradually decreases due to exhaustion of nutrients and also the accumulation of toxic products. In some cases a few cells remain viable for long time, even after death of most of the cells. These viable cells probably grow by utilising nutrients released from dead cells.

The cells attain maximum size at the end of lag phase and become smaller in log phase (exponential phase). In spore forming species, the sporulation occurs at the end of log phase (exponential phase) or in the early part of stationary phase.

REPRODUCTION

Bacteria usually reproduce by asexual methods. The asexual reproduction takes place by binary fission, conidia, budding, cysts and also through endospore formation.

1. **Binary fission**. In binary fission, single cell divides into two equal cells (Fig. 2.23). Initially the bacterial cell reaches a critical mass in its structure and cellular constituents. The circular double stranded DNA of bacteria undergoes replication, where

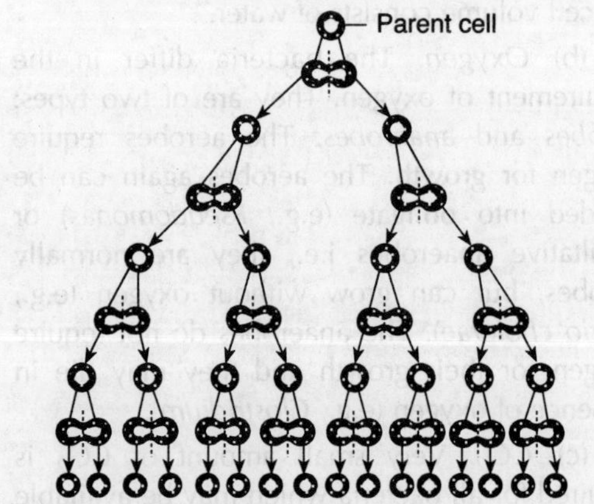

Fig. 2.23: Binary fission results in rapid multiplication of bacterial cell

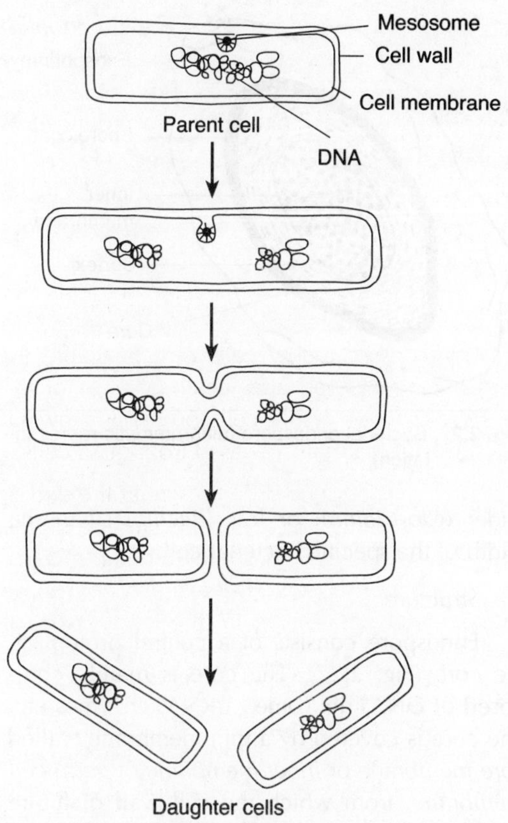

Fig. 2.24: Binary fission of a bacterium

both the strands separate and new complementary strands are formed on the original strands — results in the formation of two identical double stranded DNA (Fig. 2.24). The new double stranded DNA molecule i.e., incipient nuclei, are then distributed into two poles of the dividing cell (no spindle formation takes place like mitotic division). A transverse septum develops in the middle region of the cell, which separates the two daughter cells.

The binary fission is a rapid process and cell undergoes division at an interval of 20–30 minutes. The division becomes gradually slow after certain time due to accumulation of toxic substance and exhaustion of nutrients.

2. **Conidia**. Conidia formation takes place in filamentous bacteria like *Streptomyces* etc., by the formation of a transverse septum at

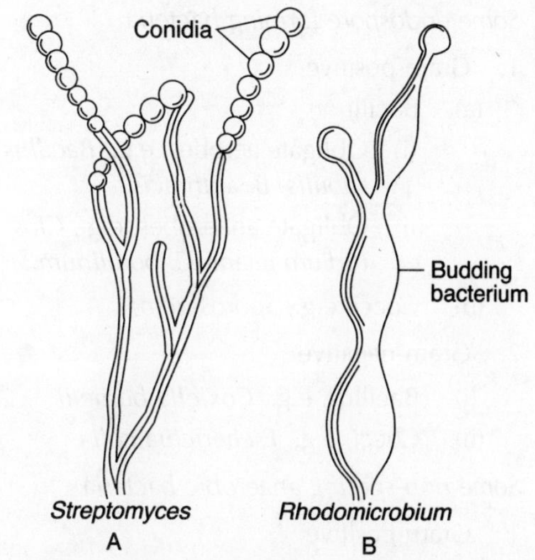

Fig. 2.25: Asexual reproduction of bacteria: A. Conidia formation, B. Budding

the apex of the filament (Fig. 2.25A). The part of this filament which bears conidia is called *conidiophore*. After detachment from the mother and getting contact with suitable substratum, the conidium germinates and gives rise to new mycelium.

3. **Budding**. The bacterial cell develops small swelling at one side which gradually increases in size (Fig. 2.25B). Simultaneously the nucleus undergoes division, where one remains with the mother and other one with some cytoplasm goes to the swelling. This outgrowth is the bud, which gets separated from the mother by partition wall, e.g., *Hyphomicrobium vulgare, Rhodomicrobium vannielia,* etc.

4. **Cysts**. Cysts are formed by the deposition of additional layer around the mother wall. These are the resting structure and during favourable condition they again behave as the mother, e.g., many members of *Azotobacter*.

5. **Endospore**. Spores are formed during unfavourable environmental condition like dessication and starvation. As the spores are formed within the cell, they are called *endospores*. Only one spore is formed in a bacterial cell. On germination, it gives rise to a bacterial cell.

A. *Some endospore forming bacteria*

1. Gram-positive

(a) Bacilli

(i) Obligate aerobes. e.g., *Bacillus subtilis, B. anthracis.*

(ii) Obligate anaerobes, e.g., *Clostridium tetani, C. botulinum.*

(b) Cocci. e.g., *Sporosarcina.*

2. Gram-negative

(i) Bacillus, e.g., *Coxiella burnetii*

(ii) Cocci, e.g., *Escherichia coli*

B. *Some non-sporing anaerobic bacteria*

1. Gram-positive

(a) Bacilli

(i) *Lactobacillus*

(ii) *Propionibacterium*

(iii) *Bifidobacterium*

(b) Cocci

(i) *Peptococcus*

(ii) *Sarcina*

(iii) *Peptostreptococcus*

2. Gram-negative

(a) Bacilli

(i) *Fusobacterium*

(ii) *Leptotrichia*

(iii) *Bacteroides*

(b) Cocci

(i) *Acidoaminococcus*

(ii) *Veillonella*

C. *Shape and position of endospore*

Spores may be oval or spherical in shape.

The position, relative size and shape remain constant in a particular species. The position of spore may be central, subterminal or terminal (Fig. 2.26). In diameter, it may be the same or

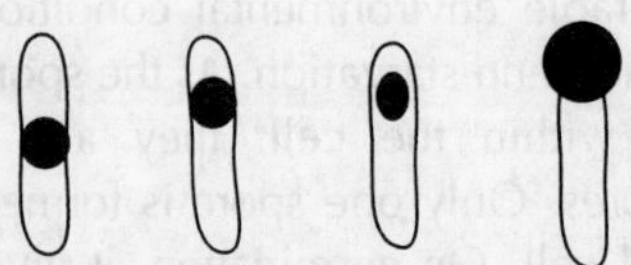

Fig. 2.26: Different position of endospore in bacterial cell

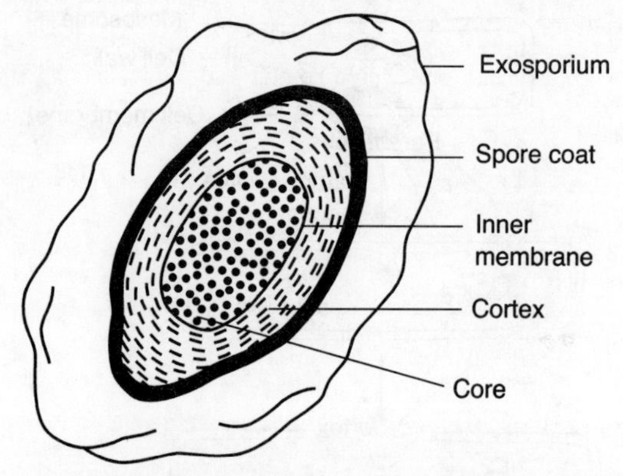

Fig. 2.27: Bacterial endospore (diagrammatic representation)

wider (*Clostridium*) or less (*Bacillus*) than the width of the specific bacterial cell.

D. *Structure*

Endospore consists of a central protoplast, the *core* (Fig. 2.27). The core is mainly composed of DNA, ribosome, t-RNA, enzymes etc. The core is covered by a thin membrane, called *core membrane* or *inner membrane* or *germ cell membrane,* from which the cell wall of future vegetative bacterium develops. It is covered by a thick layer, the *cortex* and then a multilayered thin and tough outer *spore coat,* which may be differentiated into outer and inner coat layer. In some species (*Bacillus thuringiensis*), it is covered by an additional covering, called *exosporium* or *exosporium basal layer,* which is apparently loose.

E. *Formation of endospore*

The endospore formation does not take place during active phase of growth. The sporulation starts in conditions unfavourable for the growth due to starvation, dessication, high temperature etc. The sporulation can also be induced by depleting S, C, N, Fe and PO_4 from culture medium.

During sporulation, the first detectable change is the conversion of compact nucleoid into an axial chromatin filament (Fig. 2.28). Then a transverse septum is laid down towards one pole, which separates into small and large portion. The small portion with its cytoplasm and

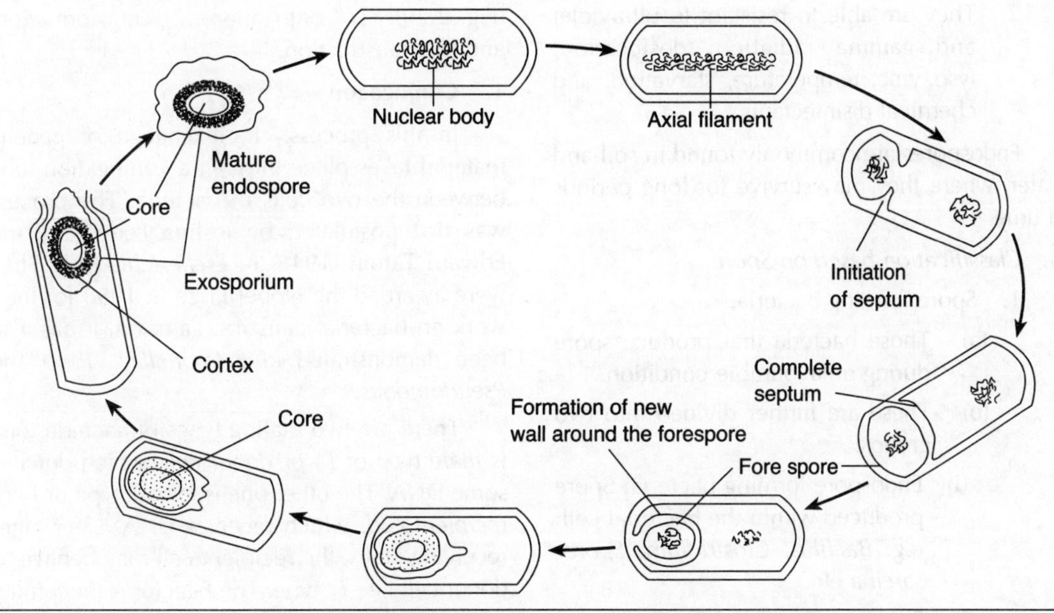

Fig. 2.28: Diagrammatic representation of endospore formation

DNA forms the *fore spore,* which later on develops into a **spore**. The membrane of large portion gradually grows around the *fore spore*. The fore spore increases in size, which becomes opaque and highly refractive, called the *endospore*. The entire process of sporulation takes place within 16 to 20 hours. The cell in which spore is formed, called *sporangium,* which remains viable for short period of time after maturation of spore. The spore is liberated by autolysis of sporangium.

The spores can survive in different adverse conditions like heat, drying, freezing, toxic chemicals and radiations. Some bacilli can resist a temperature higher than 150°C.

The endospores germinate under favourable condition which consist of three stages:

(i) *Activation* — It takes place by the induction of one or more factors such as acidic pH, heat (60°C for 1 hour), compounds containing free SH groups or through abrasion.

(ii) *Initiation* — Binding of any effector substance like L-alanine, adenosine etc. of the medium with the spore coat activates to form an *autolysin*. The autolysin destroys the peptidoglycan of the cortex. Thereafter, water is taken up and calcium dipicolonate is released. Dipicolonic acid helps to stabilise the spore protein, and both dipicolonic acid and Ca ions provide resistance to heat.

(iii) *Outgrowth* — The swelling of spore wall and disintegration of the cortex help to emerge a germ cell after breaking the spore coat, which behaves like a vegetative cell.

Each cell forms only one endospore and persists during unfavourable condition. During favourable condition, it germinates and gives rise to a single bacterial cell. *So the spore is a perennating organ and the process is called perrenation rather than multiplication*.

F. *Function of endospore*

Endospore has various functions. Some of them are given below:

1. The primary function of most endospores is to ensure the survival of a bacterium through periods of environmental stress.

2. They are able to resistant to ultraviolet and gamma radiation, desiccation, lysozyme, temperature, starvation, and chemical disinfectants.

Endospores are commonly found in soil and water, where they may survive for long periods of time.

G. *Classification based on Spore*

1. Spore forming bacteria:
   (a) Those bacteria that produce spore during unfavourable condition.
   (b) These are further divided into two groups:
      (i) Endospore forming bacteria: Spore produced within the bacterial cell. e.g. *Bacillus, Clostridium, Sporosarcina* etc.
      (ii) Exospore forming bacteria: Spore produced outside the cell, e.g. *Methylosinus*.
2. Non-sporing bacteria: Those bacteria which do not produce spore, e.g. *E. coli*, *Salmonella*.

GENETIC RECOMBINATION

In bacteria, genetic recombination takes place through interchange of genetic material. This is accomplished by three processes (Fig. 2.29): (1) Conjugation, (2) Transformation, and (3) Transduction.

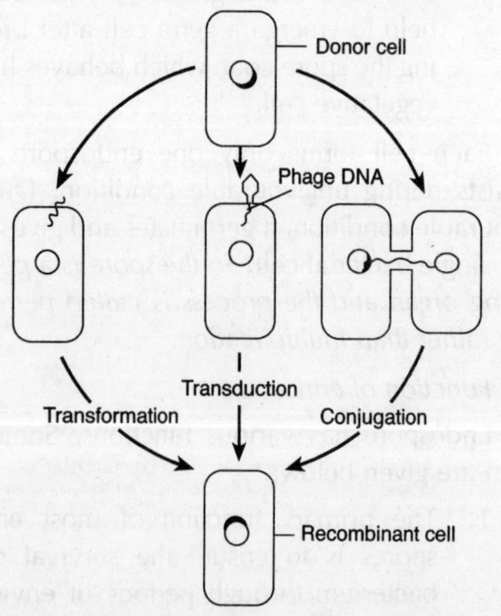

Fig. 2.29: Different types of recombination

1. Conjugation

In this process, the exchange of genetic material takes place through a conjugation tube between the two cells of bacteria. The process was first postulated by Joshua Lederberg and Edward Tatum (1946) in *Escherichia coli*. They were awarded the Nobel Prize in 1958 for their work on bacterial genetics. Later on, it has also been demonstrated in *Salmonella*, *Vibrio* and *Pseudomonas*.

There are two mating types of bacteria, one is *male type* or F^+ or *donor cell*, which donates some DNA. The other one is female type or F^- or *recipient cell*, which receives DNA. Later, after receiving DNA, the recipient cell may behave as donor cell i.e., F^+ type. The F-factor is the fertility factor, sex-factor or F-plasmid present in the cell of F^+ i.e., donor cell or male type. The plasmid takes part in conjugation is called episome.

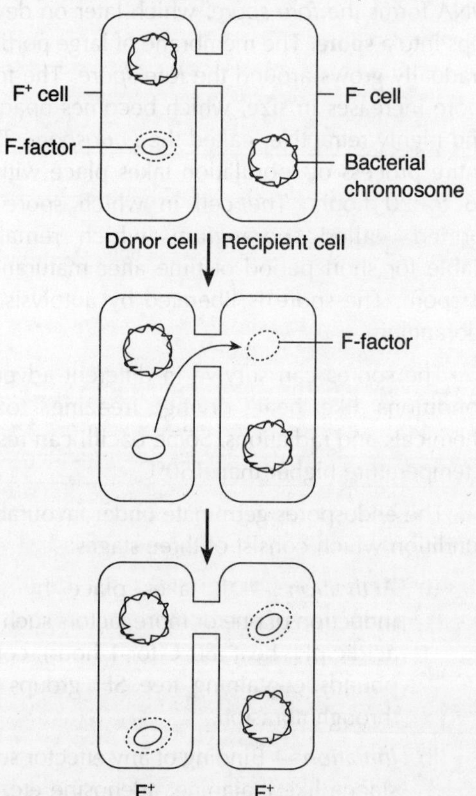

Fig. 2.30: Recombination by F-factor in conjugation

In this process, two cells of opposite mating type i.e., F^+ and F^- become temporarily attached with each other by sex pilus (Fig. 2.30). The sex pilus has a hole of 2.5 μm diameter through which DNA can pass from donor to recipient cell. The F-factor or F-plasmid is a double stranded DNA loop, present in the cytoplasm; apart from the nucleoid. The F-factor contains about 20 genes. After the establishment of conjugation tube, the F-factor prepares for replication by the *rolling circular mechanism*. The two strands of F-factor begin to separate from each other and one of them passes to the recipient i.e., F^- cell. After reaching in F^- cell, enzymes synthesise a complementary strand that forms a double helix, which bends into a loop. The conversion process is thus completed. In the donor cell i.e., in F^+, a new DNA strand also forms to complement the left over DNA strand of the F-factor.

There is another type of conjugation where passage of nucleoid DNA takes place through conjugation tube. Strains of bacteria are known as **Hfr** (high frequency of recombination) **strain**. William Hayes discovered such strains of *E. coli* in 1950s. The Hfr factor is also called episome. In Hfr strain, the F-factor is attached with the nucleoid DNA i.e., the bacterial chromosome.

In this process, Hfr and F^- cells become attached with each other by sex pilus (Fig. 2.31). At the point of attachment of F-factor, the bacterial chromosome opens and a copy of one strand is formed by the rolling circular mechanism. A portion of single stranded DNA then passes into the recipient cell through pilus. Due to agitation in medium, the conjugation tube may not survive for long time because of broken pilus. Thereby, the total length of transfer DNA may not be able to take entry to the recipient cell. *The behaviour of the transferred DNA depends on the presence and absence of F-factor*:

(i) If F-factor is indeed transferred, then it usually remains detached from the chromosome of recipient cell and enzymes synthesise a complementary DNA strand. The factor then forms a loop and exists as a plasmid, thereby the recipient cell becomes a donor.

(ii) If F-factor remains at the rear end of the transfer DNA during its entry to the recipient cell, the F-factor may not be able to take entry due to broken pilus and only a portion with new genes (Fig. 2.31) takes up the entry. Thereby, the F^- strain remains as recipient one. In F^- strain, genetic recombination takes place between donor fragment and recipient DNA.

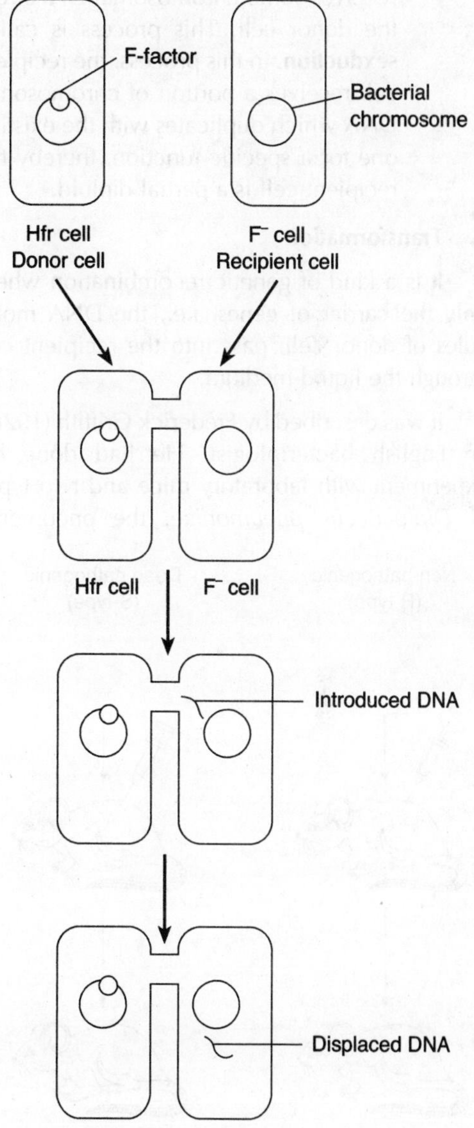

Fig. 2.31: Recombination by fragment of DNA in conjugation

(iii) Sometimes, if the F-factor gets free from the Hfr cell and maintains an independent status, then the Hfr cell converts to a F⁺ cell. Sometimes during the leaving of F-factor from the bacterial chromosome, it takes a segment of chromosomal DNA. The F-factor with segment of chromosomal DNA is called F′-factor. Later on, during conjugation, when this F′-factor is transferred, the recipient cell receives some chromosomal DNA from the donor cell. This process is called **sexduction**. In this process, the recipient cell receives a portion of chromosomal DNA which duplicates with the existing one for a specific function, thereby the recipient cell is a partial diploid.

2. Transformation

It is a kind of genetic recombination where only the carrier of genes, i.e., the DNA molecules of donor cell, pass into the recipient cell through the liquid medium.

It was described by Frederick Griffith (1928), an English bacteriologist. He had done his experiment with laboratory mice and two types of *Diplococcus pneumoniae*, the pneumonia causing organism. One type has rough (R) non-capsulated cells and another one with smooth (S) capsulated cells. The R-type is non-pathogenic, while the S-type is pathogenic. The process of transformation is mentioned below (Fig. 2.32):

(i) When live non-pathogenic (R-type) cells are injected in mice, the mice remain alive.

(ii) When dead pathogenic (S-type) cells are injected in mice, mice also remain alive.

(iii) When pathogenic (S-type) cells are injected in mice, they suffer from pneumonia and died.

(iv) When live non-pathogenic (R-type) cells are mixed with dead pathogenic (S-type) cells and are injected in mice, they also suffered from pneumonia and died. On isolation of dead tissue of mice, the smooth (S) capsulated cells are found on agar. The above experiment indicates the conversion of R-type to S-type, called *transformation*.

Later, James L. Alloway (1932), transformed the rough type cells to smooth type, by using the fragments from dead smooth-type cells and confirmed Griffith's work.

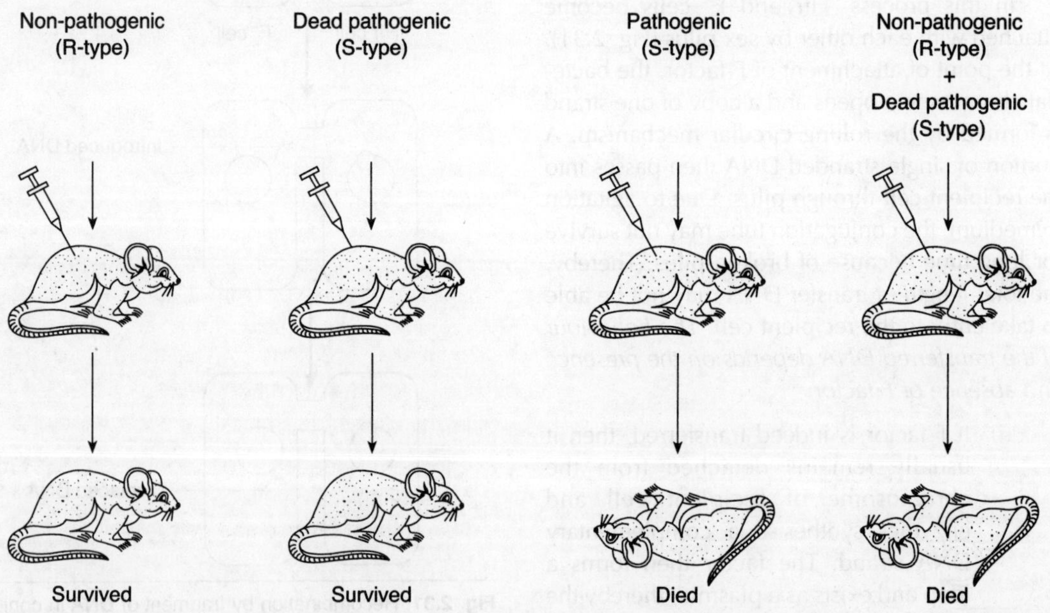

Fig. 2.32: Griffith's experiment with mice and *Diplococcus pneumoniae*

Further, Oswald T. Avery, Colin M. MacLeod and Maclyn N. McCarty (1944) also found that DNA isolated from the fragments could induce the transformation. Their experimental result was the first proof of DNA as the genetic material in living organism. The possible mechanism of transformation can be explained (Fig. 2.33). The transformation takes place in a few cell of the mixed population. It is an important method of genetic recombination. A few donor cells break apart and an explosive release and fragmentation of DNA take place. A fragment of double stranded DNA (10–20 genes) then gets attached with the recipient cell for entry (Fig. 2.33). During entry one strand of the fragment becomes dissolved by enzyme leaving the second strand, which then passes to the recipient cell through cell wall and cell membrane. After entry, a portion of single strand of double stranded DNA of recipient cell gets displaced by enzyme and then replaced by the DNA of donor cell. The displaced DNA is then dissolved by other enzyme. Thus the recipient cell becomes transformed which will display its own as well as the characters of the newly incorporated DNA.

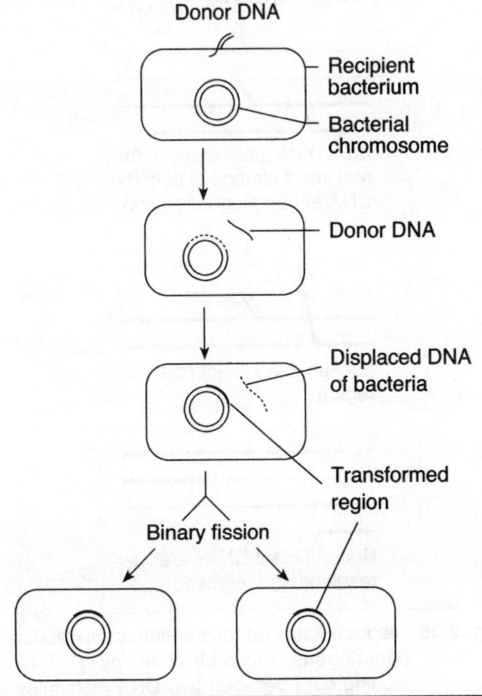

Fig. 2.33: Diagrammatic representation of Transformation

Detailed mechanism of transformation, with especial emphasis on natural and induced competence and DNA uptake

Thus the transformation takes place by horizontal gene transfer through uptake of free DNA by other bacteria. This transformation takes place either spontaneously by taking DNA from the environment, i.e., **Natural**, or by forced uptake under laboratory condition i.e., **Artificial process.**

A. Natural transformation

During natural transformation, free naked fragments of double stranded DNA of donor cell become attached to the surface of the recipient cell. The free double stranded DNA molecules may be available in the medium by lysis or natural decay of bacteria (Fig. 2.34).

After attachment of donor double stranded DNA with the surface of recipient bacterium, one strand is digested by the bacterial nuclease and the remaining one strand is then taken in by an energy-requiring transport system. This uptake of DNA takes place during late logarithmic phase of growth.

During this process, Rec A type of protein plays an important role. The Rec A protein binds with the single stranded DNA and forms a coating around the DNA (Fig. 2.34). The coated single stranded DNA and DNA of recipient cell then move close to each other to get homologous sequence. After reaching at proper place, the Rec A protein actively displaces one strand of chromosomal DNA of recipient cell. The process requires hydrolysis of ATP to get energy. The incoming DNA strand is then integrated with one strand of bacterial DNA by base pairing and ligation takes place by DNA ligase. The displaced DNA strand of recipient cell is then digested by cellular DNase activity. Any mismatch between the two strands of new region is corrected by them. Thus the transformation is completed.

If the introduced single stranded DNA fails to recombine with the recipient DNA, it is digested by cellular DNase and gets lost.

B. Artificial transformation

The *E. coli*, an ideal material for research is not transformed naturally. Later, it has been

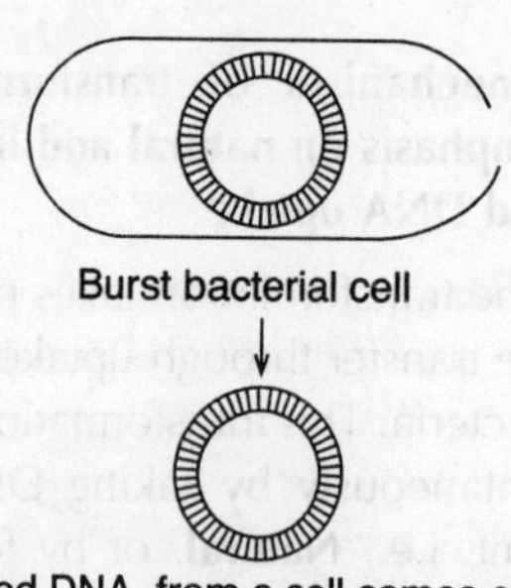

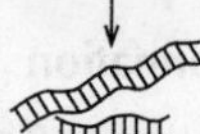

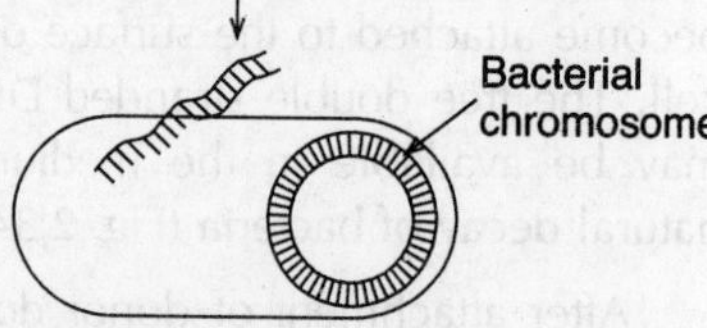

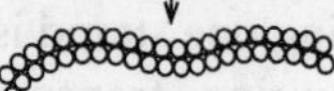

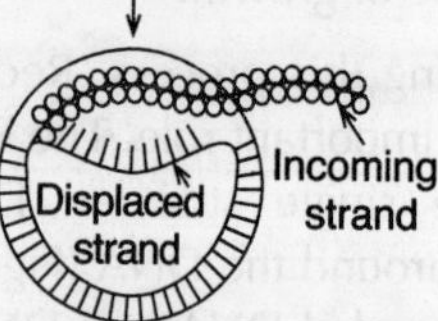

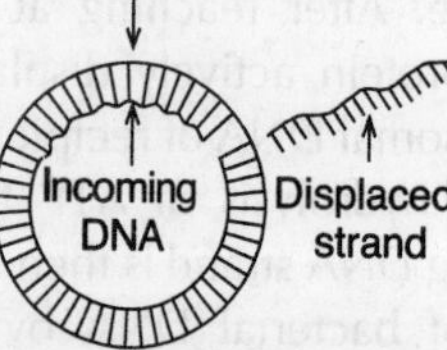

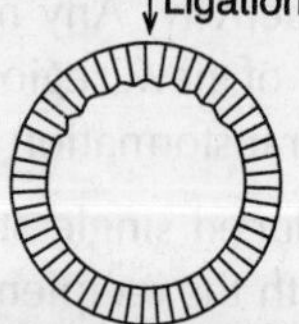

Fig. 2.34: Diagrammatic representation of the natural transformation in bacteria

discovered that the transformation in *E. coli* can be done by special physical and chemical treatments. This can be done by exposure of *E. coli* to high voltage electric field and also by high concentration of $CaCl_2$. Under such condition, the bacterial cells are forced to take up foreign DNA. This type of transformation is called artificial.

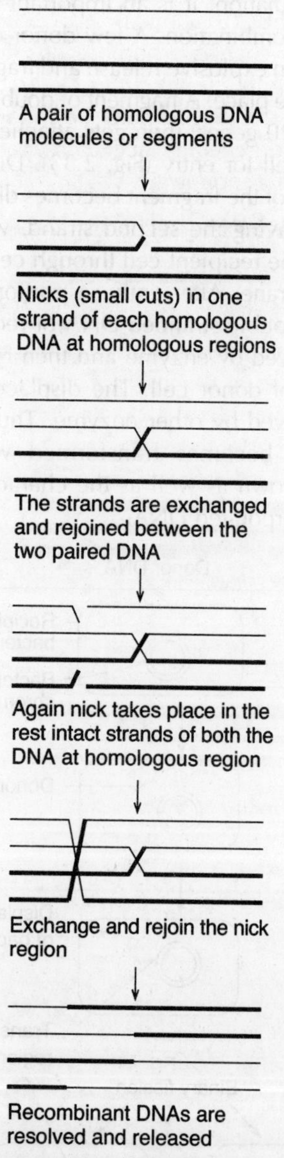

Fig. 2.35: Diagrammatic representation of a model of homologous recombination by reciprocal crosing over between two DNA molecules or DNA segments

During this process, the recipient bacterial cells are able to take up double stranded DNA fragments.

Physical or chemical treatment forces the recipient bacterial cell to receive exogenous DNA. The foreign DNA is then integrated with the chromosome by homologous recombination, mediated by Rec A protein. The Rec A protein catalyses the annealing of two DNA segments and exchange of homologous region. This involves nick i.e., small cut of DNA strands and rejoining of exchanged parts i.e., breakage and reunion. The generally accepted model of the above phenomenon is given below (Fig. 2.35):

Difference between conjugation and transformation

Conjugation	Transformation
1. It takes place between the members of same species and also between the members of two different species of *Escherichia* and *Salmonella* or *Shigella*; *Salmonella* and *Serratia*.	It takes place between the members of same species.
2. Genetic recombination is temporary.	Genetic recombination is permanent.

3. Transduction

It is a special method of genetic recombination where genetic material is transferred from the donor to the recipient cell through a non-replicating bacteriophage — temperate bacteriophage. This was discovered by Joshua Leaderberg and Norton Zinder (1952) during their research with *Salmonella typhimurium*. In this process, a small fragment of bacterial DNA is incorporated into an attacking bacteriophage (i.e., virus which infect bacteria) and when this bacteriophage infects a new bacterial cell, it transfers the genetic material into it, and thus genetic recombination takes place.

Transduction are of two types: A. Specialised transduction, and B. Generalized transduction.

A. **Specialised transduction**. In this process, the bacteriophage gets attached to a bacterial cell wall at the receptor site and the nucleic acid of bacteriophage is transferred into the cytoplasm of the host cell (Fig. 2.36A). The phage does not cause the lysis of the host bacterium. In the bacterial cell, the phage nucleic acid codes for the synthesis of specific proteins, the repressor proteins. The repressor proteins prevent the virus to produce the material require for its replication. In the bacterial cell, the viral DNA may exist as a fragment in the cytoplasm or it may attach itself to the chromosome, known as prophage (Fig. 2.36B). The bacterial cell which carries the prophage is called lysogenic and the phenomenon where the phage DNA and bacterium exist together is called lysogeny. The bacterial cell may remain lysogenic for many generations and during this period the viral DNA replicates many times together with the bacterial chromosome. However, in course of time, the phage stops the synthesis of repressor proteins in the bacterial cell, and then the synthesis of phage components starts. Now the phage DNA separates from the bacterial chromosome and starts the synthesis of phage proteins (Fig. 2.36C). During this separation, a number of genes of the bacterium get attached to it. These attached genes keep on replicating along with the phage DNA (Fig. 2.36D) and later on it develops into phage particles, those come out from the bacterial cell by bursting (Fig. 2.36E). When the new phage particle (Fig. 2.36F) infects a new bacterial cell (Fig. 2.36G, H), the attached bacterial genes present along with phage particle enters in the chromosome of the new bacterium and causes recombination (Fig. 2.36I). Thus the new bacterial cell contains its own genes and several genes from the parent bacterial cell. This type of transduction is known as specialised transduction, which is an extremely rare event.

B. **Generalised transduction**. This process of transduction is more common than specialized transduction. Here the prophage particle is present in the cytoplasm of the infected bacterial cell (Fig. 2.36J). In this process, the phage DNA starts synthesising new

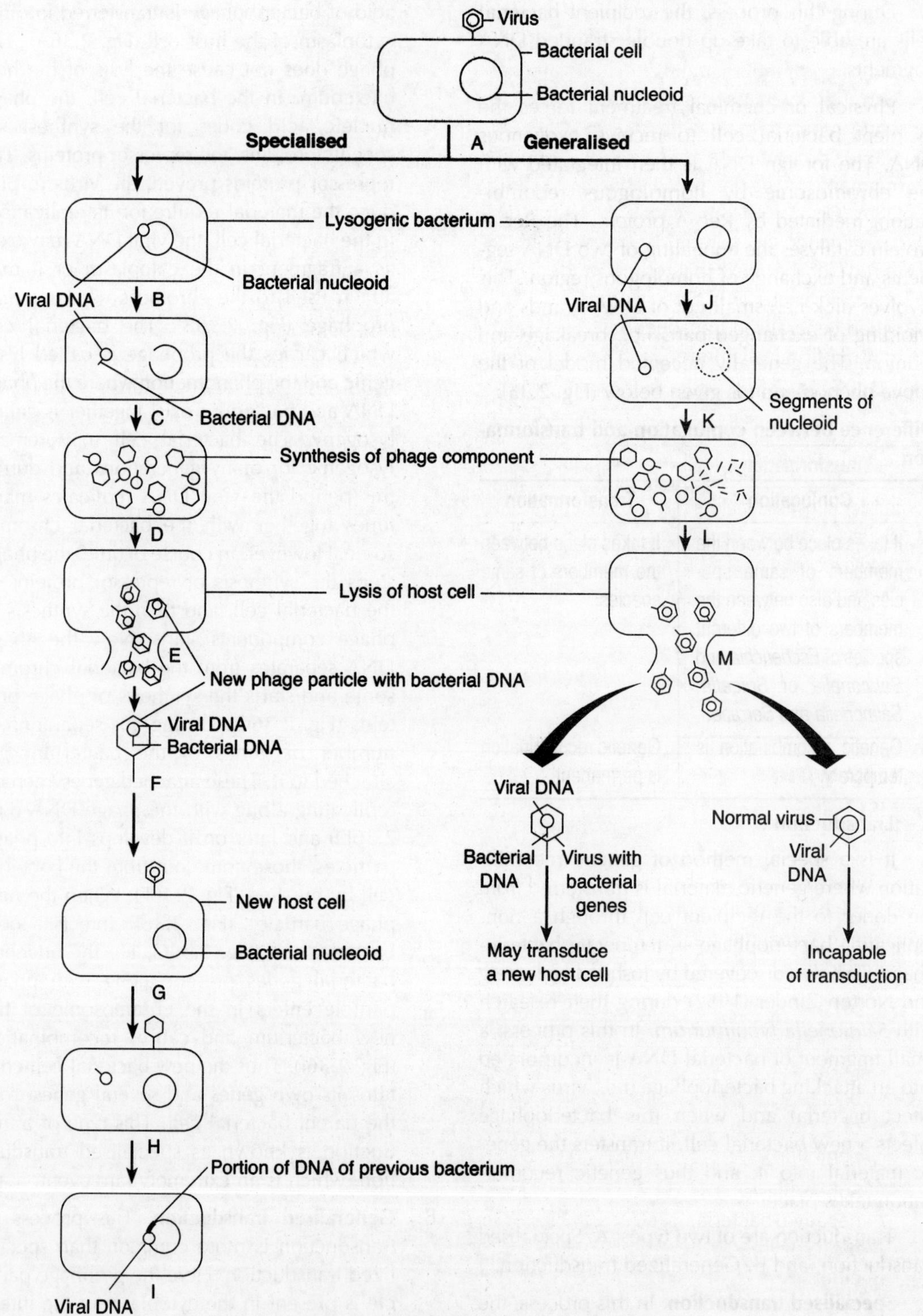

Fig. 2.36: Diagrammatic representation of Transduction

phages. During this process chromosome of bacterial cell gets fragmented (Fig. 2.36K) and some of the fragments become attached with the DNA of some new phage particle, while others remain with phase DNA (Fig. 2.36L). When the newly formed phage with fragment of bacterial chromosome in its DNA (Fig. 2.36M) attacks a new bacterium, the gene of the parent bacterium is transferred to the new bacterium and causes recombination. This type of transduction is called generalised transduction. This type of transduction is also rare.

Difference between transformation and transduction

Transformation	Transduction
1. Transfer of genetic material takes place from donor to recipient bacterium through the liquid medium.	Transfer of genetic material takes place from donor to recipient bacterium through a Bacteriophage.
2. The enzyme deoxyribonuclease can completely check the process.	The deoxyribonuclease has no effect on transduction.

BACTERIA WITHOUT CELL WALL (WALL-LESS FORMS)

There are four forms of bacteria available in nature, where cell wall is absent. These forms are either available naturally or may develop later by penicillin or lysozyme treatment in artificial or natural condition. These are:

1. **Mycoplasma**: It is a genus of naturally occurring bacteria which lack cell wall.
2. **L-forms**: It is cell-wall-deficient forms of bacteria, usually produced in the body of a patient treated with the antibiotic, penicillin.
3. **Spheroplasts**: It is derived from Gram-negative bacteria; produced artificially by lysozyme or by growing with penicillin or by any other agent able to break down the peptidoglycan layer.
4. **Protoplasts**: It is derived from Gram-positive bacteria and totally lacking cell wall; produced artificially by lysozyme in hypertonic medium.

MYCOPLASMAS

Mycoplasmas are the smallest among the known aerobic prokaryotes (Fig. 2.37). They were first discovered by Pasteur in 1843, during his work on the possible causal agent of pleuropneumonia of cattle. Thus they were called pleuropneumonia-like organism (PPLO). Pasteur was unable to isolate them in pure culture. Later, Nocard and Roux (1898), the French microbiologists, were successful in growing them in pure culture medium containing serum and confirmed by inoculation and subsequent expression of disease in healthy cattle.

Mycoplasmas are commonly found in soil, hot spring, sewage water and also in plants and animals including man. Borrel (1910) named these organisms *Asterococcus mycoides*. Later, in 1929, Nowak placed them under genus Mycoplasma.

Classification. Based on nutritional requirement, mycoplasmas are divided into the following three genera:

1. **Mycoplasma**. They require cholesterol for their growth. They parasitise on animals including man by causing damage to the mucous membranes and different joints of the body.
2. **Acholeplasma**. They do not require cholesterol for their growth. They are available in

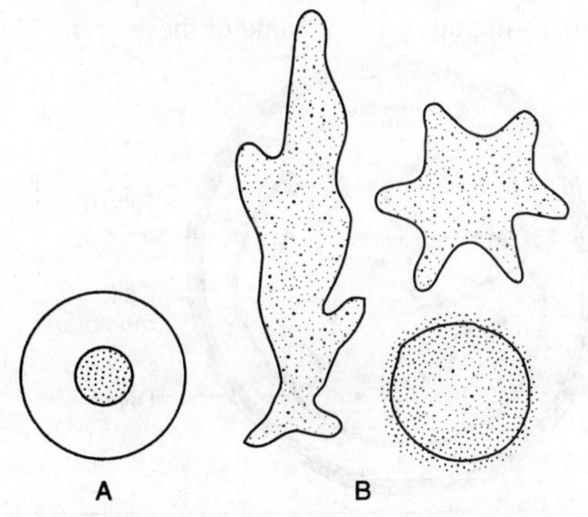

Fig. 2.37: Mycoplasmas: A. Single colony, B. Cells of various shapes

sewage water and soil as saprophytes and in vertebrates and also in plants as parasites.

3. **Thermoplasma**. They also do not require cholesterol for their growth. They are aerobic microorganisms showing good growth in acidic pH between 0.96–3.0, with an optimum temperature of 59°C.

Structure. The cell is devoid of cell wall which makes them readily deformable showing irregular and variable shapes. They may be ring-like, granular, coccoid, pear-shaped, filamentous, etc. (Fig. 2.37). The filaments are of two types: unbranched or branched. The cells are very small and measure 0.3–0.9 µm in diameter.

The cells are covered by cytoplasmic (lipoprotein) membrane (Fig. 2.38). Cell membrane covers the cytoplasm which contains nucleoplasm-like structure and ribosomes. The genetic material is composed of DNA and RNA. It is about less than 50%, the amount present in other prokaryotic organisms. The amount of RNA (8%) is more than DNA (4%).

They are usually non-motile, but some forms show gliding movements. They reproduce by vegetative means i.e., by binary fission and budding.

They are sensitive to antibiotics like chloramphenicol, streptomycin, erythromycin etc., but are insensitive to penicillin, ampicillin etc., due to the absence of cell wall.

Some Mycoplasmal diseases. Mycoplasmas cause different serious diseases in plants and animals including man. Some of these are:

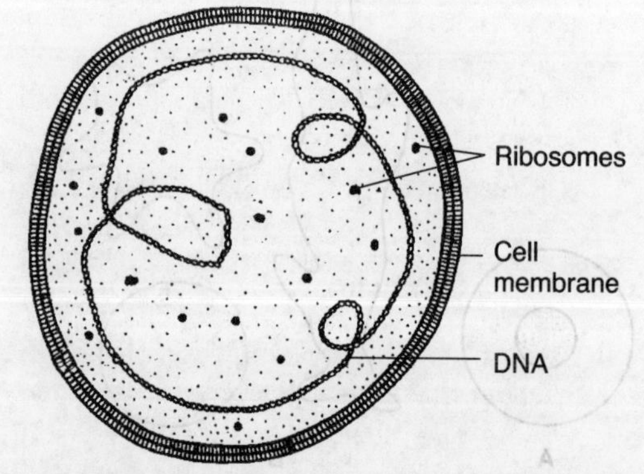

Fig. 2.38: Mycoplasmas: Structure of a cell (Electron micrograph).

(a) **Plant diseases**: (i) Little leaf disease of brinjal, (ii) Bunchy top of papaya, (iii) Big bud of tomato, (iv) Witches broom of legumes, (v) Yellow dwarf of tobacco, (vi) Strip disease of sugarcane, (vii) Clover dwarf, (viii) Cotton virescence.

(b) **Human diseases**: (i) Primary atypical pneumonia (PAP) by *Mycoplasma pneumoniae*. (ii) *Mycoplasma hominis* causes pleuropneumonia, prostatitis, inflammations of genitals etc. (iii) Mycoplasma *fermentants* causes infertility in man.

(c) **Animal diseases**: (i) *Mycoplasma agalactia* causes agalactia of goat and sheep, (ii) *Mycoplasma mycoides* causes pleuropneumonia of cattle, (iii) *M. bovigenitalium* causes inflammation of genitals of different animals.

Uses of Mycoplasma

Mycoplasma is used in various purpuses like, antibiotic discovery, patch clamping, cell lysis, and transfection. These are given below:

1. Antibiotic discovery

From 1960s to 1990s, **Merck and Co**. used a spheroplast screen as a primary method for discovery of antibiotics that inhibit cell wall biosynthesis. In this screen devised by **Eugene Dulaney**, growing bacteria were exposed to test substances under hypertonic conditions. Inhibitors of cell wall synthesis caused growing bacteria to form spheroplasts. This screen enabled the discovery of antibiotics like fosfomycin, cephamycin C, thienamycin and several carbapenems.

2. Patch clamping

Specially prepared giant spheroplasts of Gram-negative bacteria can be used to study the function of bacterial ion channels through a technique called **patch clamp**. This was originally designed for characterizing the behavior of neurons and other excitable cells. To prepare the giant spheroplasts, bacteria are grown in a medium containing chemicals that prevent the cells from dividing completely. This causes bacteria to form long "snakes" that share a single membrane and cytoplasm. After a period of time, the cell walls of the "snakes" are digested, and the bacteria collapse into very large spheres, surrounded by a single lipid bilayer. The membrane can then be analyzed on a patch clamp apparatus to determine the phenotype of the ion channels embedded in it. It is also common to overexpress a particular channel to amplify its effect and make it easier to characterize.

The technique of patch clamping giant *E. coli* spheroplasts has been **used extensively** to study the native mechano-sensitive channels (MscL, MscS, and MscM) of *E. coli* since 1987. Recently, it has been extended to study other heterologously expressed ion channels and it has been shown that the giant *E. coli* spheroplast can be used as an ion-channel expression system comparable to the Xenopus oocyte.

3. Cell lysis

Yeast cells are normally protected by a thick cell wall which makes extraction of cellular proteins to be difficult. Enzymatic digestion of the cell wall with zymolyase, creating spheroplasts, renders the cells vulnerable to easy lysis with detergents or rapid osmolytic pressure changes.

Gram-negative bacteria, such as *E. coli* are a subject to outer cell lysis by osmotic shock, lysozyme and ethylenediaminetetraacetate (EDTA), the chelating agent, with subsequent release of periplasmic gap content and spheroplasts containing inner cell content into the sucrose medium. Lysozyme catalyzes the hydrolysis of the peptidoglycan layer, while EDTA destroys the outer membrane facilitating access of enzyme to the inner layers of the cell wall. The periplasmic enzymes released during the rupture of *E. coli* outer cell wall include: an alkaline phosphatase, a cyclic phospho-diesterase, an acid phosphatase, and a 5′-nucleotidase. Sucrose-tris (hydroxymethyl) aminomethane-HCl medium is essential to the preservation of the integrity of the enzymes upon lysis. EDTA aids cell hydrolysis by binds with divalent ions, such as Ca^{2+}, and removing them from the wall thus softening the wall for further lysozyme action.

4. Transfection

Bacterial spheroplasts, with suitable recombinant DNA inserted into it, can be used to transfect animal cells. Spheroplasts with recombinant DNA are introduced into the media containing animal cells and are fused by polyethylene glycol (PEG). With this methodology, nearly 100% of the animal cells may take up the foreign DNA.

SPHAEROPLAST

A **spheroplast** is a gram-negative Bacterial cell from which the *cell wall has been almost completely removed*, by the action of **penicillin** (an antibiotic). The name was credited from the fact that after a microbe's cell wall is digested, due to membrane tension the cell acquired a characteristic *spherical shape*. Spheroplasts are **osmotically fragile**, and will lyse if transferred to a hypotonic solution.

Economic importance of Bacteria

Bacteria have great economic value in our daily life. Some of them are harmful and others are useful to the living world. They cause diseases of plants and animals, including human beings; which at times become so severe that it may threaten the existence of particular plants and animals. On the other hand, some of them are useful in maintaining the carbon and nitrogen balance in the nature and also have uses in different industries. Some of the useful and harmful activities of bacteria are given below:

A. Harmful activities. Some bacteria are harmful to human affairs in different ways:

1. *Pathogenic bacteria*: These bacteria cause great losses to plant and animal population by causing several diseases. **Some important bacterial diseases of plants and animals, including human beings** is given below (Table 2.13–2.15 and 2.18).

2. *Reduction of soil fertility*: Some facultative anaerobic bacteria are available mostly in the oxygen deficient soil which reverse the nitrifying process, thereby causing the loss of a part of its combined nitrogen. They break down

the nitrates in a stepwise manner either to ammonia through **assimilatory nitrate reduction** or to N_2 through **dissimilatory nitrate reduction**. The two pathways have a common intermediate, the nitric oxide:

Common pathway

$$NO_3^{-3} \longrightarrow NO_2^{-3} \longrightarrow NO$$

Assimilatory nitrate reduction

$$NO \begin{cases} NH_2OH \longrightarrow NH_3 \\ N_2O \longrightarrow N_2 \end{cases}$$

Dissimilatory nitrate reduction

The free nitrogen goes to the atmosphere, resulting in the reduction of soil fertility by loosing valuable N_2 source, the nitrate. The bacteria those cause denitrification are called **denitrifying bacteria**. e.g., *Bacillus cereus, Paracoccus denitrificans*, and *Pseudomonas aeruginosa*.

3. *Spoilage of food*: In favourable temperature and humidity, bacteria can grow luxuriantly in many food materials. They change the flavour, appearance and smell of food.

Table 2.13: Some important plant diseases

Name of disease	Causal organism
Wilt of potato, tomato and tobacco	*Pseudomonas solanacearum*
Citrus canker	*Xanthomonas axonopodis pv. citri*
Blight of bean	*X. phaseoli*
Black rot of crucifer	*X. campestris*
Crown gall of apple, rose and tomato	*Agrobacterium tumefaciens*
Ring rot of potato	*Corynebacterium sepedonicum*
Canker of tomato	*C. michiganese*
Scab of potato	*Streptomyces scabies*
Wilt of cucurbits	*Erwinia tracheiphila*
Fire blight of apple	*E. amylovora*
Soft rot of mango	*Bacterium cartovorus*
Wilt of tobacco	*Phytobacterium solanacearum*
Leaf spot of cherry	*Coccomyces hiemates*
Bacterial leaf blight of rice	*Xanthomonas campestris pv. oryzae*

Table 2.14: Some important human diseases

Name of disease	Causal organism
Cholera	*Vibrio cholerae*
Tuberculosis	*Mycobacterium tuberculosis*
B. coli	*Escherichia coli*
Typhoid	*Salmonella typhi*
Tetanus	*Clostridium tetani*
Leprosy	*Mycobacterium leprae*
Gonorrhoea	*Neisseria gonorrhoeae*
Pneumonia	*Streptococcus pneumoniae*
Diphtheria	*Corynebacterium diphtheriae*
Syphilis	*Treponema pallidum*
Plague	*Yersinia pestis*
Bacillus dysentery	*Shigella dysenteriae*
Jaundice	*Leptospira ictero-haemorrhagiae*
Meningitis	*Neisseria meningitidis*
Whooping cough	*Bordetella pertussis*

Table 2.15: Some important animal diseases

Name of disease	Causal organism
Anthrax of sheep and elephants	*Anthrax bacilli*
Actinomycosis of cattle	*Mycobacterium bovis*
Black leg of cattle	*Clostridium chanvei*
Vibriosis	*Vibrio tetus*
Brucellosis of domestic animals	*Brucella* sp.

Different species of *Streptococcus, Micrococcus* and *Lactobacillus* cause spoilage of milk and different milk products. The exotoxin produced by *Clostridium botulinum* causes botulism disease showing the symptoms like swelling of tongue, double vision and respiratory disturbances.

4. *Pollution of water*: Some bacteria pollute water and make it unsuitable for drinking. Diseases like cholera (*Vibrio cholerae*), typhoid fever (*Salmonella typhi*) and bacillus dysentery (*Shigella dysenteriae*) are commonly transmitted by drinking water.

5. *Abortion*: Some species of *Salmonella* cause abortion of sheep, horse, goat and other animals.

6. *Biological warfare*: Bacteria causing diseases like anthrax, black-leg, tuberculosis, etc. are used as secret war-agents.

B. Useful activities

1. *Increase soil fertility*: Plants do not have the efficiency of direct utilisation of huge amount (78%) of atmospheric nitrogen. However, some free-living and symbiotic bacteria are able to fix free nitrogen into nitrogenous compounds. These bacteria are of the following three categories:

   (i) *Nitrifying bacteria*. This group of bacteria oxidises ammonia to nitrate in two steps. The first step involves the oxidation of ammonia to nitrous acid by bacteria like *Nitrosomonas*, etc. and the second step involves oxidation of nitrous acid to nitric acid by *Nitrobacter*.

Step I: $NH_3 + 1\frac{1}{2}O_2 \longrightarrow HNO_2 + H_2O + \text{Energy}$

Step II: $2HNO_2 + O_2 \longrightarrow 2HNO_3 + \text{Energy}$

   (ii) *Ammonifying bacteria*. Some saprophytic bacteria hydrolyse the proteins of dead animals and plants into amino acids and other organic nitrogenous substances. Some ammonifying bacteria like *Bacillus ramosus, Proteus vulgaris, Clostridium* sp. etc., convert amino acids into ammonia, which combines with water and carbon dioxide to form ammonium carbonate, used by many crops as nitrogen source.

   (iii) *Nitrogen fixing bacteria*. These bacteria can fix atmospheric nitrogen. They are of two types:

   (a) *Symbiotic nitrogen fixers* like *Rhizobium*, found in the root nodules of leguminous plants which fix N_2 about 100–400 kg/hectare/year. The filamentous Actinomycete, *Frankia* found in root nodules of *Alnus rugosa* can fix N_2 about 150 kg/ hectare/year.

   (b) *Free-living fixers* like *Clostridium* and *Azotobacter* can fix N_2 25–50 kg/hectare/year.

2. *Industrial importance*: From ancient time men have used microorganisms in the preparation of their food, drinks etc. Some of the uses are:

   (i) *Dairy industry*. The lactic acid bacteria are used in the preparation of

Table 2.16: Some Nitrogen fixing bacteria of Free-living and symbiotic types

Organism	Nature	Habitat
A. **Symbiotic**		
Gram-negative		
(i) *Rhizobium leguminosarum*	Rods	Root nodules of Pea (*Pisum sativum*)
(ii) *R. sesbaniae*	–do–	Stem nodules of *Sesbania*
(iii) *R. trifolii*	–do–	Root nodules of Clover
(iv) *Bradyrhizobium japonicum*	–do–	Nodules of Soybean
Gram-positive		
(v) *Frankia* spp.	Mycelial	Nodules of *Casuarina, Alnus*, etc.
B. **Free living**		
Gram-negative		
(i) *Azotobacter agills*	Cocci	Aerobic and soil inhabiting
(ii) *A. chroococcum*	–do–	–do–
(iii) *Klebsiella pneumoniae*	Rods	Can fix nitrogen in anaerobic condition
Gram-positive		
(iv) *Clostridium pasteurianum*	Rods	Obligate anaerobe and soil inhabiting.

butter, cheese, curd etc. The curd is prepared from milk by bacteria in two steps. They convert lactose sugar to glucose and then glucose to lactic acid, which sours the milk and coagulates the milk protein (casein) forming curd. Some bacteria used in dairy are *Lactobacillus plantarum, L. acidophilus, L. bulgaricus, Streptococcus lactis, S. thermophilus* etc.

(ii) *Vinegar industry*. *Clostridium acetobutylicum* and *Acetobacter aceti*, the acetic acid bacteria are used in the conversion of alcohol to acetic acid, the vinegar. This was done in wooden vats and takes several weeks to complete the process.

(iii) *Butanol and acetone production*. Butanol and acetone are produced by the action of species of *Clostridium* on molasses. They are used as solvent in different industry.

(iv) *Fibre retting*. Butyric acid bacteria such as *Clostridium tetani, C. botulinum* etc., are used in the retting of jute, hemp and flax fibres, thereby the bast-fibres become loosen and can be extracted easily.

(v) *Curing of tea and tobacco*. The curing of tea and tobacco is done by some bacteria which give particular taste, flavour or smell.

3. *Biological control of insect*. Many bacteria like *Bacillus thuringiensis* etc., secrete proteins which are highly toxic to caterpillars and insects belonging to Lepidoptera. They are, however, non-toxic to other animals and also plants, thereby the bacteria are used as ideal agent for biocontrol of insect.

4. *Degradation of petroleum*. Many bacteria like *Pseudomonas, Micrococcus, Candida* and *Achromobacter* are able to degrade petroleum hydrocarbon in water from different vehicles, thereby they check pollution.

5. *Decomposition of dead animals*. Some bacteria are able to decompose the dead organic matters into inorganic form which mix with soil and increase the soil fertility. These inorganic substances are then absorbed by the plants as nutrients.

6. *Bacteria in medicine*. Bacteria are the sources of different antibiotics and they are also used as serum and vaccine.

(i) *Sources of antibiotic*. Antibiotics are the chemical substances secreted by microorganisms, which inhibit the growth and development of other microorganisms. Antibiotics like bacitracin, polymyxin B, terramycin, streptomycin etc., are of bacterial origin (Table 2.17).

Table 2.17: Bacteria as useful sources of antibiotics

Name of Antibiotic	Producing organism
Bacitracin	*Bacillus subtilis*
Polymyxin	*B. polymyxa*
Terramycin	*Streptomyces numosus*
Streptomycin	*S. griseus*
Neomycin	*S. fradiae*
Chloromycetin	*S. venezualae*

(ii) *Serum and vaccine*. Sera (sing. serum) are used by men as therapeutic measure before infection or after development of diseases like pneumonia, diphtheria etc. Vaccines are commonly used for immunisation against diseases like cholera, typhoid etc.

(iii) *Production of vitamins*. Some bacteria produce vitamins of commercial importance. *Pseudomonas denitrificans* is used to produce Cobalamin (Vitamin B_{12}) and *Clostridium butylicum* is used in the production of Riboflavin (Vitamin B_2).

Table 2.18: List of some human bacterial pathogens and their important characters

Name of Pathogen	Disease produced	Shape	Response to Gram stain	Response to acid fast stain	Spore forming or not	Motile or non-motile	Aerobic or anaerobic	Average size (μm)
1. *Corynebacterium diphtheriae*	Diphtheria	Bacillus	Gram+ve	Acid fast +	Non-sporing	Non-motile	Aerobic	2–6 × 0.3 – 0.5
2. *Mycobacterium tuberculosis*	Pulmonary tuberculosis	Bacillus	Gram+ve	Acid fast +	Non-sporing	Non-motile	Aerobic	2–10 × 0.2 – 0.4
3. *M. leprae*	Leprosy	Bacillus	Gram+ve	Acid fast +	Non-sporing	Non-motile	Aerobic	1–8 × 0.2 – 0.5
4. *Salmonella typhi*	Typhoid fever	Bacillus	Gram+ve		Non-sporing	Motile (peritrichous flagella)	Anaerobic	2–4 × 0.6
5. *Shigella dysenteriae*	Bacillary dysentery	Bacillus (non-capsulated)	Gram–ve		Non-sporing	Non-motile	Aerobic	1–3 × 0.5
6. *Haemophilus influenzae*	Influenza	Bacillus	Gram–ve		Non-sporing	Non-motile	Aerobic	1.5 × 0.3
7. *Yersinia pestis*	Plague	Ovoid bacillus	Gram–ve		Non-sporing	Non-motile	Aerobic and facultative anaerobic	1.5 × 0.7
8. *Clostridium tetani*	Tetanus	Bacillus	Gram+ve		Spore forming	Motile	Anaerobic	2–5 × 0.4 – 0.5
9. *Peptococcus*	Urinary tract infection, Puerperal sepsis	Coccus	Gram+ve		Non-sporing	Non-motile	Aerobic	0.3–1
10. *Coxiella burnetii*	Q fever	Coccus (pleomorphic)	Gram–ve		Spore forming	Non-motile	Aerobic	0.3 – 0.4
11. *Bordetella pertussis*	Whooping cough	Cocco-bacilli	Gram–ve		Non-sporing	Non-motile	Aerobic	0.25–0.5 × 0.5–2.0
12. *Neisseria gonorrhoeae*	Gonorrhoea	Oval cocci in pairs	Gram–ve		Non-sporing	Non-motile	Aerobic	0.6–1

7. *In sewage work*. Putrifying bacteria are used to remove the solid and semi-solid constituent of sewage under anaerobic conditions. After treatment the constituents get decayed and liquefied. Those are filtered and the liquid is drained out in the river.

8. *Preservation of green fodder*. Some bacteria are used in the preservation of green fodder in pits by the **ensilage process.**

CYANOBACTERIA

Cyanobacteria are also known as blue green bacteria. They are the largest group of photosynthetic prokaryotes. Since their recognition as a biological group from the early part of 19th century, the cyanobacteria have been considered as a division or class of algae, the cyanophyta (blue green algae). **The detailed description of the group is available under algae (Chapter 3).**

Role of microorganisms in biogeochemical cycling

Soil fertility is maintained either artificially through the application of fertilizer or naturally through biogeochemical cycling. Thus, addition of fertilizer through natural process is very much

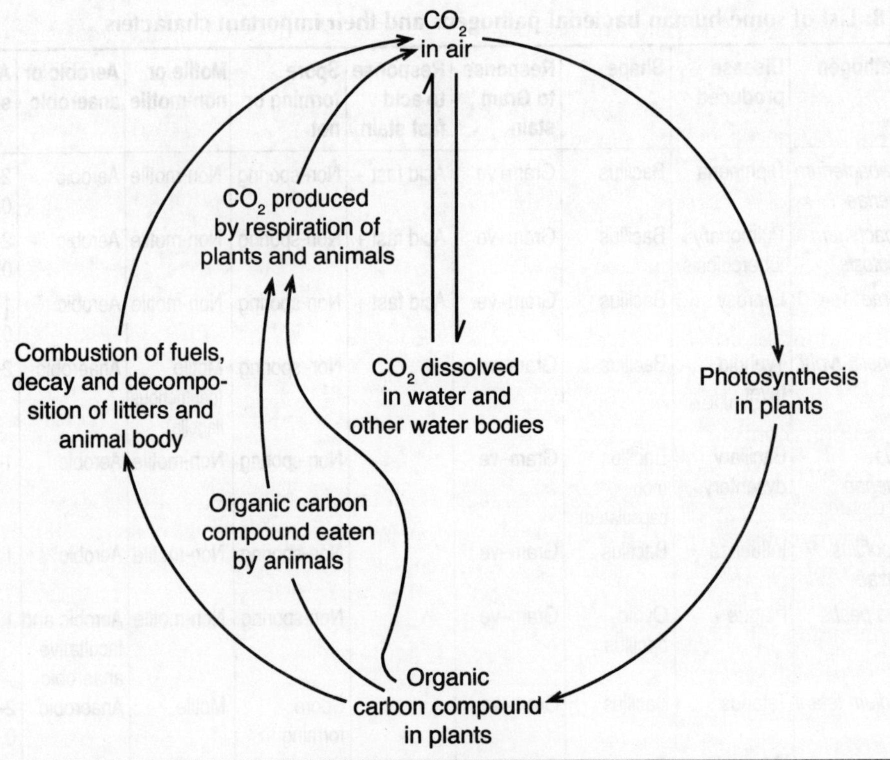

Fig. 2.39: Carbon cycle

useful for the plant growth. This can be done by microorganisms of different groups through their enzymatic action. The major elements available in living biomass are carbon, hydrogen, nitrogen, oxygen, sulphur, phosphorus etc. Out of the above elements, only carbon cycle and nitrogen cycle are briefly discussed:

Carbon cycle

Carbon is an essential element that can be cycled from inorganic to organic form. It is present in nature as inorganic CO_2 and in living and dead organisms as various organic matters. The cyclic conversion is maintained between inorganic and organic form (Fig. 2.39).

CO_2 becomes available in nature by microbial decomposition of organic substance, burning of fuels such as coal and petroleum; and respiration in plants and animals. Carbonates present in rocks and sediments also contribute in this respect.

The inorganic CO_2 are converted to organic carbon by different types of autotrophs i.e., photo-lithotrops and chemolithotrophs. After the fixation of carbon into organic compounds, it can move in heterotrophic organisms. Later on, the degradation of organic compounds takes place in soil and CO_2 goes back in the atmosphere.

The decomposition of organic matters takes place by both aerobic and anaerobic condition. During aerobic degradation, 60-90% of carbon of organic residue is utilised as a source of energy, releasing CO_2 in atmosphere. But in anaerobic condition, about 95-98% carbon is released as methane (CH_4). Methane again goes back to nature as CO_2 through oxidation.

Decomposition of organic matters in soil

Organisms like bacteria including actinomycetes and also fungi of different groups such as Zygomycotina, Ascomycotina, Basidiomycotina and Deuteromycotina play a major role in the decomposition of organic matters in the soil.

The decomposition process starts while the plants are at the senescent stage and it continues throughout their life-cycle. The different types of microbes decompose at different level of decomposition. Most of the work in this respect has been done on fungi. Different steps of decomposition — as done by fungi — are given:

1. The primary colonisers are weak parasites and/or saprophytes of sugar fungi i.e., the members of Mucorales etc., which colonise in senescent tissues.
2. The secondary colonisers grow in dead tissues. In the first phase, members of Ascomycotina and their imperfect forms decompose cellulose and in this stage sugar fungi also become active. Finally, members of Basidiomycotina appear and decompose lignin.

The plant parts comprise chiefly of cellulose, hemicellulose, pectic substances and lignin. These are degraded by the following fungi:

1. *Cellulose*: Cellulose is broken down into simple sugars. The cellulose decomposers are comprised of Ascomycotina and Deuteromycotina, represented by various species of the genera *Aspergillus, Penicillium, Fusarium, Chaetomium, Trichoderma, Phoma, Cladosporium* etc.
2. *Hemicellulose*: The fungi involved in the breakdown of hemicelluloses are the species of different genera such as *Aspergillus, Penicillium, Chaetomium, Fusarium, Trichoderma, Glomerella, Alternaria* etc.
3. *Pectic substances*: Microbial breakdown of pectic substances are caused by pectinases, produced by different microbes like *Aspergillus, Penicillium, Rhizoctonia, Rhizopus, Monilia* etc.
4. *Lignin*: It is the hardest chemical of woody parts of the plants, degraded by different members of Agaricales and Aphylloporales under Basidiomycotina.

In addition to fungi, both aerobic (bacilli and pseumonads) and anaerobic (methanogens) bacteria and also Actinomycetes are involved in the decomposition of organic matters.

Nitrogen cycle

Nitrogen is one of the important elements in biological compounds, mainly of nucleic acid and protein and, therefore, it is essential for life. Atmosphere contains about 78% nitrogen, but free nitrogen cannot be utilised by most of the organisms, except a few blue green algae and some bacteria.

The fixation of nitrogen takes place by physical, chemical and biological means. Small amounts of N_2 are converted to ammonia by electrical discharges in atmosphere i.e., by physical force, and settle down on earth by rain. About 30 million metric tonnes of N_2 are produced by industry, however, about 200 million metric tonnes of nitrogen are fixed every year by biological organisms. The biological N_2 fixation takes place by a few organisms like:

1. Symbiotic bacteria: *Rhizobium, Bradyrhizobium, Frankia.*
2. Free-living bacteria: *Azotobacter, Azomonas, Dersia* etc.
3. Blue green algae: *Nostoc, Anabaena,* etc., fix nitrogen inside heterocyst where oxygenic photosynthesis does not occur.

In nitrogen cycle, free N_2 gas of atmosphere is converted into ammonia or oxidised to nitrate at different stages. Blue green algae and N_2 fixing bacteria play a significant role in converting the atmospheric gaseous nitrogen into organic nitrogenous compounds and, finally, to nitrate, which is soluble in water. Nitrates are absorbed and utilised by the plants for the synthesis of amino acids vis-a-vis proteins. The plants are consumed by herbivores. When the plants and animals die, they are decomposed by bacteria, thus the N_2 becomes released in the atmosphere.

The N_2 cycle (Fig. 2.40) thus consists of the following steps:

1. Nitrogen fixation
2. Ammonification
3. Nitrification

4. Denitrification, and finally
5. Release of gases in the atmosphere.

Biological nitrogen fixation: The major share of nitrogen fixation is occupied by biological N_2 fixation. The biological N_2 fixation of atmospheric nitrogen depends on the nitrogenase enzyme system, composed of nitrogenase and nitrogenase reductase. Nitrogenase is very much sensitive to O_2 and it becomes irreversibly inactivated on exposure even at low concentration of O_2. In leguminous plants, N_2 fixation takes place in root nodules, where the enzyme is protected by red pigment leghemoglobin.

The plants are eaten by herbivores and fixed nitrogen goes into their body and after their death, it becomes decomposed and mixed with the soil.

The nitrogen may be fixed in the soil by the following means:

(a) Physical fixation takes place by lightning and thereby N_2 comes down in soil through rain.

(b) There is huge industrial production of N_2 fertiliser, these are mixed with the soil during cultivation.

(c) Farmyard manure is one of the major sources of N_2.

(d) Major share of N_2 is contributed by the biological fixers.

Ammonification: In this process nitrogen in organic matter of dead plants and animals is converted to ammonia and amino acids.

Urea is applied in the field as additional nitrogen source. The microbial decomposition of urea causes the release of ammonia which is returned to atmosphere or may go to neutral aqueous environments as ammonium ions.

Nitrification: This is the process of conversion of ammonia to nitrate. It takes place in two steps. Ammonia is first oxidised to nitrous acid by *Nitrosomonas, Nitrospira* and *Nitrococcus*. The molecular O_2 acts as electron acceptor in this step. In the second step, nitrous acid becomes oxidised and converted into nitric acid,

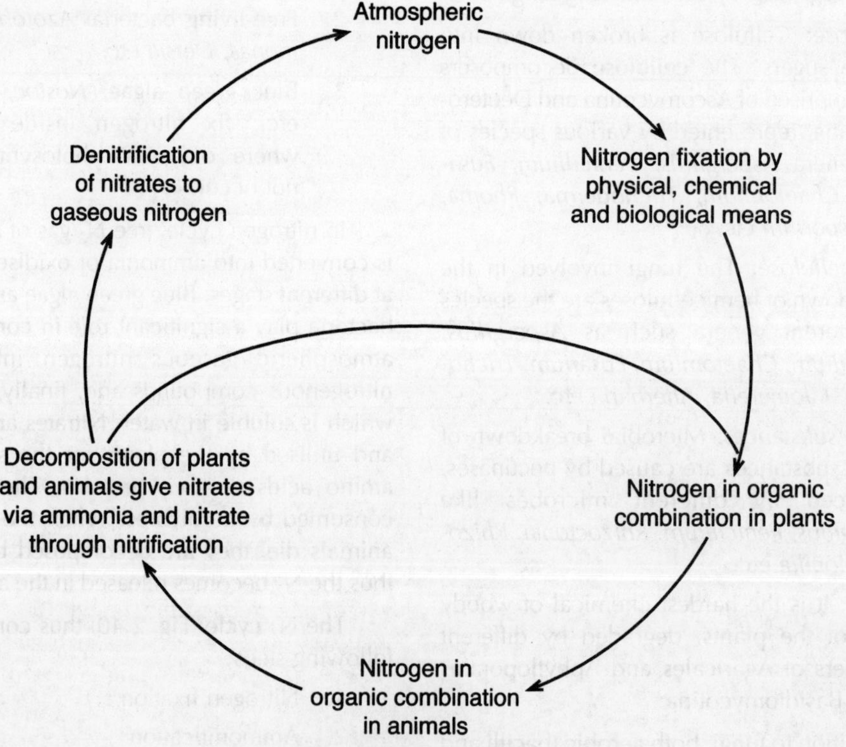

Fig. 2.40: Nitrogen cycle

mostly by *Nitrobacter*. Proper aeration in the soil is essential for the availability of oxygen.

Denitrification: During this process, nitrate is converted to molecular nitrogen by the different bacteria like *Bacillus cereus, Pseudomonas aeruginosa* etc. and thus, N_2 goes back into the atmosphere. This is called dissimilatory nitrate reduction. This takes place during depletion of oxygen in the medium.

In assimilatory nitrate reduction process nitrite is converted into ammonia.

The nitrate added to the soil is reduced to NH_3 by plants and fermentative bacteria rather than to nitrogen by denitrifying bacteria.

Industrial application of microbes

Microbes are used in industry for the production of different types of fermented foods.

Fermented foods

In microbial production of food, enzymatic activities of microbes are used in food processing. Proper substrates for fermentation, appropriate microbial populations and proper environmental conditions are essential to get the desired end products. In addition to the above, microorganisms are considered as an important source of protein to meet the demand of food for ever-growing population of human beings.

The important fermented foods are:

1. Vinegar
2. Fermented milk products
   (i) Sour cream
   (ii) Yogurt (also Yughurt/curd)
   (iii) Cheese
3. Bread
4. Single cell protein
5. Fermented vegetables
   (i) Pickles
   (ii) Soya sauce
   (iii) Poi
   (iv) Tempeh
   (v) Tofu
   (vi) Idly
6. Alcoholic beverages.

The brief informations about the above materials are given:

1. Vinegar

Different substances like fruits (oranges, grapes, apples, pear etc.), vegetables (potatoes), malted cereals (barley, wheat, rye etc.), sugary syrups (honey, molasses etc.) and many other substances are used in the commercial production of vinegar (Fig. 2.41).

The production takes place in two steps: The **first step** involves the initial anaerobic fermentation of the carbohydrate foodstuffs into alcohol by *Saccharomyces cerevisiae*, followed by **second step**, the secondary transformation of alcohol to acetic acid by Acetobacter and Glucobacter.

Process of industrial production

The commercial production of vinegar can be done by the following two methods:

(i) *Slow* or *"let home" method*

During this process, initially alcohol is synthesised at a concentration of 11-13% due to natural alcoholic fermentation. Then acetic acid bacteria is seeded into the alcoholic liquid and alcohol becomes slowly converted into acetic acid. A film of vinegar bacteria, i.e., mother of vinegar, grows on the surface of the liquid, indicates that ethanol becomes oxidised into acetic acid. Inferior quality of vinegar is produced in this process.

(ii) *Quick* or *"French Orleans" process:*

In this proces, a barrel is filled about $\frac{1}{4}$th of the total volume with raw vinegar from previous run to provide active inoculum. The barrel is then filled up to half of the volume with malt liquor, hard cider or wine and temperature is maintained between 24° to 29°C. Air is left in the barrel and acetic acid bacterium is grown as a layer on the top of the liquid. The complete conversion to acetic acid takes place within a few weeks to several months. The recovery of vinegar is made by replacing it from the barrel and the barrel is again filled with equal amount of alcohol to continue the process.

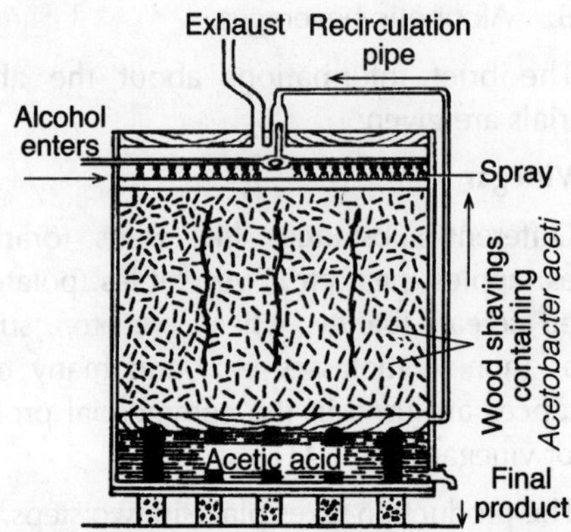

Fig. 2.41: Diagram of a commercial vinegar generator

The production of vinegar becomes retarded due to dropping of film of vinegar bacteria. Thus a raft of floating framework is provided to continue the filter.

To increase the rate of production, vinegar generator can be used where alcohol containing liquid is trickled over a surface film of vinegar bacteria. The bacterial film is maintained on wood clips. Alcoholic liquid is sprinkled on the wood clips and during the slow trickling, the liquid is poured down through the generator, thus the alcohol is gradually converted to acetic acid. Air is supplied from the bottom of the generator (Fig. 2.41).

Pests like mites, fruit flies and vinegar eel (*Anguillula aceti*) attached with the film of vinegar bacteria, cause to sink the film, deteriorate vinegar and thus reduce the yield. Some bacteria like *Lactobacillus* and *Leuconostoc* may contaminate the vinegar and thus reduce its quality. In addition to the above, by storing in copper, tin, steel or iron container, the vinegar becomes cloudy.

2. **Fermented milk products**

(i) **Sour cream**. To prepare the sour cream, butter is soured by lactic acid bacteria. Bacteria like *Streptococcus lactis* or *S. cremoris* are used to produce lactic acid and *Leuconostoc citrovorum* is used to develop necessary flavour. The cream is inoculated and incubated until the desires acidity develops.

(ii) **Yogurt/Curd**. Initial substrate used in the preparation of yogurt is the artificially concentrated milk. The fermentation of the concentrated milk is done by mixing it with *Lactobacillus bulgaricus* and *Streptococcus thermophilis* at a temperature of 42°C for 4 hours, followed by fermentation with yeast. Other organisms like *Lactococcus cremosis, L. lacties* and *L. diactotylactis* are also used in this process.

Flavour of the yogurt appears due to accumulation of lactic acid and acetaldehyde.

Sweetness can be done by addition sugar in the concentrated milk, otherwise it will be sour in taste.

For preparing good quality of yogurt, recently prepared yogurt of satisfactory quantity is added with the concentrated milk.

(iii) **Cheese**. The cheese is produced from the curd by proper shaking and separating it from whey (the liquid portion).

Natural production involves lactic acid fermentation with mixtures of different species of *Lactobacillus* and *Streptococcus*. The flavour of cheese varies with the microbial starter cultures, incubation time and inclusion of secondary microbial species at the later part. During this process, unripe cheese are inoculated with fungal spores and incubated for different period of time.

Cheese may be categorised as:

(i) **Unripe**. This type of cheese is prepared from single step fermentation, e.g., cream and cottage cheese. It is also soft in nature and contains 50-80% water and called soft cheese.

(ii) **Ripe**. Unripe cheese are added with different microorganisms and incubated for different period of time to achieve different texture, taste and aroma. The ripe cheese may be **soft**, contains about 50-80% water, e.g., Camembert and Brie, which require 1-5 months ripening. The **semisoft** type cheese contains about 45% water such as Roquifort, Blue etc. which require 1-12 months ripening. The **hard** type cheese contains less than 45% water, e.g., Colby and Cheddar which require 3-12 months' ripening. The Camembert cheese is produced by *Penicillium camemberti*, Roquefort cheese by *P. roqueforti* and Swiss cheese by *Propionibacterium shermanii*.

The process of cheese making involves nine steps:

Preparing the milk
↓
Formation of curd
↓
Cutting
↓
Cooking
↓
Separating the whey
↓
Salting the residue
↓
Applying microbes
↓
Pressing the curd
↓
Ripening the young cheese

3. **Bread**

The baker's yeast used in bread baking is the different strains of *Saccharomyces cerevisiae*. The *S. cerevisiae* is mixed with bread dough to ferment sugar, thereby CO_2 is produced. The leaven lightens the bread by forming gas bubbles in it and also gives characteristic flavour. The leavening is caused by the different enzymes secreted by the *S. cerevisiae*. The enzymes are:

(i) Amylase, that hydrolyses starch to maltose,

(ii) Maltase, that degrades maltose to glucose, and

(iii) Zymase, causes fermentation of glucose and produces ethyl alcohol and CO_2. Both the substances evaporate and make the bread spongy and soft.

4. **Single cell protein (SCP)**

The term single cell protein refers not only to the isolated cell protein, but also to any microbial biomass, from uni- or multi-cellular microorganisms that can be used as food or feed additives. These are called single cell proteins, because a single type of microorganism is artificially cultivated to get such proteins. The organisms involved in SCP production include both prokaryotic (Bacteria, Cyanobacteria) and eukaryotic members (algae, yeasts and some filamentous fungi). The mushrooms, where fruit bodies are eaten as food, are also considered in this group. These group of organisms came into consideration as animal food, because of:

(a) High protein content,

(b) High growth rate,

(c) Ability to utilise various organic substances available at cheaper rate, and

(d) Ability to utilise agricultural and industrial wastes.

For the above, the microbes get the importance and thus SCP industries have developed in different corners of the world.

The traditional system of crop production is based on season, but the production of SCP is independent on climatic and seasonal variation. The microbial biomass contains about 45-55% protein and other essential nutrition and some of them contain all essential amino acids required for human beings. Regarding the rate of production, it is shown that 50 kg of yeast is able to produce 5,000 kg protein in 24 hours which is much higher than the traditional crop production through agriculture. The production of SCP was initially started during 1960s by an oil company, but now-a-days, it is more popular and attractive.

During SCP production, both **autotrophs** and **heterotrophs** are used.

Autotrophs: Autotrophs like algae such as *Spirulina, Chlorella, Scenedesmus* etc., have been grown in different warm ponds. The organisms use the energy source from the sun instead of fuel, thus reduce the cost of SCP production. In India, Central Food Technology Research Institute (CFTRI), Mysore, Karnataka, has been working on *Spirulina*. They cultured *Spirulina* and the harvested algae are dried, blended and sold as tablets, which contain 60% protein along with unsaturated fatty acids and vitamins. The Imperial Chemical Industries (ICI) has been researching on the use of a bacterium, *Methylophilus methylotrophus*, capable of utilizing methanol for production of SCP in the trade name of '**Pruteen**' to be used as animal feed. Thus, by feeding animals with SCP, the animals could supply proteins for human consumption.

Heterotrophs: Mushrooms are being cultivated almost throughout the world for their high protein, less fat and having an essential amino acid lysine by utilising agricultural wastes (straw etc.). Besides this, other renewable carbon sources like molasses, effluents of distilleries and confectionaries, whey, etc., are used for the production of SCP.

5. **Fermented vegetables**

(i) **Pickles**. Natural microflora associated with cucumbers is used to ferment cucumbers under controlled condition, with appropriate temperature and salt concentration. This can also be achieved by the bacteria like *Lactobacillus plantarum* and *Pediococcus cerevisiae* for six to nine weeks. Gradually salt (NaCl) concentration has to increase to 15-16%. The colonisation of microbial organisms varies with the gradual increase in salt concentration:

(a) During low NaCl concentration, bacteria like *Pseudomonas* and *Bacillus* are able to grow.

(b) With the increase in salt concentration, *Streptococcus faecalis* and *Leuconostoc mesenteroides* are found to grow.

(c) With more increase in salt concentration, bacteria like *Lactobacillus plantarum* becomes dominant.

(d) At the final fermentation stage, *Candida, Pichia* and *Debaryomyces* appear and the concentration of lactic acid becomes reduced. The sour taste of pickles is due to lactic acid present in the medium.

(ii) **Soya sauce.** It is a brown, salty and tangy sauce produced from the mixed substrate consisting of soya beans, wheat and wheat bran. Initially, the mixture of soya beans and wheat is inoculated with spores of *Aspergillus oryzae* and undergo koji fermentation (dry fermentation). The fungus *A. oryzae* grows on the surface and accumulates various enzymes like proteinases and amylases. Different lactic acid bacteria also dominate at this stage. The starter is then dried and extracted. The extract is then mixed with a mixture containing sterile soya beans, steamed wheat bran and autoclaved and crushed wheat. The mixture is then incubated at 30°C for some days and then soaked with concentrated brine. Now, the mixture is called maromi, incubated for 10 weeks to 1 year. Maturation begins with lactic acid bacteria and later by alcoholic fermentation with yeasts like *Saccharomyces rouxii* and *Torulopsis* spp. Important organisms used in this process are *Aspergillus oryzae* and different species of *Lactobacillus*.

(iii) **Poi.** Poi is a type of fermented food of Hawaiian Islands. It is prepared from the Taro plant. The plants are steamed, ground and kept for 1-6 days for fermentation. In the first few hours, *Pseudomonas* and other microbes dominate, but they are successively replaced by *Streptococcus, Lactobacillus* and *Leuconostoc* and finally, *Geotrichum candidum* and different yeasts dominate. Proper flavour, texture and aroma become available due to the presence of ethanol and different acids like acetic acid, lactic acid and formic acid.

(iv) **Tempeh**. Tempeh is a type of food of Indonesia produced from soya bean seeds. Water soaked seeds are ground and inoculated with the spores of *Rhizopus*. The mixture is then incubated for 20-24 hours at 32°C. After incubation, the product is added with salt and fried in oil before eating.

(v) **Tofu**. It is a Japanese food prepared from soya bean seeds. In China, it is called *Sofu*. It is a cheese-like substance developed by fermentation of soya bean seeds by *Mucor* spp. The soaked soya bean seeds are ground and added with Mg or Ca salts and the mass is incubated at 14°C for one month after inoculating with *Mucor*.

(vi) **Idly**. Idly is prepared from rice and black gram (*Vigna mungo*) seeds in the proportion 1 : 1, especially in the southern part of India. The seeds are soaked in water for two hours and then ground properly. The mixture is kept overnight. A small amount of salt and baking powder is added to it.

The mixture is baked under steam.

The organism involved in this process is *Leuconostoc mesenteroides*.

6. **Alcoholic beverages**

The different alcoholic beverages are now becoming popular and added as an item in daily diet. This part has been discussed in the part "Different microbial products of economic value".

Different Microbial products of Economic value (Primary and secondary metabolites)

Different microbial products have high economic value and promoted to develop industry for the commercial production. The different commercial products of different groups are:

1. Medicines:
   (i) Antibiotics, e.g.
   Streptomycin
   Actinomycin-C
   Penicillin
   Griseofulvin
   Cephalosporins
   (ii) Steroids
   (iii) Vaccines
   (iv) Vitamins
2. Enzymes
   (i) Amylases
   (ii) Proteases
3. Amino acids
   (i) Lysine
   (ii) Glutamic acid
4. Organic acids
5. Organic solvent
6. Alcohol
7. Dextran
8. Synthetic fuels.

1. **Medicines**:

(i) ***Antibiotics***

Streptomycin. Streptomycin is produced by *Streptomyces griseus*, first isolated by Schatz, Bugie and Waksman from soil, they reported its antibiotic activities in 1940. Streptomycin acts on many Gram-negative bacteria, some organisms of *Salmonella* group and several species of *Mycobacterium* including *M. tuberculosis* which causes tuberculosis in human beings. **Waksman** was awarded **the Nobel Prize in 1952** for the *discovery of streptomycin and its use in the treatment of tuberculosis*. (The disease was discovered by Robert Koch in 1882).

Process of commercial production: Streptomycin, a broad-spectrum antibiotic is produced by different strains of *Streptomyces griseus*. Initially, lump of mycelial biomass is prepared by inoculating the spores into the selective medium.

The medium in the production tank contains soya bean meal (as nitrogen source), glucose (as carbon source) and NaCl. The medium is then inoculated with mycelial biomass. The process continues about 10 days with high agitation and aeration at 28°C. Maximum production is achieved at pH 7.6–8.0.

The process involves three phases (Fig. 2.42)

1. **First phase**. During this phase, there is rapid growth of microbe that causes production of huge mycelium. NH_3 is released from the soya bean meal by the action of proteolytic activity of *Streptomyces griseus* and thus the pH of the medium is increased. Small amount of streptomycin is produced in the medium.

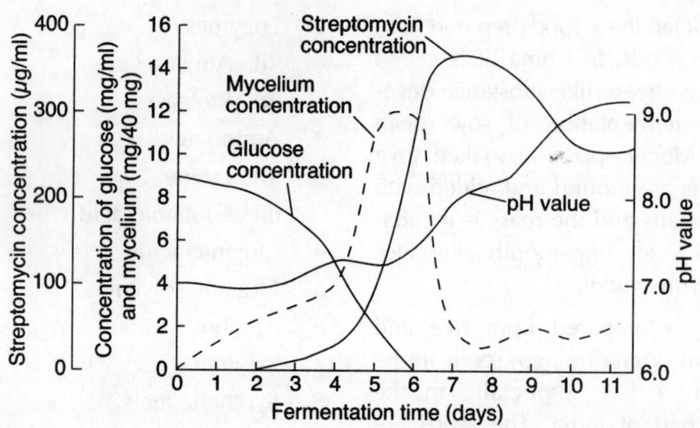

Fig. 2.42: Sequential changes occur in the medium during streptomycin production by *Streptomyces griseus*

2. **Second phase**. During this phase, there is little addition of mycelial biomass. Streptomycin, the secondary metabolite, accumulates in the medium, glucose and released NH_3 are consumed in this phase. pH of the medium remains constant ranging from 7.6–8.0.

3. **Third phase** (Final phase). In this phase, carbohydrate level depletes, streptomycin production stops and biomass declines due to autolysis. The pH level increases to about 9.0 due to release of NH_3 from the lysed mycelium, and the process normally ends at that time.

Recovery of antibiotic: After completion of the process, mycelia are separated by filtration. The streptomycin is then adsorbed onto activated charcoal and then eluted with acid alcohol. It is then precipitated with acetone and then purified by column chromatography.

Streptomycin is an *aminoglycoside* antibiotic, which inhibits protein synthesis by combining irreversibly with 30S subunit r-RNA and thus normal amino acid sequences become hampered.

Actinomycin-C. Actinomycin-C (syn. Cactinomycin) is a chromopeptide antineoplastic antibiotic, isolated from *Streptomyces chrysomallus*. It is a mixture of antibiotics containing actinomycin C_1-5%, actinomycin C_2-30% and actinomycin C_3-65%.

Actinomycin-C binds to DNA by intercalating between guanine and cytosine, forming stable antibiotic-DNA complexes that inhibit RNA and protein synthesis. It is soluble in methanol and has to store at +4°C for long term storage.

Penicillin. Penicillin, the wonder drug of the world still is one of the most useful antibiotics. The antimicrobial activity was first noted in contaminated culture plate by Alexander Fleming in 1929 and the associated fungus is *Penicillium notatum*. Later in 1940, Harold W. Florey and E. B. Chan of Oxford University, developed methods of industrial production of Penicillin and isolated the active ingredient that is used clinically. The commercial production was begun in 1941 in USA. **Fleming, Florey and Chan together shared the Nobel Prize in 1945** for the discovery of Penicillin and its production. It is also isolated from *P. chrysogenum* and other species of molds. *Penicillium notatum* has been replaced by *P. crysogenum* for the commercial production. Penicillin is active against Gram-positive bacteria, Gram-negative diplococci like *Neisseria* and also against some spirochetes. At present, the mutated strain of *P. crysogenum* yields 85,000 units/ml of the medium, which is much higher than the original strain that yields 60 units/ml.

Process of commercial production: Initially, *P. crysogenum* is multiplied by growing it in wheat bran-nutrient medium at 24°C for 7-8 days.

The mass medium consists of the following ingredients:

(i) Molasses or glucose	10%
(ii) Cornsteep liquors	4-5%

(iii) Phenyl acetic acid 0.5-0.8%

(iv) Vegetable oil 0.5%

The pH of the medium after sterilisation is maintained at 6.0.

The medium is kept in the production tank and then inoculated with the above culture. The tank is aerated by force for 1-2 days for heavy mycelial growth. The total mixture is then kept for fermentation for about 7 days.

After fermentation, the liquid medium is separated from the fungal mycelium by rotary vacuum filter. The fungal mycelium is used for animal feed supplement and the penicillin is separated from the filtrate by different extraction methods. The penicillin G is obtained from the filtrate. It is further modified into different penicillin derivatives by different methods.

The modification can be done by microbial enzymes or by chemicals. The 6-aminopenicillanic acid is derived from penicillin G by bacterial acylase. Chemical coversion is lengthy and costly.

Griseofulvin. Griseofulvin is an antifungal antibiotic, obtained from *Penicillium griseofulvum*. Synonyms of griseofulvin are grisovin, fulvicin and fulcin. It is used in the treatment of infections on skin and also of some systemic infections caused by different fungi. It causes excessive branching, abnormal swelling, stunting and distortion of fungal hyphae. The activity is rapidly lost in sunlight. The drug is orally applied.

Cephalosporins. It is a group of antibiotics produced by *Cephalosporium acremonium*, a marine fungus. They are highly effective against Gram (+ve) and Gram (–ve) bacteria. They are highly effective therapeutically with low toxicity. They have anti-bacterial property similar to semi-synthetic penicillins. The cephalosporins inhibit the cross-linking transpeptidase, thus they kill the growing cells.

(ii) ***Steroids***

Steroids are used in the treatment of various disorders in humans including regulation of sexuality. Chemical synthesis of steroids is lengthy and a costly process, thereby products also become costly. Later on, this was achieved through microbial biotransformation, thereby cost becomes lowered and the drug is available in the market at a much cheaper rate.

Different microorganisms are used in the biotransformation of different steroids. *Rhizopus nigricans* hydroxylates progesterone and forms another steroid 11-α-hydroxyprogestrone by the introduction of oxygen at the number 11 position. Similarly *Cunninghamella blakesleeana* can hydroxylate cortexolone and forms hydrocortisone.

(iii) ***Vaccines***

The antigenic properties of microbes are used to elicit a primary immune response. The inactivated virulent pathogens are used for producing vaccines. Different viral and bacterial diseases can be controlled only by vaccines. The rabies vaccine is produced in human fibroblast tissues grown in tissue culture which is friendlier than the earlier one that was grown in embryonated duck egg with painful experiences as well as with side-effects.

(iv) ***Vitamins***

Though many vitamins like riboflavin, Vitamin B_{12} etc., are produced through microbial fermentation, Vitamin B_{12} (5-deoxyadenosylcobalamino phosphate) is the only vitamin that is produced commercially through fermentation. During vitamin B_{12} production, glucose is taken as substrate and *Pseudomonas denitrificans*, the most efficient producer is taken into account.

Vitamin B_{12} can also be produced by direct fermentation, using the microorganisms like *Propionibacterium shermanii*.

It can also be produced as by-product of antibiotic fermentation by *Streptomyces*. During this fermentation cobalt salt is added to the medium.

2. **Enzymes:**

(i) ***Amylases***

Amylases break the starch to glucose. They are of various types: α–, β– and glucamylases. Fungi like *Aspergillus niger* and *A. oryzae*; and bacteria like *Bacillus subtilis* and *B. diasticus* are commonly used in this process. The amylases are used in various purposes such as:

(a) Removal of spots on clothes,

(b) Sizing agent in textile industry,

(c) Preparation of starch sizing pastes, used in paper industry,

(d) In syrup industry, for the conversion of sucrose to glucose i.e., in producing sweeteners,

(e) In bread, chocolate and brewing industry.

(ii) ***Proteases***

Proteases are the enzymes which act on peptide linkages of proteins. The enzymes are commercially available from the organisms like *Bacillus licheniformis*, and different species of *Bacillus* and *Aspergillus*. The enzymes are heat-stable and are used in modern detergent formulations and they are used in large scale in modern laundry for removal of spots of blood, milk, tea etc. The proteases are used in bakery to reduce the mixing time and improvement of the quality of loaf. They are used in leather industry for bating of hides and as meat tenderiser. Certain strains of *Bacillus* developed through recombinant DNA technology are GX 6644 and GX 6638. The GX 6644 strain is active for milk casein and GX 6638 produces an enzyme *kerazyme*, used to dissolve hairs and also to open the drains blocked by hairs.

3. **Amino acids**:

Nowadays, animal feeds are deficient in amino acid, thus it has to be added with animal feeds. Microorganisms are able to produce amino acids at commercial level. Thus the production of amino acids is increasing day by day throughout the world. The L-isomers are desired for this purpose, but not the D-isomers. (*To get only the L-isomers, the D-isomers might be separated from the mixture or new genetically-engineered strains of microbes developed.*)

(i) ***Lysine***

The cereal proteins are deficient of lysine, an essential amino acid for the humans. Thus the bread and other foodstuffs are supplemented with lysine.

Organism like homoserine requiring auxotroph of *Corynebacterium glutamicum* is used in the direct production of L-lysine. The blockage of homoserine dehydrogenase results from feedback inhibition of the enzyme that leads to the accumulation of L-lysine. In this process, cane molasses is commonly used as a substrate. About 50g lysine per litre can be produced in 2-3 days.

In other way, commercial production of L-lysine comprises of two steps. In the first step, *E. coli* is inoculated in the medium containing glycerol, corn-steep liquor and $(NH_4)_2HPO_4$ under controlled condition for three days, and thus diaminopimelic acid (DAP) is produced. In the second step, the medium is inoculated with *Enterobacter aerogenes*, which secretes DAP decarboxylase and thus DAP becomes converted into L-lysine.

(ii) ***Glutamic acid***

Different strains of the organisms like *Corynebacterium glutamicum and Brevibacterium flavum* are used in direct fermentation to produce *L-glutamic acid* and *monosodium glutamate*. Medium having glucose and mineral salts with periodic addition of urea is used for fermentation. pH 6-8, temperature 30°C and well aeration are essential for fermentation. The leaky membrane for the release of the product from the cell can be developed in different ways. The *C. glutamicum* is grown in a medium with sub-optimal concentration of biotin or by adding detergents or fatty acids that helps to develop leaky membrane through several methods.

4. **Organic acids**:

(i) ***Citric acid***

Fungi like *Aspergillus niger* is commercially used to produce citric acid.

It is a major fermentation product and has several uses in medicines, as food additive, especially in soft drinks; as metal chelating and sequestering agent. In addition, it is also used in the manufacture of ink, dyeing and engraving. Annual world production of citric acid is about 250,000 tonnes.

The fermentive medium contains molasses, ammonium nitrate, $MgSO_4$ and K_2SO_4.

Deficiency of trace element or phosphate in growth medium is required to limit the growth of the fungus, which promotes maximum citric acid accumulation. Acid is added to lower the pH, which should be maintained at 5.0 to 6.0. Iron content should be maintained at low level. After clearing the mycelial mat, citric acid is precipitated as a calcium salt by adding lime.

5. **Organic solvents**:

Out of many, a few organic solvents are produced commercially by microbial fermentation. These are:

(i) *Ethanol*. Although it develops through fermentation by yeast for beverages, but alcohol for industrial uses is also produced chemically.

(ii) *Acetone* and *Butanol*. Both the solvents are produced by microbial anaerobic fermentation with different species of *Clostridium. C. acetobutylicum* converts starch to acetone and *C. saccharoacetobutylicum* converts carbohydrates to acetone and butanol. The solvents are recovered through distillation.

(iii) *Glycerol*. Both fungus (*Saccharomyces cerevisiae*) and bacterium (*Bacillus subtilis*) are required in this process. Sodium sulphite is added in the yeast-ethanol fermentation process. The sodium sulphite reacts with CO_2 and produces sodium bisulphide, which prevents the production of ethanol from acetaldehyde by reduction.

Glycerol is used in flavouring and food colouring and is also used in the preparation of explosives.

6. **Alcohols**:

The alcohol is produced by the fermentation of different substrates with the help of yeasts.

There are two types of yeasts: Baker's yeasts and brewer's yeasts.

In brewing industry, different types of alcohols are produced depending upon the medium and the fermentative agent:

(i) **Beer**. Prepared from barley (*Hordeum vulgare*) which contains 3-6% alcohol.

(ii) Wine. Prepared from grape (*Vitis vinifera*) containing 10-20% alcohol.

(iii) Brandy. Prepared after distillation of wine, containing 60-70% alcohol.

(iv) Rum. Prepared from sugarcane (*Saccharum officinarum*) which contains 40% alcohol.

(v) Gin. Prepared from fermented rye (*Secale cereale*) which contains 3-6% alcohol.

Different species of yeasts are used in this process. These are *Saccharomyces cerevisiae, S. sake, S. ellipsoides and S. pereformus.*

The fermentation process involves the enzymatic activity of yeast that produces alcohol by the following reaction:

$$\underset{\text{Carbohydrate}}{C_6H_{12}O_6} \xrightarrow[\text{Yeast}]{\text{Zymase}} \underset{\substack{\text{Ethyl}\\\text{alcohol}}}{2C_2H_5OH} + \underset{\substack{\text{Carbon}\\\text{dioxide}}}{2CO_2} + \text{Energy}$$

7. **Dextrans**:

Dextran is a polysaccharide (glucose polymer) of very high molecular weight, ranging from 15 to 20,000 KDa. It is produced by a wide range of microorganisms like *Leuconostoc* spp., lactic acid bacteria etc.

These are used in physical fractionation of solutes of different molecular weight in chromatography.

8. **Synthetic fuels**:

Synthetic fuels produced by microbes are ethanol, methane, hydrogen and hydrocarbons. Thus the microbes may help in the energy crisis of the developing world. Different waste materials such as municipal garbage etc., are used to produce fuels by microorganisms.

(a) ***Ethanol***. It is used as fuel in automobiles mixed with gasoline (1 : 9) in USA. It is also used in Brazil as fuel in automobiles. Commercially, it is produced in two steps. In 1st step, conversion of cellulose to sugar takes place with the help of *Clostridium* spp., followed by the 2nd step where conversion of these sugar to ethanol is taking place by *Zymomonas* *mobilis* (more efficient than yeast) or *Thermoanaerobacter ethanolicus*.

(b) ***Methane***. It is used for the generation of energy like mechanical heat and electric power. Methanogenic bacteria are obligate anaerobes that produce methane (CH_4) by reducing CO_2 and/or acetate. Methane is produced at significant level from plant residue, animal waste, municipal residues, etc. This can also be achieved by growing the algae in ponds.

(c) ***Biogas***. It is a mixture of different gases like CH_4, CO_2, N_2 and O_2, dominated by methane. This can be achieved from agricultural waste and also from dungs of herbivorous animals. Gobor gas plants are very much popular in village areas. After its use in gobor gas plants, the dung residue is used again in cultivated land as fertiliser. The animal wastes are initially hydrolysed by hydrolytic bacteria, followed by acid formation by acetogenic bacteria, those convert monomers into CH_4, CO_2 and H_2. At the final stage, methanogens reduce the CO_2 and/or acetate into CH_4.

ROLE OF BACTERIA IN HUMAN WELFARE

1. As Natures' Scavengers

A scavenger is an organism that mostly consumes decaying biomass, such as meat or rotting plant material. Bacteria can acts as scavenger in two ways. These are:

(i) Saprophytic bacteria obtain food from organic remains such as animal excreta, fallen leaves, meat etc. They decompose these substances by action of digestive enzymes aerobically or anaerobically. Thus they help in sanitation of nature, known as scavengers, e.g., *Pseudomonas*.

(ii) Likewise, due to the presence of heavy metal pollutants such as lead, mercury, copper etc. leads to several toxic hazards in human beings cause cancer etc. Certain bacteria can take up these heavy and toxic metals from the waste waters and ponds and acts as natural bio-scavengers. These bacteria can accumulate metals on their cell walls are: *Escherichia coli* take up mercury, while *Bacillus circulans* can accumulate copper. These bacteria take up the metal pollutants through absorption or adsorption and in a relatively complex process. Thus they clean the nature from such pollutants and reduce its effect.

2. Biofertilizers

Some bacteria have the capacity to fix atmospheric nitrogen. They may live either symbiotically (*Rhizobium, Bradyrhizobium* etc.) or free-living (*Azotobacter, Azomonas* etc.). The importance of *Rhizobium* and *Azotobacter*, along with others is given below:

(i) **Symbiotic nitrogen fixers**:

Rhizobium: *Rhizobium leguninosarum* (syn. *Bacillus radicicola*) is a symbiotic nitrogen fixing bacterium which multiplies and lives inside the root nodules of different **leguminous plants** like Pea, Bean, Soybean, Medicago and many others.

Millions of bacteria (*Rhizobium*) are developed inside the nodules and lived as if in a home. They have the ability to absorb free nitrogen compounds. A portion of the fixed nitrogen passes into the plant, a portion diffuses out in the surrounding soil and rest remains in the nodule. On the other hand, bacteria receive some amount of carbohydrate from the host plant. Thus they show an excellent example of symbiosis. Addition of bacteria in the rhizosphere increases the number of nodules and also the amount of fixed nitrogen. Thus it can be used as a good biofertilizer.

In addition to *R. leguminosarum*, many other *Rhizobium* species are present which form association with different legume roots. These are *R. lupini, R. trifolii, R. meliloti, R. phaseoli* etc.

These bacteria are also called **rhizobia**, live freely in the soil but cannot fix nitrogen except for a strain of Cowpea *Rhizobium* (Me Comb *et. al.*, 1975). They develop the ability to fix nitrogen only when they are present inside the root nodules.

In addition to rhizobia, another nitrogen fixing mycelial bacterium, ***Frankia*** (actinomycete), is associated symbiotically with the root nodules of several **nonlegume plants** like *Casuarina, Alnus* (Alder), *Myrica, Rubus* etc.

Leaves of a few plants (e.g., *Ardisia*) develop special internal cavities for providing space to symbiotic nitrogen fixing bacteria, *Xanthomonas* and *Mycobacterium*. Such leaves are a constant source of nitrogen fertilizer to the soil.

(ii) **Fre-living nitrogen fixers**

Azotobacter: *Azotobacter beijerinkia*, an aerobic free-living nitrogen fixing bacteria is able to fix nitrogen from the air present between the soil particles. The fixed nitrogen, along with other elements, forms the organic nitrogenous compounds. After the death of bacteria, the nitrogenous compounds become decompoded by other bacteria and thus ammonia is produced. The nitrifying bacteria convert ammonia to nitrites and then into nitrates. It is then utilized by green plants. Like *Rhizobium*, addition of *Azotobacter* in soil increases the amount of nitrogen in soil, and thus increases its fertility.

In addition to the above, certain other *bacteria are also used as biofertilizer. These are Rhodopseudomonas, Rhodospirillum* and *Chromatium*. All the three photoautotrophic bacteria are able to fix atmospheric nitrogen. Inoculation of soil with these bacteria as biofertilizer helps in increasing yield and saving the use of nitrogen fertilizers.

3. Industrial Application

A. **Curing of tea**

The crude tea leaves are acted upon by bacteria, *Micrococcus condisans*. The process is known as curing, which is employed to impart a peculiar taste and flavour to the leaves. For this purpose alcohol is added to tea leaves.

B. **Curing of tobacco**

The crude dry tobacco leaves pass through curing and ripening processes before it is ready for use. Bacteria, *Bacillus megatherium* are employed in both these processes and the peculiar taste and flavour in the tobacco is due to the bacterial activity. For this purpose molasses and alcohol are added to tobacco.

C. **Tanning of leather**

The hides and skins of animals after drying, salting and clearing are steeped in fluids containing specific bacteria. The process of fermentation goes on for some time and then they are transferred to tan-pits and are allowed to be fermented further. This whole process is known as tanning and the bacteria employed in the process are obtained from cow dung and the excreta of dogs and poultry.

D. **Retting of fibres**

Retting is a process by which commercial plant fibers like jute, flax, hemp etc. are separated by the bacterial decomposition of pectin substances that hold the fibers together. In anaerobic retting, fibrous plant body immersed in water tanks where butyric acid bacteria (*Clostridium butylicum*) decompose the middle lamella and separate the fibers. In dew retting, periodic sprinkling of water is done and *Pseudomonas fluorescence* does the retting process.

E. **The sewage work**

In order to remove solid and semisolid constituents of sewage it is allowed to purify. Purifying bacteria are allowed to act upon sewage under anaerobic conditions. It gets decayed and liquified. It is now filtered and the liquid is either drained out to the river or used as manure in fields. For this purpose, in the soak pits the horse dung is filled up.

4. Biopesticides

Pesticides are the chemicals used in plant protection. Depending on the target organism, they are known as fungicides, bactericides, viricides, insecticides, nematicides, herbicides and so on. The synthetic pesticides are highly resistant to natural biodegradation leading to their accumulation in soil and water, that cause pollution. Thus people becomes interested in alternative process, i.e., the biopesticides.

The **biopesticides** *are the microorganisms or their products and are used to control pests of both plants and animals.* Several microbes like bacteria, fungi and viruses are used to control pests. Several products of the microbes are patented and registered for commercial production and used in crop field to protect the plants from different pests. Microbial pesticides should be harmless to humans, plants and other valuable animals.

They are of different types:

1. **Bio-insecticides:**

Microorganisms like bacteria, fungi and viruses show greatest commercial importance as biocontrol agents than the commercial synthetic pesticides. With the more progress of scientific achievement the bioinsecticide becomes popular in different countries.

(i) *Bacterial*

Several bacterial pathogens of different insects are being used as insecticides. These are *Bacillus, Clostridium, Pseudomonas, Enterobacter, Proteus, Serratia* etc. Out of these, *Bacillus thuringiensis* has been used extensively. The bacterium has been shown to be successful against a wide range of more than 150 insects. It shows insecticidal activity against larvae of Lepidoptera. *B. thuringiensis var. israelensis* (BTI) is highly active against larvae of mosquito vectors of malaria. Unlike DDT, the pathogen is environmentally safe in use and mosquito does not show any resistance against the bacterium. It also shows excellent result to control black fly — the carrier of widespread river blindness in Africa.

Four different toxins are produced by the *B. thuringiensis* and about 16 formulations have been prepared based on the above toxins. These are used in different countries like USA, Russia, France, Germany etc. Some of the registered products like **Sporcine, Condor, Cutlass, Thuricide, Foil** etc. are commonly used. In India, it has been found that 0.4% thuricide is more effective than malathione, endrine and DDT to control insect pests of crucifer, lac and sugarcane. In USA, different registered formulations prepared from *B. thuringiensis* are used to control pests of different crops like Alfalfa caterpillar (Alfalfa), Bollworm (Cotton), Cabbage worm (Cabbage and Cauliflower), Orange dog (Orange), Grape leaf folder (Grapes) etc.

With the help of recombinant DNA technology, the gene having insecticidal properties of *Bacillus thuringiensis* has been transferred to the crops plans like tomato in 1987, with the help of bacterium, *Agrobacterium tumifaciens*. Later, similar success has been achieved in different crop plants like tobacco, cotton etc., by using the similar technique. Now-a-days, Bt cotton is very popular among the farmers.

(ii) *Fungal*

Different entomogenous fungi have been used to protect different crops from insect damage. Species of the different genera like *Aschersonia, Beauveria, Coelomomyces, Entomophthora, Hirsutella* etc., are in common use. Some registered products are also produced and used by different countries like Australia, Brazil, China, France, Japan, USA, UK etc. List of some of the products, producing fungi in parenthesis and insects on which it acts are:

- Aseronija (*Aschersonia aleyrodis*) — Whitefly of many crops.
- Boverin and Boverol (*Beauveria brassiana*) — Pine caterpillar, Green leaf hipper, Colorado potato beetle etc.
- Mycotal and Vertalec (*Verticillium lecanii*) — Whitefly and Aphids of glasshouse crop.

(iii) *Viral*

Viruses are also very much effectively used as bioinsecticide. There are three major groups of viruses that can infect different insects. These are: (a) Nuclear polyhedrosis viruses (NPV), (b) Granulosis viruses (GV) of Baculoviridae, and (c) Cytoplasmic polyhedrosis viruses (CPV) of Reoviridae. The NPVs are effective against moths and butterflies, while CPVs are effective mainly on caterpillars. These are used in different countries like USA, UK, Canada, Japan, Germany etc. List of some viruses, registered trade names in parenthesis and target insects are given below:

- NPV (ELCAR) — Tobacco budworm and Cotton bollworm.
- NPV (GYPCHEK) — Gypsy moth.
- NPV (VIROX) — European sawfly.
- CPV (MATSUKEMIN) — Pine caterpillar.
- GV (MATEX) — Insects of different food crops like Codling moths.

2. **Bio-nematicides:**

Different fungi are known to act as nematicide. Fungi of different genera like *Arthrobotrys, Dactylella, Dactylaria* and *Monacrosporium* are used to control different members of genera like *Heterodera, Meloidogyne* and *Rotylenchulus*, cause diseases of different crop plants. The fungi damage nematode in four different ways:

(i) *By haustoria*: Fungi penetrate haustoria in the body of the nematode, digest the cell contents and draw the nutrients.

(ii) *Catching by loop*: The fungal mycelium forms loops at intervals. As a nematode passes through the loop, it constricts and thus the nematode is trapped.

(iii) *Production of adhesive hyphae*: The fungal mycelium produces some adhesive branch which may stick with the body of nematode on accidental contact.

(iv) *Formation of hyphal mesh*: The mycelium forms mesh-like cobweb and is able to catch nematodes.

The other groups of fungi are commonly present in the soil and can act as nematicides. These are *Verticillium chlamydosporium, Paecilomyces lilacinus and Dactylella oviparasitica* — they often attack nematodes as well as their eggs.

3. **Bio-herbicides:**

Fungi are found to be much more suitable as herbicide than bacteria and viruses. Fungus like *Phragmidium violacerum* is used to control European blackberry in Chile and *Puccinia chondrillina* to control rush skeleton weed in Australia. Others, like *Phytophthora palmivora*, has been developed as herbicide at commercial level. The name of the registered product, producing fungi in parenthesis and the target weeds are:

- DE VINE (*Phytophthora palmivora*) — Milk-weed vine in USA,
- LUBOA (*Colletotrichum gloeosporiodes* f. sp. *cuscutae*) — Dodders in China,
- BIOMAL (*Colletotrichum gloeosporioides* f. sp. malvae) — Round-leaved mallow in Canada and USA.

BACTERIOCIN

The bacteriocin is a form of protein, produced by bacteria. It inhibits or kills the growth of other kinds of bacteria. Actually it is a proteinaceous toxin that is produced as a very small molecule by the bacteria. It reduces the growth of bacterial strains that are similar or closely associated. The toxin is regarded to be an antibiotic of narrow spectrum and produced by a wide array of bacteria.

This toxin was first discovered by Gratia (1925), during his work on methods to destroy bacteria. The search led him to the discovery of Bacteriocin.

Structure: The structure of this toxin is usually like the shape of a protein or peptide. It consists of amino acid chain of different sizes.

Types: These preotein types are known by a host of different names. These are: Aureocin, Agrocin, Alveicin, Lactococin, Lacticin, Leucoccin etc.

Uses: These are used in the following ways:

1. These proteins are produced by non-pathogenic bacteria, those usually settle inside the human body. This is the reason of using these proteins in medicines. Use of antibiotic in the body lead to the removal of these harmless bacteria as side effect, thereby pathogenic bacteria can attack the body and give rise to ailments.

2. These proteins are also considered as a possible treatment for cancer. But more clarification is needed regarding the therapeutic capability.

Some informations about Bacteria

Largest bacterium	*Bacillus butschli* (*800 mm*). *Spirillum volutans* (*13-15 mm in length*)
Largest species of bacterium ever discovered	*Thiomargarita namibiensis*, which grows to 0.75 mm (0.03 in) in diameter, making it visible to the naked eye and up to a million times the size of more typical bacteria
Largest sulphur bacterium	*Thiophysa volutans* (*18 mm*)
Oldest microbe	*Bacillus form of bacteria. 250 million years old*
Smallest bacterium	*Dialister pneumosintes* (*0.15-0.3 mm in length*). *In naso-pharynx of man.*

VIRUSES

INTRODUCTION

The name 'virus' came from a Latin word *virus* which means venom or poisonous fluid. Although plant diseases like leaf roll of potato and human diseases like yellow fever, small pox etc., were known for long time, but the nature of causative agent was known to us quite later. Adolph Meyer (1886), an agriculture chemist of Holland, observed a diseased tobacco plant showing mottling of leaf and named it the mosaic. He was able to demonstrate the infectious nature of the sap of infected plant by grinding, filtering through double filter paper and then applying the sap to the healthy plants. The infective capacity of the fluid (sap) was lost by heating at 80°C and he concluded that certain microbes are the causative agent of tobacco mosaic. D. Iwanowski (1892), a Russian scientist, was the first to demonstrate the transmission of tobacco mosaic virus disease from infected to healthy plant through sap, even the sap was filtered through Chamberland filter candle, which is sufficient enough to remove bacteria. W. M. Beijerinck (1898), a bacteriologist of Holland, demonstrated that the invisible, filterable and non-cultivable infectious sap could diffuse through an agar gel, like a fluid. The bacteria and other organism cannot do so. From the above, Beijerinck believed that the fluid itself was alive and he called it as *"contagium vivum fluidum"* i.e., infectious living fluid.

Loeffler and Frosch (1898) observed that the agent of mouth and foot disease of cattle could pass through bacteriofilter. The agent could not be grown in artificial culture medium and was not visible under microscope. Walter Reed and his associates (1900) discovered the agent of yellow fever, the virus, which is the first discovered viral disease of man. Till date many viral diseases of vertebrates are known.

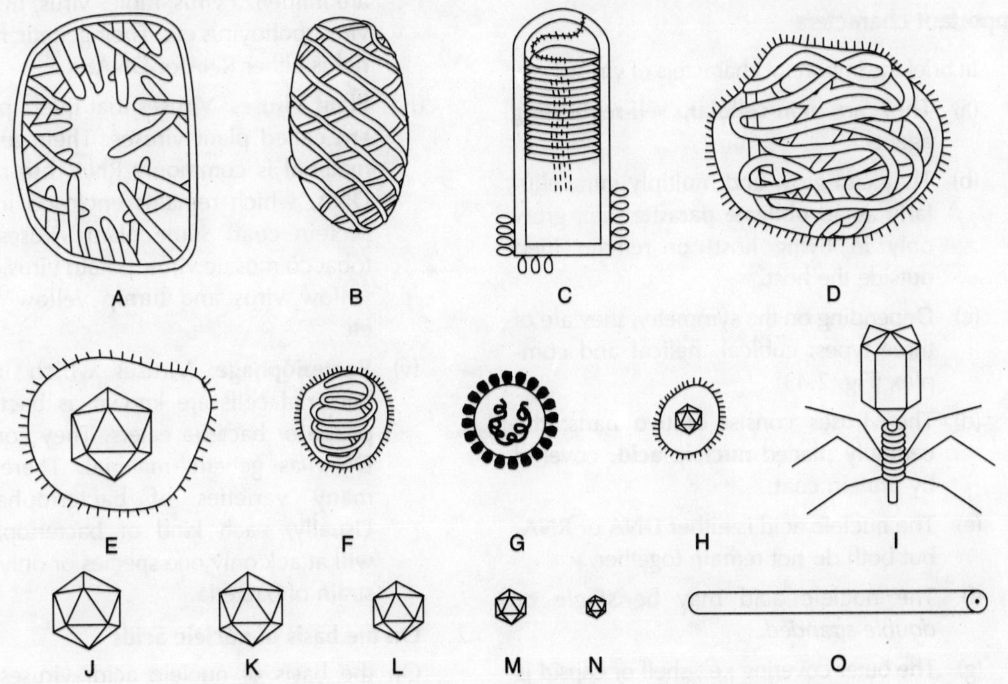

Fig. 2.43: Morphology of Viruses—A. Orthopoxvirus, B. Parapoxvirus, C. Rhabdovirus, D. Paramyxovirus, E. Herpesvirus, F. Orthomyxovirus, G. Coronavirus, H. Togavirus, I. T-even coliphage, J. Adenovirus, K. Reovirus, L. Papovavirus, M. Picorna-virus, N. Parvovirus, O. Tobacco mosaic virus

F. W. Twort (1915), a British scientist, was the first to describe some viruses that attack bacteria. Two years later, Felix d'Herelle (1917), a French scientist, independently observed and studied in detail about the virus attacking bacteria. He named the virus *bacteriophage* (bacteria-eater), commonly called phage.

Schlessinger (1933) was the first to purify virus by using differential centrifugation. Later, in 1935, Wendell M. Stanley, an organic chemist, observed that the virus could be crystallised and consisted largely of proteins. Later, it has been established that virus contains a small amount of RNA or DNA in addition to protein. Therefore, virus is nucleoprotein and the nucleic acid is the infective agent rather than protein. A few years later, in **1946, W. M. Stanley was awarded the Nobel Prize** for the above discovery.

Based on the nature of host, viruses cannot be categorised either in the animal or in the plant kingdom. The branch that deals with the viruses is called **virology**. The scientists who work in this branch are called **virologists**.

Important characters

In brief the important characters of viruses are:

(a) They are **non-cellular, self-replicating** agents.

(b) They can grow and multiply intracellularly as an **obligate parasite** (i.e., grow only in living host) or remain inert outside the host.

(c) Depending on the symmetry, they are of three types: cubical, helical and complex (Fig. 2.43).

(d) The viruses consist of two parts: the centrally placed **nucleic acid**, covered by **protein coat**.

(e) The nucleic acid is either **DNA** or **RNA,** but both do not remain together.

(f) The nucleic acid may be *single* or *double stranded.*

(g) The outer covering i.e., shell or **capsid** is made up of protein units, called **capsomeres**; except some animal viruses which are with additional polysaccharides.

(h) They have no machinery of their own for protein synthesis and thereby they use host machinery for the synthesis of protein.

(i) During replication, their nucleic acid directs the host cell to make different parts of virus and when these parts assemble together, they form a complete infectious particle, the **virion.**

(j) They are transmitted very easily from one organism to another organism.

Types of Viruses

The virus can be classified based on the the type of host and genetic material.

1. Based on the host

On the basis of host, the viruses are of three types. These are: Animal viruses, Plant viruses and Bacteriophages.

(i) **Animal Viruses**: Viruses which infect and live inside the animal cell including man are called animal viruses. These are influenza virus, rabies virus, mumps virus, poliovirus etc. Their genetic material is either RNA or DNA.

(iv) **Plant Viruses**: Viruses that infect plants are called plant viruses. Their genetic material is commonly RNA, but rarely DNA, which remains enclosed in the protein coat. Some plant viruses are tobacco mosaic virus, potato virus, beet yellow virus and turnip yellow virus etc.

(v) **Bacteriophage**: Viruses which infect bacterial cells are known as bacteriophage or bacteria eaters. They contain DNA as genetic material. There are many varieties of bacteriophages. Usually, each kind of bacteriophage will attack only one species or only one strain of bacteria.

2. On the basis of nucleic acids

On the basis of nucleic acid, viruses are classified into following **three types**. These are DNA viruses, RNA viruses and DNA-RNA viruses (Fig. 2.44 and 2.45).

DNA viruses

dsDNA

Adenoviridae

Polyomaviridae

Papillomaviridae

Herpesviridae

Iridoviridae

ssDNA

Circoviridae

Parvoviridae

dsDNA (RT)

Hepadnaviridae

Poxviridae

Asfarviridae

Fig. 2.44: Single and double stranded DNA viruses

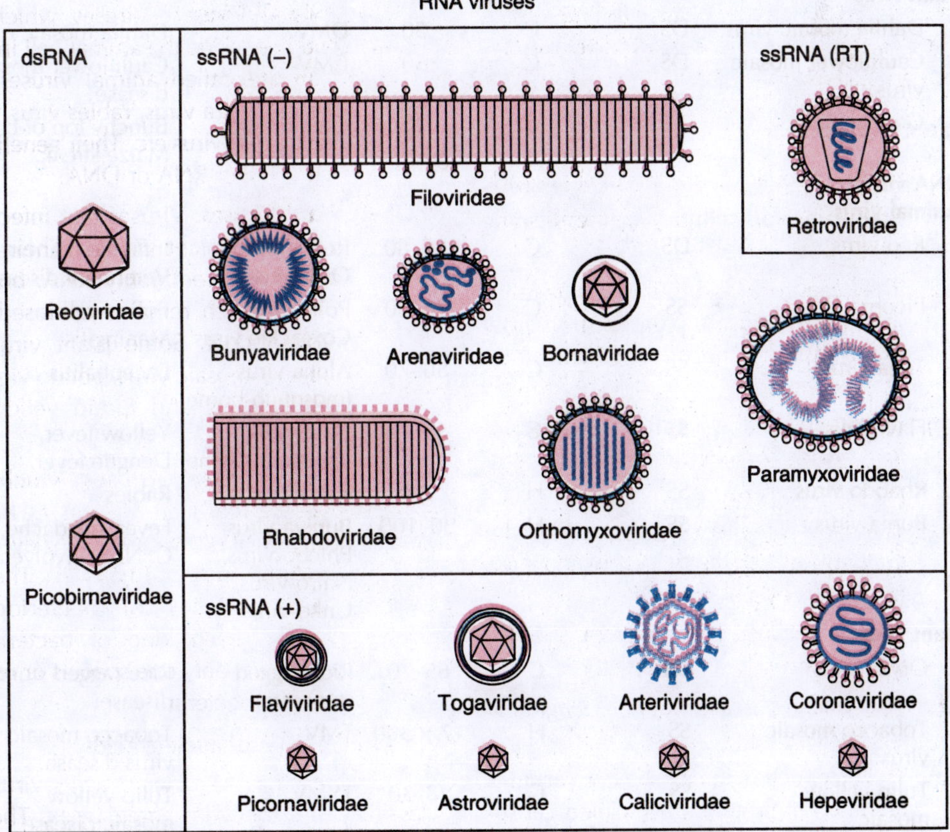

Fig. 2.45: Single and double stranded RNA viruses

(i) **DNA Viruses**: The viruses that possess, DNA as genetic material is called DNA viruses. Bacteriophages, cyanophages and most of the animal viruses contain DNA as genetic material.

(ii) **RNA Viruses**: The virus that possesses RNA as genetic material is called RNA viruses. These are plant viruses.

(iii) **DNA-RNA Viruses**: This type of viruses contains both DNA and RNA as genetic material. Rous's viruses contain both DNA and RNA as genetic material.

Table 2.19: Salient features of some common viruses

Group	Nucleic acid configuration	Symmetry	Size (nm)	Representative viruses	Diseases
A. DNA virus					
Animal virus					
1. Adeno virus	DS	C	70–90	Adeno virus (7 groups, and many serotypes)	Sore throat, Conjunctivitis
2. Pox virus	DS	B	200–400	Variola Cow pox	Small pox Cow pox
3. Hepanda virus	DS	C	42	Hepatitis B	Serum hepatitis
4. Parvo virus	SS	C	18–26	Polyoma virus	Papilloma of dog, rabbits etc.
Plant virus					
1. Dahlia mosaic virus	DS	C	50	DMV	Dahlia mosaic disease
2. Cauliflower mosaic virus	DS	C	50	CMV	Cauliflower mosaic disease
3. Gemini virus	SS	C	20	Gemini virus	Bunchy top of banana, Maize streak
B. RNA virus					
Animal virus					
1. Reo virus	DS	C	60–80	Rota virus Orbi virus	Infantile diarrhoea Veterinary disease
2. Picorna virus	SS	C	20–30	Polio virus Coxsackie virus	Poliomyelitis Meningitis
3. Toga virus	SS	C	50–70	Alpha virus (mosquito-borne)	Encephalitis
4. Flavi virus	SS	C	50–70	Flavi virus (mosquito-borne)	Yellow fever, Dengue fever
5. Rhabdo virus	SS	H	75	Rabies virus	Rabies
6. Bunia virus	SS	H	90–100	Buniya-virus, Phlebo-virus, Nairo-virus, Unko-virus	Fever, headache, C. N. S. involvement
Plant virus					
1. Oryza virus	DS	C	65–70	Rice ragged stunt virus	Rice ragged stunt virus disease
2. Tobacco mosaic virus	SS	H	17 × 300	TMV	Tobacco mosaic virus disease
3. Tulip yellow mosaic	SS	C	28–30	TYMV	Tulip yellow mosaic disease

SS = single stranded, DS = double stranded, C = cubical, B = brick-shaped, H = helical

3. **On the basis of presence of number of strands**

On the basis of number of strands of nucleic acid, the viruses are of **four types**. These are: Double-stranded DNA, Single stranded DNA, Double stranded RNA and Single stranded RNA.

(i) **Double stranded DNA**: Here viral DNA is double stranded. It is found in pox viruses, the bacteriophages like T_2, T_4, T_6, T_3, T_7 and Lamda virus, herpes viruses, adenoviruses etc.

(ii) **Single-stranded DNA**: Here viral DNA is single stranded. It is found in bacteriophages ϕX174 bacteriophages.

(iii) **Double-stranded RNA**: Here viral RNA is double stranded. It has been found within viral capsid in the reoviruses of animals and in the wound tumour virus and rice dwarf viruses of plants.

(iv) **Single-stranded RNA**: Here viral RNA is single stranded. It is found in most of the RNA viruses e.g., tobacco mosaic virus, influenza virus, poliomyelitis, bacteriophage MS-2, and avian leukemia virus.

Nature of Viral genome

The viral genome is very small. It contains just a few genes that enable viruses to infect their respective host cells and code for the synthesis of few proteins for the capsid. Viruses use host biosynthetic machinery (ribosomes, RNA, and enzymes) and raw materials (nucleotides and amino acids) to synthesize copies of their own nucleic acid and proteins.

CLASSIFICATION OF VIRUSES

So many classifications have been proposed earlier by different scientists based on different parameters, but no one is satisfactory in this respect. One of such methods is the classification of viruses based on natural hosts i.e., Bacterial viruses, Plant viruses and Animal viruses. The above method is not satisfactory, because a few viruses are able to infect both plants and animals.

Two classification systems exist:

1. The Hierarchical Virus Classification System and 2. The Baltimore Classification System.

1. The Hierarchical Virus Classification System

The classification of Andre Lwoff, Robert Horne and Paul Tournier (1962), based primarily on the nucleic acid content, got much importance.

The classification of Lwoff *et. al.* (1962) is based on the following parameters:

(i) Nature of nucleic acid i.e., DNA or RNA

(ii) Number of strands of nucleic acid

(iii) Host range

(iv) Morphology of capsid (symmetry, size and number of capsomere)

(v) Presence or absence of envelop.

Brief outline of the system of Lwoff *et. al.* (1962) is given below:

Phylum: VIRA

A. Subphylum. **Deoxyvira** (DNA viruses).

(a) Class ***Deoxyhelica*** (helical symmetry).

1. Order Chitovirales (enveloped).

(i) Family Family Poxviridae.

(b) Class ***Deoxycubica*** (cubical symmetry).

1. Order Haplovirales (no envelop).

(i) Family Microviridae (12 capsomeres).

(ii) Family Parvoviridae (32 capsomeres).

(iii) Family Papilloviridae (72 capsomeres).

(iv) Family Adenoviridae (252 capsomeres).

(v) Family Iridoviridae (812 capsomeres).

(vi) Family Inoviridae.

(c) Class ***Deoxybinala*** (viruses with head and tail).

1. Order Urovirales.

(i) Family Phagoviridae (bacteriophages).

B. Subphylum. **Ribovira** (RNA viruses).

(a) Class ***Ribohelica*** (helical symmetry).

1. Order Rhabdovirales.

Sub-order Rigidovirales (Plant viruses).

(i) Family Delichoviridae (12–13 nm)

(ii) Family Protoviridae (15 nm)

(iii) Family Pachyviridae.

Sub-order Flexiviridales (Plant viruses).

(i) Family Leptoviridae (10–11 nm)

(ii) Family Mesoviridae (12–13 nm)

(iii) Family Advoviridae (15 nm).

2. Order Sagovirales.

(i) Family Myxoviridae (9 nm)

(ii) Family Paramyxoviridae (18 nm)

(iii) Family Stomatoviridae.

(b) Class Ribocubica (cuboidal symmetry).

1. Order Gymnovirales

(i) Family Napoviridae (32 capsomeres).

(i) Family Reoviridae (92 capsomeres).

2. Order Togavirales

(i) Family Arboviridae.

2. The Baltimore Classification System

The Baltimore system of virus classification (first defined in 1971) provides a useful guide with regard to the various mechanisms of viral genome replication. The central theme here is that all viruses must generate positive strand mRNAs from their genomes in order to produce proteins and replicate themselves. The precise mechanisms whereby this is achieved differ for each virus family. These various types of virus genomes can be broken down into seven fundamentally different groups, which obviously require different basic strategies for their replication. The Nobel-prize winner **David Baltimore** originally designed the scheme. Thus, to honour Baltimore, the classification is named as "Baltimore Classification" of virus genomes. By convention, the top strand of coding DNA written in the 5′–3′ direction is + sense. The mRNA sequence is also + sense. The replication strategy of the virus depends on the nature of its genome.

Viruses can be classified into **seven groups**:

1. Classification by genome type

1.1. DNA viruses

1.1.1. **Group I**: dsDNA viruses (double stranded DNA)

1.1.2. **Group II**: ssDNA viruses (single stranded DNA)

1.2. RNA viruses

1.2.1. **Group III**: dsRNA viruses (double stranded RNA)

1.2.2. **Group IV**: (+)ssRNA viruses (positive single stranded RNA or mRNA like)

1.2.3. **Group V**: (–)ssRNA viruses (negative single-stranded RNA)

1.3. DNA and RNA Reverse Transcribing viruses

1.3.1. **Group VI**: ssRNA–RT viruses (single stranded RNA)

1.3.2. **Group VII**: dsDNA–RT viruses (double stranded DNA)

2. Subviral agents

2.1. Viroids

2.2. Satellites

2.3. Prions

The *taxonomy* of viruses is similar to that of cellular organisms:

Order (*–variales*)

Family (*–viridae*)

Subfamily (*–virinae*)

Genus (*–virus*)

Species

However, the code of nomenclature regulated by the **International Committee on Taxonomy of Viruses** differs from the others on several points. Most notably, names of orders and families are italicised, and **species names are not binomial**; instead, they generally take the form of [*Disease*] *Virus*. The recognition of orders is very recent and has been deliberately done in a leisurely manner; to date, only three orders have been named, and most families remain unplaced. Approximately 80 families and 4,000 species of virus are known.

CLASSIFICATION BY GENOME TYPE

DNA viruses

Group I: *dsDNA viruses* (double stranded DNA)

Order *Caudovirales* (tailed bacteriophages)

Family *Myoviridae* e.g., *Enterobacteria phage* T4
Family *Podoviridae*
Family *Siphoviridae* e.g., *Enterobacteria phage* λ

Unassigned

Family *Ascoviridae*
Family *Adenoviridae*
Family *Asfiviridae*
Family *Baculoviridae*
Family *Corticoviridae*
Family *Fuselloviridae*
Family *Guttaviridae*
Family *Herpesviridae* e.g., Human herpse viruses
Family *Iridoviridae*
Family *Lipothrixviridae*
Family *Nimaviridae*
Family *Papillomaviridae*
Family *Phycodnaviridae*
Family *Plasmaviridae*
Family *Polyomaviridae* e.g., Simian virus 40
Family *Poxviridae* e.g., *Cowpox virus*, *Variola virus* (small pox)
Family *Rudiviridae*
Family *Tectiviridae*

Unassigned genera

Mimivirus; type species: *Acanthamoeba polyphaga mimivirus*

Group II: *ssDNA* viruses (single stranded DNA)

Unassigned bacteriophages

Family *Inoviridae*
Family *Microviridae*

Unassigned viruses

Family *Geminiviridae*
Family *Circoviridae*
Family *Nanoviridae*
Family *Parvoviridae* e.g., *Parvovirus* B19 (most depend on coinfection with *adenoviruses* for growth)

Unassigned genera

Anellovirus; type species: Torque teno virus.

RNA viruses

Group III: *dsRNA* viruses (double stranded RNA)

Family *Birnaviridae*
Family *Chrysoviridae*
Family *Cystoviridae*
Family *Hypoviridae*
Family *Partitiviridae*
Family *Reoviridae* e.g., *Rotavirus*
Family *Totiviridae*

Unassigned genera

Endornavirus; type species: *Vicia fava endornavirus*

Group IV: (+)*ssRNA* viruses (positive single stranded RNA or mRNA like)

Order *Nidovirales* ("Nested" Viruses)

Family *Arteriviridae*
Family *Coronaviridae* e.g., *Coronavirus*

Unassigned

Family *Astroviridae*
Family *Barnaviridae*
Family *Bromoviridae*
Family *Caliciviridae* e.g., *Norwalk virus*
Family *Closteroviridae*
Family *Comoviridae*
Family *Dicistroviridae*
Family *Flaviviridae* e.g., Yellow fever virus, West Nile virus, Hepatitis C virus
Family *Flexiviridae*
Family *Hepeviridae* e.g., Hepatitis E virus

Family *Leviviridae*

Family *Luteoviridae*

Family *Marnaviridae*

Family *Narnaviridae* – naked RNA viruses

Family *Nodaviridae*

Family *Picornaviridae* e.g., *Poliovirus, Rhinovirus, Hepatitis A virus*

Family *Potyviridae*

Family *Sequiviridae*

Family *Tetraviridae*

Family *Togaviridae* e.g., *Rubella virus*

Family *Tombusviridae*

Family *Tymoviridae*

Unassigned genera:

Genus *Benyvirus*; type species: *Beet necrotic yellow vein virus*

Genus Cheravirus; type species: *Cherry rasp leaf virus*

Genus Furovirus; type species: *Soil-borne wheat mosaic virus*

Genus *Hordeivirus*; type species: *Barley stripe mosaic virus*

Genus *Idaeovirus*; type species: *Raspberry bushy dwarf virus*

Genus *Machlomovirus*; type species: *Maize chlorotic mottle virus*

Genus *Ourmiavirus*; type species: *Ourmia melon virus*

Genus *Pecluvirus*; type species: *Peanut clump virus*

Genus *Pomovirus*; type species: *Potato mop-top virus*

Genus *Sadwavirus*; type species: *Satsuma dwarf vius*

Genus *Sobemovirus*; type species: *Southern bean mosaic virus*

Genus *Tobamovirus*; type species: *Tobacco mosaic virus*

Genus *Tobravirus*; type species: *Tobacco rattle virus*

Genus *Umbravirus*; type species: *Carrot mottle virus*

Group V: (–)*ssRNA* viruses (negative single-stranded RNA)

Order *Mononegavirales* (non-segmented negative stranded viruses)

Family *Bornaviridae* – *Borna disease virus*

Family *Filoviridae* – *Ebola viruses, Marburg virus*

Family *Paramyxoviridae* e.g., *Measles virus, Mumps virus*

Family *Rhabdoviridae* e.g., *Rabies virus*

Segmented negative stranded viruses

Family *Arenaviridae*

Family *Bunyaviridae* e.g., *Hantavirus*

Family *Orthomyxoviridae* – *Influenza viruses*

Unassigned genera:

Genus *Deltavirus*; type species: *Hepatitis delta virus*

Genus *Ophiovirus*; type species: *Citrus psorosis virus*

Genus *Tenuivirus*; type species: *Rice stipe virus*

Genus *Varicosavirus*; type species: *Lettuce big-vein associated virus*

DNA and RNA Reverse Transcribing viruses

Group VI: *ssRNA-RT* viruses (single stranded RNA)

Family *Metaviridae*

Family *Pseudoviridae*

Family *Retroviridae* – *Retroviruses*, e.g., HIV 1

Group VII: *dsDNA-RT* viruses (double stranded DNA)

Family *Hepadnaviridae* e.g., *Hepatitis* B virus

Family *Caulimoviridae* e.g., *Cauliflower mosaic virus*

Subviral agents

The following agents are smaller than viruses but have some of their viral properties:

Viroids

Family *Pospiviroidae*

Genus *Pospiviroid*; type species: *Potato spindle tuber viroid*

Genus *Hostuviroid*; type species: *Hop stunt viroid*

Genus *Cocadviroid*; type species: *Coconut cadang-cadang viroid*

Genus *Apscaviroid*; type species: *Apple scar skin viroid*

Genus *Coleviroid*; type species: *Coleus blumei viroid* 1

Family *Avsunviroidae*

Genus *Avsunviroid*; type species: *Avocado sunblotch viroid*

Genus *Pelamoviroid*; type species: *Peach latent mosaic viroid*

Satellites

Satellite viruses

Single-stranded RNA satellite viruses

Subgroup 1: *Chronic bee paralysis satellite virus*

Subgroup 2: *Tobacco necrosis satellite virus*

Satellite nucleic acids

Single-stranded satellite DNAs

Doube-stranded satellite RNAs

Single-stranded satellite RNAs

Subgroup 1: Large satellite RNAs

Subgroup 2: Small linear satellite RNAs

Subgroup 3: Circular satellite RNAs

Prions

Fungal prions

Mamalian prions

GENERAL CHARACTERS

The viruses are non-cellular, self-replicating, obligate, intracellular parasitic agents — essentially composed of a protein that covers a central nucleic acid molecule, either RNA or *DNA*. The amount of protein varies from 60 to 95% and the rest is nucleic acid. A simple virus particle is often called a **virion**. They grow and multiply only in living cells. They cause diseases of animals including man, and plants of different groups, except bryophytes and gymnosperms. All known viruses are the pathogen of either plants or animals. They can exist outside the host cell for long period, but remain inert in free state. Again they become active after entering in a living host cell. They cannot grow or multiply in nutrient medium, supplemented with the extract of host cell in which it grows, but can grow only in intact living cell, thus they are **obligate parasite**. They are the smallest among the infective agents, even much smaller than the smallest bacteria and varying over a wide range from 18–400 nm (Parvo virus – 18-26 nm, Tobacco mosaic virus – 17.5 nm x 300 nm, Tulip mosaic virus – 28 nm, Polio virus – 27 nm x 30 nm, Influenza virus – 80 nm x 120 nm, Small pox virus – 400 nm, etc.). According to electron microscopic observation they are of different forms (Fig. 2.43) such as rod shaped, spherical, cubical etc. Within the host cell, the virus can grow, multiply and undergo mutation, but it does not respire. They depend completely on the enzyme system of the host cell for their activity.

Structure

The virus consists of two parts: (i) *Nucleic acid* (centrally placed) and (ii) *Protein coat*, sometimes with additional envelope.

(i) *Nucleic acid*: Viruses contain only one type of nucleic acid i.e., either DNA or RNA. The DNA containing viruses are called *Deoxyviruses*, whereas viruses having RNA are called *Riboviruses*. They vary in the structure of their nucleic acid. Most of the plant viruses have RNA either single (TMV) or double stranded (Rice ragged stunt viruses), except a few have DNA either single (Gemini viruses) or double stranded (Dahlia mosaic virus). Animal viruses have mostly double stranded DNA or either single (Polio virus) or double (Reo

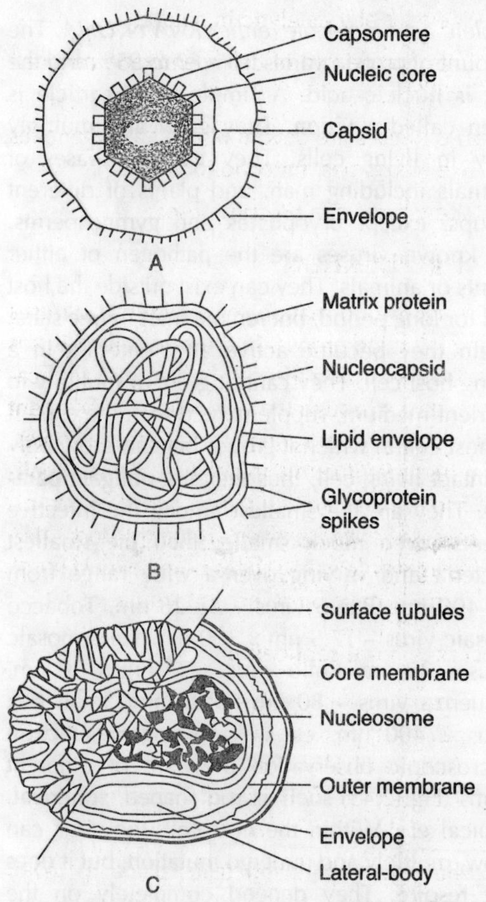

Fig. 2.46: Diagram of virus particles. A. Enveloped virus with icosahedral symmetry, B. Virus with helical symmetry, and C. Complex capsid

virus) stranded RNA and bacteriophages contain mostly double stranded DNA, but they also have single stranded RNA (f_2, R_{17}, fr) or single stranded DNA (f_1, fd, M_{13}) (Table 2.20). Each virion contains only one molecule of nucleic acid, called **genome**, consisting of nucleotide pairs whose number ranges from 1000–250,000 pairs. The amount of nucleic acid of a virion usually depends on its size. The number of genes per virion ranges from 4–8 for small viruses and 100–200 for the large viruses.

(ii) *Protein coat*: The protein coat surrounding the genome is called **capsid** and the capsid together with the enclosed nucleic acid is called **nucleocapsid**. The capsid is made up of a large number of protein subunits, called **capsomeres** (Fig. 2.47A). Many mammalian viruses have **envelope** made up of a bilayered lipoprotein, mainly of host cell origin that surrounds the nucleocapsid (Fig. 2.47B).

Symmetry

The capsid is symmetrically arranged around the central nucleic acid. Based on symmetry of capsid, the viruses are grouped into three categories: (a) *Cubical (icosahedral)*, (b) *Helical*, and (c) *Complex*.

(a) *Cubical* (*icosahedral*) capsids (Fig. 2.46A): They have a polygon with 12 corners (vertices), 20 sides (facets) and 30 edges. Each side is an equilateral triangle. They are of two types — Pentons (pentagonal capsomeres at the corners) and Hexons (hexagonal capsomeres at the corners),

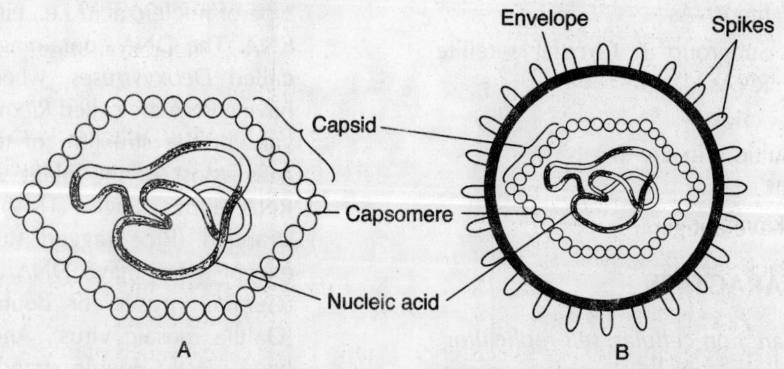

Fig. 2.47: A. Naked Viruses and B. Enveloped Virus with spikes

e.g., Herpes and Toga viruses are enveloped and Papova and Adeno viruses are naked.

(b) *Helical capsids* (Fig. 2.46B): Both nucleic acid and capsomeres are coiled together and form a spiral or helical tube. All the helical types are RNA viruses and most of them are enveloped, e.g., Tobacco mosaic virus (TMV), Influenza virus, etc.

(c) *Complex capsids* (Fig. 2.46C): Viruses which do not conform to either of the above two types due to complexity of their structure are called complex capsids, e.g., Pox virus and Bacteriophages like T_2, T_4, and T_6.

The **viruses differ from bacteria** in the following points:

(a) The viruses are very much smaller.
(b) They lack the machinery for protein synthesis.
(c) They do not have cellular organisation.
(d) They neither grow in artificial culture medium nor divide by binary fission.
(e) They have only one kind of nucleic acid.
(f) The viruses are resistant to antibiotic.

Viruses have both living and non-living characters.

Living characters of viruses

(a) They have the nucleic acid (DNA or RNA) i.e., the genetic material that can replicate.
(b) Mutation is well-established by the availability of mutant forms in some viruses.
(c) They are sensitive to stimulants like radiation, chemical substances etc.
(d) They can multiply in the living cells of the host.
(e) The viruses have antigenic property.
(f) They can attack specific host.

Non-living characters of viruses

(a) The viruses remain as inert material ouside their host.
(b) They are autocatalytic in nature.
(c) They are devoid of cell membrane and cell wall.
(d) The viruses are devoid of cellular organelles like ribosomes, mitochondria etc.
(e) The viruses can be crystallised.

VIRAL REPLICATION

It is the formation of biological viruses during infection process in the host cells. Viruses must first get inside the cell before viral replication to start. Through the generation of abundant copies of its genome and packaging these copies, the virus continues to infect new hosts. Replication of viruses greatly varies and depends on the type of genes involved in them. Most DNA viruses assemble in the nucleus while most of the RNA viruses develop solely in cytoplasm.

Process of Viral replication

Viruses multiply only in the living cells. The host cell must provide the energy and synthetic machinery and also the low molecular-weight precursors for the synthesis of viral proteins and nucleic acids.

The virus replication takes place in seven steps. These are:

1. Adsorption,
2. Entry,
3. Uncoating,
4. Transcription i.e. mRNA production,
5. Synthesis of virus components,
6. Virion assembly and
7. Release (liberation Stage).

1. Adsorption

Initially the virus attaches to the membrane of the host cell. It then injects it nucleic acid into the host cell.

2. Entry

After adsorption either the virus injects its DNA or RNA into the cell to initiate infection, but in other case the cell membrane of the host cell invaginate the virus particle, enclosing it in a

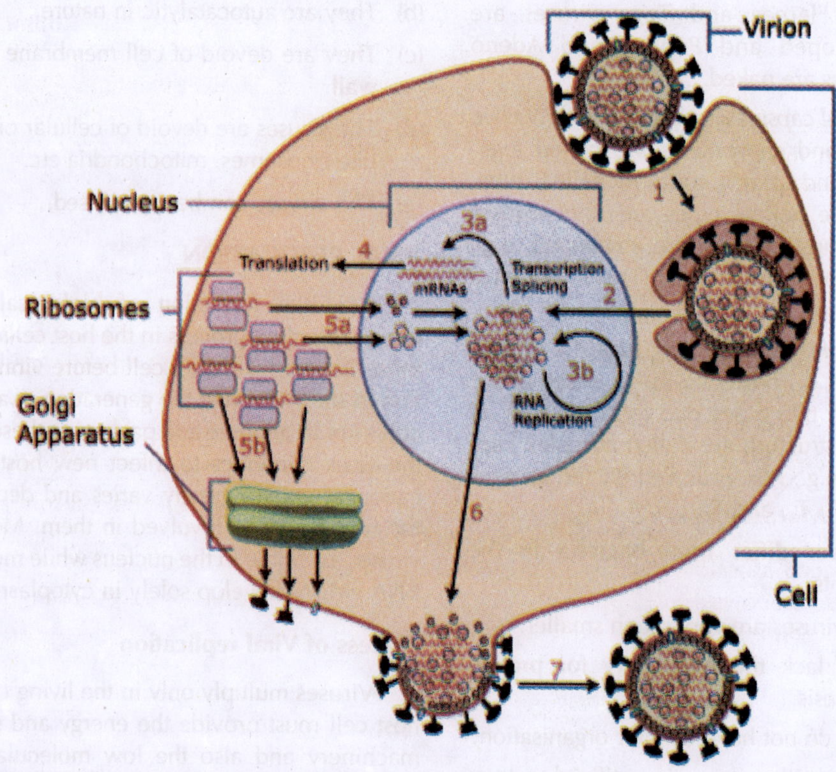

Fig. 2.48: Life cycle of Influenza virus

pinocytotic vacuole. This protects the cell from antibodies as in the case of the HIV virus.

3. Uncoating

Cellular enzymes (from lysosomes) strip off the protein coat of virus. Thus the released nucleic acid or genome becomes accessable.

4. Transcription i.e. mRNA production

In some RNA virus, the infecting RNA produces messenger RNA i.e. mRNA. From this mRNA, protein produces through translation. In others with negative stranded RNA and DNA, viruses are produced by transcription then translation (Fig. 2.48).

The mRNA is used to instruct the host cell to make virus components. The virus takes advantage from the existing cell structures to replicate of its own.

5. Synthesis of virus components

The following components are produced by the virus through existing organelles of the host:

(i) **Viral protein synthesis**: The virus mRNA is translated on the ribosomes of cell int. two types of virus protein.

(a) Structural protein: The proteins which make up the virus particle are produced and assembled.

(b) Non-structural protein: It is not found in particle, but produced mainly by enzymes for virus genome replication.

(ii) **Viral nucleic acid synthesis** (genome replication): New virus genomes are synthesized, where the templates are either the parental genome or with single stranded nucleic acid genomes, newly formed complementary strands.

(iii) **Virion Assembly**: Virion is simply an active or intact virus particle. In this stage, newly synthesized genome (nucleic acid), and proteins are assembled togather and form new virus particles.

This may take place in the cell's nucleus, cytoplasm, or at the plasma membrane in most viruses.

(iv) **Release (Liberation Stage)**: After maturation, the viruses are released by either sudden rupture of the cell, or through cell membrane by gradual extrusion (budding) in enveloped viruses.

The new viruses may invade or attack other cells, or remain dormant inside the cell. In the case of bacterial viruses, the release of progeny virions takes place by lysis of the infected bacterium. However, in the case of animal viruses, release usually occurs without cell lysis.

TOBACCO MOSAIC VIRUS (TMV)

The Tobacco Mosaic Virus (TMV) was discovered by D. Iwanowski in 1892. But later, in 1935, W. M. Stanley isolated it from infected plants and crystallised it. He found that the crystals were highly infectious.

Structure

The structure of TMV (Fig. 2.49) has been studied extensively among the plant virus so far known.

It is a rod shaped structure with rough surface, looks like a *bunch of banana*. Each rod is about 3,000 Å (300 nm) long and 170 Å (17 nm) in diameter. Each rod consists of two parts, the protein coat or capsid and nucleic acid. The capsid consists of 2,130 protein subunits, called capsomeres. The nucleic acid is a single stranded RNA molecule. The capsomeres are closely packed and arranged helically around the RNA helix, thereby it forms a hollow cylinder. The central hole (axial core) of the cylinder is about 40 Å (4 nm) in diameter which is present throughout the rod. The RNA remains deeply embedded in the protein subunits (5 nm from the outside), thereby it does not occupy the hole. The RNA thread extends the entire length of the particle in the form of a long helix.

There is about 130 helical turns and each turn consists of about $16\frac{1}{3}$ protein subunits (capsomeres) i.e., 49 subunits in three turns. Each capsomere has a molecular weight of about 17,300 and arranged with a pitch of 23 Å. Each capsomere is formed by 158 amino acids present in the form of a long chain. The RNA molecule consists of 6,000 nucleotide pairs.

The amount of protein in a rod is about 95% and rest 5% is the RNA. The molecular weight of the protein is about 40,000,000 (40 million), whereas the RNA is about 2,000,000 (2 million).

The TMV particle infects the tobacco plant and develops tobacco mosaic virus disease. The

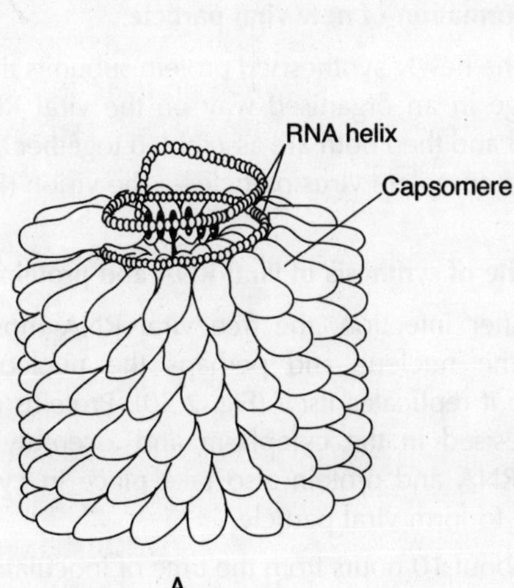

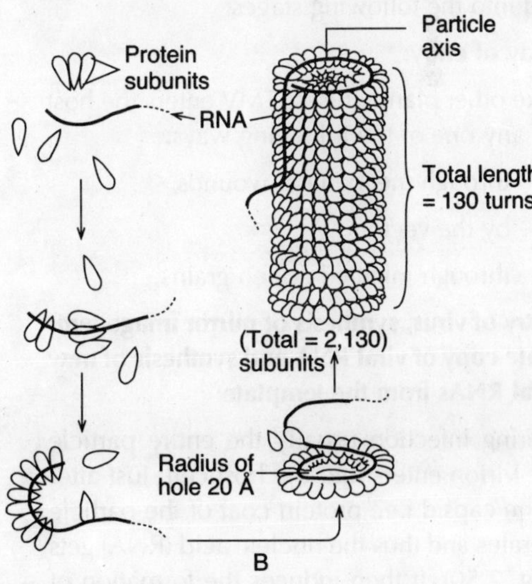

Fig. 2.49: Tobacco mosaic virus (TMV):

virus moves from one host to another (i.e., transmission) by vectors like *Microsiphon solanifolii, Myzus pseudosolani* and *M. persica*. The mechanical transmission can be done through transplanting, handling and by rubbing. During infection usually the entire particle i.e., virion, enters the host cell. Just after entering the host cell, the capsid (protein coat) of the particle degenerates and RNA starts the synthesis of m-RNA (messenger RNA) which eventually synthesises the protein subunits. The replication of RNA molecule also takes place inside the host cell. In both the cases, they utilise the raw material from the host cell. So the RNA of the virus serves both the functions: (i) synthesis of virus specific protein, and (ii) self-replication. Thus new virus particles are formed inside the host cell.

The naked RNA is also capable of infecting host cell and able to form its new generation.

TMV can infect other Solanaceous plants like tomato etc., and also to some other hosts. The particle may remain infective for about 50 years and can withstand boiling temperature for about 5–10 minutes.

Mechanism of TMV multiplication

Detailed mechanism of TMV multiplication is given below:

The multiplication of TMV in host can be divided into the following stages:

1. **Way of entry**:

Like other plant viruses, TMV enters the host cell by any one of the following ways:

(i) through mechanical wounds,

(ii) by the vectors,

(iii) through infected pollen grains.

2. **Entry of virus, synthesis of mirror image template copy of viral RNA and synthesis of new viral RNAs from the template**:

During infection usually the entire particle i.e., the virion enters into the host cell. Just after entry, the capsid i.e., protein coat of the particle degenerates and thus the nucleic acid (RNA) gets free (Fig. 2.50). It then induces the formation of enzyme RNA-polymerase by the host cell. The RNA-polymerases are also called RNA-replicases or RNA-synthetases. The viral RNA then acts as a template and with the help of RNA-polymerase, it synthesises additional RNA, which is not the copy of the viral RNA, but a *mirror image*, which remains temporarily connected to the viral strand. Thus, the two RNA threads form a double stranded RNA that soon separates into the original virus RNA and the newly formed strand i.e., the *mirror image strand. This strand acts as a template for the synthesis of more viral RNA*. From this mirror image template strand more viral RNAs are produced (Fig. 2.50).

3. **Synthesis of viral protein**:

As soon as the new viral nucleic acids (RNA) are produced, they induce the host cell to synthesise the protein molecules that will be the protein subunits required to develop its own capsid. Only a portion of viral RNA strand is required to form the viral protein (Fig. 2.50). The RNA of TMV consists of 6,400 nucleotides and its protein consists of 158 amino acids, thus it requires 158 triplate codes i.e., 474 nucleotides to code the sequence of amino acids in protein subunits.

During synthesis of viral protein, the portion of viral RNA coding for viral protein acts as m-RNA, which utilises the amino acids, t-RNAs and ribosomes of host. Some enzymes may activate or initiate the chemical reaction and thereby affect the cell physiology.

4. **Formation of new viral particle**:

The newly synthesised protein subunits then arrange in an organised way on the viral RNA strand and then both are assembled together and form a complete virus particle — the virion (Fig. 2.50).

5. **Site of synthesis of viral RNA and proteins**:

After infection, the free viral RNA moves into the nucleus and perhaps the nucleolus where it replicates itself (Fig. 2.50). Proteins are synthesised in the cytoplasm and assembly of viral RNA and protein also take place in cytoplasm to form viral particle.

About 10 hours from the time of inoculation are required for the formation of new virus particle. The particles may exist either singly or

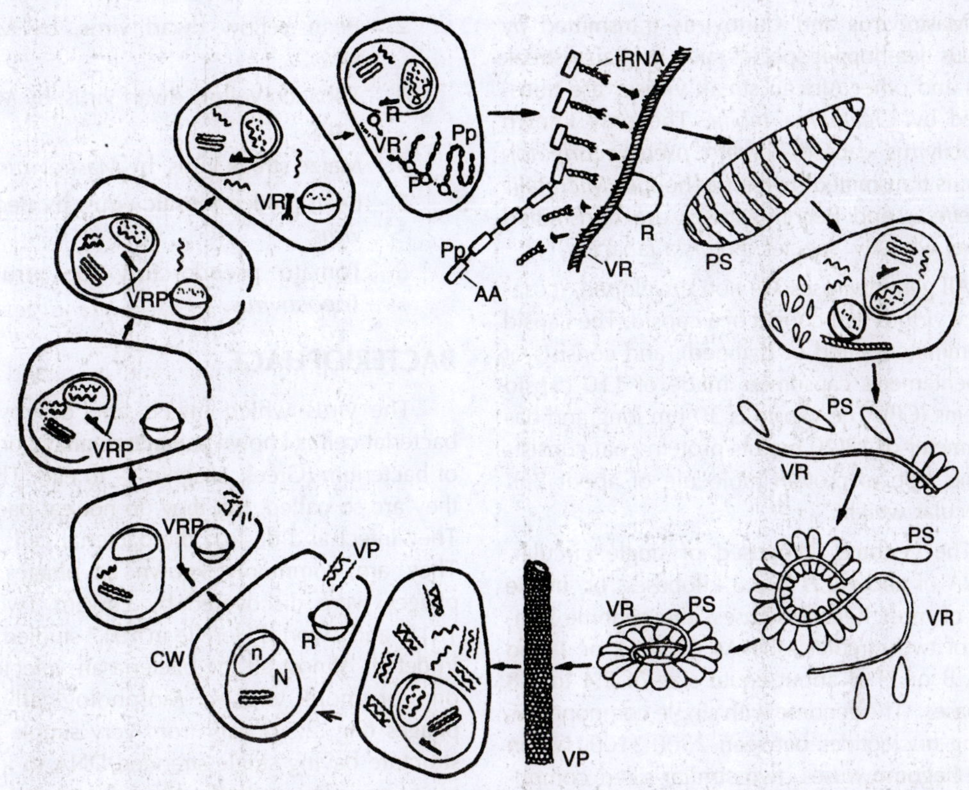

Fig. 2.50: Sequential events in the replication of TMV (Tobacco mosaic virus): CW—cell wall; N—nucleus; n—nucleolus; VP—viral particle; R—ribosome; VRP—viral RNA polymerase; VR—viral RNA; P—polyribosome; tRNA—transfer RNA, AA—amino acid and PS—protein subunit.

in groups and may form crystalline or amorphous inclusion bodies within the different regions of cell (nucleus, nucleolus and cytoplasm).

GEMINIVIRUSES

Geminiviruses are the plant viruses (Fig. 2.51), includes in the family Geminiviridae. It includes 9 genera and more than 360 species. The common genera are *Mastrevirus, Begomovirus, Topocurvirus,* and *Curtovirus*. They cause diseases like bright yellow mosaic, yellow mosaic, yellow mottle, leaf curling, stunting, streaks etc. Geminiviruses can wipe out entire crops of many different plants, such as tomatos, beans, corn, cotton etc., making them a serious threat to farming industry worldwide.

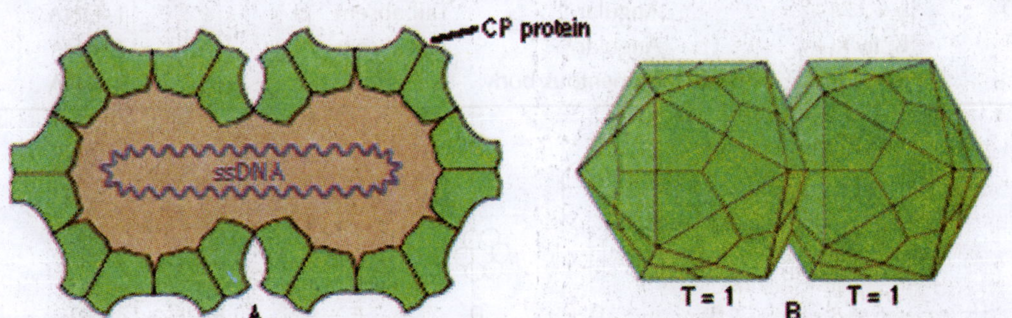

Fig. 2.51: Geminivirus: A. Sectional view showing inner ssDNA and outert CP protein and B. Twin particles of Virus

Mastrevirus and Curtovirus transmitted by various leafhopper species such as maize streak virus and other African streak viruses are transmitted by *Cicadulina mbila*. The only known Topocuvirus caused *Tomato pseudo-curly top virus*, is transmitted by the treehopper *Micrutalis malleifera*, and Begomoviruses are transmitted by the whitefly species, *Bemisia tabaci*.

All geminiviruses are non-enveloped, icosahedral virions that consist of a capsid. The capsid is geminate (paired or twinned), and consists of 22 pentameric capsomers made of 110 capsid proteins (CP). The capsid is 30 nm long and has a diameter of 18-20 nm. Its protein coat consists of one type of protein molecule of about 28K molecular weight.

The genome composed of single circular, ssDNA of about 2.6 to 2.8 kilobases, or, in the case of most Begomoviruses, the genome consists of two circular ssDNAs (DNA A or 1 and DNA B or 2) of about equal size of 2.4 to 2.8 kilobases. The genome with single component is having nucleotides between 2500-3100, but in most Begomoviruses, two similar-sized components (i.e. bipartite) having 2600 and 2800 nucleotides.

Common diseases caused the viruses are:

1. Curly top of sugar beets, by *Curtovirus*,
2. Bean yellow dwarf virus, by *Mastrevirus*,
3. Tobacco yellow dwarf virus, by *Mastrevirus*,
4. Maize streak virus, by *Mastrevirus*,
5. Bean golden mosaic virus, by *Begomovirus*,
6. Tomato pseudo-curly top virus, by *Topocuvirus*.

BACTERIOPHAGE

The virus which infects and destroys the bacterial cell is known as **bacteriophage** or eater of bacterium (Greek *phagein* — to eat). Though they are so called, but they do not eat bacteria. They infect and destroy the bacterial cells only. They are commonly known as **phages**. The phages were discovered by Edward Twort in 1915, and Twort d'Herelle in 1917 studied them in detail. Almost all the bacteria are infected by one or more viruses. Morphologically, the phages (Fig. 2.52) vary from very simple small structure having single stranded DNA ($\phi \times 174$) to a complex tadpole like structure having double stranded DNA, the T-even phages (T_2, T_4) (Table 2.20). The phage which infects the colon bacterium (*E. coli*) is called **Coliphage**. They are mostly T-even phages like T_2, T_4 etc. The viruses

Table 2.20: Important properties of some bacteriophages

Type	Example (Coliphage)	Head	Tail	Nucleic acid type
A	T-even	Angular	Contractile tail	2–DNA
B	T_1, T_5	Angular	Long non-contractile	2–DNA
C	T_3, T_7	Angular	Short non-contractile	2–DNA
D	$\phi \times 174$	Angular	Tail absent	1–DNA
E	f_2, fr, R_{17}	Angular	Tail absent	1–RNA
F	fd, f_1, M_{13}	Flexible filamentous body		1–DNA

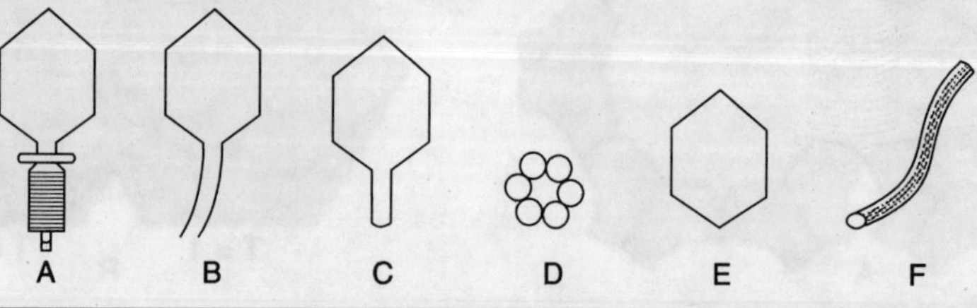

Fig. 2.52: Different types (A-F) of phages as described in Table 2.20

also infect and parasitise on blue green algae, fungi, yeasts and actinomycetes, and accordingly they are called as **Cyanophages, Mycophages, Zymophages** and **Actinophages,** respectively. Important properties of some bacteriophages are:

Ultrastructure of Bacteriophage

The detailed morphological structure of bacteriophages are known with the help of electron microscopic studies of some larger particles of T-even types which infect the bacterium *E. coli*. The structure of T-even phages are given below (Fig. 2.53):

T-even phages are spermatozoid or tadpole-like structure with a **head** and a **tail**. The head is a hexagonal, bipyramidal prismoid structure, while the tail is contractile and cylindrical. The tail is attached to the lower side of the head. The extended portion between the head and tail is called disc or **collar**. The head of T_2 phage is approximately 950 Å × 650 Å in size. The wall of the head is a proteinaceous membrane, consists of 2,000 similar protein subunits, enclosing a single, double-stranded DNA thread of about 50 nm in length. The molecular weight of DNA is 2,500,000 and the amount of nucleic acid is approximately 6×10^3 micrograms. The length of the tail is almost equal to the length of the head i.e., 950 Å and its diameter is 80 Å.

The tail consist of four parts (Fig. 2.53):

(i) A central hollow **tube** or **core** (through which DNA passes to the host cell during infection),

(ii) The proteinaceous **sheath** surrounding the core can contract longitudinally,

(iii) Hexagonal **tail plate** or **basal plate** (200 Å thick), has a pin at every corner which is attached with the distal end of the sheath,

(iv) Six long thin **tail fibres** are attached to the tail plate, one at each corner. The tail fibres are about 1,500 Å in length. The tail fibres are the organ of attachment to the host cell wall.

LIFE CYCLE

Being non-cellular the viral particle has no power of independent metabolism and reproduction like other organisms. The replication of virus takes place by two types of cycle: 1. **Lytic cycle**, and 2. **Lysogenic cycle**. Based on interac-

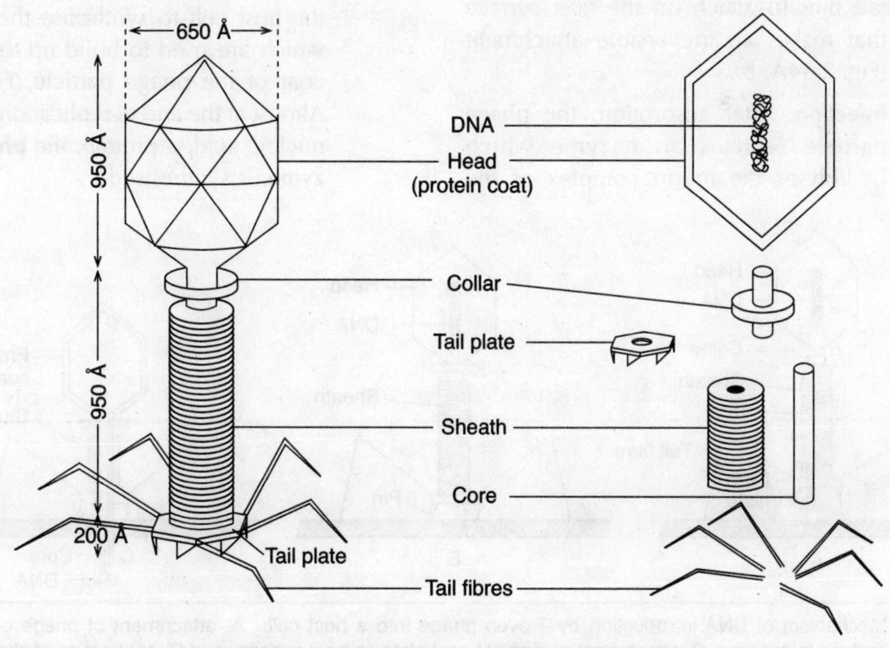

Fig. 2.53: Structure of bacteriophage (T_2) and its components

tion of phages with the bacterial cell, they have been distinguished into two types — **virulent** or **lytic**, and **avirulent** or **temperate** phages. The virulent phages perform the lytic cycle and the avirulent or temperate phages perform the lysogenic cycle. A portion of lysogenic cycle is like the lytic cycle.

1. **Lytic cycle**: This type of cycle is seen in T-even phages (T_2, T_4 etc.) which attack *Escherichia coli*. The lytic cycle consists of five steps (Fig. 2.55): (a) Adsorption, (b) Infection (Fig. 2.54), (c) Synthesis of phage components in host cell, (d) Formation of new phage particle, and (e) Liberation of phages from the host cell.

(a) **Adsorption**. The interaction between the phage specific organelle — the tail and the receptor site of the host cell is called the adsorption. The adsorption is facilitated by the negatively charged carboxyl groups on the host surface and the positively charged amino-group of protein present at the tip of the phage tail. In T-even phages, the tip of the tail fibre first attaches to the cell surface. The tail fibre then bends and allows the tail pins to attach on the host surface that makes an irreversible attachment (Fig. 2.54A, B).

(b) **Infection**. After adsorption, the phage particle secretes an enzyme which hydrolyses the murin complex of the host cell wall and forms a pore. The sheath of the tail then contracts and pushes the central tubular part, i.e., **core** of the tail, into the host wall, like an injection needle (Fig. 2.54). The nucleic acid of the phage then passes through the core and enters the host bacterium.

The empty protein shell of the phage is called **ghost**, which may remain attached even after release of nucleic acid. Once the bacterial cell receives the nucleic acid of a phage, it becomes resistant to the other phages.

(c) **Synthesis of phage components in host cell**. Once the phage nucleic acid takes the entry inside the bacterial cell, it suppresses the synthesis of bacterial protein and directs to synthesise the proteins of the phage particle (Fig. 2.55).

The DNA of phage replicates following the **semi-conservative** process. Majority of the DNA acts as a template for its own synthesis and the rest is used as template for the synthesis of viral specific m-RNA by utilizing the RNA-polymerase of the host. The newly formed m-RNA directs the host cell to synthesise the proteins which are used to build up the protein coat of the phage particle (Fig. 2.55). Almost at the end of replication of phage nucleic acid, a protein, the **phage lysozyme**, is synthesised.

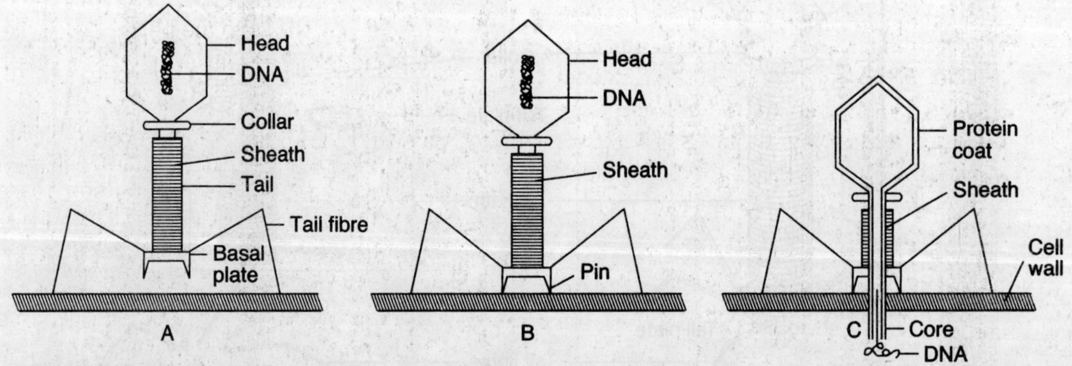

Fig. 2.54: Mechanism of DNA introduction by T-even phage into a host cell : A. attachment of phage on the host surface by tail fibre. B. attachment of pins of basal plate on host surface, and C. contraction of sheath, pushing of core and releasing of DNA inside the host

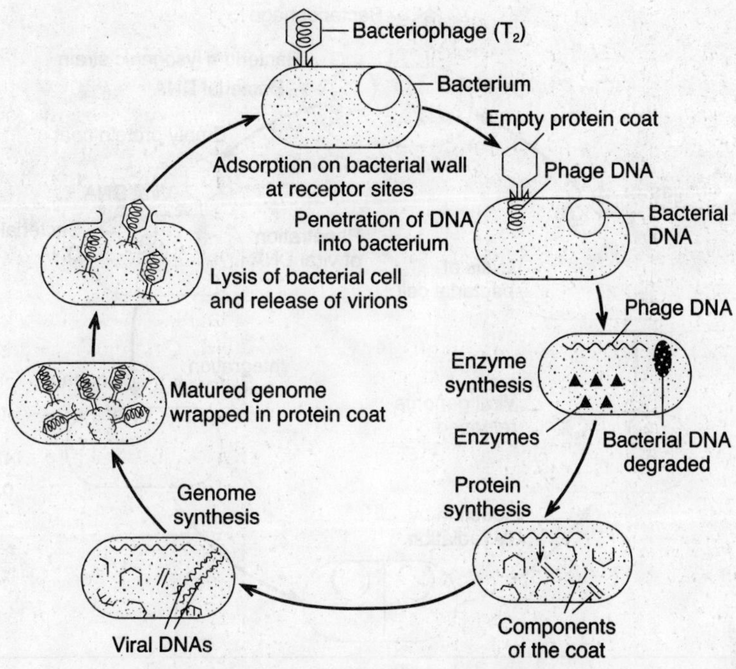

Fig. 2.55: Lytic life cycle of bacteriophage (T_2)

(d) **Formation of new phage particle**. The new phage particles are formed by the assemblage of nucleic acid and protein. This process is called **maturation**, which is controlled by viral genome (Fig. 2.55). In this process, initially the condensation of nucleic acid molecule takes place. The protein sub-units then aggregate around the nucleic acid molecule and form the head of the phage. By this time the tail formation starts. Initially the core tube is attached with the basal plate and then sheath becomes assembled around the core tube. In this stage, the tail becomes attached to the base of the head taking a collar in between. At last, the tail fibres are attached to the basal plate.

(e) **Liberation of phages from the host cell**. In a cycle of phage development, about 200 phages are formed which take about 30–90 minutes. In the host cell, the phage DNA secretes lysozyme (an enzyme) which causes the lysis of host cell wall. As a result of lysis the phage particles are liberated (Fig. 2.55).

2. **Lysogenic cycle**: A. Lwoff (1953) discovered this type of cycle in Lambda (λ) phages that attack *E. coli*. The phage involved in this cycle is called **temperate phage**, the bacterium is the **lysogenic strain** and the entire process is called **lysogeny** (Fig. 2.56).

At first the phage is adsorbed on the wall of the host bacterium and its DNA becomes injected into the host cell. Here the phage DNA, like the lytic cycle, does not take over the protein synthesis machinery of the host cell, instead, it becomes integrated with the nucleoid of the host genome. This integrated phage DNA is called a **prophage** (Fig. 2.55). Thus the new composite genome replicates as one unit. The composite genome then multiplies for indefinite number of times and produces daughter lysogenic bacteria. After a number of generations the viral genome gets detached from the composite genome and releases in the cytoplasm. This dissociation is called **induction** (Fig. 2.55). The viral genome then enters the lytic cycle and forms temperate

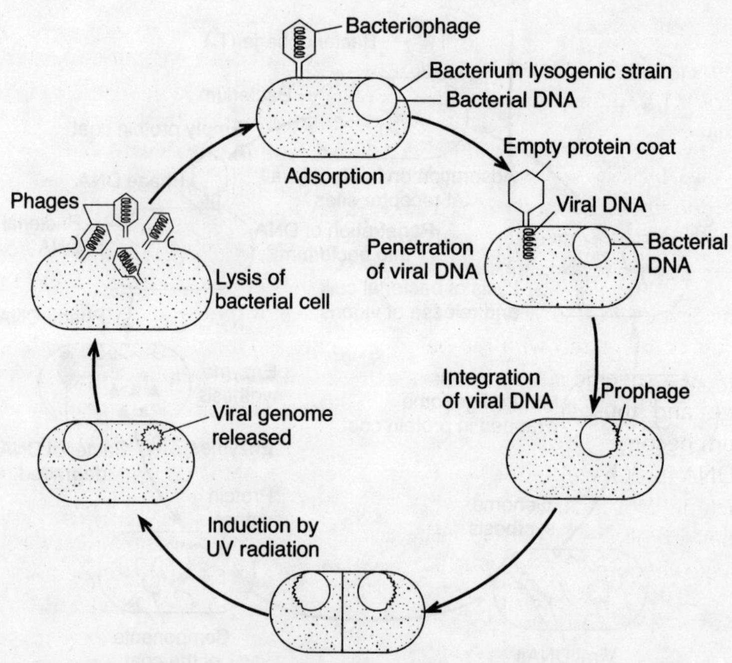

Fig. 2.56: Lysogenic life-cycle of bacteriophage (λ-phage)

phages that are released by lysis of wall of the host bacterium.

Detailed molecular mechanism of lysogeny and induction by lambda (λ) phage

Being non-cellular the viral particles have no capacity of independent metabolism and reproduction like other organisms. Thus, they need others' help for their multiplication almost like the activity of a terrorist of present day. Multiplication of viruses commonly takes place by infection, multiplication and lysis of the host bacterium, called **lytic cycle**. But in many others, the mechanism is different, those show different modes of parasitism including infection, integration with genome, multiplication along with host genome; and, later, it gets separated from the host genome, causes multiplication and comes out through lysis of host cell, called as **lysogenic cycle**. The phage involved in this cycle is called **temperate phage**, the bacterium as lysogenic strain, the injected phage DNA integrated with the host genome as **prophage** and the entire process is called **lysogeny**.

The temperate phage, after infection, may undergo any one option of multiplication i.e., either **lytic cycle** like a virulent phage, or **lysogenic cycle**. Artificial disruption of lysogenic host cell does not show the presence of infectious phages. In other way, it indicates that phages must be present in non-infectious state.

Before going through the cycle one has to know the structure and circularisation of genome of λ-phage.

Structure of λ-phage and circularisation of its genome

A λ-phage is a double stranded DNA virus with an icosahedral head of about 55 nm in diameter and a long tail (180 nm) without any contractile sheath (Fig. 2.57). The tail has a thin tail fibre at its distal end. The double stranded DNA of virus is a linear thread-like structure with 12 nucleotides long, single stranded cohesive ends at the 5′-(p) ends. The cohesive ends are complementary to each other. The single stranded complementary regions of both ends come to each other and form a circular double stranded DNA. The 5′-(p) and 3′-(OH) ends of both the thread then rejoin in vivo by DNA ligase. The circularisaion of DNA takes place after its entry inside the host bacterium (Fig. 2.57).

Activity of the circular λ-DNA

After circularisaion of λ-DNA inside the bacerial cell (*E. coli* K12), it functions in either of two alternative ways:

(i) Lytic cycle: In this way the DNA of λ-phage undergoes transcription, translation, assembly of progeny and the release by lysis of host bacterium.

(ii) Lysogenic cycle: In this way, the DNA of λ-phage is integrated with the bacterial DNA at a specific site to become a **prophage** and thus the infected host bacterium becomes lysogenic. Thus the phage DNA (prophage) replicates along with bacterial DNA. During this process the genes of phage DNA controlling lytic cycle becomes inhibited by a repressor — the λ-**repressor**.

Lytic cycle

During this process, initially the λ-phage gets attached to the bacterium with the help of tail fibre. The λ-phage then injects its DNA thread into the host bacterium (*E. coli* K12). After entry, the ds-DNA thread is converted into a circular DNA (described earlier, Fig. 2.57). The circular DNA of phage then gets attached to the membrane at the specific site and starts replication. Replication is initiated near the origin and progress in both direction symmetrically and then terminates when the two replicating forks meet each other and form a typical theta (θ)-like structure. The daughter DNA molecules synthesised and developed from the parental DNA undergo replication by 'rolling-circular model' and develop long thread-like **concate-meters**. These concate-meters contain several λ-genomes. The proper lengths with single stranded cohesive ends are cut off by enzymes and are packaged into the heads of λ-phages.

Transcription of the phage DNA develops coding messages that help to form the capsid and other proteins of phage. It is initiated by host polymerase which binds with the two promoters on λ-DNA that transcribes two different strands in opposite directions. More than 40 genes have been mapped in VDNA, having specific functions such as synthesis of viral DNA, head protein, tail protein, etc.

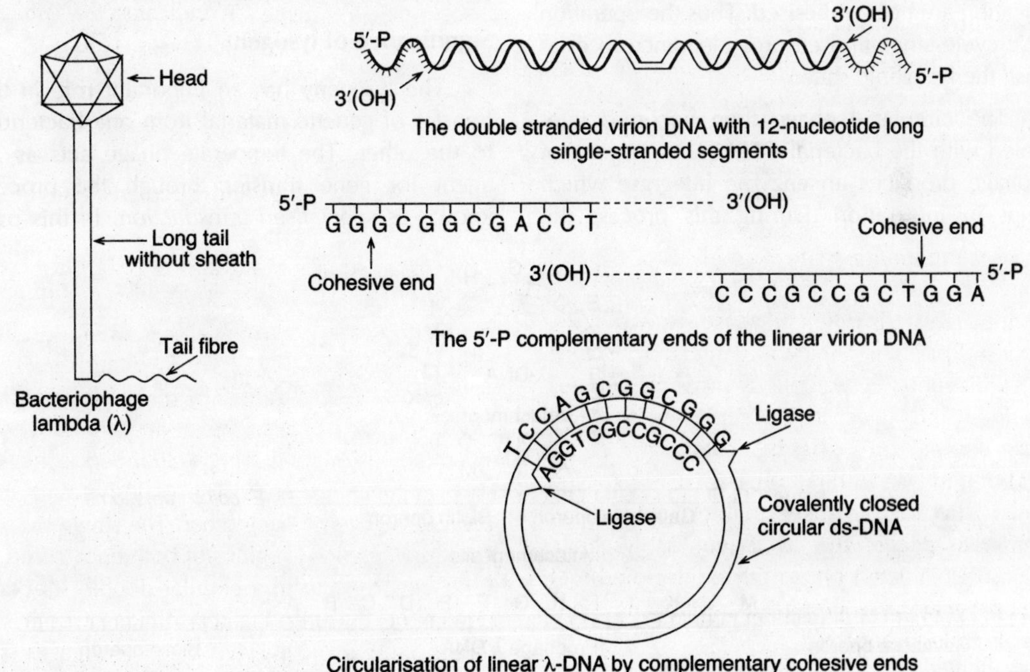

Fig. 2.57: Structure of λ-phage and circularisation of its genome

After the synthesis of sufficient number of virion components, they assemble and release by lysis of host bacterium. The released λ-phages then infect new bacterium and continue another lytic cycle or may enter into a lysogenic cycle.

Generally, the virus continues lytic cycle with a few numbers of infected cells, but major portion enters into lysogenic relationship and continues the lysogenic cycle.

Lysogenic cycle

After attachment of λ-phage with the host bacterium (*E. coli* K12), the phage pushes its DNA inside the cytoplasm of host cell. The phage DNA is then circularised in the usual pattern. The *cI* gene of λ-phage then produces the λ-repressor, an acidic protein (made up of 20 amino acid residues of molecular weight 26,000 dalton) which inhibits the action of gene controlling phage multiplication and lysis. The λ-repressor binds with the two different operators of its own genome, the OL and OR, those are involved in the initiation of transcription of phage controlling the phage multiplication. So the two essential proteins required for the initiation of phage multiplication are not synthesised. Thus the operation lytic cycle stops and ensures clearance to establish the lysogenic state.

The circular λ-phage then becomes integrated with the bacterial DNA. The *int* gene of λ-phage produces an enzyme **integrase** which helps in integration. During this process the circular DNA of λ-phage is inserted as a linear DNA into the bacterial DNA at a specific site between galactose and biotin operons (Fig. 2.58). The inserted DNA of λ-phage is called **prophage**. The λ-DNA remains with bacterial DNA for long time and replicates along with bacterial DNA.

In course of time, the lysogenic bacteria are able to produce phage particles through the process of induction either spontaneously at a very low frequency (one out of 102 or more) or due to action of different agents like X-ray, γ-ray, UV ray etc. (Fig. 2.56).

Due to induction, the prophage becomes released from bacterial chromosome and again becomes circular. The excision of prophage is catalysed by an enzyme **excisionase** produced by the xix gene of λ-phage. The rec A gene of the host bacterium produces a proteolytic enzyme which degrades the λ-repressor. At the same time, the cro gene becomes activated and produces **croprotein**, which inhibits the synthesis of λ-repressor. The circular λ-DNA then proceeds through lytic cycle and develops new crop of phage particle.

Significance of lysogeny

The lysogeny has an important role in the transfer of genetic material from one bacterium to the other. The temperate phage acts as an agent for gene transfer through the process known as *specialised transduction*. In this pro-

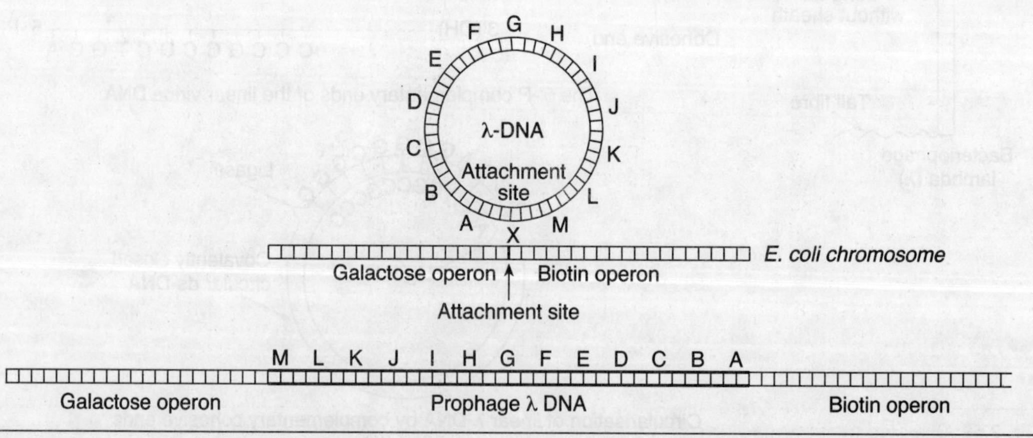

Fig. 2.58: Incorporation of λ-DNA in *E. coli* chromosome as a linear prophage between galactose and biotin operon of the host cell

cess, when a temperate phage comes out from host bacterium as prophage, it may include a part of bacterial DNA along with its DNA by mistake. After lysis, the newly developed phage can infect a new host bacterium and thus transfers the portion of previous bacterial DNA to the newly infected bacterium, and, thereby, recombination takes place.

ONE-STEP GROWTH CURVE

One-step growth curve, developed by Max Delbrück and Emory Ellis (1939), marks the starting of modern bacteriophage research.

In this experiment, susceptible bacterium (*E. coli*) is mixed with bacteriophage (T_2) and is allowed for a short period to get attached with the host cells. The culture is then diluted much, thereby any virus particle — if released by lysis of host cell — will not be able to infect new cells. The number of phage particle released from bacteria is subsequently determined at different time intervals by plaque count.

The result is plotted as *bacteriophages released from bacteria versus time*, indicating several distinct phages. Growth curve of this type consists of two periods:

1. **Latent period**. This period immediately follows phage addition and thus there is no release of virions. This is the shortest period required for virus reproduction and release.

   In the first part of this phage, the bacteria do not contain any complete infective virions. This can be shown by lysing them with chloroform. The initial stage of the latent period is called **eclipse period**. It is so named because the virions are not detectable even by disrupting the bacteria by treating with chloroform. At the end of the eclipse period, the phage virions increase in number within the host bacterium and the host bacterium is prepared for lysis.

2. **Rise period**. After the latent period, the host cells rapidly lyse and release all infective phages, called **rise period** or **burst**. At the end, no more viruses are liberated.

The term burst size is the number of viruses released from a single infected bacterium. The total number bacteriophages released can be used to calculate the **burst size**.

The experiment is carried out in enriched medium at 37°C and growth plateau is reached in approximately 30 minutes. Thus a plot of plaque forming unit (PFU) per ml against time in minutes produces a sigmoid curve (Fig. 2.59).

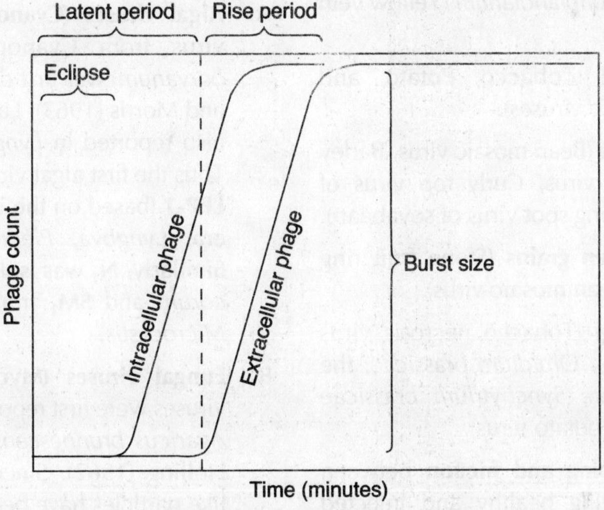

Fig. 2.59: One-step growth curve (In *eclipse period*, host cell does not contain any infective virion; in the *remaining part of latent period* shows an increasing number of infective virion, but none are released and latent period ends with lysis of host cell and rapid release of virions in rise period)

TRANSLOCATION AND TRANSMISSION OF VIRUS

Translocation of Virus

The movement of virus from one cell to the neighbouring cell of a plant is called translocation. It takes place through plasmodesmata. During movement, they multiply in most cells. Initially they infect parenchyma cells where they multiply and then pass in the neighbouring cells. In leaf parenchyma cells, the virus moves approximately 1 mm (about 8–10 cells/day). They can move more distance through the phloem (about 15 cm in first 6 minutes). After entering the phloem they can move in both upward and downward directions, thereby they can reach easily the apical meristem as well as the storage organs like rhizome, tuber etc.

Transmission of Virus

The movement of virus from one host to other host is called transmission. The transmission is dependent upon the external agents. The transmission of plant viruses takes place by the following means:

(i) **by vegetative propagating organs** like tuber (potato spindle tuber virus), bulbs (Cattle disease virus), etc.

(ii) **through collateral hosts** like lady's finger and *Croton bonplandianum* (Yellow vein mosaic virus).

(iii) **through soil** (Tobacco, Potato, and Wheat mosaic viruses).

(iv) **through seeds** (Bean mosaic virus, Barley strip mosaic virus, Curly top virus of sugar beet, Ring spot virus of soyabean).

(v) **through pollen grains** (Stone fruit ring spot virus, Bean mosaic virus).

(vi) **through fungi** (Tobacco necrosis virus transmitted by *Olpidium brassicae*, the root pathogen; *Synchytrium brassicae* transmits the potato virus).

(vii) **through rubbing and friction** between closely growing healthy and infected plants (Potato virus x, Tobacco mosaic virus, etc.).

(viii) **through insects** (Yellow mosaic of mung by white fly, *Bemisia*; Maize mosaic virus by an aphid, *Myzus*; Maize dwarf mosaic virus by an aphid, *Brachycaudus*, etc.).

Some common viral diseases

A. **Plants**: (i) Tobacco mosaic, (ii) Sugarcane mosaic, (iii) Yellow vein mosaic of bhindi, (iv) Leaf curl of papaya, (v) Leaf mosaic of curcurbits, (vi) Bunchy top of banana, (vii) Chirke of cardamom, (viii) Potato leaf roll, (ix) Tulip break, (x) Tomato black-ring, (xi) Yellow mosaic of mung.

B. **Humans**: (i) Small pox (Variola), (ii) Poliomyelitis (Poliovirus), (iii) Rabies (Rabies virus), (iv) Encephalitis (Alpha virus — mosquito-borne), (v) Common cold (Rhino virus and Human corona virus), (vi) Dengue fever and Yellow fever (Flavi virus — mosquito bone), (vii) Hepatitis (Hepatitis A, B, C, D, E, F, G), (viii) Measles (Measles virus).

C. **Animals**: (i) Hog cholera, (ii) Virus pneumonia of pig, (iii) Rabies, (iv) African horse sickness, (v) Myxomatosis of Rabbits, (vi) Granulosis of Butterflies and Moths.

Viruses or virus-like particles in Cryptogams and Gymnosperms

A. **Algal viruses (Cyanophages)**: The ds-DNA virus from Cyanophyceae (*Plectonema boryanum*) was first discovered by Safferman and Morris (1963). Later, the same virus was also reported in *Lyngbya* and *Phormidium*. Thus the first algal virus was named as strain **LPP-1** (based on the initial name of the genera, *Lyngbya, Plectonema, Phormidium*). Similarly, $\mathbf{N_1}$ was isolated from *Nostoc muscorum* and $\mathbf{SM_1}$ from *Synechococcus* and *Microcystis*.

B. **Fungal viruses (Mycophages)**: The fungal viruses were first reported in the fruit body of *Agaricus brunnescens* (syn. *A. bisporus*) by Holling (1962). Since then, virus or virus-like particles have been reported from many species of fungi (Hollings, 1978; Ghabrial, 1980) belonging to Mastigomycotina

(*Plasmodiophora brassicae*), Zygomycotina (*Rhizopus, Mucor*), Ascomycotina (*Saccharomyces ludwigii, S. cerevisiae*), Basidiomycotina (*Agaricus brunnescens, Puccinia graminis, Ustilago maydis* etc.) and Deuteromycotina (*Helminthosporium maydis, Cephalosporium crysogenum* etc.).

C. **Viruses of Pteridophytes**: Since the first isolation of virus in fern (1953), the virus-like symptoms are also observed in many ferns like *Phylitis scolopendrium, Polypodium vulgare*, etc.

D. **Viruses of Gymnosperms**: Some virus-like diseases have also been reported in pine trees.

ECONOMIC IMPORTANCE OF VIRUSES

(With special references to their role in causal organism of plant diseases, vaccine production, and role in research, medicines and diagnostics)

The viruses have both harmful and useful activities:

A. Harmful activities:

1. Viruses are responsible for various **diseases of both plants** (tobacco mosaic, yellow vein mosaic of lady's finger, leaf roll of potato, leaf curl of papaya etc.) and **animals** (small pox, meningitis, pneumonia, mumps, bronchitis etc.). The plant viruses cause damage to different parts like root, leaf, fruit, seed etc. and cause economic losses by reducing the quality and quantity of the plant products.
2. Phages often **kill the beneficial microorganisms** during commercial production of antibiotics and milk products.

B. Useful activities:

1. **In space research**: In space research, lysogenic phage cultures are used as **radiation detector** by Russians in the space ship (Vostok 2).
2. **As scavenger**: Phages are used as **scavengers** to eradicate the bacteria present in the polluted water.
3. **In therapy and prophylaxis**: to a limited extent phages are used in **therapy and prophylaxis** of some bacterial diseases.
4. **Detecting origin of life**: By holding both the living and non-living characters, viruses got the importance in determining the **origin of life.**
5. **Vaccine production**: Vaccine is an antigenic substance prepared from the causative agent of a disease or a synthetic substitute, used to provide immunity against one or several diseases.

Vaccine production has been done through several stages. These are:

(i) **Generation of antigen**: First of all the antigen is to be generated. Viruses are grown either on primary cells such as *chicken eggs* (e.g., for influenza) or on continuous cell lines such as *cultured human cells* (e.g., for hepatitis A). Bacteria are grown in **bioreactors** (e.g., *Haemophilus influenzae* type b). Likewise, a recombinant protein derived from the viruses or bacteria can be generated in yeast, bacteria, or cell cultures.

(ii) **Isolation**: After the antigen is generated, it is isolated from the cells used to generate it. A virus may need to be inactivated; possibly no further purification is required. Recombinant proteins need many operations involving ultrafiltration and column chromatography.

(iii) **Formulation**: Finally, the vaccine is formulated by adding **adjuvant, stabilizers,** and **preservatives** as needed. The **adjuvant** enhances the immune response of the antigen, **stabilizers** increase the storage life, and **preservatives** allow the use of multidose vials. **Combination vaccines** are harder to produce, because of potential incompatibilities and interactions among the antigens and other ingredients involved.

Modern techniques: Vaccine production techniques are gradually evolving. **Cultured mammalian cells** are expected to become increasingly important, compared to conventional options such as chicken eggs, due to greater productivity and low incidence of problems with contamination. **Recombination technology** that produces genetically detoxified vaccine is expected to grow in popularity for the production of bacterial vaccines that use toxoids. Combination vaccines are expected to reduce the quantities of antigens they contain, and thereby decrease undesirable interactions, by using pathogen-associated molecular patterns.

During 2010, India produced 60 percent of the world's vaccine demand worth about 900 million dolars.

6. **Role in research**: There are many ways in which viruses have been used for research. Some of the applications are mentioned below.

(a) *Recombinant DNA technology*: Viruses are commonly used as vector in recombinant DNA technology because they are able to insert their content into host cells. A vector is an autonomously replicating DNA fragment into which genes of interest can be integrated for cloning. Vector should be the DNA molecules which are amenable for manipulation, i.e., it is easy to isolate and insert into the host. They should carry target sites for restriction endonucleases and also selected markers for easy identification.

(b) *Genetic engineering*: Bacteriophages are the routinely used as viral vectors in genetic engineering. Phage vector have some advantages than the bacterial plasmid vector. These are:

(i) Phages are naturally infective,

(ii) It can accommodate large DNA fragments for cloning and

(iii) The identification of recombinant vector is comparatively easier.

The lambda (λ) and M13 phage are the most commonly used E.coli phages:

- *Lambda (λ) bacteriophages*: These are extensively used in DNA cloning research. The lambda (λ) genome contains an origin of replication genes for DNA replication etc. This is a temperate phage and can enter into both lytic and lysogenic cycle. In the host *E. coli*, the linear lambda genome becomes circular. The lambda genome has been modified to accommodate larger DNA insert, thus generating different types of λ vectors such as λgt10, λgt11 etc.
- *M13 viral vectors*: These are used for obtaining single strands of cloned DNA. These single stranded DNA are routinely used for DNA sequencing. M13 genome is used to produce a variety of modified vectors like M13mp1, M13mp8, M13mp9 etc. The M13 viral vector can be used to clone very large sized DNA inserts. Other viral vectors used in recombinant DNA technology include cosmid vectors (containing the cos sites and pac sites of λ-genome) and phagemid vectors which has an origin of replication from phage.

(c) *Genetics and Molecular biology*: These bacterial viruses have played a major role in development of genetics and molecular biology. Besides, viruses are also used for elucidation of various biochemical events. Microbiological research extensively uses viruses as both tools and model system.

(d) Molecular biology and Genetic engineering: Viruses have been sources of number of enzymes that are routinely used in molecular biology and genetic engineering. Reverse transcriptase an enzyme coded by the retroviruses is used for the generation of cDNA which is extensively required for construction of genomic libraries etc.

(e) *Plant and Animal biotechnology*: Viruses are routinely used in research in the field of plant and animal biotechnology for production of transgenics. These are as follows:

(i) Research on **plant** genomes also uses plant viruses as vectors. The virus vectors are used to produce specific products from plants. Baculoviruses are used to design vectors for production of recombinant proteins in insect cell lines e.g. BmNPV, AcMNPV.

(ii) Gene of interest from animal is initially cloned in bacterium *E. coli* and then expressed in **animal** cell culture or in embroys using shuttle vectors derived from viruses e.g. pc DNA1.1/Amp. The animal cell cultures are transfected with viruses for research on gene therapy and vaccine production.

7. **Medicine and Diagnostics**

(a) *Vaccine production*: Viral vaccines develop immunity against infection by the pathogenic strains of the same viruses. Edward Jenner (1796) started initial research on vaccines with a virus. Later, Jenner (1798, 99) used the cowpox virus (now known as vaccinia virus) as a live vaccine for preventing smallpox.

Conventional vaccines: The conventional vaccines synthesized using live attenuated viruses or killed viruses are easy to produce and economical but have a number of disadvantages. These are:

(i) Reversion of the *live attenuated viral vaccine to the initial virulent form* has also been observed in some vaccinated individual e.g. live polio vaccines.

(ii) Besides, live vaccines might also *contain certain undesirable and undefined components*, as they are usually produced in animal cell culture systems.

(iii) In addition *some of the viral vaccines are difficult to produce*, because they do not grow very well in culture system e.g. Hepatitis B virus.

(iv) The *inactivated viral vaccines are ineffective in some disease* thus live attenuated vaccines serve as a better option.

However, the attenuated vaccines can cause disease in comprised hosts.

The Recombinant – vector vaccines: In addition to inactivated and live attenuated viral vaccines, recombinant – vector vaccines are nowadays being made for prevention of a number of viral infections. In this process the genes coding for the major antigenic proteins of a virus are isolated and inserted into a vector, which can either be an avirulent virus or bacteria. This vector is inserted into human host where the vector replicates and expresses the protein products of the recombinant gene, generating either humoral or cell-mediated immune response. Viruses commonly used as vectors include vaccinia virus, adeno virus, attenuated poliovirus etc. Out of these viruses, **vaccinia virus** is most commonly used as a vector.

Vaccinia virus (Fig. 2.60) has a broad host range and remains stable for a very long time period. However, the main drawback with the virus is that:

(i) It is large in size and

(ii) Lack of unique restriction sites,

Thus foreign genes can be inserted only by homologous recombination.

A number of genes coding for major antigen of pathogenic viruses and even bacteria have been inserted into vaccinia virus DNA and expressed in animal cell culture. The important viral antigens include hepatitis B surface antigen, influenza virus hemagglutinin proteins and *herpes simplex* virus glycoproteins. It is also possible to clone genes simultaneously for 2-3 viral antigens in a given *vaccinia virus*, thus opening the possibility of vaccination with multiple antigens.

DNA vaccines: Recently, viruses have also been used to produce DNA vaccines, however DNA vaccines are still in its preliminary stage and research is underway. In a DNA vaccination, a plasmid preparation with genes of pathogenic virus and a strong viral promoter are injected into the muscles of human being as host. The antigenic proteins of the pathogens are now produced by the muscle cell evoking both humoral and cell mediated immune response. DNA vaccines against viral cancers, herpes, AIDS, hepatitis B virus etc., is now being synthesized and tested.

(b) *Gene therapy*: The introduction of functional genes into human cells to correct defective genes by replacing them is known as **Gene therapy**. Initially gene therapy was used for treatment of patients with genetic disorders, however the applications have now increased in wide areas.

Gene therapy is effective in cases where introduction of either a single gene or limited numbers of genes are required to correct a given disorder. Besides these, other important requirements of gene therapy include availability of a functional copy of defective genes and a method of gene delivery. Gene therapy is largely used in cancer treatment. However, lately it is also being used in the treatment of some infectious diseases.

Basically there are three types of gene therapy procedures are available. These are:

(i) **Ex-vivo somatic gene therapy**: It involves removal of cells with defective genes from the body, followed by introduction of therapeutic gene outside the body. The treated cells are then reintroduced into the patient.

(ii) **In-vivo somatic gene therapy**: It involves introduction of therapeutic gene directly into the target cells of the patient.

(iii) **Germ line gene therapy**: It includes the introduction of therapeutic genes into germ cells, aimed at correcting defects in the next generation. Germ line gene therapy is currently not being used in humans and is restricted only in laboratory.

In gene therapy vectors are used to carry the genes of interest into target cells. Both viral and non-viral approaches (micro-injection, Gene gun etc) are used for gene delivery. Viruses act as a very specific and efficient gene carrier systems for both ex-vivo and in-vivo gene therapy modules. There are a number of viruses which are being used as vectors in gene therapy. The important viruses are Adenovirus, Adeno-associated viruses, Vaccinia virus, Poxvirus etc.

Adenoviruses are the most plausible candidates for gene therapy because of their high specificity and easy manipulation. They show efficient nuclear invasion and heightened expression. These viruses are especially suited for treatment of respiratory disorders because of their high affinity for respiratory cells and can be used in treatment of cystic fibrosis. However some types of adenovirus may be cytotoxic with narrow host range.

Adeno associated viruses, of *Parvoviridae* family are another important group of vectors used for gene therapy. These viruses are easy to manipulate and have a single stranded genome and low immunogenicity. They require a co-infection with a helper virus for replication. Gene therapy trials for pancreatitis are currently carried out using this virus as a vector. The only disadvantage of these viruses appears to be the fact that they can carry only a limited size of DNA insert.

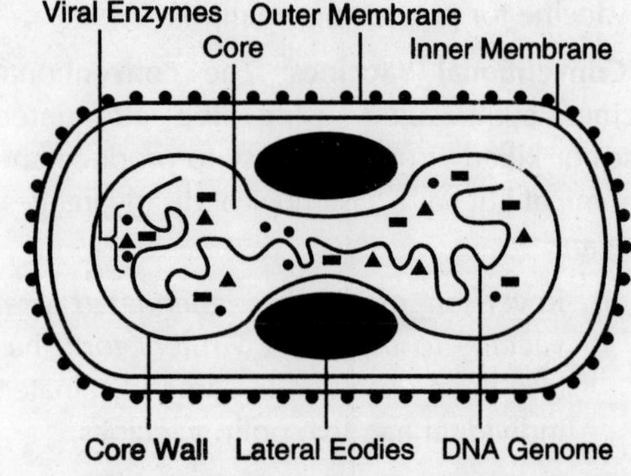

Fig. 2.[illegible]: Morphology of a Vaccinia virus

Gene therapy requires frequent repetition of the process because of elimination of target cells by routine immune processes. Thus, in future germ line gene therapy if possible appears to be an answer to our current maladies.

(c) *Cancer therapy*: Viruses can be directly used to prevent cancer by being the source of *anti-cancer vaccines* e.g. Vaccines against hepatitis B virus (causes hepatic cancer) and human papillomavirus (cervical cancer) are commercially available.

During carcinogenesis six fundamental properties are altered in cells to give rise to the destructive phenotype of cancer as illustrated in given anti-viral vaccines are prophylactic in nature and stimulate immune system to synthesize and also recruit specific antiviral molecules and cells. Cancerous cells express characteristic viral specific proteins, thus these could be targeted using vector to enhance the specificity of the vaccines. Such **therapeutic vaccines** are under clinical trials and are still not approved for human use.

Viruses, especially the RNA viruses have been found to have an inherent anti-cancer effect. They have a direct cytotoxic effect on cancer cells without involving the immune system. The direct therapeutic approach of cancer treatment using viruses is called **virotherapy**. The RNA viruses like reovirus, New castle disease virus etc., is being evaluated as viro therapeutic agents. These studies are at clinical trial stage. Genetically altered DNA viruses with enhanced specificity and cytotoxicity are also being investigated for virotherapy. Adenoviruses, Herpes viruses and Vaccinia virus are being studied for their anticancer effects.

(d) **Virus-directed enzyme prodrug therapy (VDEPT)**: In this process viruses are used to deliver an enzyme into target cancer cell. This enzyme can alter an inactive precursor of a cytotoxic drug into an active form. This treatment module is in research stages and is being evaluated for its therapeutic efficacy.

8. **Viruses in weapons and biological warfare**

Viruses may be tiny but have the capacity to cause death and devastation to large populations in epidemics and pandemics. This has led to the concern that viruses could be used for biological warfare.

9. **Viruses in agriculture**

Modification and genetic engineering methods can be used to make modified genomes that can be carried into plants by viruses acting as vectors or vehicles. This method can lead to more productive transgenics in plants.

10. **Viruses as biopesticides**

The bacterium, *Bacillus thuringensis*, is the most commonly used microbial biopesticide. Likewise, viruses are also used as a pesticide for killing a number of insect species like bollworms, fruitworms, moths etc. The Biopesticidal agents can be either prey on pests, be parasitic on them or are insect pathogens. Viruses are usually pathogenic to various insects.

Baculoviruses is a large viral group which infects insects and many other arthropods. These are extremely host specific. They form occlusion bodies that are thick protein shells which are formed outside the nucleocapsid. The **occlusion bodies** are in turn surrounded by a carbohydrate covering which makes them resistant to denaturation. Baculoviruses includes two genera. These are Nuclearpolyhedrosis virus (NPV) and Granulosis virus (GV).

In NPV, a single occlusion body contains *a number of nucleocapsids*. These nucleocapsids may be packed either singly or in groups. The occlusion body in NPV is composed of virus polyhedrin protein.

In GV, the occlusion body contains *only one nucleocapsid*. The occlusion body in GV is composed of viral granulin protein.

The insect larvae ingest baculoviruses and after reaching their gut, it proliferates inside. The viruses then spread from gut to the entire body of the insect and destroying it. The death of the insect usually occurs within 4-5 days. A number of progeny virions are released, which can further infect new insect larva.

Some other viruses like cytoplasmic polyhedrosis viruses of the family *reoviridae* and entomopox viruses of the family *poxviridae* can also be used as biopesticides. However, limited informations are available till date.

VIRUS AS CAUSAL AGENT OF PLANT DISEASES

Plant diseases cause by viruses lead to massive economic losses. Viruses caused a number of plant diseases, however there are limited studies have done in this field because it is difficult to cultivate and purify viruses. Some plant viruses like tobacco mosaic virus can be grown on plants protoplasts but most of the other viruses have to be cultured on either whole plant or tissue preparations. Thus it is difficult to study their life cycle or disease cycle.

Most of the viruses infecting plants are rod shaped or spherical in shape. They are usually

RNA viruses, containing single stranded RNA inside a protein coat. However, there are a few exceptions like cauliflower mosaic virus, which is a double stranded DNA virus. Penetration of viruses into plant cell is difficult because of the presence of complex and thick cell wall. Therefore mechanical damage of plant cell wall is important which can be caused by either insects or animals. Thus the most common mode of transmission of plants viruses is by insects like aphids, leafhoppers etc. They pick up viruses on their mouth parts and transmit it to other plants. There are basically two kinds of infections caused by plant viruses are the mosaic disease and the curling and dwarfing disease.

The mosaic diseases are the most prevalent type of diseases caused by plant viruses. This is characterized by the production of either yellowish or necrotic spots on leaves. Tobacco mosaic disease caused by Tobacco mosaic virus is the most common example of such type of disease pattern. The curling and dwarfing disease of plants also include symptoms like tumor formation, yellowing etc.

A few diseases of plants caused by viruses have been studied in detail are: Tobacco mosaic virus disease (TMV), Potato virus disease, Cauliflower mosaic virus disease.

1. **Tobacco mosaic virus (TMV)**

TMV is a thermal resistant virus which infects tobacco and other plants. It is transmitted through cell sap injury caused during harvesting through clipping of the shoots. It is a **single stranded RNA virus** and codes for protein like RNA polymerase, replication proteins and movement protein etc.

The symptoms of TMV infection are mosaic and even blisters on leaves.

The disease can be controlled through crop rotation, sanitation and use of disease tolerant varieties.

2. **Potato virus**

This plant virus is distributed worldwide and has a wide host range viz. potato, tomato etc. The symptoms include mosaic and necrosis associated with dwarfing. Genome is composed of **single stranded RNA**. The virus is transmitted through infected tubers and insects like aphids. Infection of crops with this virus cause highly significant economic losses.

3. **Cauliflower mosaic virus**

This is a plant virus with **double stranded DNA** as genetic material. This virus infects a number of plant species. The symptoms of this plant disease include leaf mottling and curling of edges. The fruits also show mottling with white and yellow patches. Cauliflower mosaic virus is transmitted by insects.

Besides viruses, viriods are cause number of plant disease. Viroids are smaller, circular RNA molecule with a rod like secondary structure **where protein coat is absent**. They do not code for any protein and use the host cell enzymatic machinery.

The "**Potato spindle tuber viroid disease**" has been discussed in the plant pathology (Chapter no. 5, page no. 698).

VIROIDS

While working with potato spindle tuber disease, T. O. Diener (1971) was the first to isolate and characterise the causal agent, *Potato spindle tuber viroid* (PSTVd) from Potato spindle tuber disease (PSTV). These are smaller than viruses and he coined the term **viroid** to those sub-viral particles. Since then only **twenty plant diseases** have been shown to cause by viroids. These are *Chrysanthemum stunt, Chrysanthemum chlorotic mottle, Citrus exocortis, Cucumber pale fruit, Cadang-cadang disease of coconut, Avocado sunblotch, Apple scar skin, Hop stunt, Cucumber pale fruit,* etc. Only **two animal diseases** such as **Scrapie** disease of sheep and Kuru disease of human are confirmed to be caused by viroids.

Structure. Viroids are small, low-molecular weight, **circular, single stranded RNA** molecule containing about 250–390 bases (which is insufficient to code even for one enzyme that may require to replicate itself and some even lack AUG — the initiation codon). The viroid **do not have capsid** (i.e., protein coat) around the RNA molecule. The molecular weight is 75,000 to 120,000 Da. The length of viroid in Potato spindle tuber disease is about 50 nm.

Replication. The mechanism of viroid replication is not yet clearly known. Two hypotheses have been proposed in this respect:

(i) **RNA directed replication**. This view suggests that in healthy plant, an enzyme replicase is present which accepts various RNA as templates. The RNA-directed RNA polymerase is available in uninfected tobacco plant.

(ii) **DNA-directed replication**. According to this hypothesis, the viroids might be replicated on DNA templates. This DNA templates are either available in healthy plants or are synthesised after infection.

Viroids differ from viruses:

1. The virus RNA is enclosed in protein coat, while viroids lack any protein coat and remain as free RNA.
2. The size of RNA is 25–370 bases in viroids as compared to 3.2–20 kb in viruses.
   - The genome of the **smallest known virus, Hepadna virus** is 3.2 kilobases in size.
   - **Virusoid** is the synonym of **viroid**.

Plant diseases. Only twenty plant diseases have so far been reported in plants caused by viroids. The viroid diseases do not significantly differ from virus diseases in the morphological expression i.e., symptoms. Like virus diseases, the symptoms like stunting, epinasty, mottling of leaves vein clearing, localised necrosis and chlorosis etc. are found in viroid diseases. The transmission of disease from one plant to another takes place generally by mechanical means. The transmission of viroid in potato spindle tuber takes place through pollen grains and seeds. **Insect vector have not been identified with any viroid transmission**. There is no recovery if the plant becomes infected by viroid (i.e., persistent) and the pathogen can be isolated only from living plants.

VIRUSOID

Virusoids are circular single-stranded RNAs depend on plant viruses for replication and encapsidation. The genome of **virusoids** consists of several hundred nucleotides and does not code for any proteins. **Virusoids** are essentially viroids that have been encapsulated by a helper virus coat protein.

This type of pathogenic RNA can infect some commercially important agricultural crops. These are subviral particles described as non-self-replicating ssRNAs. RNA replication of virusoids is similar to that of viroid, but unlike viroids, it requires that the cell also be infected with a specific "helper virus". At present, only 5 types of virusoid and their associated "helper virus" have been described. All the helper virus belongs to the family of **Sobemoviruses**. As example of a helper virus is the subterranean clover mottle virus, which has an associated virusoid packaged inside the viral capsid. Once the helper virus enters the host cell, the virusoids are released and can be found free in plant cell cytoplasm, where they possess ribozyme activity. The helper virus undergoes typical viral replication independent of the activity of virusoid. The virusoid genome is small, only 220 to 388 nucleotides long. A virusoid genome does not code for any proteins, but instead serves only to replicate virusoid RNA.

Virusoids are circular single-stranded RNAs dependent on plant viruses for replication and encapsidation. The genome of virusoids consists of several hundred nucleotides and does not code for any proteins.

Virusoids are essentially viroids that have been encapsulated by a helper virus coat protein. They are thus similar to viroids in their means of replication (rolling circle replication), but they differ in that viroids do not possess a protein coat.

Virusoids, while being studied in virology, are subviral particles rather than viruses. Since they depend on helper viruses, they are classified as **satellites**. In the virological taxonomy they appear as Satellites or Satellite nucleic acids or Subgroup 3: Circular satellite RNAs.

The term *virusoid* is also sometimes used more generally to refer to all satellites.

How do viruses, viroids and virusoids differ?

Although **viruses, viroids and virusoids** are similar, there are distinct differences also. While viruses can have a DNA or RNA genome covered by a protein capsid; viroids and virusoids consist of naked RNA. While viruses are of various shapes and sizes, viroids and virusoids are covalently closed circular RNA molecules. Viruses can infect plants, animals and human beings, while viroids infect plants and virusoids generally infect animals and human beings. While virusoids require a helper virus to establish infection, viruses and viroids do not. These are the basic criteria to differentiate these three as simple, acellular, sub-microscopic pathogenic entities.

PRIONS

The **prion** has been defined as *"small proteinaceous infectious particles which resist inactivation by procedures that modify nucleic acids"*. In brief it can be defined as "a proteinaceous infectious particle lacking nuclic acid". This fact comes as a considerable surprise to the scientific community. These diseases are often called **spongiform encephalopathies.**

Characteristic of Prions

The Prions have the following characters as mentioned below:

1. Prions are macromolecules, smaller than any known viruses, with a molecular weight near 30,000 Da.
2. Prions are smaller in size (<100 nm), hence filterable.
3. Lack of DNA and RNA genome.
4. Extreme resistance to physical and chemical agents such as heat, irradiation and formalin. But are susceptible to high concentration of phenol, sodium hypochlorite and Sodium hydroxide.
5. Slow replication, cause diseases with long incubation period (may be as long as about 35 years in human being).
6. It cannot be grown in laboratory.
7. Lack of inflammatory and immune response.
8. It can be transmitted orally to experimental animals.
9. The diseases caused by Prion are characterized by loss of motor control, dementia, paralysis, wasting and eventually death, typically following pneumonia.

Table 2.21: List of Prion diseases or Transmissible spongiform encephalopathied (TSEs) in animals and human beings

Name of disease	Natural host	Affects human
Animal diseases		
1. Scrapie	Sheep	No
2. Transmissible mink encephalopathy (TME)	Mink	No
3. Chronic wasting disease (CWD)	Deer, Mule, Elk	No
4. Bovine spongioform encephalopathy (BSE)	Cattle	Yes
Human diseases		
1. Creutzfeldt-Jakob disease (CJD)	Humans	Yes
2. Gerstmann-Straussler-Scheinker syndrome (GSS)	Humans	Yes
3. Fatal Familial Insomnia (FFI)	Humans	Yes
4. Kuru	Humans	Yes
5. Alpers Syndrome	Humans	Yes

Mechanism of cell death in prion disease

Several investigations have been done to analyse the role of PrP in neurodegeneration. Special interest are to find out the function of PrPC in neuroprotection, and the mechanism of cell death induced by, or during, the conversion from PrP^C into PrP^{Sc}.

Despite much progress on the molecular pathways involved in the activation of cell death, the mechanism of prion-induced cell death still remains obscure. **Apoptotic** cell death has been observed and described in various cell culture systems and in "in vivo", and several hypotheses have been put forward to explain the neurotoxicity that leads to **apoptosis**. Among them oxidative stress, microglia-mediated damage, and even the involvement of copper, leading to increased levels of caspase 3, Fas activation and up-regulation of the transcription factor c-jun.

The role of PrP^C as a protein with anti-apoptotic function has been highlighted in recent past. It could be speculated that alteration of PrP function (e.g. by interaction with, or sequestration by, PrP^{Sc}) also affects a homo- or hetero-dimerization of Bcl-2, or their expression, resulting in **neuronal apoptosis**. It has also been stated that binding and sequestration of ubiquitous members of the Bcl-2 family should also trigger cell death in non-neuronal cells. Since these show little or no alteration, further brain-specific factors, other than the Bcl-2 family members, may be more relevant. However, while loss of neurons can hardly be compensated, replacement of non-neuronal cells may happen unnoticed, particularly in the lymphoreticular compartment.

In brief it is to be mentioned that, *overall loss of neurons* may not be the most relevant event in prion disease. Patients may even be symptomatic through *loss of neuronal connectivity* and *neurite degeneration*, causing cell death only at the later stage.

[1. PrP = Prion protein

2. PrP^c = Normal form of protein

3. PrP^{Sc} = Infections form of protein.]

Important diseases

The important TSEs (Transmissible spongiform encephalopathied) occurring in animals are scarpie in sheep, Bovine spongioform encephalopathy (BSE) in cattle and Chronic wasting disease (CWD) in Deer, Muler and Elk. The important TSEs occurring in humans are Creutzfeldt-Jakob disease (CJD), Gerstmann-Straussler-Scheinker syndrome (GSS), Fatal familial insomnia and Kuru (Table 2.21).

Animal diseases

1. Scrapie (in sheep)

It is a prototype prion disease and the first animal transmissible spongiform encephalopathy (TSE) to be described. It is an endemic disease of sheeps and goats, present in the European continent for about two centuries.

Transmission of the disease may occur vertically from ewe to lamb. The natural transmission of the disease is believed to occur during parturition, by scarification or via the oral route. The agents first replicate in regional lymph node and later it replicate in brain and other lymphoid organs throughout the body. B lymphocytes are probably required for the spread through circulation to the brain. The incubation period of the pathogen is about 2 years. The infected animals become wasted and ataxic, and irritable. They develop intense prutitus and often rub their bodies on rocks and trees, hence the name was given. In extreme cases the animal develops paralysis and later dies. Placenta of ewe is an important source of infection, may contaminate the environment of the firm.

Once the scrapie agent has been adapted, it can infect hamsters and mice. In experimental animals, the intracerebral inoculation of infected brain tissue brings about most rapid onset of the disease. Infected hamsters have scrape-associated fibrils in the brain. The infectious fibrils contain TSE agent, i.e. the **Prion**. There is no effect on interferon and immunosuppression in this disease.

2. Bovine spongiform encephalopathy (BSE)

The disease is commonly known as **mad cow disease** emerged in cattle in UK during

1986. The disease spread very quickly and by 1995 it affects 150,000 cases. The outbreak of the disease resulted from oral ingestion of scrapie-agent through contaminated bone meal from scrapie infected sheep's dead bodies and recycling of BSE infected cattle dead bodies processed in a way that failed to destroy the scrapie agent. Such cattle feed was prohibited since 1988. The epidemic of the mad cow disease reaches its climax in 1993; caused more than 1 million i.e. 10 lacks cattle get infected.

The disease has also been found in cattles born in Austria, Belgium, Czech Republic, Denmark, Finland, France, Germany, Italy, Ireland, Israel, Japan, Liechtenstein, Luxembourg, the Netherlands, Poland, Portugal, Slovakia, Slovenia, Spain, Switzerland, and the U.K. It is also found in Canada.

Symptoms in Cattle

The disease is a transmissible, slowly progressive, degenerative, and fatal disease affecting the central nervous system of adult cattle. The infectious agent that causes **mad cow disease** is an abnormal version of a protein normally found on cell surfaces, called a **Prion.**

Treatment

There are currently no treatments for prion diseases, brain-wasting diseases that are invariably fatal. The most common human prion disease is Creutzfeldt-Jakob disease (CJD), better known as mad cow disease.

A human version of mad cow disease called variant Creutzfeldt-Jakob disease (vCJD) is believed to be caused by eating beef products contaminated with central nervous system tissue, such as brain and spinal cord, from cattle infected with mad cow disease.

3. **Chronic wasting disease (CWD)**

This is a scrape-like endemic disease if different animals like Deer, Mule and Elk, found in USA, called as Chronic wasting disease. The origin of the disease, a prion disease, is uncertain. No evidence is available that it is transmitted in human being.

Human diseases

1. **Creutzfeldt-Jakob disease (CJD)**

Creutzfeldt-Jakob disease (CJD) is a chronic degenerative disease of the central nervous system. The disease affects typically to middle-aged adults, although both young and old persons may be infected. Infected persons develop progressive dementia in less than one year, showing myoclonus, extrapyramidal motor neuron signs and also abnormal movements. Most of the infected persons die within one year. Both inherited and sporadic forms of Creutzfeldt-Jakob disease (CJD) have been seen. Inherited or familial CJD (fCJD) occurs in about 15% cases whereas sporadic CJD (sCJD) occur in about 85%. Familial CJD (fCJD) is inherited due to mutation in the PrP gene. However, majority of the CJD cases are sporadic of unknown etiology.

The disease has also been observed in iatrogenic disorder (disorder caused by medical examination and treatment) following accidental transmission from one person to another, due to the following:

(i) Corneal transplantation,

(ii) Injection of pituitary growth hormone, possibly from infected donor, and

(iii) Due to use of inadequately decontaminated intra-cerebral electrodes.

The disease is distributed in a sporadic form throughout the world, affecting 1 per million in distribution.

In 1996, a new variant of CJD (vCJD) has been recorded in Britain affecting younger persons below 45 years, which raised the possibility that Bovine spongioform encephalopathy (BSE) prions have been **transmitted to human beings through consumption of infected beef.** As per record, more than 90% of 190 cases of vCJD have occurred in Britain and rest from other European countries (France, Italy, Netherlands, Spain etc.), USA, Japan, Canada and Saudi Arabia. This report creates a panic regarding export of British beef.

2. **Gerstmann-Straussler-Scheinker syndrome (GSS)**

This disease is an inherited human transmissible spongiform encephalopathy, transmitted as a dominant autosomal trait. It affects mostly the middle-aged adult persons. Prominent clinical features of the disease include nystagmus, cerebellar ataxia and demensia. The disease is associated with the mutation in PrP gene, a codon 109 proline-to-leucine mutation. The mutation was first identified in 1989 and subsequently GSS families have been recorded. The disease can be induced in animals by using brain extracts from patients with GSS syndrome. Further this genetic disease may also cause by an infectious TSE agent.

3. **Fatal familial insomnia (FFI)**

This is an inherited disorder found as very rare case. It shows progressively severe insomnia, disturbances of autonomic function, ataxia, and myoclonus and also with other signs resembling Creutzfeldt-Jakob disease (CJD). The disease generally occurs in middle age. Fatal familial insomnia (FFI) is associated with PrP genotype, having codon 178 asparagine and 129 methionine.

4. **Kuru**

The disease was present only among the members of a cannibalistic Force tribe of Papua, New Guinea, because of their association with eating brains of relatives after a non-sterilizing ritual cooking as tribal practice.

The term Kuru means shivering tremors in local dialect, reflecting progressive cerebellar syndrome as cerebellar ataxia and tremors, caused ending with fatality within 3-6 months.

The disease is now extinct since 1950 as the practices have been abandoned because of the valuable work of Carlton Gajdusek. For the above work, he was awarded Nobel Prize in 1976.

SATELLITE VIRUSES

Satellites are sub-viral agents, composed of nucleic acid molecules that depend for their productive multiplication on co-infection of a host cell with a helper virus. Satellite nucleic acids have subsequently distinct nucleotide sequences from those of the genomes of either their helper virus or host. When a satellite encodes the coat protein in which its nucleic acid is encapsidated, it is referred to as a **satellite virus**.

In brief, it can be said that when a virus is dependent on another virus for its proper replication in a host cell, it is called *satellite virus* and such a situation is called *satellism*. The virus which helps the satellite virus to successfully infect and replicate in host cell is called *activator*.

Here the nucleic acid is not encapsidated and the virus relies for encapsidation on a helper virus.

Morphology. Capsid is round and exhibits icosahedral symmetry.

Nucleic acid. The genome is single-stranded, or double-stranded DNA, or RNA.

Translation. Virions are associated with helper virus and dependent on co-infection of helper virus during replication, or are associated with helper virus, but independent from its functions during replication.

Host range. Hosts belong to the Domin Eukaryota and includes Kingdoms Plantae and Animalia.

The first satellite virus (SV) in plant was reported to be associated with the tobacco necrosis virus (TNV) — it measures about 30 nm in diameter. Quite often the satellite virus is found independently in roots of tobacco and able to replicate normally. Sometimes both satellite virus and TNV are found to be associated together within the roots of host plant. Satellite viruses are unable to multiply in the host cell and depend on the assistance of TNV for their replication.

Categories of Satellites

Group. Satellite Viruses

Type. Single-Stranded RNA Satellite Viruses

Subgroup 1. *Chronic Bee-Paralysis Virus Associated Satellite.*

List of species:

(i) Chronic bee-paralysis satellite virus.

Subgroup 2. *Tobacco Necrotic Virus Satellite*.

List of species:

(i) Maize white line mosaic satellite virus.
(ii) Panicum mosaic satellite virus.
(iii) Tobacco mosaic satellite virus.
(iv) Tobacco necrosis satellite virus.

Group. Satellite Nucleic acids

Type. Single-Stranded Satellite DNAs

List of species :

(i) Tomato leaf curl virus satellite DNA.

Type. Double-Stranded Satellite RNAs

List of species:

(ii) Satellite of Saccharomyces cerevisiae M virus.
(iii) Satellite of Trichomonas vaginalis T1 virus.

List of tentative species:

(i) Satellite of *Ustilago maydis* killer M virus.

Type. Single-Stranded Satellite RNAs

Subgroup 1. *Large Satellite RNAs*.

List of species:

(i) Arabis mosaic virus large satellite RNA.
(ii) Bamboo mosaic virus satellite RNA.
(iii) Chicory yellow mottle virus large satellite RNA.
(iv) Grapevine Bulgarian latent virus satellite RNA.
(v) Grapevine fan-leaf virus satellite RNA.
(vi) Myrobalan latent ringspot virus satellite RNA.
(vii) Strawberry latent ringspot virus satellite RNA.
(viii) Tomato black ring virus satellite RNA.
(ix) Beet ringspot virus satellite RNA.

List of tentative species:

(i) Beet necrotic yellow vein virus satellite.

Subgroup 2. *Small Linear Satellite RNAs*

List of species:

(ii) Cucumber mosaic virus satellite
(iii) Cymbidium ring spot virus satellite RNA.
(iv) Pea enation mosaic virus satellite RNA
(v) Groundnut rosette virus satellite RNA.
(vi) Panicum mosaic virus small satellite RNA.
(vii) Peanut stunt virus satellite RNA.
(viii) Turnip crinkle virus satellite RNA.
(ix) Tomato bushy stunt virus satellite RNA, B10.
(x) Tomato bushy stunt virus satellite RNA, B1.

List of tentative species:

(i) Tobacco necrosis virus small satellite RNA.
(ii) Robinia mosaic virus satellite RNA.

Subgroup 3. *Circular Satellite RNAs*

List of species:

(i) Arabis mosaic virus small satellite RNA.
(ii) Cereal yellow dwarf virus-RPV satellite RNA.
(iii) Chicory yellow mottle virus satellite RNA.
(iv) Lucerne transient streak virus satellite RNA.
(v) *Solanum nodiflorum* mottle virus satellite RNA.
(vi) Tobacco ringspot virus satellite RNA
(vii) Velvet tobacco mottle virus satellite RNA.

AIDS

(Acquired Immunodeficiency Syndrome)

AIDS is a fatal disease, characterised by underlying cellular immunodeficiency. The disease was first detected at the 'Centre for Disease Control and Prevention', Atlanta, USA in June 1981. The probable site of origin of AIDS is thought to be Central Africa. Initially the disease appears to be common in the *homosexuals, male prostitutes* and *drug addicts*, where the incidence of *Kaposis sarcoma* (KS) and *Pneumocystis carinii* pneumonia were alarmingly high. But later, it was found that huge number of others also suffer from the disease. In India, the disease was first reported in January 1986 from the then Madras (Chennai). By 2005, more than 50 million people (including children) of the world became infected by the disease. Later, the disease became so severe throughout the world that it was considered as the first great *pandemic* disease of 20th century — loss of thousands of human lives throughout the world.

Causative agent

The causative agent is a virus. It was first isolated in 1983, by Luc Montagnaier of Pasteur Institute, Paris, from a West African patient suffering from persistent generalised lymphadenopathy and he named the virus as LAV (*Lymph Adenopathy Associated Virus*). Later, in 1984, Robert Gallo and his associates of National Institute of Health, USA, isolated the virus from AIDS patient and named the virus as HTLV-III (*Human T-cell lymphotropic virus-III*). Since then, the virus was named differently by different workers. Later, the International Committee of Virus Nomenclature gave the generic name as ***Human Immunodeficiency Virus* (HIV).**

HIV is a retro virus; the causative agent of AIDS is available in *two main strains*: (i) **HIV-1**, isolated from Central Africa, America and Europe and (ii) **HIV-2**, isolated from the prostitutes of West Africa.

Morphology of HIV

The HIV are icosahedral in shape and 90-120 nm in diameter. It can be differentiated into the following regions (Fig. 2.61):

(a) Envelop: It is the outer covering, composed of lipoprotein (lipid and viral protein) with spikes on the outer side.

(b) Core: It is an electron dense central helical core of viral capsid proteins, those surround the genome. The major inner protein is the P24 which encircles the RNA and the other one is P18 which remains at the outer side of P24, below the outer envelop.

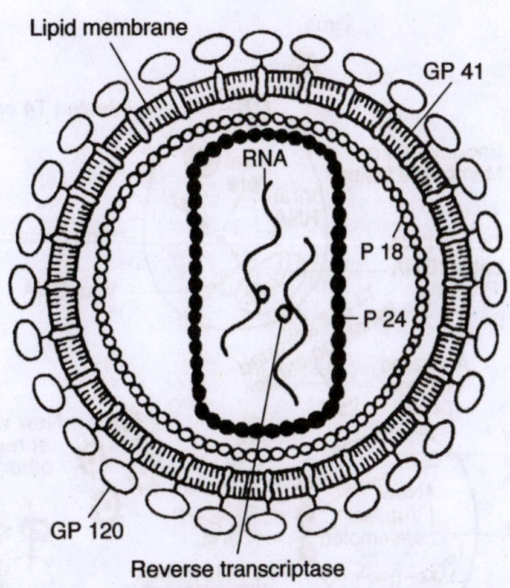

Fig. 2.61: Structure of HIV

(c) Genome: The genome consists of two identical copies of single stranded RNA (unusual in viruses) and reverse transcriptase (DNA polymerase) enzyme.

Genetic properties

Genome of HIV contains nine (9) genes, out of which three (3) are structural genes (*gag, pol* and *env*) and six (6) non-structural genes. Both HIV-1 and HIV-2 have five (5) common non-structural genes like *tat, rev, nef, vif* and *vpr*. They differ in the remaining non-structural gene. HIV-1 has *vpu*, but HIV-2 contains *vpx*. HIV-2 has approximately 40% genetic similarity with HIV-1, but differs in it's envelop. The major difference lies in the non-structural gene that HIV-1 lacks *vpx* gene that present in HIV-2, similarly HIV-2, lacks vpu gene, that present in HIV-1.

Mechanism of infection

The major role in maintaining the body's response to infection is done by T4 lymphocytes. After entering into the blood stream, HIV makes the T4 lymphocyte the primary target and gets attached to it. Once the HIV enters in a cell, it releases RNA, which becomes transcribed by reverse transcriptase into DNA i.e., the provirus. The provirus then becomes integrated into the DNA of host cell causing a latent infection (Fig. 2.62).

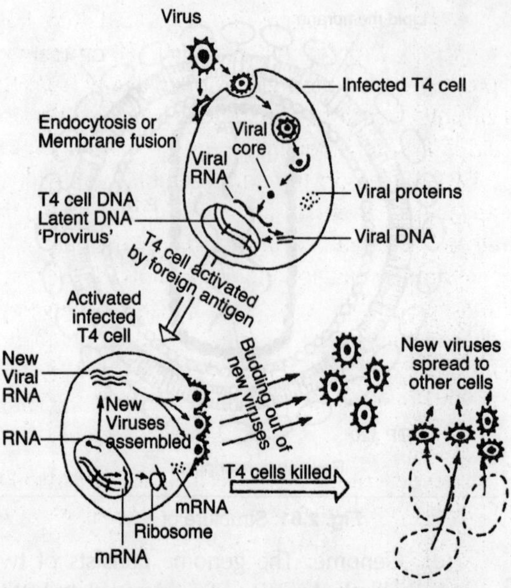

Fig. 2.62: Life cycle of HIV showing infection of T4 lymphocytes and development of new virus particles

The HIV remains dormant in a cell from 6 months to 5 years or more — the incubation period. The HIV converts T4 lymphocytes to HIV factory and thus infected T4 cells are unable to stimulate B-cells. Due to this, the B-cells are unable to multiply and produce antibodies for the newly infected virus. Thus the virus causes immunodeficiency, which becomes so severe that the infected person is unable to fight the infectious agents.

In resting cell, there is little transcription of virus. When the cell becomes activated, the DNA sequence transcribes many viral RNA. The *prophase* also develops mRNA to synthesise viral protein. Later on, viral RNA becomes covered by viral protein and forms new HIV particles; those may come out either by budding or by death of lymphocytes (Fig. 2.62). The new virus particles then proceed to infect other lymphocytes, monocytes, macrophages, glial cells of brain and also the central nervous system (CNS).

Symptoms

The early symptoms of AIDS include fever, diarrhea, enlargement of lymph glands and loss in weight. All the sufferers may not develop full-blown illness. When the individuals develop mild condition, it is known as AIDS-related complex (ARC).

In severe case, the infected person may develop *pneumonia, lymphoma* (cancer of the lymph system) and *kaposi's sarcoma* (a skin cancer). Although HIV takes long incubation period i.e., 6 months to 5 years or more, once it becomes active it causes rapid deterioration of health and may lead to death within a year.

Transmission

The HIV can be transmitted from infected to the healthy person through the following processes:

(i) *Through sexual and related contact*: Transmission occurs among hetero-sexual and homosexual individuals. During sexual contact exchange of body fluids takes place. The virus has been isolated from vaginal secretion, endocervical secretion, breast milk and also in both fluid and cells of semen. Large volume of saliva may also help in HIV transmission.

(ii) *Through blood transmission*: Transmission of HIV from infected to healthy person takes place through blood transfusion.

(iii) *Through sharing of needles*: Person uses intravenous drugs through sharing of needles and syringes without proper sterilisation — this is an important way of tansmission of HIV. In addition to that, use of any skin-piercing instrument without proper sterilisation may spread the virus from diseased to healthy persons.

(iv) *Through infected mother*: The transmission may take place from infected mother to child — transplacentally or perinatally. The placental blood or fluid may contaminate the infant, and, post-natally, infant may get infection through breast feeding.

Diagnosis of AIDS

History and medical examination

Initially doctors diagnose AIDS — infected patients from the case history and the visible symptoms. The uses of intravenous drug, transfusion recipient, polygamic and homosexual habit along with the visible symptoms of the patient make the doctor confirm that the patient is suffering from AIDS or ARC (AIDS-related complex). In addition to that, analysis of blood sample shows anaemia, lymphopenia (less number of T-helper cell), leucopenia, high immuno-globulin level, reversal of T4 : T8 ratio i.e., 0.9, etc.

Screening tests

To confirm whether the patient is suffering from AIDS or not, two types of test are commonly employed: 1. **ELISA test**, and 2. **Western blot test**.

1. **ELISA (Enzyme-linked immunosorbent assay) test**: It is the most widely used direct test for which test kits are available commercially. During this test, antigen is prepared from HIV cultivated in T-lymphocyte cell line or by recombinant technique. Antigen is coated on surface of microtitre wells and then test serum is added to it. The antibody of serum binds with the viral antigen. The unbound serum is washed out. Then antihuman goat immunoglobulin linked to a suitable enzyme (horse radish peroxidase, beta-galactosidase or alkaline phosphatase) is added and incubated for 1 hour at 37°C. Again, the well is washed out.

A suitable solution of substrate is added and left at room temperature till the control (HIV positive) develops a colour. It is read out by ELISA reader or by visual observation. The result is scored as positive (highly reactive), negative (non-reactive) and indeterminate (partially reactive).

Commercial ELISA kit used by most laboratories contains test for both HIV-1 and HIV-2 and thus any blood sample can be detected in routine examination. It is a good test with more than 99.5% sensitivity.

Though the ELISA test is highly sensitive, it is not optimal with regard to specificity. So it should be confirmed by more specific assay. The most commonly selected test for confirmation is the Western blot test.

2. **Western blot test**: This test is more specific and confirmatory assay to detect HIV infection.

The HIV proteins are separated into fragments in a polyacrylamide gel. The separated proteins are blotted electrophoretically from the gel into a strip of nitrocellulose sheet. Then the strips are incubated in test sera. The antibodies to HIV proteins (in test sera) combine with all or any fragment of HIV antigens. After washing, the strips are treated with enzyme coagulated anti-human gammaglobulin and allowed to react. Then a suitable substrate is added that develops colour bands. The position of the colour bands indicates the position of protein fraction. Like ELISA, it also shows positive, negative and intermediate result.

A patient showing positive or intermediate ELISA, but negative in Western blot indicates that ELISA test was false positive.

In case of intermediate result in Western blot, the diagnosis should be confirmed by PCR.

Prevention

Following are the preventive measures that should be taken into consideration to escape from the disease:

1. Avoidance of homosexual intercourse.
2. Avoidance of drug abuse.
3. Avoid maintaining multiple sexual partners.
4. Don't share any item like razor, injection needle, toothbrush etc., which may help contamination with blood.

Treatment

The infection is notorious and at present there is no way to get relief from AIDS, thereby the rate of mortality is 100%. There is no vaccine or drug which can save the life. But some recently introduced drugs show promising effect. These are:

1. *Azidothymidine* (AZT). It is also known as Zydovudine. This drug prevents reverse transcriptase and after taking the medicine the patient shows the following improvements:
   (a) Increase the numbers of circulating T-cells.

(b) Increase the immunological capacity.

(c) Gain in body weight.

(d) General improvement of the body, etc.

2. *Ribavirin*. The drug stops the progression of infection associated with the AIDS.

3. *Other drugs*. Drugs like *dideoxynosine* and *dideoxycytidine* are under evaluation and their action is almost similar to AZT.

SARS

SARS (Severe acute respiratory syndrome) is a **serious form of pneumonia** (Fig. 2.63). It first appeared in China in November, 2002 and was first identified in February, 2003. Now the disease is available in more than 24 countries. The disease was so severe that in between November, 2002 to July 2003, there were 8,098 cases have been reported from different corners of the world, with 774 deaths of human beings.

Infection with the **SARS** virus causes acute respiratory troubles (severe breathing difficulty) and sometimes death.

It was the first coronavirus caused severe symptoms in human beings.

Causal agent

The SARS coronavirus (SARS-CoV) causes SARS (Fig. 2.63A-B). It is a common form of virus that causes upper-respiratory tract illnesses. The common cold results from a kind of coronavirus.

Although, six different kinds of coronavirus are known to infect human beings, four of these are common, and most people will suffer at least one of them during their life time. Two other types cause **SARS** (Severe acute respiratory syndrome) and **MERS** (Middle East Respiratory Syndrome). These are less common but far more deadly.

Before SARS appeared, coronaviruses had not been particularly dangerous to human beings, but they had been known to cause severe diseases in animals.

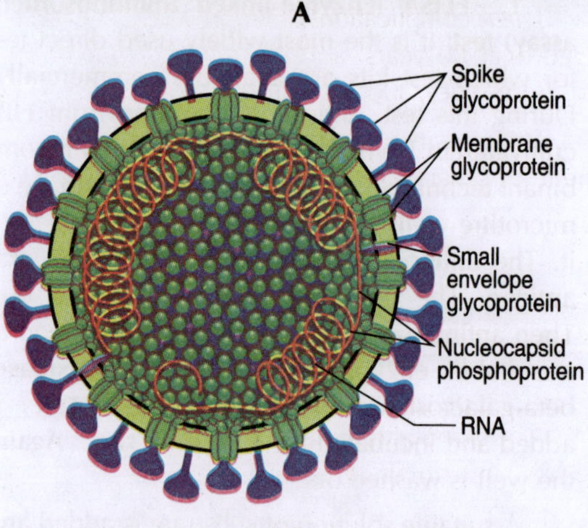

Fig. 2.63: A. SARS coronavirus (SARS-CoV) and B. Schematic drawing of SARS coronavirus

As a result, scientists first thought that animals transmitted SARS-CoV to human beings. Now it is believed that an animal virus become changed into a new, more deadly strain.

Symptoms

Droplets from coughing and sneezing and close human contact likely transmit the causal agent, the SARS virus. The respiratory droplets are probably absorbed into the body through the mucous membranes of the mouth, nose, and also eyes.

This could be through:

(i) Hugging and also by kissing,

(ii) Sharing of utensils for eating and drinking,

(iii) Speaking to someone within a distance of 3 feet,

(iv) Touching someone directly.

A person with the virus can spread the infection by leaving respiratory droplets on objects, such as door handles, doorbells, and telephones. These are then may be picked up by someone else.

Patients may go on to develop pneumonia, an infection of the lungs.

Serious complications, such as respiratory failure, heart failure, and liver failure may occur.

These complications are more likely in people aged over 60 years, and those with pre-existing co-morbidities, such as diabetes or hepatitis.

The virus may remain active in the environment for several days.

Diagnosis

The WHO advised that, to be diagnosed with SARS, a person must have all of the following:

(i) A fever of at least 100.4°F,

(ii) One or more symptoms of lower respiratory tract illness: a cough, difficulty in breathing, shortness of breath,

(iii) Radiographic evidence to suggest a diagnosis of pneumonia,

(iv) No alternative diagnosis to fully explain the illness.

SARS is very rare, and the symptoms overlap with those of flu and pneumonia. A physician will not suspect it unless the person has been in an area where there is an outbreak.

Tests

Tests can help to confirm the diagnosis of SARS by detecting the virus and antibodies.

(i) Laboratory tests are helpful to identify the SARS-CoV.

(ii) **Reverse transcription-polymerase chain reaction (RT-PCR)** testing can be used to detect the virus in blood, stool, and nasal secretions.

(iii) **Serologic** testing can detect SARS-CoV antibodies in the blood. If a person has antibodies in blood, it is likely to have the infection.

(iv) Doctors may also use a **viral culture**. This involves putting a small body tissue or fluid into a container with some cells in which the virus can grow. If the virus grows, the cells will change.

These tests may not be reliable if used in the early stages of infection.

Prevention and Treatment

Prevention:

Some procedures are to be undertaken to prevent the disease for further spread. These are:

(i) It is better not to allow any person to come from the SARS epidemic areas. Urgent medical care is needed.

(ii) The WHO recommends in isolating patients and using barrier techniques to prevent the spread of the virus, including filter masks and goggles. No drug, including antibiotics, appears to be effective against SARS.

(iv) Good and effective personal hygiene practices can also help to restrict the spread of the virus.

These include:

(a) Frequent washing of hands.

(b) Avoiding to touch the eyes, mouth or nose with unclean hands,

(c) Covering mouth and nose with a tissue paper when coughing or sneezing,

(d) Encouraging others to follow the above procedure.

(v) SARS is contagious disease and once its symptoms have developed, anyone with the disease should limit his interactions with other people till the symptoms have improved. SARS is at its most contagious stage during the second week of illness.

Treatment:

Supportive care is recommended, including the use of medications to relieve symptoms such as fever and a cough.

BIRD FLU

Bird flu or avian influenza is a viral disease of birds — it spreads in different countries by ignoring the different laws through the migratory birds. It is a highly contagious poultry disease, causes high mortality in poultry birds. The disease is known to us for more than a century. The most recent outbreak of the disease was in February 2006, from India where the poultry farms were highly affected in some parts of the country like Maharashtra (Nandurbar and Dhule district), etc. The severity was so much that 50,000 chickens were reported to have died in a week and about 900,000 chickens were slaughtered and buried. But no human infection was confirmed, though it was suspected at that time.

There have been a number of small outbreaks of bird flu in human beings since 1997 and showed a high mortality rate. The disease in humans was first reported in 1997 in Hong Kong, where 18 people were infected and 6 people died. Since then it was also found in China, Vietnam, Thailand and other countries. More attention has been given since some strains infect humans. In spite of that, in 2003 in Hong Kong two people were infected and one died. Recently, in 2004, in East Asia, as many as 10 deaths have been linked to an outbreak of the disease.

Symptoms:

On bird: The infected bird shows the following symptoms:

1. Comb becomes black in colour,
2. Release of water from nose and eye,
3. Swelling of head, eyelids, combs, neck and leg,
4. Lack of energy and appetite,
5. Irregular arrangement of feathers.
6. Movement in a circular pattern,
7. Sudden death of the bird.

On man: The infected human shows the following symptoms:

Coughing, fever, sore throat and aching muscles to eye infection and more serious pneumonia and other respiratory problems — ultimately threaten the life of humans.

Causative agent

The disease is caused by Avian Influenza A, a DNA virus. It is also called "H5N1 virus", a subtype of the above.

There are many different subtypes of type A influenza viruses. These subtypes differ because of certain proteins on the surface of the influenza A virus [hemagglutinin (HA) and neura-minidase (NA) proteins]. There are 16 known HA subtypes and 9 known NA subtypes of influenza viruses. Many different combinations of HA and NA proteins are possible. Each combination represents a subtype. All known subtypes of influenza A viruses can be found in birds.

"Human influenza virus" usually refers to those subtypes that spread widely among humans. There are only three known A subtypes of influenza viruses (H1N1, H1N2 and H3N2) currently circulating among humans. It is likely that some genetic parts of current human influenza A viruses came from birds originally. Influenza A viruses are constantly changing and they might adapt over time to infect and spread among humans.

Prevention primer

While handling poultry

1. Wear gloves, shoes and masks,
2. Use disinfectants afterwards,
3. Avoid mixing of old and new stocks of birds,
4. Avoid accumulation of foul matters (filth).

Handling migratory birds

1. Monitor all the birds in the sanctuary,
2. Segregate birds found drooping and test their blood,
3. Death of birds in sanctuary must be reported to the National Animal Disease Emergency Committee (NADEC).

Dissemination of disease

The viruses are able to infect human beings by the following ways:

1. Consuming infected birds.
2. Consuming infected eggs.

3. Viruses can disseminate through feathers, excreta etc., of infected birds.
4. Virus can disseminate through direct contact with infected birds.
5. Virus can disseminate through air.
6. The virus can disseminate in different countries through migratory birds.

Persons in danger

The following persons may be infected very easily:

1. Those who maintain poultry in domestic house,
2. Those who live near poultry farm, or
3. Those who maintain other domestic birds.

Caution from WHO

According to World Health Organization (WHO) "the Avian Influenza will continue to pose a risk to human health as the disease has proven itself capable of crossing the species barrier". The larger problem, however, is if the avian influenza strain mixes with a human influenza strain, producing a new strain that could spark a human influenza pandemic.

Alternately, the currently circulating avian influenza strain could mutate into a form that is easily spread among human beings, which could set off an influenza pandemic. Pandemic typically cause great numbers of disease and death, and are accompanied by significant social and economic disruption.

How chicken and eggs can be consumed?

1. The virus gets killed in heat. Chicken should be cooked until the juices run clear and it reaches an internal temperature of 82.2°C (180°F).
2. The egg should be boiled for more than five minutes after proper washing the outer side.

Precautions before consumption

1. Avoid of consuming frozen fowl. Because infected birds shed the virus in their droppings, saliva and nasal secretions and it spreads through droplet infection (from coughs and sneezes).
2. Don't consume the dead birds.
3. Avoid consuming different preparations like poach, half boiled eggs, etc.

Available drugs

1. The primary treatment option is the flu drug **Oseltamavir** (Tamiflu), a neuraminidase inhibitor that works by preventing the virus from escaping its host cell.
2. Other antiviral flu drugs — **Relenza** and **Zanamavir** — are also used as an alternative.

Instances of Avian Influenza Infections in human beings

List of the confirmed instances of avian influenza in human beings since 1997 up to early 2006 is given below:

1. 1997: H5N1 subtype is reported from Hong Kong in both poultry and humans. 18 people were infected and **6 people died.**
2. 1999: H9N2 subtype is low pathogenic, reported from Hong Kong and China. Two cases are confirmed, but later recovered.
3. 2002: H7N2 subtype was reported from Virginia and infection proved in one person, but recovered.
4. 2003: H5N1 subtype was reported from Hong Kong and China. Two cases were reported and **one died.**
5. 2003: H7N7 subtype was reported from Netherlands with 89 infected persons showing different symptoms, but **only one died.**
6. 2003: H9N2 subtype was reported from Hong Kong. One child becomes infected but recovered.
7. 2003: H7N2 subtype was reported from New York. The patient becomes seriously ill but recovered later.
8. 2004: H7N3 subtype was reported from Canada. This highly pathogenic subtype caused severe infections among poultry associates, but all recovered from the disease.
9. 2004: H5N1 subtype was reported from Thailand and Vietnam. It is the highly pathogenic species, first reported from Asia.

SWINE FLU

Swine influenza is also called as swine flu, hog flu, pig flu and H1N1 flu. It is caused by any one of different types of swine influenza virus (SIV). As per report of 2009, the SIV strains include:

A. Influenza C and

B. Influenza A subtypes:

(a) H1N1

(b) H1N2

(c) H3N1

(d) H3N2 and

(e) H2N3.

The SIV is commonly found in the pig population through out the world. Transmission of SIV from pig to human beings is not common and does not always lead to influenza, often results the production of antibodies in blood. If influenza caused by transmission from pig, it is called **zoonotic swine flu**. Risk of swine flu infection is common with the people with regular exposure to pigs. These strains of swine flu rarely pass from human to human. Symptoms of **zoonotic swine flu** in human beings are similar to those of influenza and influenza-like illness such as:

(i) Fever,

(ii) Chills,

(iii) Sore throats,

(iv) Muscle pain,

(v) Coughing,

(vi) Weakness,

(vii) Headache and

(viii) General discomfort.

Types of influenza virus

Out of three influenza viruses A, B & C; two cause influenza in pigs, with influenza C is rare and influenza A (Fig. 2.64) is very common. Within influenza A & C, the strains found in pigs and humans are largely distinct, although *there have been transfers of genes among strains due to reassortment crossing the boundaries of host i.e. Swine, Avian* and *Human beings*.

A. Influenza C: It infects both pigs and human beings but do not infect birds. Influenza C caused small out break in mild form among the children of Japan and California during 1996 & 1998. Due to lack of genetic diversity it does not spread in Pandemic form.

B. Influenza A: Swine influenza is known to cause by 5 subtypes, out of which 3 subtypes H1N1 (Fig. 2.64), H3N2 & H1N2 are most common and found throughout the world.

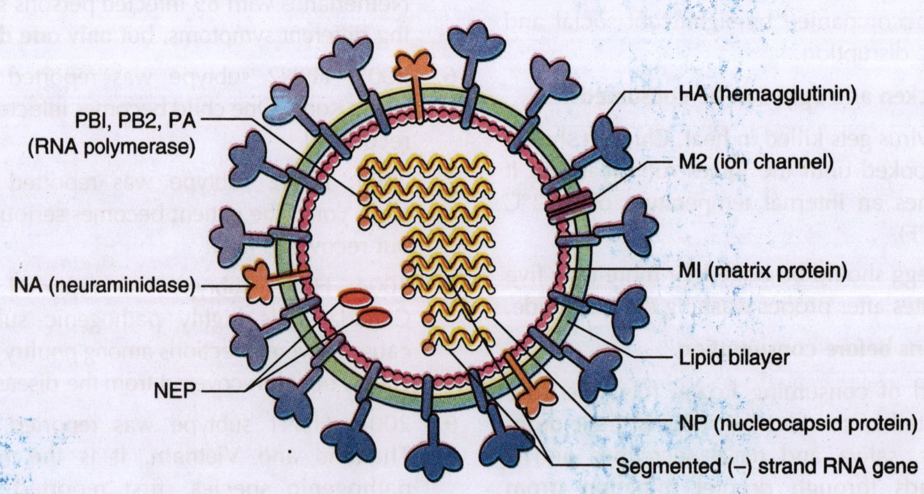

Fig. 2.64: Structure of Swine influenza virus (SIV): Influenza A subtype H1N1

History of the disease

Swine flu was first reported to be a disease related to human influenza in 1918, when pigs become sick at the same time as human. The disease appears as pandemic in human was associated with H1N1 and influenza also appears in pig. This disease was only noted as a new disease in 1918, after the outbreak in human beings. It was estimated that more than 50 million people were died world wide.

Later, in 1976 the outbreak of the disease also takes place in USA and caused small outbreak caused by H1N1 and also by H3N2.

Later vaccination program was started in USA in 1st and 11th October,1976, about 40 million people (about 24% of population) received but was not successful and setback on 16th December,1976 for the death and side effect of some persons. It was reported that vaccine killed more Americans than the disease did.

In USA, the outbreak of the disease also took place in 1988 and later in 1988.

In Philippines, the outbreak took place in swine and caused about 10% mortality.

Most recent outbreak in human took place in 2009 by H1N1 viral strain. The outbreak is due to new strain of H1N1 not previously reported in pigs. With through investigation it was proved that the new strain develop due to reassortment of just two strains, both found in swine. In India, 20 people died by swine flu in different parts of the country.

Transmission

Transmission of swine flue takes place in two ways. These are:

1. Transmission between pigs

Transmission of virus takes place through direct contact between infected and uninfected animals. It takes place during animal transport and intensive farming. The direct transfer of virus possibly takes place either by:

(a) Pigs touching noses,

(b) Through dried mucus.

(c) Airborne transmission through the aerosols produced by pigs coughing or sneezing is also an important.

(d) Through wild boar and other wild animals, those can spread the disease between farms.

2. Transmission to human

Transmission takes place through the workers of poultry and swine farm, especially peoples with intense exposure. Zoonotic infection and reassortment can co-occur and caused severity of the disease. It can transport through veterinarians and meat processing workers, although the risk of infection for both of these group of workers is lower than the persons working in farm with close contact with swine farm.

Signs and Symptoms

Disease symptoms are seen both in Swine and also in Human beings.

(i) **In Swine**

Swine influenza infection produces fever, coughing, sneezing, difficulty in breathing, lethargy, decrease in appetite, poor growth and caused economic loss due to loss in weight up to 12 lbs within 3-4 weeks. In some cases it caused abortion. But the rate of mortality is very low and ranges between 1-4%.

(ii) **In Human beings**

In 2009, "swine flu" symptoms caused by H1N1 virus is similar to those of influenza and influenza-like illness. The symptoms include fever, sore throat, cough, body aches, chills, headache, diarrhea, vomiting, and fatigue. The outbreak of 2009 in India, shows an increased rate of vomiting and diarrhea and the virus does not cause zoonotic swine flu, as it is transmitted from person to person but not from pig to human.

Common cause of death by H1N1 infection is:

(i) Respiratory failure,

(ii) Pneumonia (leading to sepsis),

(iii) High fever (leading to neurological problem),

(iv) Dehydration (due to diarrhea and vomiting) and

(v) Electrolyte imbalance.

Fatalities are more in children than the elder persons.

Diagnosis of confirmed swine flu requires laboratory testing of a respiratory sample (a simple nose and throat swab).

Prevention

Prevention of swine flu has three components:

A. Prevention in swine

B. Prevention of transmission from pig to human

C. Prevention of transmission from human to human.

A. **Prevention in swine**

Prevention of spread of disease among swine includes: (i) Facility management, (ii) Herd management and (iii) Vaccination.

(i) *Facility management*: It includes ambient temperature and disinfectants. The virus can not survive for more than two weeks outside host except in cold (but above freezing) condition.

(ii) *Herd management*: Avoid carrying the pigs, carrying influenza to non-infected herds. Introduction of new pigs to non-infected herds and to new country should be quarantined.

(iii) *Vaccination*: As illness and death is associated with swine flu involves secondary infection by other pathogens, the control strategies through vaccination may be insufficient.

Control of swine flu by vaccination becomes more difficult in recent years, as the evolution of the virus resulted in consistent responses to traditional vaccines. But the standard commercial swine flu vaccines are effective in controlling the infection when the virus strains match enough to have significant cross-protection. The pig vaccination keep pigs from getting sick, but it does not block infection or shedding of the virus.

B. **Prevention of transmission from pig to human**

Swine can be infected by both avian influenza virus and strains of human influenza virus, where antigenic shifts can occur that create new strains.

To avoid transmission from swine to human, the following procedures have to be undertaken:

(i) Encourage to use a face mask by farmers and veterinarians when dealing with infected animals.

(ii) Avoid smoking during handling of swine in farm and

(iii) Wearing of gloves is essential during handling of sick animals.

C. **Prevention of transmission from human to human**

The disease spreads among human by the following ways:

(i) Through coughing,

(ii) Through sneezing,

(iii) Through touching between infected region and then with their own nose or mouth.

Swine flu in human is most contiguous for the 1st five days of illness although it remains contiguous for ten days in children and in some peoples. Samples are collected for testing during the 1st five days.

Advice to control transmission

Following are the procedures advised to control transmission:

(i) Washing of hands with soap and water or with alcohol-based hand sanitizers after coming back from public place.

(ii) Chance of transmission is also reduced by disinfecting household surfaces like table top, telephones and other surfaces from where it is transferred via the fingers to the eye, nose or to the mouth; which can be done effectively with a dilute chlorine bleach solution.

(iii) Infected person should avoid large gathering and better to stay at home.

Treatment

In Swine:

The influenza is rarely fatal. Preventive measure should be undertaken to avoid spread of virus throughout the farm or to other farms by restricting the movement of infected pigs.

In Human:

Persons infected by swine flu are treated with antiviral drugs like Tamiflu (oseltamivir) or Relenza (zanamavir) within 48 hours of symptoms to make the illness milder. Beside antiviral drug, supportive care mainly fluid balance, relieving pain etc., in home or in hospital should be undertaken.

Precautions in brief:

Precautions to be taken to avoid infection from swine flue are given below:

1. Wash your hands frequently with soap and water. Alcohol based hand cleaners are also effective.
2. Avoid touching your eyes, nose or mouth as far as possible.
3. Avoid close contact with sick peoples.
4. Maintain high level of personal hygiene.
5. Stay at home when you are sick. Take medicine strictly as per your doctor's advice.
6. Cover your nose and mouth with a handkerchief when cough and sneeze.
7. Keep your body fit and strong by exercising, avoiding junk food and having good sleep.
8. Avoid smoking and alcohol intake.
9. Drink 8-10 glasses of water each day to help flush out toxins from your system.
10. Avoid crowded areas as far as possible.

Comparison between Common flue and Swine flue is given below (Table 2.22) to avoid anxiety.

Table 2.22: Comparison between Common flue and Swine flue

	Common Flu	Swine Flu
1. Reason	Common influenza virus	H1N1 (a subtype of Influenza virus)
2. Where it is available	In air	Multiply in the body of Pig.
3. When it becomes complicated	Fever with Pneumonia	Fever with Pneumonia and other respiratory trouble.
4. Way of transmission	Coughing, Sneezing, contact with infected person.	Same as common flu.
5. Symptoms	Fever, Tiredness, Pain inside throat, cough, vomiting tendency, diarrhea.	Same as common flu.
6. Days of curing	6-7 days.	6-7 days.
7. Persons suffers more	Old person, Child, Pregnant mother, Lung infected person.	Same as common flu.
8. Case becomes complicated	4 out of thousand.	4 out of thousand.
9. Rate of mortality	2.5 to 5.0 lacks in the world.	Up to July, 2009: 816 persons and 28 in India.
10. Mode of treatment	Proper medicine with rest in home in separate room.	Same as common flu.
11. Precaution	Cover the mouth with handkerchief during cough and cold, avoid contact with infected person.	Same as common flu.

NIPAH VIRUS (NiV)

Nipah virus (NiV) infection is newly emerging zoonosis* that causes severe disease in animals including human beings (Fig. 2.65). The natural host of NiV is the fruit bats, *Pteropus* spp. of the family Pterodidae.

NiV was first identified during an outbreak of disease that took place in Kampung Sungai Nipah, Malaysia in 1998 and the virus was isolated in 1999. The name NiV was given after the name of the village, Sungai Nipah, where the disease appears first. At that time, pigs were the intermediate hosts. At that time pigs also become infected and millions were killed in 1999 to control the spread of the disease. However, in subsequent NiV outbreaks, there were no intermediate hosts. Later the disease appears in Bangladesh in 2004 and human beings became infected with NiV as a result of consuming date palm sap contaminated by infected fruit bats. In **May 2018, an outbreak of the disease resulted in at least 18 deaths in Kerala, India**. Human-to-human transmission has also been documented, including in a hospital setting in India.

The Nipah virus is a type of RNA virus in the genus **Henipavirus**. It is a highly pathogenic paramyxovirus. NiV is morphology similar to other paramyxoviruses with pleomorphic structure, herringbone nucleocapsids, and 10 nm long surface protein spikes on the envelope (Fig. 2.65).

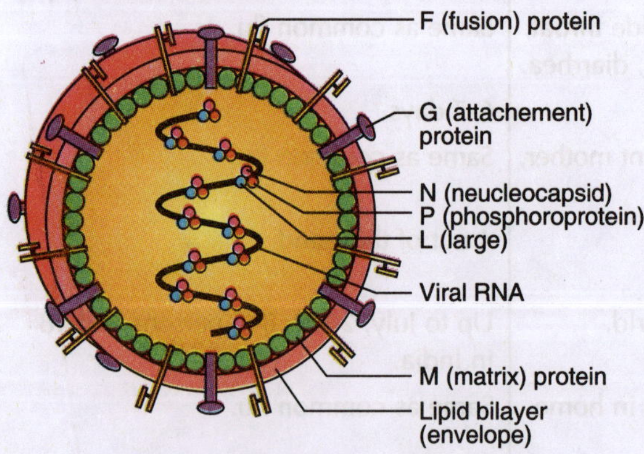

Fig. 2.65: Structure of a Henipavirus

* Zoonosis means a disease that can be transmitted from animals to human beings.

Signs and symptoms

Nipah Virus is an airborne transmission infection and can affect those who come in direct contact with contaminated bodies.

The virus is commonly associated with the inflammation of brain; thereby several days of fever can often lead to a state of confusion, disorientation and even persistent drowsiness. These symptoms can even cause a coma in a span of 24-48 hours, without proper care. Patients also show neurological, respiratory and pulmonary signs.

Transmission

Transmission from infected Bats to human beings takes place either directly through infected date palm sap (in Kerala, India) or through infected fruits via infected pig (Fig. 2.66), the intermediate host (in Malaysia).

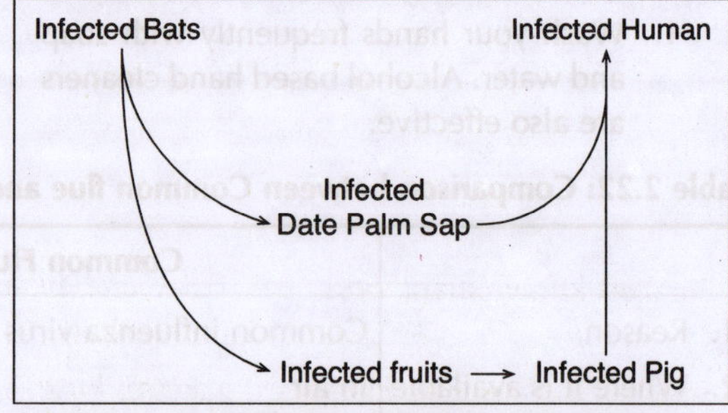

Fig. 2.66: Transmission of Nipah Virus (NiV)

Some common signs and symptoms due to NiV are headache, fever, nausea, dizziness, drowsiness and mental issues such as confusion. Symptoms may last up to 7-10 days. Watching out for respiratory illness during the early stages is also noted.

The symptoms appear from 4 to 14 days after infection. The early symptoms are not very clear and can easily be confused with that of viral fever and common cold. The virus can kill about 40-100% of infected persons. More than 60 per cent of this infection in humans comes from animals.

Prevention and cure

As there is no particular vaccine available particularly for the treatment of Nipah Virus, the only way to control the disease is through intensive supportive care.

Since drinking of raw date palm sap bitten by a bat can also cause NiV, it is safe to say that one should not cosume the date palm sap for time being. Hospitals also need to raise awareness about symptoms and transmission to avoid human-to-human infections. Detection is another issue with NiV and anyone who feels the symptoms should get tested immediately and thoroughly from a recognized laboratory.

Avoid from direct contact with infected pigs, bats and human beings in the endemic regions. Health professionals attending to such patients should take precautionary measures, such as wearing masks and gloves.

Some informations about Viruses

Largest (longest) plant virus	Citrus tristeza (rod shaped) 20,000 nm in size (0.02 mm)
Largest animal virus	Parrot fever Virus (450 μ)
Largest virus	TMV (300 nm), Pox virus (280 nm)
Largest virus	Mimivirus (the Acanthamoeba polyphaga mimivirus), with mature particles of 400 nm in diameter (icosahedral capsid), 800,000 bases and 900 genes. Later research suggested that it could be up to 800 nm long, 1.2 Mbp and 1,260 genes
Smallest animal virus	Foot and Mouth virus (10 nm size), Polio virus (27 nm in diameter)
Smallest DNA bacteriophage	Phi-X174 phage, thought to be larger than Hepatitis B, at about 4 kb. It is important to consider other self replicating genetic elements such as satelliviruses and Viroids.
Smallest DNA viruses	Hepadnaviruses such as Hepatitis B, at 3.2 kb and 42 nm; parvoviruses have smaller capsids, at 18-26 nm, but larger genomes, at 5 kb.
Smallest known plant virus	Satellite tobacco necrosis virus (Alfalfa mosaic virus; 17 nm in diameter)
Smallest RNA viruses in terms of genome size	Retroviruses such as rous sarcoma virus with genomes of 3.5 kb and particle diameters of 80 nm
Smallest virus	Virus of foot and mouth disease (20 nm)
Smallest infectious material	Prion (100 to 1000 times smaller than tobacco necrosis virus)
Smallest organism	PPLO (Pleuro Pneumonia-like Organism) or Mycoplasmagenitalium (200 to 300 nm in size)
Smallest self replicating organism	Mycoplasma

3 Algae

Introduction

The term algae (Latin word *algere*, seaweeds) was first introduced by Linnaeus in 1753. The algae comprise of a large heterogeneous assemblage of plants which are diverse in habitat, size, organisation, physiology, biochemistry, and reproduction. It is an important group of ***Thallophyta*** (Gr. *Thallos*, a sprout; *phyton*, a plant), the primitive and simplest division of the plant kingdom. The orderly systematic study of algae is called *Phycology* (Gr. *phycos*, seaweeds; *logos*, study or discourse).

The algae are chlorophyll-containing primitive plants, both prokaryotic and eukaryotic, with wide range of thalli starting from unicellular to multicellular organisations. Autophytic (which can manufacture their own food) and thalloid plant bodies are also found in Bryophytes. But the sharp demarcation between the two groups can be drawn by the following characters:

1. The sex organs, especially of female sex organ in algae are unicellular.
2. There is no embryo formation in algae.

However, the reproductive structures of some groups of algae (e.g., Chlorophyceae) are apparently multicellular and the sterile tissue is generally considered as vegetative. Bryophytes onwards in the scale of evolution have the uniform multicellular sex organs, the archegonia, which are not found in algae. For that reason bryophytes are usually called archegoniate plants.

Definitions: The definitions of algae as given by some phycologists are:

1. ***Fritsch, F. E.* (1935)** defined algae as the holophytic organisms (as well as their numerous colourless derivatives) that fail to reach the higher level of differentiation characteristic of the archegoniate plants.
2. ***Smith, G. M.* (1955)** defined algae as simple plants with an autotrophic mode of nutrition.
3. ***Chapman, V. J.* (1962)** defined algae (seaweeds of the seashore and green skeins in stagnant fresh water, ponds and pools) as among the simplest in the plant kingdom.
4. ***Prescott, G. W.* (1969)** defined algae as those chlorophyll-bearing organisms (and their colourless relatives) which are thalloid, i.e., having no true roots, stems and leaves or leaf-like organs.
5. ***Singh, R. N.* (1974)** defined that the algae are by and large simple plants which display a spectrum of photosynthetic pigments and evolve oxygen during the process of photosynthesis.

Important characteristics

1. Algae are chlorophyll-bearing autotrophic thalloid plant body.
2. Almost all the algae are aquatic.
3. The plant body may be unicellular to large robust multicellular structure.
4. The multicellular complex thalli lack vascular tissue and also show little differentiation of tissues.

5. The sex organs are generally unicellular but, when multicellular, all cells are fertile and in most cases the entire structure does not have any protective jacket.
6. The zygote undergoes further development either by mitosis or meiosis, but not through embryo formation.
7. Plants having distinct alternation of generations. Both gametophyte and sporophyte generations — when present in the life cycle are independent.

ECOLOGY AND DISTRIBUTION

Different species of algae may grow in different places. Their growth depends mainly on the type and concentration of accessary pigments possessed by them. The different accessary pigments need different intensity and quality of light. Thus the change of light wavelength greatly affects the distribution of the algae. In this respect it can be mentioned that in deep ocean and darker areas, phycocyanobilin and phycoerythrin can absorb blue light more effectively than chlorophyll a. Thus the green algae, which contain mainly chlorophyll a and β-carotene, normally grow in shallow water regions. The zonation in algal growth clearly observed in the coastal areas. Green algae are found in upper littoral zone, brown algae mostly found in the middle littoral zone and the red algae are commonly found in the lower littoral zone or in the deep oceans.

In addition to the above, substrate, topography, salinity, humidity, temperature, tides, waves, wind, water current and pollution level also affect the growth and distribution of the algae. Some algae prefer to grow in areas with stronger waves; others prefer calm water and sea bays. Some can tolerate and prefer drier conditions, thus they can grow in the supralittoral fringe; other can survive and grow in extraordinal environmental changes i.e. wet and dry, in the littoral zone, where low tide appears twice daily. The algal growth varies on the variation of substrate, such as mud, shells, stone blocks, coral reefs and lithotherm. Growth of certain types of algae accelerated with other marine organisms. In general, it can be pointed out that algae tend to grow on lithotherm or coral reefs in the coastal regions, especially in tidal pools or reef grooves. Thus these areas are best for collection of algae.

The marine macroalgae grow in the littoral zone has obvious seasonal changes. Here algae usually flourish in winter and spring. But in summer and autumn, the algae decreases significantly in species and also in richness. Mostly in summer, due to strong sunlight and high temperature becomes difficult for algae to survive and often become absent. Only some algae can be found in deeper waters.

This can be exemplified by the macroalgal flora of Taiwan. Here the Kuroshio Current with warm water and high salinity flows through the coastlines of southern region, while coastlines of northern region are influenced by the cold China Coastal Current. For the above condition, the marine macroalgal flora in southern region is quite different from the northern region. Thus the southern region is rich with tropical algae like *Helimeda opuntia, Caulerpa brachypus* f. sp. *parviflora, Cordium arabicum* and *Boergesenia forbesii,* and the northern region is known as the home of temperate marine algae like *Porphyra crispate, Bangia atropurpurea* and *Gelidium amansii,* those are absent in the southern coastal region.

General characteristics

Occurrence

The algae are ubiquitous (present everywhere) in distribution, i.e., they are found in fresh water as well as marine water, on soil, on rock, as epiphytes or parasites on plants and animals, in hot springs, in desert, on permanent snow-fields etc. But they mainly dwell in aquatic environments. Based on habitat the algae may be categorized as:

1. Aquatic algae.
2. Terrestrial algae, and
3. Algae of remarkable habitats.

1. Aquatic algae

Aquatic algae may be **fresh water** (when salinity is as low as 10 ppm) or **marine** (when salinity is 33-40%). Again, certain algae grow in **brackish water** which is unpalatable for drinking, but less salty than sea water. The fresh water algae usually grow in ponds, lakes, tanks, ditches etc. The very common ***fresh water algae*** are *Chlamydomonas, Volvox, Ulothrix, Chara, Oedogonium, Spirogyra, Nostoc, Oscillatoria* etc. Some of the very common ***marine algae*** are *Sargassum, Laminaria, Ectocarpus, Polysiphonia, Caulerpa, Bangia, Padina* etc.

Fresh water algae may be termed as **planktonic** when they grow and remain suspended on the upper part of water (e.g., *Volvox*, diatom), while the **benthic** algae are bottom-dwellers. The algae that grow at air-water interface are called **neustonic**. The benthic algae may be **epilithic**, that grow on stones; **epipelic**, attached to sand or mud; **epiphytic**, growing on plants; and **epizoic**, growing on animal body surface.

The marine algae may be **supralittoral** or **subaerial**, as they grow above the water level and in the spray zone. The **intertidal** algae grow in such a depth so that they are exposed periodically due to tides. Other marine algae are **sublittoral**, meaning that they are constantly submerged at depths as great as 30-60 metres (100-200 ft). Again, the **supralittoral** algae may be ***edaphic***, that grow in and on the soil; ***epilithic***, growing on stones; ***epiphytic***, growing on plants, ***epizoic*** — growing on animal body surface, and ***corticolous*** — growing on tree barks and parasitic on plants and animals. Some algae (e.g., *Chlorella*) live ***endozoically*** in various protozoa, coelenterates, molasses etc.

2. Terrestrial algae

Some algae are found to grow in terrestrial habitats like soils, rocks, logs etc. The algae that grow on the surface of soil are known as ***saprophytes***. Many blue-greens, on the other hand, grow under the surface of the soil, and are called ***cryptophytes***.

The algae growing in the desert soil may be typified as **endedaphic** (living in soil), **epidaphic** (living on the soil surface), **hypolithic** (growing on the lower surface of the stones on soil), **chasmolithic** (living in rock fissures) and **endolithic** algae (which are rock penetrating).

The common terrestrial members are *Oscillatoria sancta, Vaucheria geminata, Chlorella lichina, Euglena* sp., *Fritschiella* sp. and *Phormidium* sp.

3. Algae of remarkable habitats

In addition to above mentioned habitats, some algae also occur in uncommon habitats and termed as:

1. **Halophytic algae** (or Eurhaline). They grow in the highly concentrated salt lakes, and include *Chlamydomonas ehrenbergii, Dunaliella* and *Stephanoptera* sp.

2. **Symbiotic algae**. They *grow in association* with fungi, bryophytes, gymnosperms or angiosperms. The best examples of symbiotic algae found **in association with fungi** are *Nostoc, Gloeocapsa, Rivularia*; the members of Cyanophyceae and *Chlorella, Cytococcus, Pleurococcus*; the members of Chlorophyceae. This symbiotic association consisting of algae and fungi is called **lichen**. *Nostoc may* also associate with *Anthoceros* and *Anabaena* associate with the roots of *Cycas* to form **coralloid roots**.

3. **Cryophytic algae**. This group of algae growing on *ice* or *snow* provide attractive colours to snow-covered mountains. The alpine and arctic mountains become red due to the growth of the *Haematococcus nivalis*; green snow in Europe is due to the growth of *Chlamydomonas yellowstonensis*. *Scotiella nivalis* and *Raphidonema brevirostri* cause black colouration of snow, whereas *Ancyclonema nordenskioldii* is responsible for brownish purple colouration.

4. **Thermophytes** or **Thermal algae**. This group of algae occurs in *hot water springs* (50-70°C) where normal life is not possible. Many blue-greens (e.g., *Oscillatoria brevis, Synechococcus elongates, Heterohormogonium* sp.) are grown in such hot springs.

5. **Lithophytes**. They grow on the moist surface of stones and rocks, e.g., *Nostoc, Gloeocapsa, Enteromorpha, Batrachospermum* etc.

6. **Epiphytic algae**. They grow on *other plants* including other algal members. These are:

(a) **Algae on algae**. (i) *Ptilota plumosa* and *Rhodymenia pseudopalmatta* on *Laminaria hyperborean*, (ii) Diatoms on *Oedogonium, Spirogyra* etc.

(b) **Algae on bryophytes**. Blue-green algae like *Nostoc, Oscillatoria,* diatoms like *Achnanthes* etc. grow on different bryophytes.

(c) **Algae on angiosperms**. Algae like *Cocconis, Achnanthes* etc. grow epiphytically on *Lemna*, an aquatic angiosperm. Alga like *Trentepohlia* grows on the barks of different angiospermic plants, and is very common in Darjeeling (India).

7. **Epizoic algae**. The algae growing on animals like fish, snail etc. are called as epizoic, e.g., *Stigeoclonium* are found in the gills of fishes.

8. **Endozoic algae**. They grow in the tissues of animals, e.g., *Zoochlorella* sp. is found in *Hydra viridis*.

9. **Parasitic algae**. Some algae grow parasitically on different plants and animals. These are:

(a) *Cephaleuros* (Chlorophyceae) is parasitic and grows on the leaves of various angiosperms, such as tea (*Camellia sinensis*), coffee (*Coffea arabica*), *Rhododendron, Magnolia* and pepper (*Piper nigrum*). The most important one is *Cephaleuros* virescens, which causes ***Red rust of tea***.

(b) *Rhodochytrium* (Chlorophyceae) grows on ragweed (*Ambrosia*) leaves.

(c) *Phyllosiphon* (Chlorophyceae) grows on the leaves of *Arisarum vulgare*.

(d) *Ceratocolax* (Rhodophyceae) grows in *Phyllophora* thallus.

10. **Psammon**. The algae which grow in sandy beaches are called psammon, e.g., *Vaucheria, Phormidium* etc.

Thallus organisation

Thalli of algae show a ***range of organisation*** starting from unicellular form to highly organised multicellular habit where the plant body is differentiated into root-like, stem-like, and leaf-like structures giving a higher plant-like appearance. Their size ranges from a few micron to several metres. The algal thalli are grouped into the following, based on their organisation:

A. **Unicellular algae**
- (a) Motile.
  - (i) Rhizopodial.
  - (ii) Flagellated.
- (b) Non-motile.
  - (i) Spiral filamentous.
  - (ii) Coccoidal.

B. **Multicellular algae**
1. ***Colonial***:
   - (a) *Coenobium.*
     - (i) Motile.
     - (ii) Non-motile.
   - (b) *Aggregated form.*
     - (i) Palmelloid.
     - (ii) Dendroid.
     - (iii) Rhizopodial.
2. ***Filamentous***:
   - (a) *Unbranched filaments.*
   - (b) *Branched filaments.*
     - (i) Falsely branched.
     - (ii) Truly branched.
       - Simple filament.
       - Heterotrichous habit.
       - Pseudoparenchymatous habit.
3. ***Siphonaceous forms.***
4. ***Parenchymatous forms.***

A. Unicellular algae

Unicellular forms of algae functions as complete living organisms. Unicellular forms are common in all the groups of algae except Rhodophyceae, Phaeophyceae and Charophyceae. The unicells may be motile or non-motile.

(a) The **motile** unicells are either rhizopodial or flagellated.

The **rhizopodial forms** lack rigid cell wall and have cytoplasmic projections that

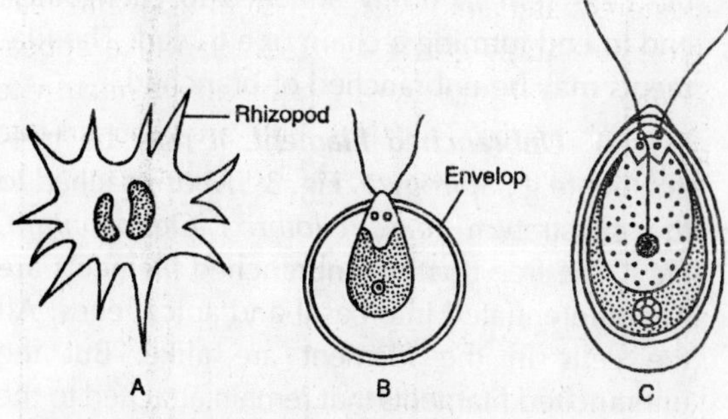

Fig. 3.1: Unicellular motile algae: A. *Chrysamoeba*, B. *Phacotus*, and C. *Chlamydomonas*

help them in amoeboid movement, e.g., *Chrysamoeba* (Chrysophyceae, Fig. 3.1A), *Rhizochloris* (Xanthophyceae).

The **flagellated** unicells resemble the motile gametes and zoospores. The flagella function as the organ of locomotion varying in number and type in different groups. The flagellated unicells are found in many groups of algae, e.g., *Phacotus* (Fig. 3.1B) and *Chlamydomonas* (Fig. 3.1C), of Chlorophyceae. *Euglena* of Euglenophyceae etc.

(b) The **non-motile** cells may be spiral filament as found in *Spirulina* (Cyanophyceae) (Fig. 3.2A). The coccoid unicellular algae are the simplest forms of algae found in Cyanophyceae, Chlorophyceae etc., e.g., *Gloeocapsa, Chlorella* (Fig. 3.2B).

B. Multicellular algae

1. **Colonial**: The colonial habit is achieved by loose aggregation of cells within a common mucilaginous investment. The cells of these usually remain connected with each other by cytoplasmic threads.

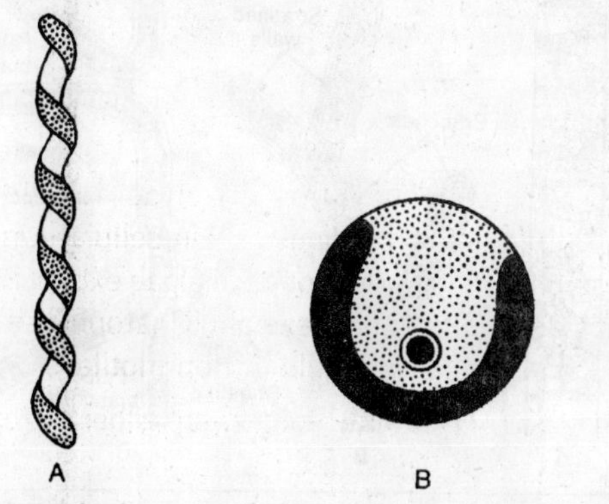

Fig. 3.2: Unicellular non-motile algae, A. *Spirulina*, and B. *Chlorella*

(a) ***Coenobium***. When a colony has a definite number of cells with a definite shape and arrangement, it is called **coenobium**.

This may be (i) motile, or (ii) non-motile.

(i) In **motile** form, cells are flagellated and whole coenobium can move by the organised beating action of flagella, e.g., *Volvox* (Fig. 3.3A), *Pandorina* (Fig. 3.3B), *Eudorina* etc. In *Volvox* the coenobium is a hollow sphere.

(ii) In **non-motile** form, the cells are without flagella, thereby the coenobium is non-motile, e.g., *Scenedesmus* (Fig. 3.3C), *Hydrodictyon* (Fig. 3.3D).

(b) ***Aggregated form***. Unlike coenobium the cells are aggregated irregularly showing a colonial mass of various size and shape. It is of three types: (i) Palmelloid, (ii) Dendroid, and (iii) Rhizopodial.

(i) **Palmelloid**. In this type the non-motile cells remain embedded in an amorphous gelatinous or muci-

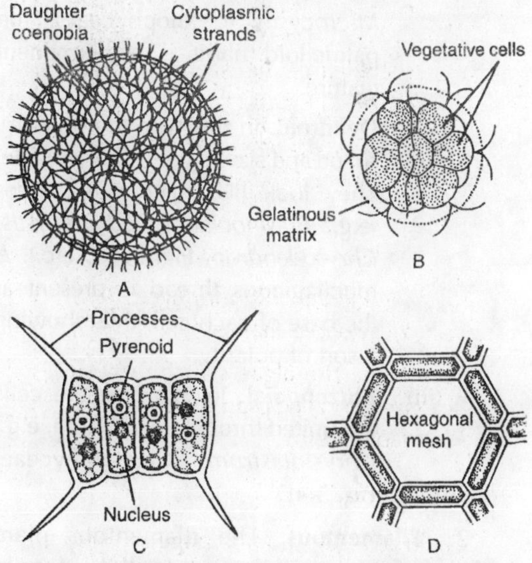

Fig. 3.3: Colonial algae: A. *Volvox*, B. *Pandorina*, C. *Scenedesmus*, and D. *Hydrodictyon*

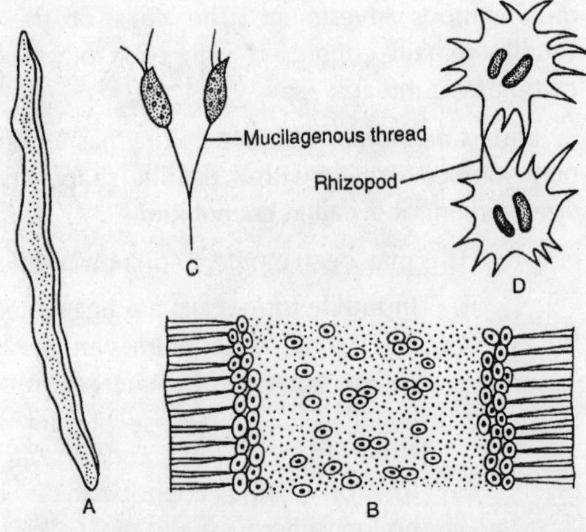

Fig. 3.4: Aggregated form: A & B. *Tetraspora*, (portion of a colony in T.S.), C. *Chrysodendron* and D. *Chrysidiastrum*

laginous matrix. Each and every cell of the organisation is independent and can perform all the functions as an individual. *Chlamydomonas* and *Chromulina* represent palmelloid as a temporary feature in their life cycle. But in *Tetraspora* (Fig. 3.4A, B) and *Palmodictyon* (Chlorophyceae), *Gleochloris* and *Chlorosaccus* (Xanthophyceae), *Phaeocystis* (Chrysophyceae) and *Microcystis* (Cyanophyceae), the palmelloid habit is a permanent feature.

(ii) **Dendroid.** In this type the number, shape and size of the cell is variable. They look like microscopic trees (e.g., *Prasinocladus, Ecballocystis, Chrysodendron*, Fig. 3.4C, etc.). A mucilaginous thread is present at the base of each cell, thus showing a sort of polarity.

(iii) **Rhizopodial.** In this type the cells are united through rhizopodia. e.g., *Chrysidiastrum* (Chrysophyceae, Fig. 3.4D).

2. **Filamentous**: The filamentous plant body is formed through repeated cell divisions in a single plane and in a single direction, where the cells remain firmly attached to each other, end to end forming a chain or a thread. The filaments may be unbranched or branched.

(a) ***Unbranched filament***. It may be free-floating (e.g., *Spirogyra*, Fig. 3.5A) or attached to the substratum (e.g., *Ulothrix, Oedogonium*, etc.). The free-floating unbranched filaments are not differentiated into basal and apical ends. All the cells in the filament are alike. But the unbranched filaments that remain attached to the substratum are differentiated into base and apex. All the cells of the filament are similar except the basal attachment cell that is specially modified for the purpose called **holdfast** or **rhizoidal cell**. The cell is devoid of chloroplast and only perform the function of anchorage. So certain degree of division of labour among the cells of the filament is established as rest of the cells perform photosynthetic and reproductive functions.

(b) ***Branched filament***. It is formed when a filament occasionally starts division in a second plane. It is of two types: (i) Falsely branched, and (ii) Truly branched.

(i) **Falsely branched.** The trichomes of blue greens may break either due to death or decay of the intercalary cells. The broken ends emerge out of the mucilaginous sheath in the form of a branch. They do not arise as lateral outgrowths, e.g., *Scytonema* (Fig. 3.5C).

(ii) **Truly branched.** When a cell in the filament occasionally starts division in a

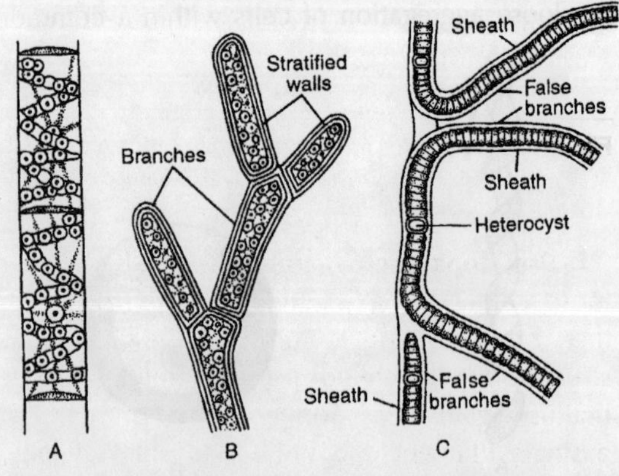

Fig. 3.5: Filamentous type: A. *Spirogyra*, B. *Cladophora* and C. *Scytonema*

second plane, true branch is formed. Thus true branches arise as lateral outgrowths of the main filament. True branches are of the following three types: Simple filament, Heterotrichous habit, and Pseudoparenchymatous habit.

Simple filament. In this branching system the whole thallus remain attached to the substratum by a basal cell and the branches may arise from any cell of the filament except the basal cell. e.g., *Cladophora* (Fig. 3.5B).

Heterotrichous habit. In this branching system the whole thallus is differentiated into prostrate and erect system. Both the prostrate and erect systems may be well-developed (e.g., *Fritschiella, Ectocarpus*, Fig. 3.6A). Progressive elimination of the prostrate system is observed in *Draparnaldiopsis* (Fig. 3.6B), *Stigoclonium*, or of the erect system as in *Coleochaete* (Fig. 3.6C).

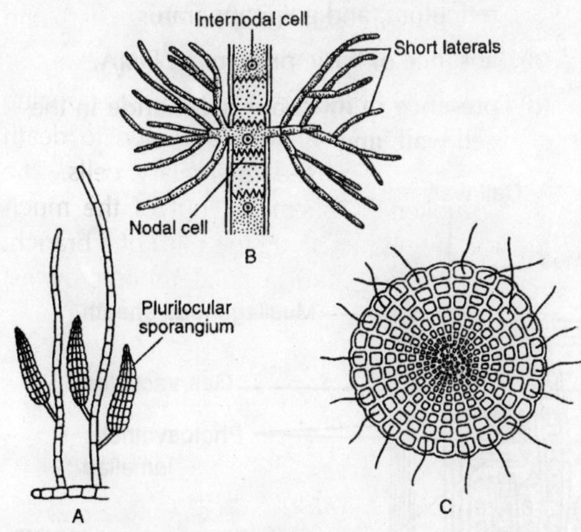

Fig. 3.6: Branched filament showing heterotrichous habit: A. *Ectocarpus*, B. *Draparnaldiopsis*, and C. *Coleochaete*

Pseudoparenchymatous habit. If one or more central or axial filaments together with their branches fuse to form a parenchymatous structure, it is called pseudoparenchymatous thallus. Again, if it is formed by the branches of a single filament it is known as uniaxial (e.g., *Batrachospermum*, Fig. 3.7A, B), or it may be multiaxial where more than one filament are involved (e.g., *Polysiphonia*, Fig. 3.7C).

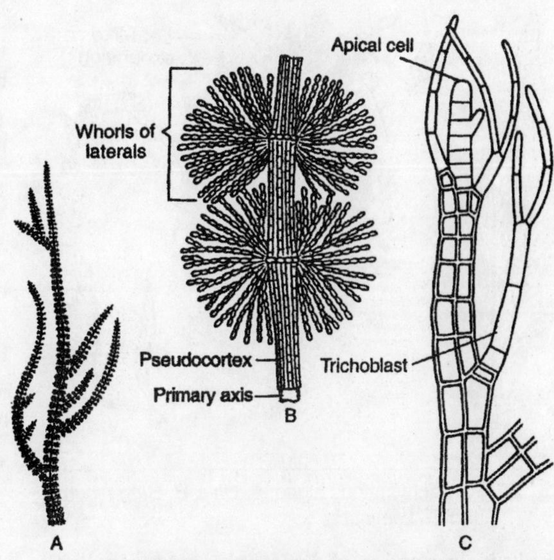

Fig. 3.7: Pseudoparenchymatous habit: A. *Batrachospermum*, B. *Batrachospermum* (portion of the plant body enlarged), and C. *Polysiphonia*

3. ***Siphonaceous forms***. In this form the thallus is aseptate and multinucleate i.e., a coenocyte. It may be simple branched (e.g., *Vaucheria*, Fig. 3.8A) or may be very elaborate with clear division of labour, differentiated into aerial and subterranean portions (e.g., *Botrydium*, Fig. 3.8B).

4. ***Parenchymatous forms***. When the cells of a filament divide in multidirectional planes, it

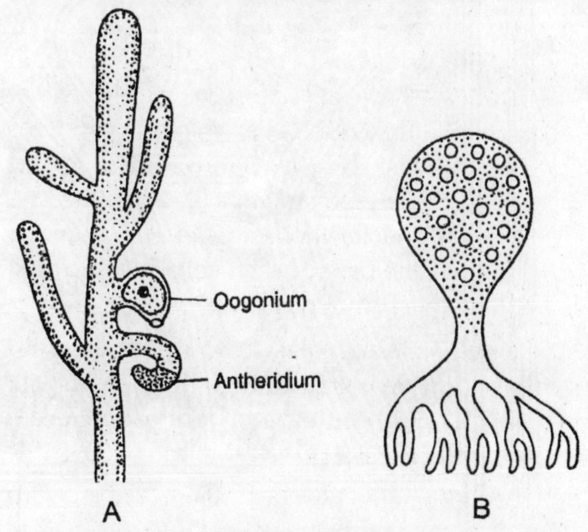

Fig. 3.8: Siphonaceous algae: A. *Vaucheria* and B. *Botrydium*

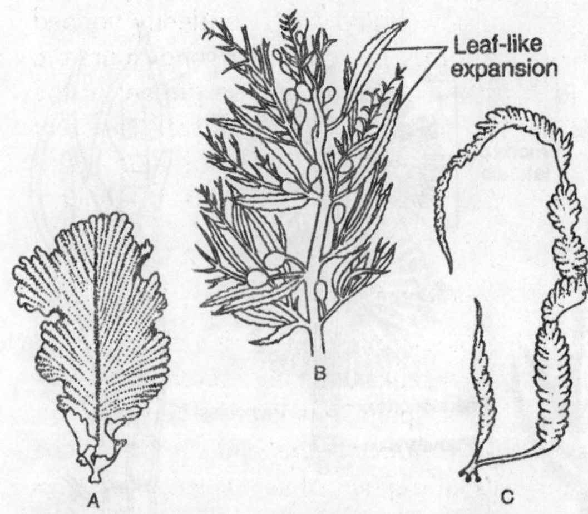

Fig. 3.9: Parenchymatous algae: A. *Ulva*, B. *Sargassum* and C. *Laminaria*

results the formation of a parenchymatous thallus and ultimately becoming foliose and flat (e.g., *Ulva*, Fig. 3.9A), tubular (e.g., *Enteromorpha, Scytosiphon*) or complex (e.g., *Sargassum*, Fig. 3.9B) structure. Growth of the parenchymatous thalli may be diffused (when all the cells can divide), intercalary (when the dividing region remain in the intercalary position) e.g., *Laminaria* (Fig. 3.9C), trichothallic (growth by a specialised intercalary meristem at the base of a terminal hair) e.g., *Porphyra* or apical (when one or more well-defined apical cells divide to produce the remainder of the thallus), e.g., *Fucus*.

CELL STRUCTURES IN ALGAE

It is difficult to generalise the structural organisation of algal cells as there is great variation in their cytological features. However, on the basis of cytological characteristics, the algal cells are categorized into: 1. **prokaryotic**, 2. **eukaryotic**, and 3. **mesokaryotic** types.

1. **Prokaryotic**. The prokaryotic cell organisation is found in Cyanophyceae (Fig. 3.10). It is characterised by:
   - (a) presence of incipient nucleus,
   - (b) lack of any membrane-bound organelles, i.e., mitochondria, plastids, endoplasmic reticulum, and golgi apparatus,
   - (c) absence of basic proteins in DNA,
   - (d) presence of mucopolysaccharide in the cell wall, and

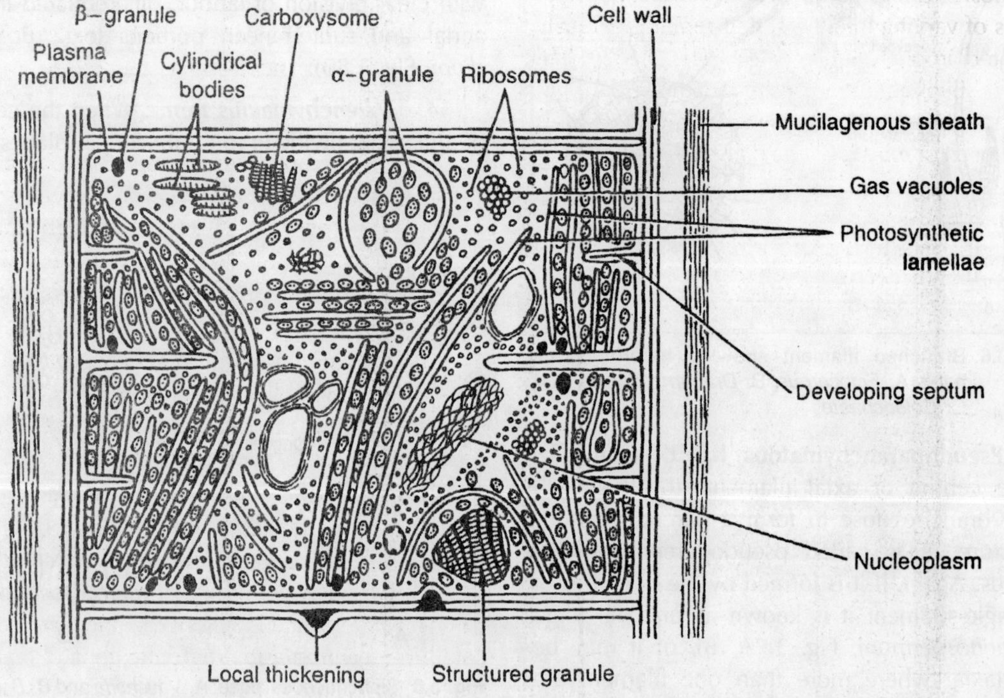

Fig. 3.10: Structure of a prokaryotic cell (blue green alga) under electron microscope

(e) the absence of mitotic figures during cell division.

2. **Eukaryotic**. In contrast, an eukaryotic algal cell (Fig. 3.11) possesses a definite nucleus and membrane bound organelles with cellulosic cell wall.

3. **Mesokaryotic**. Majority of the algae shows this type of cell organisation where the nucleus has a nuclear membrane and the mitotic figures are observed during cell division (eukaryotic characters), but the basic proteins are absent in the chromosomes (prokaryotic characters). They are called mesokaryotic algae.

The algal cells show the following structures (Fig. 3.10 and 3.11):

1. Cell Wall

Cell wall of most algae is ***cellulosic***. It also contains hemicellulose, mucilage, pectin and other substances like alginic acid, fucoidin, fucin, calcium carbonate, silica etc. in different combinations in different groups of algae. Electron microscopic studies reveal that the cellulosic wall is composed of cellulose microfibrils of varying thickness, that remain variously oriented in a granular matrix. The diatom cell wall is silicified and shows characteristic secondary structures. In **Cyanophyceae**, the cell wall is composed of ***mucopeptide*** consisting mainly of a peptide of few amino acids covalently bonded to amino-sugars, glucosamine and muramic acid. This mucocomplex is also present in the cell wall of bacteria. A true cell wall is absent in certain algae like *Gymnodinium* and *Pyramimonas*. Instead they possess a boundary membrane known as ***pellicle***.

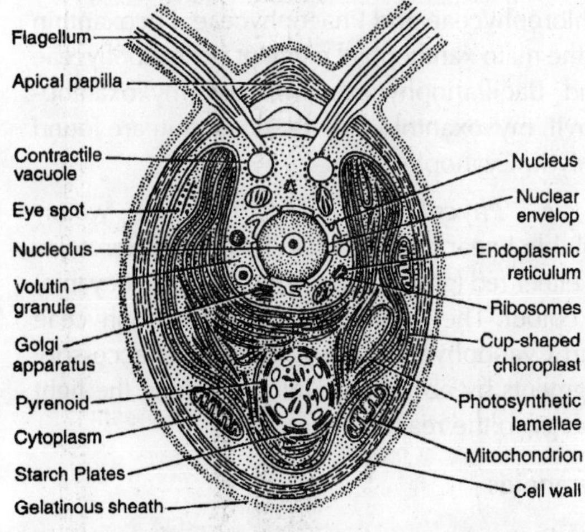

Fig. 3.11: Structure of an eukaryotic cell (*Chlamydomonas*) under electron microscope

2. The Protoplast

The protoplasmic content of a cell is called protoplast. The eukaryotic algal protoplast is surrounded by a lipoproteinaceous external boundary, called ***cell membrane***, and consists of one or more usually spherical or ellipsoidal ***nucleus*** and ***cytoplasm***. The cell membrane is made up of lipid and protein and is ***fluid mosaic*** in nature like all other biological membrane systems. It is very thin and elastic and selectively permeable. It controls the passage of materials in and out of the cells.

In all **eukaryotic algae** (Chlorophyceae etc.), the nucleus is a well-organised spherical or elliptical body. It remains surrounded by a distinct nuclear membrane. The inner side is occupied by a chromatin reticulum embedded in a matrix called karyolymph. The nuclear membrane is double layered. The outer membrane is continuous with the endoplasmic reticulum. Each nucleus contains one or more nucleoli or endosomes. The number of nucleoli varies in different algae. It may be one, two or more. The chromosomes may have a localised or diffused centromere. The number of chromosomes vary from species to species — the lowest number being n=2 (e.g., *Porphyra linearis*), while the highest is n=592 (*Netrium digitali*).

In **eukaryotic algal cell** there are membrane bound cell organelles like ***chloroplasts, mitochondria, golgi apparatus, endoplasmic reticulum*** and, in some cases, eye spot or stigma.

In **prokaryotic algal cell** (Cyanophycean members), the nucleus is not bounded by any membrane. Instead, the protoplast is differentiated into the outer peripheral ***chromoplasm*** containing photosynthetic pigments and an inner colourless ***centroplasm*** where the genetic material is not found within the membrane-bound nucleus and the DNA strands do not combine with histones to form chromosomes. Thus the centroplasm represents the ***incipient nucleus*** in Cyanophyceae. The nucleus in Dinophyceae is also not truly eukaryotic, although it is membrane-bound, but chromosomes and mitotic apparatus are absent. The cytoplasm of algal cell is divided into ***cell organelles*** and ***cytosol***.

The chloroplast

Chloroplasts are the very prominent feature of algal cells. They bear the photosynthetic pigments. It is a double-membrane structure. **Various forms of chloroplasts** are known to occur in different types of algae, of which eight main types are usually recognised: **cup-shaped** (e.g., *Chlamydomonas* and *Volvox*), ***discoid*** (e.g., *Chara*, *Vaucheria* and centric diatoms), ***parietal*** (e.g., Chaetophorales, Phaeophyceae, Rhodophyceae, many Chrysophyceae and pinnate diatoms), ***girdle shaped*** or ***C-shaped*** (e.g., *Ulothrix*), ***spiral*** (e.g., *Spirogyra*), ***reticulate*** (e.g., *Oedogonium*, *Hydrodictyon* and *Cladophora*), **stellate** (e.g., *Zygnema*), and **ribbed** (e.g., Volvocales). The basic structure of chloroplast is almost similar throughout the plant kingdom.

There are three major structural regions in the chloroplast:

1. an ***envelope*** consisting of two membranes with an enclosed space.
2. the mobile ***stroma*** containing the soluble enzymes for metabolism, protein synthesis and starch storage, and
3. the highly organised internal lamellar membranes containing pigments and involved in energy capture and transduction.

The internal lamellar system forms discs which are stacked together like piles of coins to form ***grana***. Each disc is a sac or vesicle and termed as ***thylakoid***. Each thylakoid encloses an interthylakoid space. The thylakoid system constitutes a single, complex cavity, separated from stroma by the thylakoid membrane. Built into the thylakoid membranes are pigment systems and electron carriers, which carry out the light phase of photosynthesis.

In Cyanophyceae, the thylakoids are not enclosed in membrane bound groups to form chloroplasts, instead they lie free in the cytoplasm. The thylakoids are the site of chlorophyll a, and the accessory pigments also occur on their surface in the form of small vesicles called the ***phycobilisomes***.

Pigmentation: The pigments that provide the actual colour of the thallus are of various types:

- ***Chlorophylls***: There are five types of chlorophylls found in algae, Chl a, b, c, d, and e. Of them, chlorophyll a is present in all groups of algae. Chlorophyll b is found only in Chlorophyceae, Chlorophyll c in Phaeophyceae, Cryptophyceae, Bacillariophyceae and Chrysophyceae, Chlorophyll d in some red algae, and chlorophyll e in certain Xanthophyceae.

- ***Caroteinoids***: Carotenes and xanthophylls together constitute the carotenoids. They are accessory photosynthetic pigments. Five types of carotenes are found in algae: α-carotene in Chlorophyceae, Cryptophyceae and Rhodophyceae; β-carotene in all algal groups, except Cryptophyceae; c-carotene in Chlorophyceae; e-carotene in Bacillariophyceae, Cryptophyceae, Phaeophyceae and Cyanophyceae and flavacene in members of Cyanophyceae.

- ***Xanthophylls***: Several types of xanthophylls are found in algae. Of them lutein, violaxanthin and neoxanthin are found in the members of Chlorophyceae and Phaeophyceae. Fucoxanthin is the main xanthophyll pigment in Phaeophyceae and Bacillariophyceae, whereas myxoxanthophyll, myxoxanthin and oscilloxanthin are found only in Cyanophyceae.

- ***Phycobilins***: Phycobilins are water-soluble linear tetrapyrroles. They are biliproteins of either red (phycoerythrin) or blue (phycocyanin) in colour. They are found only in Rhodophyceae and Cyanophyceae. They function as accessory pigments by absorbing and transferring the light energy to the reaction centre.

Pyrenoids

Pyrenoids are proteinaceous bodies present in chloroplasts or chromatophores — the very characteristic of algal chloroplasts. They are usually associated with the synthesis and storage of starch. In Bacillariophyceae they accumulate lipid. The number of pyrenoid may be one (e.g., *Chlamydomonas*) or more than one (e.g., *Oedogonium*) per chromatophore.

Mitochondria

Mitochondria are found in all algal cells **except Cyanophyceae**. Each mitochondrion is surrounded by a double membrane envelope. The inner membrane of plant mitochondria encloses an aqueous matrix of solutes, soluble enzymes and the mitochondrial glucose. The inner membrane is larger than the outer membrane and undergoes invagination producing sac-like **cristae** of variable shape and number — usually with a narrow neck. The whole mitochondrion is again encircled by an outer membrane lying close to the inner one, leaving an intermembrane space which is continuous with the intercristal space. The matrix is finely granular and highly proteinaceous. The organelle is semiautonomous in nature as it contains a circular DNA and ribosomes of its own, with the help of which it can synthesise some of its proteins. There is usually more than one mitochondrion per cell, but in *Micromonas* (**Chlorophyceae**) **each cell contains a single mitochondrion**.

The cells of blue green algae lack mitochondria. The cytoplasmic membrane is the site of biochemical functions normally associated with the well-defined membranous organelles in eukaryotic cells. Like bacteria, the cell membrane invaginates to form a structure called the ***mesosome*** where the respiratory enzymes are localized.

Endoplasmic reticulum(ER)

Electron microscopic studies indicate that the algal cells contain an extensive membrane network of interconnecting tubules and cisternae (flattened sac), called the endoplasmic reticulum. The ER membranes traverse the entire cytoplasm. The reticulum consists of interconnected parallel cisternae associated with ribosome, attached to the cytoplasmic face of the membrane. This form of ER is known as ***rough endoplasmic reticulum (RER)*** which is a major site of protein synthesis. On the contrary the ER membranes that do not bear ribosome are called ***smooth endoplasmic reticulum (SER)***.

Dictyosomes or Golgi apparatus

Dictyosomes or Golgi bodies are found in all algal cells except blue-green algae, and can be seen under the electron microscope. The Golgi apparatus is a component of the endomembrane system of the cell and appears to serve as an intermediate between the endoplasmic reticulum and plasma membrane. Golgi bodies may be found in the region of the nucleus (e.g., *Chlamydomonas*), near plastids (e.g., diatom and *Bulbochaete*), or it may be found anywhere in the cell. Golgi bodies are composed of 2-20 flat vesicles which are arranged in stacks. Each stack is called ***dictyosome***. All dictyosomes collectively form the Golgi apparatus. It functions in the packaging of materials for export to the cell's exterior. It is also responsible for the formation of new plasma membrane to support growth or to replace the lost one.

Eyespot or stigma

The motile vegetative and reproductive cells of algae have a pigmented spots in the anterior, middle or posterior part of the cell, known as eyespot or stigma (Fig. 3.12). It is involved directly or indirectly in light perception. The

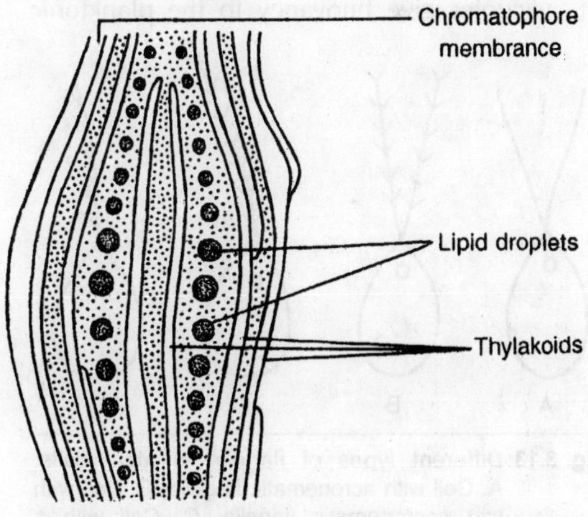

Fig. 3.12: An eyespot of *Chlamydomonas reinhardtii* (after Dodge)

stigma is usually found within the thylakoids run longitudinally through the eyespot in between two rows of granules.

Vacuoles

Almost all the algal cells, except the members of Cyanophyceae, possess one or more vacuoles. Each vacuole is bounded by a distinct membrane called **tonoplast**. Three types of vacuoles are found in motile forms:

1. **Simple vacuole**. They are very small in size and show periodic contraction and expansion. They are also called **contractile vacuoles**. They throw out the metabolic wastes of the cells. They also regulate the water content of the cell by discharging the excess amount at short interval. So they are secretory in function.
2. **Complex vacuole**. This is the characteristics of Dinophyceae and Euglenophyceae. It consists of a tube-like cytopharynx, a large reservoir and a group of vacuoles of varying sizes. The vacuoles perform the function of osmoregulation inside the cell. Sometimes, the vacuoles also store reserve food materials such as laminarin and chrysolaminarin.
3. **Gas vacuoles**. In the cells of the members of Cyanophyceae there are gas containing cavities occurring as stacks of small transparent cylinders of uniform diameter. Their walls are freely permeable to gases. The gas vacuoles give buoyancy to the planktonic forms and also serve as protective screens against incident bright light.

Flagella

Motile vegetative or reproductive cells are present in all groups of algae except Cyanophyceae and Rhodophyceae. The movement is achieved by the beating action of small filiform or thread-like protoplasmic appendages, called ***flagella***. They vary in number, length, position and presence or absence of hairs in different numbers. The number varies from one to four or many. They are mainly of two types:

1. **Whiplash** or **acronematic**. These are hairless smooth surfaced flagella (Fig. 3.13A, Fig. 3.14A), and
2. **Tinsel** or **pleuronematic**. They are having one or more rows of lateral fine filamentous hairs known as mastigonemes or flimmers.

   They are further categorised into:

   (a) *Pantonematic*. Where the mastigonemes are arranged in two opposite rows (Fig. 3.13B).

   (b) *Pantoacronematic*. Pantonematic flagellum with a terminal fibril is called panto-acronematic (Fig. 3.13C).

   (c) *Stichonematic*. It has one-sided mastigonemes (Fig. 3.13D).

If the number of flagella per cell is more than one and identical, it is known as ***isokont*** and when dissimilar, it is known as ***heterokont***.

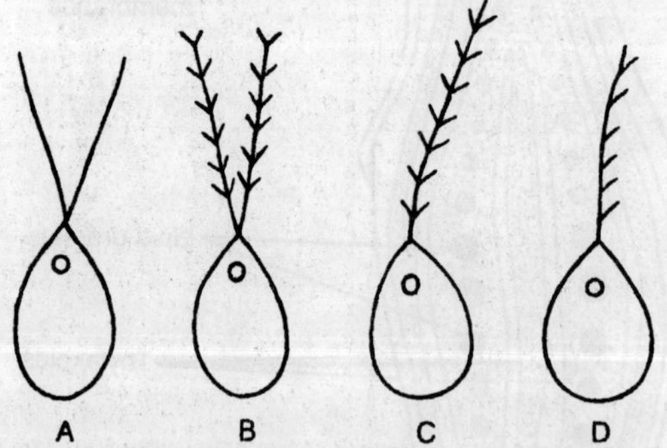

Fig. 3.13: Different types of flagella in algal cells: A. Cell with acronematic flagella, B. Cell with two pantonematic flagella, C. Cell with a panto-acronematic flagellum, and D. Cell with a sticho-nematic flagellum

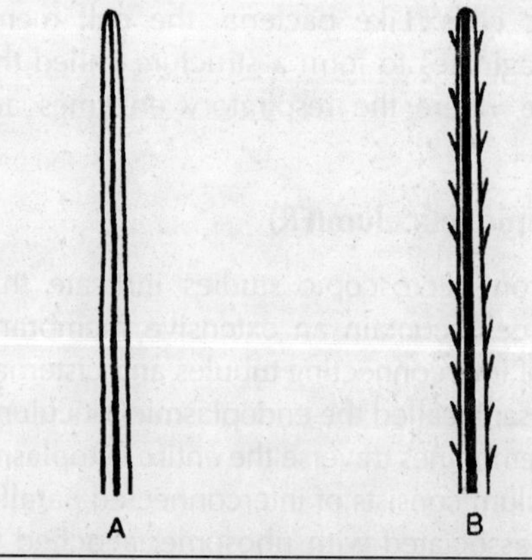

Fig. 3.14: Magnified view of flagella: A. Single whiplash flagellum, and B. Single tinsel flagellum

Flagellar roots in algae

Flagella are the extremely fine, hyaline emergence of cytoplasm. Commonly there is a single granule at the base of each flagellum (Fig. 3.15A), the **blepharoplast** or **basal body**. If algal cell has a firm wall, the flagellum emerges through a pore.

Each flagellum has a central or axial thin filament, the **axoneme**. The axoneme is surrounded by a cytoplasmic membrane or sheath formed by an extension of the cell or plasma membrane. The sheath ends just short of the apex of the flagella. The apical naked portion of the axoneme is called **end-piece**.

In transverse section, the flagellum (Fig. 3.15B) shows two central singlet fibrils surrounded by nine peripheral doublet fibrils. Each fibril is covered by a membrane and the two central ones are further covered with an additional membrane.

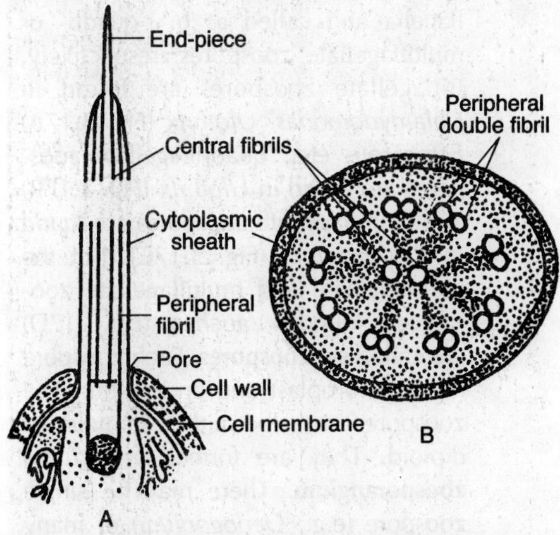

Fig. 3.15: Flagella: A. Structure of a flagellum with basal structure, and B. T.S. of single flagellum

NUTRITION

Algae are ***autotrophic*** in their mode of nutrition. They are able to manufacture their own food by photosynthesis on account of the presence of chlorophyll.

Some algae like *Cephaleuros* are parasitic, which grow on the leaves of various angiosperms, such as tea (*Camelia sinensis*), coffee (*Coffea arabica*), *Rhododendron*, *Magnolia* and pepper (*Piper nigrum*). The most important one is *Cephaleuros virescens*, which causes ***Red rust of tea***.

REPRODUCTION

The members of algae reproduce by all the three means: 1. **Vegetative**, 2. **Asexual**, and 3. **Sexual**.

1. **Vegetative reproduction**. In this type, any vegetative part of the thallus develops into new individual. It does not involve any spore formation and there is no alternation of generations. It is the most common method of reproduction in algae. The vegetative reproduction in algae is of the following types:

   (a) **Cell division** or **fission**. It is the simplest method of reproduction. The unicellular forms of algae commonly reproduce by this simple process, often called binary fission as found in *Chlamydomonas*, *Synechococcus* (Fig. 3.16A), diatoms etc. In this method the vegetative cell divides mitotically into two daughter cells, those finally behave as new individual.

   (b) **Fragmentation**. In this method, the multicellular filamentous thallus breaks into many-celled fragments, each of which gives rise to a new individual. The fragmentation may be accidental or by the formation of separation discs or by some other mechanical force or injury. It is found in *Spirogyra, Ulothrix, Oedogonium, Zygnema, Cylindrospermum* (Fig. 3.16B) etc.

   (c) **Hormogonia**. This method of vegetative reproduction is found in blue-green algae. The trichomes of blue-green algae break up within the sheath into many-celled segments called hormogonia or hormogones. They remain delimited by the formation of heterocysts, separation discs or necridia or by the death and decay of intercalary cells of the trichome. Hormogonia are commonly found in

Nostoc, Oscillatoria, Cylindrosporium etc.

(d) **Formation of adventitious branches**. Adventitious branches are formed in different large thalloid algae, which, when detached from the plant body, develop into new individuals (e.g., *Fucus, Dictyota*). Protonema-like adventitious branches are formed from the internodes of *Chara*, stolons of *Cladophora glomerata* etc.

(e) **Bulbils**. Tuber-like outgrowths are developed due to storage of food at the tip of rhizoids and on the lower nodes of *Chara*, called bulbils (Fig. 3.16C). After detachment from the plant body, bulbils grow into new plants.

(f) **Amylum stars**. Star-shaped aggregation of starch containing cells develops on the lower nodes of *Chara*. These structures are called amylum stars (Fig. 3.16D). When detached from the plant body, they grow into new plants.

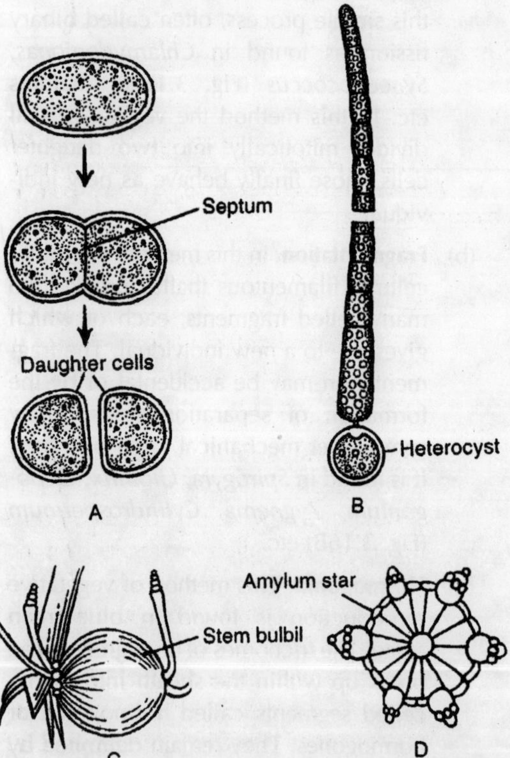

Fig. 3.16: Vegetative reproduction in algae: A. Cell division (*Synechococcus* sp.) B. Fragmentation of filament (*Cylindrospermum* sp.) C. Stem bulbil (*Chara* sp.) and D. Amylum star (*Chara* sp.)

(g) **Budding**. In *Protosiphon* bud-like structures are formed due to proliferation of vesicles delimited from the parental body by a septum, which, after detachment, grow into a new plant.

2. **Asexual reproduction**. Asexual reproduction involves the formation of certain type of spores — either naked or newly walled. It is a process of rejuvenation of the protoplast without any sexual fusion. Each and every spore germinates into a new plant. In this method, there is no alternation of generations. The asexual spores may be of various types:

(a) **Zoospores**. These are motile naked spores provided with two, four or many flagella and called as bi-, quadri- or multiflagellate zoospores, respectively. Biflagellate zoospores are found in *Chlamydomonas, Ulothrix* (Fig. 3.17A) *Ectocarpus* etc., quadriflagellate zoospores are found in *Ulothrix* (Fig. 3.17B) and multiflagellate zoospores are found in *Oedogonium* (Fig. 3.17C). But the multinucleate and multiflagellate zoospores as found in *Vaucheria* (Fig. 3.17D) are called **synzoospores**. Each zoospore has a chloroplast and an eyespot. The zoospores may be either haploid or diploid. They are formed within the **zoosporangium**. There may be single zoospore (e.g., *Oedogonium*) or many zoospores (e.g., *Cladophora*) per zoosporangium. Zoospores are either haploid or diploid depending on the nature of plant body, gametophytic or sporophytic on which it develops. The zoospores are liberated either by the disintegration of the zoosporangial wall or by the formation of an apical pore on the zoosporangium. After liberation the zoospores swim for a while, then withdraw their flagella, encyst and ultimately germinate into new plants.

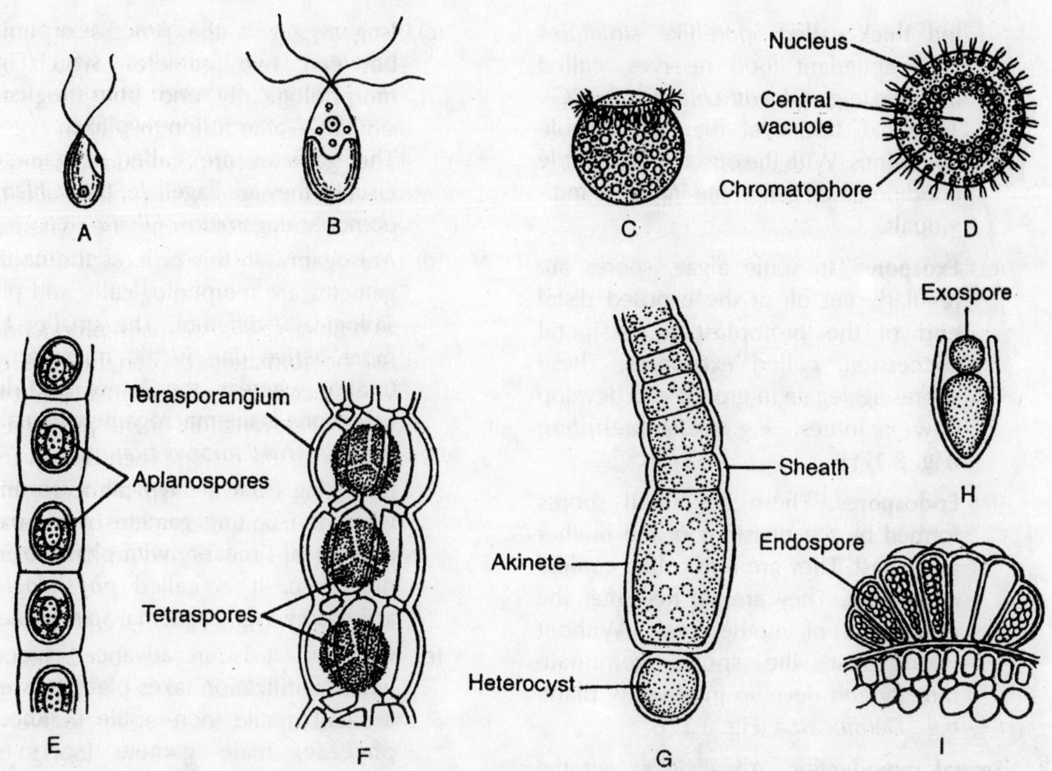

Fig. 3.17: Asexual spores in algae: A. Biflagellate microzoospore, and B. Quadriflagellate microzoospore of *Ulothrix* sp., C. Multiflagellate zoospore of *Oedogonium* sp., D. Synzoospore of *Vaucheria* sp., E. Aplanospores of *Ulothrix* sp., F. Tetraspores of *Polysiphonia* sp., G. Akinete of *Gloeotrichia* sp., H. Exospore of *Chamaesiphon incrustans*, and I. Endospores of *Dermocarpa prasina*

(b) **Aplanospores**. Aplanospores are non-motile spores. These spores are formed either singly or its protoplast may divide to form many aplanospores inside sporangium during unfavourable conditions, especially in drought (e.g., *Ulothrix* (Fig. 3.17E), *Microspora*). The aplanospores may also be formed in certain algae of semiaquatic habitat. When they appear identical to the parent cell, they are referred to as autospores (e.g., *Scenedesmus, Chlorella* etc.). Aplanospores with thickened wall and abundant food reserve are known as **hypnospores** (e.g., *Pediastram, Sphaerella* etc.). They are formed to overcome prolonged period of desiccation. With the onset of favourable condition the hypnospores either directly germinate into a new individual or their protoplasts may form zoospores. Due to deposition of haematochrome pigment in their walls, the hypnospores of *Chlamydomonas nivalis* are red in colour.

(c) **Tetraspores**. Diploid plants of some algae (e.g., *Polysiphonia*, Fig. 3.17F) produce a special type of haploid aplanospores, called **tetraspores**, formed within tetrasporangium. The diploid nucleus of a **tetrasporangium** divides meiotically to form four haploid nuclei which — with little amount of protoplasm — are developed into four tetraspores. After liberation, the tetraspores germinate to form male and female gametophytes.

(d) **Akinetes**. The vegetative cells of certain filamentous algae develop into elongated thick-walled spore-like structures with abundant food reserves, called **akinetes** (e.g., *Gloeotrichia*, Fig. 3.17G). They can tide over the unfavourable conditions. With the onset of favourable condition they germinate into new individuals.

(e) **Exospores**. In some algae, spores are regularly cut off at the exposed distal end of the protoplast in basipetal succession, called ***exospores***. These spores aggregate in groups and develop new colonies. e.g., *Chamaesiphon* (Fig. 3.17H).

(f) **Endospores**. These are small spores formed by the divisions of the mother protoplast. They are also called **conidia** or **gonidia**. They are set free after the dissolution of mother wall. Without taking rest, the spores germinate directly and develop into a new plant, e.g., *Dermocarpa* (Fig. 3.17I).

3. **Sexual reproduction**. All algae except the *members* of the class Cyanophyceae reproduce sexually. During sexual reproduction gametes fuse to form zygote (Fig. 3.18). The new genetic set up can develop by fusion of gametes coming from the different parents. Depending on the structure, physiological behaviour and complexity of sex organs, sexual reproductions are of the following five types:

(a) **Autogamy**. In this process the fusing gametes are developed from the same mother cell and after fusion they form zygote. For the above, plant developed through autogamy does not show the introduction of any new characteristic, e.g., Diatom (*Amphora normani*).

(b) **Hologamy**. In some unicellular member, the vegetative cells of different strains (+ and –) behave as gametes and after fusion they form zygote. It is an inefficient process considering the point of multiplication, but new genetic combinations are developed by this process, e.g., *Chlamydomonas*.

(c) **Isogamy**. It is the process of union between two gametes which are morphologically and physiologically similar — after fusion they form zygote. The gametes are called isogametes. Usually they are flagellate, e.g., *Chlamydomonas eugametos, Ulothrix etc.*

(d) **Anisogamy**. In this process the uniting gametes are morphologically and physiologically different. The smaller and more active one is the microgamete (male), whereas the larger and less active one is the macrogamete (female). e.g., *Chlamydomonas braunii*.

Deviating from the typical anisogamy, when the uniting gametes show morphological similarity with physiological difference, it is called ***physiological anisogamy***, e.g., *Zygnema, Spirogyra* etc.

(e) **Oogamy**. It is an advanced process where fertilization takes place between a small motile (non-motile in Rhodophyceae) male gamete (sperm or antherozoides) with a large non-motile female gamete (egg or ovum). Male gametes develop within antheridium, whereas the female gamete within the oogonium, e.g., *Oedogonium, Vaucheria, Chara, Laminaria, Sargassum, Polysiphonia, Batrachospermum* etc.

ORIGIN AND EVOLUTION OF SEX

Algae, like most of the other plants, reproduce by all the three means: vegetative, asexual, and sexual. The sexual reproduction is absent in the class Myxophyceae (Cyanophyceae) but they can reproduce by both vegetative and asexual means. In other groups the reproduction takes place by all the above three means, out of which asexual and sexual methods are very common. Many plants multiply vegetatively, but they do not involve rejuvenation of the protoplasm. The asexual reproduction takes place by means of specialised motile or non-motile sex cells, the spores, which do not undergo fusion and, on germination, they give rise to new individuals. Lastly, sexual reproduction involves the union of sex cells, the gametes, and the result of union of

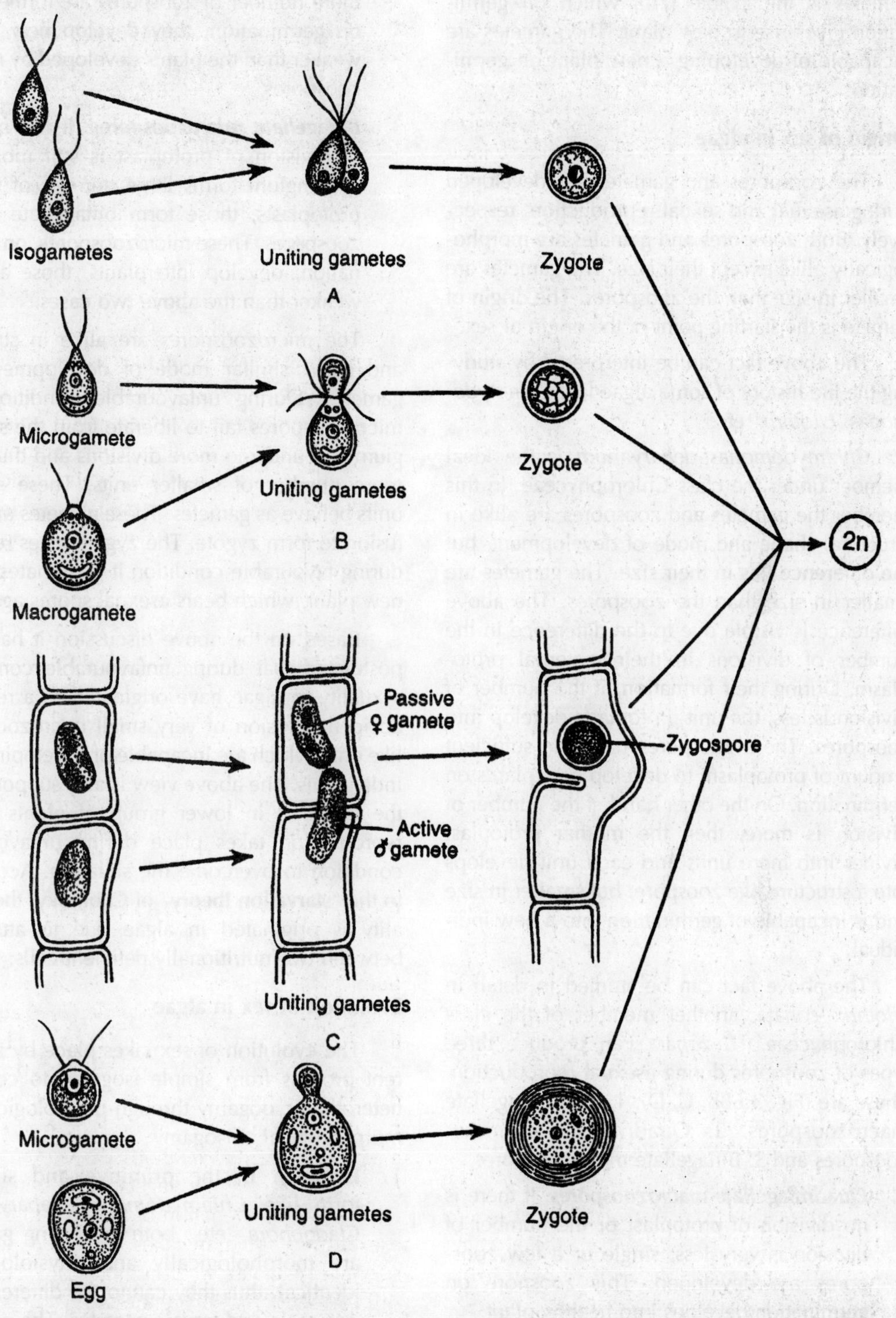

Fig. 3.18: Types of sexual reproduction in algae: A. Isogamy in *Chlamydomonas* sp., B. Anisogamy in *Ectocarpus* sp. C. Physiological anisogamy in *Spirogyra* sp., and D. Oogamy in *Chlamydomonas* sp.

gametes is the zygote (2n), which on germination gives rise to new plant. The gametes are incapable of developing a new plant on germination.

Origin of sex in algae

The zoospores and gametes are developed during asexual and sexual reproduction, respectively. Both zoospores and gametes are morphologically alike except their size. The gametes are smaller in size than the zoospores. The origin of gamete is the starting point of the origin of sex.

The above fact can be interpreted by studying the life history of some algae like *Chlamydomonas, Ulothrix* etc.

Chlamydomonas debaryanum is the ideal member under the class Chlorophyceae. In this member the gametes and zoospores are alike in structure, shape and mode of development, but the difference lies in their size. The gametes are smaller in size than the zoospores. The above difference is visible due to the difference in the number of divisions in their maternal protoplasm. During their formation, if the number of division is less, the unit protoplasts develop into zoospores. These zoospores have the sufficient amount of protoplasm to develop new plants on germination. On the other hand, if the number of division is more, then the mother protoplast divides into more units and each unit develops into a structure like zoospore, but smaller in size and is incapable of germination into a new individual.

The above fact can be studied in detail in *Ulothrix zonata*, another member of the class Chlorophyceae. *U. zonata* can produce three types of zoospores during asexual reproduction. These are (Fig. 3.68B, C, D): 1. Quadriflagellate macrozoospores, 2. Quadriflagellate microzoospores and 3. Biflagellate microzoospores.

1. ***Quadriflagellate macrozoospores***. If there is no division of protoplast or the number of division is very less, single or a few zoospores are developed. This zoospore on germination develops into healthy plant.
2. ***Quadriflagellate microzoospores***. If the number of division of protoplast is more, more number of zoospores are formed and, on germination, they develop new plants, weaker than the plants developed by macrozoospores.
3. ***Biflagellate microzoospores***. If the number of division of protoplast is still more, the sporangium forms large number of unit of protoplasts, those form biflagellate microzoospores. These microzoospores, on germination, develop into plants, those are still weaker than the above two cases.

The microzoospores are alike in structure and show similar mode of development like gametes. During unfavourable condition, the microzoospores fail to liberate from the sporangium and undergo more divisions and thus form more number of smaller units. These smaller units behave as gametes. These gametes undergo fusion to form zygote. The zygote takes rest and during favourable condition it germinates into a new plant, which bears asexual spores again.

Based on the above discussion it has been postulated that during unfavourable condition, sexuality in algae have originated as a result of accidental fusion of very small microzoospore-like units which are incapable of developing new individuals. The above view is also supported by the fact that in lower group of plants sexual reproduction takes place during unfavourable condition to overcome the situation. According to the "**starvation theory**" of Cholnoky, the sexuality is originated in algae due to attraction between two nutritionally deficient cells.

Evolution of sex in algae

The evolution of sex takes place by a different process from simple isogamy to complex heterothallic oogamy through physiological and morphological anisogamy.

1. **Isogamy**. In the primitive and simplest form like *Chlamydomonas debaryanum, Cladophora* etc., both the fusing gametes are morphologically and physiologically identical, thus they cannot be differentiated into male and female gametes. The gametes are called isogametes and the process is called isogamy (Fig. 3.18A).

2. **Physiological anisogamy**. In some algae, the gametes are morphologically alike, but differ in their physiological behaviour.

   (a) In *Ulothrix*, the gametes thus produced are morphologically identical, but the fusion takes place between gametes originating from the different filaments indicate the difference in their physiological characteristics and can be designated as + and – gametes.

   (b) In *Spirogyra*, the gametes are non-motile and identical in shape and size; those develop singly within the cell. At the time of conjugation the two filaments come very close to each other and some of the cells are connected by conjugation tube. Out of the two fusing gametes one becomes passive and remain within the cell and behave as female gamete. On the other hand, other gamete though non-motile becomes active and passes to the female through conjugation tube and behave as male gamete (Fig. 3.18C).

   Thus, though the gametes are morphologically identical, they show difference in their behaviour i.e., the ***physiological anisogamy***.

3. **Anisogamy**. In *Ectocarpus, Pandorina, Clodium* and *Chlamydomonas braunii*, the anisogamy is directly visible. Here both the gametes are ciliated i.e., motile, but unequal in size. The gametes are called anisogametes. The smaller one is called microgamete which behaves as male and the larger one is called macrogamete which behaves as female. The micro- and macrogamete are produced within the micro- and macrogametangium, respectively (Fig. 3.18B).

4. **Oogamy**. This type of sexual union is visible in *Chlamydomonas oögamum, C. coccifera* etc. Here male and female gametes are produced within antheridia and oogonia, respectively. The smaller one is active and called male gamete or antherozoid but the relatively larger one is inactive and called female gamete or egg (Fig. 3.18D).

   Usually single egg is formed within oogonium except in *Fucus* and *Sphaeroplea*. Many male gametes are formed within the antheridium.

   In Phaeophyceae, both male and female gametes are discharged from the antheridium and oogonium, respectively, and their union occurs in water. This type is called ***primitive oogamy***.

   In *Oedogonium*, the male gametes i.e., antherozoids, are smaller, flagellated and develop in pair within unicellular antheridium, but the female gamete i.e., egg, develops singly within oogonium. The fertilization takes place within oogonium.

   In *Chara*, the sex organs are further specialised. The round male sex organ is the globule containing huge number of antherozoids and the more or less oval, much protected structure is called nucule containing only one egg. The protection of egg and zygote is much more, indicating an advanced characteristic.

   In *Fucus*, separate male, female and mixed conceptacles are formed on receptacles. Out of eight (8) eggs developed in oogonium, seven (7) degenerate. Till now all the species are homothallic.

   The ***evolution of sex reaches its climax*** in the heterothallic species of Rhodophyceae. Spermatia, the male gametes, are non-motile and developed singly in spermatangium, those are carried by water current to the trichogyne, the receptive region of the female sex organ — the carpogonium.

   In *Polysiphonia* and *Oedogonium*, out of four tetraspores or zoospores developed (by meiosis) from tetrasporangia or directly from zygote, two produce female plants and other two male plants.

   From the above discussion a progressive monophyletic line of evolution can be traced from Isogamy to heterothallic oogamy through physiological anisogamy, morphological anisogamy and homothallic oogamy.

GENERAL CONCEPT OF ENDOSYMBIOSIS

Endosymbiosis is the phenomenon of the coexistence of living matter where one organism lives within the body or within the cells of another organism. This is facilitated by the fact that in most cases the relationship benefits both the organisms without a parasitic relationship where one feeds off the other. In brief, it can be mentioned for better understanding, that

1. **Endo** = within;
2. **Endocytosis** = (cyto = cell) a process of 'cell eating' – cells are engulfed, but then usually digested as food ....;
3. **Endosymbiosis** = cells are engulfed, but not digested ... cells live together is a mutually benefitting relationship, or symbiosis.

In eukaryotic cells, development of mitochondria and chloroplasts are an example of endosymbiosis. The mitochondria and chloroplasts have a structure that would have allowed their independent existence.

The endosymbiotic theory was proposed by Lynn Margulis in the 1960's and officially in her book "Symbiosis in cell Evolution" in 1981. Margulis have done a lot of work on the origin of eukaryotic cells. She looked at all the data about prokaryotes, eukaryotes, and organelles and proposed that the similarities between prokaryotes and organelles, together with their appearance in the fossil record, could best be explained by "**endosymbiosis**".

Her **endosymbiosis hypothesis** originally proposed the following:

(i) Mitochondria are the result of **endocytosis of aerobic bacteria**,

(ii) Chloroplasts are the result of **endocytosis of photosynthetic bacteria**,

(iii) In both cases by large **anaerobic bacteria** who would not otherwise be able to exist in an aerobic environment.

(iv) This arrangement became a mutually beneficial relationship for both cells (**symbiotic**).

Margulis' original hypothesis proposed that aerobic **bacteria** (those require oxygen) were ingested by **anaerobic bacteria** (grows in absence of oxygen), and may each have had a **survival advantage** as long as they continued their partnership.

Each would have performed **mutually benefiting functions** from their symbiotic relationship. The **aerobic bacteria** would have handled the toxic oxygen for the **anaerobic bacteria**, and the **anaerobic bacteria** would ingested food and protected the **aerobic "symbiote"**.

Results, "the formation of a cell with a double-membrane bound organelle". The inner lipid bilayer would have been the bacterial cell's plasma membrane, and the ouler lipid bilayer came from the cell that engulfed it (Fig. 3.19).

Other evidence that supports the endosymbiosis hypothesis

Additional evidences through the following line of studies give much support to establish the

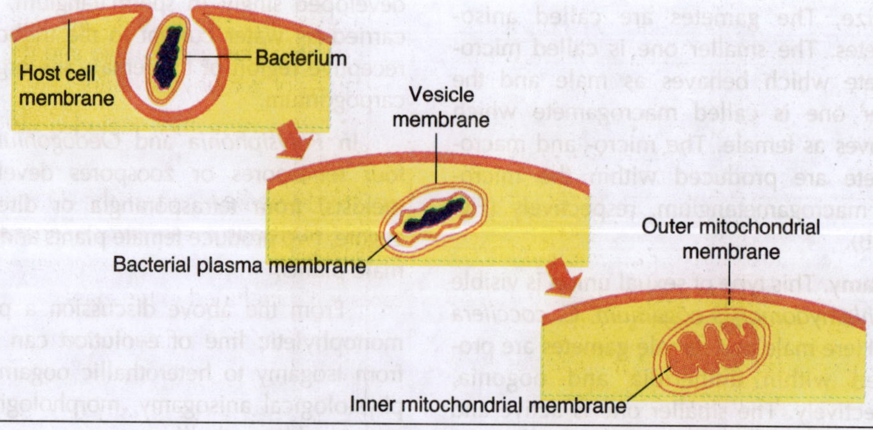

Fig. 3.19: Process of endosymbiosis

endosymbiotic theory. These are: (1) Timeline of life on the Earth and (2) Organelles have their own DNA, and divide independently of the cell they live in.

1. **Timeline of life on the Earth**

The timeline of anaerobic bacteria, photosynthetic bacteria, and aerobic cells on earth is strong evidence in support of the endosymbiotic hypothesis.

(i) **Anaerobic bacteria**: Fossil evidence of bacterial life on Earth recorded during ~3.8 billion years ago. During that time, the atmosphere of the Earth *did not contain oxygen*, and all life (bacterial cells) was **anaerobic**.

(ii) **Photosynthetic bacteria**: Fossil evidence of **photosynthetic bacteria**, or **cyanobacteria**, appears, about ~3.2 billion years ago. These bacteria use the solar energy to make sugar. Oxygen, released as a byproduct, began to accumulate in the atmosphere. However, oxygen is actually pretty toxic to cells, even our cells. Due to increase in the level of oxygen, **anaerobic** cells were now in a disadvantage position, in an oxygen-containing atmosphere, and started to die out.

(iii) **Aerobic** cells appear in the fossil record shortly after that (~2.5 billion years ago). Their cells were able to use that 'toxic' oxygen and convert it into energy (ATP) and water. Organisms that could thrive in an oxygen-containing atmosphere were now 'best suited to the environment'.

Years ago (In Billion)	Important events
4.5	Origin of the Earth.
3.5	Prokaryote Bacteria dominate.
2.5	Oxygen accumulates in atmosphere
1.5	Eukaryotes-First nucleated cell.
0.5	Cambrian explosion of multicellular Eukaryotes.

2. **Organelles have their own DNA, and divide independently in the cell they live in**

When Margulis initially proposed the Symbiotic Theory; predicted that, if the organelles were really bacterial (prokaryotic) symbionts, they would have their own DNA. If her theory did best explain the origin of eukaryotic cells, she reasoned, *organelles would have DNA that resembled bacterial DNA and be different from the cell's DNA (located in the nucleus membrane)*. Amazingly, in the 1980's this was proven to be the case for two classes of organelles, the mitochondria and chloroplasts.

The Endosymbiotic Theory provides the most plausible explanation for the development of organelles within the eukaryotic cell.

CLASSIFICATION

Before the 20th century the algae were classified into four major groups. Though the chlorophyll is commonly present in all groups, the differentiation was made on their pigments which dominate over chlorophyll and mask the green colour. The group and their differentiation were made as follows:

1. Chlorophyceae: Common name is green algae and the dominant pigments are Chlorophyll a and b.
2. Phaeophyceae: Common name is brown algae and the dominant pigment is fucoxanthin.
3. Rhodophyceae: Common name is red algae and the dominant pigment is r-phycoerythrin.
4. Cyanophyceae or Myxophyceae: Common name is blue green algae and the dominant pigment is c-phycocyanin.

Diatoms were then included under Phaeophyceae.

In the earliest part of the 20th century the class Xanthophyceae was separated from Chlorophyceae and certain pigmented flagellate types were included in the class.

Some important classifications are described below:

F. E. Fritsch (1935, 45) was the first who proposed a most comprehensive and authorative classification of algae in his book *"The Structure and Reproduction of the Algae"* based on the following characteristics:

(i) Pigmentation.

(ii) The metabolic or assimilatory food products (reserve food).

(iii) Types of flagella.

(iv) Thallus orgaization and

(v) Reproduction.

He classified the algae into eleven (11) classes. These are as follows:

1. Chlorophyceae	7. Chloromonadineae
2. Xanthophyceae	8. Euglenineae
3. Chrysophyceae	9. Phaeophyceae
4. Bacillariophyceae	10. Rhodophyceae,
5. Cryptophyceae	and
6. Dinophyceae	11. Myxophyceae

1. **Chlorophyceae** (Isokontae, green algae)

Important characteristics

(a) Cells are eukaryotic and cell wall is mainly composed of cellulose.

(b) Pigments. Presence of Chlorophyll a and b, carotenoid and xanthophylls.

(c) Chloroplast usually has pyrenoids.

(d) Reserve food is starch.

(e) Flagella if present are equal in length and are situated anteriorly. One is whiplash and other one is tinsel type.

(f) Reproduction takes place by vegetative, asexual and sexual methods. Sexual reproduction may be iso-, aniso- and oogamous type.

The class has been divided into **nine (9)** orders:

(i) **Volvocales**. e.g., *Chlamydomonas.*

(ii) **Chlorococcales**. e.g., *Chlorella, Hydrodictyon.*

(iii) **Ulotrichales**. e.g., *Ulothrix.*

(iv) **Cladophorales**. e.g., *Cladophora, Pithophora.*

(v) **Chaetophorales**. e.g., *Ulvella, Coleochaete, Trentepohlia.*

(vi) **Oedogoniales**. e.g., *Oedogonium.*

(vii) **Conjugales**. e.g., *Spirogyra, Zygnema.*

(viii) **Siphonales**. e.g., *Vaucheria.*

(ix) **Charales**. e.g., *Chara, Nitella.*

2. **Xanthophyceae** (Heterokontae, yellow green algae)

Important characteristics

(a) Cells are eukaryotic and cell wall is composed of pectic substance.

(b) The main pigments are Chlorophyll a and e, β-carotene and xanthophylls.

(c) Chloroplasts are without pyrenoid.

(d) Reserve food is oil.

(e) Flagella are two, unequal in length and are situated anteriorly. The longer one tinsel and shorter one is whiplash type.

(f) Reproduction. It takes place by all the three means — vegetative, asexual and sexual (isogamous type).

The class has been divided into **four (4)** orders:

(i) **Heterochloridales** e.g, *Heterochloris.*

(ii) **Heterococcales** e.g., *Myxochloris.*

(iii) **Heterotrichales** e.g., *Microspora.*

(iv) **Heterosiphonales** e.g., *Botrydium.*

3. **Chrysophyceae**

Important characteristics

(a) Cells are eukaryotic and cell wall is silicified or calcified (cellulose absent).

(b) Pigments. The dominating pigment is phycochrysin.

(c) Chromatophores have naked pyrenoid-like bodies.

(d) The reserve food is crysolaminarin and lucosin.

(e) The motile cells have two flagella (either equal or unequal) inserted anteriorly.

(f) Reproduction. It takes place by vegetative and sexual means. Sexual reproduction if present is of isogamous type.

The class has been divided into **three (3)** orders:

(i) **Chrysomonadales** e.g., *Crysodendron.*

(ii) **Chrysosphaerales** e.g., *Chrysosphaera.*

(iii) **Chrysotrichales** e.g., *Chrysoclonium.*

4. **Bacillariophyceae** (Diatom, yellow or golden brown algae)

Important characteristics

(a) The cells are eukaryotic and consist of two halves, epitheca and hypotheca; which are bilaterally or radially symmetrical.

(b) Cell wall is pectic and silicified, and ornamented variously.

(c) Pigments. The dominant pigments are fucoxanthin, diatoxanthin and diadinoxanthin.

(d) Chromatophore has pyrenoids.

(e) Reserve food is fat and volutin.

(f) Motile cells have single flagellum.

(g) Reproduction takes place by cell division and auxospore formation.

The class has been divided into **two (2)** orders:

(i) **Centrales** e.g., *Chaetoceras, Cyclotella.*

(ii) **Pennales** e.g., *Pinnularia, Denticula.*

5. **Cryptophyceae**

Important characteristics

(a) The cells are eukaryotic in nature.

(b) Pigments. The dominant pigments are xanthophylls – give red or brown colour.

(c) Pyrenoid-like bodies are present but they are independent of chromatophores.

(d) Reserve food is starch and/or oil.

(e) Sexual reproduction is rare and isogamous type only.

The class has been divided into **two (2)** orders:

(i) **Cryptomonadales** e.g., *Cryptomonas, Cyanomonas.*

(ii) **Cryptococcales** e.g., *Tetragonidium.*

6. **Dinophyceae**

Important characteristics

(a) The cells are eukaryotic in nature.

(b) Pigments. The dominant pigments are xanthophylls which give red or brown colouration.

(c) Cells contain many discoid chromatophores.

(d) Reserve food is starch and fat.

(e) Most members are unicellular and motile with two flagella.

(f) Some members show holozoic mode of nutrition.

(g) Sexual reproduction is rare and isogamous type only.

The class has been divided into **six (6)** orders:

(i) **Desmomonadales** e.g., *Desmocapsa, Desmomastrix.*

(ii) **Thecatales** e.g., *Prorocentrum, Exuviaella.*

(iii) **Dinophysiales** e.g., *Dinophysis, Phalacroma.*

(iv) **Dinoflagellata** e.g., *Elastodinium, Heterocapsa.*

(v) **Dinococcales** e.g., *Dinastridium, Dissodinium.*

(vi) **Dinotrichales** e.g., *Dinothrix, Dinoclonium.*

7. **Chloromonadineae**

Important characteristics

(a) Cells are eukaryotic in nature.

(a) Pigments. Dominant pigment is xanthophylls, present in excess amount.

(c) Chromatophore. Cells have numerous discoid chromatophores without pyrenoid.

(d) Reserve food is fat and oil.

(e) Flagella. Motile forms are with two equal flagella.

(f) Reproduction. It takes place by cell division. **Sexual reproduction is absent.**

The class has only **one (1)** order:

Chloromonadales e.g., *Vacuolaria, Trentonia.*

8. **Euglenoineae**

Important characteristics

(a) Plant body is unicellular and eukaryotic in nature.

(b) The dominant pigment is chlorophyll.

(c) Cells have many chromatophores.

(d) Pyrenoid-like bodies are present in some members.

(d) Reserve food is paramylon, a polysaccharide.

(f) Flagella. Most of them have one or two flagella.

(g) They possess complex vacuolar system.

(h) Reproduction. It takes place by cell division. Sexual reproduction if present is of isogamous type.

The class has been divided into **three (3)** families:

(i) **Euglenaceae** e.g., *Euglena, Ascoglena.*

(ii) **Astasiaceae** e.g., *Astasia, Menoidium.*

(iii) **Peranemaceae** e.g., *Petalomonas, Anisonema.*

9. **Phaeophyceae** (brown algae)

Important characteristics

(a) Most of the members are marine.

(b) Cell wall is composed of cellulose along with alginic and fucinic acid.

(c) Pigments. Fucoxanthin is the dominant pigment, in addition to chlorophyll a, c and carotene.

(d) Reserve food is laminarin (polysaccharide), manitol (alcohol) and fats.

(e) Some lower members have naked pyrenoid like bodies.

(f) Zoospores are flagellated, one of which is tinsel type.

(g) Sexual reproduction ranges from isogamy to oogamy.

The class has been divided into **nine (9)** orders:

(i) **Ectocarpales** e.g., *Ectocarpus, Punctaria.*

(ii) **Tilopteridales** e.g., *Tilopteris.*

(iii) **Cuteriales** e.g., *Cutleria.*

(iv) **Sporochnales** e.g., *Carpomitra, Sporochnus.*

(v) **Desmarestiales** e.g., *Desmarestia.*

(vi) **Laminariales** e.g., *Laminaria, Macrocystis.*

(vii) **Sphacelariales** e.g., *Sphacelaria.*

(vii) **Dictyotales** e.g., *Dictyota, Padina.*

(ix) **Fucales** e.g., *Fucus, Sargassum.*

10. **Rhodophyceae (Rod algae)**

Important characteristics

(a) Plant body is multicellular (either uniaxial or multiaxial) and cells are eukaryotic in nature.

(b) Cell wall is composed of outer pectic and inner cellulose.

(c) Chromatophores are present. Pyrenoid-like bodies are present in lower forms.

(d) The dominant pigment is r-phycoerythrin ($C_{34}H_{46}O_8N_4$) which gives red colour to the algae. In addition, other pigments like r-phycocyanin, chlorophyll a, d, carotene and xanthophylls are also present.

(e) Storage food is floridean starch, a polysaccharide.

(f) Reproductive cells are non-motile. Sexual reproduction is of advanced oogamous type.

The class has been divided into **seven (7)** orders:

(i) **Bangiales** e.g., *Bangia, Porphyra.*

(ii) **Nemalionales** e.g., *Nemalion, Naccaria.*

(iii) **Gelidiales** e.g., *Gelidium.*

(iv) **Cryptonemiales** e.g., *Gloeopeltis, Dumontia.*

(v) **Gigartinales** e.g., *Gigartina, Gracilaria.*

(vi) **Rhodymeniales** e.g., *Rhodymenia, Champia.*

(vii) **Ceramiales** e.g., *Polysiphonia, Ceramium.*

11. **Myxophyceae** (Cyanophyceae, blue green algae)

Important characteristics

(a) They are mostly fresh water.

(b) Plant body may be unicellular or multicellular and the **cells are prokaryotic** in nature.

(c) The cell wall is composed of mucopeptide, together with amino acids, fatty acids and carbohydrate.

(d) The dominant pigment is c-phycocyanin which gives blue green colouration to the cells. In addition, other pigments like chlorophyll a, β-carotene and c-phycoerythrin are also present.

(e) The reserve food is cyanophycean starch and cyanophycean granule (protein).

(f) There is no motile stage in the members of this class.

(g) Reproduction takes place by both vegetative (cell division, fragmentation etc.) and asexual means (exospore, endospore, heterocysts etc.). **Sexual reproduction is absent** (genetic recombination is reported in a few members).

The class has been divided into **five (5)** orders:

(i) **Chroococcales** e.g., *Gloeocapsa, Chroococcus.*

(ii) **Chamaesiphonales** e.g., *Chamaesiphon.*

(iii) **Pleurocapsales** e.g., *Pleurocapsa.*

(iv) **Nostocales** e.g., Nostoc, *Oscillatoria.*

(v) **Stigonematales** e.g., *Nostochopsis, Stigonema.*

G. M. Smith (1955) proposed the classification of algae, based on the physiological characters of vegetative cells and the morphology of motile reproductive cell. He divided the algae into seven (7) divisions. The outline of the classification is:

1. Division. CHLOROPHYTA

   (a) Class. **Chlorophyceae** e.g., *Volvox, Oedogonium.*

   (b) Class. **Charophyceae** e.g., *Chara.*

2. Division. EUGLENOPHYTA

   (a) Class. **Euglenophyceae** e.g., *Euglena, Colacium.*

3. Division. PYRROPHYTA

   (a) Class. **Desmophyceae** e.g., *Exuviaella.*

   (b) Class. **Dinophyceae** e.g., *Dinastridium.*

4. Division. CHRYSOPHYTA

   (a) Class. **Chrysophyceae** (golden brown algae) e.g., *Chromulina, Chrysoclonium.*

   (b) Class. **Xanthophyceae** (yellow-green algae) e.g., *Botrydium.*

   (c) Class. **Bacillariophyceae** (diatom) e.g., *Pinnularia, Cyclotella.*

5. Division. PHAEOPHYTA (brown algae)

   (a) Class. **Isogeneratae** e.g., *Dictyota, Ectocarpus.*

   (b) Class. **Heterogeneratae** e.g., *Laminaria, Myrionema.*

6. Division. CYANOPHYTA (blue-green algae)

   (a) Class. **Cyanophyceae** or **Myxophyceae**. e.g., *Nostoc, Anabaena, Gloeocapsa, Gloeotrichia* etc.

7. Division. RHODOPHYTA (red algae)

   (a) Class. **Rhodophyceae** e.g., *Polysiphonia, Batrachospermum.*

Papenfuss. G. F. (1955) classified algae based on the phylogeny into eight (8) phyla and twelve (12) classes. A brief outline of the classification is:

1. Phylum. Chlorophycophyta

   (a) Class. **Chlorophyceae**

2. Phylum. Charophycophyta

   (a) Class. **Charophyceae**

3. Phylum. Euglenophycophyta

   (a) Class. **Euglenophyceae**

4. Phylum. Chrysophycophyta

   (a) Class. **Chrysophyceae**

   (b) Class. **Xanthophyceae**

   (c) Class. **Bacillariophyceae**

5. Phylum. Pyrrophycophyta

   (a) Class. **Cryptophyceae**

(b) Class. **Chloromonadophyceae**

(c) Class. **Dinophyceae**

6. Phylum. Phaeophycophyta

(a) Class. **Phaeophyceae**

7. Phylum. Rhodophycophyta

(a) Class. **Rhodophyceae**

8. Phylum. Schizophycophyta

(a) Class. **Schizophyceae.**

Chapman. V. J. (1962) classified the algae into four (4) phyla and ten (10) classes, based on the morphology, pigmentation and biochemical characteristics.

1. Phylum. Euphycophyta

(a) Class. **Charophyceae**

(b) Class. **Chlorophyceae**

(c) Class. **Phaeophyceae**

(d) Class. **Rhodophyceae**

2. Phylum. Myxomycophyta

(a) Class. **Myxophyceae** (Cyanophyceae)

3. Phylum. Chrysophycophyta

(a) Class. **Chrysophyceae**

(b) Class. **Xanthophyceae**

(c) Class. **Bacillariophyceae**

4. Phylum. Pyrrhophycophyta

(a) Class. **Cryptophyceae**

(b) Class. **Dinophyceae**

Prescott. G. W. (1969) classified algae based on the following characteristics, such as photosynthetic pigments, photosynthetic products, flagella, cell wall, nature of nuclear material, life cycle and reproduction. He presented the classification in his book 'The Algae', where he classified the algae into nine (9) phyla. These are:

1. Phylum. Chlorophyta (green algae),
2. Phylum. Euglenophyta,
3. Phylum. Chrysophyta (yellow-green algae),
4. Phylum. Pyrrhophyta,
5. Phylum. Phaeophyta (brown algae),
6. Phylum. Rhodophyta (red algae),
7. Phylum. Cyanophyta (blue-green algae),
8. Phylum. Cryptophyta and
9. Phylum. Chloromonadophyta.

Round. F. E. (1973) classified algae into two groups:

1. ***Prokaryota*** (the cells are prokaryotic in nature i.e., absence of nuclear membrane and nucleolus; and absence of membrane bound organelles), and
2. ***Eukaryota*** (the cells are eukaryotic in nature).

The outline of the classification is given below:

Groups:

A. **Prokaryota**

1. Phylum. Cyanophyta

B. **Eukaryota**

It is divided into twelve (12) Phylla.

1.	Euglenophyta	7.	Dinophyta
2.	Chlorophyta	8.	Bacillariophyta
3.	Charophyta	9.	Chrysophyta
4.	Prasinophyta	10.	Phaeophyta
5.	Xanthophyta	11.	Rhodophyta
6.	Haptophyta	12.	Cryptophyta

Chapman. V. J. and **Chapman. D. J. (1973)** also classified algae into two main groups — **Prokaryota,** which includes the division Cyanophyta, and **Eukaryota** which includes ten divisions.

The outline of their classification is given:

Groups:

Prokaryota

1. Division. CYANOPHYTA

1. Class. **Cyanophyceae**

Eukaryota

1. Division. RHODOPHYTA

1. Class. **Rhodophyceae**

2. Division. CHLOROPHYTA

1. Class. **Chlorophyceae**
2. Class. **Prasinophyceae**
3. Class. **Charophyceae**

3. Division. **EUGLENOPHYTA**
   1. Class. **Euglenophyceae**
4. Division. **CHLOROMONADOPHYTA**
   1. Class. **Chloromonadophyceae**
5. Division. **XANTHOPHYTA**
   1. Class. **Xanthophyceae**
6. Division. **BACILLARIOPHYTA**
   1. Class. **Bacillariophyceae**
7. Division. **CHRYSOPHYTA**
   1. Class. **Chrysophyceae**
   2. Class. **Haptophyceae**
8. Division. **PHAEOPHYTA**
   1. Class. **Phaeophyceae**
9. Division. **PYRRHOPHYTA**
   1. Class. **Dinophyceae**
   2. Class. **Desmophyceae**
10. Division. **CRYPTOPHYTA**
    1. Class. **Cryptophyceae**

H. C. Bold and **M. J. Wynne (1978)** classified the algae into nine (9) divisions. They followed the Papenfuss (1946) idea of using the word 'phyco' before 'phyta' in the classification.

The outline of their classification is given below:

1. Division. **CYANOCHLORONTA**
2. Division. **CHLOROPHYCOPHYTA**
3. Division. **CHAROPHYTA**
4. Division. **EUGLENOPHYCOPHYTA**
5. Division. **PHAEOPHYCOPHYTA**
6. Division. **CHRYSOPHYCOPHYTA**
7. Division. **PYRRHOPHYCOPHYTA**
8. Division. **CRYPTOPHYCOPHYTA**
9. Division. **RHODOPHYCOPHYTA**

Robert Edward Lee (1989) classified the algae into 4 evolutionary groups, 15 divisions and classes (*"Phycology"* by R. E. Lee, 1989, 2nd Edn.)

The classification is given below:

Group 1. Prokaryotic algae

1. Division. **CYANOPHYTA** (blue green algae): Chlorophyll a; phycobiliproteins.
2. Division. **PROCHLOROPHYTA** (prochlorophytes): Chlorophyll a and b; no phycobiliproteins.

Group 2. Eukaryotic algae with chloroplast surrounded only by the two membranes of the chloroplast envelope.

1. Division. **GLAUCOPHYTA**: Algae that represent an intermediate position in the evolution of chloroplast; photosynthesis is carried out by modified endosymbiotic blue-green algae.
2. Division. **RHODOPHYTA** (red algae): Chlorophyll a and d; phycobiliproteins; no flagellated cells; storage product is floridean starch.
3. Division. **CHLOROPHYTA** (green algae): Chlorophyll a and b; storage product starch is found inside the chloroplast.

Group 3. Eukaryotic algae with chloroplast surrounded only by one membrane of chloroplast endoplasmic reticulum.

1. Division. **EUGLENOPHYTA** (euglenoid): Chlorophylls a and b, one flagellum with a spiralled row of fibrillar hairs; proteinaceous pellicle in strips under the plasma membrane; storage product is paramylon; characteristic type of cell division.
2. Division **DINOPHYTA** (dinoflagellates): Mesokaryotic nucleus; chlorophyll a and c_1; cell commonly divides into an epicone and a hypocone by a girdle; helical transverse flagellum; thecal plates in vesicles under the plasma membrane.

Group 4. Eukaryotic algae with chloroplast surrounded by two membranes of chloroplast endoplasmic reticulum.

1. Division. **CRYPTOPHYTA** (cryptophytes): Nucleomorph present between inner and outer membrane of chloroplast endoplasmic reticulum; starch is formed in grains between inner membrane of chloroplast endoplasmic reticulum and chloroplast envelop; chlorophyll a and c; phycobiliproteins; periplast is found inside plasma membrane.
2. Division. **CRYSOPHYTA** (golden-brown algae): Chlorophyll a and c_1 and sometimes c_2; fucoxanthin; anterior tinsel and posterior

whiplash flagella; storage product chrysolaminarin is found in vesicles in cytoplasm.

3. Division. **PRYMNESIOPHYTA** (haptophytes): Haptonema, chlorophyll a, c_1 and c_2, fucoxanthin; two whiplash flagella; scales common outside cell; storage product chrysolaminarin is found in vesicles in cytoplasm.
4. Division. **BACILLARIOPHYTA** (diatom): Silica in cell walls; chlorophyll a, c_1 and c_2; fucoxanthin; one posterior tinsel flagellum on sperm; storage product chrysolaminarin is found in vesicles in cytoplasm.
5. Division. **XANTHOPHYTA** (yellow green algae): Eyespot in chloroplast; chlorophyll a and c; anterior tinsel flagellum and posterior whiplash flagellum.
6. Division. **EUSTIGMATOPHYTA**: Eyespot large and outside chloroplast; chlorophyll a; posterior whiplash flagellum (commonly reduced) and anterior tinsel flagellum.
7. Division. **RAPHIDOPHYTA** (chloromonads): Chlorophyll a and c; anterior tinsel flagellum and posterior whiplash flagellum.
8. Division. **PHAEOPHYTA** (brown algae): Unilocular and plurilocular sporangia, chlorophyll a, c_1 and c_2; fucoxanthin; storage product chrysolaminarin is found in vesicles in cytoplasm; anterior tinsel flagellum and posterior whiplash flagellum.

A brief outline of the classification up to class (with characteristics) is given below:

Group 1. Prokaryotic algae. The prokaryotic algae have an outer plasma membrane enclosing protoplasm containing photosynthetic thylakoids, 70S ribosomes and DNA fibrils not enclosed within a separate membrane. Chlorophyll a, is the main photosynthetic pigment.

1. Division. **CYANOPHYTA** (blue green algae). Presence of chlorophyll a and phycobiliprotein, e.g., *Anabaena, Gloeotrichia* etc.
2. Division. **PROCHLOROPHYTA** (prochlorophytes). Presence of chlorophyll a and b, phycobiliproteins absent, e.g., *Prochloron*.

Group 2. Eukaryotic algae with chloroplast surrounded only by the two membranes of the chloroplast envelope.

1. Division. **GLAUCOPHYTA**

The algae have endosymbiotic blue-green algae (BGA) in the cytoplasm instead of chloroplasts. Pasteur (1914) coined the terms *Cyanelles* to endosymbiotic BGA, *Cyanone* for host and *Syncyanosis* for the association. Chlorophyll a and phycobiliproteins are present like BGA but BGA like carotenoids, myxoxanthophyll and echinenone are absent e.g., *Cyanophora paradosa, Glaucocystis* etc.

2. Division. **RHODOPHYTA** (red algae)

The Division Rhodophyta has a single class Rhodophyceae.

Class. **Rhodophyceae:**

1. The plants are generally of marine habitat called sea weeds, grows in various depth. (Out of about 4000 species, about 200 species are fresh water.)
2. Pit connections between the cells in all of the orders except Porphyridales, Erythropeltidales and haploid phase of the Bangiales.
3. Chloroplast with one thylakoid per band and no chloroplast endoplasmic reticulum.
4. Presence of Chlorophyll a and d, Phycobiliproteins includes r-phycocyanin, allophycocyanin and three forms of phycoerythrins (b, r and c).
5. Floridean starch as a storage product.
6. Lack of flagellate cells.

The class Rhodophyceae is divided into 12 orders:

1. Porphyridales. e.g., *Porphyridium, Rhodosorus* etc.
2. Rhodochaetales e.g., *Rhodochaete.*
3. Acrochaetiales. e.g., *Audouinella, Rhodochorton* etc.
4. Bangiales. e.g., *Bangia, Porphyra* etc.
5. Erythropeltidales. e.g., *Erythrocladia, Porphyropsis* etc.
6. Lemaneales. e.g., *Batrachospermum, Lemanea* etc.
7. Nemaliales. e.g., *Nemalion, Liagora* etc.

8. Gelidiales. e.g., *Harveyella, Ceratocolax* etc.
9. Gigertinales. e.g., *Gigertina, Chondrus,* etc.
10. Rhodymeniales e.g., *Rhodymenia* etc.
11. Palmariales. e.g., *Palmaria* etc.
12. Ceramiales e.g., *Delesseria, Phycodrys* etc.

3. Division. **CHLOROPHYTA** (green algae)
It is divided into 4 classes:

A. Class. **Micromonadophyceae**:

Important characteristics

1. Scaly or naked flagellates with interzonal spindles those are persistent during cytokinesis,
2. Primitive green algae, some of which gave rise to other classes in the Chlorophyta.

B. Class. **Charophyceae** (flagella are attached in a lateral position of the cell).

Important characteristics

1. Motile cells asymmetrical, two flagella are attached in a lateral position of the cell.
2. Flagellar root consist of a broad band of microtubules and a second smaller microtubular root.
3. Multilayered structure (MLS) may be present; rhizoplast absent.
4. Scales common outside the motile cells.
5. Persistant interzonal mitotic spindle in telophase.
6. Phragmoplast produces new cross-wall after cell division.
7. Eye spots usually not present.
8. Sexual reproduction involves the formation of a dominant zygote. Meiosis occurs during zygote germination.

The class Charophyceae is further divided into 4 orders:

1. Klebsormidiales. e.g., *Stichococcus, Raphidonema, Klebsormidium* etc.
2. Zygnematales. e.g., *Spirogyra, Zygnema, Cosmerium, Cylindrocystis, Mesotaenium* etc.
3. Coleochaetales e.g., *Coleochaete.*
4. Charales e.g., *Chara, Nitella.*

C. Class. **Ulvophyceae** (flagella are attached at the anterior end of the cell).

Important characteristics

1. Plants are predominantly marine.
2. Flagella are attached at anterior end of cell.
3. Externally, the motile cells have near-radial symmetry.
4. Flagella roots consist of 4 cruciately arranged microtubular roots and sometimes a rhizoplast.
5. Multilayered structure (MLS) is absent.
6. Scales may be present on motile cells.
7. Presence of persistent interzonal spindle in telophase.
8. Eyespots are common.
9. Dormant zygote is absent.
10. Alternation of generation is common.

Common genus: *Ulva.*

D. Class. **Chlorophyceae** (flagella are attached at the anterior end of the cell).

Important characteristics

1. Plants are predominantly fresh water.
2. Motile cells are with radial or near-radial external symmetry.
3. Flagella are attached at the anterior end of cell.
4. Flagella roots consists of 4 cruciately arranged microtubular roots and sometimes a rhizoplast.
5. Multilayered structure (MLS) is absent.
6. Theca is common in motile cells.
7. In telophase, the interzonal spindle collapses; phycoplast produces the new cross wall in cell division.
8. Eyespots are common.
9. Zygote undergoes a dormant period; meiosis occurs when the zygote germinates.

The class Chlorophyceae is further divided into 8 orders:

1. Volvocales. e.g., *Chlamydomonas, Volvox.*
2. Tetrasporales. e.g., *Tetraspora, Palmella.*
3. Schizogoniales. e.g., *Parasiola.*
4. Chlorococcales. e.g., *Chlorococcum, Hydrodictyon, Scenedesmus, Chlorella.*
5. Sphaeropleales. e.g., *Microspora, Sphaeroplea.*
6. Chlorosarcinales e.g., *Chlorosarcina.*
7. Chaetophorales. e.g., *Stigeoclonium, Chaetophora.*
8. Oedogoniales. e.g., *Oedogonium, Oedocladium, Bulbochaete.*

Group 3. Eukaryotic algae with chloroplast surrounded by one membrane of chloroplast endoplasmic reticulum.

1. Division. EUGLENOPHYTA (euglenoid)

It consists of only one Class — Euglenaceae.

Class. **Euglenaceae**

Important characteristics

1. They are unicellular.
2. They have chlorophyll a and b.
3. One membrane of chloroplast and endoplasmic reticulum.
4. The nucleus is mesokaryotic in nature.
5. Flagella with fibrillar hairs in one row.
6. Sexual reproduction is absent.
7. The storage product is paramylon, formed in the cytoplasm.

Common genus: *Euglena.*

2. Division. DINOPHYTA (dinoflagellates)

It consists of only one Class — Dinophyceae.

Class. **Dinophyceae**

Important characteristics

1. A typical motile dinoflagellate consists of an epicone and hypocone divided by a transverse girdle or cingulum.
2. The epicone and hypocone are normally divided into a number of thecal plates.
3. The cells can be photosynthetic or colourless and heterotrophic.
4. Photosynthetic organisms have chloroplasts surrounded by one membrane of chloroplast endoplasmic reticulum; which is not continuous with the outer membrane of the nuclear envelop.
5. Chlorophyll a and c_2 are present in the chloroplasts, with **peridinin** and **neoperidinin** being the main carotenoids.
6. Presence of pyrenoid in chloroplast.
7. Storage product is starch.
8. Eyespot is present.
9. Nucleus is dinokaryotic or mesokaryotic in nature.

Common genera: *Oxyrrhis, Heterocapsa, Peridinium, Ceatium, Amphinidium* etc.

Group 4. Eukaryotic algae with chloroplast surrounded by two membranes of chloroplast endoplasmic reticulum.

1. Division. CRYPTOPHYTA (cryptophytes)

It consists of only one Class Cryptophyceae

Class. **Chryptophyceae**

Important characteristics

1. Plants are found in both marine and fresh water habitat.
2. The plant body is dorsiventral in shape, with the cells flattened in one plane.
3. Cells contain chlorophyll a and c_2.
4. A few members lack chloroplast and are heterotrophic in nature, but most have a single-lobed chloroplast with a central pyrenoid.
5. Phycobiliproteins occur inside the thylakoids of the choloroplasts.

Common genera: *Cyathomonas, Chilomonas* etc.

2. Division. CRYSOPHYTA (golden-brown algae)

It consists only of two Classes: Chrysophyceae and Synurophyceae.

Class. **Chrysophyceae**

Important characteristics

1. Presence of chlorophyll c_1 and c_2.
2. Flagella inserted in the cell perpendicular to one another.
3. One photoreceptor as a swelling on the short flagellum.
4. Eyespot commonly present.
5. Choloroplast endoplasmic reticulum present.
6. Radially or bilaterally symmetrical silica scales may occur.
7. Contractile vacuole anterior.

Common genera: *Pedinella, Phaeaster* etc.

Class. **Synurophyceae**

Important characteristics

1. Presence of chlorophyll c_1.
2. Flagella inserted into the cell approximately parallel to each other.
3. Presence of two photoreceptors, one on each flagellum.
4. Eyespot absent.
5. Chloroplast endoplasmic reticulum usually absent.
6. Plant body bilaterally symmetrical.
7. Silica scales are common.
8. Contractile vacuole is posterior.

Common genera: *Dictyocta, Synura* etc.

3. Division. **PRYMNESIOPHYTA** (haptophytes)

It consists of only one Class — Prymnesiophyceae.

Class. **Prymnesiophyceae**

Important characteristics

1. It is a group of unicellular flagellates characterised by the possession of a haptonema between two smooth flagella.
2. Chloroplasts have chlorophyll a, c_1 and c_2 with fucoxanthin as the major carotenoid.
3. Diadinoxanthin and 19′ – hexanoyl fucoxanthin also occur in the group.
4. Two membranes of the chloroplast endoplasmic reticulum are present.
5. The anterior end of the cell has a large golgi apparatus and sometimes contractile vacuole.
6. The cells are commonly covered with scales.

Common genera: *Prymnesium, Diacronema* etc.

4. Division. **BACILLARIOPHYTA** (diatom)

It consists of only one Class — Bacillariophyceae.

Class. **Bacillariophyceae**

Important characteristics

1. Plant body is unicellular and free-living. Sometimes they are present in colony.
2. Depending on the mode of nutrition they may be **photosynthetic autotrophs, photosynthetic symbionts** or **colourless heterotrophs.**
3. Vegetative cells are diploid (2n).
4. The cells are surrounded by a rigid two-part box-like cell wall (upper part is epitheca and lower part is hypotheca) composed of silica called the frustule.
5. The cells are bilaterally or radially symmetrical.
6. The chloroplast contains chlorophylls a, c_1 and c_2 with the major carotenoid being the golden-brown fucoxanthin.
7. Reserve food is fat and volutin.
8. Motile cells have single flagellum.
9. Reproduction takes place by cell division and auxospore formation.

Common genera: *Cyclotella, Biddulphia, Rutilaria* etc.

5. Division. **XANTHOPHYTA** (yellow green algae)

It consists of only one Class. Xanthophyceae

Class. **Xanthophyceae**

Important characteristics

1. Cells are eukaryotic and cell wall is composed of pectic substance.

Table 3.1: Important features of some major classes of algae

Characteristics	Chlorophyceae (Green)	Bacillariophyceae (Diatom)	Xanthophyceae (Yellow green)	Phaeophyceae (Brown)	Rhodophyceae (Red)	Cyanophyceae (Blue green)
1. Habitat	Commonly fresh-water, some are marine or terrestrial.	Like Chlorophyceae	Mostly fresh-water, a few are marine.	Mostly marine except four genera.	Mostly marine	Like Chlorophyceae
2. Plant body	Different types: unicellular, coenobial, colonial, filamentous, siphonaceous or thalloid.	Unicellular (free or in colony)	Unicellular, filamentous, siphonaceous etc.	Multicellular, filamentous and branched.	Unicellular, filamentous or pseudoparenchymatous.	Unicellular or filamentous.
3. Cell structure	Eukaryotic	Eukaryotic	Eukaryotic	Eukaryotic	Eukaryotic	Prokaryotic
4. Cell wall	Mainly cellulose and pectic material.	Mainly silica and pectic material.	Cellulose	Inner layer cellulosic and outer layer consists of alginic and fusinic acid.	Inner layer cellulosic and outer pectic.	Mainly mucopeptide and muramic acid.
5. Flagella	Motile cells have 2, 4 or more flagella, equal in length, whiplash type (acronematic) and anteriorly inserted.	Motile cells have 1 or sometimes 2 tinsel (pantonematic) type and anteriorly inserted.	Motile forms have 2 flagella, unequal, one whiplash and other tinsel type.	Motile cells have 2 flagella, unequal in length, one tinsel and other whiplash type, laterally inserted.	Non-flagellate	Non-flagellate
6. Eye-spot	Attached on one side of the chloroplast and it has no relation with flagella.	Absent	Absent	Present in male gamete, attached with chromatophore, and related with flagella.	Absent	Absent
7. Green pigments	Chlorophylls a and b	Chlorophylls a and c	Chlorophylls a and e	Chlorophylls a and c	Chlorophylls a and d	Chlorophyll a

Contd. →

Contd. Table 3.1 →

Characteristics	Chlorophyceae (Green)	Bacillariophyceae (Diatom)	Xanthophyceae (Yellow green)	Phaeophyceae (Brown)	Rhodophyceae (Red)	Cyanophyceae (Blue green)
8. Other pigments	α and β-carotene and xanthophylls.	β-carotene, xanthophyll (fucoxanthin).	β-carotene and xanthophylls.	β-carotene, xanthophylls (fucoxantnin).	r-phycoerythrin, r-phycocyanin, α and β-carotene xanthophylls (lutein, teraxanthin etc.).	c-phycocyanin (blue), c-phycoerythrin (red), β-carotene, myxoxanthin, myxoxanthophyll.
9. Nature of nucleus	Nucleus well organised and chromosome formed during cell division.	Like Chlorophyceae	Like Chlorophyceae	Like Chlorophyceae	Like Chlorophyceae	Nuclear membrane and nucleolus absent. DNA is not associated with histones, thereby chromosomes are not formed.
10. Stored food	Starch and fat.	Oil, volutin, chrysolaminarin (leucosin).	Oil, lipid and lucosin.	Laminarin, mannitol and sucrose.	Floridean starch	Cyanophycean starch and cyanophycean protein.
11. Sexuality	Isogamy to oogamy	Iso, aniso- and oogamy.	Isogamy, rarely anisogamy and oogamy.	Iso-, aniso- and oogamy.	Advanced oogamy.	Absent (multiplication by vegetative and asexual means). Parasexuality is observed.

2. The main pigments are chlorophyll a and e, β-carotene and xanthophylls.
3. Chloroplasts are without pyrenoids.
4. Reserve food is oil.
5. Flagella are two, equal in length and are situated anteriorly. The longer one tinsel and shorter one is whiplash type.
6. Reproduction takes place by all the three means — vegetative, asexual, and sexual (isogamous type).

Common genera: *Botrydium, Ophiocytium, Tribonema, Vaucheria* etc.

6. Division. **EUSTIGMATOPHYTA**

It consists of only one Class — Eustigmatophyceae.

Class. **Eustigmatophyceae**

Important characteristics

1. Plant body is unicellular and live in fresh water or in soil.
2. Chloroplasts contain chlorophyll a, β-carotene with two major xanthophylls being violaxanthin and vaucheriaxanthin.
3. Vegetative cells usually have a characteristic polygonal pyrenoids (absent in zoospores).
4. The pyrenoid projects from the inner face of the plastid on a narrow stalk and in the cytoplasm are surrounded by flat plates of the unidentified photosynthate.

Common genera: *Pleurochloris, Polyedriella.*

7. Division. **RAPHIDOPHYTA** (chloromonads)

It consists of only one Class. Raphidophyceae

Class. **Raphidophyceae**

Important characteristics

1. They are commonly marine.
2. Chloroplasts contain chlorophyll a and c.
3. Two membrane of chloroplast endoplasmic reticulum.
4. Anterior flagella is commonly tinsel, whereas the posterior flagella is naked.

Common genera: *Heterosigma, Chattonella* etc.

8. Division. **PHAEOPHYTA** (brown algae)

It consists of only one class. Phaeophyceae.

Class. **Phaeophyceae**

Important characteristics

1. Most of the members are marine.
2. Cell wall composed of cellulose along with alginic and fucinic acid.
3. Pigments. Fucoxanthin is the dominant pigment, in addition to chlorophylls a, c and carotene.
4. Reserve food is laminarin (polysaccharide), manitol (alcohol) and fat.
5. Some lower members have naked pyrenoid-like bodies.
6. Zoospores are flagellated, one of which is tinsel type.
7. Sexual reproduction ranges from isogamy to oogamy.

Common genera: *Ectocarpus, Laminaria, Dictyota* etc.

Later on **Robert Edward Lee (1999)** revised the classification of algae ("*Phycology*", R. E. Lee, 1999, 3rd Edn.) with some modification of his earlier classification proposed in 1989 in his book "*Phycology*".

Robert Edward Lee (1999) classified the algae into four evolutionary groups, nine phyla and different classes.

The 1st group consists of **prokaryotic** algae only, the others are eukaryotic. The **eukaryotic** algae are further divided into three groups: (a) chloroplast surrounded only by the *two membranes of the chloroplast envelope*, (b) chloroplast surrounded by *one membrane of chloroplast endoplasmic reticulum* (E.R.) and (c) chloroplast surrounded by *two membranes of chloroplast endoplasmic reticulum* (E.R).

The standard botanical classification system used (suffix) in the systematics of the algae:

Phylum — phyta
Class — phyceae
Order — ales
Family — aceae
Genus
Species

The brief outline of classification (as per Lee, 1999) is given below:

Group 1. Prokaryotic algae

1. Phylum. **Cyanophyta** (cyanobacteria): Chlorophyll a; phycobiliproteins.

Group 2. Eukaryotic algae with chloroplast surrounded only by the two membranes of the chloroplast envelope.

1. Phylum. **Glaucophyta**: Algae that represent an intermediate position in the evolution of chloroplast; photosynthesis is carried out by modified endosymbiotic blue-green algae.
2. Phylum. **Rhodophyta** (red algae): Chlorophylls a and d; phycobiliproteins; no flagellated cells; storage product is floridean starch.
3. Phylum. **Chlorophyta** (green algae): Chlorophyll a and b; storage products starch is found inside the chloroplast.

Group 3. Eukaryotic algae with chloroplast surrounded by one membrane of chloroplast endoplasmic reticulum.

1. Phylum. **Euglenophyta** (euglenoid): Chlorophylls a and b, one flagellum with a spiral row of fibrillar hairs; proteinaceous pellicle in strips under the plasma membrane; storage product paramylon; characteristic type of cell division.
2. Phylum. **Dinophyta** (dinoflagellates): Mesokaryotic nucleus; chlorophylls a and c_1; cell commonly divided into an epicone and a hypocone by a girdle; helical transverse flagellum; thecal plates in vesicles under the plasma membrane.

Group 4. Eukaryotic algae with chloroplast surrounded by two membranes of chloroplast endoplasmic reticulum.

1. Phylum. **Cryptophyta** (cryptophytes): Nucleomorph is present between inner and outer membranes of chloroplast endoplasmic reticulum; Starch is formed in grains between inner membrane of chloroplast endoplasmic reticulum and chloroplast envelope; chlorophylls a and c; phycobiliproteins; periplast is found inside plasma membrane.
2. Phylum. **Heterokontophyta** (heterokonts): Anterior tinsel and posterior whiplash flagella, chlorophylls a and c_1, fucoxanthin, storage product usually chrysolaminarin occurs in vesicles in cytoplasm.

The phylum is further divided into **nine (9)** classes. These are as follows:

1. Class. **Crysophyceae** (golden-brown algae): Chlorophyll a and c_1 and sometimes c_2; fucoxanthin; anterior tinsel and posterior whiplash flagella; storage product chrysolaminarin, found in vesicles in cytoplasm.
2. Class. **Synurophyceae**: Plant body bilaterally symmetrical. Flagella inserted into the cell approximately parallel to each other. Chloroplast endoplasmic reticulum usually absent. Flagella inserted into the cell approximately parallel to each other. Presence of chlorophyll c_1.
3. Class. **Dictyochophyceae** (silicoflagellates): Golden brown algae, characterized by tentacles or rhizopodia on basically amoeboid vegetative cells. Amoeboid cells are mostly restricted to this group.
4. Class. **Pelagophyceae**: Basically unicellular algae. Cells are very small (3-5 μm) and appear as small spheres with indistinct protoplasm.
5. Class. **Bacillariophyceae** (diatoms): Silica in cell walls; chlorophylls a, c_1 and c_2; fucoxanthin; one posterior tinsel flagellum on sperm; storage product chrysolaminarin found in vesicles in cytoplasm.
6. Class. **Raphidophyceae** (chloromonads): Chlorophylls a and c; anterior tinsel flagellum and posterior whiplash flagellum.
7. Class. **Xanthophyceae** (yellow green algae): Eye spot in chloroplast; chlorophylls a and e; presence of anterior tinsel flagellum and posterior whiplash flagellum.
8. Class. **Eustigmatophyceae**: Eye spot large, remains out side the chloroplast;

chlorophyll a; presence of posterior whiplash flagellum (commonly reduced) and anterior tinsel flagellum.

9. Class. **Phaeophyceae** (brown algae): Presence of unilocular and plurilocular sporangia, chlorophylls a, c_1 and c_2, fucoxanthin; storage product chrysolaminarin found in vesicles in cytoplasm; presence of anterior tinsel flagellum and posterior whiplash flagellum.

3. Phylum. PRYMNESIOPHYTA (haptophytes): Haptonema, chlorophylls a and c_1 and c_2, fucoxanthin; two whiplash flagella; scales common outside cell; storage product chrysolaminarin, found in vesicles in cytoplasm.

A brief outline of the classification up to class (with characters) is given below:

Group 1. Procaryotic algae

1. Phylum. CYANOPHYTA

Class. **Cyanophyceae**

Important characteristics

1. The individual cells are prokaryotic in nature. The nucleus is incipient type and lacks membrane bound organelles.
2. Both vegetative and reproductive cells are non-flagellated.
3. Cell wall is made up of microfibrils and is differentiated into four (4) layers. The cell wall is composed of mucopeptide, along with carbohydrates, amino acids and fatty acids.
4. Locomotion is generally absent, but where occurs, it is of gliding or jerky type.
5. The principal pigments are chlorophyll a (green), c-phycocyanin (blue) and c-phycoerythrin (red). In addition, other pigments like β-carotene and different xanthophylls like myxoxanthin and myxoxanthophyll are also present.
6. Membrane-bound chromatophores are absent. Pigments are found embedded in thylakoids.
7. The reserve foods are cyanophycean starch and cyanophycin granules (protein).
8. Many filamentous members possess specialized cells of disputed function (supposed to be the centre of N_2 fixation) known as heterocyst.
9. Reproduction takes place by vegetative and asexual methods. Vegetative reproduction takes place by cell division, fragmentation etc. Asexual reproduction takes place by endospores, exospores, akinetes, nannospores etc.
10. Sexual reproduction is completely absent. Genetic recombination is reported in two cases.

Common genus: *Nostoc, Oscillatoria, Gloeocapsa, Gloeotrichia* etc.

Group 2. Eukaryotic algae with chloroplast surrounded only by the two membranes of the chloroplast envelope.

1. Phylum. GLAUCOPHYTA

Important characteristics

1. Members of this group have endosymbiotic cyanobacteria (cyanelles) in the cytoplasm instead of chloroplast.
2. Most of the cyanelles in this phylum lack a wall and are surrounded by two membranes – the old food vesicle membrane of the cyanome (host) and the plasma membrane of the cyanelle. In course of evolution these two membranes became the chloroplast envelop.
3. Various inclusions of the cyanobacteria like polyhedral bodies, cyanophycin granules, polyphosphate bodies are lost. The cyanome cytoplasm took over the formation of the storage product.
4. A pyrenoid, being differentiated from proteinaceous bodies, is present in the cyanelle.
5. Pigments like chlorophyll a and the phycobiliproteins are present.

Common genus: *Cyanophora, Glaucocystis, Geosiphon* etc.

2. Phylum. **RHODOPHYTA** (red algae)

The phylum Rhodophyta has a single class Rhodophyceae.

Class. **Rhodophyceae**

Important characteristics

1. Plants are generally marine habitat, called sea weeds, grow in various depth (Out of about 4,000 species, about 200 species are of fresh water).
2. Pit connections between the cells in all of the orders except Porphyridales, Erythropeltidales and haploid phase of the Bangiales.
3. Chloroplast with one thylakoid per band and absence of chloroplast endoplasmic reticulum.
4. Presence of chlorophylls a and d, phycobiliproteins include r-phycocyanin, allophycocyanin, and three forms of phycoerythrins (b, r and c).
5. Floridean starch as a storage product.
6. Lack of flagellated cells.

The class Rhodophyceae is divided into 10 orders:

1. Porphyridiales. e.g., *Porphyridium, Rhodosorus* etc.
2. Rhodochaetales. e.g., *Rhodochaete*.
3. Acrochaetiales. e.g., *Audouinella, Rhodochorton* etc.
4. Bangiales. e.g., *Bangia, Porphyra* etc.
5. Batrachospermales. e.g., *Batrachospermum*.
6. Nemaliales. e.g., *Nemalion* etc.
7. Corallinales. e.g., *Corallina, Melobesia* etc.
8. Gelidiales. e.g., *Gelidium*.
9. Gracilariales. e.g., *Gracilaria*.
10. Ceramiales. e.g., *Polysiphonia* etc.

3. Phylum. **CHLOROPHYTA** (green algae)

It is divided into 4 classes:

A. Class. **Micromonadophyceae**:

Important characteristics

Scaly or naked flagellates with interzonal spindles are persistent during cytokinesis; primitive green algae, some of which gave rise to the other classes in the Chlorophyta.

B. Class. **Charophyceae** (flagella are attached in a lateral position in the cell):

Important characteristics

1. Motile cells asymmetrical, two flagella attached in a lateral position in cell.
2. Flagellar root consisting of a broad band of microtubules and a second smaller microtubular root.
3. Multilayered structure (MLS) may be present, rhizoplast absent.
4. Scales are common, outside the motile cells.
5. Persistant interzonal mitotic spindle in telophase.
6. Phragmoplast produces new cross-walls after cell division.
7. Eyespots usually not present.
8. Sexual reproduction involves the formation of a dominant zygote. Meiosis occurs during zygote germination.

The class Charophyceae is further divided into 4 orders:

1. Klebsormidiales. e.g., *Stichococcus, Raphidonema, Klebsormidium* etc.
2. Zygnematales. e.g., *Spirogyra, Zygnema, Cosmerium, Cylindrocystis, Mesotaenium* etc.
3. Coleochaetales e.g., *Coleochaete*.
4. Charales e.g., *Chara, Nitella*.

C. Class. **Ulvophyceae** (flagella are attached at the anterior end of the cell).

Important characteristics

1. Plants are predominantly marine.
2. Flagella attached at anterior end of cell.
3. Externally, the motile cells have near-radial symmetry.
4. Flagella roots consist of 4 cruciately arranged microtubular roots and sometimes a rhizoplast.
5. Multilayered structure (MLS) is absent.
6. Scales may be present on motile cells.

7. Presence of persistent interzonal spindle in telophase.
8. Eyespots are common.
9. Dormant zygote is absent.
10. Alternation of generation is common.

The class Ulvophyceae is further divided into 6 orders:

1. Ulotrichales. e.g., *Ulothrix*.
2. Ulvales. e.g., *Ulva, Monostroma*.
3. Cladophorales. e.g., *Cladophora, Chaetomorpha*.
4. Dasycladales. e.g., *Acetabularia, Neomeris* etc.
5. Caulerpales. e.g., *Derbesia, Caulerpa, Codium* etc.
6. Siphonocladales. e.g., *Siphonocladus, Valonia* etc.

Common genera: *Ulva, Ulothrix, Cladophora* etc.

D. Class. **Chlorophyceae** (flagella are attached at the anterior end of the cell)

Important characteristics

1. Plants are predominantly fresh water.
2. Motile cells are with radial or near-radial external symmetry.
3. Flagella are attached at the anterior end of cell.
4. Flagella roots consist of 4 cruciately arranged microtubular roots and sometimes a rhizoplast.
5. Multilayered structure (MLS) is absent.
6. Theca is common in motile cells.
7. In telophase, the interzonal spindle collapses; phycoplast produces the new cross wall in cell division.
8. Eyespots are common.
9. Zygote undergoes a dormant period; meiosis occurs when the zygote germinates.

The class Chlorophyceae is further divided into 8 orders:

1. Volvocales. e.g., *Chlamydomonas, Volvox*.
2. Tetrasporales. e.g., *Tetraspora, Palmella*.
3. Schizogoniales. e.g., *Parasiola*.
4. Chlorococcales. e.g., *Chlorococcum, Hydrodictyon, Scenedesmus, Chlorella*.
5. Sphaeropleales. e.g., *Microspora, Sphaeroplea*.
6. Chlorosarcinales e.g., *Chlorosarcina*.
7. Chaetophorales. e.g., *Stigeoclonium, Chaetophora*.
8. Oedogoniales. e.g., *Oedogonium, Oedocladium, Bulbochaete*.

Group 3. Eukaryotic algae with chloroplast surrounded by one membrane of chloroplast endoplasmic reticulum.

1. Phylum. **Euglenophyta** (euglenoid)

It consists of only one Class Euglenaceae.

Class **Euglenaceae**

Important characteristics

1. They are unicellular.
2. They have chlorophyll a and b.
3. One membrane of chloroplast and endoplasmic reticulum.
4. The nucleus is mesokaryotic in nature.
5. Flagella with fibrillar hairs in one row.
6. Sexual reproduction is absent.
7. The storage product is paramylon, formed in the cytoplasm.

The class Euglenophyceae is further divided into 3 orders:

1. Heteronematales. e.g., *Peranema*, etc.
2. Eutreptiales. e.g., *Eutreptia, Eutreptiella* etc.
3. Euglenales. e.g., *Euglena, Trachelomonas, Phacus* etc.

2. Phylum. **Dinophyta** (dinoflagellates)

It consists of only one Class Dinophyceae.

Class **Dinophyceae**

Important characteristics

1. A typical motile dinoflagellate consists of an epicone and hypocone divided by a transverse girdle or cingulum.

2. The epicone and hypocone are normally divided into a number of thecal plates.
3. The cells can be photosynthetic or colourless and heterotrophic.
4. Photosynthetic organisms have chloroplasts surrounded by one membrane of chloroplast endoplasmic reticulum, which is not continuous with the outer membrane of the nuclear envelop.
5. Chlorophyll a and c_2 are present in the chloroplasts, with peridinin and neoperidinin being the main carotenoids.
6. Presence of pyrenoid in chloroplast.
7. Storage product is starch.
8. Eyespot is present.
9. Nucleus is dinokaryoytic or mesokaryotic in nature.

The class Dinophyceae is further divided into 4 orders:

1. Prorocentrales. e.g., *Prorocentrum*.
2. Dinophysales. e.g., *Ornithocercus*.
3. Peridiniales. e.g., *Ceratium*.
4. Gymnodiales. e.g., *Gymnodinium*.

Group 4. Eukaryotic algae with chloroplast surrounded by two membranes of chloroplast endoplasmic reticulum.

1. Phylum. **CRYPTOPHYTA** (cryptophytes)

It consists of only one Class Cryptophyceae

Class **Cryptophyceae**

Important characteristics

1. Plants are found in both marine and fresh water habitat.
2. The plant body is dorsiventral in shape, with the cells flattened in one plane.
3. Cells contain chlorophylls a and c_2.
4. A few members lack chloroplast and are heterotrophic in nature, but most have a single-lobed chloroplast with a central pyrenoid.
5. Phycobiliproteins occur inside the thylakoids of the choloroplasts.

Common genera: *Cyathomonas, Chilomonas* etc.

1. Phylum. **HETEROKONTOPHYTA** (heterokonts):

The phylum is further divided into **nine (9)** classes. These are:

A. Class. **Crysophyceae** (golden-brown algae)

Important characteristics

1. Presence of chlorophyll c_1 and c_2.
2. Flagella inserted in the cell perpendicular to one another.
3. One photoreceptor as a swelling on the short flagellum.
4. Eyespot commonly present.
5. Chloroplast endoplasmic reticulum present.
6. Radially or bilaterally symmetrical silica scales may occur.
7. Contractile vacuole anterior.

Common genera: *Ochromonas, Paraphysomonas, Dinobryon, Anthophysa* etc.

B. Class. **Synurophyceae**

Important characteristics

1. Presence of chlorophyll c_1.
2. Flagella inserted into the cell approximately parallel to each other.
3. Presence of two photoreceptors, one on each flagellum.
4. Eyespot absent.
5. Chloroplast endoplasmic reticulum usually absent.
6. Plant body bilaterally symmetrical.
7. Silica scales are common.
8. Contractile vacuole posterior.

Common genera: *Sinura, Mallomonas, Tessellaria* etc.

The class is divided into 3 orders (Clay *et. al.*, 1999):

1. Goniomonadales. e.g., *Goniomonas*.
2. Cryptomonadales. e.g., *Cryptomonas*.
3. Chroomonadales. e.g., *Chroomonas*.

C. Class. **Dictyochophyceae** (silicoflagellates).

Important characteristics

1. These are golden-brown algae.

2. They have tentacles or rhizopodia on basically amoeboid vegetative cells.
3. Amoeboid vegetative cells are mostly restricted in this group (and Xanthophyceae).

Dictyochophyceae is divided into 3 orders (Preisig, 1995):

1. Rhizochromulinales. e.g., *Rhizochromulina, Phaeaster, Chrysoamoeba* etc.
2. Pedinellales. e.g., *Pedinella, Apedinella, Pseudopedinella,* etc.
3. Dictyocales. e.g., *Dictyocha* etc.

D. Class. **Pelagophyceae**

Important characteristics

1. Basically unicellular algae.
2. The cells are very small (3-5 μm) members of the ultraplankton and appear as small spheres with indistinct protoplasm under the light microscope.

Common genus: *Pelagomonas*.

E. Class. **Bacillariophyceae** (diatoms)

Important characteristics

1. Plant body is unicellular and free-living. Sometimes they are present in colony.
2. Depending on the mode of nutrition, they may be photosynthetic autotrophs, photosynthetic symbionts or colourless heterotrophs.
3. Vegetative cells are diploid (2n).
4. The cells are surrounded by a rigid two-part box-like cell wall (upper part is epitheca and lower part is hypotheca), composed of silica, called the frustule.
5. The cells are bilaterally or radially symmetrical.
6. The chloroplast contains chlorophylls a, c_1 and c_2 with the major carotenoid being the golden-brown fucoxanthin.
7. Reserve food is fat and volutin.
8. Motile cells have single flagellum.
9. Reproduction takes place by cell division and auxospore formation.

Bacillariophyceae is divided into 2 orders (Preisig, 1995):

1. Biddulphiales. e.g., *Melosira, Chaetoceros* etc.
2. Bacillariales. e.g., *Amphora, Navicula, Pinnularia* etc.

F. Class **Raphidophyceae** (chloromonads)

Important characteristics

1. They are commonly marine.
2. Chloroplasts contain chlorophyll a and c.
3. Two membrane of chloroplast endoplasmic reticulum.
4. Anterior flagellum is commonly tinsel, whereas the posterior flagellum is naked.

Common genera: *Heterosigma, Chattonella*.

G. Class. **Xanthophyceae** (yellow-green algae)

Important characteristics

1. Cells are eukaryotic and cell wall is composed of pectic substance.
2. The main pigments are Chlorophyll a and e, β-carotene and xanthophylls.
3. Chloroplasts are without pyrenoids.
4. Reserve food is oil.
5. Flagella are two in number, unequal in length and are situated anteriorly. The longer one tinsel and shorter one whiplash type.
6. Reproduction takes place by all the three means vegetative, asexual and sexual (isogamous type).

Common genera: *Botrydium, Ophiocytium, Tribonema* etc.

The class is divided into 3 orders:

1. Tribonematales. e.g., *Tribonema*.
2. Botrydiales. e.g., *Botrydium*.
3. Vaucheriales. e.g., *Vaucheria*.

H. Class. **Eustigmatophyceae**

Important characteristics

1. Plant body is unicellular and live in fresh water or in soil.

2. Chloroplasts contain chlorophylls a, β-carotene with two major xanthophylls being violaxanthin and vaucherioxanthin.
3. Vegetative cells usually have characteristic polygonal pyrenoids (absent in zoospores).
4. The pyrenoid projects from the inner face of the plastid on a narrow stalk and cytoplasm is surrounded by flat plates of the unidentified photosynthate.

Common genera: *Pleurochloris, Polyedriella.*

I. Class. **Phaeophyceae**

Important characteristics

1. Most of the members are marine.
2. Cell wall is composed of cellulose along with alginic and fucinic acid.
3. Pigments. Fucoxanthin is the dominant pigment, in addition to chlorophylls a, c and carotene.
4. Reserve foods are laminarin (polysaccharide), manitol (alcohol) and fat.
5. Some lower members have naked pyrenoid like bodies.
6. Zoospores are flagellated, one of which is tinsel type.
7. Sexual reproduction ranges from isogamy to oogamy.

Phaeophyceae is divided into 7 orders:

1. Ectocarpales. e.g., *Ectocarpus, Ralfsia* etc.
2. Desmarestiales. e.g., *Desmarestia.*
3. Cutleriales. e.g., *Cutleria, Aglaozonia,* etc.
4. Laminariales. e.g., *Laminaria, Chorda* etc.
5. Sphacelariales. e.g., *Sphacelaria.*
6. Dictyotales. e.g., *Dictyota.*
7. Fucales. e.g., *Fucus, Ascophyllum, Sargassum* etc.

Common genera: *Ectocarpus, Laminaria, Dictyota* etc.

3. Phylum. **Prymnesiophyta** (haptophytes)

It consists of only one Class — Prymnesiophyceae.

Class. **Prymnesiophyceae**

Important characteristics

1. It is a group of unicellular flagellates, characterized by the possession of a haptonema between two smooth flagella.
2. Chloroplasts have chlorophylls a, c_1, c_2 with fucoxanthin as the major carotenoid.
3. Diadinoxanthin and β-hexanoyl fucoxanthin also occur in the group.
4. Two membranes of the chloroplast endoplasmic reticulum are present.
5. The anterior end of the cell has a large golgi apparatus, and sometimes contractile vacuole.
6. The cells are commonly covered with scales.

The class Prymnesiophyceae is divided into two orders (Green and Jordan, 1994):

1. Prymnesiales. e.g., *Prymnesium, Emiliania, Hymenomonas* etc.
2. Pavlovales. e.g., *Pavlova* etc.

CLASSIFICATION OF ALGAE: LEE (2008)

Later, **Robert Edward Lee (2008)** further revised his classification of Algae ("Phycology", R. E. Lee, 2008, 4th Edn.) with some modification of his earlier classification proposed in 1999 in his book "Phycology".

Robert Edward Lee (2008) classified the algae into 4-evolutionary groups, 10-phyla, with the addition of **Phylum Apicomplexa** in the 3rd evolutionary groups and different classes with the addition of 3-new classes, such as **Pinguiophyceae, Boledophyceae** and **Phaeothamniophyceae** under Phylum Heterokontophyta in the 4th evolutionary groups.

The 1st group consists of **Prokaryotic algae** only, the others are Eukaryotic. The **Eukaryotic algae** are further divided into three groups: (a) chloroplast surrounded only by the *two membranes of the chloroplast envelope,*

(b) chloroplast surrounded by *one membrane of chloroplast endoplasmic reticulum* (ER) and (c) chloroplast surrounded by *two membrane of chloroplast endoplasmic reticulum* (ER).

The standard botanical classification system is used in the systematics of the algae:

Phylum – phyta
Class – phyceae
Order – ales
Family – aceae
Genus
Species

The brief outline of classification (as per Lee, 2008) is given below:

Group 1. Prokaryotic algae

Phylum. **CYANOPHYTA** (Cyanobacteria): The Cyanophyta is the only prokaryotic algal group. It contains Chlorophyll a; phycobiliproteins.

Group 2. Eukaryotic algae with chloroplast surrounded only by the two membranes of the chloroplast envelope.

1. Phylum **GLAUCOPHYTA**: Algae that represent an intermediate position in the evolution of chloroplast; photosynthesis is carried out by modified endosymbiotic cyanobacteria.
2. Phylum **RHODOPHYTA** (red algae): Chlorophyll a; phycobiliproteins; no flagellated cells; storage product is floridean starch.
3. Phylum **CHLOROPHYTA** (green algae): Chlorophyll a and b; storage product starch is found inside the chloroplast.

Group 3. Eukaryotic algae with chloroplast surrounded by one membrane of chloroplast endoplasmic reticulum (ER).

1. Phylum **EUGLENOPHYTA** (euglenoid): Chlorophyll a and b, one flagellum with a spiraled row of fibrillar hairs; protenaceous pellicle in strips under the plasma membrane; storage product is paramylon; characteristic type of cell division.
2. Phylum **DINOPHYTA** (dinoflagellates): Mesokaryotic nucleus; chlorophylls a and c_1; cell commonly divided into an epicone and a hypocone by a girdle; helical transverse flagellum; thecal plates in vesicles under the plasma membrane.
3. Phylum **APICOMPLEXA**: Heterotrophic flagellates with colorless plastids.

Group 4. Eukaryotic algae with chloroplast surrounded by two membrane of chloroplast endoplasmic reticulum (ER).

1. Phylum **CRYPTOPHYTA** (cryptophytes): Nucleomorph present between inner and outer membrane of chloroplast endoplasmic reticulum; starch formed in grains between inner membrane of chloroplast endoplasmic reticulum and chloroplast envelope; chlorophyll a and c; phycobiliproteins; periplast inside plasma membrane.
2. Phylum **HETEROKONTOPHYTA** (heterokonts): Anterior tinsel and posterior whiplash flagellum, chlorophyll a and c; fucoxanthin, storage product usually chrysolaminarin occurring in vesicles.

The phylum is further divided into **twelve (12)** classes. These are as follows:

1. Class. **Crysophyceae** (golden-brown algae): Chlorophyll a, c_1 and c_2, and structurally by 2 flagella inserted into the cell perpendicular to each other, one photoreceptor on the short flagellum that is usually shaded by an eyespot in the anterior portion of the chloroplast, contractile vacuoles in the anterior portion of the cell, chloroplast endoplasmic reticulum, and rapidly or biradially symmetrical silica (if they are present). The storage product is chrysolaminarin. Many members produce statospores enclosed in a silicified wall with a terminal pore. Mostly grows in freshwater and in soft waters (low in calcium).
2. Class. **Synurophyceae**: Presence of chlorophyll a and c_1; flagella inserted into the cell approximately parallel to one another, there is photoreceptor near the base of each flagellum, there is no eyespot, and the contractile vacuole is in the posterior portion of the cell. Chloroplast endoplasmic reticulum is present in few species, but absent in most. The cells usually are covered by bilaterally symmetrical scales.

3. Class. **Eustigmatophyceae**: Plants are yellow-green unicells that occur in freshwater, brackish water and sea water as well as in soil. Eyespot is present outside the chloroplast. Basal swelling of the tinsel flagellum adjacent to the eyespot, only chlorophyll a, chloroplasts without girdle lamellae and no peripheral ring of DNA, and chloroplast endoplasmic reticulum not connected to the nuclear envelop.

   Eyespot is a large orange-red body at the anterior of the motile cell and is completely independent of the chloroplast. Pigments like chlorophyll a and β-carotene is present, alongwith two major xanthophylls being violaxanthin and vaucheriaxanthin.

4. Class. **Pinguiophyceae**: Marine planktonic algae like "chrysophytes". Unusually high concentrations of polyunsaturated acids, especially 20:5 (n-3) (EPAeicosapentaenoic acid) in the cells. The high percentage of unsaturated fatty acids, and lack of a cell wall, makes these algae desirable as a source of unsaturated fatty acids and of animal feed.

5. Class. **Dictyochophyceae**: Golden brown algae, characterized by tentacles or rhizopodia on basically amoeboid vegetative cells. Amoeboid cells are mostly restricted to this group.

6. Class. **Pelagophyceae**: Basically unicellular algae. Cells are very small (3-5 μm) and appear as small spheres with indistinct protoplasm.

7. Class. **Bolidophyceae**: Small group of flagellated marine picophytoplankton that is related to the diatoms (Bacillariophyceae); cells contain chlorophylls a, c_1, c_2, c_3, β-carotene, diatoxanthin, and fucoxanthin, like diatoms.

   Bolidomonas is unicell with a long tinsel flagella bearing tripartite tubular hairs and a shorter smooth flagellum. Cell is only 1.2 μm in diameter and has a simple internal organization; one plastid, mitochondrion, and golgi apparatus; two membranes of chloroplast endoplasmic reticulum, outer membrane continuous with the outer membrane of the nuclear envelope; DNA contained within a ring-shaped genophore and there is no eyespot.

8. Class. **Bacillariophyceae** (diatoms): Unicellular, sometimes colonial algae, free-living aquatic photosynthetic autotrophs, colourless heterotrophs, or photosynthetic symbiotes. Cells are surrounded by a rigid two-part box-like cell wall composed of silica, called the **frustule**. Chloroplasts contain chlorophylls a, c_1 and c_2 with the major carotenoid being the golden-brown fucoxanthin.

   Two-part frustule surrounds protoplasm that has a more or less central nucleus suspended in a system of protoplasmic threads. The chloroplasts occupy most of the cell, usually as two parietal plastids although sometimes as numerous discoid plastids. The storage product chrysolaminarin occurs in vesicles in the protoplasm.

9. Class. **Raphidophyceae** (chloromonads): They have chlorophyll a and c, and two membranes of chloroplast endoplasmic reticulum. Cells are biflagellate with anterior tinsel flagellum and posterior whiplash flagellum. Freshwater members are green and marine forms are yellowish and contain carotenoid fucoxanthin.

10. Class. **Xanthophyceae** (yellow green algae): Primarily freshwater and terrestrial algae with a few marine representatives. Motile cells with anterior tinsel flagellum and posterior whiplash flagellum. Eyespot in motile cells is always in the chloroplast and the chloroplasts are surrounded by two membranes of chloroplast endoplasmic reticulum. The outer membrane of chloroplast ER is usually continuous with the outer membrane of the nucleus. In most non-motile cells the wall is composed of two overlapping halves.

    Chloroplast contains chlorophyll a and c, lack fucoxanthin with the major carotenoids being diadinoxanthin, heteroxanthin and vaucheriaxanthin ester.

    Mannitol and glucose accumulate during photosynthesis in the plastids. The principle storage product is probably a β-1,3 linked glucan similar to paramylon, although lipids have been suggested as also being important.

Zoospores are biflagellate, presence of long anterior tinsel flagellum and short posterior whiplash flagellum.

11. Class. **Phaeothamniophyceae**: The class that is most closely related to the Xanthophyceae and Phaeophyceae, based on nucleotide sequencing.

Cells have two membranes of chloroplast endoplasmic reticulum with the outer membrane of chloroplast ER continuous with the outer membrane of the nuclear envelope; chloroplasts have a ring-shaped genophore and girdle lamellae; flagella are inserted laterally into the motile cells. Anterior tinsel flagellum has tripartite hairs that lack lateral filaments. Posterior flagellum lacks hairs. New daughter cells are formed by eleutheroschisis (parent cell wall is completely cast off and new daughter cell walls are formed). Vesicles under the plasma membrane appear similar to the physodes that occur in the Phaeophyceae. Fucoxanthin and heteroxanthin occur together (only in this class). Endogenous siliceous cysts (statospores) are not produced by these algae.

Phaeothamnion is a filamentous brown alga that produces zoospores that settle to produce new filaments.

Tetrachrysis occurs in environments such as peat ponds and has cells embedded in common mucilage.

Tetrasporopsis is a colonial freshwater alga that consists of a brown, gelatinous, bladdery sac.

Phaeoschizochlamys occurs among detritus or suspended between other freshwater algae. The cells occur in mucilage of different shapes up to 0.5 mm in diameter. The cells divide to form four autospores containing two chloroplasts.

12. Class. **Phaeophyceae** (brown algae): Members are exclusively marine, except *Heribaudiella, Pleorocladia, Bodanella* and *Sphacelaria*.

They have large amounts of the carotenoid fucoxanthin in their chloroplasts as well as form any phaeophycean tannins that might be present; the plastids have chlorophyll a, c_1 and c_2, with the major carotenoid being fucoxanthin. Two membranes of chloroplast ER, which are usually continuous with the outer membrane of nuclear envelop. The long-term storage product is laminarin found in vesicles in the cytoplasm.

Cell walls are generally composed of at least two layers, with cellulose making up the main structural skeleton.

Zoospores or gametes generally have a long anterior tinsel flagellum and shorter posteriorly directed whiplash flagellum.

3. Phylum **PRYMNESIOPHYTA** (haptophytes): It includes only one class Prymnesiophyceae.

Class Prymnesiophyceae. This group includes uninucleate flagellates having **haptonema**, between two smooth flagella. Cells are commonly covered with scales. Chloroplasts lack girdle lamellae and most contain chlorophyll a and c_1 or c_2, β-carotene, diadinoxanthin and diatoxanthin. Most of the members have two whiplash flagella of approximately of same length; storage product chrysolaminarin (leucosin), found in vesicles in the posterior end of the cell. The anterior end of the cell has a large golgi apparatus and sometimes a contractile vacuole.

A brief outline of the classification up to classes (with characters) is given below:

Group 1. Prokaryotic algae

1. Phylum. **CYANOPHYTA**

Class. **Cyanophyceae**

Important characteristics

1. The individual cells are prokaryotic in nature. The nucleus is incipient type and lacks membrane bound organelles.
2. Both vegetative and reproductive cells are non-flagellated.
3. Cell wall is made up of microfibrils and is differentiated into four (4) layers. The cell wall is composed of mucopeptide, along with carbohydrates, amino acids and fatty acids.

4. Location is generally absent, but where occurs, it is of gliding or jerky type.
5. The principal pigments are chlorophyll a (green), c-phycocyanin (blue) and c-phycoerythrin (red). In addition, other pigments like β-carotene and different xanthophylls like mycoxanthin and mycoxanthophylls are also present.
6. Membrane-bound chromatophores are absent. Pigments are found embedded in thylakoids.
7. The reserve foods are cyanophycin granules (protein).
8. Many filamentous members possess specialized cells known as heterocyst that functions as the site of N_2^- fixations.
9. Reproduction takes place by vegetative and asexual methods. Vegetative reproduction takes place by cell division, fragmentation etc. Asexual reproduction takes place by endospores, exospores, akinetes, nannospores etc.
10. Sexual reproduction is completely absent. Genetic recombination is reported in two cases.

Common genus: *Nostoc. Oscillatoria, Gloeocapsa, Gloeotrichia* etc.

The class Cyanophyceae is divided into three orders:

1. Chroococcales: Single cell or cells loosely bound in gelatinous irregular colonies e.g. *Microcystis, Synechococeus*.
2. Oscillatoriales: Filamentous Cyanobacteria e.g. *Oscillatoria, Trichodesmium*.
3. Nostocales: Filamentous Cyanobacteria with heterocyst e.g. *Nostoc, Anabaena*.

Group 2. Eukaryotic algae with chloroplast surrounded only by the two membranes of the chloroplast envelop.

1. Phylum. **GLAUCOPHYTA**

Important characteristics

1. Members of this group have endosymbiotic cyanobacteria (cyanelles) in the cytoplasm instead of chloroplast.
2. Most of the cyanelles in this phylum lack a wall and are surrounded by two membranes – the old food vesicle membrane of the cyanome (host) and the plasma membrane of the cyanelle. In course of evolution these two membranes become the chloroplast envelop.
3. Various inclusions of the cyanobacteria like polyhedral bodies, cyanophycin granules, polyphosphate bodies are lost. The cyanome cytoplasm took over the formation of the storage product.
4. A pyrenoid, being differentiated from proteinaceous bodies, is present in the cyanelle.
5. Pigments like chlorophyll a and the phycobiliproteins are present.

Common genus: *Cyanophora, Glaucocystis, Goesiphon* etc.

2. Phylum. **RHODOPHYTA**

The phylum Rhodophyta has a single class Rhodophyceae.

Important characteristics

1. Plants are generally marine habitat, called sea weeds, grow in various depths. (Out of about 4,000 species, about 200 species are fresh water).
2. Pit connections between the cells in all of the orders except Porphyridales, Erythropeltidales and haploid phase of Bangiales.
3. Chloroplast with one thylakoid per band and absence of chloroplast endoplasmic reticulum.
4. Presence of chlorophyll a, phycobiliproteins include r-phycocyanin, allophycocyanin, and three forms of phycoerythrins (b, r and c).
5. Floridean starch as a storage product.
6. Lack of flagellated cells.

The class Rhodophyceae is divided into 10 orders:

1. Cyanidiales. Unicells that inhabit volcanic areas with pH values ranging from

0.5 to 3.0 e.g. *Cyanidium, Cyanidioschyson* etc.

2. Porphyridiales. Unicells or multicellular algae that are held togather by mucilage e.g. *Porphyridium, Rhodosirus* etc.
3. Bangiales. Plants having a filamentous phase with pit connections and a macroscopic phase without pit connections e.g. *Bangia, Porphyra* etc.
4. Acrochaetiales. Algae with a uniseriate filamentous gametophyte and tetrasporophyte (if both are present) e.g. *Audouinella, Rhodochorton* etc.
5. Batrachospermales. Uniaxial (one apical cell per branch); gonimoblast usually develops from the carpogonium or hypogenous cell e.g. *Batrachospermum.*
6. Nemaliales. Multiaxial (more than one apical cell per branch); usually the gonimoblast develops from the carpogonium or the hypogenous cells e.g. *Nemalion* etc.
7. Corallinales. Heavily calcified algae with reproductive organs in conceptacles e.g. *Corallina, Melobesia* etc.
8. Gelidiales. Fleshy agarophytes, carpogonial branch consisting of a single cell, the carpogonium, no differentiated auxiliary cells e.g. *Gelidium.*
9. Gracillariales. Fleshy agarophytes, two celled carpogonial branch, no auxiliary cells, or connecting cells e.g. *Gracilaria.*
10. Ceramiales. Relatively delicate or filamentous forms with an auxiliary cell cut off after fertilization and borne on the supporting cell of a four-celled carpogonial filament e.g. *Polysiphonia* etc.

3. Phylum. **CHLOROPHYTA** (green algae)

It is divided into 4 classes.

A. Class. **Prasinophyceae** (Micromonadophyceae)

Important characteristics

1. Scaly or naked flagellates with interzonal spindles that are persistent during cytokinesis.
2. Primitive green algae, some of which gave rise to the other classes in the Chlorophyta. e.g. *Ostreococcus, Pyramimonas.*

B. Class. **Charophyceae** (flagella are attached in a lateral position in the cell):

Important characteristics

1. Motile cells asymmetrical, two flagella attached in a lateral position in cell.
2. Flagellar root consisting of a broad band of microtubules and a second smaller microtubular root.
3. Multilayered structure (MLS) may be present, rhizoplast absent.
4. Scales are common, outside the motile cells.
5. Persistent interzonal motile spindle in telophase.
6. Phragmoplast produces new cross-walls after cell division.
7. Eyespots usually not present.
8. Sexual reproduction involves the formation of a dominant zygote. Meiosis occurs during zygote germination.

The class Charophyceae is further divided into 4 orders:

1. Klebsormidiales. Unbranched filaments without holdfasts; plasmodesmata absent; zoospores naked and released through a pore in the wall e.g., *Stichococcus, Raphidonema, Klebsormidium* etc.
2. Zygnematales. Sexual reproduction by conjugation; unicells or unbranched filaments without holdfasts; plasmodesmata absent in filamentous forms; flagellated cells not produced e.g., *Spirogyra, Zygnema, Cosmerium, Cylindrocystis, Mesotaenium* etc.
3. Coleochaetales. Oogamous sexual reproduction; motile cells have a covering of scales, sheathed setae present; unicells, branched filaments or discoid thalli e.g., *Coleochaete.*
4. Charales. Oogamous sexual reproduction; sterile cells surround antheridia

and oogonia; male gametes covered with scales; zoospores not produced; complex plant body with apical growth and differentiated into nodes and internodes; plasmodesmata present. e.g., *Chara, Nitella*.

C. Class. **Ulvophyceae** (flagella are attached at an anterior end of the cell).

Important characteristics

1. Plants are predominantly marine.
2. Flagella attached at the anterior end of cell.
3. Externally, motile cells have near radial symmetry.
4. Flagella roots consists of 4 crucially arranged microtubular roots and sometimes a rhizoplast.
5. Multilayered structure (MLS) is absent.
6. Scales may be present on motile cells.
7. Presence of persistent interzonal spindle in telophase.
8. Eyespots are common.
9. Dormant zygote is absent.
10. Alternation of generation is common.

The class Ulvophyceae is further divided into 6 orders (Watanabe *et al.*, 2001):

1. Utrichales. Uninuceate filamentous algae with a parietal chloroplast e.g., *Ulothrix*.
2. Ulvales.Uninucleate cells with a parietal chloroplast; thallus is a hollow cylinder or a sheet, one or two cells thick e.g., *Ulva, Monostroma*.
3. Cladophorales. Multinucleate filamentous algae with a parietal perforate or reticulate chloroplast e.g., *Cladophora, Chaetomorpha*.
4. Dasycladales. Thallus has radial symmetry composed of an erect axis bearing branches; thallus uninucleate but multinucleate just before reproduction; gametes formed in operculate cysts e.g., *Acetabularia, Neomeris* etc.
5. Caulerpales. Coenocytic algae lacking cellulose in the walls; siphonoxanthin and siphonein usually present e.g., *Derbesia, Caulerpa, Codium* etc.
6. Siphonocladales. Algae with segregative cell division; siphonoxanthin present. e.g., *Siphonocladus, Valonia* etc.

Common genera: *Ulva, Ulothrix, Cladophora* etc.

D. Class. **Chlorophyceae** (flagella are attached at an anterior end of the cell).

Important characteristics

1. Plants are predominantly fresh water.
2. Motile cells are with radial or near-radial external symmetry.
3. Flagella are attached at the anterior end of cell.
4. Flagella root consist of 4 cruciately arranged microtubular roots and sometimes with rhizoplast.
5. Multilayered structure (MLS) is absent.
6. Theca is common in motile cells.
7. In telophase, the interzonal spindle collapses; phycoplast produces the new cross wall in cell division.
8. Eyespot is common.
9. Zygote undergoes a dormant period; meiosis occurs when the zygote germinates.

The class Chlorophyceae is further divided into 9 orders:

1. Volvocales. Vegetative cells flagellated and motile e.g., *Chlamydomonas, Volvox*.
2. Tetrasporales. Non-filamentous colonies with immobile vegetative cells capable of cell division; pseudocilia may be present e.g., *Tetraspora, Palmella*.
3. Prasiolales. Marine, freshwater or terrestrial algae composed of unicells, short filaments or small sheet like thalli e.g., *Prasiola*.

4. Chlorellales. Unicells or nonfilamentous colonial algae; vegetative cells nonmotile e.g., *Chlorococcum, Hydrodictyon*.
5. Trebouxiales. Green algae involved in a lichen symbiosis e.g., *Ttrebouxia*.
6. Sphaeropleales. Unbranched filaments with new walls formed inside the old filament walls, resulting in H-shaped wall pieces e.g., *Microspora, Sphaeroplea*.
7. Chlorosarcinales. Daughter cells retained within parent cell wall; no plasmodesmata present e.g., *Chlorosarcina*.
8. Chaetophorales. Branched or unbranched filaments; plasmodesmata present e.g., *Stigeoclonium, Chaetophora*.
9. Oedogoniales.Uninucleate filamentous freshwater algae with a unique type of cell division; motile spores and gametes with a whorl of flagella at one pole e.g., *Oedogonium, Oedocladium, Bulbochaete*.

Group 3. Eukaryotic algae with chloroplast surrounded by one membrane of chloroplast endoplasmic reticulum.

1. Phylum. **EUGLENOPHYTA** (euglenoid): It consists of only one class Euglenaceae.

Class **Euglenaceae**

Important characteristics

1. They are unicellular.
2. They have chlorophyll a and b.
3. One membrane of chloroplast and endoplasmic reticulum.
4. The nucleus is mesokaryotic in nature.
5. Flagellum with fibrillar hairs in one row.
6. Sexual reproduction is absent.
7. The storage product is paramylon, formed in cytoplasm.

The class Euglenophyceae is further divided into 3 orders:

1. Heteronematales. Two emergent flagella, the longer flagellum directed anteriorly and the shorter one directed posteriorly during swimming; special ingestion organelle present e.g., *Peranema* etc.
2. Eutreptiales. Two emergent flagella, one directed anteriorly and the other laterally or posteriorly during swimming; no special ingestion organelle e.g., *Eutreptia, Eutreptiella* etc.
3. Euglenales. Two flagella, only one of which emerges from the canal; no special ingestion organelle e.g., *Euglena, Trachelomonas, Phacus* etc.

2. Phylum. **DINOPHYTA** (dinoflagellates): It consists of only one class Dinophyceae.

Class. **Dinophyceae**

Important characteristics

1. A typical motile dinoflagellate consists of an epicone and hypocone divided by a trasverse girdle or cingulum.
2. The epicone and hypocone are normally divided into a number of thecal plates.
3. The cells can be photosynthetic or colourless and heterotrophic.
4. Photosynthetic organisms have chloroplasts surrounded by one membrane of chloroplast endoplasmic reticulum, which is not continuous with the outer membrane of the nuclear envelop.
5. Chlorophyll a and c_2 are present in the chloroplasts, with peridium and neoperidinin being the main carotenoids.
6. Presence of pyrenoid in chloroplast.
7. Storage product is starch.
8. Eyespot is present.
9. Nucleus is dinokayotic and mesokaryotic in nature.

The class Dinophyceae is further divided into 4 orders:

1. Prorocentrales. Cell wall divided vertically into two halves; no girdle; two flagella borne at cell apex e.g., *Porocentrum*.
2. Dinophysales. Cell wall divided vertically into two halves, cells with elaborate extensions of the theca e.g., *Ornithocercus*.
3. Peridiniales. Motile cells with an epicone

and hypocone separated by a girdle, relatively thick theca. e.g., *Ceratium*.

4. Gymnodiales. Motile cells are with an epicone and hypocone separated by a girdle; theca thin or reduced to empty vesicles. e.g., *Gymnodinium*.

3. Phylum. **APICOMPLEXA**

Important characteristic

1. These are algae having apicoplasts (a reduced colourless plastid).
2. The typical apicomplexans are unicellular endoparasites that cause some of the most significant tropical diseases. Malaria in humans is produced by the apicomplexan *Plasmodium*.
3. The typical apicomplexan vegetative cell (merozoite) has an **apicoplast** surrounded by 4 membranes. The inner 2 membranes are the inner and outer plastid membrane while the outer membranes are derived from the vacuole membrane and the plasma membrane of endosymbiotic red alga.
4. The apical complex consists of a polar ring and conoid formed by spirally coiled microtubules.
5. The apicomplexan has laminin polysaccharide on its surface.
6. The apicomplexan has a layer of flattened membrane sacs or **alveoli** beneath the plasma membrane.

Common genus: *Kalablepharis*.

Group 4 Eukaryotic algae with chloroplast surrounded by two membrane of chloroplast endoplasmic reticulum.

1. Phylum. **CRYPTOPHYTA** (cryptophytes): It consists of only one class Cryptophyceae.

Class. **Cryptophyceae**

Important characteristics

1. Plants are found in both marine and fresh water habitat.
2. The plant body is dorsiventral in shape, with the cells flattened in one plane.
3. Cells contain chlorophylls a and c_2.
4. A few members lack chloroplast and are heterotrophic in nature, but most have a single-lobed chloroplast with a central pyrenoid.
5. Phycobiliproteins occur inside the thylakoids of the chloroplasts.

Common genera: *Cyathomonas, Chilomonas* etc.

The class is divided into three orders (Martin *et al.*, 1998; Deane *et al.*, 2002):

1. Gonimonadales. Colourless cells with no plastids. e.g., *Goniomonas*.
2. Cryptomonadales. Cells usually reddish in color with chloroplasts containing the phycobiliprotein Cr-phycoerythrin. e.g., *Cryptomonas*.
3. Chroomonadales.The remainder of the cryptophyte algae, often blue-green in color due to chloroplasts containing the phycobiliprotein Cr-phycocyanin. e.g., *Chroomonas*.

2. Phylum. **HETEROKONTOPHYTA** (heterokónts)

A. Class. **Crysophyceae** (golden-brown algae)

Important characteristics

1. Presence of chlorophyll a, c_1 and c_2.
2. Two flagella inserted in the cell perpendicular to one another.
3. One photoreceptor as a swelling on the short flagellum.
4. Eyespot commonly present.
5. Chloroplast endoplasmic reticulum present.
6. Radially or bilaterally symmetrical silica scales may occur.
7. Contractile vacuole anterior.

Common genera: *Ochromonas, Paraphysomonas, Dinobryon, Anthophysa* etc.

The class Crysophyceae is further divided into three orders (Preisig, 1995):

1. Chromulinales. Cells with the flagella inserted into the anterior prtion of the cell, e.g. *Ochromonas, Dinobryon* etc.

2. Parmales. Cells with siliceouswalls composed of five or eight plates, e.g. *Triparma*.
3. Chrysomeridales. Motile naked cells with laterally inserted flagella and an eyespot, motile cells similar to those in the brown algae, e.g. *Giraudyopsis*.

B. Class. **Synurophyceae**

Important characteristics

1. Presence of chrolophyll c_1.
2. Flagella inserted into the cell approximately parallel to each other.
3. Presence of two photoreceptors, one on each flagellum.
4. Eyespot absent.
5. Chloroplast endoplasmic reticulum usually absent.
6. Plant body bilaterally symmetrical.
7. Silica scales are common.
8. Contractile vacuole posterior.

Common genera: *Sinura, Tessellaria* etc.

The class includes only one order Sinurales.

C. Class. **Eustigomatophyceae**

Important characteristics

1. Plant body is unicellular and live in fresh water or in soil.
2. Chloroplasts contain chlorophyll a, β-carotene with two major xanthophylls being violaxanthin and vaucherioxanthin.
3. Vegetative cells usually have characteristic polygonal pyrenoids (absent in zoospores).
4. The pyrenoid projects from the inner face of the plastid on a narrow stalk and cytoplasm is surrounded by flat plates of the unidentified photosynthates.

Common genera: *Pleurochloris, Polyedriella*.

D. Class. **Pinguiophyceae**

Important characteristics

1. Marine planktonic algae like "chrysophytes".
2. Unusually high concentrations of polyunsaturated acids, especially (EPA, eicosapentaenoic acid) in the cells.
3. The high percentage of unsaturated fatty acids, and the lack of a cell wall, makes these algae desirable as a source of unsaturated fatty acids and of animal feed.

Common genera: *Glossomastix, Phaeomonas, Pinguiochrysis, Pinguiococcus* etc.

E. Class. **Dictyochophyceae** (silicoflagellates)

Important characteristics

1. These are golden-brown algae.
2. They have tentacles or rhizopodia on basically amoeboid vegetative cells.
3. Amoeboid vegetative cells are mostly restricted in this group (and Xanthophyceae).

Dictyochophyceae is divided into 3 orders (Preisig, 1995).

1. Rhizochromulinales. Marine and freshwater unicells with tentacles. e.g., *Rhizochromulina, Phaeaster, Chrysoamoeba* etc.
2. Pedinellales. Unicells with a long anterior flagellum and a second flagellum reduced to a basal body, usually three to six chloroplasts (if chloroplasts are present), marine and freshwater. e.g., *Pedinella, Apendinella, Pseudopedinella* etc.
3. Dictyocales. Marine unicells with an external silicified skeleton. e.g., *Dictyocha* etc.

F. Class. **Pelagophyceae**

Important characteristics

1. Basically unicellular algae.
2. The cells are very small (3-5 μm) members of the ultraplankton and appear as small spheres with indistinct protoplasm under light microscope.

Common genera: *Pelagomonas*.

G. Class. **Bolidophyceae**

Important characteristics

1. Small group of flagellated marine picophytoplankton that is related to the diatoms (Bacillariophyceae).

2. Cells contain chlorophylls a, c_1, c_2, c_3, carotene, diatoxanthin, and fucoxanthin, like diatoms.
3. Unicell with long tinsel flagella bearing tripartite tubular hairs and a shorter smooth flagellum.
4. Cell is only 1.2 mm in diameter and has a simple internal organization; one plastid, mitochondrion, and Golgi apparatus.
5. Two membranes of chloroplast endoplasmic reticulum, outer membrane continuous with the outer membrane of the nuclear envelope.
6. DNA is contained within a ring-shaped genophore and there is no eyespot.

Common genera: *Bolidomonas, Tetraperma, Triperma* etc.

H. Class. **Bacillariophyceae** (diatoms)

Important characteristics

1. Plant body is unicellular and free-living. Sometimes they are present in colony.
2. Depending on the mode of nutrition, they may be photosynthetic autotrophs, photosynthetic symbionts or colourless heterotrophs.
3. Vegetative cells are diploid (2n).
4. The cells are surrounded by a rigid two-part box-like cell wall (upper part is epitheca and lower part is hypotheca), composed of silica, called the frustule.
5. The cells are bilaterally or radially symmetrical.
6. The chloroplast contains chlorophyll a, c_1 and c_2 with the major carotenoid being the golden-brown fucoxanthin.
7. Reserve food is fat and volutin.
8. Motile cells have single flagellum.
9. Reproduction takes place by cell division and auxospore formation.

Bacillariophyceae is divided into 2 orders (Preisig, 1995).

1. Biddulphiales. Radial or gonoid ornamentation; many chloroplasts; no raphe; resting spores formed; motile spermatozoids with a single tinsel flagellum; oogamous sexual reproduction. e.g., *Melosira, Chaetoceros* etc.
2. Bacillariales. Pennate or trellisoid ornamentation; one or two chloroplasts; raphes possibly present with gliding; no flagellated spermatozoids; sexual reproduction by conjugation.e.g., *Amphora, Navicula* etc.

I. Class. **Raphidophyceae**

Important characteristics

1. They are commonly marine.
2. Chloroplasts contain chlorophyll a and c.
3. Two membrane of chloroplast endoplasmic reticulum.
4. Anterior flagellum is commonly tinsel, whereas the posterior flagellum is naked.

Common genera: *Heterosigma, Chattonella.*

J. Class. **Xanthophyceae** (yellow-brown algae)

Important characteristics

1. Cells are eukaryotic and cell wall is composed of pectic substance.
2. The main pigments are chlorophyll a and e, β-carotene and xanthophylls.
3. Chloroplasts are without pyrenoids.
4. Reserve food is oil.
5. Flagella are two in number, equal in length and are situated anteriorly. The longer one tinsel and shorter one whiplash type.
6. Reproduction takes place by all the three means vegetative, asexual and sexual (isogamous type).

Common genera: *Botrydium, Ophiocytium, Tribonema* etc.

The class is divided into 4 orders:

1. Mischococcales. Small coccoid cells. e.g., *Mischococcus, Seudobumilleriopsis.*
2. Tribonematales. Filamentous organisms, not coenocytic. e.g., *Tribonema.*

3. Botrydiales. Globose multinucleate thallus with colourless rhizoids. e.g., *Botrydium*.
4. Vaucheriales. Filamentous coenocyte. e.g., *Vaucheria*.

K. Class. **Phaeothamniophyceae**

Important characteristics

1. The class that is most closely related to the Xanthophyceae and Phaeophyceae, based on nucleotide sequencing.
2. Cells have two membranes of chloroplast endoplasmic reticulum with the outer membrane of chloroplast ER continuous with the outer membrane of the nuclear envelope.
3. Chloroplasts have a ring-shaped genophore and girdle lamellae.
4. Flagella are inserted laterally into the motile cells. Anterior tinsel flagellum has tripartite hairs that lack lateral filaments. Posterior flagellum lacks hairs.
5. New daughter cells are formed by eleutheroschisis (parent cell wall is completely cast off and new daughter cell walls are formed).
6. Vesicles under the plasma membrane appear (similar to the physodes that occur in the Phaeophyceae).
7. Fucoxanthin and heteroxanthin occur together (only in this class).
8. Endogenous siliceous cysts (statospores) are not produced by these algae.

Common genera: *Phaeothamnion* is a filamentous brown alga that produces zoospores that settle to produce new filaments.

Tetrachrysis occurs in environments such as peat ponds and has cells embedded in common mucilage.

Tetrasporopsis is a colonial freshwater alga that consists of a brown, gelatinous, bladdery sac.

Phaeoschizochlamys occurs among detritus or suspended between other freshwater algae. The cells occur in mucilage of different shapes up to 0.5 mm in diameter. The cells divide to form four autospores containing two chloroplasts.

L. Class. **Phaeophyceae**

Important characteristics

1. Most of the members are marine.
2. Cell wall is composed of cellulose along with alginic and fucinic acid.
3. Pigments. Fucoxanthin is the dominant pigment, in addition to chlorophyll a, c and carotene.
4. Reserve foods are laminarin (polysaccharide), manitol (alcohol) and fat.
5. Some lower members have naked pyrenoid like bodies.
6. Zoospores are flagellated, one of which is tinsel type.
7. Sexual reproduction ranges from isogamy to oogamy.

Phaeophyceae is divided into 7 orders:

The orders considered here are presented in an evolutionary sequence with the Dictyotales and Sphacelariales being the most ancient and the Ectocarpales and Laminariales the most recent (de Reviers and Rousseau, 1999; Draisma *et al.*, 2001).

1. Dictyotales. Growth by an apical cell; meiosis occurring in the production of four to eight non-motile spores; oogamous sexual reproduction. e.g., *Dictyota.*
2. Sphacelariales. Growth by an apical cell; daughter cells divided longitudinally to give a polysiphonous structure; isogamous sexual reproduction. e.g., *Sphacelaria.*
3. *Cutleriales.* Trichothallic growth forming a fan-like thallus in at least one generation; anisogamous sexual reproduction e.g., *Cutleria, Aglaozonia.*
4. *Desmarestiales*. Trichothallic growth forming axial cells; oogamous sexual reproduction. e.g., *Desmarestia.*
5. Ectocarpales. Thallus consisting of filaments or filaments compacted together;

reproduction isogamous or anisogamous e.g., *Ectocarpus, Ralfsia* etc.

6. Laminariales. Diploid thallus parenchymatous resulting from an intercalary meristem between the stipe and blade; reproduction oogamous. e.g., *Laminaria, Chorda* etc.
7. Fucales. Growth primarily by a promeristem; gametophyte reduced to egg and sperm; oogamous sexual reproduction. e.g., *Fucus, Ascophyllum, Sargassum* etc.

3. Phylum. **PRYMNESIOPHYTES** (haplophytes): It consists of only one class Prymnesiophyceae.

Class. **Prymnesiophyceae**

Important characteristics

1. It is group of unicellular flagellates, characterized by the possession of a haptonema between two smooth flagella.
2. Cells are commonly covered with scales.
3. Two membranes of the chloroplast endoplasmic reticulum present.
4. Chloroplasts lack girdle lamellae and most contain chlorophylls a, c_1/c_2; β-carotene, diadinoxanthin and diatoxanthin.
5. The anterior end of the cell has a large golgi apparatus, and sometimes a contractile vacuole.
6. The storage product is chrysolaminarin (leucosin) in vesicles in the posterior end of the cell.

The class Prymnesiophyceae is divided into two orders (Edvardsen *et al.*, 2000; Fujiwara *et al.*, 2001):

1. Prymnesiales. Cells with 2 equal smooth flagella, no eyespot, scales commonly covering the cell body. e.g., *Prymnesium, Emiliania, Hymenomonas* etc.
2. Pavlovales. Cells with two unequal flagella often covered with hairs and deposits, eyespots may be present. e.g., *Pavlova* etc.

LIFE CYCLE PATTERNS

Sequential changes of the different phages through which an organism complete the life process, starting from zygote to the zygote of next generation is called the **life cycle**. The dominant plant body may be gametophyte (haploid i.e., n) or sporophyte (diploid i.e., 2n). The gametophytic plant body (n) by mitotic division forms gametes. The union between two gametes results in the formation of zygote which undergoes meiotic division and forms haploid plant again. On the other hand, sporophytic plant body first undergoes meiosis and forms gametes. The gametes by union form zygote, which on germination produces the sporophytic plant again.

The sexually reproducing members of algae show various patterns of life cycles. It can be distinguished into the following four types:

1. **Haplontic life cycle**
2. **Diplontic life cycle**
3. **Diplohaplontic life cycle**
   (i) *Isomorphic type*
   (ii) *Heteromorphic type*
4. **Triphasic life cycle**
   (i) *Haplobiontic type*
   (ii) *Diplobiontic type*

1. Haplontic life cycle

The plant body is gametophyte (haploid) and sporophyte (diploid) stage is represented only by zygote. The gametophytic plant develops haploid gametes in the gametangium. The fusion between gametes results the formation of zygote, the only diploid stage i.e., sporophytic phase of the life cycle. The zygote undergoes meiotic division and forms four meiospores. These meiospores develop into haploid plants. The alternation of generations can be interpreted by chromosome number (Fig. 3.20).

This life cycle is also known as **monogenic life cycle**. This type of life cycle is found in majority of Chlorophyceae like *Chlamydomonas, Ulothrix, Oedogonium, Spirogyra, Chara* etc. and all members of Xanthophyceae.

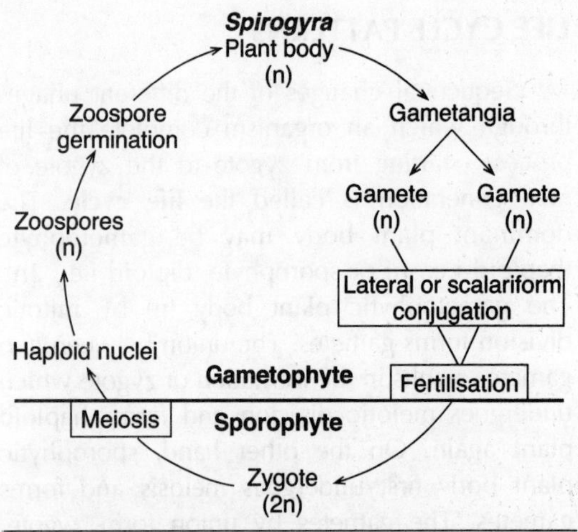

Fig. 3.20: Haplontic life cycle (*Spirogyra*)

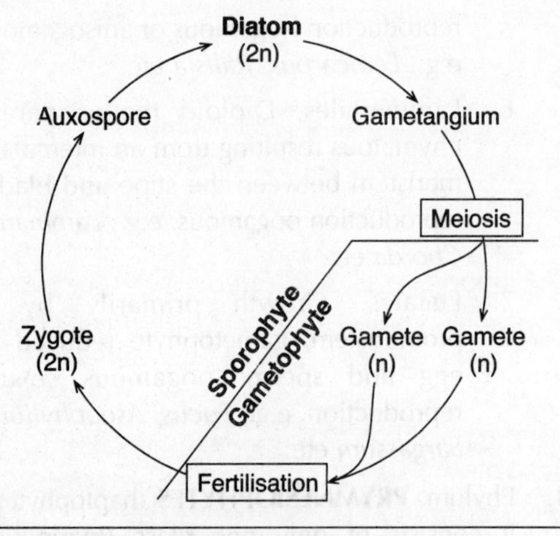

Fig. 3.21: Diplontic life cycle (Diatom)

2. Diplontic life cycle

The plant body is sporophyte and develops sex organs. Sex organs produce gametes by meiosis. The gamete only represents the gametophytic stage. The gametes undergo fertilization immediately and form zygote. The zygote does not undergo meiosis and give rise to new sporophytic plant body (Fig. 3.21).

This type of life cycle is found in majority of the members of Bacillariophyceae, some members of Chlorophyceae like *Cladophora glomerata*. *Fucus* and *Sargassum* of Phaeophyceae also show this type of life cycle.

3. Diplohaplontic life cycle

In this type the haploid and diploid phases are equally prominent and are represented by two distinct vegetative individuals. They differ only in chromosome number and function. The haploid gametophytic plant reproduces by sexual method, while diploid sporophytic plant by asexual process. In this life cycle alternation of two vegetative individuals occurs by sporogenic meiosis and fusion of gametes. It is of ***two types***:

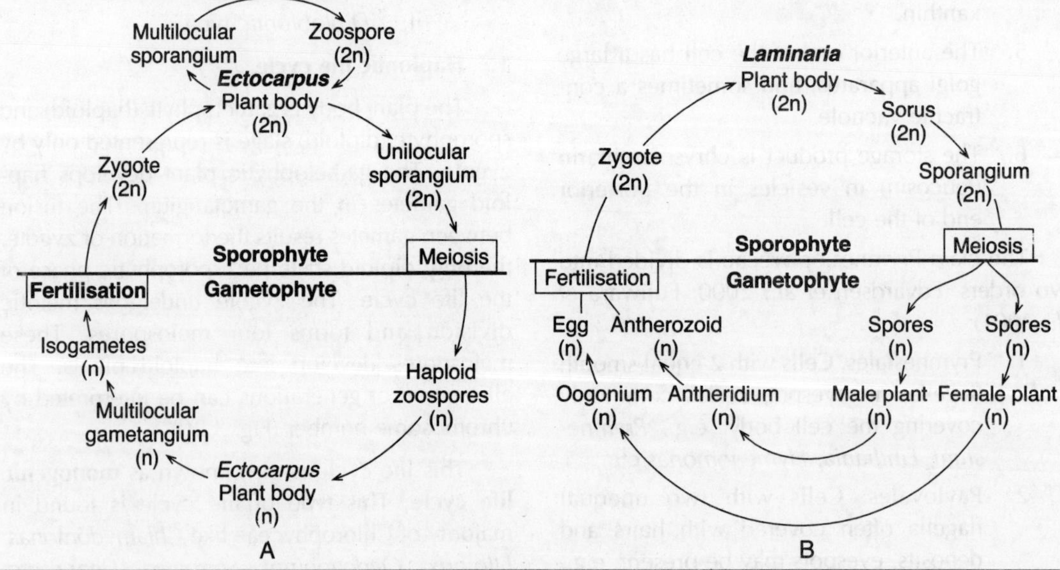

Fig. 3.22: Diplohaplontic life cycle: A. Isomorphic type (*Ectocarpus*), and B. Heteromorphic type (*Laminaria*)

(i) *Isomorphic* or *homologous diplohaplontic type*

In this type both sporophytic and gametophytic plants are morphologically similar and free living. The gametophytic plant (haploid) produces gametes, undergo sexual reproduction and form zygote. The zygote germinates directly into a sporophytic (diploid) plant. The sporophytic plant forms haploid zoospores by meiosis. These zoospores can develop new gametophytic plant.

This type of life cycle is found in *Cladophora, Ulva, Draparnaldiopsis* of Chlorophyceae and *Ectocarpus* of Phaeophyceae (Fig. 3.22A).

(ii) *Heteromorphic* or *Heterologous diplohaplontic type*

In this type both sporophytic (diploid) and gametophytic (haploid) plants are morphologically dissimilar.

Generally the sporophyte is complicated and much elaborate, but the gametophyte is simple and small as found in *Laminaria* of Phaeophyceae. In some cases like *Cutlaria* etc., gametophyte is dominant over sporophyte.

In *Laminaria* the gametophytic plant body is made up of minute filaments which produce gametes. The gametes undergo fusion and form zygote, which germinates directly into a sporophytic plant. The sporophytic plant body is macroscopic and several meters in length. The sporophytic plant bears zoosporangia and produce zoospores after meiotic division. The haploid zoospores on germination produce haploid gametophytic plant (Fig. 3.22B).

4. Triphasic life cycle

In this type, there is succession of three distinct generations. It is of **two types:**

(i) *Haplobiontic type*

In this cycle the gametophytic (haploid) phase is elaborate, dominant and persists for long time than sporophytic (diploid) phase which is represented only by zygote i.e., haplobiontic type and two successive haploid generations are interrupted only by diploid zygote stage indicate its triphasic nature.

This type of life cycle is found in the primitive members of Rhodophyceae like *Batrachospermum* (Fig. 3.23A) and *Nemalion*.

In *Batrachospermum* the gametophytic plant body develops sex organs and produces male (spermatium) and female (egg) gametes. The gametes by fusion form zygote. The zygote immediately undergoes meiosis and produces another haploid gametophytic plant, the carposporophyte. The carposporophyte develops carposporangium which produces haploid

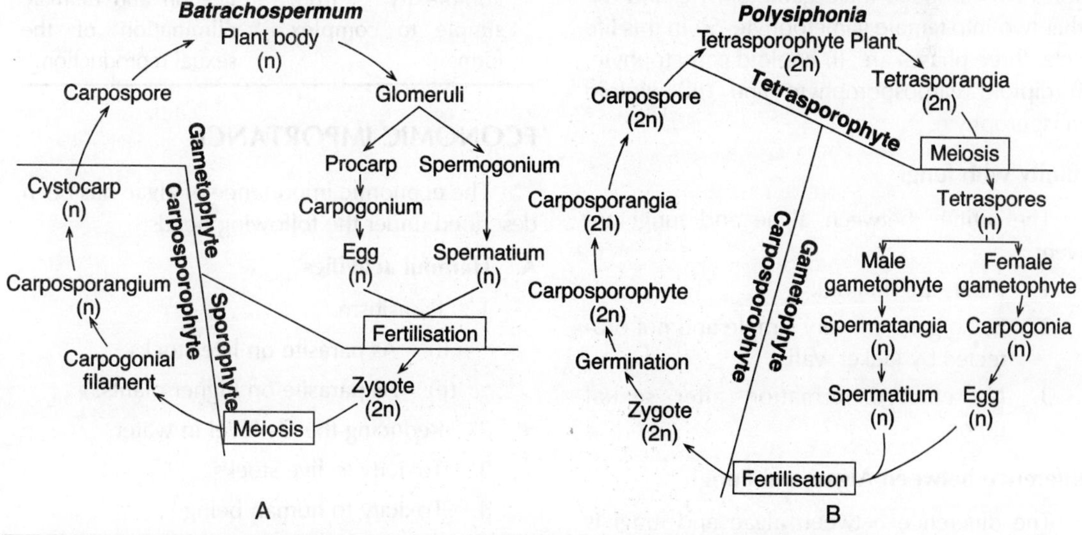

Fig. 3.23: Triphasic life cycle: A. Haplobiontic type (*Batrachospermum*), and B. Diplobiontic type (*Polysiphonia*)

carpospores. The carpospores germinate and develop new free-living gametophytic plant. So in this cycle, three phases are: (i) haploid carposporophyte, (ii) haploid gametophyte and (iii) diploid zygote.

(ii) *Diplobiontic type*

In this type there is one gametophytic phase and two sporophytic phases indicate its triphasic nature and the sporophytic phase is more elaborate and persists for long duration than the gametophyte i.e., diplobiontic type (Fig. 3.23B).

This type of life cycle is found in *Polysiphonia*, a member of Rhodophyceae.

In *Polysiphonia* (Fig. 3.23B), the gametophytic phase is represented by two types of gametophytic plant i.e., male and female plant, those bear spermatangium and carpogonium respectively. Later, the spermatangium and carpogonium develop sperms and egg respectively. The male and female gamete i.e., sperm and egg undergo fusion and form zygote. The zygote (2n) develops into a diploid carpospophytic phase. The diploid carpospores are formed in the carposporophyte. The carpospores on germination develop the diploid tetrasporophytic plants. The tetrasporophytic plant develops diploid tetrasporangia each of which produces four tetraspores (n) by meiotic division. They are liberated by splitting of sporangial wall. Out of four tetraspores two produce male gametophyte and the other two into female gametophyte. So in this life cycle, three phases are: (i) haploid gametophyte, (ii) diploid carposporophyte and (iii) diploid tetrasporophyte.

Affinity with fungi

The affinity between algae and fungi are given:

1. Plants are thalloid.
2. Sex organs are very simple and not protected by jacket wall.
3. No embryo formation after sexual union.

Difference between Algae and Fungi

The difference between algae and fungi is given in Table 3.2.

Table 3.2: Differences between Algae and Fungi

Algae	Fungi
1. Presence of chlorophyll.	1. Absence of chlorophyll.
2. Algae can manufacture food, by photosynthesis i.e., they are **autophyte**.	2. Fungi are unable to produce their own food i.e. **heterophyte** (either as parasite or as saprophyte).
3. They are mostly aquatic, only a few are terrestrial.	3. They are mostly terrestrial, but only a few are aquatic.
4. They may be unicellular to complex parenchymatous thallus.	4. They may be unicellular to thread-like mycelium which may aggregate to form plectenchyma.
5. The cell wall is mainly made up of cellulose except Cyanophyceae.	5. The cell wall is made up of chitin.
6. The reserve food is mainly starch.	6. The reserve food accumulates in the form of glycogen.
7. They grow in habitat where light is available.	7. They prefer to grow in darkness.
8. The process of reproduction proceeds in complexity from simple to complex forms.	8. There is a gradual and progressive simplification and ultimate elimination of the sexual reproduction.

ECONOMIC IMPORTANCE

The economic importance of algae has been described under the following heads:

A. Harmful activities

1. Parasitism.
   (i) As parasite on live stocks.
   (ii) As parasite on higher plants.
2. Reducing the O_2 level in water.
3. Toxicity to live stocks.
4. Toxicity to human being.
5. Fouling of ship.

6. Contamination in drinking water.
7. Blocking of photosynthesis.
8. Destruction of exposed fibres.

B. Useful activities

1. Food
2. Fodder
3. Industry
   (i) Agar agar
   (ii) Carrageenan
   (iii) Alginic acid
   iv) Diatomite
   (v) Funori
   (vi) Medicine
4. Biofertilizer
5. Minerals
6. Disposal of sewage
7. Used in biological experiments
8. Used in the production of H_2 fuel.

Introduction

The algae have enormous value to the human beings. It becomes economically important because of its both harmful and useful activities. Some of the harmful and useful activities of algae are given:

A. Harmful activities

1. **Parasitism**

(i) ***As parasite on live stocks.*** The algae of different groups like Volvocales, Chlorococcales, Cyanophyceae and several others occur in such a great abundance which makes the water colour either green or blue green and causes the blockage of gills of fishes. As a result during night time(when photosynthesis stops) the respiration of fishes is hampered causing death.

(ii) ***As parasite on higher plants.*** Out of many parasitic algae, *Cephaleuros* is the important genus. Different species of *Cephaleuros* grow on the leaves of different angiosperms like *Rhododendron, Magnolia, Camellia sinensis* (tea), *Coffea arabica* (coffee) and *Piper nigrun* (Pepper). *Cephaleuros virescens* causes **Red rust of tea** plants, causing economic losses mainly during slow growth.

2. **Reducing the O_2 level in water.** Algae like *Microcystis, Anabaena, Aphanisomenon* form water blooms and can grow well in O_2-deficient water. The continuous respiration by submerged plants and animals during night time (when the photosynthesis does not take place) causes the depletion of O_2 to almost zero level. At that condition mortality of both animals and other submerged plants takes place due to suffocation.

3. **Toxicity to live stocks** (Domestic animals like cow, pig, horse etc.). With the increase in algal population, the water becomes difficult to drink by the live stocks. Toxic substances released by some algae make the water poisonous and cause the death of livestocks after drinking that water.

4. **Toxicity to human being.** Toxins secreted by *Gymnodinium* affect the muscular and nervous systems of some fishes. *Gonyaulax catenella* secretes an endotoxin, tetraodontoxin ($C_{16}H_{31}O_{16}$), which accumulates in the body of shell fish. Consumption of such shell fish by human being causes paralysis or even death within a day.

5. **Fouling of ship.** Some algae become attached and grow on the outer surface of ship called fouling, which retards the speed. To overcome such problem the ships are dried at intervals and painted with copper paints.

6. **Contamination in drinking water.** Algae of different groups mainly blue green, green and some others contaminate the water of reservoirs. They develop a foul odour in water and make it unhygienic for human being. Some mucilaginous substances are secreted by algae on which many pathogenic bacteria can grow nicely and cause several human and plant diseases.

Different diseases may appear by drinking that water: (i) *Chlorella* and *Lyngbya* cause different skin diseases, (ii) *Gymnodinium brevis* cause respiratory disease and (iii) *Microcystis, Anabaena* cause gastric troubles.

7. **Blocking of photosynthesis**. The epiphytic algae growing on the leaf surface of other plants hamper photosynthesis, thereby the growth of the mother plant becomes reduced.

8. **Destruction of exposed fibres**. Some blue green algae grow well on the wet exposed fibres, marks by black spots. The algal growth usually follows bacterial growth and the fibres get damaged severely.

B. Useful activities

1. Food. Algae are used by man from ancient times, but due to population boom in the last decades the rate of consumption of edible algae has been increased and it will definitely increase much more in near future. The algal species is becoming a popular food to the mankind because of its high nutritive value and more yield per unit area than the conventional crops.

In Japan, the food production by marine algae is about eight times more than that of land plants.

Some algae commonly used as food are *Nostoc* and *Spirulina* (Cyanophyceae), and *Chlorella, Chondrus, Codium, Porphyra, Rhodymenia, Ulva* etc. (other groups).

(i) ***Nostoc***. *N. commune* is boiled and used as soup in China.

(ii) ***Spirulina***. It is a blue green alga, rich in protein (60%), vitamin and unsaturated fatty acids. In India it is available as tablet prepared by CFTRI (Central Food Technological Research Institute), Mysore.

(iii) ***Chlorella***. It has become popularly known to human beings after its use in the space research and nuclear submarines for the generation of oxygen.

It is also important for its nutritive value and can be compared with soyabeans. It grows very fast in controlled condition and yields about 13 metric tonnes/year/acre. It contains, carbohydrate 30%, protein 30%, lipid 15% etc. But the digestion of its cell wall is a problem to the human being and researches are going on to solve it. According to Witsch (1959) the vitamin B content of young culture is equivalent to the lemon juice.

(iv) ***Chondrus crispus***. It is commonly known as **Irish moss**. The alga is cooked with milk and with the addition of vanilla, it makes a highly popular dish, the **blancmanges**.

The gelatinous carbohydrate obtained from this algae is used in pudding. It is used as stabilizer and cleaning agent in beer industry.

(v) ***Codium*** and ***Ulva***. These are used as **salad** in Japan.

(vi) ***Porphyra***. It is a seaweed, belongs to Rhodophyceae. It contains carbohydrate 40-45%, protein 30-35% and vitamin B and C. The common name of food item is **laver or nori** in Japan, **tsats'ai** in China, **sloke** in Britain. It is very popular in different countries including Japan.

(vii) ***Laminaria***. *L. saccharina* is rich in carbohydrate (57%) and the commonly used food is called **'kombu'**.

(viii) ***Rhodymenia***. *R. palmata* is used to prepare a salty confection, commonly known as **'dulse'**.

(ix) ***Monostroma***. In Japan, it is used in the preparation of common food, known as **'aonori'**.

(x) ***Scenedesmus***. It is rich in protein and threonine and is equivalent to the skimmed milk.

(xi) ***Spirogyra*** and ***Oedogonium***. In South India, Green Laver, a kind of food is prepared from *Spirogyra* and *Oedogonium*.

2. Fodder. Many algae become popular as fodder due to their vitamin and micronutrient, in addition to their protein and carbohydrate content. The algae commonly used as fodder in different countries are *Fucus, Laminaria, Sargassum, Alaria, Rhodymenia, Ascophyllum, Macrocystis* etc.

The fat content of milk becomes increased with the addition of seaweed. The seaweeds like

Rhodymenia palmata and *Ascophyllum esculenta* are used to feed the cattle in Scotland, Ireland etc. *Macrocystis*, the Pacific-coast kelp is used directly or indirectly in the form of powder to feed cattle, hogs and poultry birds in USA because of high mineral and vitamin content. The algae also become popular as fodder in other countries like France, Norway, Germany, Denmark etc. *Macrocystis* sp. is rich in vitamins A and E — used as fodder in France. Dried *Pelvetia*, a seaweed used as cow feed increase the milk yielding capacity of cattles. The iodine and carotene content in egg-yolk increases by feeding the processed sea weeds used as food. The egg lying capacity of the poultry birds also increases by feeding the processed sea weeds like *Fucus, Laminaria* etc. *Sargassum* is used as fodder in China.

3. Industry. Algae have been used to develop many products of commercial and pharmaceutical importance. These are Agar-agar, Carrageenan, Diatomite, Alginate, Funori, Medicine etc.

(i) ***Agar-agar***. Commercially it is obtained from *Gelidium nudifrons, G. pusillum, G. robustrum, Gracilaria verrucosa* and also from different species of *Ahnfeltia, Chondrus, Gigartina, Acanthopeltis* and *Pterocladia*.

It is composed of two polysaccharides, agarose and agaropectin. The agar-agar is readily soluble in hot water but not in cold water. At 1.5% concentration it forms a clear, solid and elastic gel on cooling the melted agar to 32-39°C and does not melt again below 85°C.

Southeast Asia and Japan are the main centres of agar-agar production. For the commercial production of agar-agar following procedure is followed: Collection of plants from sea → bleached in the sun → boiled the material for few hours → the extract is acidified → material is then frozen and thawed → it is then dried. It is available in the market in the form of cakes, flakes or powder.

Use: The agar-agar is used in food, pharmaceutical, cosmetic industries and scientific laboratories.

(a) *Food*: It is used in processed cheese, jam, jellies, cream and pudding etc. It is also used as gelling and thickening agent in the preservation and canning of meat and fish.

(b) *Pharmaceutical*: It is used as **laxative**, pills, different ointments and also used in drug for slow release when requires.

(c) *Cosmetics*: It is used in cosmetics like lotions etc.

(d) *Scientific laboratories*: In scientific laboratories it is used for stiffening the culture medium used at 1.5-2.0 percent concentration.

(ii) ***Carrageenan***. It is obtained from the cell wall of *Gigartina stellata, Chondrus crispus*, the Irish moss and *Eucheuma*.

It is a phycocolloid, almost similar to agar-agar, but the ash content is higher and requires higher concentration to form solid gels. The phycocolloid consists of κ-carrageenan and λ-carrageenan.

The carrageenan acts as a blood coagulant. It is also used to stabilise emulsions and to cure cough. It is used as a component of deodorants, cosmetics, toothpastes, paints etc.

(iii) ***Alginic acid***. It is extracted from the cell wall of some brown algae like *Ascophyllum, Fucus, Ecklonia, Macrocystis, Laminaria, Durvillea* and *Lessonia*. The content of alginic acid varies in different genera and ranges from 10-40%. The salts of alginic acid found in the cell wall are the alginates.

It is non-toxic, insoluble in water, viscous, but becomes hard when dry.

It is used as:

(a) *Thickeners*. In food industries, it is used in the preparation of sauce, soup, cream etc; in textile industry as printing pastes and cosmetics.

(b) *Emulsifiers*. It is used in polish, emulsion paints etc.

(c) *Gelling agent*. It is used in confectionary, powders, paints, ice-cream etc.

(d) *Other uses*. It is used in the production of artificial fibres, plastics, rubbers etc.

(iv) ***Diatomite***. After the death of diatom cells the outer covering i.e., the silicified wall becomes accumulated at the bottom of water. The accumulation may be thicker during favourable condition. These deposits are called **diatomaceous earth** or **diatomite**. It is very suitable for use in different industries:

(a) *As filter*. It is used as filter in different industries like sugar (to filter microorganism), oil and chemical industry. Diatomite is also used as filter for battery boxes.

(b) *As insulator*. It is used as insulator in boilers and blast furnaces for its heat resistant ability.

(c) *As absorbent*. It is used as absorbent of liquid nitroglycerine.

(d) *Other uses*. Diatomite is used as abrasive (i.e., capable of rubbing or grinding down) substance for manufacture of metal paints, polish, varnish and toothpaste etc. It is also used with bakelite for fuse and switch boxes.

(v) ***Funori***. It is a type of glue, obtained from *Gloiopeltis furcata*, used as sizing agent in paper and textile industry. In cosmetic industry it is used for curling of hairs and preparation of dyeing.

(vi) ***Medicine***. Chlorellin, an ***antibiotic*** is extracted from *Chlorella*. It is used to control microorganisms like Gram –ve bacteria.

In addition to above, other algae such as *Cladophora, Laminaria, Lyngbya* and *Halidrys* are able to synthesize antibiotics.

Both Gram +ve and Gram –ve bacteria can be controlled by the extract of *Ascophyllum nodosum*. Extracts of *Cordium, Corallina* and *Durvillea* etc. are used as vermifuge. In China, an antihelminthics drug **'Tse-ko-Tsoi'**, is prepared from *Digenia simplex*, a red alga.

Different sea weeds are used as good sources of ***vitamins***:

(a) Riboflavin from *Porphyra tenera, Gelidium amansii*,

(b) Thiamine from *Rhodomela subfusca*,

(c) Vitamin C from *Porphyra lacinata* and

(d) Vitamin A from *Nitzschia*, a diatom.

Various brown algae like *Ecklonia, Eisenia, Fucus* and *Laminaria* are the richest source of **iodine**, used to control goiter. About 100 tonnes of iodine are produced per year in Japan from brown sea weeds.

The *Gelidium* is used to control stomach disorders.

The cultivation of *Nitella* in the pond greatly reduces the population of **Mosquito**, indirectly controls malaria.

4. Biofertilizer. Members of Cyanophyceae like *Nostoc, Anabaena* etc. can ***fix atmospheric nitrogen*** and form nitrogenous compounds. These compounds are further absorbed by the plant for their metabolic activity and increase yield.

5. Minerals. Minerals like Copper, Cobalt, Chromium, Boron, Iron, Zinc, Vanadium and Manganese are present in high amount in sea weeds. Hence the seaweeds are used as stock feed as well as natural fertiliser.

The *Polysiphonia* and *Rhodomela*, the members of Rhodophyceae are richest source of **bromine**.

Various brown algae like *Ecklonia, Eisenia, Fucus* and *Laminaria* are the richest source of **iodine**.

Soda and Potash are present at high percentage in brown algae, thus the ash of those algae are used in the manufacture of alum, glass wares and soap.

6. Disposal of sewage. Water-borne wastes of industry and domestic house are called 'sewage'. It is rich in sulphur, nitrogen, phosphorus and potassium. The anaerobic breakdown of sewage, gives out bad odour. So the aerobic breakdown is preferred, which does not give out bad odour and the products can also serve as fertilizer. Bacteria are carried out in a container and the required oxygen is collected from algae. Unicellular algae like *Chlamydomonas, Chlorella,*

Scenedesmus, Euglena etc. are used in this process. In turn, ammonia and nitrogen compounds become available to the algae as nutrient.

7. Use in biological experiments. Algae are used as experimental material for different research works. *Chlorella* has been used to study the path of carbon in photosynthesis and *Acetabularia* in genetical researches. *Halicystis* is used to study membrane permeability.

8. Use in the production of H_2 fuel. Production of H_2 is an important area of biotechnology for the production of non-conventional energy. Different algae like *Chlamydomonas, Dunaliella, Porphyridium, Oscillatoria* etc. are used in this process. Normally photosynthesis takes place in the above plants and the photosynthetic principal is utilized to gain the goal. In the first part of photosynthesis, photolysis of water takes place by which water splits into, hydrogen ion (H^+), O_2 and electrons. The hydrogenase enzyme of the above algae generates hydrogen gas (H_2) from hydrogen ion (H^+). Some Cyanophycean members such as *Anabaena* etc. are used in this process, but here nitrogenase is utilised as the main hydrogen producing enzyme.

ROLE OF ALGAE: IN ENVIRONMENT, AGRICULTURE, BIOTECHNOLOGY & INDUSTRY

1. Environment

Algae are the autophytes i.e. they are able to produce their own food. They grow mostly in different aquatic environments from fresh water to saline water in marine habitat. They form the base of the fresh water and marine food chain. They are most important organism in our green planet because they produce oxygen, more oxygen than anything else does. As these **algae** die, they are consumed by organisms called decomposers (mostly fungi and bacteria). The decomposers feed on decaying plants and release **important** minerals that are used by other organisms in the food web.

Single-celled Blue-green algae i.e. Cyanobacteria, that photosynthesized carbon dioxide (CO_2) from Earth's early atmosphere and released oxygen (O_2) in the atmosphere some 2.45 million years ago. This process made the world habitable by more complex organisms, like modern human beings who showed up some 200,000 years ago.

Instead of placing algae at the bottom of the food chain, we should be recognized them as a **vital key** to supporting all of life, with the unique ability to turn light energy from the Sun into carbon-based molecules that society harnesses and uses in countless ways. We rely on algae for their nutrients and proteins, but most importantly their lipids.

Lipids are molecules that are made up of carbon and hydrogen atoms and serve as the building blocks of countless other molecules. Lipids have a range of functions within algae and other organisms, such as structural features of the cell membrane (e.g., phospholipid fatty acids), energy storage (e.g., triglycerides), hormone precursor molecules (e.g., sterols), pigments (e.g., chlorophyll), and toxins – such as those involved in harmful algal blooms.

In addition to the above, algae also acts as binding agent to **bind the soil particles**. The disturbed or burnt soils are soon covered with a growth of green and blue-green algae thus reducing the danger of soil erosion. The role of Cyanophycean members as a pioneer in colony formation and thus in soil formation is well known.

2. Agriculture

Algae are used in agriculture in various ways. These are:

A. Reclamation of Alkaline 'Usar' Land

In India, due to high alkalinity in vast tracts of land, it cannot be used for cultivation of crops, commonly known as **'usar'** soil. This type of lands would be cultivable by lowering the pH and organic contents and increasing the water holding capacity of the soil. All these functions are carried out by the blue-green algae. In this soil, during rainy season the blue-green algae like *Nostoc, Anabaena, Scytonema* and *Aulosira* are grown in plenty. The process involves a series of successive growth of the algal crop in a water-logged environment.

After one year of algal growth, following changes takes place. These are:

(i) The pH changes from 9.5 to 7.6,

(ii) Organic content increased from 36.5% to 59.7%,

(iii) Nitrogen content increased from 30% to 38.4%,

(iv) Exchangeable calcium from 20% to 33% and

(v) Water holding capacity of the soil is also increased by 40%.

In such a **'reclaimed'** land, a crop like paddy grows well with a yield of 1576-2000 lbs/acre. The process is now being practiced widely.

B. Fertilizer

Various types of algae of different groups, such as blue-green algae, brown and red algae are used as fertilizer in the soil. The blue-green algae are used to enrich the N_2 content of the soil and brown and red algae are used as organic fertilizers.

(i) **Blue-green algae**: The blue-green algae like *Nostoc, Anabaena* etc. have the greatest utility, as a friend of the farmers, for their capacity to fix atmospheric nitrogen and thus enriching the soil. In paddy field they have been seen to produce an effect similar to the application of 30 kg ammonium sulphate/acre (Watanabe, 1959).

Similarly *Aulosira fertilissima* is a very common Blue-green algae in rice fields of Indian found to add about 21.6 kg nitrogen fixed/acre/crop (Singh, 1962) and also the increase of the total organic matter in the soil.

(ii) **Brown and Red algae**: The brown and red algae are used as organic fertilizers, especially on lands of coastal areas. The weed is used either directly or as a seaweed meal. Concentrated extract of different sea weeds are also sold as liquid fertilizer. Coralline algae such as *Lithothamnion calcareum* and *Lithophyllum* sp., are used profusely for liming the soil. Similarly *Chara* is also used as calcium carbonate source.

C. Pollution due to agricultural water drains

Pollution of agricultural water drains is a man-made phenomenon, developed either when the concentrations of naturally occurring substances are increased or when non-natural synthetic compounds (**xenobiotics**) are released into the environment. Organic substances released into the environment as a result of domestic, agricultural and industrial activities leads to an inorganic pollution (Mouchet, 1986). This can be controlled by the algae like *Chlorella* and *Dunaliella*.

D. Iodine level in cow milk

Seaweeds are an important source of iodine. Iodine levels in cow milk depend on the type of feeding. Feeding milk cattle with seaweeds can increase the quantity of iodine in milk. Similarly rate of egg-laying capacity in hen is also increased by algae used as feeding additive.

3. Biotechnology

In recent years, microalgae are important biological resources that have a wide range of biotechnological applications. Due to their high nutritional value, microalgae such as *Spirulina* and *Chlorella* are being mass cultured for healthy food. A variety of high-value products including polyunsaturated fatty acids (PUFA), pigments such as carotenoids and phycobiliproteins, and bioactive compounds are useful in pharmaceuticals, as well as for industrial applications. In terms of environmental biotechnology, microalgae are very useful for bioremediation of agro-industrial wastewater, and as a biological tool for assessment and monitoring of environmental toxicants such as heavy metals, pesticides and also pharmaceuticals. In recent years, microalgae have attracted much interest due to their potential use as feedstock and also for biodiesel production. There has been active research on micro-algal biotechnology in different countries, tapping into the potential of our rich micro-algal resources for high-value products and applications in wastewater treatment and assessment of environmental toxicants.

4. Industry

The role of algae in industry have been discussed in Page no. 227.

SIGNIFICANT CONTRIBUTIONS OF IMPORTANT PHYCOLOGISTS

F. E. Fritsch (1879-1954)

His full name is Felix Eugen Fritsch.

Prof. Fritsch was one of the greatest and eminent Phycologists. He obtained his Ph.D. from Munich University, Germany in 1902. Later he went to London and worked in the Department of Botany in various capacities and finally became Professor of Botany in 1948.

In 1938, he came to India to attend the "Indian Science Congress" and delivered the lecture on the "algal problems peculiar to the tropics". For his valuable contributions, he was awarded the Gold Medal from the Linnaean Society of London. He worked mainly on the fresh water algae. In 1902, he published a paper on the germination of zoospores in *Oedogonium*.

Later, in 1935 and 1945, he published the famous books "Structure and Reproductions of the Algae" in two volumes (Vol. I and Vol. II). These books are used even now for the study of algae.

G. M. Smith (1885-1959)

His full name is Gilbert Morgan Smith.

He was a botanist, who worked mainly on the algae. He was best known for his books, particularly the *Freshwater Algae of the United States,* the *Marine Algae of the Monterey Peninsula* and the two volumes of *Cryptogamic Botany.*

M. O. P. Iyengar (1886-1963)

His full name is Mandayam Osuri Parthasarathy Iyengar.

He was an outstanding Indian Phycologist and became famous internationally. He is considered as the "**Father of Indian Phycology**". He obtained his Ph.D. from London in 1909. He worked in the University of Madras and has done a lot of researches on algae. From 1929 to 1950, h e published various papers in reputed Journals such as Journal of Linnean Society, London, Journal of Indian Botanical Society and the Proceedings of Indian Science Congress.

He also published Volvocales which is a monograph. He was especially interested in Chlorophyceae. The topics of his publications include the colonial Volvocales of south India, fertilization of *Eudorina elegans,* lateral conjugation in *Spirogyra jogensis,* conjugations in *Caulerpra,* morphology and cytology of *Polysiphonia platycarpa,* life cycle of *Cylindrocapsa, Dictyosphaerium, Microdictyon, Hydrodictyon indicum* and *Botrydium.* He discovered and studied many new genera of algae such as *Dendrocystis, Fritschielle tuberosa, Pseudovalonia, Charaoisiphon, Cladospongia* etc. The well-known phycologists such as Desikachary and Sundarlingam were associated with and trained by Iyengar.

He was awarded the Birbal Sahani Gold medal from Indian Botanical Society and Sunder Lal Hora Gold Medal of the National Institute of Sciences, India. To honour and remember him, certain algae are named after him — these are *Iyengaria, Iyengariella, Anabaena Iyengarii, Chlamydomonas Iyengarii* and *Scytonema Iyengarii.*

Among his other important of papers, "Some interesting green algae" is fascinating in which he recorded the occurrence of *Chlamydomonas volvocinae* — a new species inhabiting the coenobium of *Volvox* and a few new species such as *Physicocytium indicum, Chlamydomonas eudorineae, C. oocysticola, C. ulotrichae, C. dictyosphaeriae, Paulschulzia indica, Gloeococcus pyriformis, Charadosiphon rivularis.* Iyengar and Desikachary established a new genus *Mastigocladopsis*, having both lateral and terminal heterocysts.

Iyengar was the president of Phycological Society of India. His huge collections of specimens and herberium sheets were handed over to the Department of Botany, University of Madras.

T. V. Desikachary

His full name is Thamarapu Vedanta Desikachary.

He was a student of Prof. M.O.P Iyengar. Since 1940 onwards, he worked in the University of Madras and published papers on *Gloeotricha, Nostochopsis, Camptyloneus* and *Iyengaariella tirupatiensis*. In 1969, he wrote a monograph on Cyanophyta, published by the ICAR. This book gives detail information on blue green algae (BGA) of India and neighboring countries. It comprises of neat drawings of 139 plates of more than 150 genera of BGA.

R. N. Singh

His full name is Rabindra Narain Singh.

He has done a lot of works on varied branches of algae such as Cyanophyceae, Zygnemataceae, Oedogoniaceae and Chaetophorales. He studied the life histories of *Fritschillea tuberosa and Draparnaldiopsis indica*. He has suggested the cultivation of BGA for the reclamation of **usarland** in India and has also published a monograph entitled *"Role of the BGA in the economy of Indian agriculture"*. He has contributed on N_2 fixation by BGA and established that *Aulosira fertilissima* tops the list of Indian algae.

H. D. Kumar

His full name is Har Darshan Kumar.

He worked on parasexuality and metabolic behavior of BGA, mainly *Anacystis nidulans*. He has written several books on various topics. Because the Indian subcontinent is fringed with oceanic water, studies on marine algae have tremendous prospects from fundamental and economic point of view. He wrote the Book *'Text Book of Algae'*.

M. S. Randhawa (1932-1960)

His full name is Mohinder Singh Randhawa.

He published a series of papers especially on Zygnemaceae, Oedogoniales and Vaucheriaceae from Punjab and also from Uttar Pradesh. He recorded seventy species in Zygnemaceae out of which eleven are the new species of *Spirogyra*. In 1939, he identified and described a new species, the *Sirocladium kumaoense*. He described certain terrestrial green algae including *Zygnema terrestre* and two new species of *Oedogonium*. He has reported a new kind of akinete formation in *Vaucheria geminata* and six species of *Vaucheria* from Punjab. It includes *Vaucheria amphibian*, a new species.

B. N. Prasad

His full name is Badri Nath Prasad.

He is well known for his contribution on cytology of algae, especially of *Schizomeris, Zygnema* and *Mougeotia*. His other work of general interest includes the systematic description of algae; some Nostocaceae from U.P.; algae of alkaline usar soil; thermal algae of Himalyan hot springs; notes on conjugation in *Zygnema;* loop formation in *Heterothrix* and some interesting morphological features in *Tolypothorix*. He has a credit of first time collection of certain forms in India such as *Coleochaete nitellarum, Uronema gigas, Heterothrix trichodes, Botrydiopsis arrhiza,* etc. He described six genera and many species as new records from India.

MYXOPHYCEAE

(CYANOPHYCEAE/CYANOBACTERIA)

It is a primitive group of algae, consists of 150 genera and about 2,500 species. In India, the division is represented by 98 genera and about 833 species. Members of the class Myxophyceae (Cyanophyceae) are commonly known as ***blue green algae***. The name blue green algae is given because of the presence of a dominant pigment c-phycocyanin, the blue green pigment. In addition, other pigments like chlorophyll a (green), c-phycoerythrin (red), β-carotene and different xanthophylls are also present. The members of this class are the simplest living autotrophic prokaryotes.

They have the following important characteristics:

(a) Nucleus is of prokaryotic nature i.e., devoid of nuclear membrane and nucleolus,

(b) Absence of well-organised cell organelles, and

(c) Pigments are distributed throughout the chromoplasm (the outer part of protoplasm).

Depending on the above prokaryotic characteristics many microbiologists consider the members of Cyanophyceae as bacteria. Based on prokaryotic cell structure like bacteria, Christensen (1962) placed both Cyanophyta and bacteria under a common phylum Prokaryota. Cyanophyta or blue green algae have also been named as **cyanobacteria**.

Important characteristics

The important characteristics of the division are as follows:

1. The individual cells are **prokaryotic** in nature. The nucleus is incipient type and they lack membrane bound organelles.
2. Both vegetative and reproductive cells are **non-flagellated**.
3. Cell wall is made up of microfibrils and is differentiated into four (4) layers. The cell wall composed of **mucopeptide**, along with carbohydrates, amino acids and fatty acids.
4. Locomotion is generally absent, but when occurs, it is of gliding or jerky type.
5. The principal pigments are chlorophyll a (green), **c-phycocyanin** (blue) and **c-phycoerythrin** (red). In addition, other pigments like β-carotene and different xanthophylls like myxoxanthin and myxoxanthophyll are also present.
6. Membrane bound chromatophore are absent. Pigments are found embedded in thylakoids.
7. The reserve foods are **cyanophycean starch** and cyanophycean granules (protein).
8. Many filamentous members possess specialized cells of disputed function (supposed to be the centre of N_2 fixation) known as **heterocysts**.
9. Reproduction takes place by vegetative and asexual methods. Vegetative reproduction takes place by cell division, fragmentation etc. Asexual reproduction takes place by **endospores, exospores, akinetes, nannospores** etc.
10. **Sexual reproduction is completely absent.** Genetic recombination is reported in 2 cases.

Ecology and Distribution

Cyanophyceae(Blue green algae) are generally found in almost every terrestrial and aquatic habitat like oceans, fresh water, damp soil, temporarily moistened rocks in deserts, bare rock and soil, and even Antarctic rocks. They can occur as planktonic cells or form phototrophic biofilms. They are found in almost every endolithic ecosystem. A few are endosymbionts in lichens, plants and many protists, provide energy to the host. Some live in the fur of sloths, providing a form of camouflage.

Aquatic members of Cyanophyceae form extensive and highly visible blooms in both freshwater and marine environments. The blooms appear as blue-green paint or scum. These blooms can be toxic, and frequently lead to the closure of recreational waters when spotted.

Cyanophycean growth is favoured in ponds and lakes where waters are calm and have less turbulent mixing. Their life cycles are disrupted when the water naturally or artificially mixes by the flowing water of streams. For this reason blooms of Cyanophyceae seldom occur in rivers unless the water is flowing slowly. Growth is also favoured at higher temperatures, but excessive increase due to global warming becomes problematic. At higher temperatures *Microcystis* species are able to outcompete diatoms and green algae, because of the toxins produced by them.

Based on environmental trends, models and observations suggest that Cyanophyceae will likely increase their dominance in aquatic environments. This can lead to serious consequences, particularly the contamination of sources of drinking water. Cyanophyceae can interfere with water treatment in various ways, primarily by plugging filters, may often large beds of sand and similar media, and by producing cyanotoxins which have the potential to cause serious illness if consumed. Consequences may also lie within fisheries and waste management practices. Anthropogenic eutrophication, rising temperatures, vertical stratification and increased atmospheric carbon dioxide are contributors to Cyanophyceae increasing dominance in aquatic ecosystems.

Cyanophyceae have been found to play an important role in terrestrial habitats. It has been widely reported that cyanophycean soil crusts help to stabilize soil to prevent erosion and retain water. An example of a Cyanophycean species that does so is *Microcoleus vaginatus*, which stabilizes soil by using a polysaccharide sheath that binds to sand particles and absorbs water.

Some of these organisms contribute significantly to global ecology and the oxygen cycle. The tiny marine Cyanophyceae *Prochlorococcus,* accounts for more than half of the photosynthesis in the open ocean. Many Cyanophyceae even display the circadian rhythms that were once thought to exist only in eukaryotic cells.

Cyanophyceae are arguably the most successful group of microorganisms on earth. They are the most genetically diverse; occupy a broad range of habitats across all latitudes, widespread in freshwater, marine, and terrestrial ecosystems, and they are found in the most extreme niches such as in hot springs, salt water, and hypersaline bays. Photoautotrophic, oxygen-producing Cyanophyceae created the conditions in the planet's early atmosphere that directed the evolution of aerobic metabolism and eukaryotic photosynthesis. Cyanophyceae fulfill vital ecological functions in the world's oceans, being important contributors in carbon and nitrogen amount in the world.

General account

Occurrence

Members of Cyanophyceae are available in different habitats. Most of the species are fresh water (e.g., *Oscillatoria, Rivularia*), a few are marine (e.g., *Trichodesmium, Darmocarpa*), and some species of *Oscillatoria* and *Nostoc* are grown on terrestrial habitat. Species of some members like *Anabaena* grow as endophytes in thallus of *Anthoceros* (Bryophyta) and in leaves of *Azolla* (Pteridophyta) and *Nostoc* in the root of *Cycas* (Gymnosperm).

Species of *Nostoc, Scytonema, Gloeocapsa,* and *Chroococcus* grow symbiotically with different fungi and form lichen. Some free-living members like *Nostoc, Anabaena* etc. can fix atmospheric nitrogen and increase soil fertility.

Thallus

Plants of this group show much variation in their thallus organisation (**range of thallus**). The thallus may be of unicellular or colonial forms:

1. **Unicellular form**: In unicellular form, the cells may be oval or spherical. Common members are *Gloeocapsa* (Fig. 3.24A), *Chroococcus* and *Synechococcus*.
2. **Colonial form**: In most of the members the cells after division remain attached by their cell wall or remain together in a common

gelatinous matrix, called a colony. The colonies may be of two types: (a) non-filamentous and (b) filamentous:

(a) ***Non-filamentous type***. The cells of this type divide either alternately or in three planes, thereby they form spherical (*Gomphosphaera, Coelosphaerum*), cubical (*Eucapsis alpine*, Fig. 3.24C), squarish (*Merismopedia*) or irregular (*Microcystis*, Fig. 3.24B) colony.

(b) ***Filamentous type.*** By the repeated cell division in one plane, single row of cells are formed, known as **trichome**. e.g., *Oscillatoria* (Fig. 3.24D), *Spirulina,*

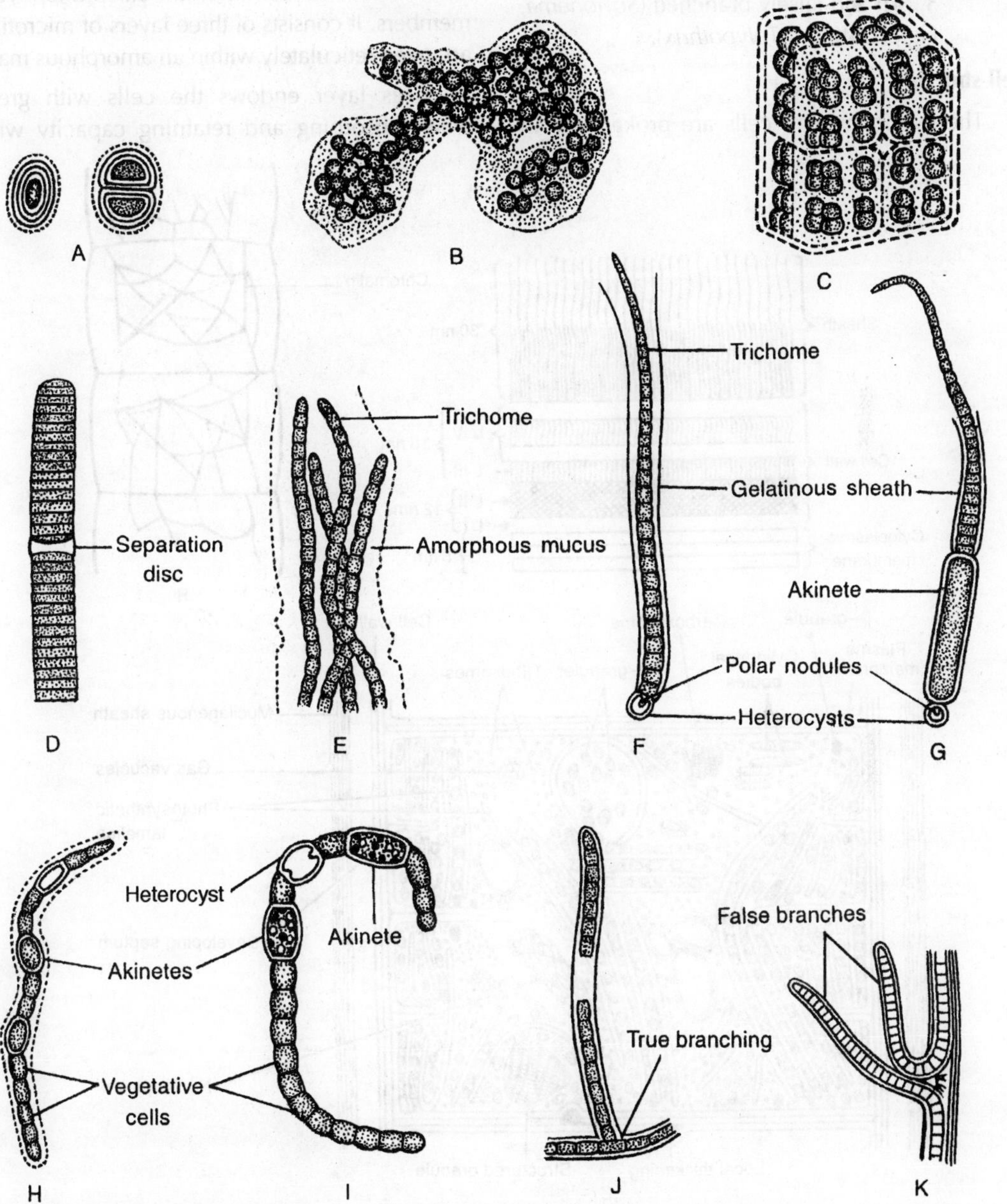

Fig. 3.24: A few members of Cyanophyceae showing thallus organization: A. *Gloeocapsa* sp., B. *Microcystis* sp., C. *Eucapsis alpina*; D. *Oscillatoria* sp., E. *Microcoleus* sp., F. *Rivularia poliotis*, G. *Gloeotrichia pisum*, H. *Nostoc* sp., I. *Anabaena* sp., J. *Mastigocladus limilosus*, and K. *Scytonema* sp.

Arthosporia etc. The trichome when covered by mucilaginous sheath is called a **filament**. The filament may contain single trichome (*Oscillatoria, Lyngbya*) or several trichomes (*Hydrocoleus, Microcoleus*, Fig. 3.24E). The trichomes may be unbranched (*Oscillatoria, Lyngbya*), branched (*Mastigocladus limilosus*, Fig. 3.24J) and falsely branched (*Scytonema*, Fig. 3.24K and *Tolypothrix*).

Cell structure

The cyanophycean cells are prokaryotic in nature, which rarely exceeds 10 μm in diameter. Each cell consists of outer covering of cell envelop which surrounds the membrane covered protoplast (Fig. 3.25C).

Envelop. It consists of outer mucilaginous sheath and inner cell wall.

Mucilagenous sheath. Presence of mucilaginous sheath is common in all cyanophycean members. It consists of three layers of microfibrils arranged reticulately within an amorphous matrix.

This layer endows the cells with greater water absorbing and retaining capacity which

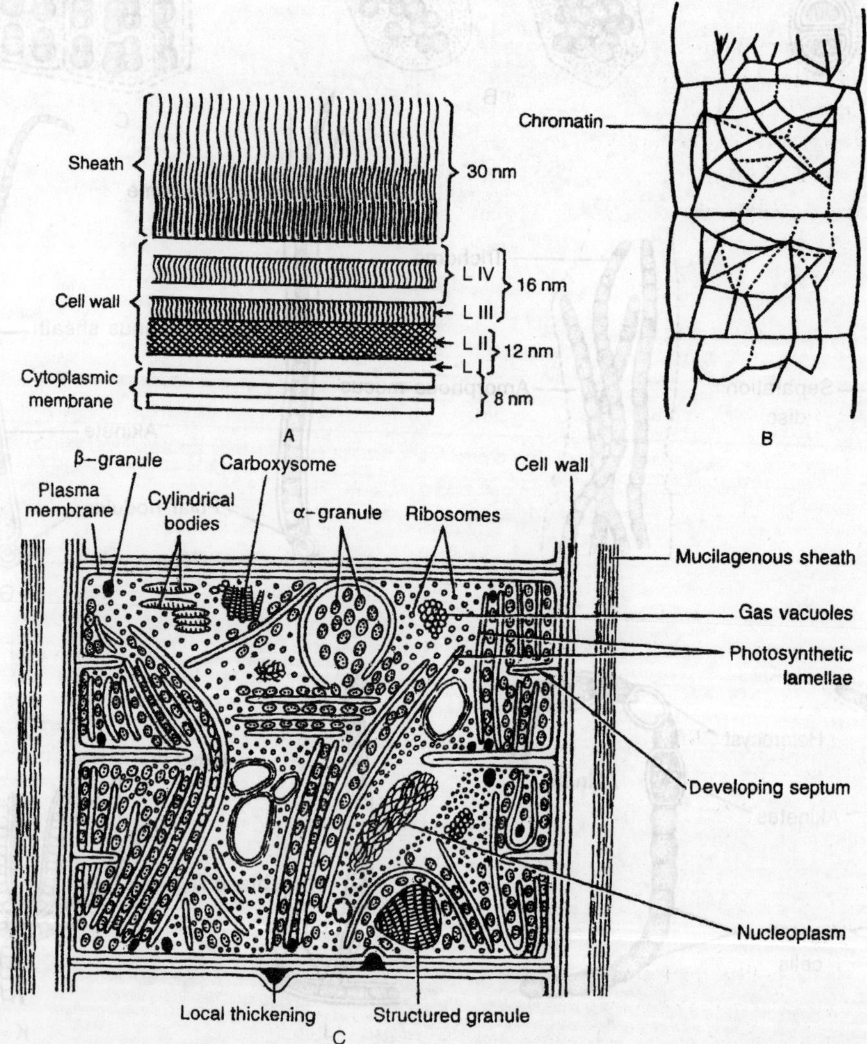

Fig. 3.25: Cell structure of *Oscillatoria*: A. Cell wall as seen under electron microscope, B. Cell structure under light microscope, and C. Cell structure under electron microscope

helps to survive during desiccation and forms a barrier for parasites.

Cell wall. The cell wall consists of four (4) (Fig. 3.25A) layers (under E.M.) named as LI, LII, LIII and LIV by Carr and Whitton (1973). Each layer is about 10 μμ in thickness. The LI is the layer situated near cell membrane and LIV is the outer-most. Cell wall is composed of mucopeptide together with carbohydrates, amino acids and fatty acids like Gram-positive bacteria. The LI and LIII layers are electron transparent, but the LII and LIV layers are electron opaque (impervious).

Cytoplasmic membrane. The cytoplasmic membrane is also known as plasmalemma present just inner to the cell wall. It consists of two electron opaque layers separated by a translucent layer (Fig. 3.25A).

This layer invaginates at different points inside the protoplast and is the site of different biochemical functions that normally takes place in different organelle like mitochondria, endoplasmic reticulum and golgi bodies in the cells of eukaryote.

Protoplast. Studies with Electron Microscope by Wilden and Mercer, Pankrats and Bowen and many others show that the protoplast consists of thylakoids, cytoplasmic inclusion and nucleoplasm.

Thylakoids. These are the complex lamellar system, which functions like the protoplasts of eukaryotes. Thylakoids are not bounded by membrane instead they appear as elongated and flattened sacs composed of two unit membranes. Each membrane is about 75 Å thick. The neighbouring thylakoids are separated by a flattened space of 50 nm. The space is occupied by rows of discoid phycobilisomes, which contain the photosynthetic pigments chlorophyll a, c-phycocyanin and c-phycoerythrin.

Cytoplasmic inclusions. The cytoplasmic inclusions present in the cyanophycean cell are ribosomes, cyanophycean granules, polyhedral bodies, polyphosphate bodies, polyglucoside bodies, α-granules, β-granules and gas vacuoles (Fig. 3.25C). The function of the above inclusions is not well-known. Cyanophycean granules are regarded as reserve food, polyhedral bodies, i.e., the carboxysomes contain enzyme ribulose bis phosphate carboxylase oxygenase. Ribosomes take part in protein synthesis. Some workers think α-granule as reserve food, a glycogen-like substance. Many prokaryotic members contain many small gas vacuoles also called pseudovacuoles. They help to float in water. They refract light and the cells become brown, purplish or black under microscope.

Nucleoplasm. The nucleoplasm is usually centrally located and contains numerous fine randomly oriented fibres of DNA. It is differentiated from the eukaryotic nucleus in absence of nucleolus and nuclear membrane. They are not differentiated from the cytoplasm by any membrane and are present in close association (Fig. 3.25C). The region is with low electron density than the surrounding cytoplasm. The DNA does not organise into chromosomes due to the absence of protein like histones or protamines.

Heterocysts

These are specialized cells of the filament distinguished from others by their thick wall, polar nodule(s) and homogenous contents.

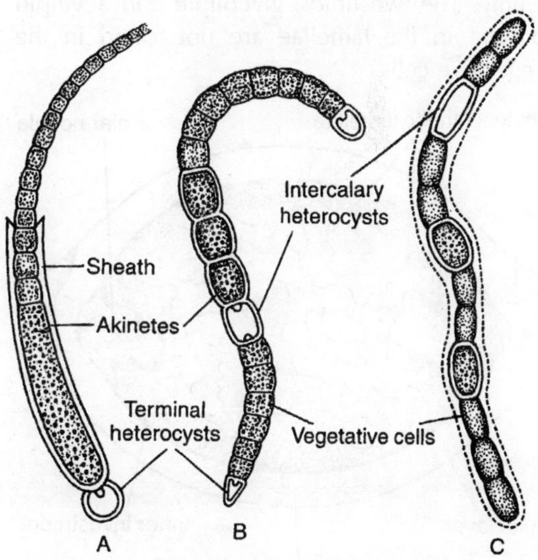

Fig. 3.26: Heterocysts at different position on the plant body: A. Terminal (*Gloeotrichia*), B. Both terminal and intercalary (*Anabaena desikacharyiensis*) and C. Intercalary (*Nostoc*)

Heterocysts are commonly found in the members of Stigonematales and Nostocales (except Oscillatoriaceae). They grow in the filament either terminally (*Gloeotrichia*) or intercalary (*Nostoc*) or both terminal and intercalary (*Anabaena desikacharyiensis*) (Fig. 3.26).

Structure. Pale yellow, thick-walled specialized cells, larger than vegetative cells are called heterocysts. The wall of the heterocyst is differentiated into three regions, an outer fibrous, middle homogenous and inner lamellar layer (Fig. 3.27). It has pore(s) at the pole(s) where it remains attached with the vegetative cell. The pores are plugged with polar nodule(s) or polar granules. The wall of the heterocyst is thicker towards the polar regions.

The inner content is dense and appears to be homogeneous. The thylakoides i.e., the photosynthetic lamellae are present, but the ribosomes are less in number. Other granular inclusions are appears to be absent. Due to absence of phycobillins and chlorophyll a, photosynthesis does not take place. The thylakoids are tightly packed due to reduction of interlamellar space and are present more towards the periphery than the central region. Due to low concentration or absence of pigment, the mature heterocyst appears to be empty. The two lipids, glycolipid and acyllipid present in the lamellae are not found in the vegetative cells.

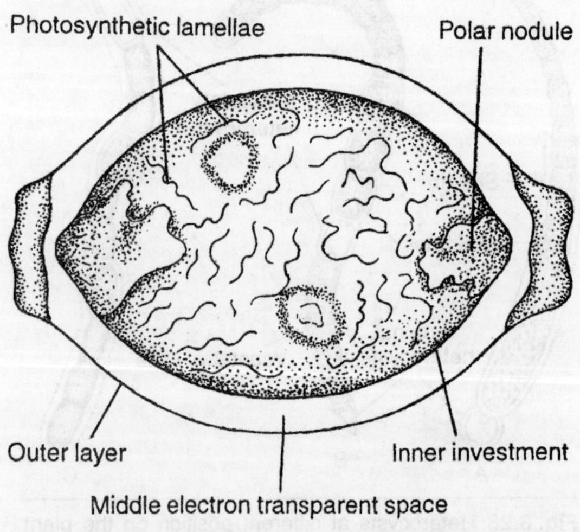

Fig. 3.27: Structure of Heterocyst (diagrammatic) as seen under electron microscope (based on Lang)

Development. *It develops either singly or in pair from single or both the daughter cells of a recently divided vegetative cell. The daughter cell may behave as proheterocyst, which later on develop into a heterocyst. The developmental changes are described below:*

(i) The inner content of protoplasm becomes paler in colour.

(ii) The wall of the polar region becomes rounded.

(iii) The connecting region(s) between vegetative cell and heterocyst becomes more constricted.

(iv) Formation of an inner polysaccharide wall external to the cell membrane.

(v) Formation of pores at one (terminal) or both (intercalary) ends.

(vi) Protoplasmic connection being established with the adjacent vegetative cell(s) through the pores.

(vii) The pores become plugged by mucilage towards maturity, which are called nodule.

(viii) The content of the cell becomes homogeneous with different chemical changes.

(ix) Gradual breakdown of thylakoids and dissolution of storage granules.

Function. The function of heterocyst is not clearly known. Many scientists have put forth different suggestions regarding the function of heterocyst:

1. Some considered it as the store-house of food materials.
2. Some others considered that they help in hormogonia formation.
3. According to Brand (1903), it is of spore-like structure.
4. Smith (1951) pointed out it as vestigial spore.
5. Fritsch (1951) told that during vegetative period it secretes certain substances which promote growth and cell division.
6. According to Fay (1968) it is the site of N_2 fixation.

7. Singh and Kumar (1970) pointed out that in addition to N_2 fixation; it also helps in growth and development of the filament.
8. Later, Steward (1970) suggested the possible roles of heterocyst such as: (i) involvement in N_2 metabolism, (ii) growth and differentiation, (iii) control of sporulation and (iv) antiquated reproduction unit.

Nitrogen fixation

Cyanophyceae (Blue-green algae/Cyanobacteria) capable of transforming atmospheric nitrogen into fixed nitrogen (inorganic compounds usable by plants). Nitrogen fixation by these organisms, play an important role in the nitrogen cycle of the biosphere.

Desikachary (1959) has listed 22 species of Cyanophyceae having the ability to fix atmospheric nitrogen. These are *Anabaena, Anabaenopsis, Aulosira, Calothrix, Cylindrospermum, Mastigocladus, Nostoc, Tolypothrix* etc.

Bold and Wynne (1978) have mentioned about the three categories of Cyanophyceae that are able to fix atmospheric nitrogen. These are:

(i) Filamentous forms with heterocyst,

(ii) Filamentous forms without heterocyst, and

(iii) Unicellular forms without heterocyst.

Singh (1961) mentioned that genera like *Anabaena, Tolypothrix* and *Aulosira fertilissima* play a very significant role in enriching the fields of rice with atmospheric nitrogen. Rippka *et al.* (1971) also shows the nitrogen fixing capacity of *Gloeocapsa*. Van Gorkom and Donze (1971) reported that the heterocyst in filamentous forms, reduce the elemental nitrogen and transfer it to the adjacent vegetative cells. Heterocyst is the main site of nitrogen fixation (Fleming and Haselkorn, 1973; Bradley and Carr, 1976). Because of this ability they are quite independent of other combined sources of nitrogen (Fogg, 1974).

According to some workers, *Tolypothrix tenuis*, able to fix atmospheric nitrogen @ 960 kg/acre/year and thus increase the soil fertility at a significant level. Nitrogen fixation by cyanobacteria in coral reefs can fix twice as much nitrogen as on land i.e. around 1.8 kg of nitrogen is fixed per hectare per day (about 660 kg/ha/year).

The mechanism of N_2 fixation by heterocyst is given in page no. 382.

Reproduction

The blue green algae reproduce by both vegetative and asexual means. **Sexual reproduction is absent.**

The vegetative reproduction performs through fission (*Synechococcus*), fragmentation (*Oscillatoria, Cylindrospermum muscicola*), hormogonia formation (*Oscillatoria, Nostoc*), hormospores (*Westiella lanosa*), planococci and Palmelloid stage.

During asexual reproduction various types of asexual spores are formed. These are akinetes (*Anabaena sphaerica, Gloeotrichia natans, Calothrix fusca*), endospores (*Dermocarpa*), exospores (*Chamaesiphon*) and nannocyte (*Microcystis*) (Fig. 3.28).

Genetic recombination

Sexual reproduction does not occur in the members of Cyanophyceae, although some evidences have well-documented regarding the gene transfer between certain members. Genetic recombination has however been reported by Kumar (1962), Bazin (1968) and Shestakov and Khyen (1970). Kumar (1962) reported genetic recombination between streptomycin- and penicillin-resistant mutants of *Anacystis nidulans*. Singh and Sinha (1965) also got the similar result in *Cylindrospermum majus* with streptomycin and penicillin. Later, Bazin (1968) pointed out that the strains of *Anacystis nidulans* resistance to either streptomycin or polymyxin could produce a double resistant strain, when grown together.

The mechanism involve here is the diffusion of DNA in the medium and its absorption by the recipient cells. Later gene transfer has also been reported by Steward and Singh (1975) and Padhy and Singh (1978). Devilly and Houghton (1977) and Stevens and Porter (1981) incubated members of Cyanophyceae in extracted DNA and confirmed the donation of functional genes.

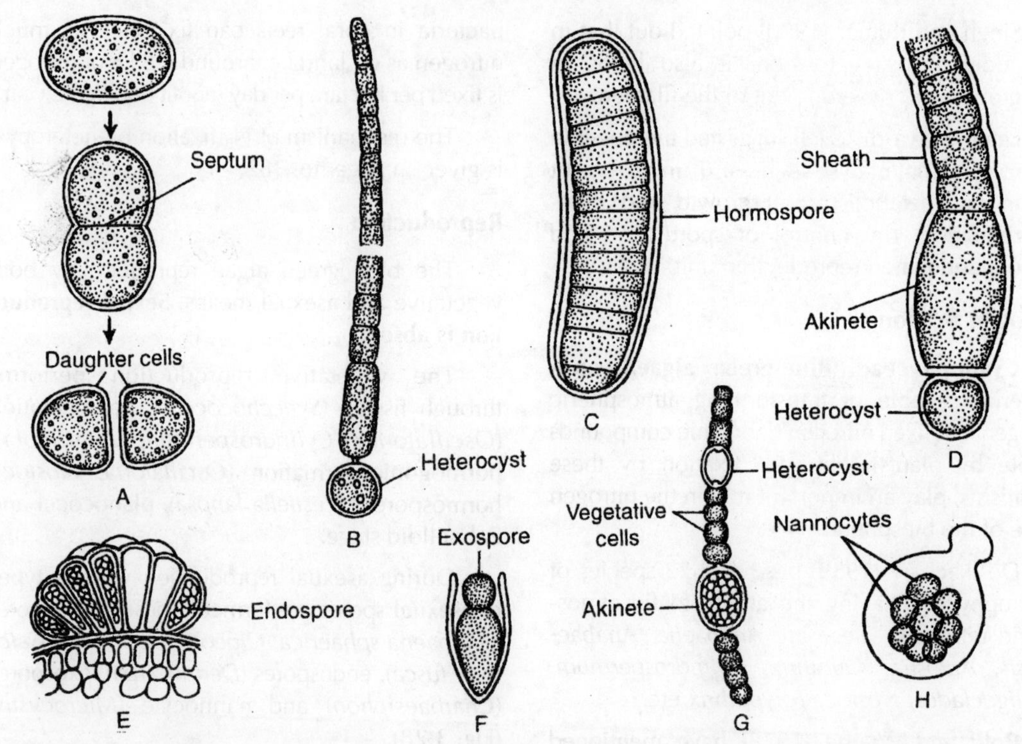

Fig. 3.28: Vegetative and asexual reproduction in Cyanophyceae: A. Cell division (*Synechococcus* sp.), B. Fragmentation of filament (*Cylindrospermum muscicola*), C. Hormospore (*Westiella lanosa*), D. Akinete (*Gloeotrichia natans*). E Endospore (*Dermocarpa prasina*), F. Exospore (*Chamaesiphon incrustans*), G. Akinete (*Anabaena* sp.) and H. Nannocytes (*Aphanothece*)

Although the role of Cyanophages have been studied extensively but no conclusive evidence of transduction in Cyanophyceae is found, but Singh and Singh (1972) reported phage-mediated streptomycin resistance. In Cyanophyceae, conjugation between cells is not known, although filament anastomosis and consequent formation of different clones have been reported by Delaney *et al.* (1976).

Gaidukov phenomenon or complementary chromatic adaptation

The efficiency to change the pigment composition, to absorb maximum light for photosynthesis, with the variation of the incident light is called **complementary chromatic adaptation.**

Many members of Cyanophyceae have the capacity to change their colour in relation to the wave length of incident light. Due to variation of the wavelength of incident light they can change their pigment composition. It may appear blue green in yellow light, green in red light and reddish in green light. Gaidukov (1903) first invented the phenomenon and according to his name it is also known as **Gaidukov phenomenon.**

Origin and affinity

This group is considered to be the most primitive because of the presence of some important features. These are

(a) presence of unorganised nucleus,

(b) absence of chromatophores,

(c) absence of flagella, and

(d) lack of sexual reproduction.

They are found in all habitats where life is possible and distributed throughout the world. Fossil records indicate that they have originated in early Pre-Cambrian period. But their ancestry is not known. Absence of flagella and the prokaryotic nature of cells lead to believe that possibly they have ***originated from unicellular aflagellate cells***. Presence of most of the members in terrestrial habitat leads to believe by most of the investigators that the Cyanophyceae have ***originated from terrestrial members***.

The members of Cyanophyceae show some relationship with both bacteria and Rhodophyceae.

Similarities with bacteria

1. Cell structure is prokaryotic in both the group, having unorganised nucleus and devoid of membrane bound organelle.
2. The capsule of bacteria (if present) and mucilaginous sheath of blue green algal cells are made up of fine fibrils.
3. Cell wall composed of mucopeptide (murein).
4. *Oscillatoria* (blue green alga) shows similarity with *Beggiatoa* (sulphur bacterium), both in shape and movement.
5. Both are sensitive to antibiotics.
6. Both the groups show similarity in many metabolic processes like nitrogen and sulphur metabolism.
7. Absence of sexual reproduction.
8. Genetic recombination has been reported in *Anacystis nidulans*, a member of Cyanophyceae, showing similarity with bacteria.

Similarities with Rhodophyceae (red algae)

1. Both the groups resemble in the absence of motile cells.
2. The Cyanophycean pigments, c-phycocyanin (blue) and c-phycoerythrin (red) are chemically similar to the Rhodophycean pigment r-phycocyanin and r-phycoerythrin.
3. *Stigonema* and some other members of Cyanophyceae have pit connections, and show relationship by having similar structures as found in the members of Rhodophyceae.

Stromatolites

Stromatolites (Gr. for 'layered rock') are the **microbial reefs** created by members of Cyanophyceae (also called Cyanobacteria) i.e. Blue-green algae. These deposits are formed by sediment trapping and binding, and/or by precipitation activities of the microbial communities (Awramik, 1976). The microbes are active on the surface layer of the Stromatolites, while the underlying build-up is a lithified remnant of earlier microbial surface communities that could be interpreted as a **trace-fossil**.

These deposits built up very very slowly: a single 1 m *structure may be* 2,000 *to* 3,000 *years old*. But the tiny microbes that make up modern Stromatolites are similar to organism that existed 3.5 billion years ago. Consider that the Earth itself has been around for 4.5 billion years, and that Human beings (*Homo sapiens*) have only been on Earth for 195,000 years.

Stromatolites are the reason why we remain alive today. Before Cyanophyceae (Cyanobacteria) the air contain only 1% O_2. Then, for 2 billion years, our **photosynthesising Stromatolites** pumped O_2 into the oceans (like trees in deep sea). When the ocean waters became saturated, the O_2 was released into the air, and with around 20% of oxygen in the air, life was able to flourish and evolve.

Even today you can see Stromatolites 'fizzing' underwater, releasing oxygen.

Economic importance

The Cyanophycean members show both beneficial and harmful activities:

Beneficial activities

The beneficial activities include their role as food, reclamation of alkaline 'usar' land, biofertilizer and pioneer of plant succession. These are discussed below:

1. AS FOOD

Blue-green algae or Cyanophyceae is used as human food and animal food supplements (Belay *et al.*, 1966) from long back. It contains small amount of protein, vitamins, β-carotene and also some minerals. The role of Cyanophyceae as an antibacterial, antiviral, antitumor and food additive have been well established (Singh *et al.*, 2005). They have antioxidant activity. It has been used for weight loss and as a nutritional supplement. It has also been used for boosting the immune system and for controlling cholesterol levels.

The important algae of this group used as food are the *Nostoc, Spirulina* and many others.

In China, *Nostoc* has been used as soup from long back (Gao and Ye, 2003). It is called "hair vegetable" (Facai) because of its hair-like appearance.

Spirulina is rich in protein (55-68%). It is high in vitamins A, E, K, C and B complex and minerals like calcium, magnesium and iron. It is high in antioxidants to fight free radicals, high in Gama linoleic acid (FLA) that keeps the heart healthy, high in lutein for assisting eye health, higher β-carotene than carrot, detoxes and balances the immune and digestive system, improve mental health, encourages weight loss and enhances the ability to generate new blood cells.

In China, the *Spirulina* is also used in commercial basis having more than 90 factories with a total production of 400 tons of *Spirulina* dry powder. It is used as pills and capsules. In addition, it is also used in pastries and chocolates. *Spirulina* is readily available at health-food stores in the developed countries. In India, it is also produced in mass scale by NBRI, Lucknow (U.P.) and CFTRI, Mysore (Karnataka).

In Japan, *Aphanothece sacrum, Nostoc verrucosum, N. commune* and *Brachytrichia* have been used as side dishes from long back.

Aphanizomenon flos-aquae has been harvested from Klamath Lake (California) since early 1980s and sold as a food and health supplement. In 1988, 10 symbol × 10^6 kg (dry weight) was marketed with a value of 100 million dollar (Carmichael *et al.*, 2000).

Incorporation of *Porphyridium* sp. biomass at 10% diet in chicken fed, lowered the serum cholesterol levels by 28%, while *egg yolk tended to have reduced cholesterol levels by* 10% and increased the linoleic acid and arachidonic acid level by 29% and 24%, respectively. In addition, the colour of egg yolk became darker as a result of the higher carotenoid levels about 2.4 times (Ginzberg *et al.*, 2000).

Spirulina prebiotic causes at least a 10 fold increase in the growth rate of the lactobacilli compared with the control (Pulz and Gross, 2004).

Microalgal biomass was predominately utilized in the health food market. Still incorporation of microalgae along with existing traditional food and some alterations are required to be attractive for the general people.

2. RECLAMATION OF ALKALINE 'USAR' LAND

In India, due to high alkalinity in vast tracts of land cannot be used for cultivation of crops, commonly known as **'usar'** soil. This type of lands would be cultivable by lowering the pH and organic contents and increasing the water holding capacity of the soil. All these functions are carried out by the Blue-green algae. In this soil, during the rainy season the Blue-green algae like *Nostoc, Anabaena, Scytonema* and *Aulosira* are grown in plenty. The process involves a series of successive growth of the algal crop in a water-logged environment.

After one year of algal growth, following changes take place. These are:

(i) The pH fell from 9.5 to 7.6,
(ii) Organic content increased from 36.5% to 59.7%,
(iii) Nitrogen content increased from 30% to 38.4%,
(iv) Exchangeable calcium from 20% to 33% and
(v) Water holding capacity of the soil is also increased by 40%.

In such a **'reclaimed'** land, crops like paddy grew well with a yield of 1576-2000 lbs/acre. The process is now being practiced widely.

3. BIOFERTILIZER

Fertilizers are expensive agricultural inputs especially for small-scale farmers, but technologies being developed which can offer farmers, the less expensive ways to ensure their soils will remain fertile and healthy. In addition, it is to be noted that due to lack of soil fertility, per-capita food production is gradually declining throughout the world.

Biofertilizers can help to maintain the natural soil habitat, and also improving fertility. Bluegreen algae, is one such example of a biofertilizer, a type of organic fertilizer which contains living organisms and harnesses naturally occurring inputs like solar energy, nitrogen, and water to ensure soil fertility and plant growth. Small-scale farmers using biofertilizers have the potential to provide larger and more sustainable yields and healthier soils for themselves, as well as for their communities.

In order to ensure that test samples of Cyanophyceae are adapted to local soil conditions, test ponds placed at village sites are being seeded with Cyanophycean members which are naturally occurring in local soils. The goal is for farmers to be able to produce their biofertilizers on-site to use on their own fields and potentially sell on the local market. By meeting these criteria, agricultural developers can help to ensure that cyanophycean biofertilizers enable sustainable food production throughout the world.

The commercial production and the use of biofertlizer show that "yields are at par or exceeded outcomes from urea application". In addition, it also improves nutritional values, like zinc and β-carotene levels were more than doubled and iron increased by more than 50% as compared to crops treated with nitrogen-rich urea and chemical fertilizers.

In the above context, it can be mentioned that about twenty two (22) filamentous members of Cyanophyceae like *Nostoc, Anabaena, Aulosira, Anabinopsis, Calothrix, Scytonema* etc. can ***fix atmospheric nitrogen*** and form nitrogenous compounds. These compounds are further absorbed by the plant for their metabiloc activity and increase yield. All the above members have ***heterocyst***. But certain non-heterocystous members like *Plectonema boryanum* are able to fix atmospheric nitrogen in anaerobic condition. In recent year, another two members of cyanophycean algae like *Nostoc entophytum* and *Oscillatoria angustissima* are useful in increasing the yield of pea plants.

The above mentioned members can be divided into **two groups**. These are:

A. Free Living nitrogen Fixing Cyanophyceae

A number of free living cyanophyceae (bluegreen algae) have the property of nitrogen fixation, e.g., *Anabaena, Nostoc, Aulosira, Totypothrix, Cylindrospermum,* and *Stigonema*. The above members are photosynthetic. Therefore, they add organic matter as well as extra nitrogen in the soil.

Aulosira fertilissima is considered to be the most active nitrogen fixer of Rice fields in India (Aiyer *et al.,* 1972). *Cylindrospermum licheniformis* grows in Sugarcane and Maize fields. Cyanophyceae are an extremely low cost biofertilisers. In Tamil Nadu, the technique of cyanophyceae inoculation to rice fields is being followed. Phosphate, Molybdenum and Potassium are supplied additionally.

B. Symbiotic Nitrogen Fixing Cyanophyceae

A number of symbiotic cyanophyceae (bluegreen algae) have the property of nitrogen fixation in association with several plants, like *Cycas* roots, lichens, liverworts (*Anthoceros*) and *Azolla* (fern). Out of the above, ***Azolla-Anabaena association*** is of great importance in agriculture.

Azolla pinnata is a very small free floating fresh water fern, multiplies very rapidly. It becomes double in every 5-7 days. The fern can co-exist with rice plants in field and grow normally. In China, the rice fields are regularly provided with *Azolla*.

In *Azolla-Anabaena* association, *Anabaena azollae* resides in the leaf cavities of *Azolla. A. azollae* fix atmospheric nitrogen. A part of the nitrogen released in the cavities and becomes available to the fern. The decaying fern release the nitrogen in the medium and is utilized by the rice plants. Further when field becomes dried during harvesting of rice, the fern functions as green manure after decomposition and enriched the field for the next crop.

4. PIONEER OF PLANT SUCCESSION

The bluegreen algae is the largest assemblage, diverse and widely distributed group of prokaryotic organisms that perform oxygenic photosynthesis. These are also called as Cyanophyceae, Myxophyceae or Cyanobacteria.

Cyanophyceae are ubiquitous in their distribution but are more common in tropics. Algal members of this group are able to withstand extreme temperature and also draught. Because of their ability to withstand extreme adverse ecological conditions and the capacity to thrive well in adverse environmental conditions and also have the response in appearance of dry conditions through the development of dormant resistant state, considered them as the **pioneer of plant succession.**

Harmful activities

1. Some members of Cyanophyceae cause damage of building plasters, stones etc. It can be avoided by spraying $CuSO_4$ and sodium arcenate.
2. Some members like *Microcystis aeruginosa*, *Anabaena flos-aquae* form water blooms and can grow well in O_2 deficient water. Continuous respiration by submerged plants and animals during night time (when photosynthesis does not take place) causes the depletion of O_2 to almost zero level. At that condition mortality of both animals and other submerged plants takes place due to suffocation.
3. Bluegreen algae contaminate the water of reservoirs. They develop a foul odour in water and make it unhygienic for human being and cause several diseases.

Different diseases like gastric troubles may appear by drinking the water contaminated with *Microcystis* and *Anabaena*.

Consideration of Cyanophyta (Blue green algae) as **Cyanobacteria**

This group of plant was known from earliest part of 19th century and considered as a Division Cyanophyta (Smith, 1933, 51, 55; Chapman and Chapman, 1973; Lee, 1980, 1989), Phylum Cyanophyta (Prescott, 1969; Round, 1973; Lee, 1999) and Class Myxophyceae or Cyanophyceae i.e., Blue green algae (Fritsch, 1935, 45) under the sub-division Algae.

The fundamental difference of cellular organisation between prokaryotes and eukaryotes were fully elucidated during 1960s. Then it became clear that the blue green algal members are prokaryotic in nature, which is unlike to the other algal members. Based on their prokaryotic cell structure **Christensen** (1962) placed both bacteria and blue green algae under the common phylum Prokaryota. The blue green algae were named as **Cyanobacteria**. Cyanobacteria are also called as blue green bacteria.

The blue green algal members have photosynthetic pigments like chlorophyll a and phycobiliproteins that makes difference with the photosynthetic bacteria, having bacteriochlorophyll. The blue green algal members also prefer oxygenic photosynthesis which indicates physiological similarity with aerobic photoautotrophs and makes difference with bacteria.

Based on the above, many phycologists like Fritsch, 1945; Prescott, 1969; Morris, 1973 and many others considered that the recognition of prokaryotic members like blue green algae as bacteria is an intrusion of bacteriologist to other field. Phycologists included the plants under algae by having thalloid plant body with chlorophyll a, the photosynthetic pigment. However the blue green algae have the following **characteristics common with bacteria**. These are:

1. Cell structure is prokaryotic in both the groups, having unorganised nucleus and devoid of membrane bound organelles.
2. The capsule of bacteria (if present) and mucilaginous sheath of blue green algal cells are made up of fine fibrils.
3. *Oscillatoria* (a blue green alga) shows similarity with *Beggiatoa* (a sulphur bacterium), both in shape and movement.
4. Both are sensitive to antibiotics.
5. Both the groups show similarity in many metabolic processes like nitrogen and sulphur metabolisms.
6. Genetic recombination has been reported in *Anacystis nidulans*, a member of Cyanophyceae, similar to bacteria.

Based on the above similarity with bacteria they may be considered as a separate group Cyanobacteria.

Still now arguments are going on between the two groups of scientists regarding the inclusion of Cyanophyceae either in Algae or in Bacteria.

Classification

All algologists recognise the class Cyanophyceae under the Division Cyanophyta dividing into five (5) orders and nineteen (19) families. The differentiation is made by the nature of plant body, absence and presence of gelatinous sheath, heterocyst and method of reproduction.

The orders are as follows:

1. Order. **Chroococcales**. Unicellular or colonial; reproduction by cell division and endospores. e.g., *Chroococcus, Gloeocapsa, Microcystis* etc.
2. Order. **Chamaesiphonales**. Unicellular or colonial; epiphytes or lithophytes, exhibiting marked polarity; reproduction by exospores and endospores. e.g., *Chamaesiphon, Dermocarpa* etc.
3. Order. **Pleurocapsales**. Heterotrichous filaments; heterocysts absent; reproduction by endospores. e.g., *Pleurocapsa*.
4. Order. **Nostocales**. Plant body filamentous; rarely branched; heterocysts commonly present though absent in some members; reproduction by hormogonia, hormospores and akinetes. e.g., *Nostoc, Oscillatoria, Anabaena, Lyngbya* and *Gloeotrichia*.
5. Order. **Stigonematales**. Heterotrichous filaments, branched truly; trichomes composed of several row of cells; reproduction by hormogonia, hormospores and rarely by akinete. e.g., *Stigonema*.

The first three (3) orders are without hormogones and the rest two (2) are with hormogones.

Order. **Chroococcales**

Identifying characteristics

1. Unicellular or colonial algae.
2. They are aquatic (either fresh water or marine) or terrestrial.
3. Colonies or the plant body is without polarity.
4. Reproduction takes place by cell division and endospore formation.
5. Nannocysts are formed very rarely.
6. Hormogonia, heterocysts and exospores are absent.

The order is divided into two families: Chroococcaceae and Entophysalidaceae.

Family. **Chroococcaceae**

Identifying characteristics

1. Plant body is unicellular or colonial.
2. Cells are generally ellipsoidal or spherical in shape and rarely spindle shaped or cylindrical.
3. Cells or colonies are embedded in thick mucilaginous covering.
4. The cells divide into two or three planes, thereby they do not form filaments.
5. Nannocyts are formed in some members.

The **common genera** of the family are: *Gloeocapsa, Chroococcus, Microcystis, Merismopedia, Gomphosphaeria, Rhabdoderma* and *Coelosphaerium.*

Life history of *Gloeocapsa* is given below:

GLOEOCAPSA

Systematic position

Fritsch (1945)	**Lee** (1999, 2008)
	Phylum. Cyanophyta
Class. Myxophyceae	Class. Cyanophyceae
Order. Chroococcales	Order. Chroococcales
Family. Chroococcaceae	Family. Chroococcaceae
Genus. *Gloeocapsa*	Genus. *Gloeocapsa.*

Occurrence

Gloeocapsa (Gr. *Gloia*, glue; *capsa*, a box or case) grows in aquatic (both fresh and salty water) and also in terrestrial condition i.e., on wet soil, wet rocks, damp walls etc. Species like *G. gelatinosa* grows in the barks of *Hevea brasiliensis* (Para rubber) and some like *G. alpine* live as a partner in lichen. *G. decorticans* occur in garden soil.

Plant body

The plant body is unicellular; the cells are spherical and remain in colonies of 2-8 or rarely

16 (Fig. 3.29) cells. Each cell is surrounded by conspicuous, concentric, generally lamellated gelatinous sheath (Fig. 3.29B). Later on the sheath surrounds the entire colony. Due to the presence of different pigments the colonies appear to be red, violet or brown colourations.

The cells are surrounded by cell wall. Cell membrane is present inside the cell wall. The protoplast is differentiated into outer pigmented **chromoplasm** region and inner colourless, **centroplasm** which represents the incipient nucleus. The chromoplasm have a special pigment, **gloeocapsin**.

Photosynthetic thylakoids are generally arranged at right angles to the cell membrane and the outer thylakoids are linked with cell membrane.

Reproduction

Gloeocapsa reproduces by **fission** (vegetative) and by the formation of **nannocytes**:

1. **Fission**. During fission individual cell constricts in the middle and subsequently the protoplasm divides in two halves and finally two cells are formed (Fig. 3.29D). This process is repeated several times and thus a colony is formed. Initially the cells remain embedded in the common sheath but later on individual sheaths are formed. Due to cell division at different planes the cells of a colony remain arranged irregularly.
2. **Nannocytes**. These are small naked protoplasts formed by repeated division of the mother protoplast. Liberating from the parent cell, the nannocytes germinate immediately and form new individual. Nannocytes are formed in *G. crepidinum* and *G. granosa*.

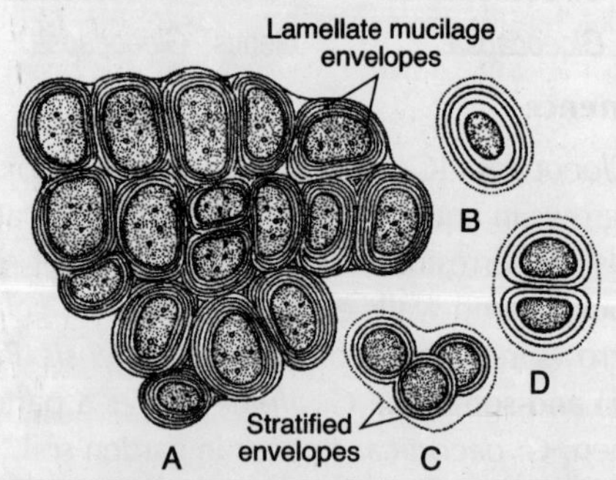

Fig. 3.29: *Gloeocapsa*: A. Cells in group with lamellate mucilage envelope. B. Single cell, C. Cells with well-defined stratified envelopes and D. Dividing cell

Economic importance

Some species of *Gloeocapsa* are able to fix atmospheric N_2 (Rippka, 1971), thereby they increase the soil fertility.

Indian species

Gloeocapsa calcarea, G. atrata, G. decorticans, G.gelatinosa.

Important features

1. Unicellular plant body with thick sheath.
2. Commonly cells remain together by mucilage.
3. Reproduction takes place by fission.

Order. **Nostocales**

Identifying characteristics

1. Multicellular and filamentous algae.
2. Filaments are generally unbranched but rarely branched.
3. Branching may be true or false type and non-heterotrichous type.
4. Heterocysts are commonly present.
5. Reproduction takes place by hormogonia, hormospores, akinetes and endospores.

Desikachary (1959) divided the Order Nostocales into five (5) families: (i) *Oscillatoriaceae,* (ii) *Microchaetaceae*, (iii) *Nostocaceae*, (iv) *Scytonemataceae* and (v) *Rivulariaceae*.

Only three families, Oscillaroriaceae, Nostocaceae and Scytonemataceae are considered below for further discussion.

Family. **Oscillatoriaceae**

Identifying characteristics

1. Plant body is uniseriate, multicellular and unbranched trichome.
2. The trichome performs oscillatory movement.
3. The individual cells are morphologically alike.
4. With some exception most of the cells are generally covered by mucilaginous sheath.

5. Growth of the filament is mostly intercalary.
6. All cells are capable of division.
7. Reproduction takes place by hormogonia.
8. Heterocysts are absent.

The common genera of the family are: *Oscillatoria, Lyngbya, Phormidium, Arthrospira, Hydrocoleum, Spirulina, Symploca, Sirocoleus, Trichodesmium* and *Microcoleus*.

The life history of *Oscillatoria* and *Spirulina* are discussed below:

OSCILLATORIA

Systematic position

Fritsch (1945)	**Lee** (1999, 2008)
	Phylum. Cyanophyta
Class. Myxophyceae	Class. Cyanophyceae
Order. Nostocales	Order. Oscillatoriales
Family. Oscillatoriaceae	Family. Oscillatoriaceae
Genus. *Oscillatoria*	Genus. *Oscillatoria*.

Occurrence

Oscillatoria is found in different habitats. They grow in fresh water of pools, ponds, streams, lakes etc. They also grow on polluted water or muddy banks of ponds and rivers. At the bottom of old ponds they grow as thick layers. They are also available in water reservoirs. The very commonly available species of *Oscillatoria* along with their habitat is given below:

1. *O. princes*. In fresh as well as in saline water.
2. *O. prolifica*. Grow as plankton in ponds.
3. *O. terebriformis*. Grow in hot springs.
4. *O. limosa* and *O. selina*. In saline water.
5. *O. sancta*. Grow on moist soil and walls.

Plant body

Plant body is a trichome. The trichomes are present either singly or matted together to form a thin or thick and spongy sheets. The number of trichomes may be thousands in a sheet. In groups they are bluish-green or black in colouration.

The trichome is long, thread-like, unbranched structure and consists of numerous cells, arranged in uniseriate fashion (Fig. 3.30A, B, C). Generally the trichomes are not surrounded by mucilaginous sheath. There is no constriction at the septal region of the adjacent cells. Morphologically the cells are alike and breadth of the cells is greater than its length except a few members like *O. splendida* (Fig. 3.30D), where the trichome is narrow and consists of cylindrical cells.

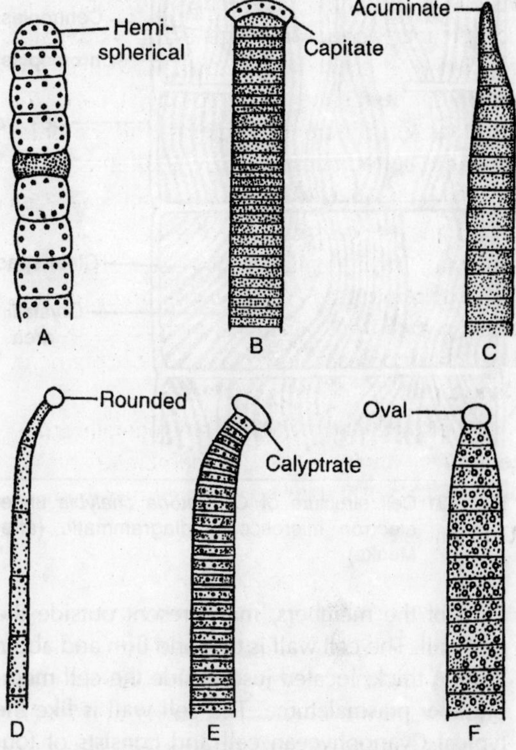

Fig. 3.30: Portion of *Oscillatoria* trichome with different types of apical cells: A. Hemispherical (*O. annae*), B. Capitate (*O. princes*), C. Acuminate (*O. salina*), D. Rounded (*O. splendida*), E. Calyptrate (*O. anguina*) and F. Oval (*O. rubescens*)

The trichomes show definite polarity and the terminal cell is of different shapes in different species. Generally the apex is hemispherical (*O. annae*) in most of the members, but some other shows variation. They may be capitate (*O. princeps*), acuminate (*O. salina*), rounded (*O. splendida*), calyptrate (*O. anguina*), oval (*O. rubescens*) etc. (Fig. 3.30).

Cell structure. Generally the cells are rectangular in shape. Mucilaginous sheath is absent in

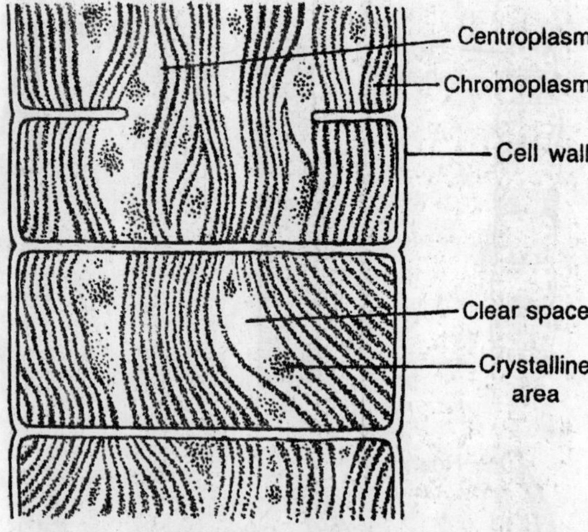

Fig. 3.31: Cell structure of *Oscillatoria chalybia* under electron microscope, diagrammatic (after Menke)

most of the members, may present outside the cell wall. The cell wall is thin and firm and about 2,000Å thick, located just outside the cell membrane or plasmalemma. The cell wall is like the typical Cyanophycean cell and consists of four layers as designated by L I, L II, L III and L IV. The L I is situated next to plasmalemma. The cell wall of *Oscillatoria* is similar to Gram +ve bacteria, both structurally as well as in composition. It is composed of mucopeptide along with carbohydrates, fatty acids and amino acids. The gliding movement takes place by contraction of fibrils of the cell wall.

The protoplast like the prokaryotic members is differentiated into outer pigmented **chromoplasm** and the central colourless **centroplasm** (Fig. 3.31). The chromoplasm region is mainly occupied by thylakoids. Photosynthetic pigments are present on the surface of the thylakoids. The pigments like c-phycocyanin and c-phycoerythrin are much higher in amounts than the other pigments. ***Plastids, endoplasmic reticulum, mitochondria and dictyosomes are absent.***

In floating members gas vacuoles are found in the cytoplasm, which provide buoyancy and possibly help in light shielding.

In addition to the above, following structures are also present in chromoplasm:

1. ***Polyhedral bodies***. They are angular in structure, which store carboxysomes. The carboxysomes contain Ribulose 1, 5-bisphosphate carboxylase/oxygenase, the CO_2 fixing enzyme.
2. ***Polyphosphate bodies***. They store phosphate molecules.
3. ***Polyglucan granules***. It is composed of 14-16 glucose units and are found between the thylakoids of actively photosynthesizing cells.

The centroplasm is also called **nucleoplasm**. It is composed of DNA reticulum where unlike eukaryotic cell it is devoid of nuclear membrane and nucleolus. The region of centroplasm is not sharply demarcated from the chromoplasm region.

Movement

Member of *Oscillatoria* shows the following two types of movements:

1. **Oscillatory movement**. The apical part of the filament of *Oscillatoria* shows pendulum-like oscillation. Such movement is called oscillatory movement (Fig. 3.32A) and hence the genus is named as "*Oscillatoria*".
2. **Gliding movement**. In this type the filament moves in forward and backward direction towards the axis of the filament. It is also called creeping or axial movement (Fig. 3.32B).

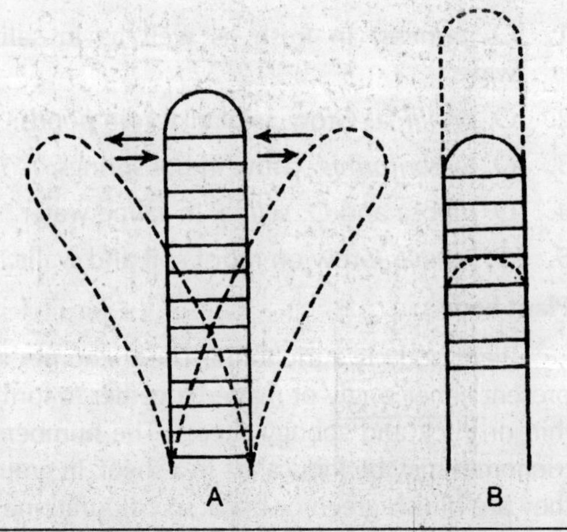

Fig. 3.32: *Oscillatoria*: A. Oscillatory movement, and B. Gliding movement

This movement is also accompanied by clockwise and anti-clockwise rotations.

Reproduction

Oscillatoria reproduces only by vegetative means. The vegetative reproduction takes place by hormogonia and fragmentation:

1. **Hormogonia**. Hormogonium is a piece of trichome, composed of a number of undifferentiated vegetative cells. A number of hormogones can develop in a filament. The hormogones are separated by separation disc.

During vegetative growth some of the cells are filled with mucilage and become dead. This dead cell is called **necredium** (pl. necredia). The necridium may form by one or more cells of the trichome. In course of development, the necredia become biconcave and function as **separation disc** (Fig. 3.33).

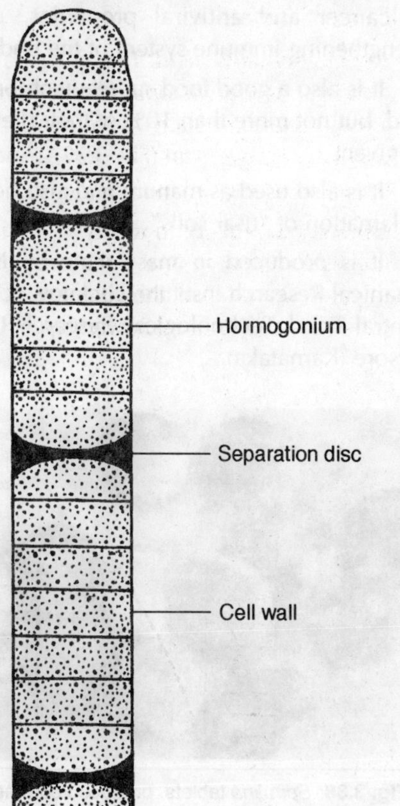

Fig. 3.33: *Oscillatoria*: A filament showing separation disc and hormogonia

Soon after separation, the hormogones keep on moving for sometimes and undergo repeated cell divisions. Thus a new trichome is formed.

2. **Fragmentation**. Sometimes, due to disturbances by insects, other animals or by any mechanical disturbances, the trichome may divide into small multicellular fragments. The cells of the fragment undergo repeated cell division to develop a new individual.

Indian species

Oscillatoria limosa, O. obscura, O. vizagapatensis, O. chilkensis and *O. sancta*.

Important features

1. Plant body filamentous, unbranched and bears extremely thin sheath.
2. Plant body exhibits oscillating or gliding movement.
3. Cells are cylindrical and breadth is greater than its length.
4. Reproduction by hormogonia and/or separation disc.

SPIRULINA

Systemic position

Fritsch (1945)		**Lee** (2008)	
Class	Cyanophyceae	Class	Cyanophyceae
Order	Nostocales	Order	Oscillatoriales
Family	Oscillatoriaceae	Family	Oscillatoriaceae
Genus	*Spirulina*	Genus	*Spirulina*

The name *Spirulina*, derived from the Latin word for 'helix' or 'spiral'. It is a bluegreen alga (Cyanobacterium). Initially it was thought to be a eukaryote and possibly a type of fungus. However, both phylogenetic and morphological analysis illustrate that these organisms are definitely bluegreen algae i.e., Cyanophyceae (Cyanobacteria). Many genera of Cyanophyceae are harmful to humans. *Spirulina*, however, are unique not only in that they are edible, but also because they provide many health benefits.

Occurrence

Spirluina includes more than 39 species are aquatic in habitat, typically inhabiting freshwater alkaline lakes. However, there are marine (brakish water) species as well. They are mesophilic, and able to survive over a wide range of temperatures; some species are even found in hot springs. It can grow both autotrophically and heterotrophically.

Plant body

Plant body of *Spirulina* is spirally coiled (Fig. 3.34) free floating trichomes with rounded ends (Fig. 3.35). Cells are arranged end to end and form a long trichome of up to 0.5 mm in length. The trichome is spiral and twisted, without mucilaginous sheath.

Fig. 3.34: *Spirulina* sp.: Trichomes under compound microscope (left) and Trichomes much enlarged (right)

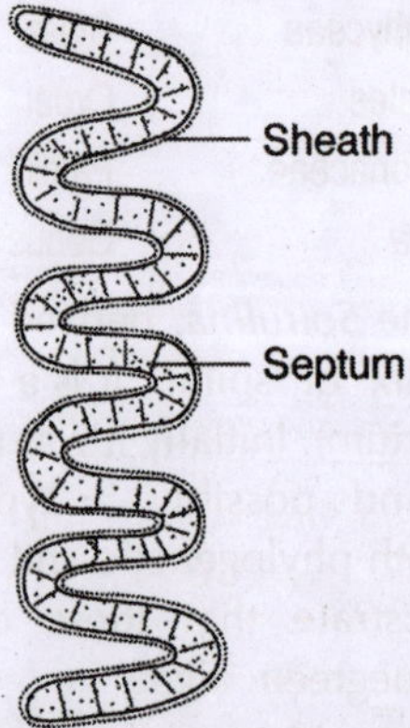

Fig. 3.35: *Spirulina* sp.: Line drawing of plant body showing septum and sheath

Cell wall is multilayered, soft; consist of complex sugars and protien but lack cellulose. Outer sheath composed of polysaccharide.

Their main photosynthetic pigment is phycocyanin, which is blue in colour. They contain chlorophyll a, and carotenoids. Some contain the pigment phycoeythrin, giving the alga a red or pink colour. *Spirulina* also have pseudovacuole, help them to float in the aquatic environments they inhabit. Reserve food is glycogen.

They are Gram-negative in nature, photosynthetic, and therefore autotrophic.

Trichomes show rotary and bending movements.

Reproduction

Reproduction takes place through fragmentation and binary fission through cell division, **true sexual reproduction is absent**.

Economic importance

Spirulina is rich in protein (55-68%). It also contains vitamins (including B_{12}), minerals and essential fatty acids. It is used as food additive for human beings in the form of pill or tablet. WHO has called it as greatest "**super food**". It has anticancer and antiviral properties, used in strengthening immune system of the body.

It is also a good food supplement for animal feed, but not more than 10% of their diet should be given.

It is also used as manure and also helpful in reclamation of "usar soils".

It is produced in mass scale by National Botanical Research Institute, Lucknow (U.P.) and Central Food Technological Research Institute, Mysore (Karnataka).

Fig. 3.36: *Spirulina* tablets, produced commercially

Indian species

Spirulina platensis, S. fusiformis etc.

Important features

1. *Spirulina* is spirally coiled free floating trichomes with rounded ends and without mucilaginous sheath.
2. Cell wall is multilayered, soft; consist of complex sugars and protien but lack cellulose.
3. Cells contain chlorophyll a, carotenoids and phycocyanin. Some members also contain phycoerythrin,
4. *Spirulina* is rich in protein (55-68%). It also contains vitamins (including B_{12}), minerals and essential fatty acids.

Differences between *Spirulina* and *Chlorella*

Spirulina and *Chlorella* are very similar in the fact that they are microalgae. ***Spirulina* is a spiral-shaped, multi-celled plant** with no true nucleus; ***Chlorella* is a spherical-shaped single-celled microorganism** with a nucleus. *Spirulina* can sometimes be up to 100 times larger than *Chlorella*.

In addition, while *Spirulinais* a blue-green alga, *Chlorella* is a green-coloured alga.

Family. **Nostocaceae**

Important characteristics

1. Cells are arranged in a single row and form a trichome.
2. Cells are structurally alike.
3. Trichomes are embedded in a common mucilaginous substance.
4. Heterocysts are present in most of the genera and they are commonly intercalary in position but rarely terminal.
5. Reproduction takes place by hormogonia.
6. The constriction between the cells is well marked and appears like a beaded necklace.

The important genera of the family are ***Nostoc***, ***Anabaena***, *Anabaenopsis*, ***Aulosira***, *Richelia*, *Cylindrospermum*, *Raphidiopsis* and *Nodularia*.

The life history of *Nostoc* and *Anabaena* are considered as follows:

NOSTOC

Systematic position

Fritsch (1945)	**Lee** (1999, 2008)
	Phylum. Cyanophyta
Class. Myxophyceae	Class. Cyanophyceae
Order. Nostocales	Order. Nostocales
Family. Nostocaceae	Family. Nostocaceae
Genus. *Nostoc*	Genus. *Nostoc*

The name *Nostoc* was first given by Paracelsus. In addition to the above, they are also called as witche's butter, mare's egg and star jelly.

Occurrence

Nostoc occurs in fresh water and also in terrestrial habitats. The fresh water members are found in ponds, pools etc. Terrestrial members grow commonly in damp soil, damp rocks and also in association with many plants like liverworts, mosses, lichens etc.

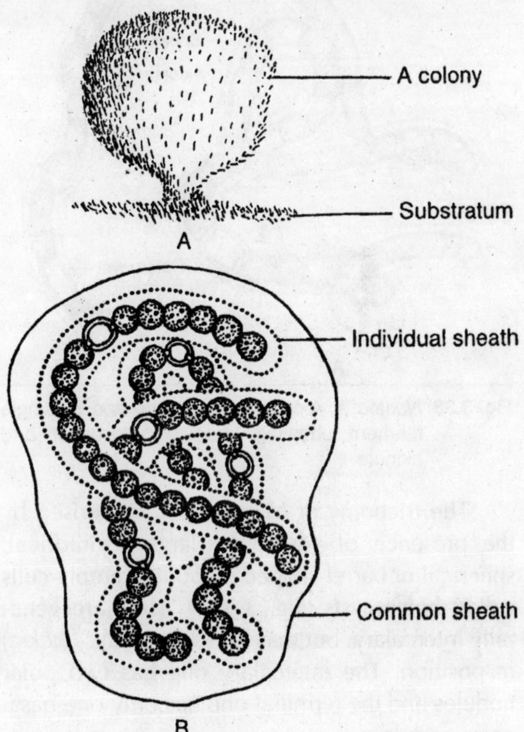

Fig. 3.37: *Nostoc*: A. Gelatinous colony. and B. Aggregated filaments

Some species like *N. punctiforme*, grow **endophytically** in the underground stem of *Gunnera manicata*. It also forms symbiotic association and grows within the thallus of *Anthoceros* and coralloid root of *Cycas*. Species like *N. collema* and *N. sphaericum* are the common phycobiont member of lichen thallus. Some species grow abundantly in rice field.

The plants are generally found in colonies embedded in gelatinous sheath in the form of ball-like colony (Fig. 3.37A) of few millimeters to even 30 cm in diameter.

Plant body

The plant body is a trichome, consists of many spherical, oval or round cells arranged end to end (Fig. 3.38B). Individual trichome is covered by a mucilaginous sheath. The sheath is hyaline or coloured. The trichome is uniseriate and unbranched (Fig. 3.37B).

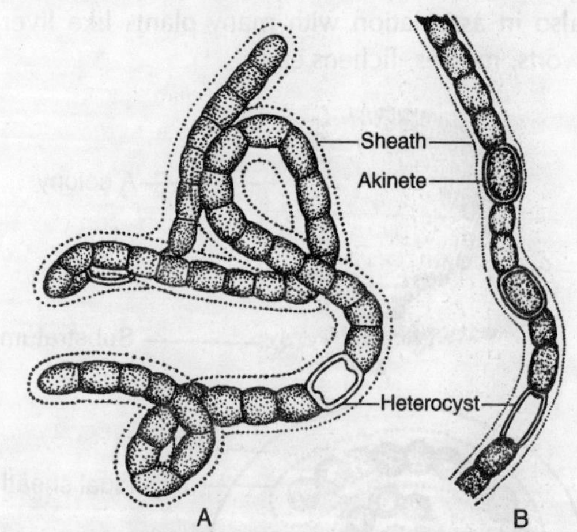

Fig. 3.38: *Nostoc*: A. A contorted filament, and B. Single filament showing sheath, heterocyst and akinete

The trichome of *Nostoc* is characterised by the presence of somewhat large, cylindrical, spherical or barrel-shaped colourless empty cells called **heterocysts** (Fig. 3.39B). They are generally intercalary, but may be terminal (*N. linckia*) in position. The intercalary one has two polar nodules and the terminal one has only one basal polar nodule.

The heterocyst generally occurs singly but rarely in chain. Some of the cells of the trichome become enlarged, which store food and becomes thick-walled. These thick-walled structures are the **akinetes**. They generally develop in chain.

The trichomes are present in aggregation within a common gelatinous sheath and form ball-like colony. Each colony consists of thousands of trichome (Fig. 3.37B). The colonies are greenish or bluish green in colour. They are globular or ellipsoidal in shape and in *N. pruniformae* it may attain the size of an egg of hen.

Cell structure. All vegetative cells are alike in structure and spherical or round in shape. Individual cells exhibit the structure of a typical cyanophycean cell.

The cell is surrounded by cell wall (Fig. 3.39A). The wall is mucopolymeric. The protoplast is differentiated into outer pigmented **chromoplasm** and inner colourless **centroplasm**. Both the regions are not separated by any membrane. The chromoplasm has different pigments, such as cyanophycean granules (protein) and cyanophycean starch. The centroplasm is the nuclear region, present as an **incipient nucleus**.

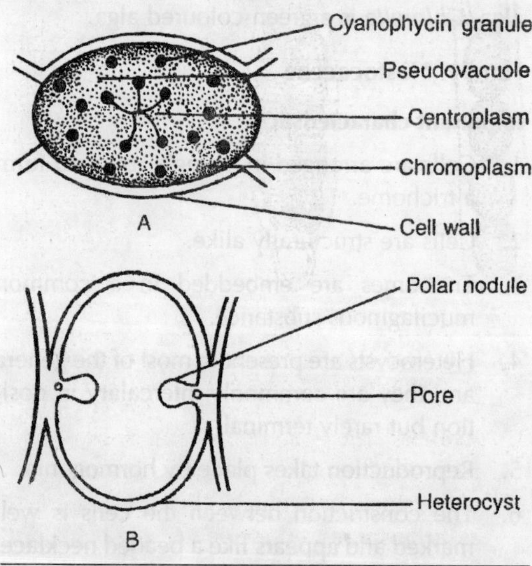

Fig. 3.39: *Nostoc*: A. Cell structure (under light microscope), and B. Single heterocyst

Reproduction

It reproduces only by vegetative means. Sexual reproduction is totally absent. The vegetative reproduction takes place by the following methods: 1. Fragmentation, 2. Hormogones, 3. Akinetes, 4. Heterocyst, and 5. Endospore formation:

1. ***Fragmentation***. The trichome may break into different fragments due to physical, physiological or other factors. Each part of the trichome is capable of developing into new trichome.

2. ***Hormogonia***. It is a very common method of reproduction. Here the trichome may break into **small multicellular fragments** due to degeneration of intercalary vegetative cells or due to the appearance of intercalary heterocyst. These small fragments are called hormogonia or hormogones (Fig. 3.40A). These fragments are further developed into new trichomes by repeated cell divisions either within the mother colony or may come out and can develop a new colony (Fig. 3.40B, C).

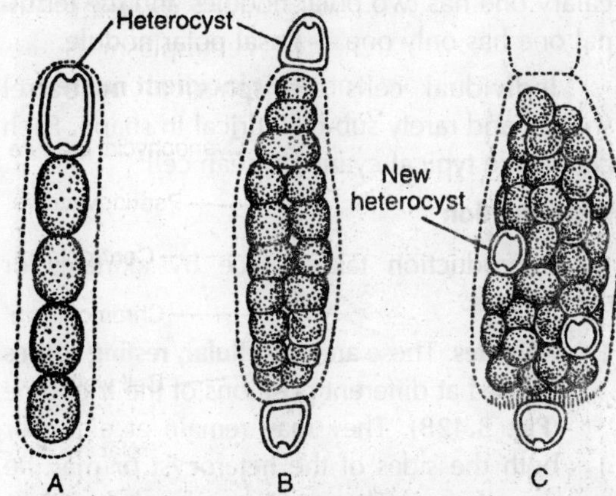

Fig. 3.40: *Nostoc*: (A-C) Different stages of hormogone germination to develop new colony.

3. ***Akinetes***. These are **unicellular, thick walled, resting spores** forms in intercalary positions either singly or in chain. They are also called as **arthospores**. Akinetes are larger in size than the normal vegetative cells (Fig. 3.38B). The akinetes store large amount of food along with cyanophycean granules. Due to highly resistant thick wall, the akinetes can survive during unfavourable conditions. During favourable condition the akinete germinates into a new trichome.

4. ***Heterocyst***. Rarely (*N. commune*) the heterocysts behave as **resting spores**. On germination it develops into new trichome (Fig. 3.41A-C). During germination its protoplast divides into two cells. Both the cells divide again and form four-celled germling, which comes out by rupturing the mother wall and develop into new filament.

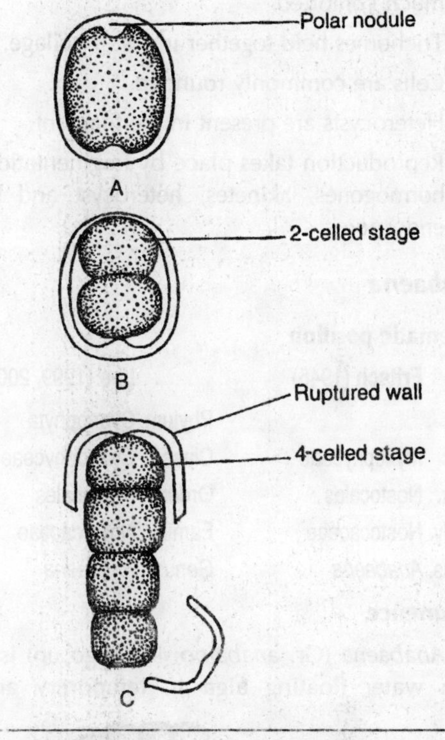

Fig. 3.41: *Nostoc*: (A-C) Germinating stages of heterocyst

5. ***Endospore***. In some species like *N. microscopicum* and *N. commune*, the content of heterocyst divides to form many spores, the endospores. The endospore after liberation and further division develops into a new filament.

Economic importance

Some species grow abundantly in rice field and increase the fertility of the soil by their efficiency to fix atmospheric nitrogen.

N. communae is boiled and used as soup in China and Japan.

Some species of *Nostoc* is used in the reclamation of alkaline soil.

Indian species

Twenty six species of *Nostoc* is reported from India, out of which *Nostoc entophytum, N. calci-cola, N. muscorum, N. punctiforme, and N. sphaericum* are very common.

Important features

1. Plant body is unbranched trichome and much contorted.
2. Trichomes held together in soft mucilage.
3. Cells are commonly rounded.
4. Heterocysts are present in the filament.
5. Reproduction takes place by fragmentation, hormogones, akinetes, heterocyst and by endospores.

Anabaena

Systematic position

Fritsch (1945)	**Lee** (1999, 2008)
	Phylum. Cyanophyta
Class. Myxophyceae	Class. Cyanophyceae
Order. Nostocales	Order. Nostocales
Family. Nostocaceae	Family. Nostocaceae
Genus. *Anabaena*	Genus. *Anabaena*

Occurrence

Anabaena (Gr. *anabaino*, I will go up) is a fresh water floating alga in temporary and permanent pools, in the rice field. Some species are terrestrial and still others are found to grow **endophytically** in the leaves of *Azolla*, the water fern (*A. azollae*) and in the coralloid root of *Cycas* (*A. cycadearum*).

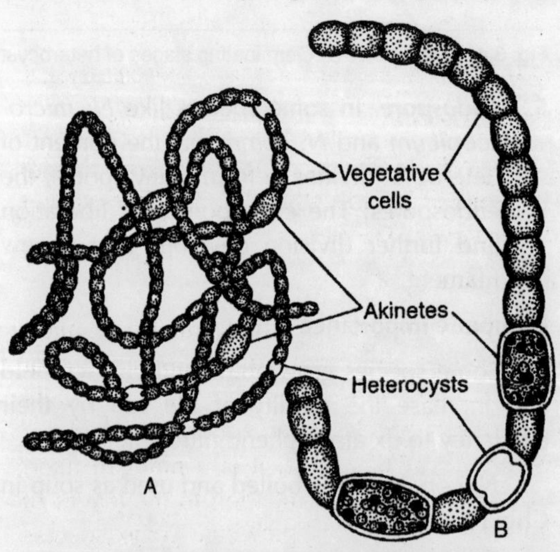

Fig. 3.42: *Anabaena*: A. Contorted filaments, and B. Filament with heterocyst and akinete

Plant body

The plant body is a trichome, consisting of many spherical or barrel-shaped and rarely sub-cylindrical cells arranged end to end (Fig. 3.42A). The trichomes are generally devoid of mucilaginous sheath, but are present in few species. The sheath is hyaline and can be seen by special treatment. The trichome is uniseriate and unbranched. In *Anabaena* the trichome shows continuing motility (Kantz and Bold, 1969).

The trichome is characterised by the presence of somewhat large, cylindrical, spherical or barrel-shaped colourless empty cells called **heterocysts**. They are generally intercalary, but may be rarely terminal in position. The intercalary one has two polar nodules and the terminal one has only one — basal polar nodule.

Individual cells are spherical or barrel shaped and rarely sub-cylindrical in shape. Each cell is like typical cyanophycean cell.

Reproduction

Reproduction takes place by akinetes or endospores.

1. **Akinetes**. These are **unicellular, resting spores** formed at different positions of the trichome (Fig. 3.42B). They may remain at either or both the sides of the heterocyst or may be away from it. They are intercalary in position, either singly or in chain. Akinetes are larger in size than the normal vegetative cells. They are generally cylindrical with rounded edges. The akinetes store large amount of food. Due to highly resistant thick wall, the akinetes can survive during unfavourable condition. During favourable condition the akinete germinates into a new trichome.
2. **Endospores**. These are small uninucleate spores, reported to develop within the heterocysts of *A. cycadearum*.

Morphologically, both *Nostoc* and *Anabaena* are almost alike but during practical handling there is no such problem to identify both the genus. The *Nostoc* trichomes are much contorted and remain in a firm gelatinous envelop, but *Anabaena* always have an extremely watery gelatinous sheath, never form colony of definite shape and with the exception of certain planktonic members they never possess contorted trichomes.

Family. **Scytonemataceae**

Important characteristics

1. Plant body filamentous with **false branching** and relatively wide sheath.
2. Trichomes generally with heterocyst (except *Plectonema*), which may be round, oval or quadrate in form.
3. Heterocysts are generally intercalary, but in *Tolypothrix* terminal heterocysts are also present.
4. Vegetative cells are usually quadrate or slightly longer than breath.
5. Filament usually consists of a single trichome (more than one in *Hydrocoryne*).
6. Reproduction is chiefly by **hormogonia**.
7. **Akinetes** are absent in most genera, but rare in others (*Scytonema*).

The common Indian genera are ***Scytonema***, *Plectonema, Tolypothrix, Hydrocoryne, Camptylonemopsis, Scytonematopsis and Pseudoscytonema*.

The life history of *Scytonema* is given below:

Economic importance

1. Certain species of *Anabaena* contaminate drinking water and cause gastric troubles. It is also reported to be toxic to cattles and other animals.
2. Certain species of *Anabaena* form water blooms and can grow well in O_2-deficient water. Due to continuous respiration by submerged plants and animals during night time (when the photosynthesis does not take place) depletion of O_2 takes place to almost zero level. At that condition, animals and other submerged plants die due to suffocation.
3. Some species of *Anabaena* are used in the production of H_2 fuel, but in this process nitrogenase is utilised as the main hydrogen producing enzyme.
4. Certain member of *Anabaena* can fix atmospheric nitrogen and form nitrogenous compounds. These compounds are further absorbed by the plants for their metabolic activity and increase yield.

Indian species

The following species are very common in India:

Anabaena fertilissima, A. spiroides, A. ambigua, A. gelatinicola, and A. sphaerica.

Important features

1. Plant body unbranched trichome.
2. Trichomes are generally devoid of mucilaginous sheath.
4. Heterocyst present in the filament.
5. Reproduction by akinete or endospores.

SCYTONEMA

Systematic position

Fritsch (1945)	**Lee** (1999, 2008)
Class. Cyanophyceae	Class. Cyanophyceae
Order. Nostocales	Order. Nostocales
Family. Scytonemataceae	Family. Scytonemataceae
Genus. *Scytonema*	Genus. *Scytonema*

Scytonema (Gr. *skytos*, leather; *nema*, thread) is a genus of photosynthetic Cyanophyceae that contains over 100 species.

Occurrence

Scytonema grows in the form of dark mats. Many species are aquatic, either free-floating or grow attached to a submerged substrate, while some others species grow on terrestrial rocks, wood, soil, or on plants. It is a nitrogen fixer, and can provide fixed nitrogen to the leaves of plants on which it is growing. Some species have a symbiotic relationship with fungi to produce lichen.

Plant body

The plant body is trichome (Fig. 3.43). The trichomes are cylindrical and isopolar. They are usually pale blue-green, olive green, brownish, or even violet in colour. The apical ends of the filament are not tapered, but the end cells may be slightly round or widened. Sometimes the cells have constricted cross walls and contain groups of long, cylindrical gas vesicles. The **heterocysts** are present throughout the filament and are rectangular and solitary, with darkly-pigmented cell walls. The trichomes are present within mucilage that is commonly tough, wide, sometimes layered, and may be brown, yellow, or orange in colour.

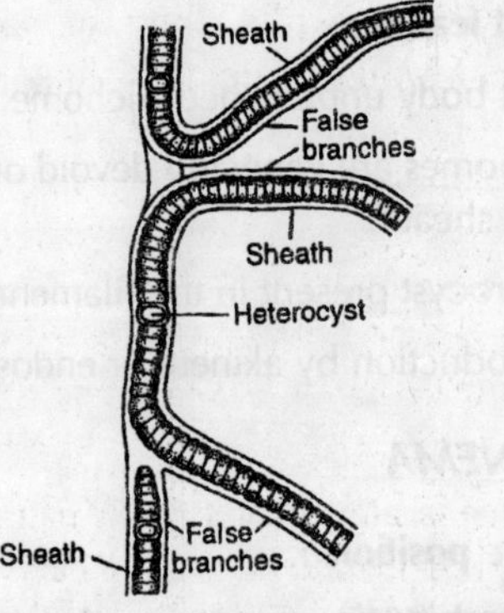

Fig. 3.43: *Scytonema hofmanii*: Plant body with false branching

Double false branching often forms when a trichome breaks apart at the site of a dead cell. Both ends of the filament then break through the mucilage sheath and continue to grow as "branches". Occasionally, false branching may also forms at a heterocyst where the filament breaks apart in a similar manner.

Reproduction

Vegetative reproduction takes place with the help of hormogones (Fig. 3.44), which are generally solitary and terminal. Pseudohormogonia, akinetes and exospores are also formed in some species.

Sexual reproduction is absent.

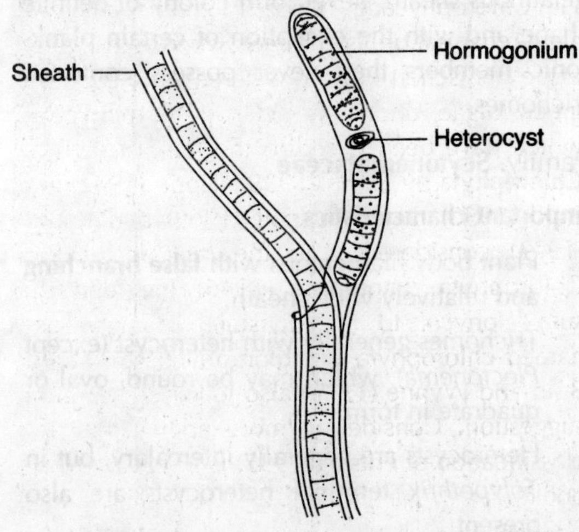

Fig. 3.44: *Scytonema* sp.: Plant body with false branching, heterocyst and hormogonium

Economic importance

Scytonema varium can be cultured artificially. It produces a broad-spectrum anti-viral protein, the **scytovirin** which can inactivate both the HIV virus and Ebola virus.

Indian species

Scytonema ampulum, S. fritschii, S. hansgirgi, S. iyengari, S. pascheri.

Important features

1. Plant body having cylindrical and isopolar trichome.
2. Heterocyst is present throughout the filament.
3. Trichomes having occasional false branching.
4. It produces a broad-spectrum anti-viral protein, the **scytovirin**.

CHLOROPHYCEAE

(GREEN ALGAE)

Chlorophyceae (*chloros*, green; *phyceae*, algal organisation) is commonly known as '***green algae***'. Fritsch (1935) considered to include the green algae under the class **Chlorophyceae**, which have been raised to the rank of division **Chlorophyta** by Smith (1938), Tippo (1942) and Bold (1950). Later Prescott (1969) and Round (1973) considered it to the rank of phylum Chlorophyta. Papenfuss (1946) included the suffix 'phyco' to the divisions of algae and named chlorophyta as Chlorophycophyta. Later Bold and Wynne (1978) also followed the same suggestion. Considering more appropriate, the classification of Fritsch (1935) is followed in this book.

This class consists of 425 genera and about 6,500 species (Alexopoulos and Bold, 1967) but, later Prescott (1969) reported that the number of species may be as many as 20,000; with more being discovered continuously. The name green alga is given because of the presence of dominant pigments like Chlorophyll a and b over the carotenoids and xanthophylls. They are all eukaryotes.

Occurrence

The members of Chlorophyceae generally grow in fresh water (about 90%) and the rest in saline water, terrestrial habitat etc. The fresh water members such as *Volvox, Oedogonium, Spirogyra* etc. grow in ponds, pools and lakes. Members of conjugales (e.g. *Spirogyra, Zygnema* etc.) and Oedogoniales (e.g., *Oedogonium* etc.) are strictly fresh water, but the members of Ulvaceae and Siphonales are predominantly marine. Some members of Volvocales, Chaetophorales and Cladophorales grow both in fresh and saline water. Some species of *Ulothrix* and *Vaucheria* are subaerial and grow on damp soil. Some members may be terrestrial and grow as epiphytes on tree trunk, leaves etc. (e.g., *Trentepohlia*); as epizoic i.e., growing on animal bodies (species of *Characium* and *Cladophora*); as endophytes (e.g., *Chlorella*), as parasites (e.g., *Cephaleuros, Rhodochytrium* and *Phyllosiphon*) and also cause diseases (detail in Parasitic algae page no. 172). They can also grow in further different habitats like hot springs (*Chlorella*), snow (*Chlamydomonas yellowstonensis*), saline water (*C. ehrenbergi*) and some remain as partners in lichen associations.

Important characteristics

1. Members of Chlorophyceae grow mostly in fresh water, a few in brackish and saline water and a few are terrestrial.
2. They show wide range of variations in their thallus structures like unicellular motile (*Chlamydomonas*) and non-motile (*Chlorella*), coenobium (*Volvox*), palmelloid (*Tetraspora*), dendroid (*Ecballocystis*), filamentous branched (*Cladophora*) and unbranched (*Spirogyra*), heterotrichous (*Coleochaete*), siphonaceous (*Vaucheria*) and parenchymatous (*Ulva*).
3. The cells are eukaryotic in nature. Usually there is only one nucleus in each cell, but in Siphonales and Cladophorales many nuclei are present in their coenocytic body. Normally the number of nucleolus is one per nucleus, but several nucleoli are present in the members of Conjugales.
4. The cell wall is mainly made up of cellulose, which comprised of hydroxyproline glycosides or xylans and mannans. In *Chara* the cell wall is encrusted with calcium and magnesium carbonate.
5. Inner to the cell wall, semipermeable cell membrane is present which encircles the protoplast. The cytoplasm contains many small vacuoles which pushes the nucleus with cytoplasm towards the periphery and called primordial utricle.
6. Flagella are 1-many, equal in size and inserted either apically or subapically. The flagella show typical 9+2 arrangement when viewed under E.M.

7. The flagellate cells have eyespot or stigma in the anterior portion, which remain inserted at one side of the chloroplast.
8. The pigments are located in the chloroplast. Chloroplast generally contains pyrenoid(s).
9. The main pigment is chlorophyll a and b; those dominate over α-and β-carotenes and xanthophylls.
10. Phycobilins are absent.
11. The reserve food is starch, composed of amylose and amylopectin.
12. They reproduce by all the three means i.e., vegetative (cell division and fragmentation), asexual (zoospore, aplanospore, akinete etc.) and sexual (isogamy to oogamy). The sexual reproduction is absent in some members of Chlorococcales.
13. The zygote or oospore is the only diploid structure in their life cycle.

Thallus organisation

The class Chlorophyceae shows a range of variation in the structure of plant body (i.e. **range of thallus**). It ranges from **unicellular** e.g., *Chlamydomonas* (Fig. 3.1C), *Chlorella, Sphaerella* etc. to **multicellular** structure. The multicellular forms may be of different types. They may have a number of cells arranged in colonies of definite shape, the **coenobium**. The number of cells in a coenobium may be definite and motile as in *Volvox* (Fig. 3.3A), *Pandorina* (Fig. 3.3B), *Pleodorina* etc. or the number may be indefinite, arranged in net-like masses and are non-motile as in *Hydrodictyon* (Fig. 3.3D). The multicells may aggregate and form a **nonmotile palmelloid** structure, where the cells remain embedded in an amorphous or gelatinous matrix as found in *Tetraspora* (Fig. 3.4A) and *Palmodictyon*. In multicellular forms the cells may be arranged in a single row to form the **filament**. The filament may be ***branched*** (e.g., *Pithophora, Cladophora* (Fig. 3.5B) etc.) or ***unbranched*** (e.g., *Oedogonium* (Fig. 3.82A), *Spirogyra* (Fig. 3.5A), *Ulothrix* (Fig. 3.67A) etc.). The multicells may aggregate and form an expanded sheet-like structure as found in *Coleochaete* (Fig. 3.6C). It shows **heterotrichous habit** where the prostrate system is well-developed. In some members the plant body is like cylindrical tube i.e., coenocytic as in *Vaucheria* (Fig. 3.8A). In some algae like *Ulva* (Fig. 3.9A), the plant body is leaf-like. The highly organised plant body in Chlorophyceae is found in *Chara* (Fig. 3.111A), where the plant is very much complicated in structure with well protected sex organs.

Economic importance

The green algae are not so economically important except a few members.

Among them *Chlorella* is very important because of its high protein content, presence of vitamins and its use in baking industry in the preparation of cake, pastries etc. It is also used in the preparation of an antibiotic, **chlorellin**; which is used to control bacteria. It is also used in different physiological experiments.

Another important member, *Chara*, is very useful to **control malaria** for its larvicidal properties. It is used as fertilizer and in the preparation of polishes.

Monostroma is used to prepare the common food 'aonori' in Japan.

In South India, Green Laver, a kind of food, is prepared from *Spirogyra* and *Oedogonium*.

Ulva and *Enteromorpha* are also eaten by some people.

Many members are used as a source of food and O_2 for many aquatic animals.

Classification

Fritsch (1935) includes all the green algae under the class Chlorophyceae. He further divided it into **nine orders**. These are Volvocales, Chlorococcales, Ulotrichales, Cladophorales, Chaetophorales, Oedogoniales, Conjugales, Siphonales and Charales.

Order. **Volvocales**

The order Volvocales includes **60 genera** and about **500 species**.

Important characteristics

1. They are commonly found in fresh water. Some are grown in brackish water, marine water and also on soil.

2. The plant body is unicellular or multicellular and the multicellular ones are colonial in habit.
3. Both unicellular and colonial members are motile, either throughout or some part of their life cycles.
4. They reproduce both asexually and sexually. Asexual reproduction takes place by zoospores, aplanospores, hypnospores etc. and sexual reproduction by iso-, aniso-, and oogamy.

Classification

Fritsch (1935) divided the order Volvocales into **3 suborders** and **7 families**. The outline of classification is given below:

Order. **Volvocales**

I. Suborder. ***Chlamydomonadineae***

Family. 1. Chlamydomonadaceae
2. Sphaerellaceae
3. Polyblepharidaceae
4. Phacotaceae

II. Suborder. ***Tetrasporineae***

Family. 1. Tetrasporaceae
2. Palmellaceae

III. Suborder. ***Chlorodendrineae***

Family. 1. Chlorodendraceae

Family. **Chlamydomonadaceae**

Important characteristics

1. Plant body is unicellular, uninucleate with definite cell wall.
2. Cells have 2-4 flagella which are equal in length.
3. Chloroplast is generally cup-shaped, but it may be H-shaped, reticulate, stellate etc.
4. Asexual reproduction takes place mainly by zoospores.
5. Sexual reproduction takes place by all the three means iso-, aniso-, and oogamy.

Life history of *Chlamydomonas* is given below:

CHLAMYDOMONAS

Systematic position

Fritsch (1945)		**Lee** (1999, 2008)	
Class.	Chlorophyceae	Phylum.	Chlorophyta
Order.	Volvocales	Class.	Chlorophyceae
Sub-order.	Chlamydomona-dineae	Order.	Volvocales
Family.	Chlamydomona-daceae	Family.	Chlamydomona-daceae
Genus.	*Chlamydomonas*	Genus.	*Chlamydomonas*

Occurrence

The genus *Chlamydomonas* (Gr. *Chlamys*, mantle; *monas*, single organism) includes about 500 species, found almost everywhere (i.e., ubiquitous). Commonly they are found in fresh water of lakes, ponds, tanks etc., but they are also available in brackish water (*C. halophila*), saline water (*C. ehrenbergii*), snow (*C. nivalis*) and some are also air borne. The snow of arctic and alpine zones becomes red due to the presence of *C. nivalis* which accumulates the red pigment haematochrome. But the snow covered mountain range of yellow stone national park (USA) becomes yellowish green due to the well-developed population of *C. yellowstonensis*.

Plant body

Plant body of *Chlamydomonas* is unicellular and motile (Fig. 3.47). The cells are usually spherical, oval or oblong in shape but other forms like ellipsoidal, pyriform etc. are also available (Fig. 3.45). Most of the species are broader towards the posterior side and pointed towards the anterior side, which gradually ends in apical papilla.

The length of the cell varies from 20-30 μm but rarely exceeds 30 μm in major diameter.

The cell wall is thin, smooth and firm. It is made up of cellulose. The major structural component of cell wall is glycoprotein. In some species (*C. gleocystiformis*), the cell wall is surrounded by mucilaginous pectin layer formed by the dissolution of pectose layer to pectin.

Inner to the cell wall, semipermeable cell membrane or plasma membrane is present which surrounds the protoplast. The protoplast contains the following:

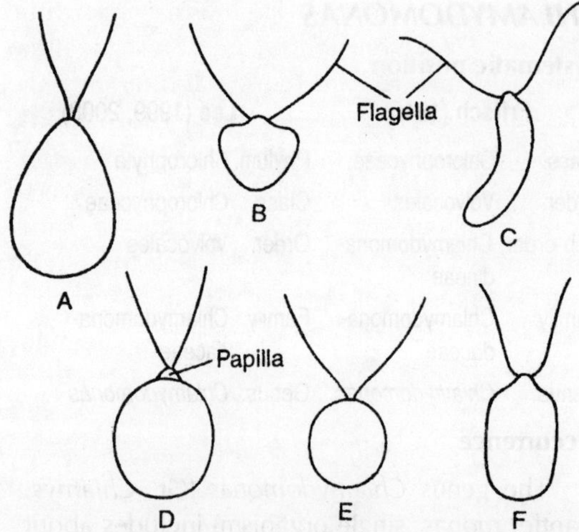

Fig. 3.45: *Chlamydomonas*: Plant body of various shapes. A. *C. polydactyla*, B. *C. gyroides*, C. *C. lunata*, D. *C. conosa*, E. *C. incerta*, and F. *C. biconvexa*

(a) **Chloroplasts**. It occupies the lower broader part. It is generally of cup-shaped

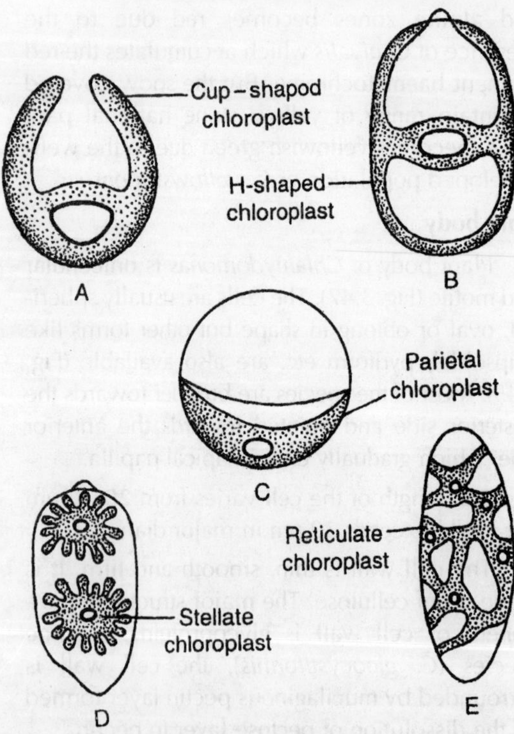

Fig. 3.46: *Chlamydomonas*: Types of chloroplasts. A. Cup-shaped (*C. eugametos*), B. H-shaped (*C. biciliata*), C. Parietal (*C. mucicola*), D. Stellate (*C. arachne*) and E. Reticulate (*C. reticulata*)

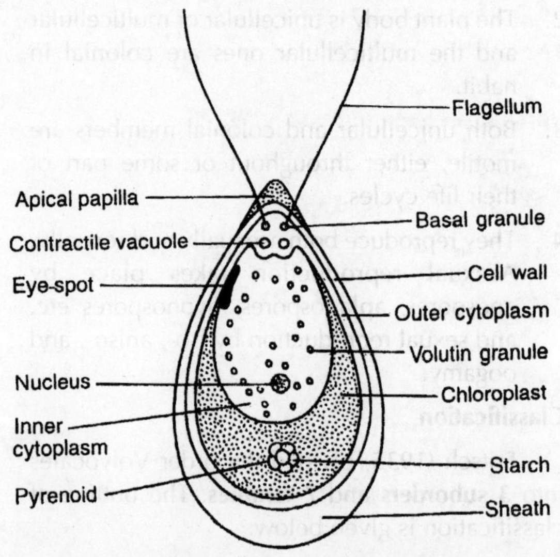

Fig. 3.47: *Chlamydomonas*: Single vegetative cell

and parietal (Fig. 3.46). (*However, the chloroplast is variable in shape, such as* H-shaped in *C. biciliata;* reticulate in *C. reticulata* and stellate in *C. arachne.*) Chloroplast has single pyrenoid with a starch sheath. The number is also variable and it may be two (*C. debaryanum*) to many (*C. gigantea*). Sometimes they may be numerous and distributed irregularly inside the chloroplast (*C. sphynicolaris*).

(b) **Eyespot** Towards the anterior end of the chloroplast at one side an circular to oval, photoreceptive organ, the stigma or eyespot is present. *The spot consists of a curved pigmented plate,* the **pigmentosa** and *a biconvex lens.*

(c) **Nucleus**. The nucleus remains suspended inside the cups and is of eukaryotic in nature.

(d) **Other inclusion**. It includes mitochondria, ER, vesicles, contractile vacuoles etc.

(e) **Flagella**. Two whiplash-type of flagella are present towards the anterior region of the cell. They are equal in length. The flagella may be very small, same size of the cell or bigger than the cell in most species.

The flagellum originates from the blepharoplast, situated towards the anterior side (Fig. 3.48). The flagella come out through very fine canals, on the outer wall (Fritsch, 1935).

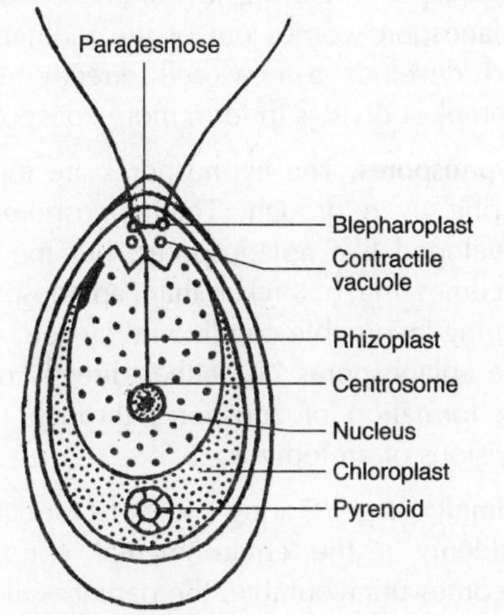

Fig. 3.48: *Chlamydomonas* showing neuromotor apparatus

In some species like *C. nasuta* **neuromotor apparatus** (Fig. 3.48) is said to be present. The neuromotor apparatus consists of two basal granules, the blepharoplasts. Both are connected by a fibre, called **paradesmose**. One of the blepharoplast is connected to the centrosome of the nucleus by a thread, the **rhizoplast**. Many fine fibrils connect the centrosome with the nucleolus.

The E.M. studies (Dodge, 1976) have not confirmed about the neuromotor apparatus.

(f) **Contractile vacuoles**. Two contractile vacuoles are present just below the blepharoplast. Possibly they regulate the water content of the cell, by discharging more water at times.

Structure as observed under E.M.

Roberts *et al.* (1972), Hills (1973) and Hills *et al.* (1973) have studied in detail the structure of *Chlamydomonas* under Electron Microscope (E.M.) (Fig. 3.49).

The cell wall is multilayered (7 layered) and it consists of proteins (Dodge, 1976). Cellulose is absent. Cell membrane like normal eukaryotic cell is lipoprotein in nature. The flagellum remains attached to the basal granule, the **blepharoplast** and shows typical 9+2 fibrillar arrangement. The chloroplast is cup-shaped (*C. eugametor*) and surrounded by double-layered unit membrane. It bears a number of band-shaped photosynthetic lamellae, the thylakoids; are lipoprotein in nature and remain dispersed in the granular matrix, the stroma. The grana-like bodies are formed by the aggregation of about 3-7 thylakoids. The chloroplast matrix contains ribosomes, microtubules, crystal-like bodies etc.

The cytoplasm contain nucleus, mitochondria, golgi apparatus, ER, ribosomes etc. The nucleus always lies in the cytoplasm, present in the cavity of chloroplast.

The Eyespot consist of 2-3 parallel rows of lipid droplets (i.e., granules) and each one measures about 75 nm in diameter. It remains embedded at one side of the chloroplast.

Reproduction

Chlamydomonas reproduces both asexually and sexually:

Asexual reproduction. Asexual reproduction takes place commonly by the zoospore formation. But some others also reproduce by aplanospores, hypnospores, palmella stage and rarely by synzoospores.

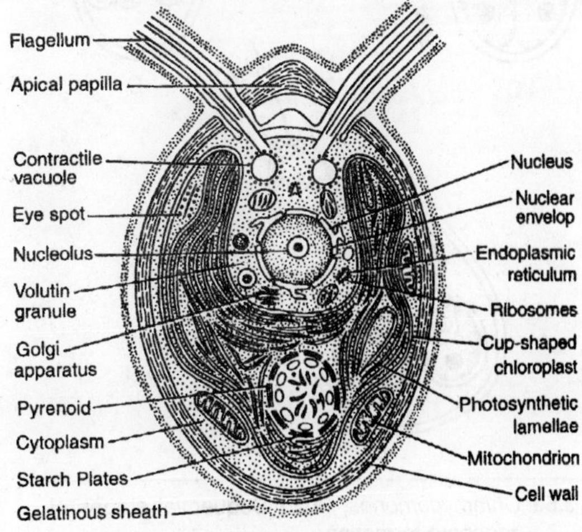

Fig. 3.49: *Chlamydomonas.* Cell structure as observed under electron microscope

1. **Zoospores**. During night with favourable environmental conditions zoospores are formed (Fig. 3.50). At the starting of zoospore formation, the parent cell withdraws its flagella and takes rest. The contractile vacuoles disappear and the protoplast slightly withdraws from the cell wall. The protoplast then undergoes repeated longitudinal division at right angles to one another and forms generally 8 but rarely up to 64 units. Each unit of protoplast secretes a wall around and develops contractile vacuoles. They are released by rupturing or gelatinisation of the mother wall. At the time of liberations they develop their flagella. These flagellated daughter units are called **zoospores**. After liberation they behave as new individuals and capable of developing new crop of zoospores after 24 hours.

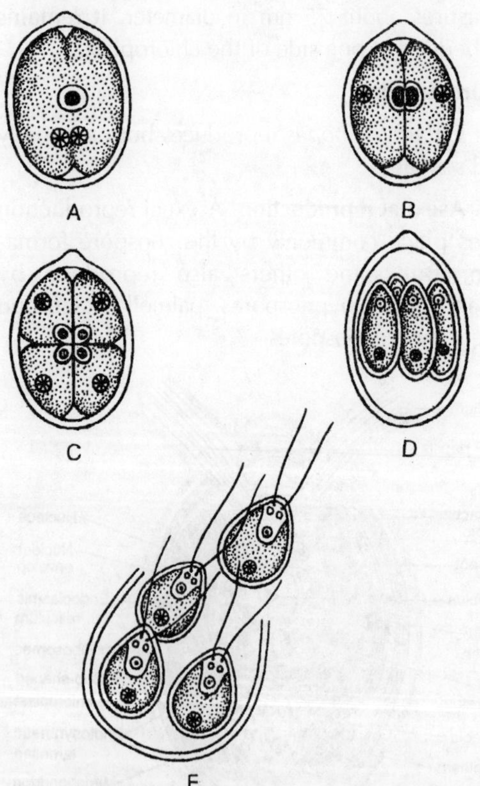

Fig. 3.50: *Chlamydomonas*: (A-E) Sequential stages of zoospore formation

2. **Aplanospores**. The aplanospores are formed during unfavourable conditions. Following the same procedure like zoospore formation, they develop 2-16 daughter protoplasts. Each one secretes thin wall around itself. These thin walled non-motile spores are called **aplanospores**. During favourable condition aplanospore comes out of the mother cell and develops a new cell directly or its protoplast divides to form more zoospores.

3. **Hypnospores**. The hypnospores are formed during severe drought. The hypnospores are developed like aplanospores, but the wall becomes much thicker than aplanospores. During favourable condition they germinate like aplanospores i.e., either directly or by the formation of zoospores through more divisions of protoplasts.

4. **Palmella stage**. During zoospore formation, suddenly if the environmental condition becomes unfavourable, the parent wall gets gelatinised. Consequently the daughter protoplasts also get gelatinised and undergo divisions. The entire structure becomes enlarged much more. This stage looks like another green alga *Palmella* of Tetrasporales and called this stage as Palmella stage (Fig. 3.51).

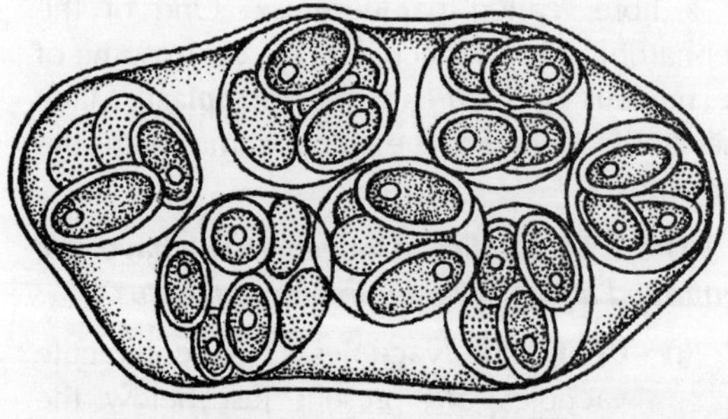

Fig. 3.51: *Chlamydomonas*: Palmella-stage

During favourable condition, the unit bodies develop into individual zoospore. Palmella stage is very common in *C. kleinii* and *C. braunii*.

5. **Synzoospore**. The multinucleate and multiflagellate zoospores are called synzoospores. They are reported (Gorbunova, 1966) to be formed in artificial culture. The mother nucleus divides into 4 nuclei, those develop a pair of flagella and finally to synzoospores.

Sexual reproduction. Sexual reproduction takes place during desiccation or deficiency of nitrogenous compound in the growing medium. It takes place very commonly through isogamy and less frequently through anisogamy and oogamy.

1. **Isogamy**. Majority of the species are isogamous. In this type ***sexual union takes place between the morphologically identical gametes*** (Fig. 3.52). In some species the vegetative cell directly develops into gametes (hologamy) or generally it divides into 8 to 64 gametes. The uniting gametes may develop from the same plant (e.g., *C. debaryanum*) i.e., homothallic or from different plants i.e., heterothallic (e.g., *C. reinhardii* and *C. moewusii*).

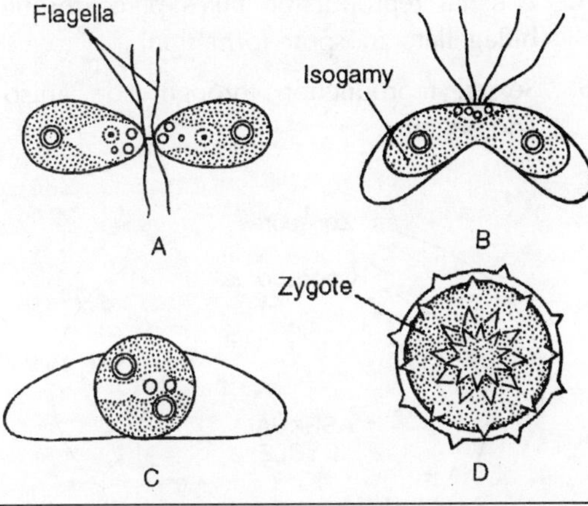

Fig. 3.52: *Chlamydomonas moewusii*: (A-D) Stages showing isogamous sexual reproduction

During union, the flagella are found to be covered by **agglutin** (not found on the flagella of vegetative cell), a chemical substance (protein) which helps in the recognition of gametes of opposite mating type.

In most of the species the gametes are unicellular, uninucleate, very small and biflagellate structures (Fig. 3.52A).

During sexual union isogametes (+ and –) come very close to each other. The wall at the point of contact dissolves resulting in the formation of diploid quadriflagellate zygote through plasmogamy followed by karyogamy (Fig. 3.52B-D).

2. **Anisogamy**. It is the **union between male and female gametes of different sizes** (Fig. 3.53A, B). The larger one is called macrogamete (female) produced only 2 or 4 in the female gametangium. The smaller one is called microgamete (male) produced 8 or 16 in the male gametangium. The microgametes are more active than macrogametes. The microgamete comes very close to the macrogamete. Both the gametes undergo fusion and form zygote (Fig. 3.53C-E). This type of sexual union is found in *C. braunii, C. suboogama* etc.

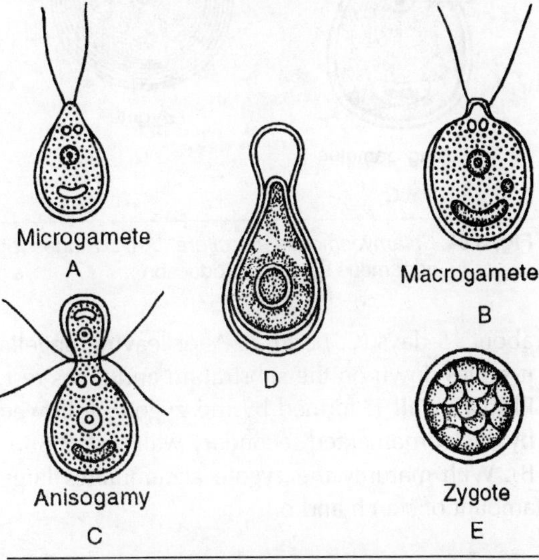

Fig. 3.53: *Chlamydomonas braunii*: A-E. Stages showing anisogamous sexual reproduction

3. **Oogamy**. It is the **union between morphologically dissimilar gametes** (Fig. 3.54A, B). The microgamete (male) is biflagellated and smaller in size than the macrogamete (female) which is non-motile and larger in size. The male cell divides repeatedly and forms 16 units, each of which is converted into male gamete. On the other hand, the female cell leaves the flagella and directly functions as female gamete. The active male gamete comes very close to non-motile female gamete and attaches itself at the anterior end. Both the gametes undergo fusion and form zygote (Fig. 3.54C, D). This type of union is found in *C. coccifera*.

Zygote. Initially the quadriflagellate zygote remains motile for several hours (*C. paradoxa*) to

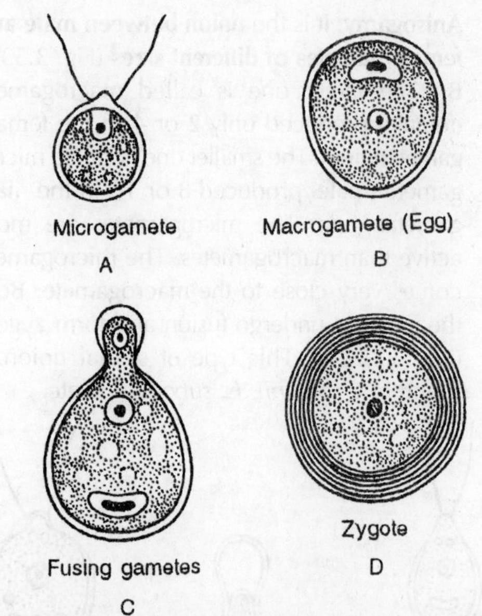

Fig. 3.54: *Chlamydomonas coccifera*: A-D. Stages of oogamous sexual reproduction

about 15 days (*C. pertusa*). After leaving flagella it settles down on the substratum and takes rest. Primary wall is formed by the zygote, followed by thick ornamented secondary wall (Fig. 3.55A, B). With maturity the zygote accumulates large amount of starch and oil.

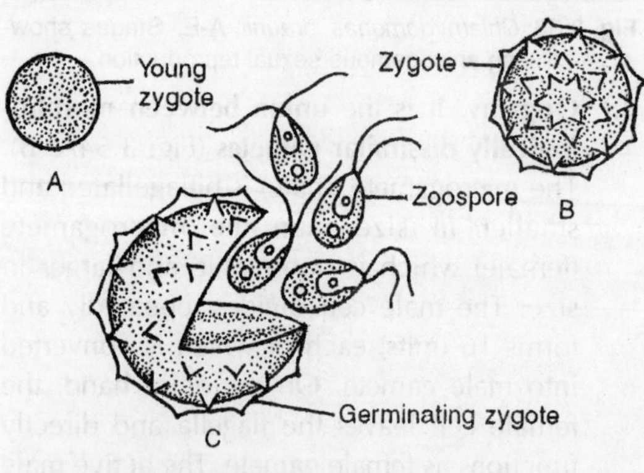

Fig. 3.55: *Chlamydomonas*: A-C. Stages of germination of zygote

Germination of zygote. During germination (Fig. 3.55C) the diploid nucleus (2n) of the zygote undergoes meiosis to form 4 (*C. chlamydogama*) or with additional mitosis forms 16-32 (*C. intermedia*), nuclei (n). The segregation of + and – strains takes place during meiosis. The inner wall dissolves and by breaking the outer wall the haploid cells are liberated. During liberation they develop flagella and behave like new individuals.

Indian species

Chlamydomonas eugametos, C. grandistigma, C. ehrenbergii.

Important features

1. Plant body is unicellular, pear-shaped and biflagellate.
2. Each cell has generally cup-shaped chloroplast, one Eyespot and two contractile vacuoles.
3. Presence of palmella-stage.
4. Asexual reproduction takes place through biflagellate zoospore formation.
5. Sexual reproduction through iso-, aniso-, and oogamy.

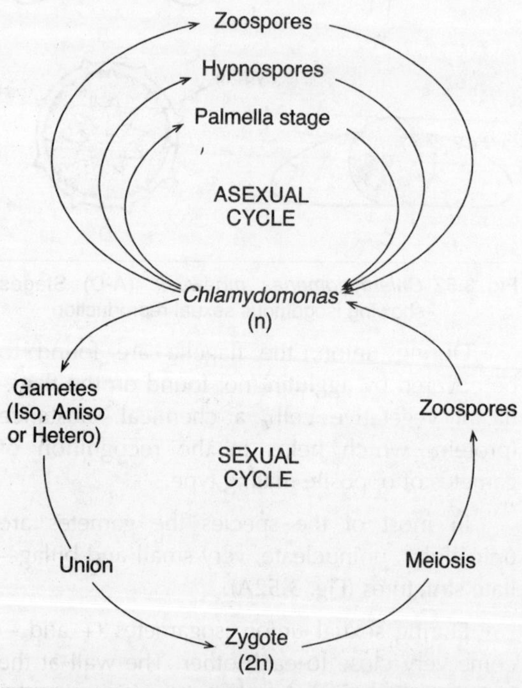

Fig. 3.56: Graphic life-cycle of *Chlamydomonas*

Life history of *Chlamydomonas* is given in Fig. 3.56 and 3.57:

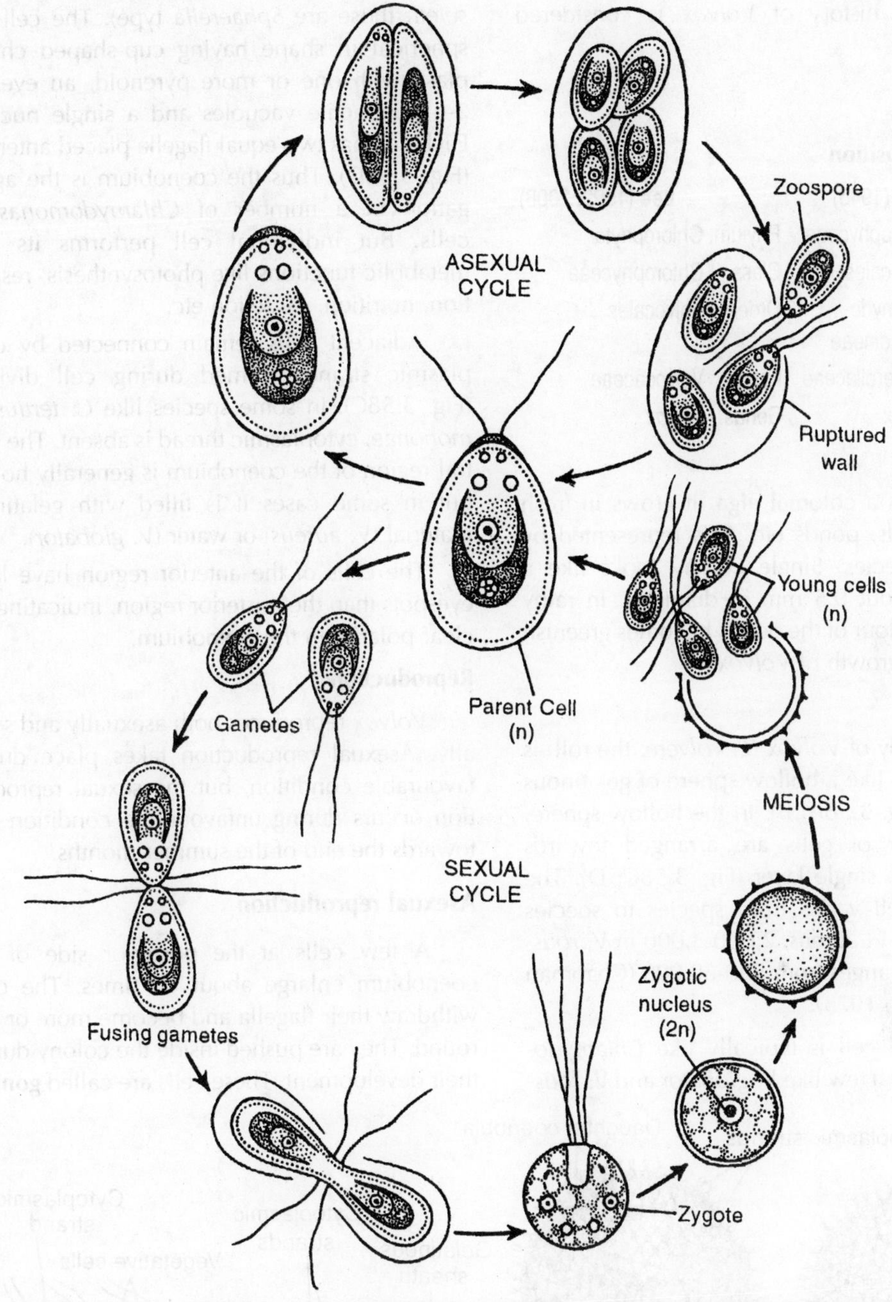

Fig. 3.57: Diagrammatic isogamous life cycle of *Chlamydomonas*

Family **Sphaerellaceae**

Important characteristics

1. The members possess unicellular thick gelatinous-walled biflagellate vegetative cells.
2. Cells with basin-shaped to more or less reticulate chloroplast.
3. Sexual reproduction is isogamous.

The important genera of the family are *Volvox*, *Sphaerella* etc.

The life history of *Volvox* is considered below:

VOLVOX

Systematic position

Fritsch (1945)		Lee (1999, 2008)	
Class.	Chlorophyceae	Phylum.	Chlorophyta
Order.	Volvocales	Class.	Chlorophyceae
Suborder.	Chlamydomonadineae	Order.	Volvocales
Family.	Sphaerellaceae	Family.	Volvocaceae
Genus.	*Volvox*	Genus.	*Volvox*

Occurrence

Volvox is a colonial alga, it grows in fresh water of pools, ponds etc. It is represented by about 20 species. Single colony looks like a small ball about 0.5 mm in diameter. In rainy season the colour of the ponds becomes greenish due to rapid growth of *Volvox*.

Plant body

Plant body of *Volvox* (*L. volvere*, the roll) is a coenobium, like a hollow sphere of gelatinous substance (Fig. 3.58A, B). In the hollow sphere, huge number of cells are arranged towards periphery in a single layer (Fig. 3.58C, D). The number of cell varies from species to species (500-1,000 in *V. aurens*, 2,000-3,000 in *V. rousseletii*) and it ranges from 500-60,000 (Chapman and Chapman, 1973).

Individual cell is typically like *Chlamydomonas* (except a few like *V. globator* and *V. rousseleti*, those are *Sphaerella* type). The cells are spherical in shape having cup-shaped chloroplast, with one or more pyrenoid, an eyespot, 2-6 contractile vacuoles and a single nucleus. Each cell has two equal flagella placed anteriorly (Fig. 3.58D). Thus the coenobium is the aggregation of a number of *Chlamydomonas*-like cells. But individual cell performs its own metabolic functions like photosynthesis, respiration, nutrition, excretion etc.

Adjacent cells remain connected by cytoplasmic strands formed during cell division (Fig. 3.58C). In some species like *C. tertius, C. mononae*, cytoplasmic thread is absent. The central region of the coenobium is generally hollow but in some cases it is filled with gelatinous material (*V. aureus*) or water (*V. globator*).

The cells of the anterior region have large eyespots than the posterior region, indicating the clear polarity in the coenobium.

Reproduction

Volvox reproduces both asexually and sexually. Asexual reproduction takes place during favourable condition, but the sexual reproduction occurs during unfavourable condition i.e., towards the end of the summer months.

Asexual reproduction

A few cells at the posterior side of the coenobium enlarge about 10 times. The cells withdraw their flagella and become more or less round. They are pushed inside the colony during their development. These cells are called **gonidia**

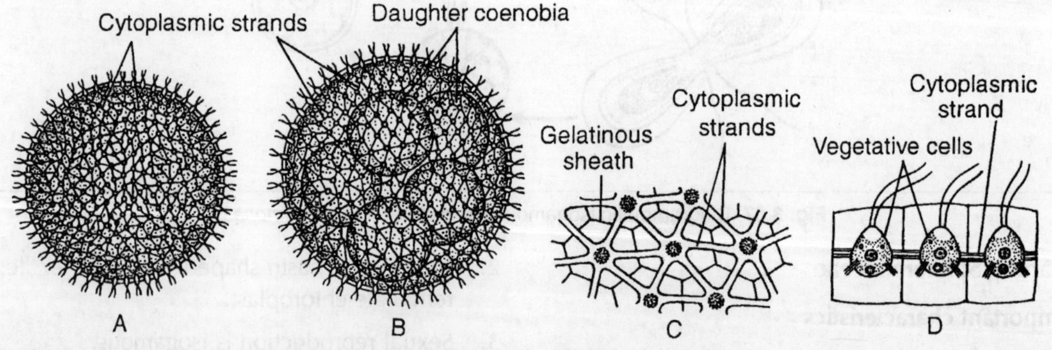

Fig. 3.58: *Volvox* sp.: A. Young daughter colony, B. Mother coenobium containing daughter coenobia, C. Surface view of a portion of colony showing gelatinous sheath and cytoplasmic strands, and D. Portion of peripheral region of colony showing position of vegetative cells and cytoplasmic strands

(Fig. 3.59A) or **parthenogonidia** or **autocolony initials**. The gonidium is separated from the vegetative cells by its position and size.

Development of daughter colony. The gonidium undergoes repeated divisions of about 15 or more times and can develop more than 3,200 cells. Those cells ultimately form a colony.

Initially the gonidium undergoes longitudinal division with respect to the colony and form 2 cells (Fig. 3.59B). The second division is at right angle to the first one and forms 4 celled stage (Fig. 3.59C). These cells again divide longitudinally (3rd division) and form 8 celled stage. The cells are arranged in such a pattern that their concave inner surface faces towards the outer side of the colony. This stage is called **plakea stage** or **cruciate plate** (Fig. 3.59D). The 4th division forms 16 celled stage (Fig. 3.59E) and that time it becomes a hollow sphere with an opening towards the outer side, called **phialopore**. The division of cells continues up to the number specific for a particular species. The cells now face towards the centre (Fig. 3.59F). This group of cells then undergoes inversion through the phialopore, by which normal pattern of the colony is achieved.

Inversion. During inversion a constriction appears at a point opposite to **phialopore**. This constricted region becomes pushed gradually towards the phialopore (Fig. 3.59G). Simultaneously the phialopore becomes enlarged, through which the lower part comes out and the edges of phialopore hang backwards. With the help of inversion the anterior side of the cells changes their position from inner to the outer side and the position of phialopore becomes reversed i.e., changes its position from outer to inner side (Fig. 3.59H). The phialopore gradually closes down and a complete hollow sphere is formed. After completion of inversion, the cells secrete their own gelatinous cell wall and each develops two flagella. Thus the daughter colony is formed.

Many such colonies may develop in a coenobium and they swim freely inside the gelatinous matrix of the mother coenobium (Fig. 3.58B). Later on the daughter coenobia come out by rupture or disintegration of the mother colony. In some species like *V. carteri* and *V. africanus* daughter colonies of 2-4 generations may remain within the mother coenobium.

Sexual reproduction

Volvox reproduces sexually during unfavourable condition i.e., towards the end of growing season (late summer). The sexual reproduction is **oogamous** (Starr, 1975). Some species (*V. globator*)

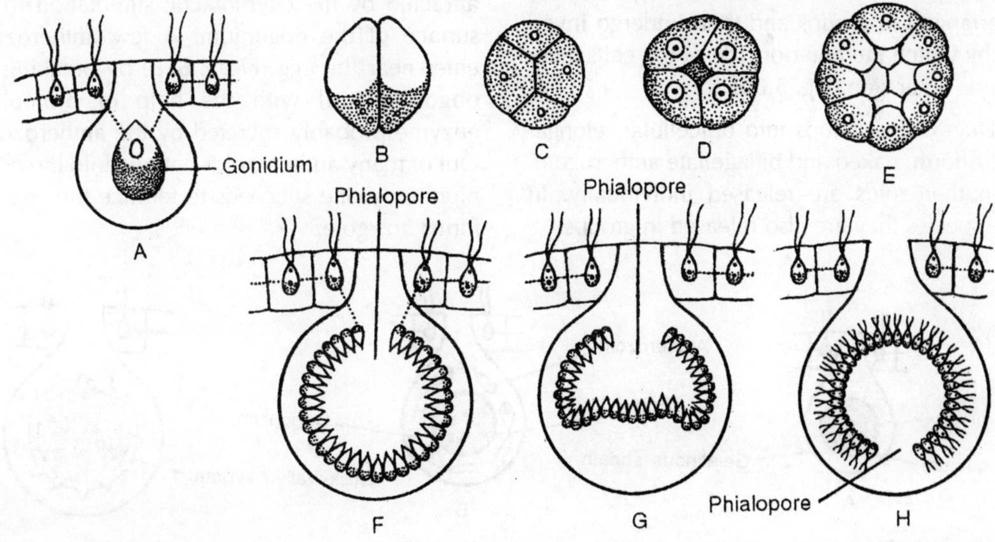

Fig. 3.59: *Volvox* sp. Stages in the development of daughter coenobium: A. Single gonidium, B–E. Stages of gonidium development. 2-, 4-, 8- and 16- celled stages, F. Final stage before inversion, G. During inversion, and H. Flagella development

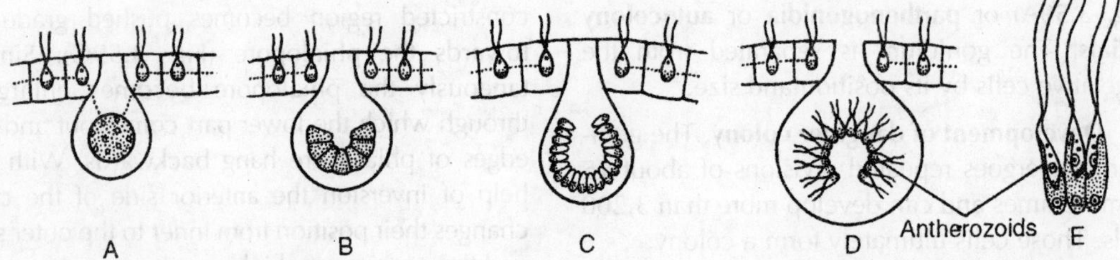

Fig. 3.60: *Volvox* sp. Stages in the development of antherozoid : A–D. Sequential stages of antherozoid development, and E. Antherozoids

is monoecious and others (*V. aureus*) are dioecious. Most of the monoecious species are of protandrous type (i.e., antheridia develop and mature earlier than oogonium). Some cells of the posterior region of the colony withdraw their flagella and develop into reproductive bodies called **gametangia**. The male gametangia are called **antheridia** and the female as **oogonia**.

Development of antheridium. During development of antheridium, an antheridial initial becomes differentiated from the colony. It is like a gonidium, which is aflagellated, larger in size than vegetative cells and contains dense cytoplasm with a single nucleus. The cell undergoes repeated longitudinal divisions like the asexual stage and forms generally about 64-128 cells (though the number varies from 16-512, depending on species). Like the asexual stage the cells are arranged in groups and then undergo inversion by which the anterior side of the cells faces towards outer side (Fig. 3.60).

Each cell develops into unicellular, elongated, fusiform, naked and biflagellate antherozoid. The antherozoids are released individually. In some species they are also released in groups.

Development of oogonium. Single vegetative cell of the colony at the posterior side withdraws its flagella, enlarges in size and become a more or less flask-shaped oogonium. The entire protoplast without undergoing any division, forms an uninucleate non-flagellated egg or female gametophyte (Fig. 3.61A). The egg or female gametophyte is spherical, uninucleate, non-flagellated, green in colour and has parietal chloroplast. It has many pyrenoids and large amount of reserve food. The mouth of the flask-shaped oogonium opens towards the outer surface of the colony.

Fertilization

After maturation, the anthrozoids (= spermatozoids) are liberated from the antheridium either singly or in mass. They move in water and get attracted by the chemotactic stimulation to the surface of the oogonium. A few antherozoids enter near the egg (Fig. 3.61B) by breaking the oogonial wall with the help of proteolytic enzyme probably secreted by the antherozoids. Out of many antherozoids entered into the oogonium only one succeeds to fertilize the egg and forms a zygote.

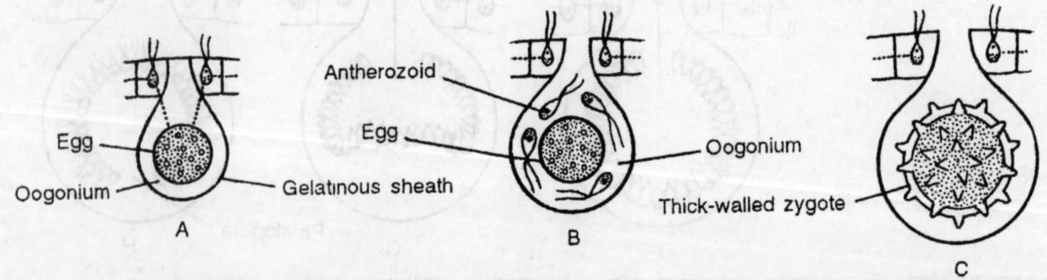

Fig. 3.61: *Volvox* sp. Stages in the development of egg and zygote: A. Portion of coenobium with an egg, B. A stage before fertilization, and C. Zygote inside the coenobium

Zygote

The zygote secretes a thick wall around itself (Fig. 3.61C). It accumulates the haematochrome and becomes red in colour. The wall of the zygote may be smooth (*V. monanae, V. globator* etc.) or spiny (*V. spermatophora* etc.). Zygote is liberated by the disintegration of the mother wall and remains dormant for a long period.

Germination of zygote

During favourable condition the zygote germinates. Before germination, the diploid (2n) nucleus (Fig. 3.62A) of the zygote undergoes meiotic division and forms 4 haploid cells (Fig. 3.62B, C). Further development of zygote varies with species:

1. In *V. minor* and *V. aureus*, after meiotic division the cells undergo repeated mitotic division and form a new colony as formed during asexual reproduction (Fig. 3.62D-F).
2. In *V. rousseletii*, out of 4 haploid cells generally only one survives. The outer wall (exospore) of the zygote breaks and the inner wall (endospore) comes out in the form of vesicle containing a single biflagellate meiospore. The meiospore is then liberated in water by breaking the inner wall i.e., endospore. The biflagellate meiospore then undergoes divisions like the development of daughter colony during asexual process and forms new coenobium.
3. In *V. campensis*, out of many zoospores formed in the oogonium by zygotic division only one survives and others degenerate. The surviving one comes out and by repeated mitotic division it forms a new colony like asexual reproduction.

Indian species

V. aureus, V. merrille, V. rousseleti, V. africanus, V. globator and *V. prolificus* are very common.

Important features

1. Plant body is coenobium and consists of large number of biflagellate, pear-shaped cells.
2. The cells of the coenobium are connected together by means of protoplasmic strands.

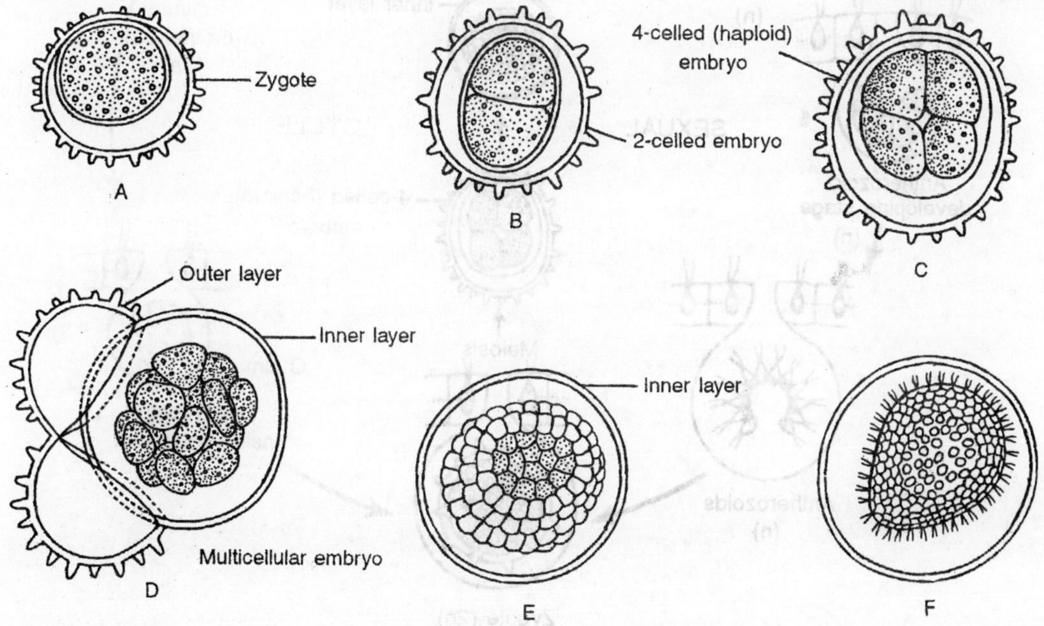

Fig. 3.62: *Volvox* sp. Stages in the development of zygote into coenobium : A. Single zygote, B–C. Reduction division and formation of 4-celled stage, D. Multicellular embryo after splitting the outer layer, E. Multicellular embryo with inner layer, and F. Young coenobium.

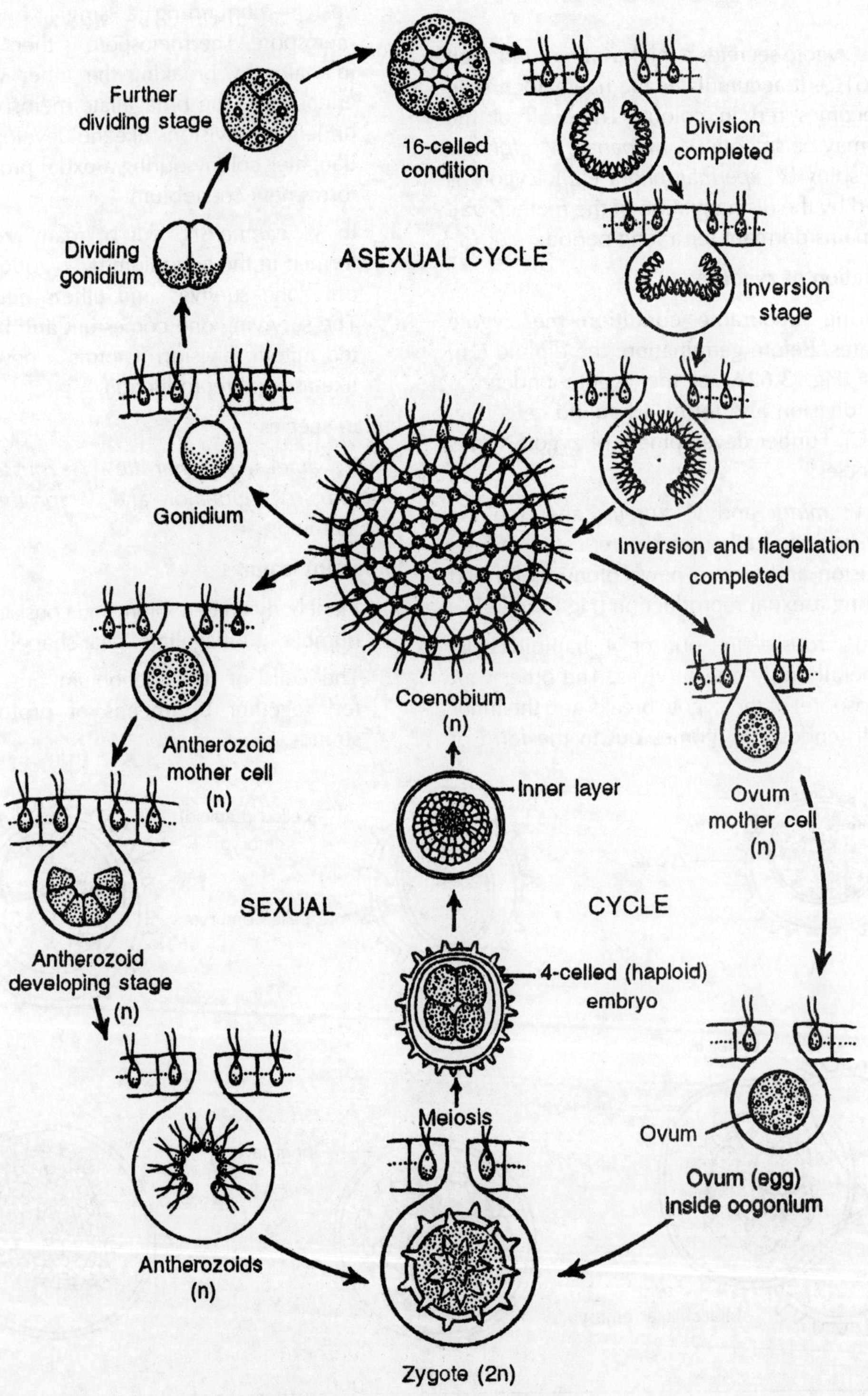

Fig. 3.63: Life cycle of *Volvox* sp.

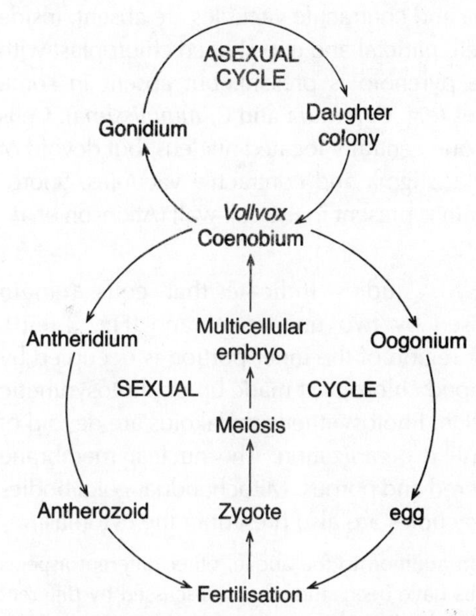

Fig. 3.64: Graphic life cycle of *Volvox* sp.

3. Young coenobia consist of only vegetative cells and are concerned with locomotion and food production.
4. Older coenobium consists of vegetative cells, daughter coenobia and antherozoid mother cells and/or ovum mother cells.
5. Sexual reproduction is oogamous and the coenobia may be monoecious or dioecious.
6. The female gametes or ova are large and non-motile, produced singly inside the oogonium.
7. The male gametes or sperms are spindle-shaped, narrow with a pair of apical cilia and are produced in bunch inside the antheridium.
8. The result of sexual union is the zygote, which on germination develop into new coenobium either directly or by the formation of single biflagellate zoospore.

Fig. 3.63 and 3.64 depict life cycle of *Volvox*.

Order. **Chlorococcales**

The Chlorococcales includes **35 genera** and about **250 species**. Most of them are fresh water, while a few are marine. The important characteristic of the order are:

1. The order includes unicellular as well as colonial members, grows mostly in fresh water.
2. In vegetative phase flagella are absent, thereby plants are non-motile.
3. Cells are devoid of eyespots and contractile vacuoles, but may appear in motile reproductive cells.
4. Cells of the colonial members do not show any polarity.
5. Commonly, cells divide during formation of reproductive bodies.
6. Plants reproduce asexually by zoospores, aplanospores, autospores etc.
7. Sexual reproduction is isogamous or anisogamous.

Classification

Fritsch (1935) divided the order Chlorococcales into **8 families**. These are:

1. Chlorococcaceae e.g. *Chlorococcum* etc.
2. Eremosphaeraceae e.g. *Eremosphaera*
3. Chlorellaceae e.g. ***Chlorella***
4. Oocystaceae e.g. *Scotielia, Oocystis*
5. Selenastraceae e.g. *Selenastrum*
6. Dictyosphaeriaceae e.g. *Dictyosphaerium*
7. Hydrodictyaceae e.g. *Hydrodictyon, Pediastrum*
8. Coelastraceae e.g. *Coelastrum* and *Scenedesmus*

Family. **Chlorellaceae**

Important characteristics

1. Vegetative body, unicellular and non-motile.
2. Reproduction takes place by non-motile spores.

Life history of *Chlorella* is given on next page:

CHLORELLA

Systematic position

Fritsch (1935)	**Lee** (1999, 2008)
	Phylum. Chlorophyta
Class. Chlorophyceae	Class. Chlorophyceae
Order. Chlorococcales	Order. Chlorellales
Family. Chlorellaceae	Family. Chlorellaceae
Genus. *Chlorella*	Genus. *Chlorella*

Occurrence

Chlorella commonly occurs both in fresh water and marine water (Kessler *et al.*, 1968), although they are found to grow in different habitats. *C. parasitica* grows symbiotically in the cells of *Paramecium* and also in *Hydra*.

Plant body

Chlorella is green, unicellular and non-motile alga. Cells are very small, spherical or globular in shape and measures 2-12 μm (Fig. 3.65A). Cells are surrounded by a thin cellulosic wall. Flagella, stigma and contractile vacuoles are absent. Inside the cell, parietal and cup-shaped chloroplast with single pyrenoid is present, but absent in some species (e.g. *C. minuta* and *C. minutissima*). Cells have one centrally located nucleus, but devoid of flagella, stigma and contractile vacuoles. Sporopollenin is present in the cell wall (Atkinson *et al.*, 1972).

E.M. studies indicate that cells remain enclosed by two unit membrane (Fig. 3.65B). Major region of the inner portion is occupied by C-shaped chloroplast made up of photosynthetic lamellae. Photosynthetic thylakoids are devoid of grana-like organization. The nuclear membrane is layered and porous. Mitochondria, golgi bodies and vacuoles are also present in the cytoplasm.

[In addition to the above, other different aspects on cells have been studied and discussed by different scientists in different time. Among those, the cellular morphology of different species (Novakova, 1969), physiological and biochemical aspects (Shihira and Krauss, 1964; Kessler, 1972, 76; Kumar and Kessler,

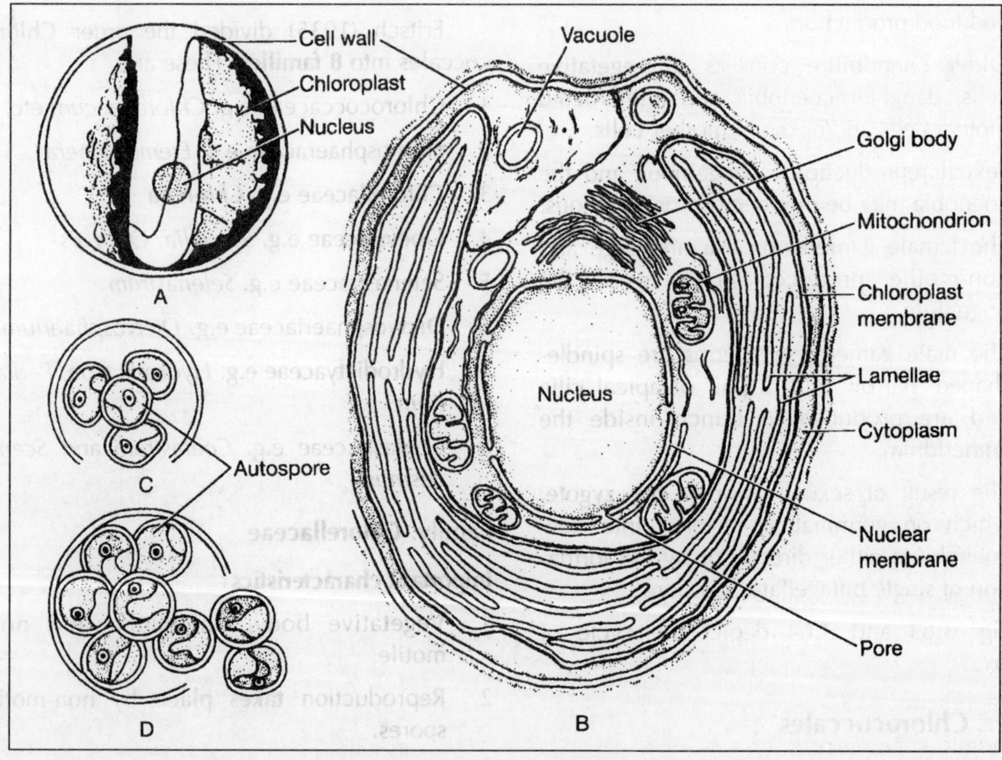

Fig. 3.65: *Chlorella*: A. Single cell under light microscope (after Grintzesco), B. Single cell under electron microscope (diagrammatic) (after Bisalputra), and C-D. Autospore formation

1975), fine structure, physiology and biochemistry (Lee *et. al.*, 1982) and concentration of polyamines (Hagewald and Kneifel, 1982) are important].

Reproduction

Chlorella reproduces only by asexual means, **vegetative and sexual reproduction is absent.**

Asexual reproduction take place by **autospores**. Autospores are the non-flagellate spores looks like parent cell. Motile asexual spore i.e. zoospores are not formed.

Autospores: During autospore formation, the content of the mature cell divides into 2, 4, 8, or, rarely 16 daughter protoplasts (Fig. 3.65 C, D). Each one becomes rounds off and develop into non-motile spores are called **autospores**. Autospores thus formed are liberated by rupturing or by the dissolution of the parent cell wall and later it becomes mature into new *Chlorella* cells. Initially although it looks like parent cell but are smaller in size.

Presence of sulphur is essential for cell division, but for photosynthesis it also requires nitrogen.

With the help of synchronous culture, Tamiya (1963) investigated 4 phases in the life cycle of *C. ellipsoidea* (Fig. 3.66).

These are as follows:

1. *Growth phase*. In this phase the autospores increase in size at the expanse of photosynthetic products.
2. *Early ripening phase*. Here the cells prepare to undergo cell division.
3. *Post-ripening phase*. Here the cells divide twice.
4. *Division phase*. In this phase the parental cell wall gelatinizes, ruptured and finally the autospores are liberated.

Indian species

The common Indian species are *Chlorella vulgaris, C. conglomerata, C. conductrix* and *C. parasitica*.

Economic importance

Economically *Chlorella* is very important. Different aspects of importances are follows:

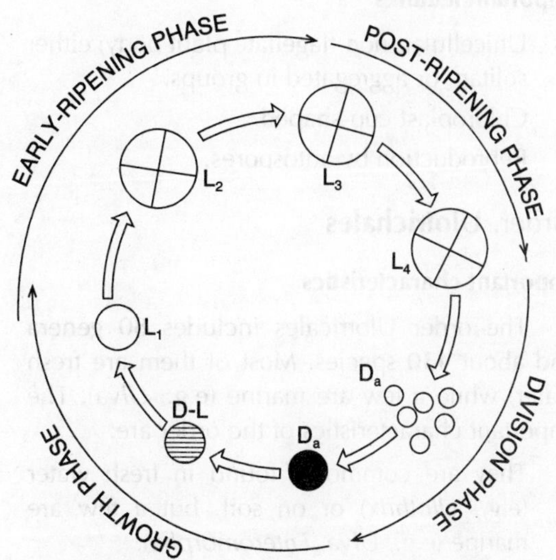

Fig. 3.66: *Chlorella*: Diagrammatic representation of the life-cycle of *C. ellipsoidea* (after Tamiya, 1963)

1. Havng similarity in photosynthetic pigments and reserve food along with higher plants and for its higher rate of growth in different conditions, it is used as a **classical material to study photosynthesis, respiration and also in other physiological aspects.**
2. *Chlorella* contains protein (about 50%), carbohydrate (about 20%), fat (about 20%) and also amino acids, vitamins and minerals. For the above it is used as a **good source of food** in different countries like USA, Germany, Japan, UK etc. It is to be noted that mass culture is costly.
3. It contains the well-known vitamins, like vitamin B_{12}, riboflavin etc.
4. It is used to **purify air in space ship and nuclear submarine** by regulating the O_2 and CO_2. Astronauts also feed on *Chlorella* soup.
5. It also produces an antibiotic, **Chlorellin**, used to control many bacterial pathogen,
6. It also have same level of inorganic content as that of Maize (*Zea mays*), shown through ash analysis.
7. It contains most of the essential amino acid as required by human beings.
8. It is also ued as efficient and cheaper agent in **cleaning sewage disposal.**

Important features

1. Unicellular, non-flagellate plant body; either solitary or aggregated in groups.
2. Chloroplast cup-shaped.
3. Reproduction by autospores.

Order. **Ulotrichales**

Important characteristics

The order Ulotricales includes 80 genera and about 430 species. Most of them are fresh water, while a few are marine (e.g., *Ulva*). The important characteristics of the order are:

1. They are commonly found in fresh water (e.g., *Ulothrix*) or on soil, but a few are marine (e.g., *Ulva, Enteromorpha*).
2. Plant body is commonly an unbranched filament; but in Ulvaceae it is parenchymatous or foliaceous.
3. Cells are uninucleate and contain chloroplast of different types like C-shaped, parietal, axial etc.
4. Each chloroplast contains one or more pyrenoids.
5. Asexual reproduction takes place by means of bi- or quadriflagellate zoospores, aplanospore and akinetes.
6. Sexual reproduction takes place by gametic union and may be iso-, aniso-, or oogamous type.

Classification

Fritsch (1935) divided the order Ulotrichales into **3 suborders** and **6 families.**

The outline of classification is given below:

Order. **Ulotrichales**

I. Suborder. ***Ulotrichineae***
   Family. 1. Ulotrichaceae
   2. Microsporaceae
   3. Cylindrocapsaceae
   4. Ulvaceae

II. Suborder. ***Prasiolineae***
   Family. 1. Prasiolaceae

III. Suborder. **Sphaeropleineae**
   Family. 1. Sphaeropleaceae

Family. **Ulotrichaceae**

Important characteristics

1. The plant body is an unbranched filament.
2. Cells of the filament are uninucleate.
3. Cells have single girdle-shaped, parietal chloroplasts.
4. Sexual reproduction is isogamous and takes place by the union of biflagellated gametes.

The life history of *Ulothrix* is given below:

ULOTHRIX

Systematic position

Fritsch (1945)	**Lee** (1999, 2008)
Class. Chlorophyceae	Phylum. Chlorophyta
Order. Ulotrichales	Class. Ulvophyceae
Subrder. Ulotrichineae	Order. Ulotricales
Family. Ulotrichaceae	Family. not mentioned
Genus. *Ulothrix*	Genus. *Ulothrix*

Occurrence

The genus *Ulothrix* (Gr. *oulos*, woolly; *thrix*, hair) includes about 30 species. Most of them are found in **fresh water** of tanks, ponds, lakes etc. The common fresh water species are *U. aequalis, U. zonata* etc. Some grow in saline water (**marine**) such as *U. pseudoflacca* and *U. flacca*. Species like *U. flacca* grows in the **intertidal region**. Another member *U. implexa* grows as **lithophytes**. They prefer to grow in cold water and are available from rainy season to spring. They disappear during summer months.

Plant body

The thallus is filamentous, long, unbranched and multicellular, where the cells are arranged in a single row (i.e., uniseriate). The filaments are bright green in colour and remain attached at one end with the substratum such as stones, rocks or some other solid objects.

The plant body (Fig. 3.67A) is differentiated into three regions: 1. Basal cell, 2. Apical cell and 3. Middle cells.

1. **Basal cell**. It is the lowermost cell of the filament, commonly long, gradually narrowed downwardly and towards the basal end it

expands and forms a disc-shaped structure. The cell is hyaline or brownish in colour, which performs the function of fixation to the substratum and called **holdfast** or **rhizoidal cell.**

2. **Apical cell**. It is the topmost cell of the filament, which is dome-shaped and green in colour.
3. **Middle cells**. All the cells in-between basal and apical cells are alike. The **cells are commonly broader than its length** i.e., rectangular in shape (The cells may be quadrangular e.g., *U. moniliformis* or it may be much longer than its breath e.g., *U. subconstricta*.)

Cell structure

All the cells except the basal holdfast can divide (Fig. 3.67B) and their wall is composed of outer pectin and inner cellulose. Inner to the cell wall is the cell membrane. The cell membrane encloses the protoplast. It consists of cytoplasm, chloroplast and nucleus. The cytoplasm forms a lining layer or **primordial utricle** adjacent to the cell membrane. Single nucleus remains embedded inside the cytoplasm. The central region is occupied by the vacuole, which contains cell sap. In the peripheral region of the cytoplasm near the cell wall, **single girdle** or **band-shaped chloroplast** is present (other type of chloroplasts are c-shaped, collar-shaped or ring-shaped.) The chloroplast contains one (*U. rorida*) to many (*U. zonata*) pyrenoids.

Reproduction

Ulothrix reproduces by all the three means: Vegetative, Asexual and Sexual:

1. **Vegetative reproduction**. It takes place by fragmentation.

   Fragmentation. In this process the filament breaks up into a number of parts. Each part is capable of developing a new plant like its parent.

2. **Asexual reproduction**. Asexual reproduction takes place in winter, during its active growth. It takes place by the formation of zoospores, akinetes and palmella stage.

(a) ***Zoospore formation***. Zoospores are formed during favourable condition with proper growth of the plant. Any cell except holdfast is capable of producing zoospores (Fig. 3.68A). During zoospore formation the protoplast becomes slightly contracted from the cell wall. The protoplast either develops into single zoospore (*U. fimbriata*) or undergoes division

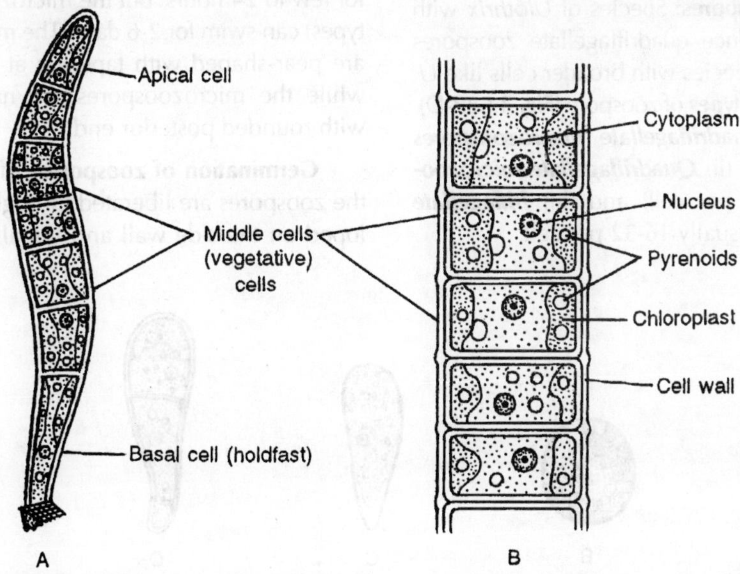

Fig. 3.67: *Ulothrix* sp.: A. Single filament with basal cell (holdfast) and apical cell, and B. Portion of filament showing vegetative cells

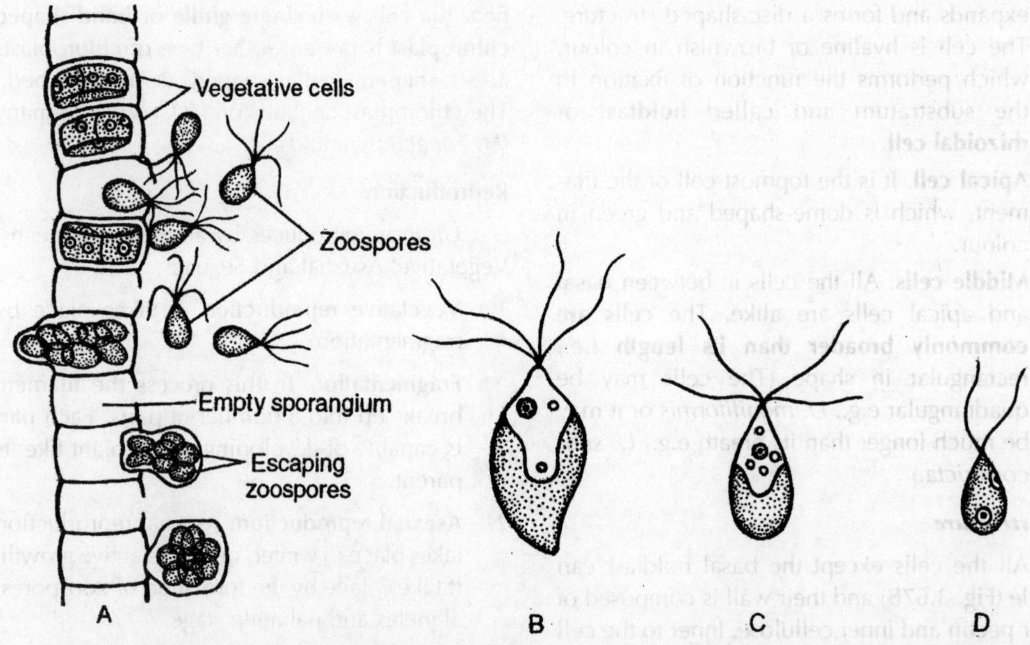

Fig. 3.68: *Ulothrix* sp.: A. Portion of filament with vegetative cells, sporangium and escaping of quadriflagellate zoospores, and B-D. Different types of zoospore (B. Quadriflagellate macrozoospore, C. Quadriflagellate micro-zoospore and D. Biflagellate microzoospore)

and form 2, 4, 8, 16 or 32 units. Each unit contains single nucleus and cytoplasm. These small units form zoospores. The zoospores are of different types:

Types of zoospores. Species of *Ulothrix* with narrow cells produce quadriflagellate zoospores of one kind. But species with broader cells like *U. zonata* produces **3 types** of zoospore (Fig. 3.68B-D). These are: (i) ***Quadriflagellate macrozoospores*** usually 4 per cell, (ii) ***Quadriflagellate microzoospores***- usually 8 per cell, and (iii) ***Biflagellate microzoospores***- usually 16-32 per cell.

The zoospores are morphologically almost alike but differ in size, number of flagella, and position of the eyespot. They also vary in their swimming period. The macrozoospores can swim for few to 24 hours, but the microzoospores (both types) can swim for 2-6 days. The macrozoospores are pear-shaped with tapering at posterior end, while the microzoospores are narrowly ovoid with rounded posterior end.

Germination of zoospores. After maturation the zoospores are liberated through a pore developed on the side wall and initially remain in a

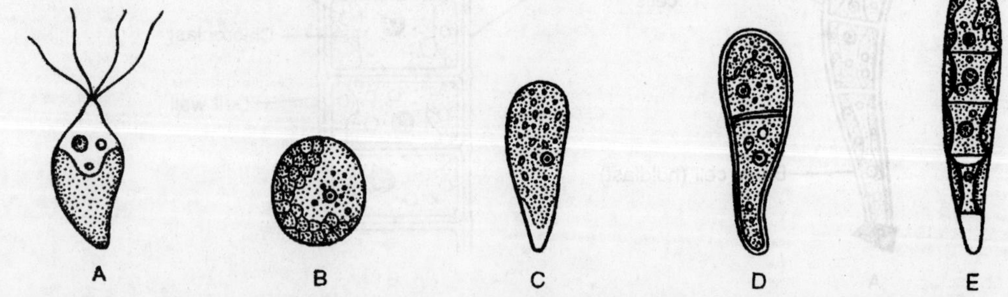

Fig. 3.69: *Ulothrix* sp.: A-E. Germination of zoospore and stages of development of new filament (A. Quadriflagellate zoospore, B. Encysted stage, and C-E. Development of new filament)

delicate mucilage vesicle (Fig. 3.68A). The vesicle soon dissolves and zoospores are liberated. The zoospores can swim for different period of time (few hours to even 6 days). Coming in contact with suitable substratum they lose their flagella and get attached. The cell divides horizontally. The lower attached cell develops into holdfast and the upper cell forms the filament (Fig. 3.69) through repeated divisions.

(b) ***Aplanospores***. With the sudden change of environment towards unfavourable condition during zoospore formation the protoplast units do not form flagella and remain inside the mother cell as non-motile units. These ***unicellular, uninucleate, thin walled non-motile units are called aplanospores*** (Fig. 3.70A). During favourable condition they germinate after or before liberation from parent cell (Dodel, 1876).

(c) ***Hypnospores***. During drought, sometimes the entire protoplast of a cell may round up and forms a **single thick walled structure, the hypnospore** (Fig. 3.70B). During favourable condition it germinates and develops into a new filament.

(d) ***Akinetes***. Akinetes are formed during extreme unfavourable condition. During this process the cell becomes enlarged, protoplast accumulates food material and then forms thick wall around itself. This **thick walled resting vegetative cell is called akinetes** (Fig. 3.70C). It is found in *U. zonata, U.acrorhiza, U. oscillarina* etc.

(e) ***Palmella stage***. Sometimes the wall of the aplanospore mother cell becomes mucilaginous. Consequently the wall of aplanospores also gets enveloped by mucilaginous substance. These coverings protect the aplanospores against desiccation. In this way many green round bodies become enclosed in a mucilaginous mass, called **palmella stage** (Fig. 3.70D). During favourable condition these green bodies come out by the dissolution of the mucilage covering and germinate into new plants.

3. **Sexual reproduction**. Towards the end of growing season they reproduce by sexual means. The sexual reproduction is **isogamous** i.e., the union between similar gametes (Fig. 3.71). The fusion takes place between the gametes developed in different filaments i.e., they are heterothallic or dioecious (*U. rorida* is monoecious).

Morphologically gametes are similar to the microzoospores. They are produced in the gametangium (Fig. 3.71A) similar to zoosporangium. Depending on species the number of gametes may be 8, 16, 32 or 64. Like microzoospores they are uninucleate and biflagellate but smaller in size. Though the gametes are morphologically similar, they are physiologically different and designated as + and – strains.

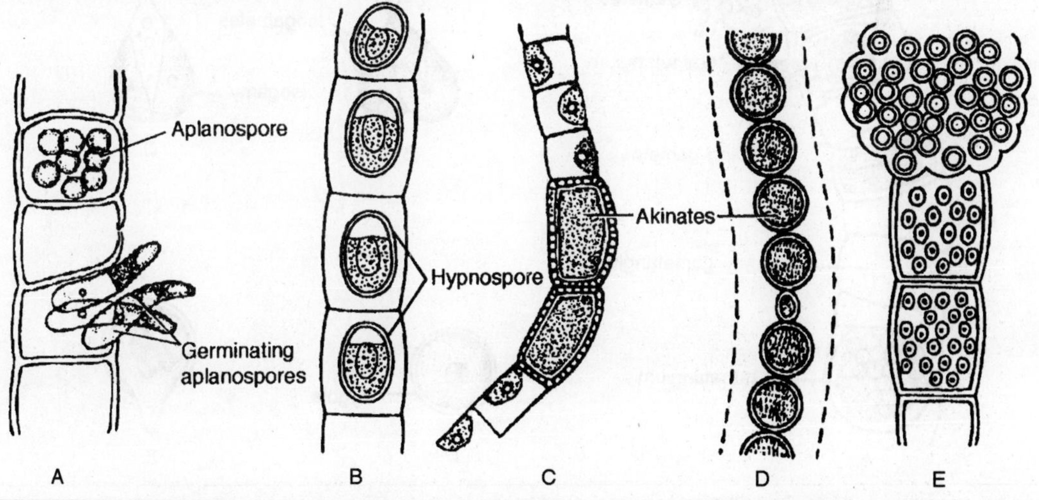

Fig. 3.70: *Ulothrix* sp.: A. Aplanospores, B. Hypnospore, C. Akinetes, and D. Palmella stage

Liberating from the gametangium, the gametes swim for some time. The gametes of opposite strain (+ and -) fuse and form a quadriflagellate spindle shaped zygote (Fig. 3.71B, C). Initially the zygote keeps on swimming but later it settles down on some substratum, withdraws its flagella and becomes round off.

The zygote takes rest for about 5-9 months. After rest the nucleus of zygote (2n) undergoes meiosis and forms 4 haploid nuclei of different strains (2+ and 2- type). The mitosis may follow the meiosis and forms 8 (Fig. 3.71E) to 16 haploid nuclei. The nuclei along with some cytoplasm form spores, called **meiospores.** The meiospores are haploid and quadriflagellate. On germination they develop into haploid *Ulothrix* filaments either + or – type.

Parthenogenesis. Sometimes, the gametes fail to fuse, lose their flagella, and secrete a thick wall around them and are called **parthenospores** or **azygospores**. After rest, each parthenospore germinates directly into a new plant.

Indian species

Ulothrix zonata, U. tenerrima, U. aequalis, U. variabilis, U. oscillarina etc.

Important features

1. This is a fresh water alga growing on substratum like sand particles, rocks etc.
2. The plant body is unbranched, filamentous and differentiated into apex and base.
3. Cells have band or girdle-shaped chloroplast containing one or more pyrenoids.
4. Reproduction takes place by vegetative, asexual and sexual means.
5. Vegetative reproduction takes place by fragmentation.
6. Asexual reproduction by takes place bi- or quadriflagellate zoospores.
7. Sexual reproduction is isogamous.
8. Zygote on germination produces 4 or more zoospores, those on germination produce new filaments.

Fig. 3.72 and 3.73 depict life cycle of *Ulothrix*.

Order. **Chaetophorales**

Chaetophorales are the plants with hair or setae.

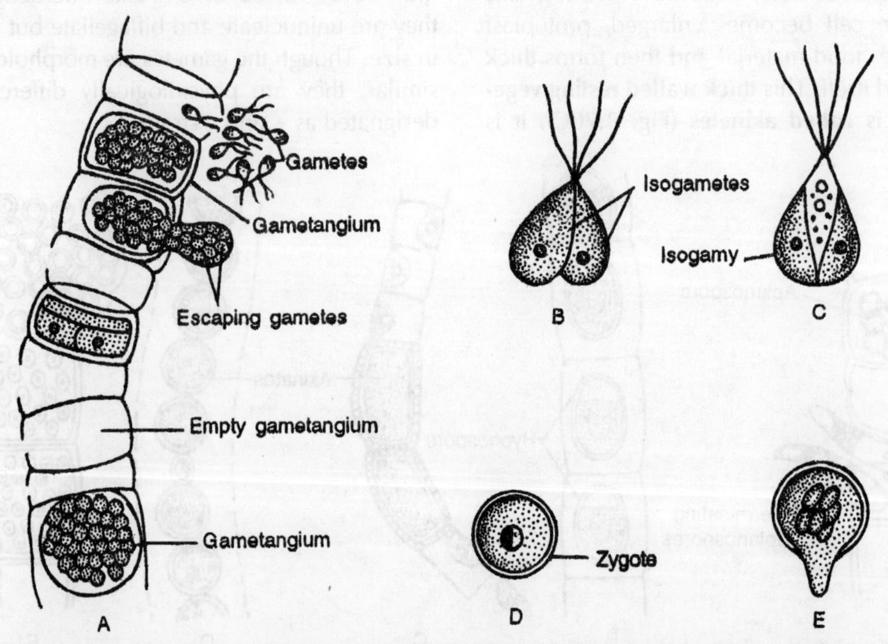

Fig. 3.71: *Ulothrix* sp.: A. Portion of filament with vegetative cells, gametangia and escaping gametes, B. Uniting gametes, C. Isogamy, D. Zygote, and E. Germination of zygote showing zoospore primordial inside

Important characteristics

1. Members are generally found in fresh water.
2. Plant body is filamentous and shows prominent heterotrichous habit; however, in Coleochaete, the prostrate system is well-developed and in *Microthamnion* the erect system is well-developed.
3. Some members have setae (*Coleochaete*) or hairs (*Stigeoclonium*) of different types. The

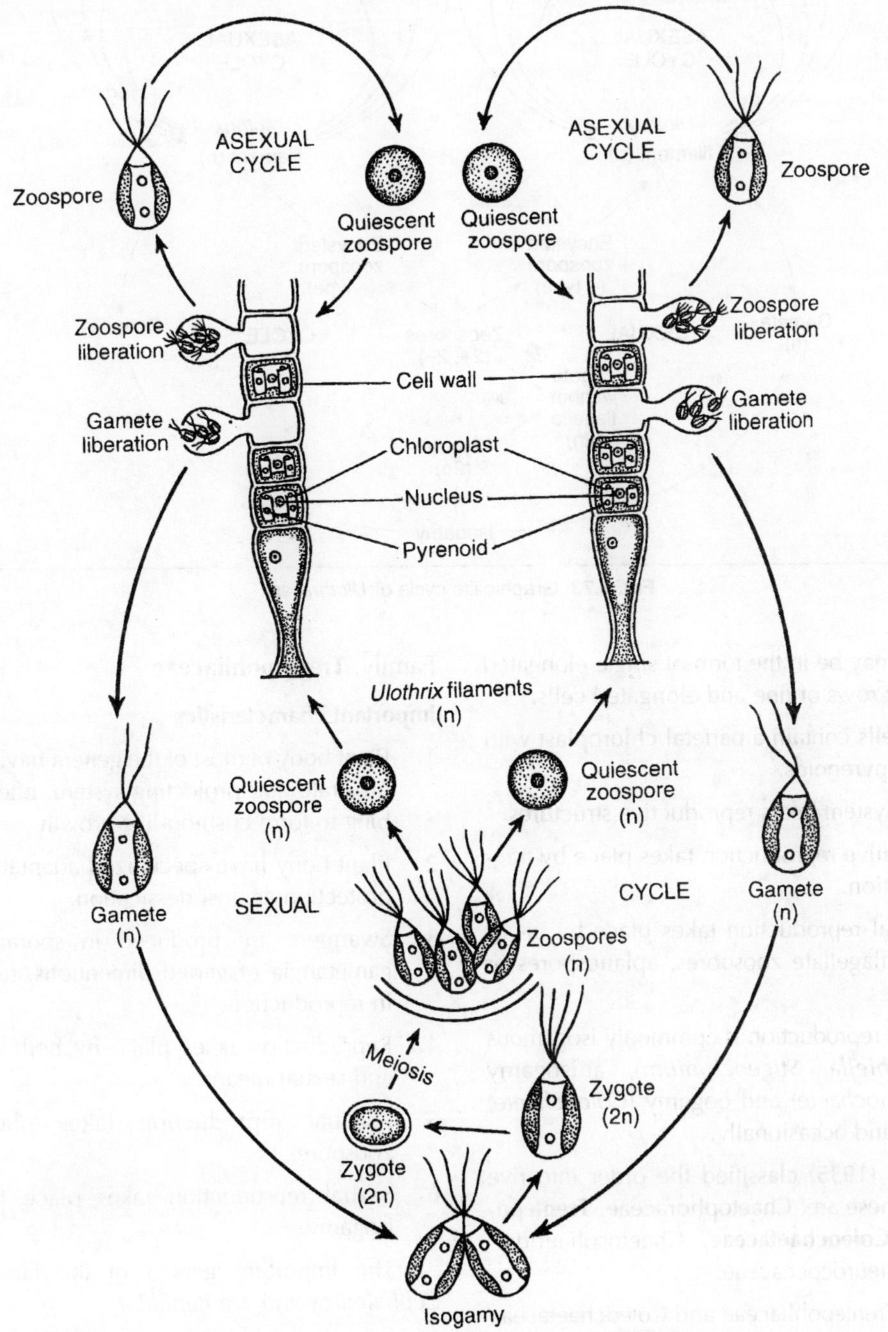

Fig. 3.72: Life cycle of *Ulothrix* sp.

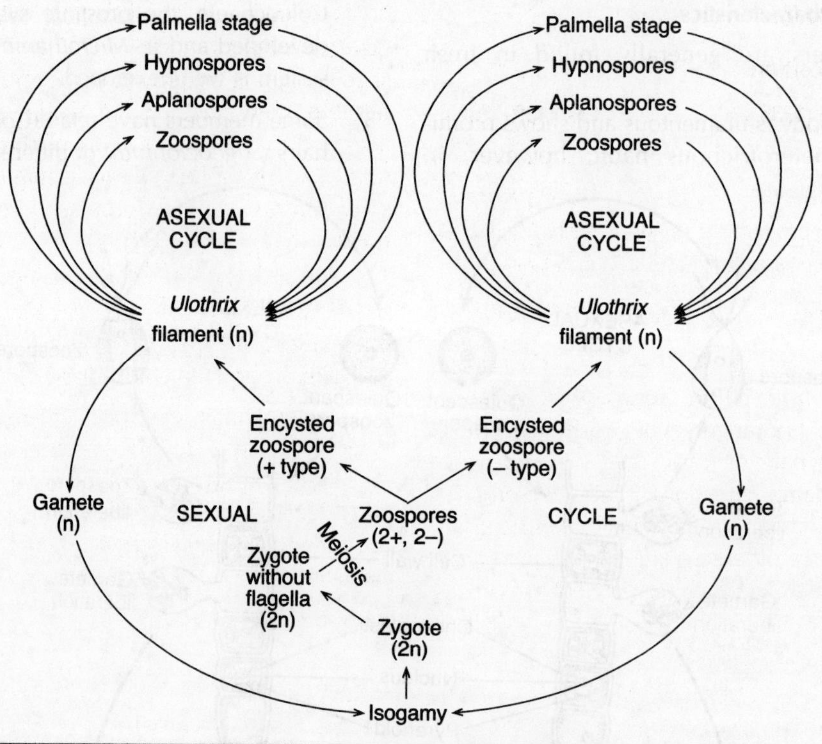

Fig. 3.73: Graphic life cycle of *Ulothrix* sp.

hairs may be in the form of single elongated cell or rows of fine and elongated cells.

4. The cells contain a parietal chloroplast with many pyrenoids.
5. Erect system bears reproductive structures.
6. Vegetative reproduction takes place by fragmentation.
7. Asexual reproduction takes place by bi- or quadriflagellate zoospores, aplanospores or akinetes.
8. Sexual reproduction is commonly isogamous (*Fritschiella, Stigeoclonium*), anisogamy (*Aphanochaete*) and oogamy (*Coleochaete*) are found occasionally.

Fritsch (1935) classified the order into five families. These are: Chaetophoraceae, **Trentepohliaceae, Coleochaetaceae**, Chaetosphaeridiaceae and Pleurococcaceae.

Only Trentepohliaceae and Coleochaetaceae are considered below for further discussion.

Family. **Trentepohliaceae**

Important characteristics

1. Plant body of most of the genera having both prostrate and projecting system, often combing to form cushion-like growth.
2. Plant body have specialized adaptations for protection against dessication.
3. Swarmers are produced in sporangia or gametangia of varied dimentions, take part in reproduction.
4. Reproduction takes place by both asexual and sexual means.
5. Asexual reproduction takes place by zoospores.
6. Sexual reproduction takes place through isogamy.

The important genera of the family are *Cephaleuros* and *Trentepohlia*.

Life history of *Trentepohlia* is given below:

TRENTEPOHLIA

Systematic position

Fritsch (1935)	Lee (1999, 2008)
	Phylum. Chlorophyta
Class. Chlorophyceae	Class. Ulvophyceae
Order. Chaetophorales	Order. not mentioned
Family. Trentepohliaceae	Family. not mentioned
Genus. *Trentepohlia*	Genus. *Trentepohlia*

Occurrence

Trentepohlia is an aerial alga growing epiphytically as orange-yellow cushions or felt-like layer on bark of trees, on rocks and walls and other damp situations. The filaments of *Trentepohlia* have a strong orange colour caused by the presence of large quantities of **carotenoid** pigments which mask the green colour of the chlorophyll.

Plants can withstand prolonged desiccation without any appreciable change. Some species also constitute the algal component of a number of crustose lichen genera such as *Graphis, Graphina, Gyalecta* and *Opegrapha*.

There are about 40 species of *Trentepohlia* mostly distributed in tropical and subtropical areas, but several species also occur in temperate environments. The genus is present in almost all continents.

Plant body

The plant body is composed of heterotrichous filament. In most of the species, thallus exhibits a differentiation into **both aerial and prostrate systems** with richly branched threads. But in some others, where both prostrate and aerial systems are not equally well developed (Fig. 3.74 D, E & L).

Branches of the aerial system are opposite, alternate, or unilateral developing either from the apex of the parent cell or sub-terminally or even from the middle and are generally slightly attenuated (Fig. 3.74 A-C). The cells are cylindrical or barrel-shaped and often length being twice than its breadth (Fig. 3.74 L), and with thick cellulose walls which are stratified by parallel, or divergent layers (Fig. 3.74 C, F).

The free ends of the apical cells are often covered by pectose caps (Fig. 3.74 F-I). In some species, during the growth of the apical cells, the **pectose caps** get pushed aside and remain attached to the lateral walls of the older cells (Fig. 3.74 J-K). The cells have numerous band-shaped or discoid chloroplasts without pyrenoids.

The characteristic orange-yellow colour of the cells is due to the presence of **orange-red haematochrome** dissolved in fat occurring as globules around the chloroplasts.

The function of this special pigment is not yet clearly known. According to some, the haematochrome is a kind of reserve food, whereas, others consider it as a light screen. Cells when young are uninucleate and later on, it becomes multinucleate.

Reproduction:

Trentepohlia reproduces by all the three means are, vegetative, asexual and sexual reproduction.

1. **Vegetative reproduction**. It takes place through fragmentation and akinete formation.

   (i) ***Fragmentation***. In this process the filament breaks up into a number of parts. Each part is capable of developing a new plant like its parent.

   (ii) ***Akinete formation***. Akinetes are thick walled structures developed during unfavourable condition from the vegetative cells of the prostrate system. It may occur in several successive cells. With the onset of favourable condition it germinates and develop new filament like the parent.

2. **Asexual reproduction**. In this process bi- or quadriflagellate swarmer's are developed in three different kinds of sporangia. These are:

   (i) ***Sessile sporangia***: The sessile sporangia may be terminal, lateral, intercalary, or rarely axillary in position (Fig. 3.75D and Fig. 3.75A-G). They are formed by enlargement of cells and liberate biflagellate swarmers without being detached from the mother plant. This

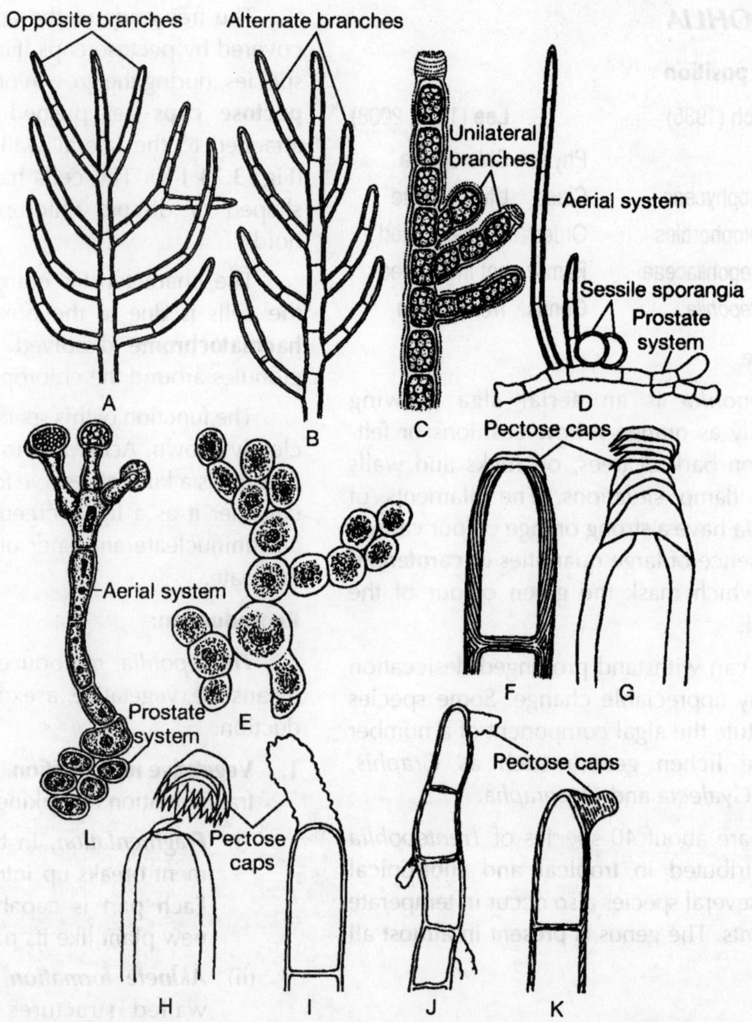

Fig. 3.74: *Trentepohlia* sp. A-C. Different types of branching in aerial branches, D. Prostrate and erect branches with sessile sporangia, E. Prostrate system, F-I. Structure and mode of formation pectose caps on free ends of apical cells with stratified wall, J-K. Pectose cap pushed aside and L. Plant body having prostrate and erect systems, where cells of aerial system having length greater than breadth and bears terminal sporangia

swarmers behaves as isogametes (Fig. 3.75 N-R). Sessile sporangia are quite commonly produced on the prostrate portion of the plant. They may behave as gametangia.

(ii) ***Stalked sporangia***: These are formed terminally or laterally by cutting off from a tubular outgrowth developed from enlarged supporting cell. The apical portion of the outgrowth swells to form sporangium and separates from the stalk cell underneath, which is often bent to assume a knee-shaped appearance (Fig. 3.75 D-E).

These sporangia produce bi- and quadriflagellate swarmers which behave as zoospores. Stalked sporangia are usually developed on the projecting system of the plant (Fig. 3.75 F-H).

(iii) ***Funnel-shaped sporangia***: These are formed from the apices of the terminal cells, become detached at maturity

(Fig. 3.75I-M). Their function has not yet been established.

Both sessile and stalked sporangia may be developed on the same plant or on separate plants. The development of sessile and stalked sporangia in the life cycle of the alga is suggestive of the possibility of alternation of generations.

3. **Sexual reproduction**. Sexual reproduction is of **isogamous type**. The gametes are biflagellate and produced in gametangia. The gametangia are terminal or intercalary in position. They are formed like sessile sporangia by the enlargement of cell. Often, they may develop on the basal region of plant body. The gametangia are solitary and spherical in structure. The gametangia often differ from the sporangia of the same species. A number of biflagellate gametes are produced in gametangia. After maturity they are liberated through a protruding terminal aperture. After sexual union they form quadriflagellate zygote. Parthenogenetic development of gametes into filaments has been observed in some species (Meyer, 1936).

Indian species

Trentepohlia monilia

Important features

1. Epiphytic alga forms orange-yellow cushions or felt-like layer on damp substrate.
2. Plants can withstand prolonged dessication.
3. Plant body heterotrichous, erect and prostrate systems with or without being equally well developed.
4. Branches of the aerial system are opposite, alternate or unilateral.
5. Cells having stratified wall, special pigment i.e. orange-red haematochrome and band-shaped or discoid chloroplasts.

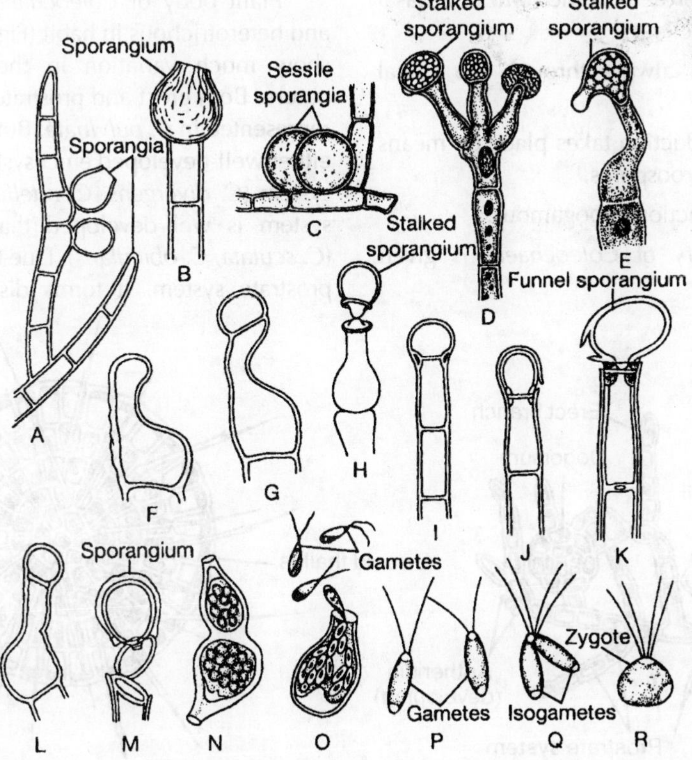

Fig. 3.75: *Trentepohlia* sp. A-C. Lateral, terminal and axillary sessile sporangia, D-E. Stalked sporangia, F-H. Formation of stalked sporangia, I-M. Development of funnel-shaped sporangia, N-O. Sporangia behaving as gametangia and liberation of gametes, P. Gametes, Q. Gametic union and R. Quadriflagellate zygote

6. Free-ends of the apical cells are covered by pectose caps.
7. Biflagellate or quadriflagellate swarmers are developed in specialized structures those are gametangia or sporangia, based on the behaviour of swarmers.
8. Biflagellate swarmers usually behave as isogametes and quadriflagellate one as zoospores.

Family. **Coleochaetaceae**

Important characteristics

1. Plant bodies possess a typical heterotrichous habit. Generally the projecting system is dominant and looks like disc.
2. Cells are uninucleate with single lamellate parietal chloroplast with one or two pyrenoids.
3. All or some cells bear a single long sheathed bristle or seta.
4. The sheath is present in the form of a basal cylinder of mucilage.
5. The growth is always through the apical region.
6. Asexual reproduction takes place by means of biflagellate zoospores.
7. Sexual reproduction is oogamous.

The life history of *Coleochaete* is given below:

COLEOCHAETE

Systematic position

Fritsch (1945)	**Lee** (1999, 2008)
	Phylum. Chlorophyta
Class. Chlorophyceae	Class. Charophyceae
Order. Chaetophorales	Order. Coleochaetales
Family. Coleochaetaceae	Family. Not mentioned
Genus. *Coleochaete*	Genus. *Coleochaete*

Occurrence

The genus *Coleochaete* (Gr. *Keleon*, sheath; *chaetos*, hair) is represented by about 10 species, out of which 3 species are found in India. They grow in fresh water either as epiphytes on different angiosperms (*Trapa, Hydrilla, Ipomea, Sagittaria, Typha* etc.) and algae (*Oedogonium, Vaucheria*) or as endophytes (*C. nitellarum*) within the cells of *Chara* and *Nitella* (Chlorophyceae).

Plant body

Plant body of *Coleochaete* is multicellular and heterotrichous in habit (Fig. 3.76 A, B). They show much variation in their heterotrichous nature. Both erect and prostrate systems are well represented in *C. pulvinata*. But others may show either well-developed erect system than prostrate system (*C. divergens, C. nitellarum*) or prostrate system is well-developed than the erect one (*C. scutata, C. obicularis*). Due to well-developed prostrate system, it forms discoid thallus and

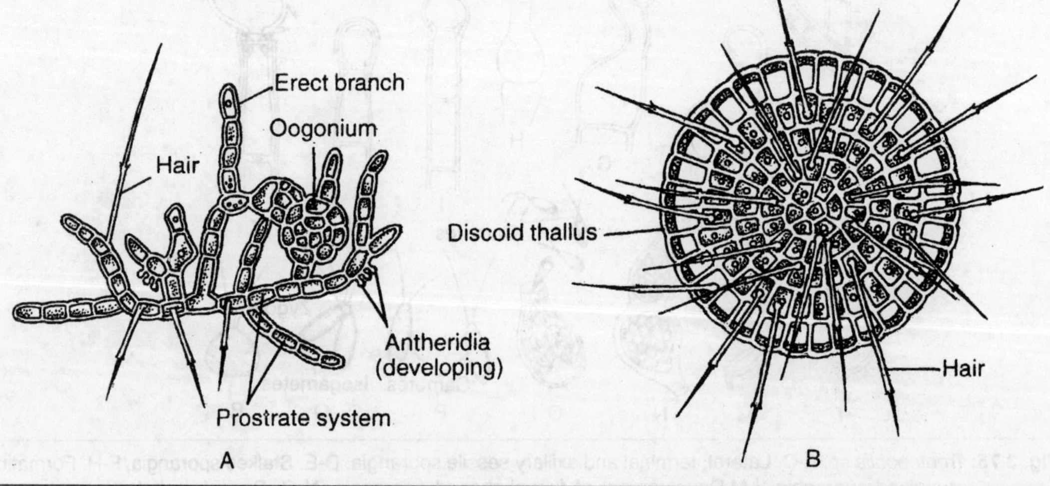

Fig. 3.76: *Coleochaete* sp.: Thalloid plant body — A. *C. divergens*, and B. *C. scutata*

looks like pseudo-parenchyma of one cell in thickness. The discoid thalli may be circular in *C. orbicularis* or irregular in *C. scutata*.

Cell structure. The cells are generally hexagonal or polygonal but may be quadrangular in some region of the thallus. The cells of the erect filament of heterotrichous habit are almost cylindrical, where the length is more than breadth. The cells are uninucleate with single large laminate chloroplast generally with one or rarely with two pyrenoids.

Reproduction

Coleochaete reproduces both asexually and sexually.

Asexual reproduction. It takes place by zoospore and aplanospore formation.

Zoospore. During spring or early summer, asexual reproduction takes place by the formation of zoospores (Fig. 3.77). Any cell of the plant body may function as zoosporangium (Fig. 3.77A). **Each zoosporangium produces single zoospore**.

Zoospores are ovoid, unicellular, uninucleate and biflagellate structure with a large parietal chloroplast. Eyespot is absent (Fig. 3.77B).

The zoospores are liberated from the zoosporangium through a pore on the sporangial wall (Fig. 3.77A). It loses flagella within a short time and secretes a wall on the periphery.

Germination. During germination the zoospore divides either transversely (Fig. 3.77C) or vertically and with subsequent divisions (Fig. 3.77D) it forms new plant body. In *C. scutata*, the first division is transverse and forms two cells. The upper cell forms seta and the lower one divides repeatedly to form discoid thallus. But in species like *C. pulvinata*, the first division is vertical and both the cells undergo repeated division to form a prostrate system. The erect system develops later from the prostrate system.

Aplanospore. Aplanospores are formed during unfavourable condition. At that condition, the zoosporangium, instead of forming zoospore, forms aplanospore. The aplanospores are non-flagellate, thick-walled and round structure. During favourable condition the aplanospores germinate and form new thalli.

Sexual reproduction. The sexual reproduction is highly specialised and **oogamous type**. The plant body may be homothallic i.e., monoecious (*C. pulvinata*) or heterothallic i.e., dioecious

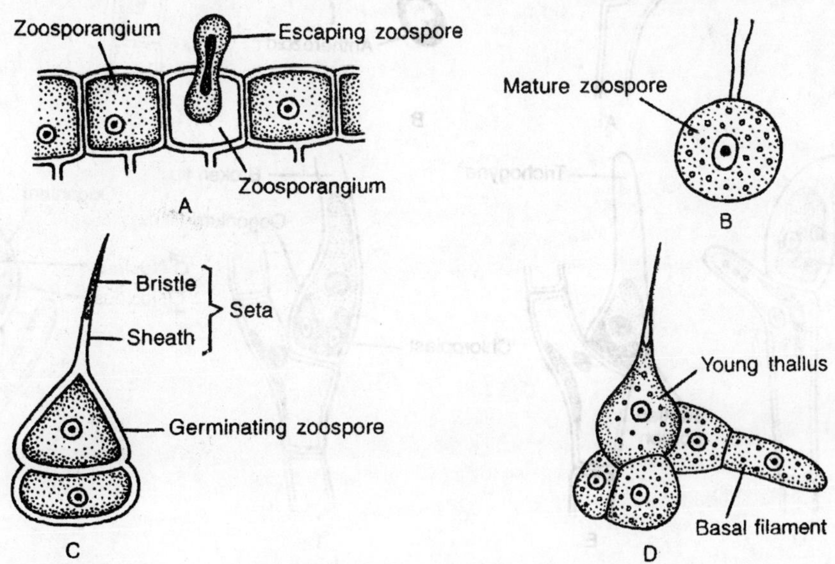

Fig. 3.77: *Coleochaete* sp.: A-D. Stages of asexual reproduction, A. Escaping of zoospore from zoosporangium, B. Mature zoospore, and C-D. Germination stages of zoospore

(*C. scurafa*). The male sex organ is called **antheridium** and female as **oogonium** or **carpogonium**.

Antheridium. In well-developed heterotrichous form like *C. pulvinata*, the antheridia are borne in groups at the apex of erect branches. But in discoid form like *C. scutata* the antheridia are developed in the half way between the centre and periphery of the thallus.

The antheridium develops as a conical outgrowth of apical cell of the erect system which is cut off from the mother cell by wall. A number of antheridia may develop at the apex of a lateral branch (Fig. 3.78A). Antheridia may form on the same filament bearing oogonia in homothallic species. **Each antheridium produces single colourless, unicellular, uninucleate, biflagellate antherozoid** (Fig. 3.78B). They are liberated by breaking the apical side of the antheridial wall.

Oogonium. The oogonium develops at the apex of an erect branch (Fig. 3.78C). Due to further growth of the lower cell, the oogonium becomes lateral in position (Fig. 3.78D, E). The oogonium is a flask-shaped structure with an elongated colourless receptive trichogyne (Fig. 3.78E). In discoid species instead of trichogyne, the neck is represented by small papilla. The basal enlarged portion contains a **single oosphere** or **egg**.

Fertilization. Before fertilization the apex of the trichogyne breaks down (Fig. 3.78F) and some colourless cytoplasm is exuded out which attracts the antherozoids. Although many antherozoids get entangled in the exuded cytoplasm, but only one enters into oogonium and reaches to the egg through broken trichogyne (Fig. 3.78G). Initially the nucleus of antherozoid is smaller in size than egg and after attaining equal size to the egg, fertilization takes place resulting in the formation of **zygote** or **oospore**.

Post-fertilization changes

After fertilization a septum is laid down which separates the basal region from the trichogyne.

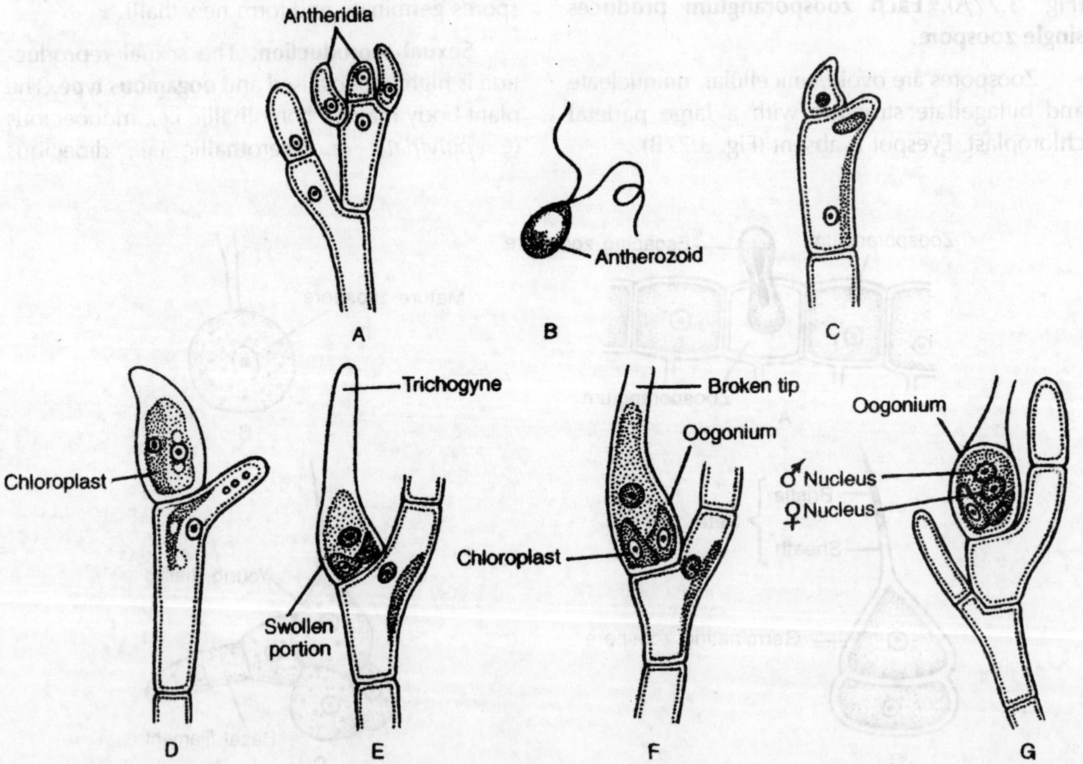

Fig. 3.78: *Coleochaete* sp.: Sexual reproduction. A. Bunch of antheridia on antheridial branch, B. Antherozoid, C-F. Oogonial development, and G. Fusion between gametes

Gradually the zygote becomes enlarged and secretes a thick wall around itself. The neighbouring cells of the oogonium get stimulated and form pseudoparenchymatous covering around it (Fig. 3.79A). The oogonium with pseudoparenchymatous sheath is called spermocarp. The **spermocarp** is red or reddish-brown in colour. It remains dormant throughout the winter.

During favourable condition the spermocarp becomes active. The colour of spermocarp changes to green and then it germinates. During germination its nucleus undergoes **meiotic division**. First it divides by a transverse division into two cells (Fig. 3.79B). This is followed by subsequent division at right angle to each other and forms octant stage. Further divisions make **16-32 celled stage** (Fig. 3.79C, D). Each cell with its protoplast metamorphoses into single biflagellate **zoospore** (Fig. 3.79E). The zoospores are liberated by breaking the zygote wall and sheath into two halves. After liberation the zoospores swim freely for some time. Later on, they germinate and form new haploid plant body of *Coleochaete*.

Indian species

C. scutata, C. orbicularis and *C. nitellarum*.

Important features

1. Plant body is thailoid, multicellular and heterotrichous in habit.
2. Cells are uninucleate with single large lamellate chloroplast generally with one or rarely two pyrenoids.
3. Asexual reproduction takes place by biflagellate ovoid zoospores and aplanospores formation.
4. Sexual reproduction is oogamous.
5. The antheridia may be terminal or intercalary. Each antheridium produces single colourless, unicellular, uninucleate, biflagellate antherozoids.
6. The oogonium is a flask-shaped structure with an elongated neck, the trichogyne.
7. Each oospore is converted into spermocarp during elaborate post-fertilization stages.
8. Oospore undergoes a period of rest and then produces zoospores (16-32) after meiosis.

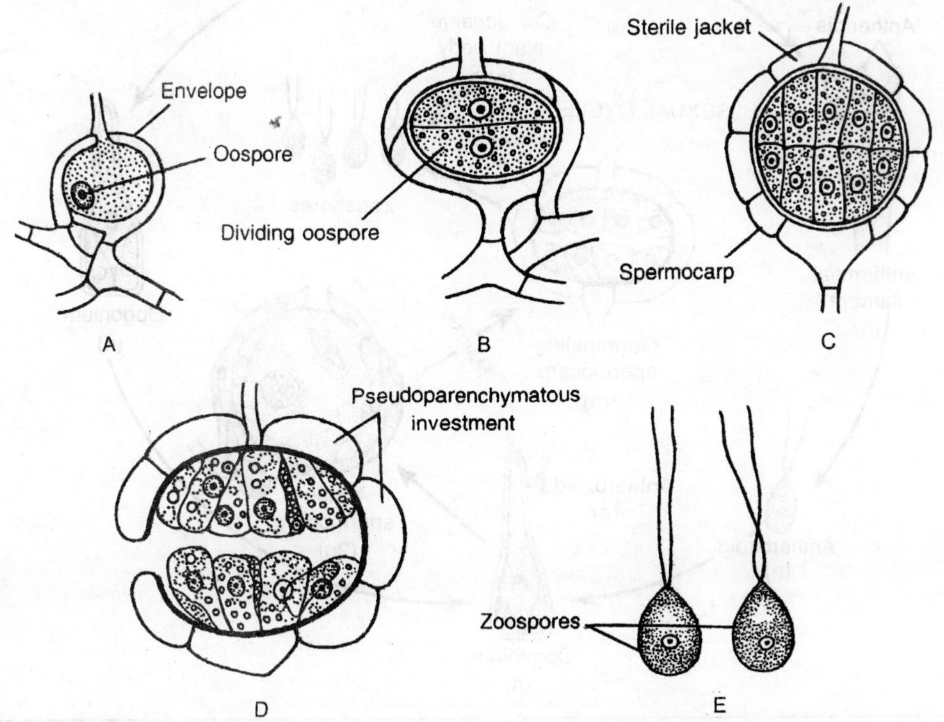

Fig. 3.79: *Coleochaete* sp.: A-E. Post fertilization changes and formation of zoospores (E)

9. The zoospores on germination gives rise to new individuals.

Fig. 3.80, 3.81 depict life cycle of *Coleochaete*.

Order. Oedogoniales

Important characteristics

1. Most of the members grow in fresh water. The order is represented by only three

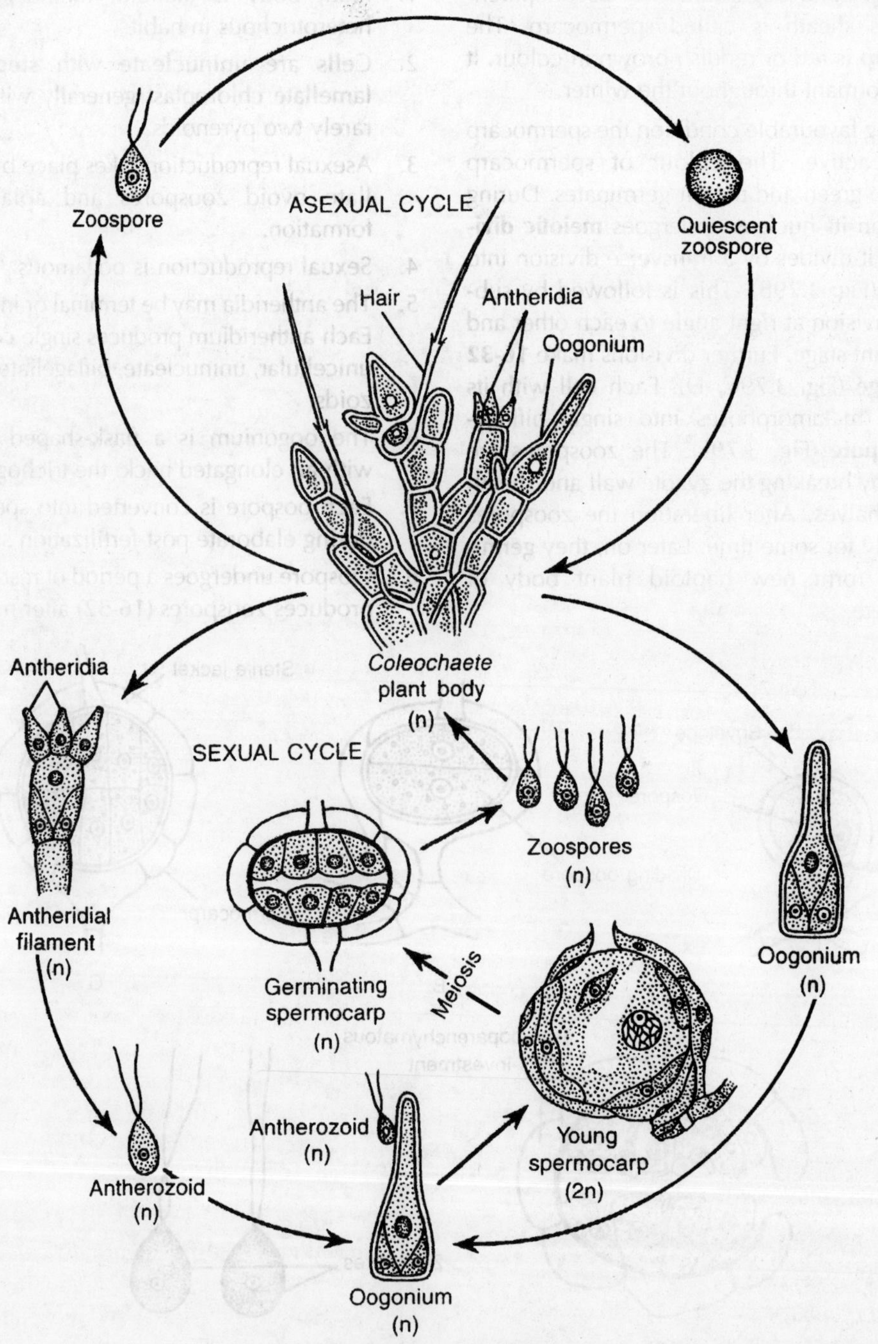

Fig. 3.80: Life cycle of *Coleochaete* sp.

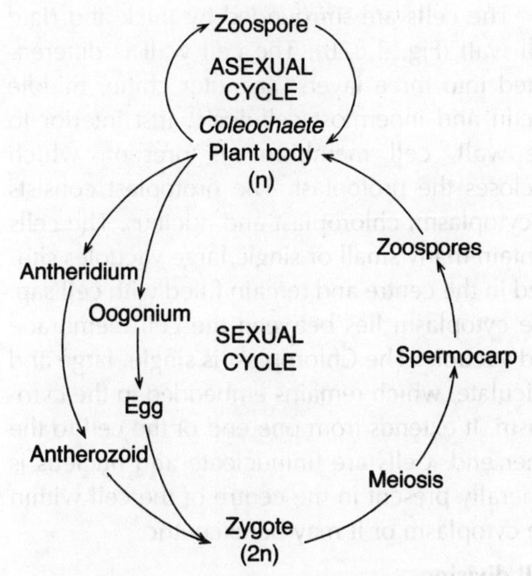

Fig. 3.81: Graphic life cycle of *Coleochaete* sp.

genera, *Oedogonium, Oedocladium* and *Bulbochaete*.

2. They are filamentous and the filaments may be branched (*Oedocladium* and *Bulbochaete*) or unbranched (*Oedogonium*).
3. The plant body is differentiated into apical and basal region.
4. It consists of cylindrical cells and the cells are longer than breadth.
5. Cells are uninucleate and have reticulate chloroplast with pyrenoids.
6. Cell division is elaborate and a cap is formed at the upper end of the daughter cell.
7. Asexual reproduction takes place by pyriform, multinucleate and multiflagellate zoospores. Flagella are arranged in a ring around the beak-like anterior end.
8. Sexual reproduction is advanced oogamous type.
9. Both androspores and antherozoides are multiflagellate.
10. Male gametes are similar to zoospore but smaller in size.
11. Heterothallic or dioecious species are of two types: **macrandrous** (where male and female filaments are of normal size) and **nannandrous** type (where male is very small i.e., dwarf male or nannandrium and the female one is of normal size).

Classification

According to Fritsch (1935) the order Oedogoniales contains only one family Oedogoniaceae. The family has only three genera: ***Oedogonium***, *Oedocladium* and *Bulbochaete*.

The life history of *Oedogonium* is given below:

OEDOGONIUM

Systematic position

Fritsch (1945)	**Lee** (1999, 2008)
	Phylum. Chlorophyta
Class. Chlorophyceae	Class. Chlorophyceae
Order. Oedogoniales	Order. Oedogoniales
Family. Oedogoniaceae	Family. Oedogoniaceae
Genus. *Oedogonium*	Genus. *Oedogonium*

Occurrence

Oedogonium (Gr. *oedos*, swelling; *gonos*, reproductive bodies) is an exclusively **fresh water alga**. Out of about 400 species more than 200 have been reported from India. They are very common in pools, ponds, lakes etc. The filamentous plant body may get attached with the stone, wood, leaves of aquatic plants, small branches of dead plant remain in water etc. by their basal cell, the holdfast. Some species like *O. terrestris* are **terrestrial**.

Plant body

The thalloid plant body is green, multicellular and filamentous. The filaments are unbranched and cells of each filament are attached end to end and form uniseriate row (Fig. 3.82A). The filament is differentiated into three types of cells: 1. Basal cell, 2. Apical cell and 3. Middle cells.

1. **Basal cell**. It is the lowermost cell of the filament. The cell is long, gradually narrowed and towards the basal end it expands to form simple, disc-like, multilobed or finger-shaped structure. The cell is generally

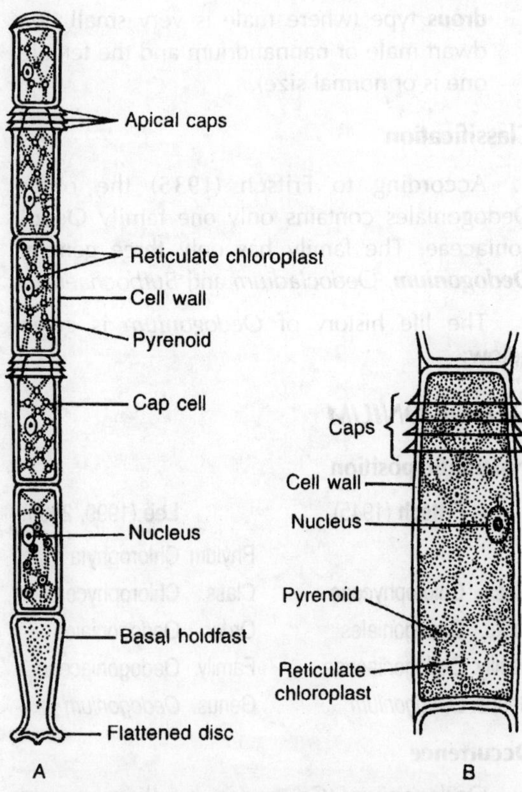

Fig. 3.82: *Oedogonium* sp.: A. Single vegetative filament with holdfast and apical cell, and B. Single vegetative cell

colourless, which performs the function of fixation to the substratum and called **holdfast**.

2. **Apical cell**. It is the topmost cell of the filament. The cell is usually rounded towards apical side and green in colour.

3. **Middle cells**. All the cells in between basal and apical cells are alike. The **cells are longer than their breadth** i.e., rectangular in shape.

Towards the upper end of some cells a ring-like structure is present known as **cap** or **apical cap** (Fig. 3.82A). The cell with cap is called **cap cell**. The number of caps on a cell indicates the number of cell divisions in that cell.

Cell structure

The intercalary cells are longer than their breadth and are cylindrical in outline.

The cells are surrounded by thick and rigid cell wall (Fig. 3.82B). The cell wall is differentiated into three layers: an outer chitin, middle pectin and innermost cellulosic. Just interior to the wall, cell membrane is present, which encloses the protoplast. The protoplast consists of cytoplasm, chloroplast and nucleus. The cells contain many small or single large vacuoles situated in the centre and remain filled with cell sap. The cytoplasm lies between the cell membrane and vacuole. The Chloroplast is single, large and reticulate, which remains embedded in the cytoplasm. It extends from one end of the cell to the other end. Cells are uninucleate and nucleus is generally present in the centre of the cell within the cytoplasm or it may be excentric.

Cell division

Growth of the filament takes place through cell division (Fig. 3.83). All cells except apical and basal ones are capable of dividing through cell division though division remains restricted in some of the cells of the filament.

The **steps of cell division** are:

1. Initially the nucleus becomes shifted from peripheral position towards the centre and then moves slightly towards the upper half of the cell (Fig. 3.83A, B).
2. Ring-like thickening develops towards the upper part of the cell wall which gradually increases in thickness (Fig. 3.83B).
3. The nucleus undergoes mitotic division and form two nuclei (Fig. 3.83C).
4. At the end of cell division (telophase), a row of microtubules develop and accumulate as a layer between the daughter nuclei (Fig. 3.83C). This layer remains in floating condition which will develop the future septum.
5. The ring-like thickening gradually elongates and splits the mother wall towards the apical region. The ring expands much more and forms a concave cylindrical structure (Fig. 3.83D). The ring material ultimately forms the cuticle of the upper daughter cell.
6. The upper part of the ruptured mother wall remains attached to the anterior end of the new daughter cell as a cap i.e., the apical

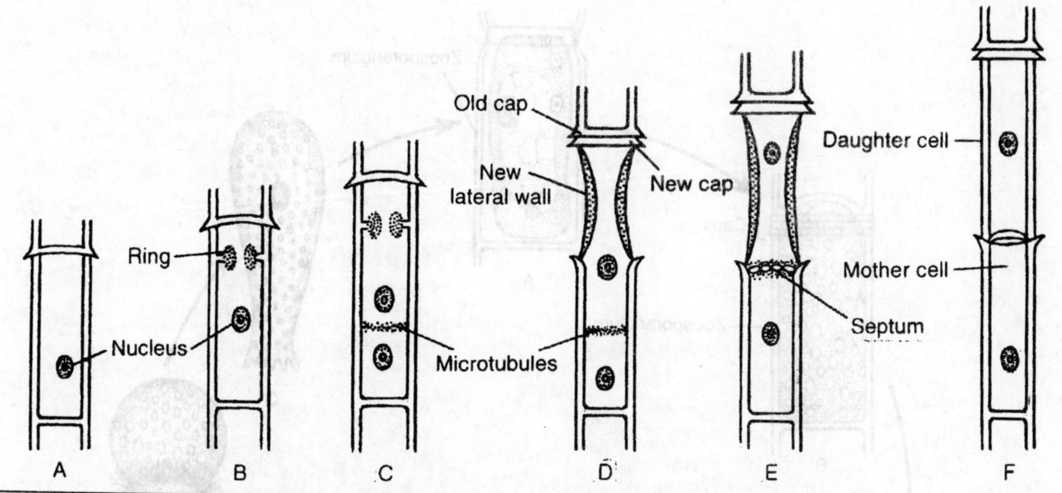

Fig. 3.83: *Oedogonium* sp.: A-F. Successive stages of cell division

cap. The other part remains towards the basal region of the daughter cell (Fig. 3.83D).

7. The floating septum gradually goes up to the base of the future daughter cell i.e., at the top of the mother cell at the ruptured end and it becomes fixed (Fig. 3.83E). Later on it develops into mature cross wall.
8. New side wall develops between the cuticle and the plasmalemma of the upper cell. Thus the two cells are formed (Fig. 3.83F). It is evident that the cell with cap is the younger one which develops between the two old cells.

Reproduction

Oedogonium reproduces by all the three means: vegetative, asexual and sexual.

Vegetative reproduction. It takes place by fragmentation and akinete formation:

1. *Fragmentation*. It takes place by accidental breakage of the filament, dying off of intercalary cells or by the formation of intercalary sporangia. The fragments are capable of developing into new filaments.
2. *Akinete*. During unfavourable condition the entire protoplast of a cell becomes a thick-walled, reddish-brown, round or oval structure, the akinete. The akinete germinates during favourable condition and develops a new filament. They generally form in chain.

Asexual reproduction. Asexual reproduction takes place by means of zoospores (Fig. 3.84A-G). Zoospores are formed singly within a cell. Comparatively younger cell i.e., the cell with cap behaves as sporangium mother cell.

The zoospores are multiflagellate and ovoid, pyriform or spherical in shape (Fig. 3.84). They are uninucleate with single chloroplast and occasionally with an eyespot.

During favourable condition, the zoospore formation begins in a cap cell of the filament. The entire protoplast of zoosporangium contracts from the wall and behave as a unit. The protoplast becomes round or oval in shape and its nucleus moves at one end. Near the nucleus a semicircular hyaline area develops. Just below the hyaline area a ring of blepharoplast granules develops, connected with each other by fibrous strands (Ringo, 1967). Later on, from each blepharoplast granule, single flagellum develops. Thus a crown of flagella is present around the colourless semicircular area.

The fully developed zoospores are liberated by breaking the zoosporangium wall. The wall of the zoosporangium breaks near the cap region and the neighbouring cells bend on one side to make way for the liberation of zoospore. During liberation, the zoospore remains as a delicate mucilaginous vesicle for 3-10 minutes. After dissolution of vesicle the zoospore gets

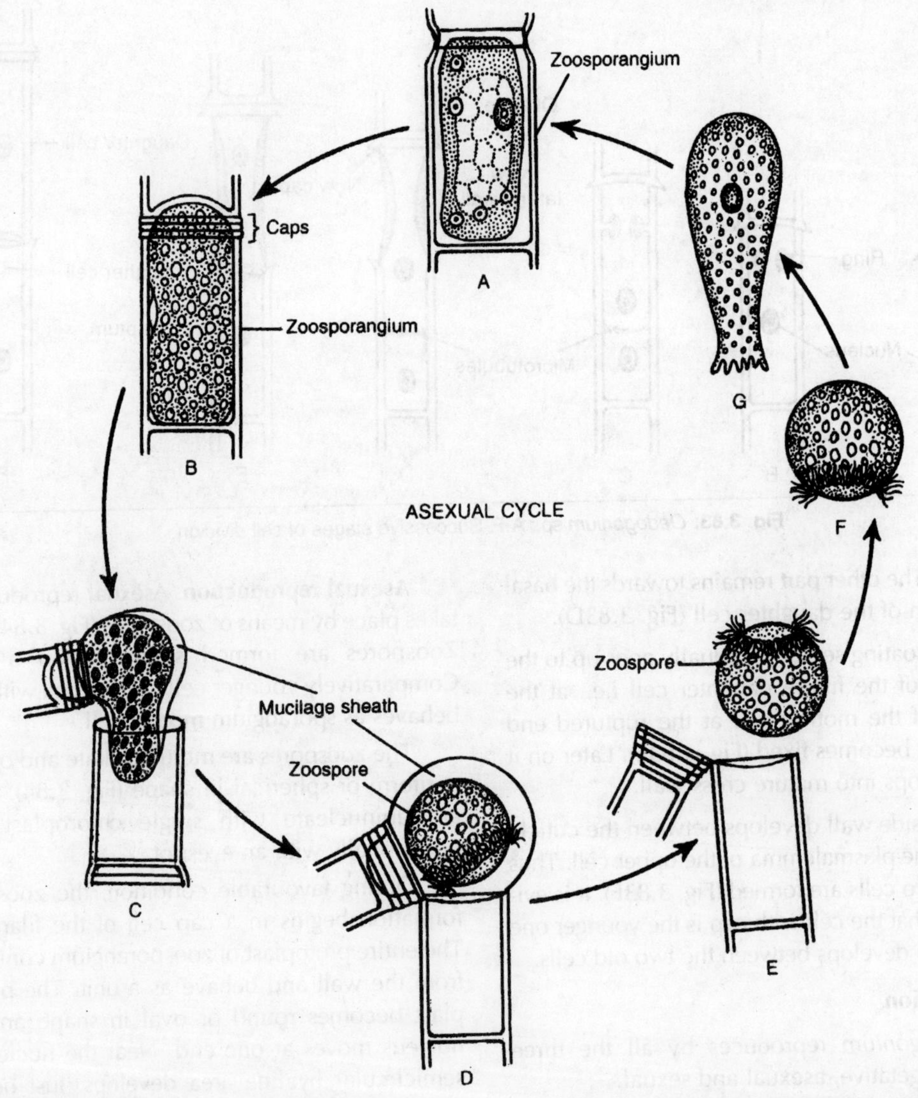

Fig. 3.84: *Oedogonium* sp. Asexual reproduction: A-E. Successive stages of zoospore formation, F. Single zoospore, and G. Germination of Zoospore

free and starts swimming in the surrounding water.

Germination. The zoospore can swim for about one hour or more. Coming in contact with substratum by the anterior end, it loses flagella and starts to elongate. The lower hyaline part becomes separated by cell wall, which forms the hold fast. Through the subsequent division and redivision in a single plane, new filament is formed.

Sexual reproduction. The sexual reproduction in *Oedogonium* is an ***advanced oogamous type***. The **male gametes** or antherozoides are produced in antheridium (Fig. 3.85) and the **female gamete** or egg is produced in oogonium (Fig. 3.86). Male and female gametes differ both morphologically and physiologically. Only one egg is produced in each oogonium and two antherozoides in each antheridium. Another motile structure, the **androspore**, is produced singly in each androsporangium. Deficiency of nitrogen and alkaline pH are the important factors for promoting sexual reproduction.

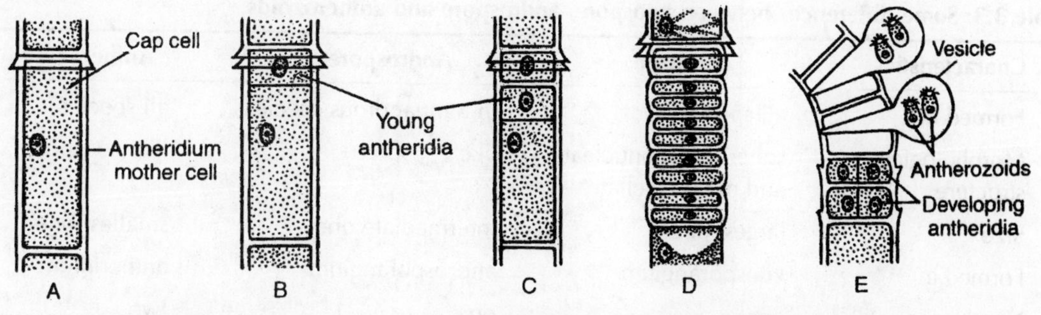

Fig. 3.85: *Oedogonium* sp.: A-E. Successive stages of development of antherozoids

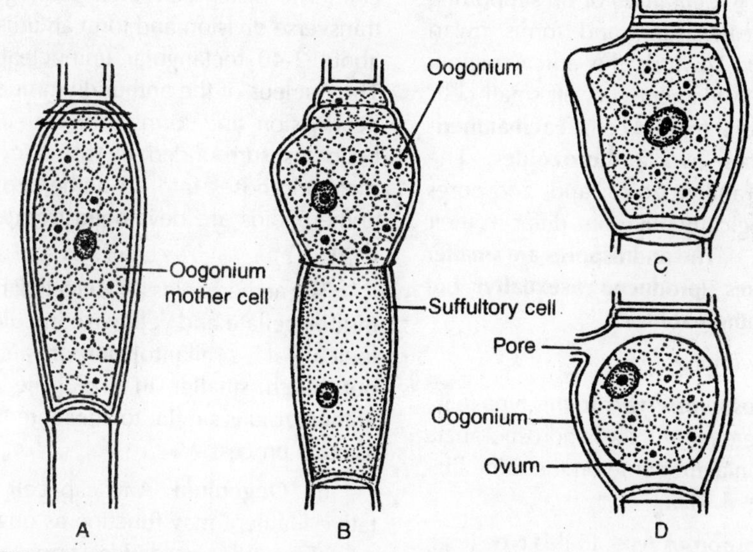

Fig. 3.86: *Oedogonium* sp.: A-D. Successive stages of development of ovum

Distribution of sex organ. Based on the size of the male (antheridial) filament the species of *Oedogonium* are divided into two groups — macrandrous and nannandrous type:

1. **Macrandrous type**. In macrandrous type the antheridium develops in the filament of normal size. It is of two types:

   (i) ***Monoecious type*** (homothallic or bisexual). In this type (e.g., *O. fragile, O. nodulosum* and *O. hirnii*) antheridia and oogonia are borne on the same filament (Fig. 3.89).

   (ii) ***Dioecious type*** (heterothallic or unisexual). In this type (e.g., *O. gracilius, O. cardiacum* and *O. aquaticum*) the antheridia and oogonia are borne on the different filaments (Fig. 3.90).

2. **Nannandrous type**. The nannandrous species are **always dioecious (heterothallic)** i.e., antheridia and oogonia are borne on different filaments. In this type the antheridia develop on a very small filament termed as **dwarf male** or nannandrium. In nannandrous type initially androsporangia are developed in series on normal sized filament. The **androspore form singly within androsporangium**. Liberating from androsporangium, the androspores swim freely in water. The androspore germinates on the

Table 3.3: Some differences between zoospore, androspore and antherozoids

	Characteristics	Zoospore	Androspore	Antherozoid
1.	Formed in	all species	in nannandrous species	all species
2.	Morphological structure	spherical, uninucleate and multiflagellate	like zoospore	like zoospore
3.	Size	largest one	intermediate one	smallest one
4.	Formed in	zoosporangium	androsporangium	antheridium
5.	Number per cell	one	one	two
6.	Further activity	develops a new plant	develops into a dwarf male	fertilizes the egg

oogonial wall (*O. ciliatum*) or on supporting cell (*O. concatenatum*) and forms dwarf male filament. Towards the apical region, the dwarf male filament cuts off small cells as the antheridial mother cells. **Each antheridium produces two antherozoides**. The androspores, antherozoids and zoospores are morphologically alike but differ in their sizes (Table 3.3). **The androspores are smaller than zoospores (produced asexually) but larger than antherozoides.**

They are of two types:

(i) ***Gynandrosporous type***. In this type (e.g., *O. concatenatum*) the androsporangia and oogonia are borne on the same filament (Fig. 3.91).

(ii) ***Idioandrosporous type***. In this type (e.g., *O. setigerum, O. confertum* and *O. iyengarii*) the androsporangia and oogonia are borne on different filaments (Fig. 3.92).

Sexual reproduction in Macrandrous species

The structure and development of antheridium and oogonium are similar in all the species belonging to either monoecious or dioecious type. They differ only in the position of sex organs. In **monoecious type** both the sex organs develop on the same filament, but in **dioecious type** they are on different filaments.

(a) **Antheridium**. Any cap cell of the vegetative filament may function as antheridial mother cell (Fig. 3.85). It divides transversely into an upper smaller antheridium and a lower larger sister cell. The sister cell then undergoes repeated transverse division and form an uniseriate row of about 2-40 rectangular uninucleate antheridia. The nucleus of the antheridium undergoes mitotic division and forms 2 nuclei. Each nucleus becomes surrounded by some cytoplasm and metamorphoses into an antherozoid. Thus two antherozoids are developed from each antheridium.

The antherozoids are unicellular, uninucleate, multiflagellate and yellowish in colour. Morphologically it is similar to zoospore and androspore, but much smaller in size. The liberation of antherozoid is similar to zoospore formed during asexual process.

(b) **Oogonium**. Any cap cell of the vegetative filament may function as oogonial mother cell (Fig. 3.86). It divides transversely into an upper oogonium and a lower supporting cell or **suffultory cell**. The lower cell may again undergoes similar divisions in repeated sequence to form two or more oogonia with a lower supporting cell. With maturity the oogonium becomes globose, which contains single egg. A receptive spot is present at one side of the egg. Before fertilisation a transverse slit or pore develops on the oogonial wall through which the antherozoids take the entry.

Sexual reproduction in Nannandrous species

The structure and development of androsporangium, antheridium and oogonium are similar in all the species either belonging to ***Gynandrosporous or Idioandrosporous type***. They differ only in the position of androsporangium. In

Gynandrosporous type the androsporangia and oogonia are borne on the same filament, whereas in *Idioandrosporous type* the androsporangia and oogonia are borne on different filaments.

(a) **Androsporangium**. The mode of development of androsporangia is alike with the antheridial development in macrandrous species. The androsporangia are larger than the antheridia of macrandrous type. The nucleus of androsporangium does not divide and the entire protoplast metamorphoses into a **single androspore**. The androspores are unicellular, uninucleate and multiflagellate. The androspores are larger than the antherozoids. The androspores are liberated by breaking the wall of androsporangium. During liberation each androspore remains in a mucilage envelop for few minutes and then becomes free to swim in water (Fig. 3.87A, B).

(b) **Germination of androspore and formation of antherozoids**. After swimming for some-

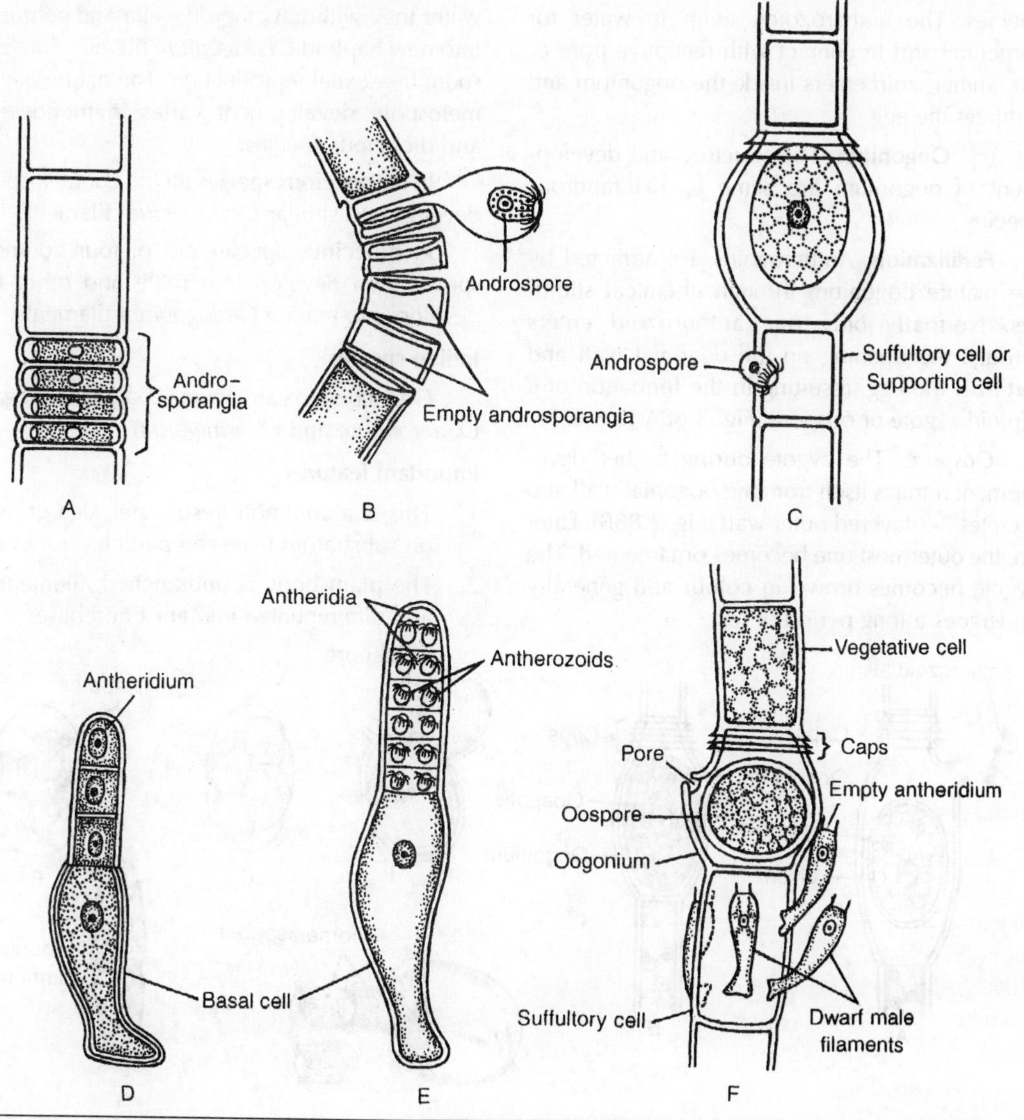

Fig. 3.87: *Oedogonium* sp.: Development of dwarf male: A-B. Development of androspore, C. Attachment of androspore on suffultory cell, D-E. Development of dwarf male and formation of antherozoid, and F. Formation of oospore after fertilization

time, it gets attached either on oogonial wall or on supporting cell (Fig. 3.87C). Then a wall develops around the andospore. The androspore elongates and cuts off a few flat cells at its apex to form the antheridia (Fig. 3.87D). The nucleus of each antheridium divides mitotically to form two nuclei. Each nucleus with some cytoplasm metamorphoses into single antherozoid. Thus **two antherozoids** are formed in each antheridium (Fig. 3.87E). The antherozoids are liberated in a similar way as found in macrandrous species. The antherozoids swim in water for sometime and in contact with receptive pore or slit, antherozoid enters inside the oogonium and fertilizes the egg.

(c) **Oogonium**. The structure and development of oogonium are same as macrandrous species.

Fertilization. Antherozoids are attracted by the mature oogonium through chemical stimulus. Normally only one antherozoid enters through the opening on the oogonial wall and fertilizes the egg, resulting in the formation of a diploid zygote or oospore (Fig. 3.88A-B; 3.87F).

Oospore. The zygote during further development retracts itself from the oogonial wall and secretes 2-3 layered outer wall (Fig. 3.88B). Later on, the outermost one becomes ornamented. The zygote becomes brown in colour and generally undergoes a long period of rest.

Germination of oospore. The oospore germinates during favourable condition (Fig. 3.88C-G). The nucleus undergoes **meiosis** and forms 4 haploid daughter nuclei. The nuclei accumulate some cytoplasm and form 4 daughter protoplasts. They liberate by rupturing the oospore wall. During liberation they develop flagella and are called **meiospores** or **zoomeiospores**. Initially they remain inside a delicate vesicle, which soon disintegrates and the zoospores get free into the environment. After swimming for some time in water they withdraw their flagella and germinate into new haploid *Oedogonium* filament like zoospore in asexual reproduction. The nature of zoomeiospore development varies in monoecious and dioecious species.

In **monoecious species** all the zoomeiospores develop into similar *Oedogonium* filament.

In **dioecious species** out of four zoomeiospores, two develop into male and other two develop into female *Oedogonium* filaments.

Indian species

Oedogonium cardiacum, O. aster, O. elegans, O. aerolatum and *O. armigerum.*

Important features

1. This is a common fresh water alga growing on substratum like sand particles, rocks etc.
2. The plant body is unbranched, filamentous and differentiated into apex and base.

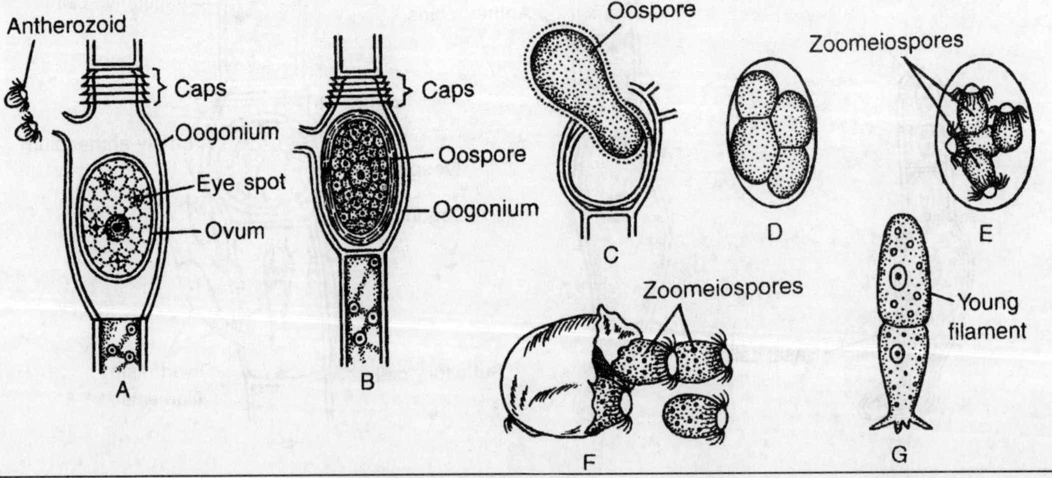

Fig. 3.88: *Oedogonium* sp.: A. A stage before fertilization, B. Oospore in oogonium, C. Liberation of oospore from oogonium, D-E. Stages of zoospore formation, F. Liberation of zoospore, and G. Young filament develops after germination of zoomeiospore.

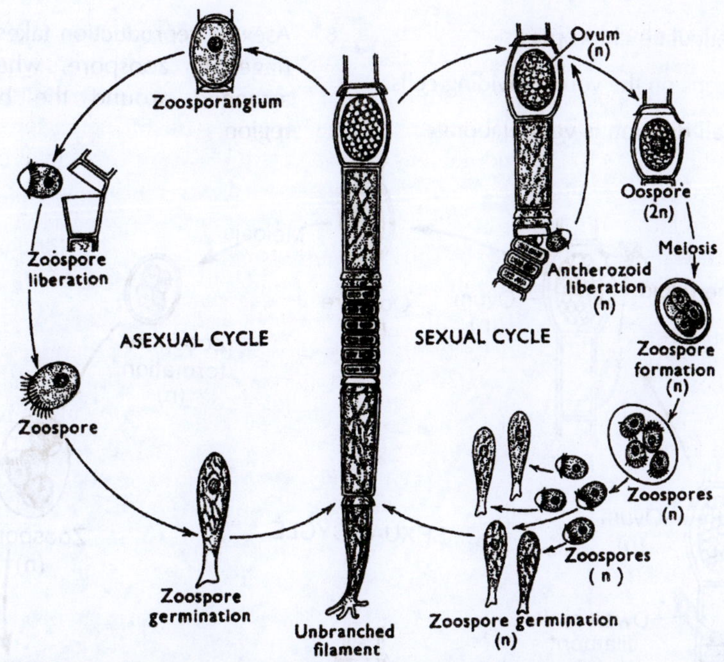

Fig. 3.89: Life cycle of macrandrous monoecious species of *Oedogonium*

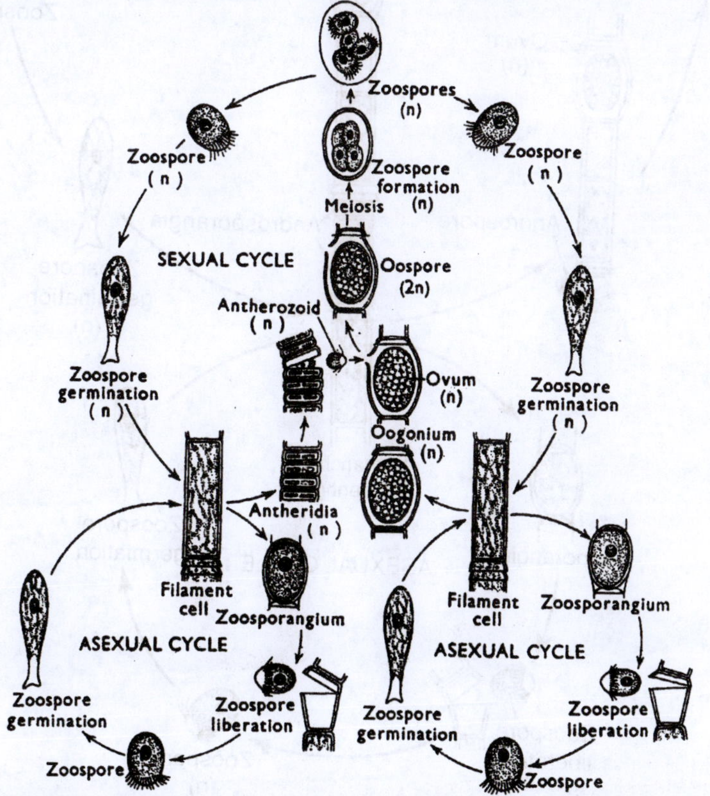

Fig. 3.90: Life cycle of macrandrous dioecious species of *Oedogonium*

3. Cells have reticulate chloroplasts.
4. Presence of caps on the young dividing cells.
5. Vegetative cell division is very elaborate.
6. Asexual reproduction takes place by multi-flagellate zoospore, where flagella are arranged around the beak-like apical region.

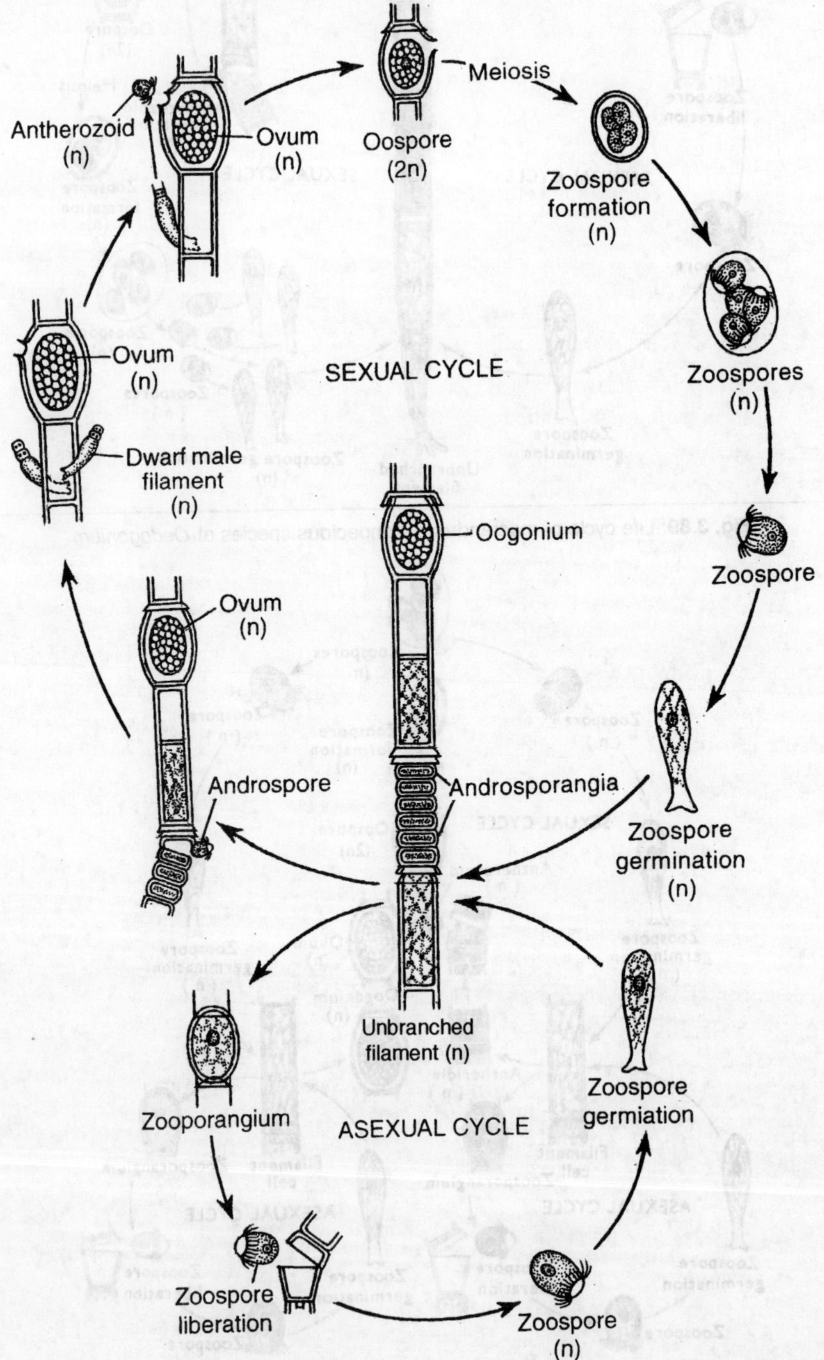

Fig. 3.91: Life cycle of nannandrous (all are dioecious) – gynandrasporus species of *Oedogonium*

7. Sexual reproduction is advanced oogamous type.
8. The female gamete i.e., ovum, is produced singly in each oogonium.
9. The male gametes i.e., antherozoids, are very much similar to zoospores but smaller in size. Two antherozoids are produced in each antheridium.
10. Based on the size of male filament the plants are divided into two groups: Macrandrous and Nannandrous type.
11. In macrandrous type the antheridia develop into the filament of normal size. But in nannandrous type the antheridia develop on small and thin male filament, the dwarf male or nannandrium (remain attached with the oogonium wall or its lower cell, the supporting cell), develop on germination of androspore. The androspore forms singly in androsporangium, develop in the normal filament.
12. The androspores are smaller than zoospores (produced asexually), but larger than antherozoids.

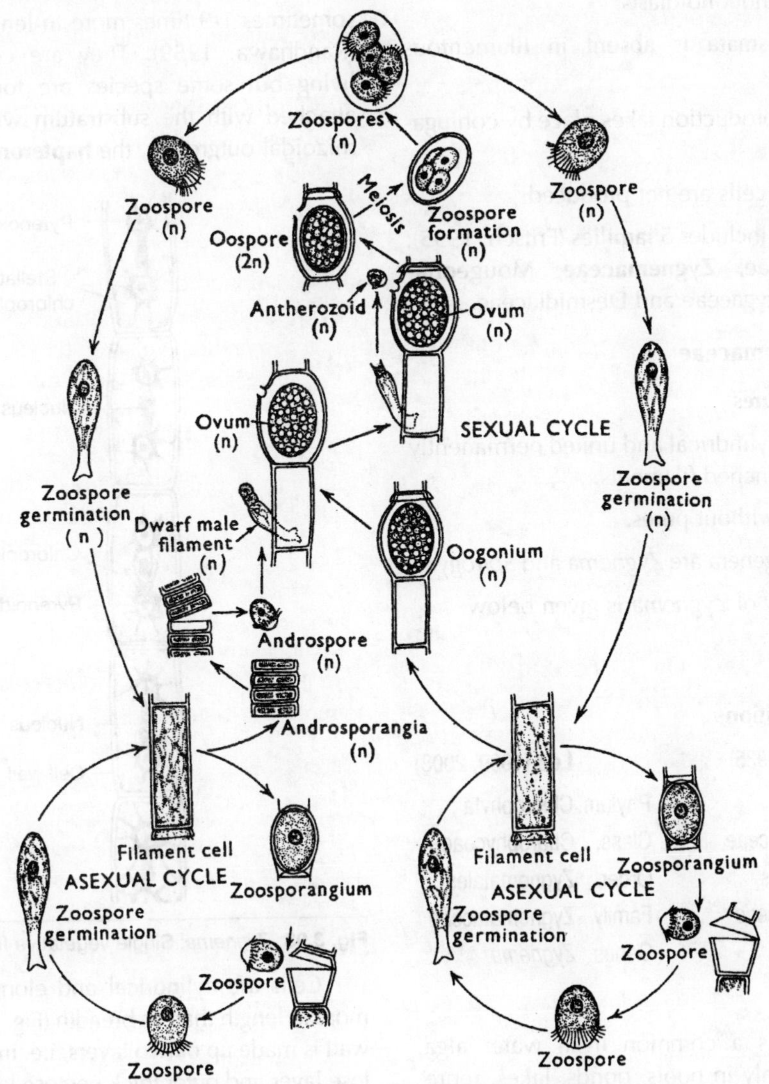

Fig. 3.92: Life cycle of nannandrous (all are dioecious) – idioandrosporus species of *Oedogonium*

13. The zygote undergoes meiotic division and produces four zoospores. In dioecious species two produces male and other two produces female plants.

Fig. 3.89 to 3.92 depict life cycle of *Oedogonium*.

Order. **Conjugales**

The order Conjugales includes about 40 genera and 3,000 species (Randhawa, 1959).

Important features

1. Plant body is unicells or unbranched filaments without holdfasts.
2. Plasmodesmata is absent in filamentous forms.
3. Sexual reproduction takes place by conjugation.
4. Flagellate cells are not produced.

The order includes 5 families (Fritsch, 1935): Mesotaeniaceae, **Zygnemaceae**, Mougeotiaceae, Gonatozygaceae and Desmidiaceae.

Family. **Zygnemaceae**

Important features

1. Cells are cylindrical and united permanently into unbranched filaments.
2. Cell wall without pores.

Common genera are *Zygnema* and *Spirogyra*.

Life history of *Zygnema* is given below:

ZYGNEMA

Systematic position

Fritsch (1935)	**Lee** (1999, 2008)
	Phylum. Chlorophyta
Class. Chlorophyceae	Class. Charophyceae
Order. Conjugales	Order. Zygnematales
Family. Zygnemaceae	Family. Zygnemataceae
Genus. *Zygnema*	Genus. *Zygnema*

Occurrence

Zygnema is a common fresh water alga, occurs commonly in pools, ponds, lakes, represented by 100 species (Randhawa, 1959), of which more than 20 species are found in India. *Z. himalayaensis* reported from Himalayas (about 7,500 ft) and *Z. terrestris* from Kashmir (6,000 to 12,000 ft).

The genus was established by Prof. Agardh (1824).

Plant body

The plant body consists of cylindrical cells, united end to end and form unbranched filaments. Cells of the filaments are almost similar in structure and the length of the individual cell is commonly more than twice of its breath, but sometimes 1-9 times more in length than breath (Randhawa, 1959). They are commonly free-living but some species are found to remain attached with the substratum with the help of rhizoidal outgrowth, the **hapteron** (*Z. terrestre*).

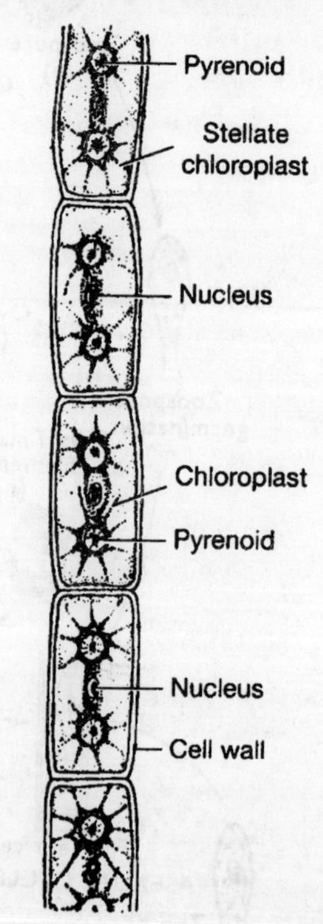

Fig. 3.93: *Zygnema*: Single vegetative filament (portion)

Cells are cylindrical and elongated and are more in length than its breadth (Fig. 3.93). The cell wall is made up of two layers, i.e. inner thin cellulose layer and outer thick pectose layer. The outer pectose mucilaginous sheath provides sliminess and thus keeps them free from epiphytic growth.

Individual cells are characterized by the presence of **two stellate or star-shaped chloroplasts**. Each chloroplast is made up of pyrenoid containing rounded body, having many irregular short outgrowths radiating in all sides. Cells are uninucleate and the nucleus remains suspended in the centre by cytoplasmic connections.

The filament grows in length as a result of cell division.

Reproduction

Zygnema reproduces by all the three means are vegetative, asexual and sexual.

1. **Vegetative reproduction**. It takes place by fragmentation and also by akinete formation.
   (i) *Fragmentation*. Due to mechanical injury or by accident, the filament divide into small fragment or sometimes even into individual cells. Each of which are capable to develop into new individual.
   (ii) *Akinete*. Akinetes are the thick walled structure, formed by the accumulation of starch and fat, and also some additional wall layers of pectose and/or cellulose in an individual cell. These are observed in species like *Z. sterile, Z. giganteum, Z. terrestre* etc. In *Z. sterile*, two akinetes are produced in each cell, whereas a chain of akinetes are formed in *Z. giganteum* (Fig. 3.94).

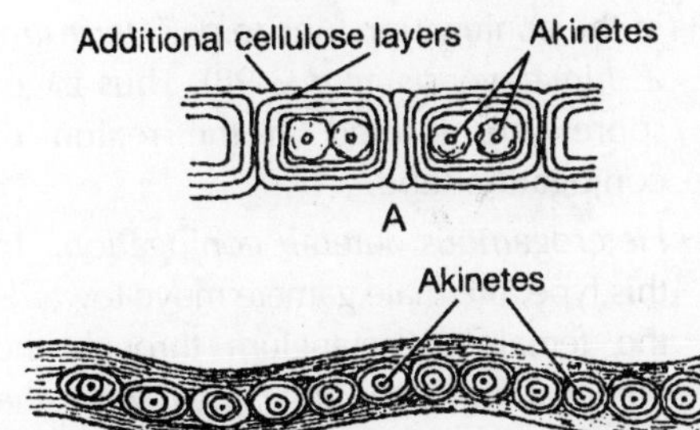

Fig. 3.94: *Zygnema*: Akinetes — A. *Z. sterile* (two in each cell) and B. *Z. giganteum* (chain of akinetes)

2. **Asexual reproduction**. No motile spores are formed. It takes place by aplanospores and parthenospores.
   (i) *Aplanospores*. These are asexual spores formed by in vegetative cells, the aplanosporangia. Spores are round or ovoid in structure. The protoplast of the cell contracts and changes into an aplanospore. These are variously smooth, coloured, or sculptured. These are formed in species like *Z. himalayensis, Z. terrestre* (Fig. 3.95), *Z. kumaoense* (Fig. 3.95) etc.
   (ii) *Parthenospores*. Due to failure of gametic union, parthenospores (azygospores) are formed in some species like *Z. collinsianum*. Here although both the gametes are formed and both of them passes inside the conjugation cannal but fusion does not take place and both the gametes are transformed into parthenospores (Fig. 3.95).

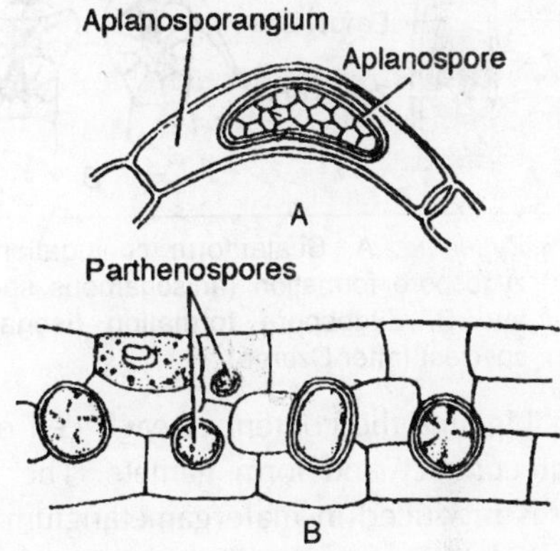

Fig. 3.95: *Zygnema*: A. Aplanospore (*Z. terrestre*) and B. Parthenospore (*Z. collinsianum*) (after Randhawa)

3. **Sexual reproduction**. Sexual reproduction takes place by **conjugation**. Some species reproduce during February to June, but others during October to November. In *Zygnema*, conjugation takes place by both the methods: *Scalariform* and *lateral type*.
   (i) ***Scalariform conjugation***. In this process, two filaments come closer, lie side by side and become embedded in mucilage secreted by them. The cells situated opposite to each other, start to form small outgrowth (Fig. 3.96), of which male cells develop longer out-

growth than the female cells. After some time, the opposite outgrowths come in contact with each other at their anterior end. Later, the opposite wall of outgrowth, at the point of contact dissolves and a tube-like passage is formed, called **conjugation canal** or **tube** (Fig. 3.96).

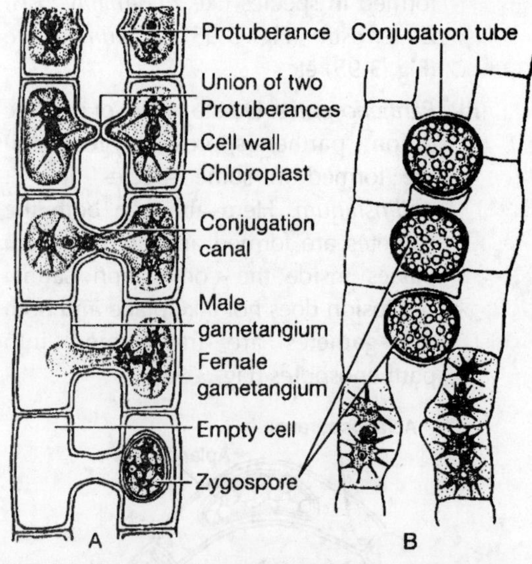

Fig. 3.96: *Zygnema*: A. Scalariform conjugation and zygospore formation (anisogamous species) and B. Zygospore formation (isogamous species) (after Czurda)

By this time the content of each cell of the filament contract and form gamete. The male gamete is produced in male gametangium (Fig. 3.96) and the female gamete in female gametangium (Fig. 3.96). The male gamete moves towards the female gametangium and fuses with the female gamete and forms **zygospore** (Fig. 3.96).

(ii) ***Lateral conjugation***. It takes place between the two cells of the same filament, of which one functions as male gametangium and the other as female gametangium. Initially, the longitudinal wall at the region of septum between two adjacent cells protrudes out. Thus two outgrowths are formed, one on either side of the septum (Fig. 3.97). Later, a passage is formed by the dissolution of the contact walls of both the outgrowths and thus a passage is formed between the 2 gametangia. This passage functions as a conjugation tube.

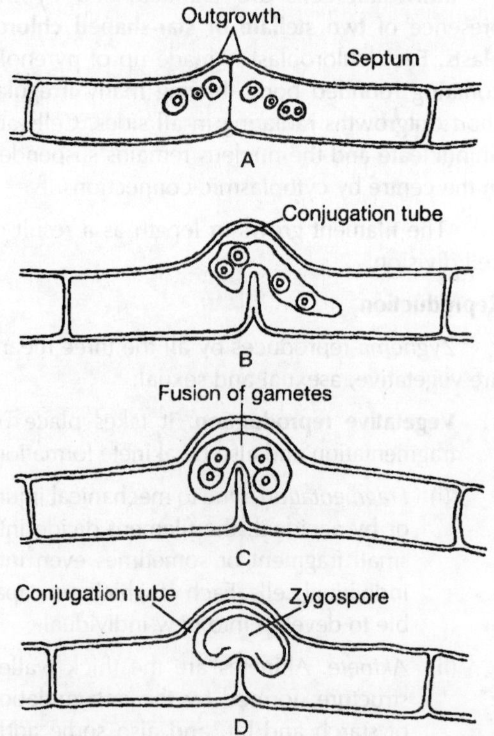

Fig. 3.97: *Zygnema*: A. Lateral conjugation in *Z. czurdae* and zygospore formation (isogamous species) (after M.O.P. Iyenger)

In *Zygnema*, lateral conjugation is of two types: Isogamous and Heterogamous.

(i) *Isogamous lateral conjugation*. In this type, both the gametes move and fuse in the conjugation tube (e.g. *Z. terrestre, Z. himalayensis*) (Fig. 3.98). Thus zygospores are formed in the region of conjugation tube.

(ii) *Heterogamous lateral conjugation*. In this type, the male gamete move towards the female gametangium through the conjugation tube and fuses with the female gamete to form zygospore (*Z. stellinum*). Thus in the filament, zygospore containg cells and empty cells are present alternately.

Zygospore. It represents the diploid (2n) phase. It contains one diploid nucleus and four stellate chloroplasts comes two from each gamete. But soon the chloroplasts of male gamete degenerate (Fig. 3.98). Later on, the zygospore becomes enveloped by a 3-layered thick wall (Fig. 3.98F-H):

(i) Outer layer i.e. **exospore** is thin and made up of cellulose or pectose.

(ii) Middle layer i.e. **mesospore** is thick and made up of cellulose, which is chitinized and variously sculptured and

(iii) Inner layer i.e. **endospore** is thin and also made up of cellulose.

The colour of zygospore varies from pale yellow to brown depends on species. After maturation of zygospore, the gametangial walls get decayed and zygospores are liberated. The zygospores sink at the bottom of the pond/lake and undergo a long period of rest.

With the onset of favourable condition i.e., with the availability of sufficient water after rain etc., the zygospore germinates. Before germination the diploid nucleus undergoes meiotic division and 4 haploid nuclei are formed (Fig. 3.98). Out of 4 nuclei, 3 degenerate and only one survives (Fig. 3.98).

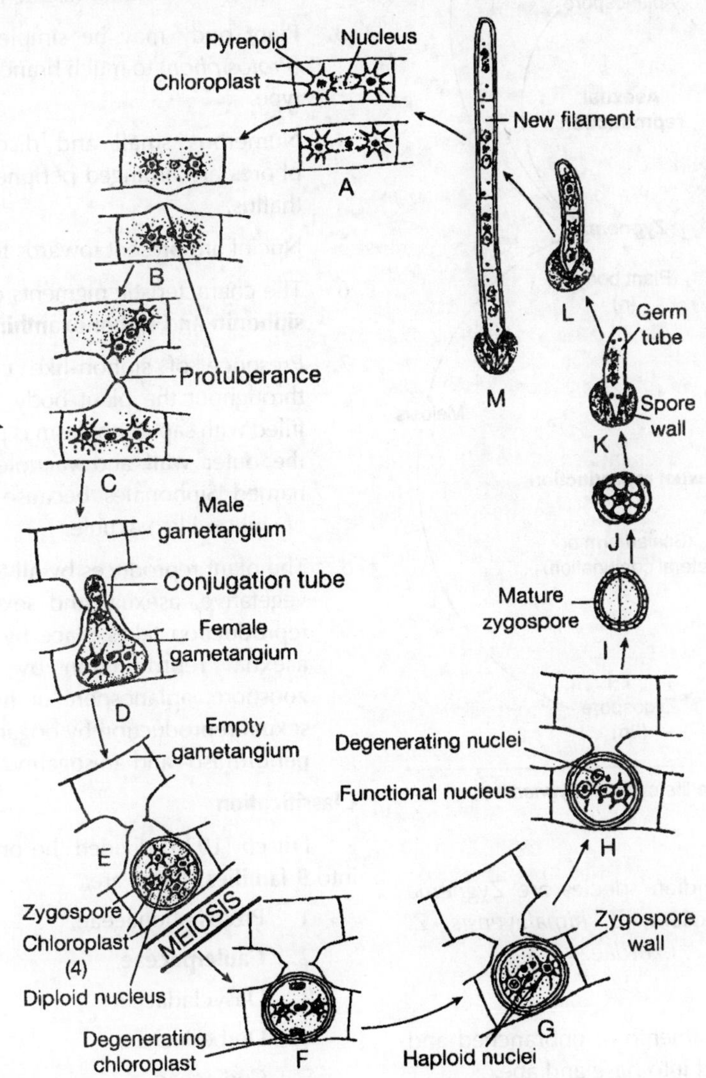

Fig. 3.98: Life cycle of *Zygnema* sp.

During germination, the exospore and mesospore wall ruptures and the inner thin wall along with the inner content grow out into a long germ-tube (Fig. 3.98K). Initially the germ-tube divides into two cells and both the cells divide and redivide transversely to form a new filament (Fig. 3.98L, M).

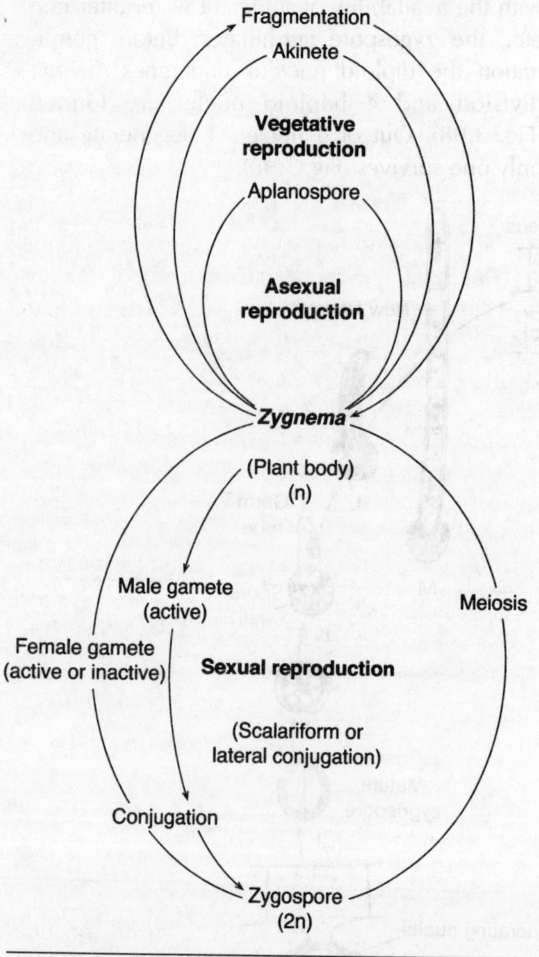

Fig. 3.99: Graphic life cycle of *Zygnema* sp.

Indian species

The common Indian species are *Zygnema giganteum, Z. globosum, Z. himalayensis, Z. gorakhporanse,* and *Z. czurdae*.

Important features

1. Plant body is filamentous, unbranched and not differentiated into base and apex.
2. Cells with stellate chloroplast.
3. Absence of flagellate gametes.
4. Sexual reproduction takes place both by scalariform and lateral conjugation.

Order. **Siphonales**

Important characteristics

1. Most of the members of Siphonales are marine. A few members are freshwater. Some members grow as epiphytes or endophytes.
2. Thalloid plant body is variously branched, aseptate and multinucleate i.e., coenocytic.
3. Plant body may be simple vesicular type (*Protosiphon*) to much branched filamentous type.
4. Numerous small and discoid chromatophores are arranged peripherally inside the thallus.
5. Nuclei are present towards the inner layer.
6. The characteristic pigments of this order are **siphonin** and **siphonoxanthin**.
7. Presence of siphon-like central vacuole throughout the plant body, which remains filled with sap. Cytoplasm is present between the outer wall and vacuole. The order is named "Siphonales" because of the presence of siphon-like vacuole.
8. The plant reproduces by all the three means vegetative, asexual and sexual. Vegetative reproduction takes place by fragmentation, asexual reproduction by multiflagellate zoospore, aplanospore or hypnospore and sexual reproduction by oogamy. Rarely they perform iso- and anisogamy.

Classification

Fritsch (1935) divided the order Siphonales into **9 families**. These are:

1. Protosiphonaceae,
2. **Caulerpaceae**,
3. Dasycladaceae,
4. Derbesiaceae,
5. Codiaceae,

6. Valoniaceae,
7. Chaetosiphonaceae,
8. Phyllosiphonaceae and
9. **Vaucheriaceae**.

This classification is also followed by M. O. P. Iyenger (1951).

Only the families Caulerpaceae and Vaucheriaceae are considered below for further discussion.

Family. **Caulerpaceae**

Important features

1. Plant body is coenocytic thallus consisting of rhizome-like creeping prostrate portion bearing rhizoids and upright shoot-like aerial portion with division of labour.
2. Sexual reproduction is isogamous and anisogamous.

The common genera of the family are *Caulerpa* and *Bryopsis*.

Life history of *Caulerpa* is given below:

CAULERPA

Systematic position

Fritsch (1935)		**Lee** (1999, 2008)	
		Phylum.	Chlorophyta
Class.	Chlorophyceae	Class.	Ulvophyceae
Order.	Siphonales	Order.	Caulerpales
Family.	Caulerpaceae	Family.	Caulerpaceae
Genus.	*Caulerpa*	Genus.	*Caulerpa*

Occurrence

Caulerpa is a marine alga represented by about 73 species. It grows both in shallow and deep sea. Plants are mainly grows in warmer seas of the tropical parts of the world. Many species have been found to grow in the Western coast of India, either in shelter, in exposed areas, on stone and also on muddy or sandy areas in the sea.

Some species like *C. prolifera, C. racemosa* and *C. crassifolia* occur as lithophytes on rocks and coral reefs. On the other hand, species like *C. verticillata* grow either as lithophyte or as epiphyte on the root of many mangrove plants. Species like *C. cupressoides* grows rooted in sand or mud at the bottom of seas.

C. racemosa is an edible alga.

Plant body

Plant body of *Caulerpa* is well developed, highly elaborate thallus and most species appears like small angiosperms. The thallus is **unicellular** and **coenocytic** i.e. containing many nuclei and without partition wall. Some species are more than 30 cm in height and some are 1 m or more in length.

The plant body consists of creeping rhizome-like part, from which many branched rhizoids are developed towards the lowers side and leaf-like erect shoots towards the upper side (Fig. 3. 100). The *rhizome-like* part is long, prostrate, branched and cylindrical, generally grown inside the sand or mud. *Rhizoids* are branched, thread-like and hyaline, meant for anchorage and absorption. In some species the rhizoids develop just below the erect shoot, but not in others. The *leaf-like erect shoots* are also called **assimilators** or **assimilatory shoots**.

Variations in the form of assimilators are also observed in different species. These are:

1. *C. chemnitzia* (Fig. 3.100A): Here the appendages are tubular in structure.
2. *C. fastigiata* (Fig. 3.100B): Assimilators having irregular branching.
3. *C. prolifera* (Fig. 3.100C): Assimilators are entire and broad leafy structure.
4. *C. racemosa* (Fig. 3.100D): Assimilators are in the form of knob-like bodies.
5. *C. taxifolia* (Fig. 3.100E): Assimilators having two rows leafy structures.
6. *C. verticillata* (Fig. 3.100F): Assimilators are cylindrical and arranged in whorl.

Internal structure

Cells are surrounded by cell wall. Wall is made up of pectin, callose, pectic acid etc., but cellulose is absent. Internally the cells contain cytoplasm, which encloses a large central

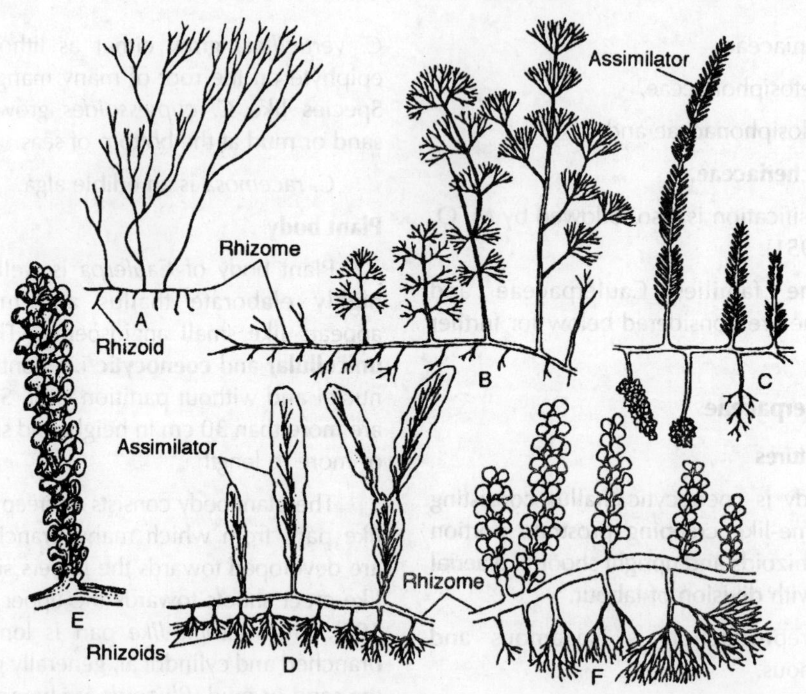

Fig. 3.100: *Caulerpa* plant body with various types of assimilators. A. *C. chemnitzia* (after Svedelius), B. *C. fastigiata* (after Reinke), C. *C. prolifera* (after Zimmermann), D. *C. racemosa* (after Boergesen), E. *C. taxifolia* (after Boergesen), and F. *C. verticillata* (after Boergesen)

siphon-like vacuole. Cytoplasm contains large number of nuclei and many **discoid chloroplasts**. Pyrenoid is absent.

Beside the normal pigments like green algae, two other pigments, characteristic of Siphonales like **siphonein** and **siphonoxanthin** are present but **α-carotene** is present instead of β-carotene.

T.S. of Rhizome: In transverse section, it is circular in outline and surrounded by outer thick covering, provides mechanical support to the plant body (Fig. 3.101A). Central region is with net work formed by transverse and longitudinal rod-like structure, the **trabeculae**, made up of callose and pectic material. Trabeculae are

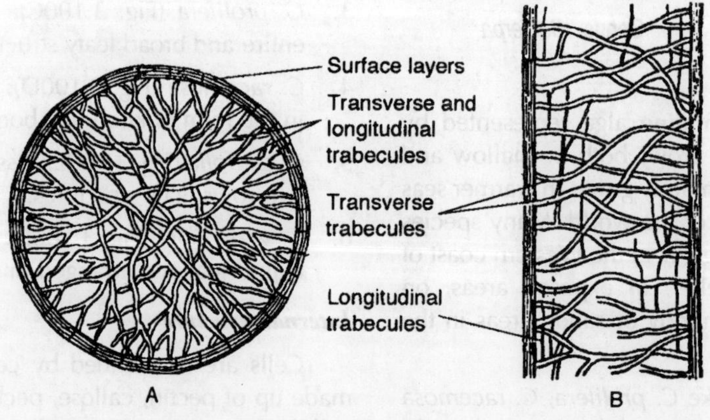

Fig. 3.101: *Caulerpa prolifera*: A. T.S. of rhizome and B. L.S. of assimilator

arranged perpendicular to the surface, possibly gives mechanical support.

L.S. of aerial erect branches: Here many transverse and longitudinal trabeculae (Fig. 3.101B) are arranged irregularly, but some of them are seen running from surface to surface.

T.S. of rhizoid: Here the trabeculae are either ill-developed or absent.

Trabeculae develop from the dense apical cytoplasm from rows of **microsomes**. It may remain attached with the inner layer either at one or at both the ends.

Reproduction

It reproduces by vegetative and sexual means. Spore formation has not yet been reported.

Vegetative reproduction: It takes place by fragmentation and 'death and decay' of older part.

(i) *Fragmentation*: During this process, the thallus gets fragmented and dispersed through water. On contact with suitable environmental condition, the fragments develop into new plant.

(ii) *'Death and decay' of older part*: Due to some environmental condition and by ageing, the older part of the rhizome undergo death and decay, separates the branches. Later new plants are developed from the separated branches.

Sexual reproduction: It has been investigated only in a few species, like *C. serrulata* and *C. prolifera*. Most of the species are dioecious but, *C. taxifolia* is monoecious. All the species are anisogamous.

Thallus is diploid (2n). During this process gametes are formed in the fertile region of the thallus, but distinct gametangia are not formed. At the time of gametogenesis, **meiosis** takes place and haploid (n) gametes are formed. Gametes are of two types. Macrogametes are slightly longer and broader than the microgametes. Gametes are uninucleate, biflagellate, and also with a stigma, but devoid of pyrenoid. Fertile region of the assimilator developed many discharge papillae which are about 2 mm or more in length.

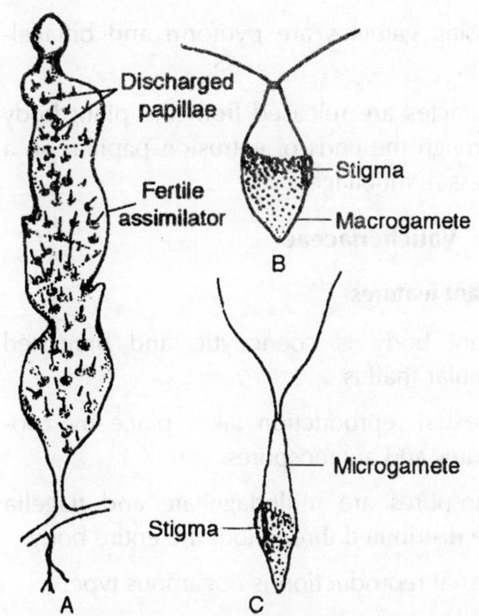

Fig. 3.102: *Caulerpa*: A. Fertile assimilator with discharge papillae, B. Macrogamete and C. Microgamete (after Schussnig).

After maturation, the gametes are liberated with some mucilage by breaking the apical region of the papillae. Fusion takes place between two anisogametes in the sea water and thus zygote (2n) is formed.

The zygote remains quadriflagellate for some time. Later, flagella are withdrawn and germinate directly into diploid plant body.

The life cycle of *Caulerpa* is therefore **diplontic type**, as the plant body is diploid and the haploid stage is represented by the gametes only. The haploid gametes soon fuse togather and form the diploid zygote, which germinate and develop the new diploid plant body.

Indian species

Caulerpa crassifolia, C. fastigiata, C. peltata, C. racemosa, C. setulariodes, and *C. taxifolia.*

Important features

1. Plant body coenocytic with remarkable differentiation of aerial and prostrate portion.
2. Development of internal skeletal strands for mechanical support of different region of plant body.
3. Sexual reproduction is both iso- and anisogamous type.

4. Fusing gametes are pyriform and biflagellate.
5. Gametes are released from the plant body through the ends of extrusion papillae in a mass of mucilage.

Family. **Vaucheriaceae**

Important features

1. Plant body is coenocytic and branched tubular thallus.
2. Asexual reproduction takes place by zoospores and aplanospores.
3. Zoospores are multiflagellate and flagella are distributed throughout the entire body.
4. Sexual reproduction is oogamous type.

The life history of *Vaucheria* is given below:

VAUCHERIA

Systematic position

Fritsch (1945)	**Lee** (1999)
	Phylum. Heterokontophyta
Class. Chlorophyceae	Class. Xanthophyceae
Order. Siphoniales	Order. Vaucheriales
Family. Vaucheriaceae	Family. Not mentioned
Genus. *Vaucheria*	Genus. *Vaucheria*

Occurrence

The genus *Vaucheria* (named after J. P. Vaucher) is represented by about 54 species, out of which 9 species are found in India.

Most of the members are terrestrial (*V. geminata, V. terrestris, V. hamata*) or fresh water (*V. uncinata*). A member like *V. sessilis* is terrestrial

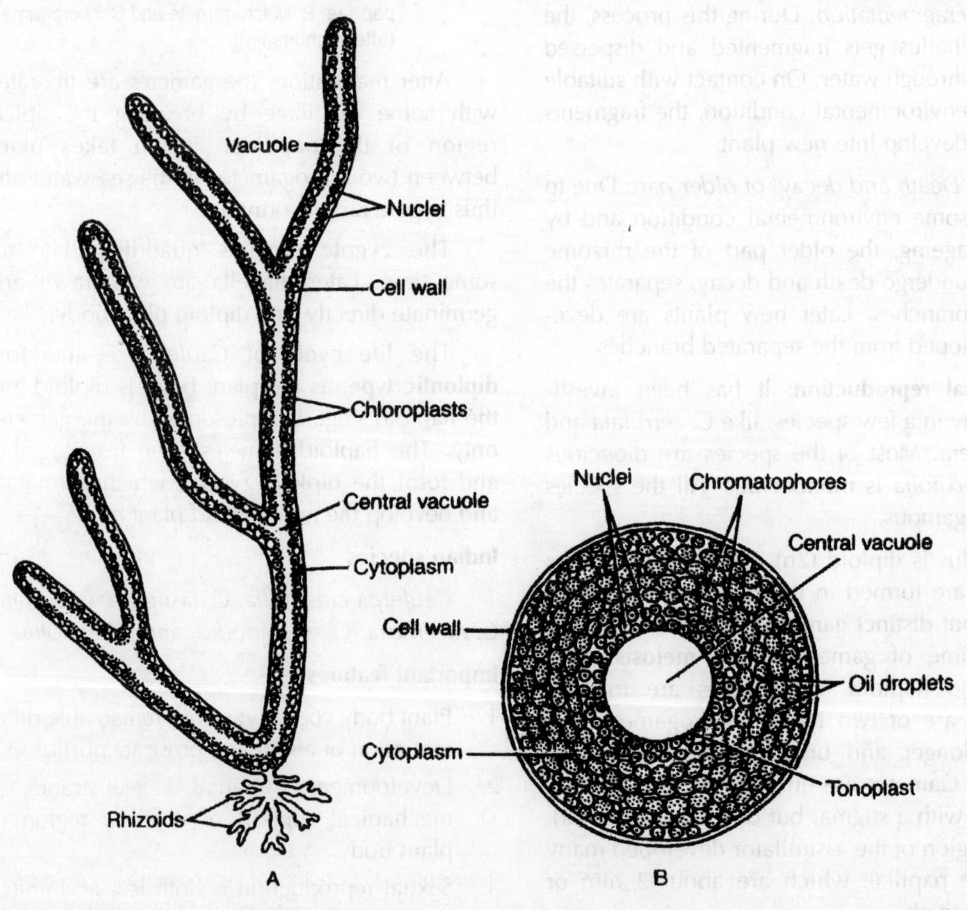

Fig. 3.103: *Vaucheria* sp.: A. Single vegetative filament with rhizoid, and B. T.S. through vegetative filament

as well as aquatic. Species like *V. amphibian* is amphibious in habitat. The terrestrial forms grow in winter in the form of green and thick layer on damp soil. A few species grow in brackish water and some are marine (*V piloboloides*). *V. jonesii* grows on snow.

Plant body

Plant body of *Vaucheria* is filamentous, much branched, coenocytic and siphonaceous thallus (Fig, 3.103A). The coenocytic body contains many nuclei. Septa may form during injury or on development of sex organ. But the filament is normally septate in *V. pseudohamata*.

In terrestrial species the plant body remains attached to the soil surface with much branched thread like structure, the **rhizoid** or **hapteron**. In floating members the rhizoids are either absent or ill-developed. Species like *V. mayyanadensis* shows well-developed rhizoid and aerial branches.

The filamentous body has a thin outer wall, which is less elastic. It is made up of outer pectic and inner cellulosic layers. In the centre of the filament a continuous vacuole is present except at the apical region, which is filled with cell sap. The protoplast is present throughout the filament between the cell wall and vacuole which contains nuclei, chromatophores and other substances. In the protoplast, nuclei remain towards vacuole and chromatophores towards periphery (Fig. 3.103B). Chromatophores are very small, circular, oval or elliptical in outline. Pyrenoids are absent.

Vaucheria contains the pigments like chlorophyll a, chlorophyll e, carotenoids and xanthophylls as in the class Xanthophyceae. The carotenoids are present in more amount than chlorophylls. The characteristic pigment of Chlorophyceae, chlorophyll b, is absent and the dominant pigment of the order Siphonales of Chlorophyceae, the siphonin and siphonoxanthin are absent in *Vaucheria*.

The reserve food material is oil, it occurs as colourless droplets in the cytoplasm. If the filaments remain exposed under continuous illumination, the assimilatory product will be starch instead of oil.

Reproduction

Vaucheria reproduces by all the three means: vegetative, asexual and sexual.

Vegetative reproduction. It takes place due to accidental breakage of the vegetative filament. Septa are formed in the injured region of the filament.

Asexual reproduction. *Vaucheria* reproduces asexually by the formation of various types of spores such as zoospores, aplanospores and akinetes.

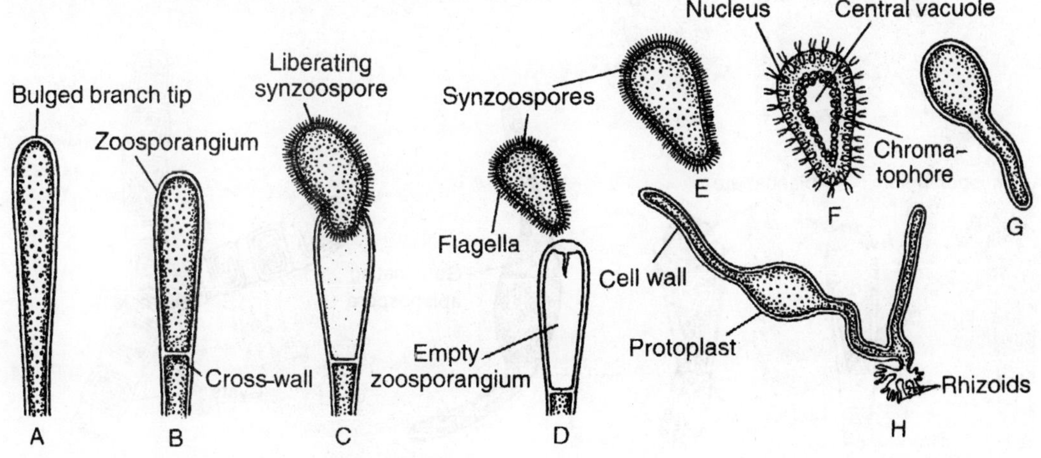

Fig. 3.104: *Vaucheria* sp.: A-E. Formation and liberation of synzoospore, F. Structure of synzoospore in T.S. showing position of chromatophore and nuclei with attached flagella, and G-H. Stages of germination of synzoospore and formation of new plant body

Zoospores. Zoospores are formed **singly** inside the zoosporangium, developed at the apical region of the filament (Fig. 104A). During formation of the zoosporangia, more amount of nuclei and cytoplasm accumulate towards the apex, thereby the vacuole becomes reduced in size. The upper portion gets separated from rest of the filament by cross-wall and forms apical sporangium (Fig. 3.104B). Consequently the sporangium becomes club-shaped and dark green in colour. After some time, the orientation of nuclei and chloroplast get reversed, thereby the nuclei present towards the centre of the vegetative filament come to the periphery and chloroplasts become shifted from periphery towards the centre.

Before liberation, entire protoplast of the zoosporangium contracts slightly. During liberation a narrow aperture develops at the apex through which the zoospore is liberated. Flagella are formed during liberation (Fig. 3.104C, D).

The zoospores are yellowish-green, ovoid, multinucleate and multiflagellate structures, having small central vacuole and many chromatophores situated between the vacuole and peripheral nuclei (Fig. 3.104F). The paired flagella are unequal in length (Greenwood *et. al.*, 1957 observed through E.M.). [Single flagellum develops from a basal granule or basal body. For each pair of flagella, a pair of basal granules is present. The basal granules remain firmly attached together at the tip of the nuclei.] Some algologists (Fritsch, 1935 and others) considered this multiflagellate and multinucleate unit as **synzoospore** or **compound zoospore** (Fig. 3.104E).

After about 15 minutes of swimming the zoospore becomes sluggish, deflagellated, rounded and then secretes a thin wall around. One or more elongated siphonaceous tubes develop at different direction. One of the tubes develops into rhizoid and other one into new thallus (Fig. 3.104G, H).

Aplanospores. Aplanospores are developed singly within aplanosporangium, during unfavourable conditions (Fig. 3.105). Like zoospores, the aplanospores are developed singly inside the **aplanosporangium** at the apical side of the aerial hyphae. During development it also cuts off from rest of the filament by transverse septum. The aplanosporangium is club-shaped, which develops single non-flagellate club-shaped or globose aplanospore (Fig. 3.105B, C). The aplanospore is liberated by rupturing the apical wall. After liberation the aplanospore germinates into new filament like the mother (Fig. 3.105D).

Rarely the protoplast of the sporangium gets separated into small units, the **microaplanospores**. On being liberated they germinate into new filaments.

Akinetes. Akinetes are formed during unfavourable condition. During akinete formation the content of the filament divides into small

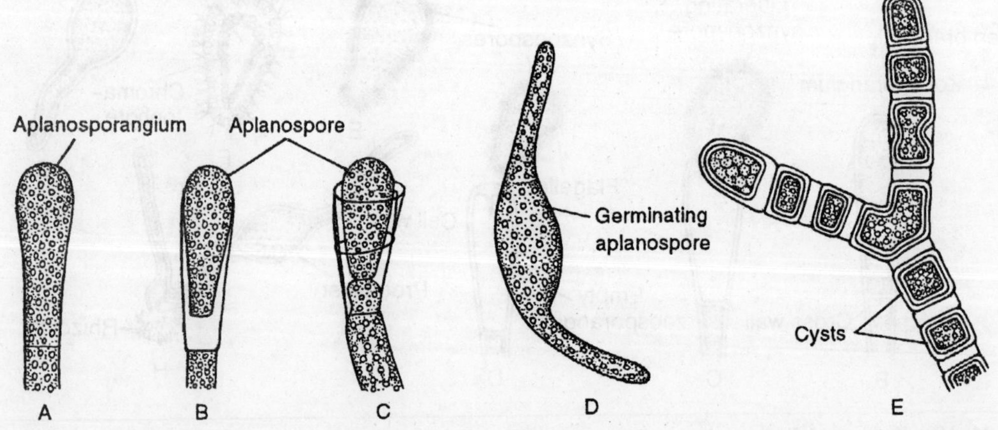

Fig. 3.105: *Vaucheria* sp.: A-C. Formation ofr aplanospore, D. Germination of aplanospore, and E. Gongrosira stage with thick-walled cysts (akinetes or hypnospores)

segments by thick gelatinous wall. These thick walled small multinucleate portions or segments are called **akinetes**, also called as **hypnospores** or **cysts**. The protoplasts of akinetes are often laden with oil. During favourable condition akinete germinates into new filament.

The above stage of akinete chain looks like another alga *Gongrosira* and therefore this stage is called **Gongrosira stage** (Fig. 3.105E).

Sometimes the protoplast of the akinete divides into many small uninucleate amoeboid segments. After liberation by breaking the akinete wall, they develop into individual filaments.

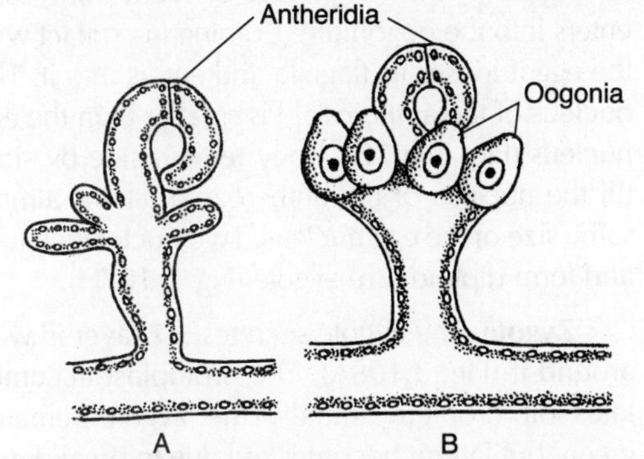

Fig. 3.106: *Vaucheria geminata*: A. Antheridium surrounded by developing oogonia, and B. Antheridium surrounded by oogonia

Sexual reproduction

The sexual reproduction in *Vaucheria* is of **Oogamous type**. It takes place by antheridium, the male sex organ and oogonium, the female sex organ (Fig. 3.107). Most of the species are homothallic or monoecious but a few (*V. litorea, V. dichotoma* and *V. mayyanadensis*) are heterothallic or dioecious.

Position. In homothallic or monoecious species (*V. sessilis, V. aversa*) antheridia and oogonia are borne very close or adjacent on a single filament or on adjacent filament. But in heterothallic or dioecious species (*V. dichotoma*), they are borne on different plants. [The sex organ shows great variation in their position not only in different species but also in the same species. The sex organs are stalked or sessile (*V. sessilis*). In *V. geminata*, a short stalk bears single terminal antheridium flanked by 3-4 oogonia (Fig. 3.106B). In *V. gardneri* single, terminal elongated antheridium is surrounded by a group of oogonium. In *V. terrestris* the stalk of antheridium and oogonium is stout, very long and well-developed. In

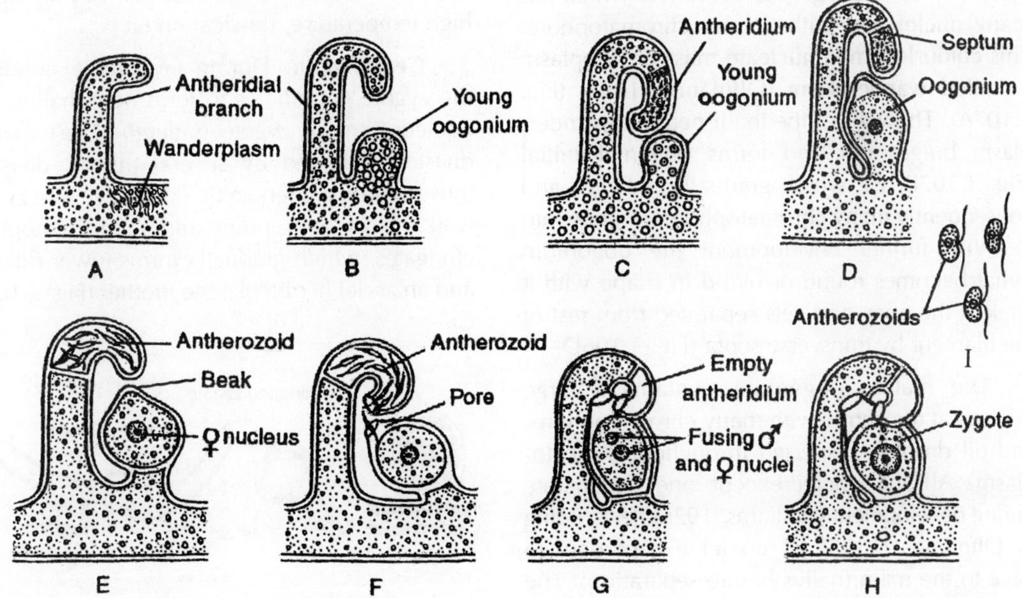

Fig. 3.107: *Vaucheria* sp.: A-F. Successive stages of development of antheridium and oogonium, G. Fusing nuclei, H. Fertilized oogonium forms zygote, and I. Antherozoids

V. bilateralis however, 2-6 oogonia are arranged bilaterally in a series and the antheridia are short-stalked and curved and open towards oogonium.]

Development

Antheridium. Initially a short lateral branch like vegetative branch projects out (Fig. 3.107A). Large number of nuclei and chromatophores accumulate towards the apical region. The apical region then bends like a horn (Fig. 3.107B, C) and a transverse septum is laid down which differentiates the apical antheridium (Fig. 3.107D). The nuclei of antheridium aggregate in the centre and divide mitotically. Each nucleus along with some cytoplasm metamorphoses into single **spindle shaped biflagellate antherozoid** (Fig. 3.107E). The flagella are unequal in length, dissimilar (one whiplash and other tinsel) and laterally inserted.

The antherozoids are generally liberated through an opening developed at the apical region of antheridium (Fig. 3.107F). Several pores are developed on anthridial wall in *V. debaryana* for the liberation of anthrozoides.

Oogonium. Initially a small protuberance develops at or near the base of antheridial branch, due to accumulation of cytoplasm. The cytoplasm of this region is colourless which has many nuclei and without any chromatophore. This colourless multinucleate mass of cytoplasm is called **wanderplasm** (Oltmanns, 1895) (Fig. 3.107A). The wall of the thalli near the wanderplasm bulges out and forms oogonial initial (Fig. 3.107B, C), which gradually enlarges and consequently many chromatophores migrate into it. With further development the oogonium initial becomes round or ovoid in shape with a beak at the apex and gets separated from rest of the filament by transverse septa (Fig. 3.106D).

The mature oogonium contains a large nucleus at the centre with many chromatophores and oil droplets dispersed throughout the cytoplasm. All the nuclei except one degenerate during development (Williams, 1926). [According to Oltmanns (1895) all nuclei except one go back to the main thallus before separation]. The protoplast along with nucleus rounds off and forms single **ovum** or **egg**. It has an hyaline area towards the anterior, known as the receptive spot (Fig. 3.107E).

Fertilization. Both the sex organs open, of almost at the same time (Fig. 3.107F). The tip of the beak of oogonium gelatinises and forms an aperture. Simultaneously an aperture is also formed at the apex of antheridium (the time of opening may vary from a few minutes to 2 hours). A small drop of colourless cytoplasm oozes out through the beak of oogonium and several antherozoids coming out from the antheridium get entangled in this drop. Out of many, only one enters into the oogonium. Coming in contact with the egg it looses its flagella and enters into it. The nucleus of the antherozoid is smaller than the egg nucleus (Fig. 3.107G). They remain side by side, till the nucleus of the antherozoid attains almost same size of the egg nucleus. Two nuclei then fuse and form diploid (2n) zygote (Fig. 3.107H).

Zygote. The zygote secretes 3-7 layered wall around it (Fig. 3.108A). The protoplast accumulates oil droplets. Initially the zygote remains green, but later it becomes red due to breakdown of chlorophyll. It remains within the oogonium and later comes out by decay of oogonial wall. It remains dormant for long period which varies with the duration of unfavourable condition like high temperature, dessication etc.

Germination. During favourable condition the zygote germinates to form new thallus. The nucleus (2n) of zygote undergoes first meiotic division followed by several mitotic divisions thus forming a coenocytic condition. The zygote wall cracks at a point and inner protoplasm elongates, which gradually forms lower rhizoids and an aerial hypha like the mother (Fig. 3.108B, C).

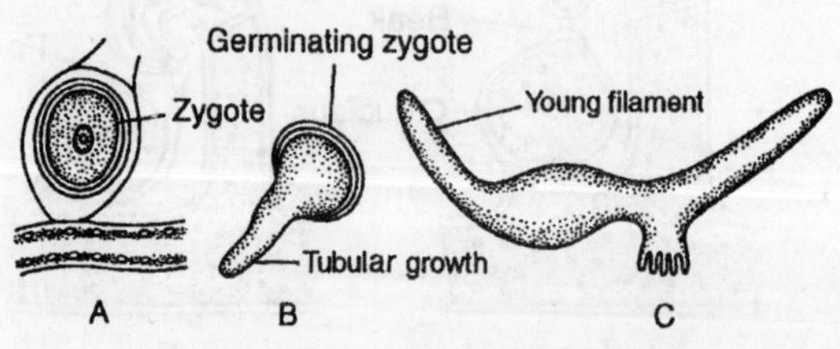

Fig. 3.108: *Vaucheria* sp.: A-C. Stages in the germination of zygote

Taxonomic status of *Vaucheria*

There has been a great controversy regarding the taxonomic status of *Vaucheria*. Fritsch 1935) included *Vaucheria* along with other three genera (*Vaucheriopsis, Dichotomosiphon* and *Pseudo-dichotomosiphon*) in the family Vaucheriaceae under the class Chlorophyceae. This was also followed by Iyenger (1951) and Venkataraman (1961). It was placed under the class Chlorophyceae, based on the following characteristics:

1. Plant body is siphonaceous filament.
2. Asexual reproduction by synzoospore or compound zoospore.
3. Sexual reproduction is oogamous type.
4. Presence of chlorophyll a.
5. Change of assimilatory product into starch, instead of oil under continuous illumination.

However, it differs from other members of Chlorophyceae, based on the following characteristics:

1. Absence of macerised cellulosic cell wall.
2. Absence of pyrenoid.
3. Paired flagella of synzoospores are unequal.
4. Absence of chlorophyll b.
5. Reserve food is oil.

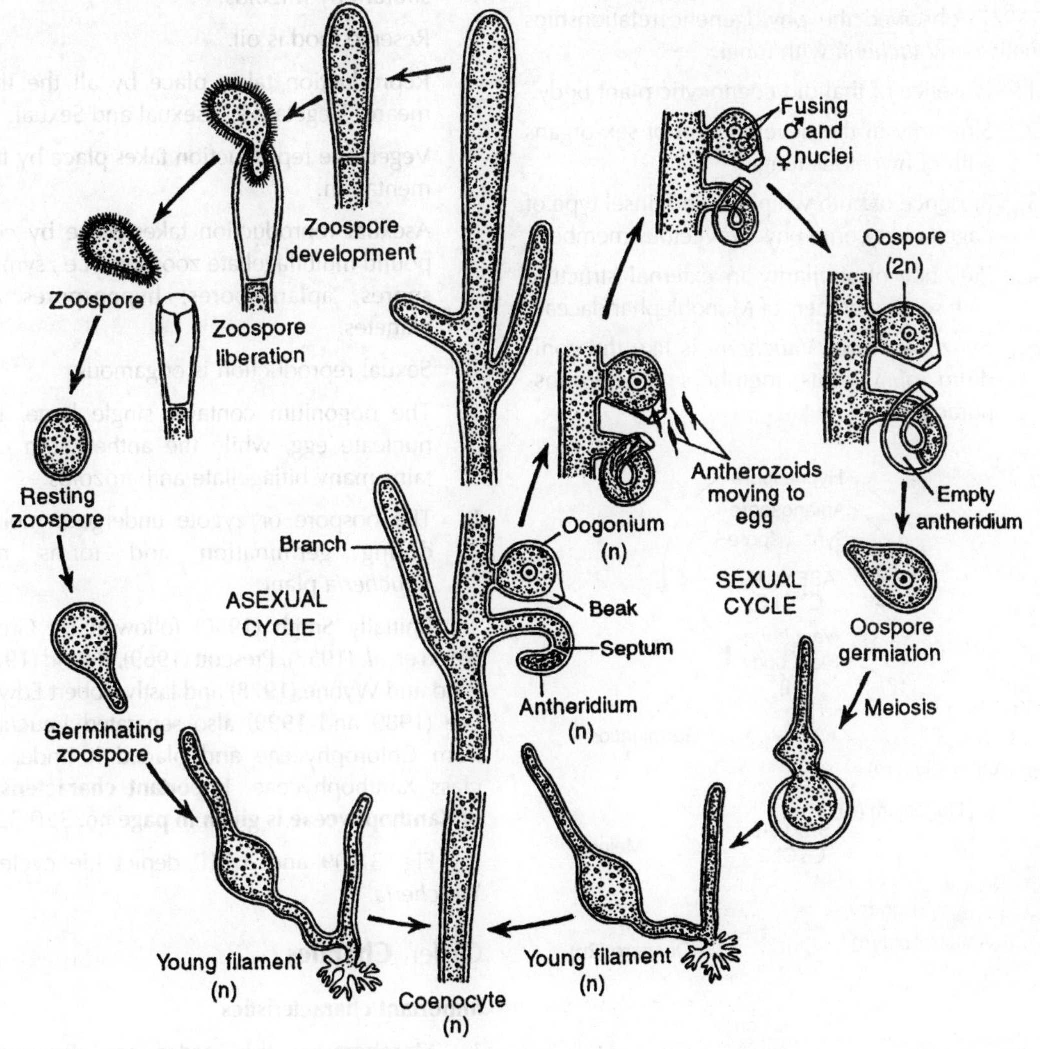

Fig. 3.109: Life cycle of *Vaucheria* sp.

Later Smith (1950) followed by Greenwood *et. al*. (1957), Prescott (1969), Round (1973), Bold and Wynne (1978) separated *Vaucheria* from Chlorophyceae and placed it under the class Xanthophyceae. They shifted *Vaucheria* to Xanthophyceae, based on the following characteristics.

1. Unequal length of flagella of antherozoids.
2. Presence of chlorophyll e, a pigment common in Xanthophyceae.
3. Absence of two xanthophylls, siphonin and siphonoxanthin.

Davis (1904), Williams (1926) and Printz (1927) observed the phylogenetic relationships between *Vaucheria* with fungi.

1. Presence of thalloid coenocytic plant body.
2. Similarity in the development of sex organs with oomycetous fungi.
3. Presence of both whiplash and tinsel type of flagella like some phycomycetous members.
4. Presence of similarity in external structure with some members of Monoblepharidaceae.
5. Synzoospore of *Vaucheria* is like the gonidium of various members of Peronosporaceae.

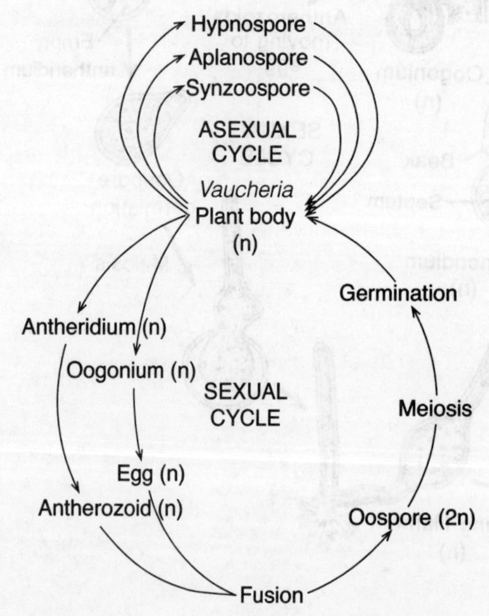

Fig. 110: Graphic life cycle of *Vaucheria* Sp.

For the above evidences someone believed that the Phycomycetes have been derived from *Vaucheria*-like ancestors. But it is very difficult to make comment on whether this similarity in characteristics is due to their real affinity or it is due to convergent evolution.

Indian species

Vaucheria sessilis, V. amphibian, V. geminata, V. uncinata, V. polysperma etc.

Important features

1. Plant body is filamentous, coenocytic and branched. It remains attached with the substratum by rhizoids.
2. Reserve food is oil.
3. Reproduction takes place by all the three means: Vegetative, Asexual and Sexual.
4. Vegetative reproduction takes place by fragmentation.
5. Asexual reproduction takes place by compound multiflagellate zoospores i.e., synzoospores, aplanospores, hypnospores and akinetes.
6. Sexual reproduction is oogamous.
7. The oogonium contains single large, uninucleate egg, while the antheridium contains many biflagellate antherozoids.
8. The oospore or zygote undergoes meiosis during germination and forms new *Vaucheria* plant.

[Initially Smith (1950) followed by Greenwood *et. al*. (1957), Prescott (1969), Round (1973), Bold and Wynne (1978) and lastly Robert Edward Lee (1989 and 1999) also separated *Vaucheria* from Chlorophyceae and placed it under the class Xanthophyceae. **Important characteristics of Xanthophyceae is given in page no.** 320-321].

Fig. 3.109 and 3.110 depict life cycle of *Vaucheria*.

Order. **Charales**

Important characteristics

1. Members of this order are distributed throughout the world.

2. Commonly they are found in fresh water with muddy or sandy bottom and also in water flowing over limestone.
3. Plants are macroscopic, much branched, and erect and commonly up to 30 cm in length.
4. The plants are differentiated into nodes and internodes. Some of the nodes bear branches of unlimited growth, those are again divided into nodes and internodes. Each node of the main axis and branch of unlimited growth bear a number of branches of limited growth.
5. Cells are very long, uninucleate and contain many discoid chloroplasts.
6. Most of the species show cortication in the internodes. The cortex consists of vertically elongated row of cells.
7. Sexual reproduction is highly advanced, oogamous type.
8. The male and female reproductive bodies are globule and nucule, respectively. Globule develops many antherozoids and nucule contains only one egg.
9. Zygote is produced after sexual reproduction. It shows very much elaborate post-fertilization changes. During germination, zygote undergoes meiosis and gradually it forms the plant body.

Classification

Fritsch (1935) placed the order Charales under the class Chlorophyceae includes only one family the Characeae having 2 sub families: 1. Nitelleae and 2. Chareae.

Bold and Wynne (1978) placed the order Charales alone under the only class Charophyceae, under the division Charophyta. The order Charales includes only one family Characeae.

Divn. Charophyta: Class. Charophyceae

H. C. Bold and M. J Wynne (1978) in their classification took out *Chara* along with some other genera like *Tolypella, Nitella, Nitellopsis, Protochara, Lamprothamnium and Lychnothamnus* from Chlorophyceae and placed them in a **separate Division Charophyta**. The Charophyta consists of single class Charophyceae; order Charales and family Characeae.

Family. **Characeae**

Important characteristics

1. The division Charophyta includes the members of green algae, commonly known as **stoneworts**.
2. Plant body shows much elaboration of vegetative structures encrusted with calcium carbonate.
3. Plant body is erect and consists of elongated, jointed, commonly green main axis bearing branches, differentiated into nodes and internodes.
4. Each node bears a whorl of lateral branchlets.
5. Asexual reproduction is absent.
6. Sexual reproduction is of oogamous type.
7. Antheridia (globule) and oogonia (nucule) show more complexity and elaboration than other Chlorophycean members.
8. Motile cells are asymmetrical and two flagella are attached in lateral position of an antherozoid.
9. Sex organs are so large that they can be visible with naked eye.
10. Zygote on germination forms protonema (*Chara, Nitella*) from which vegetative plants are developed.

The **common genera** of the family Characeae are *Chara and Nitella*.

The life history of *Chara* is given below:

CHARA

Systematic position

Fritsch (1945)	**Lee** (1999, 2008)
	Phylum. Chlorophyta
Class. Chlorophyceae	Class. Charophyceae
Order. Charales	Order. Charales
Family. Characeae	Family. Not mentioned
Genus. *Chara*	Genus. *Chara*

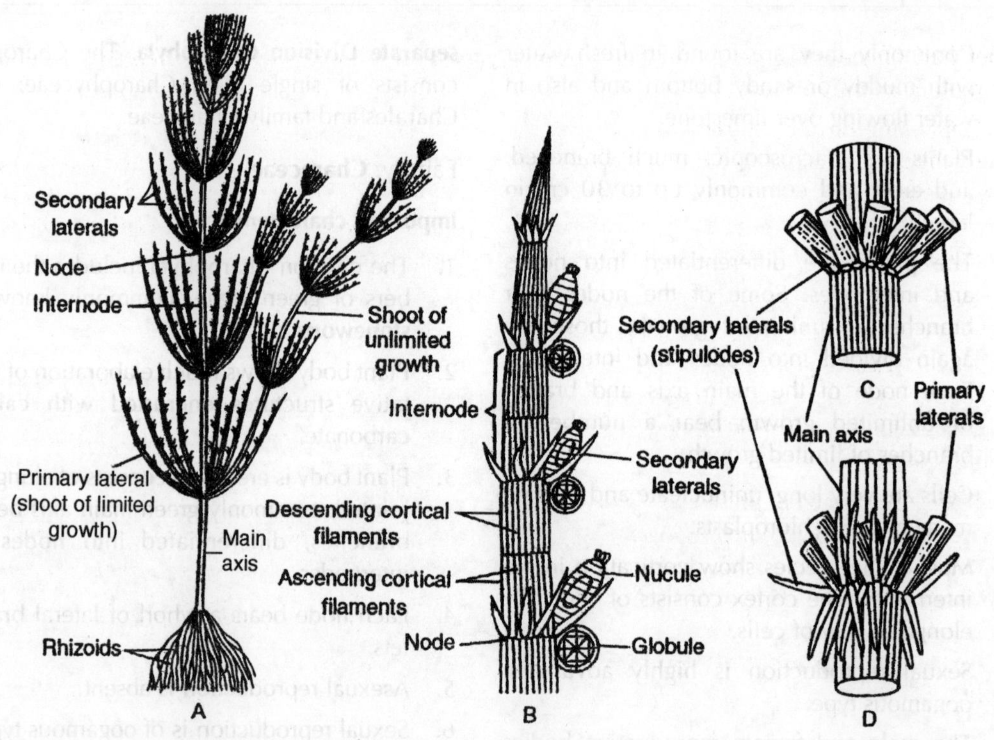

Fig. 3.111: *Chara* sp.: A. External morphology, B. Shoot (branch) of limited growth or primary lateral, and C-D. Appendages on node (C. Haplostephanous and D. Diplostephanous type)

Occurrence

Chara is represented by about 188 species (Wood and Imahori, 1959), out of which 30 species are found in India. It is commonly known as "**stonewort**". The plant body of *Chara* is encrusted with calcium and magnesium carbonate especially on the plants growing in heavy water. Thus the plants become strengthened and called stoneworts. Generally they grow in fresh water of ponds, lakes, tanks etc. in submerged condition. Some species like *C. fragilis* grows in hot *spring*, whereas *C. baltica* grows in brackish water.

Plant body

Chara is a macroscopic, multicellular, profusely branched thailoid plant body, generally attains a height of about 20-30 cm (rarely about 1 meter). It is differentiated into **rhizoid** and **main axis** (Fig. 3.111).

Rhizoid: The rhizoids are thread-like, white, multicellular, uniseriate and branched. It is an elongated branched structure having *oblique septa*. They are developed either from the base of the plant body or from peripheral cells of lower nodes of the main axis.

Main axis: It is an erect, long, branched epigeal portion of the plant body, which is differentiated into internodes and nodes.

Internodes. Generally it consists of two types of cells: (i) axial cell or internodal cell, and (ii) cortical cells.

(i) *Axial cell*. It consists of an elongated central cylindrical cell (Fig. 3.114).

(ii) *Cortical cells*. These are elongated but much smaller in diameter than axial cell and ensheathed or corticated as a layer on the outer surface of axial cell (Fig. 3.114). They originate from the node. After originating from the node, 50% of the cortical cells grow upward as the **ascending filaments** and the rest 50% grow downward as the **descending filaments** (Fig. 3.111B). The ascending filaments cover the lower half and descending filaments cover the upper half of the axial cell. Cortication is not common in all the species.

Depending on the. presence or absence of cortex, the species of *Chara* are divided into two types: **Corticate** (e.g., *C. fragilis, C, zeylanica, C. hatei* etc.) and **Ecorticate** (e.g., *C. corallina, C. succinata, C. wallichii, C. braunii* etc.).

Node: The node consists of two cells surrounded by 6-20 peripheral cells (Fig. 3.111C, D). Three types of appendages are developed from each node. These are: 1. Branches of unlimited growth, 2. Branches of limited growth and 3. Stipulodes:

1. **Branches (shoots) of unlimited growth**: They are also called **axillary branches** or **long laterals** (Fig. 3.111A) and are developed from the older nodes. These branches are also differentiated into nodes and internodes like the main axis. Each node bears branchlets like the main axis.
2. **Branches (shoots) of limited growth**: They are also called **primary laterals, branchlets** or **leaves** (Fig. 3.111B). About 6-16 branchlets develop in whorls around the node of main axis or branch of unlimited growth. It is also divided into 5-15 nodes and internodes. Each node develops some unicellular, hair-like **secondary laterals**. Sex organs are developed on lower nodes of each branchlet.
3. **Stipulodes**. These are unicellular outgrowths developed from lower nodes of branchlets i.e., branches of limited growth. The number of stipulode at each node may be equal to the number of branchlets which is called **unistipulate** (*C. nuda, C. brouni, C. coralline*) or if double it is called **bistipulate** (*C. contraria, C. tomentosa, C. baltica*).

Depending on the arrangement of stipulodes species of *Chara* are divided into **haplostephanous** (i.e., stipules are arranged in single row) e.g., *C. braunii* (Fig. 3.111C), and **diplostephanous** (i.e., stipules are arranged in two rows) e.g., *C. delicatula* (Fig. 3.111D).

Cell structure

The nodal cells are short, uninucleate, with dense and granular cytoplasm and many discoid chloroplasts without pyrenoids. Small vacuoles may be present in the cytoplasm.

The internodal cells are long, with a large central vacuole, many nuclei and many discoid chloroplasts in the cytoplasm. The cytoplasm is differentiated into outer **ectoplasm** and inner **endoplasm**. The endoplasm shows streaming movement.

Growth

Growth of *Chara* takes place by a dome-shaped **apical cell**. The cell undergoes repeated transverse divisions and forms a row of three cells (Fig. 3.112A-B). The upper one remains as **apical cell**, middle biconcave one forms the **nodal initial** and the lower one forms the **internodal initial**. The nodal initial undergoes repeated vertical divisions and ultimately forms two central cells surrounded by 6-20 peripheral cells. Branches of limited growth are developed from the peripheral cells arranged in single row. The internodal initial does not divide further and elongates much more to form long internode (Fig. 3.112).

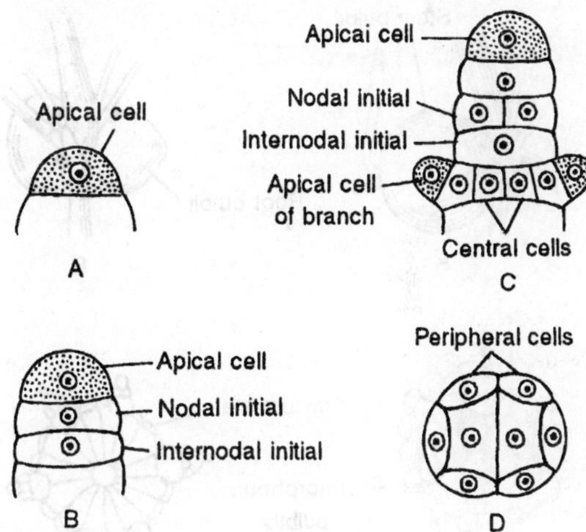

Fig. 3.112: *Chara* sp.: A-C. Stages of growth, and D. T.S. of node

Reproduction

Chara reproduces by both vegetative and sexual means. **Asexual reproduction is absent.**

Vegetative reproduction

The vegetative reproduction takes place by the formation of following structures:

1. ***Bulbils***: These are small oval or spherical bodies developed on stem or root nodes. Bulbils are formed on root of *C. aspera* and stem of *C. baltica*. After detachment, they germinate and develop into new plants (Fig. 3.113A, B).
2. ***Amorphous bulbils***: These are small cells developed and aggregated at the node, called amorphous bulbils. They are found in *C. fragilis, C. baltica* etc. On being detached from the mother plant, they germinate and develop into new plants (Fig. 3.113C).
3. ***Amylum stars***: These are multicellula aggregations of cells, looking like stars and the cells are densely filled with amylum starch; thus they are called amylum stars. The amylum stars are developed at the nodal cells of the basal region e.g., *C. stelligera* (Fig. 3.113D).
4. ***Secondary protonema***: These are thread-like structures developed from primary protonema or from the basal cell of the rhizoid. New plants are also developed from the secondary protonema.

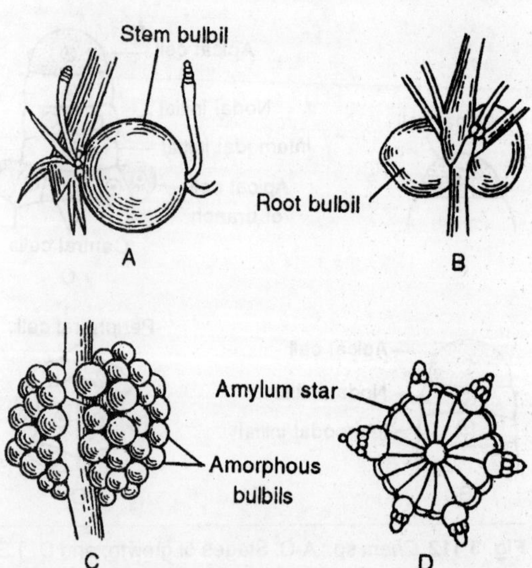

Fig. 3.113: *Chara* sp.: A. Stem bulbil, B. Root bulbil, C. Amorphous bulbil, and D. Amylum star

Sexual reproduction

Sexual reproduction of *Chara* is an **advanced oogamous type**. The sex organs are macroscopic and large. The male sex organ is spherical and yellow to red in colour, called **globule**. The female sex organ is more or less oval and green in colour, called the **nucule** or **oogonium**. They develop on the nodes of the branch of limited growth (i.e., primary lateral), intermingled with secondary laterals. **Nucule is always situated singly above the globule** (Fig. 3.111 B, 3.114).

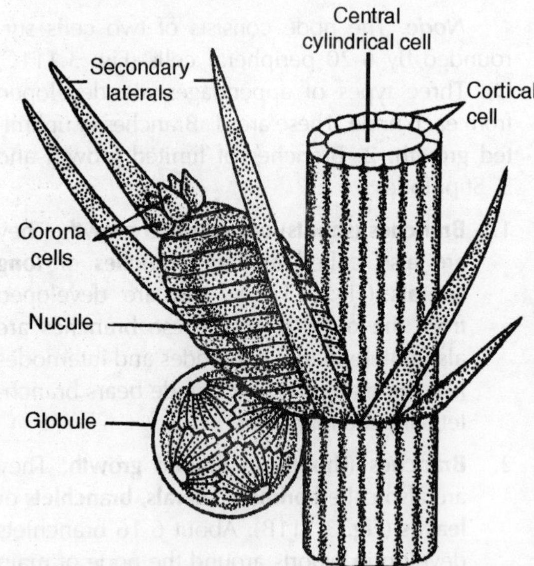

Fig. 3.114: *Chara* sp.: A portion of the branch of limited growth showing attachment of nucule, globule and secondary laterals at the node

Most of the species are **homothallic** or monoecious (i.e., male and female sex organs develop on the same plant), but some are **heterothallic** or dioecious (e.g., *C. wallichii*).

Structure of a matu.re globule: Mature globules are spherical in shape and yellow to red in colour (Fig. 3.115G). Each globule consists of eight curved plates, situated towards the outer side, which are the ***shield cells***. From the inner side of the each shield cell, a centrally placed rod shaped structure is developed, called the ***manubrium***. At the distal end of each manubrium one or more globose cells developed are called ***primary capitula***. Each primary capitulum develops two or more ***secondary capitula***. Finally each secondary capitulum develops 2-4 long ***antheridial filaments*** (Fig. 3.115 H, I). Each antheridial filament has 25-250 cells and each cell i.e., antheridium (Fig. 3.115 J, K) forms a

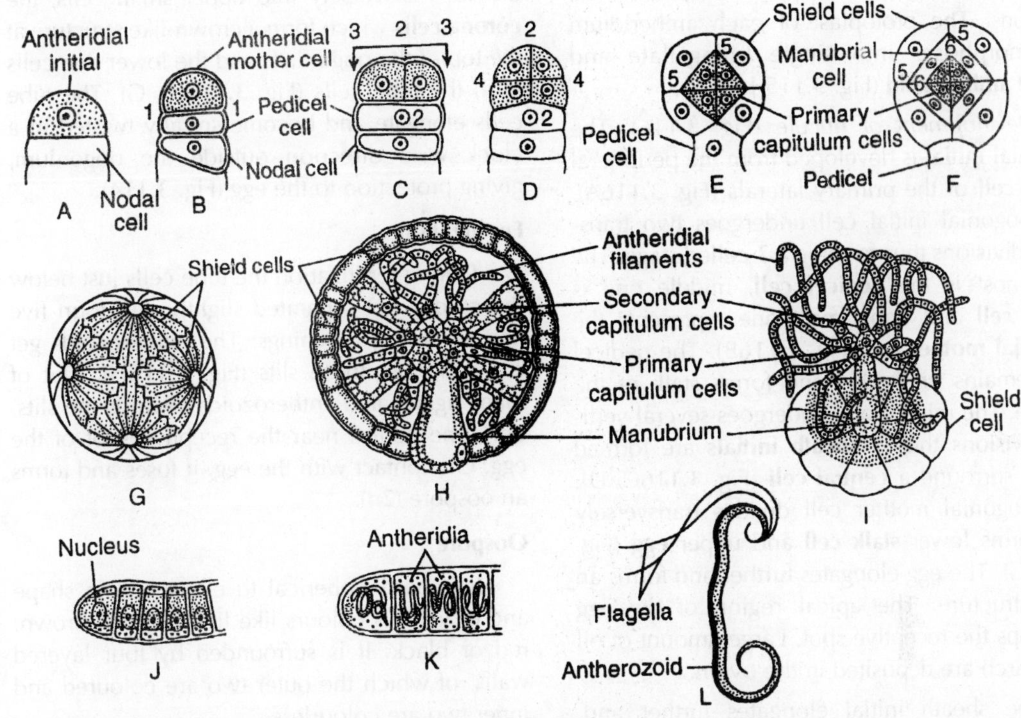

Fig. 3.115: *Chara* sp. Development of globule. A-F. Stages in the development of globule, G. Mature globule, H. Globule in longitudinal section, I. A shield cells with manubrium, primary and secondary capitulum cells and spermatogenous filaments, J-K. Stages in spermatogenesis, L. An antherozoid

biflagellate, coiled and uninucleate antherozoid (Fig. 3.115L). Thus a globule can develop as much as 20,000 to 50,000 antherozoids.

Structure of a mature nucule or oogonium: The nucule of *Chara* is oval with a short stalk. Like globule it is also developed at the node of primary laterals just above the globule in homothallic species. It **consists of** centrally placed ***one central cell, one stalk and one large egg at the top*** (Fig. 3.116H). The entire structure is covered from the base by ***five spirally twisted tube cells*** except at the apex, where they form a crown made-up of five corona cells (Fig. 3.114, 3.116H).

The jacket of nucule shows similarity with the neck cells of archegonium of Bryophyte.

Development of sex organs

Development of globule (Fig. 3.115). The globule develops at the node of branches of limited growth. Single peripheral cell of each node functions as the **antheridial initial** (Fig. 3.115A). The antheridial initial first undergoes transverse division (1-1) to form 2 cells, of which the lower one is the **pedicel cell**, which forms the stalk. The upper one is the **antheridial mother cell**. The antheridial mother cell, then undergoes two vertical divisions right angle to each other (2-2,3-3) followed by one transverse division (4-4), thus an octant (8 celled stage) is formed. Each cell of the octant stage then undergoes periclinal division (5-5) to form outer 8 and inner 8 cells. Either the outer or the inner cells then undergo another periclinal division (6-6), thus forming 3 layers of 8 cells each (Fig. 3.115 B-F).

The outer 8 cells form the 8 **shield cells**, the middle 8 cells form the **manubrium** and the inner 8 cells form **primary capitula**. The **primary capitula** further divide and form two or more **secondary capitula** (Fig. 3.115H, I). Each secondary capitulum further divides and forms 2-4 **antheridial filaments** consisting of 25 to 250 antheridial cells or antheridia, formed by repeated mitotic divisions. The protoplast of each antheridium metamorphoses into single biflagellate and coiled **antherozoid** (Fig. 3.115 J, K, L).

Development of nucule (Fig. 3.116). The **oogonial initial** is developed from the peripheral nodal cell of the primary laterals (Fig. 3.116A). The oogonial initial cell undergoes two transverse divisions thus forming a 3 celled stage. The lowermost is the **pedicel cell**, middle one is **nodal cell** and upper-most one represents the **oogonial mother cell** (Fig. 3.116B). The pedicel cell remains undivided and forms stalk of the nucule. The middle one undergoes several vertical divisions thus 5 **sheath initials** are formed which surround a **central cell** (Fig. 3.116C, D). The oogonial mother cell divides transversely and forms lower **stalk cell** and upper **egg** (Fig. 3.116G). The egg elongates further and forms an oval structure. The apical region of the egg develops the receptive spot. Large amount of oil and starch are deposited in the ovum.

The sheath initial elongates further and, divides transversely into upper small cells, the **corona cells** which form a crown-like structure at the top of the oogonium and the lower five cells form the tube cells (Fig. 3.116 F, G). The tube cells elongate and become spirally twisted in a clock wise direction outside the oogonium, giving protection to the egg (Fig. 3.114).

Fertilization

During fertilization the tube cells just below the corona get separated slightly and form five narrow slits or openings. The antherozoids get entry through these slits (Fig. 3.116H). Out of many aggregated antherozoids towards the slits, only one comes near the receptive spot of the egg. On contact with the egg, it fuses and forms an oospore (2n).

Oospore

It is hard, spherical to ellipsoidal in shape and of various colours like light yellow, brown, red or black. It is surrounded by four layered walls, of which the outer two are coloured and inner two are colourless.

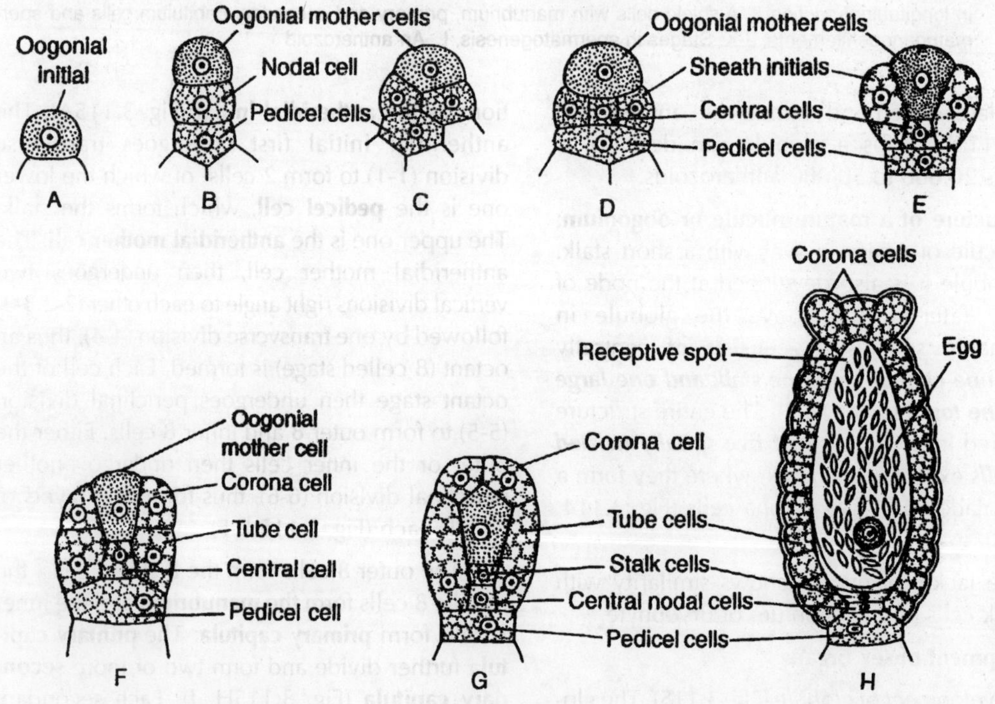

Fig. 3.116: *Chara* sp. Development of Nucule: A-G. Stages in the development of nucule, and H. L.S. of Mature nucule

Germination

During germination the nucleus of oospore , migrates towards the upper region (Fig. 3.117B). The nucleus then undergoes ***meiotic division*** to form 4 haploid nuclei (Fig. 3.117C). The oospore then divides into two unequal cells of which the upper lenticular cell contains one nucleus and lower large basal cell contains three nuclei (Fig. 3.117D). The nuclei of the basal cell gradually degenerate. The lenticular cell projects out by rupturing the oospore wall and divides ***mitotically*** by an oblique longitudinal septum to form a larger **protonemal initial** and a small **rhizoidal initial** (Fig. 3.117E). Both the initials grow in opposite direction. The protonemal initial is differentiated into nodes and internodes and form the upper part of the plant body, whereas the rhizoidal initial forms rhizoids (Fig. 3.117 F-H). Secondary rhizoids may develop from the lower node of protonemal filament (Fig. 3.117 G).

Fig. 3.118 depicts the life cycle of *Chara*.

Taxonomic status of *Chara*

The systemic position of *Chara* has been controversial for long time. **Fritsch** (1935) placed *Chara* under the order Charales of the class Chlorophyceae based on:

(i) Cellulosic nature of cell wall,

(ii) Chlorophyll a and chlorophyll b as photosynthetic pigments,

(iii) Starch as reserve food,

(iv) Numbers of flagella are two and of equal in length, and

(v) Life cycle patterns are like the typical chlorophycean members.

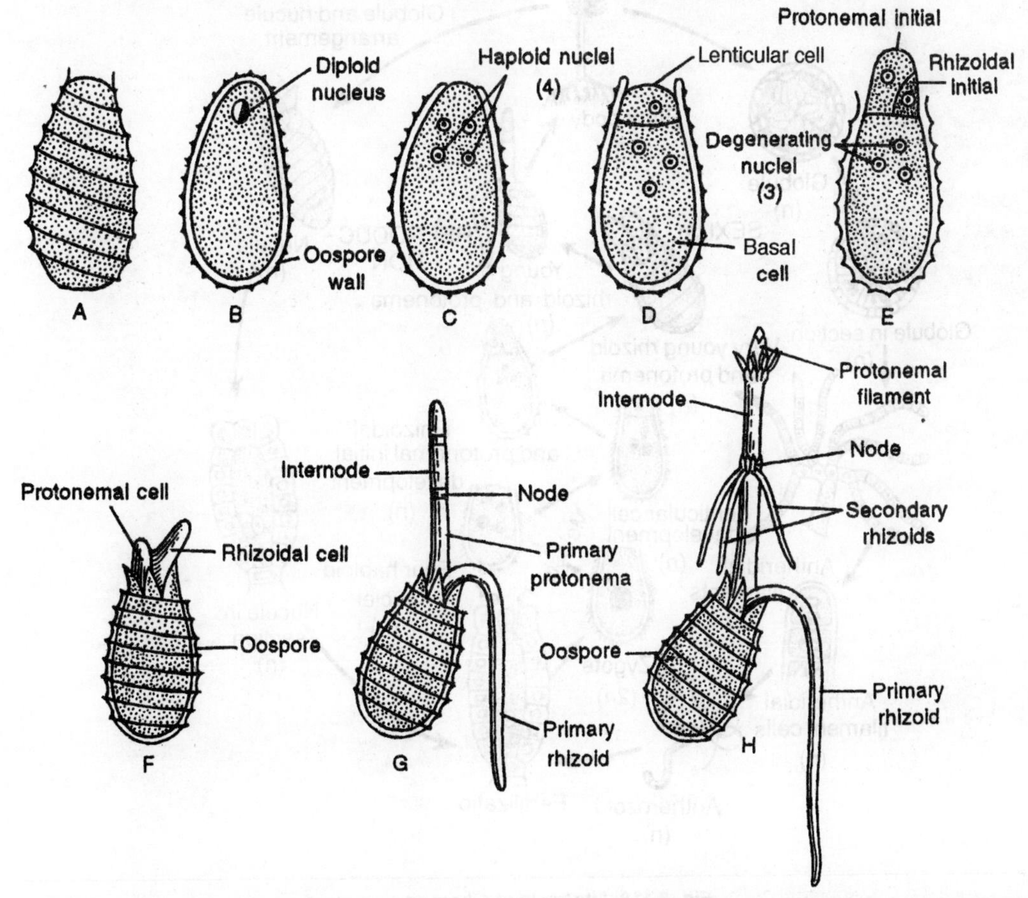

Fig. 3.117: *Chara* sp. A-H. Succesive stages in oospore germination

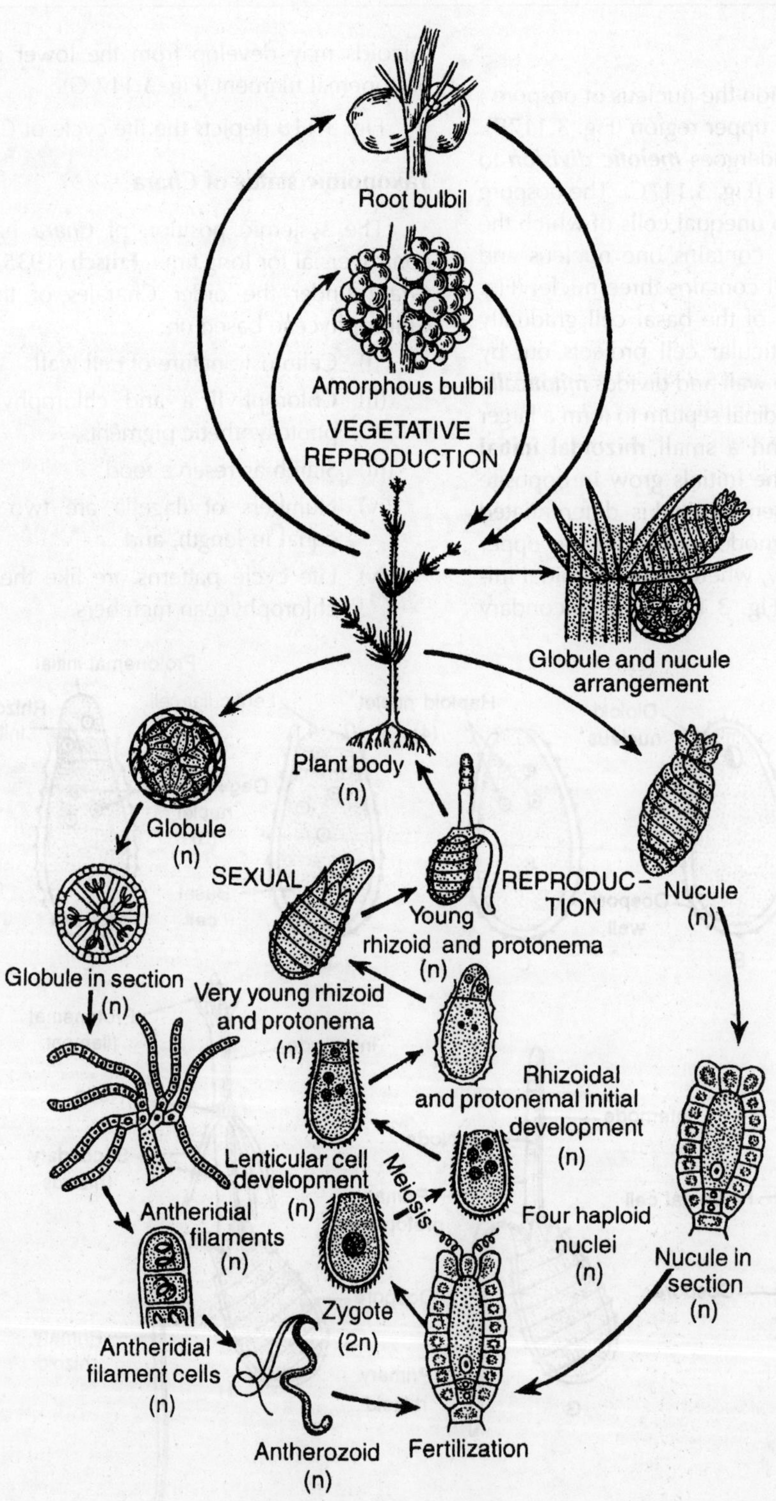

Fig. 3.118: Life cycle of *Chara* sp

Later **Smith** (1938, 55) placed the order Charales in a separate ***class Charophyceae*** under the ***division Chlorophyta***. He placed *Chara* under the class Charophyceae based on the following characteristics:

(i) Members are commonly known as stoneworts,

(ii) Plant body is an erect branched thallus differentiated into a regular succession of nodes and internodes.

(iii) Each node bears a whorl of branches of limited growth (the leaves), but branches capable of unlimited growth may arise axillary to the leaves,

(iv) The members of the class are heavily calcified,

(v) Female sexual reproductive structures are one-celled, surrounded by a sheath of sterile cells and are always borne upon the "leaves",

(vi) The antheridia are one-celled, united in uniseriate branched filaments surrounded by a common spherical envelope composed of eight cells.

Following Smith, **Prescott** (1965) also placed *Chara* under a separate ***class Charophyceae*** of the ***phylum Charophyta***.

Later, **Bold** and **Wynne** (1978) placed *Chara* and other members under the ***division Charophyta***. But instead of using the name "Charophycophyta" like other divisions, they named the division Charophyta. The authors have not included the word *phyco*, in the division because of their "uncertainty that these plants, the stoneworts and brittleworts, are, in fact, algae". So they have shown a doubt about these plants of being even the 'algae'.

Chara shows marked advancement in the following characteristics:

(i) The plant body consists of highly developed aerial portion differentiated into nodes and internodes and a prostrate portion like the root system,

(ii) Elaborate sexual reproductive organs,

(iii) Elongated biflagellate anthrozoids,

(iv) Oogamous type of sexual reproductions, and

(v) Elaborate post fertilization changes.

All the above features suggest that the status of *Chara* should be ***much higher than algae and closer to Bryophytes***. According to **Gramblast (1974) *Charophyta occupies an isolated position between green algae and Bryophytes***.

But there are certain characteristics which retain *Chara* where it is present. These are:

(i) Plant body is haploid,

(ii) Though the nucule looks like a complex structure it is actually a simple unicellular structure like any other algae,

(iii) The diploid stage is restricted in the zygote.

The difference between *Chara* and the members of green algae are strong enough to put *Chara* in a separate class under the division Chlorophyta. According to Church, *Chara* is a remnant of many probable evolutionary tendencies that have failed to attain land habit. On the basis of morphological and cytological characters and oospore ornamentation, pattern Dr. P. chatterjee, Dr. Samit Roy and Dr. Ruma Pal suggested that the genus *Chara* represents a specialised group (order Charales) and it should be placed under the class chlorophyceae.

Indian species

Chara brachypus, C. fragilis, C. gymnopitys, C. zeylanica, C. braunii, C. benthamii, etc.

Important features

1. The plant body shows very much complexity in their structure.
2. They remain attached with the substratum by rhizoids.
3. The main axis is differentiated into nodes and internodes. Each node bears a number of branches of limited growth and sometimes single branches of unlimited growth.
4. The branches of limited growth are also differentiated into nodes and internodes. Each node bears both the sex organs (Nucule i.e., female and globule i.e., male)

and secondary laterals. Nucule is situated above the globule.

5. Reproduction is of two types: Vegetative and Sexual.
6. Vegetative reproduction takes place by means of specialised star-like, tuber-like and protonema-like structures.
7. Sexual reproduction is of oogamous type. The nucule is oval-shaped and very much protected, which contains one egg and globule is round and develops many antherozoids. Zygote is produced after sexual reproduction. It shows very much elaborate post-fertilization changes.
8. During germination, zygote undergoes meiosis and gradually it forms the plant body of *Chara*.

XANTHOPHYCEAE

(YELLOW-GREEN ALGAE)

Xanthophyceae is an algal group consisting of about 76 genera and 450 species. They are commonly known as **yellow-green algae**, due to the presence of more carotenoids than chlorophylls and give the characteristic colouration. They occur in various habitats like fresh water, saline water and also in terrestrial condition on the soil.

Important characteristics

The important characteristics of the class Xanthophyceae are:

1. Members of Xanthophyceae are commonly fresh water (*Tribonema*) and most of them are free floating. [Few members are found to grow on mud (*Botrydium*, Fig. 3.119) and also on walls or tree trunks (*Characiopsis*, *Ophiocytium* etc.). A few members like *Halosphaera* are marine.]
2. Plant body is unicellular (*Heterochloris*) or multicellular. [The multicellular bodies also exhibit various forms like palmelloid (*Chlorogloea*), dendroid (*Mischococcus*), coccoid (*Chlorobotrys*), rhizopodial (*Stipitococcus*), filamentous (*Heterococcus*) and siphonaceous (*Botrydium*)].
3. The cell wall is often absent, but when present it contains more pectic compounds than the members of Chlorophyceae. Occasionally cellulose is also present. [In some non-motile forms the cell wall is silicified and made up of two halves, those overlap each other.]
4. The motile forms bear usually two flagella but rarely one. They are unequal and inserted at the anterior end. The flagella are of two types. The larger one is tinsel (or pantonematic or pleuronematic i.e., bearing hair-like appendages) and the shorter one is whiplash (or acronematic i.e., without hairs and their surface is smooth) type.
5. The chromatophores are discoid in shape and are numerous in each cell.
6. The pyrenoides are absent or rarely present.

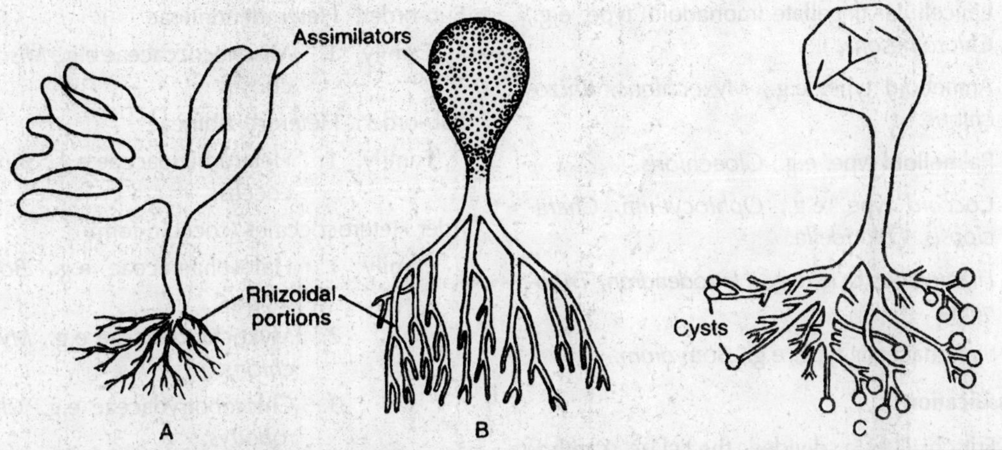

Fig. 3.119: Different species of *Botrydium*: A. *B. divisum*, B. *B. granulatum* and C. *B. tuberosum*

7. The plastids are yellow-green in colour. The photosynthetic pigments are chlorophyll a, chlorophyll e (very little), β-carotene (fairly high concentration) and xanthophylls. The chief xanthophyll is diadinoxanthin. The other xanthophylls are violaxanthin, lutein, neoxanthin, flavoxanthin and flavacin. The carotenoides are normally present in excess amount than chlorophyll. **Chlorophyll b is absent**.
8. The reserve food is oil, lipid and lucosin. **Starch is not formed**.
9. Plants reproduce commonly by vegetative and asexual means. Vegetative reproduction takes place by cell division. Asexual reproduction by zoospores, aplanospores or akinetes. Sexual reproduction, though rare, may be isogamous, anisogamous or oogamous. Isogamy is common. Both iso- and anisogamy are found in *Botrydium*.

Range of thallus

Members of Xanthophyceae show well-marked parallelism with Chlorophyceae in their thallus structure, includes motile, palmelloid, dendroids, filamentous and siphonaceous forms (Kumar and Singh, 1971). Members of this class show a wide range of forms. Like the members Chlorophyceae i.e. green algae, the plant body may be unicellular and motile (*Chlorochromonas*), palmelloid with a swimming phase (*Chlorosaccus*) or a rhizopodial type (*Chlorarachnion*). Some members also show a coccoid habit (*Botrydiopsis*). In some others, the plant body is palmelloid and non-motile (*Mischococcus*). Truly filaments are well represented by unbranched filaments (*Tribonema*), siphonaceous state represented either by coenocytic thalli (*Botrydium*). The following levels of organization are found in the class Xanthophyceae:

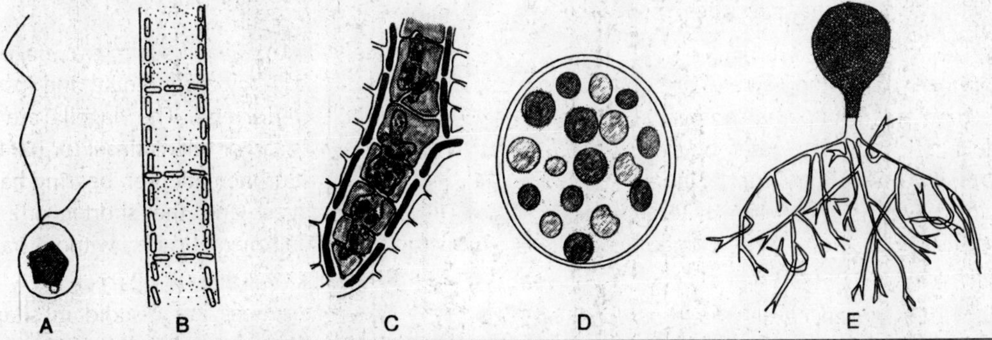

Fig. 3.120:Range of thallus organization in Xanthophyceae: A. *Chloromeson*, B. *Tribonema*, C. *Myxochloris*, D. *Botrydiopsis*, and E. *Botrydium*

1. Unicellular flagellate (monadoid) type. e.g., *Chloromeson*.
2. Amoeboid type. e.g., *Myxochloris, Rhizochloris*.
3. Palmelloid type. e.g., *Gloechloris*.
4. *Coccoid* type. e.g., *Ophiocytium, Characiopsis, Chloridella*.
5. *Filamentous* type. e.g., *Hetrodendron, Tribonema*.
6. Siphonaceous type. e.g., *Botrydium*.

Classification

Fritsch (1935) divided the class Xanthophyceae into four orders. The outline of the classification is:

Order. Heterochloridales (motile form).

Sub-order. Heterochlorineae

Family. 1. Heterochloridaceae e.g., *Heterochloris*.

Sub-order. Heterocapsineae

Family. 1. Heterocapsaceae e.g., *Chlorosaccus*.

Sub-order. Heterodendrineae

Family. 1. Mischococcaceae e.g., *Mischococcus*

Sub-order. Heterorhizidineae

Family. 1. Heterorhizidaceae e.g., *Stipitococcus*.

Order.Heterococcales (coccoid form)

Family. 1. Halosphaeraceae e.g., *Botrydiopsis*.

2. Myxochloridaceae e.g., *Myxochloris*.

3. Chlorobotrydaceae e.g., *Chlorobotrys*.

4. Chlorotheciaceae e.g., *Chlorothecium*.

5. Ophiocytiaceae e.g., *Ophiocytium*.

Order. Heterotrichales (filamentous)

Family. 1. Tribonemaceae e.g., *Tribonema*

2. Heterocloniaceae e.g., *Heterodendron*.

Order. Heterosiphonales (siphonous form)

Family. 1. Botrydiaceae e.g., *Botrydium*.

BACILLARIOPHYCEAE

(DIATOMS)

Bacillariophyceae is a large group of algae consisting of 200 genera and over 10,000 species, out of which 92 genera and about 569 species are reported from India. They are commonly known as **Diatoms**. The diatoms are the most beautiful microscopic algae due to their structure and sculpturing of their walls. They occur in various habitats like fresh water, saline water and also in terrestrial condition on or within the soil. Sometimes they also occur as epiphytes along with algae, on the leaf of forest trees, mostly in tropical rain forests. Depending on the mode of nutrition they may be photosynthetic autotrophs or photosynthetic symbionts or heterotrophs.

Important characteristics

The important characteristics of the class Bacillariophyceae are:

1. They are commonly unicellular and free-living but some members form colonies of various shapes like filaments, mucilaginous colonies etc.
2. Microscopic cells are of different shapes. They may be oval, spherical, triangular, boat-shaped etc.
3. Plant bodies are either bilateral or radial in symmetry.
4. The cells are surrounded by a rigid cell wall, called frustule, consisting of upper epitheca and lower hypotheca; arranged in the form of a box with its lid.
5. The cell wall is composed of pectic substances impregnated with high amount of siliceous substance.
6. The wall may have secondary structures like spines, bristles etc.
7. Vegetative cells are diploid (2n).
8. The cells generally have many discoid or two large plate-like chromatophores. Some cells possess stellate chromatophore.
9. The photosynthetic pigments are chlorophyll a, chlorophyll c along with xanthophylls like fucoxanthin, diatoxanthin and diadinoxathin.
10. Reserve food is oil, volutin and crysolamilarin.
11. Some vegetative cells show gliding movement.
12. Motile structure (antherozoid) has single pantonematic flagellum.
13. Vegetative multiplication takes place by cell division, which is very common. Some of the cells become very much reduced in size.
14. They produce characteristic spore, the auxospore which develops to regain the normal size.
15. Sexual reproduction takes place by isogamy and oogamy.

Classification

Fritsch (1935) and many others followed the classification of Karsten (1928) and Hustedt (1930). The outline of the classification is given:

Class. Bacillariophyceae

Order. Centrales

1. Discoideae. e.g., *Melosira*
2. Solenoideae. e.g., *Corethron*
3. Biddulphioideae e.g., *Biddulphia*
4. Rutilarioideae e.g., *Rutilaria*

Order. Pennales

(i) Araphideae

5. Fragilarioideae e.g., *Synedra.*

(ii) Raphidioideae

6. Eunotioideae e.g., *Eunotia.*

(iii) Monoraphideae

7. Achnanthoideae e.g., *Cocconeis*

(iv) Biraphideae

8. Naviculoideae e.g., *Navicula.*
9. Epithemioideae e.g., *Epiyhemia.*
10. Nitzschioideae e.g., *Bacillaria.*
11. Surirelloideae e.g., *Surirella.*

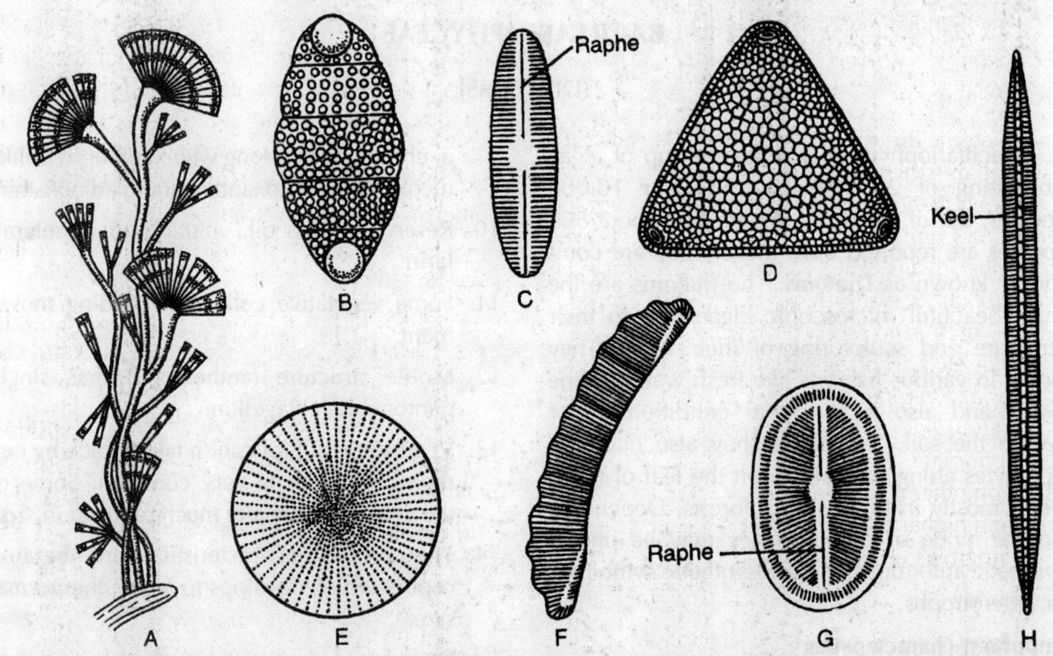

Fig. 3.121:Different species of Diatom: A. *Licmophora flabellata*, B. *Biddulphia pulchella*, C. *Achnanthes linearis*, D. *Triceratium planoconcavum*, E. *Cascinodiscus*, F. *Eunotia dp.*, G. *Cocconeis placentula* and H. *Bacillaria paradoxa*

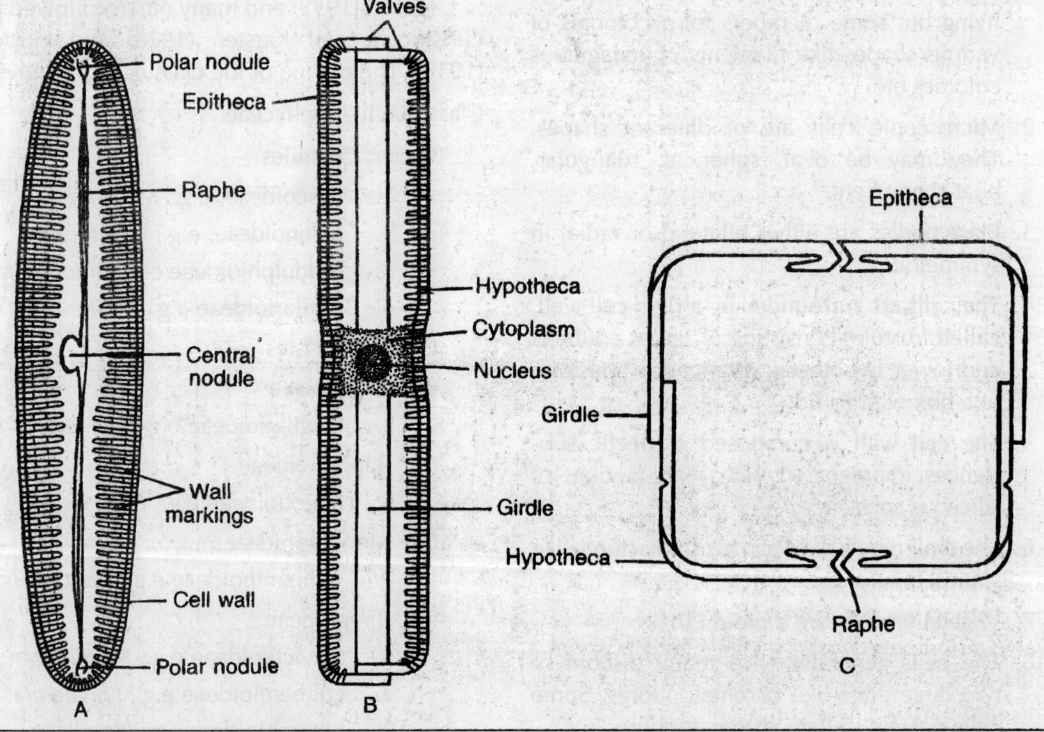

Fig. 3.122:Cell structure of *Pinnularia viridis* (Pennales): A. Frustule in valve view, B. Frustule in girdle view, and C. Frustule in transverse section.

The **General account** of Bacillariophyceae is given below:

Occurrence

Diatoms are found in all possible habitats. Commonly they are found in fresh water (*Denticula tenuis, Navicula pupula, Meridion circulare, Cymbella ventricosa, Melosira variens, Amorpha ovalis* etc.), sea water (*Corethron, Biddulphia, Sceletonema, Fragilaria, Tropidonensis* etc.) and soil (*Pinnularia, Navicula, Frustulia* etc.). The terrestrial species (*Amorpha, Navicula, Pinnularia* etc.) are able to withstand dessication for a long period. Some diatoms (*Gomphonima, Cymbella* etc.) can grow as epiphyte on other algae (*Enteromorpha, Cladophora* etc.) and higher plant. *Licmophora*, a member of diatom, grows endozoically.

Plant body

Plant body is unicellular, generally moves singly. The cells are of different shapes viz. round, oval, elongated, rod-shaped, triangular, disc-shaped etc. Sometimes they become aggregated and get embedded in a gelatinous matrix, but they do not behave like multicellular organisms. In colonial form the cells may be present as uniseriate row (e.g., *Melosira*), like a branched body (e.g., *Licmophora flabellate*) or other forms also (Fig. 3.121).

Cell structure. The cell consists of cell wall and protoplast (Fig. 3.122A, B, C).

1. ***Cell wall***. The cells are covered by a siliceous wall, the frustule. It consists of two overlapping halves, the theca. The upper one is **epitheca** and lower one is **hypotheca**. Both the theca consist of two portions:

(a) **Valve** — the upper flattened top and

(b) **Connecting band** or cingulum (pl. cingula) — the incurved region. The common region of the connecting bands, where both the theca remain fitted together, is the **girdle**. [When the diatoms are observed from the valve side i.e., valve side is uppermost, called the **valve view**, but when viewed from the connecting band, it is the **girdle view**.] Depending on symmetry, the cells are divided into two orders: **Pennales** (bilaterally symmetry) and **Centrales** (radially symmetry). In some pinnate diatoms (*Cybella cistula, Pinnularia viridis* etc.) an elongated slit is present on their valves, called **raphe**. The raphe is interrupted at its midpoint by thickening of the wall called **central nodule**. Similar thickening is also present at the ends called **polar nodules**. Some members like *Tabellaria fenestrate* etc. of the order Pennales, do not have raphe, called **pseudoraphe**.

Besides raphe or pseudoraphe, the cell walls have other types of openings, called pores and locules.

Based on electron microscopic studies, Hendey (1971) observed four basic-types of secondary structures. These are: ***Punctae*** (small perforations on valve surface), ***Canaliculi*** (tube-like narrow channels which run through the valve surface), ***Areolae*** (large box-like depressions) and ***Costae*** (rib-like structures on the valve surface).

The cell wall is mainly made up of pectic substances, impregnated with silica. The content of silica varies from 1% (*Phaeodactylum tricornutum*) to about 50% on the basis of dry weight of the cell.

2. ***Protoplast***. The entire content present inside the cell membrane is the protoplast. The cell membrane encloses a large central vacuole surrounded by cytoplasm. The cytoplasm contains single nucleus, mitochondria, golgi bodies and chloroplasts. The chloroplast may be of different shapes like stellate, H-shaped, discoid etc. In some species the chloroplasts contain pyrenoids.

The *photosynthetic pigments* are chlorophyll a, c_1 and c_2, β-carotene, fucoxanthin, diatoxanthin and diadinoxanthin. The latter two are present in small quantity (The *golden-brown colour* of *diatom cells* is due to the presence of xanthophylls like fucoxanthin, diatoxanthin and diadinoxanthin. The term **diatomin** is used for the mixture of chlorophyll and carotenoids, particularly carotene and several brown xanthophylls pigments.)

The *reserve food* of diatoms is **chrysolaminarin** and **oil droplets** (*they do not store in the form of starch*).

Locomotion. All diatoms with raphe are motile. Most of the members of the order Pennales contain raphe and perform gliding movement. The gliding movement is caused by the circulation of cytoplasm within the raphe by the release of mucilage. The rate of movement varies from 02-25 μm/sec. The locomotion is affected by temperature, light etc.

Reproduction

Diatom reproduces by vegetative and sexual means.

Vegetative reproduction

Vegetative reproduction performs with the help of cell division (Fig. 3.123). It takes place usually at midnight or in the early morning.

During cell division the protoplast of the cell enlarges slightly, thus the cell increases in volume and slightly separates both the theca (epitheca and hypotheca). Then the protoplast undergoes mitotic division and gets separated along the longitudinal axis through the median line. Thus one half of protoplast remains in **epitheca** and the other one in **hypotheca**. One side of the protoplast thus remains naked. Now both the theca i.e., epitheca and hypotheca of mother cell behave as epitheca of the daughter cells. Thus new silicious valves are deposited towards the naked sides of the protoplast and always behave as hypotheca of the daughter cells. Connecting bands are developed between the theca. Later on, the daughter cells get separated.

During cell division, both the theca i.e., epitheca and hypotheca of the mother cell behave as epitheca of the daughter cells. So at the side where the hypotheca behaves as epitheca, the cell becomes reduced in size. Thus with continuous cell division some cells gradually become reduced in size.

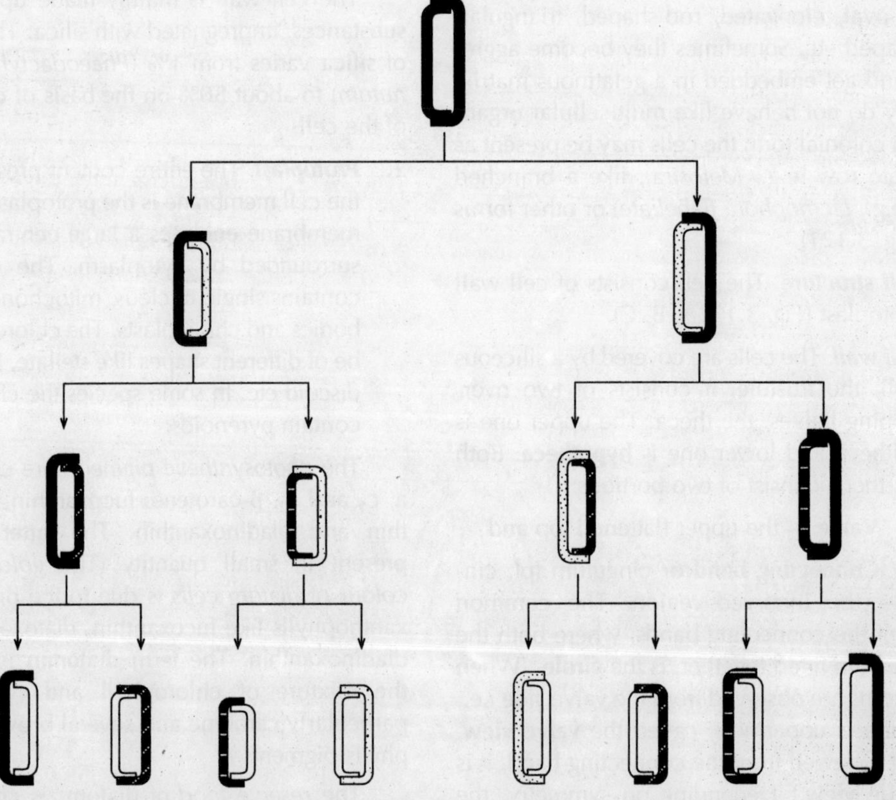

Fig. 3.123: Cell division in diatoms (diagrammatic) showing reduction in cell size in successive generations except the three (extreme right side cells)

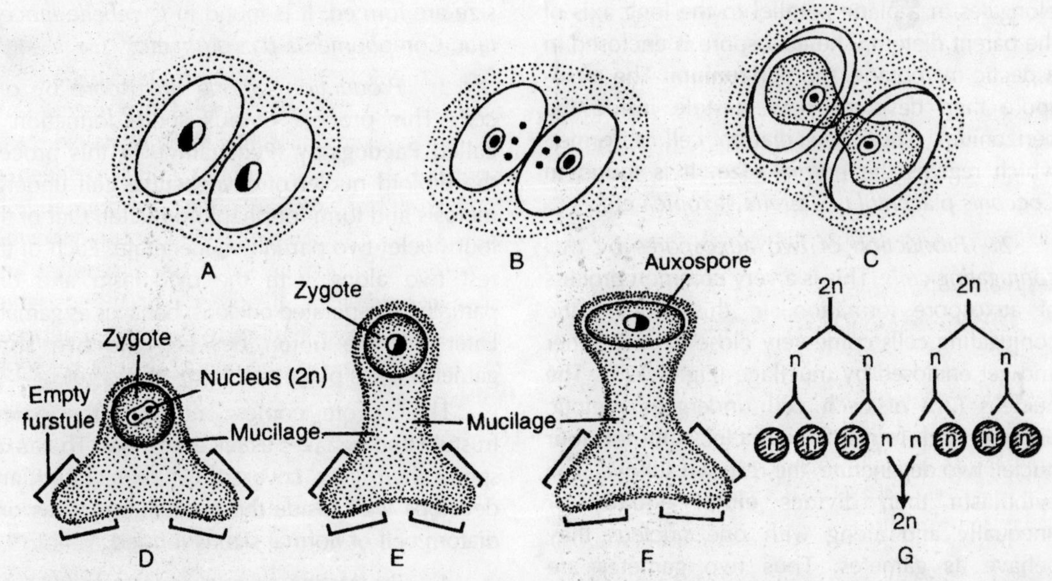

Fig. 3.124:A-F. Production of one auxospore by two conjugating cells of *Cocconeis placentula*, and G. Nuclear behaviour during reproduction

Sexual reproduction

The pattern of sexual reproduction differs in both orders — Pennales and Centrales. During this process, auxospore is formed in both the groups. *During cell division, those cells become reduced in size, are able to regain their normal size through the formation of auxospore, so it is a "restorative process" rather than multiplication*.

Auxospore formation in Pennales

It takes place through gametic union, autogamy and parthenogenesis. These are of the following types:

1. *Production of one auxospores by two conjugating cells*. In this process two uniting cells come very close to each other (Fig. 3.124) and become covered by a mucilaginous sheath. The diploid nucleus of each cell undergoes meiosis. Out of four nuclei, three degenerate and only one survives. The surviving nucleus behaves as gamete (n). The gametes come out from the parent frustules and unite together, to form a zygote (2n). After a short period of rest the zygote elongates considerably and functions as an auxospore. The auxospore projects out from the parent frustules along with mucilage and

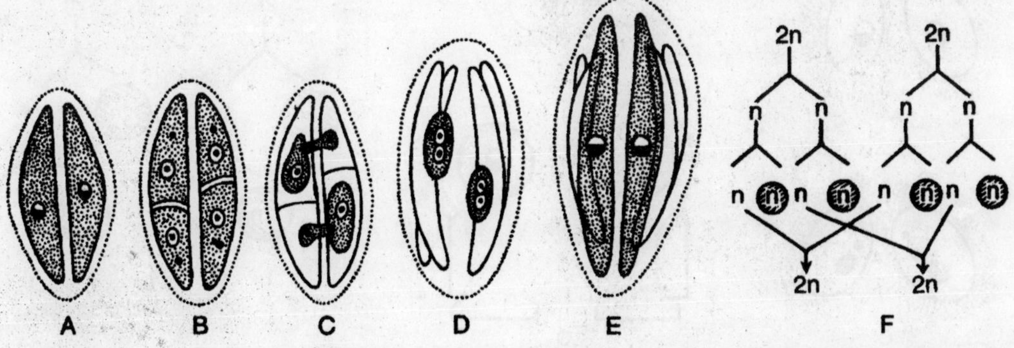

Fig. 3.125:*Cymbella lanceolata*: A-E. Production of two auxospore by two conjugating individuals, and F. Nuclear behaviour during reproduction

elongates in a plane parallel to the long axis of the parent diatom. The auxospore is enclosed in a pectic membrane, the **perizonium**. The auxospore then develops new frustule inside the perizonium. Thus new diatom cell is formed which regains the normal size. It is found in *Cocconis placentula, Surirella saxonica* etc.

2. *Production of two auxospores by two conjugating cells*. This is a very common process of auxospore formation. In this process the conjugating cells come very close to each other and get enclosed by mucilage (Fig. 3.125). The nucleus (2n) of each cell undergoes meiotic division and forms four nuclei. Out of four nuclei, two degenerate, the rest two survive. The cytoplasm then divides either equally or unequally and along with one nucleus they behave as gametes. Thus two gametes are formed in each cell.

The pattern of union between the gametes varies from species to species. Both the gametes of a cell may be active and fuse with the gametes of other cell, thus two zygotes are produced in a single cell or out of two, one becomes active and fertilizes with the opposite one and thus one zygote is produced in each cell.

The zygotes elongate and function as auxospores. The auxospores develop the **perizonium** around themselves and both of them develop new frustules on their outer sides i.e., inside the perizonium. Thus two diatom cells of normal size are formed. It is found in *Cymbella lanceolata, Comphomema parvulum* etc.

3. *Production of one auxospore by one cell*. This process of auxospore formation is called **Paedogamy** (Pedogamy). In this process the diploid nuclei of a vegetative cell undergo meiosis and form four haploid nuclei. Out of the four nuclei two partially degenerate. Each of the rest two along with the cytoplasm and one partially degenerated nucleus, behaves as gamete. Later on, the union between the two sister gametes takes place and forms the zygote.

The zygote comes out from the parent frustule and behaves as an auxospore. The auxospore then gets covered by perizonium and develops wall inside the perizonium. Thus one diatom cell of normal size is formed.

4. *Production of one auxospore by autogamy*. In this process the diploid nucleus undergoes first meiotic division. Thus two haploid nuclei are formed. The two nuclei in the protoplast come side by side, fuse together and form diploid (2n) nucleus. This is called autogamous pairing.

The protoplast along with diploid (2n) nucleus comes out from the parent frustule and behaves as an auxospore. The auxospores are then covered by perizonium. New wall develops on the auxospore inner to the perizonium. Thus a new individual of normal size is developed. This is found in *Amphora normani*.

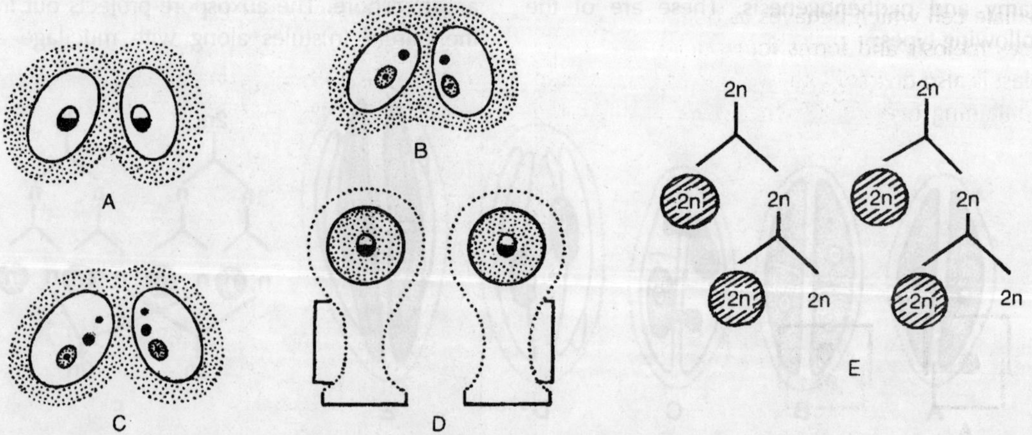

Fig. 3.126: A-D. Production of auxospore by parthenogenesis in Pennales, and E. Nuclear behaviour during reproduction

5. *Production of auxospore by parthenogenesis*. The diatom cells come together and are covered by a common mucilage envelop (Fig. 3.126). The diploid nucleus undergoes two sequential mitotic divisions. **Meiotic division does not take place here**. One nucleus in each mitotic division degenerates. Thus only one diploid (2n) nucleus along with protoplast remains, and comes out from the mother cell and behaves as an auxospore.

The auxospore is then covered by perizonium and secretes new wall around itself. Thus normal sized cell is formed.

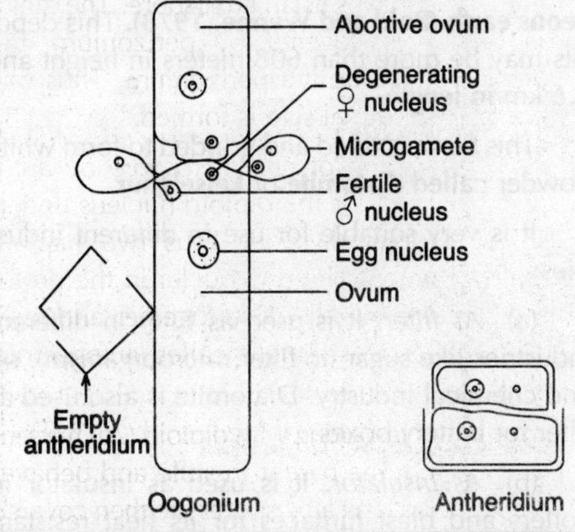

Fig. 3.127: Production of auxospore by oogamy in Pennales

6. *Production of auxospore by oogamy*. In this process (Fig. 3.127) the nucleus (2n) of female cell which behaves as oogonium, undergoes meiosis and forms four nuclei. The protoplast is also divided into two unequal parts, each containing two nuclei. The lower half is larger and behaves as functional ovum and the upper smaller one as non-functional ovum. The functional ovum contains one functional nucleus and one non-functional nucleus, which gradually degenerates at maturity.

The male cell (2n) behaves as antheridium, also undergoes meiosis and forms four nuclei. The protoplast also divides into two parts. Thus two microgametes are formed. Each of which contains two nuclei, of which one is functional and other is non-functional. The microgametes are naked, globular and non-flagellate.

After coming out, the male gamete fertilizes the egg and forms the zygote (2n). Later it functions as an auxospore and forms new individual of normal size. It is found in *Rhabdonema adriaticum*.

Auxospore formation in Centrales

It takes place by autogamy and oogamy:

1. *Auxospore formation by autogamy*. The protoplast of the vegetative cell (Fig. 3.128) secretes mucilage which separates both the theca. The nucleus (2n) then undergoes meiosis and forms four nuclei. Of the four nuclei two degenerate and the other two undergo fusion to form diploid (2n) nucleus again. This is called **autogamy**. The protoplast with 2n nucleus functions as an auxospore. The auxospore forms fresh frustule inside the perizonium covering and forms cell of normal size. It is found in *Melosira nummuloides*.

2. *Auxospore formation by oogamy*. Oogamy takes place by the fusion of egg and sperm developed inside the oogonium and antheridium respectively (Fig. 3.129).

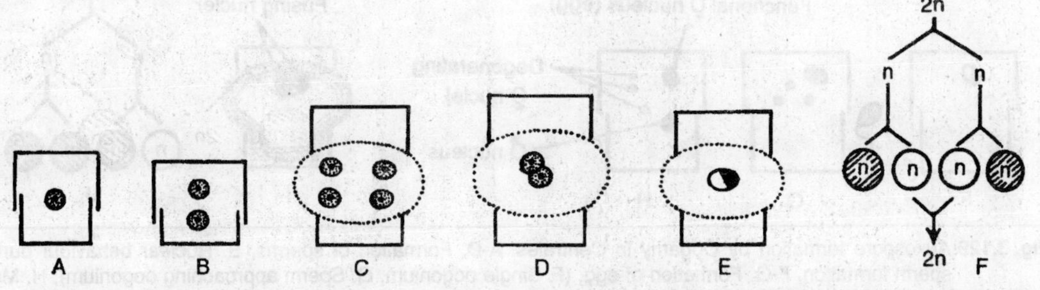

Fig. 3.128: A-E. Auxospore formation by autogamy in Contrales, and F. Nuclear behaviour during reproduction

Oogonium. Single vegetative cell behave as an oogonium. The protoplast of **oogonium** undergoes **meiotic** division and forms four nuclei. Of the four nuclei three degenerate and the remaining one functions as an **egg**.

Antheridium. The pattern of development of sperms varies in different species. In species like *Melosira* varians the protoplast undergoes **meiotic** division and forms four haploid nuclei. Each haploid nucleus with some protoplast metamorphoses into an uniflagellate (tinsel type) sperm. In others the number of sperms may go up to 8 or even 128.

Fertilization. After coming out of the antheridium only one sperm enters inside the oogonium and fertilizes the egg. The resultant zygote undergoes mitotic division but one nucleus degenerates in each division. The remaining nucleus with its protoplast behaves as an auxospore. The auxospore then develops new wall inside the perizonium covering and forms new cell of normal size like the mother. It is also called **firstling cell**.

From the above processes of sexual reproduction in both pennales and centrales, it becomes clear that the sexual process in diatom does not lead to multiplication but is to regain the normal size.

3. *Resting spores*. These spores are formed during unfavourable conditions. Some members reproduce by the formation of thick-walled resting spores, the **cysts** or **statospores**. They are formed in *Melosira*.

Economic importance

The diatoms are used in various purposes either directly or indirectly. The different uses of diatoms are:

1. **Diatomite**. After the death of diatom cells the outer coverings i.e., the silicified walls become accumulated at the bottom of water. The accumulation may be thicker during favourable conditions. These deposits are called **diatomaceous earth (Bold and Wynne, 1978)**. This deposits may be more than 608 meters in height and 1.6 km in length.

This earth is dried and grinded to form white powder called **diatomite** or **keiselghur**.

It is very suitable for use in different industries:

(a) *As filter*. It is used as filter in different industries like sugar (to filter microorganism), oil and chemical industry. Diatomite is also used as filter for battery boxes.

(b) *As insulator*. It is used as insulator in boilers and blast furnaces for its heat resistant ability.

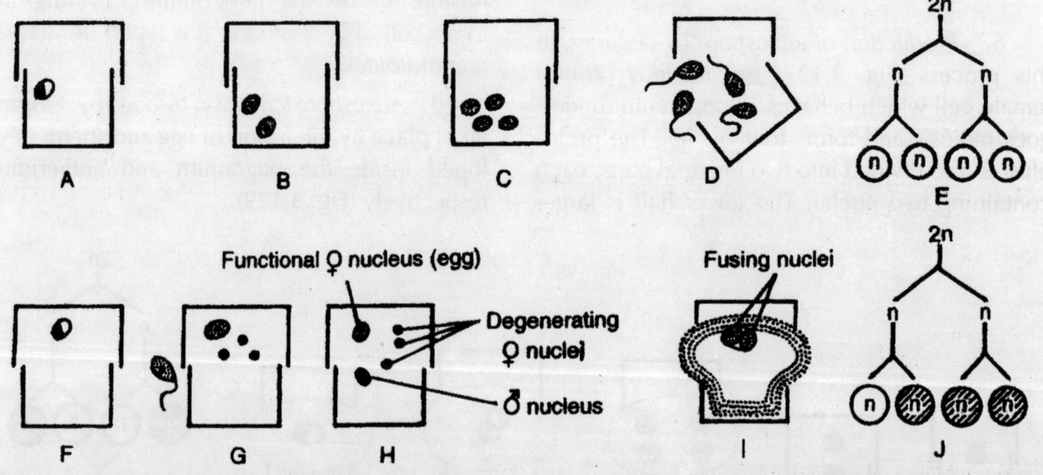

Fig. 3.129:Auxospore formation by Oogamy in Centrales: A-D. Formation of sperms, E. Nuclear behaviour during sperm formation, F-G. Formation of egg, (F. Single oogonium, G. Sperm approaching oogonium), H. Male nucleus entered inside the oogonium, I. Fusing male and female nuclei and J. Nuclear behaviour during Egg formation

(c) *As absorbent*. It is used as absorbent of liquid nitroglycerine.

(d) *Other uses*. Diatomite is used as abrasive (i.e., capable of rubbing or grinding down) substance for the manufacture of metal paints, polish, varnish, toothpaste etc. It is also used with bakelite for electrical fuse and switch boxes.

2. **Petroleum**. Much of the petroleum is considered to be of diatom origin as they are found in association with large oil deposits.

3. **Food**. Due to their great abundance in the different seas and their use as food by marine animals, they are called the 'grasses of the sea'. Those animals may be consumed as food by man and maintain the food chain.

4. **Testing of microscopic lenses**. Due to the fine markings on shell, the diatom cells are used to test microscopic lenses.

PHAEOPHYCEAE

(BROWN ALGAE)

Phaeophyceae is a large group of algae consisting of 240 genera and over 1,500 species, out of which 32 genera and 93 species are reported from India (Mishra, 1966). They are commonly known as **brown algae**, due to the presence of a golden brown xanthophyll pigment, **fucoxanthin** ($C_{40}H_{54}O_6$) in the chromatophores.

Occurrence

About 99.7% members are marine (Chapman and Chapman, 1973) and a few grow in fresh water. The **marine** members are *Macrocystis, Nercocystis, Postelsia, Laminaria, Sargassum* etc. The **fresh water** members are *Pleurocladia, Heribaudiella, Pseudobodanella, Lithoderma* and *Sphacelaria*. Members like *Pleurocladia lacustris* grow **both in fresh water and marine** habitats.

Important characteristics

The important characteristics of the class Phaeophyceae are given below:

1. Plant body is **immobile, multicellular** and highly differentiated both externally and internally. [Unicellular, colonial (motile and non-motile) and unbranched filamentous forms are completely absent).
2. They range from simple **microscopic heterotrichous filament** (*Ectocarpus*) **to largest alga** (*Macrocystis pyrifera*), which attains a length of 60-90 meters. (The largest forms are known as kelps or rockweeds. *Lessonia flavicans* reachs a length of 4 meters and looks like a miniature tree. *Nereocystis luelkeana,* the bladder kelp which attains a length of 25-30 meters. *Postelsia palmaeformis* appears like a palm tree and commonly known as Sea Palm) (Fig. 3.130).
3. Commonly the plant body is differentiated into **hold fast**, a short or elongated **stipe** and an expanded **blade**. The blade performs photosynthesis and bears reproductive structures. Many species remain afloat by having air bladders.
4. The photosynthetic pigments include **chlorophyll a, chlorophyll c, β-carotene** and **xanthophylls** like lutein, fucoxanthin, flavoxanthin and violaxanthin. The **fucoxanthin** is however present sufficiently which partially mask the chlorophyll and carotenoid, thereby giving the characteristic brown colouration.
5. The growth of the plant body may be **apical** (Fucales, Dictyotales), **intercalary** (Laminariales) or **trichothallic** (Ectocarpales).
6. The cell wall is differentiated into outer and inner layers. The outer mucilaginous layer has **fucinic** and **alginic acid**, but the inner layer is mainly **cellulosic**. [The alginic acid is used to manufacture artificial silk and adhesive, obtained commercially from *Sargassum, Laminaria* etc].
7. The cells usually have many small vesicles and white granules. The granules are called **fucosan vesicles.**
8. Pyrenoides are usually absent, but, if present, is of single stalk type.
9. Motile structures (zoospores and gametes) have **two laterally inserted unequal flagella,** of which larger one is tinsel or pantonematic and the smaller one is whiplash or acronematic type.
10. The reserve foods are commonly **laminarin** and **mannitol**. Sucrose and glycerol are also present in some members.
11. They reproduce by all the three means vegetative, asexual and sexual.
    (a) Vegetative reproduction takes place by fragmentation. Special reproductive branches, the **propagules**, are developed in some members of Sphacelariales;

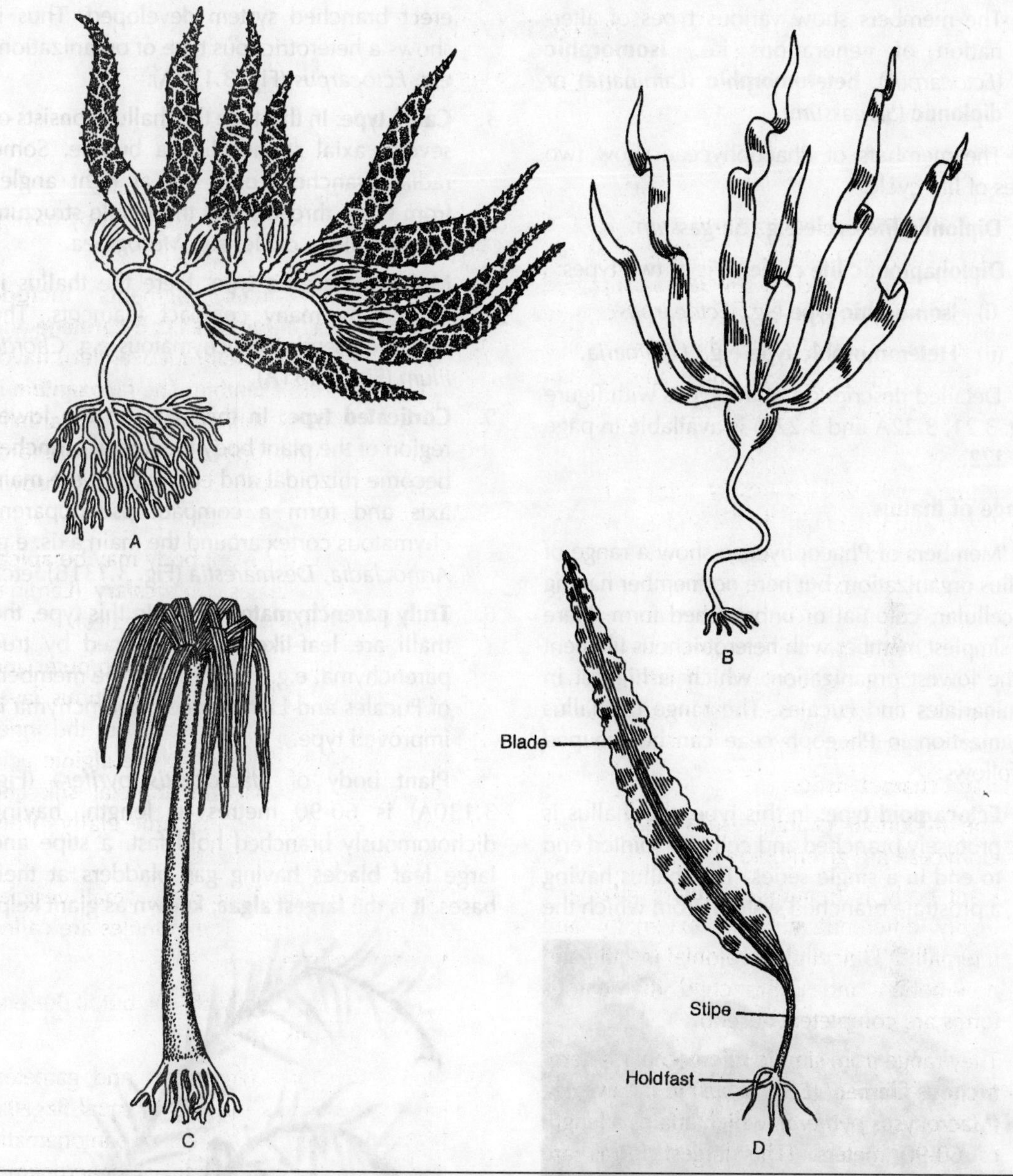

Fig. 3.130:Some marine members of Phaeophyceae: A. *Macrocystis pyrifera* (largest alga), B. *Nereocystis luelkeana* (bladder-kelp), C. *Postelsia palmaeformis* (sea-palm) and D. *Laminaria saccharina* (Devil's apron).

those develop to new plants after detachment.

(b) Asexual reproduction takes place by **zoospores** except Tilopteridales, Dictyotales and Fucales. The zoospores produced in **unilocular sporangia are haploid, while in pleurilocular sporangia they are diploid.**

(c) Sexual reproduction ranges from **isogamy** (Ectocarpales and Sphacelariales) to **oogamy** (Fucales, Dictyotales and Laminariales) through **anisogamy** (Cutleriales and Tilopteridales).

12. In most of the members fertilization is external. Zygote does not undergo meiotic division and on germination it develops diploid thallus.

13. The members show various types of alternation of generations i.e., **isomorphic** (*Ectocarpus*), **heteromorphic** (*Laminaria*) or **diplontic** (*Sargassum*).

The members of Phaeophyceae show two types of life cycle:

1. **Diplontic life cycle** e.g., *Sargassum*.
2. **Diplohaplontic life cycle**. It is of two types:
   (i) **Isomorphic type** e.g., *Ectocarpus*.
   (ii) **Heteromorphic type** e.g., *Laminaria*.

Detailed description of the types with figure (Fig. 3.21, 3.22A and 3.22B) is available in page no. 222.

Range of thallus

Members of Phaeophyceae show a range of thallus organization, but here no member having unicellular, colonial or unbranched form. Here the simplest member with heterotrichous filament is the lowest organization, which is highest in Laminariales and Fucales. The range of thallus organization in Phaeophyceae can be grouped as follows:

2. **Ectocarpoid type**: In this type, the thallus is profusely branched and cells are jointed end to end in a single series. The thallus having a prostrate branched system, from which the erect branched system developed. Thus it shows a heterotrichous type of organization. e.g. *Ectocarpus* (Fig. 3.132A).
3. **Cable type:** In this type the thallus consists of several axial strands like a bundle. Some radial branches come out at right angles from these threads, and the whole structure appears like a cable. e.g. *Mesogloea*.
4. **Multiseptatecable type**: Here the thallus is formed by many compact filaments. The cells are pseudoparenchymatous e.g. *Chorda filum* (Fig. 3.131A).
5. **Corticated type**: In this type, in the lower region of the plant body the lateral branches become rhizoidal and coil around the main axis and form a compact pseudoparenchymatous cortex around the main axis. e.g. *Arthocladia, Desmarestia* (Fig. 3.131B). etc.
6. **Truly parenchymatous type**: In this type, the thalli are leaf-like lamina formed by true parenchyma. e.g. *Punctaria*. In the members of Fucales and Laminariales, parenchyma is improved type.

Plant body of *Macrocystis pyrifera* (Fig. 3.130A) is 60-90 metres in length, having dichotomously branched hold fast, a stipe and large leaf blades having gas bladders at their bases. It is the **largest algae**, known as giant kelp.

Fig. 3.131: Range of thallus in Phaeophyceae: A. *Chorda filum* and B. *Desmarestia ligulata*

Plant body of *Nereocystis* is 20-25 meters long, having a haptera, a stipe and terminal large gas bladder having many blades on it. Other members like *Fucas, Sargassum, Laminaria* (Fig. 3.130D)., *Postelsia* etc., all having true parenchymatous structures.

Classification

Fritsch **(1935, 45)** classified the Class Phaeophyceae into nine orders. This was also followed by Mishra (1966). The outline of the classification is given below:

Class. Phaeophyceae

Order. 1. **Ectocarpales** e.g., *Ectocarpus, Halothrix*.
2. Tilopteridales e.g., *Ptilopteris*.
3. Cutleriales e.g., *Cutlria*.
4. Sporochnales e.g., *Sporochnus*.
5. Desmarestiales e.g., *Desmarestia*.
6. **Laminariales** e.g., *Laminaria*.
7. Sphacelariales e.g., *Sphacelaria*.
8. Dictyotales e.g., *Dictyota*.
9. **Fucales** e.g., *Sargassum*.

Only the orders Ectocarpales, Laminariales and Fucales are considered below for further discussion.

Order. **Ectocarpales**

Important characteristics

1. The plant body is filamentous, branched and heterotrichous, a few are pseudoparenchymatous.
2. In filamentous form the cells are arranged uniseriately.
3. The plant body elongates by intercalary growth.
4. Each cell contains chloroplasts with pyrenoids.
5. Reproduction takes place both asexually and sexually. Asexual reproduction takes place by zoospores formed in pleurilocular sporangia. Sexual reproduction takes place by both isogamy and anisogamy, though anisogamy is rare. **Oogamy is absent**.
6. They perform either isomorphic or heteromorphic alternation of generations.

Family. **Ectocarpaceae**

Important characteristics

1. The thalloid plant body is filamentous, branched and heterotrichous.
2. Sexual reproduction is isogamous.
3. Alternation of generation is isomorphic.

Life history of *Ectocarpus* is given below.

ECTOCARPUS

Systematic position

Fritsch (1945)	**Lee** (1999, 2008)
	Phylum. Heterokontophyta
Class. Phaeophyceae	Class. Phaeophyceae
Order. Ectocarpales	Order. Ectocarpales
Family. Ectocarpaceae	Family. Ectocarpaceae
Genus. *Ectocarpus*	Genus. *Ectocarpus*

Occurrence

Different species of the genus *Ectocarpus* (Gr. *ekos*, external and *kapos*, fruit) is found throughout the world, out of which 16 species are found in India (Mishra, 1966). They **grow in marine habitat**, either **free-floating, epiphytes** (on other sea plants) or **lithophytes** (on rocks). They are commonly available in both tropical and temperate seas. In India they are commonly found in the western coast. Species like *E. spongiosus* and *E. conigerare* are **free-floating**, *E. breviarticulatus* and *E. coniferus* grow on larger algae like *Laminaria* and *Fucus* as **epiphytes**. *E. dermonematis* is **endophytic** and *E. fasciculatus* is an **epizoic** species, grows on the fins of fishes. Galls are formed in *Laminaria* and *Cystoseria* due to the presence of *E. deformans* and *E. valiantei* respectively as **endophytes**. They are the **most primitive** of all the members of the class Phaeophyceae.

Plant body

The plant body is **filamentous** and **heterotrichous**, which is differentiated into prostrate and erect systems (Fig. 3.132A, B).

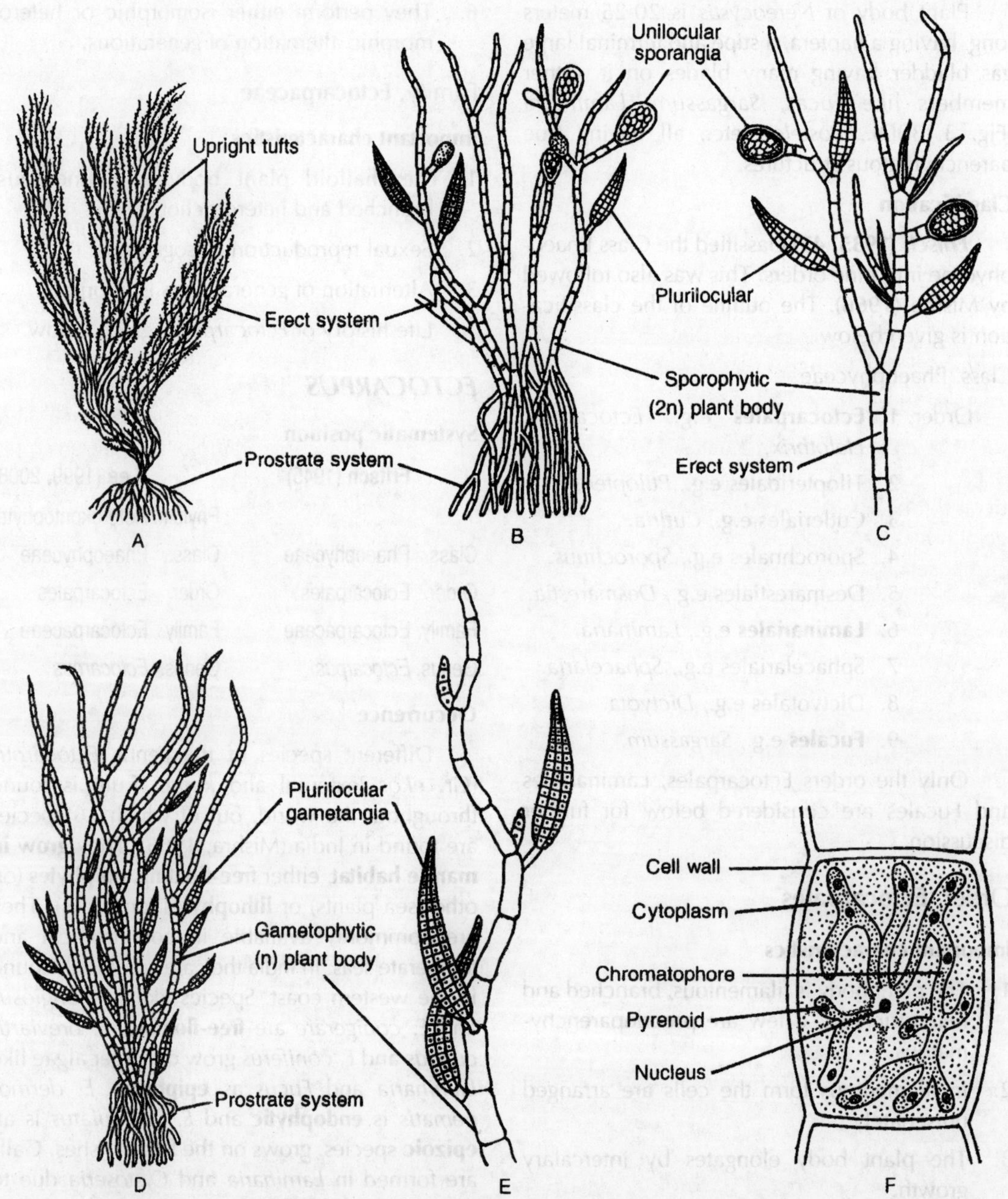

Fig. 3.132: *Ectocarpus* sp.: A. Habit, B. Sporophytic (2n) plant body with both erect and prostrate system bearing unilocular and plurilocular sporangia, C. Portion of erect filament of sporophytic (2n) plant, D. Gametophytic (n) plant with both erect and prostrate system bearing plurilocular gametangia, E. Portion of erect filament of gametophytic (n) plant and F. Cell structure

The prostrate system is profusely branched and attached with the substratum. Some species like *E. filifer* etc, develop multicellular hairs from the prostrate system.

The erect system develops from the prostrate system which remains free-floating. It is much branched and well-developed. Both the main axis and branches are uniseriate (**monosiphonous**), but the lower part may become multiseriate (**polysiphonous**) due to longitudinal division e.g., *E. geminifructus*.

The prostrate system serves the function of anchorage with the substratum or on other plants and the erect system is photosynthetic and bears reproductive organs. The apical part of each filament generally terminates into hairs.

Two types of plant bodies are differentiated genetically, one is haploid and other one is diploid. **Both haploid and diploid plants are morphogically identical** (Fig. 3.132B, D).

Cell structure

The cells are rectangular or cylindrical (Fig. 3.132F). Cell wall is differentiated into outer pectic and inner cellulosic layers. The characteristic gelatinous substance present in the cell wall is composed of **algin** and **fucoiden**. Inner to the wall, cell membrane is present which encircles the protoplast. The protoplast contains one central nucleus and many chromatophores. The number and shape of chromatophore varies with species. They may be ribbon-shaped, band-shaped, discoid etc. and are associated with pyrenoids.

The photosynthetic pigments are chlorophyll a, chlorophyll c, β-carotene and fucoxanthin. The fucoxanthin masks the chlorophyll and gives the characteristic brown colouration.

The cytoplasm contains many vacuoles, which are called **physodes**. The physodes contain polyphenols, probably function as lysosomes. Morphologically, haploid and diploid vegetative filaments are almost alike, but **the cells of haploid filaments are comparatively shorter in length than the diploid filaments**.

Growth

The nature of growth varies with the region of plant body and also with species. The erect system shows intercalary, diffuse or trichothallic growth, whereas it is apical in prostrate system.

Reproduction

Ectocarpus reproduces both asexually and sexually:

Asexual reproduction. It takes place by zoospores. The zoospores are biflagellate having one whiplash and other tinsel-type of flagella. **The diploid plant (2n) develops two types of sporangia**. These are unilocular sporangia and pleurilocular or neutral sporangia (Fig. 3.132 B, C). **The *unilocular sporangia* develop haploid zoospores i.e., zoomeiospores, but the *plurilocular* sporangia develop diploid zoospores.**

1. **Unilocular sporangia**. The unilocular sporangia develop from the apical cell of short lateral branches (Fig. 3.133A). The apical cell enlarges and functions as **sporangial initial**. The diploid nucleus of the sporangial initial first undergoes meiosis followed by several mitotic divisions, thus 32-64 haploid nuclei are formed. These nuclei accumulate some cytoplasm and develop individual units (Fig. 3.133 B-D). Each unit metamorphoses into a **pyriform, uninucleate, biflagellate zoospore also called zoomeiospore** (Fig. 3.133 E). The flagella are unequal and laterally inserted. Out of two flagella, posterior one is short and whiplash type (i.e., acronematic) and the anterior one is larger and tinsel type (i.e., pantonematic).

During liberation the apex of the sporangium wall gets dissolved and the haploid zoospores are liberated in a gelatinous mass. After some time they get free from the gelatinous mass and swim freely in water. They remain motile for about 30 minutes.

On contact with suitable substratum, the zoospore withdraws its flagella and forms a new cell wall around it (Fig. 3.133 F). Within short time a germ tube is formed, which divides many times and form prostrate filament (Fig. 3.133 G, H). Some cells of the prostrate filament become active and form erect filaments.

Thus the plants developed on germination of haploid zoospores (i.e., zoomeiospores) are gametophytic.

2. **Plurilocular sporangia**. These are elongated, multicellular body developed on diploid (2n) i.e., sporophytic plant body (Fig. 3.134). They develop initially like the unilocular sporangia, at the tip of short lateral branches. The apical cell enlarges and functions as sporangial initial. This sporangial initial becomes enlarged

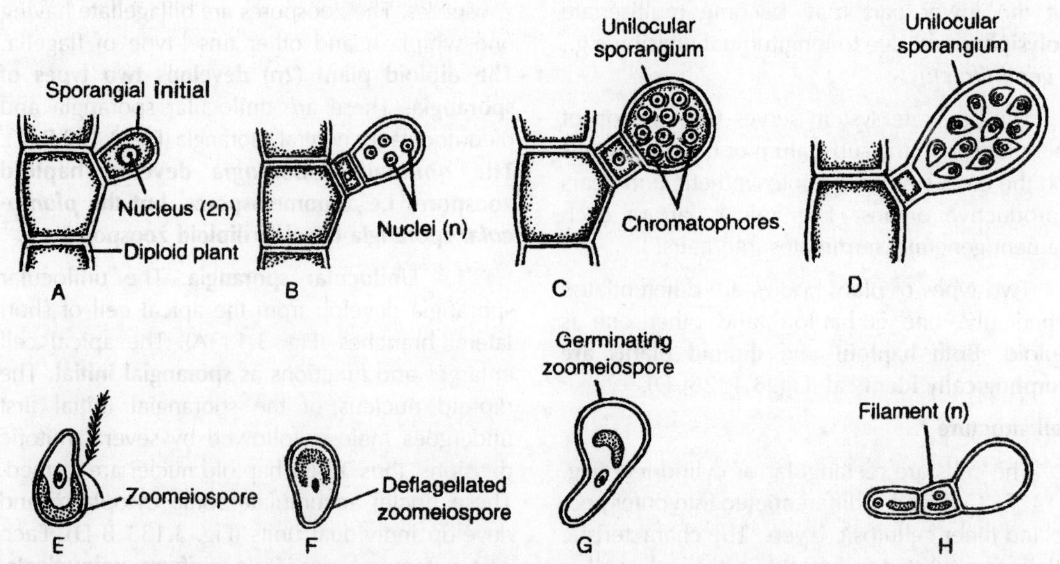

Fig. 3.133: *Ectocarpus* sp.: A-D. Development of unilocular sporangium, E. Zoomeiospore, and F-H. Germination of Zoomeiospore

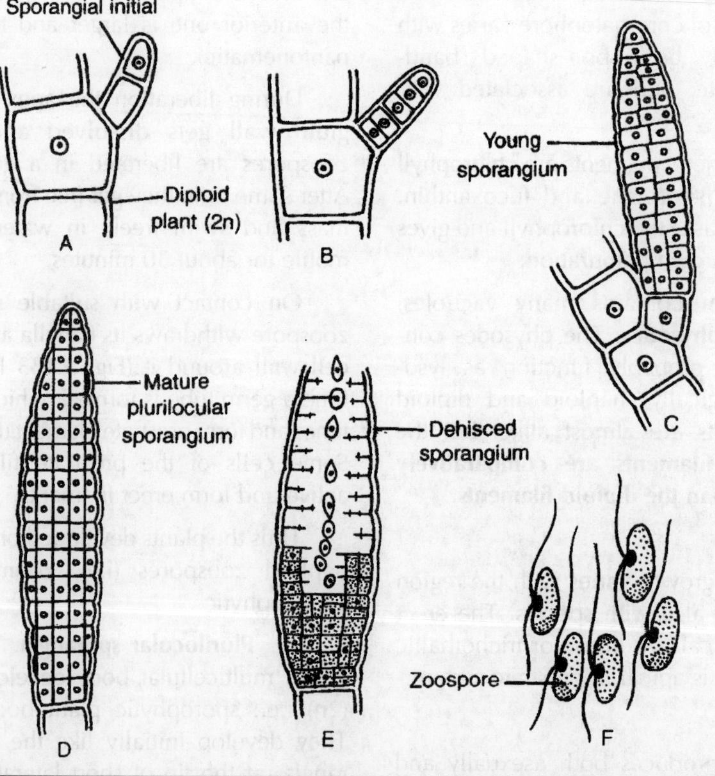

Fig. 3.134: *Ectocarpus* sp.: A-D. Development of multilocular sporangium, E. Liberation of zoospores from sporangium, and F. Zoospores

and undergoes repeated mitotic division, thus 6-12 cells are formed. The cells are arranged in vertical row (Fig. 3.134 A, B). The cells then undergo several vertical divisions thus a multi-chambered structure is formed i.e., the **plurilocular sporangium** (Fig. 3.134 C, D). There is **no reduction division during the formation of zoospore** in plurilocular sporangium. So, each small cubical cells of plurilocular sporangium contains single **diploid nucleus**. Each unit then metamorphoses into a single, uninucleate (2n) and biflagellate zoospore.

The **zoospores formed from plurilocular sporangia** are alike with the zoospores i.e., zoomeiospores developed in unilocular sporangia but are **diploid**. The zoospores of plurilocular sporangia liberate through apical (Fig. 3.134E) or lateral aperture and on germination they **produce the sporophytic (2n) plant**.

Sexual reproduction

The sexual reproduction is both ***isogamous*** and ***anisogamous*** type. **Oogamy is absent.** Anisogamy is very common. Anisogamy may be of two types: morphological anisogamy (*E. secundus*) and physiological anisogamy (*E. siliculosus*). The gametes are produced inside the plurilocular gametangia, developed on haploid plants.

Plurilocular gametangia. They are large, elongated, sessile or short stalked, multicellular structures (Fig. 3.135A). Morphologically, both plurilocular gametangia and plurilocular sporangia are alike. The plurilocular gametangia produce haploid gametes; on the other hand plurilocular sporangia produce diploid zoospores. Though both are morphologically more or less alike, the gametes are slightly smaller in size than the zoospores. The development of plurilocular gametangia is alike with the development of plurilocular sporangia.

The gametes are liberated from the gametangia following the same procedure as that of zoospore liberation from the plurilocular sporangia (Fig. 3.135).

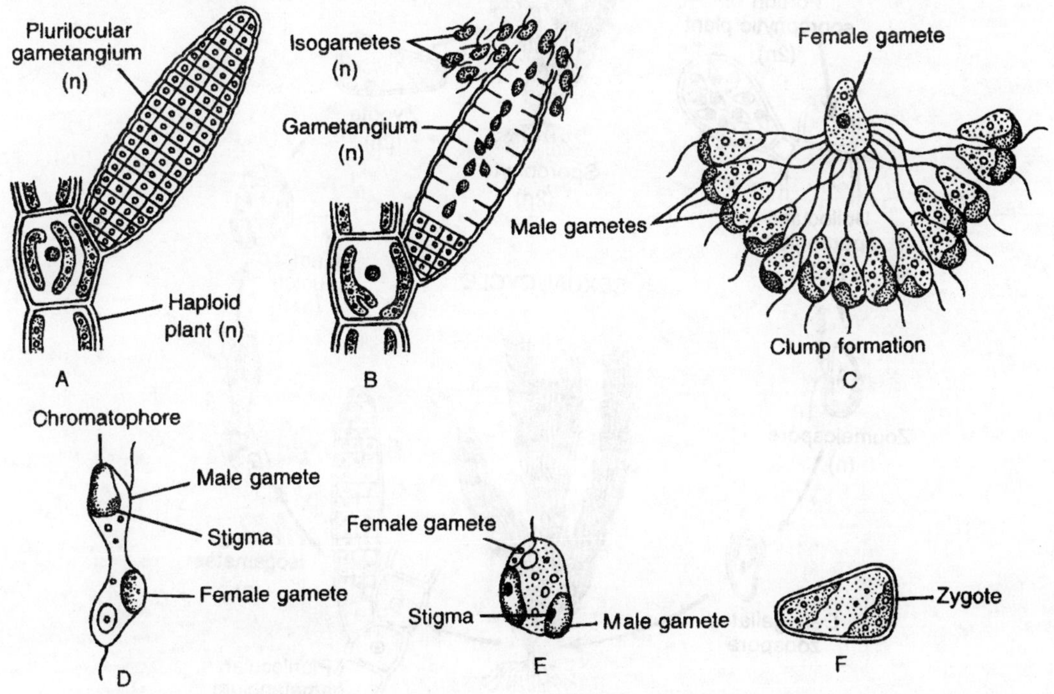

Fig. 3.135: *Ectocarpus* sp.: A. Portion of gametophytic (n) plant with plurilocular gametangium, B. Liberation of gametes from gametangium, C. Male gametes around a female gamete before fertilization (clump formation), D-E. Fusion of male and female gametes and F. Single zygote

Fertilization

Majority of the species show physiological anisogamy (Fritsch, 1945), but morphological anisogamy is observed in *E. secundus* (Sauvagean, 1896). In physiological anisogamy both the uniting gametes are morphologically similar but in morphological anisogamy female gamete is larger than the male gametes.

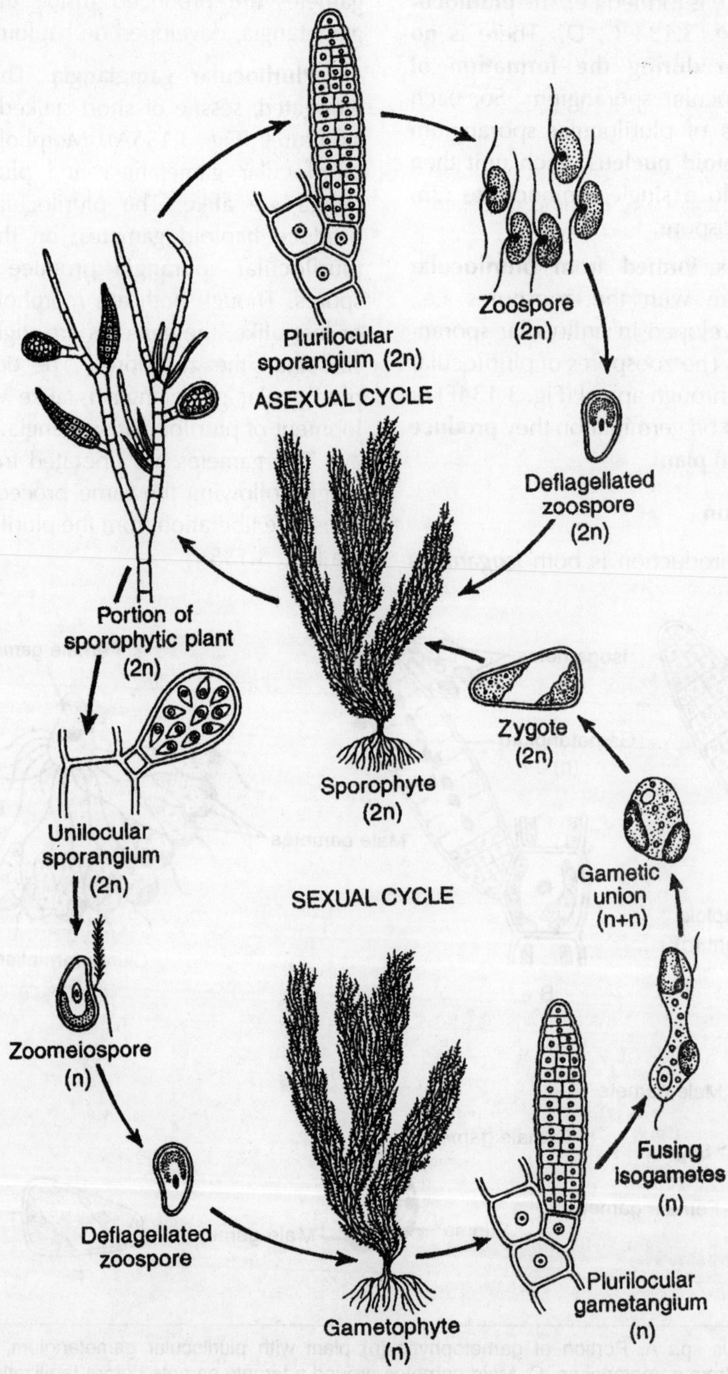

Fig. 3.136: Life cycle of *Ectocarpus* sp.

During fertilization, many male gametes encircle the female gamete and get entangled by the anterior large flagellum. This stage is called **clump formation** (Fig. 3.135C). Out of many, only one male gamete fuses with the female gamete (Fig. 3.135D, E) and. the remaining gametes go astray and gradually get destroyed.

The uniting gametes then form zygote (Fig. 3.135F), through plasmogamy and karyogamy.

Germination of zygote. The zygote undergoes germination **without any reduction division** and rest. On germination it develops into a sporophytic (2n) plant. The sporophytic plant again develops unilocular and plurilocular sporangia. The life cycle is diplontic type (Fig. 3.136 and 3.137).

Indian species

Ectocarpus arabicus, E. filife, E. enhali, E. coniger, E. zeylanicus, E. rhodochortonoides etc.

Important features

1. It is a marine brown alga, distributed throughout the temperate and tropical seas of the world.
2. Plant body is filamentous, much branched and heterotrichous, having basal rhizoids and well-developed branched erect system.
3. Both sporophytic and gametophytic plants are alike (isomorphic).
4. The sporophytic plant bears both plurilocular and unilocular sporangia. The plurilocular sporangium produces zoospores (2n) through mitosis. They germinate to produce new-diploid sporophytic plant. On the other hand, unilocular sporangium produces zoomeiospore through meiosis, followed by several mitotic divisions. Zoomeiospores on germination develop gametophytic plants.
5. The gametophytic plants bear plurilocular gametangia (look similar to plurilocular sporangia) produce gametes.
6. Sexual reproduction may be isogamous or anisogamous type. Physiological anisogamy is very common.
7. Direct germination of zygote results in the formation of a diploid sporophytic plant.

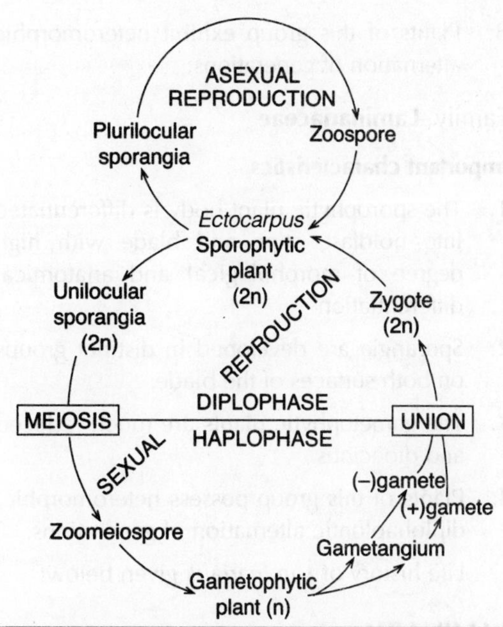

Fig. 3.137: Graphic life cycle of *Ectocarpus* sp.

Order. **Laminariales**

Important characteristics

1. Members are commonly called "kelps", because of very large size and complexity in structure. Some of the members reach up to 50-60 meters.
2. They are exclusively marine and commonly grow in sub-littoral and littoral zones of rocky coasts of temperate and polar regions.
3. Plant bodies are mostly parenchymatous and growth takes place by intercalary, medullary and superficial meristems.
4. Dominant plant body is very large and sporophytic (2n) in nature. They are usually differentiated into root-like holdfast or, hapteron, stalk-like stipe and leafy blades. But the gametophytic plant body is very small and microscopic.
5. Reproduction takes place by both asexual and sexual means.
6. Asexual reproduction generally takes place by zoospores developed in unilocular sporangia.
7. Sexual reproduction is of oogamous type. Sex organs i.e. antheridia and oogonia are developed on gametophytic plant.

8. Plants of this group exhibit heteromorphic alternation of generations.

Family. **Laminariaceae**

Important characteristics

1. The sporophytic plant body is differentiated into holdfast, stipe and blade with high degree of morphological and anatomical differentiation.
2. Sporangia are developed in distinct groups on both surfaces of the blade.
3. The gametophytic plants are much reduced and dioecious.
4. Plants of this group possess heteromorphic, diplohaplontic alternation of generations.

Life history of *Laminaria* is given below:

LAMINARIA

Systematic position

Fritsch (1945)	**Lee** (1999, 2008)
	Phylum. Heterokontophyta
Class. Phaeophyceae	Class. Phaeophyceae
Order. Laminariales	Order. Laminariales
Family. Laminariaceae	Family. Laminariaceae
Genus. *Laminaria*	Genus. *Laminaria*

Occurrence

Laminaria (Latin. a thin plate or leaf) with some 40 species are exclusively grown on marine habitat in comparatively shallow water, mainly along the rocky shores, just below the tidal region. They are distributed widely in the Atlantic and Pacific coasts, especially where the water is cool. They are popularly called as "**devil's apron**".

Some common members are *L. digitata. L. saccharina, L. japonica, L. flavicans and L. flexicaulis.*

Plant body

The plant body of *Laminaria* is diploid and attains a length of 2 to 12 metre or more.

The sporophytic plant body is differentiated into **holdfast, stipe** and **blade** (lamina) (Fig. 3.138).

Holdfast: It is an organ of attachment in variable forms. It may be like a solid disc or a cluster of profusely branched cylindrical root-like organs, also called **haptera**.

Stipe: It is an unbranched and cylindrical or slightly flattened stalk-like structure, developed on the holdfast and bears terminal blade(s).

Blade (lamina): Blade may be simple (*L. saccharina*) or a cluster of vertically divided segments (*L. cloustoni, L. digitata*) (Fig. 3.138A, B). Blades are flat, long, ribbon-shaped with wavy or smooth margin, and tough and leathery in texture. Growth is intercalary and meristematic zone is situated between the stipe and blade.

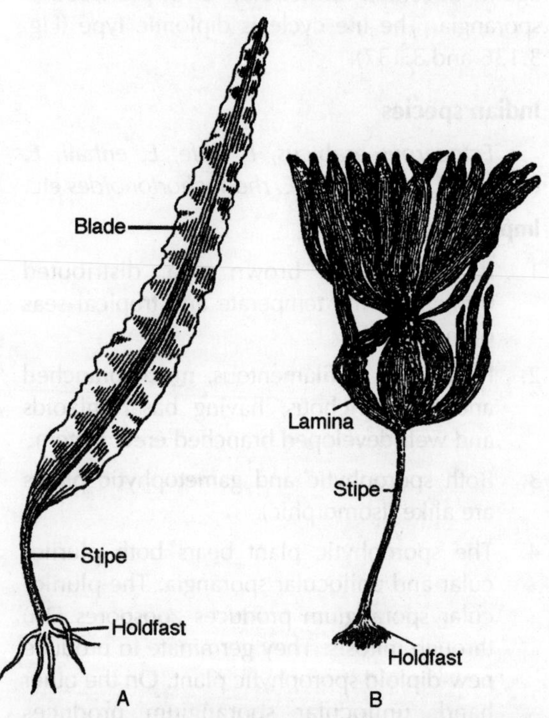

Fig. 3.138: *Laminaria* spp. plant bodies showing holdfast, stipe and blade. A. *L. saccharina* and B. *L. digitata*

Most of the members are perennial (*L. saccharina* etc.) and live about 5 years or more, but in annual member (*L. ephemera*), a new sporophyte develops in each spring.

Internal structure

1. **Stipe**. In transverse section, the stipe shows the following regions: (i) Meristoderm, (ii) Cortex and (iii) Medulla.

(i) **Meristoderm**. It is the outermost region, consists of one or two layers of small cubical cells containing chromatophores. It remains surrounded by a layer of mucilage and serves the function of assimilation.

(ii) **Cortex**. It lies just below the meristoderm (between meristoderm and central medulla). The cortex is usually differentiated into outer and inner cortex. The *outer cortex* is much wider and composed of elongated cells with pointed ends and the *inner cortex* composed of longer cells with square ends. Mucilage canals or tracts are present in the cortex. The cortical cells serve the function of both storage and mechanical support.

(iii) **Medulla**. It is the innermost layer, composed of loosely arranged cells and the spaces between them are filled with mucilage. Cross-connections are developed due to fusion of papillose outgrowths of the longitudinal walls. Due to regular cross-connections in the inner layer of cortex and also in the medulla, the layers appear as an entangled net work. The medulla is characterised by the presence of elongated hyphal structure, the '**trumpet hyphae**', formed due to swelling at the septa which is larger at one side than the other (Fig. 3.139 B). The transverse wall is perforated and shows a close resemblance with the sieve plates of sieve tubes. In older hyphae, the pores are sometimes blocked by callus pads.

2. **Blade** (lamina). Like stipe, the blade also shows the following three main regions in transverse section: (i) Meristoderm, (ii) Cortex (outer and inner) and (iii) Medulla. (Fig. 3.139A). Only the elements of the medulla are drawn out due to greater surface area.

Reproduction

The plant reproduces by all the three means: vegetative, asexual and sexual.

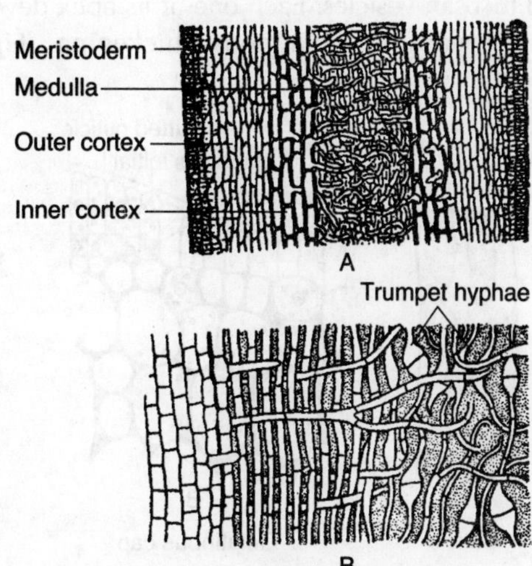

Fig. 3.139: *Laminaria*: A. T.S. of lamina showing internal structures, and B. Portion of medulla in section showing the arrangement of trumpet hyphae

Vegetative reproduction

The vegetative reproduction takes place by stolons (*L. sincelarii, L. longipes* etc.). The stolons develop horizontally from the holdfast, grow upwardly into plantlets.

Asexual reproduction

It takes place by **haploid zoospores** developed within the club-shaped unilocular sporangia, borne inside the sorus (pl. sori) (Fig. 3.140A) These haploid zoospores are also called **meiozoospores** or **gonozoospores**. Sori develop extensively on both the surface of blade and nearly cover the entire area. Numerous sterile hyphae, the **paraphyses** are present intermingled with the sporangia.

Development: Sporangia develop from the superficial cells (meristoderm) of the blade. Initially these cells elongate and divide transversely into upper *paraphysis initial* and lower *basal cell* (Fig. 3.140B).

The paraphysis initial elongates and forms an elongated thread with a club-shaped upper region. During this time, the basal cell broadens and thus the paraphyses occupy a limited part towards its basal region. The upper region of the paraphysis contains numerous chromatophores and fucosan vesicles. Each one at its apex develops a cup-like gelatinous thickening (Fig. 3.140C).

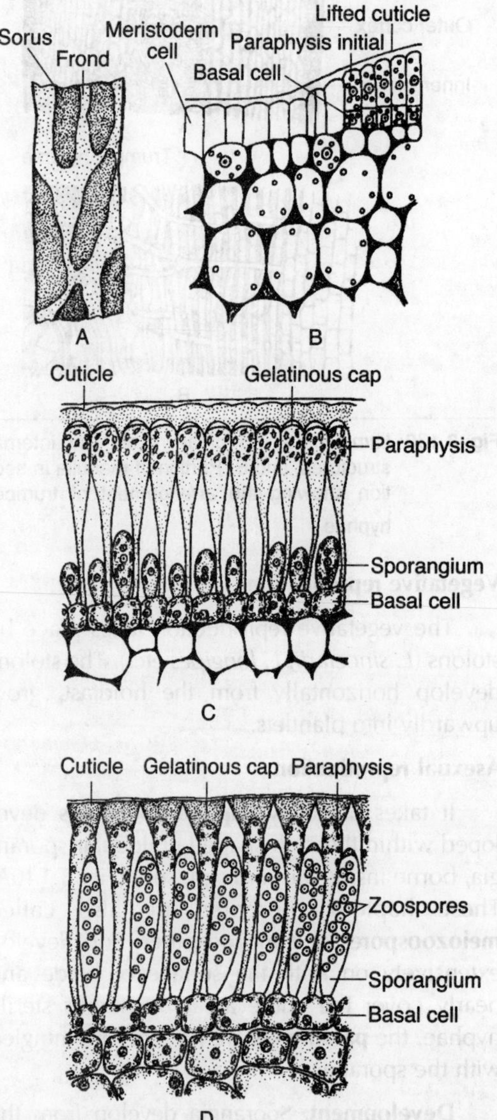

Fig. 3.140: *Laminaria*: A. Portion of lamina showing sori, and B-D. Development of zoospores in sporangium

By the side of paraphysis, the basal cells enlarge and differentiate into sporangia by transverse wall (Fig. 3.140D). During growth, the sporangia make their way and remain between the paraphyses. The nucleus of the sporangium undergoes first meiotic division, following successive mitotic divisions, thereby 32, sometimes 64 haploid nuclei are developed. Each nucleus with some cytoplasm and chromatophore gets differentiated into a unit, which later gets differentiated into zoospore. Zoospores are without eyespot (Henry and Cole, 1982), with two lateral flagella (biflagellate) of unequal length (longer one tinsel and shorter one whiplash).

Dehiscence of the sporangia takes place by the lateral pressure of paraphyses on the sporangium. After liberation, the zoospores swim actively for some time. Out of 32 or 64 zoospores, 50% develop into **male gametophyte** and the remaining 50% into **female gametophyte**.

After swimming for some time, the meiozoospores settle down to germinate. After withdrawing flagella, it becomes rounded and secretes a wall. On germination, it develops a small protuberance, which increases in length and its apex becomes bulbous and appears almost like a dumb-bell. The bulbous region develops into a filamentous gametophyte, consisting of elongated cells. The male and female gametophyte develop male and female sex organ, respectively.

Sexual reproduction

It is of **oogamous type**. Antherozoids and egg produce inside the antheridium (male sex organ) and oogonium (female sex organ) respectively. The male plants become many celled before they develop antheridia (Fig. 3.141A), but the female plants develop oogonia at 2-3 celled stage.

Antheridia develop at the tip of lateral branches of many-celled male plants. Protoplast of each one-celled antheridium metamorphoses into a single pyriform or ellipsoidal antherozoid with two flagella (biflagellate) placed laterally (Fig. 3.141B). Antherozoids are liberated through the apical region of antheridium. The male plant degenerates after the gametes are released.

Oogonia develop on 2-3 celled female plants (Fig. 3.141C). Oogonia are longer and thicker than the other cells of the female plant. The oogonial protoplast rounds off and becomes differentiated into an egg. At maturity, the egg is not extruded out, but remains attached at the apex of the oogonial wall.

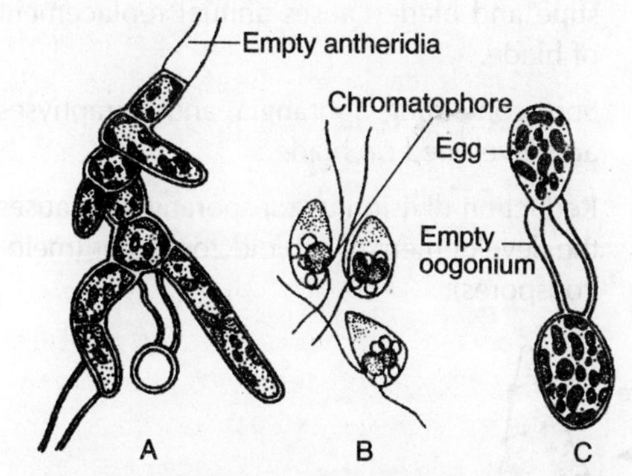

Fig. 3.141: *Laminaria*: A. Male gametophyte, B. Anthero-zoids and C. Female gametophyte

Fertilization

After liberating from the antheridium, the antherozoid swims in water. On reaching the oogonium, it unites with egg and forms zygote.

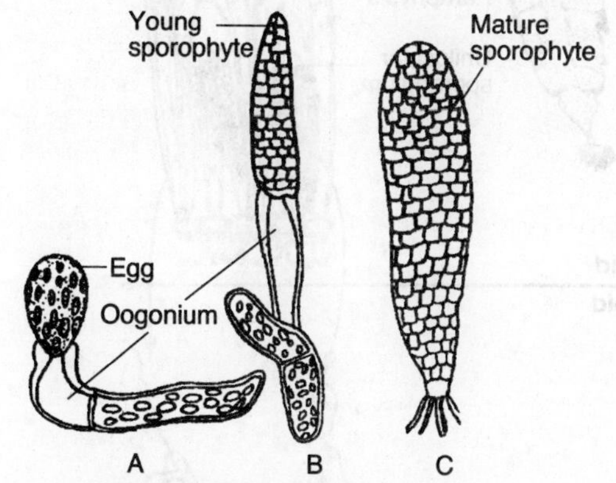

Fig. 3.142: *Laminaria*: A-C. Stages in the development of sporophyte

Germination of zygote

Soon after the formation of zygote, it develops into a sporophytic plant. Initially, the zygote undergoes successive transverse division and develops into a vertical row of 4-10 cells. Except the lowermost one, all other cells of the filament further divide by both vertical and transverse divisions, thus forming an expanded, thin and flat blade (Fig. 3.142B). The lowermost cell elongates to form the first rhizoid, but later several rhizoids are developed (Fig. 3.142C). Further division takes place in the lower part of the blade, thus forming meristematic region. With further development, the lower and upper parts of the meristem get differentiated into stipe and blade respectively.

Parthenogenetically, an egg may develop into a plant, as that of sporophytic plant, which is irregular in shape.

Alternation of generations

Laminaria exhibits a **heteromorphic** alternation of generation, showing a prominent and well-developed sporophyte and a very small gametophyte. Asexual reproduction takes place in the sporophytic plant. The zoospores are developed inside the zoosporangium through meiosis and are thus called as zoomeiospores. Half of the zoomeiospores develops multicellular male thalli, where antherozoids are produced inside the antheridium. The other half develops into female thalli, those develops egg inside oogonium. The entire process of sexual reproduction is unique of its kind. The egg does not retain within oogonium, but it still remains attached with the oogonial wall, indicating a transitional stage between the primitive and advanced nature. There is spontaneous germination of zygote without rest, while it still remains attached with the female gametophyte. This is found to be an important feature in the life cycle of *Laminaria* (Fig. 3.143 and 3.144).

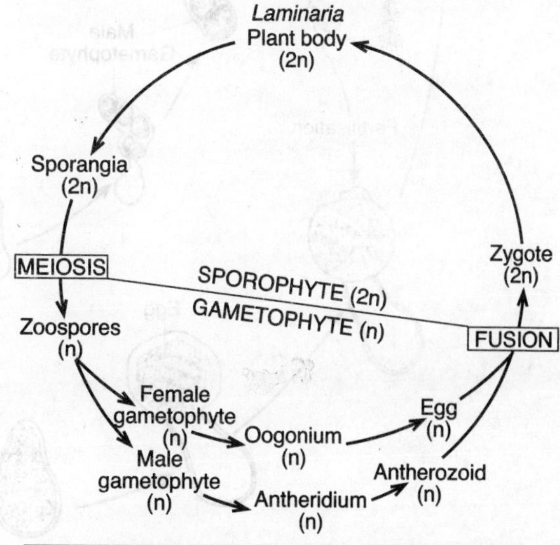

Fig. 3.143: *Laminaria*: Graphic life cycle

Important features

1. Sporophytic plant body is gigantic in size.
2. Plant body is differentiated into three regions: holdfast, stipe and blade.
3. Internal structure of the plant body is very much complex.
4. Presence of meristematic tissue between stipe and blade causes annual replacement of blade.
5. Sori containing sporangia and paraphyses are developed on blade.
6. Reduction division in zoosporangium causes the development of haploid zoospores (meio-zoospores).

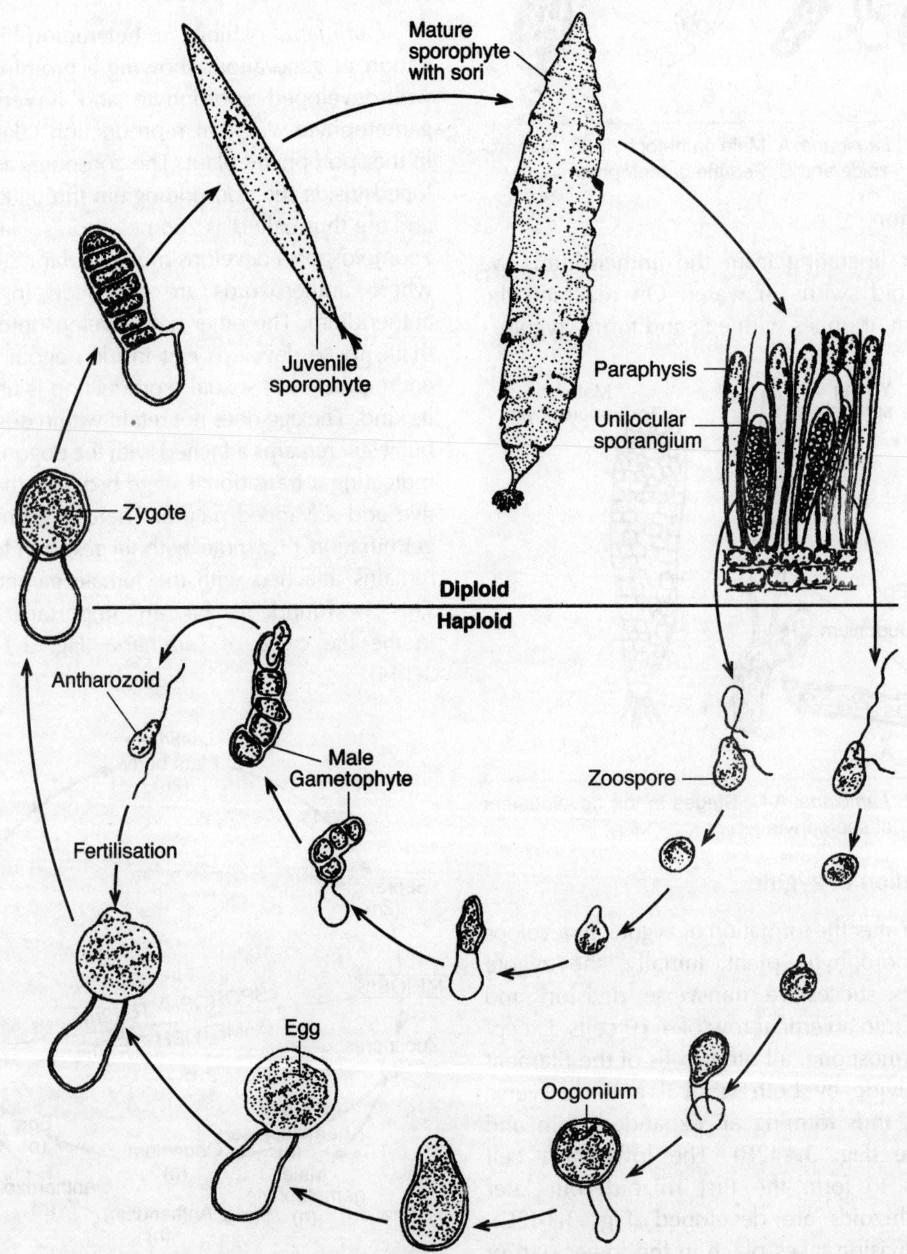

Fig. 3.144: *Laminaria*: Diagrammatic life cycle

7. Gametophytes are heterothallic.
8. Sexual reproduction is oogamous.
9. Fertilization is external.
10. Alternation of generations is heteromorphic, where predominant sporophyte alternates with the much reduced gametophyte.

Economic importance

The economic importance of *Laminaria* is given below:

The burnt ash (Europe) as well as the large brown seaweeds (America) are refered to as "**kelp**". The ash contains a large amount of **iodine** (15 kg/ton).

The industrial use of kelps is for the **alginate** which has several uses:

(i) **In making ice-cream:** Manufacturers add alginate powder before freezing their product, which prevents the water from forming coarse ice-crystals in ice-cream, thus it results in a smoother product (Smith, 1955).

(ii) **Goiter** is treated by giving kelp pills or kelp ash.

(iii) **Stipes of *Laminaria*** are used in surgery for opening fistulae (natural opening), because of their swelling properties.

(iv) Brown seaweeds have also been used as **green manure** for the soil and also as food for animals and human beings.

(v) "**Kombu**" is a food made from members of Laminariales in Japan.

(vi) In Mainland China, *Laminaria* is known as "**haidai**" (sea belt), which has been widely used as a food and medicine for over 1,500 years.

Order. **Fucales**

Important characteristics

1. The plant body is a sporophytic parenchymatous thallus, grows by single apical cell.
2. The plant body bears two types of sporangia, the microsporangia and macrosporangia.
3. Microsporangia behave as antheridia and develop antherozoids, whereas macrosporangia behave as oogonia and develop single egg within each.
4. Sexual reproduction is oogamous type.
5. The sporangia are borne within special cavities, the conceptacles.
6. Absence of alternation of generations.

The order Fucales (Fritsch, 1935) is divided into seven families: Fucaceae. Himanthaceae, Cystoseiraceae, Sargassaceae, Ormosiraceae, Durvilleaceae and Ascoseiraceae.

Only **Sargassaceae** and **Fucaceae** are considered for further discussion.

Family. **Sargassaceae**

Important characteristics

1. Plant body is thalloid, much branched and radially or bilaterally symmetrical.
2. Branching is monopodial.
3. Growth takes place by three-sided apical cell.
4. Some branches are modified into leaves, which bear axillary air-bladders.
5. Sex organs are developed in conceptacles, embedded in cylindrical or flattened receptacles.

Life history of *Sargassum* is given below:

SARGASSUM

Systematic position

Fritsch (1945)	**Lee** (1999, 2008)
	Phylum. Heterokontophyta
Class. Phaeophyceae	Class. Phaeophyceae
Order. Fucales	Order. Fucales
Family. Sargassaceae	Family. Not mentioned
Genus. *Sargassum*	Genus. *Sargassum*

Occurrence

The genus *Sargassum* (Spanish *sargazo*, seaweed) is represented by about 150 species (Nizamuddin, 1970), out of which 16 species are found in India (Mishra, 1966). It is found in temperate, subtropical and tropical regions of both northern and southern hemispheres. It is very common in Africa, South America, Australia etc.

In West Africa, a part of Atlantic Ocean becomes densely occupied by *Sargassum* and the region is called as 'Sargasso sea'. *Sargassum filipendula*, a **free-floating** large kelp found in the Sargasso Sea, was discovered by Columbus in 1492, as the ships were held fast by the sea weeds. In India it is found in Porbandar, Bombay, Okha, Lakshadweep Island etc.

Plant body

The plant body is diploid (2n), erect and branched thallus (Fig. 3.145). The thallus is differentiated into a basal **holdfast** and an expanded, leafy, cylindrical **main axis**.

The holdfast is discoid and serves the function of anchorage with the substratum. The main axis is generally of 10 to 50 cm in length. It is erect, flattened or cylindrical structure. The main axis bears many primary laterals arranged spirally in a phyllotaxy of $\frac{2}{5}$. Due to its unlimited growth, the **primary laterals** are also called **long shoots**. The main axis and primary laterals (long shoots) bear flat expanded structures, called **secondary laterals** or **leaves**.

The **leaves** are flat, simple structures with distinct midrib and dentate, serrate or entire margins, with an acute apex.

Sometimes, the leaves growing towards sunlight show many dots on both the surfaces. These dots are the ostioles i.e., openings of the sterile conceptacles. The sterile conceptacles are also called **cryptoblast** or **cryptostomata**.

On the main axis as well as on the primary laterals, the secondary laterals i.e., the leaves are replaced by many spherical, hollow bodies, called **air bladders**. The air bladders help to float them in water (Fig. 3.145).

The axils of leaves develop long much branched flattened or cylindrical structures called **receptacles**. The receptacle bears many fertile flask-shaped structures, the **conceptacles**. The conceptacles bear sex organs.

Internal structure

Axis. It is generally of circular in outline and differentiated into 3 regions: outer **meristoderm**, middle **cortex** and innermost **medulla** (Fig. 3.146A).

Meristoderm. The meristoderm is made up of single layer of closely packed cells. The cells are meristematic in nature. The cells contain chromatophores and perform photosynthesis. This layer can store food material.

Cortex. The cortex is situated next to meristoderm and occupies major part of the axis. It consists of compactly arranged parenchyma cells

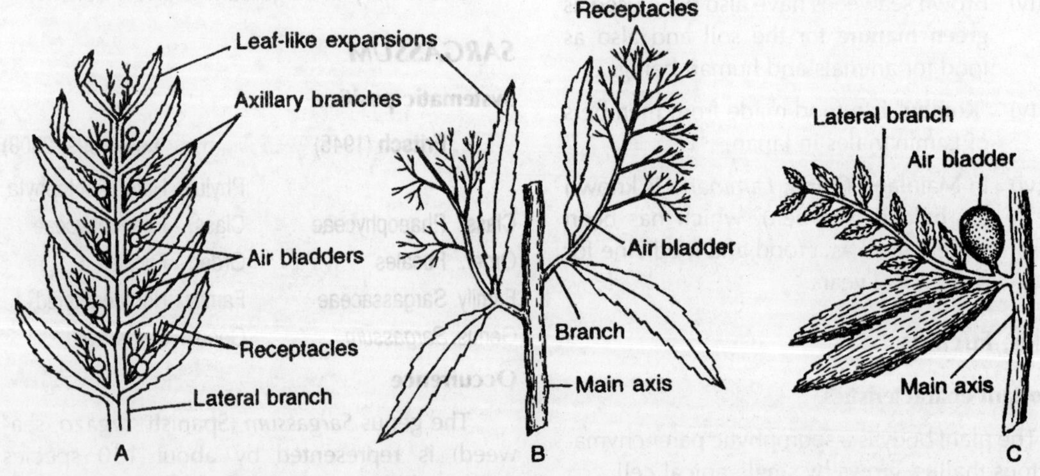

Fig. 3.145: *Sargassum* sp.: A. Portion of a branch with leaf-like expansions, air-bladders and receptacles, B. Portion of the main axis showing arrangement of branches, leaf-like expansions and receptacles, and C. Lower part of main axis with lateral branch and air bladder.

of polygonal shape, rarely with intercellular spaces. The cells are smaller in size than meristoderm. Like the outer layer this layer also stores food material.

Medulla. The medulla i.e., the inner layer consists of narrow, thick walled elongated cells. This layer possibly helps in conduction.

Leaf. It is flat and differentiated into outer meristoderm, middle cortex and inner medulla like the axis (Fig. 3.146B). The medulla is round and present in the middle region. On both surfaces of the leaf there are many sterile conceptacles, the **cryptostomata** or **cryptoblasts** (Fig. 3.146B, D). These are flask-shaped with many sterile unbranched filaments, the **paraphyses** developed from the base. The paraphyses protrude out through the opening present on the outer side, the **ostiole** (Fig. 3.136D). The cells of the wall have many chromatophores.

Air bladder. Internally it is almost alike with the axis but without medulla (Fig. 3.146C). The central region is occupied by a large hollow cavity filled with air and gases. Outer to the cavity, cortex is present; which consists of a few layers and thinner cells than axis and finally it ends with a single layered outer meristoderm.

Thus the thallus shows division of labour along with differentiation of tissues. It serves the function of anchorage, photosynthesis, storage, conduction and support.

Reproduction

It reproduces by both vegetative and sexual means. **Asexual reproduction is absent**.

1. *Vegetative reproduction*. It takes place by fragmentation. Due to death and decay of the older part, the younger region gets separated. The separated region grows and finally develops into a new individual like the mother. The **free floating members** like *S. hystrix* and *S. natans*, multiply only by this method.

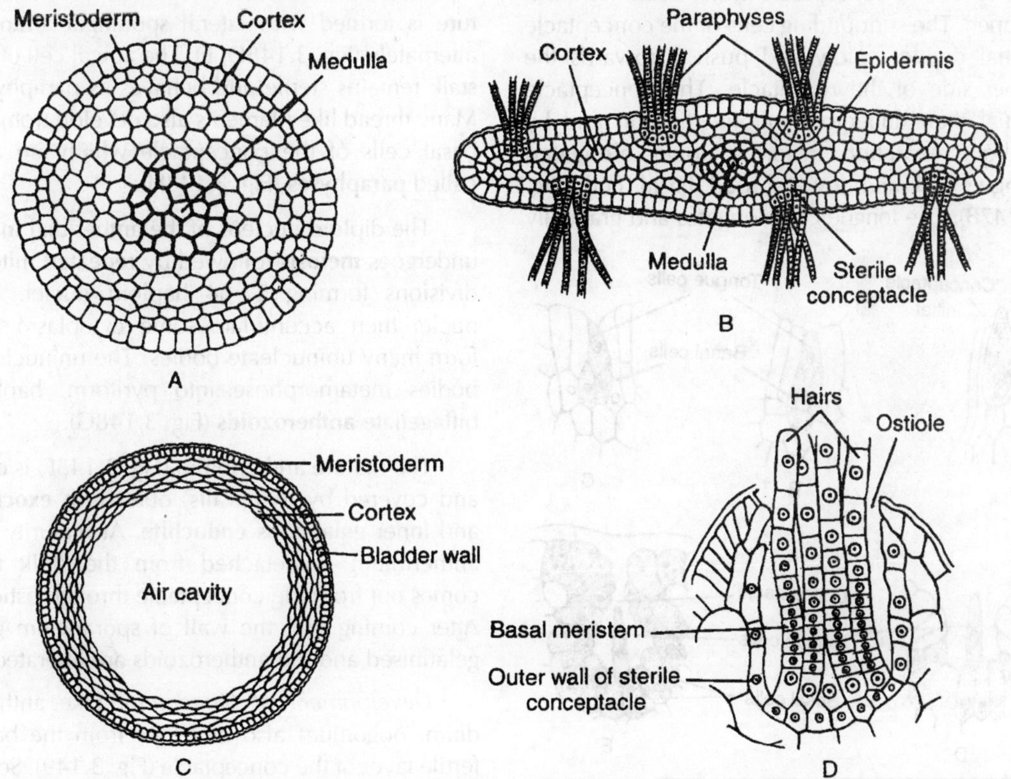

Fig. 3.146: *Sargassum* sp.: Internal structures : A. T.S. of main axis, B. T.S. of leaf, C. T.S. through air-bladder, and D. V.S. through a sterile conceptacle

2. *Sexual reproduction*. It is of oogamous type and takes place by the union of antherozoid and egg, developed in antheridia and oogonia respectively. The sex organs develop in separate flask-shaped bodies the **conceptacles**, developed on branched **receptacles**. The conceptales with antheridia or oogonia are called **male** or **female conceptacles**.

Development of conceptacle. The fertile and sterile conceptacle are almost similar. The difference lies in the activity of basal cells of the linear wall of conceptacle. In sterile conceptacle it only develops sterile hairs, the paraphyses, but in fertile conceptacle it develops either antheridia or oogonia and also paraphyses in some regions.

During development (Fig. 3.147) single superficial cell on the receptacular branch becomes enlarged and functions as conceptacle initial (Fig. 3.147A). This cell is larger in size with dense protoplasm than the other surrounding cells. The conceptacle initial becomes flask-shaped. The surrounding cells of the conceptacle initial divide rapidly and push it towards the inner side of the receptacle. The conceptacle initial then undergoes mitotic division and by oblique septation it forms upper elongated tongue cell and lower broad basal cell (Fig. 3.147B). The tongue cell elongates and gradually

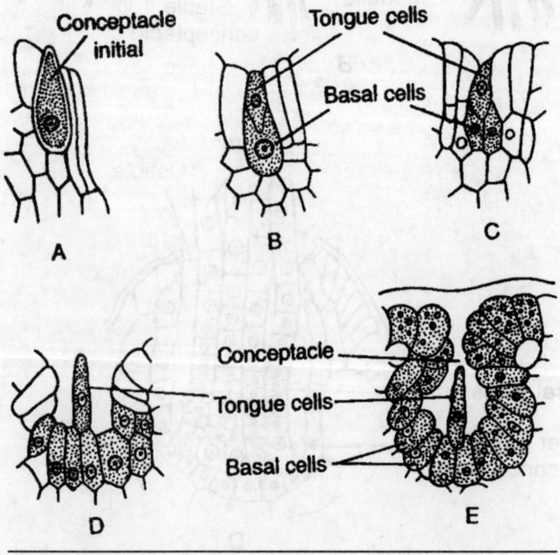

Fig. 3.147: *Sargassum* sp.: A-E. Development of conceptacle

disappears. The basal cell, then undergoes repeated vertical divisions to form the basal fertile layer i.e., the inner layer of the conceptacle. The reproductive organs are developed from this inner layer.

Development of antheridium. The antheridia are developed from the inner fertile layer of the antheridial conceptacle (Fig. 3.148). The lower basal cells of the conceptacle are the **antheridial initials** forming papilla like outgrowths (Fig. 3.148A). Each one divides by a transverse wall into two cells. The lower one remains as conceptacle wall, whereas the upper one (Fig. 3.148A) again divides transversely and forms a lower stalk cell and an upper antheridial cell (Fig. 3.148B). The antheridial cell develops into an **antheridium** (Fig. 3.148C). Due to rapid growth of the stalk cell, the antheridium becomes pushed at one side (Fig. 3.148D). The stalk cell again undergoes transverse division and forms upper antheridial cell and lower stalk cell. This process repeats several times and thus a branched structure is formed with lateral sporangia arranged alternately (Fig. 3.148D, E). The apical cell of the stalk remains sterile and behaves as paraphysis. Many thread like filaments also develop from the basal cells of the conceptacle which are also called **paraphyses** (Fig. 3.148H).

The diploid nucleus of the antheridial initial undergoes **meiosis** followed by repeated mitotic divisions forming 32-64 haploid nuclei. The nuclei then accumulate some cytoplasm and form many uninucleate bodies. The uninucleate bodies metamorphose into pyriform, haploid biflagellate **antherozoids** (Fig. 3.148G).

The mature antheridium (Fig. 3.148F) is oval and covered by two walls, outer firm **exochite** and inner gelatinous **endochite**. At maturity the antheridium is detached from the stalk and comes out from the conceptacle through ostiole. After coming out, the wall of sporangium gets gelatinised and the antherozoids are liberated.

Development of Oogonium. Like antheridium, oogonium also develops from the basal fertile layer of the conceptacle (Fig. 3.149). Some of the cells of this layer function as an **oogonial initials** (Fig. 3.149A). The oogonial initial under-

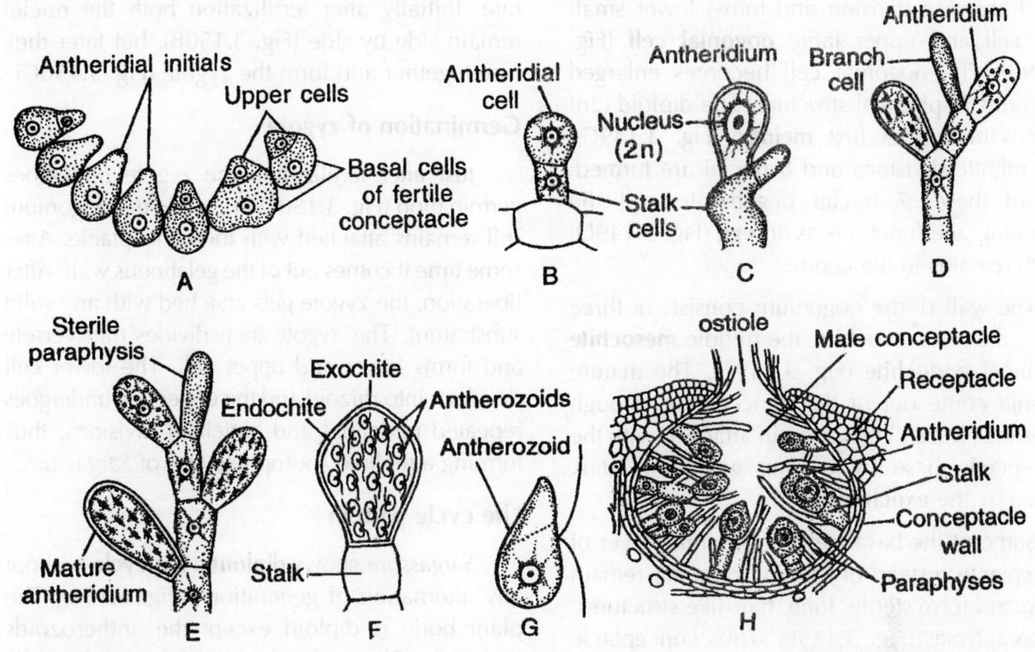

Fig. 3.148: *Sargassum* sp.: Development of antheridium: A. Cells of basal fertile layer of conceptacle, B-E. Development of antheridium, F. Mature antheridium, G. Single antherozoid, and H. V.S. of male conceptacle showing antheridium along with paraphyses

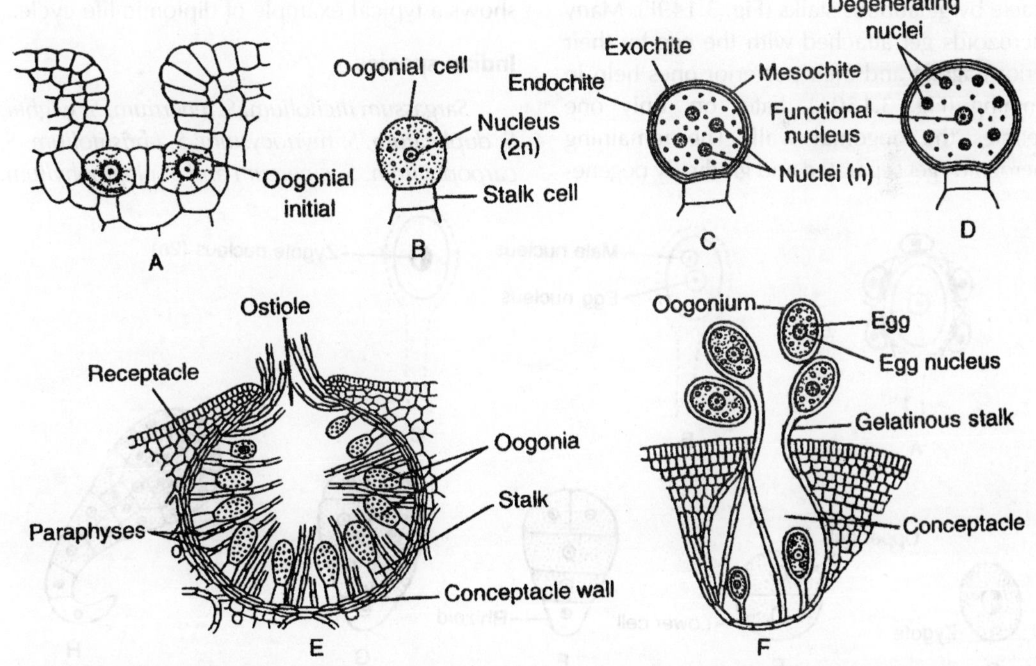

Fig. 3.149: *Sargassum* sp.: Development of oogonium: A-D. Development of oogonium, E. Vertical section of female conceptacle showing oogonia and paraphyses inside, and F. Release of oogonium from conceptacle

goes transverse division and forms lower small stalk cell and upper large **oogonial cell** (Fig. 3.149B). The oogonial cell becomes enlarged and forms a spherical structure. The diploid (2n) nucleus undergoes first **meiotic** (Fig. 3.149C), then mitotic divisions and 8 nuclei are formed. Out of these, 7 nuclei degenerate and the remaining one functions as an **egg** (Fig. 3.149D) which remains in the centre.

The wall of the oogonium consists of three layers, the outer **exochite**, the middle **mesochite** and inner **endochite** (Fig. 3.149C). The mature oogonia come out of the conceptacle through the ostiole, but still they remain attached with the conceptacle base by a long gelatinous stalk formed by the **exochite**.

Some of the basal cells of the inner layer of conceptacle instead of forming oogonia remain sterile and form sterile, long, hair-like structures, the paraphyses (Fig. 3.149E). Thus conceptacle contains oogonia intermingled with paraphyses.

Fertilization

Fertilization takes place when the eggs remain outside but still attached with the conceptacle by gelatinous stalks (Fig. 3.149F). Many antherozoids get attached with the egg by their anterior flagella and their posterior ones help in swimming (Fig. 3.150A). Later on, only one penetrates the oogonial wall. The remaining antherozoids get separated and gradually degenerate. Initially after fertilization both the nuclei remain side by side (Fig. 3.150B), but later they fuse together and form the zygote (Fig. 3.150C).

Germination of zygote

Just after fertilization the zygote undergoes germination (Fig. 3.150D-H), while the oogonium still remains attached with the conceptacle. After some time it comes out of the gelatinous wall. After liberation, the zygote gets attached with any solid substratum. The zygote then divides transversely and forms lower and upper cell. The lower cell develops into rhizoid and the upper cell undergoes repeated periclinal and anticlinal divisions, thus forming a thalloid sporophyte (2n) of *Sargassum*.

Life cycle pattern

Sargassum shows **diplontic life cycle** without any alternation of generations (Fig. 3.151). The plant body is diploid except the antherozoids and eggs. The antherozoids and eggs i.e., the gametes, represent only the haploid (n) stage. After fusion zygote is formed. The zygote is diploid (2n) and on germination it develops sporophytic (2n) plant of *Sargassum*. Thus it shows a typical example of diplontic life cycle.

Indian species

Sargassum ilicifolium, S. tenerrium, S. wightii, S. duplicatum, S. myriocystum, S. christifolium, S. carpophyllum, S. cinereum and S. plagiophyllum.

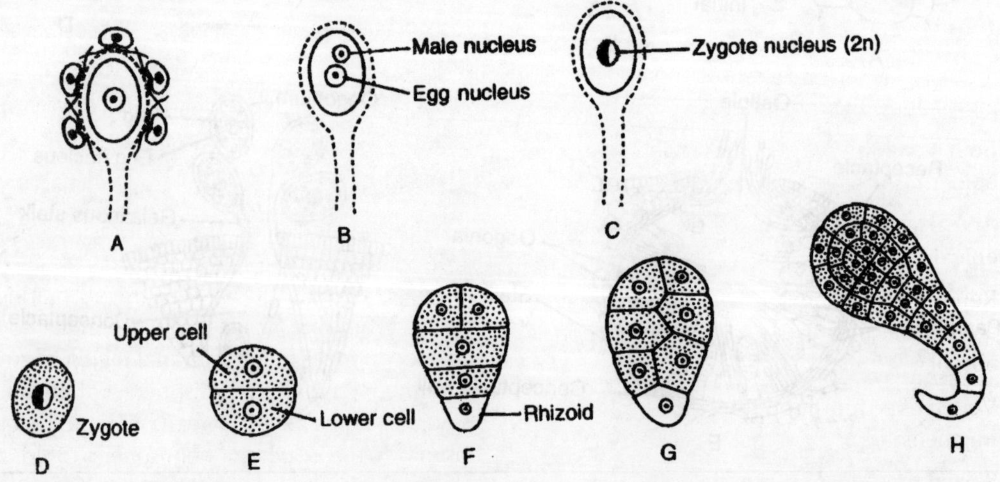

Fig. 3.150: *Sargassum* sp.: A-C. Stages of fertilization, D. Zygote, E-H. Development of Zygote

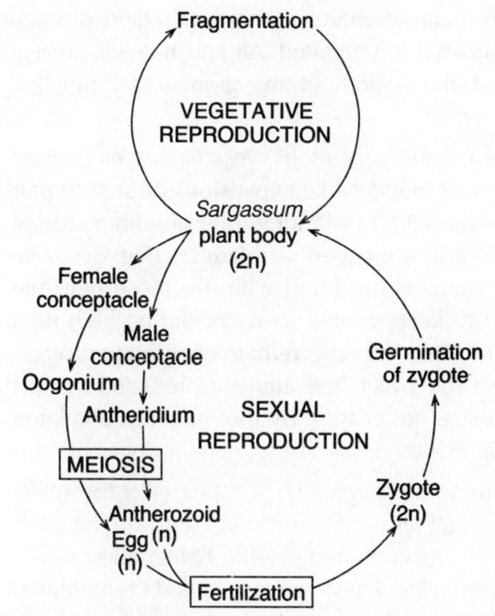

Fig. 3.151: Graphic life cycle of *Sargassum* sp.

Important features

1. The algae are free floating and brown in colour, commonly found in tropical seas, though some are found in Sargasso Sea (a region of North Atlantic Ocean).
2. The plant body is diploid and differentiated into root, stem and leaf-like structures.
3. The main axis i.e., stem is vertically elongated and differentiated into nodes and internodes.
4. It bears long shoots of unlimited growth (primary laterals), leaves (secondary laterals), air bladders and receptacles.
5. Both stem and leaves are differentiated into epidermis, cortex and medulla.
6. Apical growth takes place by a three-sided apical cell.
7. Reproduction takes place by vegetative and sexual means. *Asexual reproduction* is absent.
8. Vegetative reproduction takes place by fragmentation.
9. Sexual reproduction is oogamous.
10. Oogonia and antheridia are borne in unisexual conceptacles, those remain embedded in receptacles.
11. Oogonium produces one egg and the antheridium produces 64 biflagellate antherozoides.
12. Fertilization is internal, as the egg is not come out from the oogonium.
13. Zygote germinates directly and produces a new sporophytic (2n) plant.

Family. **Fucaceae**

Important characteristics

1. Plants are dichotomous or pinnate strap-shaped simple or with costate branches, often with piliferous cryptostomata, or may be with buoyant air bladders.
2. Receptacular portions of the thallus terminate main branches or form stalked lateral branchlets.
3. Plants are heterogamous to oogamous.
4. Each oogonium forms several eggs.

Life history of *Fucus* is given below:

FUCUS

Systematic position

Fritsch (1945)	**Lee** (1999, 2008)
	Phylum. Heterokontophyta
Class. Phaeophyceae	Class. Phaeophyceae
Order. Fucales	Order. Fucales
Family. Fucaceae	Family. Not mentioned
Genus. *Fucus*	Genus. *Fucus*

Occurrence

The genus *Fucus* (Gr. *phykos*, world of algae) is represented by more than 100 species (Bold and Wynne, 1978) is exclusively marine in habitat. They are cold sea water algae found attached to **rocks in the intertidal region** of coasts in northern temperate hemisphere. For the above they are called as **rockweed**. It is completely absent in the southern hemisphere (Bold and Wynne, 1978) and also in India (Mishra, 1966).

Plant body

The plant body is well branched, leathery, slimy and dark brown in colour, rarely exceeding 30 cm in length. The thallus is differentiated into three regions, holdfast, stipe and blade (Fig. 3.152A).

1. **Holdfast**: Holdfast is the broad, disc-shaped structure functions as an organ of attachment. It is also called **hapteron** (Fig. 3.152B). It does not penetrate inside the substratum by any means.
2. **Stipe**: It is a short cylindrical, stem-like structure develops from the holdfast (Fig. 3.152B).

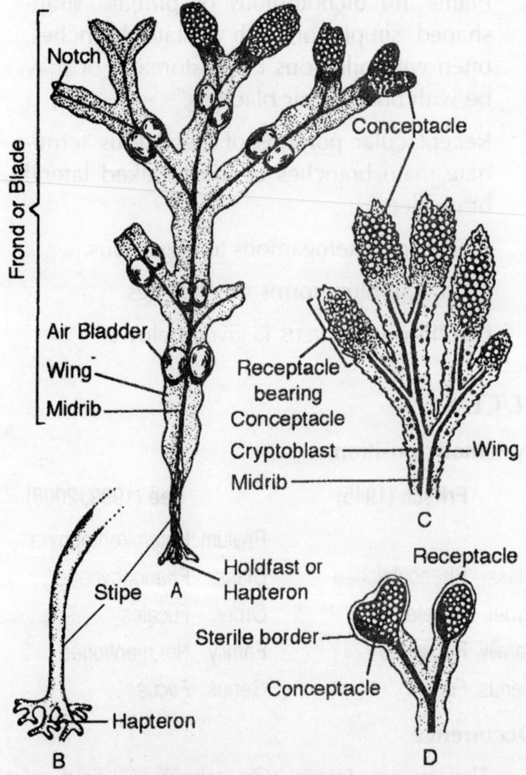

Fig. 3.152: *Fucus* sp. A. Plant of *F. vesiculosus* showing habit, B. Holdfast and stipe of *F. serratus*, C. Frond of *F. serratus*, and D. Fertile branch of *F. spiralis* bearing receptacles

3. **Blade**: It is the broad, dichotomously branched expanded region, also called **frond** (Fig. 3.152A), develops on expansion of stipe and is continuous with its midrib. The margin of the frond varies with species, may be smooth, serrate or entire. The apical region of each branch is bifurcated. An apical cell is present at the bottom of the groove of bifurcated region.

In some species (*F. vesiculosus* etc.) there are small bladder-like expansions present in pair near the region of forking, one on either side of the midrib are called **air bladders** (Fig. 3.152A). The bladders are filled with air; those help the plant to keep upright in water during high tide. At low tide, the plants remain exposed to drought above the water line and are protected against excessive desiccation by the mucilage available in the thallus.

In some species (*F. serratus* etc.) the frond bear irregularly scattered small dots are the position of pits containing sterile hairs called **sterile conceptacles**. These are also called **cryptoblasts**. During reproductive phase, sex organs are developed in **fertile conceptacles** developed in fertile frond, the **receptacl**: (Fig. 3.152C, D).

Internal structure

Internally the frond of the plant body is differentiated into meristoderm, cortex and medulla (Fig. 3.153A).

1. **Meristoderm**: It is the outermost compact layer with palisade-like cells containing plastids situated perpendicularly to the surface of the frond is called **meristoderm**. Cells of this layer are capable to divide actively and have both protective and photosynthetic function.
2. **Cortex**: It is situated in between the outer meristoderm and central medulla, consisting of few layers thin-walled polygonal parenchymatous cells may be differentiated into outer and inner cortex. The inner cortex probably serves the function of storage.
3. **Medulla**: The central region of the frond is the medulla. It consists of loose, tangled mass of elongated cells are called **medullary hyphae**. These hyphae have thick mucilaginous wall and often have pits on the transverse walls (Fig. 3.153B). The hyphae are separated by mucilaginous substance secreted by both the hyphae.

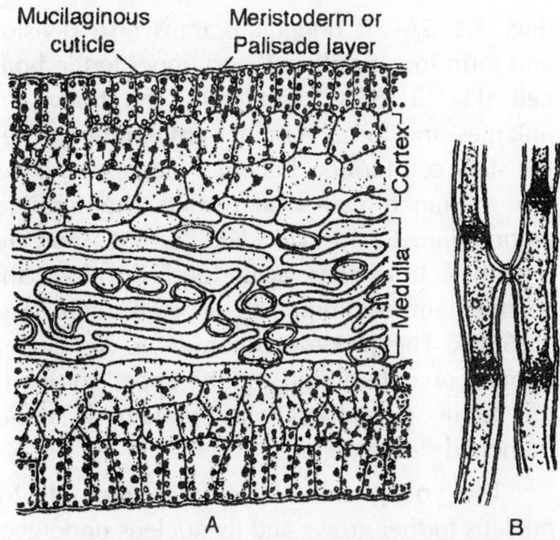

Fig. 3.153: *Fucus* sp. A. T.S. of frond showing meristoderm, cortex and medulla, and B. Medullary hyphae showing pits on the transverse wall

Reproduction

It reproduces by both vegetative and sexual means. **Asexual reproduction is absent.**

1. *Vegetative reproduction*. It takes place by fragmentation. Due to constant agitation of wave during high tide, the frond may be broken and get detached from the mother. The fragments may get attached in salt marshes and grow as independent plant.

2. *Sexual reproduction*. It is of oogamous type and takes place by the union of antherozoid and egg, developed in antheridium and oogonium respectively inside the fertile conceptacles. Species may be monoecious (*F. spiralis, F. furcatus* etc.) or dioecious (*F. vesiculosus, F. serratus* etc.). In monoecious species, both male and female sex organs are formed in the different or even in the same conceptacle i.e. bisexual conceptacle. *Development of the conceptacle is almost same as in Sargassum*.

With the onset of reproductive phase the apical end of each branch become swollen. The swollen ends are covered with small projections,

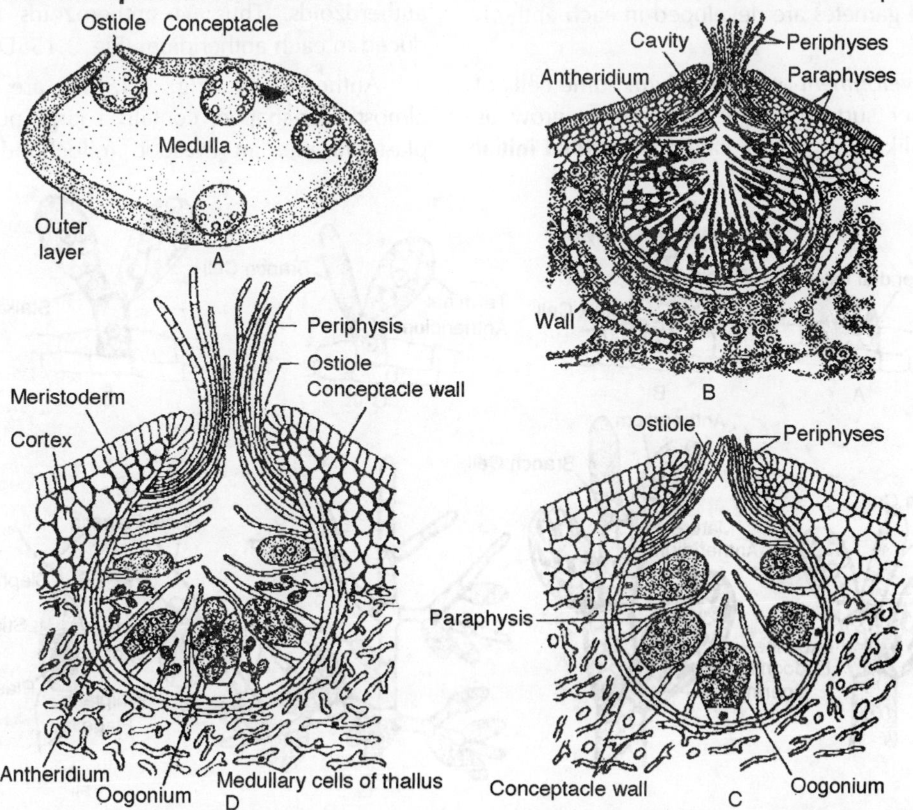

Fig. 3.154: *Fucus*: A. Section through frond showing conceptacles embedded in the upper part of medulla, and B-D. Conceptacles: Male (B), female (C) and bisexual (D)

the **papillae**. Papillae having minute pore at the apex, called **ostiole**. Each ostiole leads into an almost spherical cavity developed inside the thallus, called **conceptacle**. The conceptacles contain sex organs. The fertile frond containing the fertile conceptacles are called **receptacle**. The branches bearing receptacles are shed annually.

Male conceptacle. The male conceptacles are flask-shaped containing fertile paraphyses bearing **antheridia** intermingled with sterile hyphae, the **paraphyses** and also lined with sterile hyphae at the neck, the **periphyses**. Both paraphyses and periphyses serve the function to release the antherozoids.

The antheridia are small oval or club-shaped unicellular structures develop commonly on the lower part of fertile paraphyses at the floor of the fertile conceptacle. Mature antheridia are orange in colour having two walls. Outer wall is firm, the **exochite** and the inner gelatinous layer, the **endochite** (Fig. 3.155F). 64 haploid antherozoids or male gametes are developed in each antheridium.

Development of antheridium. Some cells of the inner surface of the conceptacle grow as papilla-like out growth, called **antheridial initial** (Fig. 3.155A). It undergoes transverse division and form lower **stalk cell** and upper fertile **body cell** (Fig. 3.155A). Gradually, the body cell enlarges and forms terminal **antheridium**. Then the stalk cell further becomes active and develop a branch cell which shifts the terminal antheridium to one side (Fig. 3.155C) and the branch cell divides again and develops the second antheridium and a stalk cell (Fig. 3.155D). The process continues to develop a number of antheridium and the upper portion of the fertile paraphysis may remain as richly branched sterile region.

The young antheridium with diploid (2n) nucleus further grows and its nucleus undergoes **meiosis**, thus 4 haploid nuclei are produced. These haploid nuclei undergo further mitosis results the formation of 32 nuclei. These nuclei accumulate some cytoplasm and develop 32 **antherozoid mother cells**. The antherozoid mother cell with further mitosis develops 2 antherozoids. Thus 64 antherozoids are produced in each antheridium (Fig. 3.155D-F).

Antherozoids (Fig. 3.155H) are minute, almost pear-shaped, contain a large nucleus, a plastochrome, an eyespot, mitochondria with

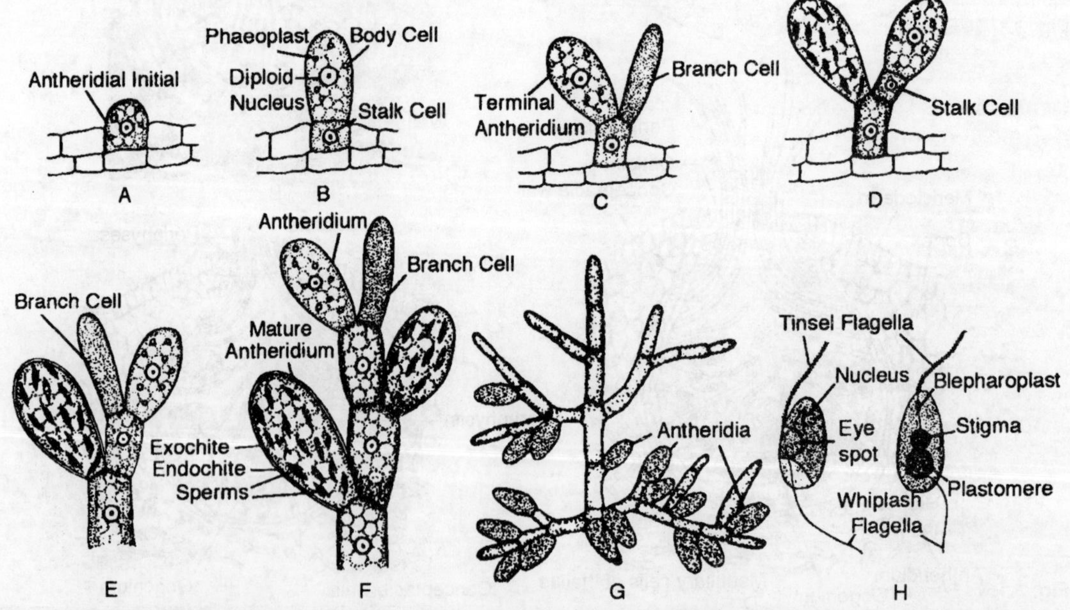

Fig. 3.155: *Fucus*. A-F. Development of antheridium, G. Antheridia on fertile paraphyses, and H. Structure of antherozoids (in different view)

two laterally attached unequal flagella. The shorter (tinsel type) one is directed forward and the longer (whiplash type) one is directed backward.

Liberation of antherozoids: The antherozoids are liberated in mass still enclosed in the inner layer, the endochite, called sperm bladders. The endochite gelatinized at one or at both ends and the sperms set free, those swim in all directions.

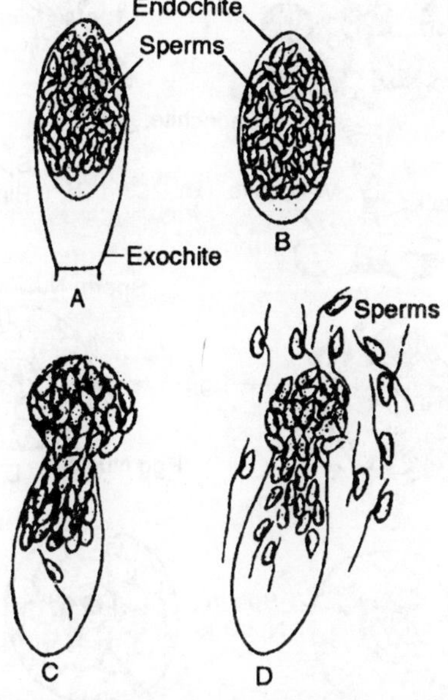

Fig. 3.156: *Fucus.* Liberation of antherozoids

Female conceptacle. The female conceptacles are also flask-shaped containing stalked **oogonia** intermingled with sterile hyphae, the **paraphyses** (Fig. 3.157) and also lined with sterile hyphae at the neck, the **periphyses**. Both the paraphyses and periphyses serve the function to release the eggs.

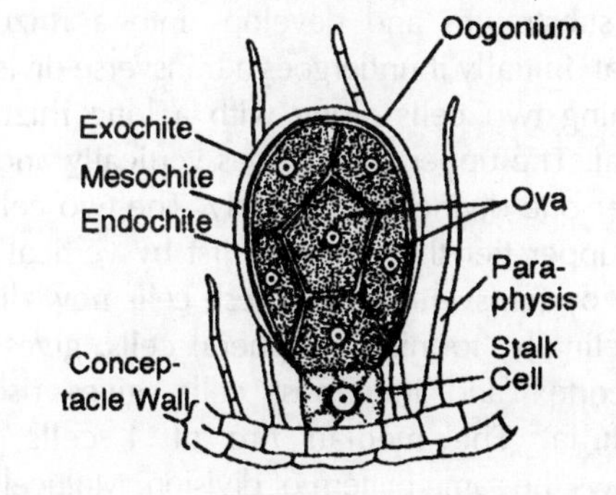

Fig. 3.157: *Fucus.* Portion of the conceptacle wall bearing mature oogonium containing ova and a few sterile hyphae, the paraphyses

The **oogonia** are small oval or globular unicellular structures develop directly at the floor of the fertile conceptacle. Mature **oogonia** have three walls. Outer wall is exochite, the middle **mesochite** and the innermost gelatinous layer, the **endochite**. The **exochite** and **endochite** are thin and the **mesochite** is thick. The oogonial wall remains thin where it is attached with the stalk. Eight (8) haploid eggs or female gametes are developed in each oogonium.

Development of oogonium (Fig. 3.158). Some cells of the inner surface of the conceptacle grow as papilla-like out growth, called

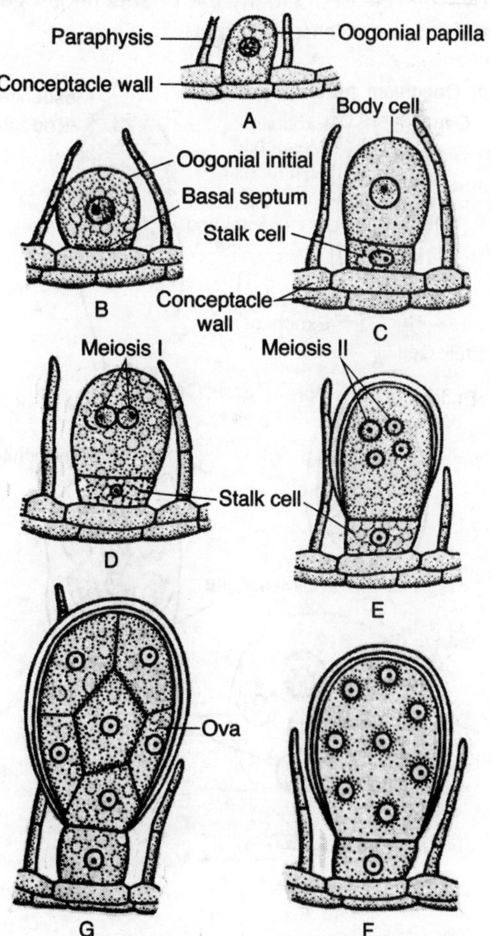

Fig. 3.158: *Fucus.* A-F. Successive stages in the development of oogonium

oogonial initial. It undergoes transverse division and form lower stalk cell and cytoplasm with numerous oil drops. The diploid (2n) nucleus of oogonium undergoes meiosis followed by mitosis results the formation of 8 haploid nuclei. Cleavage of the oogonial cytoplasm along with eggs results the formation of 8 uninucleate haploid eggs.

Liberation of eggs (Fig. 3.159). During liberation, both endochite and mesochite imbibe water and swells, which caused pressure, thereby the exochite get ruptured and the ova are released still enclosed in a bladder composed of endochite and mesochite. Gradually the mesochite get gelatinized and develop a pore at the apex through which the endochite along with the ova are released. Finally the endochite get gelatinized and releases the ova.

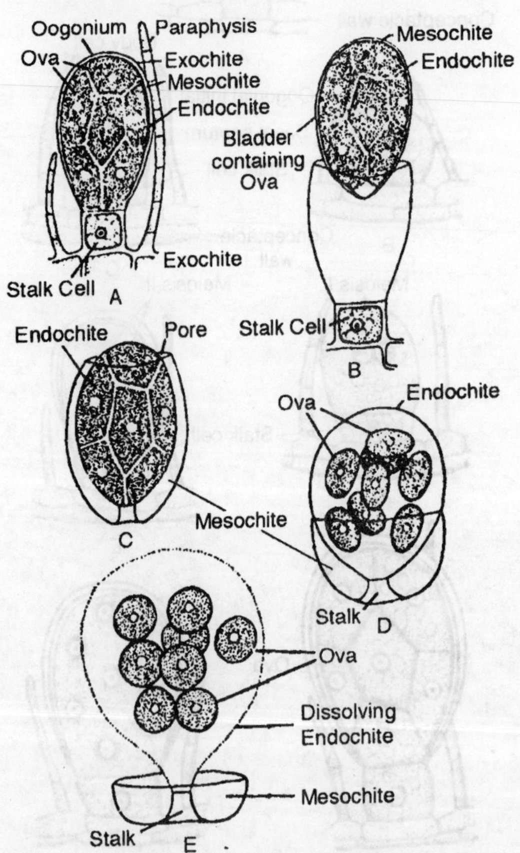

Fig. 3.159: *Fucus*. A-E. Successive stages in the liberation of egg from the oogonium

Fertilization:

Fertilization (Fig. 3.160) is external. During fertilization, the ovum released some volatile substance which attracts the antherozoids. Many antherozoids surround the egg but only one gets fertilized, thereby zygote is produced.

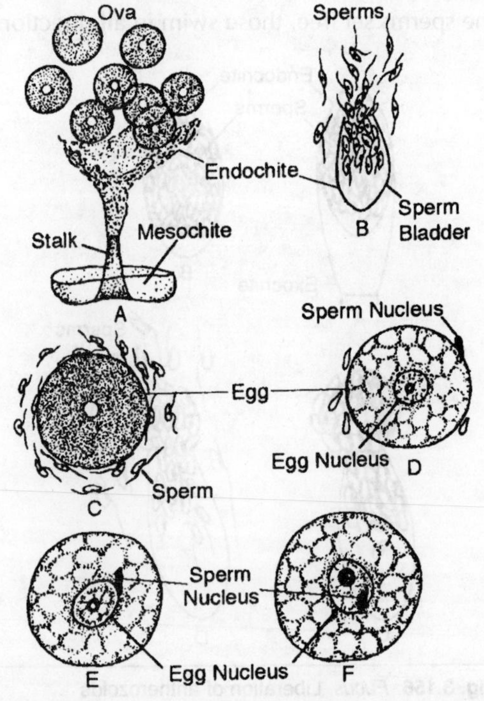

Fig. 3.160: *Fucus*. A. Liberation of ova, B. Liberation of antherozoids, C-F. Subsequent stages of fertilization

Germination of zygote (Fig. 3.161): After a short period of rest, the zygote elongates towards the substratum and develops into a rhizoidal initial. Initially it undergoes a transverse division, forming two cells along with a long rhizoidal initial. The upper one divides vertically and the lower one divides transversely. The two cells of the upper tier then divides first by vertical and then by transverse wall. These cells now divide periclinally, forming peripheral cells, gives rise to cortex and innermost cells, gives rise to medulla. The median tire of 4 cells also follows the same pattern of division. Multicellular rhizoids are developed from the lower cells of the median tire. During this development, an apical superficial cell is differentiated in the

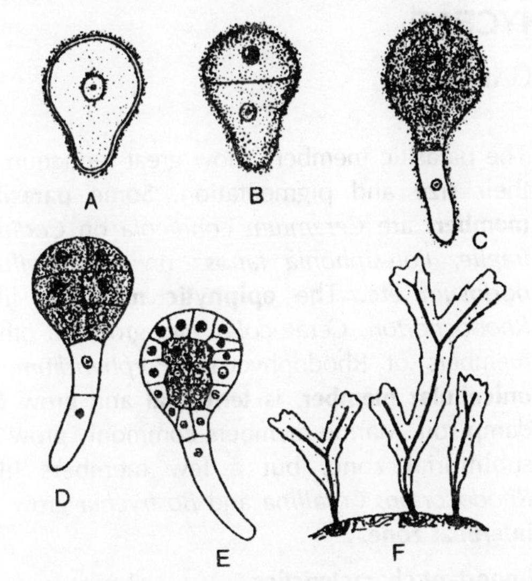

Fig. 3.161: *Fucus*. A-E.Germination of zygote, and B. Young sporophyte

multicellular embryo. In course of further development a multicellular saprophytic thallus is formed.

Life cycle is diplontic type (Fig. 3.162).

Indian species

Fucus turbinatus, F. granulatus, F. taxifolius etc.

Important features

1. The plant body is diploid and differentiated into holdfast, stipe and dichotomously branched ribbon-like frond.
2. Presence of bladder as floating device.
3. Internal structure of plant body is complex.

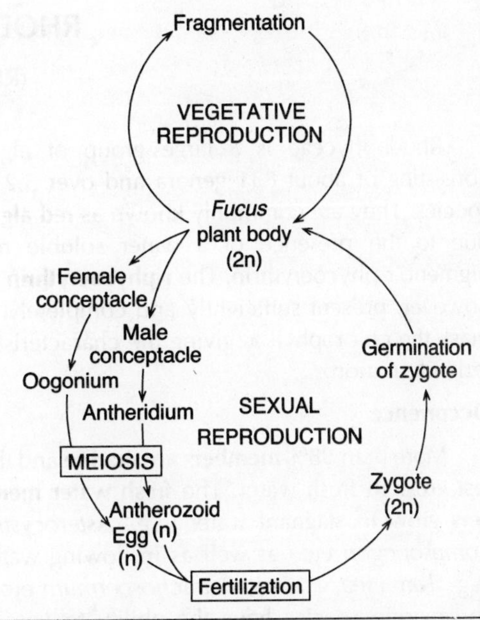

Fig. 3.162: *Fucus*. Graphical representation of the life cycle

4. Sexual reproduction is heterogamous.
5. Sex organs in the same or different conceptacles.
6. Reduction division takes place in antheridium and oogonium during formation of gametes.
7. Eight eggs are developed in each oogonium.
8. Fertilization of ova takes place in water, outside the plant body.
9. Absence of morphological alternation of generations.

RHODOPHYCEAE

(RED ALGAE)

Rhodophyceae is a large group of algae consisting of about 831 genera and over 5,250 species. They are commonly known as **red algae** due to the presence of a water soluble red pigment, r-phycoerythrin. The **r-phycoerythrin** is, however, present sufficiently and completely to mask the chlorophyll a, giving the characteristic red colouration.

Occurrence

More than 98% **members are marine** and the rest grow in fresh water. The **fresh water members** grow in stagnant water (e.g., *Asterocystis, Compsopogon* etc.) as well as in flowing water (e.g., *Lamanea, Thorea, Batrachospermum* etc.). The marine species have the ability to live at greater depth (even at 30-90 meters) than the other members of different classes. They also exhibit a high degree of parasitism and epiphytism. The parasitic members show great reduction in their size and pigmentation. Some **parasitic members** are *Ceramium condicola* on *Codium fragile, Polysiphonia lanosa* on *Ascophyllum nodosum* etc. The **epiphytic members** like *Rhodochorton, Ceratocolax* etc. grow on other members of Rhodophyceae. *Porphyridium*, **a unicellular member**, is **terrestrial** and grow on damp soil. Marine members commonly grow in sublittorial zone, but a few members like *Rhodocorton, Corallina* and *Bostrychia* grow in **intertidal zone.**

Important characteristics

1. Most of the members (more than 98%) are marine and 20 species (as per report) grow in fresh water. The members may grow either as saprophytes, parasites and also as epiphytes.

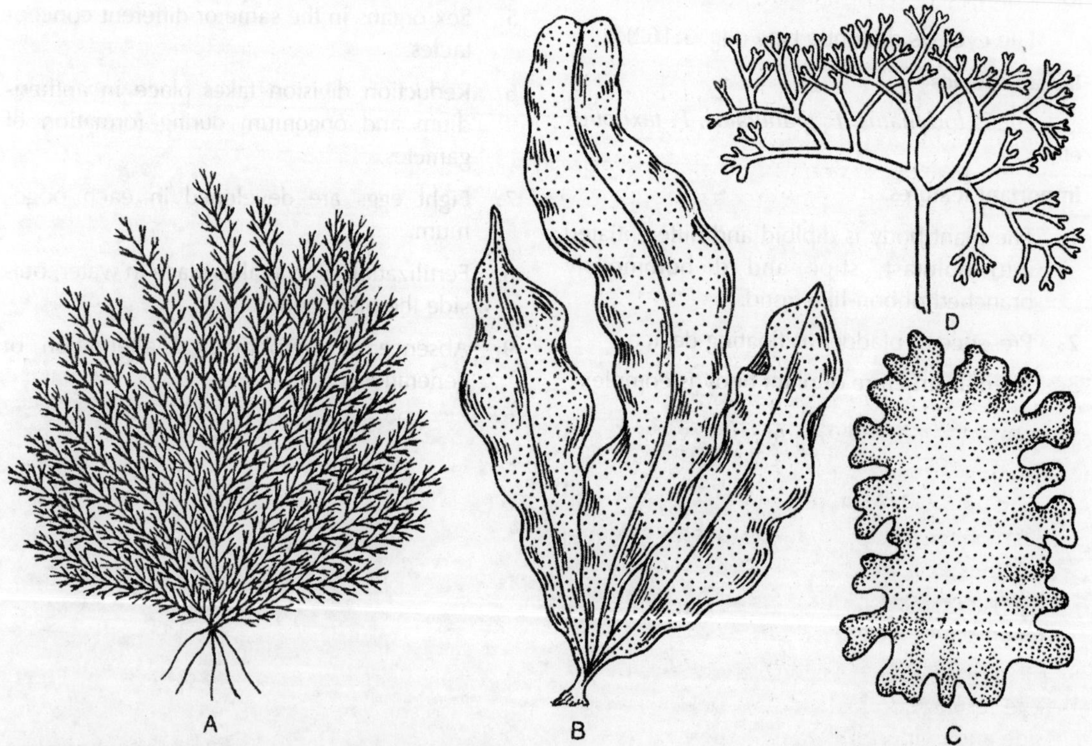

Fig. 3.163: Types of red algae : A. Feathery thallus of *Polysiphonia*, B. Leaf like thallus of *Grinnellia*, C. Flat thallus of *Porphyra* and D. Branched thallus of *Chondrus*

2. The plant body may be **unicellular** (*Porphyridium*) or multicellular. The multicellular form may be **filamentous** (*Goniotrichum*), **parenchymatous** (*Porphyra, Grinellia*), **pseudoparenchymatous** (*Helminthocladia*), **feathery** (*Polysiphonia*) or **ribbon like** (*Chondrus*) (Fig. 3.163).
3. They do not attain the size like that of the brown algae (Phaeophyceae), but may reach up to 2 meters in *Schizymenia* (Abbott, 1967).
4. The flagellated motile stages are totally absent.
5. The cell wall consists of outer pectic and inner cellulosic layer. The mucilaginous material of the outer layer mainly consists of **agar-agar** and **carrageenans** and constitute major portion of dry weight of the cell wall.
6. In multicellular forms, the cell walls have pits, through which cytoplasmic connections are maintained. These cytoplasmic threads are the so-called plasmodesmata. (E.M. studies have not confirmed it.)
7. The members of Rhodophyceae show much variation in the number of nuclei in a cell. In the subclass Bangioideae, cells are uninucleate, but in the subclass Florideae most of the members are multinucleate and the number of nuclei is 3,000-4,000 in *Griffithsia*.
8. The cells may have one chromatophore with a central pyrenoid (Bangioideae) or many discoid and parietal chromatophores with pyrenoids (Florideae).
9. The photosynthetic pigments are chlorophyll a, chlorophyll d; α- and β-carotene; xanthophylls like teraxanthin, lutein, violaxanthin, zeaxanthin, flavoxanthin and biliproteins such as r-**phycoerythrin** and r-phycocyanin. The characteristic red colouration of the algae is due to the sufficient presence of r-**phycoerythrin** which completely masks the chlorophyll a. [The chief Xanthophyll is teraxanthin and chlorophyll b is absent.]
10. The reserve food is **floridean starch**, floridiside and mannoglycerate.
11. Reproduction takes place by all the three means: vegetative, asexual and sexual
    (a) Vegetative reproduction takes place only in unicellular form.
    (b) Asexual reproduction takes place by monospore, neutral spore, carpospore, bispore, and tetraspore.
    (c) Sexual reproduction is of advanced oogamous types:
        (i) The male sex organs are known as spermatangium. Single non-flagellate male gamete is produced in each spermatangium, called **spermatium**.
        (ii) The female sex organs are called **carpogonia** or procarp. Carpogonia are flask-shaped with a long neck, the **trichogyne**.
12. During fertilization, the spermatium comes in contact with the trichogyne with the help of water current.
13. In Rhodophyceae the post-fertilization changes are highly elaborate. They develop **carposporophyte**. Carposporangia are developed from each carposporophyte and each carposporangium produces single carpospore.
14. Most of the Rhodophycean members show biphasic or triphasic life cycle patterns.

Range of thallus

Thallus structures in Rhodophyceae range from simple to complex, unicellular to much complex filamentous organization.

Unicellular form:

Unicellular forms, of course, have the simplest structure. There are few unicellular red algae (*Bulboplastis, Porphyridium, Asterocystis*), but *Porphyridium* is a common unicellular species found in many freshwater pools.

Multicellular form:

Most of the members are multicellular and less than 30 cm in length, but a few are as long as 1 meter. Multicellular forms are represented by simple filaments, branched (*Nemalion, Polysiphonia*) or unbranched; to flat (*Porphyra*), feathery (*Polysiphonia*), lace-like (*Gelidium*) or ribbon-like (*Chondrus*). Simple filaments are abundant among the red algae. Simple heterotrichous forms are represented by *Audouinella*, having the basal attachment ranges from single holdfast cell to a creeping or coalesced prostrate system. In *Asterocystis*, the thallus is pseudofilamentous type. Majority of the members grouped into two types of thallus organizations are Uniaxial and Multiaxial form.

(i) **Uniaxial form**: In Uniaxial form, growth occurs by single apical cell and the thallus contains a single axis or central filament.

In *Batrachospermum*, the axial cells produce determinate branches in whorls. In addition, the pericentral cells produce the down-growing rhizoidal filaments that cover the corticating cells. These may produce secondary whorl of branches.

In *Platysiphonia*, the uniaxial foliose forms have conspicuous apical cells throughout their growth, but in some algae like *Nitophyllum*, the apical cell is obscured in older plants.

In some members, the reduced upright filaments are heavily calcified. They either possess irregularly spreading thallus with smooth surface (*Phymatolithon*) or having short branched or unbranched uprights (*Lithothamnion*).

In Ceramiaceae, majority of the members are uniaxial type, ranging from uncorticated thallus (*Antithamnion*) to heavily corticated (*Ptilota*) one.

(ii) **Multiaxial form**: In Multiaxial form, growth occurs by a group of apical cells and the thallus contains several axial or central filaments.

Simple multiaxial form (*Liagora*) has soft dichotomous branching, gelatinous and calcified thalli. The central axial filaments are colourless and loosely arranged. In members like *Galaxura*, the apical cells of the axial filaments are grouped in apical depression. The thallus is heavily calcified, thus rigid in structure.

Comparatively more specialized multiaxial thalli are heavily calcified found in Cryptonemiales such as *Arthrocladia, Jania* and *Amphiroa*. Here the determinate branches are formed in interrupted series, between which the lightly calcified axial filaments developed from the articulating point of the thallus.

Highly complex multiaxial types include terete*, fleshy or foliose species with dense forms. Here the apical cells can be distinguished only at their young stages. Terete forms include *Agardhiella* and *Gracilaria*, while large foliage form includes *Gigartina*, which is about 1 metre in length.

In *Gastroclonium* and *Champia* of Rhodymeniales, growth occurs characteristically, where the hollow tubular thalli are chambered and the chambers are separated by transverse medullary hyphae.

Lastly, parenchymatous forms is seen in the members like *Smithora* and *Porphyra*. In *Gelidium* thallus is lace-like and in *Chondrus* it is ribbon-like.

[* Terete means cylindrical or slightly tapering, and without substantial furrows or ridges.]

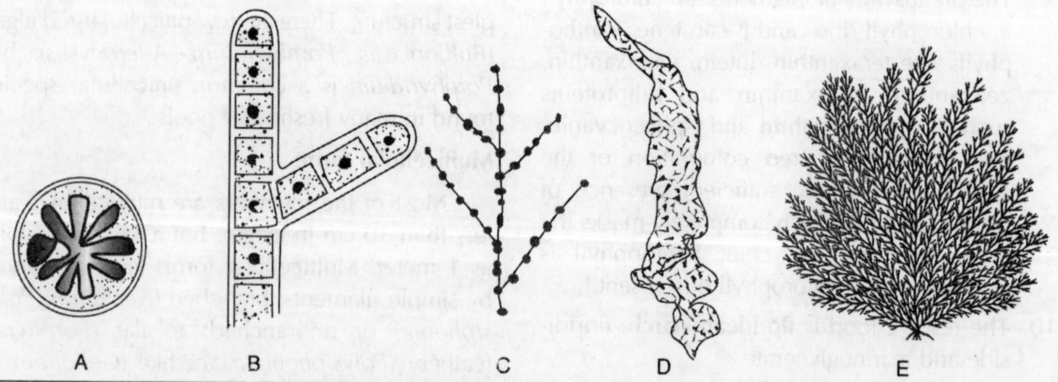

Fig. 3.164: Range of thallus in Rhodophyceae: A. *Porphyridium*, B. *Asterocystis* (pseudofilamentous type), C. *Batrachospermum*, D. *Porphyra* and E. *Gelidium*.

Classification

Fritsch (1935, 45) classified the Class Rhodophyceae into two sub-classes and seven orders. The outline of the classification is given below:

Class. Rhodophyceae

Sub-class Bangioideae

Order. 1. Bangiales

Sub-class Florideae

Order. 1. Nemalionales

2. Gelidiales
3. Cryptonemiales
4. Gigartinales
5. Rhodymeniales
6. **Ceramiales.**

Comparison between the sub-class Bangioideae and Florideae is given (Table 3.4) below. Only Ceramiales is considered below for further discussion.

Order. **Ceramiales**

Important characteristics

1. Plant body is either uniaxial or multiaxial.
2. Immediate after fertilization auxiliary cell is formed from the supporting cell.
3. Carpogonial filament is always four-celled.
4. Tetrasporophytic plants are distinct which contain tetrasporangia and by reduction division they form haploid tetraspores.
5. The tetraspores are tetrahedrally arranged.
6. They show diplobiontic life cycle pattern.

The order Ceramiales (Fritsch, 1945) includes four families such as Ceramiaceae, Delesseriaceae, Rhodomelaceae and Dasyaceae.

Family. **Rhodomelaceae**

Imtortant characteristics

1. Plant body multiaxial (polysiphonous).
2. Cells of the central axis are uninucleate.
3. Both male and female reproductive structures are developed on trichoblasts, the special branches present towards the apex of the thallus.
4. Male trichoblasts are branched but female trichoblats are unbranched.

The genera included in the family Rhodomelaceae are *Polysiphonia, Lophothalia, Colacopsis, Condria, Vidalia* etc.

Table. 3.4: Comparison between the sub-class Bangioideae and Florideae

Bangioideae	Florideae
1. Terrestrial or aquatic, either freshwater or marine.	1. Mostly marine.
2. Thallus structure simple, either unicellular or multicellular.	2. Thallus structure is relatively complex, either uniaxial or multiaxial.
3. Cells are always uninucleate.	3. Except apical and reproductive cells, all others are multinucleate.
4. Each cell contains a single stellate chromatophore in an axial position.	4. Cells contain many discoid chromatophore located peripherally in each cell (except Nemalionales).
5. Each chromatophore has single pyrenoid.	5. Chromatophores lack pyrenoid.
6. Pit connection absent.	6. Pit connections present between neighbouring cells.
7. Thallus shows difused intercalary growth.	7. Growth takes place with the help of a single apical cell.
8. Sexual reproduction is generally absent in majority of members.	8. Sexual reproduction is very common.

Life history of *Polysiphonia* is given:

POLYSIPHONIA

Systematic position

Fritsch (1945)	**Lee** (1999, 2008)
	Phylum. Rhodophyta
Class. Rhodophyceae	Class. Rhodophyceae
Order. Ceramiales	Order. Ceramiales
Family. Rhodomelaceae	Family. Not mentioned
Genus. *Polysiphonia*	Genus. *Polysiphonia*

Occurrence

The genus *Polysiphonia* (Gr. *poly*, many; *siphon*, tube) is represented by more than 150 species (Bold and Wynne, 1978), out of which about 16 species are reported from India. They grow in marine habitat and are cosmopolitan in distribution. Commonly they are found in littoral and sublittoral zones. In India they are found in western and southern coasts. Commonly, they grow as lithophytes i.e., on rocks or stones (e.g., *P. elongata*), some species like *P. urceola*, *P. ferulacea* grow as epiphytes on *Laminaria*. *P. fastigiata* grows as semiparasite on *Ascophyllum nodosum*.

Plant body

Plant body is multiaxial well branched thallus of dark brown, reddish or bluish red colouration appearing as a very small bush (Fig. 3.165A). The height of the bush varies from a few to several centimeters.

Most of the species are heterotrichous in habit, consisting of **prostrate** and **erect systems.**

Prostrate system. It may be multiaxial and well developed. From the lower side of the prostrate system many unicellular rhizoids are developed (Fig. 3.165 B, C). The rhizoids are much lobed at the apex and form definite **attachment discs** (e.g., *P. urceolata, P. nigrescens*). [In *P. elongata* and *P. violacea*, the prostrate system is absent and many rhizoids develop from the lowermost cells of the erect system and by aggregation they form **massive attachment disc** (Fig. 3.165 C).

Erect system. The erect filaments develop from the prostrate system. The erect system consists of main axis and many branches (Fig. 3.166). The branches are of two types: **long branch** and **short branch.**

The long branches are called **branches of unlimited growth** or **long lateral branches** and the short branches i.e., **branches of limited growth** are called **trichoblasts** (Fig. 3.166A). The long branches develop in a spiral or radial symmetry. The trichoblasts are spirally arranged, dichotomously branched, colourless and mostly annual structures bearing sex organs. The trichoblasts may develop both from main axis and long branches.

The main axis and long branches consist of a **central siphon** of many elongated cylindrical

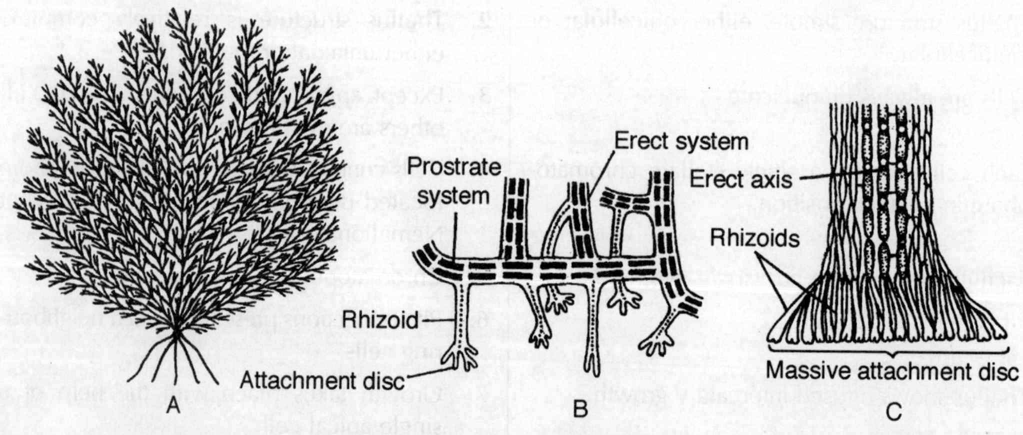

Fig. 3.165: *Polysiphonia* sp.: A. Habit of Plant body, B. Lower portion of thalus with attaching disc and some erect filament developed on it, and C. Lower portion of *P. violacea* with a massive attaching disc

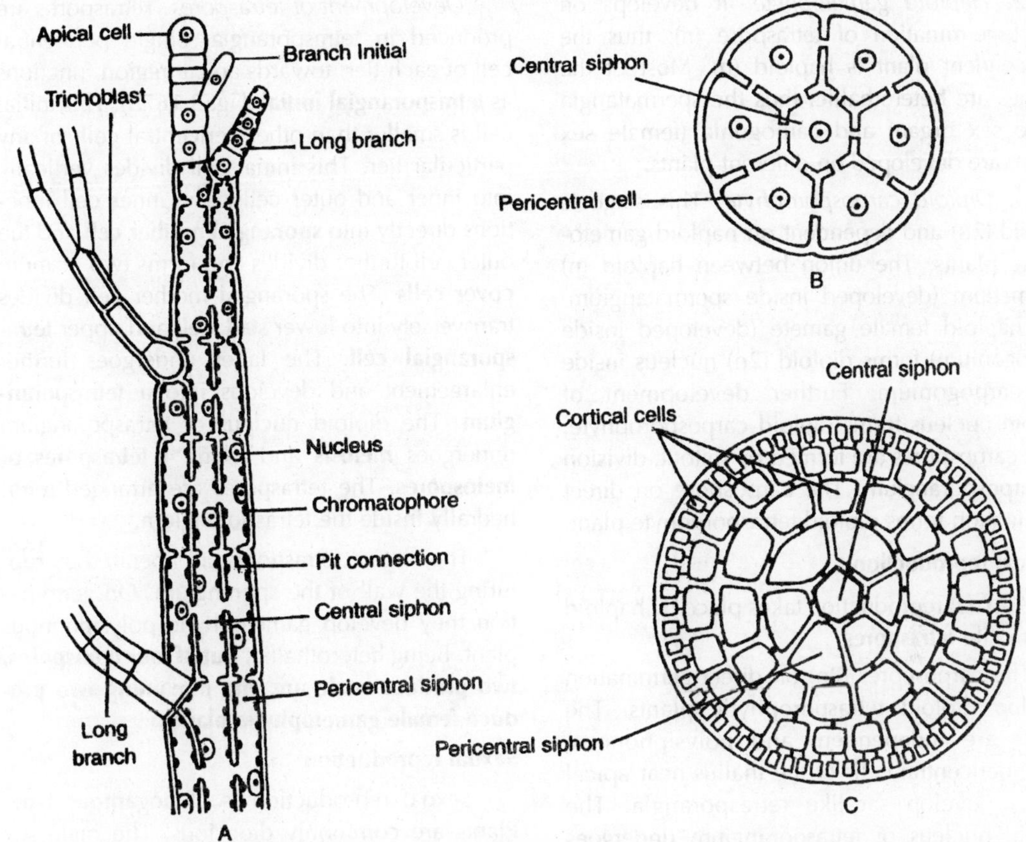

Fig. 3.166: *Polysiphonia* sp.: A. Apical portion of a plant body showing central siphon, pericentral siphon, long branch and trichobiast, B. T.S. of thallus showing central siphon, surrounded by pericentral cells, and C. T.S. of old thallus showing cortical cells in addition to central and pericentral siphons

cells situated in vertical row (Fig. 3.166 B, C). It is surrounded by 4-20 **peripheral siphons**. So the plant body is **polysiphonous** and named *Polysiphonia*. Only the central siphon is present at the apical region of both main axis and the long branches. In most of the species, pericentral siphon is covered by 3 layers of cortical cells formed due to periclinal and anticlinal divisions of the cells of pericentral siphon. All cells of the plant body are connected with each other by pit connections (cytoplasmic connections). The short branches or **trichoblasts are monosiphonous**.

Cell structure

The cells have thick wall, differentiated into outer pectic and inner cellulosic layers. The cells are uninucleate with many discoid chromatophores without pyrenoids. Neighbouring cells are connected by pit connections. The cells contain large central vacuole. Reserve food is floridean starch.

Reproduction

Polysiphonia reproduces both asexually and sexually. Sexual reproduction is of oogamous type. In the life cycle of *Polysiphonia* three kinds of plants are recognised. These are: 1. diploid tetrasporophyte, 2. haploid gametophyte and 3. diploid carposporophyte (Fig. 3.172).

1. *Diploid tetrasporophyte*. It develops on direct germination of carpospore (2n = 40), thus the plant is diploid (2n). It is an independent plant which, instead of developing sex organs develops tetrasporangia. The diploid nucleus of tetrasporangia undergoes **meiosis** and develops four (4) haploid (n = 20) tetraspores.

2. *Haploid gametophyte*. It develops on direct germination of tetraspore (n); thus the independent plant is haploid (n). Most of the species are heterothallic, thus the spermatangia (male sex organ) and carpogonia (female sex organ) are developed on different plants.

3. *Diploid carposporophyte*. This stage is diploid (2n) and dependent on haploid gametophytic plants. The union between haploid (n) spermatium (developed inside spermatangium) and haploid female gamete (developed inside carpogonium) forms diploid (2n) nucleus inside the carpogonium. Further development of diploid nucleus forms diploid carposporophyte. Later carpospores are formed by mitotic division of carposporangium. The carpospore on direct germination forms diploid tetrasporophyte plant.

Asexual reproduction

Asexual reproduction takes place by haploid non-motile **tetraspores**.

The carpospores (2n) on direct germination develop diploid tetrasporophytic plants. The plants are independent and polysiphonous. Some pericentral cells of the thallus near apical region develop sac-like tetrasporangia. The diploid nucleus of tetrasporangium undergoes **meiosis** and forms four tetraspores. The spores are arranged tetrahedrally (Fig. 3.167A).

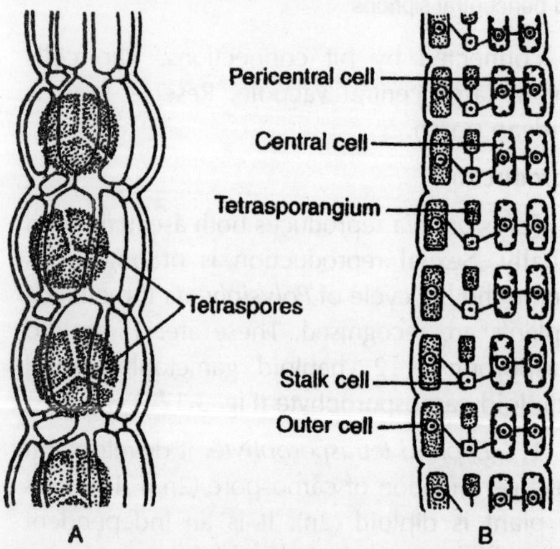

Fig. 3.167: *Polysiphonia* sp.: A. Portion of tetrasporophytic plant (2n) with tetraspores, and B. Portion of tetrasporophytic plant showing development of tetraspores.

Development of tetraspores. Tetraspores are produced in tetrasporangia. Single pericentral cell of each tier, towards apical region functions as **tetrasporangial initial** (Fig. 3.167B). This initial cell is smaller than other pericentral cells of any particular tier. This initial cell divides vertically into inner and outer cells. The inner cell functions directly into **sporangial mother cell** and the outer cell further divides and forms two or more **cover cells**. The sporangial mother cell divides transversely into lower **stalk cell** and upper **tetrasporangial cell**. The latter undergoes further enlargement and develops into a **tetrasporangium**. The diploid nucleus of tetrasporangium undergoes **meiosis** and forms **4 tetraspores** or **meiospores**. The tetraspores are arranged tetrahedrally inside the tetrasporangium.

The mature tetraspores are liberated by rupturing the wall of the sporangium. On germination they develop gametophytic polysiphonous plant. Being heterothallic, **out of four tetraspores, two produce male and the remaining two produce female gametophytic plants.**

Sexual reproduction

Sexual reproduction is of **oogamous type**. Plants are commonly dioecious. The male sex organs i.e., spermatangia and female sex organs i.e., carpogonia, are developed on male and female plants, respectively.

1. *Male reproductive organ*. It is called **spermatangium** or **antheridium**.

Initially male trichobiast develops as side branch on the plant body (Fig. 3.168A). It becomes branched. In some species both the branches become fertile, but in others only one remains fertile and the rest undergo repeated dichotomy to form dichotomous sterile structure. The monosiphonous fertile branch(es) of male trichome bears many unicellular and spherical spermatangia. Each spermatangium is a uninucleate structure which produces single **spermatium**, the male gamete.

During development of spermatangium (Fig. 3.168B-D), all cells except a few basal cells, divide periclinally and form pericentral cells on both the sides. Each **pericentral cell** undergoes several, divisions and forms **spermatangial**

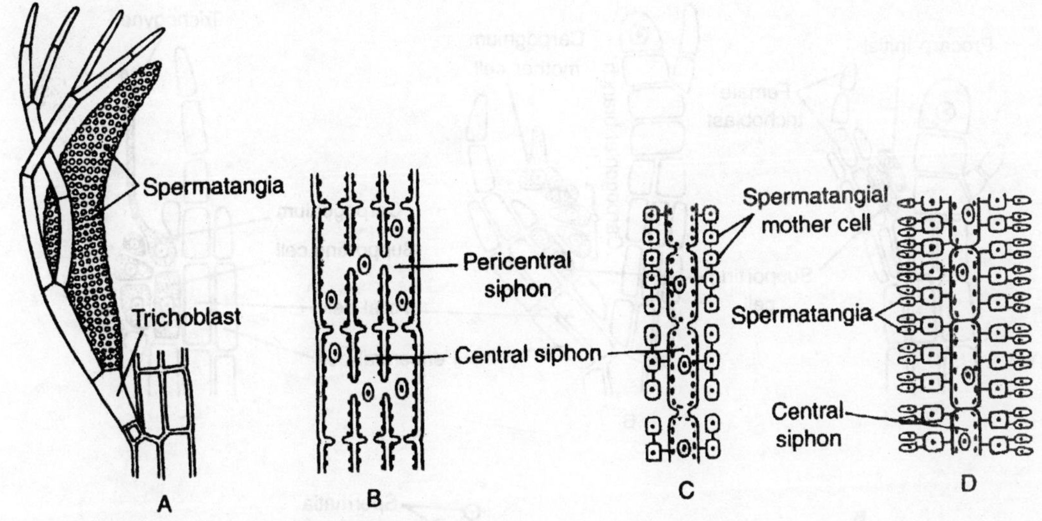

Fig. 3.168: *Polysiphonia* sp.: Development of male reproductive organ. A. Portion of thallus with antheridial branch, and B-D. Sequential development of spermatangia

mother cells. Each one cuts off 2-4 unicellular bodies, the **spermatangia**. Each spermatangium develops into a single non-motile male gamete, the **spermatium.**

The spermatia are liberated from the spermatangium, through a narrow apical slit on the wall. The spermatia are dispersed through water.

2. *Female reproductive organ*. The female reproductive organ is called **carpogonium**.

The carpogonium develops at the top of 2-5 celled carpogonial filament (Fig. 3.169). The carpogonial filament develops on the female trichoblast. The carpogonium is a flask-shaped body, with a basal swollen region containing an egg and an upper elongated neck region, the **trichogyne** (Fig. 3.169C).

During development of carpogonium, initially a **female trichoblast initial** is developed on central siphon, a few cells (3-4) below the apical cell. The female trichoblast initial, then undergoes repeated divisions and forms a **female trichoblast** of 5-7 cells. The lowermost three cells of the female trichoblast divide vertically and form three tiers of pericentral cells. Any one of the pericentral cells of the middle tier towards the mother axis becomes the **supporting cell**. The supporting cell cuts off a small initial at its outside, the **procarp initial** (Fig. 3.169A). The procarp initially undergoes repeated divisions and forms a 4-celled branch, the **procarp** or **carpogonial filament** (branch) (Fig. 3.169B). The apical cell of the carpogonial filament functions as **carpogonium mother cell**. The cell further develops into a **carpogonium**. The carpogonium has a swollen basal region containing egg and an elongated tubular region, the **trichogyne** (Fig. 3.169C).

At the later stage, the carpogonium develops two initials from the supporting cell, one at the base, the **basal sterile filament initial** (Fig. 3.169D) and another at the lateral side, the **lateral sterile filament initial**. The lateral sterile initial divides transversely and forms two-celled lateral sterile filament (Fig. 3.169E). The carpogonium is ready for fertilization at this stage. The pericentral cell adjacent to the supporting cell starts growing to cover the fertilized **carpogonium**. Later on they form sheath (the protective covering) around the fruit body, called as **pericarp**.

Fertilization

The spermatia are dispersed with the help of water. A few spermatia become attached at the tip of the receptive trichogyne. Out of many, only one becomes successful. The common wall of successful spermatium and trichogyne dissolves

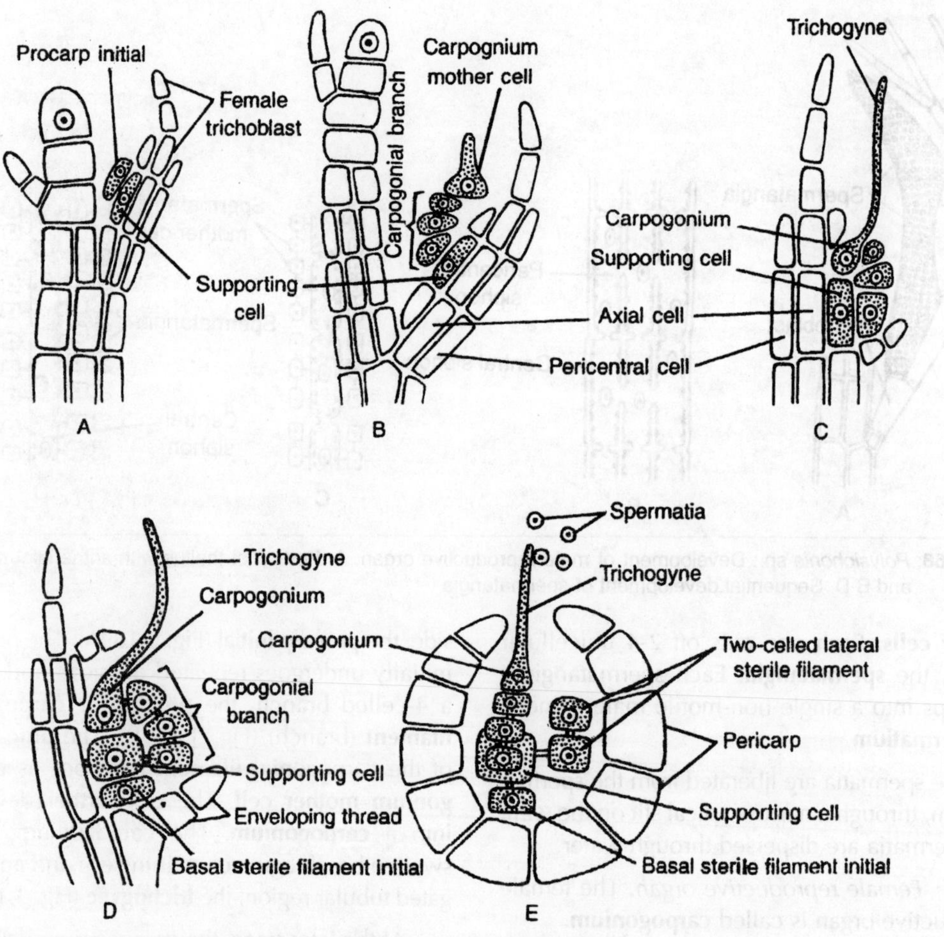

Fig. 3.169: *Polysiphonia* sp.: A-E. Development of carpogonium

at the point of contact and the male nucleus passes to the female nucleus present at the base of the carpogonium. The fusion between the nuclei results in the formation of **zygote**.

Post-fertilization changes

At the starting of this phase, an auxiliary cell is developed from the supporting cell situated just below the basal region of the carpogonium (Fig. 3.170A). Simultaneously, the lateral sterile filament increases in length (4-10 celled) by cell division as well as elongation and the basal sterile initial divides to form a two (2)-celled filament. The auxiliary cell has a single haploid nucleus. A tubular connection is then developed between the auxiliary cell and carpogonium (Fig. 3.170B). The carpogonial nucleus (2n) divides **mitotically** into two nuclei, of which one is transported to the auxiliary cell and the other one remains in the carpogonium. Thus the auxiliary cell contains one haploid and one migrated diploid nuclei. The haploid nucleus (n) is degenerated. Gradually the trichogyne shrivels (Fig. 3.170B). Many vegetative filaments then develop from the adjacent vegetative pericentral cells, which gradually develop the total covering. The diploid nucleus of auxiliary cell then divides **mitotically** and forms two nuclei. One of them then migrates into the outgrowth developed on the auxiliary cell. This outgrowth after separating by a partition wall forms **gonimoblast initial** (Fig. 3.170C). In this way many gonimoblast initials

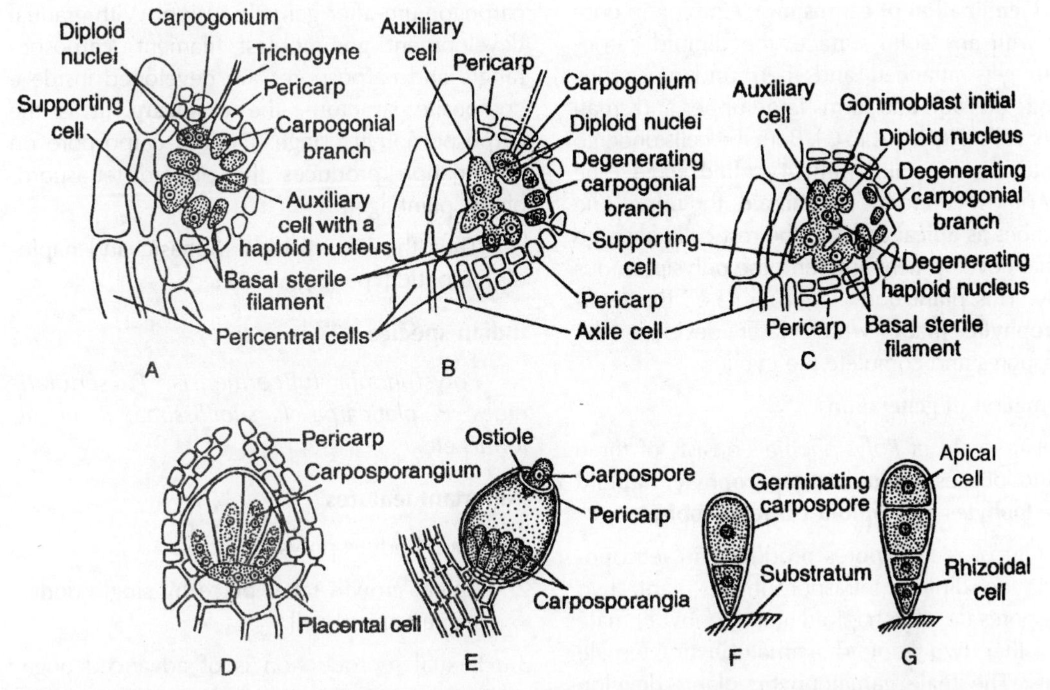

Fig. 3.170: *Polysiphonia* sp.: Post-fertilization changes : A. Fertilized carpogonium, B. Continuity between auxiliary cell and carpogonium, C. Degeneration of carpogonial filament, D. Formation of carposporophyte, E. Liberation of carpospore from mature cystocarp, and F & G. Germination of carpospore

can develop on auxiliary cell. Each initial by repeated mitotic divisions forms **gonimoblast filament**. The terminal cell of the gonimoblast filament develops into **carposporangium**, which forms single **diploid carpospore** inside (Fig. 3.170D, E).

During this development the auxiliary cell, supporting cell, carpogonium and some cells of basal and sterile filaments fuse together and form an irregular cell, the **placental cell** (Fig. 3.170D). The haploid nuclei (n) of the placental cell gradually degenerate and have simply a nutritive function.

The placental cell, gonimoblast filament and carpogonia are covered by many vegetative filaments and form an urn-shaped structure, the **cystocarp** (Fig. 3.170E, 3.171). The outer covering of cystocarp is called **pericarp**. The diploid part of the cystocarp represents the **carposporophyte**. Some cells of basal and sterile filament along with some cells of carpogonial filament gradually degenerate.

The carposporangium develops single diploid carpospore. After liberating from the carpogonium they come out through the ostiole of **cystocarp** (Fig. 3.171).

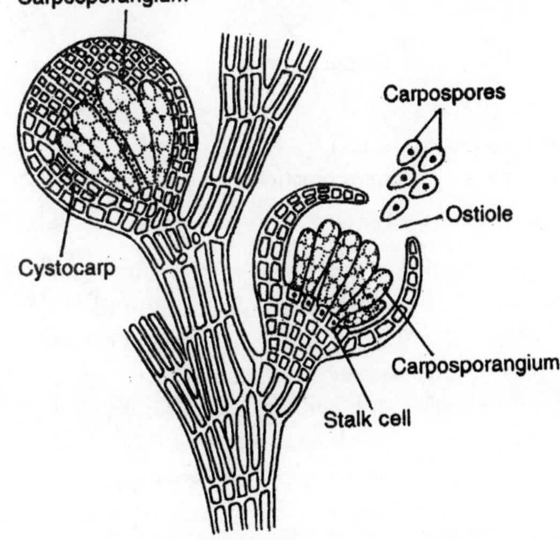

Fig. 3.171: *Polysiphonia* sp.: A branch bearing cystocarps

Germination of carpospore. Coming in contact with any solid surface, the diploid carpospore gets attached and then undergoes first **mitotic** division and forms large upper and small lower cells (Fig. 3.170F, G). Both the cells undergo **mitotic** division and form **4 celled stage**. The lower most cell forms the **rhizoid**, the upper one functions as **apical cell** and the rest cells undergo further development and form the polysiphonous body. This plant body is diploid i.e., the **tetrasporophytic plant**, which later develops the tetraspores and complete the cycle.

Alternation of generations

Life cycle of *Polysiphonia* consists of three distinct phases: **diploid tetrasporophyte, haploid gametophytes** and **diploid carposporophyte**.

Out of 4 tetraspores produced in tetrasporangia on diploid tetrasporophytic plant, two tetraspores develop haploid (gametophytic) male and other two haploid (gametophytic) female plants. The male gametophytic plants develop male gametes inside spermatangia and female gametophytic plants develop female gametes inside carpogonigia. Zygote develops inside carpogonium after gametic fusion. With gradual development gonimoblast filament, carposporangia and carpospores are developed inside a composite structure, the **cystocarp**. It is the carposporophytic stage. Diploid carpospore on germination produces the diploid tetrasporophytic plant again.

Thus the life cycle is triphasic and haplodiplobiontic type (Fig. 3.172).

Indian species

Polysiphonia tuticorinensis, P. sertularioides, P. platicarpa, P. suotilissima, P. unguiformis etc.

Important features

1. Plant body is polysiphonous.
2. Apical growth takes place by single dome-shaped apical cell.
3. Sexual reproduction is of advanced oogamous type.
4. Post-fertilization stage is much elaborate.
5. Cystocarp is well-developed.

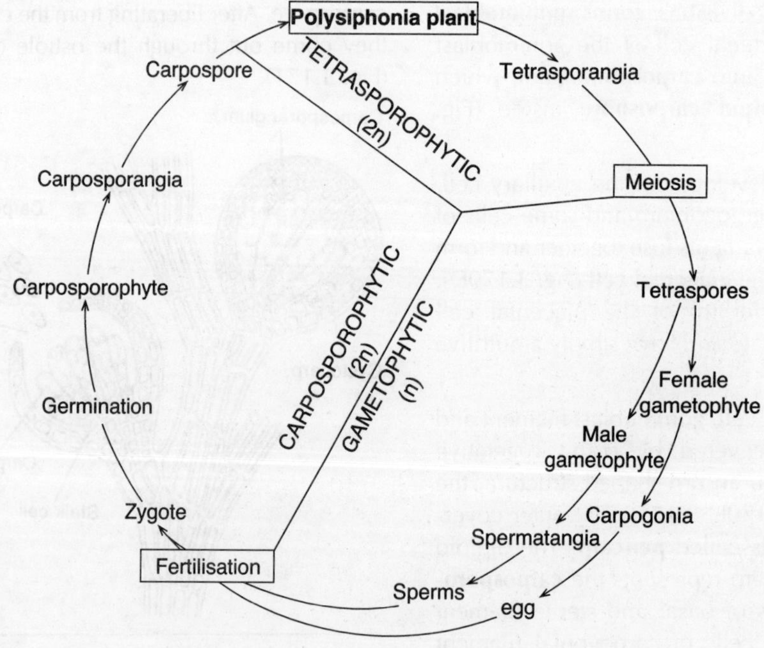

Fig. 3.172: Graphic life cycle of *Polysiphonia* sp.

PROCHLOROPHYTA

This is an unique prokaryote group of phytoplankton, with no phycobilin pigments, which were initially found in 1975 near the Great Barrier Reef and off the coast of Mexico (***Prochloron***). Prochlorophyta was soon assigned as a new algal sub-class by Ralph A. Lewin (1976).

The **Prochlorophyta** is a group of photosynthetic bacteria, an important component of picoplankton. These organisms are abundant in nutrient poor tropical waters and use an unique photosynthetic pigment, divinyl-chlorophyll, to absorb light and acquire energy. They lack red and blue phycobilin pigments and have stacked thylakoids, making them distinctly different from Cyanobacteria (or Cyanophyta), but some authors consider them as part of the Cyanobacteria, as the group **Prochlorales**.

Prochlorophytes are very small microbes generally between 0.2 and 2 μm (photosynthetic picoplankton). They morphologically resemble Cyanobacteria(formally known as blue-green Algae). Members of Prochlorophyta have been found as coccoid (spherical) i.e., Coccus shaped, as in *Prochlorococcus*, and as filaments, as in *Prochlorothrix*.

Prochloron

Prochloron (Gr. *Pro*, before; *chloros*, green) is a unicellular oxygenic photosynthetic prokaryote commonly found as an extracellular symbiont on coral reefs, particularly in didemnid ascidians (sea squirts). It was discovered by R.A. Lewin (1975).

According to **endosymbiotic theory**, *Prochloron is* the predecessor of photosynthetic components, the **chloroplast**, which is present in photosynthetic eukaryotic cells. But this theory is not supported by phylogenetic studies. According to them, *Prochloron* is not on the same line of descent that leads to chloroplast-containing algae and land plants.

Prochloron is a unicellular prokaryotic alga, contains both Chlorophyll a and Chlorophyll b. The cell wall is like Cyanophyce (Myxophyceae), the blue-green alga. The thylakoids and cytoplasm together occupy a thick peripheral band. However, the thylakoids are present not as single lamellae, as in blue-green algae, but in pairs or, sometimes, in thicker stacks. Both thylakoids and cytoplasm are absent from a large central zone which is generally electron-

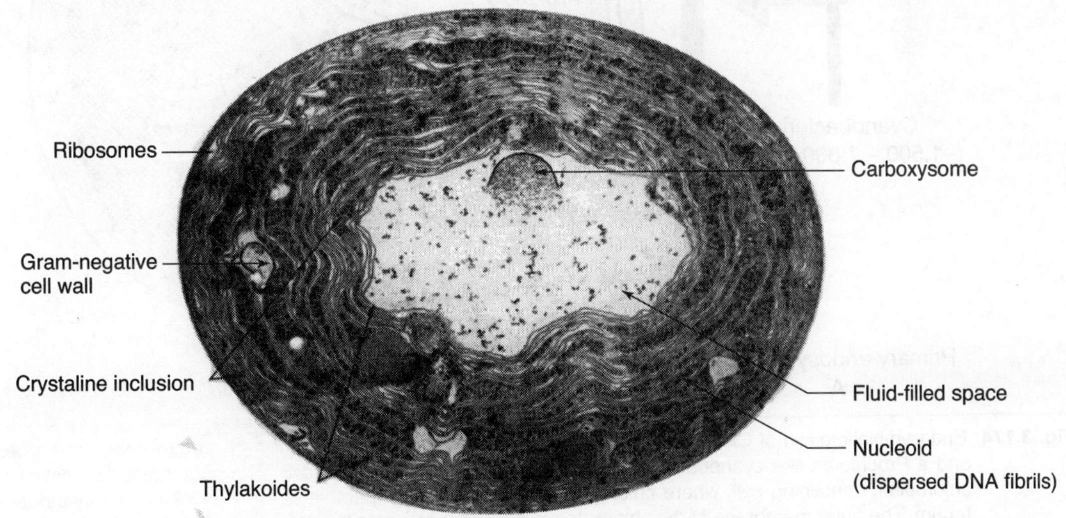

Fig. 3.173: Electron microscopic structure of *Prochloron*

transparent, but may contain electron-dense granules and fibrils. The function of the central zone is not known.

Evolutionary significance of *Prochloron*

Evolutionary pathways regarding the origin of plastids (chloroplasts) in support with the **endosymbiotic theory** is one of the most interesting and also complex problems in the area of evolution and development. It was a great finding to discover a bacterium like ***Prochloron*** which contains all the light harvesting pigments, present in chloroplasts. All **green algae** have **chl** b so the common ancestor of all green algae must be with chl b.

According to endosymbiotic theory this common ancestor should have been the **host** of the previously free-living symbiotic **bacterium** (i.e. prokaryotic alga) which ultimately developed into a chloroplast. Hence the green algae must have acquired their plastids from the **endosymbiosis** of photosynthetic bacteria having **chl** b (such as *Prochloron*). It is believed that the outer membrane of the chloroplast envelope represents the vacuolar membrane of the host cell and the inner membrane represents the cell membrane of the photosynthetic bacteria.

The unique photosynthetic machinery of *Prochloron* among cyanobacteria makes them fascinating to scientists who investigate the evolution of the characters in photosynthetic organisms. Informations about the genetics of *Prochloron* will help the scientists to determine the relationships between these bacteria and other photosynthetic organisms, especially because of their distinct functional and structural photosynthetic features.

Prochloron cells are structurally comparable to the photosynthetic machinery of green plastids, having the thylakoids placed in the cytoplasm of

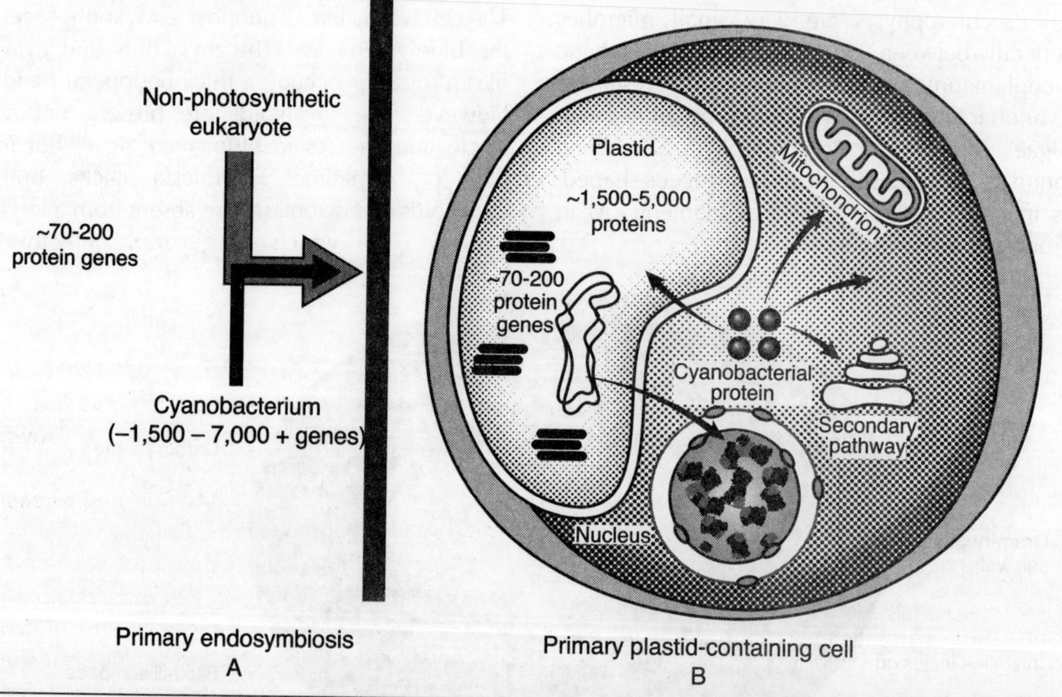

Fig. 3.174: Endosymbiotic origin of Chloroplast: A. Primary endosymbiosis involving a **non-photosynthetic eukaryote** and a Prochloron-like cyanobacterium containing Chl a and Chl b, and B. The schematic diagram of a chloroplast containing cell, where chloroplast is structurally comparable to the *Prochloron* like cyanobacterium. The outer membrane of the chloroplast envelope represents the vacuolar membrane of the host cell and the inner membrane represents the cell membrane of the cyanobacterium.

the bacterial cells (Fig. 3.174). This feature forms the basis for the assumption that *Prochloron* is similar to the **symbiotic bacteria** which evolved into the first chloroplast within the autotrophic eukaryotes. The light traping pigment composition of *Prochloron* has much more similarity with both Cyanobacteria and members of Chlorophyta. According to endosymbiotic theory, it was proposed that *Prochloron* or its ancestor is the most probable precursor of the chloroplast found in photosynthesizing eukaryotes.

Land plants like green algae also have chloroplast surrounded by a double membrane and chl *b* containing chloroplasts, help to assume to have arisen originally by the symbiosis of ***Prochloron*-like** prokaryotic cells.

CHAROPHYTA

Bold and Wynne (1978) created the Division Charophyta includes the order Charales alone under the only one class Charophyceae. Charales includes only one family Characeae. In their classification, they include *Chara* along with some other genera like *Nitella, Tolypella, Nitellopsis, Protochara, Lamprothamnium* and *Lychnothamnus* from Chlorophyceae.

The evolutionary significance of Charophyta is given below:

Evolutionary Significance of Charophyta

Based on the speculation, it can be said that approximately 450–500 million years ago, an ancient freshwater green alga was successfully colonized on land. From this important event in our planet's natural history, the spectacular diversity of plants that now occupy most of our terrestrial ecosystems that arose gradually. Consequently, due to dramatic changes, the planet's atmosphere and biogeochemistry, yield the conditions that support our biota.

Morphological, biochemical, and molecular data strongly suggest a common ancestry of land plants and Charophyceae. Although it is now widely accepted that the Charophyceae are a sister group to the land plants, there is considerable disagreement over the systematics of different charophycean taxa and whether a *Chara*- or *Coleochaete*-like alga, was the ancestral to the land plants.

Comparative analyses of complete **18S rRNA gene sequences** of charophycean algae, together with sequences from Bryophytes, Pteridophytes and higher plants **now almost confirm a common ancestry of charophytes and land plants**. These data suggest that the Charales diverged early into a separate lineage, whereas other orders within the Charophyceae show close affinities to the bryophytes. More than 500 years ago, **Punius** (1469), in his book "Historiae Naturalis", assigned the green alga *Chara* to the land plants as an aquatic horse-tail. He, for the first time, reflected upon the relationship between green algae and land plants. Since then, the systematic position of *Chara* and related algae within the plants has been controversial.

Earlier part of 20th Century, Schenck *et al.* (1917) placed the stoneworts (Characeae) in a common evolutionary lineage with the brown algae (Phaeophyceae). This lineage was independent of the green algae (Chlorophyceae) which were grouped with the red algae (Rhodophyceae) and fungi (Eumycetes, Phycomycetes). Later, Karsten *et al*. (1936) recognized a common ancestry of the 'green plants' (i.e. Chlorophyceae, Bryophyta and Pteridophyta), but the Characeae and Conjugatae were still excluded from the green algae. Instead, the Rhodophyceae were regarded as their closest relatives.

Later part of 20th Century, the studies was based only on morphological data, supplemented **by ultrastructural and biochemical characters, and more recently, by molecular data**. The most important observation by Pickett-heaps (1967, 69, 72, 75) showed that two radically different types of **microtubule organization** are involved during cytokinesis in the green algae. Species of the Volvocales, Tetrasporales, Chlorococcales, Oedogoniales and some Ulotrichales are characterized by *the collapse of the interzonal spindle apparatus after mitosis and by developing a '**phycoplast**' with the microtubules all oriented in the plane of cell division*. In contrast, Klebsormidiales, Coleochaetales, Charales and Conjugales along with the land plants have a *persistent spindle and develop a cleavage furrow* (*Klebsormidium*) or a '***phragmoplast***' in which the microtubules are oriented perpendicular to the eventual plane of cytoplasmic division. Based on the assumption that these features must be evolutionarily conserved. Pickett-heaps and Marchant (1972) proposed a new phylogenetic scheme of green algal evolution. According to them, the **phycoplast** and the **phragmoplast** containing green algae constitute two different evolutionary lineages. The latter lineage subsequently designated as Charophyceae by Stewart and Mattox (1975), were suggested to be the progenitors of the bryophytes and higher land plants. Such a relationship is further supported by **synapomorphic characters**, such as the usage of ***glycolate oxidase*** rather than glycolate dehydrogenase in photorespiration (Frederick *et al.* 1973), the possession of ***copper-zinc superoxide dismutase*** (De Jesus *et al.* 1989) and two class, the ***aldolases*** (Jacobshagen and Schnarrenberger 1990), and flagellate spores and/or gametes with a **multilayered structure (MLS)**. Whereas it is unlikely that all MLS's found so far are homologous structures (Sluiman 1983). The combined features of a persistent and open mitotic spindle, MLS, and a unique type of peroxisome structurally and biochemically similar to those of land plants (Stewart *et al.* 1972) are considered reliable indicators of a common ancestry (Graham, 1993).

To emphasize the proposed relationship of the Charophyceae to the land plants, the terms '**Streptophyta**' and '**Anthocerotophyta**' were created to combine the charophytes and the embryophytes into a single division (Bremer 1985, Swiman 1985). The classification of the green algae into two classes, Chlorophyceae and Charophyceae (Stewart and Mattox, 1975) was later extended to a total of five classes as more taxa were studied by comparative cytology (Mattox and Stewart 1984). The concept of the Charophyceae remained unchanged with the exception that the *Chlorokybales* (single genus *Chlorokybus*), were added as a 5th order with the Zygnematales, Klebsormidiales, Coleochaetales, and Charales. This classification has gained wide acceptance since then, although different views of the systematic position of various charophycean orders have been suggested, in different books. Bold and Wynne (1985) followed the view of Gramblast (1974) who, after extensive *studies on the fossil records* of charophytes, stated that these algae occupy an isolated position between green algae and bryophytes. Taking these data into consideration, together with the unique complex and reproductive structures, sperm morphology, and possession of a protonematal stage, they were classified as a separate division, **Charophyta**, whereas the remaining charophycean algae were assigned to different chlorophycean orders. Still another view is proposed by Hoshaw *et al.* (1990) as the characteristic for the classification of the Charophyceae (*phragmoplast in cell division, glycolate and urea metabolism*) have not yet been extensively studied in members of the Zygnematales, these algae were provisionally separated from the Charophyceae as a phylum Conjugaphyta. Not only the systematic position of the different charophycean taxa is controversial, but also which taxa eventually gave rise to the land plants, needs more investigations on the above parameter.

Lastly it can be mentioned that the study of charophytes is still in an early but exciting phase of discovery which needed yet to be resolved.

SOME GENERAL ASPECTS ON ALGAE

LUMINOCITY IN ALGAE

The emission of light from living organisms was not experienced until 1830 (Sweeney, 1979).

Two types of light emission take place among living organisms (Hastings, 1986):

1. **Bioluminescence (chemiluminescence):** Here energy from an exergonic chemical reaction is transformed into light energy.

2. **Photoluminescence**: This depends on *prior absorption of light.*

Luciferin is the compound responsible for bioluminescence, which is oxidised with the help of luciferase, results in the emission of light.

The basic reaction (Hastings, 1983): Luciferin becomes oxidised by luciferase, resulting in an electronically excited product (P)* which emits a photon (hv) on decomposition:

$$\text{Luciferin} + O_2 \xrightarrow{\text{Luciferase}} (P)^* \longrightarrow P^+ \text{ hv}$$

Bioluminescence is observed in many organisms of different phyla, not only in plants but also in fungi and animals. This includes bacteria to dinoflagellates to jellyfish and brittle stars to worms, fireflies, molluscs, and fish (Hastings, 1986). e.g., *Gonyaulax polyedra, Pyrocystis fusiformis, Ceratium fusus, Peridinium pentagonium, Pyrocystis fusiformis* etc.

TYPES OF PLASTIDS IN ALGAE

Plastids are the small bodies of various shapes present in many living cells containing DNA, and/or pigments and reserve food material. It is of two types: *chromoplast* (pigmented) including chloroplasts (chlorophyll containing plastids), and leucoplasts (colourless). The leucoplasts are the colourless plastid responsible for the accumulation of storage products. ***The basic type of plastid in algae is the chloroplast, responsible for photosynthesis***. Chloroplast develops from proplastid. Proplastid is a reduced plastid with few thylakoids. In heterotrophic algae, it remains as proplastid. *Plastids are absent in Cyanophyta, where pigments remain in the peripheral region of protoplast, the chromoplasm.*

Chloroplasts

Chloroplast is bounded by the double membrane, the chloroplast **envelop**. In many eukaryotic algae, the chloroplast envelop is further surrounded by one or two membrane of **chloroplast endoplasmic reticulum**, with attached ribosome on outer surface. Chloroplast consists of a series of flattened membranous vesicles, the **thylakoids** and a surrounding matrix, the **stroma**; both are responsible for photosynthesis. The thylakoids contain **chlorophylls**, the site of photochemical reaction and the chemical reaction, i.e., dark fixation of CO_2, takes place in stroma. Thylakoids are also associated with other pigments, like **phycobilisomes**. The **pyrenoid**, a well-differentiated region of chloroplast present in stroma, is frequently associated with storage products. Presence of pyrenoid indicates its primitiveness in the evolutionary point of view. Pyrenoid contains ribulose-1, 5-bisphosphate carboxylase/oxygenase (Rubisco), an enzyme which fixes CO_2. It commonly contains small and spherical lipid droplets between the thylakoids, associated with synthesis and growth of lipoprotein membrane of chloroplast. Most of the motile algae have an orange-red eyespot, consists ol a group of tightly packed carotenoid lipid globules, and acts as organ of photoreceptor.

Most of the chloroplast contain DNA of prokaryotic nature. DNA microfibrils are circular and attached with the chloroplast membrane and lack histones (the basic protein).

Pigments: Photosynthetic algae have chlorophyll. The chlorophylls are of four types: Chlorophyll a, b, c (c_1 and c_2) and d. Chlorophyll a is universally distributed in all photosynthetic algae. **Carotenoids** are yellow, red or orange pigments, occur inside the plastid (outside in some cases). They are of two types, carotenes and xanthophylls. β-carotene is commonly found in algae. **Phycobiliproteins**, the water soluble blue or red pigments, are located in Rhodophyta or Cryptophyta inside thylakoids.

Types of chloroplast

Based on the structural configuration, the chloroplasts in eukaryotic algae are of **5 types:**

Type 1. One thylakoid per band, no endoplasmic reticulum (Rhodophyta) (Fig. 3.175A).

Other characteristics:

(a) Thylakoids can be free from one another, with phycobilisomes (containing phycobiliproteins) on the surface of thylakoids.

(b) Phycobilisomes on the surface of one thylakoid alternate with the phycobilisomes present with the adjacent thylakoid.

(c) Phycobilisomes appear as discs when phycocyanin predominates, or as 35 nm when phycoerythrin predominates.

(d) In advanced members, peripheral thylakoids enclose the rest of the thylakoids, but in primitive members the thylakoids terminate close to the chloroplast envelop.

(e) The clumps of DNA are scattered throughout the plastids.

(f) Pigments: In addition to chlorophyll a, chlorophyll d is also found but not in all members.

Type 2. Two thylakoids per band, two membranes of endoplasmic reticulum (Cryptophyta) (Fig. 3.175B).

Other characteristics:

(a) Chloroplasts contain bands of two thylakoids.

(b) Phycobiliproteins are dispersed within the thylakoids.

(c) Eyespots occur as lipid droplets in the chloroplasts (like Chlorophyta and most Heterokontophyta).

(d) The clumps of DNA are scattered throughout the plastids.

(e) Pigments: In addition to chlorophyll a, chlorophyll c_2 is also found, but c_1 is absent.

Type 3. Three thylakoids per band, one membrane of chloroplast endoplasmic reticulum (Dinophyta, Euglenophyta) (Fig. 3.175C).

Other characteristics:

(a) The thylakoids are grouped in bands of three with a girdle or peripheral band running parallel to the chloroplast envelope (Euglenophyta) or without girdle band (Dinophyta).

(b) Eyespot occurs as a group of membrane-bounded lipid droplets, free of the chloroplast (like Eustigmatophyceae).

(c) The clumps of DNA are scattered throughout the plastids (except Euglenophyta, which shows a variable distribution of chloroplast DNA).

(d) Pigments: In addition to chlorophyll a, chlorophyll b is also found (except in Dinophyta, where chlorophyll c_2 is present, but c_1 is absent).

Type 4. Three thylakoids per band, two membranes of endoplasmic reticulum (Prymnesiophyta and Heterokontophyta except Eustigmatophyceae) (Fig. 3.175D).

Other characteristics:

(a) The thylakoids are grouped in bands of three with a girdle or peripheral band running parallel to the chloroplast envelope (Heterokontophyta) or without girdle band (Prymnesiophyta and Eustigmatophyceae).

(b) Eyespots occur as lipid droplets in the chloroplasts in most of the Heterokontophyta, but eyespot occurs as a group of membrane-bounded lipid droplets, free of the chloroplast in Eustigmatophyceae (like Dinophyta, Euglenophyta).

(c) The clumps of DNA are scattered throughout the plastids (in Prymnesiophyta and Eustigmatophyceae) or DNA occurs in ring just within the girdle lamellae (in Bacillariophyceae, Chrysophyceae, Raphidophyceae and Xanthophyceae except *Vaucheria*).

(d) Pigments: In addition to chlorophyll a, chlorophyll c is also found in most of the heterokontophyta and fucoxanthin is the principal xanthophyll in golden-brown algae like Bacillariophyceae, Chrysophyceae, Prymnesiophyceae and Phaeophyceae).

Type 5. Two to six thylakoids per band, no chloroplast endoplasmic reticulum (Chlorophyta) (Fig. 3.175E).

Other characteristics:

(a) Thylakoids occur in bands of 2-6, with thylakoids running from one band to the next.

(b) Eyespots occur as lipid droplets in the chloroplasts (like Cryptophyta and most Heterokontophyta).

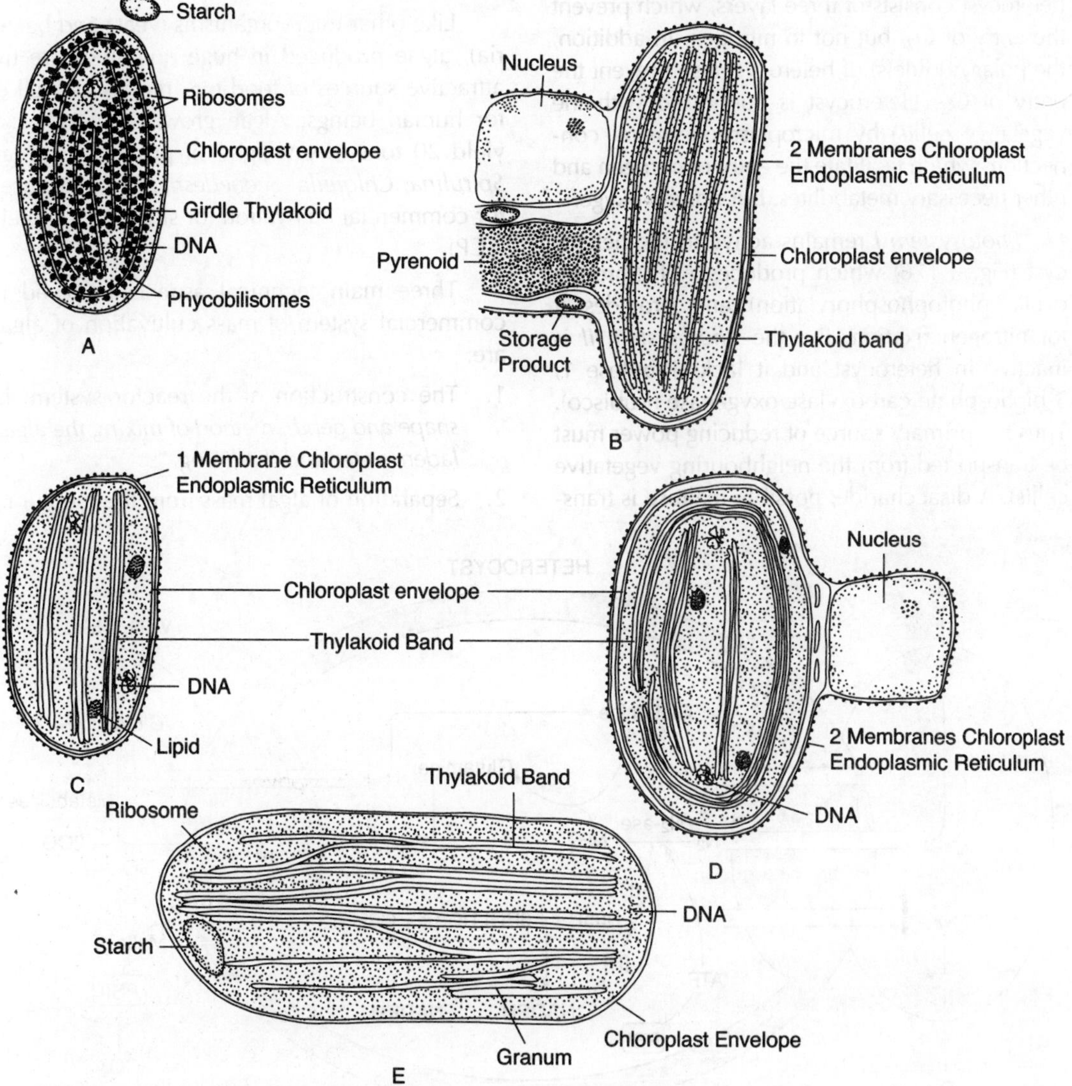

Fig. 3.175: Types of chloroplasts in eukaryotic algae: A. One thylakoid per band, no endoplasmic reticulum, B. Two thylakoids per band, two membranes of endoplasmic reticulum, C. Three thylakoids per band, one membrane of chloroplast endoplasmic reticulum, D. Three thylakoids per band, two membranes of endoplasmic reticulum and E. Two to six thylakoids per band, no membranes of endoplasmic reticulum

(c) The clumps of DNA are scattered throughout the plastids.
(d) Pigments: In addition to chlorophyll a, chlorophyll b is also found.

N_2 FIXATION BY HETEROCYST

In Cyanophyta (Cyanobacteria), due to cut off of an external source of combined nitrogen (nitrate or ammonium), vegetative cell(s) of the filament develop into thick walled specialised cells, known as heterocyst(s). The wall of the heterocyst consists of three layers, which prevent the entry of O_2, but not to nitrogen. In addition, the polar nodule(s) of heterocyst also prevent the entry of O_2. Heterocyst is connected with the vegetative cell(s) by microplasmodesmatal connections which facilitate the entry of nitrogen and other necessary metabolites, but not the oxygen.

Photosystem I remains active inside heterocyst (Fig. 3.176) which produces ATP (through cyclic photophosphorylation) and provides it for nitrogen fixation. But the *Photosystem II* is inactive in heterocyst and it lacks Ribulose 1, 5-biphosphate carboxylase/oxygenase (Rubisco). Thus the primary source of reducing power must be transported from the neighbouring vegetative cell(s). A disaccharide, possibly maltose is transported to heterocyst and then hydrolysed to form glucose. The glucose is then used to provide NADPH by oxidation through pentose phosphate pathway. *In exchange, the fixed nitrogen is transported to vegetative cell(s).* Ammonium is assimilated by Glutamine synthatase and the product glutamine is transported to the vegetative cell(s), which then converted into glutamate again by glutamate oxyglutarate aminotransferase (GOGAT).

CULTURE OF MICROALGAE FOR SCP

Like other microorganisms (yeast and bacteria), algae produced in huge quantities are the attractive sources of food for animals as well as for human beings. Algae grown in ponds can yield 20 tons (dry weight) of protein/acre/year. *Spirulina, Chlorella, Scenedesmus* etc., are used for commercial cultivation for single cell protein (SCP).

Three main technical aspects involved in commercial system of mass cultivation of algae are:

1. The construction of the reactor system: its *shape and depth, method of mixing the algal-laden water and the lining.*
2. Separation of algal mass from the medium.

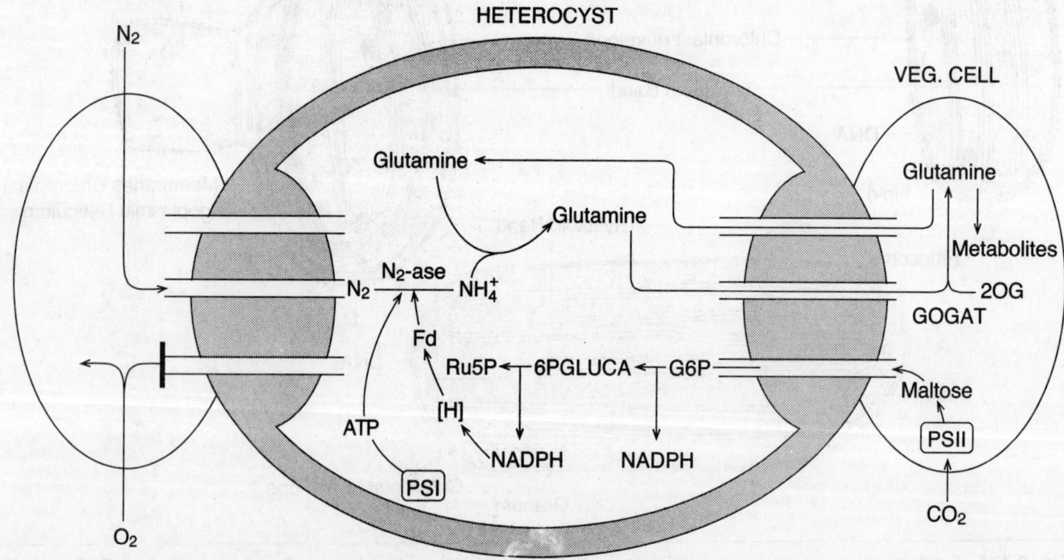

Fig. 3.176: Diagrammatic representation of carbon and nitrogen exchange between the heterocyst and vegetative cells during heterocystous nitrogen fixation in Cyanophyceae (Cyanobacteria)

3. Dehydration of algal mass to facilitate product distribution and storage.

Mass cultivation

Mass production of algae requires a less expensive, yet reliable enclosure for growing the culture. Different designs are constructed for the mass production, but two basic systems are distinguished:

(i) Horizontal turbulent flow created by various types of devices, and

(ii) Inclined baffled surfaces (cascades) where the algal suspension is constantly pumped back from the lower to the upper part of the growing region.

Designs

Horizontal Ponds: Three basic designs are used in the mass cultivation:

1. Oblong forms (race ways) are constructed either as one or several units jointed together, with agitation by a propeller, paddle wheel(s) etc.
2. Circular ponds are made with agitation by a rotating arm and
3. Closed circulation systems, where the pond is covered by transparent film or in shallow trenches roofed with plastic or the culture is circulated in plastic tubes.

Sloped cultivation units: This system designed as such to create a turbulence flow while the algal culture flows through a sloping enclosure.

Among the other systems, the **closed circulation system** is mentioned:

In a closed system the algal cultures are not exposed to the atmosphere, but are covered with a transparent material or contained within transparent tubing. Out of many, one design followed by Richmond in Israel, used a covered 100 m^2 *Spirulina* pond with a lining of 0.8 mm PVC and the cover is 0.02 mm thick polythene used for greenhouse.

Advantages:

(i) Advantages lie in preventing evaporation and elevating the temperature. This system is suitable in areas where evaporation is higher in warm sunny regions. The high rate of evaporation increases the cost of water and pumping as well as algal growth due to increase the salinity of the medium.

(ii) Another advantage is the reduction in the amount of dirt and insects contaminating the algal product.

Disadvantages:

(i) Marked reduction in the light penetration due to the covering, accumulation of dust on the outer surface and also the water condensation on its inner surface.

(ii) This system increases the initial investment of the covering.

The current high market value of different algae, such as *Spirulina*; in connection with the health food market, justify the importance of this covering system.

Pond lining: The lining of pond represents the most expensive item in pond construction. At present, commercial ponds are lined with either concrete or plastic. In tropical India, the UV-resistant polyvinylchloride (PVC) is used as the suitable lining material. The UV-resistance polyethylene would also be suitable as it is chemically and biologically inert.

Mixing and Turbulence: If the environmental conditions are not growth-limiting, a suitable amount of nutrition is mixed to create a *turbulent flow* which is found to be the most important requisite for consistently higher rate of yield of algal mass. High population density (500-1,000 mg dry wt./liter) is required to get high output/unit area and the mixing is required to prevent from sinking of cells to the bottom of the pond. Accumulation of organic matter in "dead" areas would affect cell deterioration and anaerobic decomposition would decrease the quality of the product. The maintenance of high turbulence is related to the nutritional and gaseous gradients formed around the algal cells during their metabolic activity and also with the mutual shading.

The turbulent flow in the pond may be created by paddle wheel, free propeller, combination of pump and gravity flow, injector, airlift etc.

Supply of carbon dioxide (CO_2): The supply of CO_2 is required for the proper yield in the large scale cultivation, particularly for species grown near neutral pH. To avoid loss of CO_2, the mixing can be done by "gas exchanger" as designed by Heussler *et al*. (1978).

Harvesting of the algae

Efficient method of harvesting of algae from the growth medium is essential for the good yield. Many methods used in this process are centrifugation, electrofloatation, sand bed filtration and autoflocculation.

Dehydration of algal mass

The step is very important in respect of production cost. The various methods differ in the extents of both capital investment and energy requirement and also have effect on the food value and taste of the product. Out of many, *spray drying, surface drying* and *sun drying* are effective:

(a) Despite the very attractive future of SCP as nutrient for man, there are some problems on universal global adoption.

(b) Taste of human beings of different corners greatly varies depending on their climatic condition and other factors.

(c) The high nucleic acid content of cells can produce intestinal disturbances.

(d) It is important to ascertain that the amino acids of the microbial protein would meet dietary requirements of the consumer.

CULTURE OF ALGAE FOR β-CAROTENE

Carotenoids are yellow, orange or red pigments that usually occur inside the plastid, but may be present outside the plastid in certain cases. Naturally occurring carotenoids can be divided into two groups:

(a) Oxygen-free hydrocarbons, the carotenes, and

(b) Their oxygenated derivatives, the xanthophylls.

Two types of carotenes are found in algae, **α- and β-carotene**. Out of which, β-carotene is very common in different algal groups.

Carotenoids are mostly used as natural food colour, as additive for animal feed to increase the colour of their products like egg yolk, poultry and fish meat and also to improve animal health.

Out of more than 400 known carotenoids, a few are used commercially for the above purposes. These are β-carotene, lutein, astraxanthin, bixin etc.

β-carotene is the only carotenoid which has the potential to form two molecules of vitamin A (retinol). In human beings, the conversion of β-carotene into vitamin A occurs in small intestine, where both the compound β-carotene and vitamin A are absorbed and subsequently pass on to plasma, tissue and organ of the body. The average daily carotene uptake is 1.5 mg, equivalent to 250 retinol equivalents or 2,500 IU of vitamin A.

The unicellular halotolerant alga, *Dunaliella bardawil is found to contain high concentration of β-carotene.*

Commercial production: *Dunaliella* is the most suitable and commonly used in mass cultivation in ponds. It grows well in media with high magnesium, calcium, sodium and the respective anions, sulphate and chloride. In desert areas, the high sunlight vis-a-vis access to brackish water or sea water at temperature ranges from –5°C to more than 40°C, make them suitable for commercial production.

The area with high light intensity having source of salt water (near sea or salt lakes), increases rate of water evaporation and this is considered to be a suitable place for the commercial cultivation of *Dunaliella*, in the production of β-carotene.

Four methods have been followed in the large scale cultivation of *Dunaliella* and commercial production of β-carotene:

1. Extensive cultivation.
2. Intensive cultivation.
3. Semi-intensive cultivation, and
4. Highly intensive cultivation.

Only first two types are described:

1. **Extensive cultivation.** It uses no mixing

and requires minimal control of the environment. To decrease fungal contamination and attack of predators (amoebae, ciliates etc.), very high salt concentration is employed by the cultivars. Naturally selected strain is well-adapted to nearly salt saturation condition. In this process:

(i) Productivity is low (<0.1 g β-carotene m^{-3})

(ii) Large area is needed for commercial production

(iii) Operation cost is very low.

2. **Intensive cultivation**. This process uses high biotechnology to control all factors affecting cell growth and chemistry. Ponds commonly used are oblong, lined and up to an appx. 3,000m^2 in size. This may yield 3-9 kg β-carotene on total harvest (5-15g β-carotene m^{-3}).

During large scale production of β-carotene, the rate is about 200 mg $m^{-2}d^{-1}$/year, thus in an area of 50,000 m^2, 3650 kg β-carotene is produced /year.

Dunaliella is found to grow well in sea water based media containing 1.5M NaCl, 0.4M $MgSO_4$ and 0.1 M $CaCl_2$ under pH.

Market products

Three different products are available in the tiarket: (β-carotene extracts, *Dunaliella* powder for human use (tablet or hard capsules) and dried *Dunaliella* powder for feed use (cattle, poultry, fish etc.).

ALGAL BIOFERTILIZER

Algae are used as biofertilizer either directly or indirectly. Apart from bacteria, the soil contains algae, fungi and other microorganisms; and algae occupy about 3 times than the bacteria. They belong to different groups like Cyanophyta, Chlorophyta, Phaeophyta, Rhodophyta etc. They are used as the sources of *nitrogen, potassium* and *calcium*. Seaweeds are either allowed to rot in the field or composted with other organic materials. The prepared manures when added in cultivated lands, they not only enrich the soil with mineral nutrients, but also help in soil binding.

A. **Nitrogen source**: Members of Cyanophyta play a great role in the fixation of atmospheric nitrogen. 32 genera and 152 species of blue green algae (BGA) so far reported are capable of fixing atmospheric nitrogen. They grow in aerobic, anaerobic/microaerobic condition. Some of the important nitrogen fixing members of Cyanophyta are the different species of *Nostoc, Anabaena, Spirulina, Oscillatoria* and *Cylindrospermum*. They are with *(Nostoc)* or without (*Oscillatoria*) heterocyst. Some of the algae (*Nostoc*) are multiplied in the laboratory and applied in the field as fertilizer. The algae grow well in the rice field and increase the nitrogen content of the soil. The members of Cyanophyta, which are rich in nitrogen and phosphorus, are better fertilizer than seaweeds. In tropical countries, bottom mud of dried ponds is regularly used as manure (due to high content of members of blue-green algae) in crop fields. A blended mixture of BGA and seaweed (*Macrocystis*) may serve as an ideal fertilizer and this can relieve the shortage of fertilizer in the developing countries.

The nitrogen fixing members of blue green algae are classified into the following major groups:

1. Unicellular or colonial aerobic nitrogen fixers, e.g., *Chlorogloeopsis, Gloeocapsa, Gloeothece* and *Aphanothece*.

2. Unicellular or colonial anaerobic/microaerobic nitrogen fixers, e.g., *Xenococcus, Synechococcus, Dermocarpa* etc.

Structure of β-carotene

3. Heterocystous filamentous aerobic nitrogen fixers, e.g., *Nostoc, Anaebaena, Anabaenopsis, Tolypothrix, Westiella, Aulosira* etc.
4. Non-heterocystous filamentous anaerobic nitrogen fixers, e.g., *Oscillatoria* and *Phormidium*.
5. Non-heterocystous filamentous aerobic nitrogen fixers, e.g., *Trichosdesmium*.

B. **Potassium source**: Seaweeds are rich in potassium, but poor in nitrogen and phosphorus than the farm manure. Sea weeds of Phaeophyta and Rhodophyta are used as fertilizer mainly in the coastal regions. They are used in different countries like Sri Lanka, Ireland, and France etc., for Coffee plantation, in Potato and Turnip fields. In China, they are used in ground-nut fields, whereas in Japan in rice fields.

Seaweeds are used as compost. Sometimes the weeds are dried and burnt and the ashes are scattered over the cultivated lands.

Concentrated liquid extracts of seaweeds are used as liquid fertilizer.

C. **Calcium source**: The powder prepared after grinding of different members of *Lithophyllum* and *Lithothamnion* are used in place of lime. Some species of *Chara* are also used for similar purpose.

ALGAL TOXINS

Toxins are the compounds produced by microorganisms and being toxic to plants and/animals.

The toxins produced by different members of algae are called algal toxins. The members are mostly Cyanophyta, Dinophyta and Prymnesiophyceae.

Different toxins such as *Anatoxin, Saxitoxin* and its derivatives are produced by members of Cyanophyta and Dinophyta. Other toxins like *microcystins, hepatotoxins and lipopolysaccharides* are produced by Cyanophyta. Okadaic acid and polyether toxin are produced by Dinophyta as found in diarrhetic shellfish poisoning (DSP).

Some algal members and their toxins are named:

Cyanophyta

1. *Anabaena flos-aquae*. It contains four types of anatoxins — anatoxin A, B, C and D. Out of four, the well-known **anatoxin A** is an alkaloid, a nitrogen-containing compound of low molecular weight.
2. *Microcystis*. It contains a toxin called **microcystin**, a polypeptide of 10 amino acids. It causes death at a dose of 0.5 mg/kg body weight.
3. *Aphanizomenon flos-aquae*. It contains a mixture of **saxitoxin** and three related toxins. Saxitoxin is an alkaloid responsible for paralytic shellfish poisoning. The minimum lethal dose is 10 μg/kg body weight. Consumption through drinking water, even at low concentration, may cause diarrhea in human beings.
4. *Lyngbya majuscula*. It is a marine alga, which produces **debromoaplysiatoxin**, that causes swimmer's itching. Another toxin, lyngbyatoxin A is also produced by this alga.
5. *Oscillatoria nigrovidis*. It produces the toxin such as **oscillatoxin A, bromooscilatoxin A, dibromooscillatoxin A, bromoaplysiatoxin** and **debromoaplysiatoxin**.

Dinophyta

Gonyaulax tamarensis. It produces the alkaloid toxins such as **saxitoxin, neosaxitoxins, gonyautoxin II** and **gonyautoxin III**. They are acid stable and non-volatile, which cause the death of shellfish. The algae become lodged in the gills of the shellfish and, when the fishes are eaten by animals and human beings, poisoning results. The poisoning cause paralysis and death. This type of poisoning is called paralytic shellfish poisoning, saxitoxin poisoning etc.

Similar type of poisoning also takes place with *G. cantenella*.

Prymnesiophyceae

Prymnesium parvum. This alga produce; a potent exotoxin of glycopepticle in nature, that cause severe mortality of fish in brackish water in many European countries. The toxin is very much effective in aquatic gill-breathing animal; like molluscs and fish.

List of some other toxic algae of different groups

Cyanophyta:

1. *Aphanothece nidulans.*
2. *A. cyanea.*
3. *Anabaena circinalis.*
4. *A. limnetica.*
5. *A. scheremetievi.*
6. *A. nadsonii.*
7. *A. inequalis.*
8. *Gloeotrichia echinulata.*
9. *Lyngbya birgei.*
10. *Microcystis aeruginosa.*
11. *M. toxica.*
12. *Nodularia spumigena.*

Dinophyta:

1. *Gonyaulax monilata.*
2. *Gymnodinium brevis.*
3. *G. mikimotoi.*
4. *G. veneficum.*
5. *G. flavum.*
6. *Pyrodinium phoneus.*

4 Fungi and Allied Organisms

KINGDOM FUNGI

INTRODUCTION TO TRUE FUNGI

Fungi (singular *fungus*- mushroom, from Greek) are chlorophyll-less thallophytic plant-like organisms. Due to absence of chlorophyll; they are heterophytes i.e., depend on others for food. They grow in various habitats and show much diversity in structure, physiology and reproduction. They developed from long back in the geological time scale. Their existence was found from fossil records of Pre-Cambrian period. Information from ancient literature indicate that the fungi were used as food by human beings. At present, the fungi are used in medicine and also as food in addition to other aspects. The fungi cause diseases of crops (spots, rusts and smuts etc.) and on human beings (Aspergillosis, Blastomycosis etc.). Some of the plant diseases like Rust of coffee (c.o. *Hemileia vastatrix*) during early 1870s, almost completely destroyed the coffee cultivation in Sri Lanka (erstwhile Ceylon), Downy mildew of grapes (c.o. *Plasmopara viticola*), caused severe damage in France during 1878-1882, thus wine industry suffers a lot and Brown spot of rice (c.o. *Helminthosporium oryzae*) caused femine in Bengal (undivided), India in 1943 due to 90 per cent reduction of yield in 1942 (Ghosh *et al.*, 1960). More than 5,000 genera and 50,000 species of fungi have been recorded, but their number may be much more than the actual record. The subject which deals with fungi is known as **Mycology** (*mykes*, mushroom; *logos*, study) and the concerned scientist is called **mycologist**.

In recent year, Mycology includes Fungi, Chromista and Protozoa.

Definitions: Various definitions of fungi as proposed by mycologists are:

1. **Alexopoulos (1962)** defined fungi as "nucleated, spore-bearing, achlorophyllous organisms which generally reproduce sexually and asexually and whose usually filamentous, branched somatic structures are typically surrounded by cell walls containing cellulose or chitin or both".
2. **Alexopoulos and Mims (1979)** defined fungi as "eukaryotic spore bearing, achlorophyllous organisms that generally reproduce sexually and asexually, and whose usually filamentous, branched somatic structures are typically surrounded by cell walls containing chitin or cellulose, or both of these substances, together with many other complex organic molecules".

Important characteristics

1. Fungi are cosmopolitan in distribution i.e., they can grow in any place where life is possible.
2. They are heterotrophic in nature due to the absence of chlorophyll. On the basis of their mode of nutrition, they may be parasite, saprophyte or symbiont.
3. The somatic body may be unicellular (*Synchytrium, Saccharomyces*) or filamentous

(*Mucor, Aspergillus*). The filament is known as **hypha** (plural, hyphae) and its entangled mass is known as **mycelium**.

4. The hypha may be aseptate i.e., coenocytic (without septa, containing many nuclei) or septate. The septate mycelium in its cell may contain only one (monokaryotic), two (dikaryotic) or more nuclei.
5. The septa between the cell may have different types of **pores: micropore** (*Geotrichum*), **simple pore** (most of the Ascomycota and Mitosporic fungi i.e. earlier Deuteromycotina) or **dolipore** (Basidiomycota, except rusts and smuts).
6. The cells are surrounded by distinct cell wall, composed of fungal cellulose i.e., **chitin**; but in some lower fungi, the cell wall is composed of **cellulose** or **glucan**.
7. The cells generally contain colourless protoplasm due to absence of chlorophyll, containing nucleus, mitochondria, endoplasmic reticulum, ribosomes, vesicle, microbodies, etc.
8. The cells are haploid, dikaryotic or diploid. The diploid phase is ephemeral (short-lived).
9. In lower fungi like Chytridiomycota, the reproductive cells (zoospores and gametes) may be uni- or biflagellate, having whiplash and/or tinsel type of flagella. But in higher fungi like Zygomycota, Ascomycota, Basidiomycota and Mitosporic fungi i.e. earlier Deuteromycotina, motile cells never form at any stage.
10. In response to functional need, the fungal mycelia are modified into different types, such as: Plectenchyma, Stroma, Rhizomorph, Sclerotium, Hyphal trap, Appressorium, Haustorium, etc.
11. The unicellular fungi, where entire plant body becomes converted into reproductive unit, are known as **holocarpic fungi** (e.g., *Synchytrium*). However, in many others, only a part of the mycelial body is converted into reproductive unit, thus they are called **eucarpic fungi** (e.g., *Mucor, Penicillium*).
12. They reproduce by three means: vegetative, asexual and sexual.
    (a) **Vegetative reproduction** takes place by fragmentation (*Mucor, Penicillium, Fusarium*), budding (*Saccharomyces, Ustilago*) and fission (*Schizosaccharomyces*).
    (b) **Asexual reproduction** takes place by different types of spores. These are zoospores (*Synchytrium*), conidia (*Talaromyces, Aspergillus*), oidia (*Coprinus, Collybia*), chlamydospore (*Fusarium*), etc. The spores may be unicellular (*Aspergillus*) or multicellular (*Alternaria*).
    (c) With the exception of Mitosporic fungi i.e., earliest Deuteromycotina, **sexual reproduction** takes place by the following five processes: Gametic copulation (*Synchytrium*), Gametangial contact (*Penicillium*), Gametangial copulation (*Rhizopus, Mucor*), Spermatization (*Puccinia, Podospora*) and Somatogamy (*Polyporus, Agaricus*).

Affinities with plants and animals

The affinities of fungi with both plants and animals are given below:

1. Fungi and animals share a molecule called **chitin** that is not found in plants.
2. Fungi and many invertebrate animals use the complex carbohydrate, chitin; for structural purposes. In fungi, it is the structural component of cell wall. But in animals, it forms the hard structures such as exoskeletons of insects and the beaks of octopuses and other mollusks.
3. On molecular level, chitin is similar to the plant molecule cellulose, used in plant to build up cell walls and other structures, but the chitin molecule has a modification that makes it stronger than cellulose.
4. All the three groups (fungi, plants and animals) of organisms have **mitochondria** and **nuclei** in their cells.
5. The most obvious similarity between fungi and animals is their **trophic level**, that is,

their place in the food chain. Neither fungi nor animals are producers (heterotrophs), as plants can produce (autotrophs). Both must use external food sources for energy.

General account

The fungi are chlorophyll-less plant-like or achlorophyllous non-vascular organism.

Occurrence

The fungi are cosmopolitan in distribution and occur in all possible habitats. Due to absence of chlorophyll, they depend on others for food, that is why they may be **saprophytes** (*Ascobolus*), **parasites** (*Fusarium*) or **symbionts** (*Termitoces*). Most of the fungi are terrestrial which grow in soil, on dead and decaying organic material. Some grow on both plants and animals (*Fusarium oxysporum f.* sp. *lycopersici* – on tomato and immunodipressed mice, Ortoneda *et al*, 2004). They can grow on foods like jam, bread, fruits etc. Some members are also found in water — **aquatic fungi** (*Rhizophydium planktonium* on diatom, *R. couchi* on *Spirogyra*). They are also present in the air. Thus the fungi are universal in their distribution.

Vegetative structure

Except certain **unicellular** forms (Fig. 4.1A) like *Synchytrium, Saccharomyces*; the fungi are generally thalloid and filamentous type. The filaments are known as the **hyphae** (sing. hypha). The tangled mass of these hyphae is the **mycelium**. The hypha may be aseptate which contains many nuclei and vacuolated cytoplasm. The multinucleate tubular aseptate hypha is called **coenocyte** (Fig. 4.1C). On the other hand, hypha that has partition walls dividing the hypha into many cells is called **septate** (Fig. 4.1D-F). The septate mycelium in its cell may contain only one (monokaryotic), two (dikaryotic) or many nuclei (multinucleate) with vacuolated protoplasm. The septa between the cells may have **different types of pores**. In *Geotrichum*, small

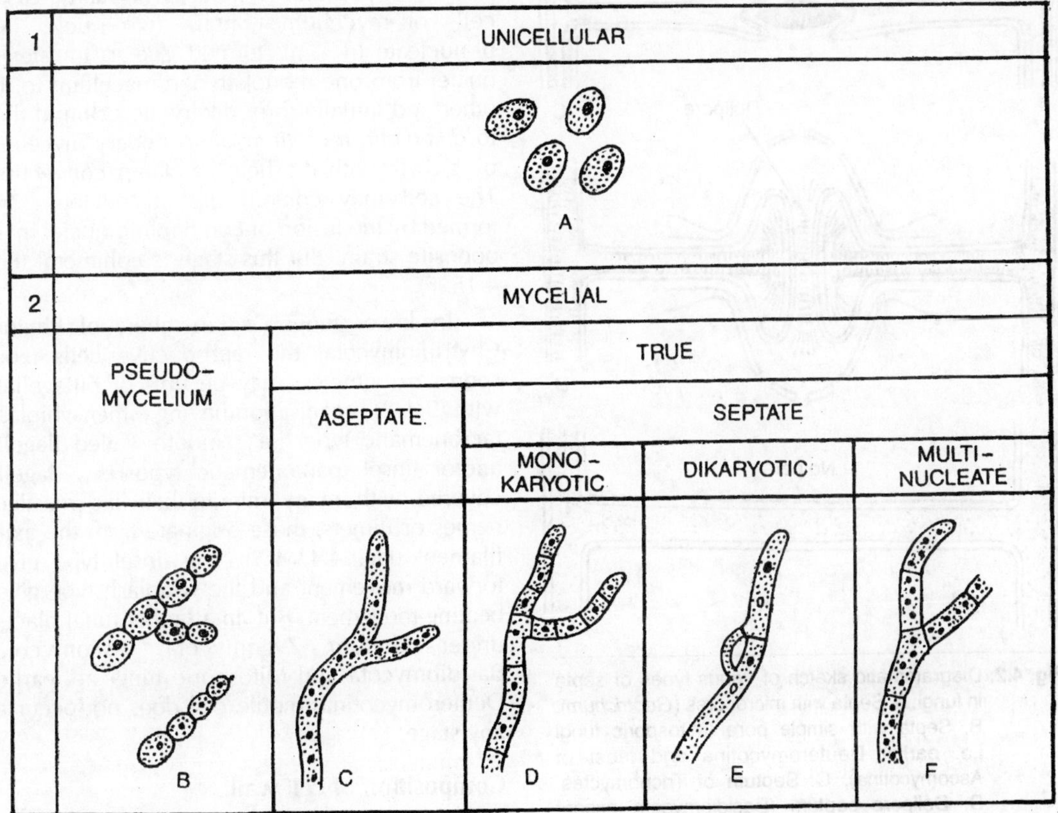

Fig. 4.1: Different types of thalli in fungi: A-B. *Saccharomyces*, C. *Mucor*, D-E. *Agaricus*, and F. *Penicillium*

pores are present on the wall, called *micropore* (Fig. 4.2A), when only one bigger sized pore is present on the wall, it is called *simple pore* (Fig. 4.2B), found in most of the Ascomycota and Mitosporic fungi i.e. earlier Deuteromycotina. In Trichomycetes and Mucorales, the pore is surrounded by an overarching bifurcation of the margin of partition wall, which looks like a bordered pit of tracheid. A biumbonate, electron dense plug of material is present between the two bifurcations (Fig. 4.2C). But in Basidiomycotina, except rust and smuts, the complicated type of pore with single opening is called *dolipore* (*dolium*, a large Jar or cash, i.e., barrel) *septum* (Fig. 4.2D). Lastly, in lower fungi, the older part and the basal region of reproductive part forming partition wall, without any pore, is called *solid septum* (Fig. 4.2E).

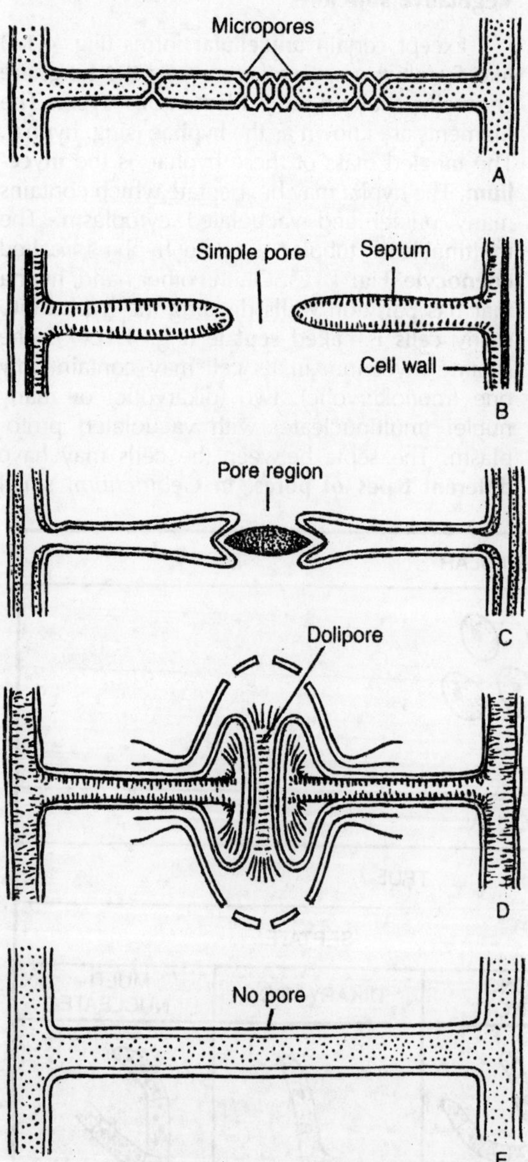

Fig. 4.2: Diagrammatic sketch of vrious types of septa in fungi: A. Septa with micropores (*Geotrichum*), B. Septa with simple pore (Mitosporic fungi i.e. earlier Deuteromycotina and most of Ascomycotina), C. Septum of Trichomyctes, D. Dolipore septum (Basidiomycota except rusts and smuts) and E. Solid septum of lower fungi.

The cells have distinct cell wall, made up of fungal cellulose i.e., chitin along with other substances. Fungal cells have chlorophyll-less vacuolated protoplasm, which contains nuclei/nucleus, mitochondria, endoplasmic reticulum, ribosomes, microbodies, vesicles, etc. (Fig. 4.3). The chief reserve foods are glycogen and oil.

Mycelium that contains only one nucleus (n) in each cell is called *monokaryotic mycelium* or primary mycelium. Whereas, in many cases, cells of mycelium contain two nuclei i.e., bi-nucleate (n + n), formed due to transfer of nuclei from one monokaryotic mycelium to the other and initially form dikaryotic cell and then to *dikaryotic mycelium* or secondary mycelium by growth with the help of clamp connection. The cell may contain diploid nucleus (2n), formed by the fusion of two haploid nuclei (n) of opposite strain. But this stage is ephimeral (Fig. 4.1E, F).

In lower fungi i.e., members of Phylum Chytridiomycota, the reproductive cells (zoospores or gametes) may be uni- or biflagellate with (9 + 2) organisation having either whiplash (acronematic type) i.e., smooth-walled flagella and/or tinsel (pantonematic type) i.e., flagella covered with many minute hair-like mastigonemes or flimers, those originate from the axile filament (Fig. 4.4, 4.5). The tinsel type gives forward movement and the whiplash type gives beating movement. But, in all other fungi placed under Phylum Zygomycota, Ascomycota, Basidiomycota and Mitosporic fungi i.e. earlier Deuteromycotina, motile cell does not occur at any stage.

Composition of cell wall

The cells are surrounded by outer rigid structure, the cell wall. Its composition varies in

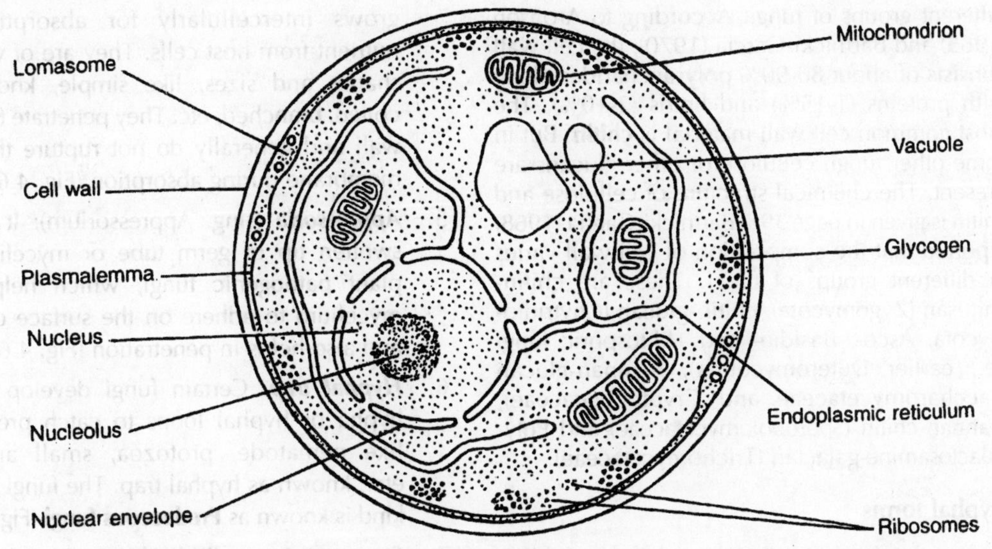

Fig. 4.3: Diagrammatic representation of the ultrastructure of a typical fungus cell in cross-section

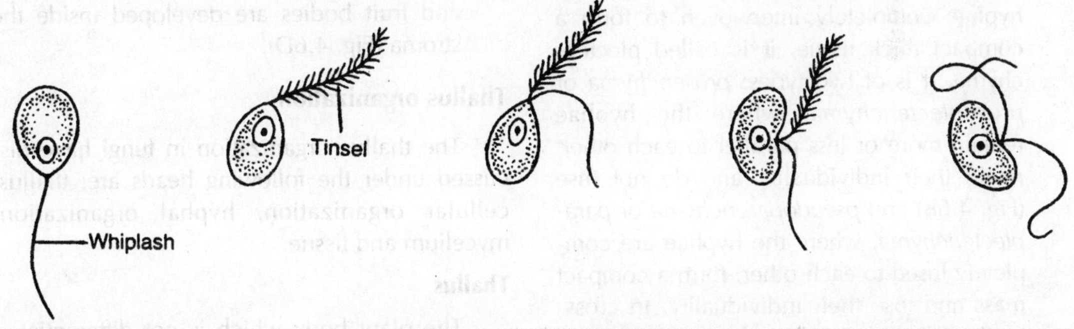

Fig. 4.4: Various kinds of zoospores in fungi

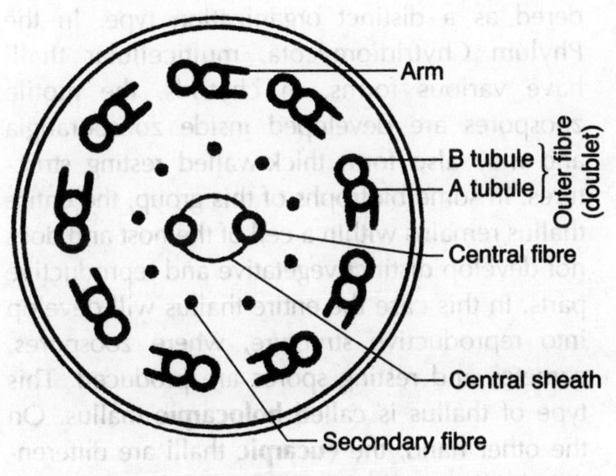

Fig. 4.5: Cross-section of a typical flagellium showing 9+2 arrangement

CH_2OH … CH_2OH

Structure of Cellulose

CH_2OH … $NHCOCH_3$

Structure of Chitin

different groups of fungi. According to Aronson (1965) and Bartnicki-Garcia (1970), the cell wall consists of about 80-90% polysaccharides along with proteins (1-15%) and lipids (2-10%). The most common cell wall material is **chitin**. But in some other fungi, cellulose or other glucans are present. The chemical structure of cellulose and chitin is given in page 395. Bartnicki-Garcia (1968) reported that the composition of cell wall varies in different groups of fungi. These are chitin-chitosan (Zygomycota), chitin-glucan (Chytridiomycota, Asco-, Basidio- and Mitosporic fungi i.e., earlier Duteromycotina), mannan-glucan (Saccharomycetaceae and Cryptococcaceae), mannan-chitin (Sporobolomycetaceae) and Polygalactosamine-galactan (Trichomycetaceae).

Hyphal forms

In response to functional need, the fungal mycelia are *modified into different types of structure*:

1. ***Plectenchyma***: When the component hyphae completely interwoven to form a compact thick tissue, it is called plectenchyma. It is of two types: *prosenchyma* or *prosoplectenchyma*, where the hyphae remain more or less parallel to each other, retain their individuality and do not fuse (Fig. 4.6B) and *pseudoparenchyma* or *paraplectenchyma*, where the hyphae are completely fused to each other, form a compact mass and lose their individuality. In cross-section the whole mass looks like parenchyma of angiospermic plant (Fig. 4.6B).
2. ***Sclerotia*** (sing. Sclerotium): It is a compact structure of different shapes and sizes, formed by the aggregation of mycelia. It may be round to elongated pod-like, very minute dots to large ball-like structure weighing approximately 14 kg (30 lbs). It may survive for many years as resting stage (Fig. 4.6A).
3. ***Rhizomorph***: In this case, hyphae are aggregated longitudinally in varying degree of complexity, where the hyphae lose their individuality and the whole structure behaves as an organised unit. They can withstand adverse environmental conditions and after few years, they can start growing (Fig. 4.6E).
4. ***Haustoria*** (sing. Haustorium): These are intracellular outgrowth of the mycelium that grows intercellularly for absorption of nutrient from host cells. They are of various shapes and sizes, like simple, knob-like, coiled, branched, etc. They penetrate the cell wall and generally do not rupture the cell membrane during absorption (Fig. 4.6F, G).
5. ***Appressoria*** (sing. Appressorium): It is the swollen tip of germ tube or mycelium of plant pathogenic fungi, which helps the mycelium to adhere on the surface of host and also helps in penetration (Fig. 4.6H).
6. ***Hyphal trap***. Certain fungi develop sticky hypha or hyphal loops to catch predators like nematode, protozoa, small animals etc., known as hyphal trap. The fungi of this kind is known as **Predaceous fungi** (Fig. 4.6I).
7. ***Stroma***. It is a solid body of various shapes and sizes, formed by the compact aggregation of mycelium. Reproductive structures and fruit bodies are developed inside the stroma (Fig. 4.6D).

Thallus organization

The thallus organization in fungi have discussed under the following heads are: thallus, cellular organization, hyphal organization, mycelium and tissue.

Thallus

The plant body which is not differentiated into roots, stems and leaves, called thallus. In chytrids, the coenocytic thallus is being considered as a distinct organization type. In the Phylum Chytridiomycota, multicellular thalli have various forms. In chytrids, the motile zoospores are developed inside zoosporangia and may also form thick-walled resting structures. In some biotrophs of this group, the entire thallus remains within a cell of the host and does not develop distinct vegetative and reproductive parts. In this case the entire thallus will develop into reproductive structure, where zoospores, gametes and resting spores are produced. This type of thallus is called **holocarpic** thallus. On the other hand, the **eucarpic** thalli are differentiated into vegetative and reproductive parts. Here the vegetative part i.e. the rhizoid, serves the function of anchorage and absorption of

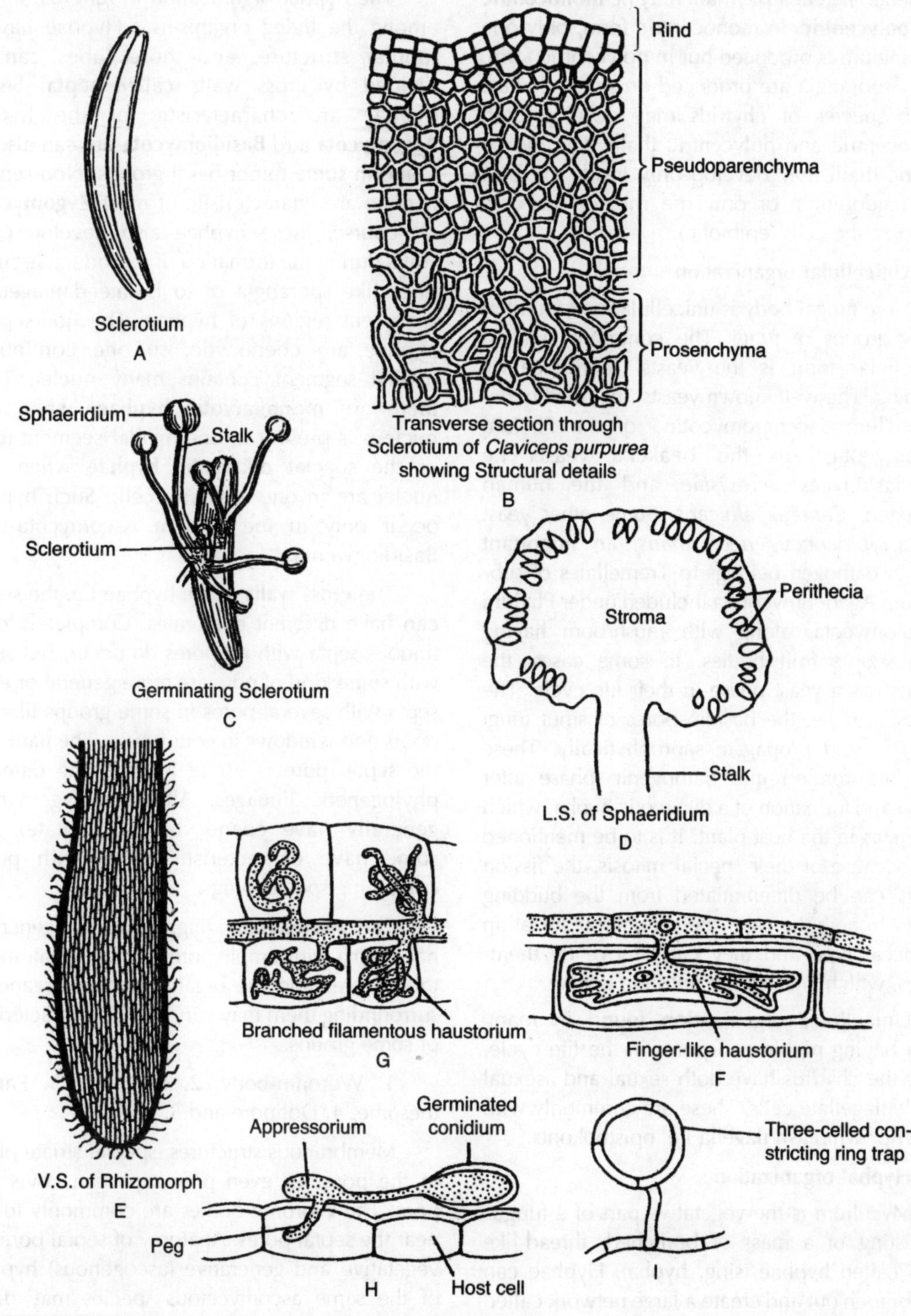

Fig. 4.6: Different type of modifications of the hyphal structure: A-D. *Claviceps purpurea*, E. *Aemillariella mellea* (= *Armillaria mellea*), F. Haustorium of *Erysiphae graminis* in epidermal cell of oat in longitudinal section. G. Haustoria of *Peronospora calotheca* in stem cells of *Asperula odorala*, H. Appressorium of *Erysiphe graminis tritici* on wheat, and I. Hyphal trap of *Monocresporium*

nutrient. The eucarpic thalli may be **monocentric** and **polycentric**. In monocentric form, only one sporangium is produced but in polycentric form, many sporangia are produced on single thallus. Some species of chytrids may develop both monocentric and polycentric thalli. The monocentric thalli may develop only inside the host cell (endobiotic) or only the rhizoids become colonize the cells (epibiotic).

1. Unicellular organization

Here fungal body is unicellular and found in many groups of fungi. The commonly known unicellular form is the yeasts, reproduce by budding. The well known yeasts, included under subphylum Saccharomycotina of the Phylum Ascomycota, are the beaker's yeasts i.e *Saccharomyces cerevisiae* and the human pathogen, *Candida albicans*. Some other yeast like *Cryptococcus neoformans,* an important human pathogen belongs to Tremellales of subphylum Agaricomycotina included under Phylum Basidiomycota, along with mushroom having macroscopic fruit bodies. In some cases, the fungus has a yeast phase in their life cycle. The meiospores i.e. the basidiospores of smut fungi can live and propagate saprophytically. These cells transformed into pathogenic phase after fusion and formation of a dikaryotic hypha, which colonizes in the host plant. It is to be mentioned that because of their special mitosis, the fission yeasts can be differentiated from the budding yeasts. In both the cases the cellular organization is coccal type and they can't move by themselves, which affects their nutrition uptake.

Unicellular organization found in many fungi having motile zoospores in the life cycle. Only the chytrids have both sexual and asexual motile flagellate cells. These are commonly with posterior whiplash flagella i.e. opisthokonts.

2. Hyphal organization

Mycelium is the vegetative part of a fungus consisting of a mass of branched, thread-like parts called hyphae (sing. hypha). Hyphae can also branch out and create a large network called a mycelium. There are two main types of hyphae, septate and coenocytic (without septa containing many nuclei).

The hyphal organization in fungi is unique among the living organisms. Hyphae have a tubular structure, and those tubes can be divided by cross walls called **septa**. Septal hyphae are characteristic of the hyphal **Ascomycota** and **Basidiomycota** but can also be found in some minor basal groups. Non-septate hyphae are characteristic of most **Zygomycota**. Of course, these hyphae also develop cross walls during the formation of reproductive structures like sporangia or to isolate damaged or senescent regions of hyphae. The non-septate hyphae are coenocytic, so one continuous hyphal segment contains many nuclei. Then there are **monokaryotic hyphae**, when one nucleus is present in one hyphal segment (cell) or the special **dikaryotic hyphae** when two nuclei are in one segment (cell). Such hyphae occur only in the Phylum Ascomycota and Basidiomycota.

The **cross walls** of the hyphae i.e. the septa, can have different anatomies. Completely continuous septa with no pores do occur, but septa with some kind of pore are more general or even septa with several pores in some groups like the doors and windows in your house. The nature of the septal pore is useful in detecting different phylogenetic lineages. Although Zygomycota generally have coenocytic (nonseptate), but some have characteristic septa with pores containing special plugs.

The **Ascomycota** (Fig. 4.7A, B) generally have septa with simple pores. But the anatomy of those pores and the organization of organelles surrounding them may vary and be characteristic of some groups.

1. Woronin-body, 2. Pore-cap, 3. Parenthesome, 4. Dolipore and 5. Wall-ER

Membranous structures, special striate plugs in the pores or even pore-caps can cover the pore. The Woronin-bodies are commonly found near the septal pores. Anatomy of septal pores of vegetative and generative (ascogenous) hyphae of the same ascomycetous species may differ significantly, and the pore organization of vegetative hyphae is generally more complex than that of generative hyphae.

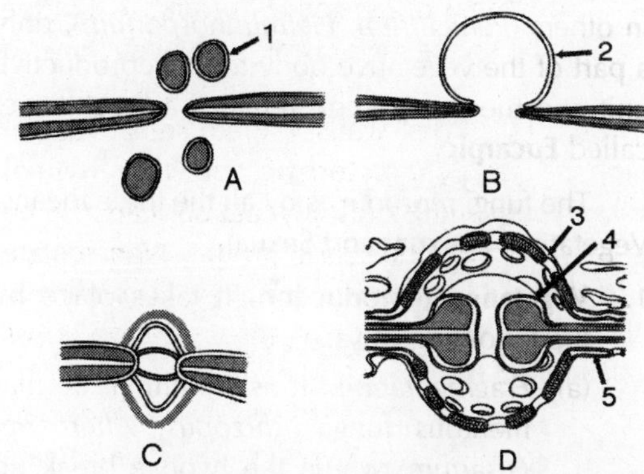

Fig 4.7: Structure of some types of septal pore: A-B: Ascomycota (A. Simple pore of vegetative hyphae with Woroning bodies; B. Simple pore with pre-cap of a generative hypha); and C-D: Basidiomycota (C. Simple pores with different structures. D. Dolipore with parenthesome)

The hyphae of Basidiomycota (Fig. 4.7C, D) have diverse septal pore organization. Some smut fungi have no septal pores. Several members of smut and rust fungi have simple pores. A special pore anatomy, the **dolipore**, is exclusive to the sub-phylum Agaricomycotina, under phylum Basidiomycota. The border of this pore is swollen like a nut, and different structures such as membrane-like bands, may occur in the tube of the pore. The dolipore can be covered by parenthesome, a membrane organization continuous with the wall endoplasmic reticulum (ER). The parenthesome can be sacculate, continuous or porate.

Bear in mind that the pores represent a direct physical connection with the neighbouring hyphal segments and their cytoplasm is continuous. Cellular organelles can migrate across the pores; even nuclei can sometimes migrate to the neighbouring segments. This is one reason why hyphal segments cannot be considered as analogous to the autonomous eukaryotic cell.

3. Mycelium and tissue

Fungal hyphae can group together to form a more or less organized multicellular structure like tissues of multicellular eukaryotic organisms. However, these are not the real tissues, but bear in mind that fungal tissues do not fulfill the definition of real tissues.

The organized hyphal mass is the mycelium. Although simple hyphae can usually be visible, the mycelial structures are the more obvious as fungal parts. The several different types, forms and levels of mycelial organization have different functions.

(i) **Rhizomorphs**. In rhizomorph, the hyphae are grouped together forms a cord-like structure. Rhizomorph help to find and exploit the nutrients of the environments for the fungi, but they can also serve as resting propagules and, being robust in structures, they provide physical integrity for the different fungi. They can grow to various distances but some are extremely effective, such as *Serpula lacrymans*, whose rhizomorphs can grow for several meters to find woody substrates, or members of the genus *Armillaria*, which includes the largest living organisms. The robust rhizomoprhs of such species helps to preserve the multihectare-sized (about 3.9 km) continuous genet of the *A. ostyae*. Differentiated rhizomorphs have a thick-walled, compact rind covering their surface and the medial part consisting of hyphae with thinner walls and bigger lumina. Some ectomycorrhizal fungi have differentiated rhizomorphs with central hyphae of large diameter and partly or totally dissolved septa, making the transportation more effective. One concept uses the term rhizomorph only for differentiated structures and hyphal cords when there are no differentiated parts, only the cord-like organization of similar to hyphae.

The hyphal mantle of **ectomycorrhizae** is another characteristic mycelial tissue that can even represent the main biomass of an ectomycorrhizae in root tip. The two main anatomical types of those mantles are *plectenchymatous*, when the hyphae forming the mantle can easily be detected, or *pseudoparenchymatous* when it is to be composed of non-tubular cells such as inflated, polygonal or isodiametric in shape. Other layer-like tissues formed by hyphae i.e., the **peridium** enclosing sporocarps. This can be divided into several subtypes, such as angular and the epidermoid, depending on if the cell walls are straight or wavy respectively.

(ii) **Sclerotia**. Sclerotia (sing. sclerotium) comprised of other important mycelial structures, the resting structure which can tide over the fungus over unfavourable condition. Their hyphal tissue is kept together by an extracellular glucan matrix as well. Sclerotia may have differentiation in tissue with a melanized compact rind and a medular tissue with thinner walled hyaline hyphae. Like rhizomorphs, sclerotia are of many different categories.

(a) The *Botrytis cinerea*, a well known plant pathogenic sclerotium-forming fungus, whose sclerotia overwinter on plant stems. The life cycle of the fungus starts from those sclerotia, in the next vegetation period.

(b) Another commonly known sclerotium-forming fungus is the *Claviceps purpurea*, caused ergot disease of rye, which infects cereals and forms sclerotia in their spikes. These **sclerotia** contain the medically important ergot alkaloids at high concentration. The sclerotia overwinter and another mycelial structure, the **stroma**, which develops from them during spring. This stroma has a stem-like part and a head in which several *perithecia develop to produce asci and ascospores*.

(iii) **Sporocarps**. Mycelial tissues form the fungal fruiting bodies are the sporocarps. The main functions of sporocarp are to support and development of spore producing structures, to hold and protect them and also to help in the spore dispersal. Sporocarps may be sexual or asexual, depending on whether they produce sexual or asexual spores. Sexual sporocarps occur in the Ascomycota are termed as **ascocarp** or **ascoma**, and in Basidiomycota as **basidiocarp** or **basidioma**. Some zygomycota also form sexual sporocarps.

Reproduction

In unicellular fungi (*Synchitrium, Saccharomyces*), entire vegetative cell is transformed into a reproductive unit, called **Holocarpic**. However, in others (*Penicillium, Helminthosporium*), only a part of the vegetative body forms reproductive unit and the rest portion remains as vegetative, called **Eucarpic**.

The fungi reproduces by all the three means: Vegetative, Asexual and Sexual.

1. **Vegetative reproduction**. It takes place by the following ways:

(a) *Fragmentation*. It is common in filamentous fungi (*Rhizopus, Alternaria, Fusarium*) where the hyphae break up into two or more fragments due to some external force and each one develops into a new individual (Fig. 4.8A, B).

(b) *Budding*. It takes place in unicellular fungi (*Saccharomyces*, Fig. 4.8C). A small outgrowth, the bud emerges out from the parent cell. Nucleus divides into two and one passes to the bud. The bud is then separated by partition wall, but continues its growth. Later on, it breaks off from the mother and grows individually. Sometimes, the process repeats very fast and the buds remain attached with the mother in chain, that looks like mycelium, called *pseudomycelium* (Gr. *pseudo*, false + mycelium) (Fig. 4.1B). This process takes place in ascospores and basidiospores of some fungi.

(c) *Fission*. Normally unicellular fungi (*Schizosaccharomyces*) reproduce by this method, where the vegetative cell elongates, and divides into two daughter cells of equal size by simple constriction in the middle with simultaneous nuclear division (Fig. 4.8D).

2. **Asexual reproduction**: It takes place by means of several types of spore generally form during favourable condition. The spores may be unicellular (*Penicillium, Aspergillus*) or multicellular (*Fusarium, Helminthosporium*). They may be exogenous, developed on conidiophore (*Penicillium*) (Fig. 4.9C) or endogenous, developed in sporangium (*Mucor*) (Fig. 4.9I) or in pycnidium (*Ascochyta pisi*). Some of the spores are:

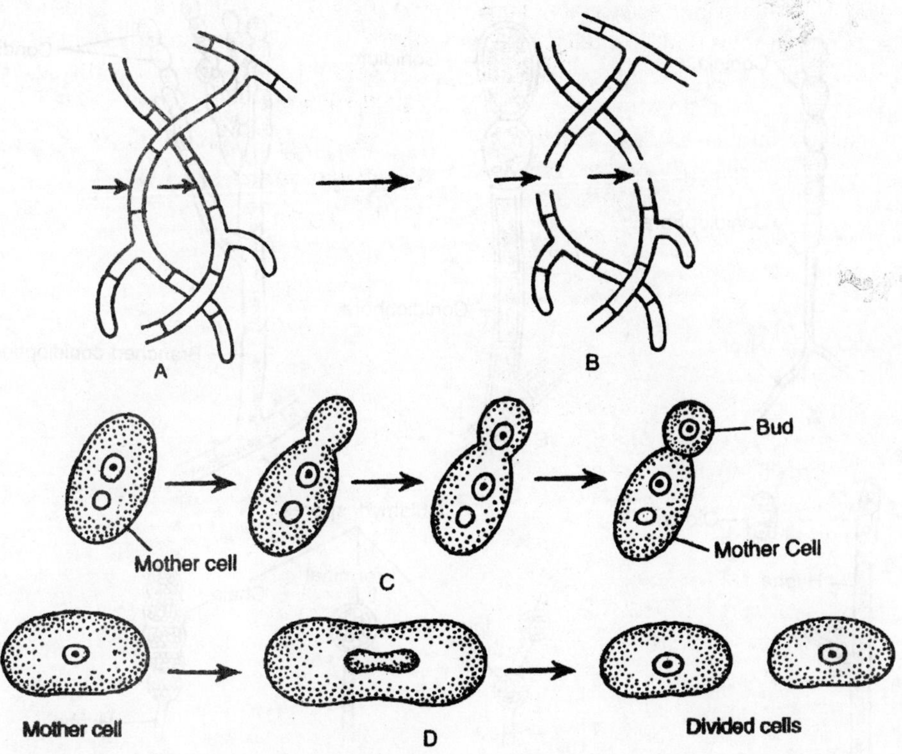

Fig 4.8: Diagrammatic representation of the types of vegetative reproduction in fungi: A-B. Before and after mycelial fragmentation, C. Stages of budding in yeasts, and D. Stages of fission in yeast.

(a) *Zoospore*. The zoospores may be uni- or biflagellate, generally pear-shaped, produced in sporangium, e.g., *Synchytrium* (Fig. 4.17A).

(b) *Conidia*. These are exogenously produced non-motile spores develop by constriction at the end of specialised hyphal branches, called conidiophores. They may produce singly (*Venturia inaequalis*) or in chain (*Penicillium, Aspergillus*) (Fig. 4.9A, B, C).

(c) *Oidia*. In some fungi (*Mucor mucedo*), the hyphal tips often divide by transverse wall into large number of small segments, may remain in chain or becomes free from each other, these are known as oidia. The oidia on germination develop into new plants (Fig. 4.9D).

(d) *Chlamydospore*. The chlamydospores are thick walled, round to oval in outline, coloured brown or black. They produce either terminally or intercalary at some intervals throughout the length of hyphae, e.g., *Fusarium* (Fig. 4.9E, F, G).

(e) *Sporangiospores*. These are globose, multinucleate, non-motile aplanospores, formed inside the sporangium. The sporangiospore germinates by producing germ tube. Later on, it develops profusely branched mycelium (Fig. 4.9H, I).

3. *Sexual reproduction*: It is the process of union between two compatible nuclei. The nuclei in some members are contributed by two well-organized gametes. The whole process of sexual reproduction consists of three phases, in the sequence of plasmogamy, karyogamy and meosis:

(a) Plasmogamy. It involves the union of two protoplasts, brings two haploid nuclei close together in the same cell.

(b) Karyogamy. It involves the fusion of two haploid nuclei brought together during

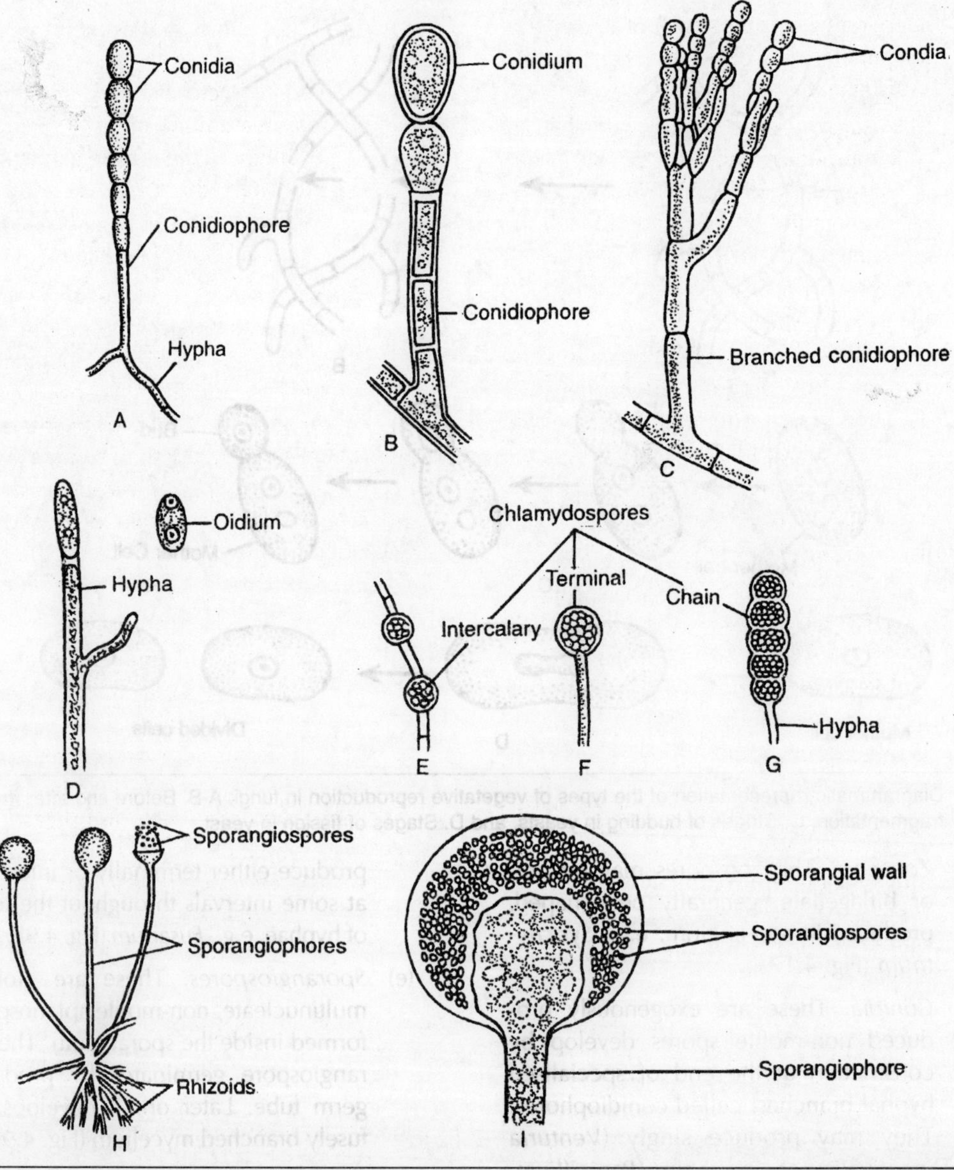

Fig 4.9: Various types of asexual spores of fungi: A. Conidia in chain borne on conidiophore in *Erysiphe*, B. Endoconidium from conidiophore in *Ceratostomella* sp., C. Conidia on branched conidiophore in *Penicillium*, D. Development of oidium, E-G. Chlamydospores, H. Portion of plant body of *Rhizopus*, showing rhizoids, sporangiophores and sporangia, and I. Portion of sporangiophore of *Rhizopus* with single sporangium.

plasmogamy. This results in the formation of diploid nucleus i.e., zygote, which is ephemeral (short-lived).

(c) *Meiosis*. It follows karyogamy and reduce the number of chromosome from diploid zygote nucleus to original haploid number in the daughter nuclei.

The plasmogamy i.e., the first phase of sexual reproduction, differs in different fungi. The **different methods of plasmogamy** are:

(a) **Planogametic copulation**. Planogametes are motile gametes. This process involves the fusion of two gametes, where either one or both are motile.

Depending on the structure and nature of gametes, it is of three types: Isogamy, Anisogamy and Oogamy:

(i) *Isogamy*. The uniting gametes are morphologically similar, but physiologically different. This process is common in primitive unicellular fungi, e.g., *Synchytrium* (Fig. 4.10A) etc.

(ii) *Anisogamy*. Both the uniting gametes are morphologically similar, but different physiologically and in size. The smaller one is more active, considered as male and the larger less active one as female, e.g., *Allomyces* (Fig. 4.10B) etc.

(iii) *Oogamy*. Both the uniting gametes are morphologically and physiologically different. The male gamate is smaller and motile, and the female gamete is larger and non-motile, e.g., *Monoblepharis* (Fig. 4.10C).

(b) **Gametangial contact**. The uniting gametes are present in different gametangium, thus the male and female gametangia are known as antheridium and Ascogonium (in Ascomycota), respectively. The gametes are never released from gametangium. Both the gametangia come in close contact and transfer male gamete to the egg through contact and dissolution of common wall. The gametangia do not lose their identity, e.g., *Ascobolus*, *Penicillium* (Fig. 4.10D, E).

(c) **Gametangial copulation**. The process involves the fusion of the entire content of the uniting gametangia. Such fusion occurs in the following two ways:

(i) The two gametangia fuse by the dissolution of their common wall resulting into the formation of a single cell in which content of both the gametangia mix with each other and their morphological identity are completely lost, e.g., *Rhizopus*, *Mucor* (Fig. 4.10F).

(ii) The entire thallus acts as gametangium. Both the gametangia come in close contact and the male gametangium transfer its entire content to the female gametangium through the pore developed in contact area e.g., *Rhizophidium* (Fig. 4.10G), *Polyphagus*.

(d) **Spermatization**. Certain fungi produce many unicellular non-motile, male cells, the spermatia. The spermatia are brought in contact by agents like wind, water and insect either to the trichogyne of the ascogonium or to somatic hyphae or even to special receptive hyphae. The wall at the point of contact dissolves and content of spermatia passes to the female organ, e.g., *Puccinia*, *Podospora* (Fig. 4.10H).

(e) **Somatogamy**. In many higher fungi belonging to Asomycota and Basidiomycota, the development of gametes and gametangia are completely lacking. In such fungi, somatic hyphae anastomose with each other to bring together the compatible nuclei. It is regarded as a reduced and efficient form of sexuality, designated as somatogamy, e.g., *Polyporus*, *Agaricus*, *Morchella* (Fig. 4.10I).

Status of fungi in living world

The status of fungi have already been discussed under the "Modern concept of classification of Living World" in Chapter 1 (Page 19).

Classification

Many scientist classified fungi in so many different ways from time to time. Earlier systems of classification were based mainly on: (i) Thallus organisation and composition of cell wall, (ii) Life cycle patterns, (iii) Type of asexual spores, (iv) Presence and absence of sexual reproduction and its type. However, recent systems are based on physiological, biochemical, cytological, genetical, serological characteristics etc., in addition to morphological characteristics.

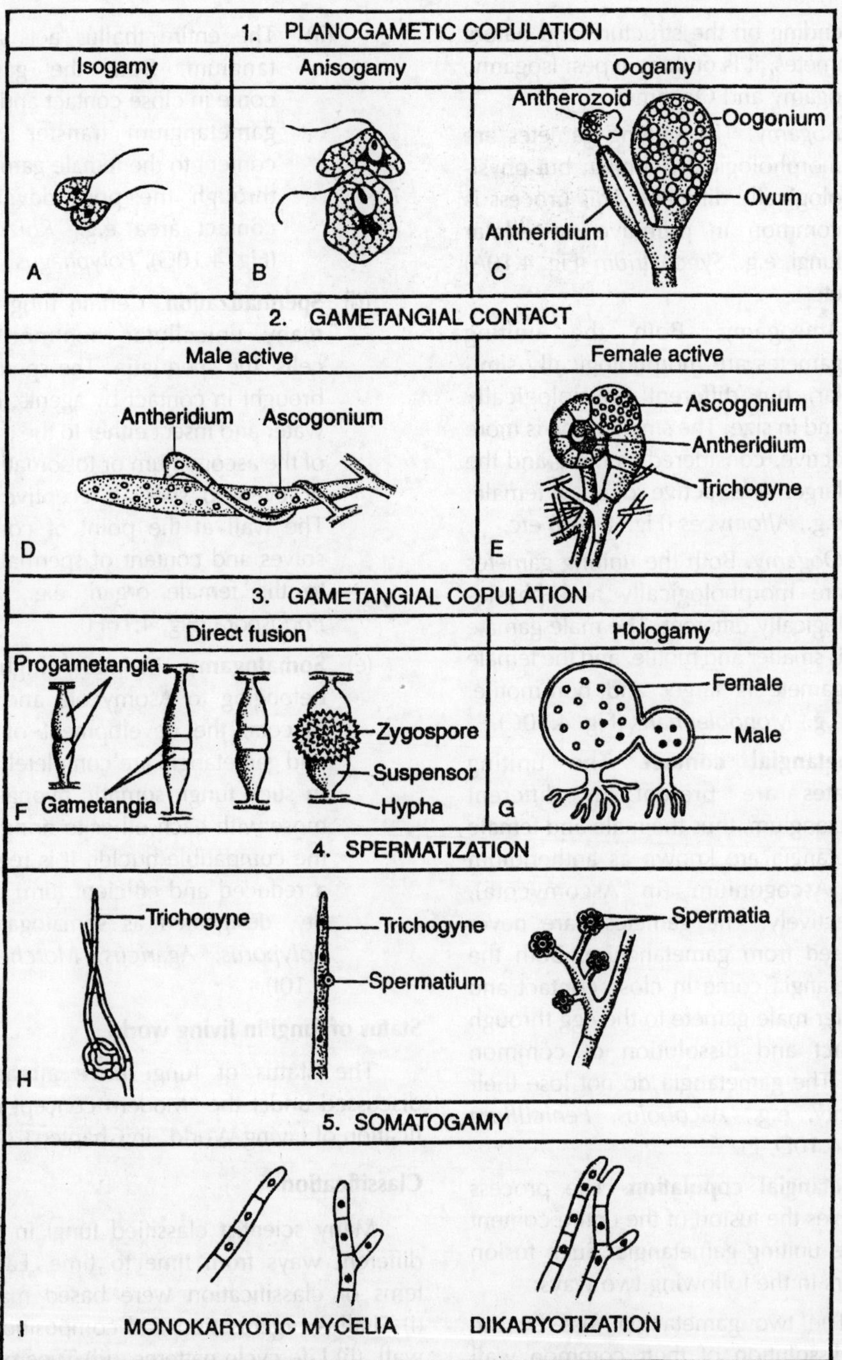

Fig. 4.10: Various methods of Plasmogamy in fungi : A. *Synchytrium*, B. *Allomyces*, C. *Monoblepharis*, D. *Pythium*, E. *Ascobolus*, F. *Rhizopus*, G. *Rhizophidium couchii*, H. *Podospora anserina* and I. *Agaricus*

Bauhin (1623) is the first who described about 100 species of fungi in his book 'Pinax Theatri Botanici'. But the detailed classification of fungi was first proposed by Elias Fries in his book "Systema Mycologicum", published in three volumes between 1821 and 1831, where he divided fungi in four classes: (i) Coniomycetes, (ii) Hyphomycetes, (iii) Gastromycetes

Schematic outline of the classification of H.C.I Gwynne-Vaughan and B. Barnes (1926):

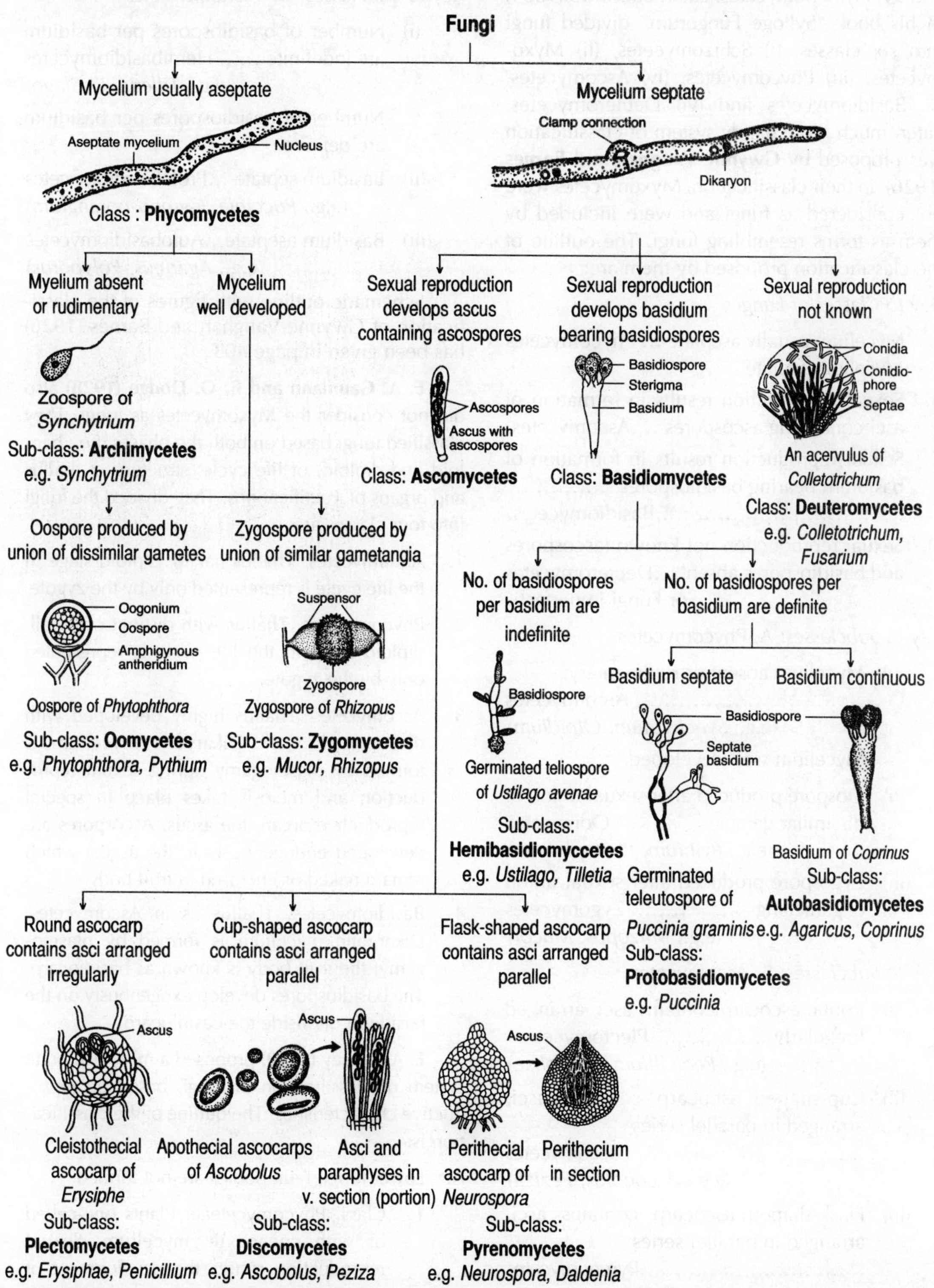

and (iv) Hymenomycetes. Later, Saccardo (1884) in his book "Sylloge Fungorum" divided fungi into six classes: (i) Schizomycetes, (ii) Myxomycetes, (iii) Phycomycetes, (iv) Ascomycetes, (v) Basidiomycetes and (vi) Deuteromycetes. Later, much more handy system of classification was proposed by **Gwynne-Vaughan** and **Barnes** (1926). In their classification, Myxomycetes were not considered as fungi and were included by them as forms resembling fungi. The outline of the classification proposed by them are:

Key to Classes of Fungi

A. Mycelium usually aseptate ... Phycomycetes
Mycelium septate.

B. Sexual reproduction results in formation of asci containing ascospores ... Ascomycetes.

C. Sexual reproduction results in formation of basidium bearing basidiospores ....Basidiomycetes.

D. Sexual reproduction not known (ascospores and basidiospores absent)...Deuteromycetes or Fungi imperfecti.

Key to Subclasses: A. Phycomycetes

(i) Mycelium absent or rudimentary. .... Archimycetes (e.g., *Synchytrium, Olpidium*)
Mycelium well-developed.

(ii) Oospore produced after sexual union of dissimilar gametes............ Oomycetes (e.g., *Pythium, Phytophthora*)

(iii) Zygospore produced after sexual union of gametangia .... Zygomycetes (e.g., *Rhizopus, Mucor*)

Key to Subclasses: B. Ascomycetes

(i) Round ascocarp contains asci, arranged irregularly..................Plectomycetes (e.g., *Penicillium, Erysiphae*)

(ii) Cup-shaped ascocarp contains asci, arranged in parallel series .... Discomycetes (e.g., *Ascobolus, Peziza*)

(iii) Flask-shaped ascocarp contains asci, arranged in parallel series ....Pyrenomycetes (e.g., *Daldenia, Xylaria*)

Key to Subclasses: C. Basidiomycetes

(i) Number of basidiospores per basidium are indefinite ...... Hemibasidiomycetes (e.g., *Tilletia, Ustilago*)
Number of basidiospores per basidium are definite.

(ii) Basidium septate ...Protobasidiomycetes (e.g., *Puccinia, Gymnosporangium*)

(iii) Basidium aseptate...Autobasidiomycetes (e.g., *Agaricus, Polyporus*)

Schematic outline with figures of the classification of Gwynne-Vaughan and Barnes (1926) has been given in page 403.

E. A. Gaumann and B. O. Dodge (1928) also did not consider the Myxomycetes as fungi. They classified fungi based on both the phases (i.e., haploid and diploid) of life cycle; structure of thallus and organs of fructifications. They divided the fungi into four classes are:

1. Archimycetes. Thallus naked; diploid stage in the life cycle is represented only by the zygote.
2. Phycomycetes. Thallus with distinct cell wall; diploid stage in the life cycle is represented only by the zygote.
3. Ascomycetes. Thallus highly developed with distinct cell wall. Dikaryotic mycelium is formed by plasmogamy during sexual reproduction and meiosis takes place in special reproductive organ, the ascus. Ascospores are developed endogenously in the ascus, which remain naked or enclosed in fruit body.
4. Basidiomycetes. Thallus as in Ascomycetes; Dikaryotic mycelium is formed by plasmogamy; the fruit body is known as basidiocarp. The basidiospores develop exogenously on the basidium on/inside the basidiocarp.

E. A. Bessey (1950) proposed a more elaborate system of classification of fungi, based on reproductive characteristics. The outline of the classification is:

A. Lower fungi: Fruit bodies are not formed.

1. Class. Phycomyceteae: Plants one-celled or with coenocytic mycelium. Sexual reproduction results the development of oospore or zygospore.

B. Higher fungi (Phylum. Carpomyceteae): Fruit bodies are formed.

1. Class. Ascomyceteae: Fruit body contains asci, where ascospores are formed endogenously.
2. Class. Basidiomyceteae: Fruit body contains basidia, where basidiospores are formed exogenously.
3. Class. Fungi imperfecti: Sexual stage not known.

G. W. Martin (1961) included slime moulds under fungi. His classification is based on septation of hypha and characteristics of spore. The outline of the system is:

A. Myxomycetes: Plant body having naked protoplast.

B. Eumycetes: Plant body may be unicellular or filamentous and are true walled fungi. It is divided into four classes:

1. Class. Phycomycetes: Mycelium coenocytic.
2. Class. Ascomycetes: Mycelium septate and sexual spore i.e., ascospores develop endogenously.
3. Class. Basidiomycetes: Mycelium septate and sexual spore i.e., basidiospores develop exogenously.
4. Class. Deuteromycetes: Mycelium septate and sexual reproduction absent. Reproduction takes place only by asexual spores.

C. J. Alexopoulos (1962) also included slime molds under fungi and placed them under Division Mycota and it was divided into two subdivisions: Myxomycotina (wall-less form) and Eumycotina (true, walled fungi). The outline of his classification is:

Division. Mycota: Microscopic, unicellular or filamentous thallus with chitinous or cellulosic cell wall containing typical eukaryotic nucleus. Reproduction takes place both asexually and sexually.

A. Subdivision. Myxomycotina: Plant body is plasmodium.

1. Class. Myxomycetes: Vegetative phase is plasmodium. Reproduction takes place by very small multinucleate spore.

B. Subdivision. Eumycotina: Vegetative phase is either unicellular or branched mycelium with distinct cell wall. Hyphae are coenocytic (aseptate and multinucleate) or septate. Each cell may be uni-, bi- or multinucleate. Reproduction by asexual (spore) or sexual (gamete) means.

The sub-divisions has been divided into eight natural classes and one form-class. These are:

1. Class. Chytridiomycetes: Motile cells with single whiplash flagellum placed posteriorly.
2. Class. Hypochytridiomycetes: Motile cells with single tinsel flagellum placed anteriorly.
3. Class. Oomycetes: Vegetative phase consists of multinucleate and well-developed mycelium. Motile cells have two flagella i.e., biflagellate consists of one whiplash and the other tinsel type.
4. Class. Plasmodiophoromycetes: Multinucleate thallus without cell wall, remains parasitically inside the host cells. Motile cells are with two whiplash flagella of unequal in length.
5. Class. Zygomycetes: Well-developed coenocytic (multinucleate and aseptate) mycelium grows either parasitically or saprophytically. Motile cells are absent.
6. Class. Trichomycetes: Multinucleate simple or branched thallus; mostly parasite on arthopods.
7. Class. Ascomycetes: Well-developed mycelium with septate hyphae. Ascospores are produced endogenously inside ascus.
8. Class. Basidiomycetes: Well-developed mycelium with septate hyphae. Basidiospores are produced exogenously on basidium.

1. Form-class. Deuteromycetes: Well-developed mycelium with septate hyphae. Sexual phase unknown. Reproduces mainly by asexual means.

Ainsworth G. C. (1966, 71, 73) proposed a more natural system of classification of fungi. This classification is based on morphology, especially on reproductive structure. He includes fungi along with slime molds under the kingdom Mycota (Myceteae of Alexopoulos and Mims, 1979). Based on the presence or absence of plasmodium and pseudoplasmodium; the kingdom Mycota is further divided into two divisions: Myxomycota i.e., slime molds (Gymnomycota of Alexopoulos and Mim, 1979) and Eumycota or true fungi (Mastigomycota and Amastigomycota of Alexopoulos and Mims, 1979). Divisions are subsequently divided into subdivision, class, subclass, order, family and then to genus. According to his classification, division ends in mycota, subdivision in mycotina, class in mycetes, subclass in mycetidae, order in 'ales' and family in 'aceae'. A schematic outline of Ainsworth's (1973) classification is given:

Kingdom: Mycota

Important features:

(i) Free-living, parasitic or mutualistic symbionts, devoid of chlorophyll.

(ii) Cell wall composition is very variable, majority contain chitin and glucan.

(iii) Reserve food materials are oil, mannitol and glycogen.

(iv) Except some unicellular members, majority are filamentous.

A. Division. Myxomycota

Wall-less organisms possess either a plasmodium (a mass of naked multinucleate protoplasm having amoeboid movement) or a pseudoplasmodium (an aggregation of separate amoeboid cells). Both are of slimy consistency, hence **slime molds**.

1. Class. Acrasiomycetes (cellular slime molds)
2. Class. Hydromyxomycetes (net slime molds)
3. Class. Myxomycetes (true slime molds)
4. Class. Plasmodiophoromycetes (endoparasitic slime molds).

B. Division Eumycota (True fungi, all with walls)

(a) Subdivision Mastigomycotina (motile cells – zoospores present, perfect state spore-oospore).

1. Class. Chytridiomycetes (unicellular, zoospore with single *whiplash* flagellum).
2. Class. Hyphochytridiomycetes (unicellular, zoospore with single tinsel flagellum).
3. Class. Oomycetes (aseptate mycelium, zoospores with two flagella).

(b) Subdivision. Zygomycotina (mycelium aseptate, perfect state spore-zygospore).

1. Class. Zygomycetes (mycelium immersed in the host tissue).
2. Class. Trichomycetes (mycelium not immersed in the host tissue).

(c) Subdivision. Ascomycotina (yeasts or septate mycelium, perfect state spore-ascospores formed in ascus, usually within ascocarp).

1. Class. Hemiascomycetes (no ascocarp, asci naked).
2. Class. Loculoascomycetes (fruit body an ascostroma, asci bitunicate i.e., 2-v alled).
3. Class. Plectomycetes (fruit body cleistothecium, asci unitunicate i.e., 1-walled).
4. Class. Laboulbeniomycetes (fruit body perithecium, asci unitunicate, exoparasite of arthopods).
5. Class. Pyrenomycetes (fruit body perithecium, asci unitunicate, not parasitic on arthopods.
6. Class. Discomycetes (fruit body apothecium, asci unitunicate).

(d) Subdivision. Basidiomycotina (yeast or septate mycelium, perfect state spore – basidiospore formed on a basidium).

1. Class. Teliomycetes. Basidiocarp lacking, teliospores grouped in sori or scattered within the host tissue, parasitic on vascular plant.
2. Class. Hymenomycetes. Basidiocarp present. Hymenium is completely or partly exposed at maturity. Basidiospore ballistospores.
3. Class. Gasteromycetes. Basidiocarp present. Hymenium enclosed in

basidiocarp. Basidiospore not ballistospores.

(e) Subdivision. Deuteromycotina or Fungi imperfecti. Yeast or septate mycelium. Perfect state unknown.

1. Class. Blastomycetes. Budding (Yeast or Yeast like) cells with or without pseudomycelium. True mycelium lacking or not well-developed.

2. Class. Hyphomycetes. Mycelia sterile or bearing asexual spore directly or on conidiophore, in various aggregation.

3. Class. Coelomycetes. Mycelial; asexual spore formed in pycnidium or acervulus.

Schematic representation of the outline with figure, the classification of G. C. Ainsworth (1973) is given:

Schematic outline of the classification of G. C. Ainsworth (1973):

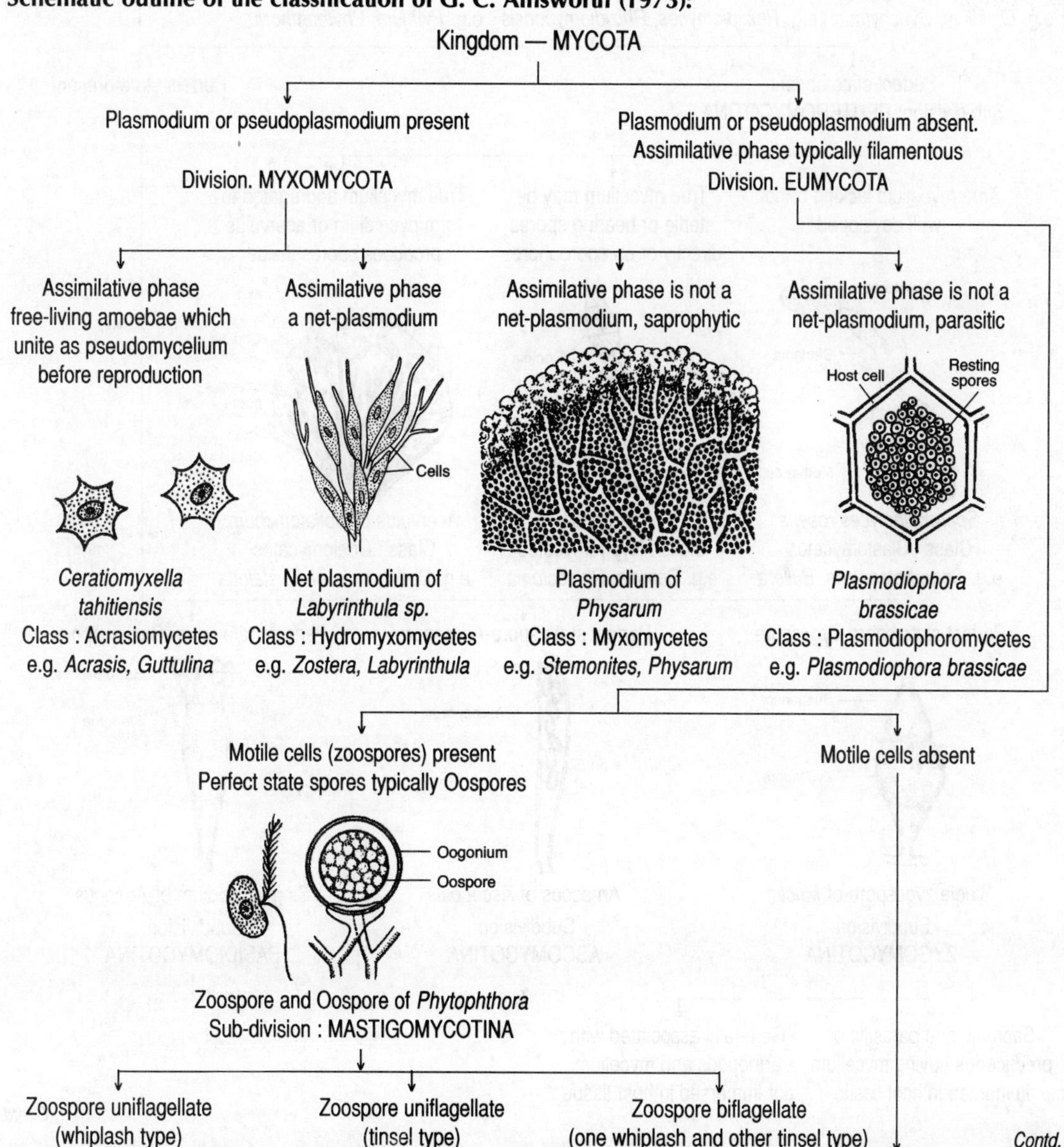

Contd.

Contd.

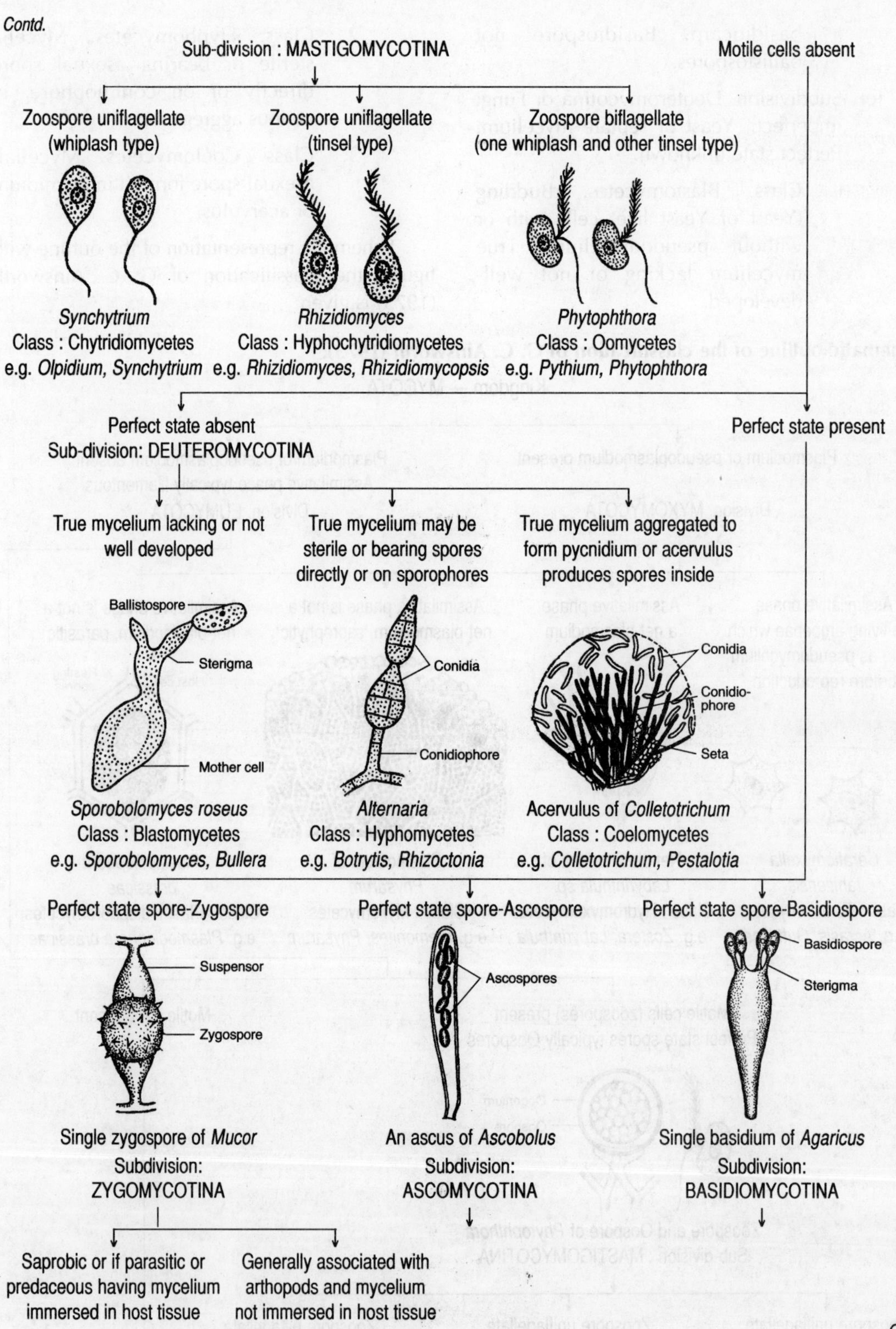
Sub-division : MASTIGOMYCOTINA
Motile cells absent
Zoospore uniflagellate
(whiplash type)
Zoospore uniflagellate
(tinsel type)
Zoospore biflagellate
(one whiplash and other tinsel type)
Synchytrium
Class : Chytridiomycetes
e.g. Olpidium, Synchytrium
Rhizidiomyces
Class : Hyphochytridiomycetes
e.g. Rhizidiomyces, Rhizidiomycopsis
Phytophthora
Class : Oomycetes
e.g. Pythium, Phytophthora
Perfect state absent
Sub-division: DEUTEROMYCOTINA
Perfect state present
True mycelium lacking or not
well developed
True mycelium may be
sterile or bearing spores
directly or on sporophores
True mycelium aggregated to
form pycnidium or acervulus
produces spores inside
Ballistospore
Sterigma
Mother cell
Conidia
Conidiophore
Conidia
Conidio-
phore
Seta
Sporobolomyces roseus
Class : Blastomycetes
e.g. Sporobolomyces, Bullera
Alternaria
Class : Hyphomycetes
e.g. Botrytis, Rhizoctonia
Acervulus of Colletotrichum
Class : Coelomycetes
e.g. Colletotrichum, Pestalotia
Perfect state spore-Zygospore
Perfect state spore-Ascospore
Perfect state spore-Basidiospore
Suspensor
Zygospore
Ascospores
Basidiospore
Sterigma
Single zygospore of Mucor
Subdivision:
ZYGOMYCOTINA
An ascus of Ascobolus
Subdivision:
ASCOMYCOTINA
Single basidium of Agaricus
Subdivision:
BASIDIOMYCOTINA
Saprobic or if parasitic or
predaceous having mycelium
immersed in host tissue
Generally associated with
arthopods and mycelium
not immersed in host tissue

Contd.

Contd.

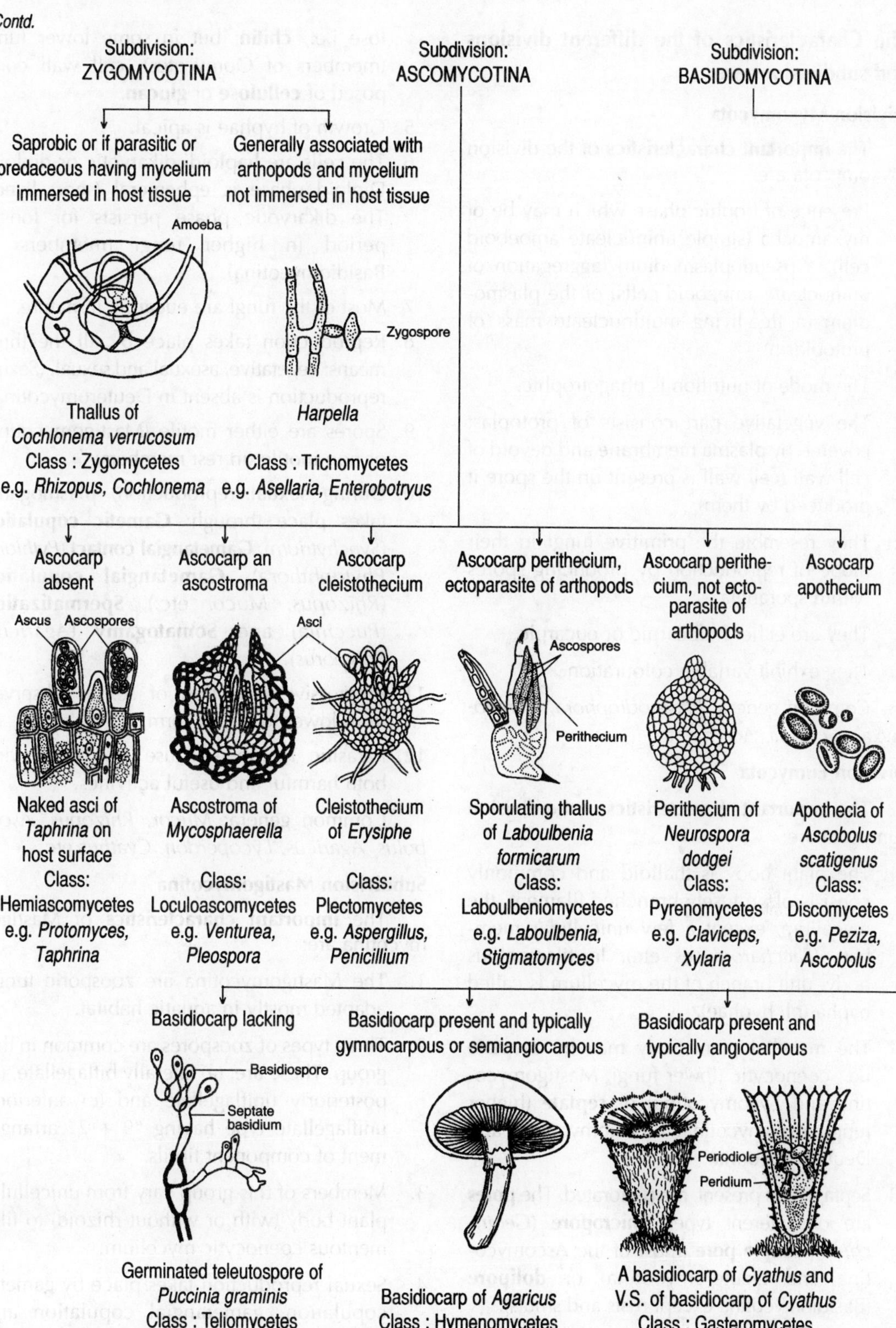
Subdivision:
ZYGOMYCOTINA
Subdivision:
ASCOMYCOTINA
Subdivision:
BASIDIOMYCOTINA
Saprobic or if parasitic or predaceous having mycelium immersed in host tissue
Generally associated with arthopods and mycelium not immersed in host tissue
Amoeba
Zygospore
Thallus of Cochlonema verrucosum
Harpella
Class : Zygomycetes
e.g. Rhizopus, Cochlonema
Class : Trichomycetes
e.g. Asellaria, Enterobotryus
Ascocarp absent
Ascocarp an ascostroma
Ascocarp cleistothecium
Ascocarp perithecium, ectoparasite of arthopods
Ascocarp perithecium, not ectoparasite of arthopods
Ascocarp apothecium
Ascus
Ascospores
Asci
Ascospores
Perithecium
Naked asci of Taphrina on host surface
Class: Hemiascomycetes
e.g. Protomyces, Taphrina
Ascostroma of Mycosphaerella
Class: Loculoascomycetes
e.g. Venturea, Pleospora
Cleistothecium of Erysiphe
Class: Plectomycetes
e.g. Aspergillus, Penicillium
Sporulating thallus of Laboulbenia formicarum
Class: Laboulbeniomycetes
e.g. Laboulbenia, Stigmatomyces
Perithecium of Neurospora dodgei
Class: Pyrenomycetes
e.g. Claviceps, Xylaria
Apothecia of Ascobolus scatigenus
Class: Discomycetes
e.g. Peziza, Ascobolus
Basidiocarp lacking
Basidiocarp present and typically gymnocarpous or semiangiocarpous
Basidiocarp present and typically angiocarpous
Basidiospore
Septate basidium
Periodiole
Peridium
Germinated teleutospore of Puccinia graminis
Class : Teliomycetes
e.g. Puccinia, Ustilago
Basidiocarp of Agaricus
Class : Hymenomycetes
e.g. Agaricus, Polyporus
A basidiocarp of Cyathus and V.S. of basidiocarp of Cyathus
Class : Gasteromycetes
e.g. Cyathus, Geastrum

The Characteristics of the different divisions and subdivisions are:

Division Myxomycota

The **important characteristics** of the division Myxomycota are:

1. Presence of trophic phase which may be of myxamoeba (simple uninucleate amoeboid cell), a pseudoplasmodium (aggregation of uninucleate amoeboid cells) or the plasmodium (a free-living multinucleate mass of protoplasm).
2. The mode of nutrition is phagotrophic.
3. The vegetative part consists of protoplast covered by plasma membrane and devoid of cell wall (cell wall is present on the spore if produced by them).
4. They resemble the primitive fungi in their mode of reproduction by producing spores within sporangia.
5. They are either holocarpic or eucarpic.
6. They exhibit variable colouration.

Common genera: *Plasmodiophora, Ceratiomyxa, Lycogala, Acryria* etc.

Division Eumycota

The **important characteristics** of the division Eumycota are:

1. The plant body is thalloid and commonly consists of profusely branched filament, the mycelium, except a few unicellular members (*Saccharomyces* etc). In filamentous body, unit branch of the mycelium is called hypha (pl. hyphae).
2. The mycelial plant body may be **aseptate** i.e., coenocytic (lower fungi, Mastigomycotina and Zygomycotina) or **septate** (higher fungi, Ascomycotina, Basidiomycotina and Deuteromycotina).
3. Septa, when present, are perforated. The **pores** are of different types: **micropore** (*Geotrichum*), **simple pore** (most of the Ascomycotina and Deuteromycotina) or **dolipore** (Basidiomycotina except rusts and smuts).
4. The hyphal wall is made up of fungal cellulose i.e., **chitin**; but in some lower fungi (members of Oomycetes), cell wall composed of **cellulose** or **glucan**.
5. Growth of hyphae is apical.
6. The cells are haploid, dikaryotic or diploid. Diploid phase is ephemeral (short lived). The dikaryotic phase persists for longer period in higher fungi (members of Basidiomycotina).
7. Most of the fungi are eucarpic in nature.
8. Reproduction takes place by all the three means: vegetative, asexual and sexual. (Sexual reproduction is absent in Deuteromycotina).
9. Spores are either motile (Mastigomycotina) or non-motile (in rest members).
10. During sexual reproduction, plasmogamy takes place through: **Gametic copulation** (*Synchytrium*), **Gametangial contact** (*Pythium, Phytophthora*), **Gametangial copulation** (*Rhizopus, Mucor* etc.), **Spermatization** (*Puccinia*) and **Somatogamy** (*Agaricus, Polyporus*).
11. Progressive reduction of sex is observed from lower to higher form.
12. Parasitic members cause diseases having both harmful and useful activities.

Common genera: *Mucor, Rhizopus, Ascobolus, Agaricus, Lycoperdon, Cyathus* etc.

Subdivision Mastigomycotina

The **important characteristics** of Mastigomycotina are:

1. The Mastigomycotina are zoosporic fungi, adapted mostly in aquatic habitat.
2. Three types of zoospores are common in this group. These are: (a) laterally biflagellate, (b) posteriorly uniflagellate and (c) anteriorly uniflagellate type having "9 + 2" arrangement of component fibrils.
3. Members of this group vary from unicellular plant body (with or without rhizoid) to filamentous coenocytic mycelium.
4. Sexual reproduction takes place by gametic copulation, gametangial copulation and gametangial contact.

Common genera: *Synchytrium, Olpidium, Physoderma, Coelomomyces* etc.

Subdivision Ascomycotina

The **important characteristics** of Ascomycotina are:

1. Vegetative body is unicellular or commonly well-developed, branched septate mycelium with uni- or multinucleate cells having perforated septa.
2. Mostly, the cell wall is composed of chitin and glucans, but in unicellular form, it is composed of glucans and mannans.
3. Vegetative reproduction takes place by fragmentation (in filamentous form), fission and budding (in unicellular form).
4. Asexual reproduction takes place by non-motile spores, such as conida, oidia and chlamydospores.
5. Sexual reproduction takes place by gametangial copulation (*Saccharomyces*), gametangial contact (*Penicillium*), somatogamy (*Morchella*) or spermatization (*Polystigma*).
6. Complete absence of motile structures.
7. The product of sexual reproduction is the ascospores grown inside a small specialised sac-like structure, called ascus.
8. The fruiting bodies (inside which asci are developed) are the ascocarps. The ascocarps may be cleistothecium (*Penicillium*), apothecium (*Ascobolus*), perithecium (*Daldenia*) or ascostroma (*Elsinoe veneta*).

Common genera: *Saccharomyces, Ascobolus, Peziza, Daldenia, Penicillium* etc.

Subdivision Basidiomycotina

The **important characteristics** of Basidiomycota are:

1. Presence of well-developed, branched and septate mycelium having simple (e.g., Ustilaginales and Uredinales) or dolipore (e.g., Auriculariaceae, Aphyllophorales and Agaricales) septum (Fig. 4.2B & 4.2D).
2. The mycelial cells may contain one nucleus, called monokaryotic i.e., primary mycelium or two nuclei, called dikaryotic i.e., secondary mycelium. The secondary mycelia may organize and form fruit body, called tertiary mycelium.
3. The cell wall is mainly composed of chitin and glucans.
4. Reproduction:
   (a) Vegetative reproduction takes place by budding and fragmentation.
   (b) Asexual reproduction takes place by *conidia, oidia,* or *chlamydospores* that are absent in some higher taxa of this subdivision.
   (c) Sex organs are absent. During sexual reproduction, the dikaryotic cell is formed by *somatogamy, spermatization* or by buller phenomenon. The dikaryotic phase persists for long period of time. Karyogamy occurs in basidium mother cell and forms diploid nucleus, which is ephemeral (short-lived). 4-haploid basidiospores are formed by meiosis. Basidiospores are developed *exogenously* on the horn-shaped structure, called sterigmata (generally 4) on the basidium.
5. Basidia are of two types: *Holobasidium* (aseptate) e.g., *Agaricus, Polyporus* etc. (Fig. 4.67A) and *Phragmobasidium* (septate) e.g., *Puccinia, Ustilago* (Fig. 4.67B) etc.
6. Except in lower forms (*Puccinia, Ustilago*), secondary mycelia by aggregation form fruit body, called basidiocarp (*Agaricus, Polyporus* etc.). It has the spore bearing structure, called basidium. The number of spores on each basidium is commonly four, but 2 or more than 4 are also reported.

Common genera: *Polyporus, Agaricus, Puccinia, Cyathus, Lycoperdon* etc.

Subdivision Deuteromycotina

The **important characteristics** of Deuteromycotina are:

1. The members of this group are either saprophytes or parasites.

2. The vegetative body is mycelial and composed of profusely branched and septate hyphae with perforated septa and their cells are usually multinucleate.
3. In parasitic species, the hyphae grow intra- or intercellularly.
4. They reproduce only by asexual methods and the sexual stages are not known.
5. The asexual reproduction takes place commonly by conidia, or by blastospores, chlamydospores and arthrospores. The conidia are developed exogenously on conidiophores. The conidiophores are either free or grouped together to form specialized structures, such as *synnemata* (sing. *synnema*) and *sporodochia* (sing. *sporodochium*) or produce in fructification known as *acervuli* (sing. *acervulus*) or *pycnidia* (sing. *pycnidium*).
6. Many genera of the class exhibit genetic variation as a result of heterokaryosis and parasexuality.

Common genera: *Fusarium, Pyricularia, Cercospora, Septoria* etc.

Consideration of fungi as Kingdom

Robrert Whittaker (1969), an American Ecologist was the first to separate the fungi from plant kingdom and given the status as Kingdom Fungi in his **Five Kingdom System** of classification which consists of Monera, Protista, Plantae, Fungi and Animalia. Mycologists now regard the fungi as being distinct from plants and classified the fungal members within their own kingdom namely the Fungi or Mycetae (Talbot, 1971, Holmes, 1983).

Later, **C. J. Alexopoulos** and **C. W. Mims (1979)** placed fungi and slime molds under the kingdom of their own, called Myceteae under the superkingdom Eukaryonta. The kingdom is divided into three divisions and further the divisions are divided into sub-division, class and form-class. The outline of the classification is given:

Kingdom. Myceteae (Fungi): Achlorophyllous, saprobic or parasitic organisms with unicellular or more typically, filamentous soma (thallus), usually surrounded by cell walls that characteristically consists of chitin and other complex carbohydrates, nutrition absorptive, except in the slime molds (Division Gymnomycota) where it is phagotrophic, propagation typically by means of spores produced by various types of sporophores; asexual and sexual reproduction usually present.

The kingdom is subdivided into three major divisions:

A. Division. Gymnomycota: Phagotrophic organisms with somatic structures devoid of cell walls:
   - (a) Subdivision. Acrasiogynomycotina
     1. Class. Acrasiomycetes
   - (b) Subdivision. Plasmodiogymnomycotina
     1. Class. Protosteliomycetes
     2. Class. Myxomycetes

B. Division. Mastigomycota: Fungi with centrioles; flagellate cells typically produced during the life cycle; nutrition typically absorptive; varying from unicellular that becomes converted into a sporangium, to an extensive, filamentous, coenocytic mycelium, asexual reproduction typically by zoospores; sexual reproduction by various means:
   - (a) Subdivision. Haplomastigomycotina
     1. Class. Chytridiomycetes
     2. Class. Hyphochytridiomycetes
     3. Class. Plasmodiophoromycetes
   - (b) Subdivision. Diplomastigomycotina
     1. Class. Oomycetes

C. Division. Amastigomycota: Fungi without centriole, no motile cells, nutrition absorptive, single-celled to mycelial with a limited or extensive, septate or aseptate mycelium, asexual reproduction by budding, fragmentation, sporangiospores or conidia; sexual reproduction, where known, by various means; haplobiontic-haploid life cycle with zygotic meiosis.
   - (a) Subdivision. Zygomycotina
     1. Class. Zygomycetes
     2. Class. Trichomycetes

(b) Subdivision. Ascomycotina
 1. Class. Ascomycetes
(c) Subdivision. Basidiomycotina
 1. Class. Basidiomycetes
(d) Subdivision. Deuteromycotina
 1. Form class. Deuteromycetes.

Later, Webster (1980), Deacon (1980) and Hudson (1986) also proposed their classifications. Among all the above classifications, Ainsworth's classification has been accepted by many mycologists, with some comments. Say for example, the Mastigomycotina is a composite grouping covering all fungi that form motile flagellated zoospores. Among others, unrelated classes like Chytridiomycetes and Oomycetes were included in the same group. Oomycetes has indeed quite a different evolutionary origin to that of most other fungi.

Hawksworth *et al.* (1983 and 1995)

Ten years after the classification of Ainsworth (1973), Hawksworth *et al.* (1983) revised Ainsworth's classification in the 7th edition of the "Dictionary of the Fungi". The changes made by them are:

1. The Division Myxomycota divided into eight classes instead of four classes.
2. The Sub-division Ascomycotina is directly divided into thirty seven (37) orders and arranged alphabetically; there is no classes in-between.
3. The subdivision Basidiomycotina is divided into four classes instead of three, where class Teliomycetes is replaced by Urediniomycetes and Ustilaginomycetes.
4. In the sub-division Deuteromycotina, the class Blastomycetes was not considered.

Later, **Hawksworth *et al.* (1995)** thoroughly revised the classification in the 8th edition of the "Dictionary of the Fungi". The classification was based on the sequence of 18s rRNA among the different members. Though members of fungi show similarity in their morphology, mode of nutrition and ecology, but the higher dissimilarity is observed in the base sequences of their 18s rRNA (or more precisely in the DNA coding for it), which is now considered as the most important parameter to determine the genetic relationship. Thus, it shows that the fungi are polyphyletic aggregation of unrelated members.

Based on the above fact and phylogenetic consideration, the entire fungal community are now segregated out and placed them under **three different Kingdoms: Fungi, Straminopila (Chromista)** and **Protozoa.** Based on comparison of several factors such as, 18s rRNA, cytoskeleton protein, chemical features like chitin, mitochondrial codon UGA coding for tryptophan instead of termination, storage of glycogen, elongation factors and morphological structure of motile cells like male gamete in animal and zoospores in fungi, there is close similarity of fungi with animals than to other two groups like Straminopila and Protozoa.

The different taxa considered in this system are:

Domain
 Kingdom
  Phylum
   Class
    Order
     Family
      Genus
       Species

Further the species are divided into varieties, biological strains, physiological races, etc.

The different **taxa** considered in this system along with their '**ending**' are:

Phylum	—	mycota
Sub-phylum	—	mycotina
Class	—	mycetes
Sub-class	—	mycetidae
Order	—	ales
Family	—	aceae

But the genera and species have no standard ending. In this system, **division** has been replaced by **phylum** and all taxa are *written* in *italics*.

All the three groups are placed under the Domain *Eukaryota*. Out of three kingdoms, kingdom Fungi includes only fungi, but the other two kingdoms include non-fungal phyla also. The earlier consideration of the sub-division Deuteromycotina representing the asexual stages of Ascomycotina and Basidiomycotina (Ainsworth 1973, Hawksworth *et al.*, 1983) was now reconsidered as a formal taxon as they are not a monophyletic group. The members are described under **Mitosporic fungi.**

The outline of the classification is:

Kingdom **Fungi**
- Phylum Chytridiomycota
  - Class Chytridiomycetes
- Phylum Zygomycota
  - Class Trichomycetes
  - Class Zygomycetes
- Phylum Ascomycota (no classes)
  - Mitosporic (Anamorphic fungi)
- Phylum Basidiomycota
  - Class Basidiomycetes
  - Class Teliomycetes
  - Class Ustomycetes

Kingdom **Straminopila (Chromista)**
- Phylum Hyphochytriomycota
- Phylum Labyrinthulomycota
- Phylum Oomycota

Kingdom **Protozoa**
- Phylum Plasmodiophoromycota
- Phylum Dictyosteliomycota
- Phylum Acrasiomycota
- Phylum Myxomycota
  - Class Myxomycetes
  - Class Protosteliomycetes

The classification of Hawksworth *et al.* (1995) has been followed in this book.

The characteristics of the different phyla are:

Kingdom **Fungi** (i.e., true fungi).

Phylum. **Chytridiomycota**

Important characteristics:

1. Vegetative body is coenocytic and thalloid, either globose or ovoid structure, either an elongated simple hypha, or well developed mycelium.
2. Cell wall is mainly made up of chitin and glucan.
3. Nuclear division is intranuclear and centric type.
4. Members of this group produce motile cells at some stage of their life cycle.
5. Motile cells (zoospores and gametes) possess single posteriorly placed, whiplash type of flagellum except a few polyflagellate cells.
6. Sexual reproduction takes place by planogametes developed in gametangia. The fused gametes form zygote. Zygote on germination develops either into a resting spore or resting sporangium except a few those develop diploid thallus.

Common genera: *Synchytrium, Monoblepharis, Rhizophidium* etc.

Class Chytridiomycetes.

Phylum **Zygomycota**

Important characteristics:

1. The thallus is normally haploid, consisting of coenocytic mycelium and its wall contains chitin and chitosan.
2. The mycelium contains cell organelles like other fungi, except typical golgi bodies and centriole.
3. Asexual reproduction takes place by aplanospores.
4. Sexual reproduction takes place by gametangial copulation results in the formation of zygospore.

Common genera: *Mucor, Rhizopus, Phycomyces, Cunninghamella* etc.

Class Trichomycetes

Class Zygomycetes

Phylum **Ascomycota**

Important characteristics:

1. Vegetative body is unicellular or commonly well developed, branched septate mycelium with uni- or multinucleate cells having perforated septa.
2. Mostly, the cell wall is composed of chitin and glucans, but in unicellular form, it is composed of glucans and mannans.
3. Vegetative reproduction takes place by fragmentation (in filamentous form), fission and budding (in unicellular form).
4. Asexual reproduction takes place by non-motile spores, such as conidia, oidia and chlamydospores.
5. Sexual reproduction takes place by gametangial copulation (*Saccharomyces*), gametangial contact (*Penicillium*), somatogamy (*Morchella*) or spermatization (*Polystigma*).
6. Complete absence of motile structures.
7. The product of sexual reproduction is the ascospores grown inside a small specialized sac-like structure, called ascus.
8. The fruit bodies (inside which ascus developed) are the ascocarps. The ascocarps may be cleistothecium (*Penicillium*), apothecium (*Ascobolus*), perithecium (*Daldenia*) or ascostroma (*Elsinoe veneta*).

Common genera: *Saccharomyces, Penicillium Daldenia, Ascobolus, Morchella* etc.

Phylum **Basidiomycota**

Important characteristics:

1. Presence of well-developed, branched and septate mycelium having simple (e.g., Ustlaginales and Uredinales) (Fig. 4.2B) or dolipore (e.g., Auriculariaceae, aphyllophorales and Agaricales) septum (Fig. 4.2D).
2. The mycelial cells contain one nucleus, called monokaryotic i.e., primary mycelium or two nuclei, called dikaryotic i.e., secondary mycelium. The secondary mycelia may organize and form fruit body, called tertiary mycelium.
3. The cell wall is mainly composed of chitin and glucans.
4. Reproduction
   (a) Vegetative reproduction takes place by budding and fragmentation.
   (b) Asexual reproduction takes place by conidia, oidia or chlamydospores. This is lacking in some higher taxa of this subdivision.
   (c) Sex organs are absent. During sexual reproduction, the dikaryotic cell is formed by somatogamy, spermatization or by buller phenomenon. The dikaryotic phase persists for long period of time. Karyogamy occurs in basidium mother cell and forms diploid nucleus, which is ephemeral (short lived). 4-haploid basidiospores are formed by meiosis. Basidiospores are developed exogenously on the horn-shaped structure, the sterigmata (generally 4) on the basidium.
5. Basidia are of two types: Holobasidium (aseptate) e.g., *Agaricus, Polyporus* etc. (Fig. 4.56A) and Phragmobasidium (septate) e.g., *Puccinia, Ustilago* (Fig. 4.56B) etc.
6. Except in lower forms (*Puccinia, Ustilago*), secondary mycelia by aggregation form fruit body, the basidiocarp (*Agaricus, Polyporus* etc.). The number of spores on each basidium is commonly 4, but 2 or more than 4 are also present.

Common genera: *Agaricus, Polyporus, Puccinia* etc.

Class Basidiomycetes

Class Teliomycetes

Class Ustomycetes

Kingdom **Straminopila** or **Chromista** (i.e., pseudo-fungi).

Phylum **Hyphochytriomycota**

It is a very small group, comprising of about 23 known species.

Important characteristics:

1. The organisms are thalloid, soil-inhabiting or aquatic, chitrid-like.
2. Cell wall contains both chitin and cellulose.
3. Thalli are either holocarpic or eucarpic. The holocarpic thalli are endobiotic and converted into a zoosporangium. In eucarpic forms, the thalli may consist of a single reproductive organ bearing a branched rhizoidal system or may be polycentric with septate and branched hyphae.
4. Motile cells (zoospores) possess single anteriorly placed flagellum, converted with flagellar hairs, which developed inside zoosporangium and are released through discharge tubes.
5. Sexual reproduction has not been demonstrated conclusively in any member. But some evidence suggestive of sexual cycle has been described for *Anisolpidium ectocarpi* parasitises on *Ectocarpus mitchellae*. Suspected zygote has been reported in some, but meiosis has not been reported.

Common genera: *Rhizidiomyces, Reessia, Hyphochytrium* etc.

Phylum **Labyrinthulomycota**
(Net Slime Molds)

Important characteristics:

1. Members are found primarily in estuarine (a wide tidal mouth of a river) and near shore habitat; associated with algae, leaves of higher plants and organic debris.
2. Members are mostly saprobic or weak parasites and exhibit nutrition through absorption.
3. Vegetative body is a net slime mold.
4. Presence of an ectoplasmic net-work of anastomosing, branched, wall-less filaments, produced by cells with a specialized cell surface organelle, the **sagenogen** or **bothrosome**.
5. Cell walls are composed of scales derived from Golgi.
6. Zoospores are flagellate, heterokont (unequal flagella) type. Flagella are laterally inserted. Larger one is tinsel, directed anteriorly and the shorter one is whiplash, directed posteriorly.

Common genera: *Labyrinthula, Thraustochytrium* etc.

Phylum **Oomycota**

Important characteristics:

1. Members of Oomycota are found to grow in both fresh water and salt water as well as in terrestrial habitat.
2. They are either unicellular or filamentous, composed of profusely branched and coenocytic hyphae.
3. Septa develop in older region and also at the base of reproductive structures.
4. Cell wall is composed primarily of β-glucans, but also contains hydroxyproline, an amino acid; and small amount of cellulose.
5. Cells contain mitochondria with tubular cristae and with various types of biochemical and molecular characteristics.
6. Cell divisions (both mitotic and meiotic) are intranuclear and centric (i.e., the nuclear envelop remains intact until the end of division and centrioles are present at the poles of the dividing nuclei).
7. Unicellular forms are holocarpic, but filamentous forms are eucarpic.
8. Asexual reproduction takes place by means of biflagellate zoospores with shorter whiplash and longer tinsel flagella.
9. Zoospore ultrastructure shows various characteristics.
10. Sexual reproduction is oogamous and

takes place by gametangial contact and which produces thick-walled sexual spore, the oospore.

11. Meiosis takes place in the developing gametangia (antheridia and oogonia).

Common genera: *Pythium, Peronospora, Albugo* etc.

Kingdom **Protozoa** (i.e., the slime moulds).
Phylum **Plasmodiophoromycota**

(Endoparasitic Slime Molds)

This group is commonly known as endoparasitic slime molds. They are obligate parasites grow on algae, aquatic fungi and higher plants (commonly in the roots).

Important characteristics:

1. Members of this class are obligate parasites (i.e., biotrophic) on fresh water algae, aquatic fungi and higher plants (commonly in the roots).
2. Vegetative body consists of a naked holocarpic plasmodium.
3. Plasmodia are of two types in their life cycle: Sporangiogenous plasmodium (forms sporangium) and cytogenous plasmodium (gives rise to cyst i.e., resting spore).
4. Zoospores biflagellate, having unequal flagella of whiplash type situated in opposite direction, the shorter one in anterior and longer one in posterior side.

Common genera: *Plasmodiophora, Octomyxa, Sorodiscus* etc.

Phylum **Dictyosteliomycota**

(Dictyostelid Cellular Slime Molds)

Important characteristics:

1. These are saprobic slime molds, grow in the middle of organic debris like dung, decaying plants and also in soil.
2. Somatic phase is microscopic and the fructifications are minute, inconspicuous (not easily noticed) and ephemeral (short-lived).
3. The somatic amoebae have filose pseudopodia and a nuclear envelope persisting up to later stage.
4. Somatic amoebae aggregate together to form a **pseudoplasmodium**.
5. The amoebae never fuse together, but retain their individuality with full cooperation as member of an association till the formation of **sorocarp** (Gr. *sorus*, heap; *karpos*, fruit).
6. The sorocarp is differentiated into two regions: stalk and spores. On germination, spores develop into myxamoeba.

Common genera: *Dictyostelium, Polysphondylium* etc.

Phylum **Acrasiomycota**

(Acrasid Cellular Slime Molds)

Members of this group are commonly known as Acrasid Cellular Slime Molds. They are found profusely in the upper layer of humus, in deciduous forests and in cultivated lands.

Important characteristics:

1. Somatic phase mainly consists of amoeboid cells or myxamoebae.
2. Myxamoebae aggregate to form a pseudoplasmodium, which develops fruit body.
3. Lack of flagellated cells except in *Pocheina rosea*.
4. Fruit bodies may be sorocarp (in *Dictyostelium*) or sporocarp (in *Protostelium*).

Common genera: *Acrasis, Protostelium* etc.

Phylum **Myxomycota**

(Plasmodial slime mold)

They are commonly known as true slime molds or plasmodial slime molds, found in damp places especially on old wood and other decomposing plant parts.

Important characteristics:

1. Somatic body is a free-living plasmodium.
2. They feed on yeast cells, protozoa, fungal spores and other substances.
3. Reproduction takes place by asexual and sexual means.

(a) Asexual reproduction takes place by fragmentation in plasmodium or by binary fission in myxamoebae.

(b) Sexual reproduction takes place by fusion between flagellated zoospores or myxamoeba to form zygote, from which multinucleate plasmodium develops by mitotic divisions. They develop different types of fructification. These are **sporangium, aethalium** and **plasmodiocarp** (Fig. 4.124D-G). Meiosis takes place during spore formation in the fructification.

Common genera: *Ceratiomyxa, Physarum* etc.

Class Myxomycetes

Class Protosteliomycetes

Key to different Phylum and other groups of Fungi

The members of the kingdom Fungi i.e., true fungi are placed under four phyla and one unnatural group. Different groups are distinguished from each other as follows:

A. Motile cells i.e. zoospores present .... Chytridiomycota

Motile cells absent 2

Perfect-state absent .... Mitospori fungi

Perfect-state present 3

B. Perfect-state spores zygospores ... Zygomycota

Perfect-state spores ascospores ... Ascomycota

Perfect-state spores basidiospores .... Basidiomycota

Alexopoulos, Mims and Blackwell, 1996

Fungi (polyphyletic group)

Kingdom Fungi

Phylum Chytridiomycota

Phylum Zygomycota

Phylum Ascomycota

Phylum Basidiomycota

Kingdom Straminopila (including some algae)

Phylum Oomycota

Phylum Hyphochytriomycota

Phylum Labyrinthulomycota

Kingdom Protista

Phylum Plasmodiophoromycota

Phylum Dictyosteliomycota

Phylum Acrasiomycota

Phylum Myxomycota

Economic importance

Fungi are very much useful in our day-to-day life. Due to absence of chlorophyll, they depends on living, dead or rotten organic substances for food. Some are useful and others are directly or indirectly harmful to the mankind. Still some like *Claviceps purpurea* have both harmful and useful activities. Considering the overall activities, this can be said safely that in one hand the fungi is our friend and on the other hand the great foe, either directly or indirectly.

Some useful and harmful activities of fungi are:

Table 4.1: Important distinguishing characters of the three Kingdoms

Characters	Fungi	Straminopila	Protist
Nutrition	Heterotrophic (absorptive or osmotrophic)	Autotrophic (photosynthetic and absorptive)	Heterotrophic (phagotrophic) or autotrophic (photosynthetic)
Flagellar mastigonemes	Absent	Tubular	Not tubular
Cell wall	Chitin and β-glucans	Often cellulose; absence of Chitin and β-glucans	Absent when trophic; various when present
Mitochondrial cristae	Flattened	Tubular	Tubular

A. **Useful activities**: Directly or indirectly fungi are beneficial to human being. Fungi is used in medicine industry, as food, in food preparation, in other industry and also in agriculture. Some of the useful activities are:

1. **Preparation of Medicine**: Various kinds of fungi are used in the production of various kinds of medicine. The most important members are *Penicillium notatum, Claviceps purpurea, Saccharomyces cerevisiae, Eremothemium ashbyii, Aspergillus proliferous* etc.

   (a) *Antibiotic*: Antibiotics are the metabolic product of some microorganisms which are active against other microorganism(s). Sir **Alexander Fleming (1929)** was the first who invented the **wonder drug Penicillin** from *Penicillium notatum*. Later, Penicillin was also commercially produced from *P. crysogenum*. Later on, other different types of antibiotics were isolated from different fungi. The list of some fungi along with their produced antibiotics and range of activity are given below.

   (b) *Alkaloid*: Several alkaloids are produced and accumulated in the sclerotium of *Claviceps purpurea* which causes Ergot disease of rye. Out of several alkaloids, Ergometrine and its semisynthetic analogues like methyl ergometrine and methyl ergometrine maleate have prominent uterine action, those control haemorrhage of mother during child's birth, having side-effect with increase in blood pressure and decreased milk secretion.

   (c) *Steroid*: Rheumatic arthritis, allergy and some other diseases are controlled by steroid. Many fungi have the capacity to synthesize different steroids. Steroid like *cortisone* is produced by *Aspergillus niger* from plant glycosides by fermentation.

   (d) *Vitamins*: Vitamins are the micronutrients required for the growth of living organisms. Vitamin B-complex, Vitamin A and Vitamin B-12 are found respectively from *Saccharomyces cerevisiae, Rhodotorula gracilis* and *Eremothemium ashbyii*.

   (e) *Statins: Aspergillus terreus*, a soil-borne fungus, produces a secondary metabolite called **lovastatin** and *Aspergillus phoma* produces **squalestatin**. Both have been used to reduce or remove low density lipoproteins from blood vessels in human beings. These are implicated in attracting stem cells to damaged tissues. The stem cells then appear to regenerate the tissue.

Name of antibiotics	Produced fungi	Range of activity
1. Penicillin	*Penicillium notatum* and *P. chrysogenum*	Bacteria (both Gram +ve and –ve)
2. Griseofulvin	*P. nigricans* *P. griseofulvum*	Fungi
3. Cephalosporin	*Acremonium* sp.	Bacteria (Gram +ve)
4. Citrinin	*P. citrinum*	Fungi
5. Palutin	*P. palutum*	Fungi and Bacteria
6. Fumagillin	*Aspergillus fumigatus*	Protozoa
7. Fusidic acid	*Fusidium coccineum*	Bacteria
8. Viridin	*Trichoderma viridi*	Fungi

(f) *Immuno suppressants*:

(i) *Cyclosporin A* is a primary metabolite of several fungi, like *Trichoderma polysporum* and *Cylindrocarpon lucidum*. It acts by inhibition of lymphocytes. The inhibition of T cells proliferation results in the suppression of the activation process associated with invasion of foreign bodies.

(ii) *Gliotoxins*, belongs to a class of compounds called epipolythiodioxopiperazines, produced by many fungi including *Aspergillus fumigatus*, also acts as an *Immuno suppressant*.

(g) *Other compounds*: An anticancerous substance, **calvacin** is present in the giant Puffball (*Clavatia gigantea*), that prevents stomach tumours. Similarly, fungi like *Ganoderma lucidum* have promising significant roles in **anticancerous, anti HIV** and **antihepatitis-B** diseases. The fungi like *Coriolus versicolor* have also **anticancerous** activity.

2. **Food**: Fungi are used as food by humans from prehistoric period. Some fungi have been used directly as food and some are used in food processing:

(a) *Direct use*: Fruit bodies of some fungi, like *Agaricus brunnescens, A. campestris, Volveriella volvacea, V. diplasia, Pleurotus sajor-caju, Lentinus edodes* etc., are used as food due to their high protein content (21–30% on dry weight) and have good amount of lysine, an amino acid; minerals like Na, Ca, K and P; Vitamins like B, C, D and K and very little amount of fat. These are recommended as ideal foods for heart and diabetic patients. The above fungi can also grow artificially at commercial level.

(b) *Used in food processing*: The nutritious food soyabean cannot be digested easily by human beings. Soyabeans are fermented by fungi like *Rhizopus oryzae, R. oligosporus* to prepare an easily digestable and tasty food known as "tempeh".

Several Indian foods such as jalebies, idli, murcha, papadam and toddy, are actually the fermented products caused by several fungi like *Saccharomyces bayanus* for jalebies; *Trichosporon pollulans and Torulopsis candida* for idli; *Rhizopus arrhizus* for murcha; *Saccharomyces cerevisiae* for papadam and toddy.

Flavours and colours: Many fungi are responsible for a range of flavours including terpenes, methanol and lactones. Some fungi produce a range of compounds that alter the colour of food. Red wine is produced by using *Monoascus purpureus*.

Fermentation by yeast, particularly by *Rhodotorula* and *Saccharomyces*, results in quick growth of the organism producing *single cell protein (SCP)*. Because of high nucleic acid content, the fermented products are not suitable for human nutrition, but can be used as feed for animal.

3. **Fungi in industry**: Many fungi are used in the production of alcohol, bread, cheese, enzyme and organic acids.

(a) **Alcohol production**: Alcoholic fermentation by fungi is the basis of brewing industry. The enzyme zymase of microorganisms like yeast is responsible for alcohol production.

The reaction is:

$$C_6H_{12}O_6 \xrightarrow[\text{Yeast}]{\text{Zymase}} 2C_2H_5OH + 2CO_2$$

Wines are produced from grapes or other fruits by *Saccharomyces ellipsoideus* with about 14% alcohol concentration. Beer is brewed from barley malt by *Saccharomyces cerevisiae* with 3–8% alcohol production.

Sake is a Japanese alcoholic product made from rice. Steamed rice is inoculated with

Aspergillus oryzae to produce koji. Further rice mash is mixed with bacteria and yeasts to produce lactic acid. The mash and koji are then mixed and fermented for about 20 days and the alcohol concentration reached about 18%. The alcohol is then filtered, pasteurized and stored before consumption.

(b) **Bread and cake production**: During alcoholic fermentation by yeast, CO_2 being released as bubbles are used in baking industry to make the breads and cakes as spongy in appearance.

(c) **Cheese production**: Some species of *Penicillium (P. roquiforti* and *P. camemberti)* are used in the production of Roquefort and Camembert cheese by hydrolysis of fats and also to develop specific *texture* and *flavour* to cheese.

(d) **Enzyme and organic acid production**: Many fungi are used in the commercial production of enzymes and different organic acids. List of some fungi along with produced enzymes and/or acids and their uses are given:

4. **Soil fertility**: Decomposition of litter and wood, mainly in the forest, takes place by the combined action of different type of fungi. Fungi like *Fusarium, Chaetomium, Chytridium, Penicillium, Aspergillus* etc., can decompose the structural polymers such as cellulose, hemicellulose, lipid, protein, starch etc. The more recalcitrant (persistant) polymers like lignin are decomposed by white rot fungi of Basidiomycotina. By decomposing the organic matters, fungi help to increase minerals and other substances, thereby the fertility of soil is increased.

5. *Plant nutrition*: Several fungal members like *Rhizoctonia, Tricholoma, Boletus, Phallus, Amanita* etc., associated with the roots of higher plants form *mycorrhizal* relationship. The fungal partner supplies water and minerals and in turn, they take nutrition from the plant.

Enzyme Organic acid	Produced fungi	Uses
Enzymes		
1. Amylase	*Aspergillus oryzae, A. niger*	Alcohol industry and in pharmaceuticals.
2. Cellulase	*Trichoderma reesli*	Production of cheese and hydrolysis of cellulose.
3. Invertase	*Saccharomyces cerevisiae*	Paper industry and confectionary.
4. Zymase	*Saccharomyces cerevisiae*	Ethyle alcohol production.
Organic acids		
1. Citric acid	*Aspergillus niger, A. wentil*	Soft drinks and other foods. Manufacture of ink and leather tanning.
2. Fumaric acid	*Rhizopus stolonifer*	Manufacture of wetting agents.
3. Gallic acid	*Aspergillus gallomyces, Penicillium glaucum*	Manufacture of ink and dyes.
4. Gluconic acid	*Aspergillus niger, Penicillium purpurogenum*	In textile, leather, food and photographic industries and also in pharmaceuticals.
5. Kojic acid	*Aspergillus oryzae*	As insecticide and antibiotic.
6. Itaconic acid	*Aspergillus terreus, A. itaconicum*	Manufacture of synthetic fibre and plasticisers.

6. *Manufacture of Phytohormone*: Gibberellin, an important phytohormone produced by *Gibberella fujikuroi*, the pathogen of Bakanae disease of rice, has been used to enhance growth of many plants.

7. *As insecticide*: Fungi like *Cordyceps melonthae, Aschersonia aleyroidis, Empusa sepulchralis* etc., are used as

insecticides to control different types of insects.

8. *Biological research*: Fungi like *Neurospora,* Yeast etc., have been used in genetical and cytological studies. *Physarum polysephalum* has been used to study DNA-synthesis and morphogenesis.

9. *Test organism*: Some strains of *Aspergillus niger* have been used to detect trace elements like Zn, Cu, and Mo, even if the substances are present in very minute quantity in the substrate. These elements when absorbed by the fungus give a particular colour to the conidia. Similarly, *Neurospora crassa* has been used to detect Vitamin B complex.

B. **Harmful activities**: Fungi are also harmful to the human beings in various ways, either directly or indirectly. They may cause diseases of plants, human beings, and animals; spoilage of food etc. Some of the harmful activities are:

1. *Fungi causing plant diseases*: Fungi cause several minor and major plant diseases. Some of them also cause famine in different parts of the world. List of some diseases along with causal organism are given:

Name of disease	Causal fungi
1. Early blight of potato	*Alternaria solani*
2. Late blight of potato **(caused famine in Ireland, 1845**; leading to death of more than 10 lakh people)	*Phytophthora infestans*
3. Wart disease of potato	*Synchytrium endobioticum*
4. Brown spot disease of rice **(caused famine in Bengal, 1943**; leading to death of more than 20 lakh people)	*Helminthosporium oryzae* **(perfect stage** – *Cochliobolus miyabeanus*)
5. Blast disease of rice	*Pyricularia oryzae*
6. Black stem rust of wheat	*Puccinia graminis tritici*
7. Loose smut of wheat	*Ustilago nuda tritici*
8. Damping off of tobacco	*Pythium debaryanum*
9. Tikka disease of ground nut	*Cercospora arachidicola*
10. Red rot of sugarcane	*Colletotrichum falcatum*
11. Whip smut of sugarcane	*Ustilago scitamini*
12. Powdery mildew of peas	*Erysiphe polygoni*
13. Downy mildew of peas	*Peronospora pisi*
14. White rust of crucifer	*Albugo candida*
15. Leaf rust of coffee	*Haemelia vastatrix.*

2. *Fungi causing human diseases*: Like plant disease, some fungi also cause human diseases. Some of the important human diseases caused by fungi are given:

Name of disease	Causal fungi
1. Mental disorder	*Lipomyces neoformis*
2. Ringworm of skin and nails	*Trichophyton, Microsporum, Epidermophyton*
3. Lung disease	
(a) Aspergillosis	*Aspergillus niger, A. flavus*
(b) Blastomycosis	*Blastomyces dermatitidis*
(c) Coccidiodomycosis	*Coccidioides immitis*
(d) Paracoccidioidomycosis	*Paracoccidioides brasiliensis*
4. Ear infection (Otomycoses)	*Aspergillus flavus, A. nidulans*
5. Allergy	*Alternaria, Phoma, Trichoderma, Aspergillus*, etc.

3. *Fungi causing animal diseases*: Like plants and human being, some animals also suffer from different diseases caused by fungi. Some important animal diseases caused by fungi are given:

Name of disease	Causal fungi
1. Athelete foot	*Tinea rubrum*
2. Mucormycosis	*Mucor, Rhizopus*
3. Aspergillosis	*Aspergillus*
4. Penicillosis	*Penicillium*

Besides above diseases, *Saprolegnia parasitica,* an aquatic fungi live as parasite on egg and gills of fishes. Members like *S. ferax, Achlya* sp. and *Ichthyophorus haferi* cause severe damage to fishes.

4. *Fungi causing disease of vegetables* and *fruits*: Different species of fungi cause

disease (post-harvest) in vegetables and fruits. List of some commonly available diseases are:

Name of disease	Causal fungi
1. Soft rot of tomato	*Rhizopus stolonifer*
2. Soft rot of sweet potato	*R. stolonifer*
3. Fruit rot of banana	*Alternaria tenuis*
4. Fruit rot of apple	*Aspergillus fumigatus*, *Alternaria* spp.
5. Fruit rot of orange	*Penicillium digitatum*

5. *Spoilage of food*: Different types of food are destroyed by different fungi like *Mucor, Aspergillus, Penicillium, Rhizopus* etc. Some of them are given below.

Fungi	Food	Condition
1. *Mucor mucedo*, *Rhizopus stolonifer*, *Cladosporium herbaceum*	Meat	in frozen condition (even at –6°C temperature).
2. *Rhizopus* spp.	Bread	in normal temperature.
3. *Aspergillus*	Jam, Jellies, Pickles etc.	in normal temperature.

6. *Destruction of timber*: Species of *Polyporus and Armillaria* are very common — cause wood rot of many trees. *Polyporus abietinus* causes decay of *Abies* wood, *P. schweinitzii* causes bud rot of mature conifers, *P. sulphureus* causes wood rot of apple, pear etc., while *Armillaria mellea* causes red rot of apple trees.

7. *Destruction of goods*: Fungi like *Alternaria, Mucor, Trichoderma, Penicillium, Chaetomium, Cephalothecium* etc., destroy paper, rubber, leather, camera lenses etc. In addition to the above, other fungi like *Chaetomium globosum, Memnoniella echinata* destroy jute products.

8. *Poisonous fungi*: Some fungi like *Amanita phalloides* produce toxins, like *α-amanitin* causes lesions of stomach cells and *phalloidin* affects on liver. Fungi, like *Aspergillus flavus, A. fumigatus* and *Penicillium islandicum* when infest ground nut and mustard cakes, produce a poisonous substance, the aflatoxin that binds with DNA and inhibits transcription, resulting into the inhibition of protein synthesis.

9. *Hallucinogenic drug*: **LSD** (d-lysergic acid diethylamide), the well-known hallucinogenic drug, is extracted from the sclerotia of *Claviceps purpurea*, the causal agent of ergot disease of rye. Other fungi like *Psilocybe mexicana* produce *Psilocin* and *Psilocybin* that have hallucinogenic properties. The hallucinogenic substances may destroy brain cells and cause distortion of perception power of human beings.

PHYLUM CHYTRIDIOMYCOTA

(Uniflagellate Zoosporic Fungi)

The Chytridiomycota (Gr. *Chytridion*, little pot) is a group of fungi characterized by the presence of zoospores, having single posteriorly directed flagellum. The thalli are typically microscopic and varied in structure. Molecular phylogenetics and zoospore ultrastructural characters show greater genetic diversity than expected. What was once the single order Chytridiales is now ten monophyletic orders. *Rozella, Olpidium*, and the Blastocladiomycota, organisms once thought to be within the Chytridiomycota, diverge in separate lineages and are no longer included in the phylum Chytridiomycota.

The Neocallimastigomycota are sister to Chytridiomycota, but they are distinct in their zoospore ultrastructure, behaviour of nuclear envelope during mitosis, and specific adaptation as anaerobes to digestive systems of herbivores. Molecular characterization of environmental samples demonstrates that zoosporic fungi are widespread and can survive in a range of habitats, from oceans to freshwater and even harsh environments including under artic snow or in exposed soils in alpine regions. The importance of these organisms is as parasites, saprotrophs of refractory materials, and components in food webs. Current classification is based on molecular and ultrastructural analyses of zoospores.

Important characteristics

Important characteristics of Chytridiomycota are given below:

1. Vegetative body is coenocytic and thalloid, either globose or ovoid structure, either an elongated simple hypha, or well developed mycelium.
2. Cell wall is mainly made up of chitin and glucan.
3. Nuclear division is intranuclear and centric type.
4. Members of this group produce motile cells at some stage of their life cycle.
5. Motile cells (zoospores and gametes) possess single posteriorly placed, whiplash type of flagellum except a few polyflagellate cells.
6. Sexual reproduction takes place by planogametes developed in gametangia. The fused gametes form zygote. Zygote on germination develops either into a resting spore or resting sporangium except a few those develop diploid thallus.

Common genera: *Synchytrium, Monoblepharis, Rhizophidium* etc.

Ainsworth (1973) placed Class Chytridiomycetes under the Subdivision Mastigomycotina. Hawksworth *et al.* (1995) included the members in the single Class **Chytridiomycetes**, under the Phylum **Chytridiomycota**. Thus general aspects are same in the class and phylum.

Ecology and significance

Chytridiomycota usually inhabit in fresh water ecosystems. They need water for the survival. However, Chytridiomycota often dwell within the host plants or animals. They live either parasitically and saprophytically. As Chytridiomycota often feed on decaying organisms, they are the important decomposers. Chytridiomycota can also have a negative impact on human produce, particularly *Synchytrium endobioticum*, causes wart disease of potato. Some others, like *Olpidium brassicae*, caused "Big Vein" in lettuce; are parasitic, but cause little damage to their host. While this is not true of all species, some members such as *Rhizophlyctis rosea* and *Allomyces anomalus* have structures that allow them to survive in draught and also on excessive heat.

Anaerobic **rumen fungi***, such as *Orpinomyces joyonii*, inhabit the guts of goats, cattle, and sheep. They live in anaerobic conditions and play an imporant part in the digestive process. They help in processing of plant matter while obtaining food. They are so effective in making some of their host animals to utilize cellulose. In addition, some species of anaerobic rumen fungi are used for other purposes. Xue *et al.* (1996) cloned microbial non-starch polysaccharide hydrolases from *Neocallimastix patriciarum*. They found that the genes for these hydrolases are valuable for the production of genetically-engineered cereal crops.

There is a well known disease that caused by the members of Chytridiomycota. ***Chytridiomycosis*** is a fungal infection of amphibians like frogs, caused by *Batrachochytrium dendrobatidis*. The disease is found in keratinised tissue, which includes the mouth parts of tadpoles. One of the main symptoms of chytridiomycosis is a thickened epidermis (hyperkeratosis). It is fatal, and has caused amphibian population decline in many continents. Treatment is possible if the infection is caught early, but not always effec-

* **Rumen** is the large first compartment of the stomach of a ruminant (herbivours, even-toed, hoofed mammal that has a complex 3- or 4-chambered stomach). **Rumen fungi are** the symbiotic anaerobic fungi those breaks down cellulose to unit of glucose.

tive. One important characteristic of chytridiomycosis is the way it behaves at different temperatures. The pathogenicity of the disease is enhanced at lower tempeartures, with higher rate of mortality. Weldon *et al.* (2004) proved that the disease initially known to occur in 1938, although exists in Africa for 23 years before the actual report and later it was introduced in other continents in the mid-1930s.

Diagonisis of the disease

Different techniques have developed to diagonize the disease. These are:

(a) Olsen *et al.* (2004) has developed a staining protocol to facilitate the often difficult diagnosis of this disease.

(b) Annis *et al.* (2004) has developed a method of detection which is a polymerase chain-reaction based assay.

(c) Boyle *et al.* (2004) has also developed a new method of detection, a real-time PCR Taquin assay.

(d) Van Ells *et al.* (2003) developed a method of immune-histochemical staining which proved to be a very effective method of diagnosis.

Treatment of the disease

Different techniques have developed to control the disease. These are:

Johnson *et al.* (2003) compared the efficacy of chemical disinfectants, UV light, dessication, and heat in treating chytridiomycosis. Most methods were effective.

Chemical treatments were highly effective, although concentration and exposure are the critical factors in their efficacy.

Thallus organization

The morphology of the members of this Class Chytridiomycetes (presently phylum Chytridiomycota) are variable in nature, even within species, and its use in the classification of chytrid has been inadequate. However, their variability has led to specialized terminology for this class (phylum). These are as follows.

Species may be **monocentric** (Fig. 4.11, 4.13 and 4.14), unicellular and composed of only a single zoosporangium or **polycentric** (Fig. 4.12), composed of two or more zoosporangia.

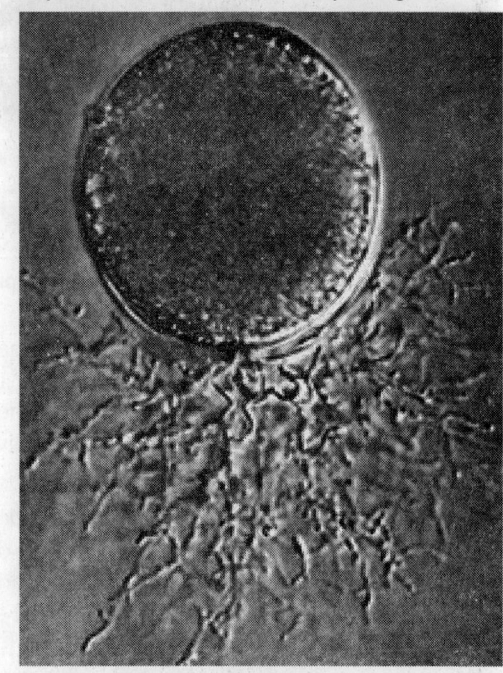

Fig. 4.11: *Spizellomyces palustris*, a monocentric, eucarpic chytrid. From Chen, *et al.* (2000).

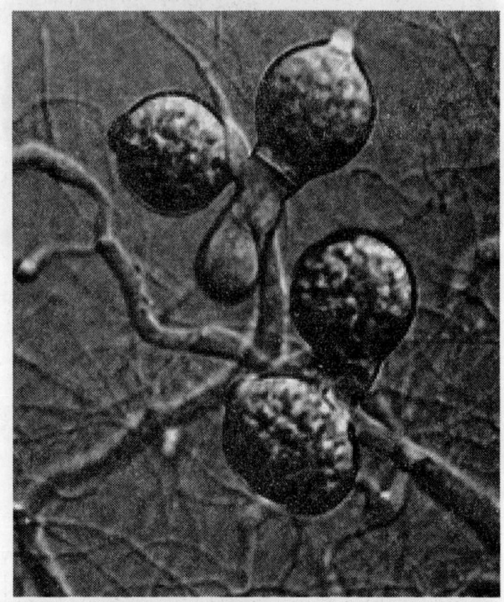

Fig. 4.12: *Polychytrium aggregatum*, a polycentric, eucarpic chytrid. From Chen, *et al.* (2000)

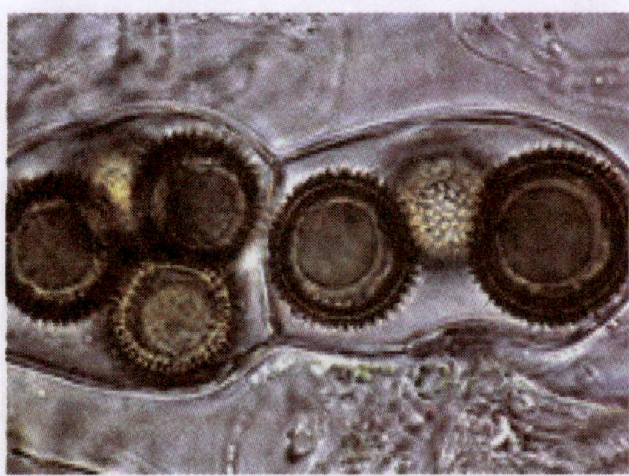

Fig. 4.13: *Rozella allomycis*, a monocentric, holocarpic, and endobiotic chytrid

They may be **Holocarpic** (Fig. 4.13), when the fungus body is entirely reproductive or **eucarpic** (Fig. 4.11, Fig. 4.12 and Fig. 4.14), when the fungus body is differentiated into reproductive as well as vegetative structures, (rhizoids).

The **parasitic members** are further divided into **epibiotic** (Fig. 4.13), those grow externally on the host, but attached internally by rhizoids, and on the other hand, some are **endobiotic** (Fig. 4.14), where the parasite grows entirely within the host.

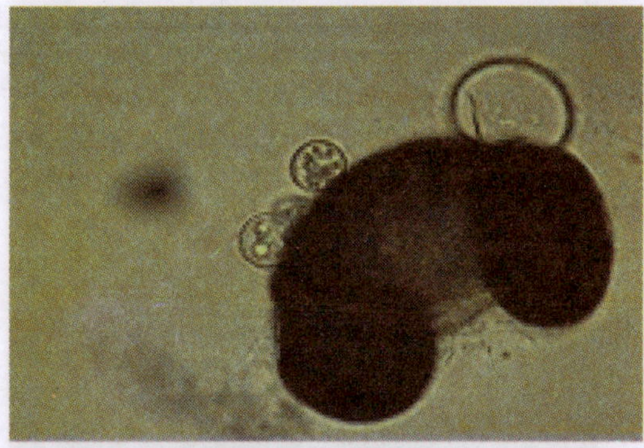

Fig. 4.14: *Rhizophydium sphaerotheca*, monocentric, eucarpic and epibiotic chytrid.

Reproduction

Fungi of this group reproduced both asexually and sexually.

Asexual reproduction

There are different forms in thallus growth and reproductive development that may occur following encystment of a uninucleate zoospore, the cyst nucleus. These are **endogenous**, *monocentric* or *polycentric type*; or **exogenous**, *monocentric* or *polycentric type*.

Asexual reproduction takes place by means of zoospores are borne in **zoosporangia**. In the early stage of development sporangia contain protoplasm with many nuclei. With further development, the protoplast undergoes cleavage into many, minute, uninucleate units, each of which develops into single posteriorly flagellate zoospore (Fig. 4.17, *Synchytrium*). Zoospores may emerge through one or more papillae during discharge of sporangium. A discharge papilla is developed in the wall of the sporangium or in the tip of a tube developed from it (e.g. *Batrachochytrium dendrobatidia*). In some species, a well-defined circular cap is formed, called operculum; through which zoospores are released. Species with operculum is called **operculate type** and others are called **inoperculate type**, where zoospores are discharged through a pore in the wall of the sporangium (e.g. *Monoblepharis*) or through discharge tube, formed when discharge papilla dissolve.

Zoospores are uninucleate, and on germination give rise to new individual like the mother.

Sexual reproduction

Sexual reproduction takes place by a variety of different methods, although it is unknown or questionable in many taxa under this phylum. With the available information, it is found that sexual reproduction takes place by the following methods, described below:

1. **Planogametic copulation**

   (a) *Conjugation between isogamous planogametes* (e.g. *Synchytrium*). In this method two morphologically similar but physiologically different planogametes unite to form a motile zygote. In some species, gametes produced from same gametangium will not fuse (e.g. *Synchytrium endobioticum*).

   (b) *Conjugation between anisogamous planogametes*. (e.g. *Allomyces macrogynus*). In this method two morphologically and physiologically different planogametes (one is larger than the other) unite to form a motile zygote.

(c) *Fertilization between egg and antherozoid* (e.g. *Monoblepharis*). In this method, motile male gametes are liberated from the male gametangium (also called antheridium) into the water and swim away. Some of the male gamete i.e. antherozoid may reach the female gametangium i.e. oogonium and one of them enters each oogonium and unite with the egg to form zygote.

2. **Gametangial copulation**

In this method, the entire protoplast of one gametangium transferred into the other (e.g. *Polyphagus euglenae*).

3. **Somatogamy**

It is the process of fusion of two somatic structures. In some members rhizoidal filaments precedes the formation of a resting spore (e.g. *Chytriomyces hyalinus*).

Chytridiales

The main distinguishing characters are:

1. True mycelium is lacking.
2. Rhizomycelium or rhizoids are present in some species.
3. Zoospores are globose, elongated when swimming, ameboid prior to encyst formation.
4. In zoospore, single large lipid globule core area of compacted ribosomes, partially enclosed by ER nucleus is to one side and not conspicuously attached to kinetosome, lipid globule is on the opposite side of the zoospore.

SYNCHYTRIUM

Systemic position

Ainsworth (1973)	Hawksworth (1995)
Kingdom Mycota	Kingdom Fungi
Division Emycota	Phylum Chytridiomycota
Subdivision Mastigomycotina	Class Chytridiomycetes
Class Chytridiomycetes	Order Chytridiales
Order Chytridiales	Family Synchytriaceae
Family Synchytriaceae	Genus *Synchytrium*
Genus *Synchytrium*	

Synchytrium is represented by about 200 species reported from all over the world, occurs as parasite on aquatic alga, bryophytes (mosses), pteridophytes (ferns) and mostly on flowering plants. About 80 species have been reported from India, of which *S. rytzii* on the members of Lamiaceae, *S. trichosanthoides* and *S. laginariae* on the members of Cucurbitaceae, *S. taraxicola* on *Taraxacum officinale, S. sisamicola* on *Sesamum indicum* and *S. endobioticum* is a well-known species which causes **wart disease** or **black wart of potato**, available in the main potato growing regions of the world, mostly in mountains with moist and cold climate. In India, this disease was first recorded at Rangbul in Darjeeling district of West Bengal in 1953 (Ganguly and Paul, 1953).

Symptoms

The disease is characterised by cauliflower-like black warty growth on tubers (Fig. 4.16), stolons and stem bases (Fig. 4.15). Sometimes, the size of the warts are more than the size of the tuber.

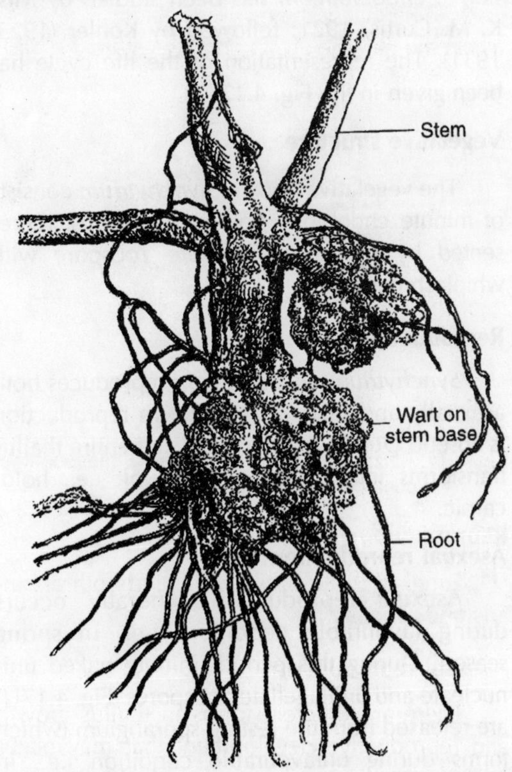

Fig. 4.15: Symptom on stem base and stolon of potato

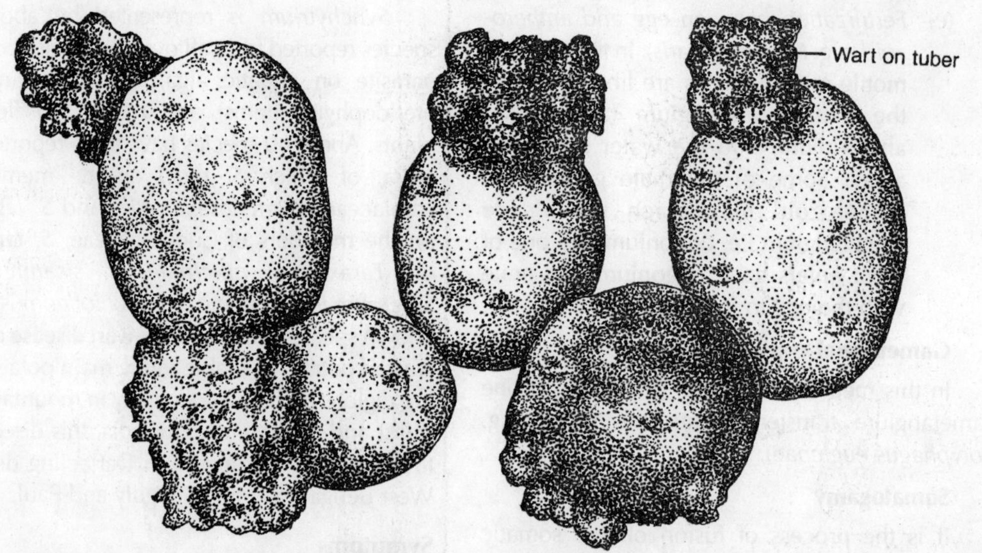

Fig. 4.16: Wart on potato tubers

Life history

The life history of the most common species i.e., *S. endobioticum* has been studied by Miss K. M. Curtis, 1921; followed by Köhler (1923, 1931). The representation of the life cycle has been given in the Fig. 4.17.

Vegetative structure

The vegetative body of *Synchytrium* consists of minute endobiotic holocarpic thallus, represented by naked uniflagellate zoospore with whiplash flagellum.

Reproduction

Synchytrium endobioticum reproduces both asexually and sexually. Vegetative reproduction is absent. During reproduction, the entire thallus transforms into a reproductive unit i.e., holocarpic.

Asexual reproduction

Asexual reproduction generally occurs during favourable condition, i.e., in spring season. During this period, minute naked uninucleate and uniflagellate zoospores (Fig. 4.17U) are released from the resting sporangium (which forms during unfavourable condition i.e., in winter season) after water soaking (Fig. 4.17T).

The zoospores are capable of swimming for about two hours. After coming in contact either with the potato 'eye' or stolon or young tuber (Fig. 4.17B), they come to rest and withdraw their flagella (Fig. 4.17C). The content of the zoospore cyst enters into the host cell through the wall by minute pore in amoeboid movement, keeping the cyst membrane outside (Fig. 4.17D).

The protoplast of zoospore, after entry in the host epidermal cell, absorb food and becomes spherical in shape. The infected host cell also enlarges in volume. The host cell surrounding the infected cell becomes stimulated and starts swelling (**hypertrophy**) resulting into the formation of tumour or wart-like structure. The infected cell dies and remains in the middle of the wart. The pathogen along with its nucleus enlarges considerably, rounds off and develops two layered walls, consisting of a thick golden brown exospore and a thin hyaline endospore. This is the **summer spore** (Fig. 4.17E).

The summer spore germinates within the infected host cell. Before germination, the nucleus enlarges and the inner wall protrudes out through a minute pore on the outer wall and forms a vesicle towards the upper portion of the infected host cell. The total content of the summer spore is transferred to the vesicle. The

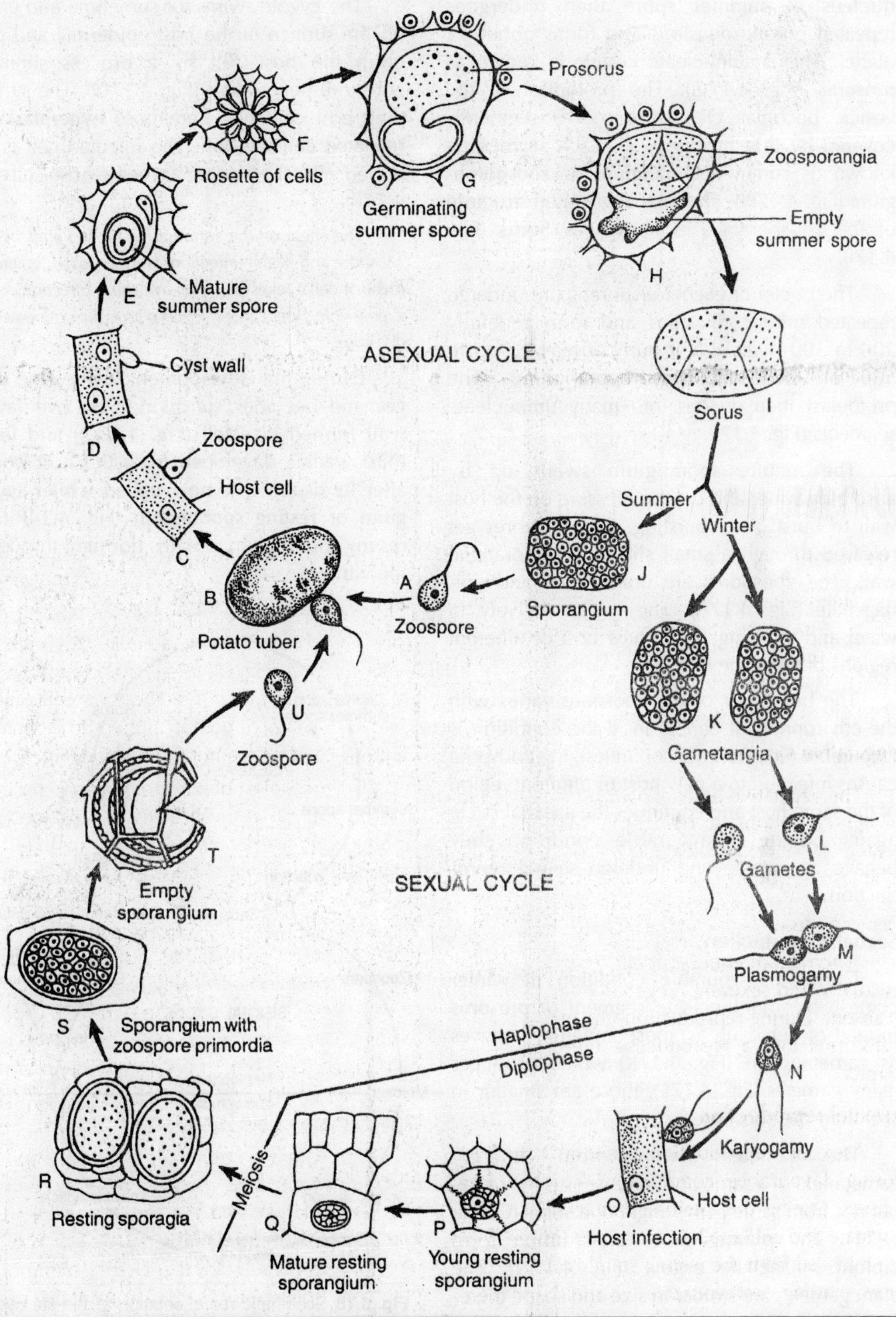

Fig. 4.17: Life cycle of *Synchytrium endobiotium*

nucleus of summer spore then undergoes repeated mitotic divisions and forms about 32 nuclei. This multinucleate vesicle is known as **prosorus** (Fig. 4.17G). The protoplast of the vesicle becomes cleaved into 4–9 segments, covered by thin hyaline wall. Each segment is known as **summer sporangium** or **zoosporangium** (Fig. 4.17H). The total aggregated structure of the zoosporangia is known as **Sorus** (Fig. 4.17I).

The nuclei of each zoosporangium undergo repeated mitotic divisions and form generally 200 to 300 nuclei (their number may go up to 500 or more in large sporangium). The protoplast then divides into many uninucleate segments (Fig. 4.17J).

The mature sporangium swells up by absorbing water and creates pressure on the host wall to burst. After bursting, the zoospores get released through a small slit on the sporangial wall. The zoospores are uninucleate and uniflagellate (Fig. 4.17A). They swim actively in water and infect again the new host or different regions of the same host.

The behaviour of the zoospore varies with the environmental condition. If the condition is favourable (i.e., summer continues), the zoospore causes infection to a new host or different region of the same host and continues the asexual cycle again. During unfavourable condition they behave as gamete and undergo sexual reproduction.

Sexual reproduction

During unfavourable condition (if winter comes), the multinucleate segment of **prosorus** instead of behaving as zoosporangium behaves as gametangium (Fig. 4.17K) which produces many gametes (Fig. 4.17L), those are smaller in size than the zoospores.

The gametes coming from different gametangia of a same or different sorus may fuse, but not from same gametangia of a sorus (Curtis, 1921). The planogametes after union form diploid biflagellate zygote (Fig. 4.17N). The planogametes are similar in size and shape therefore, copulation is **isogamous**.

The zygote swims for sometime and encysts on the surface of the host epidermis and penetrates the host cell by a process similar to zoospore penetration (Fig. 4.17O). The surrounding host cell then undergoes **hyperplasia** i.e., repeated cell division. The infected cell is then buried into the deeper layer of host cells (Fig. 4.17P).

The effect on the surrounding tissue varies between zoospore and zygote infection. *Hypertrophy* (i.e., enlargement of cells) takes place on *zoospore infection,* but the *zygote infection* causes *Hyperplasia* (i.e., repeated cell division).

During this development, the zygote enlarges and becomes surrounded by two layered wall formed by itself (Fig. 4.17Q) and then a third wall is developed from the host content after its death. It is now called **winter sporangium** or **resting sporangium** (Fig. 4.14R). The resting sporangium remains dormant throughout the winter season.

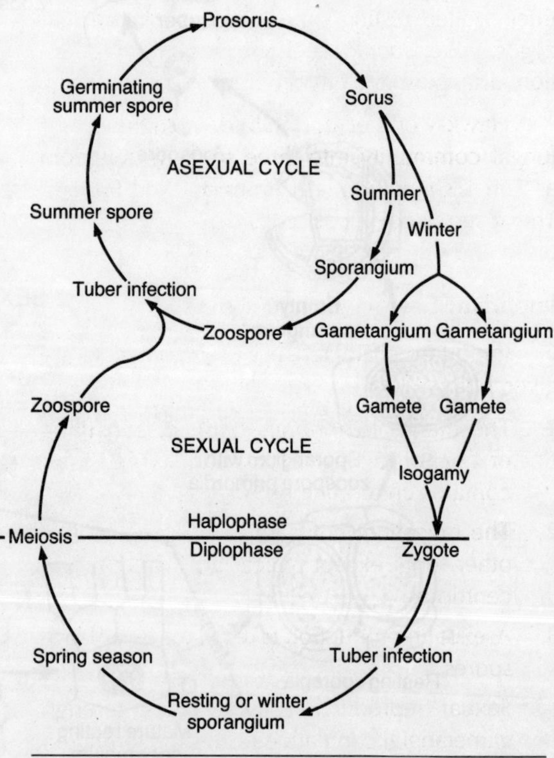

Fig. 4.18: Schematic representation of the life cycle of *Synchytrium endobioticum*

The resting sporangia are released into the soil after decaying the host tissue and are capable to germinate within about two months. With the onset of favourable condition i.e., in spring season, the resting sporangium becomes active and its nucleus undergoes repeated nuclear division of which first one is meiotic, followed by many mitotic divisions. The protoplast, along with a single nucleus, divides into many uni-nucleate segments (Fig. 4.14S).

After absorbing water, the wall of resting sporangium bursts open (Fig. 4.14T) and releases the zoospores. The zoospores are like the asexual zoospores (Fig. 4.14U), which on coming in contact with a suitable host cause infection and repeat the cycle again.

The schematic representation of life cycle of *S. endobioticum* has been given in Fig. 4.18.

Indian species:

Synchytrium endobioticum, S. brownii, S. ajrekari, S. colapsum, S. cookii.

PHYLUM ZYGOMYCOTA

(Conjugated Fungi)

The fungi belonging to this group are generally terrestrial and absence of any motile cells or zoospores in the life cycle. They produce asexually by aplanospores or conidia and developed thick walled resting spores, the **zygospore** (Gr. *zygos*, yoke; *spora*, seed) after sexual reproduction, distinguish the phylum from rest of the fungi.

Hawksworth *et al.* (1995) divided the earlier fungal community into three separate Kingdom as Fungi, Straminopila (Chromista) and Protists. The **Zygomycota** was considered as Phylum, under the Kingdom Fungi.

Important Characteristics

Important characteristics of Zygomycota are given below:

1. The thallus is normally haploid, consisting of coenocytic mycelium and its cell wall contains chitosanchitin.
2. The mycelium contains cell organelles like other fungi except typical golgi bodies and centrioles.
3. Asexual reproduction takes place by aplanospores.
4. Sexual reproduction takes place through gametangial copulation results the formation of zygospore.

Common genera: *Mucor, Rhizopus* etc.

Ecology and significance

Ecologically Phylum Zygomycota is the most diverse group of fungi. The ecological behaviour and its significance are given below:

1. They are terrestrial and live close to plants, usually in soil and on decaying plant matters. As they decompose soil, plant matter, and dung, they play a major role in the carbon cycle.
2. Some are the pathogens of plants, animals, amoeba, other fungi and also form mutualistic symbiotic relationships with plants (ectomycorrhizal, e.g. Endogonales). In addition, they form commensalistic relationships with arthropods, inhabiting in gut of the organism and feeding on unused nutrients, e.g. Harpellales and Asellariales.
3. However, members of this group can also be found in aquatic ecosystems.
4. Although, the Zygomycota are largely known to humans for the negative economic impact, but they also have some **practical use**. Certain species are used in food fermentations. In addition, people have used their pathogenic power to control insect pests. Although these are largely considered terrestrial organisms, certain species of Zygomycota also form relationships with animals.

5. Zygomycetes are known to cause serious infections, particularly for diabetics and immunocompromised individuals. These infections can also occur as a result of major burns or other tramatic injury. One such disease is **zygomycosis**. This is a rare fungal disease that occurs in humans, and can even affect the fetus. It is potentially lethal.

Thallus organization

Based on morphology and ecology, the Phylum Zygomycota is divided into two classes, Zygomycetes and Trichomycetes. The thallus organization of the two classes is given below:

1. **Zygomycetes**

The mycelium of some species of **Zygomycetes** consists of coenocytic hyphae, in which the nuclei are haploid when the organism is in the vegetative stage. The walls of which are composed of chitin, chitosan, and polyglucuronic acid. In other forms however, the mycelium is very much reduced and may possess more or less regularly spaced septa. Some species having the capacity to grow as either mycelia or yeasts are said to be **dimorphic**. Unwalled thalli also are found within some members of this group.

2. **Trichomycetes**

All members of **Trichomycetes** are obligately associated with living arthropods. Most species are microscopic with limited thalli, may be branched or unbranched. The mycelia of some have regular septations, while others lack septa except at the base of reproductive cells. Members of the orders Asellariales and Harpellales have septa with perforation, like *Linnderia pennispora*, a member of Zygomycetes. They remain attached with the host either by **holdfast**, densely aggregated in gut with hairy appearance (the class is called, hairy fungi) or does not penetrate through the chitinous gut but assumed to absorb nutrients from gut lumen.

Reproduction

Members of the Phylum Zygomycota reproduce both asexually and sexually.

Asexual reproduction

The asexual reproduction takes place through the formation of sporangia and sporangiospores. Walled sporangiospores are formed by the internal cleavage of the sporangial cytoplasm. At maturity, the sporangial wall typically disintegrates or dehisces, thereby freeing the spores that are usually dispersed by wind or water. Sporangia are formed at the ends of specialized hyphae called sporangiophores.

A unique spore dispersal strategy in the order Mucorales is exhibited by the dung fungus *Pilobolus*, the '**hat thrower**'. The entire black sporangium is explosively shot off, from the top of the sporangiophore up to distances of several meters. Phototrophic growth of the sporangiophore facilitates dispersal away from the dung onto a fresh blade of grass where it may be consumed by any herbivore, thereby completing the asexual cycle after the spores pass through the digestive system.

Some members of the Entomophthorales (e.g. *Basidiobolus, Conidiobolus*) also reproduce via forcibly discharged of asexual spores. Interestingly, species of *Conidiobolus, Basidiobolus*, and many other genera produce the second kind of spore on a long stalk that appears to have certain morphological adaptations for efficient insect dispersal.

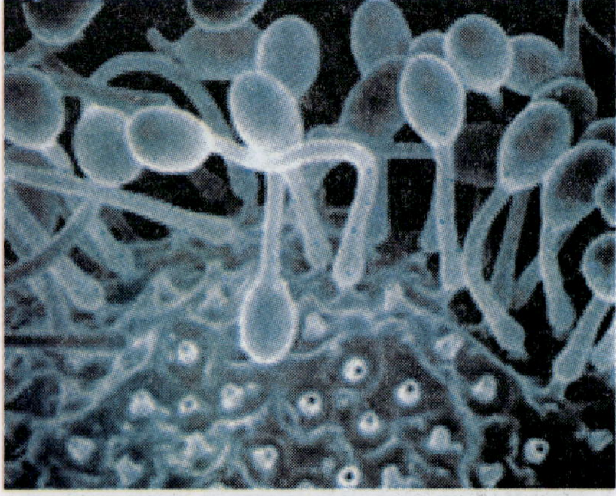

Fig. 4.19. *Benjaminiella poitrasii*: Unispored sporangia during asexual reproduction (Scanning electron micrograph)

Two variant types of sporangia include **sporangiola** and **merosporangia**. Sporangiola are simply 1-few spored sporangia, containing between 1-30 spores (Fig. 4.20). Merosporangia are elongated sporangiola with uniseriate spores usually produced from a vesicle or stalk (Fig. 4.21).

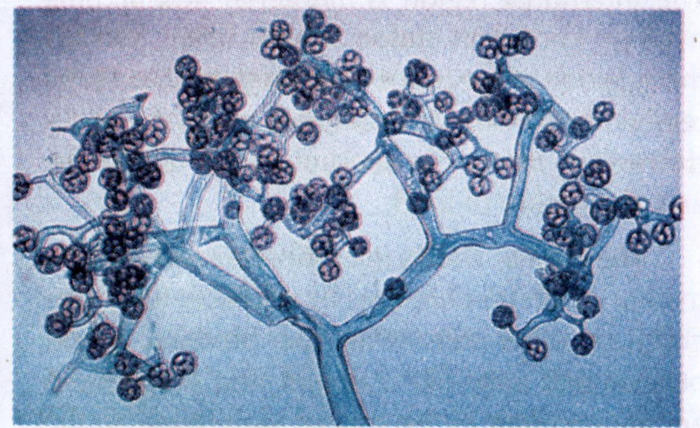

Fig. 4.20: *Thamnidium elegans* (Mucorales): Dichotomously branched sporangiophore and few-spored **sporangiola** are borne at the tips of sporangiophore branches

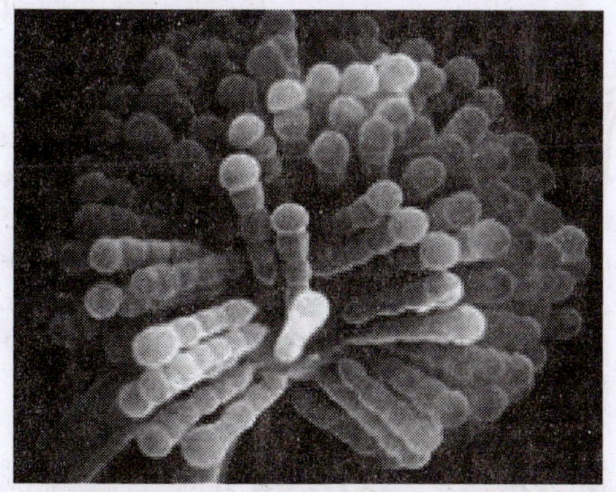

Fig. 4.21: *Syncephalastrum racemosum* (Mucorales). Scanning electron micrograph of uniseriate **merosporangia** produced on a vesicle (hidden beneath merosporangia) (From O'Donnell 1979)

Sexual reproduction

The sexual reproduction takes place following gametangial fusion with the production of Zygospore, which distinguished it from all other fungi.

A single zygospore is formed per zygosporangium. The mature zygospore is often thick-walled, and undergoes a dormant period before germination. Most Zygomycota are thought to have a zygotic or haplontic life cycle (Fig. 4.23). Thus, the only diploid phase takes place within the zygospore. Nuclei within the zygospore are believed to undergo meiosis during germination.

When hyphae from opposite mating types produce structures called progametangia that are dense and multinucleate, and they met togather. Cell walls are formed to separate the tips of the progametangia into **gametangia**, and the rear part form **suspensor**. Plasmogamy then occurs between the two gametangia to form a zygospore. Later on, karyogamy takes place within the zygospore. The cell walls of the zygospore later become thicked. Germination begins with the development of germ tube, cosiquently the diploid nuclei inside the zygospore undergoes meiosis and a germsporangium develops at the end of a germ tube. Spores are produced within the germsporangium. After liberation, the spores fall on the suitable substratum, and germinate to produce new hypha like the parent.

RHIZOPUS

(Bread-mold fungus)

Systemic position

Ainsworth (1973)	Hawksworth (1995)
Kingdom Mycota	Kingdom Fungi
Division Eumycota	Phylum Zygomycota
Subdivision Zygomycotina	Class Zygomycetes
Class Zygomycetes	Order Mucorales
Order Mucorales	Family Mucoraceae
Family Mucoraceae	Genus *Rhizopus*
Genus *Rhizopus*	

The genus *Rhizopus* (Gr. *rhiza*, root; *pous,* foot) is represented by about 50 species grows mostly as saprophyte on decaying fruits and vegetables, in soil, on various foodstuffs like bread, jellies etc. Some of the species of *Rhizopus* cause disease and some have industrial uses. *R. stolonifer* (syn. *R. nigricans*) known as "bread mold" occurs commonly on bread and it also causes soft rot of sweet potato and leak diseases of straw-berries. List of various species of *Rhizopus,* harmful and useful to mankind, are given in Table 4.2:

Table 4.2: List of the various species of *Rhizopus* and their importance

Name	Importance
Harmful	
1. *Rhizopus stolonifer*	Soft rot of sweet potato and leak disease of strawberries
2. *R. artocarpi*	Fruit drop of Jack fruit
3. *R. arrhizus*	Fruit rot of apple
4. *R. equinus, R. cohnii, R. rhizopodiformis* and *R. microsporus*	Mucormycosis in man and also in domestic animals
Useful	
1. *R. stolonifer*	For the manufacture of fumaric acid and cortisone
2. *R. nodosus, R. sinensis*	Lactic acid fermentation
3. *R. oryzae*	Alcoholic fermentation
4. *R. oligosporus*	For the preparation of delicious food "Tempeh" from soyabean.

Vegetative structure

The vegetative plant body is eucarpic and consists of white cottony, much branched mycelium. The mycelial plant body is differentiated into nodes and internodes (Fig. 4.22A). The internodal region is the aerial and arching hyphae, known as **stolon**, which when touches the substratum forms the **nodal region**. The nodal region bears much branched *rhizoid,* grows downward inside the substratum for anchorage and absorption of food.

The hyphal wall is microfibrillar and consists mainly of chitin-chitosan. In addition to chitin-chitosan, other substances like proteins, lipids, purines and salts like calcium and magnesium are also present in the hyphal wall. Inner to the cell wall, cell membrane is present which covers the protoplast (Fig. 4.22B, C). The protoplast contains many nuclei, mitochondria, endoplasmic

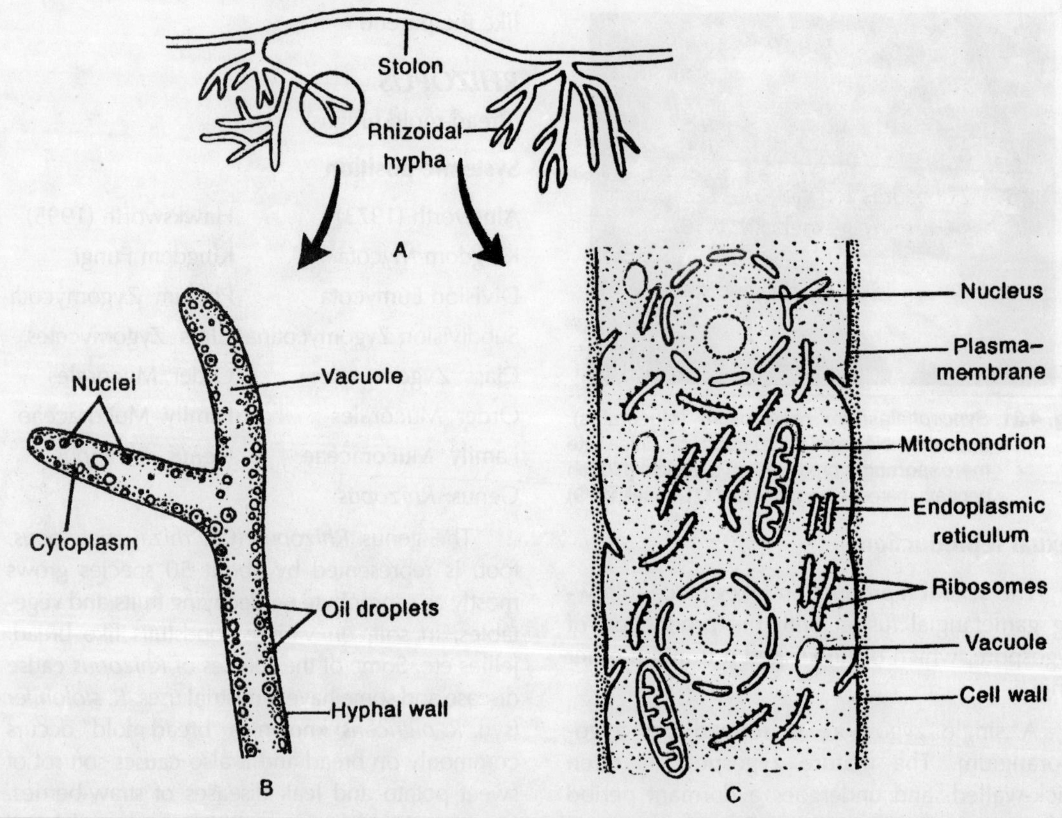

Fig. 4.22: *Rhizopus stolonifer*. A. Vegetative mycelium, B. Portion of hypha under light microscope, and C. Portion of hypha under electron microscope

reticulum, ribosome, oil droplets, vacuoles and other substances. The size of the vacuole enlarges with age by coalescence of smaller vacuoles.

Reproduction

Rhizopus reproduces by vegetative, asexual and sexual means:

1. **Vegetative reproduction**: It takes place by *fragmentation*. Due to accidental breakage the stolon may break up into two or more small units. Each unit is capable of growing as mother mycelium.

2. **Asexual reproduction** : It takes place by the formation of sporangiospore and chlamydospore.

   (a) **Sporangiospore formation**. During favourable condition, the non-motile spores such as sporangiospores or aplanspores are formed inside the sporangium. The sporangium develops singly at the apex of sporangiophore.

The sporangiophore develops in tuft from the upper side of node opposite to the rhizoidal hyphae (Fig. 4.23B). Initially, a number of elongated hyphae develop aerially from the upperside of the node which elongate upto a certain height. The nuclei and cytoplasm push more and more towards the apical side, consequently the apex of the aerial hyphae swells up (Fig. 4.23C). The swollen part enlarges and develops into a large round sporangium (Fig. 4.23D). With maturity, the protoplast inside the sporangium is differentiated into a thick dense layer of multinucleate cytoplasm towards the peripheral region just inside the sporangial wall, called the *sporoplasm* and a vacuolated portion with a few nuclei towards the centre, called *columellaplasm*. A series of small vacuoles then appears between the sporoplasm and columellaplasm (Fig. 4.23E). These vacuoles become flattened and coalesce to form a continuous clevage cavity. This is followed by the formation of a septum towards innerside of the cavity. With further development, the septum becomes dome-shaped and pushes its way into the sporangium (Fig. 4.23F, G).

Protoplast of the sporoplasm then undergoes cleavage to produce many small multinucleate (2–10 nuclei) segments. These segments are transformed into globose non-motile sporangiospores (Fig. 4.23F, G).

After the maturation of spores, the wall of sporangium dries and columella collapses like an inverted cup with irregular surface. The sporangial wall breaks in different fragments leaving a portion as collar on the sporangiophore. The powdery mass of spores are exposed to the atmosphere (Fig. 4.23H).

In ideal moisture and temperature, if the spores fall on suitable substratum, they germinate by germ tube to form new mycelia (Fig. 4.23I, J)

   (b) **Chlamydospore**. During unfavourable condition, thick-walled, nutrition-rich, intercalary mycelium segments arise by septation of mycelium, termed as chlamydospores. They get separated from each other when the connecting mycelium dries up. With the onset of favourable condition, the chlamydospore germinates and gives rise to a new mycelium.

3. **Sexual reproduction**: Sexual reproduction takes place during unfavourable condition by means of **gametangial copulation**. The gametangia looks like, but equal or unequal in size and, by conjugation, they give rise to *zygospore*. Most of the species of *Rhizopus* are heterothallic (*R. stolonifer*), but few species (*R. sexualis*) are homothallic.

In **heterothallic** species, zygospores are produced by the union of two gametangia developed from mycelia of compatible strains (Fig. 4.23K, L); whereas, in **homothallic** species, the uniting gametangia develop from mycelia that derived from a single spore.

When heterothallic species are cultured, two mycelia of compatible strain come near to each other, the mycelia produce small outgrowth, called *progametangia* (Fig. 4.23 K, L). The apical region of the two progametangia come in close contact (Fig. 4.23M). Nuclei and cytoplasm of each progametangium push more and more towards the apical region which

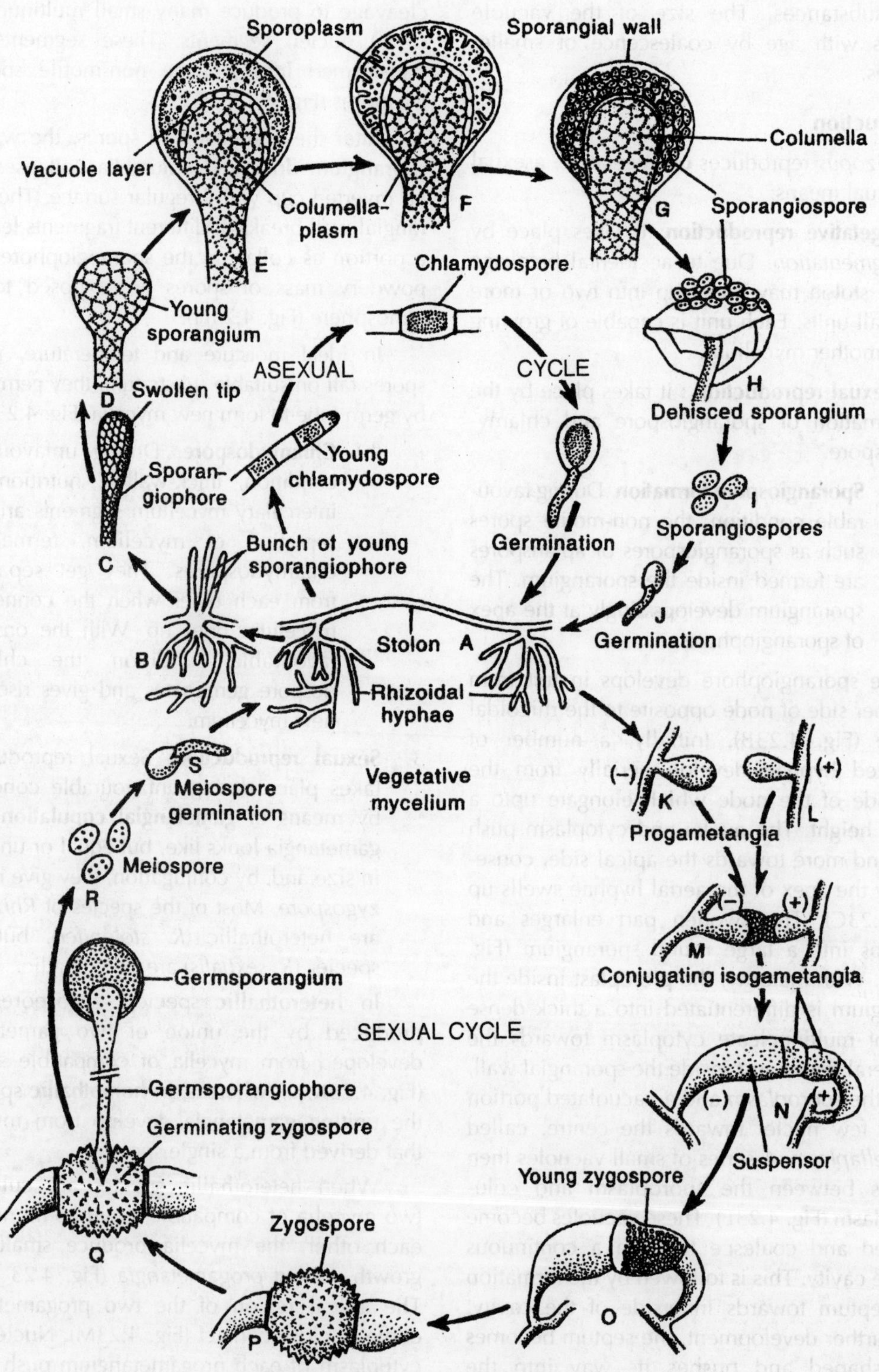

Fig. 4.23: Life cycle of *Rhizopus stolonifer*

swell up with dense protoplasm. The rear region becomes vacuolated. A septum is laid down separating the apical region, which is called *gametangium*; and the basal region, is called *suspensor*. The undifferentiated multinucleate protoplast of the gametangium is called *aplanogamete* or *coenogamete* (Fig. 4.23N).

There is much variation in size of the gametangium pairs. In some pairs, the uniting gametangia are **equal** in size, but in other pairs they are **unequal**.

After maturation of gametangia, the common wall at the point of their contact dissolves and the protoplast of both the gametangia unite to form zygospore (Fig. 4.23O, P). The nuclei of opposite gametangia fuse together to form diploid (2n) nuclei and unpaired nuclei gradually degenerate.

The young zygospore enlarges and probably secretes five layered (two in exospore and three in endospore) thick wall, which undergoes a period of rest (Fig. 4.23P).

After resting period, the zygospore germinates. On germination, the innermost layer comes out after cracking the outer walls and produces a *promycelium*. With further development, the promycelium is differentiated into a lower stalk like germsporangiophore and an upper spherical germsporangium (Fig. 4.23Q).

Meiosis occurs during **zygospore germination** and the haploid nuclei form haploid spores like sporangiospores inside the germsporangium. These spores are also known as **meiospores** (Fig. 4.23R). Each meiospore after liberation germinates like sporangiospore, and forms new mycelium like mother thallus (Fig. 4.23S).

Sometimes failure of gametangial copulation results in parthenogenic development of zygospore, by any one gametangium, called *azygospore* or *parthenospore*. It is however haploid in nature and its nuclei do not undergo meiosis before spore formation.

Indian species:

Rhizopus stolonifer, *R. artocarpi*, *R. oryzae*, *R. sexualis* etc.

Mucor

(Pin-mold fungus)

Systemic position

Ainsworth (1973)	Hawksworth (1995)
Kingdom Mycota	Kingdom Fungi
Division Eumycota	Phylum Zygomycota
Subdivision Zygomycotina	Class Zygomycetes
Class Zygomycetes	Order Mucorales
Order Mucorales	Family Mucoraceae
Family Mucoraceae	Genus *Mucor*
Genus *Mucor*	

The genus *Mucor* (*L. muceo*, be moldy) is represented by about 80 species, found throughout the world and about 17 species from India, commonly known as mold. They grow mostly as saprophytes on decaying fruits and vegetables, in soil (*Mucor strictus, M. flavus*), on various food-stuff-like bread, jellies, jams, syrups. M. mucedo, is a coprophilous species (grows on dungs of herbivorous animals like cow etc.), known as black mold. Species like *M. mucedo* and *M. racemosus* are well-known air contaminants. M. javanica is known to cause alcoholic fermentation. List of various species of Mucor, harmful and useful to the mankind, are given in Table 4.3.

Table 4.3: List of various species of *Mucor* and their importance

Name	Importance
Harmful	
1. *M. pusillus*	Bronchomycosis in man.
2. *M. circinelloides*, *M. javanicum*, *M. racemosus*, *M. spinosus*	Mucormycosis in man and also in domestic animals.
Useful	
1. *M. javanica*	Alcoholic fermentation.

Vegetative structure

The vegetative plant body is eucarpic, consists of white cottony coenocytic much-branched mycelium. The mycelia ramify all over the substratum. The hyphae are usually prostrate, but

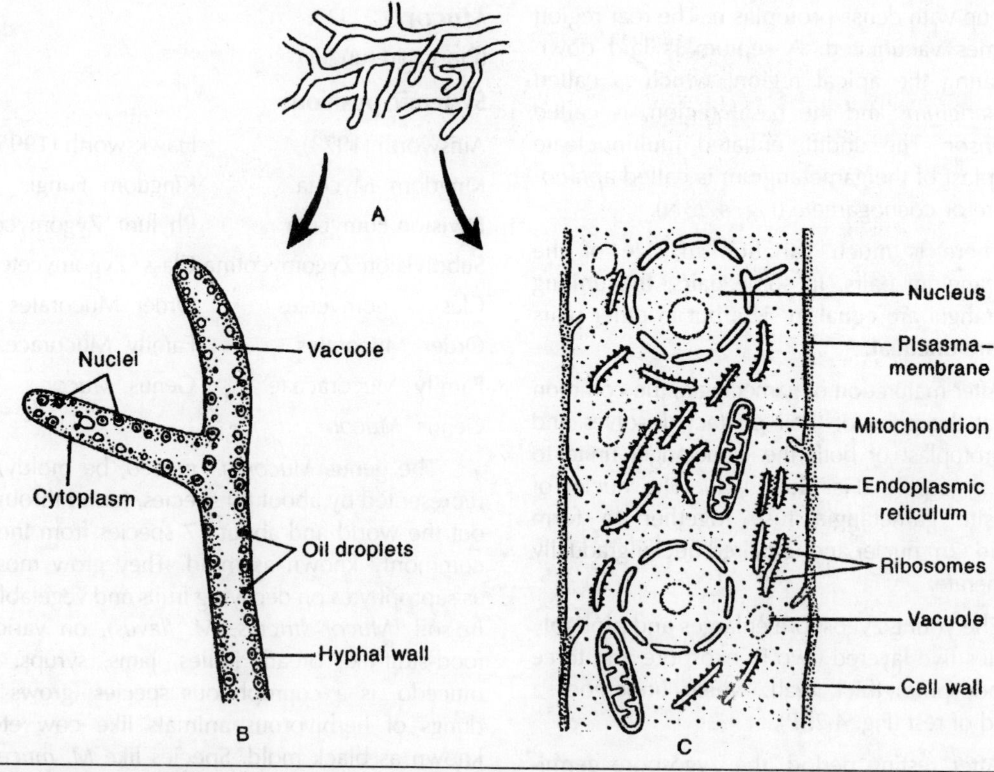

Fig. 4.24: Mucor: A. Vegetative *mycelium*, B. Portion of hypha under light microscope, C. Portion of hypha under electron microscope

some of them penetrate into the substratum and serve the function of both anchorage and absorption of nutrients (Fig. 4.24A).

The hyphal wall is microfibrillar, consist mainly of chitin-chitosan. In addition, other substances like other polysaccharides, lipids, purines, pyrimidines, protein, Ca and Mg are also present. Inner to the cell wall, cell membrane is present which covers the protoplast. The protoplast contains many nuclei, mitochondria, endoplasmic reticulum, ribosomes, oil droplets, small vacuoles and other substances (Fig. 4.24B, C).

Reproduction

Mucor reproduces by vegetative, asexual and sexual means.

1. **Vegetative reproduction**: It takes place by fragmentation. Due to accidental breakage, the mycelium may break up into two or more units. Each unit is capable to grow as mother mycelium.

2. **Asexual reproduction**: It takes place by the formation of sporangiospore, oidia and chlamydospore (Fig. 4.25).

   (a) ***Sporangiospore formation***: During favourable condition, the nonmotile spores known as sporangiospores or aplanospores are formed inside the sporangium.

The sporangiophores develop singly and scatteredly on the upper side of the superficial mycelium (Fig. 4.25B). The sporangiophore is generally unbranched, however it is branched in *M. brunneus* and *M. racemosus*. After attaining a certain height, the nuclei and cytoplasm push more and more towards the apical side, consequently the apex of the aerial hyphae swells up (Fig. 4.25C, D). The swollen part enlarges and develops into large round sporangium. With maturity, the protoplast inside the sporangium is differentiated into a thick dense layer of multinucleate cytoplasm towards the peripheral region just inside the sporangial wall, called **sporoplasm**

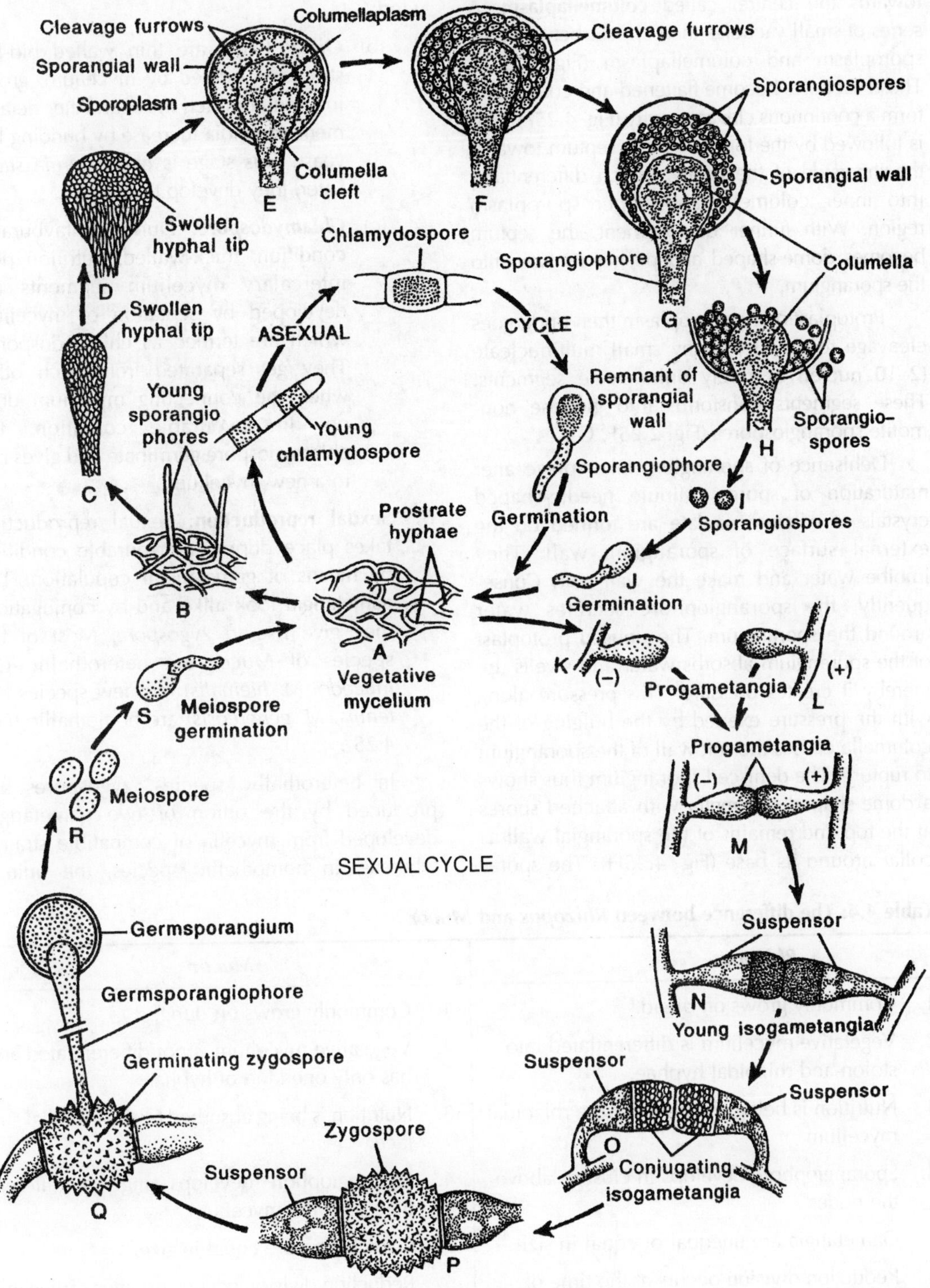

Fig. 4.25: Life cycle of *Mucor mucedo*

and a vacuolated portion with a few nuclei towards the centre, called **columellaplasm**. A series of small vacuoles then appear between the sporoplasm and columellaplasm (Fig. 4.25E). These vacuoles become flattened and coalesce to form a continuous clevage cavity (Fig. 4.25F). This is followed by the formation of a septum towards the innerside of the cavity, which differentiates into inner columella and upper sporoplasm region. With further development, the septum becomes dome-shaped and pushes its way into the sporangium.

Protoplast of the sporoplasm then undergoes cleavage to produce many small multinucleate (2–10 nuclei) or rarely uninucleate segments. These segments transforms into globose non-motile sporangiospores (Fig. 2.25F, G).

Dehisence of sporangium takes place after maturation of spores. Minute needle-shaped crystals of calcium oxalate are formed on the external surface of sporangium wall. They imbibe water and make the wall soft. Consequently, the sporangiophore secretes water around the sporangium. The unused protoplast of the sporangium absorbs water and swells up, thereby it creates pressure. This pressure along with the pressure exerted by the bulging of the columella causes the soft wall of the sporangium to rupture. The dehisced sporangium thus shows a dome-shaped columella with attached spores at the top and remains of the sporangial wall as collar around its base (Fig. 4.25H). The spores are dispersed chiefly by insect and also by wind.

(b) ***Oidia***: Oidia are thin walled bid-like structures formed by mycelium, grown in a medium rich in sugar. After detachment, the oidia increase by budding like yeast. This stage is called *torula stage*. Later, they develop to mycelia.

(c) ***Chlamydospore***: During unfavourable condition, thick-walled, nutrition rich, intercalary mycelium segments are developed by septation of mycelium which are termed as chlamydospores. They get separated from each other when the connecting mycelium dries up. In favourable condition, the chlamydospore germinate and gives rise to a new mycelium.

3. **Sexual reproduction**: Sexual reproduction takes place during unfavourable condition by means of gametangial copulation. The gametangia look alike and by conjugation, they give rise to *zygospore*. Most of the species of *Mucor* are heterothallic (*M. mucedo, M. hiemalis*), but few species (*M. tenuis, M. genereosis*) are homothallic (Fig. 4.25).

In heterothallic species, zygospores are produced by the union of two gametangia developed from mycelia of compatible strains; whereas, in homothallic species, the uniting

Table 4.4: The difference between *Rhizopus* and *Mucor*

Rhizopus	***Mucor***
1. Commonly grows on bread.	1. Commonly grows on dung.
2. Vegetative mycelium is differentiated into stolon and rhizoidal hyphae.	2. Vegetative mycelium is not differentiated and has only one kind of hypha.
3. Nutrition is being absorbed by the rhizoidal mycelium.	3. Nutrition is being absorbed by the mycelial surface.
4. Sporangiophore develops in clusters above the nodes.	4. Sporangiophore develops singly at different places on the mycelium.
5. Gametangia are unequal or equal in size.	5. Gametangia are equal in size.
6. Reduction division occurs at the time of zygospore germination after rest.	6. Reduction division occurs just after karyogamy before rest of a zygospore.

gametangia develop from mycelia that derived on germination of a single spore.

When two mycelia of compatible strains come close to each other, the mycelia produce small outgrowth, called *progametangia* (Fig. 4.25L, M). The apical region of the two progametangia come in close contact. Nuclei and cytoplasm of each progametangium push more and more towards the apical region and its tip swells up with dense protoplasm. The rear region becomes vacuolated. A septum is laid down, separating the apical region, which is called *gametangium*; and the basal region is called *suspensor* (Fig. 4.25N). The undifferentiated multinucleate protoplast of the gametangium is called *aplanogamete* or *coenogamete*.

After maturation of gametangia, the common wall at the point of their contact dissolves and the protoplast of both the gametangia unite to form *zygospore* (Fig. 4.25O, P). The nuclei of opposite gametangia fuse together to form diploid (2n) nuclei, unpaired nuclei gradually degenerate.

The diploid nuclei undergo meiosis before resting stage of zygospore. In heterothallic species normally 50% of the nuclei are of "+" strain and the other 50% of "–" strain. Accordingly, the spores are of + or – type, but in *M. mucedo* (a heterothallic species) the spores are of either + or – type. All the nuclei except one degenerate and the surviving nucleus undergo repeated mitosis to produce large number of nuclei of the same strain.

The young zygospore enlarges and probably secretes five layered (two in exospore and three in endospore) thick wall, of which the outer one is black and warty. The zygospore then undergoes a period of rest.

After resting period, the zygospore germinates and, on germination, the innermost layer comes out after cracking the outer walls and produces a *promycelium*. The content of the zygospore migrates into the tip of the promycelium which swells up and differentiates into a lower stalk like germsporangiophore and the upper spherical germsporangium (Fig. 4.25Q).

The haploid nuclei of germsporangium form haploid spores called *sporangiospores* inside the germsporangium. These spores are also known as *meiospores* (Fig. 4.28R). Each meiospore, after liberation, germinates like sporangiospore and forms a new mycelium like the mother thallus (Fig. 4.25S).

Sometimes failure of gametangial copulation results in parthenogenic development of zygospore by any one gametangium called *azygospore* or *parthenospore*. It is, however, haploid in nature and its nucleus does not undergo meiosis before spore formation.

Indian species:

Mucor indicus, M. mucedo, M. silvaticus, M. praini, M. hiemalis.

Heterothallism in Mucorales

Heterothallism is a sexual phenomenon, where two compatible partners are required to produce sexual spores. Thus in heterothallic species sexual reproduction that occurs only between two self-sterile mycelia. On the other hand homothalism is also a sexual phenomenon, where single organism is capable to produce sexual spores.

Ehrenberg (1829) was studied for the first time the production of zygospore in the order Mucorales. After long time, Blakeslee (1904) reported that in several genera of Mucorales, zygospores are nor formed at all. He supported his idea with facts and reasons. He observed and reported that two types of species are found, named as homothallic and heterothallic species.

In **homothallic species**, two hyphae of the same mycelium produced by single spore fuse with each other and zygospores are developed, e.g., *Mucor hiemalis*. But in **heterothallic species** like *Mucor mucedo* and *Mucor stolonifer*, the fusion and production of zygospore takes place only among the different strains (+ and –) of hyphae, those develop on different mycelia. In these species the zygospores are not produced by the union of to hyphae develop from same strain.

Blackeslee (1904) observed and reported that in heterothallic species, when the mycelia of + and – strains kept apart, the zygospores are not produced and only sporangia are developed. But zygospores are produced as and when mycelia of both the + and – strains grown together. Mycelia of both the strains are morphologically similar but different in their physiological behavoiur.

It has also been observed that the growth of + mycelium is comparatively faster and the gametangia are bigger in size than the – mycelium and they can be designated as female and male gametangia respectively. In recent year, many mycologists do not support this view and stated that this behavoiur takes place only because of nutrition.

Based on experimental evidence, Blakeslee proved the phemenon heterothallism. On agar medium in petridishes, he inoculated + and – strained spores and observed that the fusion i.e union takes place only at those regions where the hyphae of different strains i.e + and – strains come in contact. Here the zygospores are produced at the points of contact of hyphae of different strains and no zygospores are produced at the point of contact of hyphae of the same strains.

Based on the experimental results (Fig. 4.26) the phenomenon of heterothallism may be clearly understood. Here 5 spores of different strains are inoculated at different region on sterilized nutrient agar-agar medium in petridishes, as A+, B-, C+, D- and E+ respectively. After the experiment, it is observed that zygospores are produced at the regions where the mycelia of opposite strains developed from + and – strained spores. These are regions AB, BC, DC, DE and AD, where fusion between the mycelia takes place. As there are same strains at the regions AC and AE, no zygospores are produced.

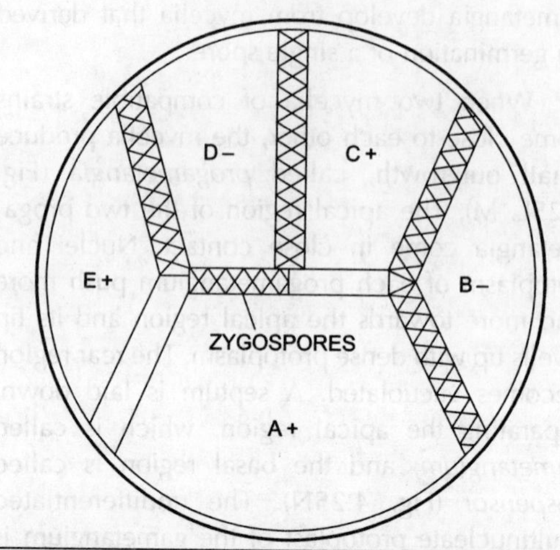

Fig. 4.26: Heterothallim in Mucorales (after Blakeslee)

He also observed that the hyphae of + and – strains of different heterothallic species develop zygospores but are imperfect zygospores and these zygospores do not mature. Further, he also observed that in heterothallic species like *Mucor mucedo*, the zygospore on germination develop germsporangia, which contain single strained spores. In this condition, the fusion is possible only when the mycelia are produced from spores of germsporangia of different strains. But in species like *Phycomyces nitens*, + and – strained spores are produced in the same germsporangium.

Although, the phenomenon of heterothallism was first reported in the order Mucorales, but now it become clearly established in many other fungi like rusts and smuts.

PHYLUM ASCOMYCOTA

(Sac Fungi)

The Phylum Ascomycota (Gr. *askos*, sac or bladder; *mykes*, fungi), commonly known as "sac fungi", is the largest group under the Kingdom Fungi, comprised of six classes and about 30,000 species. Majority of the members are terrestrial, although many members live in water. They may be saprophyte or parasite, cause diseases. The most important and interesting members are the yeasts used in bakery and alcohol industry. *Claviceps purpurea,* the causal agent of ergot disease of rye, develops sclerotia that contain a lot of alkaloids including **L.S.D.** (lysergic acid diethylamide), the well known **hallucinogenic drug** and **Ergometrine**, a popular **medicine**. Ergometrine and its semisynthetic analogue, methyl ergometrine have prominent uterine action, commercially used as methyl ergometrine maleate with trade names EMATRIN, ERGOLIN, METHERGIN, UTERGIN, UTROWIN etc., or ergometrine maleate with trade names ERGOMETRIN and LERIN as tablet (0.125 mg) or injection (0.2 mg/ml), which control haemorrhage of mother during childbirth, but the drugs raise blood pressure and decrease milk secretion. The wonder drug penicillin is an important antibiotic, produced by *Penicillium notatum,* discovered by Sir Alexander Fleming in 1929. Later, it was also isolated from *P. crysogenum* and *P. italicum*.

Important characteristics

Important characteristics of Ascomycota are:

1. Vegetative body is unicellular or commonly well-developed, branched septate mycelium with uni- or multinucleate cells having perforated septa.
2. Mostly, the cell wall is composed of chitin and glucans, but in unicellular form, it is composed of glucans and mannans.
3. Vegetative reproduction takes place by fragmentation (in filamentous form), fission and budding (in unicellular form).
4. Asexual reproduction takes place by non-motile spores, such as conida, oidia and chlamydospores.
5. Sexual reproduction takes place by gametangial copulation (*Saccharomyces*), gametangial contact (*Penicillium*), somatogamy (*Morchella*) or spermatization (*Polystigma*).
6. Complete absence of motile structures.
7. The product of sexual reproduction is the ascospores grown inside a small specialized sac-like structure, called ascus.
8. The fruiting bodies (inside which asci are developed) are the ascocarps. The ascocarps may be cleistothecium (*Penicillium*), apothecium (*Ascobolus*), perithecium (*Daldenia*) or ascostroma (*Elsinoe veneta*).

General account

The general account of the phylum *Ascomycota* is given below in different heads:

Somatic structure

The somatic body generally consists of well developed septate and profusely branched mycelium except yeasts, those are unicellular or pseudomycelial. Each cell contains one to several nuclei. Septa having single pore, develop at the centre with spherical Woronin bodies on each side of the pore. The cytoplasmic continuity is maintained through the pore. The **Woronin bodies** are the round to oval structures of unknown chemical composition, can plug the pore if hyphae get damaged (also found in many members of *Mitosporic fungi*, earlier called as *Deuteromycotina*). The pores are so wide through which nuclei and mitochondria can pass from one cell to other. Cell wall may composed of glucan and mannans (unicellular forms) or chitin and glucans (multicellular forms). Hyphal walls that are basically two-layered: A thin electron dense outer layer and a thick electron transparent inner layer. Due to aggregation of somatic mycelium, structures-like sclerotium (pl. sclerotia), rhizomorph and stroma (pl. stromata) are produced.

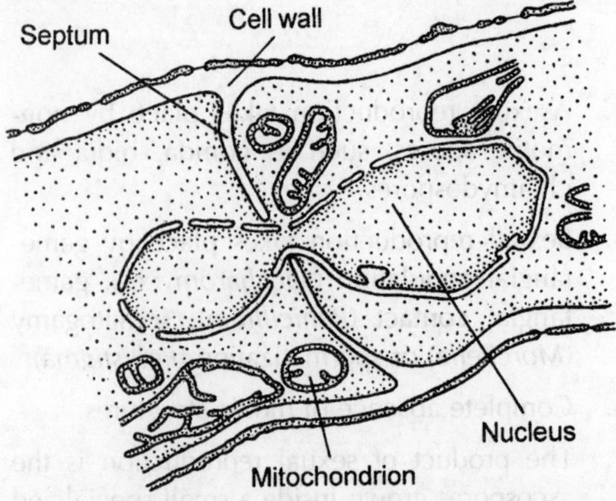

Fig. 4.27: L.S. of an Ascomycetous hyphae with perforated septum shows passing of nucleus

Reproduction

Members of this group reproduce both asexually and sexually. The asexual or conidial stage is called *imperfect stage* and the sexual or ascus producing stage is called ascigenous or *perfect stage*. Some fungi reproduce only sexually while others reproduce both asexually and sexually. **Flagellate structures are not developed in this group**. Some terminologies are used by Hennebert and Weresub (1977) in respect of reproduction are: *anamorph*, for asexual forms; *teleomorph*, for sexual forms and *holomorph*, for the whole fungus in all its forms and phases. Based on the presence of anamorphic and/or telemorphic stages members of *Ascomycotina* are of three types:

1. Members readily produce asexual spores (conidia etc.) but rarely or never form sexual spores, e.g. *Pyricularia grisea* (Perfect state *Magnaporthe grisea*).
2. Some members never or rarely produce asexual spores but produce only ascospores, e.g. *Sordariales*.
3. Other members regularly produce both asexual and sexual spores. e.g. *Fusarium solani* (Sexual stage *Nectria haematococca*).

The first case is very common with plant pathogens, reproduces through anamorphic or asexual state and placed in the artificial form-class Fungi Imperfecti or Fungi Anamorphici or Deuteromycota or Mitosporic fungi and commonly called as "**waste box fungus**". When the teleomorph of such fungus is found, it is placed under *Ascomycota* (if they develop ascospores) in different name. Thus the fungus has two names, an **anamorphic** and a **teleomorphic** name. It should be noticed, that anamorphic taxa (genera, family etc.) are not taxa in the same sence as teleomorphic taxa.

The dual system of classification has advantages on practical point of view. Many septate fungi rarely or only at certain times of the year produce sexual states on their host, however they often readily form the conidial state (anamorph) on culture media. With the help of classification based on anamorphic characters, it is possible to identify the particular fungus without having to wait for it, till the development of sexual state.

Some anamorphic genera have teleomorphs in more than one genus and on the other hand, some teleomorphic genera have anamorphs in more than one genus. The first case is exemplified by anamorphic genus *Fusarium* has teleomorphs in the genera like *Calonectria, Nectria* and *Gibberella* and in the second case the teleomorphic genus *Mycosphaerella* has anamorphs that are either Coelomycetes (genera like *Septoria, Phoma* and *Ascochyta*) or Hyphomycetes (genera like *Cladosporium* and *Pseudocercospora* (Table 4.5). Both the anamorphic and telemorphic names are considered as valid.

Asexual reproduction

Asexual reproduction may be carried on by budding, fission, fragmentation and formation of conidia, blastospores and chlamydospores, depending on the environmental condition.

1. *Budding* and *fission* are the methods of propagation normally found in yeasts and also in a few other members of *Ascomycota*.
2. *Fragmentation* is the method of breakdown of filament into two or more fragments due to natural or artificial means during

Table 4.5: Some examples of anamorphic and teleomorphic *Ascomycota*

Anamorphic state	Teleomorphic state
Cylindrocarpon radicola	*Nectria radicicola*
***Fusarium** solani*	*Nectria haematococca*
***Fusarium** episphaeria*	*Nectria episphaeria*
***Fusarium** rigidiusculum*	*Calonectria rigidiuscula*
***Fusarium** graminearum*	*Gibberella zeae*
Pseudocercospora aleuritis	***Mycosphaerella** aleuritis*
Pseudocercospora musae	***Mycosphaerella** musicola*
Septoria tritici	***Mycosphaerella** graminicola*
Phoma caricae-papayae	***Mycosphaerella** caricae*

favourable condition. Later on, the fragments behave like new individual.

3. *Chlamydospores* are thick walled spores formed during unfavourable condition, develop either singly or in chain. During favourable condition the chlamydospores on germination develop new mycelium.
4. *Blastospores* (Gr. *blastos*, a bud; *spora*, seed, spore) are thin walled spores develop at the apex of mycelium by budding.
5. *Conidia* are developed towards the apical region on simple or branched conidiophores, very common in the members of *Ascomycota*. The conidiophores are developed in various ways.
   (i) Develop scatteredly on the mycelial mat,
   (ii) Develop in **acervuli**, the fruit body;
   (iii) Develop in **pycnidia**, another type of fruit body.

Sexual reproduction

Like others, the sexual reproduction in *Ascomycota* takes place by plasmogamy, karyogamy and meiosis. The plasmogamy aims to bring two compatible nuclei together in a cell. The plasmogamy in ascomycetous members takes place by the following procedures:

Types of plasmogamy

1. **Gametangial copulation**

In this process both the gametangia come close together, coiled around or touch at the tip and fuse. Two gametangia unite together and form fusion cell, where two nuclei comes from different gametangia fuse together and form a diploid cell, which functions as an ascus. Diploid nucleus of ascus divides meiotically and form haploid ascospores. The gametangial copulation may be isogamous (*Eremascus*) or anisogamous type (*Dipodascus albidus*).

2. **Hologamy**

In this type the whole i.e. entire mature somatic cell functions as gametangium. Gametes are not formed in this process. Fusion takes place between the mature somatic cell and form zygote. The diploid nucleus undergoes meiosis and form ascospores. This type is found in *Schizosaccharomyces octosporus* (Fig. 4.28), *Saccharomyces cerevisiae* etc.

3. **Gametangial contact**

It is also called gametangy. In this type, morphologically differentiated gametangia are developed in many higher members. The male gametangia are called antheridia (sing. antheridium) and female gametangia as ascogonia (sing. ascogonium). The gametangia may be uninucleate (*Sphaerotheca*) or multinucleate (*Pyronema, Ascobolus,* etc.). A pore develops at the point of contact of antheridium and ascogonium through which the male nucleus passes inside the ascogonium. Sometimes the ascogonium develop septate or aseptate and usually unbranched or branched tubular receptive processes, called trichogyne, through which the male nucleus passes inside the ascogonium. Fertilization tube does not develop.

4. **Autogamy**

In this type, antheridia are absent or remain non-functional and only the ascogonial nuclei participate in the development of ascus and ascospores.

In *Talaromyces vermiculatus* syn. *Penicillium vermiculatum* (Fig. 4.45G, H, I) as the tip of the antheridium touches the multinucleate ascogonium, the common wall dissolve and the two protoplasts come in contact with each other (Dangeard, 1907). But nuclear migration does not take place and the antheridial nucleus exists inside the antherdium even after the formation of ascogenous hyphae.

5. **Spermatization**

In this type antheridia are not formed. The male sex cells i.e. **spermatia** are develop as minute, spherical and uninucleate bodies. In some members spermatia are develop either at the apex of special hyphae, the spermatiophore (Gr. *spermation*, little seed; *phoreus*, bearer) or develop inside the flask-shaped body, the spermogonia. After detachment, the spermatia are carried by water, wind or insect to the trichogyne (Gr. *thrix*, hair; *gyne*, woman) or other receptive part of the well developed female sex organ the ascogoniun. After contact, the common walls dissolve and the spermatia empties its content in the female organ. The process of fusion between spermatium and ascogonium is called **spermatization.** This type is found in *Mycosphaerella tulipiferae, Neurospora sitophyla* (Fig. 4.53) and many other higher *Ascomycota*.

In some cases, microconidia and oidia are also function as spermatia.

6. **Somatogamy**

In this type, antheridia and spermatia are not formed. Fusion takes place between the somatic hyphae of two compatible mycelia (*Ascobolus furfuraceous*). Nuclei of one mycelium migrate to the other, through septal perforation and reach up to the ascogonia. This is found in some higher *Ascomycota*.

Compatibility

Compatibility means fusing capability. Based on compatibility, members of *Ascomycota* are divided into two groups.

1. *Homothallic species*. In this type all individuals are self-compatible i.e. self-fertile (*Ascobolus crenulatus*). Therefore all individuals are able to form asci by themselves. Thus they are considered as bisexual one.
2. *Heterothallic species*. In this type two compatible individuals must be mated together before asci are formed (*Ascobolus furfuraceous*). Thus individuals are self-sterile and considered as unisexual one.

In *Ascomycota*, the heterothallism is of two types:

Bipolar heterothallism

In this type, the compatibility is determined by a pair of genes A_1A_2. The genes segregate at meiosis just before ascospore formation. Out of eight ascospores in an ascus, 50% will carry A_1 gene and the rest 50% with A_2 gene. Thus ascospores normally carry one gene either A_1 or A_2. Ascospores on germination develop the vegetative mycelium. Both type of sex organs i.e. male (antheridia or spermatia) and female (ascogonia) develop on same mycelium will carry either A_1 or A_2. Thus two sex organs (male and female) must be brought together from different genotype i.e. either from A_1 or from A_2 for the sexual act. The dikaryon thus formed by the mating of compatible thalli bear the genes $A_1 + A_2$ and the zygote with A_1A_2. Thus the compatibility in some *Ascomycota* controlled by gene present in one locus is of two-allele type, called as **bipolar heterothallism**.

Tetrapolar heterothallism

In some other members of *Ascomycota* the sexuality is controlled by two pairs of genes i.e. A_1A_2 and B_1B_2. These two pair of genes are located on different chromosomes. Thus the different combination of alleles form the mating types may be A_1 B_1, A_1 B_2, A_2 B_1 and A_2 B_2. The mating type with A_1B_1 is compatible with any other mating type except the one with same

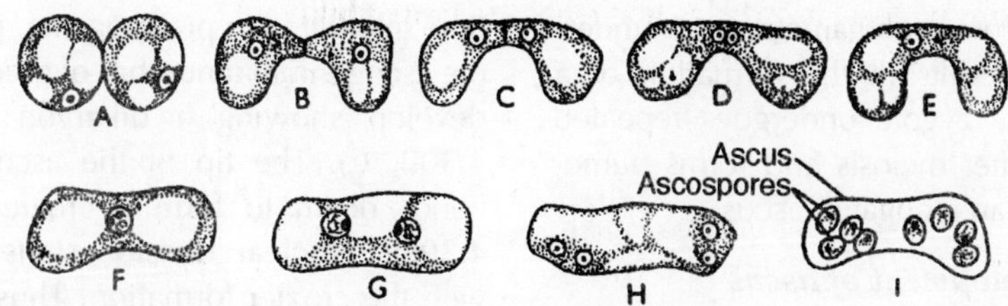

Fig. 4.28: *Schizosaccharomyces octosporus*: A-I. Stages of hologamy leading to the development of ascus containing ascospores

allele but not fully compatible with the mating types A_1B_2 and A_2B_1.

Development of Ascus

In the phylum Ascomycota, the asci develop after sexual reproduction which generally contain eight ascospores, but their number may be four (*Saccharomyces cerevisiae, Saccharomycodes ludwigii*) or many (*Dipodascus uninucleatus*). Asci may be cylindrical, globose or club-shaped (Fig. 4.31). Based on the wall structure and mode of dehiscence, the asci may be differentiated as *prototunicate, unitunicate* and *bitunicate*. The **prototunicate** asci (*Endomyces, Saccharomyces* etc.) have a thin delicate wall. Both unitunicate and bitunicate asci have two layered wall. In **unitunicate** asci (*Peziza, Ascobolus* etc.), both the layers remain together even during spore release, but in **bitunicate** asci (*Pleospora, Melaspilea* etc.) the inner wall is projected out during spore liberation, while outer wall remains in its earlier position. Asci may develop either directly or indirectly.

A. *Direct development of ascus*

In unicellular members like *Saccharomyces* and related genera, the two compatible cells come close to each other (Fig. 4.28A, B). The cells undergo plasmogamy by dissolution of their common wall and the two haploid nuclei come close to each other and undergo fusion (karyogamy) results in the formation of a diploid nucleus, the zygote (Fig. 4.28C-F). The zygote containing cell enlarges and forms an ascus mother cell. The diploid nucleus undergoes first meiotic division followed by single mitotic division, results in the formation of eight nuclei (Fig. 4.28G-I). The nuclei along with some cytoplasm form ascospores. In some cases, mitotic division does not take place after meiosis and forms four ascospores (*Saccharomyces cerevisiae, Saccharomycodes ludwigii*) in an ascus.

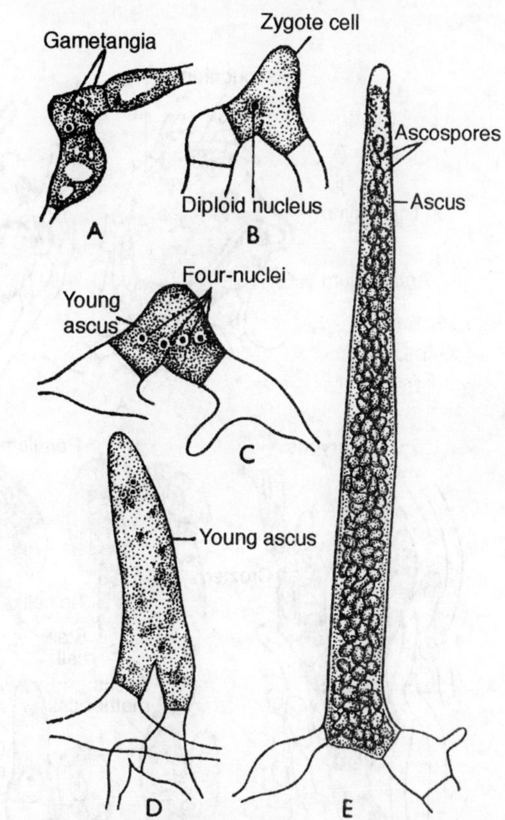

Fig. 4.29: *Dipodascus uninucleatus*: A-B. Stages of gametangial copulation leading to zygote formation, C. Four-nucleate young ascus. D. Multinucleate ascus. E. Ascus containing numerous ascospores

In filamentous members like *Dipodascus uninucleatus* (Fig. 4.29), gametangia developed from vegetative mycelium undergo plasmogamy and one nucleus from each gametangium undergoes karyogamy results in the formation of a single zygote. The zygote undergoes repeated mitotic division after meiosis and forms numerous ascospore in an elongated ascus.

B. *Indirect development of ascus*

In filamentous members like *Penicillium, Peziza, Pyronema* (Fig. 4.30) and many others, ascus develops in indirect way. The transfer of male nuclei from antheridium to ascogonium takes place by plasmogamy and both the nuclei of opposite mating arrange themselves in pairs, called dikaryon. Karyogamy maintains a long time interval with plasmogamy. From the wall of the ascogonium a number of ascogenous hyphae develop showing a dikaryon condition (Fig. 4.30B, C). The tip of the ascogenous hyphae bends down to form a crozier or hook (Fig. 4.30D). Nuclear division starts simultaneously with the crozier formation. Thus four nuclei are formed by mitosis inside the crozier. Two septa are laid down that demarcate uninucleate terminal cell, binucleate (of opposite mating) penultimate cell and uninucleate basal stalk cell (Fig. 4.30E, F). The binucleate penultimate cell behaves as ascus mother cell. The wall of ascus mother cell extends more in length than the

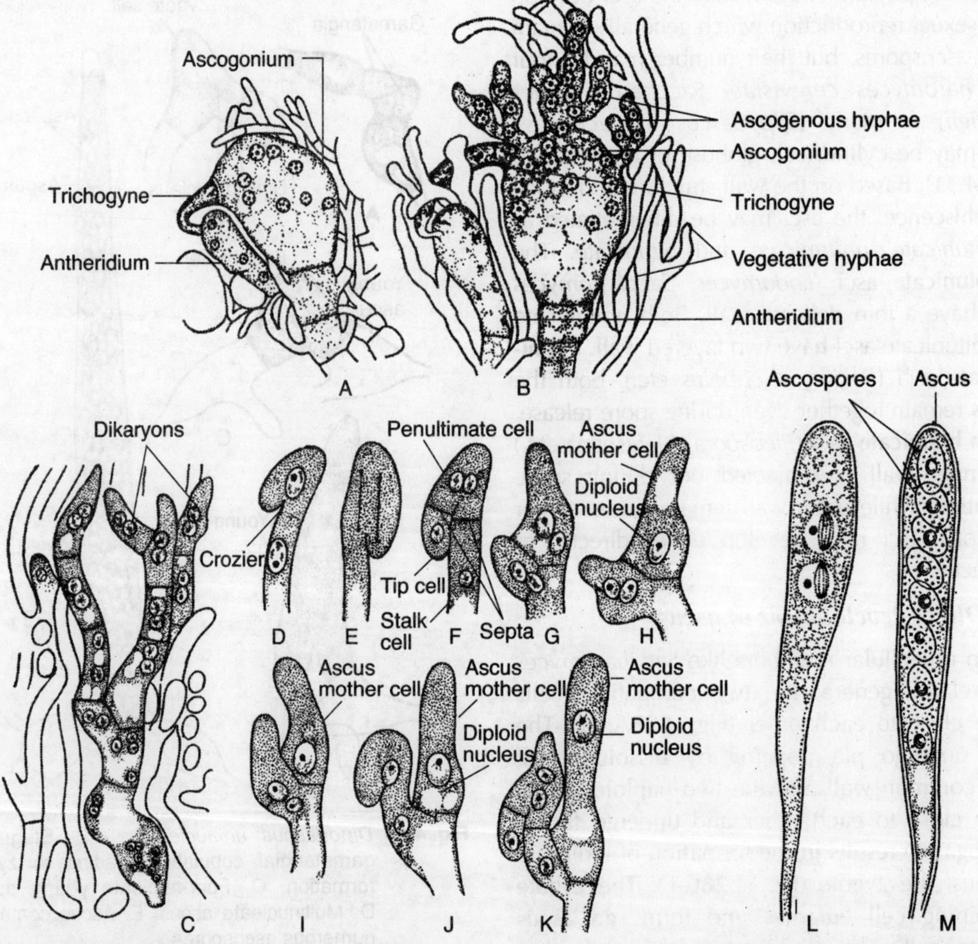

Fig. 4.30: *Pyronema confluens*: A-M stages in gametangial contact leading to the development of ascus containing ascospores. A. Contact between antheridium and ascogonium, both are multinucleate, B. Development of forked ascogenous hyphae from ascogonium, C-M. Development of ascus and ascospores from ascogenous hyphae

breadth, thus forms an elongated ascus. Simultaneously, both the nuclei of ascus mother cell fuse and form diploid nucleus (Fig. 4.30H-K), which undergoes first meiotic division and then mitotic division results in the formation of eight nuclei. The cytoplasm accumulates around each nucleus to form ascospore (Fig. 4.30L, M). Division may take place both in nucleus and cytoplasm of an ascus results in the formation of 2-celled ascospore.

The apical cell of ascogenous filament bends and may fuse with the basal cell and the region extends again and forms new binucleate penultimate cell as before (Fig. 4.30F-K). This process may repeat several times and a cluster of asci may develop.

Ascospores

The ascospores are produced in ascus (Fig. 4.31). Each ascus commonly contains **8 ascospores**, with variations having:

(a) **1 (one) ascospore** in *Monosporascus cannonballus* (Fig. 4.31A),

(b) **4 ascospores** in *Saccharomyces ludwigii* etc. (Fig. 4.31B),

(c) **8 ascospores** in *Ascobolus scatigenus* (Fig. 4.31C),

(d) **Many ascospores** in *Dipodascus uninuclestus* (Fig. 4.31D), and

(e) *Thelebolus stercoreus*, a member of the family Thelebolaceae under Pezizales, where ascoma having **single large ascus containing over 1000 ascospores** (Fig. 4.31E).

Ascocarps

The asci become naked (Fig. 4.32A, B) in Hemiascomycetes (*Saccharomyces, Schizosaccharomyces, Taphrina,* etc.), but in others they are found intermingled with mycelium and form ascocarps of different shapes such as **cleistothecium** (globose), **perithecium** (flask-shaped), **apothecium** (saucer-shaped) or **ascostroma** (looks like perithecium but lacks its distinct wall and also differs in development). In *cleistothecium* the asci may be arranged scatteredly (*Penicillium*) or are arranged in tuft at the base (*Erysiphe*, Fig. 4.32C). In *Perithecium* (Fig. 4.32D, E), the asci develop from the bottom and are intermingled with sterile mycelia called paraphyses (Members of Pyrenomycetes, *Claviceps, Xylaria* etc.). Asci are arranged in a layer on the top of the exposed hymenial surface mixed with paraphysis in *Apothecium* (Fig. 4.32G, H) fruitbody (*Ascobolus, Peziza*), and lastly *Ascostroma* (Fig. 4.32F) may contain one or more cavities, that looks like perithecium, where asci are developed from the base. This type of ascocarp is also known as *Pseudo-*

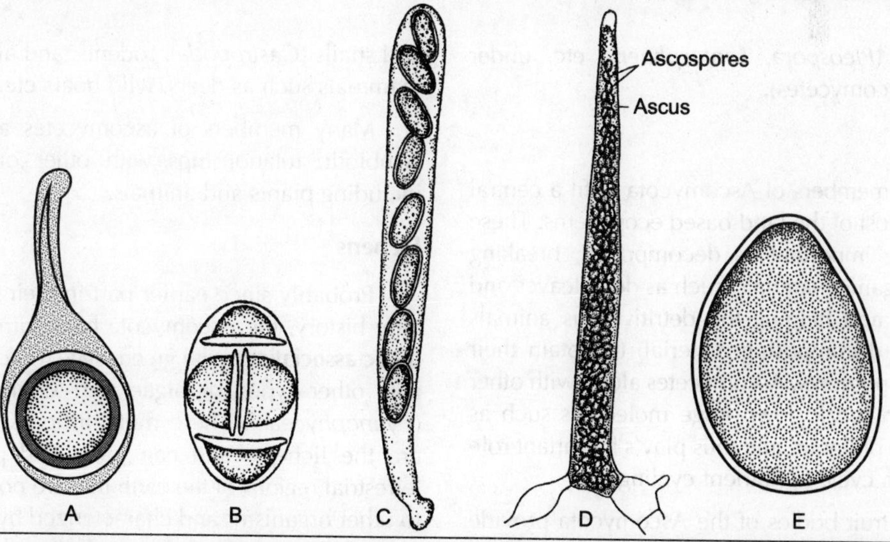

Fig. 4. 31: Various types of ascus containig ascospores (A = one ascospore, B = 4-ascospores, C = 8-ascospores, D = Many ascospore, E = A large ascus bearing over 1000 ascospores).

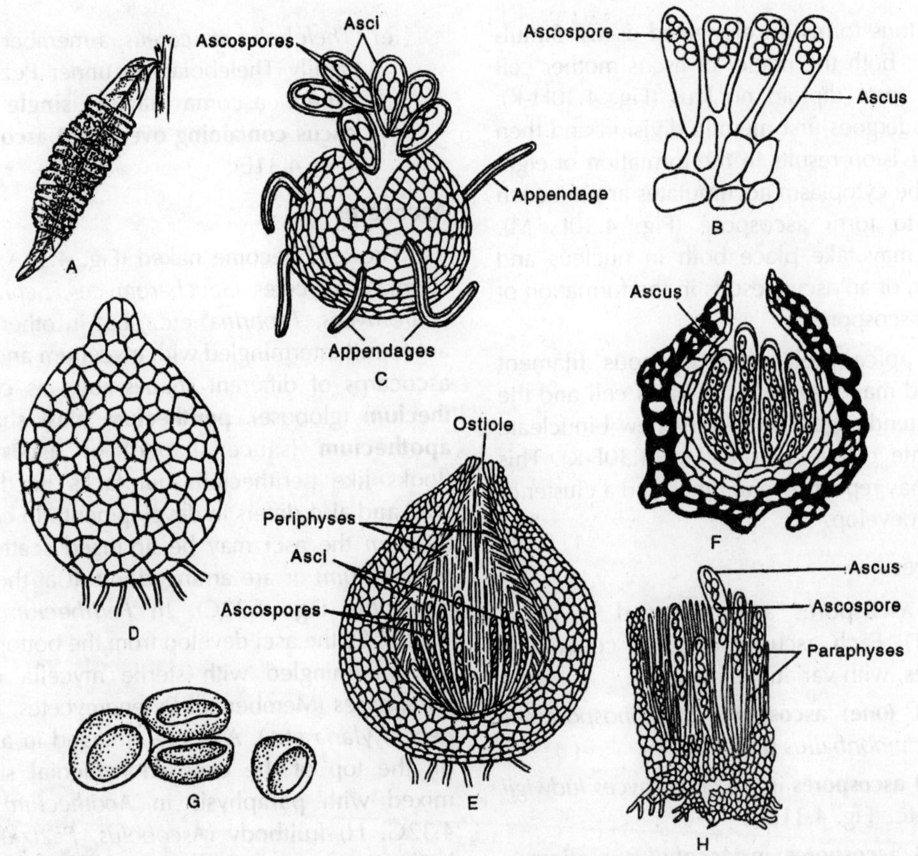

Fig. 4.32: Naked asci and asci in different types of ascocarps: A. Peach leaf curl disease (c.o. *Taphrina deformans*), B. Naked asci on peach leaf (in section), C. Cleistothecium of *Erysiphae*, D. Perithecium of *Neurospora*, E. Perithecium of *Neurospora* (in section) showing asci inside, F. Ascostroma of *Mycosphaerella tulipiferae* in section showing asci inside, G. Apothecia of *Ascobolus* and H. Portion of *apothecium of Ascobolus* (in section)

thecium (*Pleospora, Leptosphaeria* etc. under Loculoascomycetes).

Ecology

The members of Ascomycota fulfil a central role in most of the land-based ecosystems. These fungi are important as decomposers, breaking down organic materials, such as dead leaves and animals, and helping the detritivorous animals (feed on decomposing material) to obtain their required nutrients. Ascomycetes along with other fungi can break down large molecules such as cellulose or lignin, and thus play's important role in carbon cycle (a nutrient cycling).

The fruit bodies of the Ascomycota provide good food for many animals like insects, slugs and snails (*Gastropoda*), rodents; and also larger mammals such as deers, wild boars etc.

Many members of ascomycetes also form symbiotic relationships with other organisms, including plants and animals.

Lichens

Probably since earlier part in their evolutionary history, the Ascomycota have formed **symbiotic associations** with green algae (*Chlorophyta*), and other types of algae and Cyanobacteria (*Cyanophyceae*). These mutualistic associations are the lichens, and can grow and persist in terrestrial regions of the earth that are not suitable to other organisms and characterized by extreme temperature and humidity, including the Arctic, the Antarctic, deserts, and mountain tops. While the photoautotrophic algal partner generates metabolic energy through photosynthesis, the fungus offers a stable, supportive matrix and protect cells from radiation and dehydration. Around 42% of the Ascomycota (about 18,000 species) form lichens, and almost all the fungal partners of lichens belong to the Ascomycota.

Mycorrhizal fungi and endophytes

Members of the Ascomycota form two important types of relationship with plants: as mycorrhizal fungi and as endophytes.

1. **Mycorrhizae** are the symbiotic associations of fungi with the root systems of higher plants, which can be of vital importance for growth and persistence of the plant. The fine mycelial network of the fungus enables the increased uptake of mineral salts that occur at low levels in the soil. In return, the plant provides the fungus with metabolic energy, the photosynthetic products.
2. **Endophytic fungi** live inside plants, and those that form ***mutualistic*** or ***commensal associations*** with their host, without any damage. The exact nature of the relationship between endophytic fungus and host depends on the species involved. In some cases fungal colonization of plants can give a higher resistance against insects, nematodes and bacteria. In the grass, endophytic fungal symbiont produces poisonous alkaloids, which can affect the health of herbivorous mammals and discourage or kill insect herbivores.

Symbiotic relationships with animals

1. Several Ascomycetes of the genus ***Xylaria*** colonize the nests of leaf-cutter ants and other fungus-growing ants and the fungal gardens of termites. Since they do not generate fruiting bodies until the insects have left the nests.

   Bark beetles (family Scolytidae) are important symbiotic partners of ascomycetes. The female beetles transport fungal spores to new hosts in the characteristic tucks in their skin, the *mycetangia* (Gr. *mykes*, fungus; *angeion*, vessel). The beetle form tunnels into the wood and where they lay their eggs. Spores released from the mycetangia germinate into hyphae, which can break down the wood. The beetle larvae then feed on the fungal mycelium, and, on reaching maturity, carry new spores with them to renew the cycle of infection. A well-known example of this is Dutch elm disease, caused by *Ophiostoma ulmi*, which is carried by *Scolytus multistriatus*, the European elm bark beetle.

Plant disease interactions

One of the most harmful roles is as the agent of many plant diseases. For instance:

1. Dutch elm disease, caused by the closely related species *Ophiostoma ulmi* and *Ophiostoma novoulmi*, has led to the death of many elms in Europe and North America.
2. A disease of maize (*Zea mays*), which is especially prevalent in North America, is brought about by *Cochliobolus heterostrophus*.
3. *Taphrina deformans* causes leaf curl of peach.
4. *Uncinula necator* is responsible for the disease powdery mildew, which attacks grapevines.
5. Species of *Monilinia* cause brown rot of stone fruit such as sour cherries (*Prunus ceranus*) and peaches (*Prunus persica*).
6. Members of the Ascomycota such as *Stachybotrys chartarum* are responsible for fading of woolen textiles, which is a common problem especially in the tropics.
7. Blue-green, red and brown molds attack and spoil foodstuffs, e.g. *Penicillium italicum* rots oranges.
8. Cereals infected with *Fusarium graminearum* contain mycotoxins like deoxynivalenol, which causes *Fusarium* ear blight and skin lesions when eaten by pigs.

Human disease interactions

Several cases have been observed where Ascomycota are responsible to cause many human diseases. These are:

1. *Aspergillus fumigatus*, the most common cause of fungal infection in the lungs of immune compromised patients often caused death. In addition, it also causes Allergic bronchopulmonary aspergillosis, which often occurs in patients with Cystic fibrosis as well as Asthma.
2. *Candida albicans* is a kind of yeast that attacks the mucous membranes, causes infection of mouth or vagina candidiasis, and is also blamed for "yeast allergies".
3. The fungus like *Epidermophyton* cause skin infections but are not very dangerous for people with healthy immune systems. However, it can be life-threating with damaged immune system. The fungus *Pneumocystis jirovecii* is responsible for severe lung infections that occur in AIDS patients.
4. Ergot (*Claviceps purpurea*) when attacks rye inflorescence and produces highly poisonous and carcinogenic alkaloids in sclerotium, caused ergotism if consumed by human. Symptoms include hallucinations, stomach cramps, and a burning sensation in the limbs, also called "Saint Anthony's Fire".
5. *Aspergillus flavus*, which grows on peanuts and other hosts, generates aflatoxin, which damages the liver and is highly carcinogenic.

Beneficial effects for humans

The ascomycetous fungi have brought some important benefits to human society.

1. The mould fungi, *Penicillium chrysogenum* and *Penicillium notatum*, those, produces an antibiotic **penicillin**, the wounder drug; which destroy the bacteria. This antibiotic triggered a revolution in the treatment of bacterial infectious diseases in the 20th century.
2. The medical importance of *Tolypocladium niveum* as an immunosuppressor can hardly be exaggerated. It excretes **cyclosporin**, which, as well as being given during Organ transplantation to prevent rejection, is also prescribed for autoimmune diseases such as multiple sclerosis, although there is some doubt over the longterm side-effects of the treatment.
3. Some ascomycetous fungi can be altered relatively easily through genetic engineering procedures. They can then produce useful proteins such as insulin and human growth hormone, which is employed to dissolve blood clots.
4. Several species are common model organisms in biology, including *Saccharomyces cerevisiae, Schizosaccharomyces pombe*, and *Neurospora crassa*. The genomes of a number of ascomycete fungi have been fully sequenced.

   Baker's Yeast (*Saccharomyces cerevisiae*) is used to make bread, beer and wine, during which process sugars such as glucose or sucrose are fermented to make ethanol and CO_2. Bakers use the yeast for CO_2 production, causing the flupy nature of bread. On the other hand, brewers used it for ethanol production, with the CO_2 being released into the atmosphere during fermentation.
5. Enzymes produced by the fungus used in the manufacture of cheese. The fungus like *Penicillium camemberti* is used in the manufacture of the Camembert cheeses, while *Penicillium roqueforti* is used in the manufacture of the, Roquefort and stilton cheese (Fig. 4.33).
6. In Asian countries, *Aspergillus oryzae* is used to make soy sauce, by adding to a pulp of soaked soybeans, and is also used to break down starch in rice and other grains into simple sugars for fermentation to prepare alcoholic beverages.
7. Finally, some members of the Ascomycota are choice edibles like morels (*Morchella* spp.), truffles (*Tuber* spp.), and lobster mushroom (*Hypomyces lactifluorum*) are some of the most delicious food for man.

Fig. 4.33: Stilton cheese veined with *Penicillium roqueforti*

Life cycle

The Ascomycota have diversity in many features, still they possess some common unifying characters like:

1. Somatic structure, having loose mass of septate mycelium,
2. Mode of asexual reproduction, and
3. Mode of sexual reproduction

In general, they have **two reproductive phases, i.e. Asexual and Sexual stage**.

Some members reproduce by budding and others develop spores, the conidia.

The conidia are produced in huge numbers either directly from vegetative mycelium or on specialized structure, the conidiophore. The spores are readily dispersed by wind, insect etc. In favourable condition, they germinate and develop into a somatic mycelium. This is a type of asexual reproduction. The cyclic sequence in this phase is called asexual cycle.

During sexual reproduction, changes involved in a cyclic order, takes place in regular sequence, called sexual cycle.

Thus the life cycle in Ascomycota generally consists of **two cycles: Asexual** and **Sexual cycle**.

During favourable condition, the asexual stage may repeat several times, results the production of conidia in huge quantity. The conidia are very much responsive in alteration of climatic conditions. In some cases, the conidia can perennate during unfavourable condition. During development, the same mycelium from which conidia are being developed may takes part in the formation of gametangia.

Gametangial union starts through plasmogamy and in course of development asci are formed. Asci commonly contain eight (8) ascospores, are produced through karyogamy and meiosis.

In Ascomycota, the haplophase and diplophase intervened by karyogamy, may be of short or long in duration. Depending on the species, the asci may or may not enclosed in ascocarp, the fruit body of the fungus.

Members of Ascomycota may be **homothallic** or **heterothallic** in nature. In homothallic species, the asci are formed in the same thallus and on the other hand, in heterothallic species asci are developed by the interaction of two compatible thalli.

The heterothallism is controlled by single gene with two alleles, are 'A' and 'a' and because of segretation during meiosis, preceeds ascospore formation, results the formation of eight (8) ascospores present normally in an ascus. The ascospores are of two mating types, includes four (4) in each.

In some four (4)-spored members like, *Podospora anserina* and *Neurospora tetrasperma*, the ascospores have 2-nuclei i.e. binucleate; contain nuclei of both mating types. These spores on germination gives rise to fully fertile mycelia and appear as homothallic.

At times, uninucleate ascospores on germination do not produce fertile mycelia and the ascospores are thus formed by interaction between mating types. These fungi are basically heterothallic in their mating behavior.

In most Ascomycota, perennation takes place by ascospores. Its germination during favourable condition produces one or more germ tubes which grow into somatic i.e. vegetative mycelium. Thus the stage during which asci and ascospores are produced is called **sexual stage** or **ascigerous stage** or **perfect stage** and the cycle is called **sexual cycle**.

A large number of Ascomycota have both conidial or ascospore stages through which they dessiminate and propagate themselves. But many Ascomycota, have only ascogenous stage and their asexual stage is not found or does not exist at all. Usually both the asexual and sexual stages are completed either saprophytically or parasitically, depending on their existence as saprophyte or parasite.

In many of the parasitic Ascomycota, the asexual stage is completed on living tissue but the sexual stage is formed only on dead tissue of the host plant.

The life cycle of Ascomycota is presented in Fig. 4.34.

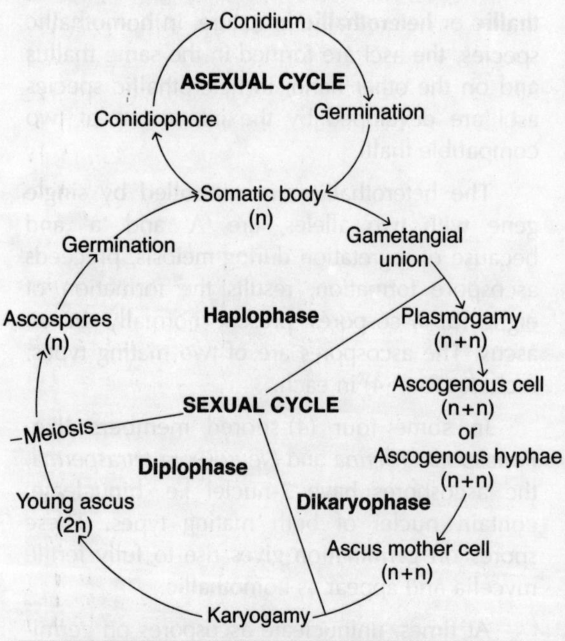

Fig. 4.34: Life cycle of Ascomycota

Heterokaryosis and Parasexuality

Many members of Ascomycota are not known to have a sexual cycle. Such asexual members may be able to undergo genetic recombination between individual through *Heterokaryosis* and *Parasexual cycle*.

Parasexuality refers to the process of *heterokaryosis*, takes place by merging of two hyphae of different individuals, through *anastomosis*, followed by a series of events results the formation of genetically different nuclei in the cell of mycelium. The merging of nuclei is not followed by meiotic events, such as formatin of gametes and an increased number of chromosomes per nuclei.

Mitotic crossing over may enable recombination, i.e., an exchange of genetic material between homologous chromosomes. The chromosome number may then be restored to its haploid state by nuclear division, thereby each daughter nuclei being genetically different from the nuclei of original parent. In other way, nuclei may lose some chromosomes, results the formation of aneuploid cells. It is found in *Candida albicans*, a member of the order Saccharomycetales.

Candida albicans has an elaborate and efficient, mating system that promotes conjugation between diploid (2n) a and α strains. The product of mating is a tetraploid (4n) a/α cell that must undergo a reductional division to return to the diploid (2n) stage. Despite the presence of several "meiosis-specific" genes in the *C. albicans* genome, meiotic division has not been observed. Instead, tetraploid (4n) products of mating can be induced to undergo efficient, random chromosome loss, often producing strains that are diploid, or close to diploid, in ploidy level. Using single nucleotide polymorphism (SNP) and comparative genome hybridization arrays for the analysis the genotypes of products from the *C. albicans* parasexual cycle, it is shown that the parasexual cycle generates progeny strains with shuffled combinations of the eight *C. albicans* chromosomes. In addition, several isolates had undergone extensive genetic recombination between homologous chromosomes, including multiple gene conversion events. Progeny strains exhibited altered colony morphologies in *in-vitro* condition, shown that the parasexual cycle generates phenotypic variants of *C. albicans*. In many fungi like *Saccharomyces cerevisiae* and *Schizosaccharomyces pombe*, the conserved *SPO*11 protein is integrated to meiotic recombination, where it is required for the formation of DNA double-strand breaks. It is shown that deletion of *SPO*11 prevented genetic recombination between homologous chromosomes during the parasexual cycle of *C. albicans*. It suggests that at least one meiosis-specific gene has been reprogrammed to mediate genetic recombination during the alternative parasexual life cycle of *C. albicans.*

SACCHAROMYCES

Systemic position

Ainsworth (1973)	Hawksworth (1995)
Kingdom Mycota	Kingdom Fungi
Division Eumycota	Phylum Ascomycota
Subdivision Ascomycotina	Class (no class)
Class Hemiascomycetes	Order Saccharomycetales
Order Endomycetales	Family Saccharomycetaceae
Family Saccharomycetaceae	Genus *Saccharomyces*
Genus *Saccharomyces*	

Leeuwenhoek (1680) first described yeast as the simplest and the most widely distributed members of Fungi. Guilliermond (1920) reported the presence of only 18 genera of yeast. They are the oldest known cultivated fungi. The genus *Saccharomyces* (Gr. *Saccharon*, sugar; *mykes*, fungus) consists of about 41 species. *S. cerevisiae*, commonly known as Brewer yeast or Baker's yeast is used widely in wine and baking industry. It produces two types of enzymes: an extracellular *invertase* and an intracellular *zymase*. The invertase hydrolyses canesugar to dextrose or invert sugar and zymase breaks invert sugar into ethyl alcohol and carbon dioxide.

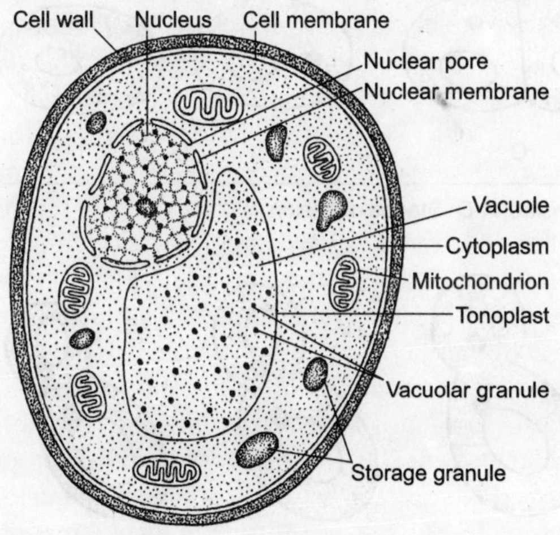

Fig. 4.35: Cell structure of *Saccharomyces cerevisiae* under electron microscope

Vegetative body

The thalloid somatic body is unicellular, but during rapid multiplication by budding the cells may remain attached in chain forming *pseudomycelium* (Fig. 4.39). The cells may be globose, elliptical, oval to even rectangular in shape and measure about 5–6 × 6–8 μm.

Electron microscopic studies (Fig. 4.35, 4.36) and chemical analysis of *Saccharomyces cerevisiae* show that the cells are surrounded by a distinct cell wall with three layers. The outermost layer mainly consists of protein-mannan and some chitin; the middle layer mainly of glucan and the innermost layer consists of protein-glucan. Some phosphate and lipids are also present (Matile *et al.*, 1969), while cellulose is absent in the cell wall. Inner to the cell wall is the cell membrane (plasmalemma), i.e., an usual unit membrane having series of shallow, elongated pits or invaginations (Fig. 4.36). In the centre, the cell having a large central vacuole,

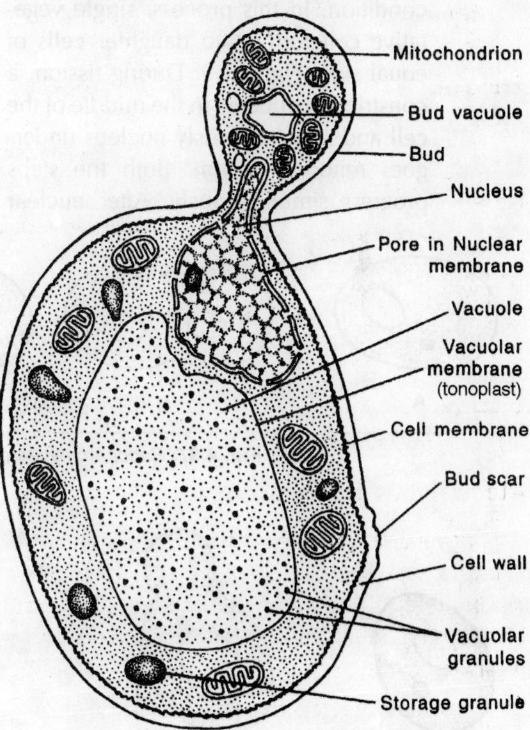

Fig. 4.36: Structure of a budding cell of *Saccharomyces cerevisiae* under Electron microscope

limited by a single membrane, the *tonoplast*; which contains a watery substance, granules of polymetaphosphate and lipid. The cytoplasm is granular and contains organelles like nucleus, mitochondria, golgi apparatus, endoplasmic reticulum, ribosome and other substances like glycogen bodies, volutin granules, oil globules etc. Some hydrolytic enzymes like proteases, esterases, ribonuclease etc., are also present in the cytoplasm. The nucleus having outer perforated double unit membrane remains by the side of the vacuole. The nucleus is bipartite in nature having major Feulgen positive and a smaller Feulgen negative regions (Moor and Muhlethaler, 1963).

Reproduction

Saccharomyces reproduces by vegetative, asexual and sexual means.

A. **Vegetative reproduction**: Vegetative reproduction takes place by fission and budding.

   (a) *Fission*: It takes place during favourable condition. In this process, single vegetative cell forms two daughter cells of equal size (Fig. 4.37). During fission, a constriction appears in the middle of the cell and simultaneously nucleus undergoes mitotic division. Both the steps progress simultaneously. After nuclear migration, one at each side, partition wall forms almost in the half way of the mother cell and, as such, two daughter cells are formed.

   (b) *Budding*: Budding also takes place during favourable condition. The protoplasm of vegetative cell swells up at one side in the form of a bud (Fig. 4.38). The nucleus undergoes mitotic division. Out of two nuclei formed by mitosis, one goes to the buds and other one remains in the mother. Bud enlarges and eventually cuts off from the mother by partition wall. The size of the bud is always smaller than the mother cell. After maturation, these buds separate from the mother and leave a convex scar on the surface, called *bud scar*. Similar scar with concave surface remains on the wall of the bud, called *birth scar*.

Sometimes due to rapid division, large number of buds develop without being detached from one another and persist in the form of branched or unbranched chain, called *pseudomycelium* (Fig. 4.39). Finally the cells get detached and grow individually.

B. **Asexual reproduction**: It takes place during unfavourable condition by the formation of thick walled spore called endospore (Fig. 4.40).

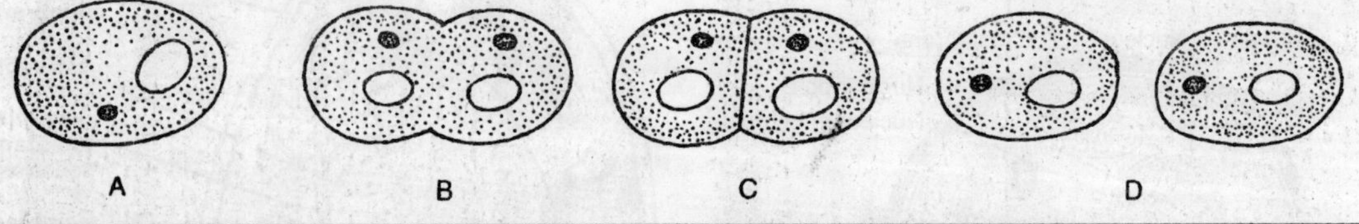

Fig. 4.37: *Saccharomyces cerevisiae*: A-D. Stages of fission

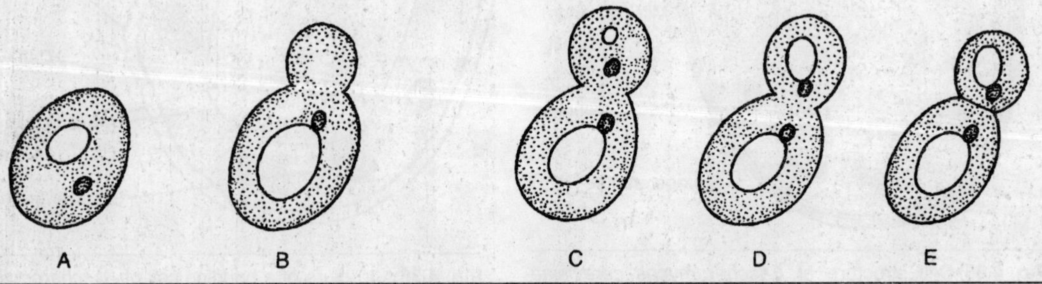

Fig. 4.38: *Saccharomyces cerevisiae*: A-D. Stages of budding

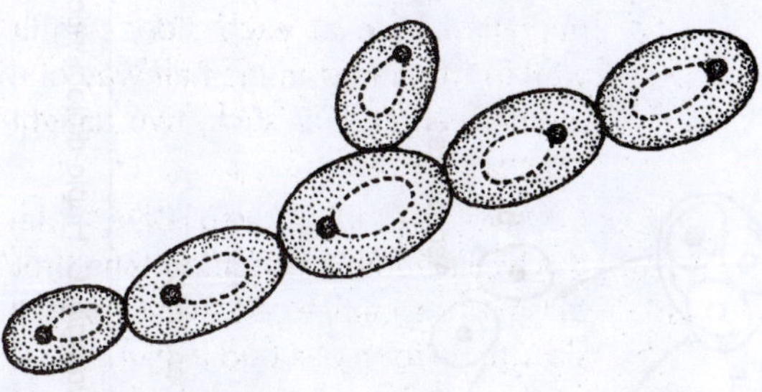

Fig. 4.39: Pseudomycelium of *S. cerevisiae*

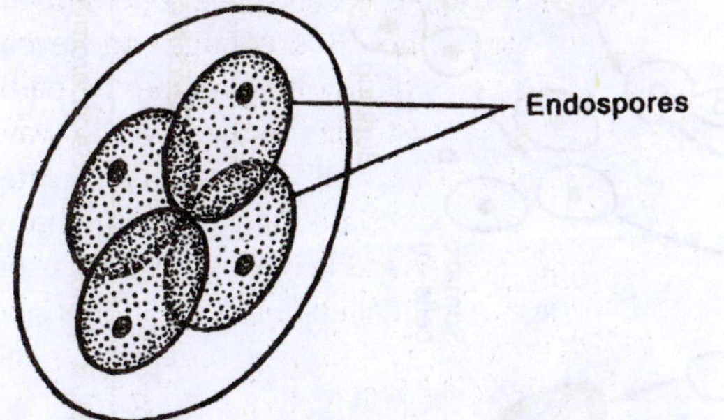

Fig. 4.40: *Schizosaccharomyces pombe*: Endospores (four) in a cell

During this process nucleus divides *mitotically* and forms four nuclei. The protoplast divides into four units, each with one nucleus and forms four endospores. During favourable condition, endospore germinates by budding and buds grow individually.

C. **Sexual reproduction**: Sexual reproduction takes place during unfavourable condition. In this process, two vegetative cells or ascospores behave as gametangia (Fig. 4.41). Two such cells come very close and develop beak-like outgrowth towards each other. Both the outgrowths come in contact and the intervening walls between them dissolve. The nuclei of both the gametangia come to the fused outgrowth (conjugation tube) and they fuse therein to form a diploid zygote. The zygote behaves as an ascus. The diploid nucleus of zygote undergoes *meiotic division* forming 4 or 8 (with additional mitosis) ascospores. The ascospores are liberated by breaking the ascus wall and behave as somatic cell.

Life cycle patterns in yeast

Three patterns of life cycle are found in yeast: Haplobiontic, diplobiontic and haplo-diplobiontic (Fig. 4.42):

1. *Haplobiontic type*: This type of life cycle is characterised by more elaborate haploid phase than the diploid phase, found in *Schizosaccharomyces octosporus* (Fig. 4.42A) The diploid phase is restricted only in the zygote. The vegetative cells are haploid and behave as gametangia. Two such gametangia fuse together and form a diploid cell. The diploid cell behaves as an ascus whose nucleus divides first meiotically, then mitotically; results in the formation of eight ascospores. After maturation, the ascospores liberate by bursting the ascus wall. The ascospores then behave as vegetative cell and continue multiplication through budding.

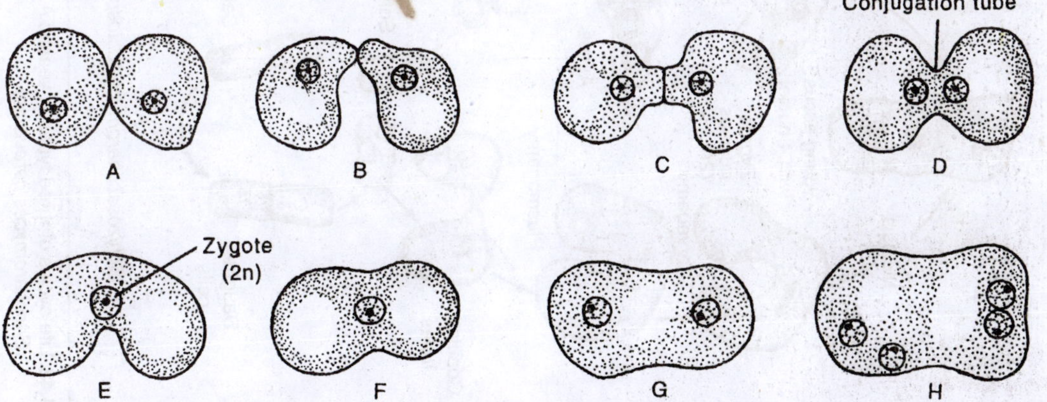

Fig. 4.41: Different stages (A-H) of sexual reproduction

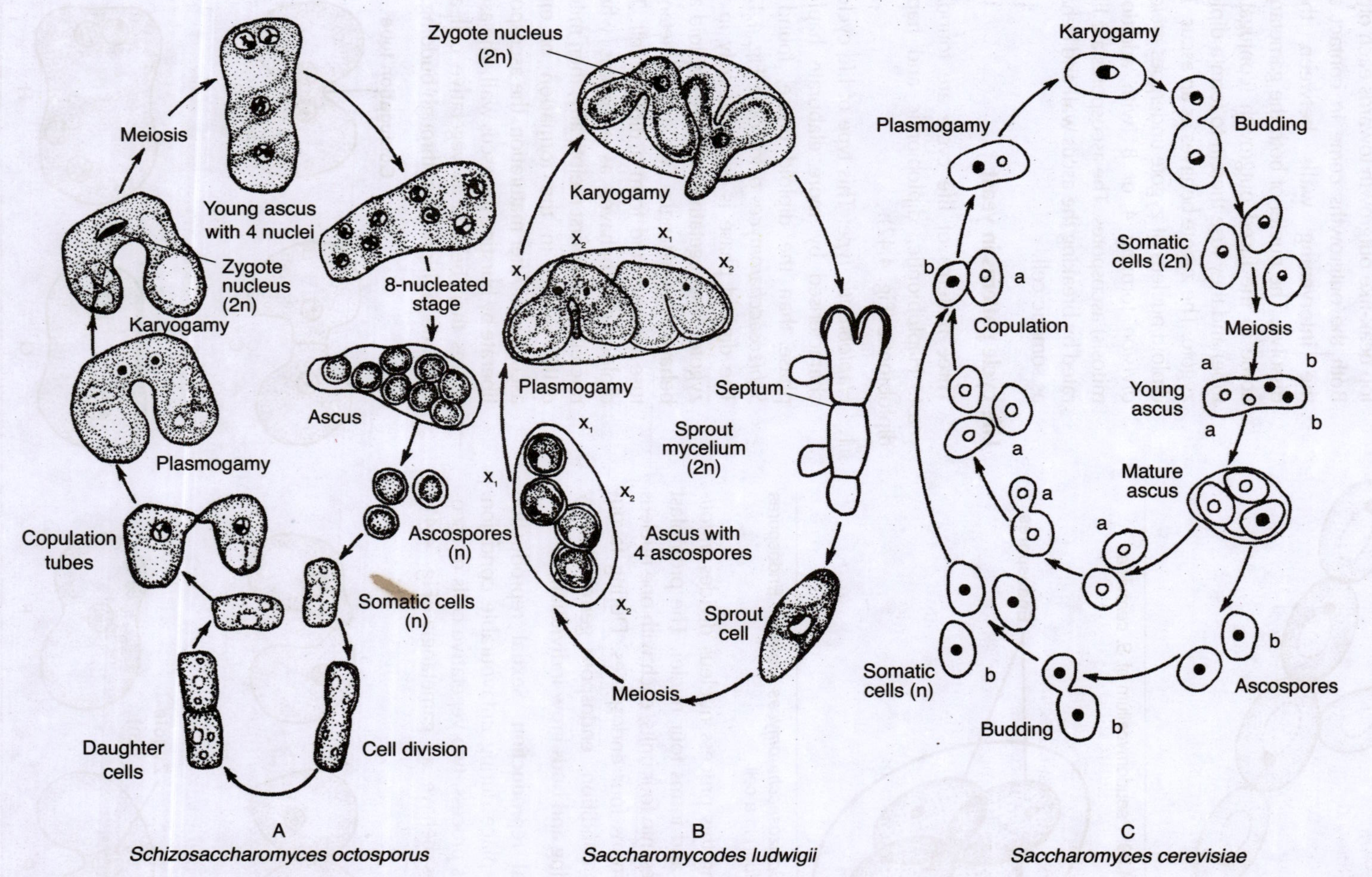

Fig. 4.42: Life cycles of three types in yeasts: A. Haplobiontic (*Schizosaccharomyces octosporus*). B. Diplobiontic (*Saccharomycodes ludwigii*), C. Haplo-diplobiontic (*Saccharomyces cerevisiae*).

2. *Diplobiontic type*: This type of life cycle is characterised by more elaborate diploid phase than the haploid phase, found in *Saccharomycodes ludwigii* (Fig. 4.42B). The haploid phase is restricted only in ascospore, with short duration. The ascospores behave as gametangia and, without liberating from ascus, they unite in pair. The paired gametangia after fusion produce diploid zygote. The zygote then germinates by producing germ tube which comes out through the ascus wall. The germ tube becomes multicellular from which diploid sprouts develop by budding. After detachment from the mother, the diploid sprouts function as asci and produce four ascospores by reduction division.
3. *Haplo-diplobiontic type*: This type of life cycle is represented by haploid and diploid phases, of more or less equal duration, found in *Saccharomyces cerevisiae* (Fig. 4.42C). The haploid cells of opposite mating type normally multiply by budding. Two such cells of opposite mating behave as gametangia and undergo fusion. The fused gametangia develop a diploid zygote. The diploid zygote like the haploid cells undergoes budding and forms many diploid cells. With the scarcity of food, the diploid cell behaves as an ascus and by reduction division it forms four haploid ascospores. After liberating from the mother wall, the ascospores undergo budding and form many haploid somatic cells.

PENICILLIUM

(Painter's brush)

(*Penicillium* is the asexual stage and the sexual stage is known as ***Talaromyces***. As the fungus is popularly known as *Penicillium*, it is described here as *Penicillium*).

Systemic position

Ainsworth (1973)	Hawksworth (1995)
Kingdom Mycota	Kingdom Fungi
Division Eumycota	Phylum Ascomycota
Subdivision Ascomycotina	Class (no class)
Class Plectomycetes	Order Eurotiales
Order Aspergillales	Family Trichocomaceae
Family Aspergillaceae	Genus *Penicillium*/
Genus *Penicillium*	Talaromyces

Penicillium is a saprophytic fungus, commonly known as **blue** or **green mold**. According to Raper and Thom (1949), the genus includes 136 species, distributed throughout the world. They are present in soil, in air, on decaying fruits, vegetables, meat, etc.

The "**wonder drug**" penicillin was first discovered by Sir Alexander Fleming at Sant Mary's Hospital, London, in 1929; during his work with a bacterium *Staphylococcus aureus* responsible for boil, carbuncle, sepsis in wounds and burns etc., get contaminated with mold spore (*Penicillium notatum*) which after proper growth causes death of *S. aureus* showing lytic zone around itself. He isolated and called this anti-microbial compound as *Penicillin*. Later Raper and Alexander (1945) selected a strain of *P. crysogenum,* more efficient than *P. notatum,* in the production of penicillin. The importance of *Penicillium* is mentioned in the Table 4.6.

Table 4.6: Importance of *Penicillium*

Species of *Penicillium*	Importance
Harmful	
1. *Penicillium divaricatum*	Causes decay of wood.
2. *P. purpurogenum*	Causes spots on papers, books, etc.
3. *P. expansum*	Rotting and spoilage of fruits like apple, orange, grapes etc.
Useful	
1. *P. notatum, P. digitatum P. italicum* and *P. crysogenum*	Source of antibiotic penicillin
2. *P. griseofulvum*	Source of antibiotic griseofulvin.
3. *P. roqueforti* and *P. camemberti*	Used in the hydrolysis of fat and flavouring of cheese.

Vegetative structure

The vegetative body is mycelial (Fig. 4.43A, B). The mycelium is profusely branched with septate hyphae, composed of thin-walled cells containing one to many nuclei (Fig. 4.43C). Each

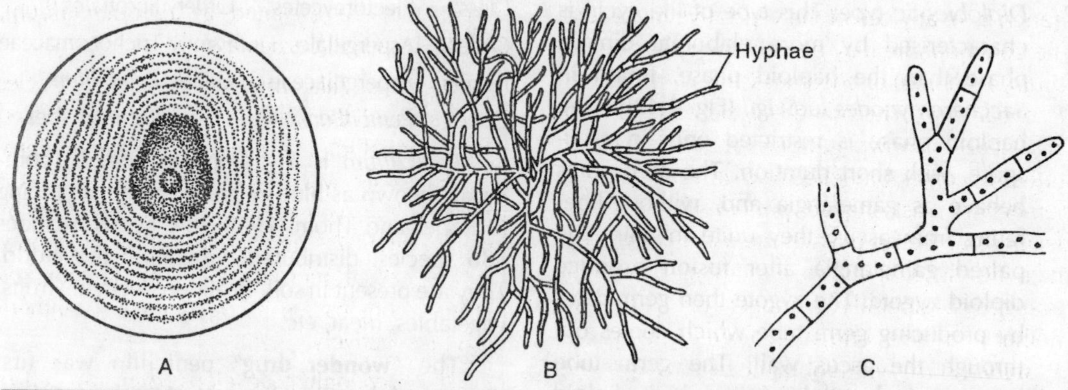

Fig. 4.43: *Penicillium*: A. A colony, B. Mycelium of *Penicillium*, C. Mycelium, enlarged view

septum has a central pore, through which cytoplasmic continuity is maintained. Some mycelia grow deeper into the substratum to absorb food material and others remain on the substrate and grow a mycelial felt. The reserve food is present in the form of oil globules.

Reproduction

Penicillium reproduces by vegetative, asexual and sexual means.

Vegetative reproduction: It takes place by accidental breaking of vegetative mycelium into two or more fragments. Each fragment then grows individually like the mother mycelium.

Asexual reproduction: Asexual reproduction takes place by unicellular, uninucleate, non-motile spores, the **conidia**; formed on conidiophore (Fig. 4.44).

The conidiophore develops as an erect

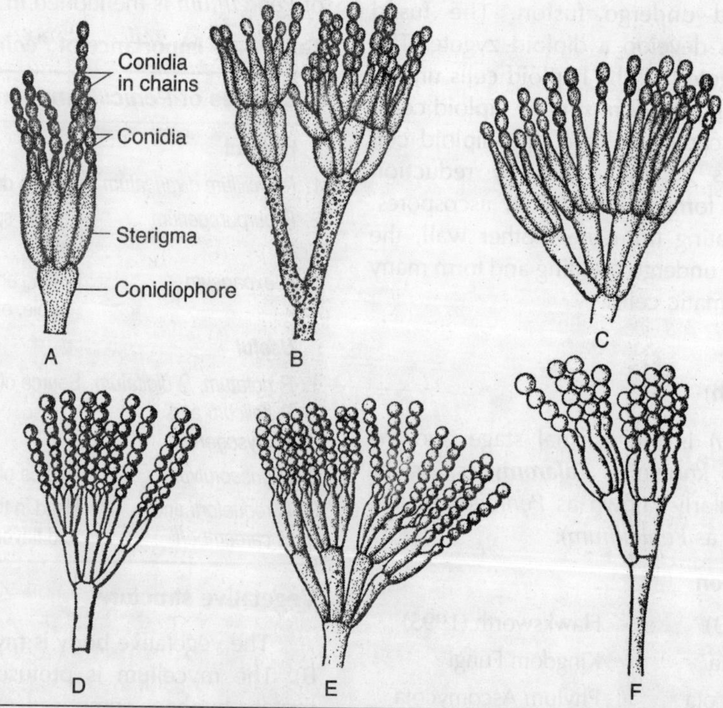

Fig. 4.44: Different types of conidiophore of *Penicillium* species: A. *P. thomii*, B. *P. lanoso-caerulium*, C. *P. vermiculatum*, D. *P. glaucum*, E. *P. herquei* and F. *P. chrysogenum*

branch from any cell of the vegetative mycelium. The conidiophore may be unbranched (*P. spinulosum, P. thomii*) or becomes variously branched (*P. expansum*). The branch of the conidiophore (Fig. 4.45B) is known as *ramus* (plural *rami*) which further becomes branched known as *metulae*. A number of flask-shaped phialid or *sterigmata* develops at the tip of each metulae. Each sterigmata develops at its tip a number of conidia arranged basipetally (younger one near the mother and older one away from it). In species (*P. spinulosum*) with unbranched conidiophore, the sterigmata develops at the tip of conidiophore. Rarely (*P. claviforme*) many conidiophores become aggregated to form a club-shaped fructification called *coremium,* which develops conidia known as *coremiospores*.

During the development of conidium, the tip of the sterigma swells up and its nucleus divides mitotically into two nuclei, of which one migrates into the swollen tip and by partition wall the swollen region cuts off from the mother and forms the uninucleate conidium. The tip of the sterigma swells up again and following the same procedure second conidium is formed, which pushes the first one towards the outer side. This process repeats several times and thus a chain of conidia is formed.

The conidia (Fig. 4.45C) are oval, elliptical or globose in structure having smooth, rough, echinulate outer surface and of various colourations like green, yellow, blue etc.

After maturation, the conidia get detached from the mother and are dispersed by wind. On suitable substratum, they germinate (Fig. 4.45D) by developing germ tube. The nucleus undergoes repeated mitotic division and all nuclei enter into the germ tube. The septa formation continues with the elongation of germ tube and finally a new septate branched mycelium develops.

Sexual reproduction: It has been studied only in few species (Fig. 4.45). It shows great variation from isogamy (*P. bacillosporum*), oogamy (*P. vermiculatum*) to somatogamy (*P. brefeldianum*). Most of the species are homothallic, except a few like *P. luteum* are heterothallic. Ascocarps are rarely formed. Based on the ascocarps, different genera can be assigned as *Europenicillium, Talaromyces* and *Carpenteles*.

The genus *Talaromyces* consists of 15 species (Pitt, 1980). All the species of *Talaromyces* studied are homothallic. The account of sexual reproduction deals with *Talaromyces vermiculatus* (= *Penicillium vermiculatum*) was described by Dangeard (1907). This species shows oogamous type of sexual reproduction. The female and male sex organs are ascogonium and antheridium, respectively.

The ascogonium develops from any cell of the vegetative filament as an erect uninucleate and unicellular body (Fig. 4.45E). The nucleus then undergoes repeated mitotic divisions and produces 32 or 64 nuclei (Fig. 4.45F).

The antheridium develops simultaneously with the ascogonium from any neighbouring hypha (Fig. 4.45F). It is also an uninucleate and unicellular branch which coils around the ascogonium. The apical region of antheridial branch cuts off by septum and forms a short, somewhat inflated unicellular and uninucleate antheridium (Fig. 4.45G).

After maturation of both ascogonium and antheridium, the tip of the antheridium bends and touches the ascogonial wall. The common wall at the point of contact dissoves and the two cytoplasm then intermixed. The nucleus of the antheridium does not migrate (Fig. 4.45H) into the ascogonium (Dangeard, 1907). Later, the pairing of nuclei into the ascogonium takes place by the ascogonial nuclei only. The ascogonium then divides by partition wall into many binucleate cells, arrange uniseriately (Fig. 4.45H).

Some of the binucleate cells of the ascogonium projects out by the formation of multicellular ascogenous branched hyphae, whose cells are also dikaryotic (Fig. 4.45I). The apical cells of the dikaryotic mycelia swell up and function as an ascus mother cells (Fig. 4.45J). Both the nuclei of ascus mother cell undergo karyogamy and form diploid (2n) nucleus (Fig. 4.45K). The nucleus then undergoes first meiosis, then mitosis, results in the formation of 8 nuclei; those after accumulating some cytoplasm form 8 ascospores (Fig. 4.45L).

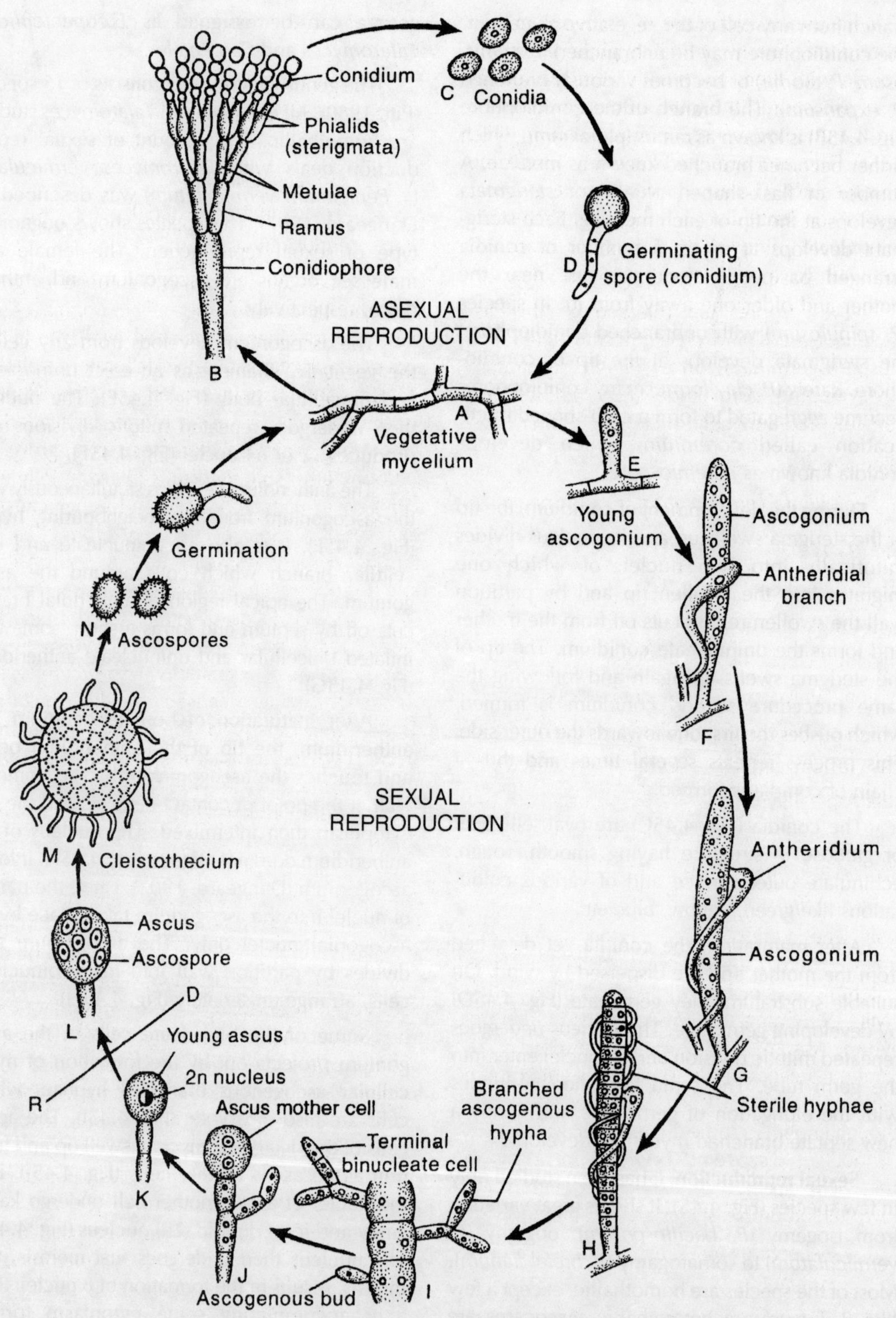

Fig. 4.45: Life cycle of *Talaromyces vermiculatus* (*Penicillium vermiculatum*)

With the development of ascogonium and antheridium, many sterile hyphae gradually entangle with them and finally after the formation of ascospores, the total structure becomes a round fruit body i.e., cleistothecium (Fig. 4.45M). The asci arrange irregularly inside the cleistothecium. The ascospores may be globose, elliptical or lenticular in shape with smooth, echinucleate, pitted (Fig. 4.45N) or branched outer wall like a pully-wheel in lateral view.

The ascospores are released by the dissolution of ascus and cleistothecium wall. The ascospore germinates on a suitable substratum by developing germ tube (Fig. 4.45O) and ultimately into a mycelium like the mother.

Indian species

Penicillium vermiculatum, P. wortmanni.

ASPERGILLUS

Systemic position

Ainsworth (1973)	Hawksworth (1995)
Kingdom Mycota	Kingdom Fungi
Division Eumycota	Phylum Ascomycota
Subdivision Ascomycotina	Class (no class)
Class Plectomycetes	Order Eurotiales
Order Aspergillales	Family Trichocomaceae
Family Aspergillaceae	Genus *Aspergillus*
Genus *Aspergillus*	

Aspergillus (L., *aspergillus*, a special type of brush), the conidial stage of the fungus was studied from long back under the class Deuteromycetes or Fungi imperfecti (now called as Mitosporic fungi), at present placed under the phylum Ascomycota with the discovery of sexual or teleomorphic stage in some of the anamorphic species. The teleomorphic stage although known in some anamorphic species but it is not yet discovered or does not exists in others. The species of *Aspergillus* with some known teleomorphic stages are given below:

1. *Aspergillus nidulans* — *Emericella*
2. *A. glaucus* — *Eurotium*
3. *A. fumigatus* — *Neosartorya*

More detail study of *Aspergillus* become essential for its importance in different aspects, such as:

1. Environmental pollution,
2. Biodegradation,
3. Biodeterioration,
4. Genetical studies,
5. Toxins in food materials, and
6. Industrial uses.

Members of *Aspergillus* are found to grow throughout the world especially in tropical countries. The members are chiefly available in air and cause contamination in laboratory cultures of fungi and also in bacteria. In this respect *Aspergillus niger* is very common and called as 'laboratory weed'. Different species grow as saprophytes in soil, foodstuffs like Jam, Jelly etc. and also in textiles, leather and paper causing deterioration. Species of *Aspergillus* like *A. halophilicus, A. restrictus* and *A. glucans* are able to grow with a Relative Humidity below 70% and able to damage the stored grains are called **'storage fungi'** (Christensen, 1965), reduce the percentage of germination and accumulation of toxin inside caused health hazards. *Aspergillus nidulans* has the historic importance in the discovery of **parasexual phenomenon** by Pontecorvo and Roper (1952). Some species are weak parasite on vegetables and fruits and some others also cause disease **aspergillosis**, in animals including human beings. *A. flavus* may infest the foods like groundnut meal and other dried foods and produces **aflatoxin**, a carcinogenic substance is well known to produce **cancer in both human and poultry**. In 1960, about 1,00,000 Turkey poults died in England by eating groundnut cake infested with *Aspergillus flavus*. Different species of *Aspergillus* are used in industry and their fermentation products yield alcohol, different acids like citric, gluconic and itaconic acid and many other useful chemicals. Species of *Aspergillus* is used for the hydrolysis of starch to glucose. *A. oryzae*, involved in the industrial production of soy sauce and *A. niger* used for the production of citric acid and enzymes such as glucose oxidase and lysozyme.

The parasexual cycle was discovered and then developed as a genetic system in

Aspergillus nidulans (= *Emericella nidulans*) by Pontecorvo and co-workers during the 1950s (Pontecorvo, 1956).

Aspergillus niger causes 'smut' of date palm and fig and black mould of cotton balls, vegetables and also on other fruits.

Somatic structure

Somatic body mycelial, which form a dense mat on the subtrate. Hyphae are multinucleate, hyaline, septate and branched.

Reproduction

The fungus reproduces by both asexual and sexual means are given below.

Asexual reproduction

Asexual reproduction takes place by the formation of conidia. During asexual reproduction

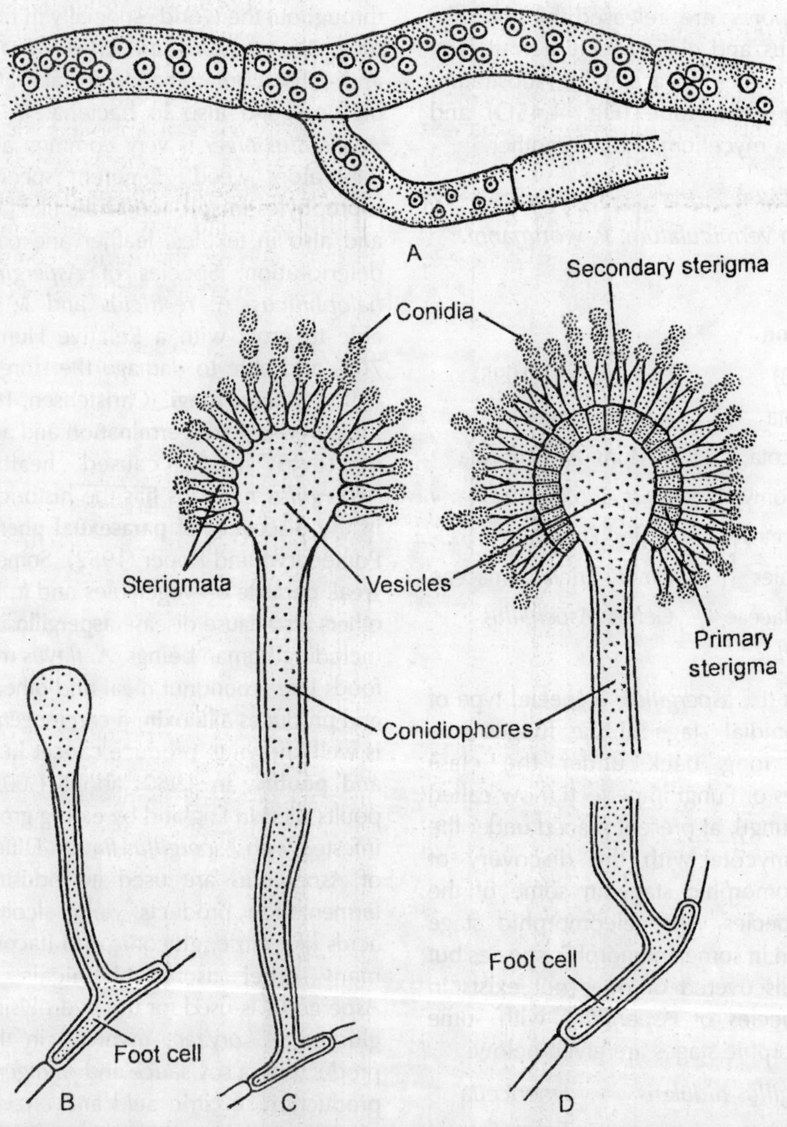

Fig. 4.46: Different conidiogenous structures in different groups of *Aspergillus*. A. Somatic mycelium, B. Foot cell with developing conidiophore, C. *A. glaucus* group (phialids borne directly on the vesicles) and D. *A. nidulans* group (phialids on a series of cylindrical cells, the **metulae**)

certain cell of the somatic hyphae enlarge and develops a thick wall, called foot cell. Foot cells are formed in series on the somatic hyphae. From the upper side of **foot cell** there arises a long and erect branch, the **conidiophore**. The conidiophore elongate and swells at the apex into a swollen structure the **vesicle**. Nuclei and cytoplasm moves towards the vesicle. Conidiophores are usually simple but branched in *A. glucus*. With further development, the vesicle enlarges and form different shapes like globose (*A. niger*), clavate (*A. clavatus*) or hemispherical to elliptical (*A. terreus, A. nidulans*). On the outer surface of the vesicle bottle-shaped structures called conidiogenous cells i.e. sterigmata or phialids developed in dense series and produce chain of conidia in basipetal (a chain of spores with the youngest at the base) sucessation. The sterigmata are developed either in one series i.e. uniseriate (*A. glaucus*) (Fig. 4.46C) or in two series i.e. biseriate (*A. nidulans*) (Fig. 4.46D). In biseriate condition, the inner layer is called **primary sterigmata** and the outer layer is the **secondary sterigmata**.

Development of conidia

The conidial ontogeny is holoblastic (Hanlin, 1976). The development of conidia in *A. niger* (Fig. 4.47) is described below.

Conidia developed from phialids may be terms as **phialoconidia**. Young phialids are more or less club-shaped and uninucleate (may be multinucleate in others). Single nucleus of the phialid divides and the tip of the phialid expands to form a spherical knob-like structure. One nucleus passes into the spherical knob. The expansion of the first conidium causes rupture of the wall of phialid near the apex which persist as a cap around the first formed conidium. Before rupture of the phialid wall a layer of wall mate-

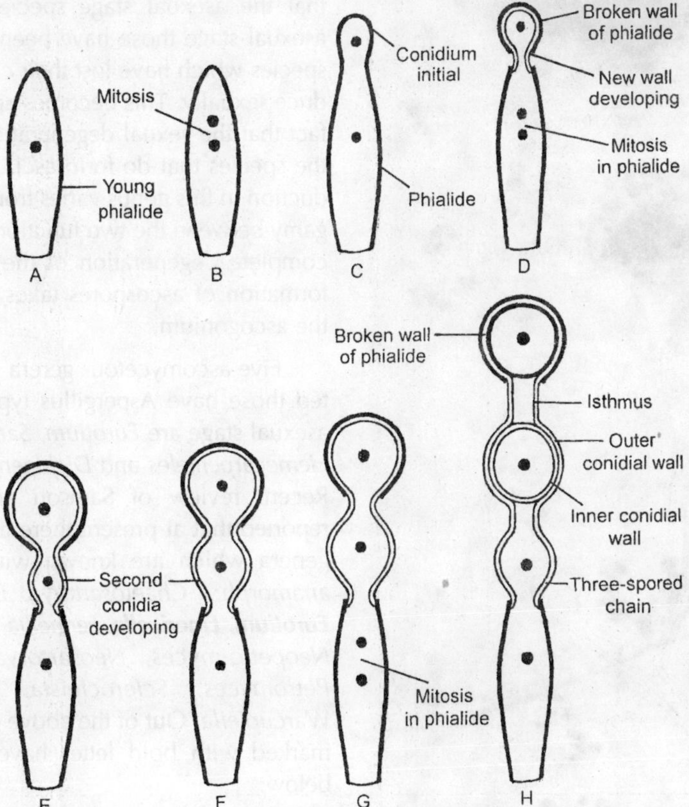

Fig. 4.47: Different stages in the development of conidium in *Aspergillus niger* (modified from C. V. Subramanian, 1971)

rial is laid down, which forms the outer layer of conidiun, inside which inner wall also develops later.

Nuclear division continues in the phialid. Nucleus with some cytoplasm extruded out from the broken tip of the phialid and wall is laid down around it and form the second conidium, which is not covered by the remnant of the broken wall of the phialid. The cytoplasmic continuity is maintained between the conidia by a cylindrical connection called **isthmus** (Fig. 4.47H). With further development cytoplasmic continuity is interrupted by the formation of the inner wall. The process continues and forms a chain of conidia where the youngest one remain at the base and oldest one at the top i.e. they arranged in a basipetal manner.

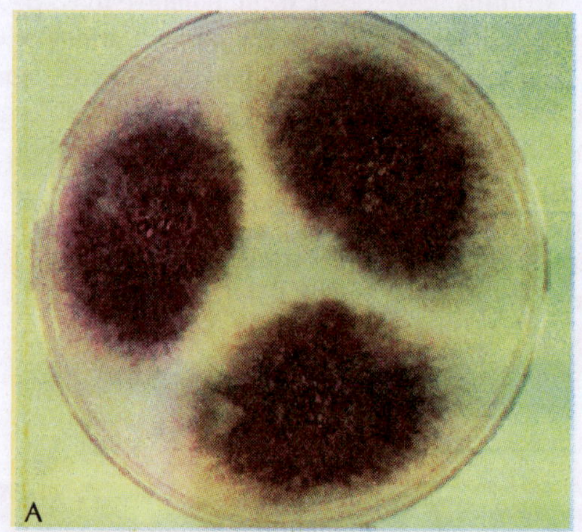

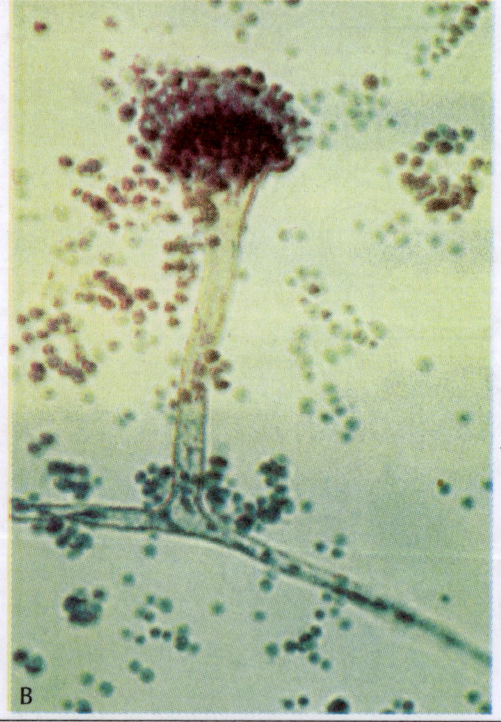

Fig. 4.48: Culture of *Aspergillus nidulans*: A. Colonies on petridishes, B.Conidial head.

The conidia are usually brown (*A. tamari*), black (*A. nige*), green (*A. flavus*), yellow (*A. ochraceus*), white (*A. candidus*) etc. in colour depending on the species. The conidia are dessiminated by air current. Coming in contact with the suitable environment the conidia germinate to form new mycelium.

Sexual reproduction

Although *Aspergillus* is commonly available with the asexual stage but perfect states are known only in a few species. It is quite possible that the asexual stage species exists with the asexual stage those have been derived from the species which have lost their capability to reproduce sexually. This becomes apparent due to the fact that the sexual degeneration occurs even in the species that do form asci. The sexual reproduction in this genus varies from normal plasmogamy between the two functional gametangia, to complete degeneration of the antheridium and formation of ascospores takes place directly by the ascogonium.

Five ascomycetous gerera so far been reported those have Aspergillus type of conidia i.e., asexual stage are *Eurotium, Sartorya, Emericella, Hemecarpenteles* and *Dichlaena* (Fennell, 1973). Recent review of Samson and Varga (2010) reported that at present there are 13 teleomorph genera which are known with an *Aspergillus anamorph: Chaetosartorya, Dichotomomyces,* ***Eurotium, Emericella****, Fennellia, Hemicarpenteles, Neopetro-myces, Neosartorya, Penicilliopsis, Petromyces, Sclerocleista, Dichlaena* and *Warcupiella*. Out of the above only 13 genera, 2 marked with bold letter have been described below:

1. *Eurotium*

The genus *Eurotium* includes 19 species

(Blaster, 1976) are widely distributed in soil. The type species is *E. herbariorum*. They cause spoilage of food with high osmotic concentration. The very common species *E. repens* commonly found to grow on jams and its imperfect stage belongs to *A. glaucus* group. The fungus develops yellow spherical ascocarps on media with high sugar content (20% sucrose).

Eurotium develop both antheridium and ascogonium close to each other on aerial somatic hyphae. Both are elongate and multinucleate. The antheridium may or may not be functional. When antheridium is functional, its nuclei enter the ascogonium and pairs with the nuclei of ascogonium, otherwise only ascogonial nucleform pairs. Simultaneously, both the structure becomes invested with the sterile hyphae developed from the stalk of ascogonium. After pairing of nuclei, the ascogonium becomes septate and develops ascogenous hyphae from binucleate cells. The ascogenous hyphae become branched and penetrate within the ascoma which dissolve the surrounding pseudoparenchyma derived from the investing hyphae. The branch of ascogenous hyphae are of different length, those develop ascus through crozier formation. Mature cleisto-

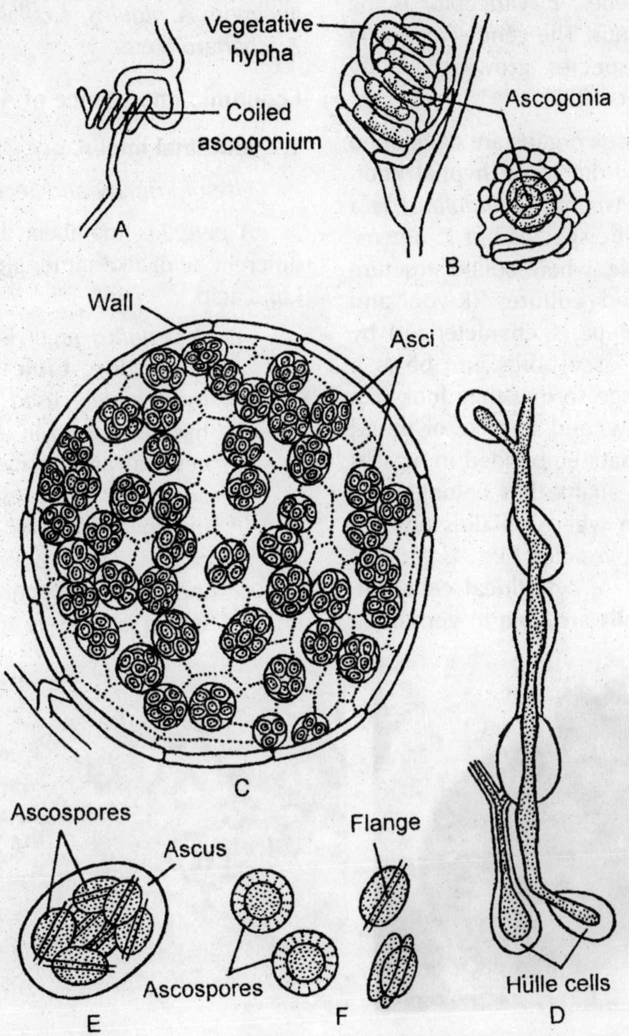

Fig. 4.49: *Aspergillus*. A-C (*Eurotium repens*). A. Coiled ascogonium, B. Ascogonium encircled by sterile hyphae, C. Asci inside the cleistothecium; D-F (*Emericella nidulans*). D. Hülle cells and E. Ascus containing ascospores and F. Ascospores in both top and lateral view.

thecium consists of cluster of globose or pear-shaped asci arranged scatteredly and surrounded by a single layered, bright yellow coloured peridium. Asci contain eight ascospores. With maturity the peridium of ascoma breaks irregularly, simultaneously the ascus wall also breaks down and ascospores are released. Ascospores are broadly lenticular, colourless and unornamented.

2. ***Emericella***

Raper and Fennell (1965) treated the species now classified as *Emericella* in the *A. nidulans* group, and distinguished eleven ascosporic species with five varieties. *E. variecolor* is the type species of this genus. The genus *Emericella* contains about 15 species growing widely (Christensen and Raper, 1978).

In *Emericella*, no sex organs are developed and ascoma develop from loose hyphal knot. *Emericella nidulans* (*Aspergillus nidulans*) is a well-known **homothallic** species but *E. heterothallicus* is **heterothallic**, where coiled structure is produced in mated cultures (Kwon and Raper, 1967). The genus is characterized by red to violet or blue ascospores and bears a double equatorial flange so that they look like pully-wheels (side view) and star-like or round (front view), by ascomata embedded in masses of **Hülle cells** and by anamorphs belonging to the *A. nidulans* group where phialids are not borne directly on the vesicles like *A. glaucus* group but on a series of cylindrical cells, the metulae. The **hülle cells** are able to germinate and form mycelium but it is doubtful about their role in propagation. Wiley and Simmons (1973) also included *Emericella nivea*, the telemorph of *A. niveus* in the genus, because of the similar ascoma structure. *E. nivea*, however, is clearly distinct from the other *Emericella* species by the hyaline to pale yellow ascospores and the anamorph belonging to the *A. flavipes* group. The species is similar to *Fennellia flavipes* (Wiley and Simmons, 1973).

Indian species

Aspergillus flavus, A. fumigatus, A. niger, A. awamori, A. nidulns, A. allahabadii, A. luchensis, A. chartarum etc.

Economic importance of *Aspergillus*

1. **Industrial importance**

Use of Primary and Secondary metabolites

Aspergillus members are able to produce different acids like citric, gluconic, itaconic and kojic acid.

(i) *Aspergillus niger* is used in citric acid production. **Citric acid** is one of the most widely used food ingredients. It has also found in using in the pharmaceutical and cosmetic industries as an acidulant. It is also used as a hardener in adhesive and for retarding the setting of concrete (Brooke, 1994) and in the industrial production of **gluconic acid**, which is used as an additive in certain

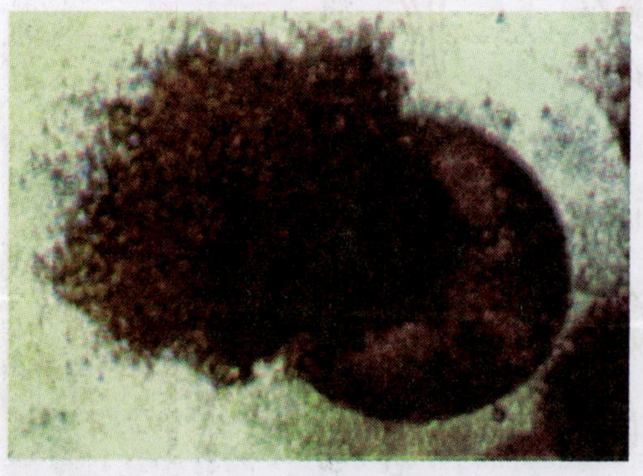

A

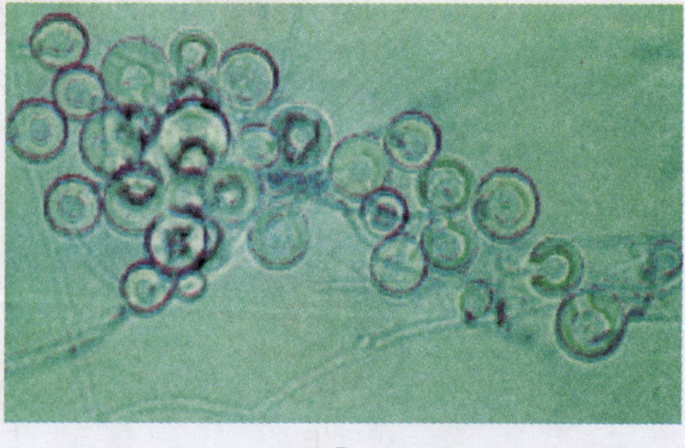

B

Fig. 4.50: *Emericella nidulans* (anamorph *A. nidulans*): A. Cleistothecium releasing ascospores, and B. Thick-walled Hülle cells.

metal cleaning applications, as well as for the therapy for calcium and iron deficiencies.

(ii) *A. oryzae* is used in the fermention for **kojic acid** production, used for skin whitening and as a precursor for synthesis of flavour enhancers (Ruijter *et al.*, 2002).

(iii) *Aspergillus terreus* produces a secondary metabolite **lovastatin** (Alberts, 1998). Lovastatin was sold under the brand named MevacorTM and became a blockbuster drug for Merck (Shu, 1998).

2. Food fermentation and enzymes

Food fermentations are the ancient technologies, used microorganisms and their enzymes in the improvement of human diet. Fermented foods enhanced textures, flavours, and aromas, and may also possess certain health benefits including superior digestibility. For vegetarians, fermented foods are used as palatable, protein-rich meat substitutes.

A. Fermentation by koji moulds

(i) **Rice wine**: *Aspergillus oryzae* and *A. sojae*, sometimes called as **koji moulds**, are used in many ways (Abe and Gomi, 2008). The hydrolytic enzymes produce by the fungi, help them for growth on starch and other carbohydrate substrates (Bennett, 2001). In the koji process, the fungal enzymes are used in the beer fermentations of western countries. The koji moulds release amylases that break down starch of rice which in turn can be fermented to make rice wine.

(ii) **Miso and Soy sauce**: The koji moulds are also effective in a variety of legume fermentations of which miso and soy sauce are best known (Reddy *et al.*, 1986). Miso is a mixture of soybeans and cereals usually used to flavour soups (Shibasaki and Hesseltine, 1962). Soy sauce is a flavourful, salty liquid sauce made from soybeans that have been fermented by koji moulds.

B. Enzymes

(i) Commercial enzymology is directed on bio-based business in which *A. niger* and *A. oryzae* are among the major producers for hydrolytic enzymes (Godfrey and West, 1996). Being high capacity of secretion, their extracellular enzymes can be used in different industries like *beverage* and *brewing industries, baking, in making animals feeds; and also in the paper pulping industry*.

A. niger has been developed as an efficient host for the production of proteins using genetic engineering techniques (Archer and Turner, 2006).

(ii) *Aspergillus* has numerous enzymes that can turn complex polymers into sugars, lipids and other simpler molecules, used for fuels and chemical synthesis.

3. Animal diseases

Airborne *Aspergillus* spores are found almost everywhere and we are routinely or almost constantly exposed to them. Such exposure is a normal part of the human condition and generally posses no adverse effect on health. Nevertheless, *Aspergillus* can and does cause animal disease in three major ways:

1. through induction of allergenic responses,
2. through the production of mycotoxins, and
3. through localized or systemic infections.

Allergies and asthma

Allergic reactions to *Aspergillus* in human being may be caused by air borne fungal spores and from fungi ingested in food. Airborne spores are readily inhaled when we breathe; they also come into contact with the eyes and other exposed parts of the body. They develop certain hypersensitivity reactions such as asthma, hay fever (allergic rhinitis), and food allergies.

Mycotoxins

Aspergillus originally was considered a serious problem in agriculture for its capacity in the biodeterioration of stored crops and also as an opportunistic pathogen of many field crops, particularly under high moisture conditions (Christensen and Kaufman, 1969). During the early 1960s, the discovery of **aflatoxins** is associated with massive deaths of poultry birds and other domestic animals throughout the world raised new awareness that posed threats to foods and feeds beyond their ability to rot plant materials (Goldblatt, 1969).

In addition to aflatoxins, other important mycotoxins produced by *Aspergillus* include **ochratoxin, patulin** and **fumigillin** (Cole and Cox, 1981; Bennett and Klich, 2005). Aflatoxins are still recognized as the most important mycotoxin, synthesized mainly by a few species, of which *A. flavus* and *A. parasiticus* are the most important.

Localized or systemic infections

The expression of aflatoxin-related diseases are influenced by factors like age, nutrition, sex, species and the possibility of concurrent exposure to other toxins. The main target organ in mammals is the liver, so **aflatoxicosis** is primarily a hepatic disease. Conditions increasing the aflatoxicosis in humans include limited availability of food, environmental conditions that favour mould growth on foodstuffs, and lack of regulatory systems for aflatoxin monitoring and control (Henry *et al.*, 1999). Aflatoxins have been isolated from all major cereal crops. Aflatoxin B1 is considered as **carcinogen** by WHO (Moy, 1998).

About 40 of the 250 species of *Aspergillus* have been reported as **human pathogens** (Klich, 2006) but the majority of cases are associated with just one species: *A. fumigatus* (Latge and Steinbach, 2008). The respiratory system is the normal portal of entry. The animal diseases caused by *Aspergillus* infection are all lumped together under the term **'aspergillosis.'** Occupational mycoses, allergies, localized mycoses, and invasive mycoses are all assembled together under the same umbrella called as 'Aspergillosis'.

Aspergillosis was first recognized as an avian disease (Ainsworth and Austwick, 1959). At present, it has been reported from all domestic and many wild species of animals. **Birds** show high susceptibility to *Aspergillus*.

ALTERNARIA

(*Alternaria* is the asexual stage and Sexual stage is named *Pleospora infectoria*, Webster, 1980)

Systemic position

Ainsworth (1973)	Hawksworth (1995)
Kingdom Mycota	Kingdom Fungi
Division Eumycota	Phylum Ascomycota
Sub-division Deuteromycotina	Class (no class)
Class Hyphomycetes	Order Pleosporales
Order Moniliales	Family Pleosporaceae
Family Dematiaceae	Genus *Pleospora* (asexual stage *Alternaria*)
Genus *Alternaria*	

Alternaria is a large genus includes 299 species, obiquitous in its distribution. They grow

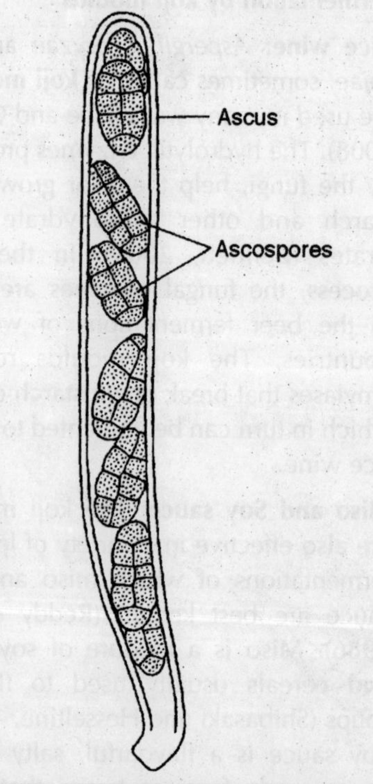

Fig. 4.51. Single ascus of *Pleospora infectoria*, the perfect stage of *Alternaria*.

either as saprophyte on soil, air and also on decaying plant parts or as facultative parasite on different hosts. In addition to the above they also cause allergies (Hyde and Williams, 1946) and **skin diseases** of human beings. Diseases caused by different species of *Alternaria* are given in table 4.7

Vegetative structure

Vegetative body consists of light brown, but becomes darker with age, septate much branched hyphae which grow both inter- and intracellularly inside the host tissue. The cells are usually multinucleate. Cells of *A. brassicicola* contain 1-many nuclei (Knox-Davis, 1979).

Table 4.7: List of some pathogenic species and diseases caused by *Alternaria*

Name of some pathogenic species	Disesaes
1. *Alternaria alternata* (*A. tenuis*)	Black-point disease of wheat.
2. *A. triticina*	Leaf blight of wheat.
3. *A. solani*	Late blight of potato.
4. *A. brassicae*	Grey leaf spot of *Brassica* sp.
5. *A. citri*	Infect roots of citrus.
6. *A. helianthi*	Infect *Helianthus annuus*

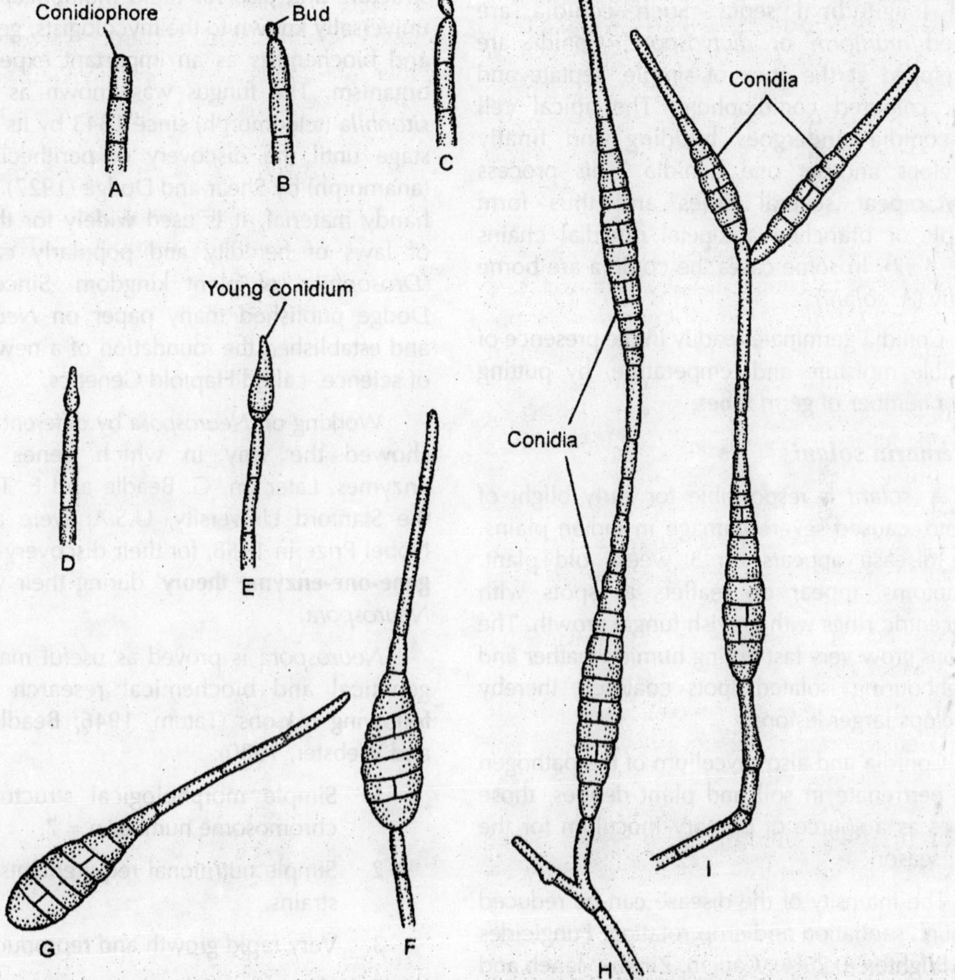

Fig. 4.52: *Alternaria.* A-F. Development of conidium of *A. solani*, G. Mature conidium of *A. solani*, H. Two conidia in chain (*A. brassicae*), and I. Branching of conidial chain in *A. brassicae*

Reproduction

Alternaria reproduces commonly by producing conidia. Formation of microsclerotia and chlamydospores has been reported in *Alternaria alternata* (Amrani and Najim, 1985). The sexual stage is named as *Pleospora infectoria*, a Loculoascomycetous fungus; develop in pseudothecia (Webster, 1980).

Asexual reproduction

Asexual reproduction takes place by developing conidia.

Conidia: Conidia are short, obclavate or beaked, dark coloured and with both transverse and longitudinal septa. Such conidia are called *'muriform'* or *'dictyospore'*. Conidia are developed at the apex of simple, septate and dark coloured conidiophore. The apical cell of conidia undergoes budding and finally develops another one conidia. This process may repeat several times and thus form simple or branched acropetal conidial chains (Fig. 4.52). In some cases the conidia are borne singly (*A. solani*).

Conidia germinate readily in the presence of suitable moisture and temperature, by putting out a number of germ tubes.

Alternaria solani

A. solani is responsible for early blight of potato, caused severe damage in Indian plains. The disease appears on 3 weeks old plant. Symptoms appear on leaflets as spots with concentric rings with grayish fungal growth. The lesions grow very fast during humid weather and neighbouring isolated spots coalesce, thereby develops larger lesions.

Conidia and also mycelium of the pathogen can perrrenate in soil and plant debries, those serves as a source of primary inoculum for the next season.

The intensity of the disease can be reduced through sanitation and crop rotation. Fungicides like Blightox (0.25%), Captan, Zineb, Maneb and also Dithane Z-78 are used as spray to control the disease.

NEUROSPORA

Systemic position

Ainsworth (1973)		Hawksworth (1995)	
Kingdom	Mycota	Kingdom	Fungi
Division	Eumycota	Phylum	Ascomycota
Subdivision	Ascomycotina	Class	(no class)
Class	Pyrenomycetes	Order	Sordariales
Order	Sphaeriales	Family	Sordariaceae
Family	Sordariaceae	Genus	*Neurospora*
Genus	*Neurospora*		

The genus *Neurospora* is very important for its simplicity in both internal and external structure and also for rapid multiplication. It is universally known to the mycologists, geneticists and biochemists as an important experimental organism. The fungus was known as *Monilia sitophila* (teleomorph) since 1843 by its conidial stage until the discovery of perithecial stage (anamorph) by Shear and Dodge (1927). Being a handy material, it is used widely for the study of laws of heredity and popularly called as '*Drosophila*' of plant kingdom. Since 1927, Dodge published many paper on *Neurospora* and established the foundation of a new branch of science, called Haploid Genetics.

Working on *Neurospora* by different workers showed the way in which genes control enzymes. Later on, G. Beadle and E. Tatum of the Stanford University, U.S.A; were awarded Nobel Prize in 1958, for their discovery of '**one-gene-one-enzyme theory**' during their work on *Neurospora*.

Neurospora is proved as useful material in genetical and biochemical research for the following reasons (Tatum, 1946; Beadle, 1959 and Webster, 1980):

1. Simple morphological structure with chromosome number n = 7,
2. Simple nutritional requirements of wild strains,
3. Very rapid growth and reproduction,
4. Very easy to develop mutants, and
5. Very easy to analyse the tetrads.

The life history (Fig. 4.53) of *N. sitophila* is described below:

N. sitophila

The name *Neurospora* (Gr. *neuron*, nerve; *spora*, seed) is given for the ornamentation of nerve-like ribs on the outer wall of ascospores. The genus contains 12 species out of which *N. sitophila, N. crassa* and *N. tetrasperma* are very common. It is a saprophytic fungus grows on rotting leaves, bread, leather, burnt ground and charred vegetation. It grows well as contaminants of laboratory cultures of bacteria and fungi. *N. crassa* is called pink bread mold and *N. sitophila* is called red bread mold.

Somatic body

The somatic body i.e. mycelium of *N. sitophila* consists of well-branched and multicellular hyphae. The aerial hyphae form mass of mycelium on the substrate. *Neurospora* reproduces both asexually and sexually.

Asexual reproduction

Asexually the mycelium reproduces by two types of conidia: Macroconidia and Microconidia.

Macroconidia: The aerial hyphae get differentiated into branched structure, called conidiophore. The macroconidia are developing in chain on branched conidiophores. Macroconidia are oligonucleate and contain about 2-8 nuclei, usually 2 or 3 and about 6-8 μm in diameter. Mature conidia are detached from the conidiophores and disseminated by wind.

Microconidia: These are smaller uninucleate, oval, sticky conidia of about 3-4 μm in diameter develop in a similar way as macroconidia but are less frequent. The conidiogenous cells are called phialids. On germination both types of conidia give rise to somatic hyphae, which develop very quickly and then start reproducing asexually.

Sexual reproduction

N. terricola and *N. dodgei* are homothallic, *N. tetrasporus* is secondary homothallic whereas *N. crassa* and *N. sitophila* are heterothallic. The *N. crassa* and *N. sitophila* are eight (8) spored whereas *N. tetrasperma* is 4 spored.

N. sitophila is an 8 spored, bisexual and heterothallic fungus (Alexopoulos and Mims, 1979). Sexual reproduction takes place by multicellular ascogonium and conidium. The conidia act as male cells, whereas the female sex organ is represented by protoperithecium (Webster, 1980). Protoperithecium is the structure where multinucleate ascogonium with long hyphal branch behaves as trichogyne surrounded by several layer of interwoven hyphae. Protoperithecium is also called bulbil or archicarp. Macroconidia or microconidia are chemically attracted by the trichogyne and unite with it. *Spermatization* takes place not only by the microconidia but also by the macroconidia and even by the conidial germ tube that can transfer nuclei to the receptive trichogyne. Although the members are bisexual i.e. develop both ascogonia and microconidia or macroconidia, but are incompatible and cannot produce ascospores themselves. After dissolution of the common wall of trichogyne and fertilizing cell one or more nuclei migrate down the trichogyne into the oogonial cell. The development of fruit body i.e. perithecium takes place, as other typical ascomycetous members. Shortly after the migration of nuclei, ascogonial cells give rise to ascogenous hyphae, the tip of which enlarge and form elongate asci. The surrounding sterile hyphal layer increases in size and organizes itself into a perithecium with elongated neck and apical ostiole (Fig. 4.53).

The young asci are binucleate. Nuclear fusion takes place in young asci, followed by meiosis and then mitosis results the formation of 8 nuclei arranged in a linear order. Nuclei with a portion surrounding cytoplasm finally differentiated into ascospores. Each ascus contains 8 ascospores of which 4 are of one mating type and the remaining 4 are of the other mating type. The centrum of the young fruit body contains many paraphyses along with asci. The paraphyses disappear at maturity.

Ascospores are darkbrown or black, having nerve-like ridges on the outer wall. The young

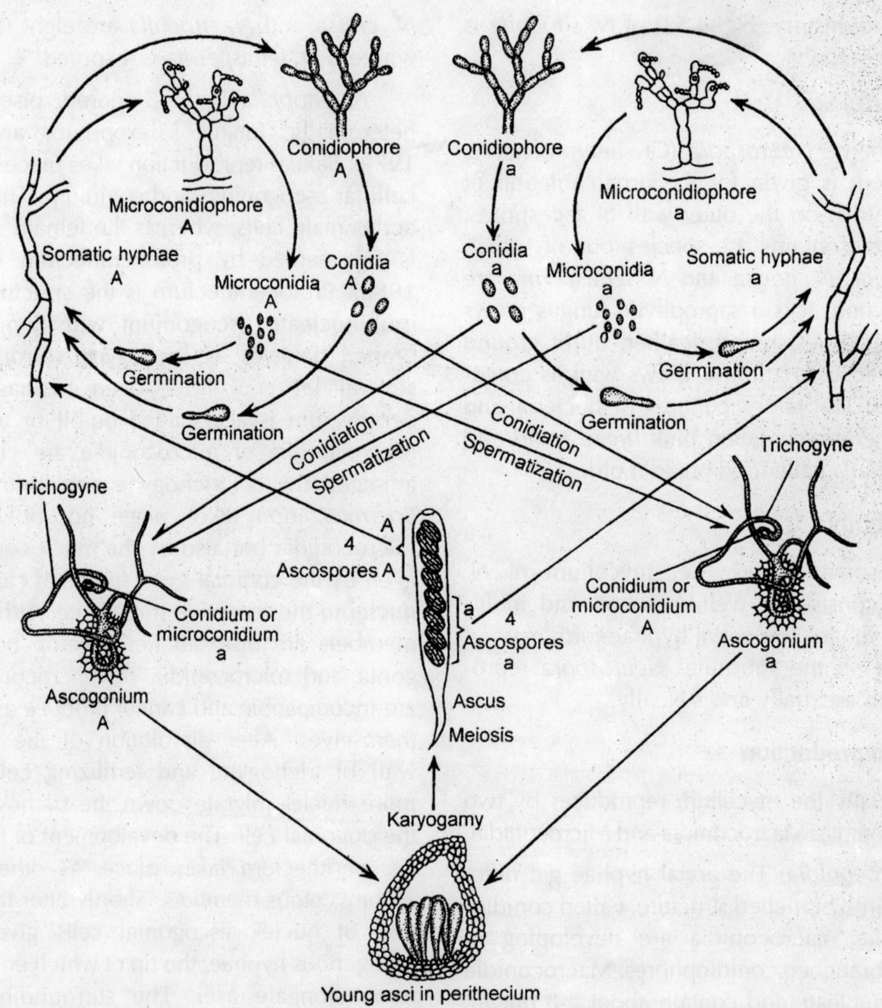

Fig. 4.53. Life cycle of *Neurospora.*

ascospores are uninucleate but become binucleate during discharge. During liberation the ascospores are shoot out from the fruit body. On germination, they produce coarse, septate, rapidly growing mycelium with multinucleate cells.

ASCOBOLUS

(Coprophilous fungi)

Systemic position

Ainsworth (1973)	Hawksworth (1995)
Kingdom Mycota	Kingdom Fungi
Division Eumycota	Phylum Ascomycota
Subsivision Ascomycotina	Class (no class)
Class Discomycetes	Order Pezizales
Ordeer Pezizales	Family Pezizaceae
Family Pezizaceae	Genus *Ascobolus*
Genus *Ascobolus*	

Most of the members of *Ascobolus* are coprophilous, grow on dungs of herbivorous animals like cow, buffalo etc., viz. *A. scatigenous* on cow dung, *A viridis* on buffalo dung and *A. minutus* on both cow and buffalo dung. *A. striisporus* grows on decaying plant materials.

Vegetative structure

Plant body is mycelial, composed of branched, septate hyphae (Fig. 4.54) with multi-

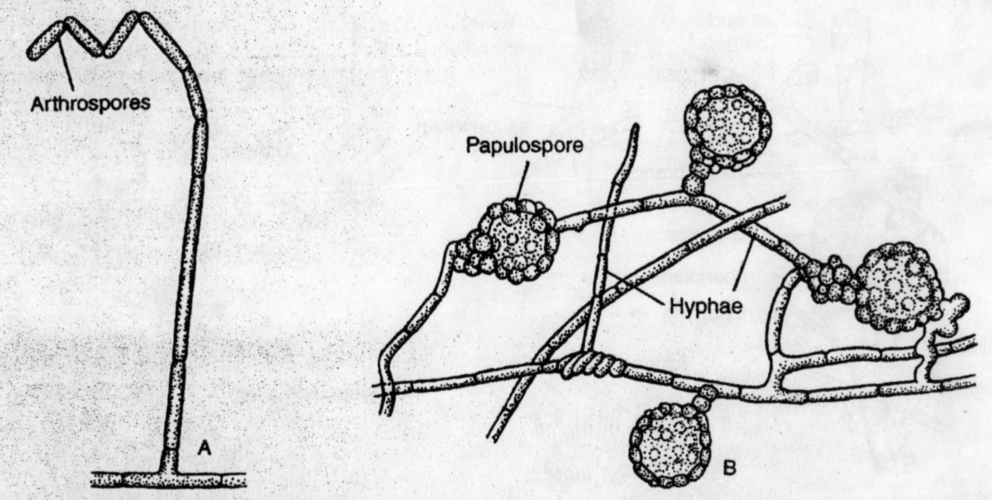

Fig. 4.54: *Ascobolus*, asexual spores: A. Arthospores (oidia) of *A. furfuraceous*; B. Papulospores of *A. scatigenus* (*A. magnificus*)

nucleate cells. The mycelia ramify profusely in the substratum and absorb nutrition by certain such hyphae.

Reproduction

Ascobolus reproduces both asexually and sexually.

Asexual reproduction: It reproduces asexually by oidia (arthospore) or papulospore formation:

Oidia (arthospores). Species like *A. stercorarius, A. denudantus* etc., produce oidia in chains on the vegetative mycelium (Fig. 4.54A) derived from an ascospore. These oidia on germination produce new mycelium, which again may develop into new oidia.

Papulospore: The papulospores are developed in species like *A. scatigenus* (Fig. 4.54B). It is the sclerotium-like structure having one or two large central storage cells covered by hyphae, and looks blister-like outgrowths on the central cell/cells.

Sexual reproduction: Sexual reproduction performs by the development of antheridia and ascogonia, the male and female sex organs, respectively (Fig. 4.55A). The species of *Ascobolus* may be heterothallic (*A. furfuraceous, A. immerses*) or homothallic (*A. crenulatus*). After sexual reproduction, it produces cup-shaped fruit body, the apothecium (Fig. 4.56).

The account of sexual reproduction (Fig. 4.55) deals with *A. scatigenus* (*A. magnificus*) was described by Dodge (1912, 1920) and Gwynne-Vaughan and Williamson (1932). *A. scatigenous* is a **monoecious** species i.e., capable of bearing both the sex organs, but are **heterothallic**. Antheridial branches are clavate or cylindrical and remain erect (Fig. 4.55A) and cut off an apical antheridium in course of time. The female branch developed by the neighbouring mycelium gradually coils the antheridium and becomes multicellular by septation (Fig. 4.55B, C, D). The multicellular ascogonial branch becomes differentiated into a terminal long septate multicellular trichogyne, a large unicellular multinucleate ascogonium and a multicellular stalk. The trichogyne elongates rapidly and coils around the antheridium (Fig. 4.55E). The apical point of the trichogyne is attached with the apical side of the antheridium. The common wall at the point of their contact dissolves and content of the antheridium passes to the ascogonium through trichogyne. The nuclei of opposite sex are situated side by side towards the peripheral region, leaving a finely granular cytoplasm towards the centre. Those nuclei are unable to pair with ascogonial nuclei and thus disintegrate in antheridium, in trichogyne or in ascogonium. After nuclear migration, the ascogonium increases in size and buds out ascogenous hyphae

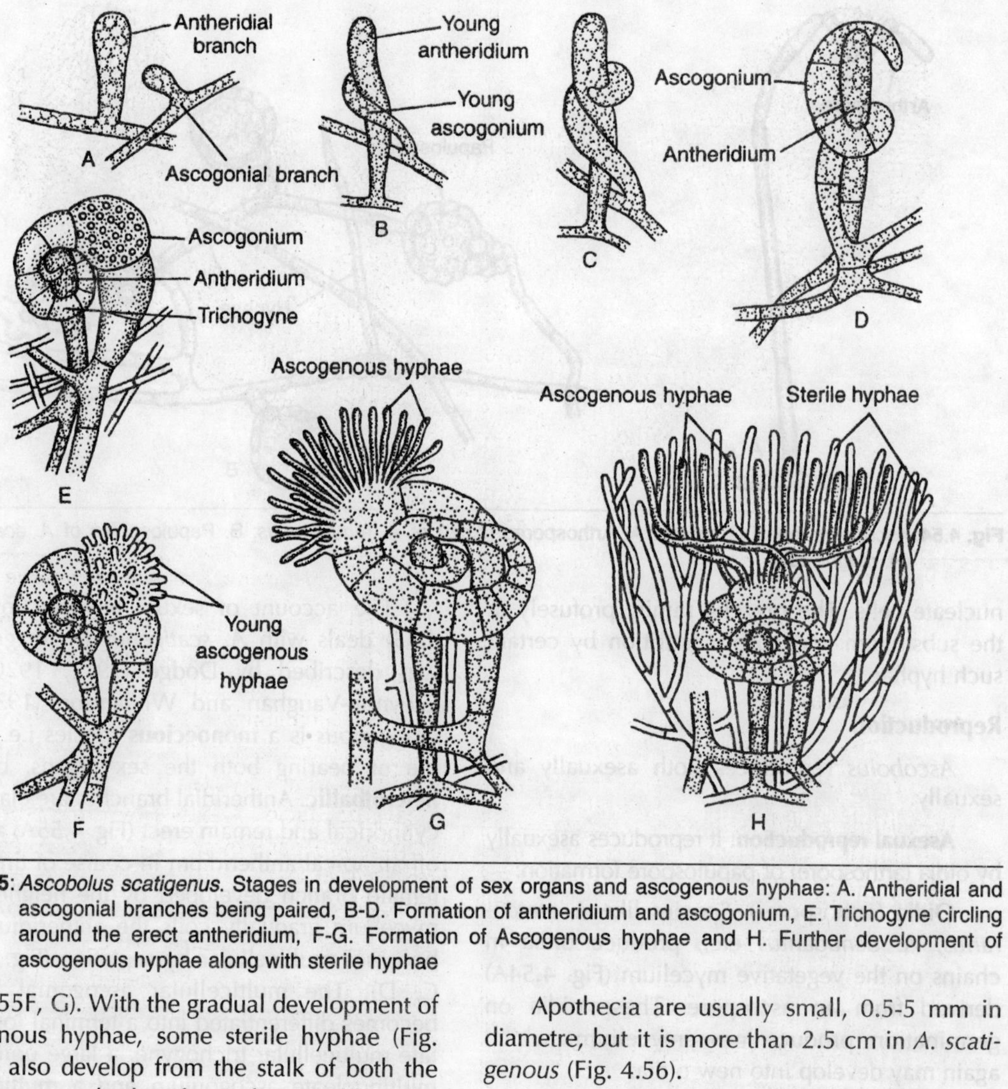

Fig. 4.55: *Ascobolus scatigenus*. Stages in development of sex organs and ascogenous hyphae: A. Antheridial and ascogonial branches being paired, B-D. Formation of antheridium and ascogonium, E. Trichogyne circling around the erect antheridium, F-G. Formation of Ascogenous hyphae and H. Further development of ascogenous hyphae along with sterile hyphae

(Fig. 4.55F, G). With the gradual development of ascogenous hyphae, some sterile hyphae (Fig. 4.55H) also develop from the stalk of both the sex organs and also from the neighbouring mycelium, and subsequently intermingle with them. Gradually they form a cup-shaped fruit body, the **apothecium** (Fig. 4.57).

Van Brummelen (1967) reported two types of ascocarp development in *Ascobolus*:

1. *Gymnohymenial* or *gymnocarpic type*. In this type, the hymenium remains exposed till the maturity of the asci, e.g., *A. magnificus*.
2. *Cleistohymenial* or *angiocarpic type*. In this type, the hymenium is enclosed at least during its early part of development, e.g., *A. immersus* and *A. furfuraceous*.

Apothecia are usually small, 0.5-5 mm in diametre, but it is more than 2.5 cm in *A. scatigenous* (Fig. 4.56).

The ascogenous hyphae grow in size and a pair of opposite nuclei migrates inside it. The ascogenous hyphae then become septate. The apical

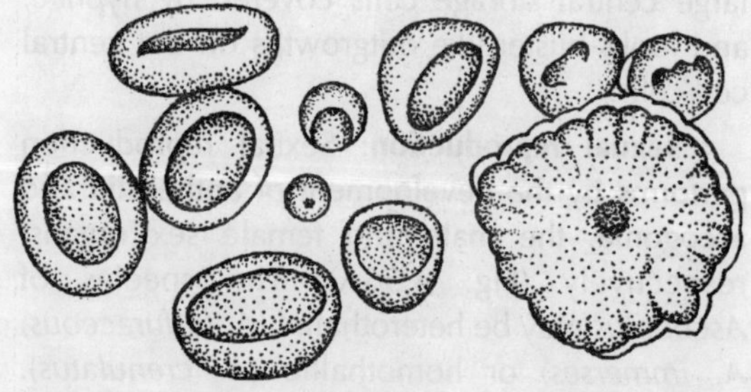

Fig. 4.56: *Ascobolus scatigenus*: Fruit bodies at different stages of maturity

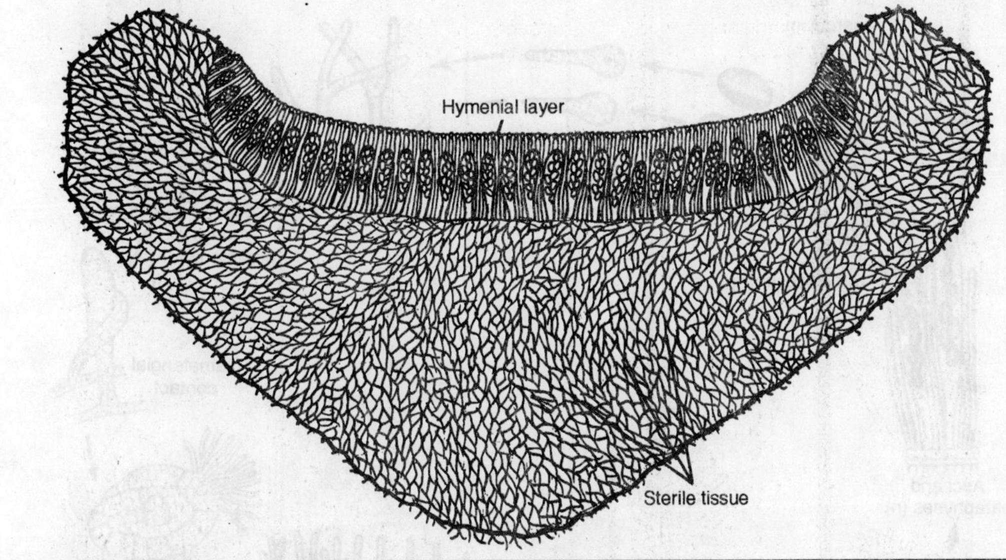

Fig. 4.57: *Ascobolus* sp.: Diagram of apothecium in section showing hymenial layer and sterile tissue

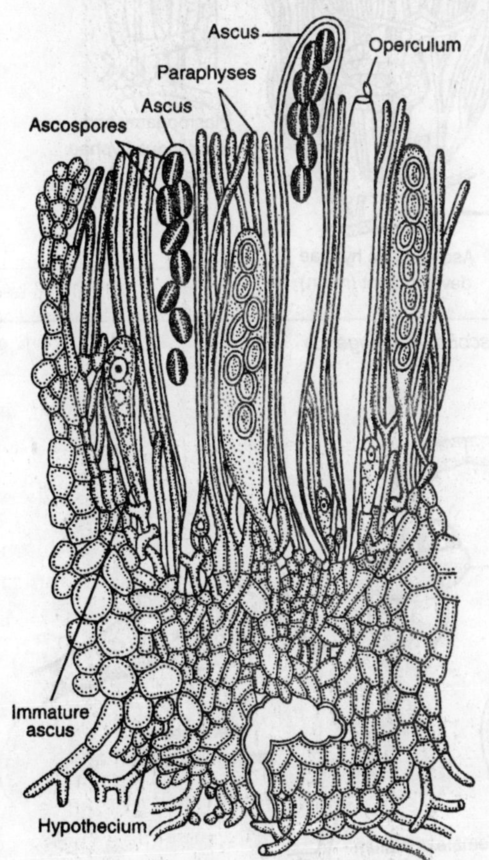

Fig. 4.58: *Ascobolus* sp.: Portion of a section of apothecium showing emergent and operculate asci, ascospores and paraphyses

cell of ascogenous hyphae forms crozier and with mitotic division of nuclei it forms three celled structure, of which the middle one, i.e. the binucleate penultimate cell, enlarges and forms an ascus which contains eight ascospores, developed sequentially by karyogamy, meiosis and mitotic division of nuclei. The asci remain intermingled with septate paraphyses in the upper region of the fruit body i.e., the hymenium (Fig. 4.57). Ascospores are liberated from ascus through the operculum. Each ascus has two cell walls.

The ascospores are one-celled, double-walled with various sculpturing on the outer wall (Fig. 4.58), consisting of warts, with longitudinal striation, ridges etc. The size of the ascospores generally ranges from 14-25 μm × 7-11 μm, but in *A. immersus* they are very large and measure approx. 70 μm × 30 μm. The ascospores are first hyaline, but becomes coloured purple brown to black.

The spores on germination (Fig. 4.59) develop new mycelium like the mother.

Indian species

Ascobolus magnificus, A. viridis, A. carbonarius, A. immersus, A. indicus, A. scatigenus etc.

Degeneration of sex

Progressive **degeneration of sex** is found in the genus *Ascobolus*. This can be traced, based on the following observations:

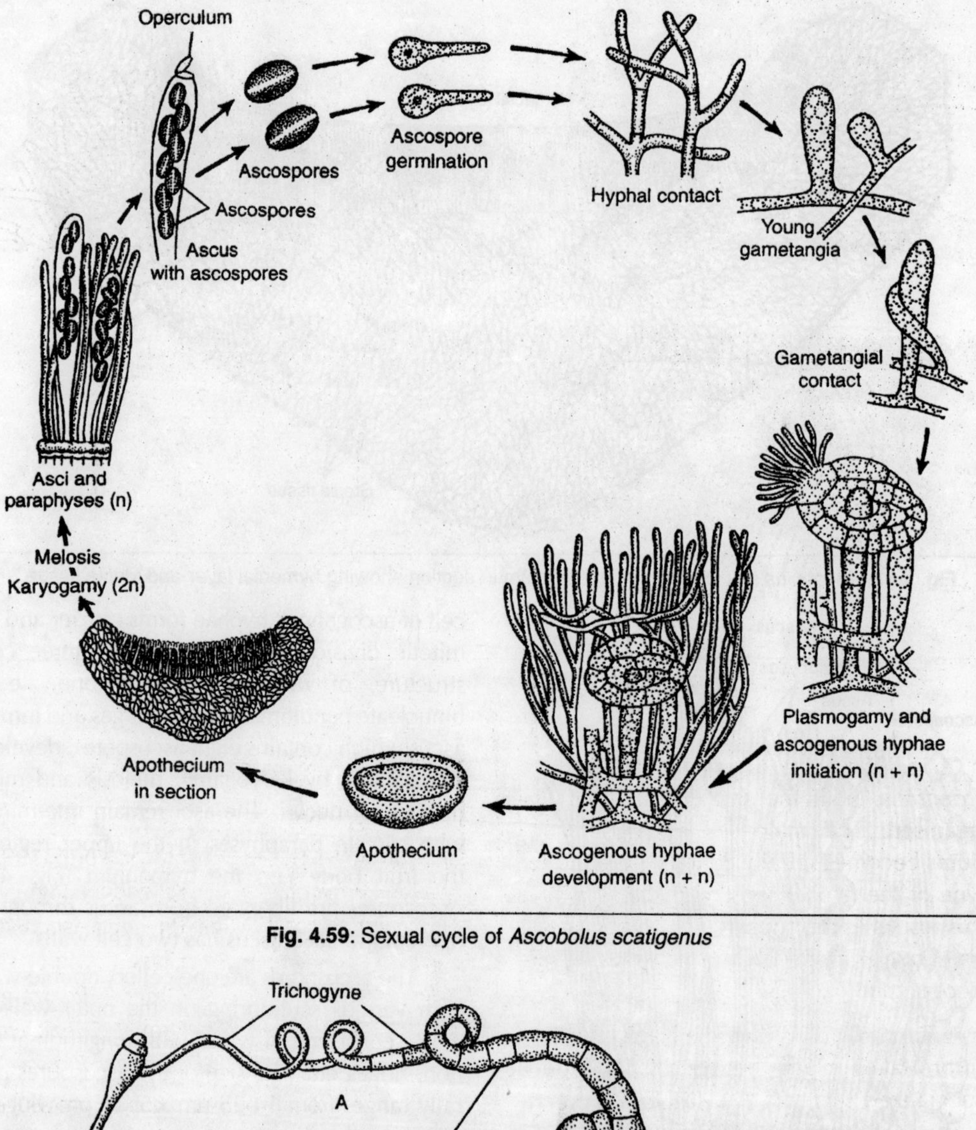

Operculum
Ascospores
Ascospore
germination
Hyphal contact
Young
gametangia
Ascospores
Ascus
with ascospores
Gametangial
contact
Asci and
paraphyses (n)
Meiosis
Karyogamy (2n)
Plasmogamy and
ascogenous hyphae
initiation (n + n)
Apothecium
in section
Apothecium
Ascogenous hyphae
development (n + n)

Fig. 4.59: Sexual cycle of *Ascobolus scatigenus*

Trichogyne
A
Ascogonium
Male conidium
Stalk
B
Female conidium
C
Fertilised
ascogonial cells

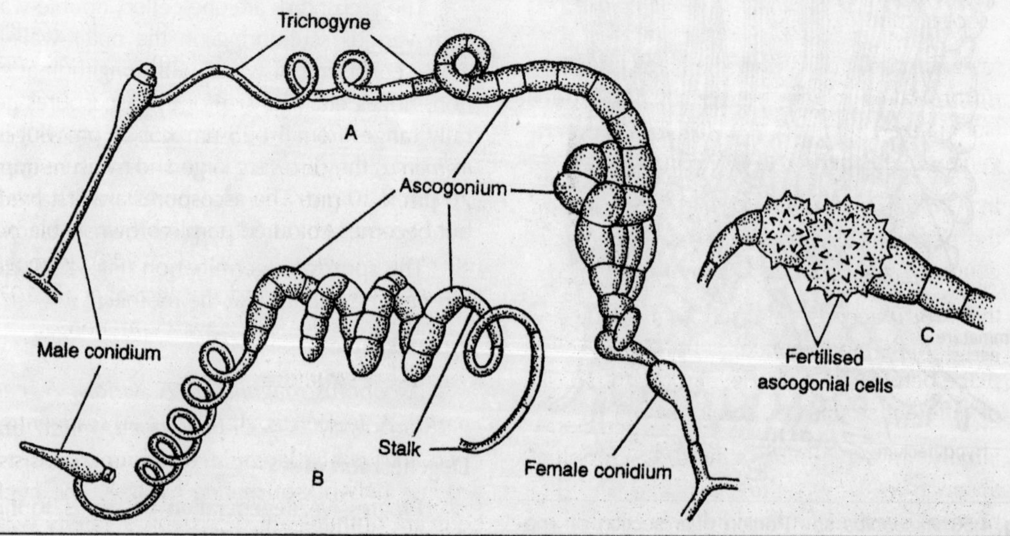

Fig. 4.60: *Ascobolus carbonarius*: A-C. Stages in sexual reproduction

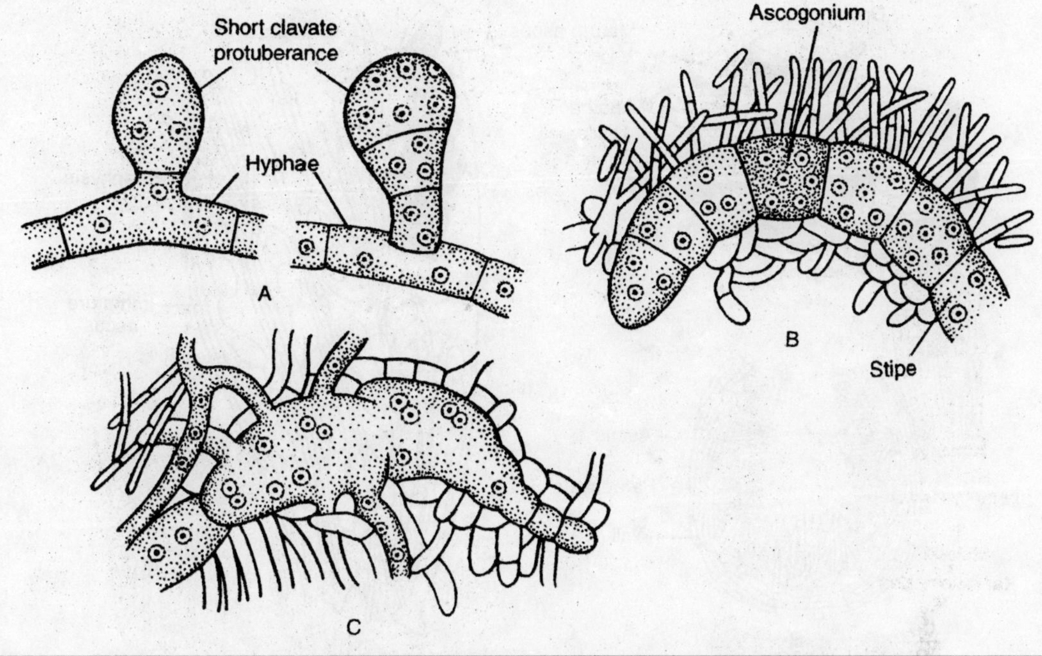

Fig. 4.61: *Ascobolus. citrinus*: A-C. Stages in sexual reproduction

1. In *A. scatigenous (A. magnificus)* and *A. strobilinus*, both the **sex organs are well-organised**. The male sex organ, antheridiums becomes spirally coiled by the trichogyne of the female sex organ, ascogonium. This results in the copulation and passage of male nuclei from the antheridium to the ascogonium (Fig. 4.55).
2. In *A. carbonarius*, antheridia are absent and union takes place between spermatium (specialized sex cell) and the tip of trichogyne, called **spermatisation** (Fig. 4.60).
3. In *A. citrinus*, the antheridia are absent and the oogonial nuclei alone pass to the ascogenous hyphae, called **autogamy** (Fig. 4.61).
4. In *A. furfuraceous,* in absence of sex organs (both male and female), copulation takes place between hyphae developed from oidia of different strains i.e., **somatogamy**.
5. Finally, in *A. equinus*, there is complete absence of any sexual act and the life process involves an haploid phase only i.e., **apomixis**.

PEZIZA

Systemic position

Ainsworth (1973)		Hawksworth (1995)	
Division	Eumycota	Kingdom	Fungi
Subdivision	Ascomycotina	Phylum	Ascomycota
Class	Discomycetes	Class	(no class)
Order	Pezizales	Order	Pezizales
Family	Pezizaceae	Family	Pezizaceae
Genus	*Peziza*	Genus	*Peziza*

Peziza is a saprophytic fungus, grows generally on dung, decaying wood and rich humus of forest soil. Some species also grow on charred or burnt wood. About 50 species has so far been reported (Hawksworth *et al.*, 1983). About 25 species are found in India (Bilgrami *et al.*, 1979), of which *P. vesiculosa* is very common.

Vegetative structure

Plant body is well-developed, much branched, frequently perennial mycelium, consists of a dense network of septate hyphae. The hyphal cells are uninucleate. The hyphae ramify within the substratum to draw nourishment.

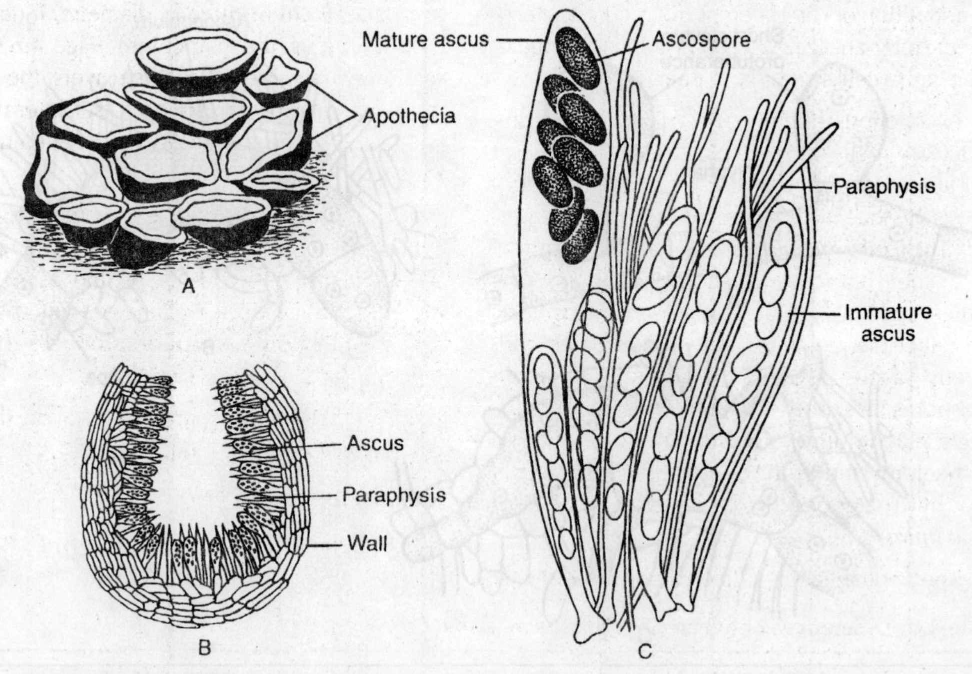

Fig. 4.62: *Peziza*: A. Apothecia in groups, B. V.S. of an apothecium, and C. Portion of hymenium showing asci and paraphyses

Reproduction

Peziza reproduces by asexual and sexual methods.

Asexual reproduction: It takes place by conidia and chlamydospores:

Conidia: Conidial stage is absent in most species, but are known to occur in *P. rapanda, P. fuckeliana, P. vesiculosa* and *P. pustulata*. Conidiophores are short and developed from the prostrate mycelium. The tip of the developing conidiophore swells up into vesicles which bears large number of conidia. Conidia are elliptical and hyaline to light coloured. Conidium germinates to form new mycelium.

Chlamydospores: Chlamydospores are thick walled resting spore, developed intercalary on the mycelium, either singly or in series. Under suitable condition, they germinate and produce new mycelia.

Sexual reproduction: Sex organs are lacking in *Peziza*. Sexual reproduction takes place either by somatogamous copulation i.e., the union between vegetative cells of the same filament

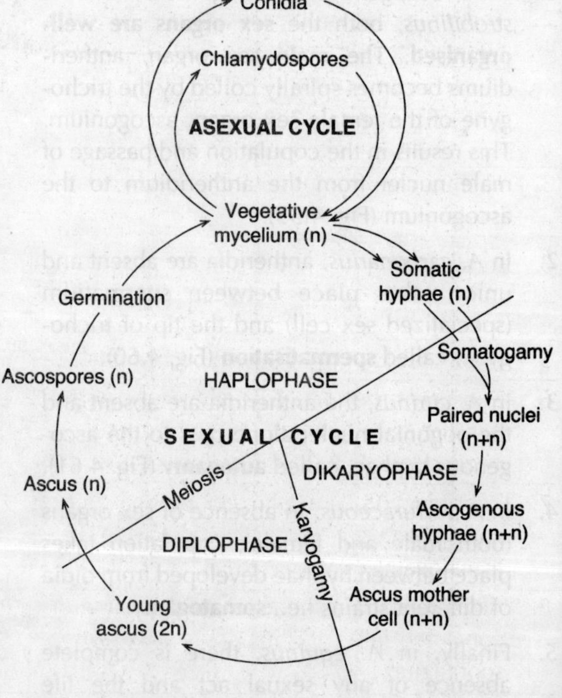

Fig. 4.63: Schematic representation of Life cycle of *Peziza*

(homothallic) or of different filaments (heterothallic) or by autogamous pairing i.e., the nuclei of the same cell arranged in pairs.

Ascogenous hyphae arise directly from the binucleate cell and develop asci directly at their apices (*P. vesiculosa, P. tectoria*). The ascogenous hyphae become many celled by septation. The cells are binucleate and the terminal cell or their branches function directly as ascus mother cell. Karyogamy, meiosis and mitosis take place sequentially in ascus mother cell, thereby ascus is formed containing eight ascospores. The surrounding vegetative hyphae weave a peridium around the asci and form an apothecium, the fruit body. Development of ascus by crozier or hook formation is reported in *P. cavinum*.

The apothecia are saucer-shaped, subsessile to sessile, 5 cm or more in diametre. Inner layer of the cup is the hymenium (Fig. 4.62A). In sectional view, it exhibits three layers: the uppermost hymenium consisting of asci intermingled with paraphyses; the middle sub-hymenium and the outer peridium (Fig. 4.62B).

The asci are elongated, containing eight spores (Fig. 4.62C). Ascospores are placed obliquely on one row inside the ascus. They are elliptical, smooth, hyaline or light coloured. On germination ascospore produces haploid mycelium.

The schematic representation of the life cycle is given in Fig. 4.63.

Indian species

Peziza repanda, P. subcupularis, P. succosa, P. vesiculosa, P. multigullulata etc.

Table 4.8: Differences between *Chytridiomycota* and *Ascomycota*

Chytridiomycota	*Ascomycota*
1. Members are smallest, simplest and most primitive fungi.	1. Members are elaborate, much complex and much advanced fungi.
2. Thallus is coenocytic with either oval multinucleate cell or small elongated hyphae.	2. Thallus is much branched and septate mycelium consists of uni- or multinucleate cells having septa with simple central pore with spherical **Woronin bodies** on each side of the pore (unicellular in yeast).
3. Thallus may be endobiotic, epibiotic or interbiotic.	3. Thallus is commonly saprophytic but many are pathogenic i.e endobiotic.
4. Cell wall is chitinous.	4. Cell wall composed of glucan and mannans (unicellular form) or chitin and glucan (multicellular form).
5. Presence of flattened mitochondrial cristae.	5. Mitochondrial cristae is absent.
6. Zoospores are motile having single posterior whiplash flagellum. Flagellar apparatus of zoospore have a pair of kinetosome.	6. Absence of motile cells in the life cycle.
7. Reproduction takes place by planogametic copulation, gametangial copulation and somatogamy.	7. Reproduction takes place by gametangial copulation, hologamy, gametangial contact, autogamy, spermatization and somatogamy.
8. After sexual reproduction, the zygote is converted into a resting spore or a resting sporangium.	8. Ascus and ascospores are developed after sexual reproduction.
9. Fruit bodies are not develop during or after sexual reproduction.	9. Commonly develop fruit body of different types towards the end of sexual reproduction.

MITOSPORIC (ANAMORPHIC) FUNGI

(Fungi without sexual reproduction, not known or does not takes place)

Mitosporic fungi are also called earlier as *Fungi imperfecti* or *Waste box fungus*, and all are placed under the class Deuteromycotina includes about 600 genera and more than 20,000 species. This group includes those fungal members where sexual or perfect stage does not takes place or not known to us and only the asexual or imperfect stage is recorded, where reproduction usually takes place by means of conidia (mitospore). Due to lack of knowledge on sexual reproduction they could not be placed under the well-defined groups of fungal phyla such as *Chytridiomycota, Neocallimastigomycota, Blastocladiomycota, Zygomycota, Ascomycota* or *Basidiomycota* and have been kept in a distinct group Anamorphic fungi (earlier Deuteromycotina). The group has been created for the temporary assemblage of fungal members, to place them in proper phylum after gaining knowledge on their perfect stage.

Possibly, almost all the fungal members kept under Anamorphic fungi may be the members of *Ascomycota*, evidenced by the similarity between the conidial stage of Anamorphic fungi and *Ascomycota*. They differ from *Chytridiomycota, Neocallimastigomycota, Blastocladiomycota* and *Zygomycota* due to the absence of aseptate mycelium and from *Basidiomycota* due to lack of clamp connection at the septa. It is thought that the members of Anamorphic fungi (earlier Deuteromycotina) are mostly the members of *Ascomycota* where either sexual reproduction did not evolve or lost in course of evolution. After gaining knowledge and finding similarity with perfect i.e. sexual stage they have to be shifted to *Ascomycota* or more rarely to *Basidiomycota*.

C. T. Ingold called the Sub division Deuteromycotina (present Anamorphic fungi) as *"dust bin"* group because of their doubtful systematic position. Alexopoulos and Mims (1979) considered Deuteromycotina (present Anamorphic) as "conidial stages of Ascomycetes or more rarely Basidiomycetes whose sexual stages have not been discovered or do not exist".

Hawksworth *et al*. (1995) pointed out that even after removal of anamorphs to their telemorphs, contain 2600 genera and about 15000 species and recommended as the second largest group of fungi. The important features of the group are given below:

Important characteristics

1. Members of this group are commonly terrestrial but many are aquatic and grow both in fresh water and marine habitat. Members are

Table 4.9: Perfect stages of some common members of Mitosporic fungi

Imperfect state	Perfect state	Phylum
Aspergillus nidulans	*Emericella nidulans*	*Ascomycota*
Fusarium moniliformae	*Gibberella fujikuroi*	"
Geotrichum candidum	*Endomyces geotrichum*	"
Helminthosporium gramineum	*Pyrenophora graminea*	"
H. oryzae	*Cochliobolus miyabeanus*	"
H. sigmoideum	*Leptosphaeria salvinii*	"
H. teres	*Pyrenophora teres*	"
Histoplasma capsulatum	*Emmonsiella capsulata*	"
Penicillium vermiculatum	*Talaromyces vermiculatus*	"
Tubercularia vulgaria	*Nectria cinnabarina*	"
Cryptococcus neoformans	*Filobasidiella neoformans*	*Basidiomycota*
Rhizoctonia solani	*Thanatephorus cucumeris*	"

either saprophytic or parasites, causing diseases of both plants and animals.

2. Except a few unicellular members, somatic body is mycelial and consists of profusely branched and septate hyphae with perforated septate cells are commonly multinucleate.
3. In parasitic members, the hyphae grow both inter- and intracellularly.
4. They reproduce only by asexual means and *sexual reproduction does not takes place or not known to us.*
5. The asexual reproduction takes place commonly by conidia or blastospores, chlamydospores or arthospores. Conidia are developed exogenously on conidiophores. Conidiophores are either free or grouped togather to form special structures like **sporodochia, synnemata, pycnidia** or **acervuli**.
6. Many genera show genetic variation, through heterokaryosis and parasexuality.

Perfect stage along with the name of some members of Mitosporic fungi is given in Table 4.9.

Classification

Ainsworth (1973) classified the subdivision Deuteromycotina into three classes: Blastomycetes, Hyphomycetes and Coelomycetes. Alexopoulos and mims (1979) preferred to name three form-subclasses under Deuteromycotina as Blastomycetidae, Hyphomycetidae and Coelomycetidae. Webster (1980) discussed the members of Deuteromycotina under three distinct ecological groups, viz. Aquatic Fungi Imperfecti, Predaceous Fungi Imperfecti, and Seed borne Fungi Imperfecti, Later, Hawksworth *et al.*(1983) recognized only two classes: Hyphomycetes and Coelomycetes.

In recent year, Hawksworth *et al.* (1995) and Kirk *et al.*(2008) discarded the traditional hierarchial divisions into orders and families as it is useless in terms of phylogenetic classification through molecular aids. The new concept is now accepted universally and they have studied the genera only in alphabetical sequence in: (i) Imperfect yeasts, (ii) Hyphomycetous Anamorphic fungi, (iii) Acervular Anamorphic fungi and (iv) Pycnidial Anamorphic fungi.

Three categories of information are considered for the description of this group of fungi. These are conidiomatal types, as mentioned in Table 4.10 and 4.11, and also Saccardo's groups and conidiogenous events.

Table 4.10: Conidiomatal types and conidial spore groups used in description of this group of fungi

Conidiomata
1. Hyphal
2. Synnematal
3. Sporodochial
4. Pycnidial
5. Pycnothyrial
6. Acervular
7. Cupulate
8. Stromatic (complex conidiomata excluded from other categories)
9. Sclerotic

Table 4.11: Conidial shape and septation (Saccardo's groups)

			Conidia hyaline or Bright (hyalo-)	Conidia pigmented or dark (phaeo-)
1.	1-celled (not E, F or G)	amerosporae	Hyalosporae	Phaeosporae
2.	2-celled	didymosporae	Hyalodidymae	Phaeodidymae
3.	2-many celled (multiseptate)	phragmosporae	Hyalophragmae	Phaeophragmae
4.	Muriform	dictyospore	Hyalodictyae	Phaeodictyae
5.	Filiform	scolecosporae		
6.	Helical	helicosporae		
7.	Branched	staurosporae		

Imperfect yeasts

Yeast cells exist as budding cells or as pseudomycelium. The buds may detach passively or may discharge explosively like ballistospores of Basidiomycotina, e.g., *Candida. Candidiasis, Cryptococcus* etc.

Hyphomycetous Anamorphic Fungi

Members of this group form spores or sporophores on vegetative mycelium, either remain free or aggregated, those form sporodochia or synnema. Members mostly grow in terrestrial habitat, but some are in aquatic environment. The group contains about 1,700 genera and 1,100 species (Hawksworth *et al.* 1995), e.g., *Alternaria, Botrytis, Cercospora, Fusarium, Helminthosporium, Monilia* etc.

Acervular Anamorphic Fungi

Acervuli are the flat or saucer-shaped fruit body, having cushion-like flat mass of hyphae from which closely packed short conidiophores develop. The acervuli are neither surrounded by any hyphal wall nor do they possess any definite ostiole, e.g., *Colletotrichum, Pestalotiopsis, Sphaceloma* etc.

Pycnidial Anamorphic Fungi

Pycnidia are globular or flask-shaped structure developed on living or dead tissues. Commonly they are glabrous, but some have brown setae. They contain many conidiophores lined on the inner side. Each pycnidium usually has an opening at the apex, the ostiole. Pycnidia of some members of Coelomycetes have many ostioles, e.g., *Ascochyta, Botryodiplodia, Phyllosticta, Phoma, Phomopsis,* etc.

Economic Importance

The mitosporic fungi have great economic importance. Some members caused different plant diseases, others cause several diseases of human beings, and still others are very useful in controlling a number of soil borne plant diseases.

1. *Plant pathogens:* A number of members causes diseases of different plants. These include leaf spot (blast of rice, *Pyricularia oryzae*; leaf spot or tikka disease of ground nut, *Cercospora personata*), blight (early blight of potato, *Alternaria solani*), blotch (leaf blotch of wheat, *Septoria tritici*), wilt (wilt of Pigeon pea, *Fusarium udum*), anthracnose (anthracnose of mango, *Colletotrichum gloeosporioides*) etc.
2. *Human Pathogens:* Animals including human beings are affected by different pathogens causing different diseases. These are *ringworm* (*Epidermophyton floccosum* and *Microsporum audowinii*), athlete's foot (*Trichophyton gypseum* and *T. purpurium*), monilasis (*candida albicans*) etc.
3. *Biological control*. Several members are able to control nematode diseases by trapping the nematodes with their sticky mycelium or loops to which nematodes adhere or trap and then the other hyphae invade and digest them. Thus, these members help in reducing the disease by killing the nematode, e.g. *Arthrobotrys oligospora, Dactylella cionopaga*.

Some common genera, *Cercospora, Fusarium* and *Colletotrichum* are discussed below:

CERCOSPORA

Systemic position

Ainsworth (1973)	Hawksworth (1995)
Kingdom Mycota	Kingdom Fungi
Division Eumycota	Mitosporic (Anamorphic)
Subdivision Deuteromycotina	Fungi
Class Hyphomycetes	Group Hyphomycetous
Order Moniliales	(Anamorphic) Fungi
Family Dematiaceae	Genus *Cercospora*
Genus *Cercospora*	

The genus includes about 2,000 species and many of them are reduced to synonyms or transferred to more appropriate genera by various authors. Vasudeva (1963) described 260 species of *Cercospora* in his monograph on Indian *Cercospora*. Mostly they are weak parasites, but occasionally they cause serious diseases of plants and a few are pathogenic to man. A list of some pathogenic species is given in Table 4.12.

Vegetative body

The mycelial plant body consists of much branched and septate hyphae. They are hyaline when young, but become greenish-brown with age. The hyphae grow intercellularly and obtain nutrition by invading branched haustoria in the mesophyll cells.

Reproduction

It reproduces only asexually by conidia.

Table 4.12: List of some pathogenic species and diseases caused by *Cercospora*

Name of some pathogenic species	Disease
Plant diseases	
1. *Cercospora personata* and *C. arachidicola*	Tikka disease of groundnut.
2. *C. musae*	Leaf spot (Sigatoka disease) of banana.
3. *C. beticola*	Leaf spot of beet and spinach.
4. *C. coffeicola*	Leaf spot of coffee.
5. *C. apii*	Leaf spot of celery.
6. *C. nicotianae*	Frog eye-spot of tobacco.
7. *C. indica*	Leaf spot of Cajanus cajan.
Human disease	
1. *C. apii*	Cercosporomycosis.

The internal vegetative mycelium becomes aggregated and forms a mass of pseudoparenchymatous tissue i.e., stroma under the epidermis (Fig. 4.64A). A tuft of short, septate, thin-walled unbranched conidiophores develops from the internal stroma (Fig. 4.64A, B). Conidia are multicellular, long, filiform, hyaline or pigmented (Fig. 4.64C, D). They are borne in acropetal succession on simple or sympodially extending conidiophores. Conidia are grey to brown in colour. They disperse by wind or rain. During germination each conidium develops one or more germ tube, which gives rise to new mycelium.

Perfect stage of some are known as *Mycosphaerella*. The perfect stage of *C. cerasella* is *M. cerasella* and for *C. personata* is *M. berkeleyii*.

FUSARIUM

Systemic position

Ainsworth (1973)	Hawksworth (1995)
Kingdom Mycota	Kingdom Fungi
Division Eumycota	Mitosporic (Anamorphic)
Subdivision Deuteromycotina	Fungi
Class Hyphomycetes	Group Hyphomycetous
Order Moniliales	(Anamorphic) Fungi
Family Yuberculeriaceae	Genus *Fusarium*
Genus *Fusarium*	

The genus includes a large number of species. Most of the species are saprophytes and grow on various substrates. Some species are facultative parasite, either in mild or severe forms. The parasitic species are commonly soil-borne. Different species cause diseases like wilt, damping off of seedling, root-rot, fruit rot, rots of vegetables etc. A list of some pathogenic species is given in Table 4.13.

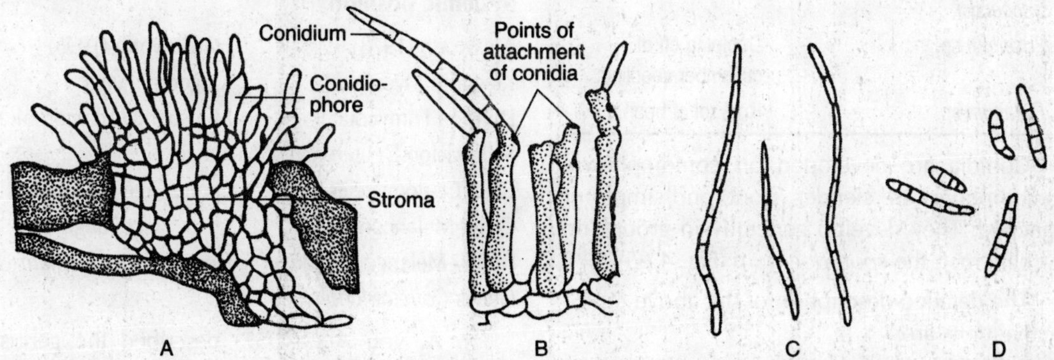

Fig. 4.64: *Cercospora*: A. V.S. of leaf showing conidiophores of *C. personata*, B. Acropetal production of conidia on sympodially extending conidiophores in *Cercospora*, C. Conidia of *C. personata*, and D. Conidia of *C. arachidicola*

Vegetative body

The plant body is mycelial and composed of much branched and septate hyphae. The hyphal cells are multinucleate. The hyphae are hyaline when young, but becomes pink, yellow or purple with maturity. In culture, they grow profusely in the form of cottony mass. In host, they grow extensively, ramify the xylem vessel and tracheid, and completely or partially plug their lumen. (As a result the conduction of water becomes hampered and the plants become wilted and died).

Reproduction

Reproduction is mainly by asexual means, by the formation of three kinds of spore. These are *microconidia, macroconidia* (Fig. 4.85) and *chlamydospores*. In addition, *sclerotia* are also formed to perform vegetative reproduction.

Table 4.13: List of some pathogenic species and diseases caused by *Fusarium*

Name of some Pathogenic species	Disease
1. *Fusarium udum* (Perfect stage *Gibberella indica*)	Wilt of pigeon pea.
2. *F. oxyporum* f. sp. *vasinfectum*	*Fusarium* wilt of cotton.
3. *F. oxysporum* f. sp. *lini*	Wilt of linseed.
4. *F. oxysporum* f. sp. *cubense*	*Fusarium* wilt of banana.
5. *F. sacchari*	Wilt of sugarcane.
6. *F. oxysporum* f. sp. *ciceri*	Wilt of gram.
7. *F. oxysporum* f. sp. *lycopersici*	Wilt of tomato.
8. *Fusarium* sp.	Damping off of cucumber seedling.
9. *Fusarium* sp.	Root rot of bean.

Conidia are developed on conidiophores. Conidiophores are slender, stout, and simple or branched. They develop generally in groups on mycelial mat, the *sporodochium* (Fig. 4.65).

The detailed description of the above reproductive units are:

1. *Microconidia*. They are very small, hyaline, oblong to ovoid, one or two celled, singly or in chain on simple or branched conidiophores. On germination they develop new mycelium like their parent.

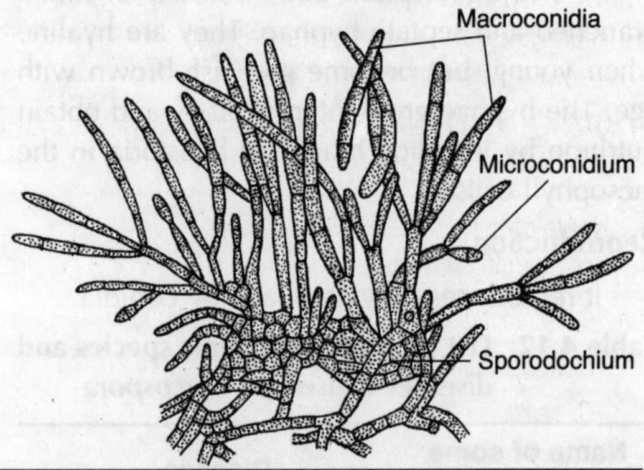

Fig. 4.65: Sporodochium of *Fusarium* sp.

2. *Macroconidia*. They are large, hyaline, slightly curved, multicellular (generally with 3-5 septate), sickle- or crescent-shaped. On germination they develop new mycelium.
3. *Chlamydospores*. They develop during unfavourable condition. The chlamydospores are round or oval, thick-walled cells formed in the hyphae. They develop either terminal or intercalary in position. During favourable condition, chlamydospore germinates and develops into a new mycelium like a parent.
4. *Sclerotia*. The mycelia often become aggregated and form compact resting bodies are called sclerotia. They function as an organ of storage and myltiply by 'vegetative propagation'.

COLLETOTRICHUM

Systemic position

Ainsworth (1973)	Hawksworth (1995)
Kingdom Mycota	Kingdom Fungi
Division Eumycota	Mitosporic (Anamorphic)
Subdivision Deuteromycotina	Fungi
Class Coelomycetes	Group Acevular Imperfect Fungi
Order Melanconiales	
Family Melanconiaceae	Genus *Colletotrichum*
Genus *Colletotrichum*	

J. A. von Arx (1957) described the genus *Colletotrichum* and based on the cultural characteristics, he recognised only 11 species. Later Sutton (1980) distinguished and described 22 species based on cultural characteristics. Different species of *Colletotrichum* cause anthracnose of many crops, therefore commonly called anthracnose fungus. A list of some pathogenic species is given in Table 4.14.

Table 4.14: List of some pathogenic species and diseases caused by *Colletotrichum*

Name of the pathogenic species	Disease
1. *Colletotrichum falcaturn*	Red rot of sugarcane.
2. *C. Coffeanum*	Coffee berry disease.
3. *C. capsici*	Ripe fruit rot & die back of chilli.
4. *C. orbiculare*	Anthracnose of *Cucumber*.
5. *C. dematium*	" " spinach.
6. *C. higginsianum*	" " *Brassica*.
7. *C. musae*	" " *Musa*.
8. *C. truncatum*	" " ground nut.
9. *C. lindemuthianum*	" " *Phaseolus*.
10. *C. gloeosporioides*	" " *Citrus*.

Vegetative body

The mycelial plant body consists of much branched septate hyphae. They are usually hyaline when young and with age, the hyphal wall becomes thicker and darker. Hyphae grow both inter- and intracellularly inside the host tissue.

Reproduction

It takes place by conidia, chlamydospores and sclerotia.

Conidia. It develops in the acervulus (Fig. 4.66A). The acervulus is saucer-shaped; either epidermal, subepidermal or subcuticular and may be separated or remains very close to each other. Conidia are developed on conidiophores. The conidiophores are short, septate, smooth, hyaline to brown coloured, unbranched or branched at the base, formed on the upper side of the stromatic surface (Fig. 4.66B). Bunch of conidiophores are surrounded or intermixed with long, dark brown, rigid, smooth, thick-walled, septate or aseptate structures, with tapered end, called *setae* (Fig. 4.66A).

The conidia are aseptate (one-celled), hyaline, cylindrical, slightly curved or lunate (sickle-shaped) with rounded ends (Fig. 4.66A, B). The conidia are dispersed by rain or wind. They are short-lived. On germination, each conidium develops one to four (generally two) germ tubes to form the mycelium (Fig. 4.66C). In contact with host surface, the mycelial tip swells to form apresorium and infects the host by infection peg developed from the lower side.

Chlamydospores. During unfavourable condition, intercalary or terminal chlamydospores are formed by septation of vegetative mycelium. The chlamydospores germinate in favourable condition.

Sclerotia. Sometimes sclerotia are produced in culture. They are brown to black in colour. On germination they develop mycelia.

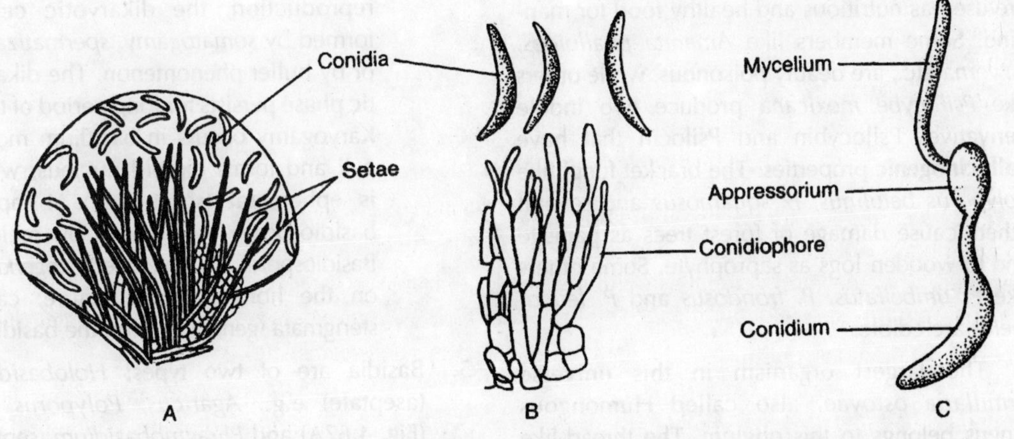

Fig. 4.66: *Colletotrichum*: A. An acervulus, B. Conidiophores and conidia, C. Germination of conidium

PHYLUM BASIDIOMYCOTA

(Club Fungi)

The members includes in the phylum Basidiomycota are characterized by the formation of basidiospores exogenously on the basidium (Pl. basidia: Gr. *basidion*, a small base) and absence of flagellated structure. The phylum included 500 genera with more than 16,000 species. The members include mushrooms, puff balls, toad stools, bracket fungi and some microscopic rusts, smuts and also some yeast.

Rust (*Puccinia* spp.) and smuts (*Ustilago* spp.) caused diseases of crop plants resulting heavy financial losses in different parts of the world. Some yeast, cause diseases of human beings. One of these yeasts, ***Sporobolomyces roseus***, is very common on moribund leaf surfaces and caused allergy because its basidiospores are **respiratory allergens**. Another one is ***Cryptococcus neoformans***, which grows commonly on old "weathered" bird droppings, and cause a fatal systemic infection of immunocompromised people. Its air-borne basidiospores initiate infection via the lungs, leading to the disease termed **cryptococcosis**. This is now the fourth most important life-threatening diseases of AIDS patients in the USA, and also common in Europen countries.

Mushrooms [*Volvariella volvacea, Agaricus brunnescens* (syn. *A. bisporus*), *A. campestris*] are used as nutritious and healthy food for mankind. Some members like *Amanita phalloides, A. verna* etc., are deadly poisonous, while others like *Psilocybe mexicana* produce two indole derivatives Psilocybin and Psilocin that have **hallucinogenic** properties. The bracket fungi like *Polyporus betulinus, P. squamosus* and several others cause damage of forest trees as parasite and to wooden logs as saprophyte. Some others like *P. umbellatus, P. frondosus* and *P. leucomelas* are edible.

The largest organism in this universe *Armillaria ostoyae*, also called Humongous fungus belongs to this phylum. The thread-like somatic body covers 2,385 acres (almost 4 square miles) of the Malheur National Forest in the Blue Mountains of Oregon, and is about 2400 years old.

The **important characteristics** of Basidiomycota are:

1. Presence of well-developed, branched and septate mycelium having simple (e.g., Ustilaginales and Uredinales) or dolipore (e.g., Auriculariaceae, Aphyllophorales and Agaricales) septum (Fig. 4.2B & 4.2D).
2. The mycelial cells may contain one nucleus, called monokaryotic i.e., primary mycelium or two nuclei, called dikaryotic i.e., secondary mycelium. The secondary mycelia may organize and form fruit body, called tertiary mycelium.
3. The cell wall is mainly composed of chitin and glucans.
4. Reproduction:

   (a) Vegetative reproduction takes place by budding and fragmentation.

   (b) Asexual reproduction takes place by *conidia, oidia*, or *chlamydospores* that are absent in some higher taxa of this subdivision.

   (c) Sex organs are absent. During sexual reproduction, the dikaryotic cell is formed by *somatogamy, spermatization* or by buller phenomenon. The dikaryotic phase persists for long period of time. Karyogamy occurs in basidium mother cell and forms diploid nucleus, which is ephemeral (short-lived). 4-haploid basidiospores are formed by meiosis. Basidiospores are developed *exogenously* on the horn-shaped structure, called sterigmata (generally 4) on the basidium.
5. Basidia are of two types: *Holobasidium* (aseptate) e.g., *Agaricus, Polyporus* etc. (Fig. 4.67A) and *Phragmobasidium* (septate) e.g., *Puccinia, Ustilago* (Fig. 4.67B) etc.

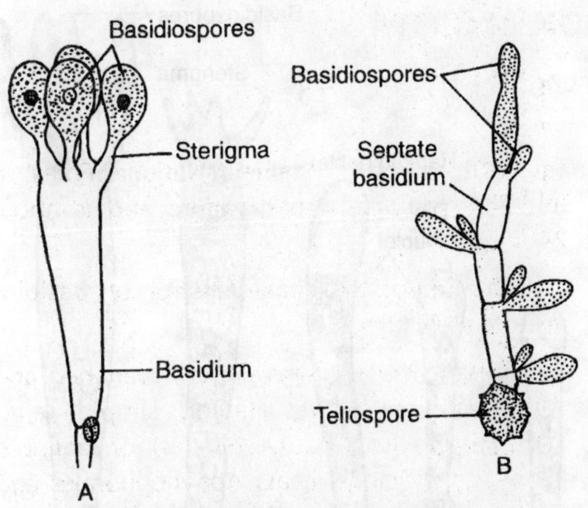

Fig. 4.67: Types of basidia: A. Holabasidium (*Agaricus*) and B. Phragmobasidium (*Ustilago*)

6. Except in lower forms (*Puccinia, Ustilago*), secondary mycelia by aggregation form fruit body, called basidiocarp (*Agaricus, Polyporus* etc.). It has the spore bearing structure, called basidium. The number of spores on each basidium is commonly four, but two or more than four are also reported.

Growth of mycelium by cell-division

(Phenomenon of dikaryotization)

The secondary mycelium has cells with two nuclei of opposite mating (Fig. 4.68A). The apical cell of secondary mycelium divides by the formation of lateral projection known as *Clamp connection*. During cell division a short backwardly directed outgrowth develops from lateral wall, known as *clamp*.

Out of two nuclei, one migrates towards the clamp (marked as hollow) and the other one (marked as solid) remains in the cell (Fig. 4.68B). Both the nuclei then undergo mitotic division (Fig. 4.68C) forming four nuclei. Out of two (marked as hollow) nuclei, one remains in clamp and the other goes back to the apical region of the cell. The remaining two (marked as solid) nuclei in the cell are distributed towards apical and basal region. Thereafter, two septa are developed in such a way that one septum separates the clamp and the other develops at the bottom level of clamp outgrowth, results in the formation of three cells; *uninucleate clamp, uninucleate*

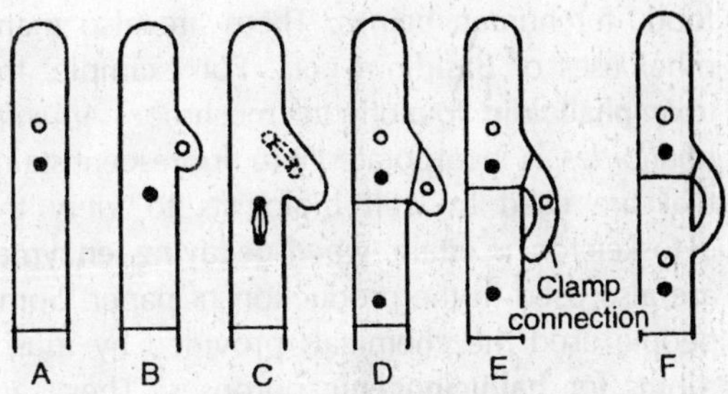

Fig. 4.68: A-F. Stages in cell division by forming clamp connection

subterminal cell and *binucleate (dikaryotic) terminal cell* (Fig. 4.68D). During this time, the clamp grows and touches the wall of subterminal cell. The wall of the subterminal cell and clamp at the point of contact dissolves and the nucleus of clamp migrates to the subterminal cell and becomes dikaryotic (Fig. 4.68E, F). Thus two dikaryotic cells are formed. This process repeats several times and thus the growth continues.

Ecology

Members of Basidiomycota are commonly found in almost every terrestrial environment. They are also found in both in freshwater and marine habitats. They caused decay of plant and waste materials and make them an important factor in the carbon cycle of the environment. However, they caused negative impacts on human beings if they attack and decay of wood structures of different type of houses and also in forest trees. The economic consequences have led to the study of wood-rotting fungi. Yang (2005) worked on different fungi after isolating from different plants like sugar maple, yellow birch, white birch wood etc., and found that the majority of the species were not host specific and can attack any type of wood. Although some species of Basidiomycota are pathogenic in both plants and animals but all the members are not harmful. Some form symbiotic relationships with the roots of vascular plants, forming mycorrhizal association. Bastidiomycota also form symbiotic relationships with insects.

While mushrooms are used for food, and considered as "**delight of the diabetic**" and is also rightly been referred to as the "ultimate health food" for human beings. There are also many other uses of Basidiomycota. For example, the toxin **phalloidin**, found in the mushroom *Amanita phalloides*, is incorporated into fluorescent stains that are used by cell biologists to view the cytoskeleton. Certain **wood-decaying enzymes** are also used in the production of paper. Some people used the chemicals provided by mushrooms for **hallucinogenic** purposes. These are often used for spiritual reasons, but also make up a part of the illegal drug.

Some members develop fruit bodies in the form of a complete or incomplete rings are called **fairy rings**. Fungi grow on dead leaves and roots in grasslands, and produce rings of fruit bodies or mushrooms. The circumference of the rings expands every year and release nitrogen into the soil, caused good growth of grasses. These rings expand every year. Folkloric legends associated these rings with fairy dances, from where they got their name. Most of them do not damage the grasses, but some fungi of this group, such as *Marasmius oreades*, the champion of fairy ring; will kill the areas under it. They accomplish this by forming the water-repellent masses of hyphae under the soil, caused dehydration of the grasses.

The fairy ring is not just of interest to scientists or landscapers. Its presence within folkloric trade has given the fairy ring representations in art and also in literature.

Development of Basidium, Basidiospore and Basidiocarp

A. Development of Basidium and Basidiospore

Basidium developed from the dikaryotic terminal cell (Fig. 4.69) of a hypha of the secondary mycelium (n+n). The terminal cell is being separated from rest of the hypha by a septum over which a clamp connection is usually found. During development the terminal cell enlarges to form young basidium, also known as **basidiole**. Initially two nuclei of the basidiole fuse to form a diploid nucleus (2n). It then produces four protuberances at the top. These protuberances are known as **sterigmata** (sing, sterigma).

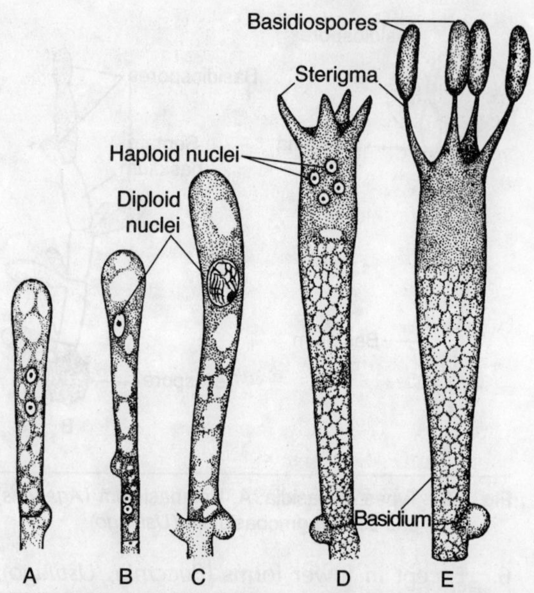

Fig. 4.69: Successive stages of development of basidium and basidiospores. A. Dikaryotic terminal cell of secondary mycelium, B-C. Cells with diploid nucleus develop through karyogamy, D. Young basidium with four sterigma and four haploid nuclei, and E. Mature basidium

Sterimata may be **lacking** (*Cyathus*) or when present, the tips of the sterigmata enlarge to produce basidiospore initials (*Agaricus, Polyporus*). The four haploid nuclei then squeeze through sterigmatal passage into each basidiospore initial. The basidiospore initial then becomes walled off to form a **basidiospore** with single nucleus.

The portion of the basidiospore in contact with the sterigma has been designated as **hilum** (Heim, 1931); and hilar appendix or apiculus becomes short, often forms sharp protrusion near the hilum at the basal end of the basidiospore.

The basidiospores may be borne symmetrically or asymmetrically on the sterigmata. Based on the presence of sterigmata on basidium, basidia are of two types:

(i) **Tetrasterigmate type**. In tetrasterigmate basidium, each basidium typically bears four basidiospores, one on each sterigma e.g. *Polyporus*.

(ii) **Bisterigmate type**. In bisterigmate basidium, basidium produces only two sterigmata and two basidiospores. e.g. *Agaricus brunnescens* (syn. *A. bisporus*).

In **bisterigmate** basidium, each of the two spores may receive two nuclei resulting in the development of binucleate basidiospores. In bisterigmatic basidia the basidiospores may also be uninucleate (*Agaricus brunnescens*, syn. *A. bisporus*), by gradual disintegration of the two unused nuclei.

The number of basidiospores may also be more than four and are without sterigmata (*Cyathus*).

The basidia are commonly slender, clavate to broadly clavate and may also be spherical to elongate. They may be septate (*Puccinia*) or aseptate (*Polyporus*). Septation is commonly transverse or vertical, and may also be oblique.

During nuclear division the orientation of the nuclear spindles may also vary in different basidia. Based on the above, basidia are of two types:

(i) **Chiastobasidial type**. In chiastobasidial type, the nuclear spindles are oriented transversely in the basidium,

(ii) **Stichobasidial type**. In stichobasidial type, basidia having nuclear spindles oriented longitudinally or obliquely.

Types of basidia

Basidia are mainly of 2 types: Homobasidia and Heterobasidia.

(i) Homobasidia

In this type, the basidia are slender, clavate to broadly clavate, or nearly globose and aseptate. The number of basidiospores may be 4 or more, with (Fig. 4.70A, B, F and H) or without sterigmata (Fig. 4.70E and G). They are also known as **autobasidia** or **holobasidia**. The homobasidia may be chiasto-, or stichobasidial type.

When the basidia are developed inside a fruit body, they are the **endobasidia**. Again basidia with basidiospores symmetrical on sterigmata are the **apobasidia** and those with basidiospores asymmetrical are the **autobasidia**.

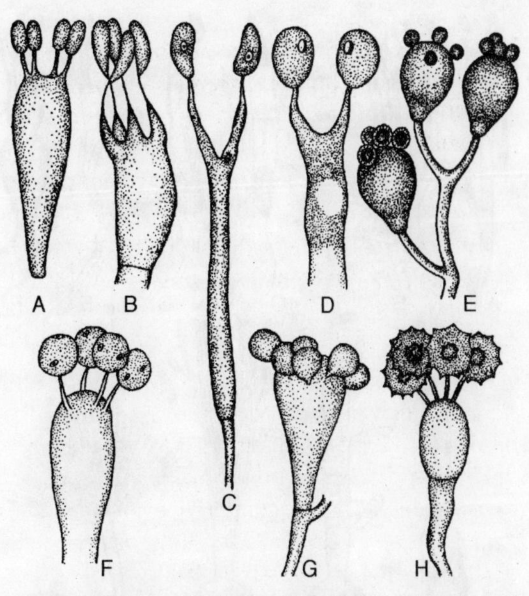

Fig. 4.70: Different types of basidia. A, B, F and H. Homobasidia with sterigmatic basidiospores, E and G. Homobasidia with asterigmatic basidiospores, and C-D. Heterobasdium with tuning fork type of basidium

(ii) Heterobasidia

In this type, the basidia are either septate or deeply lobed (Fig. 4.70C, D). They are also known as **phragmobasidia** or **metabasidia.**

In heterobasidia, the enlarged basal portion of the mature basidium in which nuclear fusion takes place has been designated as *hypobasidium* (Neuhoff, 1924) and which bears upon it the *epibasidium*, but not the sterigmata directly. Later, Linder (1940) discarded this terminology and preferred probasidium for hypobasidium and basidium for epibasidium. Heterobasidium may be of three types (Fig. 4.71): Transversely septate, longitudinally septate and deeply lobed.

B. Development of Basidiocarp

The fruit bodies of higher basidiomycota are called **basidiocarps**. These are made up of secondary mycelium i.e dikaryotic mycelium. The basidiocarps are the aerial sporophores which bear basidia. The basidiocarps are of different sizes, types, forms and textures.

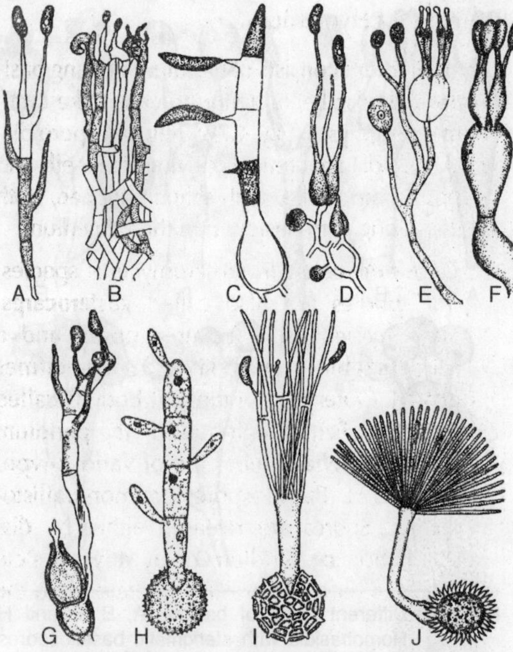

Fig. 4.71: Different types of basidia. A-C. Transversely septate heterobasidium: A. Thin-walled hypobasidium (*Auricularia*), B. Hooked epibasidium (Helicobasidium) and C. Firm-walled hypobasidium (Septobasidium); D-E. Longitudinal to obliquely septate basidium; F. Heterobasidium with four epibasidium separated from hypobasidium by a septum at the base of each. G-J. Thick-walled probasidia with tubular basidia producing basidiospores; G. Basidiospores borne on sterigmata (*Puccinia*); H-J. Basidiospores not borne on sterigmata: H. *Ustilago*, I. *Tilletia* and J. *Neovossia*

The secondary mycelia, those are present in basidiocarps are of **three morphological types**. These are:

1. **Generative hyphae**: This type of hyphae is present in all basidiocarps. Hyphae are thin walled, produce basidia and constitute hymenium. They give rise to other kinds of hyphae.
2. **Skeletal hyphae**: These hyphae are unbranched or sparsely branched, thick walled with narrow lumen. They form a rigid frame work.
3. **Binding hyphae/ligative hyphae**: These hyphae are much branched, narrow, thick walled and with limited growth. These hyphae weave with other hyphae and bind them together.

Basidiocarps may make up of hyphal types in different combinations. These are:

1. **Monomitic type**: Basidiocarps made up of only generative hyphae.
2. **Dimitic type**: Basidiocarps made of generative hyphae along with any other type of hyphae.
3. **Trimitic type**: Basidiocarps consist of all three types of hyphae.

General development of Basidiocarp

Basidiocarps develop as hyphal knots from sub-terranean mycelial strands, called rhizomorphs. The hyphal knots enlarge into round or ovoid structures which break through the surface and develop basidiocarps. The development of basidiocarps shows much variation in Hymenomycetes and Gasteromycetes.

1. **Hymenomycetes**: In many hymenomycetes species, young fruit bodies may be enveloped by a universal veil, which is broken during pileus expansion. During this a cup-like volva leaves at the base of **stipe** and **broken scales** on cap (e.g. *Amanita*).

In some other species, the hymenium is being protected during the development by a partial veil (inner veil or velum), which extends from the edge of cap to stipe (Fig. 4.72). Here the remnants of inner veil persist as a ring on the stipe is called **Annulus** (e.g. *Agaricus*).

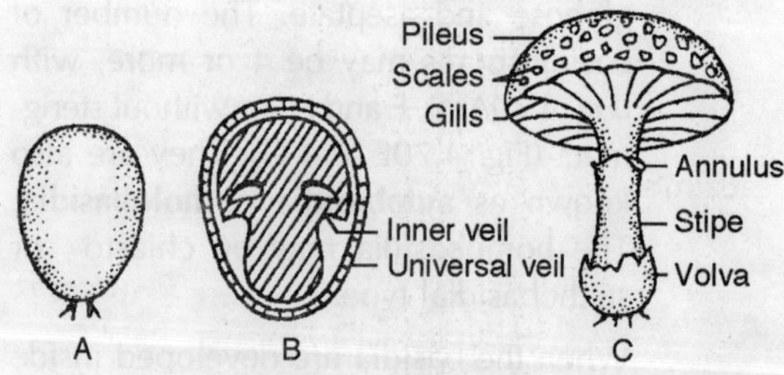

Fig. 4.72: Development of basidiocarp in Agaric: A. Button stage, B. Developing stage and C. Mature stage

In many hymenomycetes, basidiocarp has a stipe and pileus. Lower surface of the pileus has radially arranged gills and the fertile regions of basidiocarp are called **hymenium**.

Arrangement of Hymenium

The fruit bodies show much variation in hymenial arrangements. These are:

(i) *Agaricoid*: Hymenium consists of gills (Fig. 4.73A).

(ii) *Poroid*: Hymenium has pores instead of gills (Fig. 4.73B).

(iii) *Hydnoid*: Hymenium is toothed or spiny.

(iv) *Clavate*: Fruit bodies are club shaped or coralloid shaped and outside of which is covered by hymenium.

(v) *Resupinate*: Fruit bodies with flattened hymenium, appressed to underside of solid surfaces.

(vi) *Gasteroid*: Hymenium is enclosed in fruit bodies.

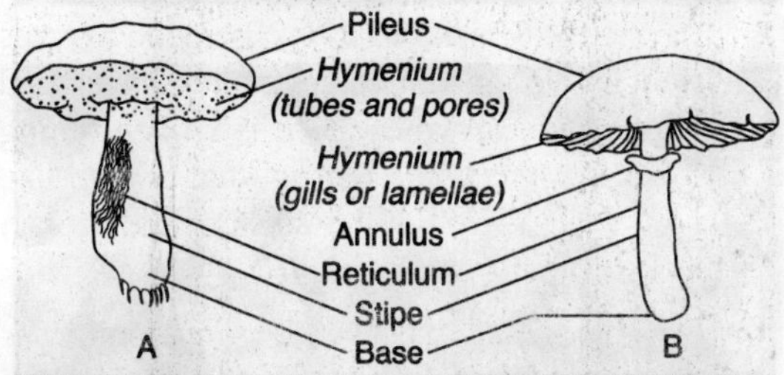

Fig. 4.73: Gills of Agaric: A. Poroid type and B. Agaricoid type

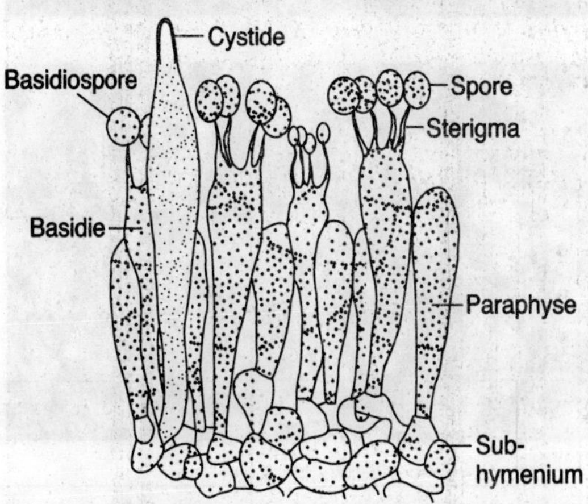

Fig. 4.74: Hymenial layer showing basidium, Cystidia and Paraphyses

Structure of Hymenium

Hymenium consists of basidium bearing basidiospores. It also bears sterile structures like cystidia and paraphyses (Fig. 4.74), helps in spore dispersal. Cystidia are conical or cylindrical cells and paraphyses are sterile, club shaped hyphae, both develop along with basidium in the hymenium.

2. **Gasteromycetes**: In gasteromycetes species, fruit bodies are also called **gasterocarps**. Here hymenium is never exposed and a recognizable fertile layer is sometimes absent. Outer wall of the fruit body is called **peridium**. Fertile region inside the peridium is called **gleba**. Glebas are of various types (Fig. 4.75). Basidiospores are non-ballistospores. Spores are released either by disintegration of peridium or by any physical forces, after the spores are liberated into the peridium.

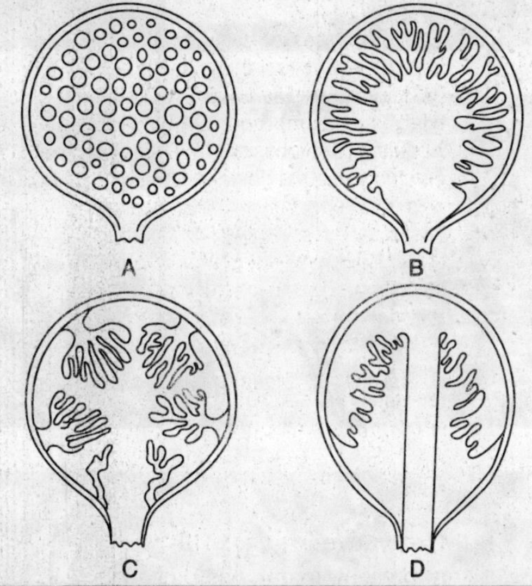

Fig. 4.75: Glebal tissue in basidiocarps (*Gasteromycetes*): A-C. Cavities lined with hymenium (A. Small cavities, B. Irregular cavity developed in upper half and C. Several cavities) and D. hymenium on the underside of the pileus-like structure

Types of Basidiocarp

Basidiocarp is the sporocarp of Basidiomycota. Basidiocarp is also called as **basidiome** or **basidioma** (plural: basidiomata). These are multicellular structure on which the spore

Fig. 4.76: Different types basidiocarps: A. Jelly fungus (*Tremella fuciformis*), B. Club fungus and Coral fungus (*Artomyces pyxidatus*), C-D. Polypore (*C. Ganoderma* sp. and D. *Polyporus sulphureus*), E. Cantharelloid fungus (*Cantharellus ignicolor*), F-H. Gasteroid fungus (F. *Lycoperdon perlatum*, G. *Simblum* sp. and H. *Cyathus olla*),

producing structure hymenium is borne. Basidiocarps are developed in hymenomycetes and gasteromycetes, but such structures are not produced in rusts and smuts. Basidiocarps are of two types:

(i) **Epigeal** (above ground) **type**: Those are visible with naked eye and having more or less agaricoid morphology, commonly called mushrooms, and

(ii) **Hypogeal** (underground) **type**: Those are not visible and are usually called false truffles.

All basidiocarps have hymenial layer, and on its outer surface basidia and basidiospores are produced, commonly along with some sterile structures like paraphyses, cystidia etc., help in spore dispersal. In the simplest form, basidiocarp consists of an undifferentiated fruiting structure with a hymenium layer on the surface. Based on the degree of differentiation into a stipe, pileus, and type of hymenophore, basidiocars are classified into various types. These are of seven (7) types: Jelly fungus, Club fungus and Coral fungus, Polypore, Cantharelloid fungus, Tooth fungus or hydnoid fungus, Gasteroid fungus, False truffle, Secotioid fungus, Agaricoid fungus (Fig. 4.76).

1. Jelly fungus

It is an informal group, considered to species having a gelatin-like consistency. The structural hyphae of these fungi have wall, that are not thin and rigid as most fungi but instead are expanded out to a rather diffuse and indefinite extent. During dry weather condition, these hyphae collapse down and become rather hard and resistant to bending, but with moist weather they expand back in their original gelatinous texture. They can exist for many months in dry state, e.g.

Tremella foliacea- Jelly leaf,

Tremella fuciformis- Snow fungus,

Ascocoryne sarcoidea- Jelly drops.

2. Club fungus and Coral fungus

These groups of fungi are called as clavarioid fungus, typically having erect, simple or branched basidiocarps, that are formed on the ground, on decaying vegetation, or on dead woods. Most of them have fruit bodies those are erect, simple or branched. Spores are borne on the sides of the clubs or branches and the spore-bearing surface is typically smooth or rigided, sometimes warted to spiny, e.g.

Artomyces sp.

Chaetotyphula sp.

Clavariachaete sp.

3. Polypore

Polypores are also called bracket fungi or pore fungi. They produce shelf or bracket-shaped or sometimes circular basidiocarps called conks that lie in a close planar grouping of separate or interconnected horizontal rows of caps that can be several hundred killograms in weight. Fruit bodies are mainly found on living or dead trees and woody debris and may resemble mushrooms. They may forn annual or perennial fruit bodies. The perennial one grow larger year after year. Bracket fungi are typically tough and produce basidiospores, within the pores that typically develop in the under surface.

e.g. *Polyporus* spp.

Ganoderma sp.

Hexagonia sp.

4. Cantharelloid fungus

In this type, fruit bodies are medium to large size and with flaring funnel shape. The fruit bodies have wavy-edged cap of bright yellow to orange in colour and usually flat in young stage and later becomes dome-shaped. The gills are well spaced, shallow, blunt-edged and fairly thick, often with connecting veins in between. The forked and thick-edged gills are the same colour as cap or paler, running down the stalk. The stalk is solid and also coloured like cap or slightly paler. The flesh is white or yellowish in colour. Veil, annular ring or volva is absent. Spores are yellowish in colour. The odour is fruity and taste is mild to peppery. Some species are poisonous,

e.g. *Cantharellus cibarius*.

5. Gasteroid fungus

In gasteroid fungi the spores are produced inside their fruit bodies, rather than on the outer surface. The members are earlier places under the obsolete class Gasteromycetes. The class is artificial, however, since species which include earthstars, puffballs, stinkhorns and false truffles are not closely related to each other.

They form diverse type of fruit bodies, but in all cases spores are formed internally and remain till maturity. The mature spores are released passibly in different ways.

In the puffballs (*Lycoperdon, Bovista* and *Calvatia*) , spores are formed inside the spherical to pestle-shaped fruit bodies and are released either by wind or by raindrops. In the later case, the basidiocarp develops an apical hole, the ostiole; through which spores are puffed out by the pressure of raindrop falling on the surface of basidiocarp. The earthstars (*Geastrum* sp.) have a hard outer layer that splits open in a star-like manner to get a puffball like spores sac.

e.g. *Lycoperdon* spp.

Geastrum sp.

In Stinkhorns and ther relative members like *Phallus, Simblum, Mutinus, Clathrus* etc., form spores within gelatinous puffball-like "eggs". On maturity, the eggs split and various strange spore-receptacles emerge out. The spores are coated with a slime of evil-smelling that attracts different flies, help in dispersal of spores.

In bird's nest fungi like *Cyathus, Crucibulum* etc., form small egg-like packets of spores within cup-shaped fruit bodies. The spore packets are ejected by rain-splash and dispersed in long distance. Gradually the packets get damaged and release the spores.

7. Secotioid fungus

Secotioid fungi included those having growth form, intermediate between mushroom-like hymenomycetes and closed bag-shaped ga teromycetes. Secotioid fungi may or may not have opening caps, but they often lack the vertical geotrophic orientation of the hymenophore needed for wind dispersal of spores and the spores are not forcibly discharged or otherwise prevented from being dispersed.

e.g. *Lentinus tigrinus*, where gills are completely enclosed and never exposed (Fig. 4.77A).

8. Agaricoid fungus

This type includes the genus *Agaricus* of both edible and poisonous species (Fig. 4.77B). The genus includes about 300 species distributed worldwide. The genus includes "button mushroom" (*Agaricus brunnescens*, syn. *A. bisporus*) and "field mushroom" (*Agaricus campestris*).

The genus *Agaricus* is characterized by a fleshy cap, the pileus; underside of which having a number of radiating gills on both side of which basidia and basidiospores are developed. Spores are chocolate-brown in colour. Cap is situated centrally on stem-like stipe and a partial veil, which protects the developing gills and later forms a annular ring, present on the stipe. Spores are dispersed by wind and insects.

Life Cycle Pattern

The general Life pattern of the Phylum Badidiomycota resembles much with the Phylum Ascomycota.

A

B

Fig. 4.77: Different basidiocarps: A. Secotioid fungi (*Lentinus tigrinus*) and B. Agaricoid fungus (*Agaricus perobscurus*)

The haploid basidiospore on germination develop primary mycelium, is the haplophasic somatic body. The developed primary mycelium often produces small unicellular reproductive structure the oidia. The oidia may behave as spore, and on *germination* it produces primary mycelium or as spermatium and on *spermatization* with compatible primary mycelia form dikaryotic secondary mycelia. The secondary mycelium may also be formed by somatogamy between two compatible primary mycelia. The apical cell of dikaryotic hypha of secondary mycelium develops into basidium.

The resting spores may also be formed from the dikaryotic hyphal cells. Each resting spore on germination gives rise to a basidium. Karyogamy takes place in the basidium produces diplophase (2n) which is followed immediately by meiosis and ultimately basidiospores (n) are produced. The basidiospore on germination produces haplophasic somatic body and completes the life cycle.

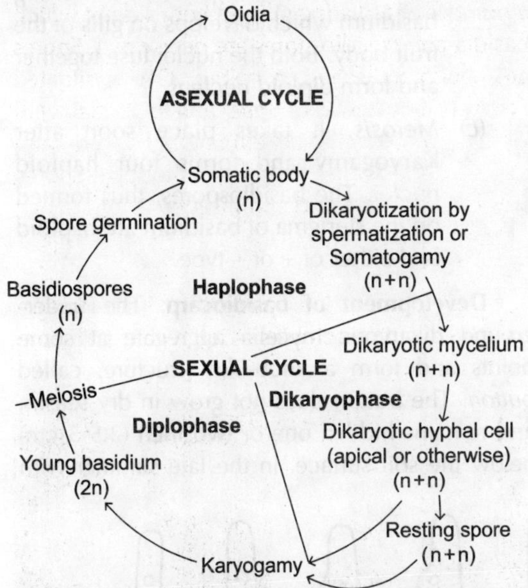

Fig. 4.78: Life cycle pattern in the Basidiomycota

Thus the life cycle of Basidiomycota members comprises of the *three phases* the haplophase (n), dikaryophase (n+n) and diplophase (2n); out of which the dikaryophase is very much prolonged and is nutritionally independent of the haplophase. The haplophase is also well represented. Just as in almost all the fungi the diplophase is very short and is followed by meiosis. The general life cycle pattern is briefly presented in Fig. 4.78).

Some common genera, *Agaricus, Cyathus, Lycoperdon, Polyporus, Puccinia* and *Ustilago* are discussed below:

AGARICUS

(Gill fungi)

Systemic position

Ainsworth (1973)	Hawksworth (1995)
Kingdom Mycota	Kingdom Fungi
Division Eumycota	Phylum Basidiomycota
Subdivision Basidiomycotina	Class Basidiomycetes
Class Hymenomycetes	Order Agaricales
Order Agaricales	Family Agaricaceae
Family Agaricaceae	Genus *Agaricus*
Genus *Agaricus*	

Agaricus is a saprophytic fungus, commonly grows on damp wood, decomposing organic matters like humus, horse dung etc. During summer with rainy weather, it is fairly common in the grassy lands. Butler and Bisbay (1958) recorded 25 species of *Agaricus* from India.

Two important **edible** species of *Agaricus*: *A. campestris*, the field mushroom and *A. brunescens* (syn. *A bisporus*), the cultivated mushroom, are cultivated commercially in various parts of India. Another two species, *A. sylvaticus* and *A. placomyces* are highly **poisonous**, those may cause gastrointestinal disturbances (Alexopoulos and Mims, 1979). *A. xanthodermus* is also found to be a poisonous species (Webster, 1980).

Vegetative body

Vegetative body is mycelial and consists of septate much branched hyphae. Spores on germination develop into *monokaryotic* or *primary mycelium,* either + or – type (Fig. 4.79A, B). The primary mycelium is short-lived and it soon transforms into *dikaryotic* or *secondary mycelium* (Fig. 4.79C) by the fusion of two cells of different monokaryotic mycelium (+ and –) following clamp connection (Fig. 4.80). The hyphae of the dikaryotic mycelia interlace and twist together to form thick white hyphal cord, called *rhizomorph* which bear the fruit bodies.

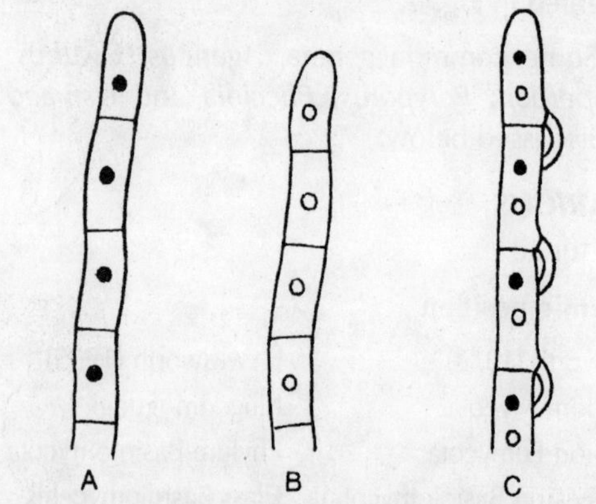

Fig. 4.79: *Agaricus*. A-B. Monokaryotic mycelia, C. Dikaryotic mycelium

Reproduction

Agaricus reproduces by all the three means : vegetative, asexual and sexual.

Vegetative reproduction : It is mostly propagated by vegetative means where dikaryotic mycelium develops *spawn,* the mushroom seed. The mass of spawn divides artificially into small blocks that are grown in soil supplemented with organic manure to obtain fruit bodies.

Asexual reproduction : It takes place by *chlamydospores,* that are formed rarely during unfavourable condition. Terminal or intercalary chlamydospores are developed on dikaryotic mycelium, which on germination during favourable condition produce dikaryotic mycelium.

Sexual reproduction : Sex organs are absent in *Agaricus* and sexual reproduction takes place by somatogamy. Most of the species including *A. campestris* are heterothallic, but *A. brunescens* is homothallic. Somatogamy includes plasmogamy, karyogamy and meiosis. Karyogamy does not take place immediately after plasmogamy, but meiosis follows soon after karyogamy:

(a) *Plasmogamy*. Two cells of monokaryotic hyphae of opposite strains (+ and –) come in contact with each other (Fig. 4.80A). The cell wall dissolves at the point of contact and a dikaryon (n + n) is formed (Fig. 4.80B, C). This dikaryotic cell develops into dikaryotic mycelium by regular cell divisions through clamp connection (Fig. 4.80D-G). The dikaryotic mycelia are subterranean and after aggregation at some points they form button which remains dormant before the rain comes during late summer. After rain, the soil becomes soft and the button develops into fruit body.

(b) *Karyogamy*. It takes place in the young basidium which develops on gills of the fruit body. Both the nuclei fuse together and form diploid nucleus.

(c) *Meiosis*. It takes place soon after Karyogamy and forms four haploid nuclei. The basidiospores, thus formed on the sterigma of basidium are haploid and either of + or – type.

Development of basidiocarp: The underground dikaryotic mycelia aggregate at some points and form a knob-like structure, called *button*. The button does not grow in dry season and remains hidden one or two inch (2.5-5 cm) below the soil surface. In the late summer with

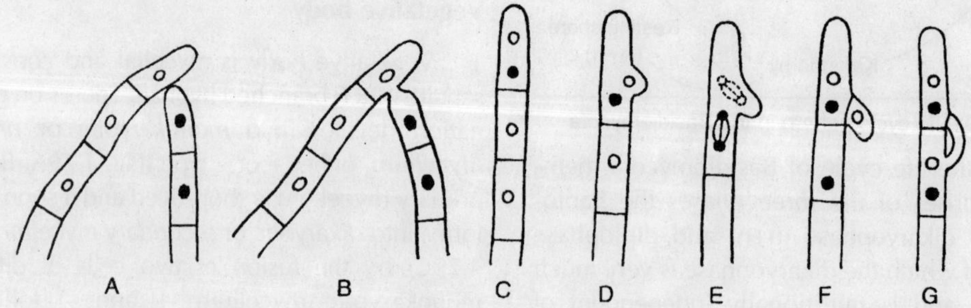

Fig. 4.80: *Agaricus*: A-B. Stages of Plasmogamy, C-G. Growth of mycelium by clamp connection

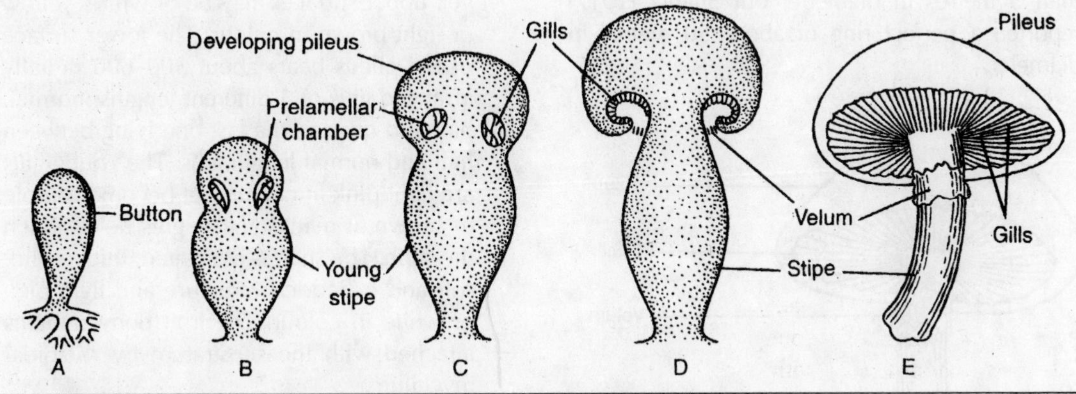

Fig. 4.81: *Agaricus*: A-E. Development of basidiocarp

heavy rain, when the soil becomes moist and soft, the button grows rapidly and develops the *basidiocarp* (Fig. 4.81).

During development, the button is differentiated into a basal bulbous part and an apical hemispherical region. The bulbous part gradually differentiates into elongated, solid, cylindrical structure, the *stipe* and the hemispherical region differentiates into a round, convex region, looks like the top of an open umbrella, the *pileus* (Fig. 4.81E). Towards the bottom of the hemispherical region some hyphae are drawn apart and form a ring-like cavity, the *prelamellar chamber* (Fig. 4.81B, C). The upper surface of prelamellar chamber becomes deeply concave and lined with alternating radial bands of slow and rapidly dividing cells. The region with rapid division forms gill-primordia, which develops into gills, that hang downwardly into the prelamellar chamber (Fig. 4.81D). The top of the hemispherical region (pileus surface) expands resulting in the increase in radial interspaces between the gills. The edge of the pileus of young basidiocarp connects with the stipe by a membranous tissue called the *velum, partial veil* or *inner veil* (Fig. 4.81E). Further expansion of the pileus causes discontinuation with velum and the pileus fully expands out like the top of an open umbrella, with numerous gills attached on its lower side (Fig. 4.81E). The velum remain attached with the upper part of the stipe in the form of a ring, the *annular ring* or *annulus.*

Fairy ring: Due to centrifugal growth (i.e., grows outward from the centre) of the dikaryotic mycelium, fruit bodies are formed in a circular ring around the spot where the fruit body formed in the last year. These fruit bodies are arranged in a ring, imagine that these circular ring marks the path of dancing fairies, called *fairy ring* (Fig. 4.82). The perfect ring of *Agaricus* is usually less

Fig. 4.82: *Agaricus*: Fairy ring formed by basidiocarp

than 5 metres in diametre, but Shantz (1917) reported a perfect ring of about 50 metres in diameter.

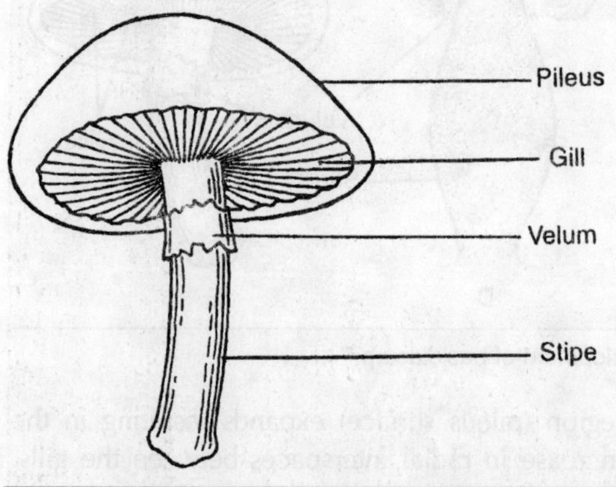

Fig. 4.83: *Agaricus*: Single fruit body

Structure of Basidiocarp

A. **External structure** : The mature basidiocarp (Fig. 4.83, 4.84) is an open umbrella shaped structure with a broad expanded *pileus* on a long massive stalk-like *stipe*. The pileus is 5–12 cm (2-5 inch) in diametre with a convex upper surface, may be of white, yellow or light brown in colour. The lower surface of the pileus bears about 300–600 radially arranged gills of 3 different lengths: normal, half and quarter; the last one lying between half and normal length gills. The young gills are light pink in colour, but becomes purple or brown at maturity. The gills never touch the stipe. The stipe is elongated, thick, solid, soft and cylindrical structure and light-pink or white in colour. The fruit body remains attached with the substratum by rhizoidal mycelium.

B. **Internal structure** :

(a) **Stipe**. It is differentiated into central *medulla,* composed of loosely interwoven hyphae and an outer *cortex,* made up of densely compacted hyphae (Fig. 4.85A).

(b) **Pileus**. Like stipe, it is also differentiated into outer compact and inner loose hyphae (Fig. 4.85B).

(c) **Gills**. It is differentiated into three regions: trama, sub-hymenium and hymenium (Fig. 4.85C-D):

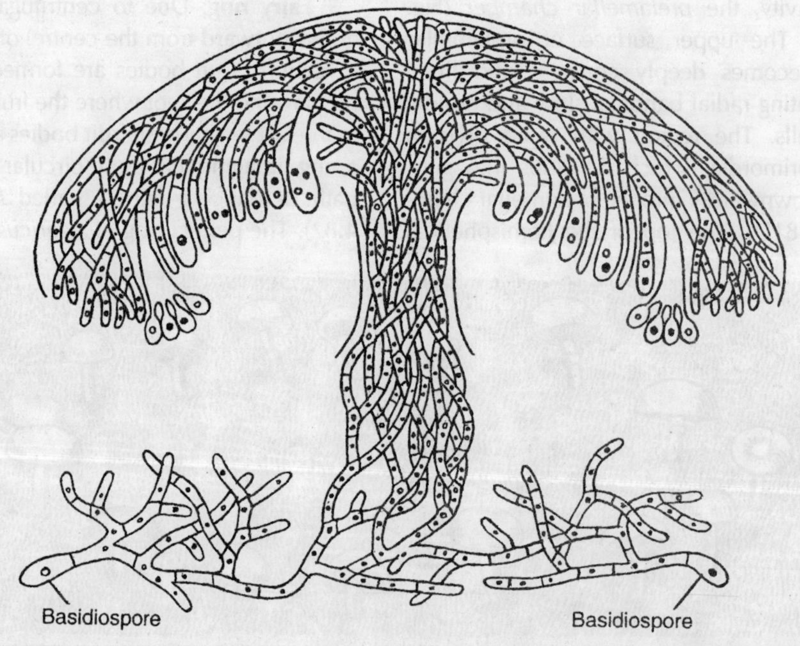

Fig. 4.84: *Agaricus*: Diagram of basidiocarp development by dikaryotic mycelia

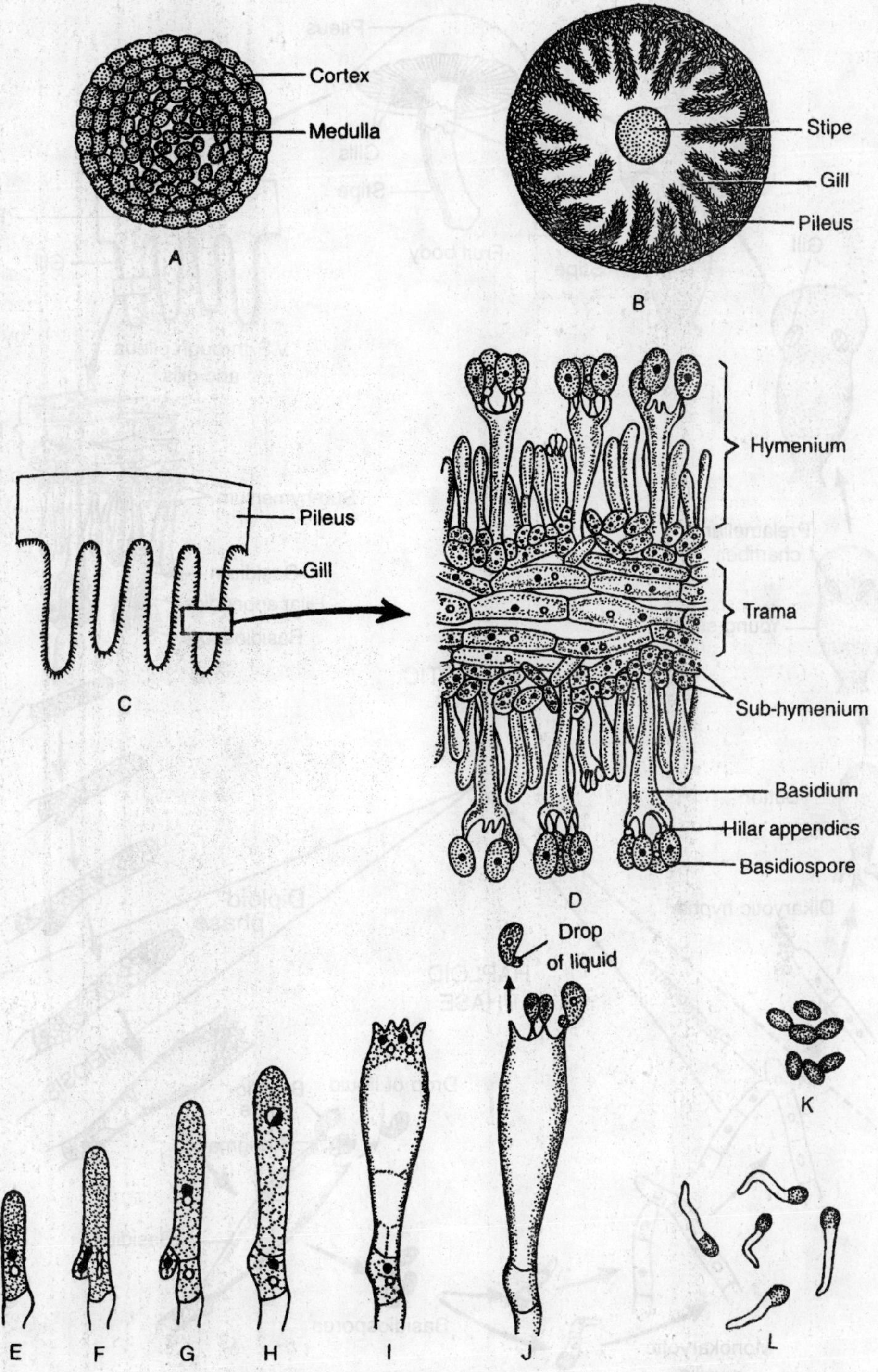

Fig. 4.85: *Agaricus*. Internal structure of basidiocarp: A. Transverse section (T.S.) of stipe, B. T.S. of pileus, C. Vertical section of pileus through gill region, D. Magnified view of a portion of gill, E-J. Development of basidium and formation of basidiospores, K. Basidiospores and L. Germination of basidiospores

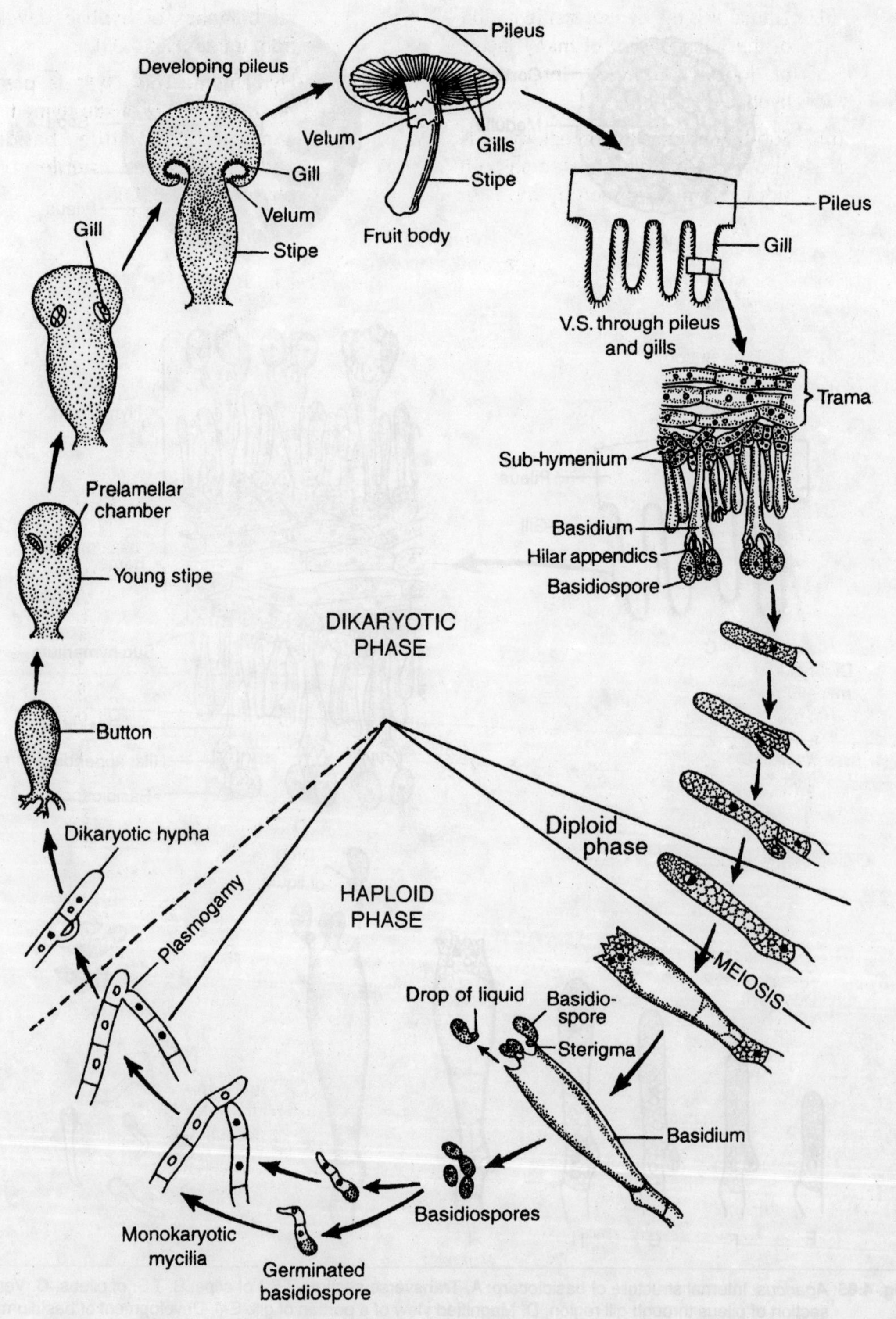

Fig. 4.86: Life cycle of *Agaricus*

(i) Trama. It is the central sterile region of the gill, consists of many layers of loosely arranged interwoven hyphae (Fig. 4.85D).

(ii) Sub-hymenium (hypothecium). It is also a sterile zone, situated on both sides of trama, formed by the lateral branches of hyphae develops from trama (Fig. 4.85D).

(iii) Hymenium. This layer is present on the outerside of sub-hymenium, composed of fertile basidium and club-shaped sterile paraphyses (Fig. 4.85D).

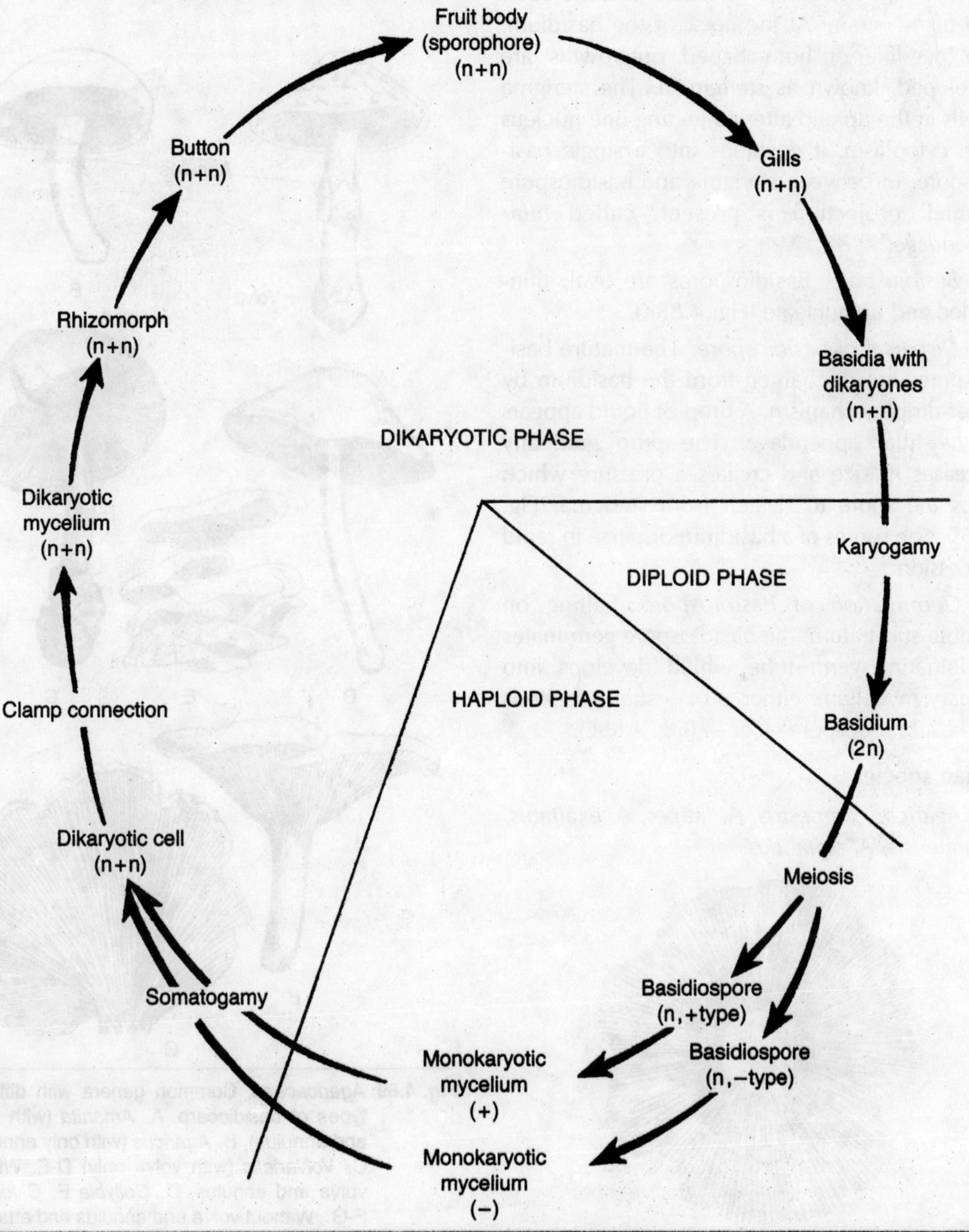

Fig. 4.87: Life cycle of *Agaricus campestris*

Development of basidium. Young basidium is aseptate, fertile dikaryotic cell present in the hymenial zone (Fig. 4.85E-J). As the basidium matures, the nuclei (+ and –) of dikaryon fuse together and form diploid nucleus (2n). This diploid stage is ephemeral. The diploid nucleus undergoes meiosis and forms four haploid nuclei (n), of which two are of "+" strain and the other two of "–" strain. At the apex of the basidium, four peg-like or horn-shaped outgrowths are developed, known as *sterigmata*. The sterigma swells at the tip and after collecting one nucleus with cytoplasm, it develops into a single basidiospore. In between sterigma and basidiospore a small projection is present, called *hilar appendage*.

Basidiospore. Basidiospores are oval, thin-walled and uninucleate (Fig. 4.85K).

Dispersal of basidiospore. The mature basidiospores are discharged from the basidium by water-drop mechanism. A drop of liquid appears at the hilar appendage. The drop gradually increases in size and creates a pressure which helps the spore to detach from sterigma (Fig. 4.85J). The spores of a basidium disperse in rapid succession.

Germination of basidiospore. Falling on suitable substratum, the basidiospore germinates by initiating germ tube which develops into primary mycelium, either + or – strain depends on the strain of spore (+ or –) (Fig. 4.85L).

Indian species

Agaricus campestris, A. latipes, A. exaltatus, A. pratensis, A. squalidus.

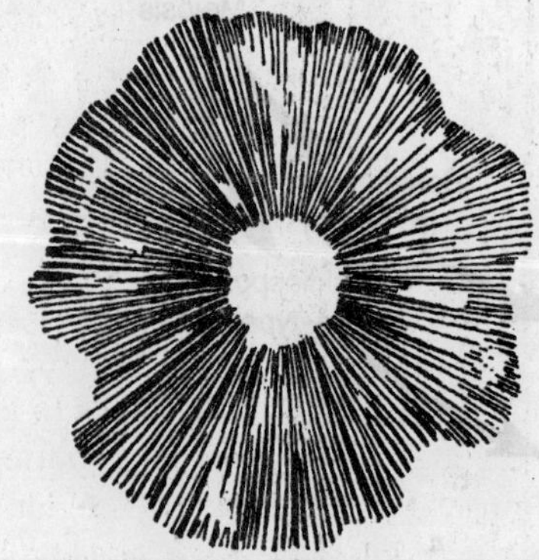

Fig. 4.88: *Agaricus.* Spore print

The diagrammatic and graphic life cycles of *Agaricus* are presented in Fig. 4.86 and 4.87 respectively.

Spore print

It is a technique to study the arrangement of gills in different way. The stipe is removed from the basidiocarp close to the pileus. The pileus is

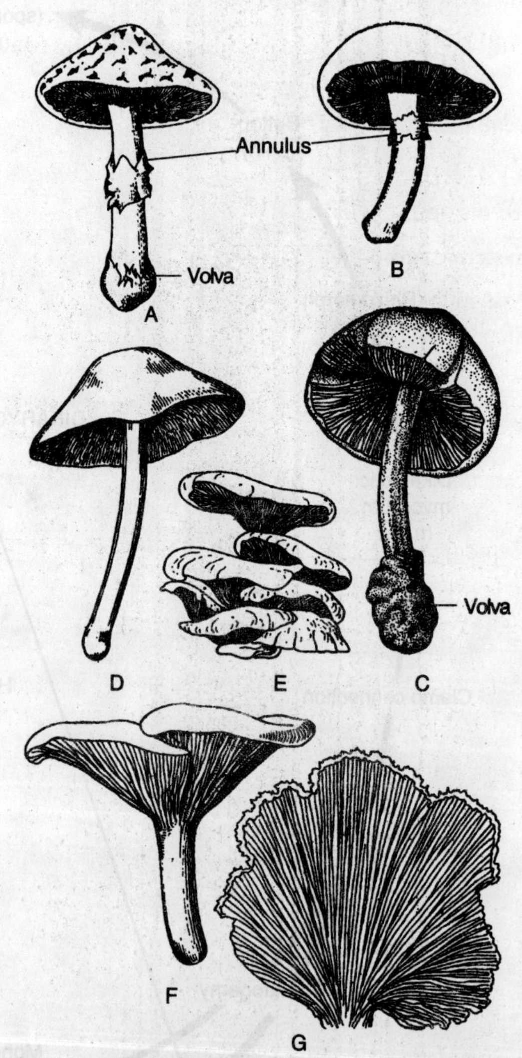

Fig. 4.89: *Agaricaceae*: Common genera with different types of basidiocarp. A. *Amanita* (with volva and annulus), B. *Agaricus* (with only annulus), C. *Volvariella* (with volva only) D-E. Without volva and annulus, D. *Collybia* E. *Clitocybe*, F-G. Without volva and annulus and attached by margin of basidiocarp, F. *Pleurotus* and G. *Schizophyllum*

then kept for a few hours on moist cotton. Then the pileus is placed, gills-side down, on a black paper and kept for overnight. Total set up is then covered with bell jar to avoid any foreign disturbance. The pileus is then carefully lifted and the spores fallen on the black paper become clearly visible. The distribution of spores in a radiating manner indicates a duplicate arrangement of gills. This is called **spore print** (Fig. 4.88).

Fruit bodies of different members of Agaricaceae

Fruit bodies of different members of Agaricaceae are shown in Fig. 4.89.

CYATHUS

(Bird's nest fungus)

Systemic position

Ainsworth (1973)	Hawksworth (1995)
Kingdom Mycota	Kingdom Fungi
Division Eumycota	Phylum Basidiomycota
Subdivision Basidiomycotina	Class Basidiomycetes
Class Gasteromycetes	Order Agaricales
Order Nidulariales	Family Agaricaceae
Family Nidulariaceae	Genus *Cyathus*
Genus *Cyathus*	

Fig. 4.90: Fruit bodies of *Cyathus* on a piece of rotten wood

Cyathus along with other members of the order Nidulariales are commonly called 'Bird's nest fungi', because of the presence of egg-like lenticular bodies inside the funnel-shaped structure, more or less like bird's nest (Fig. 4.90).

Cyathus grows commonly on the old cut stumps of bamboo, on dead logs, rotten wheat straw, dung etc.

Structure of fruit body. The fruit body or basidiocarp is funnel-shaped or inverted bell-shaped structure with flattened apical region (Fig. 4.91A). The base of the fruit body is provided with aggregated mass of hyphae, called **emplacement** (Fig. 4.91B). The funnel-shaped outer region of basidiocarp is the **peridium**, which is differentiated into three layers, a middle pseudoparenchymatous layer present in-between outer exoperidium and inner endoperidium. The exoperidium is composed of loosely woven hyphae, which bear long hairs on the outer surface, the **setae** (Fig. 4.91A, B). The endoperidium is composed of gelatinous hyphae. The inner surface of the peridium is provided with **striae** (Fig. 4.91B). The base of the basidiocarp is made up of solid mass of hyphae, forms a stalk-like structure, the **stipe** (Fig. 4.74B). A number of peridioles are present inside the basidiocarp, these remain attached with the endoperidium with the help of **sheath, middle piece** and **purse** (Fig. 4.91C).

Development of fruit body. The brown coloured mycelia of *Cyathus* become aggregated on the surface of the substratum, on which hyphal knot becomes differentiated. The peripheral region of the knot is then covered by stout hairs. At that time, the entire structure is covered with brown hairs. After that, the peridium becomes separated from rest of the hyphae. With further growth, oval to round portions are separated on the endoperidium. These round portions are the **primordia of peridioles**. With further growth, gelatinisation goes on. The apical region of the young basidiocarp is at first covered with brown hairs and with further development the hairs are pushed aside. The upper region of the cup-shaped basidiocarp is then covered by thin membrane — the **epiphragm**. Later on, the epiphragm becomes gelatinised and is torn apart in the mature basidiocarp.

The rounded portion of the endoperidium i.e., the primordia of peridiole, becomes lenticular, whereby thick-walled **tunica** is being formed. Towards the flattened side of the peridiole, a depressed region is present (Fig. 4.91C, E), its centre is occupied by a tuft of hyphae. These hyphae become closely united and run parallel to one another and form a strand-like structure with swollen middle region called **funiculus** or **purse** (Fig. 4.91C). The funiculus contains a cord-like mycelial aggregation present in a coiled fashion, the **funicular cord**. The apical region of the cord forms a hook-like structure, the **hapteron** (Fig. 4.91D). The hapteron is very sticky and can adhere easily to the solid surface. Some of the hyphae form a thin structure, the middle piece (Fig. 4.91C). The **middle piece** again becomes attached to a group of hyphae, the **sheath** (Fig. 4.91C), that develops on the inner surface i.e., the endoperidium.

Internal structure of Peridiole. In section, peridiole appears oval in outline. It is surrounded by outer compact and inner loosely inter-woven hyphae, the **tunica** (Fig. 4.91E). Inner to the tunica, a dark thick-walled zone is present i.e., the **cortex**. Inner to the cortex, a thick sterile zone is present which consists of thick-walled hyphae and inside which a thick layer is present consisting of comparatively loose thin-walled hyphae (nurse hyphae) intermingled with basidia (Fig. 4.92A). Single irregular airspace is present inside the mature peridiole. The basidia produce 4-8 basidiospores. Basidia are without sterigmata i.e., non-sterigmate (Fig. 4.92B) and develop along with nurse hyphae. The basidia are collapsed before the maturity of spores. The

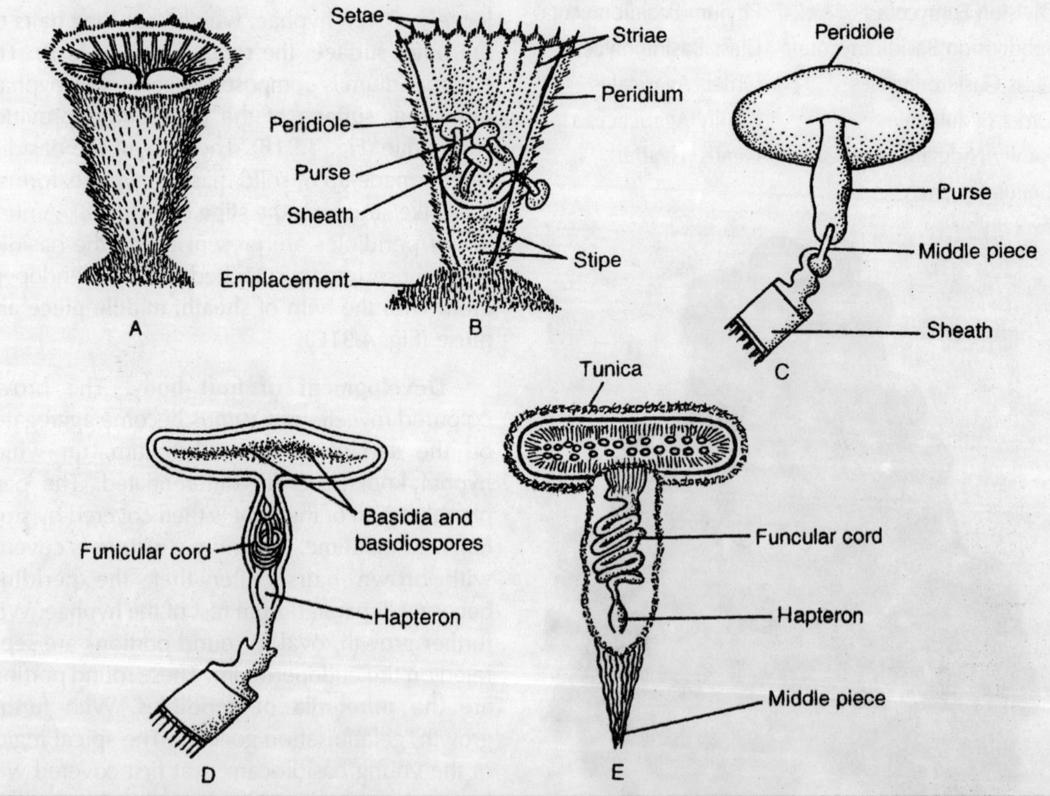

Fig. 4.91: *Cyathus* sp.: A. An entire basidiocarp, B. Sectional view of a basidiocarp showing structural details, C. An entire peridiole with purse, middle piece and sheath, D. A peridiole in section showing basidia, basidiospores, funicular cord and hapteron and E. Sectional view of a peridiole showing details of tunica, funicular cord, hapteron and middle piece

spores then remain intermingled with the nurse hyphae and by proper nursing the spores increase in volume and become thick-walled (Fig. 4.92C). Gradually the nurse hyphae disappear. After that, the spores get gelatinised. The mature spores remain suspended in the gelatinous matrix. Later on, the spores become free from the peridiole by the decay of peridiole wall. The spores on germination produce new mycelia.

Cyathus shows tetrapolar heterothallism. In *C. striatus*, the sexuality is governed by two types of sex factors Aa and Bb. Four types of basidiospores are produced: AB, Ab, aB and ab; those on germination develop the mycelium. Compatibility between the mycelium takes place only with AB × ab and Ab × aB, where there is no common factor. Nuclear fusion and meiosis take place in basidium during the development of spores (Lu and Brodie, 1964).

Dissemination of basidiospores. Brodie (1951) studied and explained clearly the mechanism of basidiospore dissemination in *C. striatus*. Initially the basidiospores are present inside the peridiole and thus the dissemination of basidiospores takes place along with the dissemination of peridiole. Brodie explained the phenomenon as '**splash-cup mechanism**'. He observed and explained that if a raindrop of about 4 mm diametre with a velocity of about 6 metres per second falls into a basidiocarp, it creates an upward thrust along the inclined side of the cup and thus the peridioles are forcibly ejected and can fly to a height of about 4-5 m (13-16.5 ft) in the air. The thrust developed during this process causes the funiculus and the purse to tear open at their attachment with middle piece. The rupture of the purse causes the expansion of **funicular cord** with its sticky attachment organ, the **hapteron**. The hapteron helps the peridiole to attach with the surrounding vegetation. The peridiole initially remains in hanging condition by funicular cord and gradu-

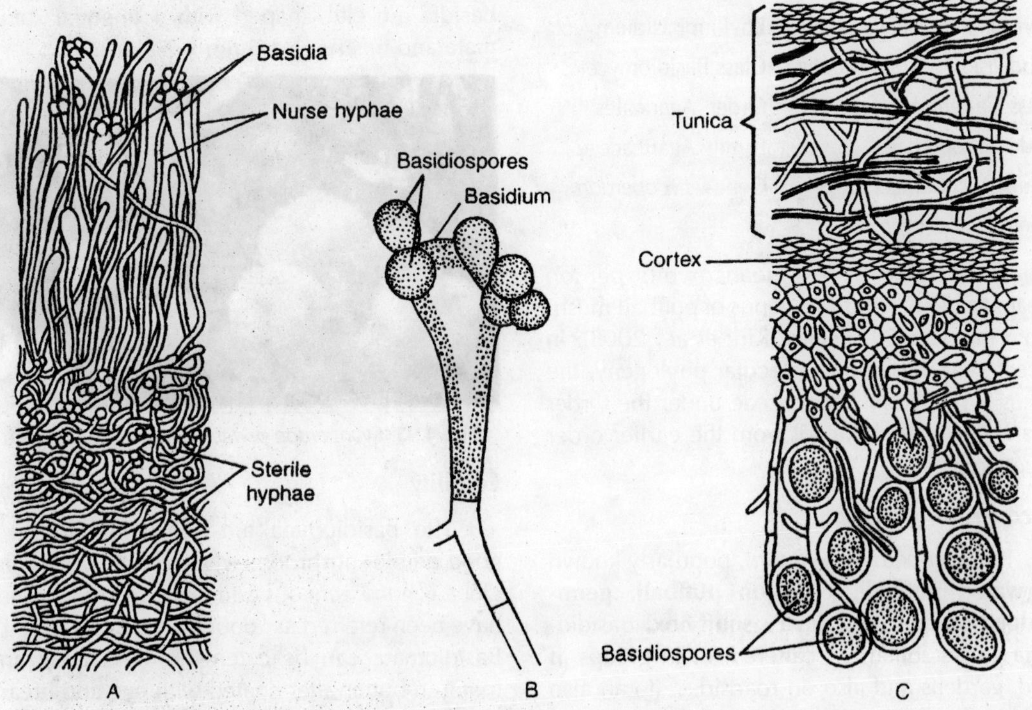

Fig. 4.92: *Cyathus* sp.: A. Portion of a section through peridiole (young), B. Basidia with basidiospores, sterigmata absent and basidiospores more than four and C. Sectional view of a portion of a peridiole (mature) showing details of tunica and cortex and basidiospores mixed with nurse hyphae

ally it coils around the object and, finally, the peridiole comes very close with the object. Later on, by dissolution of the outer wall of peridiole i.e., tunica, the spores are released in the environment.

In certain species like *C. stercoreus,* the peridioles are eaten by herbivorous animals and by incubation in body temperature, the spores are released from the peridiole and stimulated to germinate. The germinated basidiospores come out with the faeces, on which mycelium grows well and later it develops fruit body.

Indian species

Cyathus microsporus, C. hookeri, C. coleusoi, C. intermedius, C. limbatus.

LYCOPERDON

('Puff ball')

Systemic position

Ainsworth (1973)	Hawksworth (1995)
Kingdom Mycota	Kingdom Fungi
Division Eumycota	Phylum Basidiomycota
Subdivision Basidiomycotina	Class Basidiomycetes
Class Gasteromycetes	Order Agaricales
Order Lycoperdales	Family Agaricaceae
Family Lycoperdaceae	Genus *Lycoperdon*
Genus *Lycoperdon*	

Lycoperdon (*lycos* means wolf; *perdon* means to break wind) is a genus of puffball mushroom includes 50 species (Kirk *et al.*, 2008). In recent year, based on molecular phylogeny, the genus is placed in Agaricaceae under the Order Agaricales after shifting it from the earlier order Lycoperdales.

Lycoperdon perlatum

The *Lycoperdon perlatum*, popularly known as **wated puffball, common puffball, germ-studded puffball and devil's snuff-box.** Basidiomata grows solitarily, scattered, or in groups in field, gardens and also on roadsides. It can also grow in fairy rings (Dickinson and Lucas, 1982). *L. perlatum* is edible when young where the internal flesh is completely white in colour.

L. perlatum is a medium-sized puffball, ranges in shape from pear-like with a flattened top, to nearly spherical, gradually tappering to a wide stalk and measures 1.5-6 cm x 3-7 cm. It is off-white in colour where the outer surface (exoperidium) of the globose top covered with short spiny bumps or jewels; those are easily rubbed off to leave a net-like pattern. The base of the basidiomata is thick, with a thick and solid stalk (Fig. 4.93). The inner content of the upper region i.e. gleba is white and firm in young basidiomata. With maturity, the endoperidium of the top of basidiomata turns brown and powdery as the spores become mature. The exoperidium at the top of basidiomata are slugs' away, form a hole (ostiole) in the endoperidium, through which the spores are released by falling raindrops or by any external pressure. The spores are spherical, thick-walled, covered with minute spines and measures 3.5-4.5 µm in diameter. The capillitia i.e. the thread-like filaments in the gleba are yellowish-brown to brown in colour, aseptate and measures 3-7.5 µm in diameter. The basidia are club-shaped with 4 unequal sterigmata and measures 4-5 µm x 7-9 µm.

Fig. 4.93: *Lycoperdon perlatum.* Mature fruit bodies

Edibility

The basidiomata are considered to be a good edible mushroom when young, where the gleba is homogeneous and white in colour. They have been referred as "poor man's sweet-bread". Basidiomata can be eaten through slicing and frying in butter after coated with egg and bread-crumbs (Dickinson and Lucas, 1982) or used in soup (Kuo, 2007). The nutrient value of 100 grams dry basidiomata is as follows:

Nutrient value in 100 grams dry basidiomata:

1.	Energy	441.1 kcal
2.	Carbohydrate	42.0 g
3.	Fat	10.6 g
4.	Protein	44.9 g
5.	Iron	5.5 mg
6.	Manganese	0.6 mg
7.	Zinc	0.5 mg

The northern flying squirrel (*Glaucomys sabrinus*) is also used the fruit body of *L. perlatum* as food.

Other importance

In addition to edibility, the fruit body has many other importances as given below:

1. *Antifungal activity*: The extract of the fungus have antifungal acivity against other harmful fungi like *Aspergillus fumigatus, Alternaria solani, Botrytis cinerea, Verticillium dahliae, Candida albicans* and *C. tropicalis* (Pujol *et al.*, 1990).
2. *Antimicrobial activity*: Extract of the fruit body contain relatively high level of antimicrobial activity against the human pathogenic bacteria like *Staphylococcus aureus, Escherichia coli, Bacillus subtilis* and *Pseudomonas aeruginosa* having efficiency level comparable to the antibiotic ampicillin (Ramesh and Pattar, 2010).
3. *Presence of phenolic compound*: The fruit body contain a phenolic compound, cinnamic acid at a concentration @ 14 mg/kg fruit body (Barros *et al.*, 2009).

POLYPORUS

(Pore Fungi)

Systemic position

Ainsworth (1973)	Hawksworth (1995)
Kingdom Mycota	Kingdom Fungi
Division Eumycota	Phylum Basidiomycota
Subdivision Basidiomycotina	Class Basidiomycetes
Class Hymenomycetes	Order Polyporales
Order Aphylloporales	Family Polyporaceae
Family Polyporaceae	Genus *Polyporus*
Genus *Polyporus*	

Fig. 4.94: *Polyporus*. Fruit bodies on tree trunk

The genus is named *Polyporus* (Gr. *poly*, many; *poros*, pores) because of the presence of numerous fine pores on the undersurface of the fruit bodies. For the above reason, members of the family Polyporaceae are commonly called **pore fungi** and the outer margin of the fruit body looks like first brackets and thus called **bracket fungi**.

Out of many, 159 species are found in India (Bakshi, 1971). The genus has both harmful and useful members.

The harmful members cause damage to both living trees and dead logs (Fig. 4.94). Important *wood-rotting* parasites are:

1. *P. sulphureus*, the sulphur mushroom grows on oak, base of bamboo plants and can be easily recognised because of its bright yellow surface.
2. *P. abietinus*. It causes decay of *Abies* sp.
3. *P. cinnarbarinus*. It causes decay of oak branches and is easily recognised by their cinnabar red colour with corky fruit bodies.
4. *P. betulinus*. It causes heart rot of birch trees.
5. *P. squamosus*. It causes heart rot of elms and sycamore.
6. *P. glomerata*. It causes wood decay of red maple.

The *useful members are*:

1. *P. mylittae*. It is found in Australia and known as 'black fellow's bread' where sclerotia are edible. The sclerotia attain

a weight of 13.5-18 kg (30-40 lbs) or more.

2. *P. frondosus, P. leucomelas* and *P. umbellatus* are the edible members.

Vegetative structure

The vegetative body is mycelial and composed of slender, branched and septate hyphae. Initially, the mycelia are monokaryotic, those developed from germination of spore. The hyphae are much branched and soon become dikaryotic as a result of somatogamy. The dikaryotic hyphae bear clamp connection at the septa.

Reproduction

Polyporus reproduces by both asexual and sexual means:

Asexual reproduction: It is *very rare*. It takes place by **conidia,** developed either on dikaryotic mycelium or on sterile fructifications. On germination they develop dikaryotic mycelia.

Sexual reproduction: Sexual reproduction is **somatogamous**. The species are heterothallic and the fusion between two somatic and monokaryotic mycelia (somatogamy) of opposite strains results in the formation of dikaryotic mycelium. The dikaryotic or secondary mycelium is perennial, which may survive for several years. At regular interval, during favourable condition, fruit bodies or basidiocarps are developed.

Development of fruit body (basidiocarp)

The development of basidiocarp from the secondary mycelium is not clearly understood. Initially, it appears as a spherical knob-like structure which gradually comes out by bursting the bark or soil. With further development the knob may differentiate into either stalked or sessile sporophore.

The *stalked sporophore* (*P. betulinus*) has definite stalk or stipe of about 5-15 cm in height, bearing a pileus of about 2 cm in diametre. In most species, sporophores are *sessile* (*P. sulphureus, P. consor, P. adustus, P. borealis* etc.) and attached laterally with the substratum. At maturity, the fruit bodies may look like bracket, shelf or knob.

Structure of fruit body (basidiocarp)

In stipitate fruit body, the stipe bears an apical umbrella-shaped pileus. In sessile type, the fruit bodies are attached directly with the substratum and then projected outwardly and form various shapes. *On the ventral surface, the pileus is studded with many fine pores,* leading into hollow tube-like structures. The tubes are lined internally with hymenium, composed of basidium bearing basidiospores and sterile paraphyses.

The basidiocarp is made up of three types of hyphae.

These are:

1. *Generative hyphae*. Hyphae are thin-walled with dense cytoplasm and may or may not have clamp connections (Fig. 4.95A).
2. *Binding hyphae*. Hyphae are much branched, narrow and thick-walled (Fig. 4.95B).
3. *Skeletal hyphae*. Hyphae are unbranched, thick walled with narrow lumen (Fig. 4.95C), developed as lateral branch from generative hyphae.

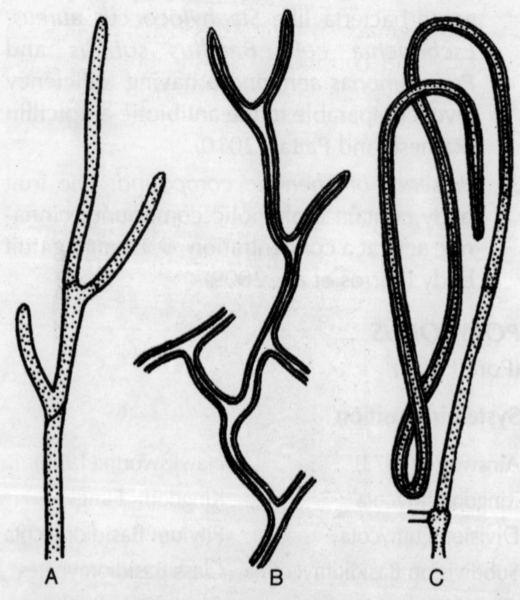

Fig. 4.95: *Polyporus.* Different types of hyphae present in the basidiocarp: A. Generative hyphae, B. Binding hyphae (Branched and thick walled) and C. Skeletal hyphae

Based on the types of hyphae present, the basidiocarps are of three types. These are **monomitic, dimitic** and **trimitic**:

1. **Monomitic**. This type consists of only generative hyphae (*P. adustus*).
2. **Dimitic**. This type consists of generative hyphae along with either binding or skeletal hyphae (*P. sulphureus*).
3. **Trimitic**. This type consists of all the three kinds of hyphae (*P. versicolor*).

V.S. of fruit body

V.S. of the fruit body shows the following five layers from upper to the lower side (Fig. 4.96):

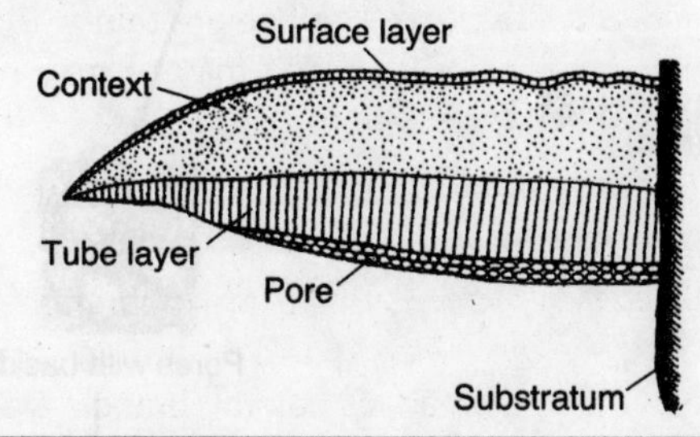

Fig. 4.96: *Polyporus*: Diagrammatic representation of a fruit body in vertical section

1. **Pileus surface**. It is the upper surface of fruit body and consists of a thin zone of thick-walled hyphae.
2. **Context**. Next to pileus is the context, it consists of very fine anastomosing hyphae with large and irregular spaces between them. Sometimes the context is differentiated into upper soft and lower hard and firm layer, called duplex.
3. **Tube layer**. Next to context is the tube layer, it consists of vertically placed tubes which vary in length according to the size of the fruit body. The tissue lying between the pore tubes consists of generative and skeletal hyphae, called dissepiment (Fig. 4.97).
4. **Pore surface**. It is the lower surface of the fruit body, where tubes open.
5. **Hymenium**. The hymenium is lined in the inner surface of the pore, consists of basidium along with paraphyses and rarely with cystedia (Fig. 4.97).

Section of pore tube

From the dissepiment tissue, short branches of hyphae develop at right angles throughout the length of the tube, those form the hymenial layer. The hymenial layer consists of the following (Fig. 4.97):

1. *Basidium*: These are fertile, clavate and single celled structures, slightly project out from the hymenial layer. The basidium bears

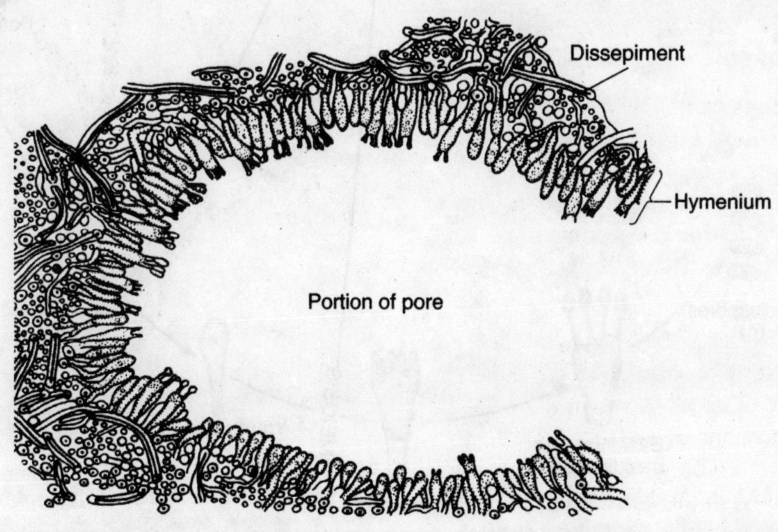

Fig. 4.97: *Polyporus*: T.S. of a pore to show hymenium

four sterigmata at its apex from which four basidiospores are abstracted.

2. *Paraphyses*: These are sterile structures, remain intermixed with basidium in the hymenial layer and help in spore dispersal.

3. *Cystedia*: These are sterile structures, usu-

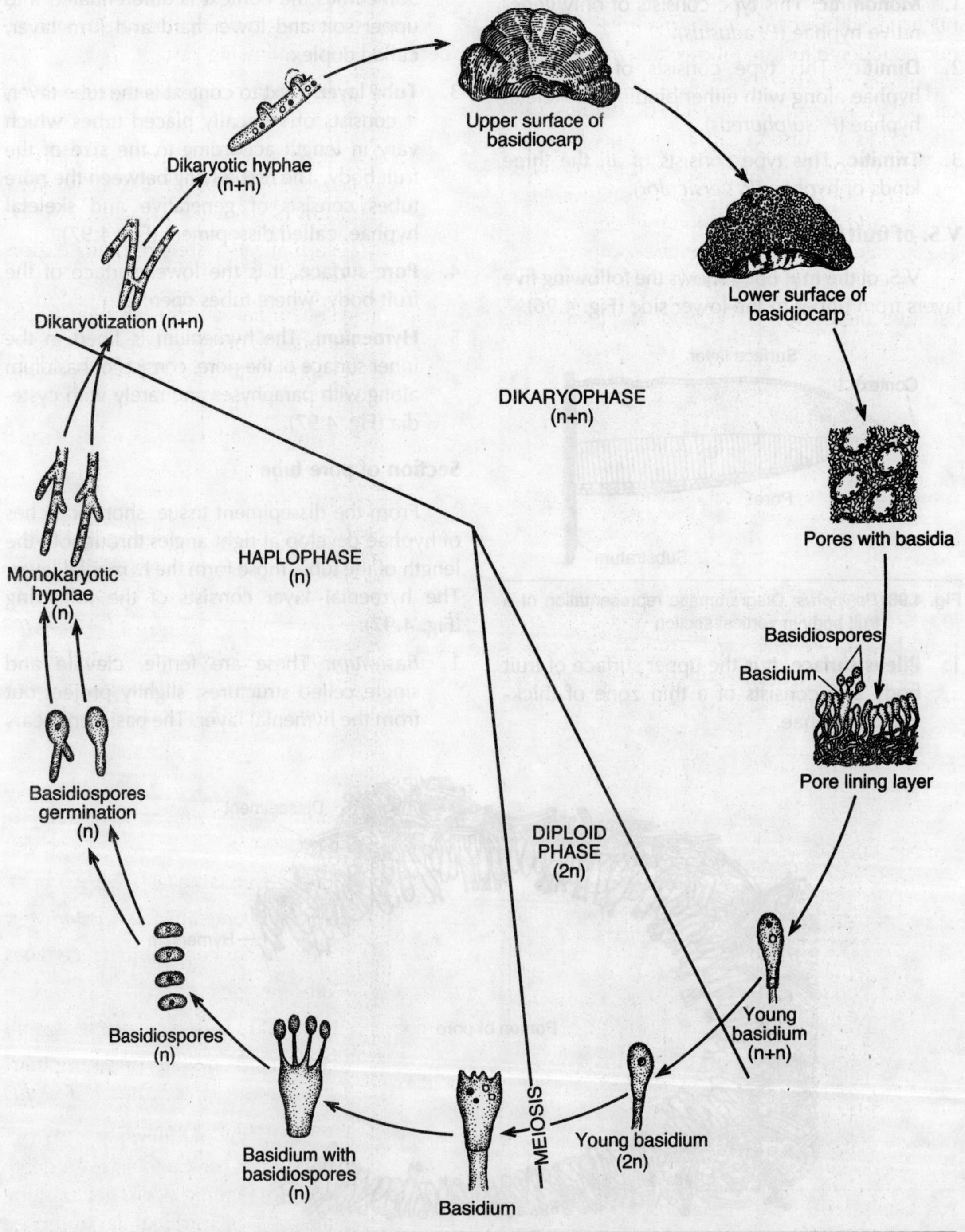

Fig. 4.98: Life cycle of *Polyporus* sp.

ally conspicuous, larger than basidium, remain intermixed with basidium in the hymenial layer and help in spore dispersal.

Young basidia are single celled and binucleate (dikaryotic). With maturity both the nuclei undergo fusion, followed by meiosis. Four sterigmata are developed at the apex of basidium, those bear single haploid basidiospores. The spores are discharged in the pore tube and gradually come out through the pore tubes. Discharge of spores continues from weeks to months and during this period millions of spores are liberated. On germination the spores develop into monokaryotic mycelium.

The life-cycle of *Polyporus* is given in the Fig. 4.98.

Indian species

Polyporus betulinus, P. agariceus, P. circinatus, P. gilvus, P. cubensis etc.

Some common genera of Indian Polyporaceae

The most common and important genera of this family are *Poria, Fomes, Daedalea, Lenzites, Polyporus, Trametes* etc. The shape of pores of this family is of different types (Fig. 4.99). These may be: (i) **Poroid** i.e., circular in outline, (ii) **Hexagonal**, (iii) **Daedaloid** i.e., intermediate between rudimentary lamellate (gills) and pores or (iv) **Lamellate** i.e., almost resembling gills.

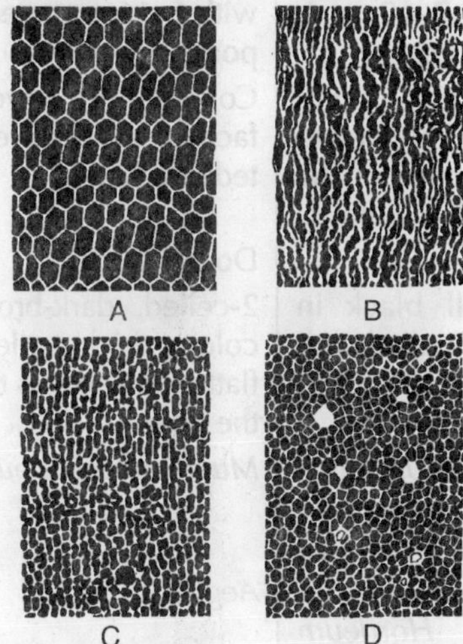

Fig. 4.99: Shape of pore surface of some members of polyporaceae: A. Hexagonal, B. Lamellate, C. Daedaloid and D. Poroid

Pores are round to angular in different genera like *Polyporus, Poria, Fomes, Ganoderma, Trametes* etc.; labyrinthiform in *Lenzites* and *Daedalea* and hexagonal in *Hexagonia*.

The common genera of this family can be identified by the following key:

A simplified key to the common genera of Polyporaceae is given:

A. Basidiocarp always entirely resupinate, completely appressed to the substratum, without any pileus .... *Poria*

AA. Basidiocarp pileate, rarely resupinate, pore tubes not separate.

B. Basidiocarp perennial, each year a tube layer added to previous layer so that pore tubes appear stratified, distinctly or indistinctly.

C. Spores smooth on inner wall...*Fomes*

CC. Spores rough on inner coloured wall, outer wall hyaline.

D. Spores ovate to cylindric-ovate with the wall not uniformly thick, apex always thicker, truncate or rounded .... *Ganoderma*

DD. Spores globose to subglobose, occasionally oblong with the wall uniformly thick ...... *Amauroderma*

BB. Basidiocarp annual, pore tubes in a single layer.

E. Pores daedaloid or approaching lamellate .... *Daedalea*

EE. Pore surface lamellate.

Lamellae concentric ... *Cyclomyces*

Lamellae not concentric ... *Lenzites*

EEE. Pores hexgonal.

Basidiocarp sessile ...... *Hexagonia*

Basidiocarp stipitate or substipitate, ... *Favolus*

EEEE. Pores circular to angular.

F. Pore tubes sunk into an even depth context forming a distinct stratum, their bases forming an unbroken straight line................*Polyporus*

FF. Pore tubes sunk to unequal depths into context not forming a distinct stratum, their bases do not form a straight line ....*Trametes*

PUCCINIA

(Rust Fungi)

Systemic position

Ainsworth (1973)	Hawksworth (1995)
Kingdom Mycota	Kingdom Fungi
Division Eumycota	Phylum Basidiomycota
Subdivision Basidiomycotina	Class Basidiomycetes
Class Teliomycetes	Order Pucciniales
Order Uredinales	Family Pucciniaceae
Family Pucciniaceae	Genus *Puccinia*
Genus *Puccinia*	

The genus *Puccinia*, with about 1,800 species, is distributed throughout the world. About 147 species is reported from India which cause serious rust disease of cereals like wheat, oat, maize and barley. The species of *Puccinia* are **obligate parasite,** grow internally, but spores are visible externally on the host surface. Being an obligate parasite they do not grow on artificial culture medium, but Williams (1966), Helfer (1989) and others have reported that the successful growth occurs in some of the species like *P. hordei,*

Table 4.15: Distinguishing features of the three rusts of wheat

Characteristics	Black rust (*P. graminis*)	Brown rust (*P. recondita*)	Yellow rust (*P. striiformis*)
1. Appear in	Last week of November in South India and in March in North India.	January in Eastern and Northern India.	January to February in Eastern and Northern India.
2. Infected region	Mainly stem, then leaf sheath, leaf blade and ear.	Mainly leaf, rarely leaf sheath and stem.	Mainly leaf but in severe cases leaf-sheath, stem and also ears.
3. Uredosorus	Uredosori are large elongated, dark brown to rusted in colour.	Uredosori are small round to oval, bright orange in colour.	Uredosori are small round to oval, lemon-yellow in colour.
4. Uredospore	Oval, brown, thick walled with 4-round equatorial germ pores.	Oval or round, bright orange with 7–10 scattered germ pores.	Oval to spherical, yellow, with 6–16 scattered germ pores.
5. Teleutosorus	Available mainly on stem, elongated and black in colour.	Develops rarely; if present, they are confined to the lower surface of the leaf and black in colour.	Confined to lower surface of the leaf, elongated and black.
	Burst on maturity.	Do not burst.	Do not burst.
6. Teleutospore	2-celled, chestnut brown to black in colour, thick walled with very thick, round or pointed apex.	2-celled, dull black in colour, thick walled with flattened and less thick at the apex.	2-celled, dark-brown in colour, thick-walled with flattened and less thick at the apex.
7. Alternate hosts	*Berberis vulgaris, Mahonia* sp. and *Mahoberberis* sp.	*Thalictrum falvum, Isopyrum fumarioides, Clematis* spp., *Anchusa* spp.	*Muehlenbeckia hugeli.*
8. Collateral hosts	*Bromus japonicus, B. mollis, B. patulus, Hordeum murimum, Lolium prenne, Aegilops squarrosa* etc.	*Bromus japonicus, B. catharticus, Hordeum murinum* etc.	*Aegilops* sp.

P. coronata, P. striiformis, P. recondita and *P. graminis tritici* on culture medium. *Puccinia* may be **autoecious** (i.e., they can complete their life cycle on one host), e.g., *P. menthae, P. asparagi, P. butleri* or **heteroecious** (i.e., they need two different hosts, to complete their life cycle), e.g., *P. coronata, P. graminis* etc. *P. graminis* grows on different graminaceous host and *P. graminis tritici* being a **heteroecious rust** needs two hosts, wheat and barberry to complete its life cycle.

Wheat crop in India that gets damaged by different rust diseases, caused by different species of **Puccinia** are:

1. *P. graminis tritici* : Black (stem) rust.
2. *P. recondita* : Brown (leaf) rust.
3. *P. striiformis* : Yellow (strip) rust.

Distinguishing features of the three different rusts of wheat are given in Table 4.15.

Puccinia graminis

Puccinia graminis is the causal organism of **black rust disease** of different cereals like wheat, barley, rye, oat etc. As it mainly develop disease on stem, also known as **stem rust**. Persoon (1797) was the first, who reported that the **black stem rust** caused by a fungus.

Puccinia graminis is an obligate parasite and heterocious fungus, having two phases in its life cycle, dikaryophase and haplophase. The pathogen completes its life cycle with the help of two different and unrelated host, wheat and barberry. The wheat is the primary host and the barberry is the secondary host. The primary host may be different but secondary host is same. Based on the primary host the name of the subspecies is different, such as *P. graminis* f.sp. *tritici* on wheat, *P. graminis* f.sp. *hordi* on barley etc. But no one able to complete its life cycle without the presence of secondary host, the barberry.

The isolate of wheat cannot infect oat or any other host and *vice versa*, which indicate their host specificity. The phenomenon where a specific pathogen infects only a specific host is called ***biological specialisation***. Depending on the above phenomenon, the species *graminis* is again divided into different *biological strains* or *biological forms* or *subspecies*. Later, Stakman and many others designated the biological strains or subspecies as *formae specialis* abbreviated as f. sp, as named by *trinomials* i.e., the addition of third name with binomials after considering the chief host of the pathogen. So the name *Puccinia graminis* which infects wheat (*Triticum aestivum*) is called as *Puccinia graminis* f. sp *tritici*.

Thus *P. graminis* is a composite species. Based on pathogenecity, size of spores and cultural variability, it has been divided into different subspecies:

Name	Host
1. *P. graminis tritici*	Wheat (*Triticum aestivum*)
2. *P. graminis secalis*	Rye (*Secale cereale*)
3. *P. graminis agrostis*	*Agrostis caninis*
4. *P. graminis poae*	*Poa compressa*
5. *P. graminis avenae*	Oat (*Avena sativa*)
6. *P. graminis phleipratensis*	*Phleum pratense*
7. *P. graminis airae*	*Aira caespitosa*

Depending on the nature of parasitism on different agronomic varieties of the host, the subspecies is again divided into ***physiological races***. Thus the Physiological specialisation can be defined that within the species of a pathogen (*P. graminis* f. sp. *tritici*) there exist certain individuals that are morphologically similar, but differs in their physiology., biochemical charaters and pathogenecity and are differentiated on the basis of their reaction on certain host genera or cultivar. All the above biological strains/forms and physiological races have the common barberry plant as their alternate i.e., **secondary host**. The differences they show are in the graminaceous host, i.e. **primary host**.

Wheat crops of Northern and Southern region of India are largely affected by *Puccinia graminis tritici*. In South India (Nilgiri ranges) the disease appears in November, thereby the vegetative growth of the crop is affected much more than North India, where the disease appears in late March or thereafter when the crop is near its maturity. The rate of photosynthesis reduces greatly, thereby causing much loss in yield.

Life history of ***Puccinia graminis tritici*****:**

Puccinia graminis tritici is a heteroecious, macrocyclic and polymorphic pathogen:

(a) *Heteroecious* i.e., it requires two unrelated hosts such as wheat (*Triticum aestivum*), the primary host and Barberry (*Berberis vulgaris*), the secondary host; to complete the life cycle.

(b) *Macrocyclic* i.e., it has extended life cycle, consists of five types of spores i.e., uredospores, teleutospores, basidiospores, pycniospores or spermatia and aeciospores. [Demicyclic rust: All but uredinial stage is absent and Microcyclic rust: The teleutospore is the only binucleate spore produced. Although three forms are present, but pycnia or spermogonia may occasionally be absent.]

(c) *Polymorphic* (*Poly*, many; *morph*, form or structure) i.e., it produces 5 different types of spores in its life cycle.

Vegetative body: The vegetative body is mycelium. Mycelia are of two types : dikaryotic and monokaryotic. Both the types are septate, much branched, grow intercellularly and produce special haustoria, which penetrate the host cell. Simple pore if present in the septum, maintains protoplasmic continuity between neighbouring cells. The cell wall is composed of chitin and glucan. The cytoplasm also contains vacuoles, oil globules, glycogen bodies etc.

The dikaryotic mycelium (n + n) occurs in wheat plant i.e., the primary host and the monokaryotic mycelium occurs on barberry plant i.e., the alternate host of the pathogen.

Different type of spores

Puccinia graminis tritici produces five different type of spores in its life cycle. These are uredospore, teleutospore, basidiospore, pycniospore and aeciospore. Uredospores and teleutospores grow on wheat, the primary host; basidiospores on soil or on dead plants upon soil that developed from teleutospore; and the pycniospores and aeciospores on barberry plant, the alternate host of the pathogen.

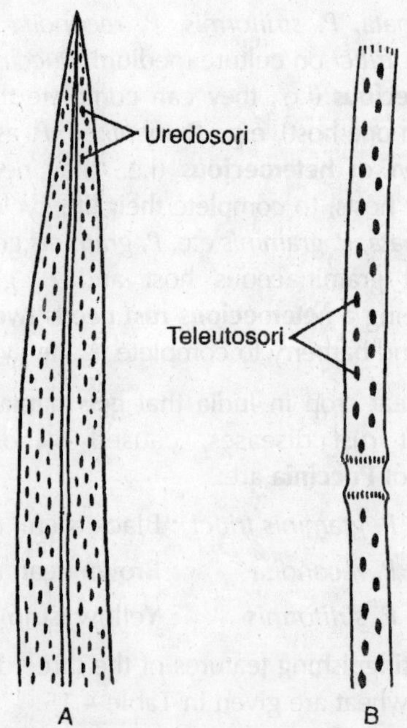

Fig. 4.100: *Puccinia graminis*: Uredosori on leaf (A) and Teleutosori on Stem (B) of wheat

Uredospore: The uredospores are borne on uredosorus (Fig. 4.100A, 4.101A). The uredosorus develops on wheat plant from the dikaryotic mycelium produced by germination of aeciospore, which comes from barberry plant generally through wind dissemination. The aeciospore after reaching the wheat plant may attack leaf, stem or glumes. On germination, the aeciospore produces dikaryotic mycelium which enters through stomata and grows intercellularly. The mycelium of the subepidermal region develops erect hyphae, those produce binucleate **uredospores** at their tips. The uredospores develop in groups under the epidermis, called **uredosorus**, which appear in the form of reddish-brown pustules. With maturity, the host epidermal wall bursts by pressure of developed spore and uredospores become exposed.

Uredospores are stalked, oval, unicellular, brown, thick walled with 4-round equatorial germ pores measuring outer layer 25-30 μm × 17-20 μm. The spore wall is thick with echinulate outer layer (Fig. 4.101B).

The uredospores in favourable condition (i.e., in winter season) again germinate (Fig. 4.101C),

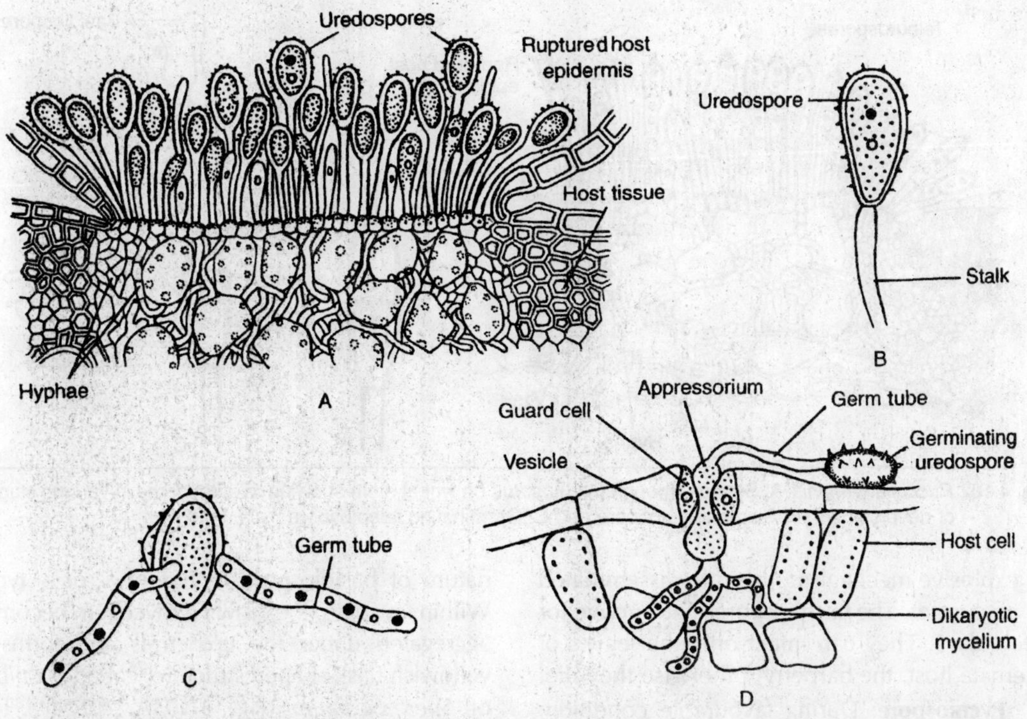

Fig. 4.101: *Puccinia graminis*: A. Section through uredosorus on wheat stem showing uredospores in various stages of development, B. Single uredospore, C. Germinating uredospore, and D. Infection of wheat leaf by germinating uredospore.

thus infect the wheat plant (Fig. 4.101D) and develop next crop of uredospore. The procedure may repeat several times in a single season.

Teleutospore/Teliospore: At the end of wheat season, the uredosori also produce teleutospore along with uredospore. At that time the sorus is known as *mixed sorus*. The mixed sorus gradually converts solely into *teleutosorus* (Fig. 4.100B, 4.102A). Teleutosorus develops exclusively by the infection of uredospore. The teleutosori look black raised streak that developed on leaf sheath and also on stem.

The teleutospores (Fig. 4.102B) are stalked, spindle-shaped, thick and smooth-walled with round or pointed apex, 2-celled and slightly constricted at the septum. Spores are chestnut brown in colour and measure about 15–20 μm × 40–46 μm. Each cell is dikaryotic (n + n) having one germ pore. The germ pore is at the top of the apical cell and it is below the septum in the lower cell. Both the nuclei in a cell fuse together and form a diploid (2n) nucleus.

The teleutospores becomes exposed by rupturing the host epidermis. It acts as a resting spore and survives during unfavourable condition. With the onset of favourable condition of low temperature with high humidity, the teleutospores germinate on soil either detached or attached condition with the dead host.

Basidiospore: It develops on germination of teleutospore. The teleutospore germinates by producing one germ tube from each cell. The germ tube has limited growth and is called *promycelium, probasidium* or *epibasidium*. The diploid nucleus passes inside the probasidium and undergoes meiosis to form four haploid nuclei, two (+) and two (–) type. The probasidium now becomes four celled by transverse wall. Each cell then develops small projection, the sterigma through which nucleus and cytoplasm push outside and develop basidiospores, of which two are (+) and two of (–) type (Fig. 4.102C).

Basidiospores are unicellular, thin-walled and very small. After discharge from basidium

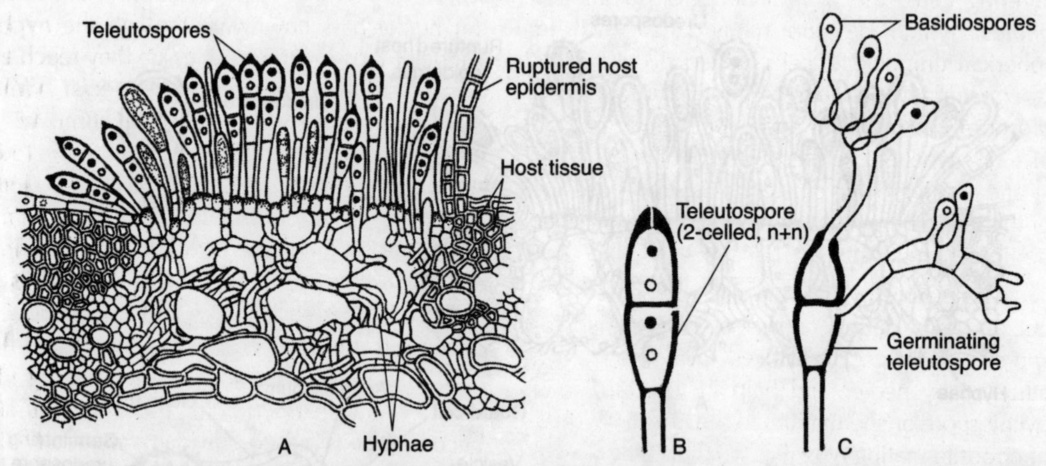

Fig. 4.102: *Puccinia graminis*: A. Section through teleutosorus on wheat stem showing teleutospores in various stages of development, B. Single teleutospore and C. Germinating teleutospore

by explosive mechanism, they are disseminated by air current. The basidiospores can survive for a few days. They can infect only the leaves of alternate host, the barberry, otherwise they die.

Pycniospore: During favourable condition, the basidiospore germinates on contact with barberry leaflet towards the upper surface by producing germ tube. The germ tube penetrates the epidermis and grows there intercellularly. The nature of the mycelium depends on the nature of basidiospore, either of + or – type. Within few days, the growing mycelium becomes aggregated under the epidermis and forms a yellowish flask-shaped structure, called *Pycnium* or *Spermogonium* (Fig. 4.103A, 4.104A). The Pycnium has small sterile mycelium at the neck, called *periphyses, which* intermingle with much larger thin-walled simple and branched *receptive* or *flexuous hyphae*. The bottom of the inner side of pycnium is lined by many uninucleate

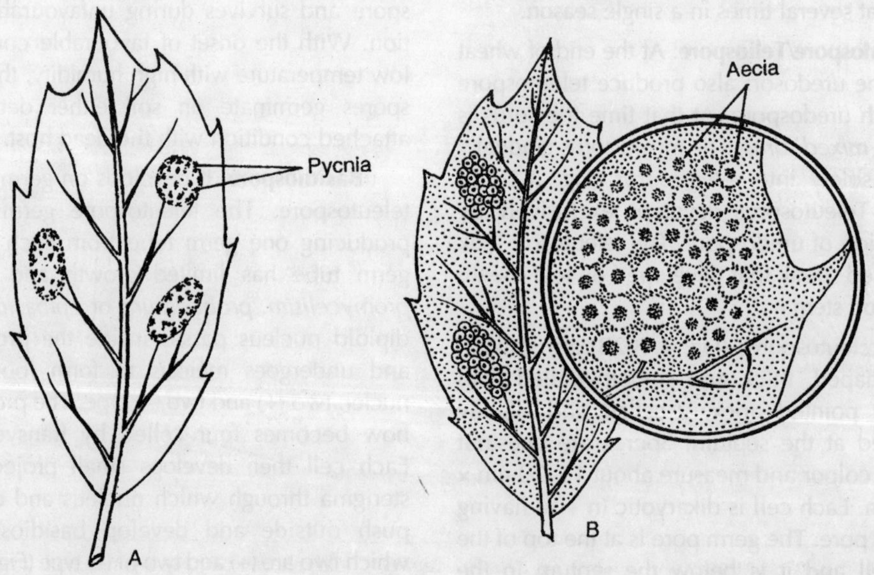

Fig. 4.103: *Puccinia graminis*: A. Upper surface of barberry leaf showing pycnia, B. Lower surface of barberry leaf showing aecia

tapering cells, the *pycniophores* or *spermatiophores,* which develop many small oval to spherical uninucleate cells, called *pycniospores (spermatia).* Depending on the nature of basidiospore, the pycnium and pycniospore may be of + or – type. Numerous pycnia of different types (+ and –) can grow in cluster on the upper surface of the leaf. The pycniospore or spermatium does not infect any host.

The mature pycnium (+ and – type) secretes nector drops during release of mature pycniospores which get intermixed. The insects get attracted by nector and help in the transfer of pycniospore or spermatium to the flexous hyphae of opposite mating type (Fig. 4.104B). The wall at the point of contact dissolves and the nucleus of pycniospore (spermatium) passes to the flexuous hyphae, thus dikaryotic condition is established. This process is known as *spermatisation.*

Aeciospore: During development of *Pycnia,* some hyphae of + and – type goes downward towards the lowerside of the leaf and forms respective *immature aecia* (Protoaecium or aecial primodium). Further growth is initiated within few days of *spermatisation.* The spermatial nuclei pass downward through the hyphae and after crossing the inner tissue they reach and dikaryotize the basal cells of the *aecial primordium.* The aecial primordium then stimulates to further growth and within few days, on the lower surface of the leaf, aecium (Pl. aecia) or cluster cup (n + n) is formed. The young aecium is buried inside the tissue below the lower epidermis, but with maturity it pushes and ruptures the epidermis, thereby spores are exposed. The aecium is inverted cup-shaped structure (Fig. 4.103B, 4.104A) with outer margin composed of short cells, called *peridium.* The basal cell after becomes dikaryotised, divided mitotically into chain of alternately arranged large and small cells (Fig. 4.104A, C). The large cells form the aeciospores (n + n) and the small one becomes sterile, called *disjunctor cell.* The disjunctor cell helps in spore dispersal.

The aeciospores are unicellular, binucleate (n + n), thin-walled and orange in colour. The young aeciospores are polyhedral in shape, but becomes globose with maturity. They are

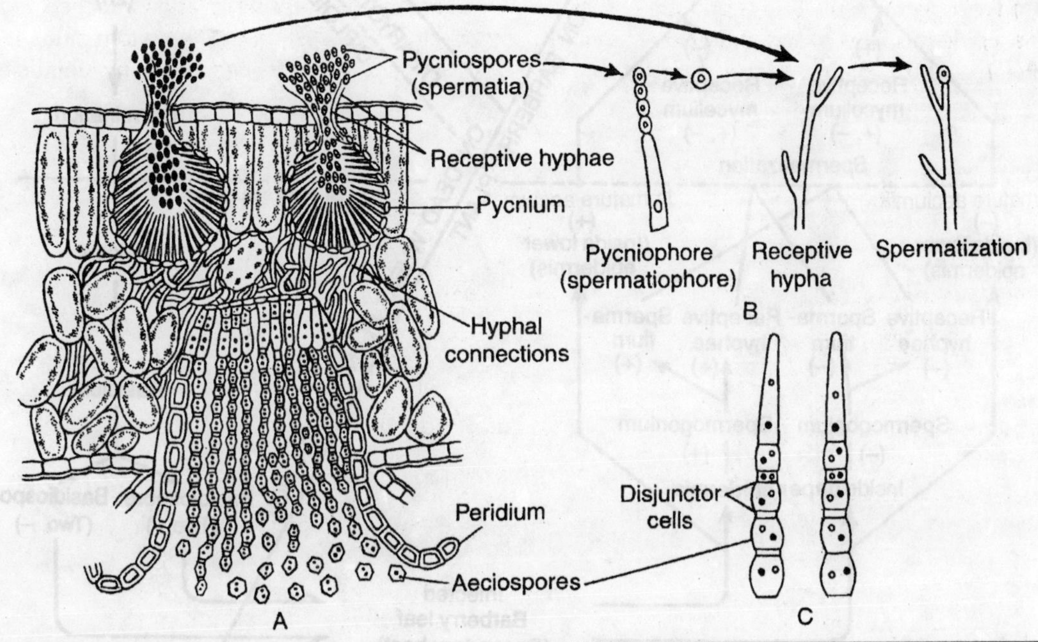

Fig. 4.104: *Puccinia graminis*: A. Section of infected barberry leaf showing upper pycnia and lower aecidium, B. Stages of spermatization, C. Two aeciospore chains showing disjunctor cells alternating gradually with young and mature aeciospore.

dispersed by air current and can infect only the graminaceous host i.e., wheat plant. After falling on wheat plant (stem and/or leaf), they germinate by producing germ tubes (n + n), that penetrate inside the host tissue and establish infection.

The graphic life cycle of *Puccinia graminis tritici* is presented in Fig. 4.105.

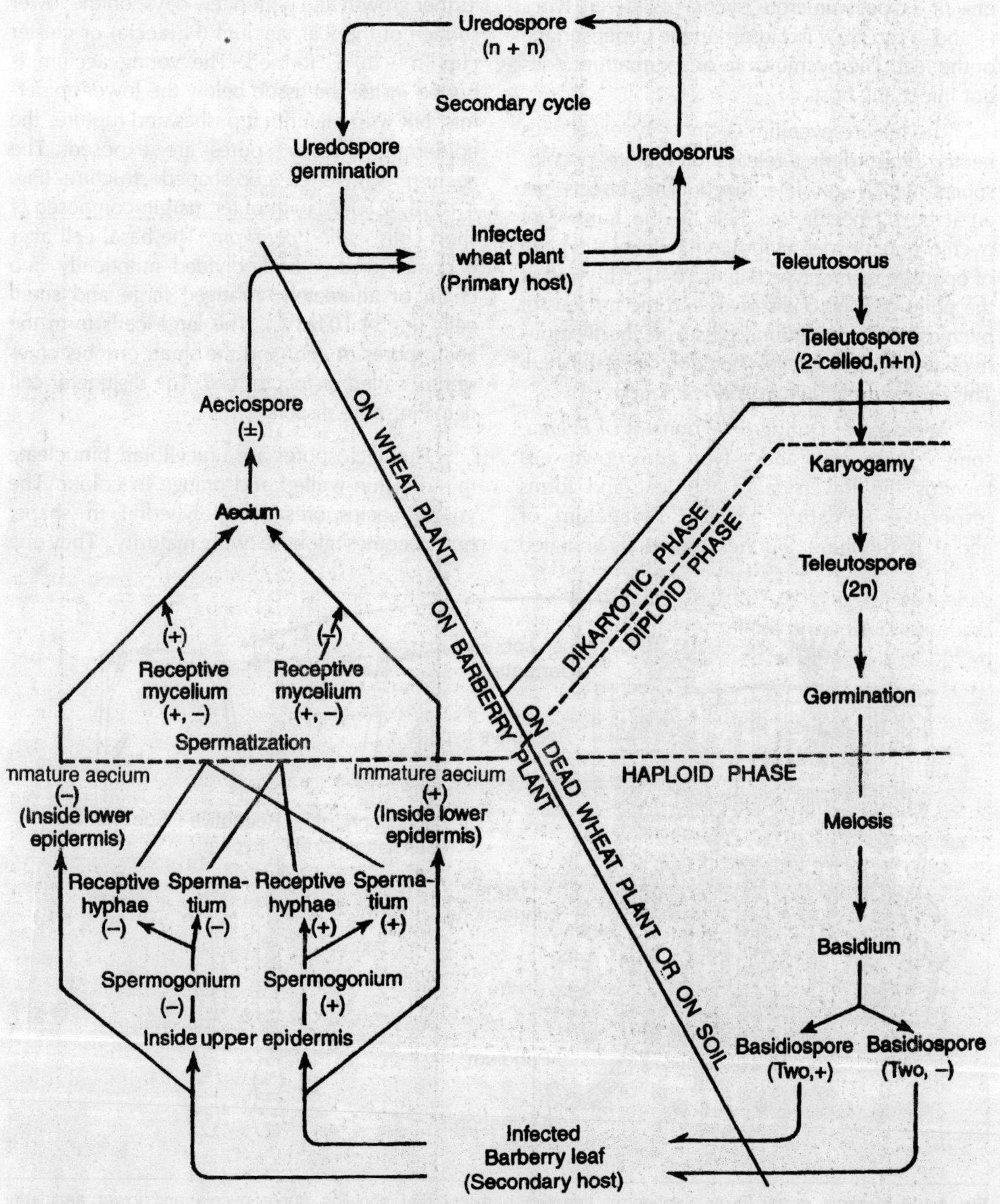

Fig. 4.105: Life cycle of *Puccinia graminis tritici*

USTILAGO

(Smut fungi)

Systemic position

Ainsworth (1973)	Hawksworth (1995)
Kingdom Mycota	Kingdom Fungi
Division Eumycota	Phylum Basidiomycota
Subdivision Basidiomycotina	Class Ustomycetes
Class Teliomycetes	Order Ustilaginales
Order Ustilaginales	Family Ustilaginaceae
Family Ustilaginaceae	Genus *Ustilago*
Genus *Ustilago*	

Ustilago (L. *ustus* = burned) includes 200 species, parasitic mostly on the members of Poaceae (Gramineae) and Cyperaceae. Sometimes they badly damaged the important crops like wheat, barley, sugarcane, oat, maize, bajra (pearl millet) etc. In addition, another smut not belongs to Poaceae or Cyperaceae is the "Anther smut of Carnation (*Dianthus caryophylium*)" and in other Caryophyllaceae caused by *U. violacea*.

In some parts of India, the disease is locally called as *kangiari* (Punjab) and *kandua* (Uttar Pradesh). In Australia, the inflorescence destroyed by loose smut is called as *snuffy ears* mainly by the farmers. Some of the important diseases caused by different species of *Ustilago* are given below (Table 4.16):

Table 4.16: List of important smut diseases caused by different species of *Ustilago*

Name of disease	Causal organism
1. Loose smut of wheat	*Ustilago tritici*
2. Loose smut of barley	*U. nuda*
3. Covered smut of barley	*U. hordei*
4. Loose smut of oats	*U. avenae*
5. Covered smut of oats	*U. kolleri*
6. Common smut of maize	*U. maydis*
7. Whip smut of sugarcane	*U. scitaminea*

Type of smut diseases

It is of two types: loose smut and covered smut disease. The difference between loose smut and covered smut are given below (Table 4.17).

Symptoms of Loose smut and Covered smut

The symptoms of smuts are usually developed as black powdery mass of spores on the inflorescence, either by replacing ovary, the loose smut (loose smut of wheat, c.o. *Ustilago segetum* var. *tritici*) or the spores remain inside ovarian wall, the covered smut (covered smut of barley, c.o. *Ustilago hordei*).

A. Loose smut of wheat

The symptom becomes visible, when the plants are in ears. Diseased ears emerge out of the flag leaf, a little earlier than the healthy ones.

Table 4.17: Differences between loose smut and covered smut

Loose smut	Covered smut
1. Pathogen perpetuate as mycelium inside the embryo of seeds.	Pathogen perpetuate as teliospores remain enclosed in a membranous covering (pericarp) of the grains.
2. Pathogen remains in the embryo till germination.	Commonly the teliospores are released at the threshing time and dusted over the grain surface.
3. Infection of seedling takes place by the active growth of perennating hyphae and remains just behind the growing point keep place with the apical meristem.	Infection of young seedling occurs through coleoptile.
4. The spores directly exposed to the air form the infected inflorescence during flowering.	The spores are not directly exposed to the air.

The remarkable symptom is the "smutted" grain heads, which contain masses of black powdery spores instead of grain on deformed spikelet. The spores completely replace the grain head so that there is no grain to be harvested on infected plants. The spores are easily separated from the deformed spikelet and disperse through wind, leaving a bare rachis. In some cultivar, partially smutted ears or only few tillers get infected leaving the others.

It may be possible to identify infected plants in the field before flowering stage; those are slightly taller and more mature earlier than the rest. Due to early maturity, pathogen got little extra time to produce and disperse spores before the florets of the healthy plants open.

B. Covered smut of Barley

Infected plants do not show symptoms until heading. Kernels of infected plants are replaced by masses of dark brown smut spores, covered by persistent membrane. Smutted heads are hard and compact, with intact awns. Infected plants may be stunted. Occasionally smut sori may also develop in leaf blades, where they appear as long streaks.

Why the green smut of rice is called false smut?

The smut pathogens (e.g. *Ustilago tritici, U. maydis* etc.) belong to the order Ustilaginales, but green smut pathogen of paddy (*Ustilaginoidea virens*) does not belong to this order. Initially it was placed under the subdivision Deuteromycotina (present Mitosporic fungi), but later it has been placed further under the subdivision Ascomycotina. So, it is not a smut in true sence, thus called false smut.

Vegetative structure

The somatic body is mycelial. Mycelium is much branched, septate, hyaline, grows intercellularly and usually inconspicuous. Based on nuclear behavious, it has two distinct stages: **primary** and **secondary**.

1. **Primary mycelium**: The primary mycelium is produced on germination of a basidiospore. The mycelium consists of slender, hyaline and septate hyphae. The hyphal cells contain a single haploid nucleus. Based on the strain of basidiospore, the mycelium is either "+" or of "-" strain. The primary mycelium is also called *monokaryotic mycelium* or *haplomycelium*.

Fig. 4.106: Smut symptom on inflorescence: A. Loose smut of Wheat, and B. Covered smut of Barley

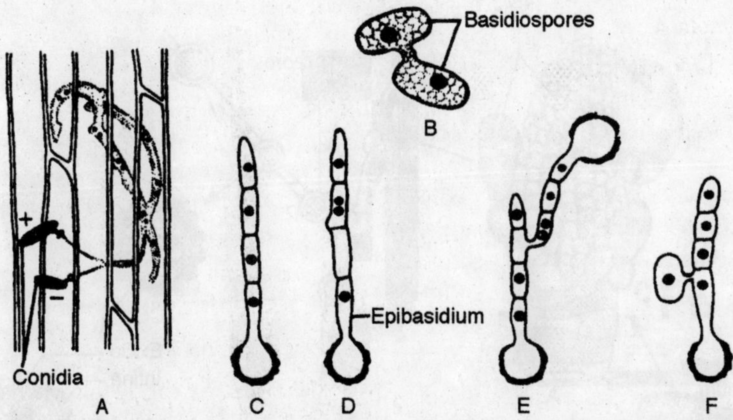

Fig. 4.107: Different methods of dikaryotization in *Ustilago*: A. *U. maydis*, B. *U. anthearum*, C and D. *U. hordei*, E. *U. nuda* and F. *U. violacea*

The mycelium is less extensively developed and either changes soon into secondary mycelium or dies.

2. **Secondary mycelium**: The secondary mycelium is long lived and dikaryotic (n + n) in nature. The mycelium consists of septate, much branched hyphae, and grows intercellularly. Each cell of the hyphae contains two haploid nuclei (n + n). This type of mycelium usually found within the host. In *U. maydis*, hyphae grows both inter- and intracellularly.

In some species, the secondary mycelium grows scatteredly throughout the different parts of the host, called as *systemic*, but insome others, it spreads only near the point of infection, called as *localized*.

Methods of diplodization

The process through which the primary mycelium converted into secondary mycelium is called **diplodization** or **dikaryotization**. In this process the two haploid cells of opposite strains come to each other and form a binucleate cell (n + n) through cytoplasmic union. Two nuclei of such cells do not fuse in the vegetative phase and continue to grow as a dikaryon. The binucleate cell so formed develops into dikaryotic hyphae. The diplodization in *Ustilago* sp. may takes place by any one of the following methods. These are:

1. By somatic fusion between the primary mycelium developed from conidia of opposite strains (Fig. 4.107A). e.g. *U. maydis*.
2. By fusion between the germ tubes of two germinating basidiospores (Fig. 4.107B). e.g. *U. anthearum*.
3. By fusion between two adjacent haploid cells of the same epibasidium (Fig. 4.107C, D). e.g. *U. hordei*.
4. By fusion between two basidia formed by germination of teliospores of opposite strains (Fig. 4.107E). e.g. *U. nuda*.
5. By fusion between basidiospore and one of the cells of the basodium of opposite strains (Fig. 4.107F). e.g. *U. violacea*.

Reproduction

Reproduction in *Ustilago* takes place by the production and germination of teliospores into basidiospores (Fig. 4.109).

Basic aspects are discussed below, based on the life history of *U. maydis*, which causes Corn smut of maize.

The dikaryotic mycelium grows intercellularly in host. Due to infection of secondary mycelium, the host cells divides more actively, results the formation of galls on cob (Fig. 4.108A).

At maturity, the growth of the secondary mycelium (n+n) stops and most of the dikaryotic cells (n+n), start to function as spore initials.

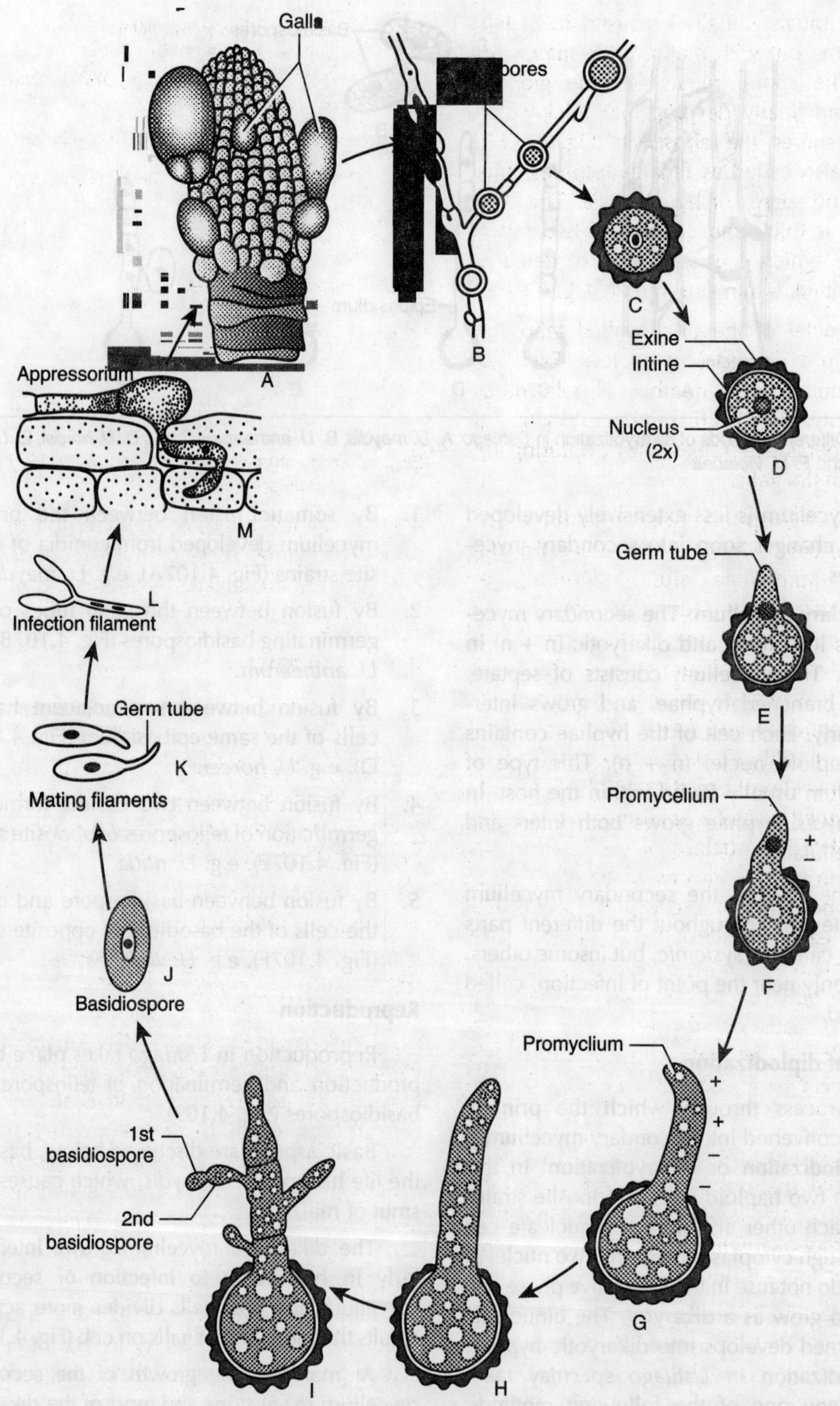

Fig. 4.108: Life cycle of *Ustilago maydis*

The spore initials initially enclosed in a gelatinous matrix, but with maturity the matrix disappears. The spore initials become globose, enlarged and finally develop into thick-walled unicellular spores, the **teliospores** (Fig. 4.108B). These are also called as **chlamydospores**, smut spores, brand spores or hypobasidia. The teliospore wall is thick and consists of two layers: outer exine, which is usually spiny or reticulate and inner intine, is thin and smooth (Fig. 4.108C).

Two nuclei of the thick-walled teliospore fuse to form a diploid (2n) nucleus. Each gall contains innumerable number of teliospores. With maturity, the galls first undergo drying and then the spores are releases by rupturing the epidermis of the galls.

Mature teliospores are unicellular and diploid (2n) (Fig. 4.108D). After a short resting period or immediate after liberation, the teliospores germinates. During germination, the spore wall cracks and a germ tube comes out in the form of a promycelium (Fig. 4.108E). The diploid nucleus migrate inside the promycelium, divides meiotically and forms four haploid (n) nuclei (Fig. 4.108F, G). Later, septa are formed in the promycelium and thus four uninucleate cells are formed, two of + and other two of - type (Fig. 4.108H). The nucleus of each cell of the promycelium divides into two daughter nuclei, of which one migrates into a bud, develops at the side of each promycelial cell and the other nucleus remain within the cell (Fig. 4108I). The uninucleate bud cell function as basidiospores (Fig. 4.108I). Later on, second basidiospore is formed following the above procedure. Thus the developed basidiospores are either + or – type.

Basidiospores are thin-walled, round or oval, haploid (n), uninucleate and unicellular (Fig. 4.108J). Each basidiospore germinates to produce a monokaryotic mycelium (Fig. 4.108K) are of either + or – strain.

On infection, primary mycelia are developed inside the host tissue. Two hyphae of opposite strains come in close contact and initially develop dikaryotic cell and then to secondary mycelium (Fig. 4.108L).

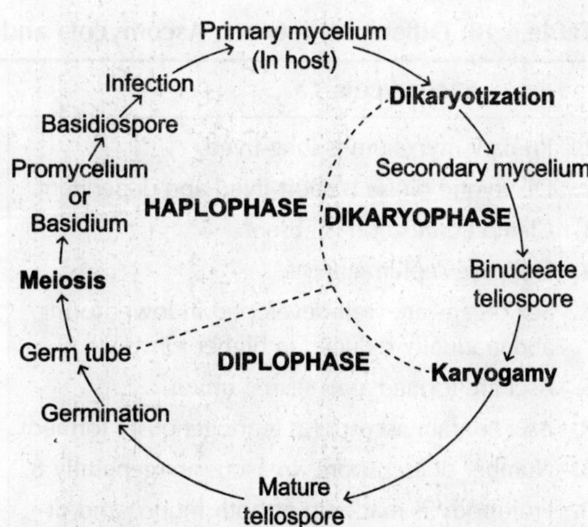

Fig. 4.109:Schematic representation of the life cycle of *Ustilago*

Sometimes, basidiospores keep on dividing through budding. These uninucleate buds of the basidiospores are called conidia (sing. conidium). These **conidia** are also behaves like basidiospore.

Bioluminescence

Bioluminescence is the phenomenon of the production and emission of light by a living organism. Bioluminescence found widely in marine animals, as well as **in some fungi**, bacteria and terrestrial invertebrates like fireflies. In some animals, the light is produced by symbiotic organisms such as *Vibrio*, a bacterium.

Aristotle (384-322 BC) was first reported a mysterious light, emanating (evolving) from the decaying wood. Later, Pliny the Elder (23-79 AD) mentioned about a *glowing, sweet fungus* found on trees in France. In 1854, Johann Florian Heller (1813-1871) identified *strands (hyphae) of fungi* as the source of light in dead wood (E. Newton Marvey, 1920). It was not until the late 19th century that bioluminescence was properly investigated. The phenomenon is *widely distributed* among marine animals (dinoflagellates) and on land it occurs *in fungi*, bacteria and some invertebrates, including insects.

Before the safety lamp to come in focus, the dried fish skins were used in coal mines of Britain and Europe as a weak source of light (Samuel, 1862). The American zoologist, E. Newton Harvey (1920) published a monograph,

Table 4.18: Difference between Ascomycota and Basidiomycota

Ascomycota	Basidiomycota
1. Primary mycelium is long-lived.	1. Primary mycelium is short-lived.
2. Dikaryotic phase is short-lived and dependent.	2. Dikaryotic phase is long-lived and independent.
3. Clamp connection is absent.	3. Clamp connection is common in secondary mycelium.
4. Dolipore-septum absent.	4. Dolipore-septum is present in most of the members.
5. Sex organs are well-developed in lower forms and gradually reduced in higher forms.	5. Sex organs are ill-developed or lacking.
6. Asci are formed after sexual union.	6. Basidia are formed after sexual union.
7. Asci contain ascospores (endogenously formed).	7. Basidium bears basidiospores (exogenously formed).
8. Number of ascospores in ascus are generally 8.	8. Number of basidiospores on basidium are generally 4.
9. Fruit body is made up of both mono- and dikaryotic hyphae.	9. Fruit body is made up of dikaryotic hyphae only.

The Nature of Animal Light, summarizing early work on bioluminescence. The above process avoided the necessity of using candles which risked sparking explosions of firedamp and he also pointed out another safe source of illumination in mines was bottles containing fireflies (Barbara, 2006).

Bioluminescence attracted the attention of the Navy (USA) during the Cold War, since submarines in some waters can create a bright enough wake to be detected. Thus a German submarine was sunk during First World War (1914-1918), having been detected in this way.

After long time, it was Osamu Shimomura, a Japanese chemist who first obtains the crystalline luciferin, by using sea firefly (*Vargula hilgendorfii*), but it was another ten years before he discovered the chemical's structure and was able to publish his 1957 paper *Crystalline Cypridina Luciferin* (P. Vincent and G. David F., 2005).

More recently, Martin Chalfie, Osamu Shimomura and Roger Y. Tsien won the 2008 Nobel Prize in Chemistry for their 1961 discovery and development of green fluorescent protein as a tool for biological research.

Till date the research is going on the curiosity of the light and how and why these mushrooms glow (including animals and microorganisms) and also the use of this mechanism for the human welfare.

A. Bioluminescence in fungi

So far more than 70 species of **bioluminescent fungi** are known, all of which belongs to the order Agaricales and are found to grow largely in temperate and tropical climates. All known bioluminescent fungi are white-spore agarics that belong to 4 distinct evolutionary lineages. These are:

(i) *Omphalotus lineage* (comprising the genera *Omphalotus* and *Neonothopanus*): It includes 12 species,

(ii) *Armillaria* lineage: It includes 5 species,

(iii) *Mycenoid linage* (*Mycena, Panellus, Prunulus, Roridomyces*): It includes 47 species, and

(iv) *Lucentipes lineage* (recently discovered): It includes 2 species, *Mycena lucentipes and Gerronema viridilucens,* which belong to a family that has not yet been formally named (Oliveira *et al.*, 2012).

Armillaria mellea is the most widely distributed among the luminescent fungi, found across Asia, Europe, North America, and South Africa (Vydryakova *et al.*, 2009).

Bioluminescent fungi emit greenish light at a wave length between 520-530 nm. The emission of light is a continuous process and occurs only in living cells (O'Kane, 1990). No correlation has been observed between the fungal bioluminescence with cell structure. The bioluminescence may occur in different regions of the fungi are:

(i) Both mycelia and fruit bodies: in *Panellus stipticus* and *Omphalotus olearius*,

(ii) Only in mycelia and young rhizomorphs: in *Armillaria mellea*.

(iii) Only in the spores: in *Roridomyces roridus*.

(iv) Only in the sclerotia: in *Collybia tuberose*.

B. Mechanism of bioluminescence

Although the biochemistry of fungal bioluminescence has not fully been characterized, but the preparation of cell-free extracts of bioluminescent regions, has allowed scientists to characterize the *in-vitro* requirements of fungal bioluminescence. Available information indicates that two-stage mechanism is involved in the process of fungal bioluminescence:

(i) *First stage*: In the first stage, a light-emitting substance (luciferin) is reduced by a soluble reductase enzyme at the expense of NADPH.

(ii) *Second stage*: In the second stage, the reduced luciferin is oxidized by an insoluble enzyme, luciferase that releases the energy in the form of bluish-green light.

Conditions that affect the growth of fungi, such as pH, light and temperature have been found to influence bioluminescence, *suggesting a link between metabolic activity and fungal bioluminescence* (Moore *et al.*, 2011).

The following list of bioluminescent mushrooms is based on a 2008 literature survey by Dennis Desjardin and his colleagues (Desjardin

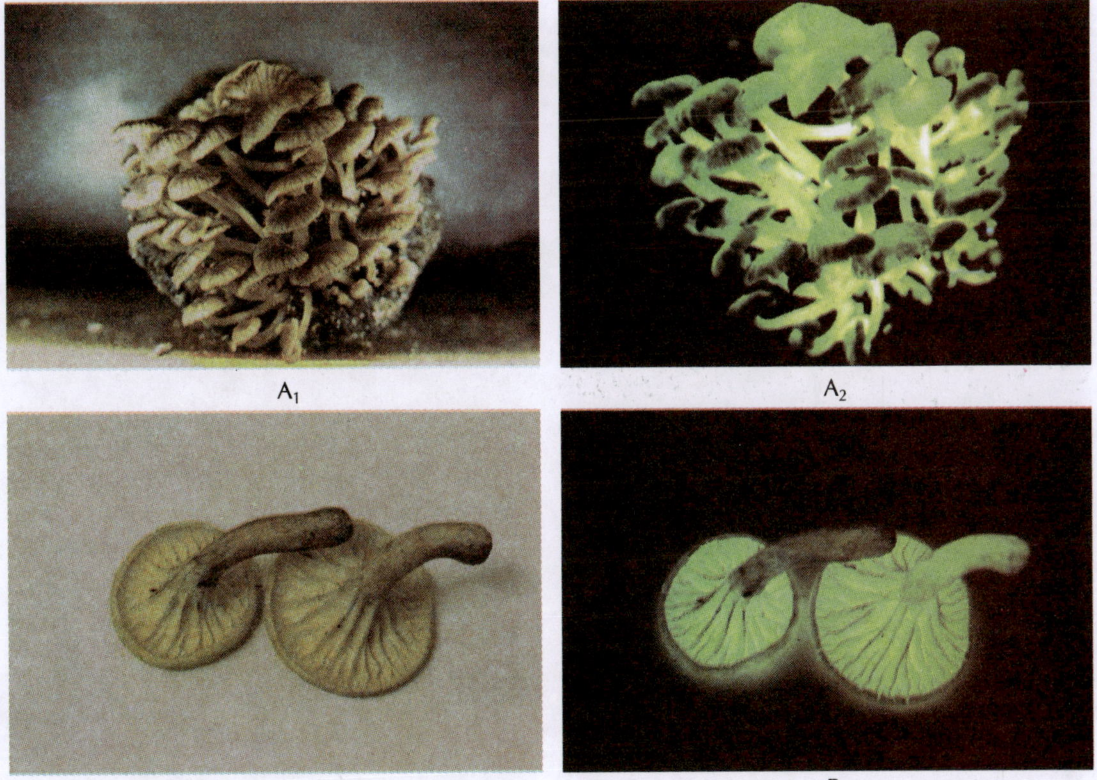

Fig. 4.110: Bioluminescence in different fungi: A. *Mycena lucentipes* (A_1 during day and A_2 glowing in the dark) and B. *Neonothopanus gardneri* (B_1 during and B_2 glowing in the dark). [Photos by: Cassius V. Stevani.]

et al., 2008), in addition to accounts of several new species published since then (Desjardin *et al.*, 2010; Capelari *et al.*, 2011; Shih Y-S, *et al.*, 2013).

Table 4.19: List of bioluminescent mushrooms available in different Geographical regions of the world along with the luminescent region*

Name of the fungus	Geographical distribution
A. Only mycelium luminescent	
1. Armillaria mellea	Europe and North America
2. Armillaria ostoyae	,,
3. Mycena deeptha	India
4. Mycena rosea	Europe
5. Roridomyces roridus	North America, South America, Europe and Japan
B. Only fruit body luminescent	
1. Filoboletus pallescens	Malaysia
2. **Mycena lux-coeli**	Japan
3. Nothopanus noctilucens	Japan
4. Omphalotus mangensis	China
5. Pleurotus decipiens	Malaysia
C. Both mycelium and fruit body luminescent	
1. Dictyopanus foliicolus	Japan
2. Mycena cahaya	Malaysia
3. Neonothopanus gardneri	South America
4. Omphalotus japonicus	Japan
5. Panellus stipticus	Australasia, Africa, Europe, North America, South America and Japan

* The list follows Desjardin et al. (2008) with some modifications along with latest information.

ALLIED FUNGI

KINGDOM CHROMISTA (STRAMENOPILA)

(Phylum Oomycota)

The **chromista** (means 'coloured') is a eukaryotic kingdom created by Cavalier-Smith in 1981. This group of organisms was included under the former Kingdom Protista. The kingdom Protista is now derecognized and placed under two separate Kingdoms the *"Chromista" and* "Protozoa". Later, Patterson (1989) introduced the term **"stramenopila"** (stramenopila means flagellar hairs; *stramen*, flagellum and *piloa*, hair). The term *Chromista* emphasized photosynthesis and *Stramenopila* indicate the presence of hairy flagella (tinsel type). As this kingdom includes the organisms having photosynthetic and non-photosynthetic characters, but all have flagellum hairs i.e. tinsel-type, thus it is better to call this group as ***"Stramenopila"*** instead of ***"Chromista"***. Thus the term Stramenopila is more appropriate, considering them all in a common group.

The kingdom *Stramenopila* includes three phyla *Hyphochytridiomycota, Labyrinthulomycota* and *Oomycota*.

Important features

1. Cell wall **having glucan-cellulose** but chitin is absent.
2. Presence of **tubular mitochondrial cristae** (shared ancestral character with algal relatives).
3. **Tripartite tubular flagellar hairs** (character exclusive to Stramenopila).

Only phylum Oomycota is discussed in the following chapter.

PHYLUM OOMYCOTA

Members of the phylum *Oomycota* are commonly aquatic, some are terrestrial and a few others are parasite on higher plants. The aquatic members may be either free-living saprophyte or parasite on algae and also on other animals etc.

The phylum is characterized by **oogamous** type of **sexual reproduction, biflagellate** (one whiplash and other tinsel type flagellum) **zoospores** and **non-flagellate gametes.**

Important features

1. Members are aquatic or terrestrial, fresh water or marine, saprobic or parasitic.
2. Members are either unicellular or well developed branched filaments consisting of coenocytic mycelium.
3. **Cell wall** is mainly **composed of Cellulose-β-Glucan** and **chitin is absent**. Cell wall having **amino acid hydroxyproline**, not found in fungi with chitinous wall.
4. Members are **diploids**, having haplobiontic B-type life cycle (Win-Tin and Dick, 1975).
5. Most of the members are eucarpic except the members of *Lagenidiales*, those are holcarpic.
6. Asexual reproduction takes place by **zoospores formed in zoosporangia**. The zoosporangia are either not well differentiated (in primitive aquatic forms) or distinct zoosporangia (in advanced formed) are formed.
7. **Zoospores are biflagellate** (one whiplash and other tinsel type), **unequal and devoid of cell wall.** The **tinsel one diverted with 2 rows of mastigonemes** and **whiplash one smooth or with fine flexous hairs placed backwardly**. Some groups (*Saprolegniales*) have dimorphic (pyriform and reniform) zoospores. **Flagella are attached anteriorly in pyriform** and **laterally in reniform type.**
8. **Sexual reproduction** is of **oogamous type**. The male and female sex organs are known as antheridium and oogonium respectively. **They develop aplanogametes.**
9. Gametic union results the formation of thick walled resting spore, the Oospore.

Ecology

Members of Oomycota are responsible for the decomposition and recycling of waste in the ecosystem, in moist and cool areas such as damp soil and freshwater habitats. Parasitic species cause destruction of the crops in many parts of the world, and steps in the history of agriculture by causing disease and remarkable damage. Some of them caused diseases of fish and mammals. Some of these are given below:

1. **Diseases of fishes**. The remarkable among them is the *Saprolegnia parasitica* on Atlantic salmon, *Salmo salar* L. parr
2. **Diseases of human beings**. *Pythium insidiosum* caused *Pythium* keratitis and Pythiosis.
3. **Diseases of plants**. The diseases caused by different groups of pathogens are:
   (i) ***Phytophthora* group**. The most famous example of this is the destruction, caused Ireland's Great Potato Famine in 1845-48, caused by *Phytophthora infestans* (**Late blight disease of potato**). In addition they also caued remarkable diseases like sudden oak death (c.o. *Phytophthora ramorum*), rhododendron root rot, (c.o. *Phytophthora connamoni, P. citriocola* and *P. cactorum*) and ink disease in the European chestnut (c.o. *Phytophthora cambivora*).

   (ii) **Paraphyletic *Pythium* group**. This group is more prevalent than *Phytophthora* and individual species have larger host ranges, although usually causing less damage. *Pythium* damping off is a very common problem in greenhouses, where the organism kills newly emerged seedlings.

   Mycoparasitic members of this group (e.g. *P. oligandrum*) parasitize on other oomycetes and fungi, and have been used as biocontrol agents.

   (iii) **Downy mildews group**. These diseases are easily identifiable by the appearance of white, brownish or olive "mildew" on the underside of leaf. They are obligate biotrophs.

   (iv) **White blister rusts group**. They belongs to Albuginales, which cause white blister disease on a variety of flowering plants. White blister rusts sporulate under the epidermis of their hosts, causing spore-filled blisters on stems, leaves and the inflorescence. They are obligate biotrophs. The Albuginales includes three pathogenic genera, are:
   (a) *Albugo* parasitic predominantly to Brassicales,
   (b) *Pustula*, parasitic predominantly to Asterales, and
   (c) *Wilsoniana*, predominantly parasitic to Caryophyllales.

Classification

The outline of the classification of Phylum Oomycota is given below:

Class *Oomycetes* (= *Peronomycetes*)
Subclass *Rhipidiomycetidae*
  Order *Rhipidiales*
Subclass *Saprolegniomycetidae*
  Order *Leptomitales*
  Order *Saprolegniales*
  Order *Olpidiopsidales*
  Order *Myzocytiopsidales*
Subclass *Peronosporomycetidae*
  Order *Pythiales*
  Order *Peronosporales*
  Order *Albuginales*

Some common genera, *Saprolegnia, Pythium, Phytophthora* and *Albugo* are discussed below:

SAPROLEGNIA

Systemic position

Ainsworth (1973)	Hawksworth (1995)
Kingdom Mycota	Kingdom Stramenopila
Subdivision Mastigomycotina	Class Oomycetes
Division Eumycota	Phylum Oomycota
Class Oomycetes	Subclass Saprolegniomycetidae
Order Saprolegniales	Order Saprolegniales
Family Saprolegniaceae	Family Saprolegniaceae
Genus *Saprolegnia*	Genus *Saprolegnia*

The genus *Saprolegnia* (Gr. *sapron*, rotton; *legnon*, border) includes 15 species (Kirk *et al.*, 2008), out of which 7 species are found in India. It is a genus of freshwater mould often called **"cotton mould"** because of the characteristics white or grey fibrous patches it form. The common Indian species are *S. ferax, S. parasitica, S. monoeca* and *S. litoralis*. Most of the members grow as saprophytes in soil and on decaying animal and plant body in fresh water ponds. But species like *S. ferax* and *S. parasitica* grow as parasite on fishes or on their eggs and cause their death. The disease is called **salmon disease**. Sometimes it causes a great problem in hatcheries of fish.

The fungus can be grown easily in laboratory by bating dead insects (Fig. 4.111), hemp seeds etc., in pond water for a few days.

Fig. 4.111: *Saprolegnia* on Herald moth

Vegetative structure

The somatic body is mycelial and consists of simple or branched, fairly wide, sought and coenocytic hyphae (Fig. 4.113A). They are eucarpic in nature. The cell wall is made up of cellulose and glucan. The reserve food material is oil globules and glycogen.

Hyphae are of two types: Rhizoidal and External hyphae.

Rhizoidal hyphae: Hyphae are shorter in size and can absorb food from the substratum or from host.

External hyphae: The hyphae grow on the surface of the host or on substratum. Sex organs are developed at the apical region of hyphae.

Reproduction

Saprolegnia reproduces by all the three means: vegetative, asexual and sexual.

Vegetative reproduction

It takes place by fragmentation and chlamydospore formation.

1. *Fragmentation*. In this process the mycelium may break into two or more segments. Each one develops into new individual.
2. *Chlamydospore* or *Gemmae*. These are unicellular, thick-walled, globose, ovoid or irregular bodies develop at the tip of hyphae or may intercalary in position. During development the apex of hyphae becomes swollen and accumulates many nuclei and large amount of food. The swollen region is cut off from rest of the filament by septa and form chlamydospore or gemmae (Fig. 4.112A). On maturation, they get detached from rest of the hyphae. They germinate to produce new mycelium.

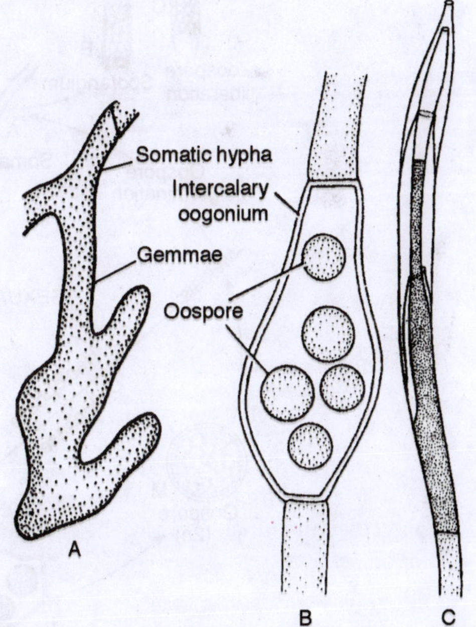

Fig. 4.112: *Saprolegnia*: A. Gemmae on somatic hyphae, B. Intercalary oogonium containing oospore and C. Sporangial proliferation.

Asexual reproduction

It takes place by zoospore formation.

Zoospores. Zoospores are produced inside zoosporangium. The zoosporangia are develop at the apex of vegetative mycelium or they may develop intercalary (*S. torulosa*). During development, the apex of the hypha becomes swollen and accumulates numerous nuclei with cytoplasm and gets separated from rest of the mycelium by basal septum (Fig. 4.113B). The mature sporangium is club-shaped (Fig. 4.113C). During development, the protoplast of the sporangium undergoes cleavage and cleaved into many uninucleate segments. Later on, they get differentiated into biflagellate and pear-shaped zoospores, called primary zoospores (Fig. 4.113D). Out of two flagella one is whiplash type and the other one tinsel type. The zoospores are released through pore developed at the sporangial apex (Fig. 4.113C, D). After released from the sporangium the zoospores swim for sometime, withdraw their flagella and develop a thick-walled around itself and are called primary cysts (Fig. 4.113E). During favourable condition, the primary cysts germinate and produce biflagellate and kidney-shaped zoospores, called **secondary zoospores** (Fig. 4.113G). Like the primary zoospores,

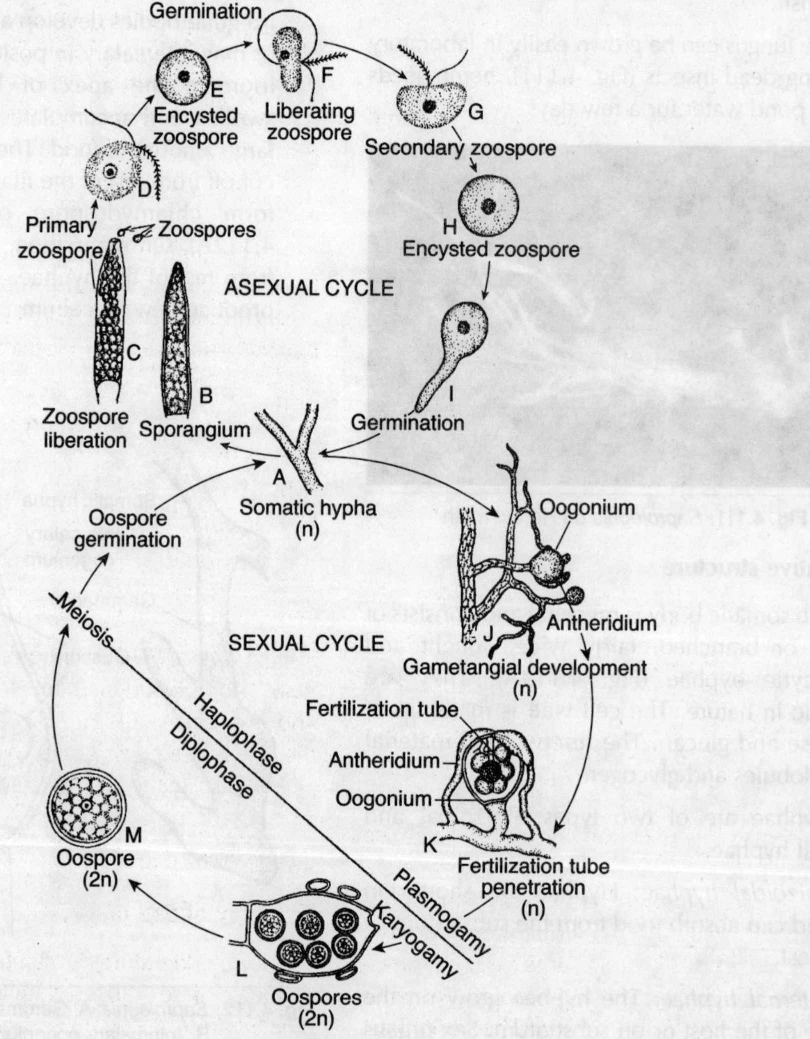

Fig. 4.113: *Saprolegnia*: Life cycle

the flagella are of two types, whiplash and tinsel type but are laterally placed. Like primary zoospores they swim for sometime, get encysted (Fig. 4.113H) and then germinate to produce new mycelium (Fig. 4.113I).

The phenomenon of the sequential production of two types of zoospores i.e. primary and secondary is called **diplanetism** and the zoospores are called diplanetic. The formation of secondary zoospores and secondary cyst may repeat several times and are called as **polyplanetism** or **repeated emergence**.

In *Saprolegnia*, after the discharge of zoospores from zoosporangium, new zoosporangium develops from the septum of the old one, thereby the new sporangium remains inside the old one. This type of development of new sporangium i.e. secondary sporangium inside the old one is called *proliferation of sporangium* (Fig. 4.112C). The zoospores of the new zoosporangium become released through the apical pore of the old one. This process may repeat several times and thus several empty sporangia are observed inside the old one.

Sexual reproduction

Sexual reproduction commonly takes place during unfavourable condition. It is of oogamous type. The male and female sex organs are antheridia and oogonia respectively. They commonly develop terminally on the apex of lateral branches or intercalary as in *S. litoralis*. They are commonly homothallic (monoecious) but a few are heterothallic (dioecious).

Oogonium. The oogonia are developed either singly or in chain at the apex of main hyphae or on lateral branches. They develop intercalary in some species (Fig. 4.113J). During development, the apex of hypha swells considerably and finally becomes spherical or globose in shape. Many nuclei and cytoplasm migrate into the swollen structure and cut off from rest of the filament by septa. With further development, the oogonium swells and a large vacuole develop in the centre; thereby its protoplast becomes pushed towards the periphery. The protoplast undergoes cleavage into many segments and transformed into eggs. Initially each segment contains many nuclei, but later on they becomes uninucleate by degeneration of all the nuclei except one. The oogonium usually contains 4-10 eggs but oogonium with 32 eggs has also been reported.

Antheridium. The antheridia are also developed at the apex of main hyphae or on lateral branches (Fig. 4.113J, K). During development, apex of the hypha swells, many nuclei and cytoplasm migrate inside it and finally separated from the rest mycelium by septa. The mature antheridium is club-shaped with many nuclei.

The antheridia may develop from various regions. When antheridia develop from the stalk of the oogonium are called **androgynous**. They may develop from the same hypha from where the oogonium develops (**monoclinous**) or from the different hypha (**diclinous**).

Fertilization. The mature antheridia become closely attached with the oogonial wall and develop simple or branched and delicate protuberances, the fertilization tube. Only one male nucleus fertilizes an egg and form oospore. Thus many oospores are developed in each oogonium (Fig. 4.113L).

Germination of oospore. After a long period of rest, oospores germinate either directly into hypha or after germination the germ tube swells and form zoosporangium. Each zoosporangium develop many biflagellate and pear-shaped zoospores. Zoospores on germination produce germ tube which develops into new mycelium.

Parthenogenesis. In pathogenic species, *S. farax*; antheridia do not develop close to oogonia and thus fertilization become impossible. Thus the unfertilized eggs act as parthenospores. They germinate directly into mycelium, like the mother.

Schematic representation of the life cycle is given below (Fig. 4.114)

Cytology of *Saprolegnia*

Much literature is available regarding the cytology of sexual reproduction of *Saprolegnia* (Dick, 1969, 1972; Dick and Win-Tin, 1973). The available literature indicates that meiosis takes place in gametangium during gametogenesis. Thus the vegetative thalli are probably diploid (2n). The diploid oospore germinates directly and develops diploid mycelium like the mother.

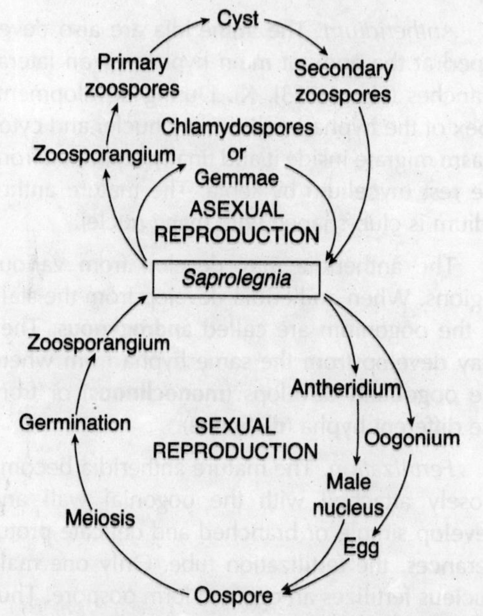

Fig. 4.114: *Saprolegnia*: Schematic representation of life cycle

PYTHIUM

Systemic position

Ainsworth (1973)	Hawksworth (1995)
Kingdom Mycota	Kingdom Stramenopila
Division Eumycota	Phylum Oomycota
Subdivision Mastigomycotina	Class Oomycetes
Class Oomycetes	Subclass Peronosporomycetidae
Order Saprolegniales	Order Pythiales
Family Pythiaceae	Family Pythiaceae
Genus *Pythium*	Genus *Pythium*

The genus *Pythium* (Gr. *pythein*, to cause to rot) is represented by 92 species (Waterhouse, 1968), of which 28 species are reported from India. Several species of *Pythium* grow as saprophytes in soil, in water; on decaying matters, but some are facultative parasites causing root rot (root tips), fruit rot (watery fruits) and damping off of both pre-emergent and post-emergent seedling of numerous angiospermic plants. *P. debaryanum, P. ultimum, P. aphanidermatum* and *P. irregulare* cause damping off of seedling, of which *P. debaryanum* is very common and causes damping-off of seedling of tobacco, chilli, tomato and mustard. *P. gracilis* grows as parasite on *Vaucheria* (alga). List of the pathogens along with diseases caused by them is given in Table 4.20.

Table 4.20: Important diseases caused by different species of *Pythium*

Causal Organism	Disease
1. *Pythium debaryanum*	"Damping-off" of tobacco, tomato, castor, chilli and cress (*Lepidium*).
2. *P. aphanidermatum*	"Rhizome rot" of ginger, "Foot rot" of papaya (*Carica papaya*). "Damping off" of potato and "Fruit rot" of cucurbits.
3. *P. myriotylium*	"Rhizome rot" of ginger.
4. *P. graminicolum*	"Rhizome and root rot" of turmeric.
5. *P. indicum*	"Fruit rot" of lady's finger (*Hibiscus esculentus*).
6. *P. butleri*	"Fruit rot" of cucurbits.

Pythium debaryanum, the very common species, is used to describe the general life history of *Pythium* (Fig. 4.117).

Vegetative structure

The mycelial plant body consists of slender, cylindrical, hyaline, coenocytic hyphae (Fig. 4.115). Septa are formed at the time of reproduction. The hyphae grow both intra- and intercellularly. No haustoria are developed.

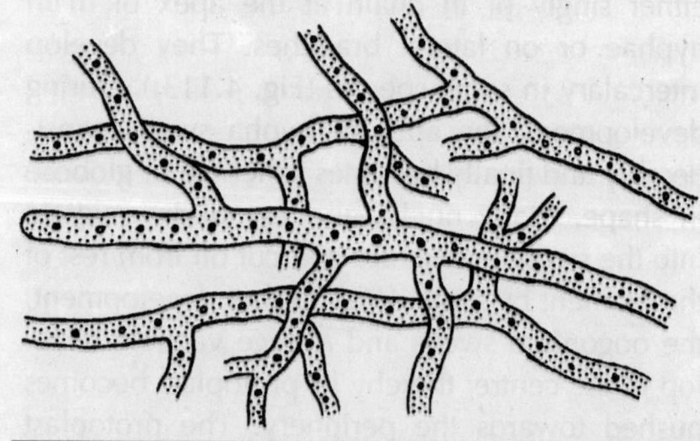

Fig. 4.115: Vegetative mycelium of *Pythium debaryanum*

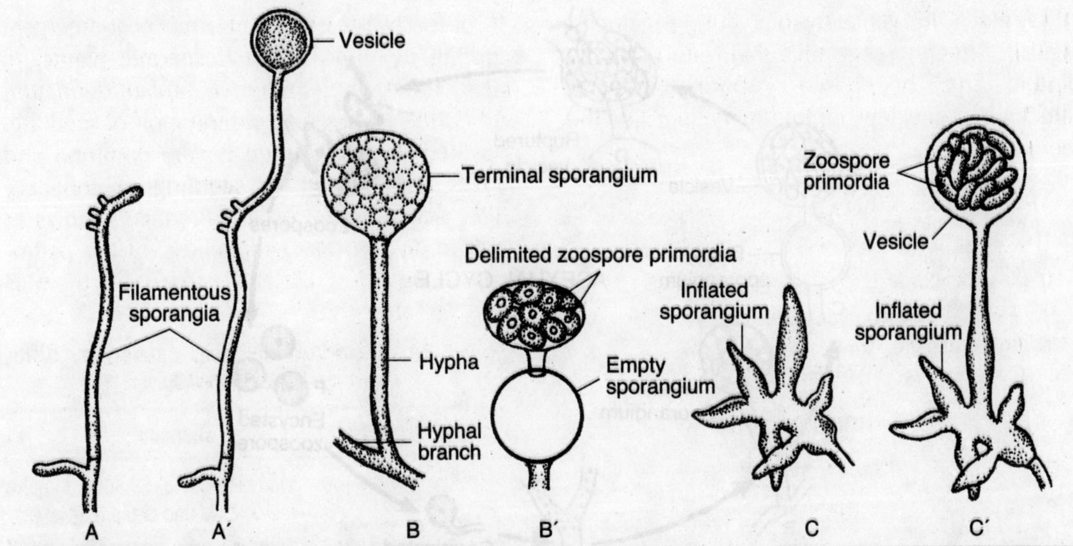

Fig. 4.116: Different types of sporangia of *Pythium* and their germination: A-A′. Filamentous sporangia, B-B′. Globose sporangia, C-C′. Lobulated sporangia

Asexual reproduction

It occurs by zoospores which are produced inside the vesicle formed by the sporangium. The sporangia are globose to oval and are either terminal or intercalary in position on the vegetative hyphae. The terminal or the portion in intercalary region of hyphae swell up and form globose to oval structure, the sporangium (Fig. 4.116B, B′, 4.117B, C). The sporangium is multinucleate. During zoospore formation, the clevage of cytoplasm starts in the thin-walled bubble like vesicle at the tip of long tube that produced from the sporangium (Fig. 4.117C). 8–20 zoospores are produced in the vesicle. After maturation, wall of the vesicle bursts and the zoospores are liberated (Fig. 4.117D).

The zoospores are of secondary type (i.e., kidney-shaped) having two lateral flagella attached at the concave region (Fig. 4.117E). After swimming for sometimes, the zoospores come to rest, encyst (Fig. 4.117F) and then germinate by germ tube to form new mycelium (Fig. 4.117G).

Sporangium of other types, like *filamentous sporangia* (Fig. 4.116A, A′) are formed in *P. gracilis, P. monospermum* and *P. papillatum*. They are scarcely distinguishable from the vegetative cell and are slender simple or branched filaments. The *lobulated sporangia* (Fig. 4.116C, C′) formed by inflated lobed hyphae are found in *P. graminicola* and *P. aphanidermatum*. In both the cases, long tube develops from the sporangium whose tip swells up and forms vesicle inside which zoospores are differentiated.

The sporangial proliferation occurs in *P. proliferum*, where new sporangia are developed by proliferation from old sporangium.

Sexual reproduction

It takes place by the development of *antheridia* and *oogonia* which are easily developed in cultures and also after death of the host when the fungus lives saprophytically. Most of the species like *P. debaryanum* are homothallic, but some like *P. heterothalicum, P. sylvaticum, P. splendens, P. intermedium* and *P. catenulatum* are heterothallic.

Oogonium develops as terminal or intercalary swelling of the hypha, which gradually becomes spherical in shape and cuts off by transverse septa from rest of the mycelium. The antheridium develops as terminal club-shaped swelling of hyphal tip, and may originate either from branches of oogonial stalk (**monoclinous**) or from neighbouring hyphae (**declinous**) (Fig. 4.117H). The antheridium is very much smaller in size than the oogonium and a number of antheridia comes in contact with the oogonial wall.

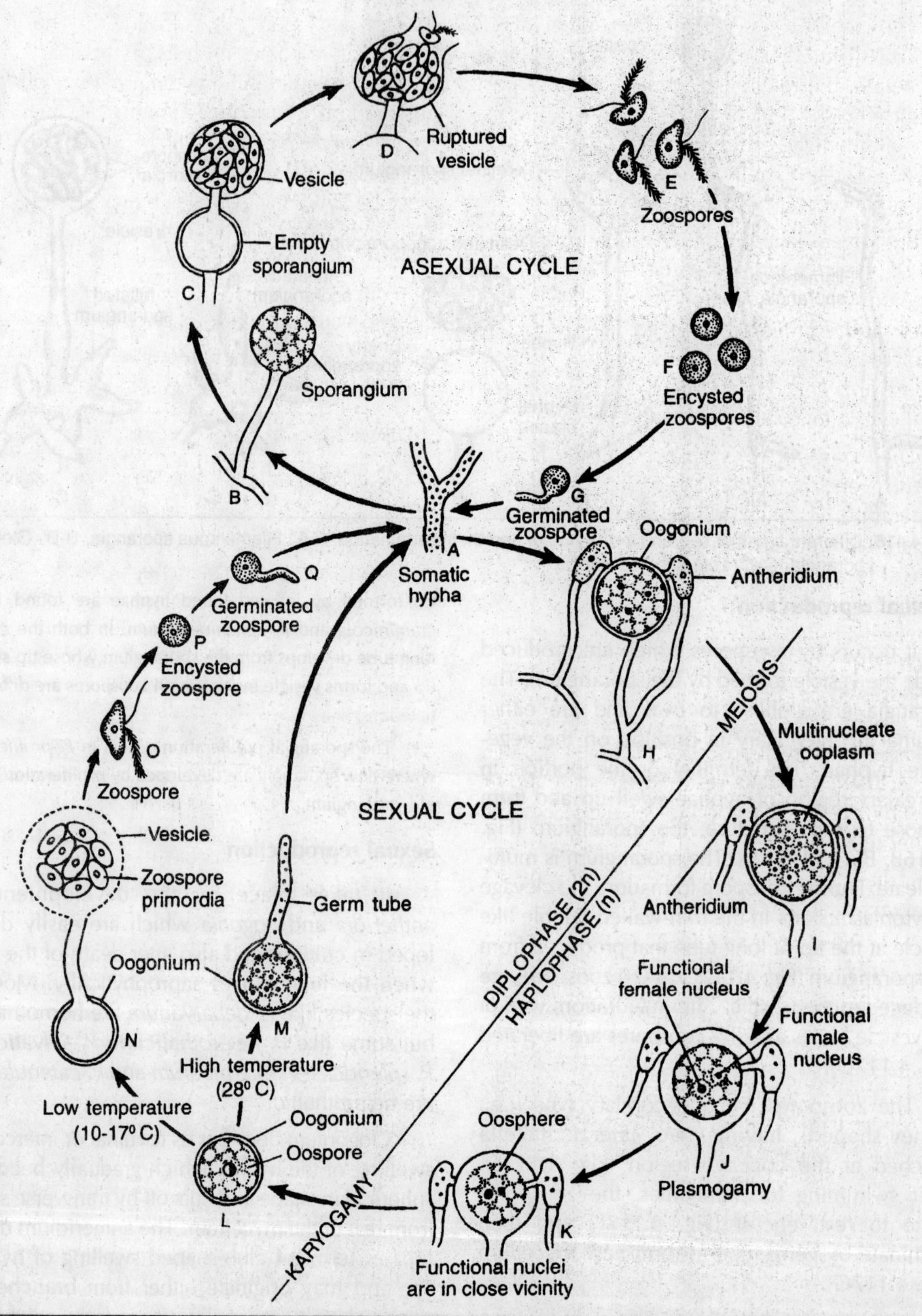

Fig. 4.117: Life cycle of *Pythium debaryanum*

The young oogonium is multinucleate and globose. Eventually, its protoplast is differentiated into central oosphere surrounded by periplasm. Both the regions are multinucleate. The number of nuclei in the oogonium gradually reduces with age. The surviving nuclei undergo meiosis and form 32 nuclei (Fig. 4.117I). All nuclei except one in the oosphere degenerate, the remaining nucleus functions as an egg (Fig. 4.117J). The young antheridium is also multinucleate and with maturity all except one degenerate. The surviving one undergoes meiosis to form 4 haploid nuclei (Fig. 4.117I).

Before fertilization, a fertilization tube develops by the antheridium at the contact wall of oogonium through which one of the male nuclei passes and fertilizes the egg. The fertilized oosphere develops into a thick walled, smooth oospore or zygote (Fig. 4.117J, K, L). The oospore germinates after a period of rest.

The pattern of oospore germination depends on the variation of environmental temperature. At high temperature (28°C), the oospore germinates by germ tube (Fig. 4.117M), while at low temperature between 10–17°C, the germ tube elongates at a length of 5–20 μm, then swells up at the tip and forms a vesicle in which zoospores develop (Fig. 4.117N). The zoospores come out of the vesicle (Fig. 4.117O) after bursting the vesicle wall. After swimming for some time, the zoospores become encysted (Fig. 4.117P) and take rest. After rest, it germinates by germ tube, which develops like the mother mycelium.

Indian species: *Pythium graminicola, P. debaryanum, P. aphanidermatum, P. butleri* etc.

PHYTOPHTHORA

Systemic position

Ainsworth (1973)	Hawksworth (1995)
Kingdom Mycota	Kingdom Stramenopila
Division Eumycota	Phylum Oomycota
Subdivision Mastigomycotina	Class Oomycetes
Class Oomycetes	Subclass Peronosporomycetidae
Order Peronosporales	Order Peronosporales
Family Pythiaceae	Family Peronosporaceae
Genus *Phytophthora*	Genus *Phytophthora*

The genus *Phytophthora* is represented by about 40 species, reported from different parts of the world. The species may be either facultative saprophyte or facultative parasites. About 27 species are reported from India. The name *Phytophthora* comes from its importance in crop destruction (Greek *phyton*, plant; *phthora*, destruction). The most important species is *Phytophthora infestans*, which causes late blight disease of potato by which the Irish famine took place in 1845. The pathogen is most destructive in low temperature and high humidity. Some other species of *Phytophthora* are also pathogenic to different hosts. A list of pathogens along with diseases caused by them is given in Table 4.21.

Table 4.21: Important diseases caused by the species of *Phytophthora*

Causal Organism	Disease
1. *Phytophthora infestans*	Late blight disease of potato.
2. *P. arecae*	Koleroga disease of *Areca* plants.
3. *P. colocasiae*	Blight of *Colocasia*.
4. *P. parasitica*	Foot rot and leaf blight of Pipper betle ("Pan"), seedling blight of castor and leaf blight of *Sesamum indicum* ("Til").
5. *P. nicotianae*	Black shank of *Nicotiana tabacum* ("Tobacco").
6. *P. palmivora*	Gumasis of citrus and bud rot of palms.
7. *P. cactorum*	Root rot, damping off and fruit rot of more than 40 angiosperms.

Vegetative structure

Vegetative body is of eucarpic thallus, consisting of mycelium. The mycelium is profusely branched and consists of aseptate hyaline, irregularly branched hyphae, where septa may develop in the older parts and at the base of sex organs. The branches arise at right angles. The mycelium grows both intra- and intercellularly. During intercellular growth, it develops finger-like haustoria inside the neighbouring host cells. The hyphae are about 4-8 μm in diameter with about 0.01 μm thick wall.

The cell wall lacks chitin and is mainly composed of glucan (about 90%). The protoplasm contains numerous nuclei, mitochondria, endoplasmic reticulum, ribosome, dictyosome and vacuoles.

Reproduction

Phytophthora reproduces asexually and sexually.

Asexual reproduction

Asexual reproduction takes place by indirect and direct germination of sporangium and also by chlamydospore formation.

Sporangium formation: The sporangia develop on sporangiophore, which come out from the infected leaf through stomata or through piercing the epidermal cell towards lower side (Fig. 4.118A). The sporangiophore is branched (*P. infestans*). The sporangium develops terminally (Fig. 4.118B-D), but it becomes shifted to lateral position by further growth of the sporangiophore (Fig. 4.118E, F) which develops again a new sporangium at its tip. Thus the sporangiophore behaves as sympodia and the branching of conidiophore of *P. infestans* is called sympodial branching.

The sporangium is thin-walled, hyaline to light yellow coloured, spherical, oval or pear-shaped with a small stalk below and thin papilla above (Fig. 4.118G). It measures 22–25 µm × 16–25 µm.

The mature sporangium is detached from the sporangiophore and is dispersed by rain splash.

Depends on the environmental condition, the germination of sporangium may be indirect or direct. Low temperature (below 15°C) and wet condition i.e., high humidity (about 100%) favour indirect germination, while high temperature (above 20°C) and dry condition i.e., low humidity (below 90%) favour direct germination of sporangium.

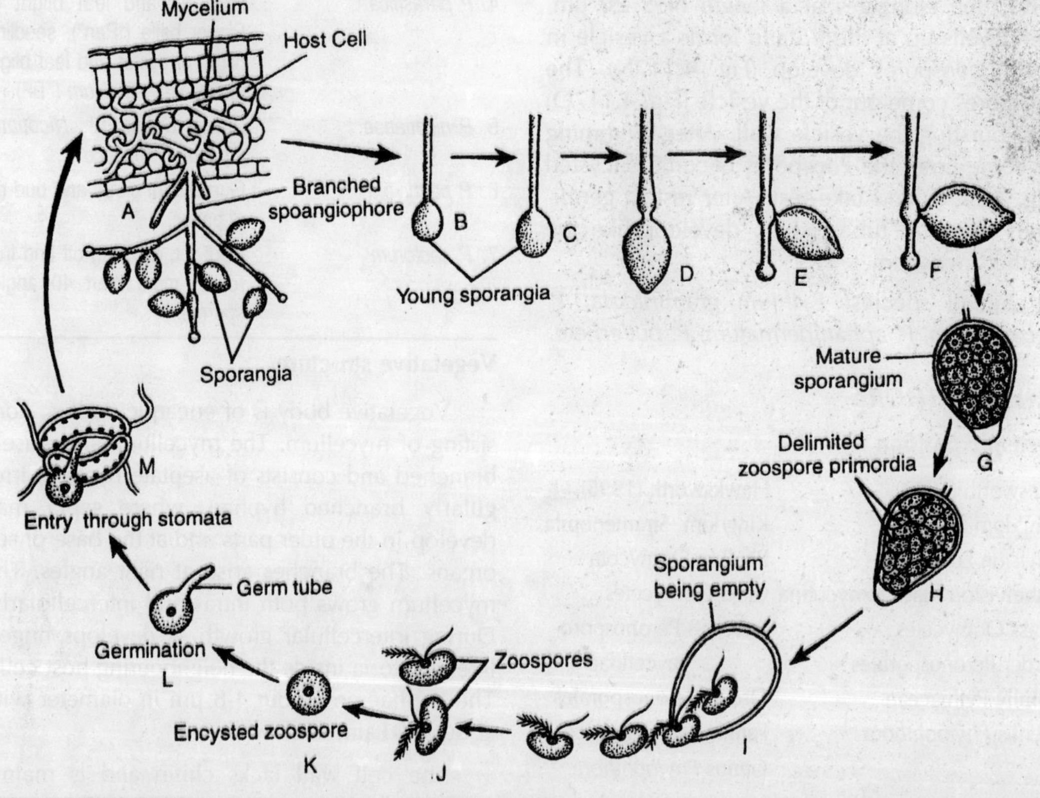

Fig. 4.118: *Phytophthora infestans*, Formation and germination of zoospores: A. Branched sporangiophore emerging through stomata. B-F. Stages of development of sporangium. G-I. Stages of formation of zoospores from zoosporangium, J. Zoospores, K. Encysted zoospore, L. Germination of zoospore and M. Entry of germ tube through stomata

During *indirect germination,* the content of sporangium breaks into 3–13 or more small uninucleate segments (Fig. 4.118H). Each segment becomes kidney-shaped zoospore and it develops two laterally placed flagella of which one being whiplash and other tinsel type (Fig. 4.118I). These zoospores can swim in waterfilm of rain or dewdrop. On coming in contact with host surface, they first lose their flagella, become encysted (Fig. 4.118K) and then germinate by producing germ tube (Fig. 4.118L). The germ tube then penetrates directly through stoma and causes infection (Fig. 4.118M) or the germ tube at its tip may produce an appressorium. The appressorium develops infection peg towards its lower side and causes further infection piercing through epidermal cell.

During *direct germination,* the sporangium behaves as conidium and germinates directly by producing multinucleate germ tube which causes infection like that of zoospores.

Sexual reproduction

The sexual reproduction is of **oogamous** type which takes place with the help of definite male and female reproductive organs, known as **antheridium** and **oogonium**, respectively. The sex organs arise at the tip of short lateral hyphae. The species of *Phytophthora* may be homothallic (*P. himalayensis*) or heterothallic (*P. infestans*).

Depending on the development of antheridium and oogonium, ***Phytophthora*** is of two types: ***Paragynous*** and ***Amphigynous***:

1. In *Paragynous species* (Fig. 4.119A) like *P. cactorum,* the antheridia are attached laterally on the oogonial wall, developed from same or different hyphae.

2. In *Amphigynous species* (Fig. 4.119B) like *P. infestans, P. erythroseptica, and P. cactorum,* the antheridium develops first and a nearby hypha penetrates through the developing antheridium and emerges out, its tip inflates to form a spherical oogonium. The antheridium remains as collar at the base of the oogonium.

The detailed process of sexual reproduction of *P. infestans* is given (Fig. 4.120):

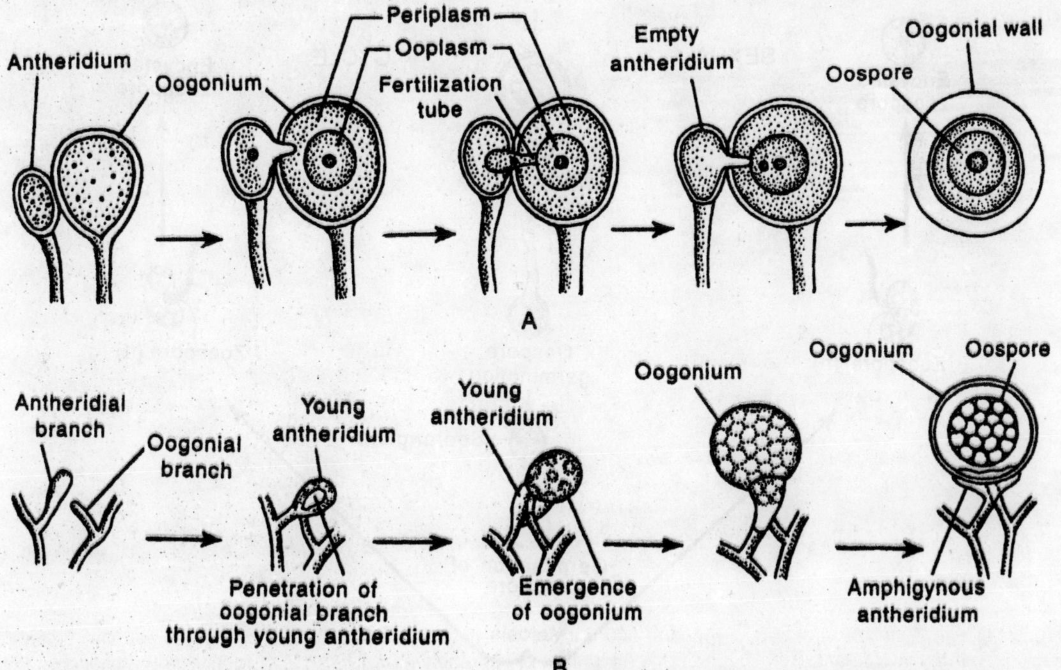

Fig. 4.119: Development of oospore in Paragynous (A) and Amphigynous (B) species of *Phytophthora*

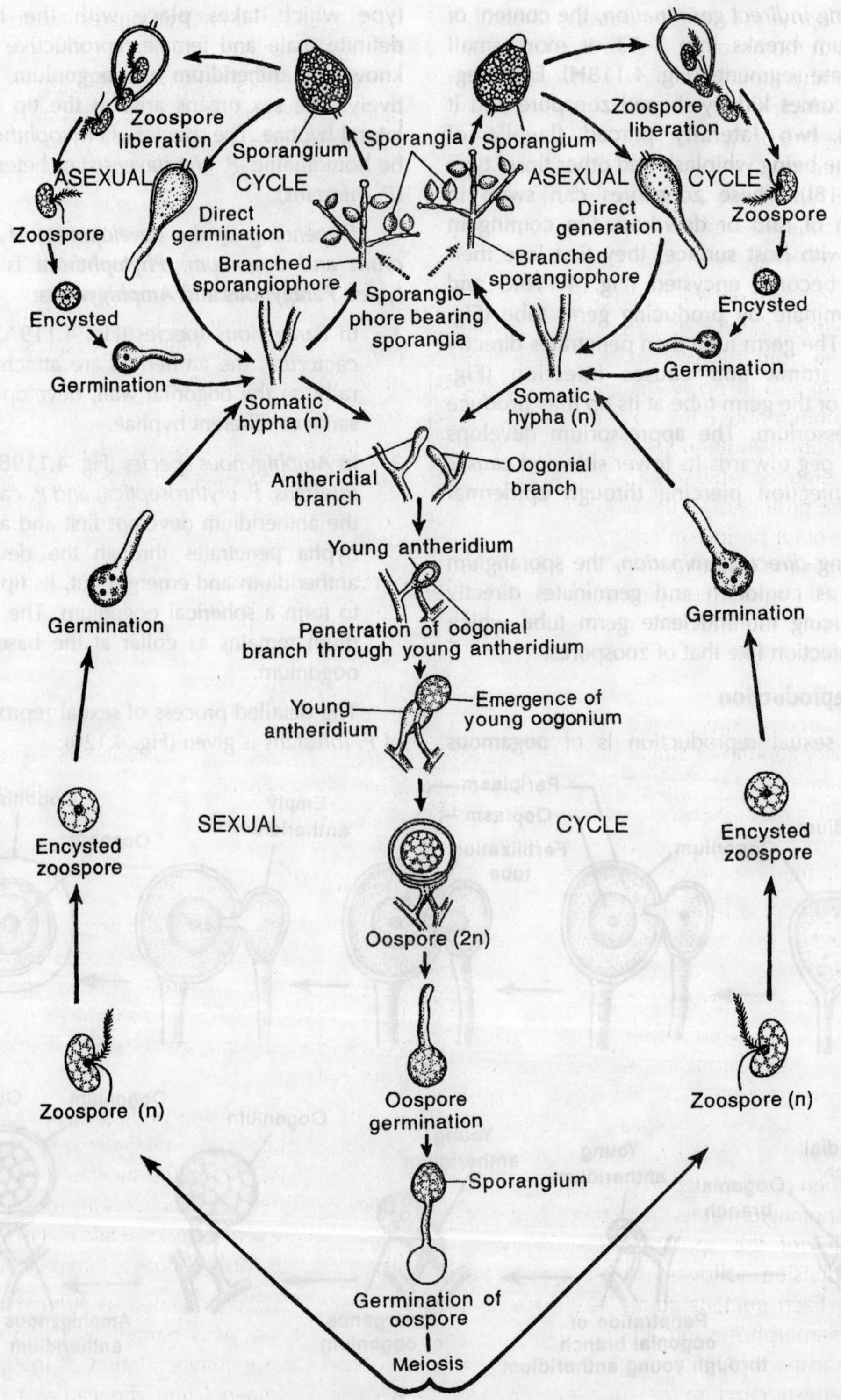

Fig. 4.120: Life cycle of *Phytophthora infestans*

Development of Antheridium: Antheridium is formed at the tip of the lateral hypha. The tip swells up and then a partition wall develops at the base which cuts off the upper antheridium containing one or two nuclei. The nuclei undergo repeated mitosis by stimulus of developing oogonium and form 8–12 nuclei. At maturity only one nucleus remains functional and the others degenerate.

Development of Oogonium: Oogonium also develops from the tip of neighbouring hypha. The hypha initially pierces the developing antheridium and its tip swells up to form a globose oogonium above the antheridium. Initially the oogonium is multinucleate and those nuclei again undergo repeated mitosis to form about 24 nuclei. The protoplast of the oogonium differentiates into outer periplasm and central ooplasm. Initially both the regions are multinucleate, later all the nuclei except one pass from ooplasm to the periplasm. The nuclei in the periplasm later degenerate and the ooplasm with its one nucleus functions as an egg.

Fertilization: After maturation of antheridium and oogonium, a single nucleus from the antheridium passes to the oogonium through fertilization tube and fertilizes the egg resulting in the formation of oospore. Karyogamy is delayed till the oospore wall matures.

The oospore is double-walled, consisting of comparatively thin exospore and thick endospore. Parthenogenic development of oospore also takes place in *P. infestans*.

Germination of Oospore: The oospore undergoes maturation for several weeks to months. Before germination, the oospore nucleus (2n) undergoes repeated mitotic division and forms a number of diploid nuclei. The oospore then produces germ tube which cuts off terminal sporangium at its tip by transverse septa. The nuclei of the sporangium first undergo *meiotic* division followed by several mitotic divisions. Each nucleus along with some cytoplasm metamorphoses into a biflagellate kidney-shaped zoospore. The zoospores after release from the zoosporangium remain motile for some time. On contact with suitable host, it becomes encysted and then germinates to form new mycelium again.

Some views regarding meiotic stage: The place of meiosis in *Phytophthora* is still not clear. According to Sansome (1963, 65, 67) and others, vegetative thallus is diploid and reduction division (meiosis) takes place in gametangium to form haploid gametes. According to Day (1974) however, the place of occurrence of meiosis in the life cycle is still a vexing question. On the other hand, Webster (1980) viewed that reduction division occurs at the time of oospore germination and the zoosporangium bears haploid zoospores.

Indian species: *Phytophthora infestans, P. parasitica, P. arecae, P. cactorum, P. colocasiae.*

ALBUGO

Systemic position

Ainsworth (1973)	Hawksworth (1995)
Kingdom Mycota	Kingdom Stramenopila
Division Eumycota	Phylum Oomycota
Subdivision Mastigomycotina	Class Oomycetes
Class Oomycetes	Subclass Saprolegniomycetidae
Order Peronosporales	Order Peronosporales
Family Albuginaceae	Family Saprolegniaceae
Genus *Albugo*	Genus *Albugo*

The genus *Albugo* is represented by about 45 species, out of which 18 species are reported from India. The common Indian species are *A. bliti, A. candida, A. evolvulii, A. portulacae, A. platensis, A. trapogonosis* and *A. ipomoeae-panduranae*.

They are obligate endo-paratites of higher plants belonging to Brassicaceae (Cruciferae), Convulaceae, Amaranthaceae, Asteraceae (Compositae) etc. *A. candida* is the most common species of the genus causes **white rust of Crucifers** like mustard, radish, cabbage, turnip etc.

The genus was named as *Albugo* in 1821 and in 1874 it was named as *Cystopus*. But based on the recommendation of International Botanical Nomenclature the earlier name i.e. *Albugo* was considered as valid.

Vegetative structure

Vegetative body is **eucarpic** thallus, consisting of **coenocytic** mycelium. The mycelium is profusely branched and consists of aseptate hyaline irregularly branched hyphae. The hyphal wall is composed of chitin and the cytoplasm is granular, contains nuclei, oil globules, glycogen and vacuoles. A system of unit plasma membrane is present towards the cytoplasm. Septa may develop in older parts and at the base of the sex organs. The hyphae grow **intercelluarly** and absorb nutrition from host cell with the development of small knob-like **haustorium** (Fig. 4.121B). Haustoria are differentiated into slender stalk (about 0.5 μm wide) and a head (about 4 μm in diameter). The haustorial head contains dense cytoplasm devoid of nucleus but packed with mitochondria, ribosomes, endoplasmic reticulum and lipids. The base of haustorium is surrounded by a collar-like sheath formed by the extension of the cell wall of the host. The haustorium is separated from the host cell membrane by an encapsulation made up of extra-haustorial matrix.

Reproduction

Albugo reproduces both asexually and sexually.

Asexual reproduction

During asexual reproduction the hyphae of intercellular mycelium form palisade like cells below the epidermis. The apices of these hyphae develop into short, erect, thick-walled and club-shaped structures, called **sporangiophores**. Sporangiophores lie perpendicularly and very closely like the palisade-like layer below the epidermis (Fig. 4.121A). The apical region of the sporangiophore is thin walled, inside which many nuclei are present in dense cytoplasm. Later on, it enlarges and consequently a constriction develops below the swollen end, results the formation of a multinucleate, oval or spherical sporangium. Following the same manner another sporangium develops at the apex of the sporangiophore. The process repeats several times and thus a chain of sporangia is formed, where the sporangia will remain in **basipetal succession** i.e. younger one at the base and oldestest at the apex of the chain. *Like, the younger brother will remain on the lap of mother and elders have to go apart.* During the successive development of sporangia, a mucilaginous pad or disc develop between the adjascent sporangia, called **disjunctor** (Fig. 4.121C).

Due to gradual development of a mass of sporangia, pressure develops inside causes swelling of the epidermis and finally it ruptures. Thereby the sporangia become visible on the host surface as white powdary mass.

The sporangia are smooth, colourless or very light in colour, multinucleate, and spherical or shortly elliptical to cylindrical, 12-25 μm in diameter (Waterhouse, 1973).

Sporangia are dispersed from the white rusty pustules, by shrinking and drying of the mucilaginous disc. Thus the sporangia get separated and initially become accumulated in the space between the ruptured epidermis of host and the sporangiophore. Later on, the sporangia are blown away by wind or rain splash. After landing on differ region of the same host or on other suitable host the sporangia start germination within a few hour under favourable condition.

Germination of sporangium: The nature of germination of sporangium depends on environmental condition:

1. **In moist condition and low temperature (about 10°C)**: In this condition the sporangium behaves as a zoosporangium. Initially, it swells by absorption of some water and its multinucleate protoplast divides to form 4-12 polyhedral daughter protoplasts. The daughter protoplast then develops into single zoospore (Fig. 4.121D). Simultaneously a papilla also protrudes from the sporangium. The papilla enlarges and the young zoospores are initially transferred inside the sessile vesicle (Venterpool, 1959). The zoospores are released in groups (Fig. 4.121D). The process of differentiation of sporangium to release of zoospores takes about 2 minutes (Webster, 1980).

Zoospores are biflagellate (Fig. 4.121E), having one short and tinsel type; and other one is long and whiplash type.

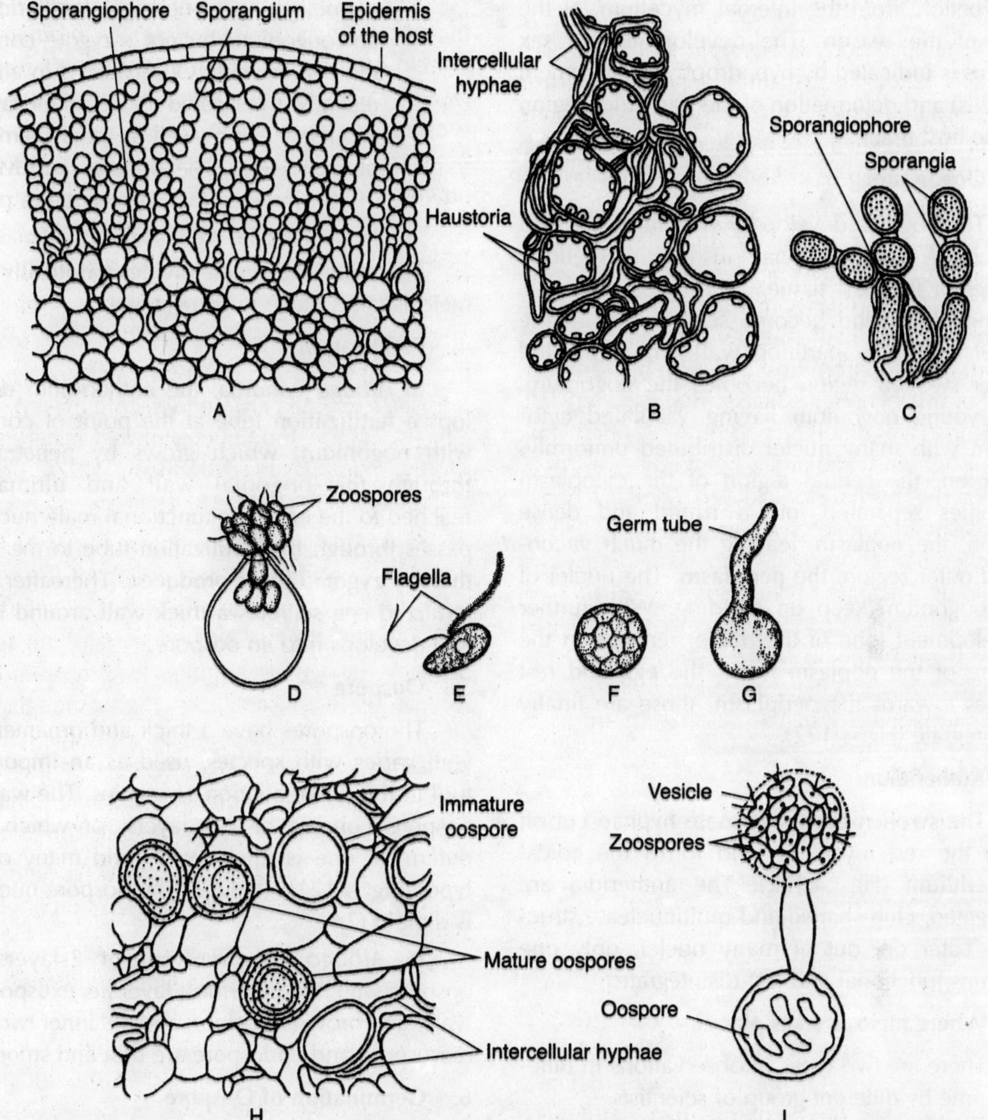

Fig. 4.121: *Albugo candida*: A. Section through a pustule showing sporangiophores and sporangia, B. Enlarged sectional view of infected leaf tissue showing intercellular hyphae and haustoria, C. Attached sporangia on sporangiophores, D. Discharge of biflagellate secondary zoospores, E. Single biflagellate secondary zoospores, F. Encysted zoospore, G. Germinating encysted zoospore, H. Portion of host tissue showng oospores at different stages of development and I. Germination of oospore

The released zoospores get in contact with the host, release their flagella (deflagellated), secrete a wall and become encysted (Fig. 4.121F). Later on, the encysted zoospore forms a germ tube (Fig. 4.121G), which penetrates the epidermis and develops into a new mycelium.

2. **In dry condition and high temperature:** In this condition the sporangium germinates directly into germ tube and directly infects the host, which is uncommon in this genus.

Sexual reproduction

Sexual reproduction is **oogamous** type. The female sex organ i.e. **oogonium** and male sex organ i.e. **antheridium**, *both are developed near each other* into the deeper layer of tissue of stem and petiole from the internal mycelium, at the end of the season. The development of sex organs is indicated by hypertrophy (enlargement of cells) and deformation of the particular region of the host tissue.

1. **Oogonoum**

The oogonia develop by swelling the apical region of female hyphae in the intercellular spaces of the host tissue. The apical region is multinucleate and become separated from the lower region by partition wall. The separated upper swollen region becomes the oogonium. The young oogonium having vaculated cytoplasm with many nuclei distributed uniformly. Later on, the central region of the cytoplasm becomes separated into a round and dense region, the **ooplasm**, leaving the much vacuolated outer region, the **periplasm**. The nuclei of the oogonium keep on dividing. With further development, one of the nuclei remains in the centre of the ooplasm forms the **egg** and rest moves towards the periplasm, those are finally disintegrate (Fig. 4.122).

2. **Antheridium**

The swollen tip of the male hyphae cut off from the rest mycelium and forms the apical **antheridium** (Fig. 4.122). The antheridia are elongated, club-shaped and multinucleate structure. Later on, out of many nuclei, only one remains functional and rest disintegrate.

3. **Where meiosis takes place?**

There are two types of observations in different time by different group of scientists.

(i) Thirumalachar *et al.* (1949) reported that in some species of *Albugo, meiosis takes place in the zygotic nucleus*. Later, Smith (1955) also reported that meiosis takes place in zygote during germination and divisions in sex organs are all mitotic.

(ii) On the other hand, Stevens (1899) stated that nuclear divisions of the gametangial nuclei represented by meiosis. Later, Sansome and Sansome (1974) based on their work on *Albugo candida*, also observed the occurrence of meiosis in gametangia (antheridium and oogonium) but not in zygote, confirm the earlier view. Observations in all the recently investigated species of the family Albuginaceae as well as other members of Oomycetes (Alexopoulos and Mims, 1979), indicate that meiosis takes place in gametangia and not in zygote.

Therefore, the zygote divides mitotically and meiosis takes place in gametangia.

4. **Fertilization**

In *Albugo candida*, the antheridium develops a fertilization tube at the point of contact with oogonium, which grows by penetrating through the oogonial wall and ultimately reached to the egg. The functional male nucleus passes through the fertilization tube to the egg; thereby zygote (2n) is produced. Thereafter, the fertilized egg secretes a thick wall around itself and develops into an oospore.

5. **Oospore**

The oospores have a thick and ornamented wall varies with species, used as an important tool in the differentiation of species. The wall of oospore consists of 2-3 layers, of which the outermost one is thick, warty and many other types (Fig. 4.121H, I; 122). The oospore nucleus is diploid (2n).

In *Albugo candida*, out of 3-layers of oospore wall, the outermost layer i.e. exospore is warty and brown in colour, and the inner two i.e. mesospore and endospore are thin and smooth.

6. **Germination of Oospore**

The oospores undergo a period of rest for several months before germination (Webster, 1980). With the onset of favourable condition, the infected host tissue initially swells and decay, thereby the oospores are released. The nucleus (2n) of the oospore undergoes repeated divisions to form as many as 32 nuclei.

Although some workers believe that nucleus of zygote divide meiotically, but Stevens (1899) and the recents workers (Sansome and Sansome, 1974; Alexopoulos and Mims, 1979, Webster, 1980) support the view that nucleus of zygote divide mitotically.

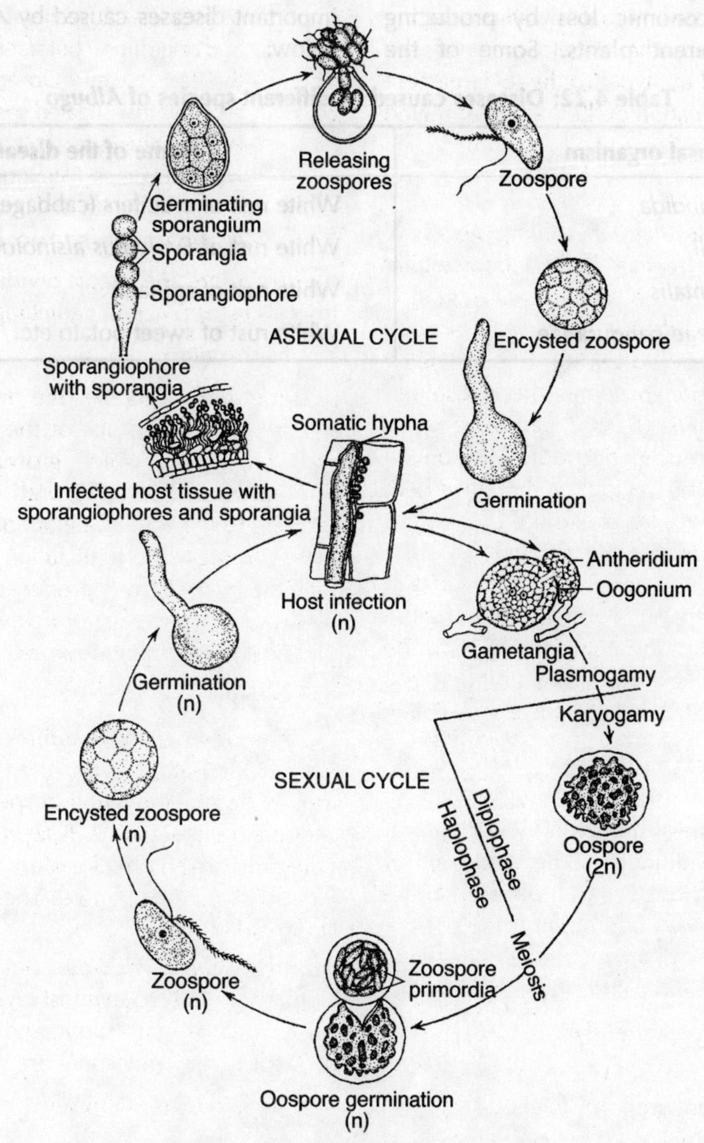

Fig. 4.122: Life cycle of *Albugo candida*

During germination in the following spring, outer wall burst and the inner wall comes out in the form of a thin spherical vesicle (Fig. 4.121I). The vesicle may be sessile or develop at the end of the wide cylindrical tube. Inside the vesicle hundreds of zoospores are formed (de Bary, 1863), but Webster (1980) pointed out that only 40-60 zoospores are formed. Later on, the vesicle dissolves, releasing the zoospore and the zoospores swim freely.

Zoospores are uninucleate, reniform and biflagellate (Fig. 4.122). Reaching on a suitable host, the zoospores release the flagella and become encysted. The encysted zoospore germinates by producing germ tubes (Fig. 4.122) and infect the new host and produce the mycelium.

Common diseases caused by *Albugo*

In India out of the available species, majority of them grow as parasite and some of them caused severe economic loss by producing diseases on different plants. Some of the important diseases caused by *Albugo* are given below:

Table 4.22: Diseases caused by different species of *Albugo*

Causal organism	Name of the disease
1. *Albugo candida*	White rust of crucifers (cabbage, radish etc.)
2. *A. evolvuli*	White rust of *Evolvulus alsinoides*
3. *A. occidentalis*	White rust of spinach
4. *A. ipomoeae-panduranae*	White rust of sweet potato etc.

ALLIED FUNGI

KINGDOM PROTOZOA

Introduction

Members of the Kingdom *Protozoa* are not the true fungi and mycologists called them as 'slime molds'. Earlier, they were placed and described separately under the division Myxomycota separating them from true fungi, those are placed under Eumycota. Anton de Bary (1887) called them as Mycetozoa (*mycete*, fungi; *zoa*, animals) and similarly zoologist have also studied this group of organisms under the phylum *Protozoa*.

The slime moulds are placed under kingdom *Protozoa*, as a new eukaryotic kingdom segregated from the kingdom Protista. The kingdom Protozoa includes 6 classes, 11 orders, 30 families, 122 genera and 1161 species. The important characteristics of the kingdom are given below:

Important characteristics

1. Members are predominantly unicellular, plasmodial or colonial phagotropic, and wall-less in the tropic state.
2. Ciliary hairs never rigid or tubular.
3. Chloroplasts, where present lacking starch and phycobilisomes, with stalked thylakoid and 3 membranes.
4. Multicellular species with minimal cells differentiation and lacking collagenous connective tissue sandwiched between dissimilar epithelia.

Present status of Slime molds

Slime mold (also spelled slime mould) is an informal name given to several kinds of unrelated eukaryotic organisms. They live freely as single cells, but can aggregate together to form multicellular reproductive structures. Slime molds were formerly classified as fungi but are no longer considered as part of that kingdom. Although they are not related to one another, but for convenience they are still sometimes placed within the paraphyletic group, the **kingdom Protozoa.**

Occurrence

The **Protozoa** are common but inconspicuous inhabitants like damp soil, rotting logs, moist dead wood, leaf mold, moist sawdust, bark of trees, decaying fleshy fungi, or on other organic matter. They often spend most of their lives within the substrate and emerge only during production of sporangia. During rainy season they may occur on leaves of grasses or on other plants.

Types of slime molds

Slime molds were earlier considered as one monophyletic group (i.e., all are closely related) within the **fungi** or perhaps within the **protozoa** called as Mycetozoa i.e. "fungus animals".

But the modern view increasingly favours the idea that there are many kinds of slime molds that are not closely related. In fact some scientists would place those different kinds, in different Phyla. Depending on observation, there are at least of three types or perhaps as many as six types of slime molds. These are:

1. **Plasmodial (or "true") slime molds**

This group includes Myxomycota or Myxogastria or Myxomycetes, examplified by *Physarum*.

These are 1-celled amoeboflagellates that combine to make *"plasmodia"* by merging cells into one huge cell with thousands of nuclei. Different kinds produce different types of fruiting bodies. The most common of which is some sort of sporangium from which spores are released. There are about 450 species, which include movable slime molds.

2. **Cellular slime molds**

This group includes Dictyostelids or Dictyosteliomycota, exemplified by *Dictyostelium*.

These are amoeboid 1-celled organisms that combine to make larger-scale single organisms in which individual cells remain distinct within the *"pseudoplasmodium"* they have formed. Due to this, they are "cellular" rather than plasmodial-type. There are about 50 species in this group.

3. **Microscopic slime moulds**

This group includes Protostelia, exemplified by *Nematostelium*. These are microscopic in structure.

These are less known than the plasmodial and cellular slime molds. Like plasmodial or "true" slime molds, they form a multinucleate plasmodium.

4. **The Acrasids or Acrasiomycota, which include *Fonticula alba***

These organisms function like cellular slime molds, but molecular and ultrastructural evidence indicate that they are evolutionarily unrelated to the other mycetozoa. They have often been classified with the Dictyostelids within the cellular slime molds, but at least one classification now considers Dictyostelids and Acrasids as separate phyla of cellular slime molds. Dictyostelids are typically soil organisms, whereas Acrasids are found on dead plants, tree bark, dung, and soil. There are 15 known species.

5. **The Labyrinthulomycota or slime nets, which include *Labyrinthuloides* and *Aplanochytrium***

These are sometime called "slime molds", but they don't seem to be closely related to other kinds of slime molds, and form a separate phylum in some classifications. They form a "net-plasmodium" and are mostly if not entirely marine.

6. **The Plasmodiophoromycota**

These are "true" or plasmodial slime molds, but differ from the Myxomycota in that the Myxomycota are free-living, whereas the Plasmodiophoromycota are parasitic on plants and fungi. They are considered by some to be an entirely different phylum.

7. **The Myxobacteria, which include *Chondromyces apiculatus, Myxococcus fulvus, Myxococcus stipitatus,* and *Stigmatella aurantiaca.***

All of the above are eukaryotes, but this group is prokaryotic or bacterial and part of the proteobacteria. In spite of that they go through a typical life cycle.

Types of plasmodia

The plasmodium is a membrane-bound single cell containing multiple nuclei. The plasmodium of' some species may have a slime sheath, which appears as a "slime track" left behind when the plasmodium migrates across a given substrate, indicating that plasmodium has been there. Plasmodial colour may be affected by pH, temperature or ingested materials.

Commonly the plasmodia are of **four types.** These are:

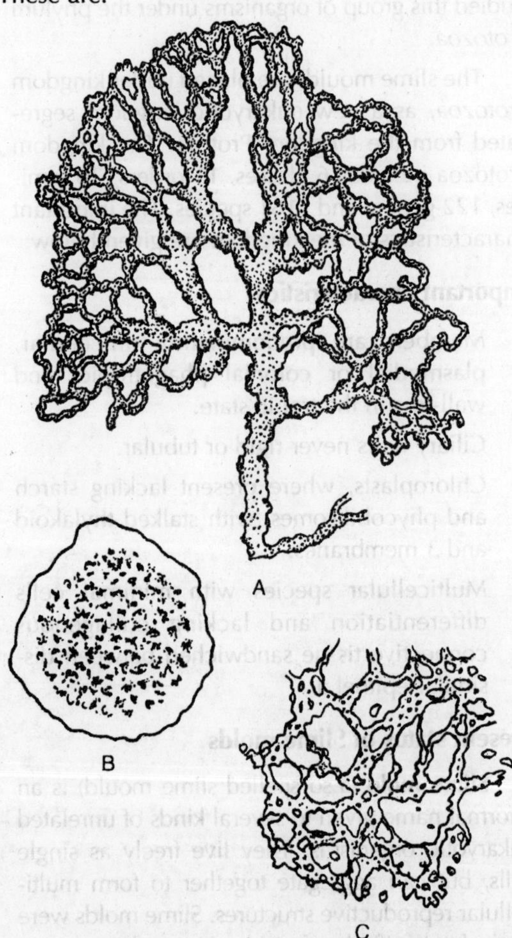

Fig. 4.123: Three types of Plasmodia in the Myxomycetes: A. Aphanoplasmodium, B. Protoplasmodium and C. Phaneroplasmodium

1. **Phaneroplasmodium**

Phaneroplasmodium (Greek *planeros,* visible) is the most common type of plasmodium. This type appears as a fan-shaped network of veins containing streaming protoplasm (Fig. 4.123C). This type is common in the order Physarales.

2. **Aphanoplasmodium**

Aphanoplasmodium (Greek *aphanes*, invisible) is rarely observed in nature. The veins are very thin and a slime sheath is lacking (Fig. 4.123A). Aphanoplasmodia are characteristic of the order Stemonitales.

3. **Protoplasmodium**

Protoplasmodium (Greek *protos,* first) is probably the most primitive type. They are always microscopic. This plasmodium has no veins; exhibits slow, irregular protoplasmic streaming and produces a single very small fruiting body (Fig. 4.123B). Protoplasmodia are found in the order Echino-steliales and Liceales.

4. **Other type**

This type of plasmodium exhibits characteristics of both phaneroplasmodia and aphanoplasmodia and occurs in the order Trichiales.

Types of Fruiting bodies

There are **four types** of fruiting bodies found in Protozoa.

1. **Sporangium**

Sporangium (Fig. 4.124) is the most common type. It is actually a small spore container which may be sessile or stalked, with wide variations in colour and shape. Sporangia usually occur in groups, since they form separate portions of the same plasmodium.

2. **Aethalium**

The **aethalium** (Fig. 4.124) is a cushion-shaped, sessile structure. Aethalia are presumed to be masses of completely fused sporangia and are relatively large, sometimes exceeding several centimeters in size.

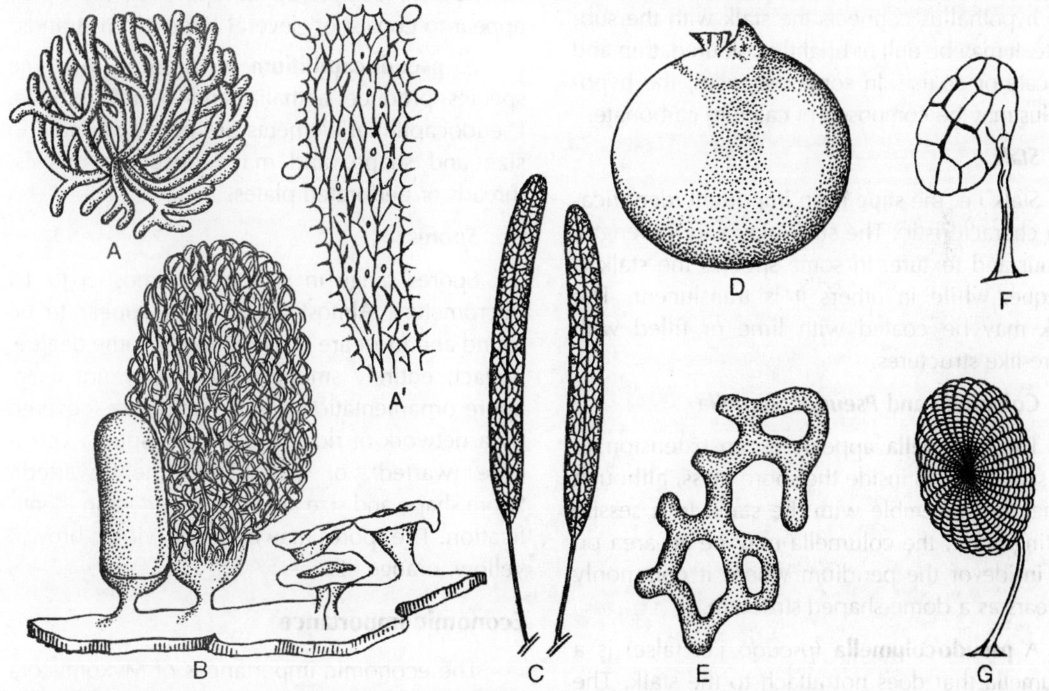

Fig. 4.124: Different types of fructification in Myxomycetes: A-A'. *Ceratiomyxa fruticulosa*. A. Columnar structures composing fructification. A'. Details of single column showing attachment of sporangia. B. Sporangia in different stages of maturity in *Arcyria* sp. C. Sporangia bearing hypothalli in *Stemonitis*. D. Aethalium of *Lycogala* sp. E. Plasmodiocarp of *Hemitrichia* sp. F-G. Sporongia in peridium on a slender stalk of *Physarum* (F) and *Dictydium* (G)

3. **Pseudoaethalium**

The **pseudoaethalium** is the false aethalium. This fruiling body is composed of sporangia closely crowded together. Pseudoaethalia are usually sessile, though a few may be stalked.

4. **Plasmodiocarp**

In this type fruit bodies are called **plasmodiocarp** (Fig. 4.124). Plasmodiocarps are usually sessile; take the form of the plasmodial veins from which they were derived. Structure of fruiting bodies is usually composed of six parts: hypothallus, stalk, columella, peridium, capillitium and spores. In some fruiting bodies a pseudocolmella or a pseudocapillitium may be present. Not all of these components are present in all fruiting body types.

The **fruiting body** is usually composed of the following **six structures**. These are as follows:

1. ***Hypothallus***

The hypothallus is a plasmodial remnant funning the base for one or more fruiting bodies. The hypothallus connects the stalk with the substrate. It may be dull or brightly coloured, thin and delicate or coarse. In some instances, the hypothallus may be composed of calcium carbonate.

2. ***Stalk***

Stalk i.e. the **stipe** is an important identification characteristic. The stalk may vary in length, colour and texture. In some species the stalk is opaque, while in others it is translucent. The stalk may be coated with lime or filled with spore-like structures.

3. ***Columella*** **and** ***Pseudocolumella***

The **columella** appears as an extension of the stalk, pushes inside the spore mass, although it may not resemble with the stalk. In a sessile fruiting body, the columella may be an area on the inside of the peridium where it commonly appears as a dome-shaped structure.

A **pseudocolumella** (*pseudo* i.e. false) is a columella that does not attach to the stalk. The pseudocolumella is found only in the order Physarales, existing as a lime mass within the spore mass. Capillitial elements may be attached to the columella or pseudocolumella.

4. ***Peridium***

Peridium (Fig. 4.124) is the covering which enclosed the spore mass. It may or may not be evident in mature fruiting bodies. In some species the peri-dium persists as a *calyculus* i.e. a cup-like structure which holds the bottom of the spore mass. The presence or absence of the calyculus may be used to identify the members along with the manner in which the fruiting body opens. The peridium may split open along the lines of dehi-scence, as a pre-formed lid, or in an irregular pattern.

In an **aethalium**, the relatively thick covering over the spore mass is referred to as a cortex rather than a peridium.

5. ***Capillitium*** **and** ***Pseudocapillitium***

The **capillitium** consists of thread-like elements within the spore mass of a fruiting body. Many myxomycetes have a capillitium, either as a single connected network, or as many free elements called **elaters**. Capillitial elements may be smooth, sculptured or spiny or they may appear to consist of' several interwoven strands.

A **pseudocapillitium** is present in some species produce aethalia and pseudoaethalia. Pseudocapillitial elements are highly variable in size and shape, and may appear as bristles, threads or perforated plates.

6. ***Spores***

Spores range in size from almost 5 to 15 micrometers. Almost all of them appear to be round and most are ornamented to some degree. In fact, entirely smooth spores may not exist. Spore ornamentation can be reticulate (covered by a network of ridges), echinate (spiny), verrucose (warted), or asperulate (finely warted). Spore shape and size are very important in identification. The spores may be black, violet, brown, yellow, orange etc.

Economic importance

The economic importances of Myxomycota are given below:

1. The slime moulds cause decay and decomposition of the organic matter in the soil.

2. They creep over the ornamental plants and make them look ugly.
3. Their attractive colours have the artistic value.
4. Plasmodia of slime moulds are the excellent material for the study of structure and physiology of protoplasm.

PLASMODIOPHORA

Systemic position

Ainsworth (1973)	Hawksworth (1995)
Kingdom Mycota	Kingdom Protozoa
Division Myxomycota	Phylum Plasmodiophoromycota
Class Plasmodiophoromycetes	Class Plasmodiophorea
Order Plasmodiophorales	Order Plasmodiophorida (= Plasmodiophorales)
Family Plasmodiophoraceae	Family Plasmodiophoraceae
Genus *Plasmodiophora*	Genus *Plasmodiophora*

Plasmodiophora is the best-known genus under the family Plasmodiophoraceae. The well-known species *P. brassicae* is an obligate endoparasite which causes **club root disease of crucifers** like cabbage, cauliflower, turnip, radish etc. The disease is also known by different common names in different countries like 'Finger and Toe' and 'Hamburg' in England; 'Club root', 'Club foot' and 'Clump foot' in U.S.A.: 'Kapoustnava Kila' in Russia : 'Kelch' and 'Kropf des Kohles' in Germany, etc.

The disease was first reported in Scotland in 1780. After many years, detailed scientific work about the pathogen and disease on cabbage was done by Woronin (1878).

Symptoms

The symptoms are visible in the root of affected plants (Fig. 4.125). The roots become abnormally increased and swell up at different regions due to hypertrophy and hyperplasia.

In Cabbage, the root bears malformation or clubs of different shapes, hence the disease is called club root disease. The development of lateral roots becomes suppressed.

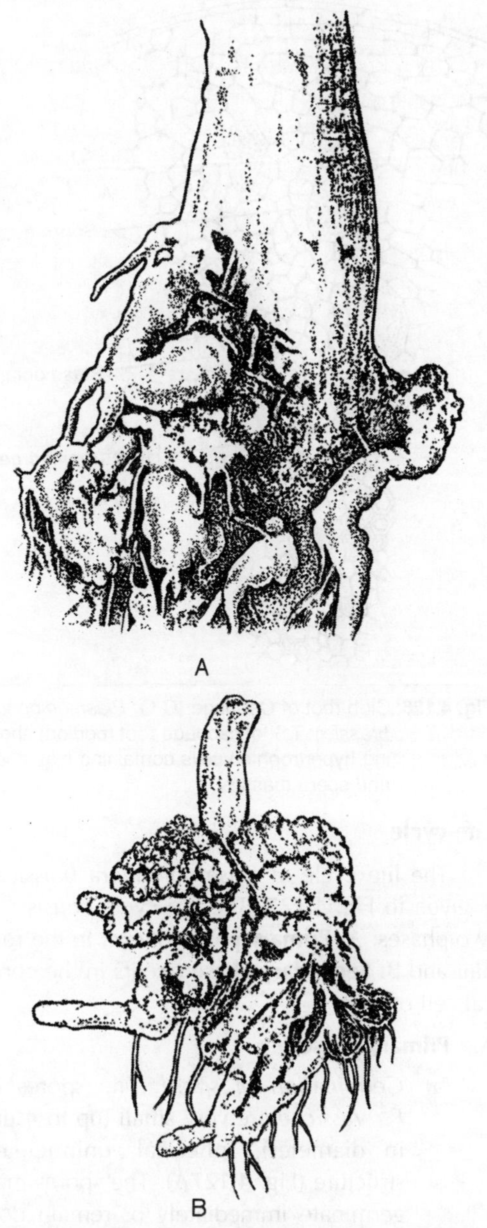

Fig. 4.125: Club root of crucifers: Club-shaped swellings of roots of different cultivars (A, B) of cabbage

The drooping of the leaves during sunny days, with gradual progress of disease, leads to permanent wilting. The disease shows some variations in symptoms in different hosts.

In section, the infected tissue shows some hypertrophied cells containing plasmodia or spore masses (Fig. 4.126).

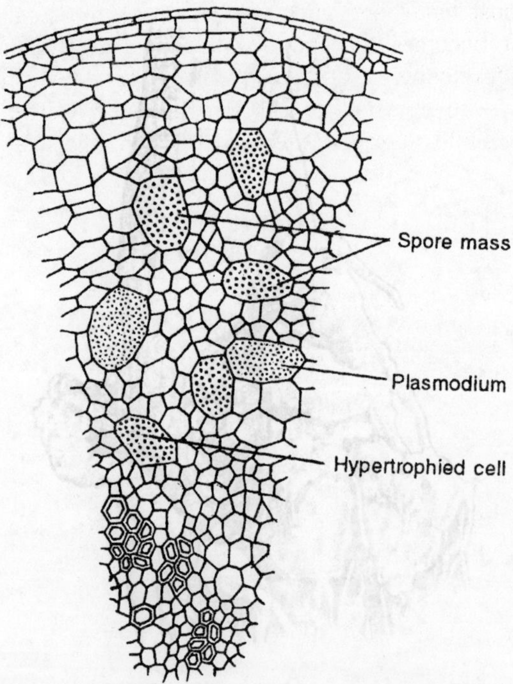

Fig. 4.126: Club root of Cabbage (C.O. *Plasmodiophora brassici*): T.S. of cabbage root (portion) showing hypertrophied cells containing plasmodia and spore masses

Life cycle

The life cycle of *Plasmodiophora brassicae* is given in Fig. 4.127. The life cycle consists of two phases: A. **Primary phase** occurs in the root hair and B. **Secondary phase** occurs in the cortical cell of root.

A. **Primary phase**

(a) *Germination of spore*. The spores of *P. brassicae* are very small (up to 4 μm in diameter), spherical, uninucleate structure (Fig. 4.127A). The spores may germinate immediately or remain dormant for several years. During favourable condition the spore swells up several times and the protoplast comes out of the spore wall through a rift (Fig. 4.127B) as a single **zoospore** or **swarm cell**. The zoospores are uninucleate and spherical or pear-shaped in structure with two unequal flagella of whiplash type, placed at the anterior end (Fig. 4.127C).

(b) *Infection to root hair*. The zoospores swim in a thin film of water. On contact with suitable host, the zoospore looses its flagella and becomes amoeboid, called **myxamoeba** (Fig. 4.127D). The myxamoeba enters into the root hair by dissolution of cell wall (Fig. 4.127E).

(c) *Gametophytic thallus* or *Gametothallus*. Within the root hair cell, the uninucleate myxamoeba enlarges to form a thallus. The protoplast enlarges in volume and its nucleus undergoes repeated mitotic divisions. Thus a small, multinucleate, naked mass of protoplasm is formed, called **plasmodium**. The plasmodium is haploid and thus it is called as **gametothallus** (Fig. 4.127F). Many gametothalli may be present in a single root hair due to multiple infection. The gametothalli do not fuse together.

(d) *Sexual reproduction*. During sexual reproduction, each nucleus of gametothallus accumulates some cytoplasm around itself and behaves as an individual unit, called **gametangium** (Fig. 4.127G). Thus the gametangium is initially very small, uninucleate spherical structure (Fig. 4.127H). The nucleus of each gametangium undergoes mitotic divisions and forms 4-8 daughter nuclei. Each daughter nucleus accumulates some cytoplasm and behaves as an individual unit (Fig. 4.127I). Each unit develops into a spindle-shaped, biflagellate gamete. Gametes are alike and are called isogametes (Fig. 4.127J). The gametes are smaller in size than the zoospores.

The gametes are liberated out by bursting the gametangial wall and root hair cells. *The gametic union (only plasmogamy) takes place in the soil*, results the formation of binucleate and quadriflagellate structure (Fig. 4.127K, L) [whether the fusing gametes come from same or different gametangia has not been established].

B. **Secondary phase**

The binucleate and quadriflagellate structure then enters again into the host tissue, initially through epiblema towards the meristematic zone. It retracts the flagella in contact with the host tissue and after entering inside host cell it becomes amoeboid, called **myxamoeba** or **plasmodium** (Fig. 4.127M). The myxamoeba can spread from cell to cell through the cell wall. Initially, it remains in cortical cell and later

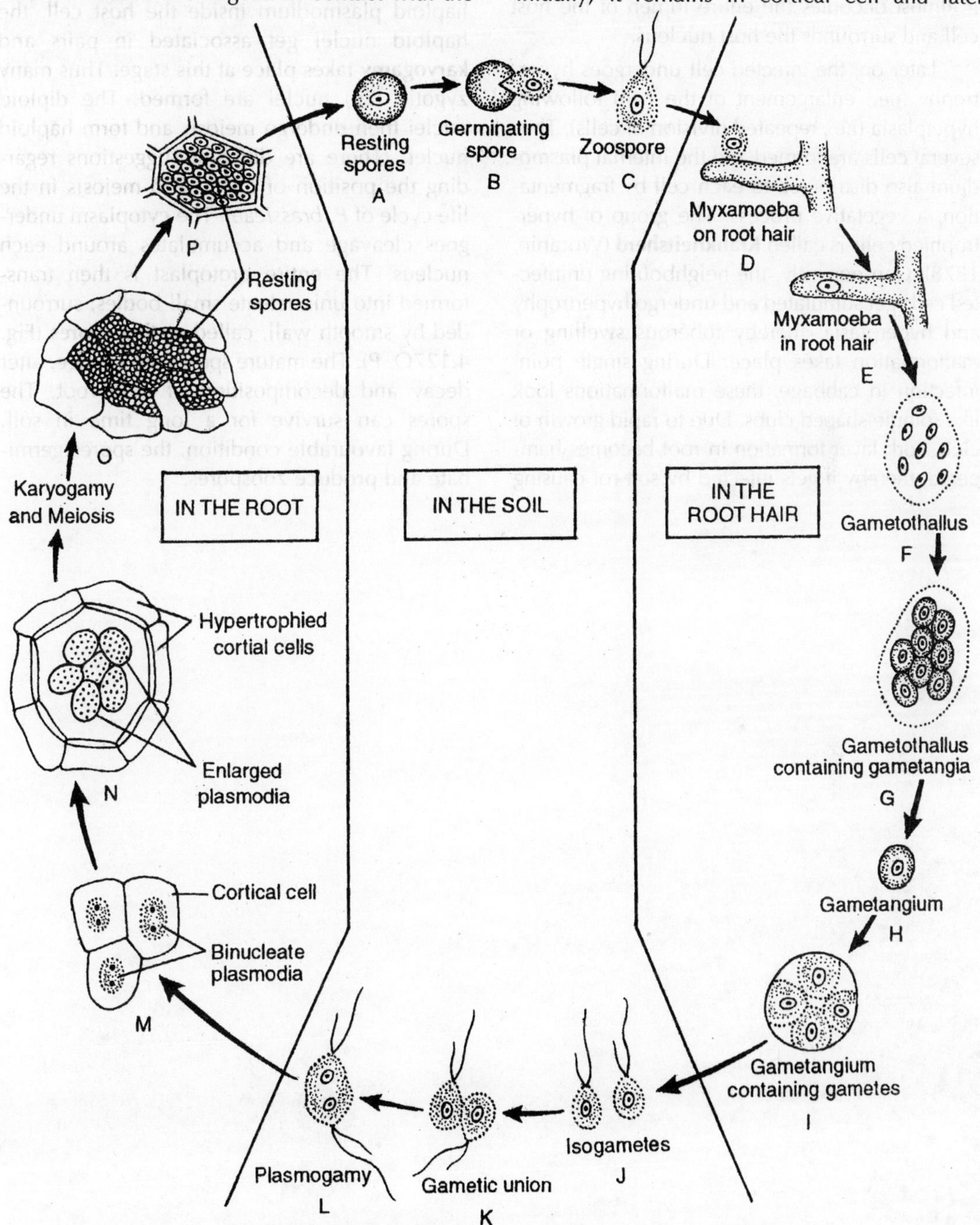

Fig. 4.127: Life cycle of *Plasmodiophora brassicae*

invades to the other tissue. Within the host cell, the **myxamoeba** enlarges and forms a naked multinucleate plasmodium by repeated division of its haploid nucleus (Fig. 4.127N). At maturity, it almost occupies the entire lumen of the host cell and surrounds the host nucleus.

Later on, the infected cell undergoes hypertrophy (i.e., enlargement of the cell) following hyperplasia (i.e., repeated division of cells). Thus several cells are formed and the internal plasmodium also distributed to each cell by fragmentation, a vegetative process. The group of hypertrophied cells is called **Krankheitsherd** (Woranin, 1878). Consequently, the neighbouring uninfected cells get stimulated and undergo hypertrophy and hyperplasia, thereby tuberous swelling or malformation takes place. During single point infection in cabbage, these malformations look like spindle-shaped clubs. Due to rapid growth of club, cork layer formation in root becomes hampered, thereby it gets infected by soft rot causing bacteria or fungi. Due to secondary infection, the decay of the club leads to the formation of toxic material, which causes wilting of the plant.

Spore formation. After maturation of the haploid plasmodium inside the host cell, the haploid nuclei get associated in pairs and **karyogamy** takes place at this stage. Thus many zygotic (2n) nuclei are formed. The diploid nuclei then undergo meiosis and form haploid nuclei. (There are different suggestions regarding the position of fusion and meiosis in the life cycle of *P. brassicae*). The cytoplasm undergoes cleavage and accumulates around each nucleus. The entire protoplast is then transformed into uninucleate small bodies, surrounded by smooth wall, called **resting spores** (Fig. 4.127O, P). The mature spores are set free, after decay and decomposition of club root. The spores can survive for a long time in soil. During favourable condition, the spores germinate and produce zoospores.

SYMBIOTIC ASSOCIATION

(LICHENS AND MYCORRHIZAE)

Introduction

Symbiosis (Gr. *sym*, together; *bios*, life) is the "living together" of two or more biological organisms. In its broadest sense, symbiotic associations include parasitic and commensal as well as mutually beneficial partnerships. The organisms, each termed a **symbiont**, may be of the same or of different species. In 1879, Heinrich Anton de Bary defined it as "the living together of unlike organisms".

Symbiosis can be obligatory, which means that one or both of the symbionts entirely depend on each other for survival, or facultative (optional) when they can generally live independently.

Lichen and Mycorrhiza are the common and ideal example of symbiosis. In **lichen**, association takes place between fungi and algae, but in **mycorrhiza** association takes place between fungi and commonly roots of higher plants.

LICHENS

Lichens are a small group of plants of composite nature, consisting of two dissimilar organisms, an alga-phycobiont (*phycos*, alga; *bios*, life) and a fungus-mycobiont (*mykes*, fungus; *bios*, life); living in a symbiotic association. Generally the fungal partner occupies the major portion of the thallus and produces its own reproductive structures. The algal partner manufactures the food through photosynthesis which probably diffuses out and is absorbed by the fungal partner.

Important characteristics

1. Lichens are a group of plants of composite thalloid nature, formed by the association of algae and fungi.
2. The algal partner-produced carbohydrate through photosynthesis is utilised by both of them and the fungal partner serves the function of absorption and retention of water.
3. Based on the morphological structure of thalli, they are of mainly four types: eprose, crustose, foliose and fruticose.
4. Lichen reproduces by all the three means: vegetative, asexual, and sexual.
   (a) *Vegetative reproduction*: It takes place by fragmentation, decaying of older parts, by soredia and isidia.
   (b) *Asexual reproduction*: By the formation of oidia.
   (c) *Sexual reproduction*: By the formation of ascospores or basidiospores. Only fungal component is involved in sexual reproduction.
5. Ascospores are produced in Ascolichen.
   (a) The male sex organ is flask-shaped spermogonium, produces unicellular spermatia.
   (b) The female sex organ is carpogonium (ascogonium), differentiates into basal coiled oogonium and elongated trichogyne.
   (c) The fruit body may be apothecial (disc-shaped) or perithecial (flask-shaped) type.
   (d) Asci develop inside the fruit body containing 8 ascospores. After liberating from the fruit body, the ascospores germinate and, in contact with suitable algae, they form new lichen.
6. Basidiospores are produced in Basidiolichen, generally look like bracket fungi and basidiospores are produced towards the lower side of the fruit body.
7. The growth of lichen is very slow, they can survive in adverse conditions with high temperature and dry condition.

Occurrence

Lichens are the nature's most unique and successful association constitutes an important component of biodiversity.

There is about 400 genera and 15,000 species of lichen found in different regions of the world. In India, they are represented by 2532 species under 324 genera and 78 families including 541 endemic (21.3%) species.

Plant body

The plant body is thalloid; generally grows on bark of trees, leaves, dead logs, bare rocks etc., in different habitat. They grow luxuriantly in the forest areas with free or less pollution and with abundant moisture. Some species like *Cladonia rangiferina* (reindeer moss) grows in the extremely cold condition of Arctic tundras and Antarctic regions. In India, they grow abundantly in Eastern Himalayan regions. They do not grow in the highly polluted regions like Industrial areas. The growth of lichen is very slow. Depending on habitat, the lichens are grouped as:

1. *Saxicolous*. Grow on stones or rocks e.g., *Dermatocarpon, Xanthoria, Verrucaria*, etc.
2. *Corticolous*. Grow on bark of trees e.g., *Grpahis, Usnea, Parmelia* etc.
3. *Terricolous*. Grow on soil (terrestrial) e.g., *Cladonia florekeana, Lecidea, Collema* etc.
4. *Lignicolous*. Grow directly on wood e.g., *Calicicum* etc.
5. *Marine*. Grow on siliceous rocky shores of sea e.g., *Caloplaca, Verrucaria* etc.
6. *Freshwater*. Grow on hard siliceous rocks in freshwater e.g. *Epheba, Hymenelia* etc.

Associated members: The composite plant body of lichen consists of algal and fungal members.

The algal members belong to Chlorophyceae (*Trebouxia, Trentepohlia, Coccomyxa* etc.), Xanthophyceae (*Heterococcus*) and also Cyanobacteria (*Nostoc, Scytonema* etc.) (Fig. 4.128).

The fungal members mainly belong to Ascomycotina and a few to Basidiomycotina. Among the members of Ascomycotina, Discomycetes are very common; producing huge

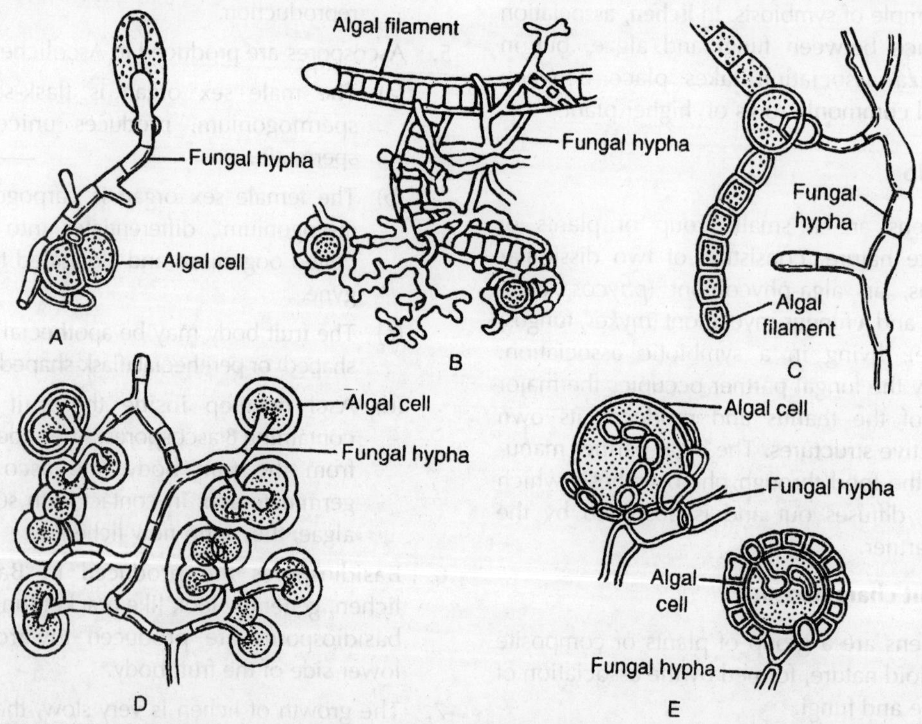

Fig. 4.128: Lichen-forming Algae: A. *Pleurococcus viridis*, B. *Scytonema* sp., C. *Nostoc* sp., D. *Gloeocapsa* sp., E. *Pleurococcus* sp.

apothecia, others belong to Pyrenomycetes or Loculoascomycetes.

The members of Basidiomycotina belong to Thelephoraceae.

Nature of association: There are three views regarding the nature of association of algal and fungal partners in lichen:

1. *According to some workers,* the fungus lives parasitically, either partially or wholly, with the algal components. This view gets support for the following evidences:
   (i) Presence of haustoria of fungus in algal cells of some lichen.
   (ii) On separation, the alga of lichen is able to live independently, but the fungus cannot survive.
2. *According to others,* they live symbiotically, where both the partners are equally benifitted. The fungal member absorbs water and mineral from atmosphere and substratum, make available to the alga and also protects algal cells from adverse conditions like temperature etc. The algal member synthesises organic food sufficient for both of them.
3. *According to another view,* though the relationship is symbiotic, the fungus shows predominance over the algal partner, which simply lives as subordinate partner. It is like a master and slave relationship, termed ***helotism***.

Classification

Natural system of classification is not available for lichens. They are classified on the nature and kinds of fruit bodies of the fungal partner. Based on the structure of fruit bodies of fungal partners, Zahlbruckner (1926) classified lichens into three sub-classses. These are:

1. ***Ascolichens***. The fungal member of this lichen belongs to Ascomycetes. Based on the structure of fruit body, they are divided into two series:
   (i) *Gynocarpeae*. The fruit body is disc-shaped i.e., apothecial type. It is also known as Discolichen (e.g., *Parmelia*).
   (ii) *Pyrenocarpeae*. The fruit body is flask-shaped i.e., perithecial type. It is also known as *Pyrenolichen* (e.g., *Dermatocarpon*).
2. ***Basidiolichens***. The fungal member of this lichen belongs to Basidiomycetes e.g., *Dictyonema, Corella*.
3. ***Hymenolichens***. The fungal member belongs to Deuteromycetes. They are also known as imperfect lichens.

Later, Alexopoulos and Mims (1979) classified lichens into three main groups:

1. Basidiolichen. The fungal partner belongs to Basidiomycetes e.g., *Dictyonema*.
2. Deuterolichen. The fungal partner belongs to Deuteromycetes.
3. Ascolichen. The fungal partner belongs to Ascomycetes e.g., *Parmelia, Cetraria*.

Structure of thallus: The plant body of lichen is thalloid with different shapes. They are usually grey or greyish green in colour, but some are red, yellow, orange or brown in colour.

External structure: Based on the external morphology, general growth and nature of attachment, *initially three main types* or *forms* of lichens (crustose, foliose and fruticose) have been recognised. Later, based on detailed structures, **Hawksworth and Hill (1984)** categorized the lichens into **five main types** or forms:

1. **Leprose**. This is the simplest type, where the fungal mycelium envelops either single or small cluster of algal cells. The algal cell does not envelop all over by fungal hyphae. The lichen appears as powdery mass on the substratum, called leprose (Fig. 4.129A), e.g., *Lepraria incana*.
2. **Crustose**. These are *encrushing lichens* where thallus is inconspicuous, flat and appears as a thin layer or crust on substratum like barks, stones, rocks etc. (Fig. 4.129B). They are either wholly or partially embedded in the substratum, e.g., *Graphis, Lecanora, Ochrolechia, Strigula, Rhizocarpon, Verrucaria, Lecidia* etc.
3. **Foliose**. These are *leaf-like lichens,* where thallus is flat, horizontally spreading and with lobes. Some parts of the thallus are attached with the substratum by means of hyphal out-

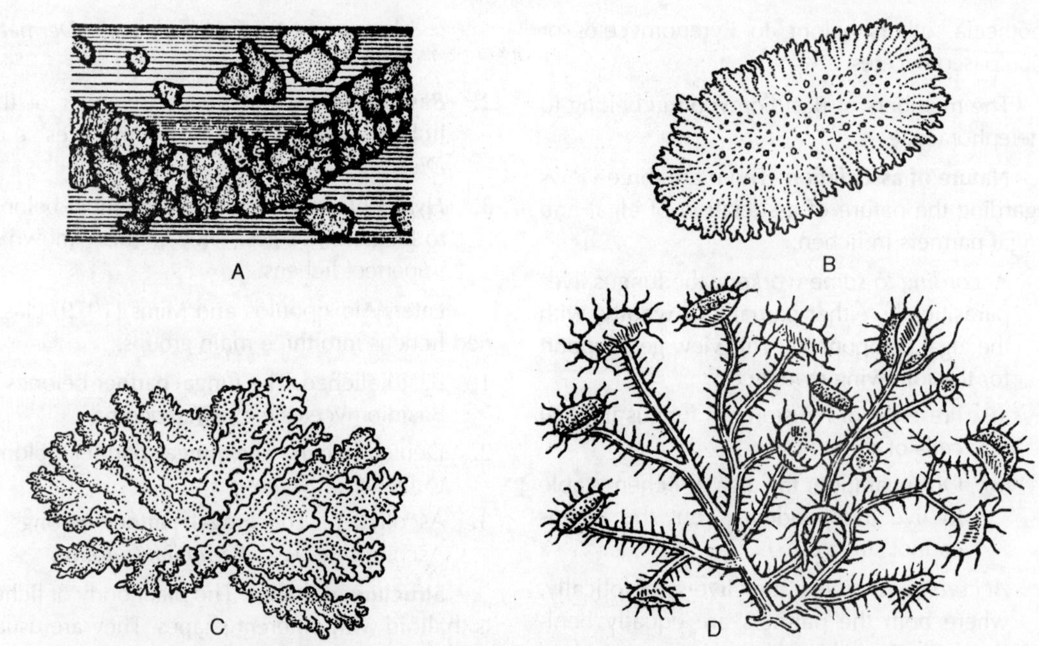

Fig. 4.129: Different forms of lichen: A. Leprose, B. Crustose, C. Foliose and D. Fruticose

growth, the *rhizines,* developed from the lower surface (Fig. 4.129C), e.g., *Parmelia, Physcia, Peltigera, Anaptychia, Hypogymnia, Xanthoria, Gyrophora, Collema, Chauduria* etc.

4. **Fruticose** (*Frutex*, shrub). These are *shrubby lichens*, where thalli are well developed, cylindrical branched, shrub-like (Fig. 4.129D), either grow erect (*Cladonia*) or hang from the substratum (*Usnea*). They are attached to the substratum by a basal disc e.g., *Cladonia, Usnea, Letharia, Alectonia* etc.
5. **Filamentous**. In this type, algal members are filamentous and well-developed. The algal filaments remain ensheathed or covered by only a few fungal hyphae. Here algal member remains as dominant partner, called filamentous type, e.g., *Racodium, Ephebe, Cystocoleus* etc.)

***Internal Structure**:* Based on the distribution of algal member inside the thallus, the lichens are divided into two types: Homoisomerous or Homomerous and Heteromerous.

1. ***Homoisomerous***. Here the fungal hyphae and the algal cells are more or less uniformly distributed throughout the thallus. The algal members belong to Cyanophyta. This type of orientation is found in crustose lichens. Both the partners intermingle and form thin outer protective layer (Fig. 4.130A), e.g., *Leptogium, Collema* etc.
2. ***Heteromerous***. Here the thallus is differentiated into four distinct layers: upper cortex, algal zone, medulla, and lower cortex. The algal members are restricted in the algal zone only. This type of orientation is found in foliose and fruticose lichens (Fig. 4.130B) e.g., *Physcia, Parmelia* etc. The detailed internal structure of this type is:
   (a) *Upper cortex*. It is a thick, outermost protective covering, made up of compactly arranged interwoven fungal hyphae located at right angle to the surface of the fruit body. Usually there is no intercellular space between the hyphae, but if present, these are filled with gelatinous substances.
   (b) *Algal zone*. The algal zone occurs just below the upper cortex. The algal cells are entangled by the loosely interwoven fungal hyphae. The common algal members may belong to Cyanophyta like *Gloeocapsa* (unicellular); *Nostoc, Rivu-*

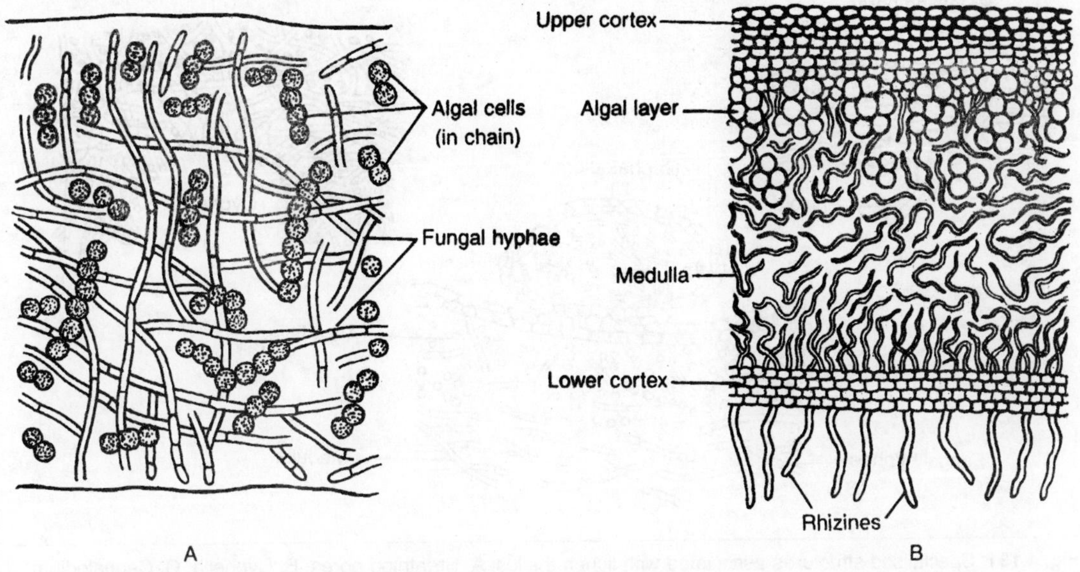

Fig. 4.130: Internal structure of lichen thallus : **A. Homoisomerous thallus, and B. Heteromerous thallus**

laria (filamentous) etc. or to Chlorophyta like *Chlorella, Cystococcus, Pleurococcus* etc. This layer is either continuous or may break into patches and serve the function of photosynthesis.

(c) *Medulla*. The medulla is situated just below the algal zone, comprised of loosely interwoven thick-walled fungal hyphae with large space between them.

(d) *Lower cortex*. It is the lowermost layer of the thallus. This layer is composed of compactly arranged hyphae, which may arrange perpendicular or parallel to the surface of the thallus. Some of the hyphae in the lower surface may extend downwards and penetrate the substratum which help in anchorage, known as **rhizines**.

The internal structure of *Usnea*, a fruticose lichen, shows different types of orientation. Being cylindrical in cross-section, the layers from outside are cortex, medulla (composed of algal cell and fungal mycelium) and central chondroid axis (composed of compactly arranged fungal mycelia).

Specialised structures

1. *Breathing pore*. In some foliose lichen (e.g., *Parmelia*), the upper cortex is interrupted by some opening, called breathing pores, which help in gaseous exchange (Fig. 4.131A).
2. *Cyphellae*. On the lower cortex of some foliose lichen (e.g., *Sticta*) small depressions develop, which appears as cup-like white spots, known as *Cyphellae* (Fig. 4.131B). Sometimes the pits that formed without any definite border are called *Pseudocyphellae*. Both the structures help in aeration.
3. *Cephalodium*. These are small warty outgrowths on the upper surface of the thallus (Fig. 4.131C). They contain fungal hyphae of the same type as the mother thallus, but the algal elements are always different. They probably help in retaining the moisture. In *Neproma*, the *Cephalodia* are endotrophic.

Reproduction

Lichen reproduces by all the three means, vegetative, asexual, and sexual.

I. ***Vegetative reproduction***

1. *Fragmentation*. It takes place by accidental injury where the thallus may be broken into fragments and each part is capable of growing normally into a thallus.

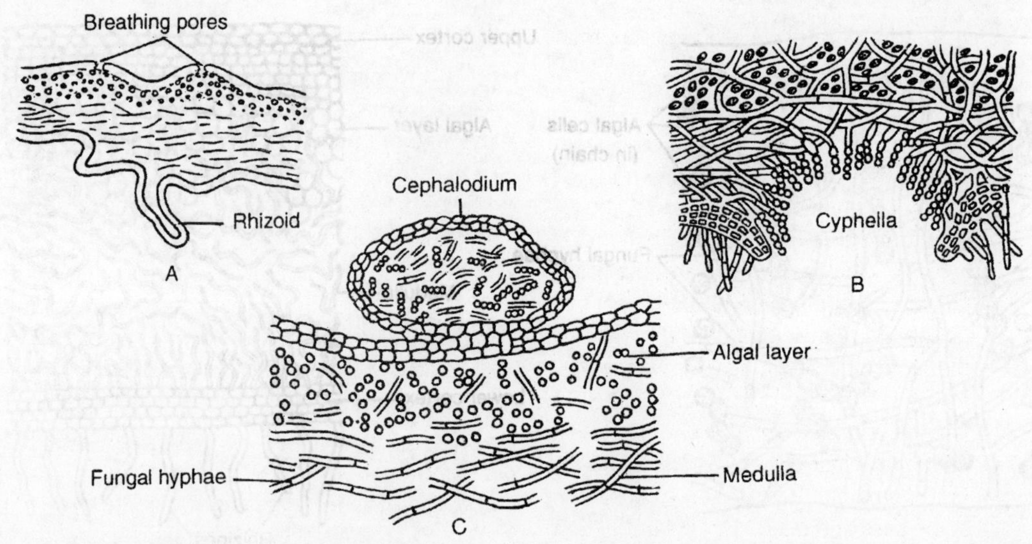

Fig. 4.131: Specialised structures associated with lichen thallus: A. Breathing pores, B. Cyphella, C. Cephalodium

2. *By death of older parts*. The older region of the basal part of the thallus dies, causing separation of some lobes or branches and each one grows normally into new thallus.

II. ***Asexual reproduction***

1. *Soredium* (pl. *Soredia*). These are small grayish white, bud-like outgrowths developed on the upper cortex of the thallus (Fig. 4.132A, B). They are composed of one or few algal cells loosely enveloped by fungal hyphae. They are detached from the thallus by rain or wind and on germination they develop new thalli.

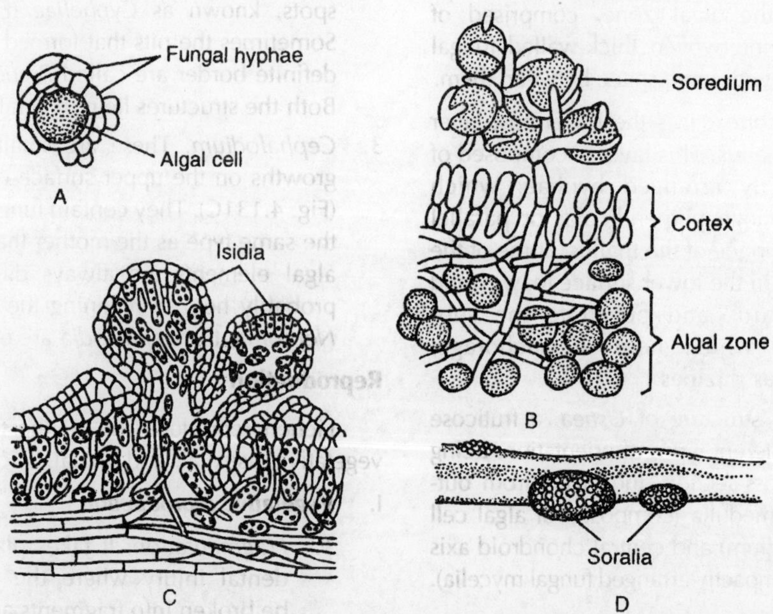

Fig. 4.132: Asexual reproductive structures: A. Soredium of *Physcia pulverulenta* with single algal cell, B. Soredium of *Parmelia* with many algal cells, C. Isidia of *Peltigera* and D. Soralia on thallus

When soredia develop in an organised manner in a special pustule-like region, they are called *Soralia* (Fig. 4.132D), e.g., *Parmelia, Physcia* etc.

2. *Isidium* (pl. *Isidia*). These are small stalked simple or branched, grayish-black, coral-like outgrowths, developed on the upper surface of the thallus (Fig. 4.132C). The isidium has an outer cortical layer continuous with the upper cortex of the mother thallus which encloses the same algal and fungal elements as the mother. They are of various shapes and may be coral-like in *Peltigera*, rod-like in *Parmelia*, cigar-like in *Usnea*, scale-like in *Collema* etc. It is generally constricted at the base and detached very easily from the parent thallus. Under favourable condition the isidium germinates and gives rise to a new thallus.

In addition to asexual reproduction, the isidia also take part in increasing the photosynthetic area of the thallus.

3. *Pycniospore*. Some lichen develop pycniospore or spermatium inside the flask-shaped pycnidium (Fig. 4.133A). They usually behave as gametes, but in certain condition they germinate and develop fungal hyphae. These fungal hyphae, when come in contact with the appropriate algal partner, develop into a new lichen thallus.

III. ***Sexual reproduction***

Only fungal partner of the lichen reproduces sexually and forms fruit bodies on the thallus. The nature of sexual reproduction in ascolichen is like that of the members of Ascomycotina, whereas in Basidiolichen is like that of Basidiomycotina members.

In Ascolichen, the female sex organ is the carpogonium and the male sex organ is called spermogonium (= pycnidium). The spermogonium (Fig. 4.133A) mostly develops close to carpogonium.

The carpogonium is multicellular and is differentiated into basal coiled ascogonium and upper elongated multicellular trichogyne (Fig. 4.133B). The ascogonium remains embedded in the algal zone, but the trichogyne projects out beyond the upper cortex.

The spermogonium is flask-shaped and develop spermatia from the inner layer (Fig. 4.133A). The spermatia behave as male gametes. The spermatium, after liberating from the spermogonium, gets attached with the trichogyne at the sticky projected part. On dissolution of the common wall, the nucleus of spermatium migrates into the carpogonium and fuses with the egg.

Many ascogenous hyphae develop from the basal region of the fertilized ascogonium. The binucleate penultimate cell of the ascogenous hyphae develops into an ascus. Both the nuclei of penultimate cell fuse and form diploid

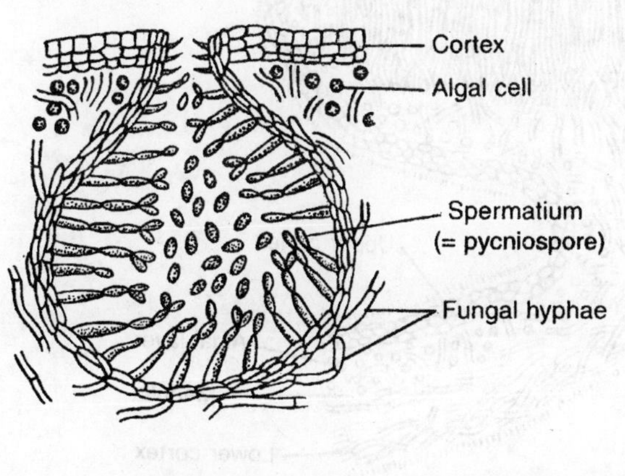

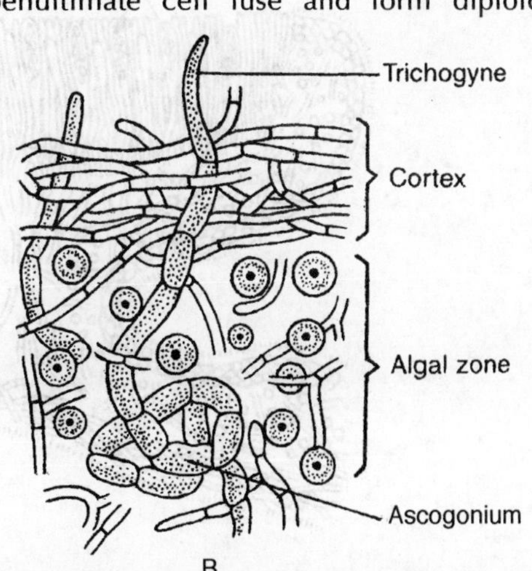

Fig. 4.133: Sexual reproductive structures: A. Spermogonium (= pycnidium), B. Carpogonium

nucleus (2n), which undergoes first meiotic and then mitotic division — results in eight haploid daughter nuclei. Each haploid nucleus with some cytoplasm metamorphoses into an ascospore.

The asci remain intermingled with some sterile hyphae — the paraphyses. With further development, asci and paraphyses become surrounded by vegetative mycelium and form fruit body. The fruit body may be *ascohymenial type* i.e., either apothecium (Fig. 4.134A) as in *Parmelia* and *Anaptychia* or perithecium as in *Verrucaria* and *Darmatocarpon* or *ascolocular type* (absence of true hymenium), which is also known as pseudothecia or ascostroma.

Internally, the cup-like (Fig. 4.134B, C) grooved region of a mature apothecium consists of three distinct parts; the middle *thecium* (= hymenium), comprising of asci and paraphyses, is the fertile zone covered by two sterile zones — the upper *epitheca* and lower *hypotheca*. The region below the cup is differentiated like the vegetative thallus into outer cortex, algal zone and central medulla (Fig. 4.134B).

Usually the asci contain eight ascospores (Fig. 4.134C), but the number may be one in *Lopadium*, two in *Endocarpon* and even more than eight in ***Acarospora***. The ascospores may be unicellular or multicellular, uninucleate or multinucleate, and are of various shapes and sizes. After liberating from the ascus, the ascospore germinates in suitable medium and produces new hypha. The new hypha, after coming in contact with proper algal partner, develops into a new thallus.

In Basidiolichen (Fig. 4.135), the result of sexual reproduction is the formation of basidiospores that developed on basidium as in

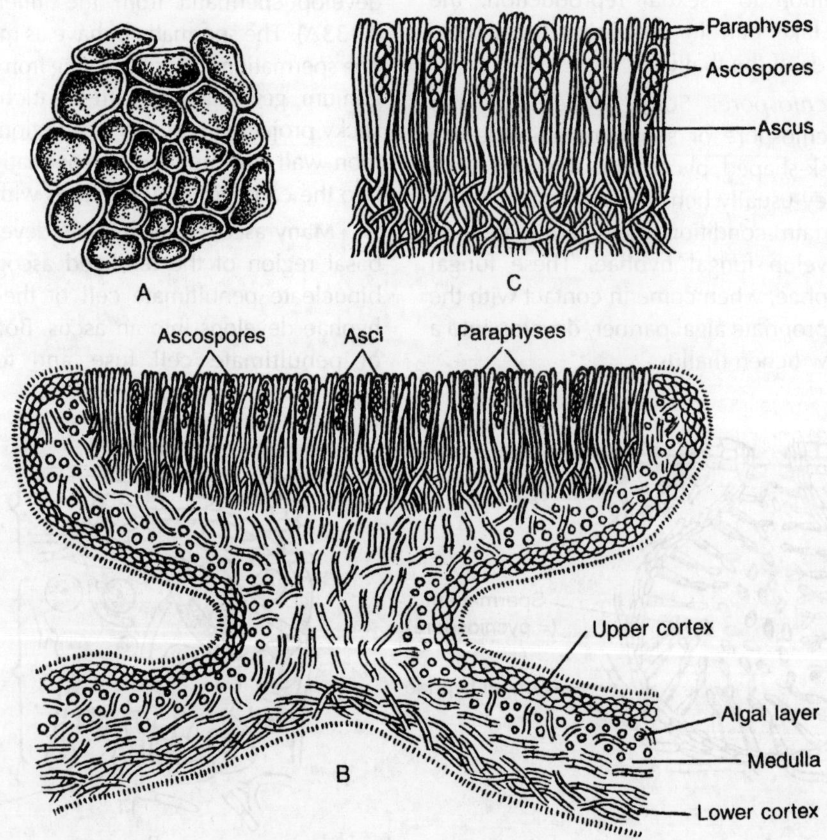

Fig. 4.134: Structural features of ascolichen: A. Apothecia, B. V.S. of apothecium and C. Portion of hymenial region

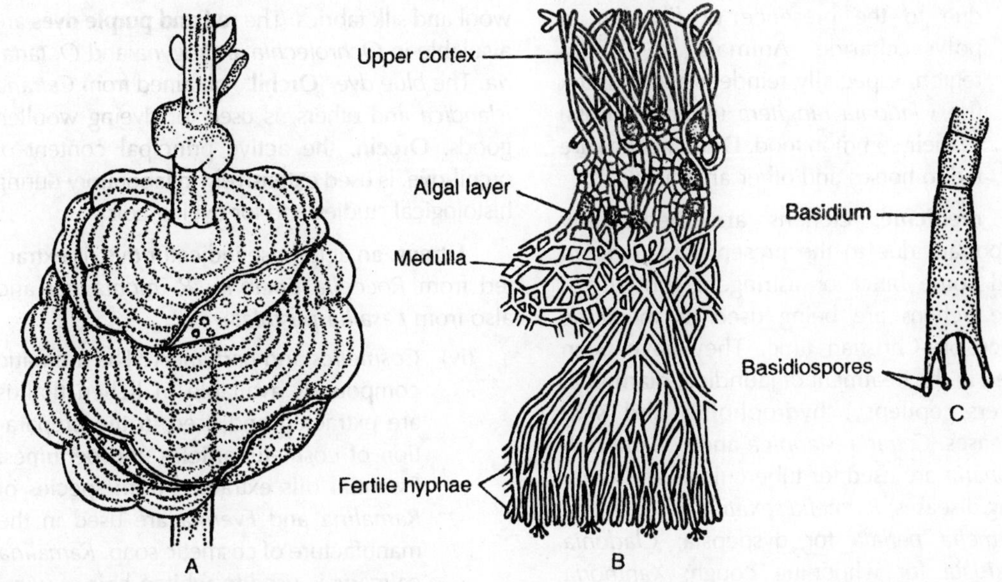

Fig. 4.135: Structural features of basidiolichen (*Cora pavonia*): A. Bracket-like thalli, B. Vertical section of thallus (portion), C. Basidium with basidiospores

typical basidiomycotina. The fungal member (belongs to Thelephoraceae) along with blue green alga, as algal partner forms the thalloid body. The thallus grown over soil produces hypothallus without rhizines, but on tree trunk it grows like bracket fungi (Fig. 4.135A) and differentiates internally into upper cortex, algal layer, medulla and lower fertile region with basidium bearing basidiospores (Fig. 4.135B, C).

Fossil Lichens

Some of the lichens have been recorded as fossils from the different geological ages. The earliest record comes from Mesozoic age — *Ophegrapha* and *Ramalinites lacerus,* of which the latter is questionable. Many lichens have also been reported from the Coenozoic age, e.g., *Lichen dichotomus, Ophegrapha thomasiana, Parmelia lacunosa, Cladonia rosea, Ramalina tertiaria* etc.

Economic importance

The lichens are useful as well as harmful to mankind. The useful activities are much more than harmful ones. They are useful to mankind in various ways: as food and fodder, as medicine and industrial uses of various kinds.

Useful activities

1. ***As food and fodder*:** Lichens are used as food by human being in many parts of the world and also by different animals like snail, catterpillars, slugs, termites etc. They contain polysaccharide, *lichenin;* cellulose, vitamin and certain enzymes.

   Some uses of lichens are:

   (i) *As food:* Some species of *Parmelia* are used as curry powder in India, *Endocarpon miniatum* is used as vegetable in Japan, *Evernia prunastri* for making bread in Egypt, and *Cetraria islandica* (Iceland moss) as food in Iceland. Others like species of *Umbillicaria, Parmelia* and *Leanora* are used as food in different parts of the world. In France, some of the lichens are used in the preparation of chocolates and pastries.

   Lichens like *Lecanora saxicola* and *Aspicilia calcarea* etc., are used as food by snails, catterpillars, termites, slugs etc.

   (ii) *As fodder: Ramalina fraxinea, R. fastigiata, Evernia prunastri, Lobaria pulmonaria* are used as fodder for animals, due to the presence of *lichenin,* a polysaccharide. Animals of Tundra region, especially reindeer and muskox use *Cladonia rangifera* (reindeer moss) as their common food. Dried lichens are fed to horses and other animals.

2. ***As medicine***: Lichens are medicinally important due to the presence of *lichenin* and some bitter or astringent substances. The lichens are being used as medicine since pre-Christian time. They have been used in the treatment of jaundice, diarrhoea, fevers, epilepsy, hydrophobia and skin diseases. *Cetraria islandica* and *Lobaria pulmonaria* are used for tuberculosis and other lung diseases; *Parmelia sexatilis* for epilepsy; *Parmelia perlata* for dyspepsia; *Cladonia pyxidata* for whooping cough; *Xanthoria parietina* for jaundice and several species of *Pertusaria, Cladonia* and *Cetraria islandica* for the treatment of intermittent fever. **Usnic acid**, a **broad spectrum antibiotic** obtained from species of *Usnea* and *Cladonia,* are used against various bacterial diseases. *Usnea and Evernia furfuracea* have been used as astringents in haemorrhages. Some lichens are used as important ingredients of many antiseptic creams, because of having spasmolytic and tumour-inhibiting properties.

3. ***Industrial uses***. Lichens of various types are used in different kinds of industries:

   (i) *Tanning industry*. Some lichens like *Lobaria pulmonaria* and *Cetraria islandica* are used in tanning leather.

   (ii) *Brewery* and *distillation*. Lichens like *Lobaria pulmonaria* are used in brewing of beer. In Russia and Sweden, *Usnea florida, Cladonia rangiferina* and *Ramalina fraxinea* are used in production of alcohol due to rich content of "lichenin", a carbohydrate.

   (iii) *Preparation of dye*. Dyes obtained from some lichens have been used since pre-Christian times for colouring fabrics etc.

Dyes may be of different colours like brown, red, purple, blue etc. The ***brown dye*** obtained from *Parmelia omphalodes* is used for dyeing of wool and silk fabrics. The **red and purple dyes** are available in *Ochrolechia androgyna* and *O. tartaria***.** The *blue dye* "**Orchil**", obtained from *Cetraria islandica* and others, is used for dyeing woollen goods. **Orcein**, the active principal content of orchil-dye, is used extensively in laboratory during histological studies and for dyeing coir.

Litmus, an acid-base indicator dye, is extracted from *Roccella tinctoria, R. montagnei* and also from *Lasallia pustulata***.**

   (iv) *Cosmetics* and *perfumery*. The aromatic compounds available in lichen thallus are extracted and used in the preparation of cosmetic articles and perfumes. Essential oils extracted from species of *Ramalina* and *Evernia* are used in the manufacture of cosmetic soap. *Ramalina calicaris* is used to whiten hair of wigs. Species of *Usnea* have the capacity of retaining scent and are commercially utilised in perfumery. *Evernia prunastri* and *Pseudevernia furfuracea* are used widely in perfumes.

Harmful activities

1. Some lichens like *Amphiloma* and *Cladonia* parasitise on mosses and cause total destruction of moss colonies.
2. Lichen like *Usnea,* with its holdfast hyphae, can penetrate deep into the cortex or deeper, and destroy the middle lamella and inner content of the cell causing total destruction.
3. Different lichens, mainly crustose type, cause serious damage to window glasses and marble stones of old buildings.
4. Lichens like *Letharia vulpina* (wolf moss) are highly poisonous. **Vulpinic acid** is the poisonous substance present in this lichen.

Ecological Importance

Lichens have some ecological importance. Some of the activities in this respect are:

1. **Pioneer of rock vegetation.** Lichens are pioneer colonizers on dry rocks. Due to their ability to grow with minimum nutrients and water, the crustose lichens colonize with luxuriant growth. The lichens secrete some

acids which disintegrate the rocks. After the death of the lichen, it mixes with the rock particles and forms thin layer of soil. The soil provides the plants like mosses to grow on it as the first successor, but, later, vascular plants begin to grow in the soil. In plant succession, *Lecanora saxicola,* a lichen, grows first; then the moss *Grimmia pulvinata,* after its death, forms a compact cushion on which *Poa compressor* grows later.

2. **Accumulation of radioactive substance.** Lichens are efficient for absorption of different substances. The *Cladonia rangiferina,* the 'reindeer moss', and *Cetraria islandica,* the 'Iceland moss' are the commonly available lichens in Tundra region. The fallout of radioactive strontium (^{90}Sr) and caesium ($^{137}C_s$) from the atomic research centres are absorbed by lichen. Thus, lichen can purify the atmosphere from radioactive substances.

The lichens are eaten by caribou and reindeer and pass on into the food-chain, especially to the Lapps and Eskimos. Thus, the radioactive substances are accumulated by the human beings.

3. **Sensitivity to air pollutants.** Lichens are very much sensitive to air pollutants like SO_2, CO, CO_2 etc.; thereby the number of lichen thalli in the polluted area is gradually reduced and, ultimately, comes down to nil. The crustose lichens can tolerate much more in polluted area than the other two types. For the above facts, the lichens are markedly absent in cities and industrial areas. Thus, lichens are used as "**pollution indicators**".

MYCORRHIZAE

Vitadini (1842) was the first to recognize the possible beneficial role of fungal mycelia which mantle the root of higher plants, and this association is named as **mycorrhiza** (pl. mycorrhizae) i.e., the fungal root, by Frank (1885).

Thus *mycorrhizae are the symbiotic associations between plant root and fungi, with bidirectional nutrient exchange between the partners.*

The autotrophic host plant acts as the carbon source for the fungus, while the fungus supplies mineral nutrients to the plant. About 90% of all land plants are associated with mycorrhiza. *The mycorrhizal association is not available in Cruciferae, Chenopodiaceae and Resedaceae.*

Scannerini (1988) briefly pointed out the common features of mutualistic symbionts. These include:

(i) Absence of any phytopathological symptoms in the partners during the active phase of mutualism,

(ii) Presence of complex interfaces between cells of the partners with a predominant type of perisymbiotic membrane, surrounding intracellular symbionts,

(iii) Presence of various types of phagocyte-like structures during establishment of symbionts and during harvesting phase to control the symbiotic population by the host.

Peterson and Farquhar (1994) classified the mycorrhizae into seven (7) distinct types. These are: (1) Ectomycorrhizae, (2) Vesicular-arbuscular mycorrhizae, (3) Ectendomycorrhizae (Arbutoid), (4) Ericoid mycorrhizae, (5) Gentianoid mycorrhizae, (6) Orchidoid mycorrhizae, and (7) Monotropoid mycorrhizae.

(1) **Ectomycorrhizae**: Ectomycorrhiza is commonly called "sheathing mycorrhiza". They occur in 3% of all seed plants in forests of temperate regions, especially on pine, beech, spruce, birch etc. Generally they cause extensive branching and growth of roots and modification of branching pattern, such as racemose type in dicots (*Fagus*) and dichotomous in gymnosperms (*Pinus*). In beech (*Fagus*) the ultimate lateral rootlets are differentiated into 'long' and 'short' roots. The long roots show indefinite growth and their branches are the short roots that are thickened, forked and mycorrhizal. They appear in various colours like white, brown, yellow, black etc., depending on the colour of the fungus. The fungus enters the cortex forming '**Hartig net**', but never goes inside the endodermis or stele. They form a mantle of varying thickness (Fig. 4.136A, B).

Majority of the fungi belong to Agaricales of

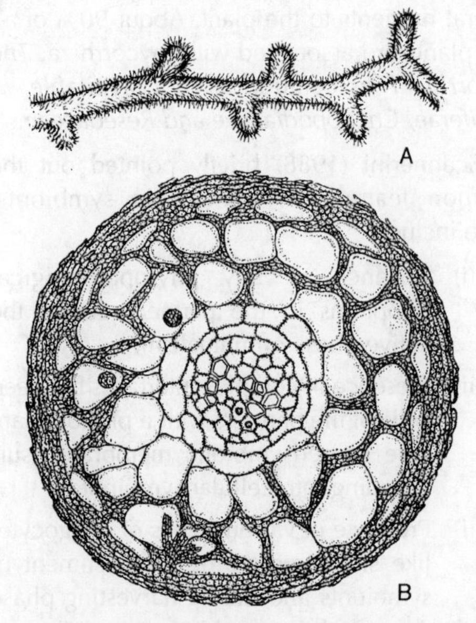

Fig. 4.136: Ectomycorrhiza: A. *Pinus* root covered with mycorrhizal fungi, B. T.S. of *Pinus* root showing ectotrophic mycorrhizal fungi forming a mantle covering and growth of fungi between cortical cells

Hymenomycetes under Basidiomycota. More than 100 species of toadstools are reported to form mycorrhiza. Most of the members are belonging to the genera *Amanita, Tricholoma, Boletus, Russula, Lactarius* etc. Members of Gasteromycetes under Basidiomycota like *Rhizopogon* and *Scleroderma* are involved in this process. Some members of Ascomycota like *Gyromitra esculenta,* all species of *Tuber* (*T. melanospora*) form mycorrhizae.

(2) **Vesicular-arbuscular mycorrhizae (VAM):** It is a type of *endomycorrhizal association*, where both vesicles and arbuscles are developed together. VAM is by far the commonest of all mycorrhizae and has been reported in more than 90% of land plants. They are found in bryophytes, pteridophytes, gymnosperm (except Pinaceae) and most of angiosperms, commonly in Leguminosae (Fabaceae), Rosaceae, Gramineae (Poaceae) and Palmae (Arecaceae). VAM is not found in Ericaceae and Orchidaceae, where other type of association is available. VAM has even been reported in Lower Devonian plant, *Rhynia*.

VAM is produced by aseptate mycelial fungi belong to Endogonaceae under Mucorales of Zygomycotina and those members produced zygospores. The important genera involved in VAM are *Glomus, Gyrospora, Acaulospora* etc. Most of the members are not culturable. The VAM is so named because of the presence of two characteristic structures i.e., vesicles and arbuscles:

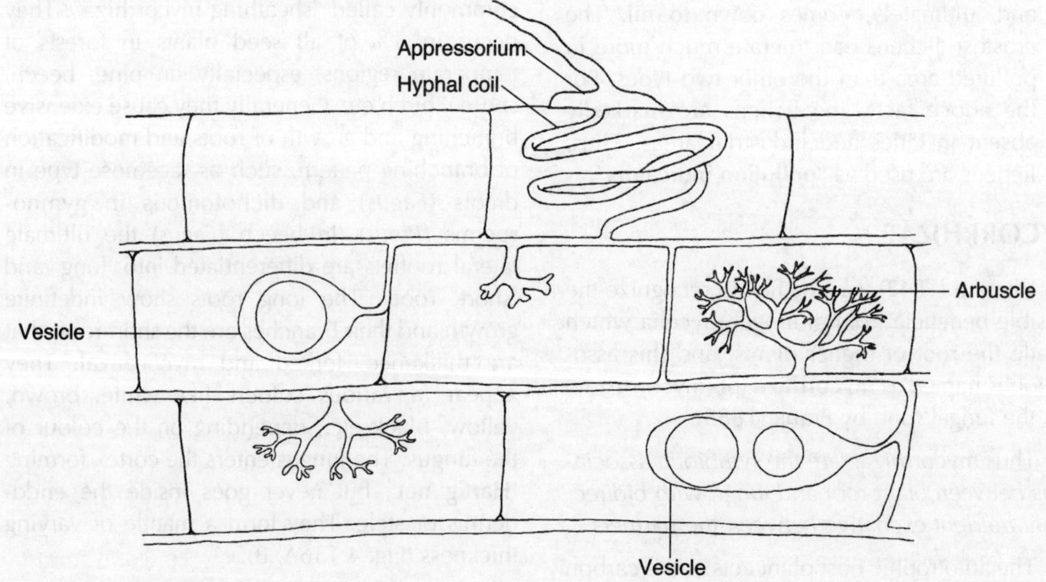

Fig. 4.137: Vesicular-arbuscular mycorrhiza showing both vesicles and arbuscles

(i) The *vesicles* are thin or thick walled vesicular structures, produced intracellularly and stored materials like polyphosphate and other minerals (Fig. 4.137).

(ii) The *arbuscles* are repeated dichotomously branched haustoria which grow intracellularly (Fig. 4.137). The arbuscles live for four days and then get lysed releasing the stored food as oil droplets, mostly polyphosphate.

There is no fungus mantle, but only a loose and very sparse network of aseptate hyphae spread into the soil. These hyphae bear different types of spores, chlamydospores, or aggregation of spores in sporocarp or zygospores. The superficial hyphae bear branches that penetrate the epidermis and then grow intercellularly only in cortex. Intercellular hyphae form arbuscles inside the parenchyma of cortex by repeated dichotomous branching of the penetrating hyphae. The cell membrane of the penetrated cell is invaginated and covers the arbuscles. The hyphae also develop both inter- and intracellular thick-walled vesicles. The chlamydospores may germinate on nutrient agar, but the hyphae stop growing when food inside the spore is used up, thus they cannot be subcultured.

This type of association was present very early in the evolution of land plants. Kidston and Lang (1921) reported the VAM-like organism with *Rhynia* and *Asteroxylon*. Later, Pyrozinski and Mallock (1975) proposed the mycorrhization/lichenisation association as a prelude to land plant evolution.

(3) **Ericoid mycorrhizae**: This is actually a type of endomycorrhiza. Ericoid mycorrhizae are found in the different members of Ericaceae like *Erica, Calluna, Vaccinum, Rhododendron* etc. The fungi are slow-growing, septate and mostly sterile. They are mostly culturable. Both *Pezizella ericae* (Ascomycotina) and *Clavaria vermiculata* (Basidiomycota) have been isolated from Rhododendrons.

During this association the rootlets of the plants are covered by very sparse, loose, dark, septate hyphae that penetrate the cortex forming intercellular coils (Fig. 4.138). After 3-4 weeks, the coils degenerate like arbuscles of vesicularasbuscular mycorrhiza (VAM).

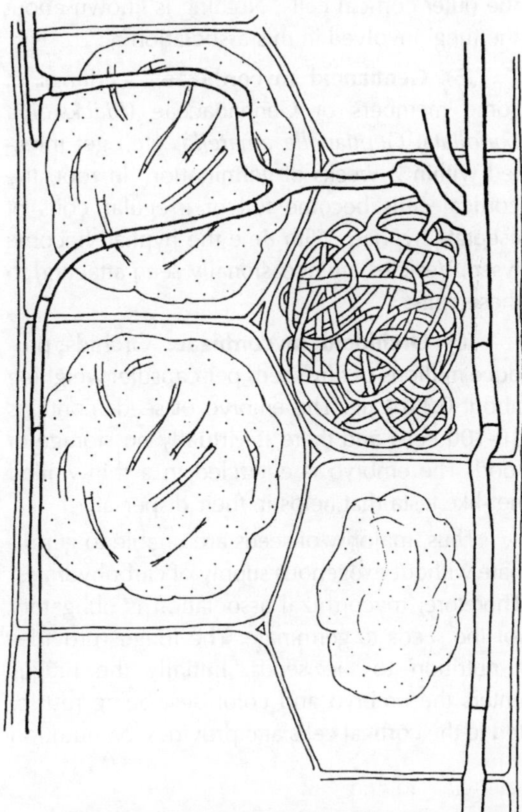

Fig. 4.138: Ericoid mycorrhiza in rootlet in *Erica* with active and lysed hyphae

Most of the members of Ericaceae grow in acid soil with less amount of P and N nutrition. The fungus gets the photosynthate from the host and improves the mineral uptake and nutrition of the host, especially P and N. Many mycotrophs of Ericaceae show high resistance to metals like Zn and Cu. The mycorrhizal plants also show high tolerance to these metals, which is totally absent in non-infected plants.

(4) **Ectendomycorrhizae (Arbutoid)**: Some members of the family Ericaceae and members of other families of Ericales have mycorrhizae intermediate in form between ecto- and endomycorrhizae types, called ectendomycorrhizae. *Arbustus* and *Arctostaphylos* of Ericaceae show this type of mycorrhizal association. In *Arbustus*, the root system is differentiated into long and

short roots. The short roots are swollen and covered by hyphal mantle. Hartig net is absent in this association, but intercellular coils develop in the outer cortical cells. Nothing is known about the fungi involved in this association.

(5) **Gentianoid mycorrhizae**: Seedlings of some members of Gentianaceae (*Blackstonia perfoliata, Gentianella amarella,* etc.) get infected within 2 weeks of germination. In root, the cortical cells become full of irregular coils of aseptate hyphae. With time the hyphae become lysed. Vesicles are occasionally seen attached to these coils.

(6) **Orchidoid mycorrhizae**: Orchids produce millions of tiny seeds per capsule, weighing about 0.3-14 µg. The embryo of seeds contains 10-100 cells and there is virtually no storage of food. The embryo is encircled in a thin-walled net-like testa that helps in their dispersal.

Thus, majority of seeds are unable to germinate without exogenous supply of carbohydrates. Therefore, mycorrhizal association is obligatory for the seeds to germinate. The fungus provides C-nutrition to the seeds. Initially the fungus enters the embryo and colonises, being restricted to the cortical cells and provides the nutrition (Fig. 4.139). For non-green orchids, this is obligatory throughout their lives. Apparently, it is a case of parasitism by orchids on the mycorrhizal fungi.

Fungi like *Rhizoctonia* (Basidiomycotina), are recognised by hyphal characteristics. *Corticium, Ceratobasidium* etc., of Aphylloporales are associated in this type of mycorrhiza.

(7) **Monotropoid mycorrhizae**: *Monotropa hypopitys* is a non-green saprophytic herb. It has short fleshy roots that are invested with a hyphal sheath and often forming Hartig net in the cortical zone. Due to absence of chlorophyll, they are unable to synthesise and supply carbohydrate to the fungus. *Boletus* is a mycorrhizal fungus associated with roots of both pine and *Monotropa*. When ^{14}C glucose was injected into the phloem of *Pinus* trees, significant amount of radioactive glucose (^{14}C) was recorded in *Monotropa* (but not in other herbs) after 5 days. This indicates that the fungus *Boletus* acts as a bridge between *Monotropa* and Pine plants. Similarly, ^{32}P injected into *Monotropa* was also detected in Pine roots within 2 hours. The above facts indicate a bidirectional flow of nutrients between the plants through the fungus *Boletus*.

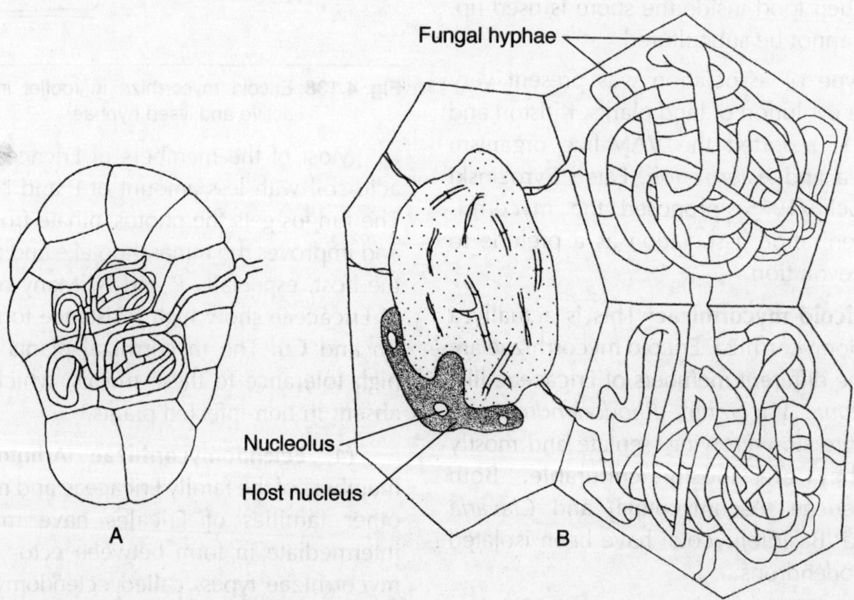

Fig. 4.139: Orchidoid mycorrhizae: A. Swollen embryo with initial infection, and B. Cortical cells of *Dactylorchis* root showing coiled active and lysed hypae

Role of Mycorrhizae in Agriculture and Foresty

Role in Agriculture

1. The mycorrhizal association helps in the formation of dichotomous branching and profuse root growth, thus enhances plant growth.
2. Ectotrophic mycorrhiza helps in uptake of mineral ions and also acts as reservoir.
3. They also help in absorption of nutrients.
4. In nutrient deficient soil, the mycelial association helps in the absorption of N, Ca, P, Zn, Fe, Na and others.
5. Mycorrhizal association is obligatory for the germination of orchid seeds.

   Mycorrhizal growth in orchids (*Rhizoctonia repens* with *Orchis militaris* tuber tissues) causes the synthesis of phytoalexins — orchinol and hirsinol. Both the compounds act as a barrier to protect infection by other pathogens.
6. Inoculation of VAM as biofertilizer provides a distinct possibility for the uptake of P in phosphorus-deficient soil.

Role in Foresty

1. Mycorrhiza plays an important role to establish forest in unfavourable location, barren land, waste lands etc.
2. Trees with facultative endomycorrhiza act as first invader in waste lands as pioneer in plant succession.
3. The application of mycorrhizal fungi in forest bed enhances the formation of mycorrhizal association that prevents the entry of fungal root pathogens. This method is very much effective in the root of *Pinus clausa* against *Phytophthora cinnamoni* infection.
4. Mycorrhiza mixed nitrogenous compounds such as nitrate, ammonia etc., is available to the plants. Thus it helps in plant growth, especially in acid soil.

APPLIED MYCOLOGY

1. Role of Fungi in Biotechnology

Biotechnology (biology + technology) is a technology that involves the use of living organisms. The term was coined in 1919 by Károly Ereky. It is not just concerned with biology, but it is a truly interdisciplinary subject involving the integration of natural and engineering sciences.

Fungi have been important in both ancient and modern biotechnological processes. Thus the term "mycotechnology" is introduced due to enormous impact of fungi in biotechnology. The processes and products that utilize fungi include baking (bread, cake etc.), brewing (beer and wine), cheese and the production of antibiotics, alcohols, enzymes, organic acids, and numerous pharmaceuticals. In textile processing the enzymatic removal of starch, sizes from woven fabrics has been in use for most of this century and the fermentation vat is probably the oldest known dyeing process. The advent of recombinant DNA technology and large scale genomics analysis has placed yeasts and filamentous fungi in the forefront of contemporary commercial applications.

Role of different fungi in biotechnology with their importance are given below in brief.

A. Beverages and Food

Alcoholic beverages

In anaerobic or with less oxygen condition, with high sugar concentration, fermentation caused by yeast and produce ethyl alcohol (ethanol) and CO_2. The alcoholic fermentation is the basis of production of different types of alcoholic beverages consumed by man and also used in other purposes. The unicellular fungus, *Saccharomyces cerevisiae* is involved in this process. The alcoholic beverages are of three types. These are wine and cider, beer, spirit and fortified wines.

Fungi as food

Fungi may be used directly or to modify as food to make it more nutritious. They are also used to produce material of nutritious value such as vitamins, amino acids and lipids. In addition, the macro-fungi are cultivated to yield fruit bodies i.e. mushrooms like *Agaricus brunnescens* (syn. *A. bisporus*), *Pleurotus sajor-caju* etc., and yeast cells and mycelial fungi grown in fermenters for food i.e. 'single cell protein'. It can develop very quickly and able to supply nutritious food in short period. They have also play a crucial role in the processing of many foods, improving the texture, digestibility, nutritional value, flavour or appearance of raw material used. The important among them are Bread, Soybean products, Soy sauce by *Aspergillus oryzae, A. tamari* or *A. sojae*.

B. Fungal secondary metabolites

Secondary metabolites are organic compounds produced by fungi which are not directly involved in the normal growth, development, or reproduction. These are as follows:

(i) *Antibiotic*: Penicillin from *Penicillium notatum*, Cephalosporin from *Cephalosporium* spp.

(ii) *Immunosupressive agent*: Cyclosporin A, an antifungal peptide produced by *Tolypocladium nivenum*.

(iii) *Cholesterol lowering drug*: Lovastatin and Compactin produced by *Penicillium citrinum* and *Monascus ruber*.

(iv) *Antitumor agent*: Taxol, a natural anti-cancerous agent, was originally discovered in plant but can also produce by fungus, *Taxomyces andreanae*. It is approved for the treatment of breast and ovarian cancer.

C. Fungal enzymes

Enzymes are catalytic proteins required to

operate a chemical reaction. Many fungi able to produce different types of enzymes and some are commercially very important. These are mentioned in the Table 4.23.

D. Biocontrol agents and other uses

Some fungi act as biocontrol agent. They act either as parasite or antagonist. Some of the commercially used biocontrol agents are given below:

(i) *Trichoderma harzianum* - White rot of onion.
(ii) *Phlebia giganteum* - Heterobasidium root rot of pine.
(iii) *Sporidesmium sclerotivorum* - Lettuce drop.
(iv) *Talaromyces flavus* - Damping off of sugar-beet.

Fungi to control insect pests

More than 400 species of fungi attack insects and mites, so many of these organisms are used as biological insecticides. Some of them are given below:

Enzyme	Produced by
1. *Aschersonia aleyrodis*	White fly
2. *Beauveria bassiana*	Colorado beetle
3. *Beauveria brongniartii*	Cockchafer
4. *Hirsutella thompsonii*	Rust mites
5. *Metarhizum anisopliae*	Beetles, bugs
6. *Nomuraea rileyi*	Caterpillars
7. *Verticillium lecanii*	Aphids, White fly

Fungus to control nematodes

Fungi that parasitise on nematodes are the nematophagus fungi. They can be divided into three groups: nematode-trapping fungi, endoparasitic nematophagous fungi and fungi that parasitise nematode eggs.

(i) *Nematode-trapping fungi*: They capture nematodes with specialized structures like constrictive and non-constrictive rings, adhesive knobs or by producing an adhesive material along the entire body surface.
(ii) *Endoparasitic nematophagous fungi*: They live in soil where they produce adhesive spores. The spores become attached to body of the nematode and on germination the germ-tube enters the body where it grows and consume the host.
(iii) *Fungi that parasitise nematode eggs*: Fungal mycelium parasitise the eggs of nematode and caused complete damage.

E. Application of Molecular Biology

Apart from the selection and conventional mutagenesis through UV, chemicals etc., the molecular techniques have opened a new avenues to produce protein and other compounds inside the yeasts and molds.

Products obtained from Recombinant DNA technology are: Pharmaceutical proteins, Commercial Recombinant Enzymes, Improved strains etc.

Table 4.23: Fungal enzymes of commercial importance

Enzyme	Produced by
1. Amylase	*Aspergillus niger, A. oryzae*
2. Asparaginase	*Aspergillus* spp. and *Penicillium* spp.
3. Catalase	*Aspergillus niger* and *Penicillium* spp.
4. Cellulase	*Aspergillus niger, Trichoderma viride, Penicillium finiculosum.*
5. Dextranse	*Penicillium* spp.
6. β-Glucanase	*Aspergillus niger, Penicillium emersonii, T. viride*
7. Glucoamylase	*Aspergillus niger, A. oryzae*
8. Glucose oxidase	*Aspergillus niger, Penicillium* spp.
9. Hemicellulase	*Aspergillus niger, Trichoderma viride*

2. Application of fungi in food industry

A. Flavour and texture

Flavour: Flavouring of food can be done by substances of fungal origin. These are:

(a) Nucleotides

Flavour potentiators have been used extensively for more than 20years. It is added with food to enhance the natural flavour of the foodstuff.

Monosodium glutamate (MSG) is a well-established compound in the potentiator market. In addition, IPM (inosine-5'-monophosphate) and GMP (guanosine-5'-monophosphate) are also useful in improving the flavour of many foods (Nelson and Richardson, 1967). Titus (1964) reported that these potentiators are beneficial in soups and gravies at levels of 25 to 100 ppm. They are also useful in improving the flavour of many products containing hydrolyzed proteins and can be used to replace beef extracts in many foods (Kuninaka, 1972). Kuninaka *et al*. (1964) reported a strong synergistic action between MSG and flavour nucleotides. Such as:

(i) 10 g MSG + 1 g IMP = 55 g MSG

(ii) 10 g MSG + 1 g GMP = 209 g MSG

Production

Monosodium glutamate (MSG) is produced by neutralization of glutamic acid which can be produced by molds and yeasts. However, glutamic acid is primarily a fermentation product of the bacterium, *Micrococcus glutamicus*. In contrast, the nucleotides are produced by treating RNA primarily from yeasts, and with nuclease from *Penicillium citrinum* (Kuninaka 1972). The enzyme cleaves the 5'-phosphodiester linkages of the nucleosides to free the flavourful 5'-nucleotides. Nuclease from *A. oryzae* and *Streptomyces griseus*, an actinomycete; has also been used in flavouring (Nelson and Richardson 1967).

(b) Yeast Cells

Yeast cells can serve as flavour enhancers in soups, gravies, sauces and snacks. Reportedly they are excellent carriers for flavourings such as cheese, barbecue, fish and chicken. In some meat products, use of yeast permits a reduction of 20 to 50% pepper and other spices (Anon, 1974).

(c) Mycelia of Mushrooms and Other Fleshy Fungi

The discovery in the late 1940s that mycelia from fleshy fungi could be efficiently produced in submerged culture led researchers to explore the growth characteristics of several mushroom genera. By manipulating medium composition, temperature, aeration and other growth factors investigators have produced mycelia with mushroom-like flavour from *Agaricus brunnescens* (syn. *A. bisporus*), *A. blazei, Coprinus comatus, Tricholoma nudum, and Lepiota rachodes*. Mycelia of *Morchella hortensis, M. crassipes and M. esculenta* have been found to have a desirable flavour and odour.

Mycelia of *Morchella* spp., are approved for sale as "morel mushroom flavouring" in USA (Litchfield, 1967). At one time, this flavouring was commercially available for use in dehydrated soup sauce and gravy formulations under the trade name Powdered Morel Mushroom Flavourings. Litchfield (1967) stated that this product was competing successfully with dried mushrooms. However, production has been discontinued (Le Duy *et al*. 1974).

(d) Blue Cheese

Nelson (1970) described a submerged fermentation procedure for the production of blue cheese flavour by *Penicillium roqueforti* on a milk-based medium. The resulting flavour contained 7-12 times the *ketone content* of good quality in commercial blue cheese and exhibited *an efficacy* 4 times that of blue cheese. The product is used commercially in salad dressing and snacks, where a blue cheese flavour is desired.

Texture

Attractive texture of different foods is established with fungi like yeasts etc., through fermentation. Fermentation not only gives food a good taste, texture and smell, but it causes changes that reduce the growth of unwanted food microbes. This improves the food's storage life and safety. Nowadays fermentations are used to make an amazingly wide range of food and drink.

(e) Bread

Yeast, *Saccharomyces cerevisiae* is mixed with sugar, flour and warm water to make bread in factory. The yeast uses the added sugar and the sugars present in the flour as its food. It breaks them down to provide the yeast with energy for growth. The yeast grows by budding. As it does this bubbles of the CO_2 gas are produced in the dough*.

The bubbles make the dough expand and rise. This is because the dough is extremely sticky and it traps the bubbles, preventing them from escaping. When the dough is baked the heat kills the yeast and the dough stops expanding.

(***Dough**: *It is a thick, malleable mixture of flour and liquid, used for baking bread or pastry*.)

B. Fermentation

Fermentation is an energy-releasing metabolic process that takes place only under anaerobic conditions.

But in the context of food production, it can be defined as "a process in which the activity of microorganisms brings about a desirable change to a foodstuff or beverage".

Fungi have been used since ancient times to make bread, cheese and wine. Food manufacturers continue to use fungi today to make a wide range of food products by a process known as **alcoholic fermentation**. Fermentation not only gives food a good taste, texture and smell, but it causes changes that reduce the growth of unwanted food microbes. This improves the food's storage life and safety. Nowadays fermentations are used to make an amazingly wide range of food and drink.

Fungi have been used as sources of food and for food processing for thousands of years. In addition to eating edible fruiting bodies, such as mushrooms, various fungi have been used to supplement and add flavour to foods. **Yeasts** are used in the fermentation of fruits to produce wines, cereals to make beer, in bread manufacture and flavouring in the form of yeast extract. **Filamentous fungi** (*Penicillium roquiforti* and *P. camemberti*) are used in the production of Roquefort and Camembert cheese by hydrolysis of fats and also to develop specific flavour to cheese.

C. Baking

Baking is a method of cooking food that uses prolonged dry heat, normally in an oven, but can also be done in hot ashes, or on hot stones. The most common baked item is bread but many other types of foods are baked. Heat is gradually transferred "from the surface of cakes, cookies, and breads to their center. As heat travels through, it transforms batters and dough's into baked goods with a firm dry crust and a softer centre".

Baking has traditionally been performed at home by women for day-to-day meals and by men in bakeries and restaurants for local consumption. When production was industrialized, baking was automated by machines in large factories. The art of baking remains a fundamental skill and is important for nutrition, as baked goods, especially breads, are a common and important food, both from an economic and cultural point of view. A person who prepares baked goods as a profession is called a **baker**.

D. Organic acids

Organic acids and their derivatives are frequently used in beverage, food, and feed production. Acidic additives may act as buffers to regulate acidity, antioxidants, preservatives, flavour enhancers etc. Beneficial effects on animal health and growth performance have been observed when using acidic substances as feed additives. Organic acids could be **classified** in groups according to their chemical structure. Each group of organic acids has its own specific properties and is used for different applications.

(i) Organic acids with low molecular weight (e.g. acetic acid, lactic acid, and citric acid), which are part of the primary metabolism, are often produced by fermentation.

(ii) Others are produced more economically by chemical synthesis based on petrochemical raw materials on an industrial scale (e.g. formic acid, propionic and benzoic acid).

(a) Citric acid

Citric acid was first discovered in lemon, contains 7-9% citric acid. It is an important intermediate of Kreb's cycle and therefore it is present in all living organisms. At present about 99% of the World's citric acid comes from microbial fermentation. Its molecular formula is $C_6H_8O_7$.

Producing Fungi

Many fungi can produce citric acid. The fungus, *Aspergillus niger* is the most important and commonly used in the industrial production of citric acid. Other less important fungi are *A. clavatus, A. wentii, Candida catenula, C. guilliermondii* and *Penicillium luteum*.

For more production, mutant strains of *A. niger* have been developed and used industrially. The strains that can tolerate high sugar concentration and low pH with reduced synthesis of undesirable byproducts like isocitric acid, gluconic acid and oxalic acid are selected for industrial use.

Applications

In food industry, Citric acid is mostly used variously. These are:

(i) **Flavouring agent**: Citric acid is pleasant in taste and palatable. Thus it is used as a flavouring agent in foods and beverages, such as jams, jellies, candies, frozen fruits, soft drinks and wine. Besides brightening the colour, it also acts as an antioxidant and preserves the flavours of the foods.

(ii) **As Stabilizer:** Citric acid can be utilized as an agent for stabilization of oils, fats or ascorbic acid. It forms a complex with metal ions like iron and copper, and prevents metal catalyzed reactions. Citric acid is also used as a stabilizer of emulsions in the preparation of cheese.

(b) Gluconic acid

Gluconic acid is mild organic acid, produced by many bacteria and fungi. Glucose on dehydrogenation forms D-gluconolactone which is then converted into gluconic acid. Its molecular formula is $C_6H_{12}O_7$.

Producing Fungi

Many fungal species of the genera, *Aspergillus, Penicillium* and *Gliocladium* produces glutamic acid.

Applications

Gluconic acid is used as an additive to foods and beverages. It is a natural constituent in fruit juices and honey, and is used in the pickling of foods. It prevents to form milk-stone in dairy industry.

E. Enzyme

Several enzymes produced by fungi are used in food industry directly or indirectly. These are: Amylase, Cellulase and Zymase. Brief information about the enzymes are given in the Table 4.24.

Mucor pusillus was found to be the best producer of rennin-like acid **protease** producing 75 U/ml of the enzyme, used in cheese production.

F. Mycoproteins

Mycoprotein is the unique and nutritious, single-cell fungal protein that can be a part of a balanced diet for human beings.

It is high in protein and fiber, low in saturated fat, and without cholesterol. This meat-free protein contributes to a balanced plant-based eating style of man.

Source

Fusarium venenatum, a natural, nutritious filamentous fungi belongs to Ascomycota, isolated from soil is the principal ingredient of mycoprotein. It is the only source of mycoprotein (trade name '**quorn**' in USA) which can produce a high percentage of protein biomass approved for market.

Synthesis

The fungus is grown in vats using glucose solution as food. Initially the fermentation vat is

Table 4.24: Fungal enzymes used in food industry

Enzyme	Producing Fungi	Application
1. Amylase (α-amylase, β-amylase and Glucamylases)	*Aspergillus oryzae* and *A. niger*	The conversion of starch to high-fructose syrup utilizes amylases i.e. producing sweeteners.
2. Cellulase	*Trichoderma reesli*	The enzyme is used in the production of cheese and also the hydrolysis of cellulose.
3. Zymase	Yeast (*Saccharomyces cerevisiae*)	(i) Alcoholic fermentation by fungi like *Saccharomyces cerevisiae* is the basis of brewing industry. Zymase of yeast is responsible for alcohol production. (ii) Wines are produced from grapes by *Saccharomyces ellipsoideus* (with about 14% alcohol concentration). (iii) Beer is brewed from barley malt by *Saccharomyces cerevisiae* (with 3-8% alcohol production). (iv) During alcoholic fermentation by yeast, CO_2 being released as bubbles are used in baking industry, gives the bread a spongy appearance.

filled with the growth medium and then inoculated with fungal spores of *F. venenatum*. As the fungus respires aerobically, so for the optimum grow rate, it is to be supplied with oxygen, and CO_2 is drawn from the vat. To make the protein, nitrogen (in the form of ammonia) is added and to support growth vitamins and minerals are also needed. The vat is kept at a constant temperature, also optimized for growth; the fungi can be double in its mass after every five hours.

When the desired amount of mycoprotein has been produced, the growth medium is drawn off from a tap present at the bottom of the fermenter. The mycoprotein is then separated and purified. It is a pale yellow solid with a faint taste of mushrooms. Different flavours and tastes can be added to the mycoprotein to be more palatable.

Once harvested, '**Quorn**' uses the age-old process of fermentation (much like that used to create beer and yogurt) to convert the carbohydrate into a mycoprotein dough, Thus it takes the texture of meat and can be used in a variety of culinary applications.

Nutrition

Mycoprotein is able to provide greater satisfaction than traditional protein sources like chicken, but rich in protein and low in caloric content. Replacing to servings of meat protein by mycoprotein can result in a daily deficit of 80 kcal. But it extends the period of satiation, which is promising for weight management of human. It is rich in fiber and protein content but very low in fat, makes the limit of fat intake.

(a) *Fusarium venenatum's* high fiber content also has potential in managing blood sugar levels.

3. Secondary metabolites (Pharmaceutical preparations)

Secondary metabolites are organic compounds produced by bacteria, fungi, or plants which are not directly involved in the normal growth, development, or reproduction of the organism. Secondary metabolites aid a host in important functions such as protection, competition, and species interactions, but are not necessary for survival. One important defining quality of secondary metabolites is their specificity.

Usually, secondary metabolites are specific to an individual species.

Secondary metabolites are small-molecule organic compounds found restricted to a particular species, genus, or family.

Different secondary metabolite produced by different fungi. Some secondary metabolites of fungi used in Pharmaceutical preparations are: Antibiotics, such as Penicillin, Griseofulvin, Fusidic acid and Cephalosporin; and Ergot alkaloids.

A. Antibiotics

Antibiotics are the metabolic products of some microorganism(s) which are active against other microorganism(s). Different types of antibiotics were isolated from different fungi used in pharmaceutical preparations are Penicillin, Griseofulvin, Fusidic acid and Cephalosporin (Table 4.25).

(i) Penicillin

Sir Alexander Fleming (1929) was the first who invented the **wonder drug penicillin** from fungus *Penicillium notatum*. Later, penicillin was also commercially produced from *P. crysogenum*. It was among the first drugs to be effective against many serious diseases, such as bacterial infections caused by staphylococci and steptococci.

Core structure of penicillin

The common core skeleton of penicillins is called "**penam**", with the molecular formula R-$C_9H_{11}N_2O_4S$, where R is the variable side chain that differentiates the penicillins.

Penicillin is a group of antibiotics including penicillin G, used intravenously; penicillin V, used orally; and procaine penicillin and benzathine penicillin are used intramuscularly. It is used effectively in controlling different bacterial infection caused by staphylococci, streptococci etc.

Penicillins are still widely used today, though misuse due to ignorance has now made many types of bacteria resistant to the antibiotic. All penicillins are β-lactam antibiotics and are used in the treatment of bacterial infections caused by susceptible, usually Gram-positive, organisms. At present many other penicillin families also exist, those are effective against many other bacteria, these include the aminopenicillins, antistaphylococcal penicillins and the more-powerful antipseudomonal penicillins.

(ii) Griseofulvin

It is an **antifungal antibiotic**, used for both in animals and human beings, to treat fungal infection of the skin (commonly ring worms), scalp and nails. Although it was first isolated in 1939 from *Penicillium griseofulvum*, but the antibiotic property was recognized by Brain *et al.*, (1946) after isolating it from *P. janczaewski* (= *P. nigricans*). Till date it was isolated from a dozen of different species, most important one is *P. patulum*. It stops fungal cell to divide (i.e. fungistatic) but does not kill them outright. Thus the treatment needs to continue for several weeks and suggested to complete the course of medicine. It also binds to keratin, in keratin precursor cells and makes them resistant to fungal infection. It is

Table 4.25: Name of antibiotic, produced fungi and range of activity

Name of antibiotic	Produced Fungi	Range of activity
1. Penicillin	*Penicillium notatum, P. crysogenum*	Bacteria (both Gram + and –)
2. Griseofulvin	*Penicillium nigricans, P. griseofulvum*	Fungi
3. Cephalosporin	*Acremonium* spp.	Bacteria (Gram +)
4. Fusidic acid	*Fusidium coccineum*	Bacteria

used orally only to control dermatophytes. It is produced commercially and is available in market in the brand name Grifulvin, S-Fulvin, Crivicin, Gris-PEG etc.

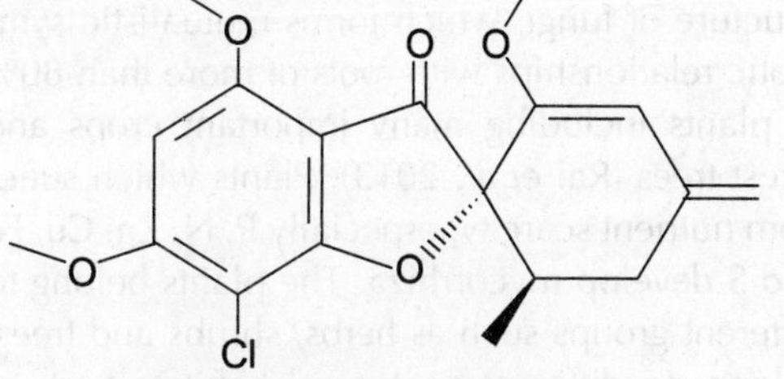

Structure of Griseofulvin

(iii) Fusidic acid (Fusidane)

Fusidic acid is a **bacteriostatic antibiotic**. It is produced by the different fungi like *Fusidium coccineum, Cephalosporium* sp., *Isaria kogana* etc.

Fusidic acid is effective primarily on gram +ve bacteria such as different species of *Staphylococcus, Streptococcus*, and *Corynebacterium*. Fusidic acid inhibits bacterial protein synthesis and replication but does not kill the bacteria, and is therefore termed "bacteriostatic".

It is often used typically in creams and eye-drops, but may also be given systemically as tablets or injections. The global problem of advancing antimicrobial resistance has led to a renewed interest in its use recently. The antibiotic is used as its sodium salt, Sodium fusidate in many countries.

The structure of fusidic acid is given below:

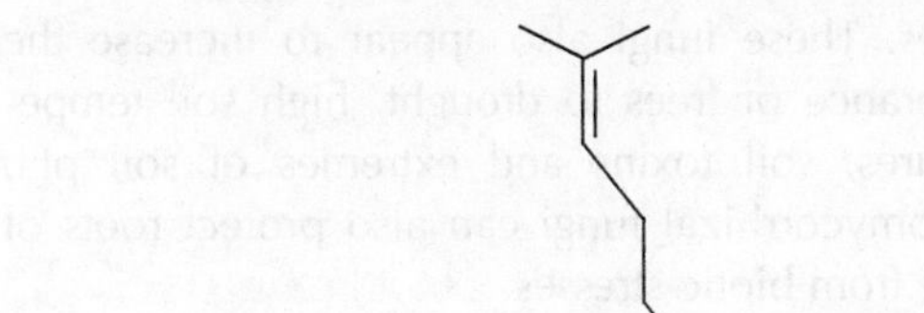

Fusidic acid

(iv) Cephalosporin

Cephalosporins are a broad group of **bactericidal antibiotic** that have the β-lactam ring and shows structural similarity and mechanism of action with other β-lactam antibiotics like penicillins, carbapenems and others.

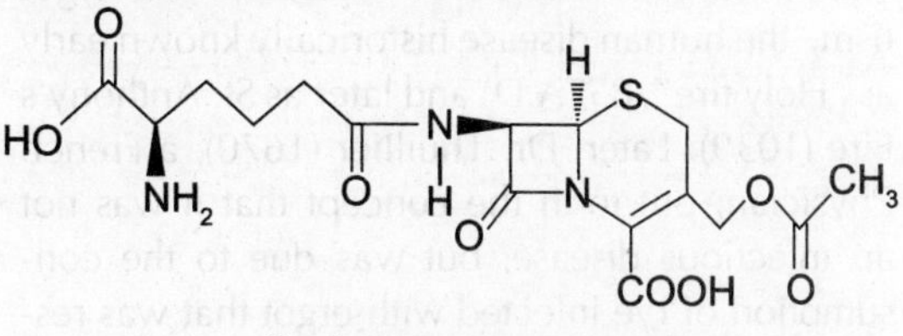

Basic core structure

Cephalosporin C

The first chemical compound of this group was isolated from *Cephalosporium acremonium*, a fungus first discovered by Giuseppe Brotzu (1948) from a sewage outfall. Crude culture filtrate of the *C. acremonium* shows new antibacterial activity, which inhibit the growth of *Staphylococcus aureus*. Later, Abraham and his co-workers at different time isolated cephalosporin P, N and C from the same fungus. These natural compounds are not potent enough but more active compounds were developed by altering the side chain. Abraham (1959) reported that his N-phenylacetyl derivative of cephalosporin C, (later named as "Cephaloram") is much more potent against *Staphylococcus aureus* strains than the parent compound. Later, many others have also developed many new derivatives.

Cephalosporins have the ability to kill bacteria by inhibiting essential steps in the synthesis of bacterial cell wall which at the end, results in osmotic lysis and death. Cephalosporins are widely used antibiotics because of their clinical efficiency and desirable safety profile. The cephalosporins are diverse in their ability on antibacterial spectrum, water solubility, acid tolerance, oral bioavailability, biological half-life and many other properties.

B. Ergot alkaloids

Ergot alkaloids are the compounds produced as a toxic mixture of alkaloids in the sclerotia of *Claviceps* spp., those are the common pathogens of various grasses. The ergot alkaloids are the derivatives of lysergic acid (ergotamine), isolysergic acid (ergocristine) or dimethyl ergoline (agoclacine). The alkaloids are synthesized through different pathways.

The ingestion of ergot sclerotia from infected cereals, commonly in the form of bread produced from contaminated flour, caused ergotism, the human disease historically known early as "Holy fire" (857A.D) and later as St. Anthony's Fire (1039). Later, Dr. Thuillier (1670), a French Physician; put forth the concept that it was not an infectious disease, but was due to the consumption of rye infected with ergot that was responsible for the outbreaks of St. Anthony's Fire.

There are two forms of ergotism: **gangrenous**, affecting blood supply to extremities, and **convulsive**, affecting the central nervous system.

Pharmacologically, these compounds are rapid acting, powerful oxytoxics, i.e., they stimulate the smooth muscle of the uterus. They are also weak vaso-constrictors. Acute ergot poisoning shows the symptoms include vomiting, diarrhoea, intense thirst, a tingling, itchy and cold skin, a rapid weak pulse, confusion, and unconciousness. Its poisoning shows constriction in the local blood vessels caused painful walking and also the development of gangrene.

4. Biofertilizer in Agriculture

What is Biofertilizer?

Biofertilizers are environment friendly, low cost, renewable sources of plant nutrients used instead of chemical fertilizers and play an important role in improving nutrient level for the crop growth and productivity. It is a live formulation of the beneficial microorganisms, which on application to seed, root or soil; mobilize the availability of nutrients by their biological activity. These are the selected strains of beneficial soil microorganisms, cultured in the laboratory and packed in a suitable carrier, which can be used either for seed treatment or soil application. Bacteria and fungi are commonly used as biofertlizer.

Fungal biofertilizers

Fungal biofertilizer is the fungal body used either alone or in combination, exerting direct or indirect benefits on plant growth and crop yield through different mechanisms. Some of the selected fungal species are used as biofertilizers (Table 4.26).

Mycorrhiza is a distinct morphological structure of fungi, which forms mutualistic symbiotic relationships with roots of more than 80% of plants including many important crops and forest trees (Rai *et al*. 2013). Plants which suffer from nutrient scarcity, especially P, N, Zn, Cu, Fe and S develop mycorrhiza. The plants belong to different groups such as herbs, shrubs and trees, belongs to different ecological habitats such as aquatic, xerophytes, epiphytes, hydrophytes or terrestrial ones (Zhu *et al*. 2008).

In recent years, use of artificially produced inoculum of two dominant types of mycorrhizal fungi has increased its significance due to its various roles in plant growth, yield and resistance against biotic and abiotic stresses. These are:

1. Ectomycorrhizal fungi

They form mutualistic symbioses with many tree species (Anderson and Cairney, 2007). They help in the growth and development of trees because the roots colonized with ectomycorrhiza are able to absorb and accumulate nitrogen, potassium, phosphorus and calcium more rapidly and over a longer period than nonmycorrhizal roots.

Ectomycorrhizal fungi help to break down the complex minerals and organic substances present in the soil and transfer nutrients to the trees. These fungi also appear to increase the tolerance of trees to drought, high soil temperatures, soil toxins and extremes of soil pH. Ectomycorrhizal fungi can also protect roots of tree from biotic stresses.

The most commonly widespread ectomycorrhizal product is inoculum of *Pisolithus tinctorius* (Schwartz *et al*. 2006) with a wide host range and their inoculum can be produced and applied as somatic mycelium in a peat or clay carrier. *Piriformospora indica* is another ectomycorrhizal fungus used as a biofertilizer with multifaceted traits of plant growth promotion, tolerance to both abiotic and biotic stresses and increased biomass (Tejesvi *et al*. 2010).

Table 4.26: Fungal biofertilizers used in Agriculture

Genus	Different species
A. Phosphorous solubilising biofertilizers	
1. *Aspergillus*	*A. tubingensis, A. niger, A. terreus, A. awamori, A. tubingensis, A. melleus* etc.
2. *Penicillium*	*P. bilaji, P. albidum, P. italicum, P. simplicissimum, P. frequentans, P. oxalicum, P. rubrum, P. expansum* and *P. citrinum*
3. *Fusarium*	*F. moniliforme* and *F. udam*
4. *Trichoderma*	*T. viridi, T. harzianum, T. virens* and *T. asperellum*
5. *Mucor*	*M. ramosissimus, M. mucedo* and *M. hiemalis*
6. *Tritirachium*	*T. album* and *T. egenum*
7. *Candida*	*C. krissii* and *C. scotti*
B. Potash solubilising biofertilizers	
1. *Aspergillus*	*A. fumigates, A. niger* and *A. terreus*
2. Ectomycorrhizal fungi	
C. Zinc solubilising biofertilizers	
1. *Saccharomyces* spp.	
2. *Ericoid mycorrhiza*	*Oidiodendron maius*
3. *Penicillium*	*P. simplicissimum*
4. *Aspergillus*	*A. niger*
D. Phosphate mobilizing biofertilizer	
1. Ectomycorrhizal fungi	*Amanita* sp., *Lactarius* sp., *Pisolithus* sp., *Rhizopogon* sp., *Cenococcum* sp., *Elaphomyces* sp., *Tuber* sp., *Pisolithus tinctorius*
2. *Glomus* spp.	*G. mosseae, G. manihotis, G. itraradices, G. aggregatum, G. cerebriforme, G. deserticola, G. globiferum, G. halonatum, G. microcarpum* etc.
3. *Acaulospora* spp.	*A. foveata, A. scrobiculata* and *A. delicata*
4. *Scutellospora* spp.	*S. erythropa, S. calospora* and *S. scutata*
5. *Gigaspora*	*G. rosea*
E. Biofertilizer enriching compost	
1. *Trichoderma*	*T. viridi, T. harzianum, T. virens* and *T. asperellum*
2. *Penicillium*	*P. albidum, P. italicum, P. oxalicum, P. rubrum,* and *P. expansum* etc.
3. *Aspergillus*	*A. tubingensis, A. niger, A. terreus, A. awamori* and *A. fumigatus*
4. *Pleurotus*	*P. ostreatus* and *P. flabellatus*
5. *Chaetomium*	*C. bostrychodes* and *C. olivaceum*
6. *Humicola*	*Humicola fuscoatra*
7. *Fusarium*	*F. solani* and *F. oxysporum*

2. **Endomycorrhizal fungi**

They form mutually symbiotic relationships between fungi and plant roots, where plant roots provide carbohydrate for the fungi and the fungi transfer nutrients and water to the plant roots (Adholeya *et al.*, 2005).

The agriculturally produced crop plants that form endomycorrhizae of the vesicular-arbuscular mycorrhiza type are now called arbuscular mycorrhizal fungi (Raja, 2006). They belongs to nine different genera are *Acaulospora, Archaeospora, Enterophospora, Gerdemannia, Geosiphon, Gigaspora, Glomus, Paraglomus* and *Scutellospora*. Arbuscular mycorrhizal fungi is a widespread group and are found from the arctic to tropics and are present in most agricultural and natural ecosystems. Arbuscular mycorrhiza are prominent P mobilizers, which facilitate mobilization of soluble phosphorus from distant places in soil, where plant roots cannot reach and thus increase the availability of P to plants.

Since mycorrhizal fungi are more efficient in the uptake of specific nutrients like P, Ca, S, N and Zn and resistant against soil-borne pathogens, interest in using these fungi as biofertilizers is increasing as they play an important role in plant growth, health and also productivity. Other fungal biofertilizers, which have been used to improve plant growth by enhancing phosphorus absorption in plants, are the phosphate solubilizing microorganisms.

The commonly widespread fungi are *Penicillium, Aspergillus, Chaetomium and Trichoderma* species. There are a number of biofertilizers available in the market (Table 4.27). However applications are based on their ability to supply and mobilize plant nutrients, control plant diseases and promote plant growth and development.

Table 4.27: List of the name of commercial products and the producing organisms:

Commercial Products	Producing organisms
1. AgBio-Endos	*Ectomycorrhizal fungi*
2. AgBio-Ectos	*Endomycorrhizal fungi*
3. Bioorganic Plus	*Trichoderma harzianum, Trichoderma hamatum*
4. Mycomax	*AM fumgi* (*Glomous intraradices*)
5. Superzyme	*Trichoderma* spp.
6. Tricho	*Trichoderma* spp.
7. Bioorganic Plus	*Trichoderma harzianum, Trichoderma hamatum*

5. Mycotoxins

Toxins are highly specific poisonous substances produced by some microorganisms. The toxins produced by fungi are called *mycotoxins*. *Mycotoxicosis* is the term applied to any disease that results from consumption of food contaminated by fungal toxin.

Many fungi produce toxins which are responsible for food poisoning and cause different types of effects on animals including human beings.

Mycotoxins are broadly classified into three groups: A. Food toxins, B. Ergot toxins, and C. Mushroom toxins.

A. **Food toxins**: These toxins are mainly responsible for poisoning of foods. The mycotoxin production occurs in many plant products, especially cereals and oilseeds. Others like peanuts, rice, corn and cotton seeds also get contaminated and store mycotoxins. The mycotoxins are produced by more than 100 mold species (Wright, 1968), mainly by *Aspergillus, Penicillium* and *Fusarium*.

According to World Health Organisation (WHO), 1969, the mycotoxins associated with human diseases are of four types. These are aflatoxins, ochratoxins, zearalenone, and trichothecenes.

(a) **Aflatoxins.** Aflatoxins are potent toxic, carcinogenic, mutagenic, immuno-suppressive agents, produced as secondary metabolites by the fungi on variety of food products. The aflatoxin problem was first recognised in 1960, when there was severe outbreak of a disease referred as "Turkey X' Disease" in U.K., in which over 100,000 turkey birds died. Aflatoxins are the most potent carcinogens, known as the suspected cause of liver cancer in human beings. The aflatoxins are produced in dried food and groundnut meal, infested with *Aspergillus flavus, A. parasiticus, A. fumigatus* or *Penicillum islandicum*.

Aflatoxins normally refers to the group of difuranocoumarins and are classified in two broad groups according to their chemical structure: the difurocoumarocyclopentenone series and the difurocoumarolactone series. About 18 types of aflatoxins have been identified, the major members are aflatoxin B_1, B_2, G_1, G_2, while M_1 and M_2 are less significant. Aflatoxins are soluble in methanol, chloroform, acetone, acetonitrile and fluoresce strongly in ultraviolet light (365 nm); B_1 and B_2 produce a blue fluorescence, while G_1 and G_2 produce green fluorescence.

Aflatoxins bind with DNA to prevent transcription, thereby protein synthesis is inhibited.

In addition to suspected carcinogenic nature, other clinical symptoms due to aflatoxin poisoning are jaundice, portal hypertension, rapidly developing ascites, etc.

Structure of aflatoxins

(i) B_1: Molecular weight, 312; melting point 268-269°C and molecular formula $C_{17}H_{12}O_6$.

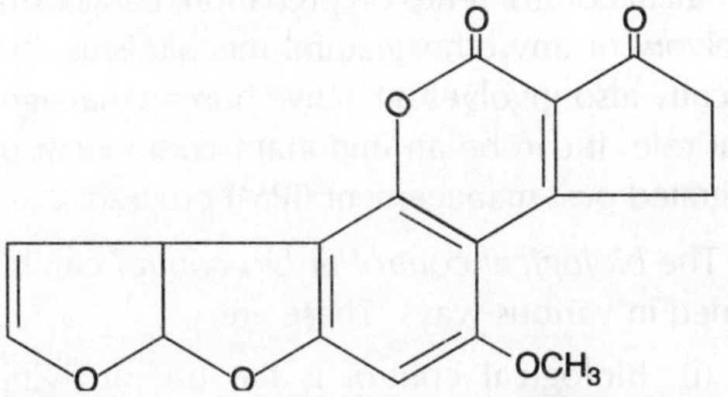

Aflatoxin B_1 isolated from *Aspergillus flavus*

(ii) B_2: It is dihydro derivative of B_1, a minor toxin with lower Rf value than B_1. Molecular weight, 314; melting point 286-289°C and molecular formula $C_{17}H_{14}O_6$.

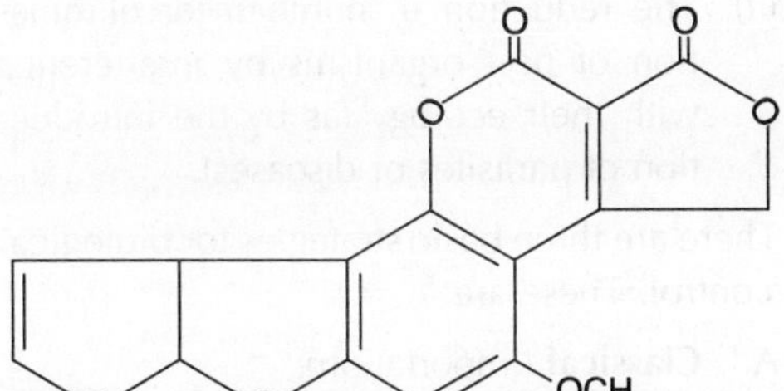

Aflatoxin G_1 isolated from *Aspergillus flavus*

(iii) G_1: Molecular weight, 328; melting point 244-246°C and molecular formula $C_{17}H_{12}O_7$.

(iv) G_2: It is the dihydro derivative of G_1. Molecular weight, 330; melting point 237-240°C and molecular formula $C_{17}H_{14}O_6$.

(v) M_1: It is the hydroxy aflatoxin B_1 with molecular weight 328; melting point 299°C and molecular formula $C_{17}H_{12}O_7$.

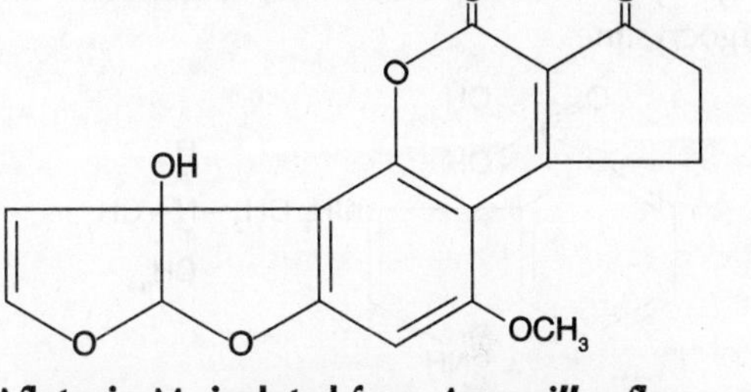

Aflatoxin M_1 isolated from *Aspergillus flavus*

(vi) M_2: It is dihydroaflatoxin M_1 with molecular eight 330; melting point 293°C and molecular formula $C_{17}H_{14}O_7$.

(vii) Other forms such as B_{2a} (Molecular weight 330, melting point 240°C, formula: $C_{17}H_{14}O_7$) and G_{2a} (Molecular weight 346, melting point 190°C, formula: $C_{17}H_{14}O_8$) are the hydroxyl derivatives of aflatoxin B_2 and G_2.

(b) **Ochratoxins**. Ochratoxins are closely related derivatives of isocoumarin, linked to an amino acid, L-β-phenylalanine. Of the nine types of ochratoxins, ochratoxin A is the most effective one. Ochratoxins are produced in maize, peanuts, beans etc., infested mainly with *Penicillium viridicatum* or *Aspergillus ochraceus*.

(c) **Zearalenone**. Zearalenone is a phenolic resorcyclic acid lactone, produced in maize and also in other cereals, infested with different members of *Fusarium* such as *F. moniliformae* and *F. graminearum*. It shows esterogenic symptoms in swine.

(d) **Trichothecenes**. Trichothecenes possess a tetracyclic 12, 13-epoxytrichothecoene skeleton. So far, more than 30 trichothecenes have been identified, of which T-2 toxin, nivalenol and deoxynivalenol, shows marked importance. Trichothecenes are produced by different species belong to the genera *Fusarium, Myrothecium, Trichoderma, Stachybotrys* and *Cephalosporium*. This group of toxins shows subepidermal haemorrhage, several local irritations and general necrosis.

B. **Ergot toxins**: These toxins are stored in the elongated sclerotia of *Claviceps purpurea*, the pathogen of ergot disease of rye. These are the poisonous alkaloids such as ergotamine, ergocrystinine, ergometrimine, ergonovin and ergocrystine.

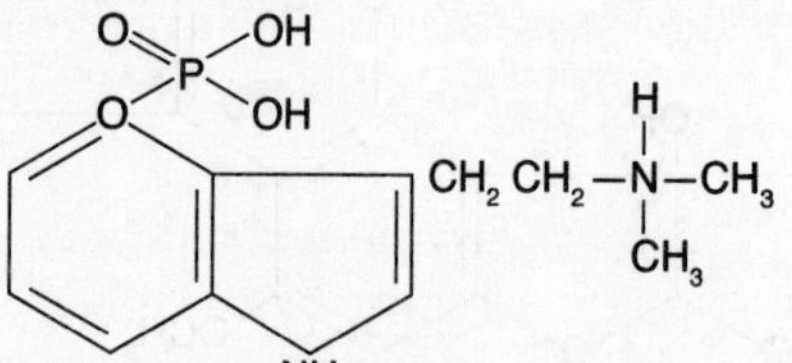

Ergometrine isolated from sclerotia of *Claviceps purpurea*

Due to consumption of sclerotia along with rye straw, the domestic animals show the following symptoms:

(i) The hoofs, legs and tail become gangrenous,

(ii) Abortion in cows.

Due to consumption of powdered sclerotia along with rye flour, following symptoms become pronounced in human being:

(i) Diarrhoea, abdominal pain and vomiting.

(ii) Effects on nerves and cause psychiatric disturbances, and

(iii) Decrease the diametre of terminal blood vessels.

C. **Mushroom toxins**: Several mushrooms produce toxins and by consuming such mushrooms as food, different abnormalities become visible in human beings like diarrhoea and vomiting within 2-3 hrs of consumption. At a later stage they cause liver damage, kidney failure, complete unconsciousness and even death. The following mushrooms are important in this respect:

(i) *Amanita phalloides* (the 'death cup'): Out of 10 toxins so far reported, the phalloidin and α-amanitin are important. The phalloidin affects the cell membrane of liver cells and α-amanitin causes lesions in stomach and intestine cells.

(ii) Different species of *Helvella* such as *H. infula, H. gigas, H. esculenta* and *H. underwoodii*, produce gyromitrin, a fatal toxin.

(iii) *Inocybe* and *Clitocybe* produce toxin muscarine that causes several abnormalities.

6. Biological control

(Mycofungicides, Mycoherbicides, Mycoinsecticides, Myconematicides)

The biological control is also called **biocontrol.** It is a method of controlling pests, such as insects, mites, aphids, weeds and disease causing organisms. Pests are the organisms which cause harms or destruction. Thus, pest can be defined as "guest without sweet packet". The biological control relies on predation, parasitism, herbivory or any other natural mechanisms, but typically also involves an active human management role. It can be an important component of integrated pest management (IPM) programs.

The *biological control* or *bio-control* can be defined in various ways. These are:

(i) Biological control is the use of living organisms to control pests.

(ii) The destruction or suppression of undesirable insects, other animals or plants by the introduction, encouragement or artificial increase of their natural enemies.

(iii) The reduction in numbers or elimination of pest organisms by interference with their ecology (as by the introduction of parasites or diseases).

There are three basic strategies for biological pest control. These are:

A. **Classical** (importation):

Where a natural enemy of a pest is introduced in the hope to achieve control,

B. **Inductive** (augmentation):

In which a large population of natural enemies are administered for quick pest control, and

C. **Inoculative** (conservation):

In which measures are taken to maintain natural enemies through regular reestablishment.

The natural enemies of insect pests are also known as biological control agents, which includes predators, parasitoids, pathogens, and competitors. The biological control agents of plant diseases are often referred to as antagonists. The biological control agents of weeds include seed predators, herbivours and plant pathogens.

Biological control can have side-effects on biodiversity through attacks on non-target species by any of the same mechanisms, especially when a species is introduced without thorough understanding of the possible consequences.

Fungi as biocontrol agent

Like other living organisms, fungi play an important role in biocontrol. Based on the target organism they are further divided into different groups. These are:

1. **Mycofungicides**

Fungi as biocontrol agents to kill fungi are called mycofungicides. This is also called *fungal fungicides* or *biocontrol agents* (BCA). There are many reports available on fungal antagonists of fungal pathogens of root and shoot systems of plants. Such interactions occur in rhizosphere (around root region) as well as phyllosphere (near leaf surface).

Microbial antagonists can suppress plant diseases and organisms that suppress pathogens can be referred to as biological control agents (BCA) (Alabouvette *et al.*, 2006).

Various fungal species can be used as biological control agents (BCA) and may provide effective activity against many pathogenic microorganisms. These are:

(i) ***Trichoderma harzianum*** – effective against *Botrytis cineria, Fusarium, Pythium* and *Rhizoctonia* (Khetan, 2001),

(ii) ***Ampelomyces quisqualis*** - a hyperparasite of powdery mildew (Liang *et al.*, 2007),

(iii) ***Chaetomium globosum*** and ***C. cupreum*** - effective against root rot disease caused by *Fusarium, Phytophthora* and *Pythium* (Soytong *et al.*, 2001),

(iv) ***Gliocladium virens*** - effective biocontrol agent of soil borne pathogens (Viterbo *et al.*, 2007),

(v) ***Coniothyrium minitans*** - a mycoparasite of *Sclerotinia* (Whipps *et al.*, 2008) and

(vi) ***Fusarium oxysporum*** (nonpathogenic species) - having biocontrol potential against rot pathogens (Fravel, 2003).

The effective biological control agent should be genetically stable, effective at low concentrations, easy to mass produce in culture on less-expensive media, and should be effective against a wide range of pathogens (Wraight *et al.*, 2001).

Several fungal taxa have been reported to be antagonist against plant pathogens and have been successfully formulated as mycofungicides or biological control products e.g. *Ampelomyces quisqualis, Aspergillus niger, Candida oleophila, Chaetomium cupreum, C. globosum, Coniohyrium minitans, Cryptococcus albidus, Gliocladium virens, G. catenulatum, Fusarium oxysporum, Phlebiosis gigantean, Pythium oligandrum, Rhodotorula glutinis, Trichoderma harzianum, T. polysporum, T. viride.*

Name of the some commercially available mycofungicide along with producing fungus and target pathogen is given in the Table 4.28.

2. **Mycoherbicides**

Fungi as biocontrol agents to kill weeds are called mycoherbicides. Although viruses, bacteria and fungi are useful as bioherbicide, but the fungi are ideal among them. Regarding application of mycoherbicides, two strategies are

Table 4.28: Some of the mycofungicide products available in the market

Products	Producing fungus	Target pathogem
1. Bioderma	*Trichoderma viride*	*Sclerotinia, Rhizoctonia*
2. Bioderma-H	*Trichoderma harzianum*	*Phytophthora, Fusarium Pythium* spp., *Cercospora, Colletotrichum, Alternaria, Ascochyta*
3. Biofox C	*Fusarium oxysporum* (nonpathogenic)	*Fusarium oxysporum, Fusarium miniliforme*
4. Ketocin	*Chaetomium cupreum*	*Fusarium oxysporum* f.sp. *lycopersici*
5. Promote	*Trichoderma harzianum* *Trichoderma viride*	*Pythium Rhizoctonia, Fusarium*
6. SoilGard (GlioGard)	*Gliocladium virens* GL21	Several plant diseases: Damping-off and root pathogens
7. Novacide	*Chaetomium cupreum*	*Fusarium oxysporum* f.sp. *lycopersici*
8. AQ10 Biofungicide	*Ampelomyces quisqualis* M-10	Powdery mildew

Table 4.29: Some of the commercial mycoherbicides products available in the market

Products	Producing fungus	Target weeds
1. Smolder	*Alternaria destruens* strain 059	*Cuscuta* spp., such as dodder, swamp dodder, large-seed dodder, field dodder, and small-seed dodder.
2. Chontrol™	*Chondrostereum purpureum* strain PFC2139	Sitka, red alder, and other hardwood species in rights-of-way and forests (used as a stump treatment)
3. Luboa	*Colletotrichum gloeosporio ides* f.sp. cuscutae	Dodder (*Cuscuta* spp.)
4. BioMal	*Colletotrichum gloeosporio ides* f.sp. malvae	Round-leaved mallow (*Malva pusiUd*) in field crops
5. Stumpout	*Cylindrobasidiom laeve*	Black wattle (*Acacia mearnsi*), Golden wattle (*A. pycnantha*)
6. DeVine	*Phytophthora palmivora* MWV	Strangler vine or milkweed (*Morrenia odorata*) in *Citrus.*

followed. These are **classical** and **inundative stratigies**. In the **classical** strategy, fungal parasite, unknown so far to that area, is introduced for attacking the target weed. But in **inundative** strategy, repeated applications of the fungal parasite are made on the weed host to eradicate completely. The fungus is not new to that locality.

The first mycoherbicide to be registered by the Environmental Protection Agency (EPA), USA as **DeVine** in 1981; it is composed of *Phytophthora palmivora* strain MWV, applied to control milkweed vine and stranglervine (*Morrenia odorata*). *P. palmivora* causes root rot, wilting of the whole plant, and also caused plant death. Single application of **DeVine** in the field was reported to reduce milkweed vine population by over 90% within 1-2 years (Woodhead 1981). Some of the common mycoherbicide along with the target herbs are given above (Table 4.29).

3. **Mycoinsecticides**

Fungi as biocontrol agents to kill insects are called mycoinsecticides or Fungal insecticides. A number of entomogenous fungi have been studied in relation to biocontrol of insect pests of different crop plants.

Entomopathogenic fungi, which cause disease in insects, include about 14 species that attack aphids. Some of the fungi act as mycoinsecticides is given below:

(i) *Beauveria bassiana* is used to manage a wide variety of insect pests like white-flies, thrips, aphids and weevils.
(ii) *Lecanicillium* spp. is deployed against white flies, thrips and aphids.
(iii) *Metarhizium* spp. is used against pests including beetles, grasshoppers, and spider mites.
(iv) *Paecilomyces fumosoroseus* is effective against white flies, thrips and aphids.
(v) *Purpureocillium lilacinus* is used against root-knot nematodes.
(vi) The fungi *Cordyceps* and *Metacordyceps* are deployed against a wide spectrum of arthropods.
(vii) *Entomophaga* is effective against pests such as the green peach aphid.
(viii) *Synchytrium solstitiale* of Chytridiomycota, is being considered as a control agent of the yellow star thistle (*Centaurea solstitialis*) in USA.

4. **Myconematicides**

Fungi as biocontrol agents to kill nematodes are called myconematicides. The fungal antagonists of nematodes consist of a great variety of organisms belonging to widely divergent orders and families of fungi. Several fungi have been reported to be effective in controlling nematodes of vegetable, fruit and cereals. Both classical nematode capturing (trap-forming) and soil fungi (opportunistic type) have been commonly observed.

(i) *Classical nematode capturing*: This group includes species of *Arthobotrys, Dactylella, Dactylaria* and *Monacrosporium*.
(ii) *Soil fungi* (*opportunistic type*): This group includes species of *Verticillium chlamydosporium, Paecilomyces lilacinus* etc.

The above fungi are effective to control nematodes of different genera such as *Heterodera, Meloidogyne, Globodera* and *Rotylenchlus* – the pests of different crops and vegetables.

7. **Medical mycology**

Medical mycology is the study of fungi that deals with diseases of animals including humans and the ecology and epidemiology of the diseases. The diagnosis, management, and prevention of fungal diseases are known as **mycoses (sing. mycosis)**. Out of more than 100,000 species of fungi, fewer than 200 cause disease in humans. Mycoses are common in human beings.

Different environmental and physiological conditions can contribute to the development of disease. Inhalation of the spores of fungi or localized colonization on skin may initiate infection on skin and also in lung. Fungal infection on skin of human was the 4th most common disease in 2010 affecting 984 million people (Hay, *et al.*, 2013).

A. **Based on the tissue level at which the infection is localized**, the mycoses are divided into the following 5 groups. These are:

(1) *Superficial mycoses*: Fungal infection is confined to the uppermost horny layer of skin, hair and also on nail;
(2) *Cutaneous mycoses*: Fungi infect the outer layers of the skin and cause an allergic or inflammatory response;
(3) *Subcutaneous mycoses*: Fungi of this group usually have low inherent virulence which has been enter in the tissues through a wound of some kind, and which remain localized or spread only by direct mycelial growth. The fungi involved are called **dermatophytes** and concerned disease is called **dermatophytosis**.

(4) *Systemic or deep mycoses*: In this type true pathogenic fungi those can establish themselves in normal hosts, or by opportunistic saprobic fungi which could not infect a healthy host, but can attack individuals whose immune system is not working. Fungi of the group can disseminate widely through the body of the infected host, and

(5) *Opportunistic fungi*: Certain saprophytic fungi and their spores can infect a person, who is sick and suffers from different diseases like severe diabetes, malignant tumers, lymphosarcoma, leukaemias etc. Fungi of this type are normally saprophyte, but based on the condition they are able to cause disease. Persons receive excessive treatment with broad spectrum antibiotic may be susceptible with opportunistic fungi.

B. **Other diseases**

(1) Allergic reactions. It takes place by the inhalation of mould spores i.e. Farmers lung *Aspergillus* sp. in grain stores.

(2) Toxicity caused by Mycotoxins, *Amanita phalloids* (death cap); Aflatoxin, *Aspergillus flavus* through peanut, caused liver cancer; Ergotamine, caused hallucination through contaminated rye grain.

COMMERCIAL USES OF FUNGI

Production of Citric acid

Citric acid is an important organic acid produced commercially by using *Aspergillus niger*. Before 1917, the citric acid was obtained only from citrus fruit. The commercial production was started as early as in 1917 using *Aspergillus niger*. About 70% of the annual production is utilised for foods, beverages and 20% in pharmaceuticals.

Commercial production

During this process selected strains of *A. niger* are grown in an open aluminum tray (2.5 m × 2 m) containing molasses.

Initially, the medium (molasses) should be free from any trace element by passing through ion-exchange resin. The medium is then added with required amount of salts of metal ions. The metal ions like Fe^{++}, Mn^{++}, Zn^{++} and phosphates are required in this procedure as co-factor, but above critical level they inhibit citric acid production and not the mycelium. Molasses solution containing inorganic nitrogenous salts are taken in the tray and seeded with *A. niger* mycelium and incubated at 30°C for 8-11 days and pH is maintained at about 2.0. Aseptic condition is not essential as low pH prevents bacterial growth.

Citric acid is produced by the fungal mycelium during idiophase of growth. At this stage, citrate synthetase activity is much more and citric acid is initially accumulated in the mycelium and then released in the medium.

After completion of incubation period, the culture fluid is taken out from the base of the tray, without disturbing the upper mycelial growth of *A. niger*. Fresh medium may be added to continue further production i.e., the next batch of fermentation. The citric acid is then recovered from culture filtrate by precipitation.

Production of ethanol

Industrial production of alcohol (Beer)

Beer is a product of fermentation of barley grains by yeasts, containing about 4% alcohol. During this process, barley grains are allowed to germinate for the conversion of starch to sugar, mainly maltose, by the naturally developed amylases. The process is called malting and the digested grains is called malt. After that, the grains were washed with water and is then (the liquid portion, called wort) removed. Hops (dried petal of *Humulus lupulus*, the vine) are then added to the wort to add colour, flavour, and stability and also to prevent contamination (due to presence of two antimicrobial substances). At that time, the fluid is filtered and yeast is mixed in large quantity. Out of many useful strains, any one strain of *Saccharomyces cerevisiae* is used. Commonly yeast is collected from previous batch culture for its further use. The mixture is then incubated for 7 days. After 7 days young beer is transferred to vats for primary and secondary ageing for 2 weeks to 6 months. Some yeast remains with the beer that is to become keg beer and the product is refrigerated for preservation. The thick wall of the keg trap produced CO_2 for continual fermentation. For bottling, the beer is either filtered to remove the yeast or pasteurised at 140°F (60°C) for 13 minutes to kill the yeasts.

Filtered yeasts are then used in different ways: 1. used to mixed with new wort, 2. pressed to tablets for consumption by human being as single cell protein (SCP).

Production of Mycoproteins

I. Enzyme: α-amylase

Both fungi (*Aspergillus niger* and *A. oryzae*) and bacteria (*Bacillus subtilis* and *B. diastaticus*) are used in the production of amylases. They are of different types like α-, β- and glucamylases. These are used in different purposes. Such as: 1. Preparation of sizing agents in textile industry, 2. Removal of spots on cloths used in laundry, 3. Production of chocolate and corn syrup, 4. Production of bread, 5. Used in alcohol industry.

II. Amino acid: L-tryptophan

L-tryptophan is an amino acid with non-polar side chains, produced commercially by a mutant (Px-115-97) of *Corynebacterium glutamicum*, 12gm/litre in the molasses medium.

Composition of medium:

(a) 10% reducing sugars as invert (as cane molasses).

(b) 0.05% KH_2PO_4.

(c) 0.05% K_2HPO_4.

(d) 0.025% $MgSO_4$.

(e) 2% $(NH_4)_2SO_4$.

(f) 1% corn-steep liquor.

(g) 2% $CaCO_3$.

pH should be maintained at 7.2.

Function: The amino acid has several uses in both plant and human beings.

In plants:

1. Tryptophan acts as precursor of IAA, which controls plant growth and development.

In human beings:

1. The L-tryptophan is converted into either Serotonin or Tryptamine. Both the compounds help in action of central nervous system and also in neurotransmission.
2. The Serotonin (5OH tryptamine) may be converted into Melatonin (Pineal gland hormone), which regulates the seasonal breeding.
3. It also helps in normal growth.

III. Vitamin: Riboflavin

Riboflavin is an important vitamin produced through microbial fermentation. It is produced by *Ashbya gossypii*, a member of Phylum Ascomycota. The growth medium contains glucose, soybean oil, glycine and inorganic salts, seeded with *A. gossypii*. The medium is incubated aerobically at 35°C which gives an yield of riboflavin at about 4.25 gms/litre.

IV. Antibiotic: Griseofulvin

Griseofulvin is an antibiotic, obtained from *Penicillium griseofulvin*.

The antibiotic is used in the treatment of superficial fungal infections and also for systemic mycoses. The drug is used orally.

V. Recombinant protein: hepatitis B vaccine

Production of recombinant vaccines requires the identification of genes for desired antigens and cloning into suitable vectors. Vectors are then introduced into the suitable host for genetic expression. Though this method has several advantages, the disadvantage lies with the low level of immunogenicity (i.e., recombinant proteins).

Production of hepatitis B vaccine

After infection, the hepatitis B virus (HBV) fails to grow in host and also in cultured cells. This character has been explained to be due to inhibition of its molecular expression and development of vaccine. Plasma of human being contains antigens at different amounts. Three types of viral proteins found to be antigenic, are: (a) Viral surface antigen (HBsAg), (b) Viral core antigen (HBcAg) and (c) e-antigen (HBeAg).

The gene HBsAg contains 6bp long sequence preceding the AUG that synthesises N-terminal methionine. During production of HBV vaccine, initially HBsAg gene of the virus is cloned and inserted into the PMA56 plasmid of yeast with the help of EcoRI. The yeast cells have the ability of secreting glycosylate protein. The gene HBsAg is inserted near the alcohol dehydrogenase (ADH) I promoter. The recombinant plasmid is inserted into the yeast cells. The transformed yeast cells are then grown in the tryptophan-free medium. The transformed cells are selected and culture of cloned yeast cells are developed. By mass culturing and isolation, it is now possible to get the antigenic material in large scale and to use it as vaccine.

The inserted gene produces particle similar to 22 μm particle of HBV as these particles are produced in serum of HBV patients. The particles that developed by HBsAg gene and those isolated from HBV-infected cells of patients have similarity in structure and high immunogenicity, which made it possible to use the recombinant product as vaccine against hepatitis B virus infection.

VI. Pharmaceuticals: Cyclosporin

Cyclosporin, is a compound of medicinal importance, available in the trade names Sandimman and Sandimmune. The compound was discovered in the early 1970s from fungi like *Tolypocladium inflatum* and *Cylindrocarpon lucidum* isolated from soil.

The cyclosporin is an extremely effective immunosuppressant agent. It is also used commonly during organ transplant.

The fungus *Tolypocladium inflatum* grows on beetle larvae and has entomopathogenic properties. Due to its insecticidal properties, they perhaps facilitate pathogenesis of insect host.

Cyclosporin might also be related to its weak antifungal activity.

It is also produced by some other fungi like *Beauveria nivea, Trichoderma polysporum*.

SOME GENERAL ASPECTS OF FUNGI AND ALLIED ORGANISMS

Mode of nutrition

Fungi are not able to synthesise their own food due to absence of chlorophyll, thus perform heterotrophic mode of nutrition. They obtain their food from external sources. Thus they live as parasites, saprophytes or in symbiotic association:

Parasites: They obtain their food from living body of both plants and animals. Parasites are divided into two groups: obligate parasites and facultative parasites.

Obligate parasites. Fungi, those can live only on living host, but neither on dead decaying material nor on artificial culture medium are the obligate parasites, e.g., *Albugo candida* causes white rust of crucifers.

Facultative parasites. Fungi, which are actually saprophytes, but on certain conditions can grow as parasites and cause diseases, e.g. *Pythium debaryanum* occasionally causes damping off of chilli seedling.

Saprophytes: They obtain their food from dead organic matters. Saprophytes are divided into two groups: obligate saprophytes and facultative saprophytes.

Obligate saprophytes. Fungi which obtain their food from dead and decaying organic matters and not from any living plants or animals are the obligate saprophytes, e.g., *Mucor*.

Facultative saprophytes. Fungi which are actually parasites, but on certain conditions can grow as saprophytes are called facultative saprophytes, e.g., *Fusarium*.

Symbiosis: This is the mutual association of fungi with other plants where both are benefitted. The partners are called *symbionts*. Lichen is an example of symbiosis where there is an association between fungi and algae.

Alexopoulos and Mims (1979) mentioned (in their book *Introductory Mycology*) that C, O, H, N, P, K, Mg, S, B, Cu, Mo, Fe and Zn are required by many fungi, probably by all. Other elements like Ca are required by some. Glucose is the best source of C and organic nitrogen compound is the best source of N. The fungi usually store excess food in the form of glycogen or lipid.

Some fungi are able to absorb larger molecules, but others must break down larger molecules into smaller molecules before these can be absorbed by them. The fungi are able to secrete extracellular enzymes which act on the substrate, digesting the food outside the fungal body and then absorb them very easily.

Hotson and Cutter (1951) were the first to develop axenic culture of *Gymnosporangium juniperi virginianae*. Williams, Scott and Kuhl (1966) were the first to grow the biotropic fungal parasite, *Puccinia graminis tritici*, as uredospores on agar medium. Later, Scott and Maclean (1969) had grown a number of rust fungi in axenic culture. This was reviewed by Scott (1976), Williams (1984) and Faster *et al.* (1993). However, the above fact created the problem of distinction between **obligate** and **facultative** biotrophic parasites.

Based on ***ecological*** and ***nutritional behaviour***, Lewis (1974) classified fungi into five groups. In this classification, he used the term symbionts in a broad sense and also included parasites. The groups are:

(i) *Ecologically obligate saprotrophs* (*obligate saprotrophs*). They are typical saprophytes, absorb nutrients from dead organic matters.

(ii) *Ecologically facultative symbiotic necrotrophs*. Nutrition is necrotrophic in the parasitic mode, otherwise they are saprophytes (facultative necrotrophic). This group includes facultative parasites and facultative saprophytes (de Bary, 1889) and also the unspecialized parasites, such as *Rhizopus stolonifer, Pythium debaryamum, Rhizoctonia* etc.

(iii) *Ecologically obligate symbiotic necrotrophs* (*obligate necrotrophs*). Fungi of this group are the extreme forms of *ecologically facultative symbionts*. This group includes vascular wilt pathogens (*Verticillium* spp. like *V. albo-atrum, V. dahliae* etc. and *Fusarium* spp. like *F. udum, F. oxysporum* f. sp. *vasinfectum* etc.), wood rotting fungi etc.

(iv) *Ecologically facultative symbionts*. Nutritionally, the fungi of this group are biotrophic in the symbiotic mode, but otherwise saprophytic (facultative biotrophs). Some facultative lichens, some facultative mycorrhizal fungi etc. are included in this group.

(v) *Ecologically obligate symbiotic biotrophs* (*obligate biotrophs*). The fungi of this group are restricted to living tissues. This group includes both parasitic and mutualistic members of sheathing mycorrhizas (SM) and vesicular-arbuscular mycorrhizas (VAM), smuts and rusts (except a few) fungi, fungi causing downy and powdery mildews, lichens etc.

[*Biotrophs:* Live on living host,

Necrotroph: A pathogenic fungus that kills host cells in advance of its hyphae and then lives on the dying and dead cells.]

Degeneration of Sex

Degeneration of sex indicates the structural degeneration of sex organs leading to the simplification of the process, but not the degeneration of sexuality. Thus, it is better to designate the phenomenon as "simplification of sex organs" rather than "degeneration of sex".

In the Phyllum Chytridiomycota (Fungi), the sexual reproduction takes place through isogamy, anisogamy to oogamy, where gametes are developed in gametangia and union of gametes results in the formation of zygote. But in different other groups of fungi like Ascomycota and Basidiomycota, progressive simplification of the process is observed in different directions with well-established sexuality. In higher members like Basidiomycota, the highest degree of sexuality is observed, though sex organs are absent.

In Coenocytic members (Oomycota and Zygomycota)

1. In Oomycota of Phylum chromista, such as *Pythium, Phytophthora* etc., show advanced type of sexual reproduction i.e., oogamy, where female (oogonium) and male (antheridium) sex organs are well-developed and after sexual reproduction they develop either oospore or zygote.
2. In members of Zygomycota such as *Mucor, Rhizopus* etc., sexual reproduction takes place through gametangial copulation, where two morphologically similar, but physiologically different (+ and –) coenogametangia are involved indicating the dioecious nature of the vegetative hyphae. This type may be considered as a degenerated state from the earlier type, i.e. oogamy. In this type, the morphological differentiation of sex organs, i.e., male and female, is abolished.

In septate members (Ascomycota and Basidiomycota)

In Ascomycota

In typical members of Ascomycota, the female sex organ is ascogonium and the male sex organ — the antheridium. Gradual degeneration is also observed in this Phylum.

1. In *Pyronema confluens* (class Pezizomycetes) and in many other members, typical oogamy is noticed. Here ascogonium and antheridium are well-developed and union between the sex organs takes place through plasmogamy. After association between the opposite nuclei in the oogonium, ascogenous hyphae develop on the outer wall containing a pair of opposite nuclei (one male and an other female). After crozier formation following mitotic division of each nucleus, an ascus is developed from the

middle (penultimate) cell. The opposite nuclei of ascus undergo meiotic and further mitotic divisions, thus forming eight ascospores.

2. In *Talaromyces vermiculatus (Penicillium vermiculatum)*, both antheridium and ascogonium develop simultaneously from the adjacent hyphae. The antheridium is uninucleate and unicellular structure, develops on antheridial branch, which coils around the multinucleate ascogonium. The tip of the antheridium is opposed to the ascogonial wall and only plasmogamy takes place without further nuclear migration (Dangeard, 1907). Further development takes place only by the ascogonial nuclei.
3. In *Aspergillus herbariorum*, the antheridium is present, but functionless. The antheridium coils round the oogonial base, but fails to establish contact with the receptive organ, the trichogyne. Pairing takes place between the two nuclei of the oogonium.
4. In *Collema pulposum* (an ascomycetous lichen), the union takes place between the oogonium and asexual spores like conidia, oidia etc. Thus the function of antheridium is replaced by asexual spores.
5. In *Humaria granulata*, the asexual spores are replaced by somatic hyphae. Here both the ascogonium and the uniting hyphae are physiologically distinct.
6. Finally, the structural degeneration is observed in *Humaria rutilens,* where union takes place between the two somatic hyphae which are sexually different. Thus, it is clear that though the greater structural simplification has been achieved, the sexuality still persists.
7. The degeneration is also clearly observed in *Ascobolus*, a genus under the class Discomycetes. These are:
   (i) In *A. scatigenous* (*Ascobolus magnificus*) and *A. strobilinus*, both the **sex organs are well-developed**. The male sex organ, antheridium, becomes spirally coiled by the trichogyne of the female sex organ, ascogonium. This results in the copulation and passage of male nuclei from the antheridium to the ascogonium (Fig. 4.55).
   (ii) In *A. carbonarius*, antheridia are absent and union takes place between spermatium (specialized sex cell) and the tip of trichogyne, called spermatization (Fig. 4.60).
   (iii) In *A. citrinus*, the antheridia are absent and the oogonial nuclei alone pass to the ascogenous hyphae, called **autogamy** (Fig. 4.61).
   (iv) In *A. furfuraceous*, in the absence of sex organs (both male and female), copulation takes place between hyphae developed from the oidia of different strains, i.e., **somatogamy**.
   (v) Finally, in *A. equines*, there is complete absence of any sexual act and the life process involves a haploid phase only, i.e., **apomixis**.

In Basidiomycota

In Basidiomycota, sexual reproduction takes place between hyphae of opposite mating types leading to plasmogamy and dikaryotic condition. Dikaryotic stage persists for different periods of time and then karyogamy takes place in the young basidium. Thus the monokaryotic mycelia function as sex organs. Though the members of Basidiomycotina show simplicity in sex organs, still they exhibit gradual degeneration among themselves.

1. In *Polyporus, Agaricus* etc., of class Hymenomycetes, the monokaryotic mycelium is well-developed and extensively branched.
2. In *Ustilago zeae* (smut fungus) of class Teliomycetes, the monokaryotic stage is not extensively developed.
3. In *Ustilago segetum* (smut fungus causes loose smut of wheat), the basidial cells develop infection thread and thus the plasmogany and dikaryotic condition are established between the infection thread of opposite mating types.
4. In *Tilletia caries* (bunt fungus causes bunt of wheat), union takes place between two

uninucleate primary sporidia developed on basidium. Thus, the mycelial involvement becomes replaced by sporidia.

Therefore, in the subdivision Basidiomycotina, the process of sexual act becomes extremely simplified due to its saprophytic to heterotrophic mode of life which causes the simplification of the process by eliminating the accessory structures.

Spore forms

Fungi reproduce both asexually and sexually by different types of spores. Some spores are the unit of dispersal, but others function as resting spores to overcome the unfavourable period, and some others have the efficiency to serve both the above functions.

The spores are mainly divided into two groups:

A. Asexual spores, and

B. Sexual spores.

A. **Asexual spores**: The spores formed during asexual reproduction are the asexual spores. Basically the asexual spores are of two types: **sporangiospores,** and **conidia.**

The sporangiospores are developed inside a sac-like structure, the sporangium. Sporangiospores are of two types, *zoospores* and *aplanospores:*

1. *Zoospores*: Zoospores (Fig. 4.4) are the characteristic spore of the subdivision Mastigomycotina. Three types of zoospores are produced in this group. These are:

   (i) Anteriorly uniflagellate. These are pyriform in shape having one whiplash (*Synchytrium*) or tinsel (*Rhizidiomyces*) type of flagellum anteriorly.

   (ii) Anteriorly biflagellate. These are pyriform in shape having two flagella (one whiplash and one tinsel type), placed anteriorly., e.g., *Saprolegnia* etc.

   (iii) Laterally biflagellate. These are kidney-shaped having two flagella (one whiplash and one tinsel type), placed laterally, e.g., *Phytophthora, Pythium* etc.

2. *Aplanospores:* The sporangiospores of this type are devoid of flagella and are called aplanospores (Fig. 4.9I). It is the characteristic spore of Mucorales under Zygomycetes, e.g., *Mucor, Rhizopus* etc.

The conidia (Fig. 4.9A, B, C) differ from aplanospores because they are not enclosed by a separate sporangial wall. They are usually developed at the apex of hyphae.

Conidia show variability in both their *form* and *development.* They are mainly divided on the basis of their mode of development. There are two basic modes of development of conidia: **blastic** and **thallic**. In *blastic type,* marked enlargement of conidium initial takes place *before* it is delimited by the septum. In *thallic type,* enlargement of the conidium initial takes place only *after* the initial has been delimited by the septum.

Blastic types

These are divided into two types:

1. Blastoconidia. They can develop from a cell of an undifferentiated hypha or conidiophore at one or more points through budding, either apically or laterally. They may develop either singly or in chains, e.g., *Aureobasidium* (Fig. 4.140A), *Cladosporium* (Fig. 4.140B), etc.

2. Phialoconidia. They develop from a bottle-shaped cell, the phialide. Single conidium reaches full size before being cut off by septation and immediately another one develops underneath. Thus generally a chain of conidia are developed, where younger one always remains near the mother, e.g., *Trichoderma* (Fig. 4.140C), *Fusarium* (Fig. 4.140D) etc.

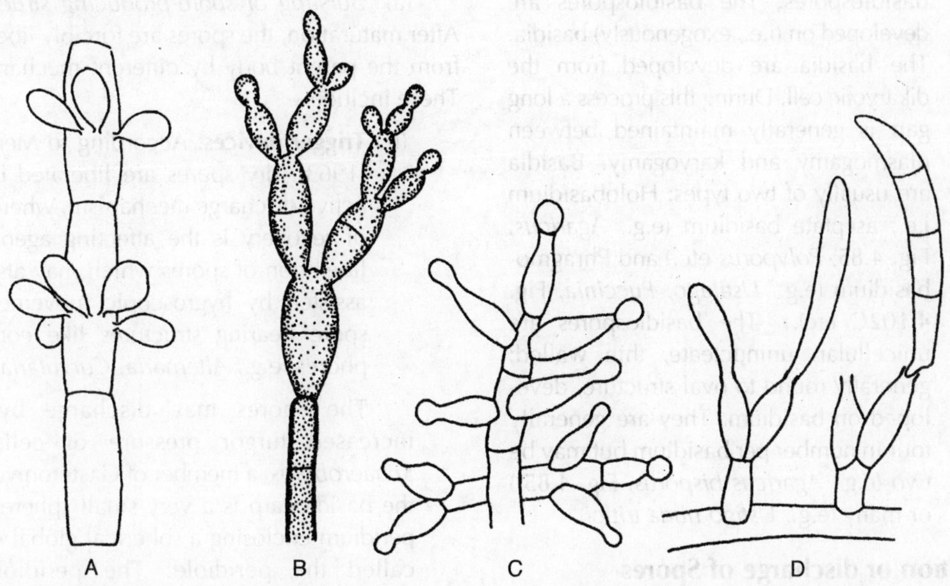

Fig. 4.140: Different types of conidia: A-B. Blastoconidia. A. *Aureobasidium*, B. *Cladosporium*; C-D. *Phialoconidia*; C. *Trichoderma* and D. *Fusarium*

Thallic types

1. Thalloconidia. In this type, the conidia are developed on hypha either singly or in short or long, branched (Fig. 4.9C) or unbranched (Fig. 4.9A) chains and enlargement of the conidium initial takes place only *after* the initial has been delimited by septum, e.g., *Erysiphe*, *Penicillium*, *Geotrichum*, *Oidium* etc.

B. **Sexual spores:** The spores formed after sexual reproduction are the sexual spores. There are four types of sexual spores in fungi. These are oospores, zygospores, ascospores and basidiospores:

1. Oospores. These are diploid (2n) spores formed by the union between egg and male nucleus. Coming in contact with the oogonium, the antheridium passes its nucleus inside the oogonium through fertilisation tube and forms oospore. It is a resting spore, e.g., members of Oomycetes under Mastigomycotina (*Pythium*, Fig. 4.117L; *Pytophthora*, Fig. 4.119B etc.).

2. Zygospores. Like Oospore, it is also a resting spore formed by the union of two gametangia into a single cell. The cell develops into a thick-walled, black and warty structure, the Zygospore, e.g., members of Zygomycetes (*Rhizopus*, Fig. 4.23P; *Mucor*, Fig. 4.25P etc.).

In majority of the members of Mastigomycotina and Zygomycotina, the sexually produced spores are developed freely and are not surrounded by sterile hyphae. But in Ascomycotina and Basidiomycotina, the spores are commonly surrounded by sterile hyphae and are called ascocarps and basidiocarps respectively.

3. Ascospores. Ascospores are developed by the union of two gametangia. The ascospores are always developed in (i.e., endogenously) a sac-like structure, the ascus, and thus the fungi containing asci are commonly called **sac fungi**. The ascospores are uninucleate and unicellular; generally oval, round or elongated structure. They are generally 8 (Fig. 4.28I) in an ascus, but may be 4 (*Saccharomyces cerevisiae*, *Saccharomycodes ludwigii*) or numerous (*Dipodascus uninucleatus*, Fig. 4.29E).

4. Basidiospores. The basidiospores are developed on (i.e., exogenously) basidia. The basidia are developed from the dikaryotic cell. During this process a long gap is generally maintained between plasmogamy and karyogamy. Basidia are usually of two types: Holobasidium i.e., aseptate basidium (e.g., *Agaricus*, Fig. 4.85; *Polyporus* etc.) and Phragmobasidium (e.g., *Ustilago, Puccinia*, Fig. 4.102C etc). The basidiospores are unicellular, uninucleate, thin walled; generally round to oval structure, developed on basidium. They are generally four in number per basidium but may be two (e.g., *Agaricus bisporus,* Fig. 4.85J) or many (e.g., *Ustigo nuda tritici*).

Liberation or discharge of Spores

The liberation or discharge of spores is the detachment of spores from the spore containing or bearing structures.

The liberation takes place either by an active or a passive process:

Active process: During active liberation, the spores are released by any one of the following ways:

(i) The hydrostatic pressure of the spore-bearing structure,

(ii) The devices that triggers off through change in the turgidity,

(iii) The development of negative pressures in cells adjacent to the spores.

Passive process: During passive liberation, the spores are released by a number of different methods which depend on the condition of spore dry, slimy, etc.

The liberation of spores depends on the length of day and night, temperature, rainfall, etc.

The different *processes of liberation* can be discussed under the following heads:

A. Active liberation: Different environmental factors like temperature, light, water, humidity greatly influence the liberation of spores. The spores of different fungi are violently discharged by the following processes:

(a) *Bursting of spore-producing structures:* After maturation, the spores are forcibly liberated from the parent body by different mechanisms. These include:

(i) **Trigger devices**: According to Meredith (1963), dry spores are liberated by an active discharge mechanism, where jerk of recovery is the affecting agency in liberation of spores which may also get assisted by hygroscopic movement of spore bearing structures like conidiophores, e.g., *Alternaria, Curvularia* etc.

The spores may discharge by the increased turgor pressure of cells. In *Sphaerobolus*, a member of Gasteromycetes, the basidiocarp is a very small sphere with peridium enclosing a spherical global mass, called the peridiole. The peridiole is enclosed in a gelatinous matrix. The peridium consists of six layers. At maturity, the peridium splits and develops 6-8 teeth. A split develops in the middle between inner three and outer three layers, except towards the apices of the teeth. The innermost layer gelatinises so that the peridiole rests in a drop of fluid. The innermost layer of the inner cup is composed of large parenchyma-like cells, initially full of glycogen, which becomes hydrolysed to glucose, thereby its suction pressure increases and takes up water. The cells tend to expand, but the outer layer resists expansion. The tension so set up is finally released by a sudden inversion of the inner cup; thereby peridioles are projected out (Fig. 4.141).

(ii) **Syringe devices**: In this type, the spores are released due to the development of *high hydrostatic pressure* within the cells. This type of spore dispersal is found in different groups of fungi.

Zygomycotina: *Pilobolus* is a coprophilous fungus, it grows on horse dung. The wall in the apex of sporangiophore is very elastic and has a circular line of weakness around the apex. With maturity, the hydrostatic pressure of the sporangiophore increases until it ruptures at the weak point and then it contracts violently and squirts

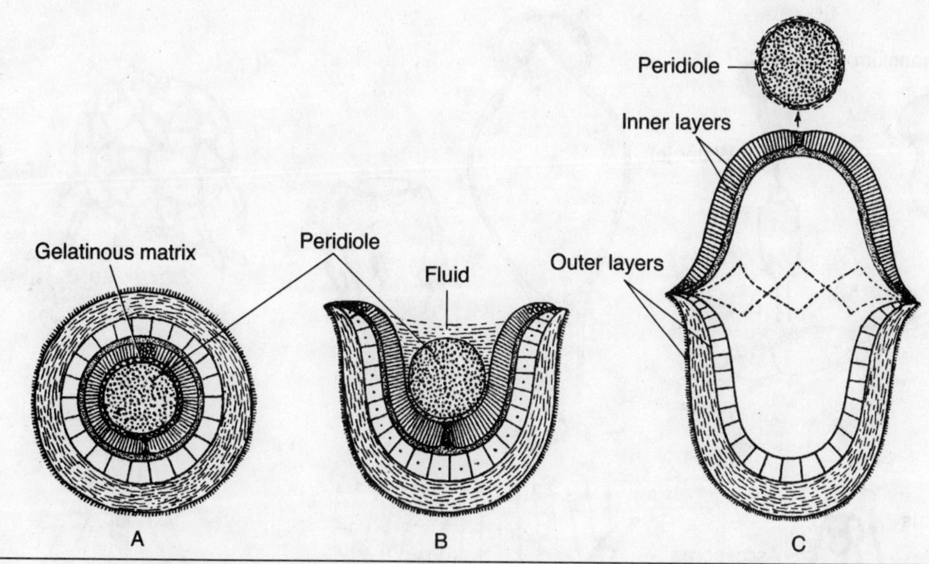

Fig. 4.141: Spore discharge mechanism: A-C. Trigger devices in *Sphaerobolus*

out a jet of water and thus projects the sporangium. The liberation of sporangium is very responsive to light. Being photoreceptive, the sporangia are directed towards the maximum light intensity (Fig. 4.142A-D).

Ascomycota with cleistothecial fruit body: *Sphaerotheca*, a member of Erysiphales develops cleistothecial type of fruit body. During this process the asci within the cleistothecium absorb water and swell up, thereby the cleistothecium ruptures and the asci protrude out and are finally ejected violently. Later on, the asci burst and disperse the ascospores (Fig. 4.142E and F).

Ascomycota with apothecial fruit body: In *Ascobolus,* the ascospores are simultaneously discharged from the upper hymenial layer of mature fruit body — this phenomenon is known as *puffing* i.e., synchronised discharge of ascospores (Fig. 4.142G).

When ascus becomes turgid, a positive hydrostatic pressure is developed which may be associated with the vacuole forces, thereby much amount of cytoplasm along with ascospores moves towards the apex of the ascus. Due to high tension, the asci become enlarged and may protrude out from the hymenial layer. The tension is released when the operculum opens and gives way to release the inner content. The elastic wall of ascus then contracts, thereby the upper half of the content is driven out as a jet which dissipates into a cloud of ascospores carrying droplets. The ascospores may reach up to a distance of 7 inch (17 cm) (Fig. 4.142G).

Ascomycota with perithecial fruit body: In Pyrenomycetes, the asci are enclosed in perithecium having a narrow opening and it is so narrowed that only one ascus can pass through it at a time. *Three types* of active discharge are found in Pyrenomycetes:

Type 1: In *Sordaria fimicola*, the asci remain attached at the base of the perithecium. Only one ascus elongates at a time through the narrow neck, until its apical region is forced through the ostiole. The gradual increase in the internal hydrostatic pressure leads to the rupture of the ascus wall and then the spores are forced out — either singly or in groups (Fig. 4.142H and I).

Type 2: In *Sporomia intermedia*, the ascus wall is 2-layered, the outer *ectoascus* is rigid and thick and inner *endoascus* is thin and extensible. As the ascus imbibes water, it swells up and due to internal pressure the *ectoascus* ruptures and the inner wall elongates 2-3 times, thereby it protrudes out through the ostiole and the spores are discharged by rupturing the *endoascus*. After

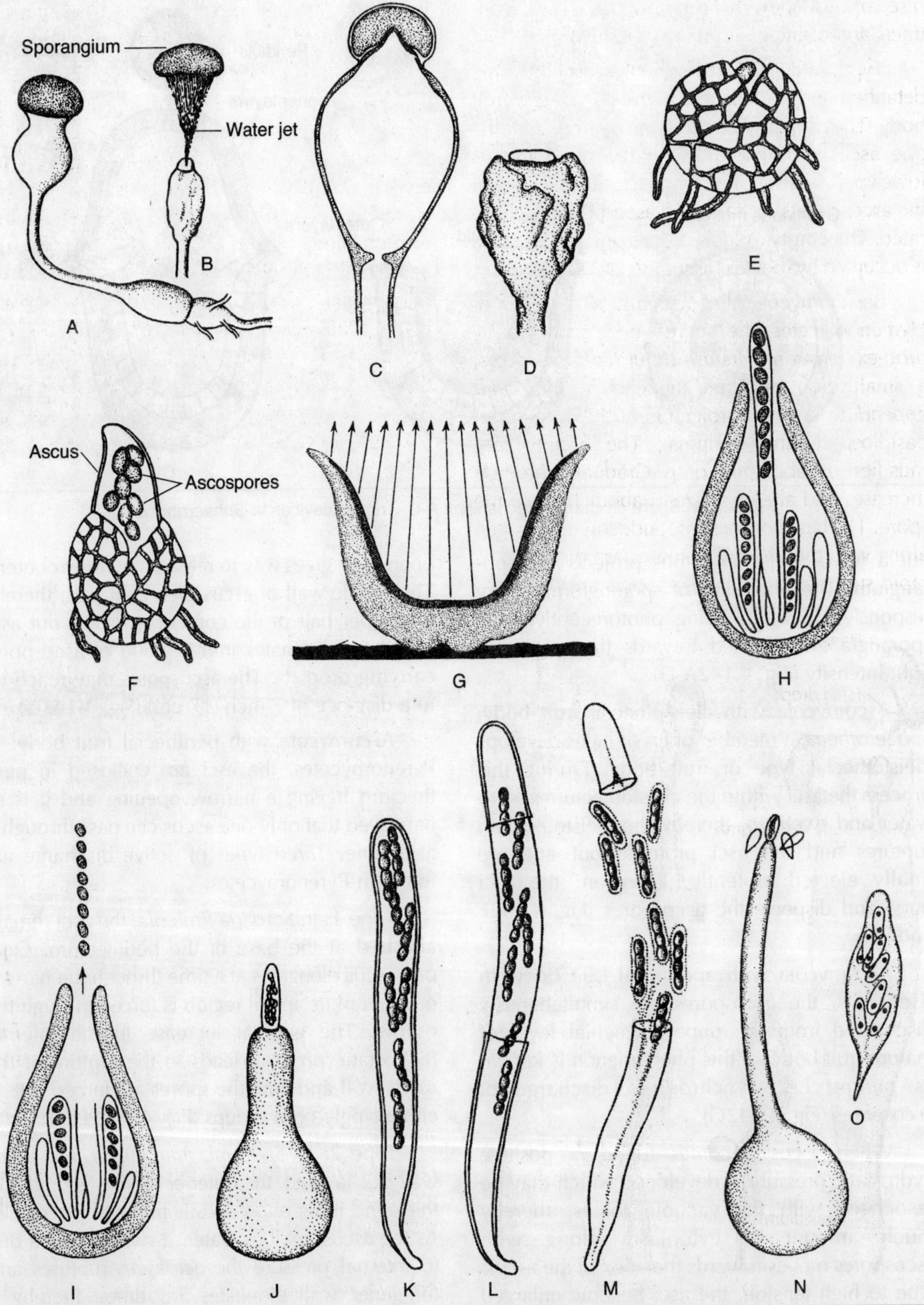

Fig. 4.142: Spore discharge mechanism: Syringe devices: A-D. In *Pilobolus* sp., E-F. In *Sphaerotheca* sp., G. In *Ascobolus* sp., H-I. In *Sordaria fimicola*, J-M. In *Sporomia intermedia* and N-O. In *Ceratocystis* sp.

discharge of spores, the empty ascus wall contracts and disintegrates (Fig. 4.142J-M).

Type 3: In *Ceratocystis*, the asci become detached and lie free within the perithecial fruit body. The asci are liberated one by one. Initially, one ascus protrudes through the neck and its lower part is firmly held by the ostiole. Later on the ascus bursts apically and ascospores are liberated. The empty ascus extrudes out and its place is occupied by the next ascus (Fig. 4.142N and O).

Basidiomycota: In *Agaricus*, a member of Hymenomycetes, the spores are discharged by *drop-excretion mechanism*. During this process, a small drop of liquid appears at the '*hilar appendics*' (a small projection at the junction of basidiospore and sterigma). The basidiospore thus lies just above the drop. Gradually the drop increases and attains a size of about 1/5th of the spore. The basidiospores are suddenly discharged along with the drop. The spores are dispersed in rapid succession. Once the spores are dispersed in air, they are free to be dispersed by air current. The sterigma does not show detectable shrinkage after discharge of spore (Fig. 4.143).

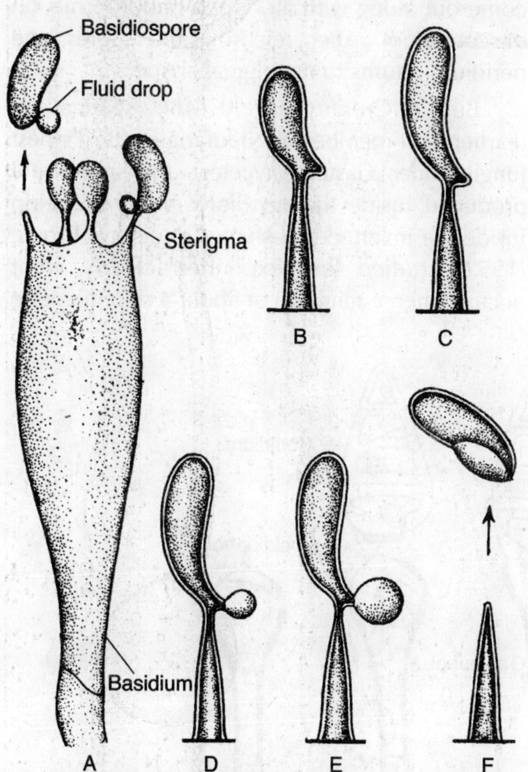

Fig. 4.143: A-F. Spore discharge by drop-excretion mechanism in *Agaricus*

(b) *Sudden changes in shape of turgid spores or turgid structures along with spores*: The liberation of spores are often associated with sudden rounding-off of turgid cells. In *Entomophthora coronata*, a member of Zygomycotina, has a single and relatively large conidium developed at the end of a straight conidiophore. The apex of the conidiophore projects into the conidium as a distinct columella. As the conidium tends to round-off, the re-entrant part bulges outward and thus the conidium bounces-off to a distance (Fig. 4.144). Similar mechanism is found in *Sclerospora* of Mastigomycotina and also in many rusts.

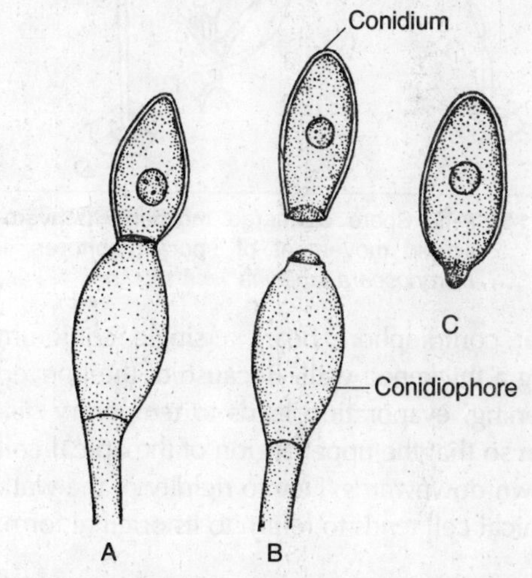

Fig. 4.144: A-C. Spore discharge mechanism: Changes in condition of turgid structures in *Entomophthora* sp.

Liberation by hygroscopic movement: This type of liberation or discharge is available in fungi with dry spores. In *Peronospora tabacina* of Mastigomycotina, the sporangiophores are dichotomously branched bearing oval sporangia. Due to hygroscopic movements of the sporangiophores the sporangia are separated and are dispersed to distant place (Fig. 4.145).

(c) *Sudden breaking of tensile water in conidia and conidiophores:* In this type, conidia are liberated by breaking of tensile water. In *Deightoniella torulosa* (Fig. 4.146), the apical

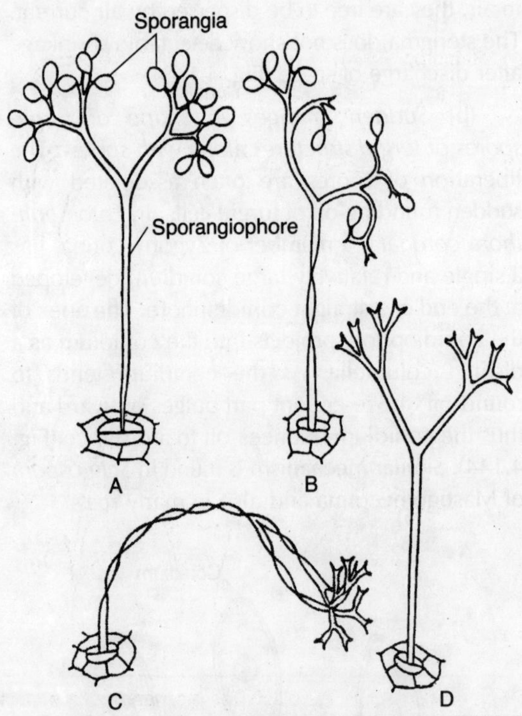

Fig. 4.145: A-D. Spore discharge mechanism: hygroscopic movement of sporangiophores in *Peronospora tabacina*

cell of conidiophore bears a single conidium having a thickened wall. Because of the uneven thickening, evaporation leads to temporary distortion so that the upper region of the apical cell is drawn downwards. Due to rigidity of the wall, the apical cell tends to return to its original form. As evaporation continues the fluid content remains under increasing tension, A break occurs either due to cohesion between the water molecules or release of adhesive force between water molecules and the cell wall. During this time a very small *gas bubble* appears and suddenly enlarges as the distorted cell returns to its original shape with a jerk that liberates the attached conidium to a moderate distance.

(d) *Liberation by impaction*: This type of spore liberation is found in puff-balls like *Lycoperdon,* earth stars like *Geastrum* and bird's nest fungi like *Cyathus* etc. — all belonging to Basidiomycotina which liberate the spores by impaction.

In *Lycoperdon* (Fig. 4.147), *Geastrum* etc. the basidiocarps contain mass of dry spores along with thread-like dry hyphae, called the capillitial threads. The basidiocarp opens at the apex by an ostiole. The dry spores may be blown out either by wind across the ostiole or may come out along with air, when raindrop hits on the peridium. After releasing the spores, the peridium returns to its original shape.

But in *Cyathus* (Fig. 4.148), Agaricaceae (earlier as a member of Nidulariales (Bird's nest fungi) under Gastero-mycetes), the spores are produced inside the peridioles which develop inside the inverted bell-shaped structure. Brodie (1951) studied and explained clearly what occurs when a raindrop of about 4 mm diametre

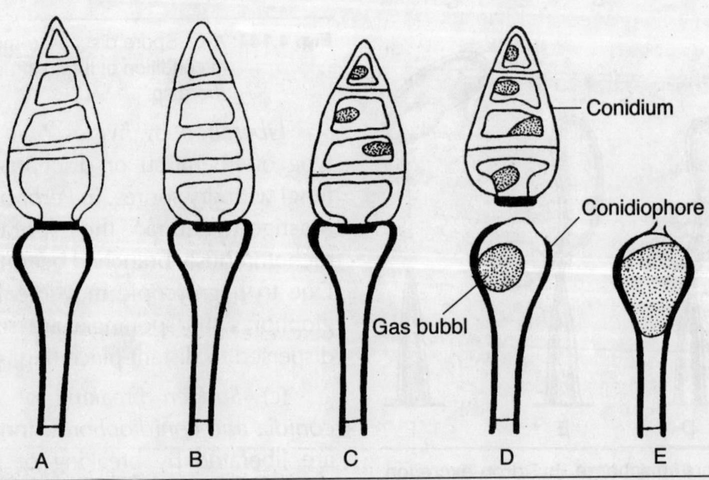

Fig. 4.146: Spore discharge mechanism: A-E. Gas bubble mechanism in *Deightoniella torulosa*

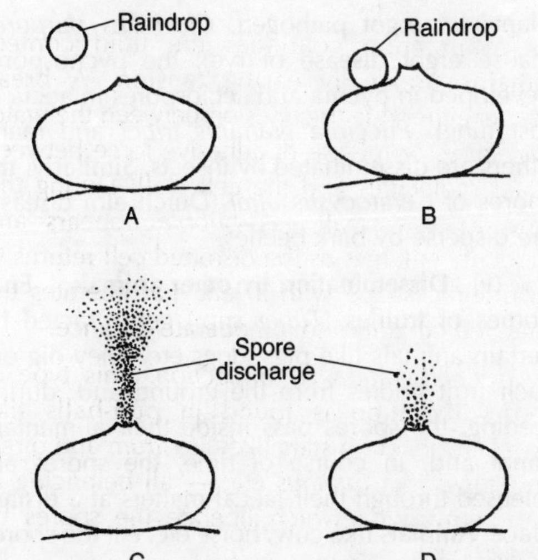

Fig. 4.147: A-D. Spore discharge mechanism: By impact rain drop in *Lycoperdon*

with a velocity of about 6 m/sec falls into the basidiocarp, creating an upward thrust along the inclined side of the cup. Thus, the peridioles are forcibly ejected and can fly to a height of about 4–5 m in the air and may attach with any substrate. This is called 'splash-cup mechanism'.

B. Passive liberation: This type of liberation is very common in a large number of fungi. The spores may be slimy or dry and, accordingly, they are dispersed by the following procedures:

(a) *Liberation of slime spores: Rhizopus* (Fig. 4.149) and *Mucor* of Mucorales and many other fungi follow this procedure. The spores are liberated as a consequence of the gelatinisation of their walls or through the autolysis of adjacent

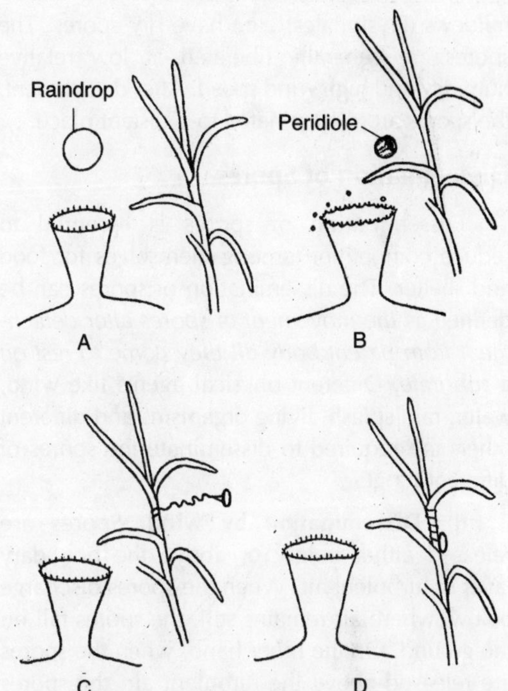

Fig. 4.148: A-D. Spore discharge mechanism: By impact rain drop in *Cyathus*

structure or cells. The sporangial wall dissolves, leaving the spores to hold together in a drop of mucilage containing water. After bursting the outer wall, the slimy mass of spores is either disseminated by animals or by air after drying.

Similar type of dispersal with some variation is also observed in different fungi like *Phallus, Dictyophora* etc.

(b) *Liberation of dry spores:* Different fungi like rusts (Uredinales), smuts (Ustilaginales), downy mildews (Peronosporales) and powdery

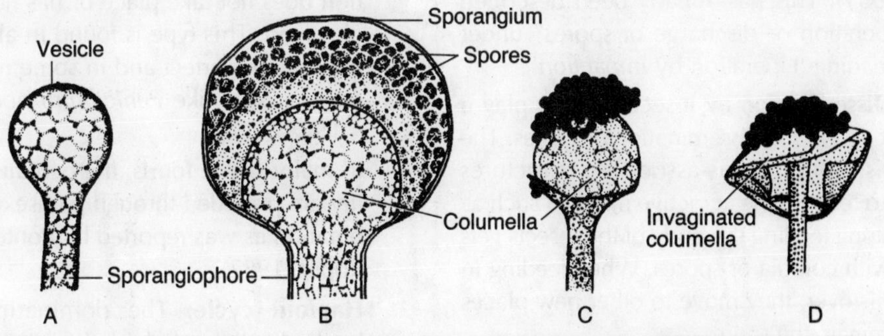

Fig. 4.149: A-D. Spore discharge mechanism: Passive liberation in *Rhizopus*

mildews (Erysiphales) etc., have dry spores. The spores are generally liberated at low relative humidity and high wind speed. After detachment, the spores are disseminated to a distant place.

Dissemination of Spores

Dissemination of spores is essential to reduce competition among themselves for food and shelter. The dissemination of spores can be defined as *the movement of spores after detachment from parent body till they come to rest on a substrate*. Different physical agents like wind, water, rain splash, living organisms and different others are required to disseminate the spores of different fungi:

(i) **Dissemination by wind.** Spores are released either below or above the boundary layer of turbulent air. When the spores discharge below, where air remains still, the spores fall on the ground. On the other hand, when the spores are released above the turbulent air, the spores are projected violently through the air and follow a quasi-parabolic curve, e.g., spores of *Erysiphae, Peronospora* etc.

(ii) **Dissemination by water.** In aquatic fungi like *Saprolegnia parasitica, S. ferax* etc., spores are disseminated by water. The dissemination is dependent upon the speed of water, chemical stimuli — if present in soil — and also on the concentration of spores.

(iii) **Dissemination by rain splash.** This type of spore dispersal is found in bird's nest fungi, the members of Nidulariales. *Cyathus,* a member of this group, shows this type of dissemination and the detailed mechanism was studied by Brodie (1951). This has already been described under "liberation or discharge of spores" under the sub-heading 'Liberation by *impaction*'.

(iv) **Dissemination by insects.** Insects play a significant role in the dissemination of spores. The insects visit the fungus-associated structures primarily to feed on the attractive material such as nectar. During feeding the body of the insects gets smeared with conidia or spores. When feeding in that place is over, they move to other new places and disseminate the spores.

The conidia develop on the ovary of rye plants by ergot pathogen, *Claviceps purpurea* (cause ergot disease of rye); the pycnospores developed in pycnia and aeciospores in aecia in rust fungi, *Puccinia graminis tritici* and many others are disseminated by insects. Similarly, the spores of *Ceratocystis ulmi* (Dutch elm disease) are disperse by bark beetles.

(v) **Dissemination by other animals.** Fruit bodies of truffles (*Tuber* spp.) get attracted by certain animals like pigs, dogs etc. They dig out such fruit bodies from the ground and, during feeding, the spores pass inside their alimentary canal and, in course of time, the spores are released through their faecal matters at a distant place. Animals like cow, horse etc. eat the spores of different fungi like *Coprinus* etc. along with herbage and thus the spore passes to their alimentary canal and afterwards it releases through the faecal matter at a distant place; thus the spores are disseminated.

(vi) **Dissemination by other ways.** Spores of different fungi are often disseminated through the contaminated soil, seeds, manure, propagating stocks, agricultural implements and also by human beings.

Life Cycle patterns

Fungi show much variation in their life cycle patterns. Some are haploid, others may be mostly dikaryotic or share both haploid and diploid equally etc. Observing the above variations, Raper (1954, 1966) recognised and reported the presence of 7 basic types of life cycle in fungi. These are (Fig. 4.150):

1. **Asexual cycle**: In this type, sexual reproduction does not take place or has not yet been observed. This type is found in all members of Fungi imperfect and in some members of other groups like *Penicillium notatum* (Fig. 4.150A).

   In many sterile forms, the benefits of sexuality are provided through parasexual recombination as was reported by Pontecorvo and Roper (1952).

2. **Haploid cycle**: The dominating part is haploid and the diploid phase is restricted to the zygote nucleus only (Fig. 4.150B). The

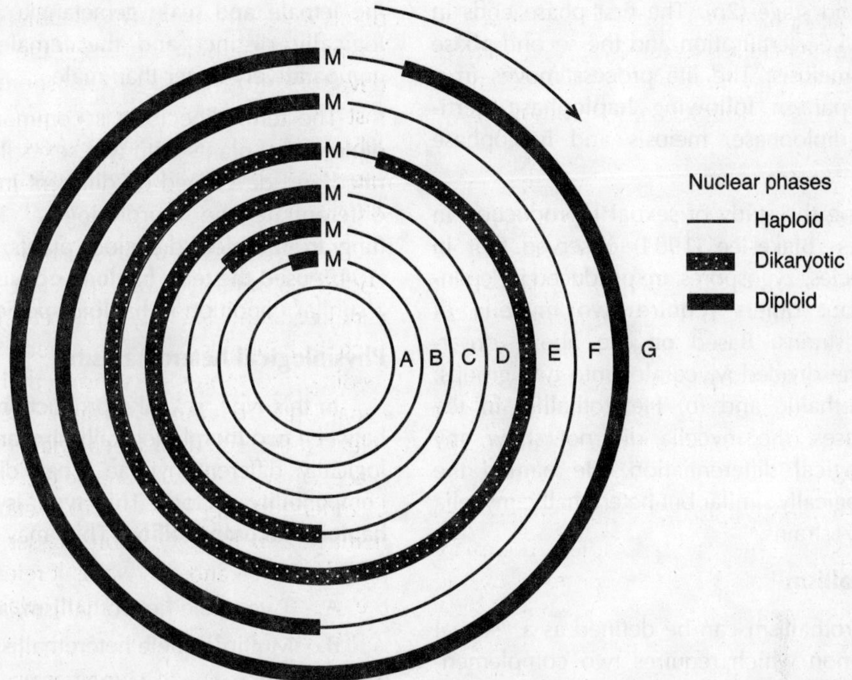

Fig. 4.150: Different types of life cycles. The cycles followed clockwise direction. "M" indicates meiotic stage: A. Sexual cycle, B. Haploid cycle, C. Haploid cycle with limited dikaryotic phase, D. Haploid dikaryotic cycle, E. Dikaryotic cycle, F. Haploid-diploid cycle and G. Diploid cycle (based on Raper, 1954).

meiosis immediately takes place after nuclear fusion i.e., zygote formation. This type of cycle is found in Mucorales, some members of Ascomycota etc.

3. **Haploid cycle with limited dikaryotic phase**: This type is almost alike to haploid cycle but paired conjugant nuclei in haploid segment (i.e., the dikaryotic phase) undergo synchronous division for variable times i.e., either lesser or greater period of time (Fig. 4.150C). This type is found in higher Ascomycotina, e.g., *Neurospora* etc.

4. **Haploid dikaryotic cycle**: This type is like that of the previous one, but once the dikaryon is formed, it continues for unrestricted period of time. Thereby, the dikaryotic phase persists as the longest phase of their life cycle (Fig. 4.150D). This type is found in Basidiomycota (e.g., *Agaricus, Polyporus* etc.), except many members of Ustilaginales i.e., smut fungi.

5. **Dikaryotic cycle**: In this type, the ascospores or basidiospores are formed after meiosis, that are fused immediately and form dikaryon. The dikaryotic phase continues till meiosis (Fig. 4.150E). This type is found in Ustilaginales (smut fungi) and sometimes in yeasts.

6. **Haploid-diploid cycle**: In this type, both haploid and diploid phase take place alternately and equally share with each other. This is found in section *Eu-Allomyces* of the genus *Allomyces, Ascocybe grovesii* of Endomycetales. This is a rare type in fungi.

7. **Diploid cycle**: In this type, haploid phase is restricted only in gametes. This is found in *Saccharomyces cerevisiae* (Winge, 1935), true slime molds (Alexopoulos, 1962), and majority of Oomycetes (Sansome, 1963).

Heterothallism and Homothallism

Sexually reproducing organisms complete their life cycle with two distinct phases: **haplophase** i.e., haploid stage (n) and **diplophase** i.e., diploid stage (2n). The first phase ends in syngamy i.e., fertilisation and the second phase ends in meiosis. The life process moves in a circular pattern following haplophase, fertilisation, diplophase, meiosis and haplophase again.

During the study of sexual reproduction in Mucorales, Blakeslee (1904) observed that in some species, zygospores are produced independently, but others require two mycelia of different strains. Based on the above observations, he divided Mycorales into two groups: (a) Homothallic and (b) Heterothallic. In the above cases, the mycelia did not show any morphological differentiation. He named the morphologically similar but heterothallic mycelia as + and – strains.

Heterothallism

Heterothallism can be defined as a "sexual phenomenon which requires two complementary strains for fertilisation". On the other hand, homothallism can be defined as a sexual phenomenon where same type of mycelia (both morphologically and physiologically) can take part in fertilisation.

Blakeslee considered the morphologically similar but genetically different strains i.e., + and – strain as equivalent to male and female sexes. With much progress in this line of research, it was found that though two mating types are required for sexual reproduction, they are morphologically different. Based on the above Whitehouse (1949) recognised two types of heterothallism:

1. Morphological heterothallism, and
2. Physiological heterothallism.

Morphological heterothallism

In this type, the thalli differ in their sex. The sexual process takes place with the association of two thalli, that differ in sexual structures like gametangia or gametes. Thus one can be described as male and other as female. This type is found in *Achlya bisexualis, A. ambisexualis, Phytophthora palmivora, Peronospora parasitica* and *Blastocladiella variabilis*. In *B. variabilis*, the female and male gemetangia are morphologically distinct and the female gamete is comparatively larger than male.

The term dioecious is commonly used for diploid (2n) plants with two sexes that are separated and developed on different individual. To differentiate the morphological heterothallic fungi from diploid dioecious plants, Whitehouse (1949) used the term **haplodioecious** to describe a similar condition in haploid species.

Physiological heterothallism

In this type, sexual reproduction takes place between two morphologically similar, but physiologically different hyphae. They differ in their compatibility factors. This type is also called **haploid incompatibility**. This may be of two types:

A. Two allele heterothallism, and
B. Multiple allele heterothallism

A. Two allele heterothallism

In this type, both mating nuclei differ in their genetic characteristics. The compatibility is governed by a pair of genetic factors, designated as 'A' and 'a'. The factors are present in two different chromosomes. Here, 'A' is dominant over 'a'. During meiosis the chromatids get separated and form four (4) spores, two of 'A' type and other two of 'a' type (Fig. 4.151). The spores on germination develop mycelia of either 'A' or 'a' type. Two complementary types are must for sexual reproduction, but mycelia with 'A' and 'A' or 'a' and 'a' are incompatible. The 'A' and 'a' are also designated as + and – type.

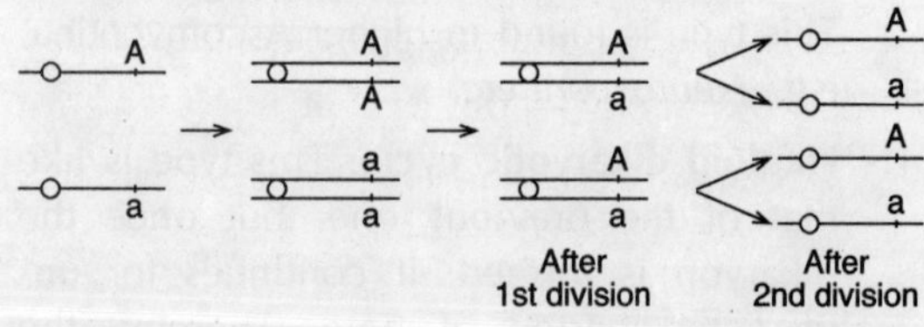

Fig. 4.151: Location of alleles on chromosomes and their meiotic behaviour

This type has been observed in different fungi like *Mucor mucedo, Neurospora sitophilla, Ascobolus magnificus, Puccinia graminis* and *Ustilago kolleri*.

B. Multiple allele heterothallism

In this type, compatibility is governed by more than two alleles. Due to the presence of large number alleles, the chances of mating of compatible strains are much more, thereby much variability in the population is observed. This may be of two types: (a) Bipolar type; (b) Tetrapolar type

(a) Bipolar type

In this type, compatibility is governed by one locus and multiple alleles are present in that locus, instead of one pair like two allele heterothallism. If the alleles A_1, A_2, A_3, A_4, ... A_n are present on locus A, then many mating types would develop after meiotic division (Fig. 4.152, Fig. 4.153). The number of mating types is directly related with the number of alleles present in a locus. The mating type with A_1 allele can unite with any other alleles, hence the chance to get complementary mating type is much more. This type of hetrothallism is found in *Coprinus comatus* and many members of Basidiomycotina, but not in rusts and smuts.

A_1 / A_2 or A_3 / A_4 or A_1 / A_3

Fig. 4.152: Location of alleles on chromosomes in different combinations.

	A_1	A_2	A_3	A_4
A_1	X	C	C	C
A_2	C	X	C	C
A_3	C	C	X	C
A_4	C	C	C	X

Fig. 4.153: Location of alleles in two loci on chromosomes in different combinations showing their meiotic behaviour (C = compatible, and X = non-compatible).

(b) Tetrapolar type

In this type, compatibility is governed by two loci instead of one as in bipolar type and multiple alleles are present in each locus like bipolar type. The alleles are present on two different loci of a chromosome. If the alleles of the locus 'A' considered as A_1, A_2, A_3, A_4, ... A_n and for locus 'B' as B_1, B_2, B_3, B_4, ... B_n, then a mating type will be A_1B_1. The mating type A_1B_1 is compatible with A_1B_2, A_2B_1 and is fully compatible with any other mating type like A_2B_2, A_3B_3 etc. but partially compatible (i.e., not fully compatible) with A_1B_2, A_2B_1 and non-compatible with the same mating type i.e., A_1B_1 (Fig. 4.154, 4.155). This type of heterothallism is found in *Ustilago maydis* and *Coprinus firmetarius*.

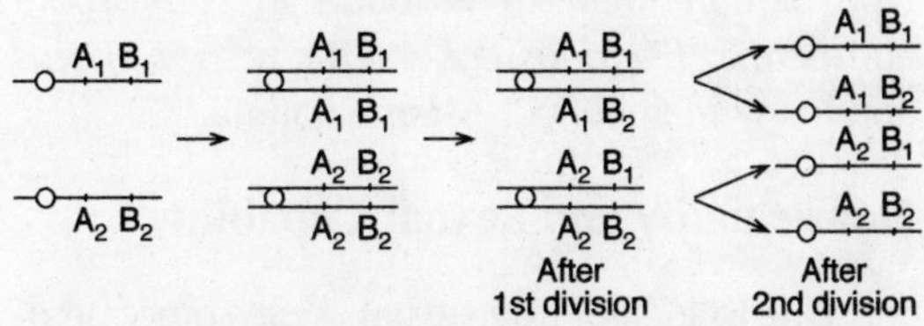

Fig. 4.154: Location of alleles in two loci on chromosomes and their meiotic behaviour

	A_1B_1	A_1B_2	A_2B_1	A_2B_2
A_1B_1	X	PC	PC	C
A_1B_2	PC	X	C	PC
A_2B_1	PC	C	X	PC
A_2B_2	C	PC	PC	X

Fig. 4.155: Chances of compatibility among different mating types in tetrapolar multiple allele heterothallism (C = Compatible, PC = Partially compatible i.e., not fully compatible and X = non-compatible)

Homothallism

As mentioned earlier that homothallism can be defined as a sexual phenomenon where same type of mycelia (both morphologically and physiologically) are involved in fertilisation.

In this group of fungi, every thallus is sexually self-fertile, i.e., self-compatible. Thus, all thalli of these fungi are of one strain. Sexual reproduction takes place in a colony that developed from a single spore. These fungi can be divided into two groups: A. Primary homothallic, and B. Secondary homothallic

A. Primary homothallic

In this type, spores are uninucleate and individual spore bears nucleus of only one genotype. This type of homothallism is found in *Allomyces javanicus, Pyronema omphalodes* and *Puccinia malvacearum.*

B. **Secondary homothallic**

These fungi are basically heterothallic type, but nuclei of compatible mating types are included in each spore during its formation. Thus sexual reproduction can occur within the culture derived from a single (heterokaryotic) spore. This type of homothallism is found in *Neurospora etrasperma, Coprinus ephemerus* f. sp. *bisporus* and many members of Ascomycotima.

Parasexuality and Sexual Capability

Due to lack of true sexual cycle some members of Deuteromycotina follow an alternative process to get the benefit of sexuality, called parasexual cycle. This phenomenon was first observed by Pontecorvo and Roper (1952) and later by Pontecorvo (1956 and 1958) and Davis (1966).

In this phenomenon, the processes of plasmogamy, karyogamy and haploidisation take place, but not at specified points and specified time in the life cycle. Thus, recombination of characteristics takes place without sexual reproduction. The basic aspects of this phenomenon include:

(i) Formation of heterokaryotic mycelium,
(ii) Karyogamy and multiplication of diploid (2n) nuclei,
(iii) Occurrence of mitotic crossing over,
(iv) Selection of diploid (2n) nuclei, and
(v) Haploidisation of some diploid (2n) nuclei.

The differences between sexual and parasexual cycle are given in the Table 4.30.

Decomposer in Forest Ecosystem

About one-third of land area of the earth is covered by forest. The living trees shed dead leaves, twigs and branches on the forest floor. All dead, fresh organic matters fallen recently to the ground is called **litter**. In a thick forest, the annual accumulation of litter ranges from about 3,500 to 5,500 kg/ha/yr. In deciduous forests (leaves fall at almost same time), a large amount of litter is accumulated every year, but in evergreen forests (leaves fall almost round the year), the accumulation does not make a rush at a time. But, such type of accumulation of organic matters is maximum in a mixed forest. Just below the fresh litter, there are the partially decomposed litters developed from the litter of the previous season, called **duff**. The gradual decomposition of organic matters by the activities of microorganisms forms a dark and finely divided organic matter called **humus.**

Forest ecosystem consists of the plants, animals and living organisms and they interact with each other. In this discussion, only fungi

Table 4.30: Comparison between sexual and parasexual cycle

Sexual cycle	Parasexual cycle
1. Nuclear fusion results the formation of heterozygous or homoygous zygotes	1. Nuclear fusion is rare in vegetative cell. Heterozygous zygotes can be identified, based on their colour and nutrition. Homozygous zygotes, if formed, are not detected
2. Zygote remains for only one nuclear generation	2. Zygote may remain active for several mitotic divisions
3. Recombination takes place by meiosis during crossing over	3. Recombination takes place by mitotic crossing over
4. Crossing over takes place simultaneously in all the chromosmes	4. Crossing over takes place only in one chromatid of a chromosome at a time
5. Meiotic products are readily recognised and can be isolated	5. Recombinants occur among vegetative cells and can be recognised by using suitable markers.

of different groups are considered as living organisms. In this respect fungi are grouped in different categories based on different aspects:

A. *Behavioural grouping based on life-history strategies and substrate grouping as done by different ecologists*. These are:

1. Fungi that utilised simple monomers: Fungi of this group can utilise only the simple monomeric nutrients such as sugars, amino acids etc., and rapidly exploit these substances. During this process, they develop many asexual spores and disperse those in more areas. Thus, the fungi are able to utilise more simple monomers present in the litters, e.g., different members of Zygomycotina like different species of *Mucor* etc.

2. Fungi that utilised complex polymers: They are able to degrade complex polymers by their enzymatic machinery. They show profuse mycelial growth which ramify inside the substrate as network and develop fruit body at the latter part of their growth, viz. many wood rotting fungi like *Polyporus, Polystictus, Fomes, Poria* etc.

B. *Based on substrate utilisation, environmental conditions for growth and interactions with other organisms*. They are divided into *five* groups:

1. *Pathogen and weak parasites*. Fungi of this group are able to *colonise on living tissue and also on the tissues with early senesce stage*. These are able to grow by overcoming the host defense factors and utilise the simple sugars and other solute materials. Viz. *Alternaria, Rhodotorula, Cladosporium herbarum, Sporobolomyces roseus* etc.

2. *Pioneer saprophytic fungi*. Fungi of this group colonise on fresh plants or animal remains. These fungi grow very rapidly and *utilise the simple soluble nutrients, but not the complex structural polymers* like cellulose etc. During both asexual and sexual stage, they develop many spores and disperse them in distant places. Viz. *Rhizopus, Mucor* and many other species of Zygomycotina.

3. *Polymer-degrading fungi*. Fungi of this group have more extended growth phase and utilise more structural polymers, like cellulose, hemicelluloses, chitin etc., and other substances like starch, protein, lipid etc. They are responsible for the major phase of decomposition of the huge amount of natural polymers.

These fungi are able to defend the resource against any invaders, either by sequestering a critically limiting nitrogen source or by secreting inhibitory substances. Viz. *Aspergillus, Penicillium, Chaetomium, Trichoderma, Fusarium, Mortierella* etc.

4. *Fungi that degrade the more recalcitrant (persistent) polymers*. Fungi of this group are able to *degrade lignin and keratin*. They become predominate in the later part of decomposition. They become active in limited nitrogen nutrients when the production of respective enzymes for degradation is stimulated. Viz. white rot fungi of Polyporaceae.

5. *Secondary or opportunistic invaders*. Fungi of this group are available at all stages of degradation of natural substrates. Some members of this group grow in close association with polymer degrading members and utilise some of the released breakdown products. Others may grow parasitically on other fungal members or on faecal pellets of microarthopods etc. The members of this group are able to tolerate the general metabolic products of many other fungi.

C. *Decomposing fungi*: The efficiency of the fungi in the process of degradation at particular environment depends on the efficiency of enzyme production, uptake of nutrients, growth efficiency and efficiency to inhibit the growth of competitors.

The decomposition takes place by both micro- and macrofungi. About 63,500 species of fungi has so far been reported from the forest area, with other 13,500 species associated with lichens. So, a total of 77,000 species have been recorded. The microfungi belong to different Phylum like Zygomycota, Ascomycota, Basidiomycota and the group Mitosporic fungi. But the macrofungi are the members of Basidiomycota, especially of Polyporaceae.

D. *Process of decomposition*: The process of decomposition depends on the various conditions of substrate and environment, such as:

(a) *Conditions of substrate*:

(i) Carbon content.

(ii) Nitrogen content.

(iii) C/N ratio.

(iv) Phosphorus content.

(v) pH.

(vi) Moisture content.

(b) *Conditions of environment*:

(i) Relative humidity.

(ii) Temperature of soil.

A single fungus can decompose certain components only, thus various species of fungi are required for the complete breakdown of the substrate. The different species of fungi are not able to act together in the degradation process, but they follow a sequential i.e., harmonic system for the complete decomposition of substrate. The rate of decomposition also varies on the efficiency of the species involved in the process.

The different elements present in the substrate do decompose together following a sequential process. These are:

(i) Sugars and amino acids: These substances are absorbed directly by the fungi and thus they disappear at the early stage of decomposition, e.g., *Alternaria, Curvularia, Cladosporium* etc.

(ii) Simple carbohydrates (sugars, starch, cellulose etc.) and some proteins: These substances are decomposed next to sugars and amino acids. Fungi become active with their cellulolytic and pectinolytic activities. They belong to Mucorales (*Rhizopus, Mucor* etc.), Agaricales (*Collybia, Mycena*), *Penicillium, Trichoderma* etc.

Rest of the substances is difficult to decompose. These are:

(iii) Cutin, fats and oils: These substances are decomposed by a few species of fungi.

(iv) Lignin, tannin and waxes: These substances are most resistant to decomposition, but gradually become decomposed by a few fungal species. Lignin is the most complex and most resistant to microbial degradation.

The complex substances that are decomposed by fungi include:

(i) Hemicelluloses: These substances are decomposed by the multiple glycan and glycoside hydrolase activities. Endoglyconases act randomly and decompose hemicellulose molecules to monomers. The processes of degradation are:

Cellulose to oligosaccharides by endo-, and exo-1,4α glucanases.

Oligosaccharides to cellobiose by exo-1,4α glucanases.

Cellobiose to glucose by α glucosidase.

(ii) Pectin: It is decomposed by the pectinase enzymes.

(iii) Lignin: It can be decomposed only by white rot fungi. The major enzymes involved in this process are laccase, manganese peroxidase, glucose oxidase and lignin peroxidase. In this process, some catalises the cleavage of aromatic ring and others generate oxidants.

In addition to lignin, the white rot fungi are able to decompose cellulose, hemicellulose and pectin and thus the wood becomes paler and fibrous. Both hard woods (lignified) and soft woods (less lignin) are decomposed, where brown rot fungi attack soft woods.

E. *Role of Fungi*: The fungi grow well in dark and shady condition and thus forest floor acts as a suitable place for fungi. The forest floor is deficient of sunlight and thus continuous flow of energy through forest ecosystem is needed to maintain the nutrient supply to autotrophic green plants and some invertebrates. The input of material from outside is very limited. Thus, a continued functioning of ecosystem is essential to maintain a balanced cycle for availability of nutrient. The fungi have a great role, because they are able to decompose the litters and woods continuously that accumulate in the forest floor. With the help of decomposition process, the carbon and other energy sources found in plants are released and recycled and become available to the plants. Thus, the recycling process is very important for the good health of the forest plants as well as millions of microfauna of forest bed.

In addition to the above process, the fungi also become a member of mycorrhizal system and supply nutrients to the higher plants.

The fungi act as a good source for humus formation during the decomposition process which helps to retain proper moisture in the forest floor. The fruit bodies developed in the forest floor serves as a good source of food for man and also for other invertebrates.

The above discussion indicates a significant role of fungi in the recycling of forest litter.

Cultivation of Mushrooms

Mushrooms are the fruit bodies of edible fungi, commonly belonging to **Basidiomycotina** (*Agaricus campestris, A. brunnescens, Pleurotus sajor-caju, Volvariella volvacea* etc.) and rarely to **Ascomycotina** (*Morchella conica, M. esculenta*).

The mushrooms were used as food since long back, probably from 3000 B.C.E. as per ancient Indian literature. Since that time, the mushrooms are being consumed in different countries like Greece, Egypt, France etc. The Greeks and Romans described mushrooms as "**food for the god**". During that period, people consumed the mushrooms after collecting them from their natural habitat. The cultivation was started in the early part of 18th century in France, but it became a thriving industry only by 1850 in Paris. In India, the first sucessful experimental cultivation of mushroom (*A. bisporus*) was initiated at Solan (Himachal Pradesh) in 1961. Later on, the cultivation of mushrooms gradually becomes popular with edible mushrooms in different parts of our country.

Assessment of nutritional value of mushrooms has been demonstrated that they complement and supplement the human diet and these are considered as "**delight of the diabetic**". Mushrooms have rightly been referred to as the "**ultimate health food**".

But there is no definite test to confirm whether a mushroom is poisonous or edible. Alexopoulos, Mims and Blackwell (1996) expressed the warning in their book "Introductory Mycology" to the Mushroom-consumers that "**you can commit many mistakes in your life, but of consuming a poisonous mushroom only once**".

Nutritive value

Mushrooms became popular for their food value. The food value of mushrooms are as follows:

1. Mushrooms are the richest source of vegetable protein.
2. The protein content varies from 1.1 to 4.98% in common cultivable mushroom (much higher than pulses, vegetables and fruits).
3. All the essential amino acids including **lysine** (550 mg/gm) are present in much higher amount than even egg.
4. Mushrooms contain sufficient quantities of mineral elements such as Na, K, Ca, P and Fe.
5. Mushrooms contain folic acid.
6. Mushrooms contain vitamins like B, C, D and K.
7. They contain little amount of fat (0.35–0.65% dry wt.) and starch (0.02% dry wt.).

Different chemical contents of mushrooms and other foods are given in Table 4.31 to 4.34:

Table 4.31: Protein content and energy value of mushrooms and other vegetables

	Name of the material	Protein content (% on dry weight basis)	Energy value (kcal)
1.	Mushrooms	26.9	16
2.	Green peas	26.1	98
3.	Green beans	21.6	35
4.	Cabbage	18.4	24
5.	Cauliflower	28.8	25
6.	Beet root	12.9	42
7.	Potato	7.6	83
8.	Brinjal	15.1	24

Table 4.32: Protein content of some cultivated mushrooms

	Name of mushroom	Protein content (% on dry weight basis)
1.	*Agaricus brunnescens*	3.94
2.	*Pleurotus ostreatus*	2.74
3.	*P. florida*	3.7
4.	*P. sajor-caju*	3.7
5.	*Volvariella volvacea*	4.98
6.	*Lentinus edodes*	1.7
7.	*Auricularia auricular*	1.1
8.	*Fammulina velutipes*	2.2

Medicinal value

Most of the mushrooms have high medicinal value to reduce blood pressure, obesity (to be fatty), constipation, atherosclerosis (fat deposition inside blood vessel) etc. Medicinal value of some mushrooms are given in Table 4.34.

Biological value

Biologically mushrooms are very much important. The biological value includes nutritive value, medicinal value and their efficiency in degradation of substrate.

Table 4.33: Chemical composition and energy value of some common mushrooms (% on fresh weight basis)

	Name of Mushroom	Water	Ash	Protein	Fat	Crude fibre	Energy Value (cal)
1.	*Agaricus brunnescens*	89.5	1.25	3.94	0.19	1.09	34.4
2.	*Pleurotus ostreatus*	92.5	0.97	2.47	0.65	1.08	23.1
3.	*Volvariella volvacea*	88.4	1.46	4.98	0.74	1.38	29.5
4.	*Termitomyces*	91.3	0.81	4.1	0.22	1.13	–

Table 4.34: Composition of vitamins and minerals of some edible mushrooms (mg/100 gm dry weight)

	Name of Mushroom	Thiamin	Niacin	Riboflavin	Ascorbic acid	Ca	K	Na	Fe	P
1.	*Agaricus brunnescens*	1.1	55.7	5.0	81.9	23	4762	–	0.2	1429
2.	*Pleurotus ostreatus*	4.8	108.7	4.7	00.0	98	–	61	8.5	479
3.	*Volvariella volvacea*	1.2	91.9	3.3	20.2	71	3455	374	15.2	677

Table 4.35: Role of some edible fungi in controlling human diseases

	Name of Mushroom	Role in controlling diseases
1.	*Agaricus brunnescens* (syn. *A. bisporus*)	Stimulating digestion, Curing hypertension, Hyperacidity, Obesity, Constipation and Atherosclerosis (fat deposition inside the blood vessel).
2.	*Auricularia polytricha* (wood car mushroom)	Strengthening health, helping blood circulation, using for the treatment of stomach ailments (sufferings), Piles, Atherosclerosis.
3.	*Pleurotus ostreatus*	Relaxing muscle, Joint pains, Antitumorous activity.
4.	*Pleurotus sajor-caju*	Reducing the rate of nephron detoriation (renal failure).
5.	*Volvariella volvacea*	Reducing blood pressure.

Cultivation procedure

It is the technique to develop the fruit bodies of edible fungi. About a dozen fungi are cultivated in 100 countries with an annual production of 2.2 million tonnes. The common four genera are *Agaricus, Lentinus, Volvariella* and *Pleurotus*. Together, they yield a major share of the total production. Out of many, *Agaricus brunnescens* (syn. *A. bisporus*), **white button** yields 56%; *Lentinus edodes*, **shiitake** yields 14%; *Volvariella volvacea*, **paddy straw** yields 8% and *Pleurotus* spp., **oyster** 7.7%. The cultivation procedure of some of the mushrooms is given:

A. Cultivation of *Agaricus brunnescens* (syn. *A. bisporus*)

The *Agaricus brunnescens* (syn. *A. bisporus*) is commonly known as **white button mushroom** (Fig. 4.156, 4.157). It contributes a major share in the mushroom production of the world. It is a temperate mushroom and can grow well in temperate conditions. *Optimum temperature, optimum moisture, proper ventilation* and *good quality of spawn* are very essential prerequisites for mushroom growth. These are:

(a) The *optimum temperature* for the mycelial growth is 24°C, while it is 14–18°C for the formation and development of fruit body.

(b) *Optimum moisture* requires nearly at the saturation point. However, direct application of excess water in bed is harmful for the growing crop.

(c) *Proper ventilation* is essential to remove toxic gases by the introduction of adequate fresh air.

(d) *Good quality of spawn* i.e., the spawn should be prepared from the tissue of

Fig. 4.156: Fruit bodies of *Agaricus brunnescens* (syn. *A. bisporus*) on mushroom bed: A. Young fruit bodies (30 days after spawning) during mass production, B. Young fruit bodies (38 days after spawning)

[Courtesy : Dr. Samir Datta]

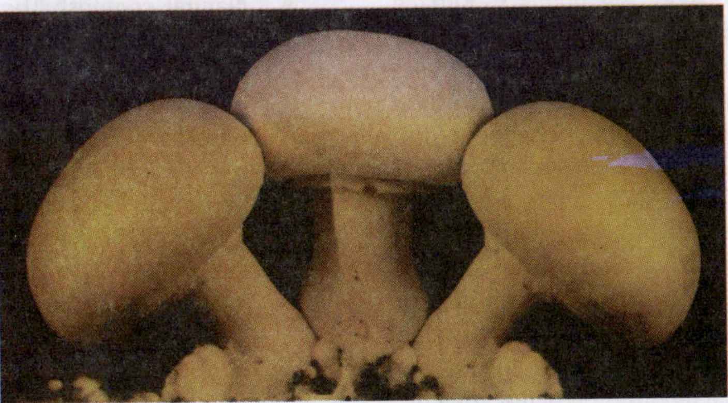

Fig. 4.157: Mature fruit bodies of *Agaricus brunnescens* (syn. *A. bisporus*) on mushroom bed

[Courtesy : Dr. Samir Datta]

single fruit body and its productive capacity should be good enough.

The cultivation procedure is:

1. Production of spawn,
2. Preparation of compost,
3. Filling of trays with compost,
4. Spawning i.e., inoculation of compost,
5. Watering of inoculated compost filled trays,
6. Casing,
7. Harvesting of mushrooms (fruit bodies), and
8. Storage of mushrooms.

1. Production of spawn

The spawn (seed of mushroom) is a pure culture of the mycelia grown on a special medium. The medium is prepared by the grains of wheat, rye, sorghum or bajra along with some ingredients. The preparation of spawn mainly consists of three steps:

(a) Preparation of substrate,
(b) Inoculation of substrate, and
(c) Incubation of inoculated substrate for spawn production.

Preparation of substrate

Take 900 gms of grains (wheat or sorghum) in 600-900 ml of water in a container and boil for 15-20 minutes. After boiling, decant the excess water and allow the grains to surface drying by spreading on polythene sheet in shade for a few hours. The grains are then mixed with chemicals like 2% calcium sulphate (gypsum) and 0.5% calcium carbonate (chalk) on dry weight basis and adjust the pH of the grain at 7–7.8. About 300–350 gms grains were then filled in milk bottles/polypropylene bags. Place a ring of tin (3.5 cm height and 3 cm diametre) towards the inner side of the open-end of polypropylene bag, tighten it with rubber band and then push the margin of the bag towards the inner side and thus a mouth is prepared. Plug the mouth of the bottle and/or polypropylene bag with non-absorbent cotton. Then cover the mouth with brown paper and tighten it with rubber band. Sterilise the substrate by autoclave at 15lb pressure for 30 minutes for 2 consecutive days. Kept the sterilised substrate in open air to cool down near to room temperature, thus making the substrate ready for inoculation.

Inoculation of substrate

The substrate is then inoculated with the mycelial culture (developed earlier, either in Potato Dextrose Agar i.e., PDA or Yeast Potato Dextrose Agar i.e., YPDA or Malt extract Agar and Rice bran decoction medium).

Incubation

Incubate the inoculated container at 20–25°C in dark for 3 weeks. Shake the container after a few days, when the mycelial growth becomes visible on the grain.

Storage of spawn

Store the spawn at 0–4°C in a refrigerator for a maximum period of 6 months, if it is not needed immediately.

The spawn can be purchased from any spawn-growing centre. (The spawn is also available in "National Centre for Mushroom Research and Training (NCMRT)", Chambaghat, Solan 173 213, Himachal Pradesh, India.

2. Preparation of Compost

The compost used in the cultivation are of two types: **Natural and synthetic**:

A. **Natural compost**: The natural compost is prepared by mixing barley or wheat straw with fresh and pure horse dung (not with the dung of other animal). *Mixed, rain wet or old dung is not suitable for the preparation of compost.* Commonly 100 kg of dung is mixed with 33 kg of straw. The mixture is then stacked a metre high heap. The heap of mixture should be kept under shade in open air. After 3–4 days, the heap was turned (to release ammonia) and stacked again. The turning process is repeated 4–5 times at an interval of 5–6 days. During this process, gypsum

($CaSO_4.2H_2O$) is added @ 25 kg/tonne (1,000 kg) dung. Finally, 40 ml nemagon is sprayed and added to the mixture. The compost was then filled in the tray of 100 ′ 50 ′ 15 cm size.

B. **Synthetic compost**: The ingredients required for the synthetic compost are:

(a)	Chopped wheat straw (3-6 cm size)	300 kg
(b)	Wheat bran	30 kg
(c)	Calcium ammonium nitrate or Ammonium sulphate	6 kg
(d)	Urea	4 kg
(e)	Potash	1.5 kg
(f)	Calcium sulphate (gypsum)	30 kg
(g)	Sawdust	10 kg

Wet the sawdust with water by spraying and mix half of the ingredient, except wheat straw and gypsum. Next day, spread the wheat straw on the cement floor and wet it thoroughly by spraying with water. The sawdust-chemical mixture is then mixed thoroughly with wetted wheat straw. This mixture is then stacked under shade into a metre high heap and covered with polythene sheet. After 5 days, the stack is scraped and rest half ingredient is thoroughly mixed with it and the entire mixture is then stacked again. This process is repeated six times. Calcium sulphate is added in the 3rd and 4th turning. Normally the compost becomes ready to cultivate after 6th turning, but 2 or more turning may be given if the smell of ammonia is yet there in the compost. During last turning, insecticide like **malathion** (10 ml dissolve in 5 l water) is added to the prepared compost. The prepared compost will be brown or dark brown in colour and is sufficient enough to fill 25 trays of 100 ′ 50 ′ 15 cm size.

3. Filling of trays with compost

Mix 3 kg of calcium carbonate with the compost prepared earlier. Fill the wooden trays with compost and compress fairly by using a wooden board (12 cm ′ 25 cm), so that a space of about 3 cm deep is left on the top of the tray.

4. Spawning i.e., inoculation of compost

Spread the spawn on the surface of compost and then cover by a thin layer of compost. Little pressure with the fingers is given to make good contact of spawn with compost. Finally the trays are covered with old newspaper. The trays are arranged one after the other in vertical stacks in such a way that sufficient aeration between the trays is maintained.

5. Watering of inoculated compost filled trays

Sprinkled water to be given on newspaper to maintain humidity. Water should be applied twice a day or less depending on the availability of moisture. The room temperature should be maintained between 24°C and 25°C for 12–15 days for the good growth of mycelium on the compost. The mycelium appears in the form of white cottony growth on the surface of bed.

6. Casing

The process of covering the mycelial mat on compost surface is made with a thin layer of soil mixed with different substances. The casing can be done with different types of mixture like :

(i) Soil : Sand : : 1 : 1;

(ii) Well-rotten cow dung : light soil : : 3 : 1;

(iii) Spent compost : Sand : Slaked lime : : 4 : 1 : 1 etc.

Casing soil should be sterilised either by chemicals like methyl bromide, formalin etc. or by heating at 70–75°C temperature for 6 hours to kill the inhabiting fungi, nematodes, insects etc.

The fruit bodies of mushroom are expected to appear after 5–20 days of casing. After casing, the room temperature should be maintained between 14–18°C for the good growth of the fruit body. The fruit bodies attain the size of button stage from pinhead within 7–8 days. Next crop appears at an interval of 8–10 days.

7. Harvesting of mushrooms i.e., fruit bodies

When the cap of the fruit body is tight with its stalk, the fruit bodies are harvested. The fruit bodies are harvested by twisting and uprooting, after holding the basal region of stalk with fingers. The lower part of the stalk is cut out where the compost remains attached.

8. Storage of mushrooms

The fruit bodies may be stored at 4°C for a few days, if it is not consumed or marketed immediately.

B. Cultivation of Paddy straw mushroom (*Volvariella volvacea*)

The paddy straw mushroom is also called tropical, straw or Chinese mushroom (Fig. 4.158). In West Bengal, it is called as 'Poal chatu'. The genus *Volvariella* belongs to the family Pluteaceae under the order Agaricales of Basidiomycotina. The common edible species under this genus are *V. volvacea, V. diplasia* and *V. esculenta*; those are grown commercially in different countries like Burma (Myanmar), China, Philippines, Malaya, India etc.

In addition to paddy straw, other substrates like water hyacinth, cotton waste, banana leaves, sawdust, sugarcane thrash (bagasse) etc., are used as substrate due to the presence cellulose, hemicellulose and lignin.

In India, the cultivation of this mushroom was first initiated in Coimbatore, Tamil Nadu, and now it is popular in different tropical parts due to the requirement of temperature ranges between 30–45°C.

The process of cultivation of straw mushroom is as follows:

1. Requirements,
2. Preparation of spawn,
3. Cultivation procedure,
4. Harvesting of fruit bodies, and
5. Preservation of fruit bodies.

1. Requirements

(i) Spawn of *V. volvacea* (600-800 gms grain spawn/bed),

(ii) Bricks,

(iii) Bamboo frame (1 m ′ 1 m),

(iv) Small water tank,

(v) Paddy straw (preferably from aman variety), apx. 36 kg,

(vi) Loose straw 5–6 kg,

(vii) Powder of Gram or Arhar seeds 200–250 gm,

(viii) Thermometer (0–100°C scale), and

(ix) White polythene sheet.

2. Preparation of spawn

The spawn can be prepared following the same procedure as adopted in *Agaricus brunnescens* (see page 611). But in addition to grains of

Fig. 4.158: Mature fruit bodies of *Volvariella volvacea*

[Courtesy : Dr. Samir Datta]

wheat or sorghum, the rice straw can also be used as substrate.

3. Cultivation procedure

Fresh paddy straw, not more than one year old and preferably from the Aman variety, should be collected from farmer or from any store. 24 straw bundles of about 1.5 kg each along with some loose straw are immersed completely in a water-filled tank by putting some weight on the bundles for about 12–15 hours. Then take out the straw bundles from the tank and keep them in stack on cement floor to drain off excess water.

4. Preparation of bed and spawning

One square bed of 1 m ' 1 m ' 1 m or 1 m ' 0.75 m ' 1 m is prepared with pre-soaked straw, keeping the butt ends (basal region) at one side, placed close to each other and arranged length-wise on a bamboo frame, supported on 4 pillars made of bricks. Same number of soaked straw bundles are placed on the previous one by keeping the butt ends in opposite direction. Inoculate the bed with spawn. The bids of spawn are placed about 8–10 cm inside the margin, maintaining a space of about 5 cm from each other. About 160–200 grams spawn is required for each layer. Powders of Gram or Arhar seeds of about 50 gms or more are spread along the line of spawning.

Second and third layers are arranged and inoculated in a similar process, but 2nd layer is placed at right angle to the 1st layer and the third layer is like the 1st layer. The spawn and seed powder on the 2nd layer will be given like the 1st layer, but on the 3rd layer those will be distributed uniformly throughout the bed. Finally, cover the top layer with loose straw. Loosely bind the bed with rope made of wheat straw at the three regions, one in the middle and one on each side. Press the bed with the help of wooden board to release the internal air and thus the spawn get compressed with the wet straw bundles. Cover the bed with polythene sheet. Watering should be done once or twice with the help of micro-sprayer. The temperature of the bed should remain 30–35°C after spawning and it should not go below 30°C during the growing season. The relative humidity should be between 80–90%. Polythene sheet should be removed after 7-10 days of spawning for the appearance of button of the mushroom. After that the buttons quickly develop into fruit bodies.

The straw once used in the mushroom cultivation can be used again. The bed should be prepared under shade away from direct sunlight and rain and also in well-aerated condition, but wind should not blow very fast.

5. Harvesting of mushroom

The fruit bodies are harvested by gentle twisting when the volva is about to rupture or is just ruptured. The production continues for 25-30 days, but in two phases. The total production per bed is approximately 3 kg. The production of second phase is comparatively less.

6. Preservation

The fruit bodies are consumed fresh or can be preserved by drying or in refrigerator for 27–48 hours. Drying can be done either in the sun or in oven at 50–60°C temperature.

C. Cultivation of Oyster mushroom (*Pleurotus*)

Species of *Pleurotus* are commonly called **Oyster mushroom** or **Dhingri** or **Wood fungus** (Fig. 4.159). It is the fourth important mushroom in the world ranking with an annual production of about 15,000 tones. It grows commercially in Japan, Taiwan, Italy, France, Thailand, Philippines and India, out of which the first three are the leading countries in its production.

The genus *Pleurotus* contains more than 50 species, of which *P. flabellatus, P. ostreatus, P. sajor-caju, P. sapidus, P. fossulatus, P. cornucopieae, P. sapathulatus* and *P. florida* have been cultivated in India.

The process of cultivation of Oyster mushroom is as follows:

1. Requirements,
2. Preparation of spawn,
3. Cultivation procedure,
4. Harvesting of fruit bodies,
5. Preservation of fruit bodies.

Fig. 4.159: Fruit bodies of *Pleurotus sajor-caju* on mushroom bed: A. Mushroom bed prepared in polythene bag and, B. Mushroom bed prepared on earthern pot

[Courtesy : Dr. Samir Datta]

Requirements

(i) Spawn of *Pleurotus* – 60 gms (30 gms grain spawn/kg of paddy straw),

(ii) Chopped and dry paddy straw (1-2 cm) – 2 kg,

(iii) Gunny or polythene bag (40″ ′ 24″) – 1 piece.

(iv) Horse gram powder – 50 gms (25 gms/kg),

(v) Thick polythene sheet (5 ft ′ 5 ft) – 1 piece,

(vi) Polythene bags (30″ ′ 18″) – 2 piece, and

(vii) Water sprayer.

Preparation of spawn

The spawn can be prepared as per method adopted in *Agaricus brunnescens* (see page 395). The spawn should not be older than 1 month.

Cultivation procedure

Take two kg of chopped straw (preferably of Aman variety) in a gunny or incised polythene bag and tighten the mouth with rope. Immerse the bag completely in water (90 l water containing 7 gm Bavistin, a fungicide along with 125 ml formaldehyde) filled tank by putting some weight on it for approximately 12-15 hours. Then take out the straw bags from the tank and keep the straw pieces in a wicker basket or a scuttle (Beng. Jhuri). Put more water in the wet straw to remove dirt, rags etc. Wait for one or more hours to drain off excess water. The wet straw pieces are then kept on polythene sheet and mixed with powder of Horse gram (20-25 gms/kg) and spawn (30 gms/kg) and if possible 10 gms fertiliser of IFCO or P.P.L (10 : 26 : 26) may be added. Take 2 polythene bags (appx. 30″ ′ 18″) and make 6-12 holes at the lower side of each bag. Then the entire mixture is put equally inside the two bags. Keep the filled bags on a bench or table in a room at 21-30°C and 65-80% humidity, with sufficient light and ventilation for 15-16 days, for spawn running. Spray water on bed twice a day by micro-sprayer. After 15-16 days, the straw pieces are covered with the mycelium and form a solid cylindrical mass. Remove the polythene bag and keep the mass on the same polythene bag in the same place. The compact mass should be watered 4-8 times throughout the day with the micro-sprayer.

The young fruit bodies will be developed after 3-4 days (i.e., 18-20 days of spawning) from all sides of the bed.

Within 2-3 days, the fruit bodies attain the size of harvesting. After harvesting the straw-mycelium mass again put inside the bag and tighten the mouth with rope. Keep it for 7 days and then again remove the mass from polythene bag and keep the polythene bag as before. Next crop of mushroom will be available within 7 days i.e., approximately 36 days after starting. Repeat the process again and the third crop will be available in 50 days.

During cropping period light should be provided for 15-20 minutes/day for better yield.

Harvesting of mushroom

The fruit bodies are harvested by gentle twisting after holding the base of the fruit bodies with fingers. The fruit bodies can be harvested generally 3 times i.e., at 22, 36 and 50 days and the total production will be 2 kg. Afterwards the bed should be destroyed.

Preservation

After harvesting, the fresh mushrooms can be sold in the market or they can be dried in sun (for three consecutive days) or in oven at 65°C. The cultivation of *Pleurotus* can also be done on earthen tray or tub.

Other modified procedure

Like the above procedure, the mixture is prepared and the tray or tub is filled with the mixture. The open end should be covered with polythene sheet. Watering should be done twice a day (varies with requirement) with the micro-sprayer. The mycelium will spread throughout the sub-stratum within 10-12 days. The fruit bodies will start to develop after 20-24 days of spawning. Within 2-3 days the fruit bodies will be the size of harvesting.

Some edible and poisonous mushrooms

Both edible and poisonous mushrooms are available in nature. List of some edible and poisonous mushrooms are given below. Common name of mushrooms are given in parenthesis.

A. Edible mushrooms:

(a) Members of Basidiomycotina:

1. *Agaricus brunnescens* (white button),
2. *A. campestris* (field mushroom),
3. *Pleurotus edodes* (shiitake),
4. *P. sajor-caju* (oyster),
5. *Volvariella volvacea* (paddy straw) etc.

(b) Members of Ascomycotina:

1. *Morchella conica,*
2. *M. esculenta.*

B. Poisonous mushrooms:

1. *Amanita phalloides* (death cap): Toxic principles are β and α-amanitin, and phalloidin.
2. *A. virosa* (destroying angles),
3. *A. verna* (fool's cap),
4. *A. muscaria* (fly-agaric) etc.

Mushroom diseases

Like higher plants, mushrooms also suffer from different diseases caused by fungi, bacteria and viruses. List of some diseases are given in Table 4.36.

Other organisms which hamper mushroom yield

In addition to diseases, mushroom yield is also hampered by molds and animal pests:

Molds: Different types of molds are associated with mushroom growing. Molds are considered as competitors for nutrients or antagonists rather than parasites. Name of different molds associated with mushroom cultivation is given:

1. Olive-green mold (*Chaetomium olivaceum*),
2. Green mold (*Trichoderma viride*),
3. Fire mold (*Neurospora crassa*),
4. Red geotrichum or lipstick mold (*Sporendonema purpurescens*),
5. Sepedonium yellow mold (*Sepedonium* spp.),
6. White plaster mold (*Scopulariopsis fimicola*),
7. Brown plaster mold (*Papulospora byssina*),
8. Black whisker mold (*Doratomyces stemonites*),
9. Inky cap (*Coprinus lagopus* and other spp.),
10. Cinnamon mold (Peziza ostracoderma, syn. *Plicaria fulva*) etc.

Animal pests: Like molds, different animal pests also hamper the yield and quality of mushrooms. These are:

Major pests:

1. Mushroom flies: Sciarids, *Lycoriella solani, L. auripila,*
2. Phorids: *Megaselia halterata,*
3. Phorids: *M. nigra,*
4. Cecids: *Heteropeza pygmaea, Mycophila speyeri* and *M. barnesi,*

5. Tarsonemid mites: *Tarsonemus myceliophagus,*
6. Red peppes: *Pygmephorus* spp.
7. Predatory mites: *Parasitus fimetorum, Digamasellus fallax, Arctoseius cetratus,*
8. Mycophagous eelworms: *Ditylenchus myceliophagus* and *Aphelenchoides composticola.*

Minor pests:

1. Collembola: *Archorutes armatus*
2. Sphaeroceridae: *Leptocera heteroneura*
3. Drosophilidae: *Drosophila funebris* etc.

Table. 4.36: List of some diseases of mushrooms

Name of diseases	Causal organisms
A. Fungal diseases	
1. Wet bubble disease	*Mycogone perniciosa*
2. Dry bubble disease	*Vertillium fungicola* var. *fungicola,* *V. fungicola* var. *aleophilum*
3. Soft mildew or Cobweb	*Hypomyces rosellus*
4. False truffle	*Diehliomyces micxrosporus* syn. *Pseudobalsamia microspore*
5. Shaggy stipe	*Mortierella bainieri*
6. Gill mildew	*Cephalosporium* spp.
7. Cap spotting	*Aphanocladium album,* etc.
B. Bacterial diseases	
1. Bacterial blotch or brown blotch	*Pseudomonas tolaasi*
2. Ginger blotch	*P. gingeri*
3. Drippy gill	*P. agarici*
C. Viral diseases	
1. Dieback disease	
2. X-disease	
3. Brown disease	
4. Water stripe disease	

5 Phytopathology

Introduction

Plant pathology is a branch of botany. It deals with the diseases of plants, helps to maintain good health and also take proper steps to increase the productivity. Plant diseases caused disasters like **famine in Ireland** (1845-1846) and **Bengal** (1943) by late blight disease of potato (c.o. *Phytophthora infestans*) and brown spot of rice (c.o. *Helminthosporium oryzae*) respectively. To overcome such problems, it is essential to carry out research on development of disease-tolerant varieties or on production of more effective pesticides in comparatively low cost or in inducing plant's own defense mechanism. Thus, plant pathologists are the plant doctors, responsible to maintain the good health of plants. Except a few cases, unlike the animal doctors, they try to look after the plant population as a whole — taking care mostly for their protection rather than cure.

Terms and definitions

During growth, plants are subjected to environmental conditions which may or may not be favourable. In or with the onset of one or more unfavourable factors of the environment, certain structural and/or physiological changes may take place. On the other hand, in contact with certain organisms, the normal metabolism and/or some structure may alter and ultimately the normal growth of the plant is affected, thus the plants with unusual growth are called **diseased plants**. Several **definitions of diseases of plants** have been proposed by different scientists from time to time. These are:

1. **Anon (1950)** defined disease as "harmful deviation from normal functioning of physiological process".
2. **Stakman and Harrar (1957)** defined plant disease as "a physiological disorder or structural abnormality that is deleterious to the plant or to any of its parts or products or that reduces their economic value".
3. **Agrios (1997)** defined plant disease as "a series of invisible and visible responses of plant cells and tissues to a pathogenic microorganism or environmental factors that result in adverse changes in the form, function, or integrity of the plant and may lead to partial impairment or death of the plant or its parts".
4. **According to the modern concept,** disease is an interaction among the host, parasite, and environment.

Plant pathology (Gr. *pathos* — suffering; *logos* — knowledge) is a branch of botany which deals with the **study of the nature, development and control of plant diseases or the study of the suffering plants.**

It is very difficult to distinguish between a healthy and a diseased plant with some abnormality. The disease of a plant may be caused by environmental factor or factors, called **non-parasitic disease** (Freezing injury of potato, water core of apple – due to high temperature) or it may be caused by microorganisms such as fungi (*Helminthosporium oryzae* — brown spot of rice), bacteria (*Xanthomonas campestris* pv. *oryzae*-Bacterial leaf blight of rice), plasmodiophorales

(*Plasmodiophora brassicae* — club root of crucifer), nematodes (*Heterodera rostochiensis* — golden-nematode disease of potato) etc. — called **parasitic disease**.

For a parasitic disease two organisms are required, one is **host**, on which disease takes place and the other one is **pathogen**, which causes disease. Roughly, we can classify the hosts into three categories: (i) **Resistant**, (ii) **Moderate resistant**, and (iii) **Susceptible**. Normally a disease takes place in a susceptible host with maximum intensity. Resistant hosts are those which are able to withstand, resist or overcome the attack of a pathogen completely or at a maximum degree. Moderate resistant hosts are able to resist the pathogen attack at a moderate level and, lastly, the susceptible plants, which are not able to resist the attack of the pathogen, produce maximum disease. The tendency of plants to escape from disease is called **klendusity**. In some cases, a variety of a plant may escape from disease, not for their resistance but for early maturity, thus they escape from natural disease incidence, called **disease escape** and the variety is called disease escaping variety. Similarly, one can classify the pathogen into three categories: (i) **Virulent**, (ii) **Moderate virulent**, and (iii) **Avirulent**. The efficiency of a pathogen to produce disease is called **virulence** and the pathogen with maximum efficiency is called virulent pathogen. Moderate virulent pathogen has the efficiency at moderate level and the avirulent pathogen is inefficient in the production of disease.

The above situation can be compared with the activity of you and your elder brother in your family to other family members. Consider your elder brother (who is in service) as a virulent pathogen and father (resistant), mother (moderate resistant) and younger brother (susceptible) as different hosts. With the same kind of mistake of others, the activity of your elder brother will be different. He can slap twice the younger brother; he may tell mother what you are doing; but to father, he has nothing to say. So in the above cases your father is resistant, mother is moderate resistant and your younger brother is susceptible to him. If you consider yourself (Student + enjoying fellowship) as a moderate virulent pathogen, you have nothing to do with both your father and mother, but you can slap once your younger brother. But if you slap him twice then he may report to your father and you may be punished. Lastly, the avirulent pathogens are not able to produce any disease in any kind of host, like a jobless inefficient uncle in your family who is a human being, yet nobody cares of him (Fig. 5.1).

An agency which incites (induces) a disease is the **incitent**. This includes both inanimate (unfavourable environmental conditions and similar situation) i.e., non-parasitic or non-infectious agencies and animate (fungi, bacteria, angiospermic plants etc.) i.e., infectious parasitic agencies. Similarly, an organism which generates a disease is known as **pathogen** or **causal organism**. For the development of disease on a host, in addition to pathogen, other environmental factors such as atmospheric humidity, temperature, wind, light, etc., are essential. Pathogen along with other environmental factors are collectively called **causal complex** (causal complex = causal organism + other environmental factors).

Considering a case in your family — father (host) brought raw materials from the market and

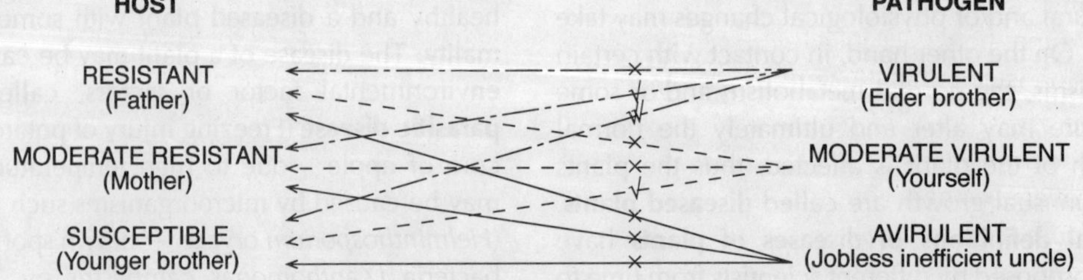

Fig. 5.1: Amount of disease produced by the pathogens to different hosts (x = no disease, √ = moderate disease and √√ = maximum disease)

mother (pathogen) is ready to prepare the food, but gas and/or water is not available; so no food can be prepared.

Casual organisms are generally parasites. Commonly the term **parasite** is applied to an organism which depends partly or wholly upon the living tissue or cell. On the other hand, a **saprophyte** is an organism which lives on organic materials or dead organic matters. In some cases, the organisms are parasites but do not play any role or may participate in disease development. Conversely, the by-products of a parasite are the major causal factors in disease development, so that parasite behaves indirectly like a pathogen. But the term pathogen is practically used for a living entity. The efficiency of a pathogen to produce disease is the **pathogenecity** and **pathogenesis** is the process or chain of events of disease development.

For the **development of disease** caused by fungi, **five steps** are followed starting from inoculum arrival to symptom expression. These are (i) **Germination** (ii) **Penetration** (iii) **Infection** (iv) **Incubation period,** and (v) **Symptom**. In bacteria it consists of four steps except the first one i.e., **germination**. The development of mycelium by the inoculum like spore is called **germination**. The initial invasion of the pathogen inside the host tissue is called **penetration** and its establishment inside the host is called **infection**. The infection may be visible externally on the host surface, called **visible infection**. On the other hand, when a pathogen has established itself inside the host tissue but is not visible from outside, it is **invisible infection**. After that, the pathogen grows (fungi) or multiplies (bacteria) inside the host tissue and ultimately the manifestation of the disease is expressed. The manifestation of disease is called **symptom**. The interval of time between infection and symptom expression is called **incubation period.**

*The above steps of development of disease can be compared with your steps to come to your college. When you are coming out from your house, consider it as the **germination**; your entry inside the bus is **penetration** and then you occupy a seat in a bus or any other vehicle, is the **infection**; then the bus or other vehicle is running along with you, is the **incubation period** and finally you reach the college, and show your performance or efficiency — the **symptom**.*

The manifestation of the disease is called **symptom**. The symptom may be visible externally, called **external symptom**, e.g., pinhead to eye-shaped, brown coloured spots on leaf, coleoptile, seeds etc., on rice indicate brown spot disease of rice by *Helminthosporium oryzae* or it may be internally invisible, called **internal symptom**, e.g., browning of woody tissue in wilt of pigeon pea, by *Fusarium oxysporum* f. sp. *udum*. When a number of symptoms are available for a particular disease, they are collectively called **syndrome**, e.g., wilt of pigeon pea, where symptoms like (i) lodging of the leaves (external), (ii) browning of the woody tissue, and (iii) presence of fungal mycelium inside the host tissue (internal) are available. If the disease is identified by observing (with naked eye) the pathogen externally on the host surface, it is called **sign**, e.g., powdery mildew of cucurbits by *Erysiphe cichoracearum*, where disease is identified by seeing the fungal spores on the host surface.

Concepts of Parasitism and Saprophytism

Various levels of parasitism of microorganisms have long been known. In some case, two organisms may live in close association with one another non-pathogenically and are essential and/or beneficial to each other's development — this is called **symbiosis**. The best examples of symbiosis are:

1. Between the **nodule forming bacteria** (*Rhizobium*) **and the leguminous plants**. Initially, the bacterium penetrates the root hairs causing abnormal growth of cells, the nodule. Up to this stage the bacterium behaves as pathogen. In the host cell, the bacterium multiplies and then starts to fix the atmospheric nitrogen and converting it into nitrogenous compounds, which are essential for protein synthesis. After the death of the bacterium the nitrogenous compounds become available for the host and also for the next crop.

Consider a case in our society with a satellite family, there are three members; husband, wife and their kid. They have some problems to prepare their daily food and also to look after their kid. Suddenly mother-in-law of husband came to their house. On that day, she was very much tired and took a good rest after taking her dinner. Up to that stage the case is pathogenic, because without giving any return she has consumed the food. But from the next morning the mother-in-law looks after the kid and sometimes she goes to the kitchen to prepare some new food, thereby the couple gets benefitted. So the mother-in-law is giving some return in exchange of taking food and shelter. This case may be considered as an ideal example of symbiosis.

2. Lichen: In lichens, an alga and a fungus are living in close association. The alga produces carbohydrate materials essential for its own development and also for the fungus. On the other hand, the fungus produces hydrolytic enzymes, which break the materials available in the substrate into simpler form.

In other case, the plants which grow on other plants without forming any parasitic or symbiotic relation are known as **epiphytes**. In tropical climate, many higher plants grow on trees or shrubs. They do not penetrate the living tissue and absorb only the essential minerals from the bark, apparently having no harmful or beneficial effect on the host plant. The epiphytes may hamper the metabolic activity by producing shades and also develop unsightly appearance of the host plants. Microorganisms may live as epiphytes but do not fall in the above category. Whereas, organisms living entirely within the plant are called **endophytes**. Such organisms may live either symbiotically or saprophytically.

Other types of parasitism are

(i) **Parasites**. The organisms which grow on living tissue are called parasites. The parasites may be of two types:

(a) **Obligate parasites**. The organisms which grow only on living tissue and not in axenic cultures (culture free from any other organism), are called obligate parasite, e.g., *Erysiphe cichoracearum* and *Sphaerotheca fuliginea* (powdery mildew of cucurbits), *Plasmopara viticola* (downy mildew of grapes), *Puccinia graminis tritici* (black stem rust of wheat), *P. recondita* (brown rust of wheat), *Synchytrium endobioticum* (black wart of potato), etc.

(b) **Facultative parasites.** The organisms which can live as saprophytes for long period in pure culture or otherwise, but under certain conditions they become pathogenic, are called facultative parasites, e.g., *Pythium debaryanum* (damping off of tobacco, tomato and chilli seedling), *Fusarium udum* (wilt of pigeon pea), etc.

(ii) **Saprophytes**. The organisms which grow on dead organic tissue or on inorganic material are called saprophytes. Like parasites, the saprophytes may be of two types:

(a) **Obligate saprophytes**. The organisms which have no relation with living host and collect their nutrition from dead organic tissue or from organic materials are called obligate saprophytes, e.g., *Mucor mucedo, Rhizopus sexuallis,* etc.

(b) **Facultative saprophytes**. These organisms are actually parasite, but, according to certain circumstances, they behave as saprophytes, e.g., *Ustilago* (smut fungus).

Other terminology

During unfavourable conditions like sudden change in temperature (high or low), effect of poisonous gas, disturbance of soil moisture relation etc., pathogen may occur inside the perennating organ or organs. With the onset of favourable condition, it becomes active and causes disease. The portion of the pathogen responsible for infection is known as **inoculum** (pl. inocula).

In presence of susceptible host, virulent pathogen and favourable environmental condition, a disease may not develop due to variation in the **amount of inoculum**. If the amount of **inoculum is less** in the diffusate, the pathogen will get more food and will germinate properly. But the amount of enzymes and toxins secreted by them become diluted in the comparatively bigger water or dew-drop and are not able to damage the host cell wall, so there will be no disease. On the other hand, if the amount of **inoculum is more** in the diffusate, the pathogen will get comparatively less amount of food in their share. Now they will fight for the food, by secreting more enzymes and toxins. For the above facts, the concentration of the enzymes and toxins will be more in the available water or dew-drop, thereby the germination and/or growth of the pathogen will be hampered and there will be no disease. So optimum concentration of inoculum is essential to develop a disease. The amount of inoculum which is responsible for successful infection is known as **inoculum potential**.

The above situation may also be noticed in your class room. Consider that all the conditions in a class room are normal, except the number of students. If the number of students present in the class is very low, then the teacher will think that he has to repeat the syllabus again and the output of the class will be less satisfactory. On the other hand, if the students present in the class is more than normal, then they are not able to sit properly and will create disturbance in the class, thereby the output of the class will be less satisfactory. So the optimum number of students in a class is essential in addition to other conditions, for the normal output of a class.

The inoculum which develops from the perenating organ or comes from the neighbouring field is the **primary inoculum**. The primary inoculum then disseminates by wind, rain, insects, or by other agencies or directly comes in contact with the host tissue and establishes infection is the **primary infection**. The primary infection that produces disease symptom is the **primary symptom**, that which produces next crop of inoculum is known as **secondary inoculum**. So the secondary inoculum may be defined as the inoculum which develops from the primary symptom. Like primary inoculum, the secondary inoculum causes **secondary infection** and **secondary symptom**. The second phase i.e., the production of secondary inoculum and development of secondary symptom may repeat several times in a growing season.

The chain of events which takes place in cyclic order to complete the life process is known as **life cycle;** the **disease cycle** is the life cycle of a pathogen including the series of events starting from the time of infection to disease expression on a host. In case of obligate parasite, disease cycle and life cycle are interdependent, but in other cases, like facultative parasite, facultative saprophytes etc., the disease cycle is dependent primarily on the parasitic part of the life cycle. Generally the disease cycle consists of two cycles, one produced by the primary infection is the **primary disease cycle** and the other one which is produced by the secondary infection is the **secondary disease cycle.** In some case, the secondary disease cycle may be absent and the pathogen produces only one cycle in its life process (wilt disease etc).

Different disease cycle: Monocyclic, Polycyclic, and Polyetic

Monocyclic plant disease: When pathogens are able to complete only one or part of disease cycle in one year, they are called monocyclic pathogens and the disease as monocyclic disease.

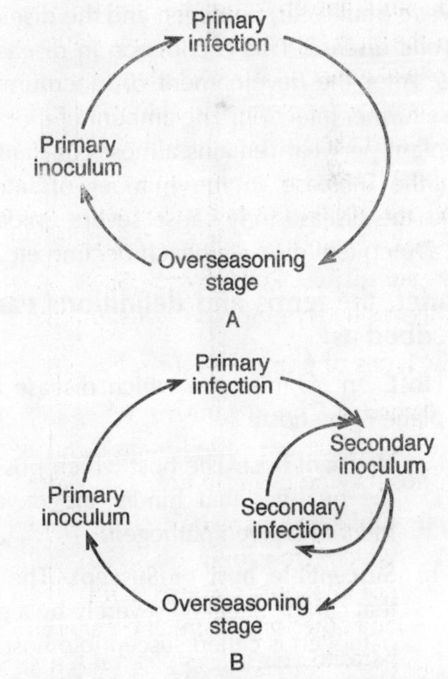

Fig. 5.2: Diagrammatic representation of disease cycle: A. Monocyclic, and B. Polycyclic

In monocyclic disease, pathogen develops primary inoculum which is the only inoculum available for the entire season. The secondary inoculum and secondary infection are totally absent in such diseases, e.g., smuts (loose smut of wheat, c.o. *Ustilago segetum*), root rot (root rot of turmeric, c.o. *Pythium aphanidermatum*) and vascular wilt (wilt of pigeon pea, c.o. *Fusarium udum*), etc.

Polycyclic plant disease: When pathogens are able to complete more than one generation (2-30) in a growing season, they are called *polycyclic pathogens* and the disease as *polycyclic disease*. The amount of inoculum in each cycle is multiplied manyfold, e.g., leaf spot (brown spot disease of rice, c.o. *Helminthosporium oryzae*), blight (late blight of potato, c.o. *Phytophthora infestans*), powdery mildews (powdery mildew disease of cucurbits, c.o. *Erysiphe cichoracearum*), etc.

Polyetic plant disease: When pathogens are not able to complete a disease cycle within one year and take several years before the inoculum is developed on plant and can be disseminated and may initiate new infection, they are called *polyetic* (multiyear) *pathogen* and the disease as *polyetic disease*. This is common in diseases of trees. After the development of inoculum they cause further infection. The amount of inoculum per plant per year remains almost constant, but with the increase in the number of infected plants the disease may cause severe epidemic, e.g., Dutch elm disease, peach decline etc.

In brief, the terms and definitions can be described as:

1. **Host**. An organism on which disease takes place is the host:
   - (a) **Resistant host**. The host which posseses the qualities that hinder the development of a given pathogen.
   - (b) **Susceptible host** or **Suscept**. The host that can be attacked severely by a given pathogen is called susceptible host.
   - (c) **Immune host**. The host which cannot be infected by a given pathogen, even under the most optimum condition.
   - (d) **Resistance**. The ability or efficiency of an organism to overcome completely or in some degree the effect of a pathogen is referred to as resistance.
   - (e) **Susceptibility**. It is the condition of a plant that makes it suitable for attack by a given pathogen or other harmful factors.
   - (f) **Immunity**. It is the **ability** of a host to resist against the infection of a given pathogen, even under the most optimum condition.
2. **Pathogen**. An organism which generates a disease is called ***pathogen***. The pathogen is also known as **causal organism**. In brief it is written as c.o.

   The efficiency of a pathogen to cause disease is the **pathogenecity,** and **pathogenesis** is the process or chain of events of disease development.
3. **Hypersensitivity**. It is the excessive sensitivity of plant tissue to some pathogens; thereby the affected cells are killed quickly and block the progress of the pathogen.
4. **Causal complex**. In addition to pathogen, other environmental factors, such as atmospheric humidity, temperature, wind speed, light etc., are essential for the development of disease on a host. Pathogen along with other environmental factors are collectively called ***causal complex***. *Causal complex = causal organism + other environmental factors*.
5. **Penetration**. The initial invasion of the pathogen inside the host tissue is called ***penetration***.
6. **Infection**. The establishment of the pathogen inside the host tissue is called ***infection***.
   - (a) **Visible infection**. The infection which is visible externally on the host surface is called ***visible infection***.
   - (b) **Invisible infection**. The infection which is not visible externally on the host surface is called ***invisible infection***.
7. **Incubation period**. The interval of time between infection and symptom expression is called ***incubation period***.

8. **Symptom**. The manifestation of the disease is called ***symptom***.

   External symptom. When the symptom becomes visible externally, it is known as ***external symptom***, e.g., brown coloured spots on leaf and on other parts of rice plant indicate that it is brown spot of rice.

   Internal symptom. When the symptom becomes internally invisible, it is called ***internal symptom***. e.g., browning of woody tissue in wilt of pigeon pea.

9. **Sign**. When the disease is identified by seeing (with naked eye) the pathogen externally on the host surface, it is called ***sign***, e.g., powdery mildew of cucurbits.

10. **Syndrome**. When a number of symptoms are available for a particular disease, they are collectively called ***syndrome***, e.g., wilt of pigeon pea, where symptoms like (i) lodging of the leaves (external), (ii) browning of the woody tissue, and (iii) presence of fungal mycelium inside the host tissue (internal) are available. These three symptoms are collectively called ***syndrome***.

11. **Symptomatology**: The symptoms are the observable and detectable effects of the pathogen on growth, development and metabolism of infected plant, and the external and internal alteration of plant or plant parts as a result of disease. The orderly study of symptoms in infected plant is called **symptomatology.**

12. **Inoculum**. The portion of the pathogen responsible for infection is called ***inoculum***. The inoculum may be spore, mycelium or any other part of the fungus, but in bacteria and virus the entire body behaves as inoculum.

13. **Inoculum potential**. The amount of inoculum responsible for successful infection is called ***inoculum potential***.

14. **Primary inoculum**. The inoculum which develops from the perenating organ or comes from the neighbouring field is the ***primary inoculum***.

15. **Primary infection**. The establishment of infection caused by primary inoculum is called ***primary infection***. The primary infection leads to the development of primary symptom.

16. **Secondary inoculum**. The inoculum which develops from the primary symptom is called ***secondary inoculum***.

17. **Secondary infection**. The establishment of infection caused by secondary inoculum is called ***secondary infection***. The secondary infection leads to the development of secondary symptom.

18. **Disease cycle**. It is the life cycle of a pathogen including series of events, starting from the time of infection to disease expression on a particular host.

19. **Parasites**. The organisms which grow on living tissue are called ***parasites***.

    (a) **Obligate parasite**. The organism which grows only on living tissue and not in axenic culture (culture free from any other organism), is called ***obligate parasite***, e.g., *Plasmopara viticola* (c.o. of downy mildew of grapes).

    (b) **Facultative parasite**. The organism which can live as saprophytes for long period in pure culture or otherwise under certain condition it becomes pathogenic is known as ***facultative parasite***. e.g., *Phythium debaryanum* (c.o. of damping off of tobacco).

20. **Saprophytes**. The organisms grow on dead organic tissue or on inorganic material are called ***saprophytes***.

    (a) **Obligate saprophyte**. The organisms which have no relation with living host and collect their nutrition from dead organic tissue or from inorganic materials are called ***obligate saprophyte***, e.g., *Rhizopus sexualis*.

    (b) **Facultative saprophyte**. The organism which is actually a parasite, in certain circumstances behaves as saprophyte, and is called ***facultative parasite***, e.g., *Ustilago* (smut fungi).

21. **Infectious disease**. It is a type of disease in

which a transmissible causal organism or particle like virus is a part of the causal complex.

22. **Symbiosis**. When two organisms live in close association with one another non-pathogenically and are essential and/or beneficial to each other's development, this phenomenon is called **symbiosis**, e.g., lichens — formed by the association of algae and fungi.

23. **Symbionts**. The organisms, taking part in symbiosis are known as **symbionts**, e.g., lichen — both algal and fungal partners are the symbionts.

24. **Epiphytes**. When one plant grows on other plant without making any parasitic or symbiotic relation, the former is known as ***epiphyte***.

25. **Endophytes**. The organisms living entirely within the host plant are called **endophytes**. Endophytes may be parasitic or symbiotic.

26. **Predisposition**. It is the effect of one or more meteorological factors which makes a plant vulnerable to attack by a pathogen.

27. **Etiology**. The determination vis-a-vis study of the cause of a disease is called etiology.

28. **Necrotroph**. The microorganisms which feed only on dead organic tissues, e.g., *Mucor, Rhizopus* etc.

29. **Biotroph**. The organism that can live and multiply only on another living organism, e.g., *Vasicular-arbuscular mycorrhizae* (VAM), powdery (*Erysiphe*) and downy (*Peronospora*) mildews.

30. **Disease triangle**. The interactions of the three components of a disease (host, pathogen and environment) have often been visualised as a triangle, commonly referred to as *disease triangle*.

The disease triangle can be discussed now:

The disease of plants is caused either by biotic or, less commonly, by abiotic means. In the first case, both host and pathogen must interact with each other and the result of their interaction will depend on the different environmental conditions like temperature, humidity etc. (Fig. 5.3). Without the requisite environmental factors pathogen will not be able to develop disease even on a susceptible host. All the three components (host, pathogen and environment) show considerable variation in nature. The **intensity of the disease** in individual plant or in plant population varies with the variation of any one component:

A. Variation in respect of host:
   1. Degree of resistance of the host.
   2. Age of the plant.
   3. Size of the plant, etc.

B. Variation in respect of pathogen:
   1. Degree of virulence.
   2. Amount of inoculum, etc.

C. Variation in respect of environment:
   1. Range of temperature,
   2. Range of humidity, etc.

The disease triangle consists of three components — host, pathogen and environment, placed at the three sides of the triangle. The length of each side is proportional to the sum total of the characteristics of each component that favours disease. The shape of the triangle varies with the variation of the factors. Such as:

1. *If the plants are resistant and/or at wrong age*, the amount of disease will be very less or zero. So in this condition, the host arm will be very small and thus the central area of the triangle will be less so that the prevalence of disease will be very little or nil.

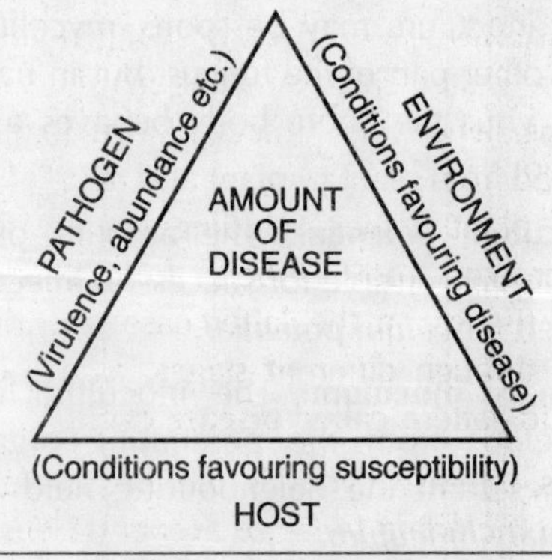

Fig. 5.3: The disease triangle

2. On the other hand, *if the plants are susceptible and at proper age* of development of disease, then the disease will be to a greater extent. So, in this condition the host arm will be comparatively larger and thus the central area of the triangle will be more, so that the prevalence of disease will be much more.
3. Similarly, *if the pathogen is more virulent, in more amounts and active* then the pathogen arm will be larger and there will be larger central area i.e., the extent of disease will be much more prevalent. The result will be reversed with less virulent pathogen if present in fewer amounts.
4. *If the environmental conditions like temperature, moisture etc., become favourable,* then larger will be the environment arm and simultaneously larger will be the central area i.e., the amount of disease will be much more. The result will be reversed with less favourable environmental condition.
5. Lastly, it can be concluded that if the extent of any one of the three components is zero, then the amount of disease will be zero i.e., no disease will develop.

Disease cycle and Environmental relation

Disease may cause by both abiotic and biotic agents. The abiotic diseases are caused by conditions external to the plant, not by living agents. They cannot spread from plant to plant, but are very common and should be considered when assessing the health of any plant. Abiotic diseases include nutritional deficiencies, soil compaction, salt injury, sun scorch etc. The **biotic diseases** are caused by living organisms. They are called **plant pathogens** i.e. **causal organisms**, when they infect plants. Pathogens can spread from plant to plant and may infect all types of plant tissue including leaves, shoots, stems, crowns, roots, tubers, fruit, seeds and vascular tissues. In the latter case, the disease progress through different stages, in a more or less cyclic pattern called **disease cycle**.

Thus, *disease cycle is the life cycle of a pathogen including series of events starting from the time of infection to disease expression on a host.* In obligate parasite, the disease cycle and life cycle are interdependent, but other cases like facultative parasite and facultative saprophytes etc., the disease cycle is dependent primarily on the parasitic part of the life cycle. Commonly for the development of a disease, a susceptible host and a virulent pathogen is required. But their presence does not always develop disease. Another one factor the **environment** is required, which control the development of disease. Thus for the development of a successful infection, susceptible host, virulent pathogen along with suitable (conductive) environment is must. These three factors together form a **disease triangle** (Fig. 5.4).

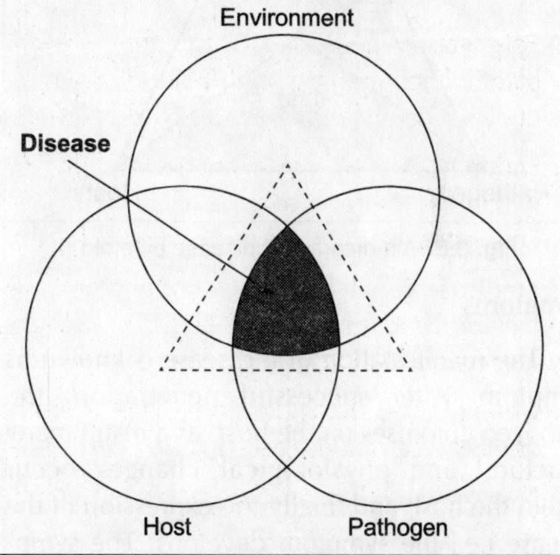

Fig. 5.4: Ven-diagram of disease triangle

Without a conducive environment, a plant disease will not occur. In most cases, a conducive environment involves high humidity, or free moisture on the leaf surface in the form of dew as well as moderate temperatures. Most fungi do not like the temperature extremes. However, the environment must remain conducive for a specific amount of time for disease to occur. **Time** is the key component lacking in the three-dimensional disease triangle and the reason for the additional fourth component since a prolonged period of exposure between the host and the pathogen in a conducive environment will be required prior to disease incidence and symptom expression.

However, in some less traditional settings a **fourth factor (time)** is included in a 4-dimensional figure to show the impact of time in addition to the host, organism, and environment. A period of time is normally required between when a host comes in contact with a pathogen and a favourable environment for disease to develop occurs. Thus, the inclusion of the fourth factor form a pyramid called **disease pyramid**.

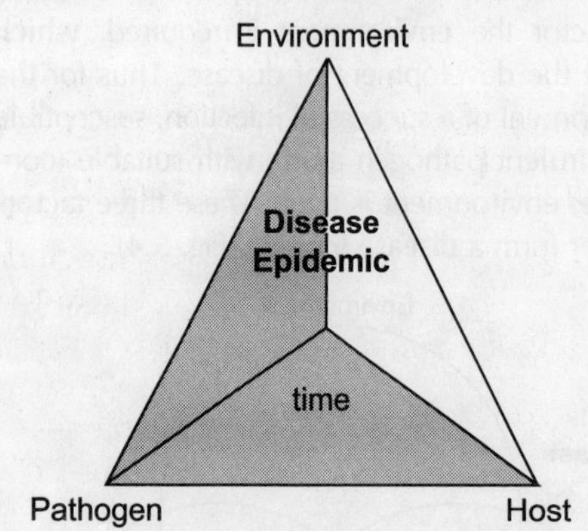

Fig. 5.5: Ven-diagram of disease pyramid

Symptoms

The manifestation of a disease is known as **symptom**. After successful penetration, the pathogen colonises on the host, as a result many structural and physiological changes occur within the host, and finally the expression of the disease i.e., the symptom develops. The symptom may be external or internal. The pathogen which is visible externally on the host surface is the **sign**. Sometimes disease caused by an organism may produce more than one type of symptom and they are collectively referred to as **syndrome** (wilt of pigeon pea, c.o. *Fusarium udum*).

Symptom: In most parasitic diseases, the growth of the pathogen takes place inside the host tissue, thus the parasite is invisible. The expression of the disease in different shapes, colours etc., are visible from outside. Symptoms are classified into three categories: 1. **Necrotic**, 2. **Atrophic** and 3. **Hypertrophic**.

1. **Necrotic symptoms**. These symptoms are evidenced by the death (necrosis) of cells, tissues or organs of the infected host due to parasitic or non-parasitic causal agencies.

The common and important necrotic symptoms produced by the different types of parasitic causal agents are: (a) **Spots**, (b) **Anthracnose**, (c) **Blight**, (d) **Scald**, (e) **Die-back**, (f) **Streak** or **Stripe**, (g) **Canker**, (h) **Damping off**, (i) **Rot**, (j) **Blotch** and (k) **Blast**.

(a) **Spots**. When necrotic region develops on the host surface, formed by killing of tissue in a limited area, it is called spot. The spots may be circular, angular or irregular in outline (Fig. 5.6A-C). Different types of leaf spot diseases are found in plants, e.g., brown spot of rice (*Helminthosporium oryzae*), ring spot of sugarcane (*Leptosphaeria sacchari*), etc.

(b) **Anthracnose**. These are more or less angular and elongated spots which appear on veins of the lower surface of leaf, and may spread on the upper surface in severe cases (Fig. 5.6E). Stem, petioles and fruits (Fig. 5.6D) may also become affected during severe infection, e.g., anthracnose of bean (*Colletorichum lindemuthianum*), anthracnose of mango (*C. gloeosporioides*) etc.

(c) **Blight**. Blight can be defined as the rapid killing of different plant parts, such as leaves, blossoms etc., and the killed tissues may become slimy and often emit pungent odour, e.g., late blight disease of potato (*Phytophthora infestans*) (Fig. 5.6F), early blight of potato (*Alternaria solani*) (Fig. 5.6G) etc.

(d) **Scald**. A white or pale appearance of leaves and/or fruits resembling that developed by scalding water is referred to as scald, e.g., leaf scald of sugarcane (*Xanthomonas albilineans*).

(e) **Die-back**. The symptom appears due to gradual dying of twigs from apex towards the base due to attack of pathogen, e.g., ripe fruit rot and die-back of chilli (*Colletotrichum capsici*).

(f) **Streak or Stripe**. It appears as narrow elongated, hydrotic, necrotic or other type of lesions on leaf veins or stems, e.g., bacterial leaf streak of rice (*Xanthomonas campestris* pv. *oryzicola*).

(g) **Canker.** It appears as sunken necrotic areas present in the bark and cortex of woody stems, on leaves and fruits (Fig. 5.6H). Sometimes the diseased portion becomes separated from the healthy region by cracks. The callus formation may take place at the healthy margin. e.g., citrus canker (*Xanthomonas axonopodis*).

(h) **Damping off.** The pathogen attacks the stem of seedlings at the ground level

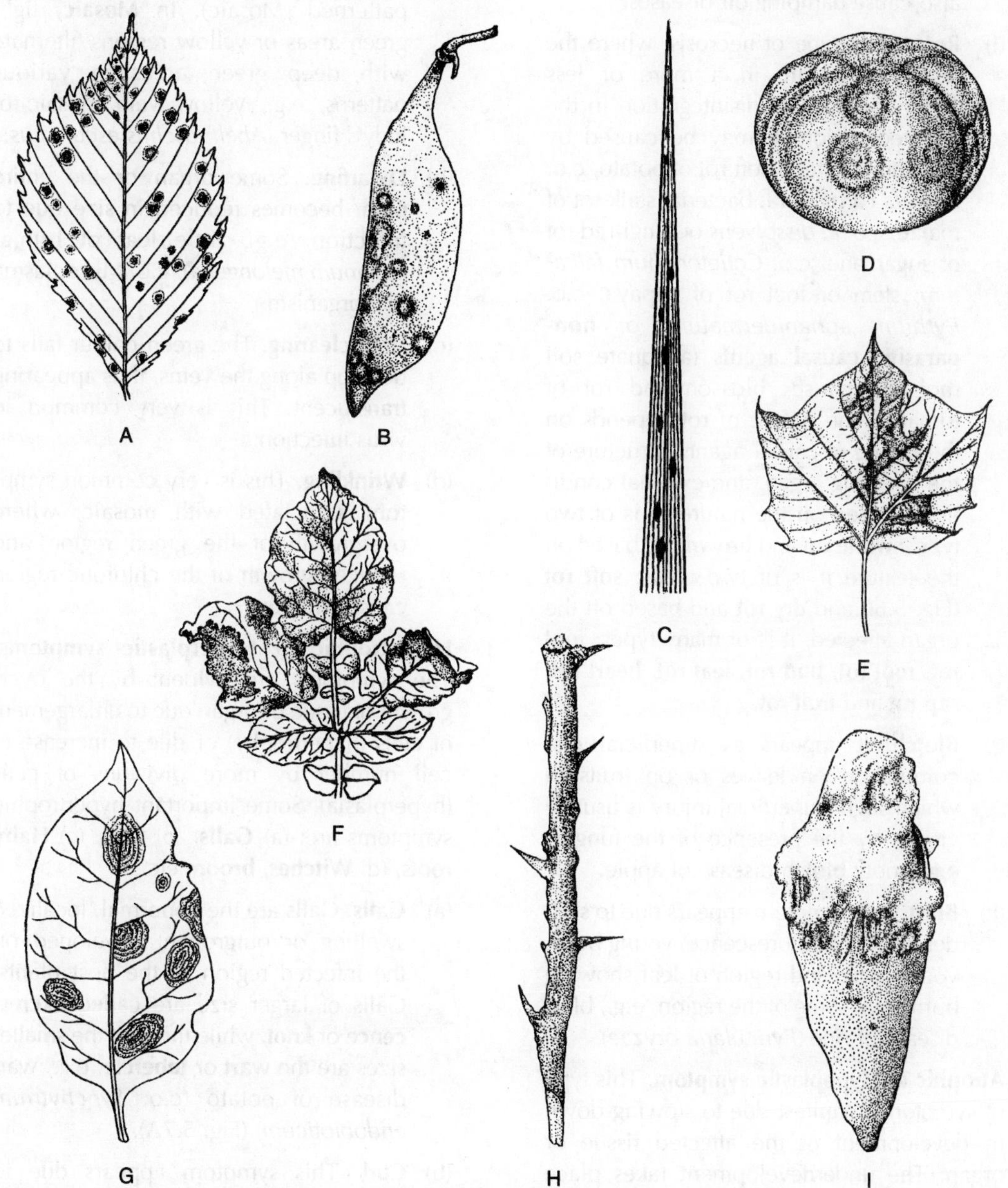

Fig. 5.6: Various types of disease symptoms: A. Leaf spots-circular to oval necrotic areas on leaf, B. Pod-spots of pea, C. Brown spots on rice leaf, D. Anthracnose of tomato, E. Anthracnose on leaf, F. Late blight of potato on leaf, G. Early blight of potato on leaf portion, H. Cirtus canker on stem, and I. Soft rot of sweet potato on storage root

and causes toppling down. Wet-rot type lesions appear at the infected region, e.g., damping off of tobacco seedling (*Pythium aphanidermatum*), damping off of jute seedling (*Rhizoctonia bataticola*). In addition to the above, other fungi like *Fusarium* and *Phytophthora* also cause damping off diseases.

(i) **Rot**. It is a type of necrosis, where the dead tissues are in a more or less advanced stage of disintegration in the infected region. It may be caused by **parasitic** (bacteria-soft rot of potato, c.o. *Erwinia carotovora*; bacterial stalk rot of maize, c.o. *E. dissolvens* or fungi-red rot of sugarcane, c.o. *Colletorichum falcatum*; stem or foot rot of papaya, c.o. *Pythium aphanidermatum*) or **non-parasitic causal agents** (adequate soil moisture causes blossom-end rot of tomato). The nature of rot depends on the nature of causal agents, structure of the infecting organ, and external conditions. Based on the nature, it is of two types: **white rot** and **brown rot**; based on the texture it is of two types: **soft rot** (Fig. 5.6l) and **dry rot** and based on the organ affected, it is of many types: **foot rot, root rot, bud rot, leaf rot, heart rot, sap rot** and **fruit rot**.

(j) **Blotch**. It appears as superficial discolouration on leaves or on fruits in which slight superficial injury is usually visible by the presence of the fungus, e.g., sooty blotch disease of apple.

(k) **Blast**. This symptom appears due to sudden death of inflorescence, young buds, young fruits and region of leaf showing burning at some of the region, e.g., blast disease of rice (*Pyricularia oryzae*).

2. **Atrophic** or **Hypoplastic symptom**. This type of symptoms manifest due to slowing down in development of the affected tissue or organ. The underdevelopment takes place due to hypoplasia i.e., subnormal cell division or rarely due to degeneration of cells. Some important types of atrophic symptoms are: (a) **Chlorosis**, (b) **Dwarfing**, (c) **Vein-clearing**, (d) **Wrinkling** etc.

(a) **Chlorosis**. It appears as yellow or lighter green in colour either due to degredation of chlorophyll or its synthesis at a slower rate. Chlorosis may be systemic (Aster-yellows) or it may be of different patterned (Mosaic). In **Mosaic,** light green areas or yellow regions alternate with deep green areas in various patterns, e.g., yellow vein mosaic of lady's finger (*Abelmoschus esculentus*).

(b) **Dwarfing**. Some organ or the entire plant becomes reduced in size due to infection, e.g., little leaf of brinjal (*Solanum melongena*), c.o. Mycoplasma like organism.

(c) **Vein-clearing**. The green colour fails to develop along the veins, thus appearing translucent. This is very common in virus infection.

(d) **Wrinkling**. This is very common symptom associated with mosaic, where overgrowth of the green region and reduced growth of the chlorotic region cause wrinkling.

3. **Hypertrophic** or **Hyperplastic symptoms**. The symptoms are evident by the overgrowth of tissue or organ due to enlargement of cells (hypertrophy) or due to increase in cell number by more divisions of cells (hyperplasia). Some important hypertrophic symptoms are: (a) **Galls**, (b) **Curl**, (c) **Hairy roots**, (d) **Witches, broom** etc.

(a) **Galls**. Galls are the abnormal, localized swelling or outgrowth, developed on the infected region of the host plants. Galls of larger size are called **excrescence** or **knot,** while those of the smaller sizes are the **wart** or **tubercle**, e.g., wart disease of potato (c.o. *Synchytrium endobioticum*) (Fig. 5.7A).

(b) **Curl**. This symptom appears due to localized overgrowth on one side or certain region of the leaves or shoots which become curled, arched or distorted, e.g., peach leaf curl disease (c.o. *Taphrina deformans*) (Fig. 5.7B).

(c) **Hairy roots**. Huge number of compactly arranged fine fibrous roots are developed due to infection.

(d) **Witches, broom**. More branches with vertical growth are developed due to infection. The branches grow vertically instead of normal horizontal pattern. Sometimes the branches and leaves become swollen.

Sign. In some parasitic diseases, the pathogen is partially or wholly visible with naked eye and the diseases are identified by seeing the structure of the pathogen. Some important signs are: (a) **Mildews**, (b) **Rusts**, (c) **Smuts,** etc.

(a) **Mildews**. Pathogen grows superficially and appears as white, violet or brown spots on young stems, leaves and also on fruits. When the fungal spores are available in the form of spreading powdery mass on the host surface, they are called **powdery mildews** (powdery mildew of cucurbits, c.o. *Erysiphe cichoracearum*). On the other hand, when the symptoms are visible in the form of tangled cottony or downy growth, they are called **downy mildews** (downy mildew of cucurbits, c.o. *Pseudoperonospora cubensis*).

(b) **Smuts**. The symptoms are usually developed as black powdery mass of spores like smut on the inflorescence, either by replacing ovary, the **loose smut** (loose smut of wheat, c.o. *Ustilago segetum* var. *tritici*) or the spores remain inside ovarian wall, the **covered smut** (covered smut of barley, c.o. *Ustilago hordei*).

(c) **Rust**. The symptoms appear as small pustules on the epidermis of leaf and stem of the host. Pustules may be powdery or compact and are of different colouration, like black (**black stem rust** of wheat, c.o. *Puccinia graminis tritici*), brown (**brown rust** of wheat, c.o. *Puccinia recondita*), yellow (**yellow rust** of wheat, c.o. *Puccinia striiformis*), white (**white rust** of crucifer, c.o. *Albugo candida*) (Fig. 5.7C) etc., depending on the species.

Concept on different types of Plant diseases

A. Systemic diseases

When a disease spreads in the entire plant to a varying extent and is associated with nearly every stage of its life cycle, it is known as systemic disease.

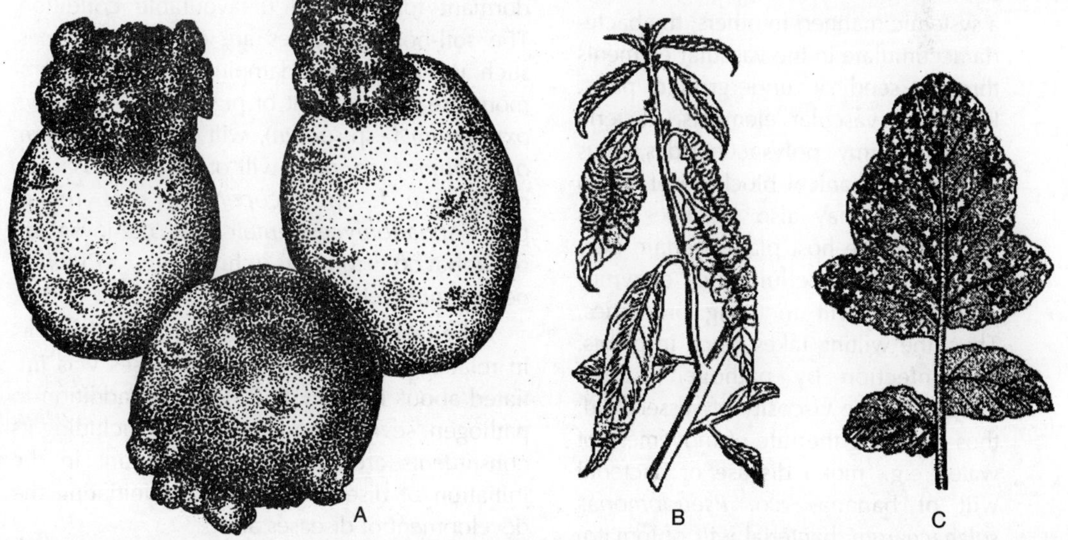

Fig. 5.7: Disease symptoms: A. Wart disease of potato (warted tuber). B. Peach leaf curl (twig with diseased leaf), and C. White rust on *Brassica* leaf

These diseases are caused by pathogens like fungi, bacteria and viruses:

1. **Fungi**. Fungi commonly enter in the root by hyphae developed on germination of different type of spores and gradually grow upwards through the xylem vessel and subsequently block the vessel by mycelium. The fungus may also produce some chemical substance (fusaric acid — a toxin, produced by *Fusarium udum* during development of wilt of pigeon pea). Due to direct blockage of vessels by mycelium and indirect effect of chemical substance, upward transport of water gets hampered and the plant becomes wilted due to continuous transpiration by the upper leafy region of plant.

   Wilting is a good example of systemic disease. Wilting causes lodging of leaves and finally death of the plant or may take place in one side of the plant (partial wilting), causes partial death with low yield, e.g., wilt of pigeon pea (*Fusarium udum*), wilt of cotton (*Fusarium oxysporum* f. sp. *vasinfectum*), wilt of tomato (*Verticillium albo-atrum*) etc.

2. **Bacteria**. Some leaf spot causing bacteria also enter the vascular system of leaf and then grow in different plant parts in a systemic manner. In others, the bacteria accumulate in the vascular elements through seed or underground parts. Inside the vascular elements, bacteria produce slimy polysaccharides, thus causing mechanical blockage of xylem vessels and may also produce toxins harmful to the host plant. Certain pectinolytic and cellulolytic enzymes possess the wilt inducing properties. Thus, the wilting takes place in plants. The infection by pathogen causes increase in the viscosity of vessel fluid, thus declining the rate of movement of water, e.g., moko disease or bacterial wilt of banana, c.o. *Pseudomonas solanacearum*; bacterial wilt of tomato, c.o. *P. solanacearum*; brown rot of potato, c.o. *P. solanacearum*; black rot of crucifers c.o. *Xanthomonas campestris* pv. *campestris*; bacterial wilt of cucurbits, ring rot of potato, etc.

3. **Viruses**. Viruses are the particles active only inside the living body. Transmission of viruses, from plant to plant (i.e., actually spread) takes place in a number of ways, like vegetative propagation, mechanical transmission through sap, seed, pollen, mites, nematodes, fungus, insects etc.

After entering the host, the virus moves from one cell to another and during movement it multiplies in most cells. They move from cell to cell through plasmodesmata. Once the virus enters the phloem, it moves towards the growing regions or food storing parts. In phloem, the virus spreads throughout the plant in systemic manner and enters the adjacent cell through plasmodesmata, e.g., tobacco mosaic virus (TMV) etc.

B. Soil borne diseases

Here the pathogen is a permanent soil inhabitant and infection occurs primarily through subterranean portions of the plant. Soil factors are thus the most influential in soil-borne diseases. The pathogens remain in the soil in active form during favourable condition, and in dormant form during unfavourable condition. The soil-borne diseases are of different types, such as wilt, root rot, damping off etc. The common diseases are: wilt of pigeon pea (*Fusarium oxysporum* f. sp. *udum*), wilt of pea (*Fusarium oxysporum* f. sp. *pisi*), wilt of tomato (*Fusarium oxysporum* f. sp. *lycopersici*); root rot of pea (*Aphanomyces euteiches*); damping off of tobacco, tomato, chilli etc. (*Pythium debaryanum*) etc.

The first comprehensive study of soil factors in relation to development of diseases was initiated about 1916 by L. R. Jones. In addition to pathogen, several conditions of soil including its constituents are very much important in the initiation of diseases. The factors initiating the development of diseases are:

Temperature. The common diseases caused by vascular fusaria are cotton wilt, tomato wilt, pea wilt, flax wilt etc. In general the growth rate of the pathogen increases with temperature to an optimum at about 27°C, above which the rate drops very quickly. In addition to soil temperature, the air temperature shows a secondary importance.

The damping off of seedling, killing of potato sprouts and canker of citrus in old stems caused by *Rhizoctonia solani* are greatly influenced by temperature between 23°C and 27°C.

Soil moisture. The soil moisture influences the development of disease. Linford (1928) found pea wilt to develop more rapidly in soil at 90% water holding capacity. Some diseases are, however, very responsive to soil moisture, e.g., *Aphanomyces* root rot of garden pea. The rate and severity of disease development increase with the soil-moisture level.

Soil pH: The pH of the soil greatly influences the development of disease. The potato scab organism *Streptomyces scabies* is not so abundant nor so severe in pH 5.2 or below. Severity of infection increases at pH 7.5, afterwards becomes reduced. The club root of crucifer's pathogen, *Plasmodiophora brassicae* grows well in acidic pH which becomes reduced in soil pH 7.0 or above.

Soil type. In the black limestone soil cotton grows well without *Fusarium* wilt, which is infected by root rot pathogen *Phymatotrichum*. The wilt of pea, by *Fusarium oxysporum* f. *pisi*, race 1, develops most extensively in silt loams, but less readily in colby silt loam, superior red clay and superior clay loam. The root-rot organism, *Aphanomyces euteiches* favours the last two type of soil due to their higher water holding capacity.

Other factors like soil fertility, etc., are important in influencing the disease.

C. Seed borne diseases

The pathogens remain dormant in seed throughout the season, safely in store house. In the next season, when the seeds are sown for the new crop, the pathogen becomes active and causes infection. The pathogens may perenate, externally (bunt of wheat, c.o. *Tilletia caries*), internally (loose smut of wheat, c.o. *Ustilago nuda tritici*- pathogen remains in the embryo) or both externally and internally (brown spot of rice, c.o. *Helminthosporium oryzae*).

Pathogens of different types that remain within seeds are fungi, bacteria, virus, nematodes etc.

Important diseases caused by different pathogens are given below (Table. 5.1).

Table 5.1: Important seed-borne diseases and their possible control

Name of disease	Name of pathogen	Location	Possible control measures
Fungal disease			
Rice			
1. Brown spot of rice	*Drechslera oryzae* (*Helminthosporium oryzae*)	Externally	Seed treatment with organo-mercurial fungicide (1% active content in the ratio 1 : 400 as dry dressing before sowing.
2. Foot rot of rice	*Fusarium moniliformae*	-do-	-do-
Wheat			
3. Foot rot of wheat	*Helminthosporium sativum* (perfect stage *Cochliobolus sativus*)	Externally and internally	Seed treatment with organo-mercurial fungicide like Agrosan G.N.
4. Loose smut of wheat	*Ustilago segetum* var. *triti*	Internally	Dry dressing with Bavistin (0.2%) before sowing or Water soaking for 72 hours gives 100% control of the disease.

Contd.

Contd.

Name of disease	Name of pathogen	Location	Possible control measures
5. Leaf blight of wheat	*Alternaria triticina*	Internally and externally	Seed dressing with mercurial fungicide.
6. Flag smut of wheat	*Urocystis tritici*	Externally	-do-
Barley			
7. Covered smut of barley	*Ustilago hodei*	Externally	Seed treatment with organomercurial fungicide.
8. Loose smut of barley	*Ustilago nuda*	Internally	Like loose smut of wheat.
Sugarcane			
9. Red rot of sugarcane	*Colletotrichum falcatum (Glomerella tucumanensis)*	Internally	Seeds are treated with Bavistin (0.1%) or Agallol.
Bajra			
10. Downy mildew of bajra	*Sclerospora graminicola*	Mycelium internally and oospores externally	Use of disease free certified seeds.
Bacterial			
Rice			
1. Bacterial blight of rice	*Xanthomonas campestris* pv. *oryzae*	Internally	Seeds soak in water at room temperature for 12 hrs., followed by hot water treatment at 53°C for 30 minutes (Sinha and Nene, 1967).
Cotton			
2. Angular leaf spot of cotton	*Xanthomonas axonopodis* pv. *malvacearum*	Externally and internally	Seeds are delinted with sulphuric acid (H_2SO_4) and then treated with vitavax or pentavax.
Potato			
3. Brown rot of potato	*Ralstonia solanaceanum*	Seed borne	Use of disease free tuber and disinfected knife. Pretreatment of tubers with 500 or 1000 ppm Streptomycin. Seeds are treated with hot water at 49°C for 30 minutes.
Cabbage			
4. Black rot of cabbage	*Xanthomonas campestris* pv. *campestris*	Seed borne	Seeds are treated with hot water at 49°C for 30 minutes.
Viral			
Potato			
1. Leaf roll of potato	Potato virus-1, or Solanum virus-14	Infected seed tuber	Seed tubers are stored at 37.5°C for 20-25 days in a humid atmosphere.
2. Rugose mosaic of potato	Potato virus-X and Potato virus-Y	Inside small tubers	Use of disease free seeds and control of aphids reduce the disease.
Bean			
3. Mosaic of bean	Phaseolus virus 1 or common bean mosaic virus	Internally	Disease free seeds.

Contd.

Contd.

Name of disease	Name of pathogen	Location	Possible control measures
Nematodes			
Rice			
1. White tip of rice	*Aphelenchoides besseyi*	Lie in a quiescent stage beneath, half of the seeds.	Seed treatment in hot water at 55°C for 10-15 minutes. Soaking of seeds in 0.1% soln., of Carbosulfan for 12 hrs.
2. Ufra disease or rice	*Ditylenchus angustus*	Externally	Application of Carbofuran, Mocap etc. in between planting and flooding.
Wheat			
3. Earcockle of wheat	*Anguina tritici*	Nematode gall along with seeds	Use of selected clean seeds. Seeds are soaked in 18% brine soln., in the floatation of galls.
Potato			
4. Potato rot nematode	*Ditylenchus destructor* and *D. dipsaci*	Infected tuber	Soaking of tubers in Chloropyriphos (0.1%) for 12 hrs.

D. Air borne diseases

The pathogen is disseminated by air, as a result primary and/or secondary infection occur from aerial pathogens. Aerial factors are thus the most important.

It is obvious that the environment becomes more complex for pathogen to maintain its life cycle in relation to the aerial portions of host plants. The interaction of temperature and humidity is often most important in this type. The development of inoculum and the local or widespread dissemination of the pathogen by wind are the important steps of the disease cycle which are commonly dependent upon environmental factors, e.g., mildews (powdery mildew of cucurbits, c.o. *Erysiphe cichoracearum*; powdery mildew of peas, c.o. *E. polygoni* etc.) and rusts (black stem rust of wheat, c.o. *Puccinia graminis* f. sp. *tritici*; brown rust of wheat, c.o. *P. recondita* etc.). List of some other air borne diseases are given below (Table 5.2):

Table 5.2: List of some air borne diseases

Name of disease	Name of pathogen
1. Early blight of potato	*Alternaria solani*
2. Leaf spot of beet	*Cercospora beticola*
3. Blast disease of rice	*Pyricularia oryzae*
4. Stripe disease of barley	*Helminthosporium gramineum*
5. Tikka disease of groundnut	*Cercospora arachidicola*
6. Fruit rot of tomato	*Phoma destructiva*
7. Blight of egg plant	*Phomopsis vexans*
8. Grey blight of tea	*Pestalotia theae*
9. Grey leaf spot of mango	*Pestalotiopsis mangiferae* (= *Pestalotia mangiferae*)
10. Scab of peach	*Cladosporium caryophylum*

Dissemination of Plant diseases

The spread of disease from one plant to other plants or from one locality to another, involves either the transport of its infective parts or the entire body is known as **dissemination.** A knowledge of dissemination is very useful to control a disease. These are of following types:

1. **Wind dissemination: Four general types** of dissemination takes place by air and wind. These are as follows:

   A. The **conidia are borne singly** or **in chain** on the apex of aerial conidiophores and are detached as well as liberated from the fruit body as a powdery mass and are subsequently carried away by air current, e.g.,

Penicillium, Aspergillus, Albugo candida, etc.

B. The **conidia are borne inside the fruit body** and are separated by some explosive mechanism which release them in the air and are subsequently carried away by the air current.

These are of **two types:**

(a) By **expulsion of the ascus**: The ascospores are forcibly ejected from ascus fruit bodies.

Again, these may be categorised into **two types:**

(i) **Successive explosion** of the spore sacs takes place from perithecium till end. Either all the ascospores shoot out at once or they are liberated in succession, e.g., apple scab fungus (*Venturia inaequalis*), powdery mildew pathogens (*Uncinula, Phyllactinia, Podosphaera, Sphaerotheca and Erysiphe*).

(ii) **Simultaneous explosion** of a number of asci takes place from apothecium, which forms visible cloud of spores ("Puffing"), e.g., *Sclerotinia* brown rot of stone fruits, *Pseudopeziza* leaf spot of clover.

(b) By the **basidium**: The forcible separation of basidiospores takes place in succession by hydrostatic pressure, e.g., rust and smut fungi.

C. The wind may help in **indirect spread** of disease by carrying the infected leaves or floral parts from one tree to another.

D. Sometimes, the wind blown rain may **indirectly** help in the transport of pathogen or its infective parts.

2. Water dissemination: Water disseminates the pathogen in **two ways:**

A. **Motile spores or organisms** may swim from one place to another in the water medium, e.g., wart disease of potato, c.o. *Synchytrium endobioticum*; club root of crucifer, c.o. *Plasmodiophora brassicae*.

B. **Non-motile spores** are transported by the flow of irrigation water or rain water:

(a) Those which have lodged on the plant by some agencies like wind, etc., and

(b) Those which are residual in origin and are embedded in mucilaginous matrix, e.g., fire blight of pome fruits.

3. Insect dissemination: Insects can disseminate disease either actively or in inactive ways. The infective bodies of fungal, bacterial and also viral diseases are disseminated by insect. It can be divided into following **five types:**

A. Infective bodies are carried externally by the insect and infection takes place without direct inoculation, e.g., late blight of potato, different rusts and smuts.

B. Infective bodies are carried externally by insect and infection takes place by transfer of inoculum through feeding, e.g., fire blight of apple, black rot of crucifers.

C. Infective bodies are not carried by insect, but infection takes place through injuries made by insect, e.g., onion blight, powdery mildew of grasses, bacterial leaf spot of cotton.

D. Infective bodies are carried internally, but mechanically i.e., without any biological relation with the insect, e.g., *Claviceps purpurea* survives in the digestive canal of some arthopods.

E. Infective bodies are carried internally and later they increase in number by multiplication inside the insect body, e.g., rice tungro virus disease.

4. Seed dissemination: Pathogens can be carried internally (*Ustilago nuda tritici*, c.o. of loose smut disease of wheat; *Ustilago hordei* c.o. of covered smut of barley) or externally (*Tilletia caries*, c.o. of bunt of wheat) or both externally and internally (*Helminthosporium oryzae* c.o. of brown spot disease of rice) and are disseminated from one locality to the other by the transfer of infected seeds.

5. **Dissemination by animals other than insects:** Pathogens can be transported to other locality by animals other than insects, such as nematodes (bacterial brown spot of potato), slugs (black rot of cabbage), sap suckers (Chestnut-tree blight) etc.
6. **Dissemination by propagating stock**: Pathogens can be transported to other locality through propagating stock and are common in cases where multiplication takes place by using propagating stocks. Propagating stock may act simply as a carrier of spores of a pathogen or with latent infection or sometimes with visible symptoms, e.g., *Fusarium* stem rot of sweet potato, Foot rot of sweet potato, virus disease of tobacco.
7. **Dissemination by shooting out of spores**: This type of dissemination is found in *Taphrina deformans*, a member of Ascomycetes; where ascospores shoot out from the asci and in contact with suitable host, it causes disease, e.g., peach leaf curl disease.
8. **Dissemination with plant debris**: Due to careless sanitary practice, infected plant debris may remain in the field and the pathogens are transferred in different places by plouging during preparation of soil before plantation. This type of dissemination is very common in late blight of potato (c.o. *Phytophthora infestans*), where the pathogen remains as oospore in the plant debris.
9. **Dissemination with manure**: Pathogens are disseminated through contaminated manures used during field preparation or at a later part of cultivation. This is very common with "damping off" of tobacco seedling, c.o. *Pythium debaryanum*.
10. **Dissemination by soil**: During field operation, contaminated soil may shift to other places through ploughing, cause infection in plants at different places in the field, e.g., club root of crucifer, c.o. *Plasmodiophora brassicae*.
11. **Dissemination by field operation**: Disease can be disseminated during field operations like watering, weeding, transplanting, e.g., tobacco mosaic virus disease.
12. **Dissemination by natural root grafting:** Disease can be disseminated during natural root grafting between healthy and diseased plants. It is found in oak wilt disease, c.o. *Endoconidiophora fagacearum*.
13. **Dissemination by export and import of commodities**: Disease can be disseminated during export and import of commodities. This may happen: (i) **with contaminated container** (ii) **with contaminated commodities,** or (iii) **both**. In this way, wart disease of potato was introduced in India. The Quarantine Law has been enforced to avoid such introduction.

Geographical distribution of diseases

Human beings have been modifying the phytogeography* since 16th century according to their need and the process has been accelerated with the increase in population. Due to increase in population of human beings, greenhouse gas concentrations in the atmosphere are being altered by human activities, and thus, causing global climate change. These activities become more intense after the **Industrial Revolution** at the end of the 18th century result from the use of natural resources such as fossil fuel burning, deforestation and other land use changes. Thus, the atmospheric concentration of carbon dioxide (CO_2) has reached levels significantly higher than in the last 650 thousand years (Siegenthaler *et al.*, 2005). Since 2000, the growth rate in CO_2 concentration is increasing more rapidly than in the previous decades (Canadell *et al.*, 2007). Similar trends have been observed for methane (CH_4), nitrous oxide (N_2O), and other greenhouse gases (Spahni *et al.*, 2005).

As per need, human beings transported entire plants or other parts like seeds, grains, root stocks etc., from one country to other by different ways like ships, vehicles etc. Along with the plant parts or plants as a whole, the pathogen may also being shifted to new other countries.

Based on the concept of disease triangle, in addition to susceptible hosts and virulent

* Phytogeography i.e. geographical distribution of plant species and their influence on earth surface.

pathogens, environmental conditions are also equally important in the successful development of disease.

The effects of climate change on plant pathogens and the diseases they cause have been examined in some pathosystems. Predicted climatic changes are expected to affect pathogen development and survival rates and modify host susceptibility, resulting changes in the impact of diseases on crops. The effects of these climatic changes will differ by pathosystem and geographical region. These changes may affect not only the optimal conditions for infection but also host specificity and mechanisms of plant infection.

Thus, the changes in the atmospheric temperature, CO_2 and ozone concentrations, precipitation, and drought, effects on the activity of pathogens and their ability to infect plants and survive in natural and agricultural environments. Changing in abiotic conditions will also affect the microclimate surrounding plants and the susceptibility of plants to infection. These changing conditions are expected to affect microbial communities in the soil and canopy pathosystems, possibly altering the currently observed beneficial effects of these communities. Because both pathogens and host plants will be affected by the changing of climate, dramatic changes in the magnitude of disease expression in a given pathosystem, the geographical distribution of particular plant diseases, the economic importance of particular diseases in a given location, and the set of diseases that challenge each crop are expected. These changes will affect the measures used by farmers to effectively control disease, as well as the feasibility of particular cropping systems in particular regions.

Diseases are one of the most important components that are having a major share in affecting agriculture. Plant diseases are the result of interaction between susceptible host, virulent pathogen and favorable environmental conditions (Garrett, 2008; Klopfenstein *et al.*, 2009; Grulke, 2011). As environment is one of the important components in disease development, changes in it can be strongly correlated with the changes in disease severity and losses due to it (Elad and Pertot, 2014). Strange and Scott (2005) estimated loss of at least 10% in global food production due to diseases. Agrios (2004) estimated that annual losses by disease cost US$220 billion. Disease control methods like chemical methods not only result in huge costs, but also increase environmental contamination threats and health issues.

Osborne (1911) pointed out that the geographic distribution of pathogens and diseases are the outcome of dynamic processes that involve:

1. Host availability, susceptibility and their abundance;
2. Availability of pathogen in active form and their survival in new climatic condition, and
3. Evolutionary changes in hosts and pathogens and suitability of climate conditions.

As environment and diseases are closely related, climate change will probably alter the geographical and temporal distribution of phytosanitary (measures for the control of plant diseases) problems. Pathogens tend to follow the host plant in its geographical distribution, but the rate at which pathogens become established in the new environment is a function of the mechanism of pathogen dispersal, suitability of the environment for dispersal, survival between seasons, and physiological and ecological changes in the host plant. Chakraborty *et al.* (2000) studied on the trend with the virulence of pathogens and their tendency of migration. The more aggressive strains of pathogen with broad host range, such as *Rhizoctonia, Sclerotinia, Sclerotium* and other necrotrophic pathogens can migrate from agroecosystems to natural vegetation, and less aggressive pathogens from natural plant communities can start causing damage in monocultures of nearby regions. In unspecialized necrotrophs, the range of hosts can be extended due to crop migration.

At the regional level of the country, the distribution of the diseases is promoted by wind direction and also by the availability and distribution of host plants. In addition, the availability of collateral hosts and volunteer plants also help in survival of pathogen that helps in disease

Table 5.3: Plant disease epidemic due to climate change

Plant disease epidemic	Climate change	Author
1. Downy mildew of grape vines (c.o. *Plasmopara viticola*), in north-west Italy	Increase in precipitation and temperature.	Salinari *et al.* (2006)
2. Swiss needle cast disease (c.o. *Phaeocryptopus gaeumannii*) in Oregon coasts	Increase in temperature during *winter and duration of leaf wet*ness during spring and autumn.	Stone *et al.* (2008)
3. Alternaria epidemic of apple (c.o. *Alternaria mali*) in Kashmir, India.	Prolonged rains followed by high temperature.	Bhat *et al.* (2015)

occurrence. The disease becomes endemic to a particular locality due to favourable climatic conditions in respect of pathogen. Due to this, pathogen develop from the perennating organ or comes from outsource through wind, survives well and cause diseases. In more favourable condition in large areas the disease may also appear as epidemic form. Some examples of epidemic diseases are given above (Table 5.3):

Briefly it can be mentioned that, the geographical distribution of plant diseases and their intensity of occurrence, depends on three factors at various levels. These are:

1. ***Host factors***: These include susceptibility, aggregation and distribution of hosts; introduction of new hosts, introduction and/ availability of collateral and alternate hosts.
2. ***Pathogen factors***: These include introduction of pathogen or presence of virulent strains of pathogen, with their high production rate, low death rate, rapid dispersal and high adaptability.
3. **Environmental factors**: These include favourable conditions in respect of temperature, rainfall, humidity, intensity and duration of light etc.

Epidemiology and Disease forecasting

Epidemiology

The term epidemic (Gr. *epi* — upon; *demos* — people) means upon people, and epidemiology is the science which deals with all the factors related to plant and pathogen involved in severe disease development, subject to the influence of the existing environmental factors. The knowledge of plant disease epidemics is of great importance in countries based on agriculture. In addition to low irrigation and deficiency of nutrition, plants also suffer from severe disease of a particular crop, sometimes collapse the associated industry and also greatly hamper the economy. Such damages were noticed by diseases like coffee rust in Sri Lanka, downy mildew of grapes in France, yellow rust of wheat in Pakistan etc. There were incidences where all the favourable factors developed synchronously, thereby disease appeared as an unpredictable magnitude lead to national disaster and caused the death of millions. The most remarkable among them are the Irish Famine (1845–46) caused by late blight of potato (c.o. *Phytophthora infestans*) and Bengal famine (1943) caused by the outbreak of brown spot of rice (c.o. *Helminthosporium oryzae*). Van der Plank (1973) provided a unique conceptual basis for the quantitative study of epidemic development and suggested the use of mathematical models to describe, interpret and predict the progress of an epidemic plant disease.

The four components of a disease pyramid are host, pathogen, environment, and time. All the factors are important and influence each other in disease development. Simultaneous occurrence of host, pathogen and favourable environmental condition are necessary to develop disease and this is found to be effective in monocyclic process. But in case of polycyclic process, time is also an important factor in disease progress. The time affects the above three factors individually, thereby the shape of the pyramid becomes altered. According to many others, **man functions** as the fifth component, and also influences in the development of diseases, particularly in crop plants. The fifth component has the capacity to control:

(a) Selection of host cultivar.

(b) Rate and time of irrigation.

(c) Amount of fertilizer.

(d) Chemical used to protect the plant from disease etc.

The above factors greatly influence the development of disease. Thus man has the capacity to reduce the disease incidence even with the presence of other four components in favourable condition. Thus man also becomes an important component.

Disease forecasting

Forecasting is the scientific way of prediction of the possible appearance of disease. It is a component of epidemiology of plant disease. The plant disease varies from season to season, both in incidence and severity due to variation of environmental and some other factors. Based on epidemiological knowledge, disease forecasting help to predict the possible outbreaks or increase of intensity of disease and, therefore, allows the farmers to determine whether, when, where and what type of management practice should be undertaken. Due to ignorance or misprediction, the farmers may apply pesticides in the field which are of no use and thus the costs of production will unnecessarily increase. The appropriate forecasting will help to determine when a disease is likely to appear so that this can decide the appropriate time of spray of pesticide. Thus this can reduce the unnecessary increase of cost of cultivation due to misuse of pesticide and also may predict use of pesticides at proper time to protect the crop and thus both farmers and the country may get the benefit.

The methods used in disease forecasting are mainly based on the following aspects:

1. Inter-crop weather and the intensity of primary inoculum

Weather during inter-crop season may help the pathogen to survive either in nature or in vector. Survival of pathogen during winter months helps the pathogen to gain success in the next crop. So many examples may be cited in this context:

(a) *Survival of more vectors*: In Stewart's wilt of corn (c.o. *Erwinia stewarti*), the pathogen survives in the winter within the body of vector, the flea battle. Assessment of vector population at the onset of spring gives a clear indication of the extent of survival of vectors during the past colder months. If the sum of mean temperature from December-February is lower than −1 °C, most vectors die, thereby the survival of pathogen is not expected. But if the winter becomes mild, it will help the vectors to survive and severe outbreak of the diseases is expected.

(b) *Survival of pathogen in resting structure(s)*: Pathogens like *Claviceps purpurea* (ergot disease of rye), *Venturea inaequalis* (apple sacb), *Monilinia fructicola* (brown rot of stone fruits), etc., may overwinter by developing resistant structures, the perfect stages. If the propagules are present in more numbers in the field to be cultivated, it may indicate the coming outbreak and the possible severity of the disease.

The soil borne pathogens, such as *Sclerotium, Verticillium* and nematode like *Heterodera* etc., can survive in the soil as sclerotia and cysts, respectively. Inoculum and the disease are directly proportionate i.e., greater the amount of propagules, more severity in the intensity of the disease. This can easily be predicted after assessment of the number of propagules per gram of soil. However, the severity of disease may vary with the fluctuation of environmental conditions during cropping season.

2. Weather condition during growing season and the production of secondary inoculum

The temperature and moisture level play a great role in the development and spread of some air borne diseases. This is highly significant for many compound interest diseases of polycyclic nature (blast disease of rice, c.o. *Pyricularia oryzae*, tikka disease of ground nut, c.o. *Cercospora arachidicola* etc.). The availability of certain environmental conditions for a minimum period may help to build up huge inoculum pressure through completion of many disease cycles resulting in severe outbreak of the disease. This can be predicted by trapping the spores daily over the cultivated field and by measuring the associated factors like favourable temperature and high humidity.

3. Disease in young crop

Spread of the disease in young crop may sometimes indicate the possible severity of the disease in the mature crop. The wheat seedling gets infected from leaf rust caused by *Puccinia recondita* in USA by overwintering local infections. The intensity of infection in the crop depends on the severity of winter and survival of pathogen in seedling. The amount of infection at the onset of spring and its periodical assessment and forecasting help the farmers to take proper measures to control the disease at different degree.

4. Amount of inoculum in planting material

The inoculum, if present in seeds and planting material, may cause disease in plants with high intensity at the onset of favourable condition in the cropping season. Thus the seed lot or the propagating material should be checked in the laboratory and the farmers warned not to cultivate such seeds or propagating stock of that locality. The inoculum of loose smut of wheat pathogen, *Ustilago segetum* var. *tritici,* remains in the embryo out of season. This can be tested by dipping the seeds in 10% KOH overnight, in which the embryo becomes soft and gets separated. The separated embryos are then tested under microscope for the possible presence of the pathogen. Based on the above result, appropriate forecasting can be done regarding the selection of seed lot by the farmers.

Thus the disease forecasting at appropriate time will help not only the farmers but also the country as a whole to check the economic loss from disease(s).

Diagnosis of plant diseases

For the diagnosis of a plant disease, it is essential to determine the cause i.e. whether it is caused by biotic or abiotic factors. In some cases, by observing the typical sign or symptom, an experienced person can determine whether the disease is caused by environmental factors (abiotic) or by pathogens (biotic). Sometimes, detailed examination of symptoms and other character become essential to diagnose a disease. After proper study of the disease, one has to go through the books having list of known diseases.

Diseases caused by the biotic agents

The biotic agents include parasitic angiosperms, nematodes, fungi, bacteria, mollicutes, protozoa etc. The agents, i.e., the pathogens, cause different types of diseases. The pathogen may be present on and/inside the host tissue. Observing the superficial pathogen (either with hand lens or with naked eye), experienced persons are able to identify the disease. But, others can identify the disease after proper microscopic examination of both external and/internal pathogen. Internally, pathogen may present adjacent to infected region in the vascular tissue, at the basal region or in the root of the plant. The pattern of appearance of disease varies with the pathogen.

1. Parasitic angiosperms

The presence of parasitic angiosperms like dodder, witchweed, mistletoe etc., on a plant reduces the growth of host plant, thus clearly indicating the disease caused by parasitic angiosperms.

2. Nematodes

If a parasitic nematode is present in or on the plant and also in rhizosphere region, it shows a few symptoms indicating the nematode as a pathogen or its involvement in disease manifestation. If the nematode is identified and was reported earlier to cause such a disease, then the diagnosis can be done very easily.

3. Fungi

If any fungus is present in or on a diseased plant, it may be a pathogen or a saprophyte.

Thus one has to study properly the nature of mycelium, character of spore and fruit bodies. After proper identification with the help of appropriate literature, one can confirm that the disease is caused by particular pathogen. If the report on the pathogen and its corresponding disease is not available, then one has to confirm about its pathogenecity, following Koch's rules of proof. In the early stage of infection, spores or any fruit body may not be available. Thus it needs to grow the fungus in selective medium for sporulation, that will help to confirm the pathogen. The sporulation can also be done by keeping the infected tissue in moist chamber made up of plastic or glass.

4. **Bacteria**

To diagnose bacterial disease and to identify the causal organism one has to study the symptom and the bacterium present around the infected region. To confirm the bacterium as a pathogen, one has to isolate and grow the bacterium in selective medium for development of pure culture. A single colony of the pure culture is inoculated to a susceptible host to develop symptom. This is compared with the symptom produced by the known species of bacterium.

Because of unicellular structure and small size (0.8 × 1-2 μm), it becomes difficult to differentiate bacteria of identical morphological nature by observing under the microscope. Due to this, several techniques have been adopted for their identification. The immunodiagnostic techniques such as agglutination and precipitation, fluorescent antibody staining and enzyme linked immunosorbant assay have been used to identify the pathogenic bacteria. The bacteria can also be identified by analysing the fatty acid profile and also by comparing the number of DNA pieces released by certain restriction enzymes or % of hybridization of DNA of an unknown bacterium with any known one.

5. **Mollicutes**

These are small, wall-less and polymorphic bacteria grow in young phloem of infected plants. They do not grow on nutrient medium (except *Spiroplasma*) and are commonly visible under electron microscope. They develop the symptoms like proliferation of root and shoot, stunting of plants, yellowing of leaves, development of abnormal flower leading to the death of host plant.

The diagnosis of such disease is based on:

(a) Symptomatology.

(b) Electron microscopy.

(c) Graft transmissibility.

(d) Sensitivity to tetracycline and moderately high temperature (32-35°C).

6. **Viruses and Viroids**

Some viruses cause specific symptoms in their host and due to this the disease and its associated virus can easily be identified by studying the symptoms. In others, where the above procedure is not applicable, the diseases are diagnosed and the associated viruses are identified by the following procedures:

(a) By transmission or sap inoculation, or by grafting, or often by vectors like insects, fungus, nematodes, mites etc.

(b) By using specific antisera. The viruses can be identified by:

 (i) ELISA (enzyme-linked immunosorbent assay).

 (ii) Microprecipitin test.

 (iii) Gel diffusion test etc.

(c) By electron microscopy technique.

7. **Involvement of more than one pathogen**

Sometimes, more than one pathogen become available in the infected region, of which one is the pathogen and the other one may be either pathogen or saprophyte. The above fact needs confirmation through the different procedures for the different groups of pathogens. The common example in this respect is the Tungro disease of rice caused by two viruses:

(a) Rice tungro bacilliform virus (RTBV), and

(b) Rice tungro spherical virus (RTSV).

Disease caused by the abiotic agents

When pathogen(s) are not found following either of the above procedures, then it must be assumed that the disease is caused by abiotic agent (s). Unlimited number of diseases has so far been reported to cause by different environmental factors by hampering the normal physiology of the plant. The interference of the physiology of plants may result due to many reasons and some of them are:

(a) Accumulation of toxic substance in soil and air,

(b) Increase or decrease of other factors like water, humidity, O_2, CO_2, temperature, mineral nutrients, light etc., and

(c) The pollutants released by the machines used during cultivation and/or for other purposes.

Common examples of such diseases are:

(i) Lightning injury of potato.

(ii) Potassium deficiency of cabbage, tobacco, tomato causes mottle and chlorosis of older leaves, particularly near the margin.

(iii) Calcium deficiency causes initially the reduction in size of leaf and shoot followed by chlorosis and necrosis of leaf tissue.

(iv) Blossom-end rot of tomato caused by high moisture in the soil.

(v) Wilting of potato, cabbage etc., may be caused due to flooding injury, and

(vi) Water core of apple is caused due to high temperature.

Plant diseases and Crop loss

Plant diseases may be caused by both abiotic and biotic factors. The abiotic factors i.e. the different environmental factors like water, temperature, irradiation etc., and biotic factors like pathogens belongs to fungi, chromistans, bacteria, viruses, nematodes etc. The diseases caused by abiotic factors are called **non-parasitic diseases** (freezing injury of potato, water core of apple-caused by high temperature, etc.) and by biotic factors are called **parasitic diseases.** The diseases caused by different biotic factors are fungi (*Helminthosporium oryzae* – brown spot of rice), chromistans (*Phytophthora infestans* – late blight of potato), bacteria (*Xanthomonas oryzae* – bacterial leaf blight of rice), viruses (tobacco mosaic virus i.e. TMV – tobacco mosaic virus disease), nematodes (*Heterodera rostochiensis* – golden nematode disease of potato) etc.

The **crop losses** may cause by abiotic (water, temperature, irradiation and nutrients) and biotic factors (weeds, animal pests and pathogens). These are mentioned briefly in Table 5.4.

Although on the basis of global scale the weeds caused major crop loss, but major losses in agricultural crops takes place due to insect damage and plant diseases. On and around the losses in crops have been estimated as follows:

1. Losses due to animal pests, 18%;
2. Microbial diseases, 16% (and 70-80% of these losses were caused by fungi);
3. Weeds, 34%;

Table 5.4: Different causes of crop losses

Causes of crop losses

Abiotic factors
1. Water
2. Temperature
3. Irradiation
4. Nutrients

Biotic factors

Weeds	Animal pests	Pathogens
1. Monocots	1. Insects	1. Fungi
2. Dicots	2. Mites	2. Chromistans
3. Parasitic weeds	3. Rodents	3. Bacteria
4. Slugs/Snails	4. Viruses etc.	
5. Birds		
6. Other animals		

4. Making a grand total of 68% average annual loss of crop production in tons (modified after Oerke, 2006).

Although plant diseases are caused by different pathogens like fungi, chromistians, bacteria, viruses, but plant pathogenic fungi are more than bacteria or viruses. From middle half of 20th century and presently, control of plant disease relies heavily on synthetic chemicals (pesticide) to increase the crop productivity. Unfortunately, despite large increases in the use of pesticide, crop losses have not significantly decreased during the last 40 years.

Diseases reduce crop productivity in various ways, examplified by:

1. **Reducing the stand** (i.e. population) of plants (pathogens that kill the host-like damping-off pathogens that kill seedlings);
2. **Reducing photosynthetic rate** (fungal, bacterial, virus diseases);
3. **Accelerating plant senescence** (most pathogens);
4. **Depleting assimilate** (nematode pathogens);
5. **Consuming tissue** (necrotrophic pathogens);

Crop losses can be quantitative and/or qualitative, and expressed in absolute terms (kg ha-1, or financial loss ha-1, for example) or in relative terms (% loss in production of tons):

1. **Quantitative losses** result from reduced productivity giving a lesser yield per unit area;
2. **Qualitative losses** from pathogens can result from:
   (i) Reduced content of a normal ingredient(s) of the crop,
   (ii) Reduced market quality (for example miss-shaped, blemished fruit and vegetables),
   (iii) Reduced storage quality,
   (iv) Contamination of the harvested product with pathogens, parts of pathogens or toxic products of the pathogens (for example, mycotoxins).

Agricultural **survey with statistics** analysis makes it clear that crop losses directly attributable to fungi are very considerable. Of course, it is changing all the time because, most of the cases, losses depend on the weather, but it appears that world agriculture sustains average losses (monetary value) of around 16% annually as a result of plant diseases. This overall average shows least and heavy disease ranges from 1-2 to 30-40 percent. Among crops, the total global **potential loss** due to all diseases varied from about 50% in wheat to more than 80% in cotton production. Other estimated **actual losses** are 26-29% for soybean, wheat and cotton, and 31, 37 and 40% for maize, rice and potatoes, respectively (Oerke, 2006).

Most important plant diseases caused heavy losses with remarkable social impacts are given below:

1. **Late blight of potato** (c.o. *Phytophthora infestans*). The disease appears in Ireland in severe form between 1840-1847 and caused major damage in potato. By 1845, the disease totally devastated the potato crop of U.S.A., Canada and Europe, mainly in Ireland. As a result, **Irish famine** occurred that caused the **death of about one million people** and more peoples migrated to other countries.
2. **Coffee rust** (c.o. *Hemileia vastatrix*). The disease appears in severe form during early 1870s, which almost completely destroyed the coffee cultivation in Sri Lanka (erstwhile Ceylon). Due to this, coffee cultivation was shifted to Westen Hemisphere, where pathogen was not available. Later, Coffee was replaced by Rubber and Tea cultivation along with associated industries.
3. **Downy mildew of grapes** (c.o. *Plasmopara viticola*). The disease caused severe damage in France during 1878-1882. The wine industry suffered a lot due to the disease.
4. **Red rot of sugarcane** (c.o. *Glomerella tucumanensis*). The disease becomes so severe in Indian states like Bihar, Punjab and Uttar Pradesh during 1938-1939; that the production was highly reduced resulting in about 70 percent reduction of the sugar production. Both farmers and associated mills became highly affected.

5. **Brown spot of rice** (c.o. *Helminthosporium oryzae*). The disease appeared as epidemic in Bengal (undivided Bengal, India) in 1942 and caused about 90 percent crop loss (Ghosh *et al*.,1960); thereby **Bengal famine** took place in **1943**. The disease was so severe that about **five million people** died of starvation.

Prevention of Plant diseases

Prevention i.e. avoidance or blockage, indicate "an action taken to stop something from happening". As with all ailments, whether human or plant, prevention is a key. Prevention is much easier than control and it is less expensive. This is true for our own health, as well as our field health. Just as eating in right way and getting regular check-ups is vital to prevent health issues, similarly providing a healthy environment for the plants is also essential for preventing from diseases and maintains good health.

Prevention is the best remedy when it comes to dealing with most plant diseases. Many diseases cannot be effectively controlled once symptoms develop or become severe. For effective prevention following things have to be maintained:

(i) Maintain a healthy field environment through proper maintenance,

(ii) Keep pruning equipment clean and in good repair, and

(iii) Do not over-fertilize, and because fungal diseases love wet environments, always watering in the morning so foliage get the time to dry before nightfall.

If you do these things consistently and take time to observe your plants regularly, chances are good that you will have very few diseases to contend with.

Some of the ways in preventing diseases are given below:

1. **Sanitation practices**

Sanitation is the process of cleaning of infected plants and plant residues from the field to reduce the source of primary inoculum and ultimately the disease. This is important for all parts of the crop field before new crop to grow. Good sanitation includes picking up of plant debris, and also keeping weeds to a minimum number. Foliage or stems leftover from diseased plants can result the same diseases for the next year. The plant debris should be burnt in outside field. Several diseases like powdery mildew of wheat, powdery and downy mildew of pea etc., can be successfully controlled by this method.

2. **Crop rotation**

The amount of inoculum in soil gradually increases by growing the same or similar crops, for many years in the same field. Soil borne diseases can be controlled by crop rotation method which helps in reducing the amount of inoculum at significant level. Soil borne pathogen with narrow host range can be controlled by this method. It is helpful to know about the different crop families, and how to rotate them in your field. Wheat bunt, rice bunt, wilt of pigeon pea and linseed can be controlled by crop rotation.

3. **Fertilize at proper dose to keep the plants healthy**

Fertilize the field at recommended dose to keep plants healthy, and not more. Over-fertilizing often leads to more problems, because plants will put out lots of weak, fresh growth that will attract the pests and diseases. The recommended amount of organic fertilizer or regular applications of compost or composted manure will help the plants to stay healthy. Healthy plants are better able to defend from diseases.

4. **Inspect plants regularly for diseases**

This is an easy way to keep the diseases out of the field and make sure that you are not bringing them with new plants. Before bringing the plants from the nursery, check to make sure, that they are healthy. If you see signs of fungal diseases or insects, or lots of yellowing or wilting foliage, do not select the plants to cultivate. In addition, observe regularly if any disease appears with the new plants. Take necessary action as per recommendation.

5. **Watering in the morning**

This is one of those old and reliable bits of advice that just makes sense about the control of the pathogen. Many fungal diseases need damp, cool environments to infect the host. So if the plants' foliage is wet overnight, that gives these diseases a chance to make a good start in the field. The easiest way to prevent this is to watering as early in the day as possible, so the plants surface can dry off before nightfall.

6. **Mulching**

Mulches are useful for maintaining soil moisture and keeping weeds down, but they are also very helpful to keep the field disease-free. It can be done with polythene sheet etc. Mulch prevents soil that is infested with soil-borne fungi from splashing up onto the plants' foliage. In rose field mulching will prevent the black spot of roses at a significant level.

7. **Provide good air circulation**

Fungal diseases like powdery mildews and black spots are more likely to be a problem if plants have poor air circulation. If they are planted too close enough airflow does not takes place. This stagnant environment is favourable for many fungal diseases. It is better to pruning out the excess branches, that provide better air circulation, thereby spread of disease becomes less; due to reduction of secondary inoculum production.

8. **Watchout for insect pests**

Insect pests, with their chewing and burrowing habit, are aggravating enough to deal with disease spread. But many of them, such as aphids, are a double-threat because they transmit diseases between plants. Regular observation and elimination of insect pests by physical and chemical means caused less damage of the crop.

Control of plant diseases

Plant diseases cause enormous damage to the plant population including important crop plants, which, at different times, snatch the food from the mouth of hungry human population and hamper medicine production causing sufferings to the human beings. In continuation of the above facts, we can exemplify the late blight disease of potato, that devastated the potato cultivation in European countries, especially in Ireland during 1840-1845 resulting in the **Irish famine** in **1845-1846**, which caused the **death of about one million people** due to scarcity of food and more than one million people migrated to other countries. Similarly, brown spot of rice devastated the rice cultivation of Bengal in 1942, caused about 90% crop loss (Ghosh *et al.*, 1960). The famous **Bengal famine** in **1943** was caused due to the destruction of rice crop by the fungi, resulting in the **death of about five million people** due to starvation. In the above cases, both potato and rice were the principal food in both the areas. So many cases can be exemplified in this regard. The above conditions threaten us not to neglect but to give more attention to control the disease of plants for the benefit of increasing human population.

To control a disease, it becomes necessary to know the cause, the name and life history of the causal organism including its process of infection and appearance of disease, etiology etc.

Various control measures of plant diseases are discussed under the following heads:

A. Prophylaxis

- I. Exclusion (legislation)
  - 1. Quarantine
  - 2. Inspection and certification
  - 3. Avoidance of pathogen
  - 4. Use of pathogen-free propagating material
  - 5. Use of epidermal coating
- II. Eradication or reduction of pathogen
  - 1. Cultural methods
    - (i) Eradication
      - (a) Plant parts
      - (b) Roguing
      - (c) Alternate hosts
      - (d) Collateral hosts
    - (ii) Crop rotation
    - (iii) Sanitation

2. Physical methods
   (i) Surgery
   (ii) Water treatment
   (iii) Hot air
   (iv) Low temperature
   v() Radiation

III. Direct protection

1. Environmental manipulation
   (i) Modification of agronomic practice
   (ii) Avoiding any injury of plants
   (iii) Biological methods
      (a) By antagonists of pathogen
      (b) Hyperparasitism
      (c) Control through trap crops
      (d) Control through antagonistic plants
      (e) Breeding for resistant varieties

2. Chemical methods
   (i) Copper compounds
   (ii) Sulphur compounds
   (iii) Heterocyclic compounds
   i(v) Quinones
   (v) Benzene compounds
   (vi) Systemic fungicides
   (vii) Antibiotics
   (viii) Growth regulators
   (ix) Soil fumigants

B. Immunization

1. Cross-protection
2. Induced resistance
3. Transgenic plants

A. Prophylaxis or **Prevention**

Prophylaxis is a preventive measure which includes the protection of the host from exposure to pathogen from infection or from the surrounding environmental factors that favours disease development. It includes a wide range of control measures which can be divided into three sub groups: **Exclusion, Eradication,** and **Direct protection**:

I. Exclusion (legislation): Exclusion is the process to keep the pathogen away from the host. This measure includes **quarantine, inspection** and **certification** etc.:

1. **Quarantine**. Quarantine is the legal restriction on export and import of propagating materials like seeds, roots, tubers etc., to prevent the spread of disease at international level. Similar measures are sometimes taken at inter-state level in the larger countries. Quarantine laws were first introduced in the USA in 1912, as Federal Quarantine Acts. **In India**, it was introduced as **Destructive Insects and Pests Act (DIP Act) in 1914** and it has been amended from time to time. The latest amendment was published in 2003. These are:

   (i) To prevent the introduction and spread of exotic pests that are destructive to the country by regulating the import of plants and plant products through adequate policy and statuary measures.

   (ii) To support India's agricultural exports through credible export certification.

   (iii) To facilitate safe global trade in agriculture by assisting producers, exporters and importers and by providing technically comprehensive and credible phytosanitary certification.

   The plant quarantine stations are:

   (i) National Plant Quarantine Station (NPQS), Rangapuri, New Delhi.

   (ii) Regional Plant Quarantine Stations: Amritsar, Chennai, Kolkata and Mumbai.

   (iii) 75 Notified points of entry at various Seaports (42), Airports (19) and Land frontiers (14)

   (iv) 65 Inland Container Depots and 11 Foreign Post Offices.

Table 5.5: List of some plant diseases introduced in India from other countries

Plant disease	Introduced from	Year of introduction
Before DIP Act (i.e., before 1914)		
1. Leaf rust of coffee (*Hemileia vestatrix*)	Sri Lanka	1879
2. Late blight of potato (*Phytophthora infestans*)	Europe	1883
3. Flag smut of wheat (*Urocystis tritici*)	Australia	1906
4. Downy mildew of grapes (*Plasmopara viticola*)	Europe	1910
5. Downy mildew of cucurbits (*Erysiphe cichoracearum*)	Sri Lanka	1910
6. Downy mildew of maize (*Sclerospora phillipinensis*)	Java	1912
After DIP Act (i.e., on or after 1914)		
1. Blast disease of rice (*Pyricularia grisea*)	South-East Asia	1918
2. Powdery mildew of rubber (*Oidium heveae*)	Malaya	1938
3. Black shank of tobacco (*Phytophthora nicotianae*)	Dutch	1938
4. Crown gall of apple (*Agrobacterium tumefaciens*)	England	1940
5. Bunchy top of banana (*Banana virus 1 & Musa virus 1*)	Sri Lanka	1940
6. Wart disease of potato (*Synchytrium endobioticum*)	Netherlands	1953
7. Onion smut (*Uromyces cepulae*)	Europe	1958
8. Golden nematode of potato (*Heterodera rostochiensis*)	Europe	1961

The centres allow the entry of propagating material when it is accompanied by the certificate of the Ministry of Agriculture of exporting country, that the propagating material is free from any disease. However, the imported materials should be thoroughly checked by the experts at the entry point. Before and even after the introduction of DIP Act in 1914; several plant diseases were introduced in India at different times (Table 5.5). Among many others, **late blight of potato *from* Europe; wart disease of potato *from* Netherlands** and **leaf rust of coffee *from* Sri Lanka** are most remarkable. **Domestic quarantine** is applicable in India for wart disease of potato, by which transportation of potato tuber from Darjeeling to other parts is totally prohibited.

2. **Inspection and certification**. Compulsory inspection and certification system of living plants, propagating stocks, etc., have been introduced in controlling the spread of the disease in many agriculturally advanced countries. Within 15 days of shipment, the plant material should be thoroughly examined and provided with a certificate mentioning data that the plant materials are found to be healthy. Many voluntary inspection and certification systems are also in effect in many areas.

3. **Avoidance of pathogen**. Some diseases can be controlled by avoiding the pathogen. Some seed-borne pathogens like *Colletorichum lindimuthianum* (bean anthracnose) and *Pseudomonas phaseolicola* (bacterial blight of bean) are unable to cause disease at low humidity. So the seeds for the next growing season should be collected from the plants growing in dry regions.

4. **Use of pathogen-free propagating materials**. Pathogen free propagating materials like tubers, bulbs, seeds etc., are usually obtained by growing the crops in an area free from pathogen or not suitable for the vector of the pathogen (for viral disease).

5. **Use of epidermal coating**. Epidermal coating of plant surface with chemicals is useful in exclusion of pathogens. Water emulsion of dodecyl alcohol, when sprayed on plants like rice, cucumber etc., forms a coating which persist for about 15 days and protects the plant from stem blights, powdery mildew, etc.

II. Eradication or reduction of pathogen: Eradication of the pathogen can be done by cultural and physical methods:

1. **Cultural methods**. This includes different practices taken by man to control the different plant diseases. These methods are aiming either to eliminate the pathogen from host or creating conditions unfavourable to the pathogen or to increase the vigour of the host. The different methods are **Eradication, Crop rotation, Sanitation** etc.

(i) **Eradication**. Eradication means complete destruction of pathogen. It can be done by partial or complete destruction of host, destruction of alternate and collateral hosts. etc. Eradication should be done at regular intervals in the field, greenhouses and also in nurseries.

(a) ***Plant parts***. Host plants infected with pathogen or suspected to carry pathogen should be **removed partially** or **entirely** and should be **burnt**. Citrus canker (c.o. *Xanthomonas axonopodis* pv. *citri*) can be controlled by eradication of infected plant parts. This process is equally effective in other canker diseases.

(b) ***Roguing***. Removal and destruction of affected plants from the fields i.e., **Roguing** at an early stage, prevents wide dissemination of the pathogen. Roguing is an effective practice in controlling the virus diseases like yellow vein mosaic of lady's finger (*Hibiscus esculentus*), cattle disease of cardamom etc. The loose smut of wheat (c.o. *Ustilago segetum var. tritici*), covered smut of barley (c.o. *U. hordei*) are being effectively controlled by this process.

(c) ***Alternate host***. For heteroecious pathogen, **eradication of alternate host** is very effective to control the disease. Eradication of barberry is effective in controlling black stem rust of wheat (but is not successful in India).

(d) ***Collateral hosts***. Similar effect is found through eradication of colateral wild hosts and weeds. Destruction of wild collateral hosts, like *Hibiscus tetraphylus* and *Croton bonplandianum* is found to be very effective in controlling yellow vein mosaic of lady's finger. Similarly, sugarcane mosaic can be controlled by complete removal of maize cultivation from the neighbouring field, because the latter helps in perpetuation of virus.

(ii) **Crop rotation**. The amount of inoculum in soil gradually increases by growing the same or similar type of crops, for many years in the same field. Soil borne diseases can be controlled by crop rotation method which helps in reducing the amount of inoculum at significant level. Soil borne pathogen with narrow host range can be controlled by this method. Wheat bunt, rice bunt, wilt of pigeon pea and linseed can be controlled by crop rotation.

(iii) **Sanitation**. Sanitation is the process of cleaning of infected plants and plant residues from the field to reduce the source of primary inoculum and ultimately the disease.

The plant debris should be burnt in outside field. Several diseases like powdery mildew of wheat, powdery and downy mildew of pea etc., can be successfully controlled by this method.

2. **Physical methods**. The most commonly used physical methods are **Surgery, Water treatment, Hot air, Low temperature, Moisture modification, Radiation**, etc.

(i) **Surgery**. It is the process of removal of infected tissue to prevent further damage, applicable for localized disease. The cut end should be covered by thin layer of Bordeaux mixture or **Boroline** (commercial) or **Nail-polish** (commercial) to prevent further infection.

(ii) **Water treatment**. Water can be used in different ways: **hot water, water at room temperature** and **steam** to control the diseases.

Hot water. It is used to control the perennating pathogen in seeds, bulbs, soil etc. Seeds treated with hot water at 54°C for 10 minutes are effective to control brown spot of rice, loose smut of wheat etc. Propagating stocks, bulbs, etc., become free from nematodes that are present inside them by treatment with hot water. Citrus root stocks and bulbs of many ornamental plants become free from *Radopholus smiles* and *Ditylenchus dipsaci*, respectively, by hot water treatment. Application of hot water (85°C or above) in soil for 30 minutes or more kills most soil borne pathogens.

Water at room temperature. Soaking the infected seeds of wheat and barley in water at room temperature for 56-64 hours effectively control the smut diseases, like loose smut of wheat and covered smut of barley, respectively (Tyner, 1953). 100% control of loose smut of wheat is possible by 48 hours water (irrespective of the quality and pH) soaking treatment with seeds having about 60% infection.

Steam. Soil can be sterilized by steam under pressure in a container or passing steam through the perforated pipe present under the soil in greenhouses and nursery seed beds for 30 minutes. This procedure is effective in controlling nematode diseases.

(iii) **Hot air**. Hot air is used for curing the post-harvest diseases. The soft rot of sweet potato (*Rhizopus stolonifer*) can be controlled by exposure of tuberous roots at 28–32°C for two weeks that helps in healing of wounds and prevents infection. Similarly, the black chaff (*Xanthomonas campestris* pv. *translucens*) can be controlled by dry heat treatment of barley seeds at 72°C for 7-10 days.

(iv) **Low temperature**. Possibly this is the most effective and widely used method to control many post-harvest diseases of soft and fleshy plant products. Low temperature effectively control the spread of the existing pathogen and thereby reduces the disease incidence.

(v) **Radiation**. UV-light, X-ray and γ-rays have been used to control post-harvest diseases of many fruits and vegetables by killing the pathogen. Post-harvest diseases of tomato, strawberry etc., can be controlled by application of γ-ray.

III. Direct protection

1. **Environmental manipulation**

(i) **Modification of agronomic practices.** Control of disease by modifying the agronomic practices can be done in two ways: **Creating conditions to increase vigour of the host** and **creating conditions to discourage the pathogen**:

Creating conditions to increase vigour of the host. Improved cultural practices often help to increase resistance against pathogen by increasing host vigour. Thus proper manuring, proper date and depth of sowing, proper irrigation and drainage, avoiding injury to plants during weeding etc., help to increase host vigour. Application of proper dose of nitrogen helps in reduction of brown spot (*Helminthosporium oryzae*) and blast (*Pyricularia oryzae*) disease of rice. Nitrogen fertilizer increases the susceptibility of wheat crop to rust diseases, but potassium reduces it. Reduction of nitrogen in the NPK fertiliser helps in reducing the rust incidence of wheat.

Creating conditions to discourage the pathogen. Conditions which discourage the pathogen include (a) **proper spacing** to discourage the stem rot pathogen, *Macrophomina phaseoli* to Jute plant; (b) **proper soil drainage** to discourage damping off pathogen, *Pythium* to chilli, tomato etc; (c) **change of planting season** from June-July to September-October of sugarcane discourage the leaf rust pathogen *Puccinia sacchari*. (d) **Mixed cropping** of moth (crop of Family Leguminosae) between the rows of cotton plant reduces the root rot incidence of cotton (*Rhizoctonia bataticola*) by reducing the soil temperature. Proper aeration of stored products, flooding or drying for long time, soil amendments to change soil pH etc., also discourage pathogens.

(ii) **Avoiding any injury of plants**. Injury of plants should be avoided during field operations, like watering, weeding, transplanting etc. The tobacco mosaic virus disease can be transmitted very easily with minimum injury. Bacterial leaf blight of rice pathogen (*Xanthomonas campestris* pv. *oryzae*) can easily be transmitted through cut ends of the apical portion of the leaves which is the normal practice adopted in many places during transplantation.

(iii) **Biological methods**. The disease can be controlled by different types of biological methods: **by antagonists of pathogen, hyperparasitism, control through trap crops, control through antagonistic plants** and **breeding for resistant varieties.**

(a) ***By antagonists of pathogen***. Soil with pure culture or with antagonistic species of *Streptomyces* is called **suppressive soil**. Soil containing different pathogens such as *Phytophthora cinnamoni* (root rot of several fruit trees), *Pythium debaryanum* (damping off), *Fusarium oxysporum* (vascular wilt) etc., are called **conductive soil**.

The cut ends of the pine stumps are infected by *Heterobasidion annosus* (= *Fomes annosus*) and then spread into roots of the same plant as well as to the roots of the standing plants and thereby causes huge economic loss. This can be protected by the application of oidia of *Peniophora gigantia* at the cut end immediately after the felling of the trees, either as a water suspension or as powder. The *P. gigantia* successfully competes with *H. annosus* and protects the standing plants. So, many examples can be cited in this regard. *Trichoderma harzianum* and *Gliocladium virens* are used in soil or as seed priming treatment, proved to be effective against damping off pathogens like *Pythium, Sclerotium* etc. Similarly, *Pythium nunn* and *P. oligodandrum*, protect vegetables from pathogenic species of *Pythium*.

(b) ***Hyperparasitism***. This is the process of controlling the pathogenic microbes with the help of other microbes or viruses. The pathogenic microbes parasitise or antagonize the pathogen. Bacteriophage, microparasites and nematophagous fungi are the pathogenic microorganisms which take part in hyperparasitism. The bacteriophages are effective in laboratory, but not in practical field. The uredospore of *Melampsora* spp. and *Puccinia graminis* *tritici* are attacked by fungi *Cladosporium* and *Trichothecium roseum*, respectively. *Fusarium roseum*, the carnation stem rot pathogen, can be controlled by the application of *Bacillus subtilis* in the soil or to the stem cuttings.

(c) ***Control through trap crops***. Several plants may get infected by different pathogens, but after infection the host secretes some substances, which are inhibitory to the pathogen or any one of its growth stage. Such hosts are called trap crops. The root knot nematode, *Meloidogyne incognita* and the golden cyst nematode, *Heterodera rostochinensis* can be controlled by the trap plants like *Crotalaria* spp. and *Solanum nigrum*, respectively. Trap plants are grown in rotation with the required crop plants.

(d) ***Control through antagonistic plants***. Certain plants like marigold, asparagus, etc., release certain toxic substances in the soil which are effective to control several soil borne nematodes. Plantation of marigold around the tomato field causes the reduction of population of nematode, like *Helicoverpa armigera*.

(e) ***Breeding for resistant varieties***. Resistant varieties are the handy, highly productive, disease resistant plants. So many different high yielding resistant varieties of plants have been developed by proper breeding techniques. Some economically important rice and wheat, resistant to different diseases, are given below:

Rice: (i) Blast resistant varieties: CO 4, CO 29, CO 30, K 332, Jaya, Rasi, Vani, ARC 6087, ARC 6633, A 67, A 90, IR 579.

(ii) Bacterial leaf blight resistant varieties: IET 4141, BJ 1, TKM 6.

(iii) Brown spot resistant varieties: T 141, ARC 11253, ARC 11073, ADT 31.

Wheat: (i) Smut resistant varieties: NP 710, NP 770.

(ii) Rust resistant varieties: UP 262, Kalyan sona, HD 2009 (Arjun).

2. **Chemical methods**. Chemicals used in plant protection are called **Pesticides**. Pesticides include fungicides, bactericides, herbicides, insecticides etc. The pesticides are used in different regions of the plants or plant products to control different types of diseases.

Fungicides are the substances which kill the fungi without causing any damage to the host. It belongs to different groups of chemicals like inorganic compounds, organic compounds, benzene compounds etc. Horstall (1956) recognized the fungicides into two categories: **(i) Contact fungicides,** and **(ii) Residual fungicides**.

Contact fungicides: These groups of fungicide are aimed at the fungus itself either before or after it has been found on the host.

Residual fungicides: These groups of fungicide are applied to the host before the fungus reaches it; thereby it forms a protective layer on the host surface.

Later, Martin (1964) divided the fungicides into two groups: (i) **Direct**, and (ii) **Protective**.

Direct: This group of fungicide kills fungi on the host surface i.e., they act as therapeutic or eradicants.

Protective: These fungicides are applied before the reaching of the pathogen to the host.

Another group of fungicide is called systemic fungicide. The **systemic fungicide** is taken up in one part and translocated within the host plant by which the host becomes fungitoxic. This group of fungicide may be both therapeutic and protective in their mode of action.

These chemicals are applied on soil, seeds, different plant parts, entire plant and also on different plant products, by foliar spray, seed treatment (both dry dressing and wet treatment), soil treatment etc.

Chemicals used in plant disease control

I. **Copper compounds**. The commonly used copper compounds are:

(a) **Bordeaux mixture**. It is a mixture of copper sulphate: quick lime (or hydrated lime): water in the proportion 8 lb (One lb = 453.6 grams): 8 lb or 12 lb: 100 gallon (1 gallon = 4.55 lt). During preparation, copper sulphate is put in water, lime is dissolved in water in another container, and both the solutions are then mixed together. The prepared fungicide should be used immediately as it deteriorates on storage. This fungicide is effective specifically on white rusts, blights, downy mildews etc. This fungicide has several advantages, such as: comparatively less expensive, safe and easy to handle, very easy to prepare, and adhesive property.

This fungicide was discovered by Pierre Marie Alexis Millardet (P. M. A. Millardet) in 1882 while working in the University of Bordeaux in France.

(b) **Burgundy mixture**. This is another type of fungicide having the composition like Bordeaux mixture with some modification, comprising of washing soda (Na_2CO_3) in place of quicklime or hydrated lime. It is less phytotoxic than Bordeaux mixture and is effectively used to control leaf blights etc.

(c) **Insoluble copper**. It consists of either basic copper chloride or basic copper sulphate (basicop) or copper oxides (perenox), etc. It is less effective and less phytotoxic than Bordeaux mixture. Commonly, they are used as spray.

II. **Sulphur compounds**. The sulphur is used both in organic and inorganic forms. This group of chemicals acts as efficient fungicide to control different plant diseases:

(a) **Organic compounds**. This group of fungicides is mostly derived from dithiocarbamic acid and is widely used in the control of plant diseases:

Thiram (Tetramethyl thiuram disulphide). It is available in the market under the trade names Arasan, Tulisan etc. This fungicide is used as foliar spray, soil drench etc., to control seedling blight, damping off and some rusts.

Ferbam (Ferric dimethyl dithiocarbamate). It is sold in the market under the trade name Coromet or Fermate. This is a protectant fungicide, commonly used as foliar spray and effectively control the early and late blight of potato in addition to various fungal diseases of fruits and vegetables.

Ziram (Zinc dimethyl dithiocarbamate). It is found in the market under the trade name Zerlate. The fungicide is used to control the foliar diseases of vegetables and ornamental plants.

Zineb (Zinc ethylene bisdithiocarbamate). Its trade name is **Dithane Z-78**. The fungicide is highly effective and safe, used as foliar spray and soil treatment to control leaf spots (blast of rice), leaf blights (early and late blight of potato), fruit rots etc.

Maneb (Manganese ethylene bisdithiocarbamate). It is available in the market as Manzate and **Dithane M-22**. It is a very useful broad-spectrum fungicide applied as foliar spray to control fruit diseases and especially foliar diseases of many vegetables.

Nabam (Disodium ethylene bisdithiocarbamate). The available trade name is Dithane D-14 and Dithane A-40. It is very effective to control seedling blights and leaf blights.

(b) **Inorganic compounds**. This group includes sulphur and lime sulphur:

Sulphur. Elemental sulphur is used as dust, wettable powder or liquid to control leaf blight, fruit rot and rusts. Sulphur dust is very effective to control powdery mildews.

Lime sulphur. It is produced by mixing calcium polysulphide and calcium thiosulphide. It is less phytotoxic than elemental sulphur and is commonly used against powdery mildews, apple scab, anthracnose etc.

III. **Heterocyclic compounds**. This group includes **Captan** which is used to control blights, leaf spots, fruit rots etc. **Difolatan** is used to control early and late blight of potato, **Folpet** is used against powdery mildews and **Captafol** against apple scab etc.

IV. **Quinones**. The quinine like **Chloranil** (trade name Spergon) is used for the treatment of bulbs, seeds, flowers and vegetables through soil drench, dipping treatment, spraying and dusting. **Dichlone** (trade name Phygon) is used to control blights, cankers of fruits, fruit rots etc., by seed treatment and spraying.

V. **Benzene compounds**. **Tetrachlor** (PCNB) is used to control various soil borne diseases of vegetables and also primarily against *Plasmodiophora, Rhizoctonia* and *Sclerotinia*. **Diazoben** (Dexon) is used to control damping off and root rots caused by *Pythium* and *Phytophthora*., **Dinocap** (trade name Mildex, Karathane) is used to control powdery mildews, and **Chlorothalonil** is used against blights, downy mildews, scabs, leaf sports, rusts, anthracnoses etc.

VI. **Systemic fungicides**. This group of chemicals, introduced first in 1966, was an excellent step in the history of plant disease control. The fungicides are absorbed by roots or leaves and translocated within the plant through the xylem, by which the plant becomes fungitoxic. They are very much effective when applied for seed treatment root dipping, soil drench etc.

Till date, many systemic fungicides have been discovered and available in the market. They are of different chemical nature, such as (a) **Oxathiins**, (b) **Benzimidazoles**, (c) **Pyrimidines**, (d) **Thiazoles**.

(a) **Oxathiins**. This group includes Carboxin and Oxycarboxin which are effective to control rust and smut diseases and also the diseases caused by *Rhizoctonia*. **Carboxin** (trade name **vitavax**) is used as seed treatment to control damping off diseases. **Oxycarboxin** (trade name **pantavax**) is used as seed treatment and foliar spray to control rust diseases.

(b) **Benzimidazoles**. This group includes several important fungicides such as **Benomyl** (methyl-n (butyl carbamyol)-2-benzimidazole carbamate) sold in the market as **benlate**. It is a safe, broad-spectrum fungicide applied as seed treatment, foliar spray, root-dip treatment etc., and is highly effective against *Fusarium, Verticillium, Ceratocystis, Rhizoctonia* and *Thielaviopsis*. It is commonly used to control blast of rice, powdery mildews, apple scab, loose smut of wheat etc. **Bavistin** (2-(methoxy-carbamoyl)-benzimidazole) is another highly active fungicide, effective against many members of ascomycotina and fungi imperfecti. **Thiabendazole** is a broad-spectrum fungicide, used to control leaf spots, bulb diseases and is commonly used for post-harvest diseases.

(c) **Pyrimidines**. This group includes mainly dimethirimol, ethirimol and triarimol, used as seed and soil treatment and are effective mainly against powdery mildews.

(d) **Thiazoles**. This group includes different fungicides such as triadimenol (Bayton), triadimefon (Bayleton), biertanol (Baycor) etc. They are very much effective against leaf spots, rusts, smuts, blights and powdery mildews etc.

VII. **Antibiotics**. Many antibiotics are useful to control plant diseases. Generally, they are absorbed and translocated systematically within the plant body. They are mostly active against bacteria, mollicutes and also some fungi. They are applied in the form of seed treatment, spray and dusting. The commonly used antibiotics are: (a) **Streptomycin**, (b) **Griseofulvin**, (c) **Cycloheximide** and (d) **Tetracycline**.

(a) **Streptomycin**. It is available in the market as phytomycin, agrimycin etc. and used as spray. It is effective against bacteria, like *Xanthomonas* spp. *Pseudomonas phaseoliculosis* and fungi, like *Phytophthora infestans* and *P. parasitica*.

(b) **Griseofulvin**. It is effective against the fungal pathogens, like *Alternaria solani, Sclerotiana fructigena, Botrytis fabae* etc., and used to control the diseases caused by them.

(c) **Cycloheximide**. It is a highly effective antifungal antibiotic, commonly used to control diseases caused by *Erysiphe polygoni, Ustilago hordei, Coccomyces heimalis, Sclerotinia fructicola* etc.

(d) **Tetracycline**. It is available in the market as Terramycin, Aureomycin etc., and used to control the diseases caused by bacteria and mycoplasmas.

VIII. **Growth regulators**. Growth regulators like Kingson, Gibberellic acid etc., are used to control the diseases of potato and tomato caused by *Phytophthora* and *Fusarium*, respectively, through increasing the disease resistance capability of the host.

IX. **Soil fumigants**. These chemicals are used mainly to control plant parasitic nematodes. Some of the nematicides are Nemagon, Furasan, Temik, Zinophos etc. — used before and after plantation.

B. Immunization

Plants cannot be immunized by vaccination like animals and human beings due to the absence of capability to produce antibodies. But, with the help of genetic engineering, scientists are able to introduce and express the genes from mice coding for the synthesis of antibodies against viruses in plants. Thus, those plants are now able to produce antibody like proteins, called **plantibodies**, against particular plant pathogen; but it is not yet known whether such plantibodies would effectively protect plants from the action of that pathogen.

The immunization can be done by various ways like 1. **Cross-protection**, 2. **Induced resistance**, and 3. **Transgenic plant production.**

1. **Cross-protection.** This procedure is used to protect plants from infection of a pathogen by prior inoculation with avirulent or mild strain of the same pathogen. This has been successfully applied to viral diseases of citrus by mild strains of *Citrus tristeza* virus, viral diseases of tomato with mild strain of TMV, etc. The protection of rice plant from brown spot pathogen, *Helminthosporium oryzae* can be done by prior (before 48hr.) inoculation with avirulent strain of the same pathogen (Sinha and Trivedi, 1969). But due to non-availability of milder strain, involvement of extensive labour, and non-persistence of strain, the procedure was not accepted commercially.

2. **Induced resistance**. The induction of resistance can be done by pathogen, through naturally occurring compounds obtained from the pathogen or by using chemicals of different groups, such as metal salts, reducing agents, amino acids, metabolic inhibitors, etc.

Plants infected with one pathogen become much more resistant to subsequent infection by other pathogens. Thus, TMV in tobacco induces a systemic resistance not only to itself, but also to other viruses, bacteria (*Pseudomonas tabaci*) and fungi (*Phytophthora nicotianae*). On the other hand, resistance induced in tobacco by fungus (*Thielaviopsis basicola*) shows resistance to TMV.

Resistance to pathogen can be induced by treating the host plant with the naturally occurring compounds, such as protein coat of TMV, glycoprotein fraction of the pathogen *Pseudomonas solanacearum*, etc.

Resistance in rice seedling against *Helminthosporium oryzae* can be induced by various methods, like seed treatment (Sinha and Giri, 1979; Sinha and Hait, 1982), root dip treatment and foliar spray (Sinha and Giri, 1979; Trivedi and Sinha, 1980) with metal salts ($HgCl_2$, $BaCl_2$), reducing agents (Sodium selenite, Cysteine), amino acids (DL-Methionine), metabolic inhibitors (Cycloheximide, Sodium azide) etc., at a very dilute concentration ranges from 10^{-4} to 10^{-6} M.

3. **Transgenic plant production**. The plants that acquired genetic make up induced in them through genetic engineering are the transgenic plant. With the help of this procedure the quality and/or quantity of the plant products can be improved as well as the plants become resistant to the pathogens and insects. Bt crops can be developed by introduction of genes from *Bacillus thuringensis* (Bt) which produces a crystal protein, toxic to a range of pathogens. Bt cotton, Bt corn, Bt rice can be developed following the above procedure.

Biological control of Plant diseases

This is the method of protection, **directly** by total or partial destruction of pathogen population by microorganism and/or **indirectly** by immunization i.e., improving the resistance of plant.

Though the scientists were interested on biological control during 1920s, it came to be implemented after the first International Symposium on "Ecology of Soil Borne Plant Pathogenes – Prelude to Biological Control" held at Barkeley, U.S.A.; in 1963.

The direct protection through biological control is applicable in both rhizosphere (the soil near the living root) and phyllosphere, but there is **limited economic success** in phylloplane (the surface of the leaf) in comparison to rhizosphere due to:

1. Certain fluctuations in temperature.
2. Intense radiation.
3. Relatively low and variable availability of nutrients.
4. Variable relative humidity.
5. Availability of antimicrobial substance of host origin, such as gallic acid etc., and
6. Variable action of wind and rain on the phyllosphere.

The **direct biological methods** of controlling plant diseases have already been discussed in page 649.

The **indirect biological method** of controlling plant diseases i.e., **Immunisation** have also been discussed in page 653.

Integrated Plant disease management

In a disease pyramid, three principal components are considered. These are **host, pathogen**, and the **environment**. All the three components act together during disease development. Thus, to protect a plant from disease(s) one has to tackle all the three parameters at the same time. The management programme should cover the integration of all the methods; those favour the host, discourage the pathogen and modify the environment. This type of management programme to control plant diseases is called **Integrated Disease Management** i.e., IDM or **Integrated Pest Management** i.e., IPM.

The procedure of management is different in respect of the principal component:

A. **Management of host**. It includes:
   - (a) Procedure to improve plant vigour,
   - (b) Induction of disease resistance, and
   - (c) Breeding for disease resistance.

B. **Management of pathogen**. It includes:
   - (a) Eradication or reduction of inoculum,
   - (b) Application of chemicals on plant surface, to discourage the pathogen, and
   - (c) Prevention through legislation (i.e., keeping the pathogen away from the host) by quarantine, etc.

C. **Management of environment**. It includes:
   - (a) Crop management,
   - (b) Soil management, and
   - (c) Water management.

According to G. N. Agrios (1997), the main goals of an integrated plant disease control program are:

1. To eliminate or reduce the amount of initial inoculum,
2. To reduce the effectiveness of initial inoculum,
3. To increase the resistance of the host,

4. To delay the onset of disease, and
5. To slow down the secondary cycle.

The programme of Integrated Plant Disease Control can be taken against a particular common disease, such as late blight disease of potato or against all diseases affecting a particular crop such, as Potato, Apple, *Citrus* etc.

Integrated management in an Annual crop

The following programmes must be undertaken for annual plants like **potato** against different diseases like late blight (fungus), brown rot and ring rot (bacteria), leaf roll of potato (virus – potato virus-1), etc.:

1. Stock tuber should be free from any disease i.e., a healthy tuber is needed.
2. Field should be cleaned from previous years plant debris and tubers, if any, where pathogen may perpetuate for the next season.
3. Crop rotation must be done with legume or any other crop plant, except Solanaceous members.
4. Plant should be protected from disease through foliar spray at regular intervals.
5. Make the soil loose by ploughing and expose it to sunlight for few days in summer months and/or fumigate the soil with chemicals.

Integrated management in a Perennial crop

For perennial crop like orchard crop (peach, apple, pear etc.); following programme should be undertaken:

1. Nursery-stock should be fumigated (to eradicate the nematode, if present).
2. The stock should be fumigated (to eradicate the nematode, if present).
3. The growing field should be fumigated before planting to eradicate any pathogen like *Armillaria, Phytophthora,* if present.
4. Proper drainage in the field should be maintained.
5. Stock should not grow in old field near the old plants that may carry different pathogens like canker, etc.
6. The field should be properly irrigated, and supplemented by fertilizers.
7. Plants should be sprayed regularly to protect them from different diseases and insect pests.
8. Usually the fruits become susceptible to several fruit rotting and fruit spotting fungi that attack at different stages till harvest or even in storage. So the fruits should be sprayed at 10-14 days' interval. Sometimes insects can damage at any stage, thereby the fungi infect the fruit through wound. Thus, the insecticides should be applied at intervals.

So the integrated control programme includes several control methods like **legislation** (quarantine, inspection etc.); **cultural practices** (sanitation, crop rotation etc.); **chemical controls** (spraying of plants, fumigation of soil, surface sterilization of planting materials etc.). All the required procedures for a particular crop should be followed to overcome the disease(s) and/or insect pests.

Plant disease clinics

Plant disease clinics identify and recommend the possible control measure to rescue the plant or mostly the population as a whole from disease(s). Individual care is to be taken for ornamental plants, fruit trees etc.; but for crop plants, it is taken as a whole.

According to L. W. Barnes (1994), the plant disease clinic develops from a perceived need for an organized systematic professional effort to assist in the identification of plant disease problem. In the USA, most of the states operate one or more plant disease clinic(s) through their department of Plant Pathology for many years (G. N. Agrios, 1997). Plant growers, homeowners, country extension agents would send diseased plants, soil along with disease plants, sometimes insects to the clinic. The experts, i.e., plant pathologists, identify the disease and recommend the possible control measures to the growers, free of cost. In early days of plant disease clinic, it was looked after by the extension plant pathologists and the diagnoses were made by the advanced plant pathology graduate students assisted significantly by more junior graduate students. In early days, plant disease clinics were equipped with surface sterilants, dissecting microscopes, culture tubes, nutrient media for culturing bacteria and fungi. With time, according to importance and need, the charge of the clinic was given to M.S. or Ph.D plant pathologists, where nematode isolation and identification were taken into consideration. Identification of virus was made by host symptomatology alone.

Since 1970, much more importance was given to the **plant disease clinic**. Every state of the USA has minimum one plant disease clinic (4 plant disease clinics in Florida). Some states also have one or more **private clinic(s)**. Now-a-days, plant disease clinics often have one scientist, one or more laboratory assistants; they are also equipped and trained for diagnosis of even viral disease through host range tests, serological tests, cell inclusion, identification, electron microscopy of plant sap and dot-blot assays of radioactive or colour-producing DNA probes. The modern clinics also have modern computers with databases expert systems for pathogen and disease identification, such as CD-ROM video capabilities and e-mail facilities for transmitting the result of diagnosis and the recommendations for disease control against payments.

Plant disease clinics have now come up in India in almost all Agricultural Universities, started in Punjab Agricultural University, Ludhiana, a few years back.

In some states of the USA, some persons with varying level of education and training in Plant Pathology (B.S.; M.S. or Ph.D.) have their **own plant disease clinic** (private) and offer their services as private doctors to individual or group of growers. At times, the growers get information from salesmen of pesticides (it includes fungicides, herbicides, insecticides, nematicides etc.), seed or fertilizer etc.; who have little or no education and training in plant pathology. From them the growers received limited or wrong information, thereby wrong pesticides or excess dose of pesticides were applied. The limited or wrong information causes crop loss as well as contamination of environment with the pesticide which may cause direct misuse and ecological disturbance. The above problem may be solved by establishment of more well-equipped clinics with efficient pathologists, the **Plant Doctor**.

Agrios (1997) hoped that some day in the not too distance future **"Plant Doctors" will take their place among professionals in the way general practitioner physicians and veterinarians have done already**. Until then, it is an optimistic hope for the plant growers to expect effective diagnosis and control of pests.

With the help of plant doctors following recommendations and I.P.M. coordination, a nation can save important crops and earn millions of money per annum.

Classification of Plant diseases

Plant diseases can be classified in different patterns. Before 1850, classification was based on the effect rather than cause. But, after the acceptance of microorganisms as a causal factor, classification was done according to causal factors. The plant diseases may be classified in different ways, based on:

1. the location of the disease, such as localized and systemic disease.
2. the affected plant parts, such as root, stem, leaf, etc.
3. the hosts affected, such as vegetable diseases, cereal diseases, timber diseases, diseases of ornamental plants, diseases of shade trees, etc.
4. the symptoms, such as necrotic (death of affected tissue), atrophic (slow down the growth of the plant) and hypertrophic (overgrowth of plant or plant parts).
5. the spread and severity, such as endemic, epiphytotic or epidemic, sporadic and pandemic.
6. the major causal agents, such as non-parasitic, parasitic and viral diseases (Table 5.6).

Major types of classification of plant diseases are

1. **Based on the location of the disease**

On the basis of location, the diseases can be classified into *two groups*: **Localized disease** and **Systemic disease**:

Localized disease. When symptoms of a disease develop on a particular area of a plant body, it is known as localized disease. The infective bodies may be spores, mycelium or any other part of the pathogen, e.g., brown spot disease of rice, c.o. *Helminthosporium oryzae*; blast disease of rice , c.o. *Pyricularia oryzae*.

Systemic disease. When a disease affects the entire plant, it is known as systemic disease. It may be due to pathological or physiological causes, and ultimately, death occurs. Sometimes, wilting may take place in one side of the plant, thus causing partial death with low yield.

Pathological disease. The diseases are caused by pathogens, like fungi, bacteria and viruses, e.g., wilt disease of tomato either by fungi like *Fusarium oxysorum* f. sp. *lycopersici* (*Fusarium* wilt), *Verticillium albo-atrum* (*Verticillium* wilt) or by bacteria like *Pseudomonas solanacearum* (bacterial wilt). The tobacco mosaic virus disease is caused by virus TMV.

Physiological disease. Flooding injury of potato takes place due to soil moisture disturbance. If rain water stands in the field for 2-3 days, potato plants undergo permanent wilting due to decay of root system. Soft rot also takes place in the storage organ.

2. **Based on the spread and severity**

On the basis of spread and severity, diseases can be classified into *four groups*: **Endemic, Epiphytotic** or **Epidemic, Sporadic** and **Pandemic**:

Endemic diseases. When diseases are **constantly present** year after year in moderate to severe form in a particular area, they are called as **endemic disease.**

This is due to the fact that pathogens can survive year after year, from one crop season to the next, **in soil** (onion smuts, c.o. *Urocystis cepulae*), **in plant debris** (stem rot disease of jute, c.o. *Macrophomina phaseoli*), or **in wild plants** (cucumber mosaic virus on wild weed plants, blast disease of rice, c.o. *Pyricularia oryzae*; on *Setaria intermedia, Digitaria marginata, Panicum repens* etc.), or may come regularly from diseased area through propagating stocks, seeds or by winds.

Epiphytotic or **epidemic diseases**. When diseases occur **widely, but periodically** in severe form, they are called **epiphytotic diseases**.

These diseases are very responsive to the variations of environmental conditions, such as temperature, humidity etc. Like endemic disease, the pathogens of epidemic diseases are constantly present, but the disease manifestation takes place only with the onset of favourable conditions, e.g., powdery mildews, rust of cereals etc.

On the other hand, the disease may be epiphytotic due to the irregularity of the occurrence of inoculum, because of inefficiency of the pathogen to survive in soil, plant debris etc., from one cropping season to the next. Black rot of crucifers depends on introduction of inoculum with the seed.

Sporadic diseases. When diseases occur in **small areas** and **at very irregular intervals**, they are known as **sporadic diseases**, e.g., blotch disease and angular leaf spot of cucumber.

Pandemic diseases. When diseases occur in **vast** and **much extensive areas**, they are known as **pandemic diseases**, e.g., late blight disease of potato, which occurs throughout the world and causes great economic loss.

3. **According to major causal agents**

On the basis of major causal agents, J. C. Walker (1969) classified plant diseases into *three distinct groups*: 1. **non-parasitic diseases**, 2. **parasitic diseases**, and 3. **viral diseases**. Further, on the basis of the causal agency or organism, he subdivided the groups are:

1. **Non-parasitic disease**

(i) **Low-temperature**: Freezing injury of potato, alfalfa and garden pea.

(ii) **High-temperature**: Heat injury of apple fruits (above 38°C), water core of apple, sunscald of vegetables, heat canker of flax etc.

(iii) **Unfavourable oxygen relations**: Black heart of potato.

(iv) **Unfavourable soil moisture relations**: Wilting of potato, cabbage and turnip (by water-logging for 48-72 hours).

(v) **Injurious atmospheric impurities**: Apple scald, discolouration of onion in storage (due to leakage of ammonia in cooling plant).

(vi) **Lightning injury**: Lightning injury of potato and cabbage.

(vii) **Mineral excess**: Crinkle leaf of cotton (due to Mn toxicity), internal bark necrosis of delicious apple.

(viii) **Mineral deficiencies**: Blossom-end rot of tomato fruit (Ca deficiency), heart and dry rot of sugar beet (Bo deficiency).

2. **Parasitic diseases**

Diseases incited by:

(i) **Bacteria**. Bacterial blight of paddy, citrus canker etc.

(ii) **Plasmodiophorales**. Club root of crucifers, powdery scab of potato etc.

(iii) **Phycomycetes**. Black wart of potato, late blight of potato, etc.

(iv) **Fungi Imperfecti**. Early blight of potato, white rot of onion etc.

(v) **Ascomycetes**. Dutch elm disease, apple scab etc.

(vi) **Basidiomycetes**. Corn smut, loose smut of wheat etc.

(vii) **Phanerogams**. Dwarf mistletoe (*Arceuthobium camplyopodum*) on *Pinus*.

(viii) **Animal parasites**

(a) **Nematodes**. Golden nematode disease of potato, ear cockle of wheat etc..

(b) **Insects**. Insect galls of various trees.

3. **Virus diseases**. Tobacco mosaic, potato mosaic, bean mosaic etc.

Koch's postulates

Robert Koch (1843-1910) was the first to discover that typical *Bacillus* is responsible for the Anthrax disease of cattle. This was the first report stating that a bacterium can cause an animal disease. His sequential observations lead to the establishment of postulates known as **Koch's postulates** or Rules of Proof. He proposed his postulates in **1876**. With the help of these postulates or **Rules of Proof,** a particular suspected pathogen could be finally confirmed and be accepted as a pathogen for a particular disease. The Koch's postulates or the Rules of Proof are:

1. The organism must be found associated with all the diseased plants or animals so far examined.
2. The organism must be isolated from diseased plant or animal and should be grown in sterilized nutrient medium to develop pure culture and must be described properly.
3. The isolated organism, when inoculated to healthy plants or animals of the same variety, must develop identical symptom(s) as observed earlier.
4. The organism must be re-isolated from the newly infected plant or animal in sterilized medium, and must be identical with the first isolated organism.

If all the four postulates have been followed and found true, then one can confirm that the isolated pathogen is responsible for the particular disease. The above procedure of confirmation is applicable for fungi, bacteria, nematodes, etc., those can easily be isolated, grown and then inoculated with the plants or animals. But the procedure is not applicable for those which cannot grow on sterilized medium, like viruses, phytoplasmas, protozoa, and some phloem-inhabiting bacteria.

Importance of Plant diseases

Human beings depend on plants for food, clothing, shelter, medicine etc. But due to disease at different times, the plants suffer a lot and their productivity becomes reduced. Due to reduction of productivity the human beings get affected directly. For the benefit of mankind pathologists (plant doctors) are trying to cure the infected plants and protect the healthy plants against diseases. The plant diseases are mostly harmful, but a few are useful also.

A. **Importance due to harmful effect**

1. **Late blight of potato** (c.o. *Phytophthora infestans*). The disease appears in severe form between 1840-1847, caused major damage of potato in Ireland. By 1845, it totally devastated the potato cultivation of U.S.A., Canada and Europe, mainly in Ireland. As a result **Irish famine** occurred that caused the **death of about one million people** and more than one million migrated to other countries.

2. **Coffee rust** (c.o. *Hemileia vastatrix*). It appeared in severe form during early 1870s which almost completely destroyed the coffee cultivation of Sri Lanka (erstwhile Ceylon). Due to this, the coffee cultivation was shifted to Western Hemisphere, where *H. vastatrix* was not available. Later, Rubber and Tea cultivation replaced coffee along with associated industries.

   The disease also hampered a lot the coffee production of Brazil during 1974-1975. The disease had crossed the Atlantic Ocean to Brazil, where it had spread severely.

3. **Downy mildew of grapes** (c.o. *Plasmopara viticola*). It caused severe damage in France during 1878 to 1882. The wine industry suffered a lot due to the above disease.

4. **Sigatoga disease of banana** (c.o. *Mycosphaerella musicola*). The disease caused major damage of banana production and threatened banana-based industries of South and Central America in 1930s. Later, in 1940, the disease was checked by the large scale application of Bordeaux mixture.

5. **Red rot of sugarcane** (c.o. *Glomerella tucumanensis*). It became so severe in Indian states like Punjab, Bihar and Uttar Pradesh during 1938-1939; that the production was highly reduced resulting in about 70% reduction of sugar production. Both farmers and associated mills became highly affected.

6. **Brown spot of rice** (c.o. *Helminthosporium oryzae*). The disease appeared as epidemics in Bengal (India) in 1942 and caused about 90% crop loss (Ghosh *et al.*, 1960); thereby **Bengal famine** took place in **1943**. It was so severe that about **five million people** died of starvation.

7. **Bacterial blight of rice** (c.o. *Xanthomonas campestris* pv. *oryzae*). This disease was first observed in 1884 by farmers of the Fukuoka in Japan. By 1926, it became widespread in different regions of Japan. In India, it was first reported from Maharashtra (Bombay, now Mumbai) in 1951 (Srinivasan *et al.*, 1959). The disease became severe in Shahabad (Bihar) in 1963, spread widely throughout the country in 1966, after the introduction of TN-1 (Taichung Native-1) variety. Now the disease is found to be a threatening stage in the yield of rice throughout the country (Srivastava, 1967).

So many other diseases are in record (Klinkowshi, 1970), those also caused severe damage of crops in different countries and thereby the associated industry becomes threatened.

B. **Importance due to useful effect**

1. **Ergot disease of Rye** (c.o. *Claviceps purpurea*). Several alkaloids are produced and accumulated in the sclerotium of *Claviceps purpurea* which cause ergot disease of rye. Out of several alkaloids, ergometrine and its semisynthetic analogue methyl ergometrine and methyl ergometrine meleate have prominent uterine action, those control haemorrhage of mother during childbirth, having side-effects with increase in blood pressure and decreased milk secretion.

2. **Tulip mosaic virus disease** (c.o. Tulip mosaic virus). The disease was first

Table 5.6: List of plant diseases of national and international importance (Pathak *et al.*, 1996)

Crop	Disease	Pathogen
Rice	Tungro	Viral
	Bacterial blight/wilt	*Xanthomonas campestris* pv. *oryzae*
	Blast	*Pyricularia oryzae*
	Brown leaf spot	*Helminthosporium orzyae*
	Sheath blight	*Rhizoctonia solani* (Perfect state—*Thanatephorus cucumeris*)
	False smut	*Ustilaginoidea virens* (Perfect state—*Claviceps oryzae sativae*)
	Nematodes pest	Nematodes
Wheat	Rusts	*Puccina graminis* f. sp. *tritici*, *P. striiformis* f. sp. *tritici*, *P. recondita*
	Smut of wheat	*Ustilago tritici*
	Karnal bunt	*Neovossia indica*
	Foliar blights	*Helminthosporium sativum* and *Alternaria triticina*
	Nematode pests	
	Molya	*Heterodera avenae*
	Ear cockle	*Corynebacterium michiganense* var. *tritici* and *Angunia tritici*
Pearl Millet	Downy mildew	*Sclerospora graminicola*
	Ergot	*Claviceps fusiformis*
	Smut	*Tolyposporium penicillariae*
Corn or Maize	stalk rot	Fungal and bacterial
Pigeon pea	Sterility mosaic	Viral
	Phytophthora blight	*Phytophthora drechsleri* f. sp. *cajani*
	Fusarium wilt	*Fusarium udum* (perfect state *Gibberella indica*)
Chick pea	*Ascochyta* blight	*Ascochyta rabie*
Soybean	Yellow mosaic	Viral
	Bud blight	Viral
Pea	Powdery mildew	*Erysiphe polygoni*
Potato	Viral diseases	Viral
	Bacterial wilt	*Pseudomonas solanacearum*
	Late blight	*Phytophthora infestans*
	Nematode pests	*Globoderma rostochiensis* (*Heterodera rostochiensis*) and several species of *Meloidogyne*.
Ground nut	*Aspergillus* root rot	*Aspergillus* spp.
	Rust	*Puccinia arachidis*
	Tikka disease	*Cercospora arachidicola* and *Cercosporidium personatum* (Perfect state *Mycosphaerella berkeleyii*)
Mustard	White rust	*Albugo candida*
	Alternaria blight	*Alternaria* spp.

[*Contd.*]

[Contd. Table 5.6]

Crop	Disease	Pathogen
Mango	Mango malformation	A complex caused by *Fusarium moniliforme* var. *subglutinans* and mites
	Powdery mildew	*Oidium mangiferae*
	Diplodia stem and other rots of fruits	*Diplodia* sp.
Banana	Bunchy top	Mycoplasmal
	Bacterial wilt	*Pseudomonas solanacearum*
	Nematode pests	Nematodes
Guava	Wilt	*Fusarium oxysporum* f. sp. *psidi*
Apple	Scab	*Venturia inaequalis*
Tea	Blister blight	*Exobasidium vexans*
Coconut	Root wilt	Mycoplasmal
	Phytophthora bud rot	*Phytophthora palmivora*
	Ganoderma wilt	*Ganoderma lucidum*
Areca-Nut	Fruit drop	*Colletotrichum* sp.
Sugarcane	Red rot	*Colletotrichum falcatum*
	Smut	*Ustilago scitaminea*
Cotton	New cotton wilt	*Fusarium oxysporum* f. sp. *vasinfectum*
Tobacoo	Leaf curl	Viral
	Phytophthora black leg	*Phytophthora nicotianae* var. *nicotianae*
	Bacterial wilt	Bacterial
Tapioca	Mosaic	Viral
Ginger	Rhizome rot	*Pythium aphanidermatum* and *Pythium myriotylum*
Cardamom	Katte disease	Viral
Coffee	Die back	*Colletotrichum coffeanum*
	Rust	*Hemileia vastatrix*
Rubber	*Phytophthora* leaf fall	*Phytophthora palmivora*
	Pink disease	*Pellicularia salmonicolor*
Jute	*Macrophomina* stalk rot	*Macrophomina phaseoli*
	Bacterial wilt	*Pseudomonas solanacearum* var. *asiaticum* along with *Rhizoctonia bataticola* and *Fusarium solani*.
Black Pepper	*Phytophthora* fruit fall and *Phytophthora* wilt	*Phytophthora capsici*
	Nematode pests	Nematodes
Brinjal	Little leaf	Mycoplasmal
Okra	Yellow vein mosaic	Viral
Cauliflower and Cabbage	*Sclerotinia* rot	*Sclerotinia sclerotiorum*
	Curd rot of cauliflower	Bacterial
Tomato	Leaf curl complex	Viral
	Bacterial wilt	*Pseudomonas solanacearum*

reported by Carolus Causius (1576) in tulip. It shows variegation or breaking in flower colour and became popular as ornamental plant. Similarly, variegated plant of *Abutilon,* probably *A. striatum* var. *Thompsonii*, became popular in Europe during 1868 as ornamental plant.

Host-pathogen relationships

Nutrients are indispensable elements required for the growth of all living organisms including host plants and pathogens (Fig. 5.8). Host plants are those on which disease occurs and pathogen causes disease. Pathogens are the living organisms or biotic agents. Based on the way of collection of nutrition, the pathogens are broadly divided into three groups: Necrotrophs, Biotrophs and Hemibiotrophs.

1. **Necrotrophs**. A pathogen that kills host cells in advance of its growth and then lives on the dying and dead cells. They acquire nutrients released from dead cells.

In this case, host and pathogen does not coexist in a harmonious pattern. Thus, during disease development, incomplete cellular relationship is being observed between host and the pathogen. Based on involvement of a number of host species, the pathogens are of two types:

(i) *Having broad host range*. Pathogen secretes toxins that act on metabolism, common to many plants.

(ii) *Having narrow host range*. Pathogen secretes specific toxin acts on host with no genes to interact with the toxin, thus host being susceptible.

2. **Biotrophs**. A pathogen that lives on living host tissues, ensure prolonged supply of the carbon and other nutrients to them.

Normally they infect through natural openings like stomata etc., or by direct penetration through the cell wall. They grow intercellularly and only penetrate the cell wall of neighbouring cells. They grow without eliciting the host defense responses. But if host cells become die in advance of their invasion, host plant become resistant as the pathogen unable to establish a parasitic relationship with the host.

3. **Hemibiotrophs**. Pathogens those obtain nutrients from the host cell by exhibiting the biotrophic phase in early stages of the infection process and then switching to the necrotrophic mode during later stages of their life cycle (Glazebrook, 2005).

Host response influences on the parasitic relationship. In some case, pathogen and the host lead to infection and establishment of the pathogen but not in others. The hosts those allow them to infect and establish inside, are the **susceptible hosts** and those restrict or retards the development are the **resistant hosts**.

Thus, resistance and susceptibility is based on the host-pathogen relationship.

Environmental factors like temperature, humidity, light etc., also influence host-pathogen relationship. All the above factors at favourable amount or intensity help in successful relationship, otherwise it retards the process.

Time is also an important factor to establish a successful host-pathogen establishment. With optimum temperature and humidity and dew drop, if spores fall in the morning on the dew drop accumulated on leaf surface, may not cause disease, as the dew drop dries due to sunshine and the spores will not get the time to germinate. Thereby no disease will establish. But spores fall on dew drop in midnight help in successful infection and disease become established. Thus, time is an important factor in host-pathogen relationship.

In presence other favourable conditions, the relationship of host plant and pathogen varies based on nutrient requirements. As per nutrient requirement, these are of **five** types:

(a) *Plant antagonistic to pathogen*. In this case plant secretes some inhibitory compound, which inhibits pathogen to develop, thereby no disease takes place. Thus, host is the resistant one.

(b) *Pathogen antagonistic to plant.* In this case pathogen secretes some compounds and the host is sensitive to the secreted compound(s) i.e. toxin, thus disease develop in host. Thus, host is the **susceptible** one.

(c) *Mutual antagonism*. In this case, inhibition or death of both the pathogen and host tissue takes place. Thus, host plant behaves as **resistant** one. This is the case of incompatible interaction between host and pathogen. This can be seen in incompatible interaction between *Puccinia graminis* f. sp. *tritici* (stem rust pathogen) and resistant wheat cultivars, where death of pathogen and host cells takes place.

(d) *Mutual adjustment*. In this case, compatible cellular relationship develops between host and pathogen in symbiotic mode. Thus, both become benefitted in this relationship. In mycorrhizal association both the partners live in a symbiotic mode.

In other case, as with biotrophs like rusts, mildew etc., pathogen grow and reproduce well on living host cells by taking shelter and nutrients, thereby it affects the growth of host but host cells are not killed.

(e) *No relationship*. In this case, plant and pathogen ignore each other or plant does not respond to the secreted chemical or toxin of the pathogen. Thus the plant is a **non-host**.

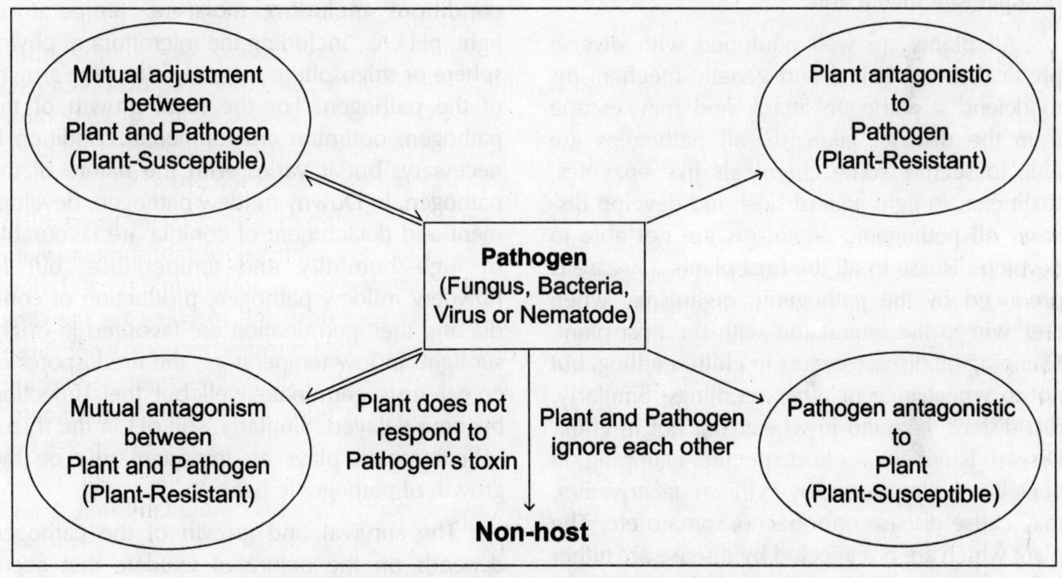

Fig. 5.8: Relationships between plant and pathogenic organism

HOST-PARASITE INTERACTION

Host and pathogen interact variously, when a pathogen comes in contact with the host, either on the root or on any aerial parts, thereby disease may develop. The host plant tries to prevent the entry of the pathogen and pathogen tries to enter and establish in the host. Thus, interaction takes place between host and pathogen. If pathogen wins in the game of interaction, plant becomes diseased. The interaction between host and pathogen during disease development is called **host-parasite interaction.**

All plants are well equipped with diverse physical, biochemical and genetic mechanisms to defend a pathogen attack and may escape from the disease. Likewise, all pathogens are able to secrete some chemicals like enzymes, toxin etc., to fight against host and develop disease. All pathogenic organisms are not able to develop disease to all the host plants. Disease is produced by the pathogenic organisms, when they win in the interaction with the host plant. Damping off disease occurs in chilli seedling, but not to wheat, rice or other seedlings. Similarly, rust disease is found in wheat, but not in chilli. Disease is not always host specific. Damping off of chilli seedling caused by *Pythium debaryanum*, may cause disease on tobacco, tomato etc. The plant which are not affected by disease are either resistant or may be a non-host plant.

To develop a disease on host by any biotic agent, infection is must and the pathogen follows different mechanisms to develop disease.

For the disease to occur, initially contact between host and pathogen is must, then recognition between host and pathogen, penetration and entry of pathogen, infection, colonization, finally expression of the disease i.e. manifestation, showing symptoms.

In general, the mechanism of infection is given below.

Mechanism of infection

(Pre-penetration, Penetration & Post-penetration)

The mechanism of infection can be discussed in the following steps.

A. Pre-penetration

During pre-penetratin the inoculum (spore, mycelium etc.) of the pathogen after arrival on the host surface sharply interact with the host and also with the surrounding environmental conditions including moisture, temperature, light, pH etc., including the microflora of phylosphere or rhizosphere that may affect the growth of the pathogen. For the ideal growth of the pathogen, optimum environmental condition is necessary, but it varies with the nature of the pathogen. In Downy mildew pathogen, development and detachment of conidia are favourable in high humidity and temperature, but in Powdery mildew pathogen, production of conidia and theit germination are favoured in bright sunlight. In low temperature, the uredospores of cereal rusts germinate well but their infection become delayed. Similarly, soil pH in the rhizosphere region plays an important role on the growth of pathogenic bacteria.

The survival and growth of the pathogen depends on the nature of exudate that accumulates both on leaf and root surface of host, along with the microbial population present on it. The exudates of leaf and root may favour or inhibit the growth of the pathogen.

Leaf exudate. Leaves exudate contain substances which may go in favour or inhibit the growth of the pathogen. The glands of gram leaf hairs contain malic acid which is antifungal and arrests the growth of *Uromyces ciceris arietini*.

Root exudate. Root exudates contain mainly sugars and amino acids that promote the growth of fungi and bacteria, but hydrocyanic acid, antibiotics and different organic acids inhibit their growth. Amino acids like proline in the root exudate enhance the spore germination of *Rhizopus arrhizus*, whereas protocatechuic and cathecol in onion root exudate inhibit spore germination of *Colletotrichum circinaus*. Besides these, the rhizosphere region contains microbial population which is antagonistic to the growth of the pathogen.

During this process, pathogen has to neutralize or resist in the exudates for their survival. Thus, the pathogen has to overcome the above mentioned barriers during pre-penetration stage before it can survive for host penetration. Once a favourable relationship is being established with the host, multiplication or growth of the pathogen begins. The fungal spore germinates, produces mycelium and progress for penetration or may give rise to an apperssorium which anchors the fungus to the host surface. Later, penetration peg is produced from the appersorium with which the pathogen causes host penetration. In bacterial pathogen rapid proliferation of cells takes place in a relatively short time. As the bacterial cells are deliate in structures may be easily killed during unfavourable conditions, but they may survive under slime layer. But multiplication of viruses takes place only in the living host cells.

B. **Penetration**

Successful penetration depends on the combined effect of various factors. These are:

(i) The nature and activity of pathogen,

(ii) Favourable physical condition, and

(iii) Host susceptibility.

Out of the above three factors, the *nature and activity* of the pathogen is most important, which control the disease development. In addition, *inoculum potential* is also important as biological energy for successful colonization in host plant. It performs with the availability of the following:

(i) *Density of inoculum*. It refers to the active propagule available per unit area of aerial organ (leaf or stem) or per unit volume of soil,

(ii) *Availability of nutrients*. Availability of nutrients for the pathogen to germinate or grow,

(iii) The virulence of the pathogen, and

(iv) Host susceptibility.

In addition, *physical and physiological condition of the host plant* has an effect on the activity or efficiency of the pathogen for the progress of the disease. The concept that encircles the phenomenon is the **predisposition**. Different factors involved in predisposition, are:

(i) Age of the host plant at the time of infection,

(ii) Genetics of the host plant,

(iii) Genetics of the pathogen,

(iv) Environmental conditions (light, temperature, humidity etc.) surrounding the host surface (leaf, stem or root), and

(v) Single versus mixed infection.

Generally adverse conditions predispose a plant to greater degree of susceptibility to attack by pathogens. The conditions involve in the above phenomenon are temperature, free moisture etc. The effect of temperature in nature is extremely variable depending on the nature of host and pathogen. Free moisture is necessary for pathogen development, in most fungal and bacterial diseases. Due to change of predisposing factors, genetically resistant may behave as susceptible or vice versa. Some fungal pathogens have specifity in part of host infection. Some of them restricted in cortical tissue, vascular tissue and some grows only in heart wood.

Under favourable predisposition, the virulent pathogen with inoculum potential, enter the host tissue called **penetration**. After entry, the pathogen established in cells is called **infection**. The pathogen secretes offensive chemical weapons to break host barriers. These are enzymes, toxins, antibiotic etc., and progress further in host tissue by causing damage. Parasitism and host resistance run parallel and are inseparable.

Some enzymes those produced normally by pathogen are called **constitutive enzymes** and others are adaptive or **inductive enzymes**, whose formation is stimulated by the presence of appropriate substrate. Enzymes catalyse many of the biochemical processes of living host and thus play a fundamental role in host-pathogen interaction. During this interaction pathogen later also degrade the host tissue into metabolite and used by pathogen. Enzymes break down cell walls and connective cell layers.

The pathogen also affects the respiratory processes, growth, ionic exchange of the host cell causing interference with the cell permeability and other normal functions.

Viruses, however, on account of their basic nature cannot produce any of these substances. They multiply by directing the host cell to manufacture viral nucleic acid and viral protein.

Different ways penetration

Bacteria enter inside the host tissue mainly through wounds and natural opening. Viruses enter mainly through wounds. Whereas, the fungal pathogens enter through natural openings, wounds, direct penetration through cuticle and outer wall of the surface cells, root hairs etc. Direct penetration process is more complicated than the entry through natural openings or by wounds. It is performed by breaking the host structural barriers through different chemical reactions. This can be discussed under the following heads:

1. **Entry through natural openings**

Both bacteria and fungi enter inside the host through natural openings like stomata, lenticels, hydathodes, nectaries etc. In this process the pathogens enter easily inside the host except in some cases where sub-stomatal hairs may resist against the entry.

2. **Entry through wounds**

Wound on host tissues caused by natural calamities like storm, fire, etc.; during field operations; by insects, by accidental breaking of parts etc.; offer easy entry of pathogens inside the host. Viruses enter inside the host mainly through wounds.

3. **Entry by direct penetration of surface cells edit**

The entry of pathogen inside the host caused by direct penetration of the outer wall of surface cells and it is rather a difficult process which usually requires high moisture or free water supply. It is even more difficult in leaves with waxy covering which allows water to drain off freely.

Fungal pathogens penetrate into host either by boring of outer wall of the surface cell or by pressure and enzymes secreted by pathogen.

After the hypha made contact with a suitable host surface, but not able to penetrate directly, follows the development of an appressorium and later from its lower surface infection peg and penetration tube develops. The infection peg penetrates through the cuticle and cell wall at a point softened by enzymatic action and followed by mechanical pressure. Later penetration tube develops and grows inside the same cell and also the neighbouring cells.

In some cases, germ tube produced by spore germination passes down between the radial walls of the adjoining cells without actually entering the cells. Other fungi send only haustoria in the epidermal cell through the outer cutinized wall. In certain others, the penetration hypha penetrating the cuticle induces the formation of a local internal swelling of the underlying cellulose wall, so that a small papilla projects from the underside of the wall into the cell.

This papilla is penetrated by the penetration hypha which enters the host cell and ultimately develops into a haustorium.

Bacteria are mostly weak parasites which cannot employ force to effect penetration. Their penetration is affected by chemical action. The plant parasitic nematodes pierce the host surface with spears or stylet.

Direct penetration depends on:

(i) Nature of cell walls and efficiency of the enzymes of pathogen, and

(ii) Potential (mechanical) force of the pathogen.

(i) Nature of cell walls and efficiency of the enzymes of pathogen

Cell wall mainly composed of pectin, cellulose, hemicellulose and lignin. The outer layer of host plant cell may have a layer of wax which is followed by cuticle impregnated with wax.

Fungi secrete three main substrates softening and solubilizing enzymes. These are pectolytic, cellulolytic and lignolytic.

Wax and **Cutin**. No pathogen secretes enzymes that would degrade wax. Only entry can enable a pathogen to break this layer by *potential (mechanical) force*. Cutin layer is penetrated either by pressure exerting force or by the action of degrading enzymes: cutinase, cutin esterase, and carboxyl cutin esterase. Cutinase breaks cutin into fatty acid and hydroxyfatty acid.

Pectin. Pectic substances which form the basic material of middle lamella, primary and secondary walls are degraded by pectinolytic enzymes. The pectinolytic enzymes act against pectin and pectic substances in many steps, are:

(i) The enzymes pectin methyl esterases (PME) hydrolyse the pectic substances into methanol and pectic acid;

(ii) The enzymes pectin glycosidases, polygalacturonases (PG) and polymethyl galacturonases (PMG) degrade the pectic acid and methylated chains of pectin.

The degradation of pectic substances provides nutrients for many fungal pathogens and due to weakening of the cell wall facilitates inter- and intra-cellular invasion by the hyphae.

Subsequently in the secondary wall of the cell, pectin is replaced by cellulose. Cellulose layers contain appreciable amounts of protein.

Cellulose. Cellulose forms the structural framework of cell walls. Cellulolytic enzymes act upon the cellulose and break it to simple compounds (glucose) and make the way for easy penetration of the pathogen into the host cell.

Hemicelluloses. In the spaces between the cellulose microfibrils of the cell wall, **hemicelluloses** are present which are degraded by hydrolytic cleavage by enzymes hemicellulases. The degradation of hemicelluloses by the enzymes hemicellulases produces simpler sugars. In vascular wilt diseases, the molecules released by cellulose degradation can cause plugging of the vessels.

Lingin. The most complex chemical compound in the plant cell wall is **lignin** which occurs chiefly in the matrix surrounding the cellulose fibrils. It is one of the most structurally complex biopolymers whose degradation is caused by polyphenol oxidases produced by wood-rotting fungi. But fungi cannot utilize the degradation products.

Nematodes. The attack of nematodes is both internal as well as external. Both endophytic and ectophytic nematodes use force as well as enzymes to dissolve or break the cell wall to come in contact with cell protoplasm through their stylet. But endophytic nematodes enter bodily in the host cell, whereas the ectophytic ones remain outside the host.

4. **Entry through specific parts or organs**

The pathogens have specificity of host organs or parts during their entry. The organs are stem, leaf, root hairs, floral parts, coleoptile, buds, etc. These are exemplified by:

(a) *Claviceps purpurea* infects the ovaries of rye flower,

(b) *Erysiphe graminis* infects the epidermal cells of leaves of cereals and grasses,

(c) *Plasmodiophora brassicae*. Soil inhabiting pathogens enter through root hairs.

(d) *Synchytrium endobiuticum* infects and remains confined in the epidermal cells of potato tuber,

(e) *Ustilago nuda* causes infection only in the flowers of wheat,

(f) Certain wood rotting fungi are – *Merulius lacrimans* and *Fomes annosus*.

C. **Post-penetration stage**

After the successful penetration, successful post-penetration normally takes place. But successful entry of pathogen may not always lead to immediate infection and disease development. The disease development may be delayed or failure. The process of disease development becomes delayed due to prolonged incubation period. Success of the post-penetration process mainly depends upon the competition of pathogen for nutrition, and production of enzymes and toxic substances and their effects on the metabolic activities of host plant.

Further, due to presence of toxic substance(s) in host cytoplasm, the pathogen may fail to develop further. Host-pathogen interaction may also result **hypersensitivity** of the host tissue, where rapid death of the affected cells prevents further spread of the pathogen. The spots produced during hypersensitivity are called **hypersensitive spots**.

But in most plant diseases, host infection is followed by invasion, a condition when a pathogen grows rapidly in the host tissue. Mode of invasion varies with the nature of pathogen.

(i) **Bacteria** invade host tissues intracellularly and destroy them,

(ii) **Fugal mycelia** invade both inter- or intracellularly but may or may not cause destruction immediately after invasion.

(iii) **Viruses** always invade host tissues intracellularly. They multiply in the living host cells by directing them to manufacture viral nucleic acid and viral protein, their movement from cell to cell are through plasmodesmata.

Further fungal hyphae and spores, and bacterial cells may move through vascular tissues once they gain entrance in them. Successful host invasion of the pathogen is invariably associated with disease syndromes of various types and varying degrees.

After penetrating the host cell walls, the pathogen comes in contact with the host cytoplasm from which it gets its required nutrition. In response to the activities of pathogen, the host metabolic processes, including osmo-regulation, respiration, photosynthesis, etc., get disturbed.

Pathotoxin

(Definition, criteria and example), HV toxin and Wild-fire toxin

The process of pathogenesis in plants is simplified by different factors like susceptibility of host, virulence of pathogen, presence of favourable environmental conditions, enzymes and metabolites of pathogen. In relation to disease development, the term toxin is commonly used for those substances, usually but not invariably of pathogen origin, those are injurious to host plants and indirectly or directly have a role in disease development.

Definition

Toxins have been defined by many in different ways. Some of these are given below:

(i) The pathogen metabolites which cause rapid death of host cells are regarded as toxins.

(ii) Any substance produced by a pathogen which is injurious to host plant and it directly or indirectly play a crucial role in pathogenesis.

(iii) Leopold (1960) defined toxin as a product of microorganism or microorganism-host interaction which acts directly on living host protoplasts to influence either the cause of disease development or symptom expression.

(iv) Wheeler and Luke (1963) have recognized pathotoxins as those that have a definite role to play in the causation of diseases.

(v) Wood (1967) defined toxin as "substances produced by parasites other than microorganisms that damage or kill protoplasts of plants irrespective of their origin or nature".

Criteria to be a toxin

Wheeler and Luke (1963) have proposed the following criteria for pathotoxins. These are:

(i) It should be applied at concentrations which could be reasonably expected in or around the diseases plant.

(ii) It should produce in a susceptible host and all the symptoms characteristic of the disease.

(iii) The pathogen and its toxin should exhibit similar host-specifity.

(iv) The ability of the pathogen to produce the toxins should vary directly with its ability to cause disease, and

(v) A disease should involve a single toxin.

Characters of toxin

The diseases incited by toxins (produced by pathogens) are **characterized** by:

(i) Rapidly spreading tissue necrosis with sharp differentiation from healthy tissue or separated from it by yellowish halo zone.

(ii) Dark colour, and

(iii) Mostly leathery structure of the affected tissue.

In chemical nature, the toxins are of different types. These are peptides, glycoproteins, amino acid derivatives, terpenoids, polysaccharides and organic acids.

Based on the source of origin (Wheeler and Luke, 1963), the microbial metabolites having toxic effects on host plants i.e. toxins are broadly classified into three broad groups are: 1. pathotoxin, 2. phytotoxins and 3. vivotoxin.

1. **Pathotoxins.** These are the toxins, play a major role in disease production and all or most of the symptoms characteristic of the disease are produced in susceptible plants. Most of these toxins are produced by pathogens during pathogenesis, e.g. **Victorin**: *Cochliobolus victoriae*, the causal agent of Victoria blight of oats. This is a *host specific toxin*.
2. **Phytotoxins.** These are toxic substances produced in the host plant during host-pathogen interactions for which a causal role in disease is merely suspected rather than established. These are the products of parasites which induce few or none of the symptoms caused by the living pathogen. They are non-specific and there is no relationship between toxin production and pathogenicity of disease-causing pathogen (Table 5.7). e.g. **Alternaric acid** – *Alternaria solani*, the causal agent of early blight of potato; **Piricularin** – *Pyricularia oryzae*, the *causal* agent of Blast disease of rice. (Wheeler and Luke, 1963).
3. **Vivotoxins.** These are the toxic substances produced in the infected host by the pathogen and/or its host which functions in the production of the disease, but is not itself the initial inciting agent of the disease. e.g. **Fusaric acid** – *Fusarium* sp. (Table 5.7), the wilt causing agent. (Dimond and Waggoner, 1953).

Two other terms are generally used in this context are **exotoxin** and **endotoxin.** Exotoxin is the extracellular toxin which diffuses from living bacterial cell and endotoxin is the internal toxin formed inside bacterial cell, liberated after death.

Mode of action of toxins

Toxins act as poisons on protoplasm of host cells. As the effect of toxins on plant cells is variable in nature, thus it is quite possible that they act in many different ways. Some of them follow more than one mechanism, of which one may be of greater importance in particular host-parasite interaction. They may alter permeability of cell membrane; may inhibit the activity of enzyme; due to structural homology may form complex with some essential metabolite and thus normal metabolism of host may hamper, and rate of respiration mostly increased in infected host tissue in in-vivo condition.

Based on the variation in the action of toxin in different hosts during host-parasite interaction, the toxins are divided into two groups. These are **general toxins** and **specific toxins.**

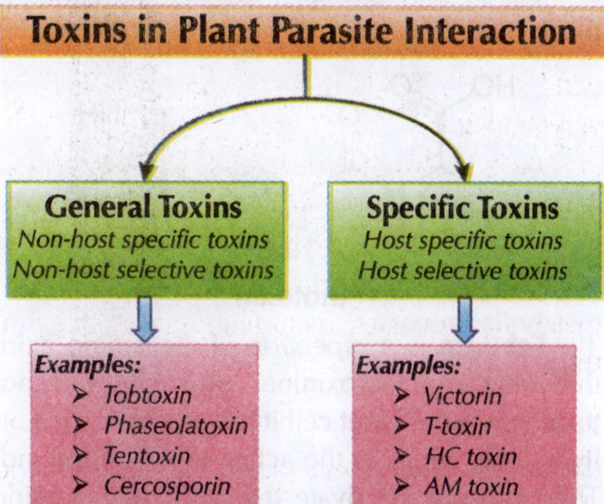

(i) **General toxins.** General toxins are those that act as general protoplasmic poisoning and they effect on many species of plants of different families. They are also called **non-host specific** or **non-host-selective toxins.**

(ii) **Specific-toxins**. Specific toxins are toxic to only a few plant species or variety of plants and harmless to others. They are also called **host specific** or **host-selective toxins.**

A. General toxins or non-host-specific toxins or non-host-selective toxins

These are produced some or the entire disease syndrome, not only on host plant but also on other non-host species. These are regarded as general protoplasmic poisons. These have been also called *"secondary determinants of disease"*, as they are non-host specific in toxin action. Some of those are given below:

1. Tabtoxin or Wildfire toxin

Tabtoxin is produced by *Pseudomonas syringae* pv. *tabaci*, a bacterium; the causal agent of wildfire disease of tobacco. The bacteria produced a dialyzable, thermostable toxin in culture called as **tabtoxon** (Clayton, 1933). Application of culture filtrate on leaves of different species, produce the typical netrotic lesion with a chlorotic hallow around it. It is produced not only on tabacco leaves, but also on the leaves of other some non-host plants. The bacterium is specific on tobacco, but toxin is not. The mutant of the pathogen unable to produce the toxin and also lost its pathogenicity, thereby no disease is produced.

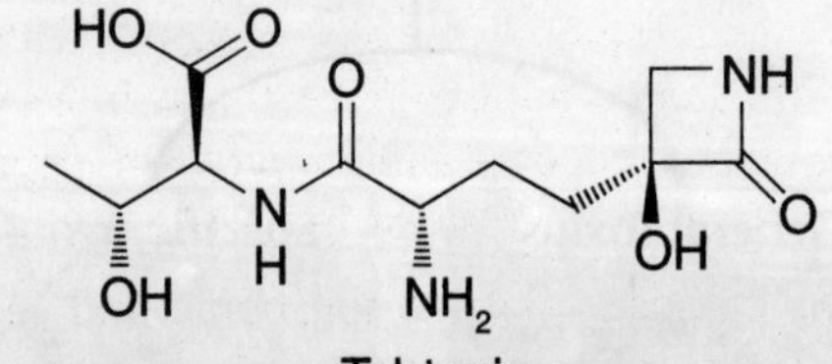

Tabtoxin

Tabtoxin is a *dipeptide* of the amino acid, threonine and tabtoxinine. As such, it is not toxic, but in the plant cell it releases tabtoxine on hydrolysis, which is the active toxic compound. The tabtoxine inactivate the enzyme glutamine synthetase, causes depletion of glutamine levels and as a consequence accumulation of ammonia. Ammonia destroys the thylakoid membrane of chloroplast, caused *chlorosis* and even *necrosis*.

In addition to the above some other strains of *P. syriggae* pv. *tabaci* infect beans and soybean and other pathovars of *P. syringae* that infect maize, coffee and oat also produce tabtoxin.

2. Phaseolotoxin

This toxin is produced by *Pseudomonas syringae* pv. *phaseolicola*, the causal agent of halo blight of bean and some other legume, which on application to the host, produces localized necrotic symptom with outer chlorotic halo. Both infected plants and those treated with purified toxin also show reduced growth of new leaves, disruption of apical dominance, and accumulation of ornithine, an amino acid.

Phaseolotoxin is a *tripeptide* with amino acids, ornithine, alanine and arginine; carrying a phosphosulfamyl group. Just after its production by the bacterium in host plant, the plant enzymes cleave the peptide bonds to release alanine, arginine and also phospho-sulfamylornithine i.e. PSMO. The last one is the actual biologically active moity. The phaseolotoxin binds with the active site of cell and it inactivates the enzyme ornithine carbomayl transferase, which catalyses the inter-conversion of ornithine and citrulline, thus leads to an accumulation of ornithine and depleted level of arginine.

Inhibition of the enzyme has been observed in both *in vivo* and in *vitro* condition. The phaseolotoxin induced chlorosis is suppressed by citrulline. The mechanism of chlorosis induction in host plants is not yet clearly elucidated.

3. Other non-host specific toxins

In addition to the above two toxins, other toxins involve in pathogenesis are Tentoxin, Cercosporin, Hull tor toxin, *Pyricularia* toxins and toxins in wilt diseases are also important.

B. Specific-toxins or host specific toxins or host-selective toxins

These toxins are host specific and acts on cells/tissues of susceptible host varieties. They are unable to act on non-host plants and also on resistant varieties. Thus, they are closely related with disease development. They remain involved directly in pathogenesis from the beginning of infection to later stages and cause natural symptoms when applied to healthy susceptible host plants and thus they are regarded as primary

Table 5.7: Toxins produced by pathogenic fungi and bacteria and respective diseases

Pathogen	Produced Toxin	Diseases
Fungi		
1. *Alternaria* spp.	Alternaric acid, Alternariol, Zinniol	Leaf spot of different plants
2. *Ceratocystis ulmi*	Ceratoulmin	Dutch elm disease
3. *Cercospora* spp.	Cercosporin	Leaf spot of beet
4. *Cochliobolus* spp.	Ophiobolins	Blight of cereals
5. *Fusarium oxysporum* f.sp. *lycopersici* and other *Fusarium* sp.	Fusaric acid, Lycomarasnin	Wilt of different plants
6. *Pyricularia oryzae*	Pyricularin	Blast of rice
Bacteria		
1. *Pseudomonas syringae* pv. *syringae*	Syringotoxin	Leaf spot of citrus
2. *Pseudomonas syringae* pv. *tagetis*	Tagetitoxin	Leaf spot of marigold

determinant of the disease. The distinctive characters are as follows:

(a) They produce all the normal disease symptoms in susceptible hosts as found in natural infection.

(b) Pathogenecity of a race of pathogen is directly proportional to its ability in toxin production.

(c) Toxin affects only in susceptible varieties.

(d) Toxin has the same host range as the pathogen, and

(e) Toxin is involved in all stages of disease development; thus, they are considered as the primary determinant of disease.

The well-known host specific toxins produced by the pathogens are:

(i) *Helminthosporium victoriae* on oat (Victorin or HV-toxin).

(ii) *Cochliobolus heterostrophus* race T on maize (T-toxin).

(iii) *Cochliobolus carbonum* on maize (HC-toxin).

(iv) *Alternaria alternata* on apple (AM-toxin) etc.

Some of them are discussed below:

1. **Victorin or HV (*Helminthosporium victorie*) toxin**

The toxin is produced by the fungus *Cochliobolus (Helminthosporium) victoriae*, the causal agent of blight disease of oat. The oat variety "Victoria" and other derived varieties were extensively cultivated as fodder in USA. The fungal disease, blight caused by *Helminthosporium victoriae* appear in 1940 and in 1944-45, the disease become a major problem, devastated the crop in many areas by 1947.

The pathogen is both soil borne and seed borne. The pathogen attacks the basal portion of oat plant but causes blight on leaves and ultimately destroy the entire plant. The pathogen remains confined to the base of stem but initially develops symptoms on leaves. Meehan and Murphy (1947) suggested the production of a translocatable toxin by the pathogen caused symptoms on leaf. Culture filtrate of pathogen also produces all the characteristic symptoms in these cultivars. The toxin produced by the fungus is controlled by a single gene. The toxin produces all external symptoms like the pathogen, with similar histological and biochemical changes like:

(i) Changes in cell wall structure,
(ii) Loss of electrolytes from cell,
(iii) Increased respiration, and
(iv) Decrease protein synthesis and growth.

Victorin

Chemically it is a complex chlorinated, partially cyclic pentapeptide. The primary target is the cell membrane where it binds itself. But it is not yet clear, how it affects cell metabolism in susceptible host.

2. **HC-toxin**

The toxin is produced by the fungus *Cochliobolus (Helminthosporium) carbonum*, the causal agent of leaf spot disease of maize. The toxin acts on and toxic to specific maize lines. In general HC-toxin has the same relation with corn as HV-toxin on oat, but not so effective like the latter.

HC-toxin is a cyclic substituted polyamide with the formula $C_{32}H_{50}N_6O_{10}$ (Pringle, 1970). On complete hydrolysis it gives alanine: proline: an unsaturated analogue of leucine: an unknown hydroxy amino acid :: 2:1:1:1.

HC-toxin

HC-toxin induces little increase in electrolyte release and respiration, and also in dark fixation of CO_2. Electrolyte release is observed nearly 24h after infection or toxin application. But the toxin does not show its effect on isolated chloroplast or mitochondria (Yoder, 1971). Application of HC-toxin on cell protoplast causes an increase in nitrate uptake, but its significance is not yet clear.

Recent studies indicate that the resistant cultivars have a gene HM1, which codes for HC-toxin reductase, which detoxify and reduces the toxin, is absent in susceptible cultivar.

Table 5.8: Other host specific toxins

Pathogen	Toxin	Host plant
1. *Alternaria alternata* f.sp. *lycopersici*	AL-toxin	Tomato
2. *A. citri* (lemon race)	ACL-toxin	Rough lemon
3. *A. citri* (tangerine race)	ACT-toxin	Dancy tangerine
4. *Alternaria* spp.	AK-toxin	
5. *Cochliobolus (Helminthosporium) heterostrophus*	T-toxin	Maize
6. *Cochliobolus sacchari*	HS-toxin	Sugarcane
7. *Corynespora cassicola*	CC-toxin	Tomato
8. *Perconia circinata*	PC-toxin	Sorghum
9. *Phyllosticta maydis*	PM-toxin	Maize, with male sterile cytoplasm.

3. **Other host specific toxins**

Many other host-specific toxins are known in different fungi. List of other host-specific toxins are given in above page (Table 5.8).

Host defense mechanism with special reference to phytoalexin

Every organism wants to live normally with proper nutrition. In host parasite infection, host want to defend the pathogen and pathogen want to produce disease. Plant defends themselves from pathogenic attack mainly by:

(i) **Structural defence**, that acts as a physical barrier and inhibits the infection by pathogen, and

(ii) **Biochemical defence** i.e. the biochemical reactions take place in cells/tissue as a whole.

Both the above types of defence may be used by plants against pathogen at any stage of infection and they may be **pre-existing** i.e. exists before infection, or may be induced by infection, i.e. **post-infectional**. Both the structural and biochemical defence mechanisms of plants in pre-existing and post-infectional forms are briefly discussed below:

I. Structural defence

1. Pre-existing defense structures

Structures present on the surface of the plant, acts to defend the infection by pathogen. These structures include:

(i) *Wax on cuticle*. Presence of waxes on the surface of aerial parts of the plant form a hydrophobic surface, thus do not allows in forming water film on it. Thereby pathogens are not able to deposite on and fail to germinate (fungus) or multiply (bacteria).

In addition, a negative charge usually develops on the leaf surface due to the presence of fatty acids, the main component of the cuticle. The negative charge prevents or reduces the chance of infection by many pathogens.

(ii) *Thickness of the cuticle*. Thick cuticle of the host offers adequate resistance against direct penetration by the pathogen, those try to enter the host through the leaf surface. The cuticle thickness obstructs the entry of pathogen. In addition, a thick cuticle also reduces the exit of the pathogen from inside the host (if present), thereby the secondary infection become reduced.

(iii) *Structure of the epidermal cell wall*. Thicker and stouter outer wall of epidermal cells acts as a competent barrier against direct penetration by fungal pathogens.

(iv) *Size, shape and opening time of stomata and structure of lenticels etc*. Both fungi and bacteria enter in plants through stomata, but their structure and behavior also decide the fate of entry.

The stem rust of wheat pathogen takes the entry, if **stomata** are open. Cultivars of wheat, where stomata open late in the day, behave as resistant. In Citrus (Szincum variety), the stomata are small and possess very narrow openings, that surrounded by broad lipped raised structures which prevent the entry of water drops containing citrus canker bacterium.

Similarly, the size and internal structures of **lenticels** may play a defensive role against the pathogens. Varieties of apple having small lenticels on fruits prevent the entry of the pathogen while those having large openings easily allow the pathogen to take entry.

Nectaries provide openings in the epidermis of host and may play a defensive role due to high osmotic concentration of the nectar. In apple, the resistant varieties have abundant hairs in the nectaries acts to defend infection while susceptible varieties are with less hair.

(v) *Presence of thick-walled cells*. Due to the presence of thick walls, pathogens (fungi, bacteria, nematodes etc.) are not able to progress through or beyond that structure. In many cereals, due the presence of bundles of extended area with sclerenchyma, the stem rust pathogen may stop to spread further. In angular leaf spot diseases, the Leaf veins effectively obstruct the spread of the pathogen, *Pseudomonas* sp.

2. Post-infectional defense structures

After the establishment of physical contact with the host plant, **early recognition** of the pathogen is very important. After recognition the pathogen gets the entry and infects the plant. But later, to check the further growth and invasion by the pathogen in host, the host plants develop some structures or mechanisms by which they protect the growth of the pathogen. Sometimes death of the invaded cell or tissue as takes place, called necrosis and the tissue as a whole called necrotic tissue. These are briefly discussed below:

(i) **Defense reactions in the cytoplasm**

After hyphal invasion, the host cells cytoplasm and nucleus gets stretched to break into two, or in some cases both cytoplasm and nucleus enlarges. In some other case, the cytoplasm becomes granular and dense and develops granular particles. These result in the disintegration of mycelium and thus the invasion stops. Such cytoplasmic defence mechanisms are seen in some weak pathogens. e.g. some mycorrhizal fungi.

(ii) **Defense structures in cell wall**

Defense structures in cell wall have limited importance in host defense. It includes morphological changes in cell wall. These are as follows:

(a) In response to pathogen, cell walls become thickened by producing **callose material**, thus prevent the entry.

(b) On invasion of bacterial cells, the outer layer of cell walls of parenchyma cells produce an **amorphous fibrillar material**, that traps the bacteria and prevents multliplication.

(c) Similarly, on invasion of fungal hyphae, **callose papillae** being deposited in the inner layers of parenchyma cells and thus prevent the growth of pathogen. In some cases, fungal hyphae after entering into the cell lumen get **enveloped by callose material**, that later become infused with phenolic substances and forms a sheath around the hyphae.

(iii) **Defense structures developed by host tissues**

Four types of defence structures are developed in the host tissues after successful penetration. These are:

(a) **Deposition of gum.** After infection, host plant produces gummy substances around the spots, which inhibit the further progress of pathogen (Fig. 5.9A). This is commonly found in stone fruits.

(b) **Abscission layers**. Abcission layers (Fig. 5.9B) are commonly formed in plants to separate the older leaves and ripen fruits. But this layer may develop in young leaves in response to infection by pathogen, thus a gap is formed between the healthy and infected tissue by dissolution of middle lamella in one or two layers of cells and inhibits further progress of the pathogen.

(c) **Tyloses.** Tyloses are the out growths of protoplast of adjacent living parenchyma cells of xylem vessels those protrudes into xylem vessels through pits in response to attack by vascular pathogens (Fig. 5.9C). In resistant plants, the tyloses block the further flow or growth of the pathogen and prevent the pathogen to grow further. It also hampers the flow of water and nutrients, thereby wilting of plant may take place.

(d) **Formation of cork layers**. Due to stimulation of host cells by certain substanes secreted by pathogen (some fungi, bacteria, virus etc.) during infection, host forms multilayered cork cells below the infection zone (Fig. 5.9E, F). Cork cells are dead cells, which stop the

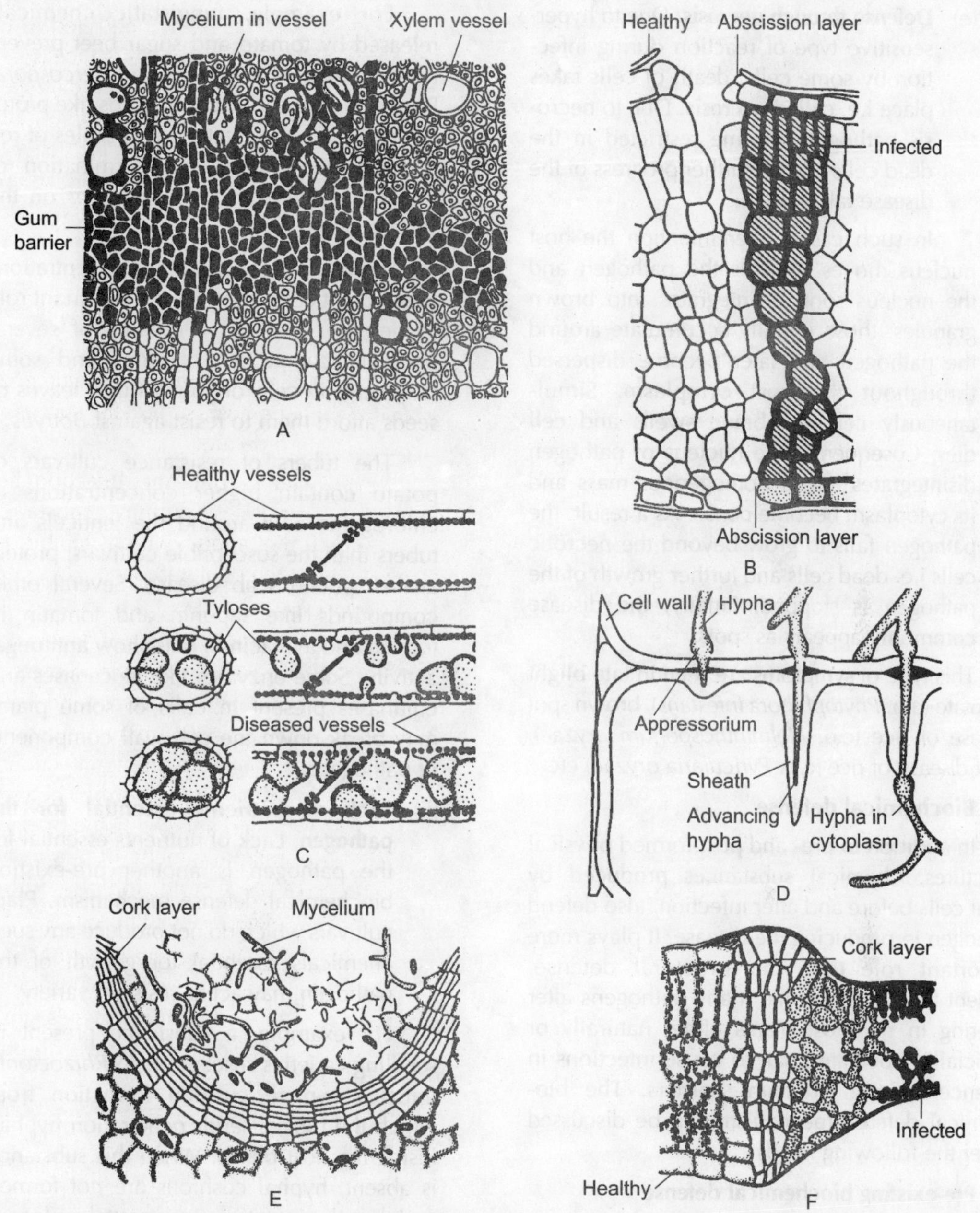

Fig. 5.9: Different defense structures developed after infection: A Gum deposition in cells, B. Abscission layer, C. Tyloses, D. Formation of sheath around developing hyphae, and E-F. Development of cork layer in tubers (E) and leaf (F).

flow of nutrients beyond that layers, deprive the pathogen from nutrition and thus these layers inhibit further invasion by the pathogen inside the living tissue. This is found in soft rot of potato caused by *Rhizopus* sp., Damping off of jute seedling (after 20 days old seedling) by *Rhizoctonia bataticola*, scab disease of potato caused by *Streptomyces scabis*, and necrotic lesions of tobacco caused by tocacco mosaic virus etc.

(e) **Defense through necrosis**. Due to hypersensitive type of reaction during infection by some cells, death of cells takes place i.e. called **necrosis**. Due to necrosis pathogen become restricted in the dead cells and no further progress of the disease takes place.

In such cases, after infection the host nucleus moves towards the pathogen and the nucleus soon disintegrates into brown granules, those initially accumulate around the pathogen and later become dispersed throughout the host cytoplasm. Simultaneously cell membrane swells and cell dies. Cosequently the nucleus of pathogen disintegrates into a homogenous mass and its cytoplasm become dense. As a result, the pathogen fails to grow beyond the necrotic cells i.e. dead cells and further growth of the pathogen is stopped. Finally, the disease commonly appears as spots.

This type of symptoms are seen in late blight of potato (c.o. *Phytophthora infestans*), brown spot disease of rice (c.o. *Helminthosporium oryzae*), blast disease of rice (c.o. *Pyricularia oryzae*) etc.

II. Biochemical defense

In addition to pre- and post-formed physical structures, chemical substances produced by plant cells before and after infection, also defend pathogen in producing the disease. It plays more important role than the structural defense, evident from the fact that many pathogens after entering in non-host plants either naturally or artificially inoculated, fail to cause infections in absence of any structural barriers. The biochemical defense mechanism can be discussed under the following heads.

1. Pre-existing biochemical defense

(i) **Presence of protective chemicals in cells**. Plant generally exudes many organic substances present in cells through the above ground parts (phylosphere) and roots (rhizosphere). Some of the compounds released by some plants are known to have an inhibitory effect on certain pathogens during the pre-penetration stage.

For example, fungistatic chemicals released by tomato and sugar beet prevent the germination of *Botrytis* and *Cercospora*. Presence of phenolic compounds like protocatechuic acid and catechol in scales of red onion variety inhibit the germination of conidia of *Colletotrichum circinans* on the surface of red onion.

Inhibitors present at high concentrations in the plant cells also play an important role in defense of plants. Presence of several phenolic compounds, tannins and some fatty acids in cells of young fruits, leaves or seeds afford them to resist against *Botrytis*.

The tubers of resistance cultivars of potato contain higher concentrations of chlorogenic acid around the lenticels and tubers than the susceptible cultivars; protect against potato scab disease. Several other compounds like saponin and tomatin in tomato and avinacin in oats show antifungal activity. Some enzymes like glucanases and chitinases present in cells of some plants may break down the cell wall components of pathogens.

(ii) **Lack of nutrients essential for the pathogen**. Lack of nutrients essential for the pathogen is another pre-existing biochemical defense mechanism. Plant cultivars which do not produce any such chemicals essential for growth of the pathogen may act as resistant variety.

For example, a substance present in seedling varieties susceptible to *Rhizoctonia* initiates hyphae cushion formation from which the fungus sends penetration hyphae inside the host plants. When this substance is absent, hyphal cushions are not formed and thus the infection does not take place.

(iii) **Absence of common antigen**. It is now well known, that the presence of a common protein (antigen) in both the host and pathogen determines the disease occurrence in the host. But if the antigen is present or absent in the host or vice-versa, it makes the host resistant to the pathogen.

Varieties of linseed which have an antigen common to their pathogen are susceptible to the disease, rust of linseed caused by *Melampsora lini*. In contrast, the absence of antigen in linseed varieties but occurring in the pathogen is resistant to the pathogen. Similarly, cotton plants also behave as resistant or susceptible against leaf spot pathogen, *Xanthomonas campestris* pv. *malvacearum*, in absence or presence of common antigen.

2. Post-infectional biochemical defense

Due to infection caused by pathogen or injuries by any means, the plant cells or tissue as a whole, synthesize many chemical substances, those inhibit the growth of the pathogen. These substances are generally produced around the site of infection, to overcome the problem. These are phenolic compounds phytoalexins, and others.

(i) **Phenolic compounds**. These are the most common compounds produced by plants in response to injury or, infection by the pathogen. In host plants, the synthesis of phenolic compounds takes place either through "acetic acid pathway" or "Shikimic acid pathway".

Some common phenolic compounds toxic to pathogens are chlorgenic acid, caffeic acid and ferulic acid. These phenolic compounds are produced at a much faster rate in resistant cultivars than the susceptible one. Probably the combined effect of all phenolic compounds present in host plants is responsible for inhibiting the growth of the pathogen.

(ii) **Phytoalexins**. Phytoalexins are antimicrobial low-molecular-weight compounds (secondary metabolites) that are synthesized *de novo* following .pathogenic attack (Bailey & Mansfield, 1982). This can also be induced in plant by injury or physiological stimuli.

The concept of phytoalexin was first postulated by Müller and Börger in 1941 from their studies of the reaction of potato varieties to virulent and avirulent strains of *Phytophthora*. Cruickshank and Perrin (1964) were able to show that pisatin in pea fulfulled all the criteria required by Müller and Börger's theory and this substance remains one of the most fully investigated phytoalexins known today.

The basic response in both resistant and susceptible cells is the same, but the basis of differentiation between resistant and susceptible hosts being the speed of formation of the phytoalexin. All phytoalexins are lipophilic compounds. These are broad spectrum inhibitors and are chemically diverse with different types of characteristic of particular plant species.

Bill (1981) stated that, the peak concentration of phytoalexins almost always coincides with the development of necrosis. Although the exact mechanism of production of phytoalexin has not been properly understood, it is considered that a metabolite of the host plant interacts with specific receptor on the membrane of pathogen, resulting in the secretion of "phytoalexin elicitor" which enters the cells of host plant and stimulates the phytoalexin synthesis.

Phytoalexins are considered to stop the growth of pathogens by altering the plasma membrane and inhibiting the oxidative phosphorylation.

They are widespread in Fabaceae, Solanaceae and 15 other families, but they are not cosmopolitan in availability, as not been reported from Cucurbitaceae and among Poaceae, they occur in oat, rice and sorghum, but not in maize. Phytoalexins tend to fall into several classes including terpenoids, glycosteroids and alkaloids. Some common phytoalexins are Ipomeamarone, Orchinol, Pistatin, Phaseolin, Medicarpin, Rishitin, Isocoumarin, 'Gossypol' Cicerin, Glyceolin, Capsidiol etc.

List of some phytoalexins, chemical nature, the host and the pathogens are given

Table 5.9: List of phytoalexin, producing host, pathogen and their chemical nature

Phytoalexin	Host	Pathogen	Chemical nature of the phytoalexin
1. Ipomeamarone	*Ipomoea batata*	*Ceratocystis fimbriata*	Furanosesquiterepene ketone
2. Orchinol	*Orchid malitaris*	*Rhizoctonia repens*	Phenanthrene
3. Pistatin	*Pisum sativum*	*Ascochyta pisi*	Kievitone
4. Phaseolin	*Phaseolus* sp.	*Monilia fruticola*	
5. Medicaripin	*Medicago lupulina*	*Colletotrichum phomoides*	Dimethyl-homoptero-carpin
6. Rishitin	*Solanum tuberosum* var. *rishiri*	*Phytophthora infestans*	Non-sequiter-pene
7. Isocoumarin	*Daucus carrota*	*Fusarium oxysporum* f. *lycopersici, Rhizopus solonifer*	–
8. *Cicerin*	*Cicer arietinum*	*Ascochyta rabie*	
9. Gossypol	*Gossypium* sp.	*Verticillium alboatrum*	
10. Capisidiol	*Capsicum fruitescens*	*Peronospora trifolium*	
11. Wyerone acid	Broad bean	*Botrytis fabae*	

in the Table 5.9. Their chemical structures are given below.

(iii) **Substances produced by host to resist enzymes of pathogen.** Some host plants produce chemicals which neutralize the enzymes produced by the pathogen, thus defend themselves from the attack of pathogen.

Bean plants, infected with *Rhizoctonia solani* cause necrosis. In resistant bean culti-

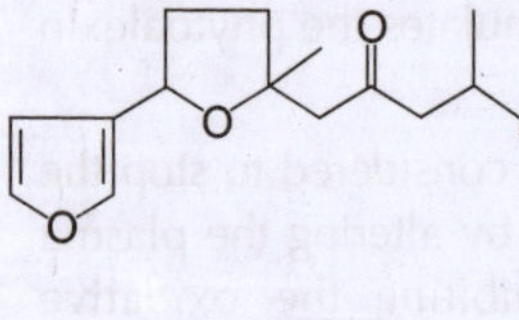

Ipomeamone 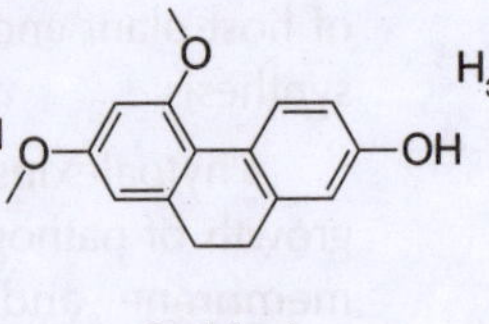 Orchinol 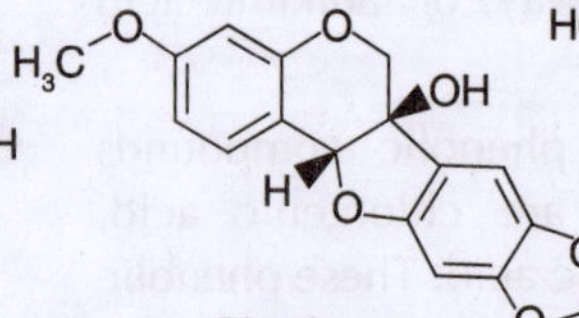 Pisatin 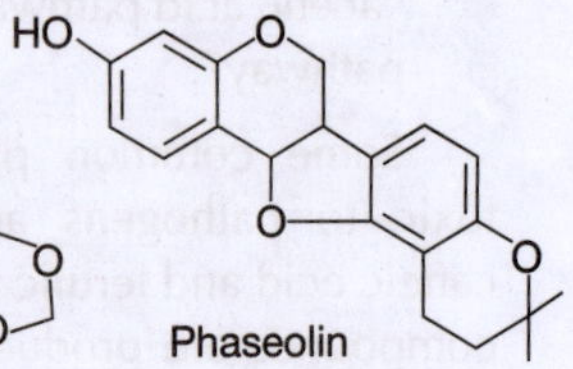 Phaseolin

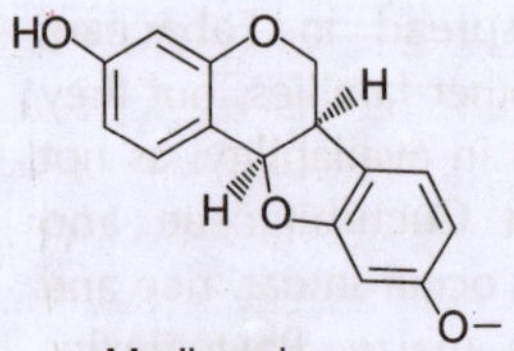

Medicarpin 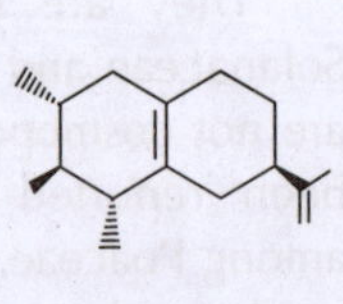 Rishitin 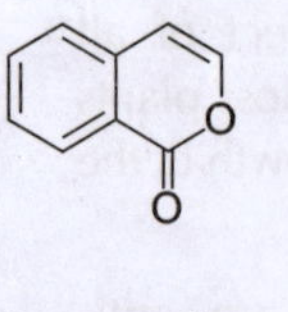 Isocoumarin 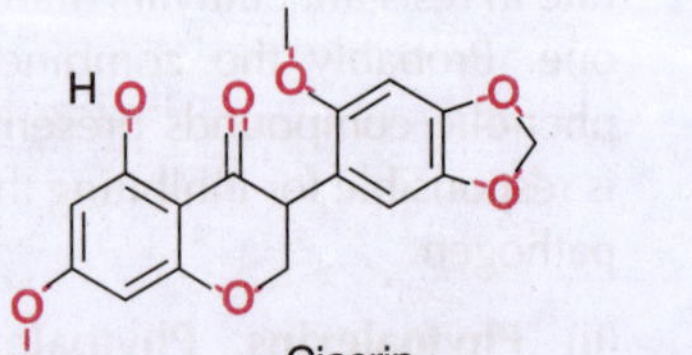 Cicerin

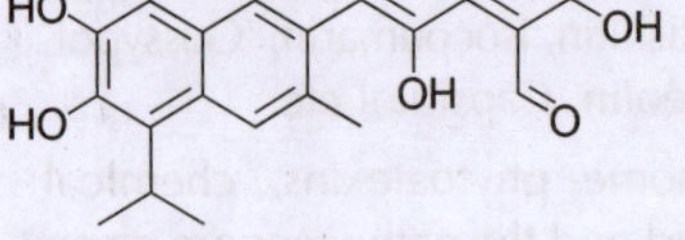

Gossypol 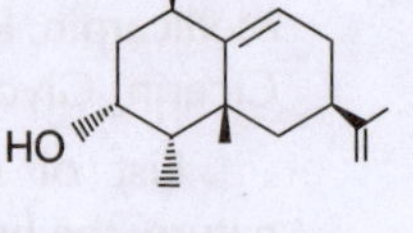 Capsidiol 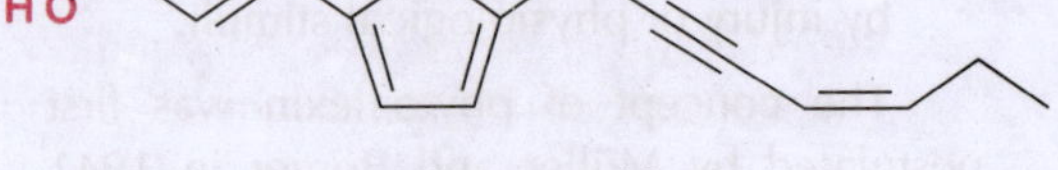 Wyerone acid

vars, the entry of pathogen causes the separation of methyl group from methylated pectic substances and forms polyvalent cations of pectic salts which contain calcium. The calcium ions accumulate in infected as well as neighbouring healthy tissues and because of the calcium accumulation, the pathogen fails to disintegrate middle lamella by its polygalacturonase enzymes. In susceptible bean cultivars, these are known to dissolve the middle lamella of healthy tissue.

(iv) **Detoxification of pathogen enzymes and toxins**. In some cases, the host plants produce chemicals which deactivate the toxins produced by the pathogens.

Pyricularia oryzae, the pathogen of blast of rice; produces the toxins, picolinic acid and pyricularin. Although resistant varieties convert these toxins into N-methyl picolininic acid and pyrecularin into other compounds, the susceptible cultivars become affected by these toxins.

Similarly, in cotton and tomato wilts, the toxin fusaric acid produced by the pathogen gets converted into non-toxic N-methyl-fusaric acid amide in resistant cultivars.

As in case of detoxification of toxins, the toxic enzymes produced by the pathogen is deactivated by phenolic compounds or their oxidation products. Some varieties of cider apple are resistant to brown rot disease caused by *Sclereotinia fructigena*. It may be, because the resistant cultivars produce pheolic oxidation products which inactivate the pectinolytic enzymes produced by the pathogen.

(v) **Biochemical alterations.** It has been observed that infection of the host by the pathogen brings about biochemical changes in the host which may prove toxic to the pathogenic microorganisms and cause resistance to the pathogen. Production of certain new enzymes and other compounds are synthesized and accumulated in higher concentration. This may also add to the resistance of the plant by being toxic to pathogenic microorganisms.

Induced systemic resistance (ISR) and Systemic acquired resistance (SAR)

The Induced systemic resistance (ISR) and Systemic acquired resistance (SAR) are the two different phenomena but both are involved in the active plant defense responses to plant pathogenic attack. ISR is similar to hypersensitive response while as SAR is similar inherent immunity system of plants.

The terms were first time coined by **Ross** (1961) during his studies on interactions between tobacco and its virus disease, the Tobacco mosaic virus (TMV). Induction of ISR is the function of **non-pathogenic** plant growth promoting *Rhizobacteria* (PGPR). However, SAR is triggered by infection of a pathogen.

Induced systemic resistance (ISR)

ISR is the systemic resistance which is naturally present in plant but can be induced or enhanced by plant associated non-pathogenic *Rhizobacteria* or by plant growth promoting *Rhizobacteria*. It is independent of salicylic acid and hence no pathogenesis related proteins are produced but it is plant specific and depends on genotype of the plant. Here plant specific i.e. the host specificity as shown by *Rhizobacteria* during activating resistance response. Actually, these bacteria, first 'recognize' their host plant before setting on ISR. Plant hormones, Jasmonic acid and ethylene are required during Induced systemic resistance (ISR). The Jasmonic acid regulates plants' responses to abiotic and biotic stresses including pathogenic attack. Sometimes it is produced as volatile compound which can reach to plant parts, as well as to the nearby plants and gives warning regarding pathogenic attack and triggers the plant defense responses. In addition, the wounding or pathogen attack also stimulates the production of ethylene which then onset the defense responses in favour of plant. Typical ethylene induced visual defense responses include rapid senescence, ripening and abscission of infected tissue like sudden fall of leaves or the fruits.

Systemic acquired resistance (SAR)

SAR is the resistance which is once acquired remains systemically in the plant body. It is the inherent resistance capability of a plant which is activated on exposure to the pathogen. It is similar to human immune response followed by vaccination.

In this type, plants cellular defense system, use recognition receptor proteins to identify pathogen which may have been plant's pathogen in the past.

Induction of SAR requires pathogenic necrosis in the host tissue. SAR once induced remains active against many pathogens for long time and not only resist pathogen attack but also able to cure diseases.

SAR develops primarily from necrotic tissue caused by the pathogen. Necrosis is followed by release and accumulation of salicylic acid in phloem tissue which triggers initially the hypersensitive response and induction of SAR. Salicylic acid is a plant hormone, required for the production of PR proteins, which inhibits induction of virulence factors and also known to control another plant hormone, the ethylene. These proteins are the products of pathogen activated PR genes and are also necessary to induce plant defense, some of them function as antimicrobials degrading cell wall of the pathogen. Some PRs are specifically antibacterial, antifungal and antiviral. They also possess lytic chitinase, lysozyme and peroxidase enzyme activities which are functional against different plant pathogens. They also act as messenger to signal the attack of the pathogen. This signaling process activates lignin formation and its deposition creates efficient barrier to infecting agents. Biochemical genetics of signaling pathway is still not clearly known. The onset of SAR determines plant's defending ability, which is found to be very high and elaborated as various plant parts including distantly present leaves, branches, stems and roots, also acquires immunity to pathogen. All plant parts need not be a site of infection or necrosis or pathogenic contact.

Induction of ISR and SAR by elicitors

Elicitors are compounds, which activate chemical defense in plants. Various biosynthetic pathways are activated in treated plants depending on the compound used. Commonly tested chemical elicitors are salicylic acid, methyl salicylate, benzothiadiazole, benzoic acid, chitosan, and so forth which affect production of phenolic compounds and activation of various defense-related enzymes in plants. Both ISR and SAR can be induced by both natural and artificial eli-citors. These are as follows:

1. Natural elicitors

Non-pathogenic and plant growth promoting *Rhizobacteria* isolated from soil and/or rhizosphere are the potential natural elicitors. Apart from *Rhizobacteria*, many fungi, viruses and nematodes are also able to induce ISR. The residence of PGPR either in symbiotic condition or free-living state in plant rhizosphere is generally enough to induce ISR in host plant naturally and such plants are naturally resistant to their pathogens. This is because antagonistic but plant growth promoting properties such as production of hormone, antibiotic etc., of these bacteria are also found to be the inducers of ISR. Thus, PGPR not only promote plant growth but also as biocontrol agent. Three PGPR i.e. *Bacillus*, *Pseudomonas* and *Rhizobium* are ideal among natural elicitors. Salicylic acid and pathogenic organisms are the only known natural chemical and biological elicitors of SAR.

2. Artificial elicitors

Elicitors or activators of ISR and SAR are successfully used in bio-control. In conventional agriculture both biological and chemical form of elicitors are used. They are either injected into the stems or applied as leaf sprays on plant body. Biological elicitors are formulations (inoculum) of cell extracts, living bacteria, viruses and also fungi, of which bacteria or precisely *Rhizobacteria* are routinely utilized to induce ISR. Chemical elicitors which are commonly used are

Beta-aminobutyric acid, Benzothiadiazole, 2, 6-dichloroisonicotinic acid, Silicon dioxide, polyacrylate, salicylate etc. which induce SAR. They have been used successfully in integrated pest and disease management either alone (only chemical or biological elicitor) or in combination of both.

Comparison between ISR and SAR

Both ISR and SAR are dissimilar with respect to their induction, signaling pathways and their co-ordination, level of defense elicited in host and also in protein expression.

ISR is induced by jasmonic acid and ethylene molecules through coordinated signaling pathway; proteins other than pathogenesis related are too expressed. Level of defense induced is lesser and also depends on host plant genotype and presence of non-pathogenic bacteria.

SAR is induced by salicylic acid. SAR and PR proteins are induced together. The defense level produced by SAR is high, long lasting, broad spectrum and independent of plant genotype but requires necrotic pathogens or presence of inducer analogs. It has been indicated that ISR and SAR show phenotypic similarity and together can produce protection at higher level than they produce alone.

SELECTED PLANT DISEASES

A. BACTERIAL DISEASES

1. *BACTERIAL LEAF BLIGHT OF RICE*

The bacterial leaf blight of rice was first observed in 1884 by farmers of Fukuoka in Japan. By 1926, the disease became wide-spread in the different regions of Japan, mainly in the south west and north east regions. In Japan, the disease is known as **'white withering disease'**. Ishiyama (1922) established that the disease was caused by bacteria. In India, the disease was first recorded from Khopoli of Maharastra in 1951. Later, Srinivashan *et al.,* (1959) reported that the disease was caused by *Xanthomonas oryzae.* Due to introduction of high yielding varieties and frequent use of nitrogen fertilizer the disease was commonly observed in different parts of India. In India, Srivastava and Rao (1966) estimated the loss ranges from 6-60% depending on the variety of rice. Now the disease is commonly recorded in Asian countries like China, India, Indonesia, Japan, Sri Lanka, Thailand and Philippines. Mew *et al.,* (1993) reported that it is one of the most destructive diseases of rice in Asia.

Host. *Oryza sativa* L.

Symptoms

The symptoms are visible at all stages of development and in all parts of rice, except root.

The symptoms appear as water soaked tiny spots on the margin of a mature lower leaf of seedling. With maturity, the leaves become yellow and gradually dry out. On leaf blade, lesions generally appear on the margin or any other region and a few are also formed at the apex as water soaked stripes. The lesions gradually enlarge both in length and width and the leaf becomes yellow within few days (Fig. 5.10A, B).

Glumes of immature grains also become infected showing discoloured spots surrounded by water-soaked tiny spots. The infected grains mature earlier than the healthy one.

The most harmful phase of disease is the **Kresek** or wilting symptom, which was first described in Indonesia. In tropical countries, the apex of the leaves of seedlings is trimmed off before transplanting to reduce the size for proper handling. The infection starts through the cut ends and within one or two weeks, the symptom appears as water-soaked green spots on the leaf.

A

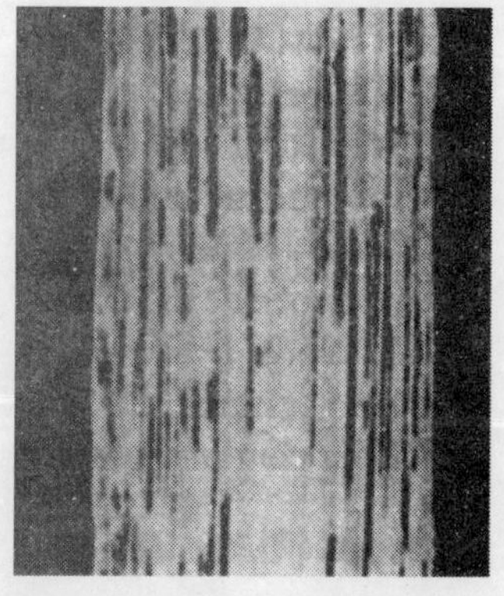

B

Fig. 5.10: Bacterial leaf blight (BLB) of rice: A. Symptoms on leaves of infected plants, and B. Magnified portion of a blighted leaf

Gradually the leaves become greyish-green, roll along the midrib and then withering of entire leaf (including leaf sheath) takes place. The bacterium can move through xylem vessel and infect the other leaves, thereby the entire plant becomes dead. (When a few leaves wilted and are seen to float on water, it is called **Kresek** or wilting symptom; but when complete death of entire plant takes place it is called **Hama lodoh**.)

In comparatively resistant varieties, just inside the margin of the leaf blade, the symptom appears as yellow stripes without showing any necrosis. Milky or opaque bacterial drops may be observed in the early morning on the young lesions.

Collateral hosts

Leersia oryzoides and *Leersia oryzoides* var *japonica* are well-known collateral hosts, effective only in Japan.

Causal organism

The disease is caused by *Xathomonas campestris* pv. *oryzae* (Ishiyama) Dye. Earlier it was known as *X. oryzae, Bacillus oryzae* or *Pseudomonas oryzae*. It is a **Gram-negative, non-spore forming, rod-shaped bacterium with round ends measuring 1-2 µm × 0.8-1 µm with monotrichous flagellum of 6-8 µm** (Ishiyama, 1922). The bacterial cells are capsulated and grow in mass. In *vitro* culture, the colonies show yellowish in colour. The optimum temperature for growth of bacterium is 25-30°C and the thermal death point is 53°C.

Disease cycle

The disease cycle of bacterial leaf blight of rice (BLB) is not clearly known. The source of primary inoculum and the initiation of disease also vary in different countries and even in different places of the same country. The various predispositions regarding disease cycle are based on the different informations published by the plant-scientists of different countries (Fig. 5.11).

The BLB is a **vascular disease**. The bacterium can perpetuate in different ways, such as through soil, seeds, wild grasses, diseased stubbles and straw.

Soil. The bacterium may survive in soil for 1–3 months, depending on soil acidity and moisture (Tagami and Mizukami,1962). Due to

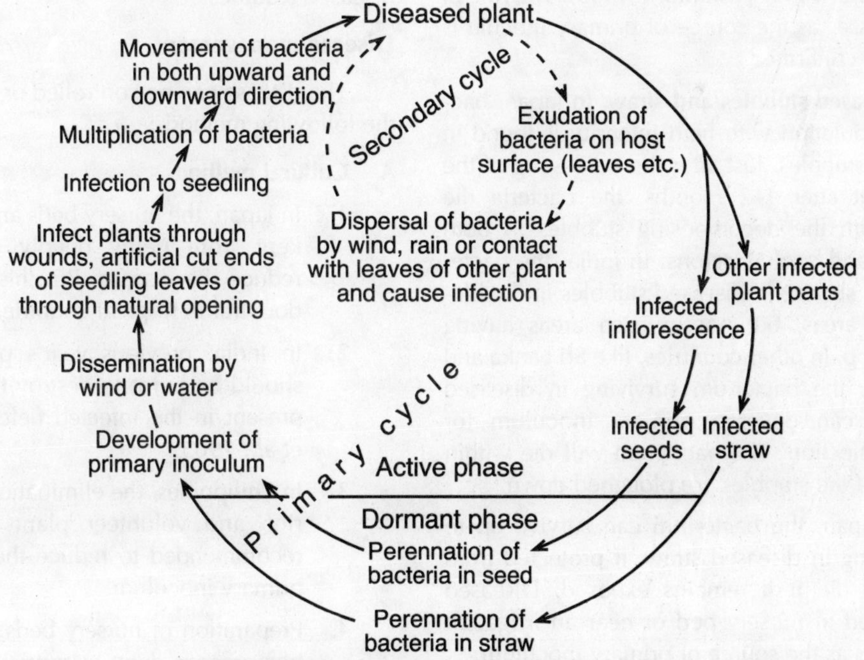

Fig. 5.11: Disease cycle of Bacterial leaf blight of rice

short period of survival, the soil inoculum does not play any role in the transmission of disease to the next season.

Seeds. The bacterium can survive for varying periods in infected seed and is transmitted to the next crop. According to Kauffman and Reddy (1975), the bacterium remains viable in infected seed only for two months. More recently, Debnath (1992) reported that, in India, the bacteria can propagate from season to season through infected seeds. After sowing of infected seeds, the bacteria grow on roots and shoot surfaces which act as primary inocula. The bacteria present in palea and lemma ooze out into soil water during germination and through flooding, the bacterium is transmitted by submergence of seedlings in nursery beds. The bacterium then establishes itself in new host, increases in number by multiplication and initiates further infection during favourable condition.

Wild grasses. The role of several weeds as the source of primary inocula suggests a new line of interest. Out of several weed hosts tested in Japan, *Leersia oryzoides* and *Leersia oryzoides* var *japonica* are considered as active natural hosts (Goto, 1953; Yoshimura, 1959). The role of wild grasses as the source of primary inocula is yet to be confirmed.

Diseased stubbles and straw. In Japan, bacterial population with high intensity is found in infected stubbles, just after the harvesting of the crop. But after 1–2 months, the bacteria die along with the decomposing stubbles in both warmer and cooler regions. In India, the bacterium can survive in diseased stubbles in double-cropped areas, but not in such areas having single crop. In other countries, like Sri Lanka and Thailand, the bacterium surviving in diseased stubbles can develop primary inoculum for further infection. The bacterium will die within 20 days if the stubbles are ploughed down.

In Japan, the bacterium can survive up to next spring in diseased straw, if protected from rain; but die if it remains exposed. Diseased straw used in nursery bed or near an irrigation canal acts as the source of primary inoculum.

In some regions of India, ratoon in lowlands with optimum soil moisture is the source of primary infection.

The bacterium enters the host through wounds made by the chafing and striking of leaves by wind, either through the cut ends (by artificial cutting during transplanting) of seedling leaves, or through broken roots during transplanting or through natural opening like hydathodes and stomata.

After entering the leaf, the bacterium multiplies and moves in both upward and downward direction. The symptom is visible within about 10 days of artificial inoculation. The exudates of bacteria that develop on the leaf surface are dispersed by rain. The dispersed bacteria in contact with other leaves may cause infection or they may fall in water and carried to the other fields during irrigation and may cause further infection. In course of time, the inflorescence gets infected and will develop the **infected seeds**. The infected plants are cut out for harvesting, whose upper part, after drying, becomes the **diseased straw,** and lower cut end is the **diseased stubbles**. Then the disease may continue through the infected seeds, disease straw and/or through diseased stubbles.

Disease management

The disease can be controlled or reduced by the following methods:

A. Cultural methods

1. In Japan, the nursery beds and fields are kept from being deeply flooded to reduce the disease. But this method is doubtful in tropical countries.
2. In India, previous year's plant debris should be burnt to destroy the bacteria present in the infected field (Sulaiman *et al.*, 1965).
3. In Philippines, the elimination of ratoon rice and volunteer plants has been recommended to reduce the source of primary inoculum.
4. Preparation of nursery beds in disease-free areas is done to restrict the inflow of primary inoculum into the main field

which might reduce the disease incidence (Srivastava, 1969 and Kauffman, 1971).

5. Injury is a prerequisite for entry of the bacterium in the host. As pruning of the apical region is a common practice in many parts of India, it is imperative that pruning should be avoided to resist the entry of the pathogen.

6. The time of application of nitrogen fertilizer influences the disease incidence. Application of nitrogen fertilizer at tillering stage increase the disease incidence, but similar application — at the time of transplanting or during flower initiation — reduces the disease incidence (Have and Kauffman, 1972).

B. Physical methods

1. Soaking the seeds in water for 12 hours, followed by hot water treatment for 30 minutes at 52–54°C, reduce the disease incidence.

C. Chemical methods

(a) Seed treatments. As the disease is seed-borne, various seed treatments are followed to reduce the disease incidence:

1. Soaking the seeds for 12 hours in a mixed solution of 0.05% Ceresan and 0.02% Agrimycin, followed by hot water treatment for 30 minutes at 52–54°C.
2. Soaking the seeds for 8 hours in 0.01% Ceresan and 0.6 gram of Streptocycline in 5 gallons (1 gallon = 4.55 lt) water.
3. Soaking the seeds for overnight in Streptocycline solution at 100 ppm concentration.
4. Soaking the seeds for 16–24 hours in decinormal nitric acid solution, followed by washing in tap water and finally drying in sun.

(b) Foliar spray. Spraying of different chemicals is also effective to control the disease:

1. Foliar spray of 60–120 ppm Chloramphenicol, an antibiotic, in combination with phenyl mercury acetate at 10–20 ppm has been found effective in Japan.
2. Foliar spray of a mixture of 100 ppm each of Cellucidin and Dithianone is also very effective.
3. According to Singh (1975, 76), 5 sprays of Agrimycin mixed with Copper oxychloride at 12 days' interval is effective to control the disease.

(c) Other methods

1. Chlorination of irrigation water help in reducing the disease (Padmanabhan and Jain, 1966).

D. Disease tolerant varieties. In India, many disease tolerant varieties are recommended for cultivation. These are W 259 and W 348 (Padmanabhan, 1965) Tainan 3 and Chainung 292 etc. (Srivastava *et al.*, 1967).

2. *CITRUS CANKER*

Citrus canker is stated to have originated from China and now the disease becomes widespread in all the citrus growing regions of the world. Sometimes, the disease becomes so severe in USA that mass eradication of diseased plants or the entire orchards has to be undertaken to control the disease. It also causes serious damage in the citrus fruit growing regions of India, China, Java and Japan.

Host. *Citrus aurantium* L. (sweet orange) and other *Citrus* spp.

Symptoms

The disease affects all parts of the plant, except the root, but the most susceptible regions are the leaves, fruits and twigs:

(i) **On leaf**: Lesions on leaf first appear as small, watery, round, translucent spots. The lesions are raised and become yellowish brown in colour. They appear first on the lower surface and then gradually spread on both the surfaces of the leaf (Fig. 5.12B). With the progress of disease, the surface of the spots becomes greyish and finally ruptures in the centre showing rough, corky and crater-like appearance. The size of the

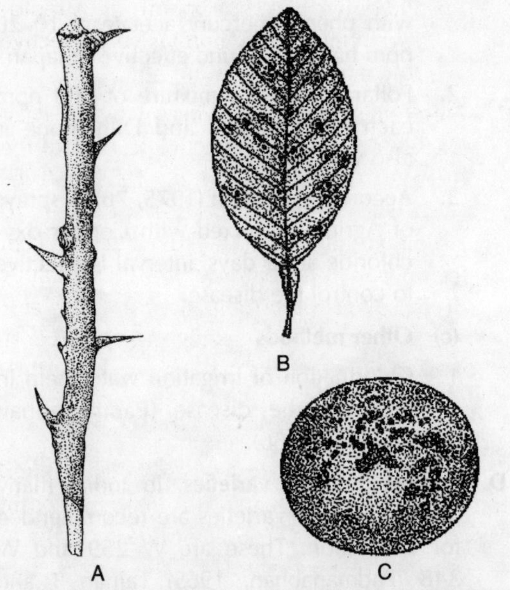

Fig. 5.12: Citrus canker: A–C: Disease symptoms on different parts. A. on stem, B. on leaf, C. on fruit

spots ranges from 1 mm to 1 cm in diameter. Lesions occurring on petioles cause premature defoliation.

(ii) **On fruits**: Lesions on fruits are almost similar to those of leaves, except that the yellow halo and crater-like depression in the central region are more prominent. Lesions often coalesce and form very irregular patches of rough and scabby raised areas (Fig. 5.12C). The injury takes place only on the skin and causes no effect on the pulp or juice.

(iii) **On twigs**: Lesions on stem are like those of leaves, but they coalesce to form elongated, rough and scabby regions (Fig. 5.12A). On larger branches the lesions are irregular, rough and more distinct.

Causal organism. The disease is caused by *Xanthomonas axonopodis* pv. *citri*. Earlier, it was known as *Pseudomonas citri, Xanthomonas citri* or *X. campestris* pv. *citri*.

The bacterium is Gram-negative, rod-shaped, aerobic and capsulated, having single polar flagellum (monotrichous). They grow in chains and measure 1.5–2.0 μm × 0.5–0.75 μm. Colonies on beef extract agar are circular, slightly raised, glistering and amber yellow to straw yellow in colour.

Disease cycle

The bacterium enters the host through stomata, lenticels and also through wounds

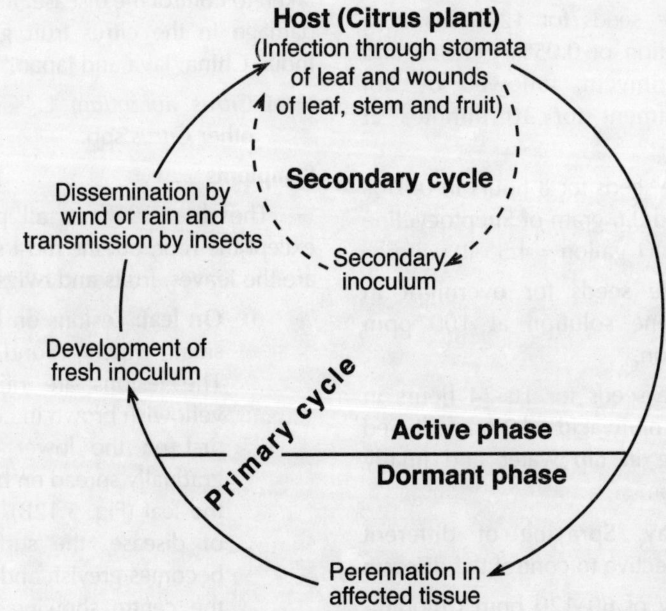

Fig. 5.13: Disease cycle of Citrus canker

caused by insects, movement of thorns etc (Fig. 5.13). The bacterium multiplies rapidly in the intercellular spaces, thereby dissolves the middle lamella and establishes themselves in the cortical region. Mild temperature (20–30°C) and wet weather with uniform rain are most suitable conditions for their infection.

The bacterium does not survive in the soil or on infected plant parts fallen on the ground, due to antagonistic action of other soil-microorganisms. Infected twigs with old lesions on the tree are the main source of perennation of the bacterium. The bacterium then disseminates by wind, rain and/or by insects like Citrus leaf miners (*Thosconyrsa citri* and *Phyllocnistis citrella*) and causes infection to the new healthy plant. However, the disease is commonly disseminated into the new locality through the infected nursery stock.

Disease management

The disease can be controlled or reduced by the following methods:

A. Cultural methods

1. **Sanitation**. Pruning of the infected parts during dry season and destruction of the affected plants by burning reduce the source of primary inoculum.
2. **Quarantine measure**. Strict quarantine measures should be adopted to restrict the dissemination of bacteria from one country to other, even from one locality to others of the same country.
3. **Disease-free nursery stock**. Disease can be controlled at a significant level by using disease-free nursery stock.
4. **Vigour of the plant**. The vigour of the plant should be maintained by application of proper amount of fertilizer and irrigation.

B. Chemical methods

1. **Fungicides**. Spraying fungicides like Lime-sulphur and Bordeaux mixture is often very useful to control the disease.
2. **Antibiotics**. Spraying of antibiotics like Phytomycin (2500 ppm) or Streptomycin (500–1000 ppm) is very effective to control the disease.

C. Other substances. Spraying of Neem cake (160 lb/acre) is very effective to control the disease (Vaheeddudin *et al.*, 1959). 50 lbs (1 lb = 453.6 grams) cake is soaked in 20 gallons (1 gallon = 4.55 lt) of water and kept for a week to decompose. It is then sprayed on the plant without filtration. 2 sprays at 2 weeks' interval during August-September is very effective.

3. *ANGULAR LEAF SPOT OF COTTON*

The disease is also known as **Bacterial blight** and **Black arm of cotton**. It was first reported in Alabama (U.S.A) in 1891. The disease occurs in all cotton growing regions of the world, like Russia, Pakistan, China, Australia, Sudan, and Egypt and also in many countries. *In India*, the disease was first reported from Madras (Tamil Nadu) in 1918, and now occurs in Maharashtra, Uttar Pradesh, Andhra Pradesh, Madhya Pradesh and Tamil Nadu. The disease appears as epiphytotic in India, during 1948 to 1952 and become well established in 1960s, due to the replacement of rain-fed indigenous diploid cotton by high yielding tetraploid *hirsutum* and *barbadense*, those were highly susceptible to the disease.

The disease has declined in importance during 1990s due to the wide availability of resistant varieties. However, it remains as a potentially important disease because of the appearance of new races. In Australia during 1980, it was reported that bacterial blight was of increasing importance due to the proliferation of new races of the pathogen *X. axonopodis* pv. *malvacearum* (Anonymous, 1980).

In India, the disease causes an annual loss range from 5% to 25%.

Host. *Gossypium hirsutum* L. and *Gossypium barbadense* L.

Symptoms

The Angular leaf spot can affect the host plant throughout its life and symptoms vary to

A

B

Fig. 5.14: Angular leaf spot of cotton: Symptoms showing, A. Angular leaf spot on leaf and B. Boll rot on fruit

some extent with the plant age and organ affected. All the organs such as leaf, seedling, older plants and also the bolls are affected by the pathogen.

1. **On leaf**: The most common symptom is angular leaf spot which begins with dark-green, water-soaked spots, initially more clearly visible on the lower side of the leaf blade. The spots are angular in shape, being delimited by the smaller veins. Older spots become dark-brown or black and are visible on the upper side of the leaves. The angular spots may be few in number in more resistant host material, but on susceptible cultivars, they can cover much of the leaf, causing chlorosis, followed by necrosis and lamina gets deformed.

Sometimes infection on the leaf occurs as water-soaked tissue, which later turns black, on either side of the main veins. This is referred to as **vein blight**.

2. **On young seedling**: Similar spots may be found on the cotyledons of young seedlings where infection occurs from the soil or seed during germination. Under favourable conditions, infection may spread from the seedling cotyledon or the leaf onto the petiole and then to the main stem, leads to seedling mortality in susceptible variety.

3. **On old plant**: In older plants, the lesions can girdle the main branches and may break, with the loss of leaves and fruiting branches. This phase of the syndrome is known as **black-arm** because of the blackened appearance of the affected petioles and also the branches.

4. **On developing boll**: Water-soaked lesions can occur on the epicalyx and more commonly on developing boll. The boll rot begins as roughly spherical water-soaked spots on the boll surface which can expand to >1 cm in diameter on susceptible varieties, becoming black with age. The pathogen penetrates the boll cortex to cause internal boll rot. The blight bacterium can also be introduced into the young boll during feeding on the seed by the cotton stainer insects (*Dysdercus suturellus, D. peruvianus, D. koenigii* etc.). This causes the lint to become stained yellow or brown and sometimes leads to internal boll rot.

Causal organism

The disease is caused by the bacterium, *Xanthomonas axonopodis* pv. *malvacearum*. The bacterium is gram-negative, rod-shaped and measures 0.9 μm x 1.2 μm, with single polar flagellum. The bacterium develops singly or in pairs or may develop in chain. They are capsulated, non-spore forming, facultative aerobe, and also acid fast negative and non-fermentative.

Optimum temperature for the growth is 31-32°C, with maximum and minimum temperature being 42°C and 6°C respectively. Thermal death point of the pathogen is 50°C. Relative humidity of 85%, early sowing, delay sowing, delayed thinning, poor tillage, late irrigation and potassium deficiency in soil are favourable for the growth of the pathogen. Frequent rain followed by bright sunshine during October-november is highly favourable.

Disease cycle

In soil, the bacteria survive on infected dried plant debris for several years. The bacteria are also seed borne both externally and internally and remains in the form of slimy mass on the fuzz of seed coat. After sowing the bacteria multiplies and infect the seedling through the micropyle. Plants that develop from the bolls falling off prematurely also provide a source of primary inoculum. They also attacks other hosts like *Jatropha curcas, Thunbergia thespesioides* and *Eriodendron anfrutuosum*. Here the primary infection starts mainly from the seed borne bacteria. The secondary spread of the bacterium may be through irrigation water, wind, wind-blown rain splash, insects and other implements (Fig. 5.15). The bacteria enters through the natural opening or wound caused by insects.

Disease management

The disease can be controlled or reduced by the following procedures:

A. Cultural methods

1. *Crop rotation*. The pathogen cannot survive in the soil outside of crop residues and is therefore readily controlled with crop rotations by non-host crops. One crop season without cotton is usually sufficient to virtually eliminate crop residues as a source of primary inoculum.

2. *Other cultural practices*. Early thinning, good tillage, early irrigation, early earthing and addition of potash reduces the intensity of infection.

3. *Seed treatment*. Pathogen may survive on seed both externally and internally.

   (i) *Seed dressing*. The seeds (with internal inoculum) are dressed with **oxycarboxin** and **carboxin** @ 2 gr/kg seeds, cause elimination of the pathogen.

   (ii) *Seed treatment with antibiotic*, such as Streptomycin sulphate, results in complete eradication of both external and internal inoculum.

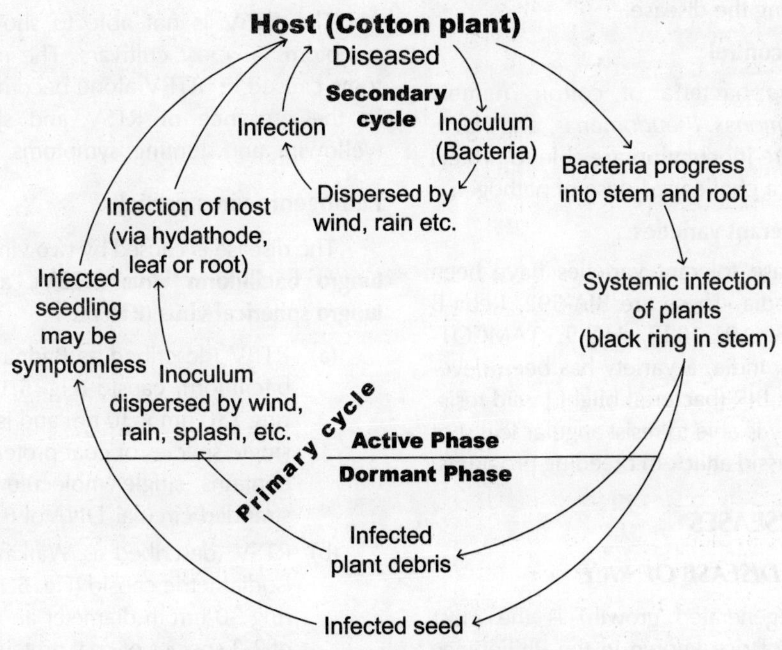

Fig. 5.15: Disease cycle of Angular leaf spot of cotton

(iii) *Hot water treatment*. Hot water treatment of seeds at 56°C for 10 minutes eliminate completely both external and internal pathogen, without affecting the germination of seeds (Verma *et al*., 1977).

(iv) *Other methods*. The seeds (with external inoculum) are immersed initially in **sulphuric acid** for 10-15 minutes, then washing by suspending in water and finally drying and treating them with **Agrosan GN** or **Ceresin**.

4. *Sanitary practice*: Removal and burning of the diseased plant debris from the field help in reducing the soil borne inoculum.

B. Chemical methods

The secondary spread of the disease can be controlled by spraying of Vitavax, Pentavax or copper-oxychloride alone or in combination with Streptomycin sulphate is very much effective in controlling the disease.

C. Bio-pesticides

Neem-based ecofriendly bio-pesticides such as Agricare and Plantotype are very much effective in controlling the disease.

D. Biological control

Phylloplane bacteria of cotton, namely species of *Aeromonas, Pseudomonas*, and *Flavobacterium* on pre-inoculation are able to protect the leaf against a challenged dose of pathogen.

E. Disease tolerant varieties

Many disease tolerant varieties have been developed in India. These are BJA-592, Reba-B 50, P-14-T-128, 101-102B, HG-9, TAMCOT-CAMD-E etc. In India, a variety has been developed named as BJR (bacterial blight jassid resistant). This variety is able to resist angular leaf spot pathogen and jassid attack in breeding programs.

B. VIRAL DISEASES

1. *TUNGRO DISEASE OF RICE*

Tungro (degenerated growth) is the most serious disease of rice known in the Philippines since middle of 20th century. It is of unknown etiology. In 1965, Rivera and Ou was the first to report that the causal agent is a virus. Now, the disease occurs severely in South and South East Asia, from India to Philippines, along with some neighbouring countries. In India, the outbreak of the disease occurred in 1965 in the eastern region and most recent outbreak of the disease took place in West Bengal in 1981. Earlier, the disease was known as **penyakit merah** (Malaysia) and **yellow orange leaf** (Indonesia).

Host. *Oryza sativa* L.

Symptoms

The symptoms of the disease are visible as stunting of plants and discolouration of leaves (Fig. 5.16A). The symptoms may vary in different cultivars. Some cultivars show prominent clearing of veins in discoloured infected region of the leaf, but in others it remains indistinct. The cultivars also show various shades of orange or yellow colouration. Some small necrotic and rusty spots may develop in the discoloured zone of older leaves. It shows the symptoms like stunting growth, less number of tillers, reduction in the length and number of panicle, number of spikelets, weight and starch content of grains — thereby the total yield becomes much reduced.

The RTSV is not able to show any clear symptoms in most cultivars. The milder symptoms caused by RTBV alone become intensified in the presence of RTSV and show typical (yellowing and stunting) symptoms.

Pathogenic virus particle

The disease is caused by two viruses: (a) **Rice tungro bacilliform virus** (**RTBV**), and (b) **Rice tungro spherical virus** (**RTSV**).

(a) RTBV (described as Badnavirus). It has bacilliform capsid (Fig. 5.16B), measuring 130 nm × 30 nm and is made up of single species of coat protein of 36 K. It contains single molecule of double-stranded circular DNA of 8.3 kbp.

(b) RTSV (described as Waikavirus). It has isodiametric capsid (Fig. 5.16B), measuring 30 nm in diameter and is made up of 2-3 species of coat protein. It contains single-stranded RNA of about 10 kbp.

A

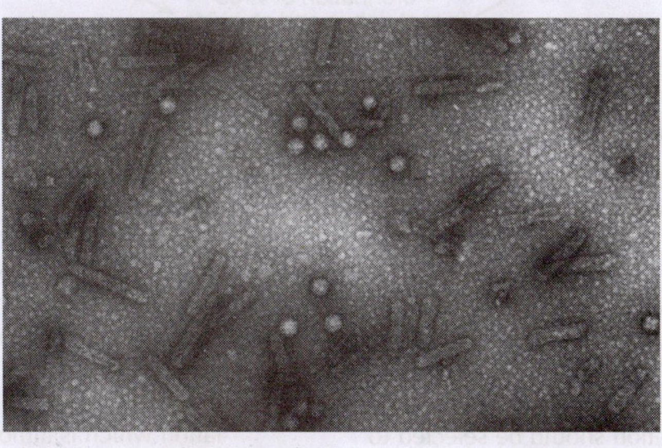

B

Fig. 5.16: Tungro disease of rice: A. Patch of infected plants showing stunting and yellow orange discolouration of leaves (Photo courtesy A. Choudhury), and B. Purified particles of the bacilliform (Badnavirus) and Spherical (Waikavirus) viruses together cause the rice tungro disease (After Plant Pathology by G. N. Agrios; Academic Press, London, 1997)

Disease cycle

The disease is caused by both the viruses RTBV and RTSV (Fig. 5.17). They are able to multiply independently and are transmitted by the green leaf hoppers (*Nephotettix virescens* and *N. nigropictus*) and other leafhoppers. The *N. virescens* is the principal vector of the disease where adults are more efficient than the youngers. It can transmit both viruses together after feeding on infected plant. The RTSV is easily transmitted by both the green leafhoppers (*N. virescens* and *N. nigropictus*), but RTBV is transmitted by *N. virescens* only in presence of RTSV. The vectors move during night, but remain with the plant during day.

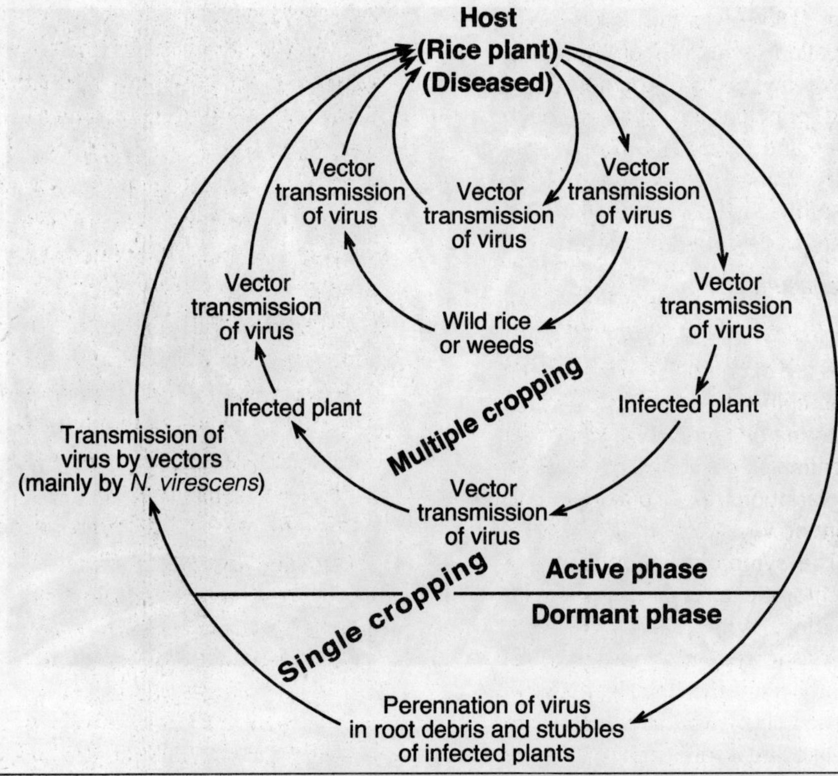

Fig. 5.17: Disease cycle of Tungro disease of rice

Disease management

The disease can be controlled or reduced by the following methods:

A. Cultural methods

1. **Sanitation.** The stubbles should be removed by ploughing to reduce the amount of inoculum in the field.
2. **Transplantation time**. The proper time of transplantation should be selected to avoid vector pressure during early stage of growth of seedlings. The vector pressure becomes more in the rainy days of April and May.
3. **Location of seed bed**. The location of seed bed should be at a distance of more than 10 m from standing crop.

B. Chemical control

1. Just after emergence of seedling, the application of Foratox 10G @ 1.75 kg/ha or Furadan 3G @ 1.5 kg/ha in the seedbed keeps the area free from vectors.
2. Spraying of insecticides like Carbofuran and Cypermethrin at 0.1% conc., gives significant reduction of vector population.
3. Application of Carbofuran @ 1 kg/ha in the root zone causes much reduction of vector population and is better than the spray treatment. Application of BPMC and Bendiocarb in the root region also gives significant control of vector population which is almost equivalent to the spraying of Carbofuran.

2. *TOBACCO MOSAIC VIRUS*

This is the best known of all virus diseases of plants and distributed throughout the world. The virus affects more than 150 genera of dicotyledonous plants of which tobacco, tomato and cucurbits are very important. The virus does not affect any monocotyledonous plants. The disease causes great economic loss by reducing both the quantity and quality of tobacco. It does not develop any external symptom on the affected grapes and apples.

Adolph Mayer (1886) was the first to point out the mosaic pattern symptom on leaves of tobacco plants. Iwanowski (1892) demonstrated that the infecting principal can pass through bacteria-proof filter and able to induce mosaic disease in healthy plants. A few years later, Beijerinck (1898) proved the existence of a virus in the affected plant as described by Mayer.

Host. *Nicotiana tabacum* L.

Symptoms

The symptoms are of different types showing various degree of chlorosis, mottling, curling, dwarfing and blistering of leaves (Fig. 5.18A, B); dwarfing of the entire plant and discolouration of flowers. The variation takes place due to variation of strain of virus and the variety of plants affected. The symptoms on young leaf appear as faint chlorotic circular lesions with gradual vein clearing. With gradual development, the characteristic systemic mosaic symptom appears. With maturity, the leaf shows abnormal dark-green spots which develop into crumpled blister-like areas of irregular outline and rest of the tissue becomes more or less chlorotic. During severe infection, the leaves become narrow, wrinkled and thin with severe malformation. The infected young plants become much more stunted than the older one. The disease causes partial pollen sterility but is rarely fatal to the host.

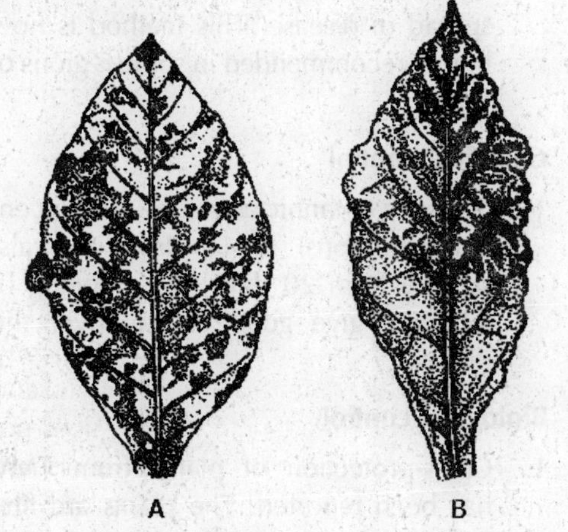

Fig. 5.18: Tobacco mosaic virus: A-B. Disease symptoms on tobacco leaves

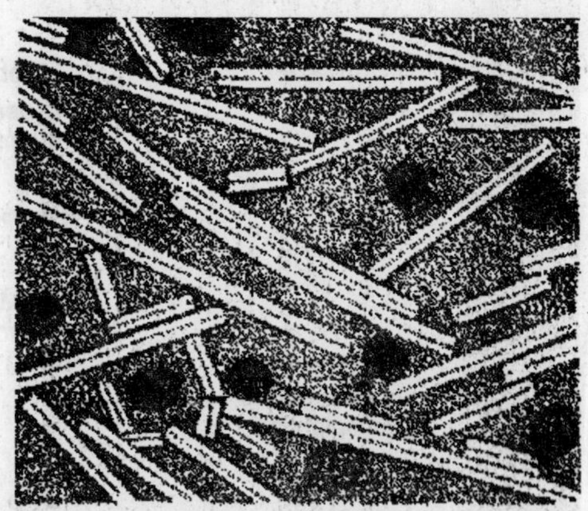

Fig. 5.19: Tobacco mosaic virus: Virus particles (stained with phosphotungstic acid)

Causal virus

The pathogen of this disease is Tobacco Mosaic Virus (TMV), recently designated by the generic name Tobamovirus, consisting of a number of viruses including Tobacco Mosaic Virus, Tomato Mosaic Virus etc.

For details of TMV structure see Fig. 2.50.

Disease cycle

The TMV survives in infected plant debris, on surfaces of contaminated seeds, on contaminated seedbed cloths, tobacco refuses from warehouses, cigars, cigarettes, pipe and chewing tobacco which form the source of primary inoculum (Fig. 5.19).

The virus is dispersed from plant to plant by mechanical transfer during transplanting, through field operations and implements and contact by man during cultivation (Fig. 5.20). The virus initially infects wounded tissue of the seedling in seedbed or the transplanted plant in the field. It develops systemic infection in all plants. After invading wounds it infects the parenchyma cell. The virus then multiplies and gradually moves from cell to cell through the plasmodesmata and, after reaching the phloem, it moves systematically and infects the whole plant. In the cytoplasm of the host the virus remains as a crystalline body. The infected plant again serves as the source as the primary inoculum, either directly or indirectly, for the next season.

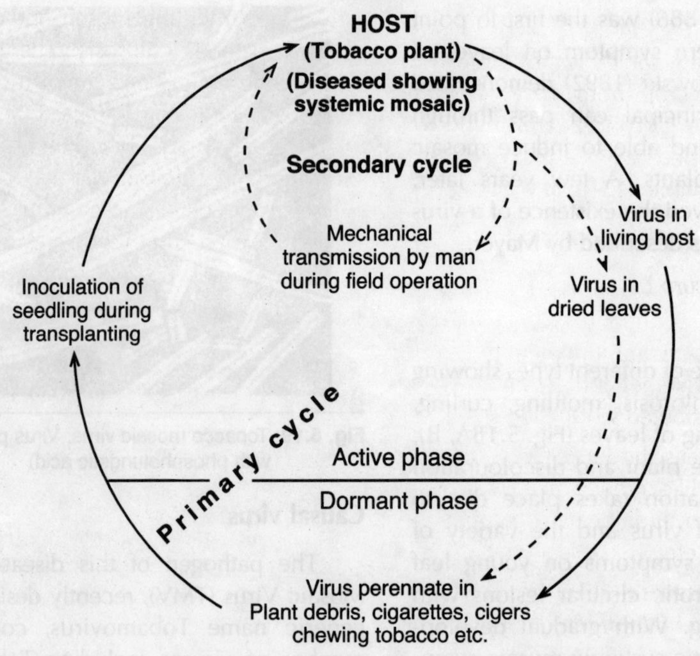

Fig. 5.20: Disease cycle of Tobacco mosaic virus

Disease management

The disease can be controlled or reduced by the following methods:

A. Cultural methods

1. Previous year's plant debris should be collected, removed and burnt outside, in the field.
2. Infected plants should be eradicated and burnt to check further spread of the disease.
3. Seed beds should be located far away from the tobacco warehouses.
4. Seed beds should be free from tobacco refuse.
5. Proper care should be taken to avoid the chance of contamination through cultivation implements.
6. After weeding, hands should be thoroughly washed with soap in running water.
7. Susceptible hosts and weeds should be destroyed thoroughly.
8. Chewing or smoking of tobacco of any kind should be avoided to reduce the source of primary inoculum.

B. Physical control

1. Soil of seed bed should be sterilized by hot water or steam.
2. According to Hare and Lucas (1959), infection by TMV is inhibited by whole milk. For the above, whole milk used by spraying on seedling before transplanting and washing of hands by dipping in milk — both reduce the spread of disease. This method is now highly recommended in some regions of USA.

C. Chemical control

1. Spraying of tannic acid (0.5–1.0 percent concentrations) at 10 days intervals, starting from 20 days after planting of tobacco, give good control over the disease.

D. Biological control

1. Cross-protection of plants from TMV has been reported. The plants are first inoculated with a milder strain of TMV, thereby the plants are able to resist the action of the virulent strain.

E. Disease tolerant variety

1. Use of tolerant variety is the best way to protect the crop from TMV. Tolerant varieties like TMVRR-2 and TMVRR-3 are developed by Central Tobacco Research Institute (CTRI) at Rajamundry.

3. *YELLOW MOSAIC OF BEANS*

The yellow mosaic of bean is a viral disease. The disease was first reported in 1925 in infecting garden peas (*Pisum sativum*) in the North-eastern United States and in Netherland. Today the virus is believed to be distributed worldwide.

In addition to peas the virus cause diseases of several legumes like French bean, soybean, mung bean, urd bean, lima bean, broad bean or faba beans, peanuts etc.

In addition to several legumes, it is also known to infect several non-leguminous plants including *Gladiolus* sp., *Fressia* sp., Opium Poppy (*Papaver somniferum*), *Canna* spp. and *Eustoma russellianum*.

The virus is spread by a number of aphid species non-persistently as well as being seed and mechanically transmitted.

Host

The virus cause diseases of several legumes like French bean (*Phaseolus vulgaris*), soybean (*Glycine max*), mung bean (*Vigna radiata*), urd bean (*Vinga mungo*), lima bean (*Phaseolus lunatus*), broad bean or faba beans (*Vicia faba*), peanuts (*Arachis hypogaea*) etc.

Symptoms

Symptoms of the disease vary greatly based on the type of virus, crop variety, the stage of the host and also the climatic condition.

On leaf: The symptoms appear as tip necrosis, streaks, mottling, mosaic and development of yellow patches chlorosis. The most attracting sign is the yellow and green mottling (to show various colours) on foliage. Often the tissue of dark green areas becomes raised over the surrounding chlorotic tissue. The clearing of veins also occurs in some crops. Due to uneven growth on infection the leaves may deform and margin can curl downwardly.

Fig. 5.21: Yellow mosaic of bean: Symptoms on leaflet

On other parts: Although the pods do not seem to be affected, but are less developed or deformed with fewer seeds.

Overall, plants become stunted.

Causal virus

The disease is caused by ***Bean yellow mosaic virus* (BYMV)**. It is a plant pathogenic virus in the genus Potyvirus and the virus family Potyviridae. The BYMV is a monopartite strand of positive-sense, single-stranded RNA surrounded by a capsid made for a single viral encoded protein. The virus is a filamentous particle that measures about 750 nm in length.

List of common yellow mosaic diseases of different beans along with causal virus is given below:

Diseases of different beans	Causal virus
1. Soybean yellow mosaic	Bean yellow mosaic virus (BYMV)
2. Mung bean yellow mosaic	Mung bean yellow mosaic virus (MBYMV)
3. Urd bean yellow mosaic	Mung bean yellow mosaic virus (MBYMV)
4. Bean yellow mosaic	Bean yellow mosaic virus (BYMV)
5. Lima bean yellow mosaic	Bean green mosaic virus (BGMV)

Disease cycle

This virus is transmitted by species of aphids and also by mechanical inoculation. With the help of the following ways the virus completes their disease cycle:

1. BYMV is transmitted through seeds. The virus over-seasons in one of its cultivated and wild hosts from which the aphids transmit the virus to the cultivated crop.
2. The spread of the virus with the involvement of bean plants takes place naturally in three different ways are:
   (i) For breeding of vectors,
   (ii) For providing virus-infected materials, and
   (iii) As recipients of viruses from infective vectors.
3. BYMV have a wide host range and well adapted to survive as it can perpetuate in some perennial plants.
4. In areas with tropical and subtropical environment, the cultivation practices also allow the viruses to survive throughout the year as same bean cultivars are grown in the same locality.

Disease management

There is no treatment against viruses and it is very difficult to control fully the population of aphids.

The incidence of the disease can be controlled or reduced by the following methods:

A. Cultural method

1. Disease-free seeds. Use of disease-free seeds will prevent the disease incidence.
2. Remove and destroy the disease affected leaves/plants from cultivated fields to reduce secondary spread of the disease.
3. Destroy weed hosts and inter crop with non-host crops like pearl millet, sorghum and maize.

B. Chemical control

1. The following insecticides used to control the disease by foliar spray are: Dimethate (0.03%), Metasystox (0.1%), monocropophs (1.6%) or Triazophos (1.25 ml/l) at the early stage of the disease are proves effective. Secondary spray may also be given at 7-10 days interval as and when required.

There are no treatments against **viruses** and full **control** of the population of aphids is very difficult. In fact, aphids are not killed rapidly enough to prevent virus spreading. Mineral oil (1%), either used alone or mixed with insecticides, considerably reduce the spread of the **virus**.

4. *YELLOW VEIN MOSAIC OF BHINDI (LADY'S FINGER)*

Vein-clearing is a symptom of some virus diseases of plants in which the veins become unnaturally clear or translucent. The common vein-clearing diseases in India are Yellow vein mosaic of bhindi and Cardamom vein clearing.

Fig. 5.22: Yellow vein mosaic of lady's finger: Symptoms on leaf.

Yellow vein mosaic of bhindi is also known as vein clearing of bhindi. The disease was first reported in bhendi i.e. okra (*Abelmoschus esculentus*) plant in 1924 in India and Sri Lanka. It is an important vegetable crop in India, where the production becomes greatly reduced by this disease in all growing regions. In case the plants get infected at early stages of development it causes 80% crop loss.

Host. *Abelmoschus esculentus* (L.) Moench (bhendi, okra)

Symptoms

The symptoms are developed both on leaves and fruits.

On leaves. The symptoms develop mainly on leaves throughout the growing period of the plant. It shows alternate **green** and **yellow** patches, vein clearing, and also vein **chlorosis** of leaves. The **yellow** network of veins is very **conspicuous**, and the veins and veinlets are **thickened**. In severe progress of the disease, the chlorosis may extend to the **interveinal** area and may result in complete **yellowing** of leaves. Infected plants become stunted.

On fruit. The size of the fruit is reduced, malformed and become yellowish green in colour.

Causal virus

This disease is caused by a complex consis-ting of the *monopartite begomovirus*, the **Bhendi yellow vein mosaic virus** (BYVMV, family: Geminiviridae) and a **small satellite DNA β component.** BYVMV can systemically infect bhendi upon agro-inoculation but produces only mild leaf curling in this host plant.

DNA β induces typical symptoms of bhendi yellow vein mosaic disease (BYVMD) when co-agroinoculated with the begomovirus to bhendi.

Disease cycle

The disease is transmitted by white fly, *Bemisia tabaci*. The population of fly is high enough during hot summer months and the crop becomes seriously affected. The virus also survives on various weeds growing along the roadside: e.g., *Croton bonplandianum, Malvastrum tricuspidatum, Ageratum* spp. etc.

Disease management

The disease can be controlled or reduced by the following procedures.

C. Cultural methods

1. *Crop rotation*. The disease can be reduced by rotation with other crops like legumes etc.
2. *Disease free seeds*. Seeds should be collected from disease free plants.
3. *Destruction of weeds*. The disease can be reduced by eradication of host weeds such as *Croton bonplandianum, Malvastrum tricuspidatum, Ageratum* spp. etc.
4. *Removal and destruction of infected plants*. Removing and destroying the disease affected plants from crop fields is practiced to avoid secondary spread.

D. Chemical methods

1. The vectors that are responsible for the spread of virus need to be controlled by spraying dimethoate 0.03 percent or monocrotophos 0.05 percent at 10 days intervals, (spraying must be done in the afternoon).
2. Foliar spray of 5 to 10 ml neem oil in a litre of water at weekly intervals reduces the disease at a significant level.
3. Application of Furatox-10 G @ 15 kg/ha in soil, followed by four foliar spray of Metasystox 25 EC @ 0.03% at an interval of two weeks from the date of sowing reduced aphid population and the disease incidence.
4. *Disease tolerant varieties*. Growing disease tolerant varieties, e.g., Akra anamia, Akra Abhay. Punjab Padmini etc., are very useful in controlling the disease.

C. VIROID DISEASES

1. *POTATO SPINDLE TUBER VIROID DISEASE*

Potato spindle tuber viroid (PSTVd) caused serious disease in potato (*Solanum tuberosum*), tomato (*Lycopersicon esculentum*), and also in brinjal (*Solanum melongena*). Natural infections have also been reported on pepino (*Solanum muricatum*), avocado (*Persea americana*), sweet potato (*Ipomoea batatas*) and also in a wide range of wild solanaceous hosts.

PSTVd was first reported as a disease of potatoes in North America. Later on, the disease is also reported in Asia, Africa, North America, South America, Europe and also in New Zealand. There have been previous detections and eradications of PSTVd of potato in Australia, including Western Australia.

Symptoms

PSTVd have both mild and severe strains. Symptoms may be confused with those of nutrient imbalance, spray and insect damage or other plant diseases caused by true viruses. Symptoms become more pronounced in warm conditions and in high light intensity.

1. On Potato plant

Foliage symptoms on potato are often difficult to recognise in early stage of infection and are rarely distinguishable before maturing. Infection with severe strains, where the stems remain upright and internodes are longer and slenderer than the normal one. Leaflets are slightly reduced in size with fluted margins. Leaves near ground level remain upright in contrast to healthy plant where leaves that often rest on the ground. Twisted leaflets along with wrinkled leaf surfaces are also observed with severe strains.

Tubers become elongated, often with pointed ends (Fig. 5.23A). Eyes are deep, more prominent and surface cracking occurs. Tubers of some cultivars develop knobs and swellings (Fig. 5.23B). With some strains, the foliar and tuber symptoms are milder than the severe strain or may not be visible.

The foliage and tuber symptoms become progressively more severe with the following generation.

2. On tomato plant

In mature tomato plants, infection with severe PSTVd strains causes purpling and yellowing of the leaflets, shortening of leaflet internodes thus it appears like a 'bunchy top'. The leaflets show down-curling and twisting, and the plant becomes stunted.

Shoot growth becomes spindly, flowers may abort and fruit becomes dark green with thicker outer wall and fail to ripen normally. With some strains, symptoms may be milder than the severe strains on plants and fruits may be infected without showing any external symptoms.

Transmission and dispersal

The disease is mechanically transmitted by

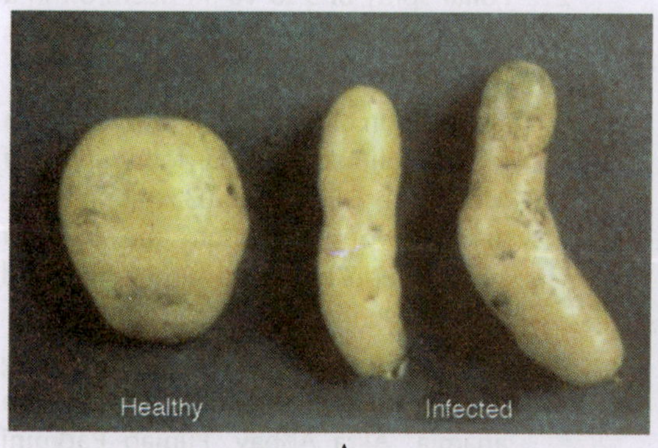

A

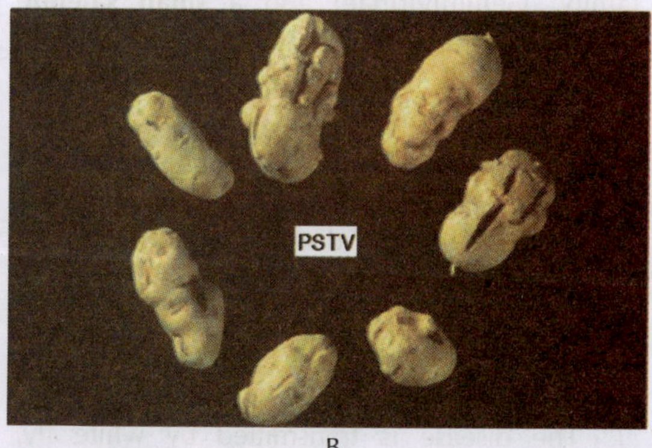

B

Fig. 5.23: Potato spindle tuber viroid disease: A. Healthy potato tuber (left), and spindle shaped tubers (right) affected by PSTVd, and B. Highly infected potatos

contact between healthy and diseased plants, wheels of tractor and also by other tools. In potato plants, it is found mosty in the upper leaves and tubers.

Transmission in true seed of potato depends on the cultivar. PSTVd can pass through both the pollen and ovule.

Long range dispersal mainly occurs through the movement of infected seed tubers and also by tubers.

PSTVd can also spread through aphids. But this only occurs in the presence of Potato leaf roll virus (PLRV).

Economic impact

PSTVd exist with both severe and mild strains. Sensitive cultivars infected with severe strains caused reduction in yield, up to 40% in individual plants, due to reduction in both the size and number of tubers. Although the quality of the tuber can be severely affected by infection but the infected potatoes are safe for consumption. Losses vary with cultivar, pathogenic strain and environmental condition, but are particularly severe under dry conditions. Mostly, degree of severity of infection caused a progressive reduction in yield.

Disease management

The chemical or biological controls are not available to control PSTVd associated with infected plants. Therefore, the control can essentially be done through the use of healthy virus-free seeds or planting material, and good sanitation practices. If infection appears, destruction of infected plants and strict hygiene are measures to be taken to prevent infection of subsequent crops is the only way to control the disease.

D. FUNGAL DISEASES

1. *BROWN SPOT DISEASE OF RICE*

Brown spot disease of rice is also known as **helminthosporiosis, sesame spot, seedling** and **leaf blight**. The disease is almost worldwide in distribution, but occurs severely in South-East Asia, Philippines, Japan etc. The disease is seed borne and due to non-selective export to other countries, it becomes distributed throughout the world.

In India, Sundarraman first reported the disease in 1922. Now this disease is commonly available in all rice growing areas, especially in heavy monsoon areas of West Bengal, Assam, Tamil Nadu, U.P. (eastern region) and a few regions of Kerala; mostly at the seedling stage.

The disease causes damages of different degree in different countries during epiphytotics (usually occurs widely, but periodically, in a destructive form). The outbreak of the disease in Bengal was in 1942, caused up to 90% loss in yield (Ghosh *et al.*, 1960); and thus caused **Bengal famine in 1943**. It was so severe that about **five million** (5,000,000) **people** died of starvation.

The loss in reduction of grain weight ranges from 4.6-29% (Bedi and Gill, 1960), Later Padmanabhan (1973) discussed in detail about the famine in his article "**Great Bengal Famine**".

Importance. The disease causes great economic loss. Three phases of damage resulting from the disease are:

1. Low rate of germination of infected seeds.
2. Leaf infection causes the reduction of photosynthetic area, resulting in the reduction of yield.
3. Due to infection and general weakening of the plant, poor setting of seeds takes place.

Host. *Oryza sativa* L. (Family–Gramineae i.e., Poaceae).

Symptoms

The pathogens usually attack all parts of the rice plant at all stages of development, excepting roots. The symptoms occur on the coleoptile, leaf blade, leaf sheath, and also on the inflorescence.

1. **On coleoptile**. Symptoms on the coleoptile appear as spots (Fig. 5.24A). The spots are brown, small, pinhead to oval in shape. Rarely the spot takes the form of long streaks. In severe cases, the coleoptile becomes blighted and causes death of the seedling.

2. **On leaf blade**. Symptoms on leaf blade appear as discrete, dark brown, elliptical to eye-shaped spots (Fig. 5.24B). A fully developed spot has a greyish brown centre surrounded by reddish brown margin; which is further surrounded by yellowish halo in early stage of development. The mature spot becomes 4-6 mm in size. In severe cases, the spots coalesce and the leaf gradually becomes dried.
3. **On leaf sheath**. Symptom on leaf sheath is very similar to the symptoms present on leaf blade.
4. **On inflorescence**. If the inflorescence becomes infected at an early stage, it does not develop any grain. Depending on the stages of seed development, the degree of damage varies causing complete sterility or partial filling up of the grain. Dark brown spots are seen on the glumes (Fig. 5.24C). The endosperm becomes brown and reduces the quality of the grain.

Collateral host

The pathogen can survive on some grasses like *Leersia hexandra, Echinochloa colonum* etc.; out of season, these plants act as collateral host.

Causal organism

The causal organism i.e., pathogen is *Drechslera oryzae* (Breda. de Hann) Subramanian and Jain, (earlier called as *Helminthosporium oryzae* Breda de Haan) which is the asexual state of the pathogen. The sexual state is *Cochliobolus miyabeanus* (Ito and Kuribay) Drechsler ex Dastur.

The fungus consists of septate, prostrate and much-branched hyphae. The hyphae grow both inter- and intracellularly. These become short segmented when they grow in the host.

The conidiophores arise as lateral branches from prostrate hyphae (Fig. 5.25A). They are erect, stout, short, unbranched except at the base, where they form branches. Towards the apical region, they bear conidia at regular intervals in the grooves which become visible clearly as scar after detachment of conidia. They are dark brown to light brown at the base and to some extent paler at the apical region. The size of conidiophores may be up to 680 μm (but in India, it is not more than 175 μm), with width ranges from 5-7 μm.

The conidia are cylindrical to ellipsoid, straight or slightly curved and slightly tapering towards the rounded end (Fig. 5.25B), smooth-walled, light brown in colour (5-10 celled) and

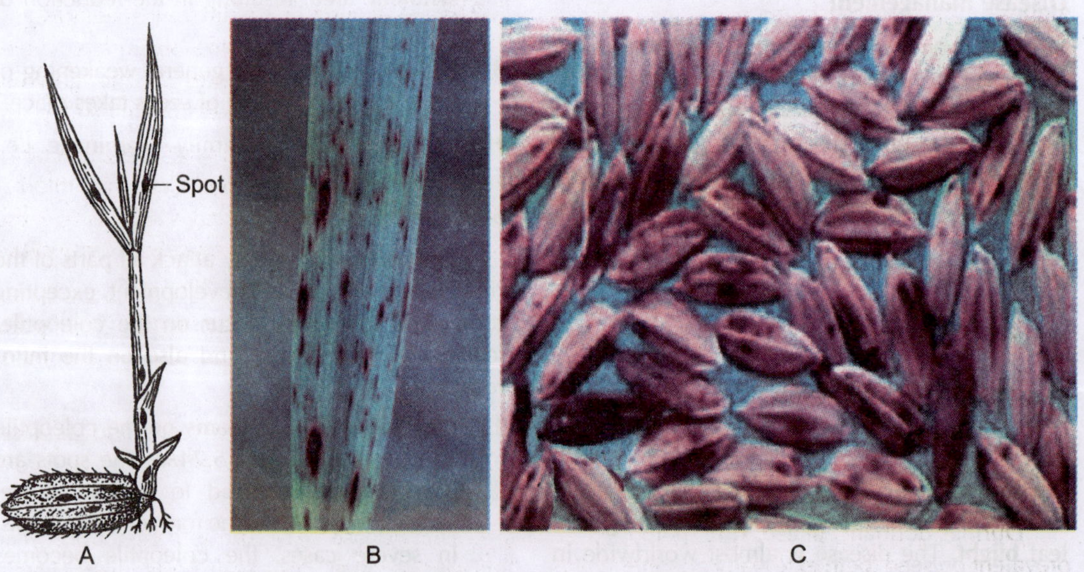

Fig. 5.24: Brown spot of rice: A-C Disease symptoms: A. On seedling, B. On leaves, and C. On grains

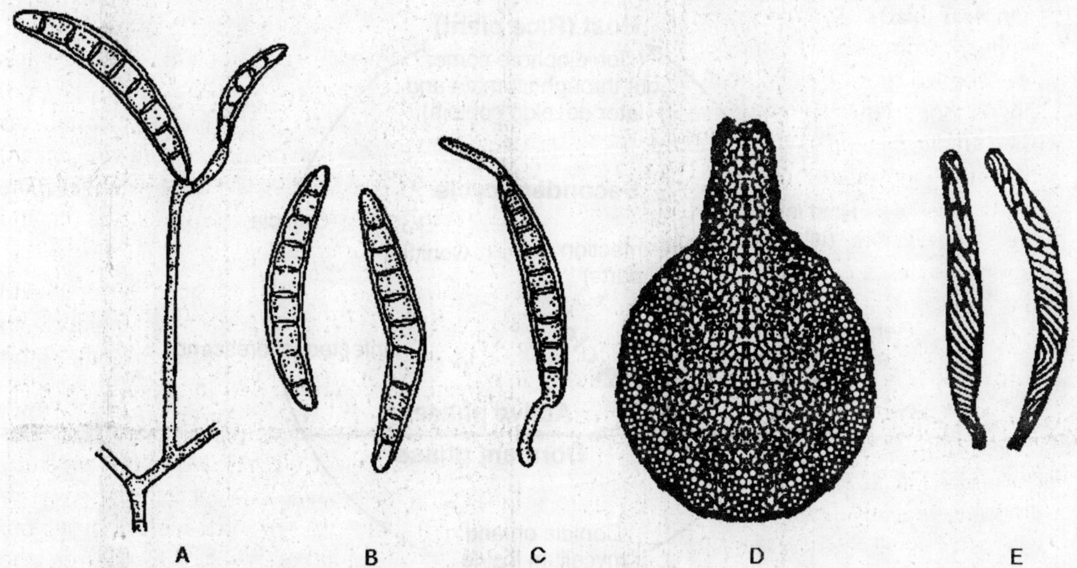

Fig. 5.25: Brown spots of rice: A. Conidiophore bearing conidia, B. Conidia, C. Germinated conidium, D. Single perithecium, and E. Asci (D and E from Ito and Kuribay, 1927)

vary greatly in size from 10-19 mμ x 45-105 mμ. The wall of the apical region is moderately thin and germination takes place by both the ends (bipolar) (Fig. 5.25C).

The perfect stage is *Cochliobolus miyabeanus*, obtained in culture (Japan) and disease stubbles (Mexico). They develop the **perithecial** fruit body (Fig. 5.25D). The perithecia are black, globose, pseudoparenchymatous, with a beak having ostiole. Asci are long, fusiform, cylindrical, slightly curved (Fig. 5.25E) and measure 21-36 μm × 145-235 μm. Asci contain mostly 4-6 filiform ascospores present in the form of a close helix. Ascospores are hyaline or pale olive-green, 6-15 septate, 250-469 μm × 6-9 μm in size.

Disease cycle

Brown spot of rice is a localized disease i.e., the pathogen is restricted to the infected zone and cannot pass long distance through the inner tissue (Fig. 5.26). The built up inoculum from the infected region disseminates further and causes infection.

During dormant phase, the pathogen is prevalent on seed (both externally and internally) and also on infected plant debris.

1. **From seed**. During germination of infected seeds, the coleoptile gets infected and then gradually it affects the seedling, showing blight symptom.

The infected regions of the seedling i.e., coleoptile, leaves, etc., then produce a large number of conidia. These conidia, after dissemination by wind, fall on the different regions of the same seedling or on different seedlings and cause infection. This is the secondary cycle, which repeats several times in the growing season and enhances progress of the disease.

Towards the end of the season, the inoculum from the infected leaf may go to the inflorescence and develops infected seeds; where the pathogen may present as mycelium (inside the seed) and/or conidia (on seed).

2. **From plant debris**. Previous year's infected plant debris also act as source of primary inoculum.

Thus, through infected seeds and infected plant debris, disease may appear in the next season and repeat the cycle.

The disease may also progress from year to year through collateral hosts, like *Leersia hexandra, Echinochloa colonum* and *Setaria* sp.

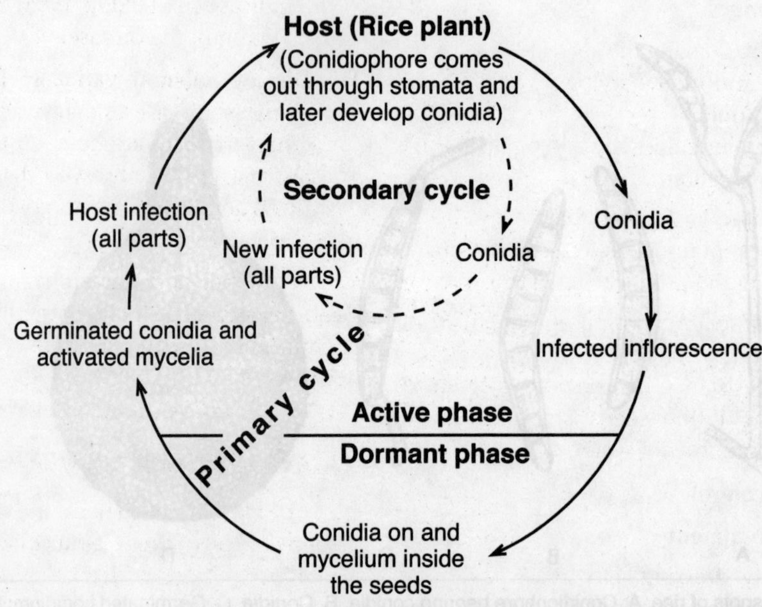

Fig. 5.26: Disease cycle of Brown spot of rice

Predisposing factors

1. A relative humidity of 92.5% or above favours conidial production (Chattopadhyay and Dasgupta, 1965).
2. Cloudy weather and higher day temperature (28-30°C) during the latter part of the growing season (Sept.–Oct.) are favourable for the rapid spread and development of the disease.
3. Optimum temperature for germination of conidia is between 25-30°C, minimum at 2°C and humidity of about 92% or more is the best for spore germination.
4. The best development of spots occurs in shade and high humidity. Sunlight retards the growth of spots.

Disease management

The disease can be controlled or reduced by the following procedures:

A. Cultural methods: The following cultural methods are useful in controlling the disease:

1. **Eradication of collateral hosts**. Collateral hosts like *Leersia hexandra*, *Echinochloa colonum* and *Setaria* sp. should be eradicated, where the pathogen can survive out of season.
2. **Sanitation**. Cleaning and burning of the previous year's infected plant debris are effective to reduce the source of primary inoculum.
3. **Irrigation water**. During irrigation, the water should not pass from infested field to the field without disease.
4. **Leaching of metals**. Leaching down of metallic elements like Fe, Mn and K from soil during rain should be checked, otherwise the plants become more susceptible to the disease.
5. **Application of metals**. Addition of metals like K in the field during growth period, reduce the severity of the disease.
6. **Spacing**. Proper spacing during transplanting also reduces the disease incidence.
7. **Nitrogen fertilizer**. Optimum nitrogen fertilizer (Urea) application, transplanting at optimum dates proved to be helpful for the reduction of disease.

B. Physical control

1. **Sunlight.** Soaking the seeds for 24 hrs in water and then drying (52-54°C) in bright sunlight reduces the source of primary inoculum by killing the activated mycelium.

2. **Hot water.** Hot water treatment at 55°C for 10 minutes gives better control by reducing the primary inoculum.

3. **Cold water.** According to Central Rice Research Institute (C.R.R.I.), Cuttack, some resistance can develop inside the seedling if the seeds are allowed to germinate in cold water.

C. Chemical control

1. **Seed treatment**

(a) **Dry dressing.** Dry dressing of the infected seeds with Ceresan or Agrosan GN. at 2-2.5% reduces the disease.

(b) **Wet treatment.** In Japan, seed treatment with Silver nitrate, Copper sulphate, Mercuric chloride, Phenol, Formaldehyde and Calcium hypochlorite have been found very effective in controlling the disease.

2. **Foliage treatment.** Spraying or dusting of fungicide, two to three times at regular intervals, gives effective control of the disease. Dithane Z-78 (0.2%), Dithane M-45 (0.3%), Hinosan (0.1%), Blitox (0.3%), Benlate (0.2%) and Bordeaux mixtute (5: 5: 50) are used to control the disease.

D. Disease tolerant varieties: It is better to cultivate disease tolerant varieties than the others, which are less costly. Singh and Sharma (1975) observed that varieties like IR-24 and Padma are comparatively resistant and can be grown in the epiphytotic areas.

A number of disease tolerant varieties have been developed in India, particularly in C.R.R.I. (Central Rice Research Institute), Cuttack, Orissa. These are CH 13, CH 20,CH 45, T 141, T 2114, IET 13238, CR 84-30, JB 83 etc.

2. *BLAST DISEASE OF RICE*

The blast or 'rotten neck' is a most destructive disease reported from almost 80 rice growing countries of the world. The crop suffers heavy losses in humid areas with high rain fall. The disease was recognized as a major problem of rice production from long back in Japan, USA and also in many other countries. In India the disease is present in most destructive form in Jammu and Kashmir, Uttar Pradesh, Bihar, Orissa, Andhra Pradesh and Tamil Nadu. The disease is also available in other parts with less frequency.

Host. *Oryza sativa* L. (Family–Gramineae i.e., Poaceae).

Symptoms

The symptoms of the disease occur on leaf, leaf sheath, columns, nodes (column joints), panicle, branch bases and also on grains.

Fig. 5.27: Blast disease of rice: A. Symptoms on leaves, and B. Node blast

On leaf. The lesions appear on leaf as small greyish or bluish, water soaked areas of 1-3 mm in diameter (Fig. 5.27A, B). During humid weather the lesions enlarge very rapidly and become 1-15 cm long and approx. 0.5 cm broad. The mature spots are spindle-shaped showing brown or reddish brown margin surrounding a grey or whitish centre and the entire spot is surrounded by yellowish halo.

In humid weather, spots on susceptible varieties enlarge very quickly and kill the entire leaf within a few days showing burnt appearance in the field, but remain as pinhead sized brown spots in resistant varieties. In moderate resistant varieties, lesions appear as small, round or elliptical in shape surrounded by brown margin and are a few mm in length.

Leaf sheath. More or less similar spots are observed on leaf sheath as found on leaf.

Culms. Brown to black spots are formed on the culms.

[Culm: central axis of the mature grass stem, consists of nodes and internodes.]

Rachis. Rachis of the maturing inforescence shows brown to black spots or rings.

Ears. Ears also show similar small spots.

If the infection occurs before grain formation, the grains are not formed and the panicles hang down. However, due to damage by necrosis of the neck tissue, the ear tends to break and fall off. Maximum damage takes place at this stage.

Causal organism

The causal organism i.e., pathogen of the blast disease of rice is *Pyricularia grisea* (Cooke) Sacc., named earlier as *Trichothecium griseum, Pyricularia oryzae* Cav. or *Dactylaria oryzae*. The perfect stage of the pathogen is *Magnaporthe grisea* (Herbert) Yaegashi and Uddagawa, an ascomycetous fungus.

The mycelium is septate, uni- or multinucleate, branched and hyaline to olivaceous in colour.

The conidiophores develop singly or in fascicles through the stomata (Fig. 5.28). They are simple or rarely branched, 2-4 septate, with swollen base and show truncated denticles along its length (in mature stage).

Conidia are pyriform (pear-shaped) to obclavate in shape with rounded base and narrowed tip, 2-septate, rarely 1-3 septate; hyaline or pale olive, 14-40 μm × 6-13 μm, with a distinctly protruding basal hilum (Fig. 5.28).

A typical lesion on leaf can develop 4,000 to 6,000 conidia/night for more than two weeks.

During germination, appressoria are developed at the tip of germ tube, either on host surface, on glass slide or on cellophane paper. They are globose, oblong or ovoid and measure 5-15 μm in diameter.

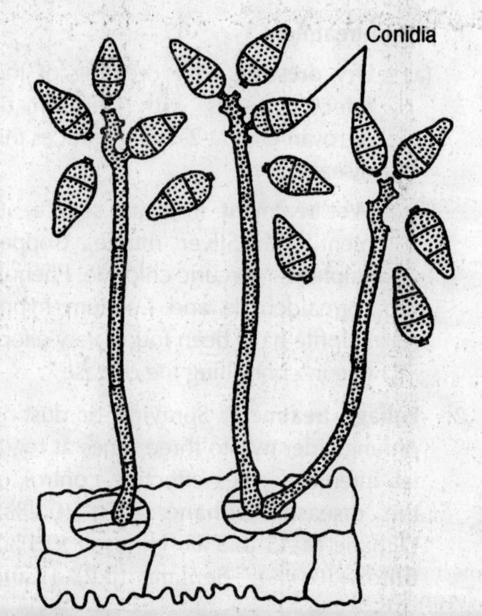

Fig. 5.28: Blast disease of rice (c.o. *Pyricularia grisea*): Conidiophores with attached and detached conidia

The fungus often produces chlamydospores of about 5-12 μm in diameter.

Pathogen can develop perithecia in culture (Fig. 5.29A, B). It is named as *Magnaporthe grisea*, and also called as *Ceratosphaeria grisea* and *Phragmoporthe grisea*. Peritheca are dark brown to black, glabrous, elongated, with spherical to subspherical base, mostly 100-180 μm in diameter with a long neck 500-1,200 μm in

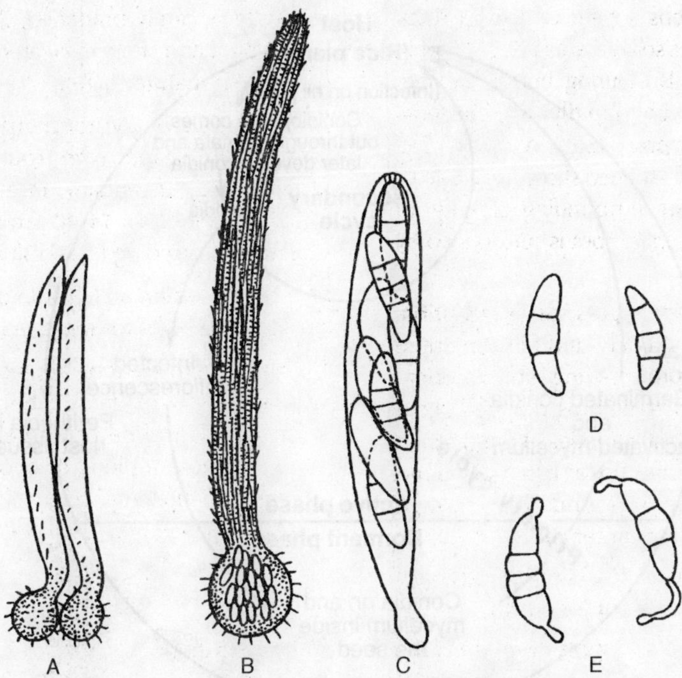

Fig. 5.29: Blast disease of rice (c.o. *Pyricularia grisea*, perfect state *Magnaporthe grisea*): A. Perithecia, B. Perithecium much enlarged, C. Single ascus containing ascospores, D. Ascospores and E. Germinating ascospores

length. The neck is cylindrical of more than half of its length i.e., up to 1,100 µm. Asci (Fig. 5.29C) are cylindrical to clavate, hyaline, unitunicate (ascus has single wall), with short stipe measuring 60-80 µm × 10-12 µm. It contains 8 ascospores. The asci are intermingled with paraphyses. Ascospores (Fig. 5.29D) are hyaline, biseriate, fusiform, curved with rounded ends, 1-4 septate (generally 3 septate), slightly constricted at the septa measuring 18-23 µm × 5-7 µm. Germination is bipolar (Fig. 5.29E).

Collateral hosts. *Saccharum officinarum* (De Gutirrez, 1954), *Setaria intermedia* (Pawar and Kulkarni, 1954) *Echinochloa crusgalli, Stenotaphrum secondatum, Digitaria sanguinalis and Eremo chloa ophiuroides* (Johnson, 1954); *Arundo donax, Brachiaria mutica, Dinebra retroflexa, Digitaria marginata, Panicum repens* (Padmanabhan, 1965).

Disease cycle

The pathogen can perpetuate in grain both externally and internally, on straw piles and also on host other than rice (Fig. 5.30). The inocula, i.e., the conidia, developed from all or any of the above sources can infect the leaf of the rice plant. The infected regions develop crops of conidia; those are disseminated and infect the different regions of the same plant and/or the other rice plants. Towards the end of the season, the pathogen may perennate as conidia on seed and also on straw piles. Next season, perennating conidia may serve as a source of primary inoculum. Likewise, in some areas they develop perithecia containing asci and ascospores. The perithecia remain with the plant debris or straw piles. Next season, the ascospores may cause infection.

Disease management

The disease can be controlled or reduced by the following methods:

A. Cultural methods. The following practices are used to reduce the disease incidence:

1. **Early planting**. Early planting shows less prevalence of disease than late planting.
2. **Sanitation**. Sanitation i.e., cleaning of the plant debris from the field reduces the source of primary inoculum, thereby reduces the disease incidence.

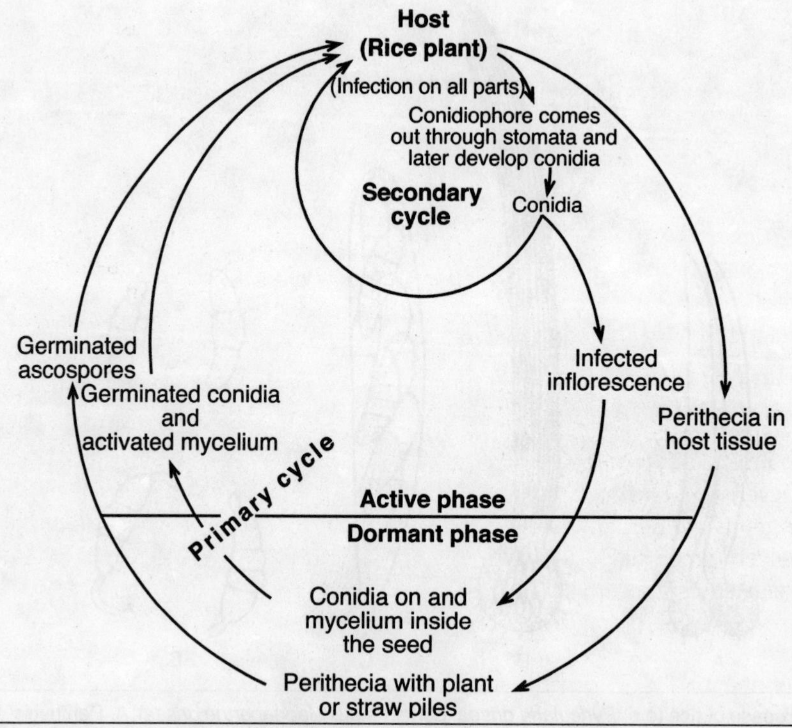

Fig. 5.30: Disease cycle of Blast disease of rice

3. **Destruction of collateral hosts**. Destruction of collateral hosts like *Panicum repens, Leersia hexandra* etc., reduces the disease incidence.

B. Chemical methods. Chemical of different groups are used in different ways to control the disease. These are:

1. **Seed treatment**. The pathogen is seed borne, thereby seed treatment is essential to reduce the disease. The seeds are treated with different chemicals to reduce the disease incidence.
   (i) Seed treatment with Agrosan GN (organomercurial) is effective to eliminate the seed borne inoculum present externally.
   (ii) Copper sulphate (20 ppm) mixed with Aureofungin (20 ppm) is effective to control seed borne inoculum (Thirumalachar, 1967).
   (iii) Kapoor and Singh (1982) reported that seed treatment with Benomyl (1: 400 w/w) gives protection to the seedling by inhibiting spore germination and appresorium formation of the pathogen.
   (iv) Fungorene (8g/kg seed) is found to be highly effective in controlling the seed borne inoculum.
2. **Foliar spray**. Cu-fungicides are not effective during severe appearance of the disease. It causes phytotoxicity, thereby reduces yield. Copper fungicide mixed with mercury compound and Phenyle mercuric acetate is highly effective as foliar spray.

C. Disease tolerant varieties. The following cultivars are recommended as disease tolerant varieties: A-67, A-90 and A-200 (in Maharastra); T-141 and T-603 (in Orissa); Co-4 (in Tamil Nadu); Kamala and Kukulu (in Bihar); Jaya and Pankaj (in Madhya Pradesh), etc.

3. *GREEN SMUT OR FALSE SMUT OF PADDY*

The smut pathogens belong to the order Ustilaginales, but green smut pathogen of paddy does not belong to this order. Initially it was placed under the subdivision Deuteromycotina, but later it has been placed under the subdivision Ascomycotina. So, it is not a smut in true sense, thus called **false smut**.

Cooke (1818) was the first to report the disease from Trinnevally in South India. Since then, disease gradually proceed in different regions of India and also in other countries, like America, Burma (Myanmar), Ceylon, Philippines etc.

Host. *Oryza sativa* L.

Symptoms

The symptom is confined to the grains of ears. Due to development of fructification of the pathogen, the grains are partially or completely covered by velvety green masses or balls (Fig. 5.31). The kernels enlarge several times than the normal one. Their colour is orange-yellow on the periphery which encircles central white region. The surface of the ball cracks with maturity. The glumes remain unchanged and are found sticking closely to the centre of the mass which burst out above. The green ball consists of lightly interwoven mycelia along with the glumes and neighbouring host tissue. At different stages of development the ball shows 3 layers:

1. *Innermost layer*: It is yellowish in colour, with radiating mycelium and spores are in the developing stage,
2. *Middle layer*: It is orange in colour and composed of mycelium and spores,
3. *Outermost layer*: It is green in colour, consists of mature spores.

Causal organism

The pathogen *Ustilaginoidea virens* (Cke.) Takahashi is a member under the Sub-division. Deuteromycotina; Order: Moniliales; Family: Dematiaceae. Later, with the development of knowledge on sexual reproduction, its perfect stage was placed under the Sub-division Ascomycotina; Order: Sphaeriales; Family: Clavicepitaceae and named as *Claviceps oryza-sativa* Hashioka.

The green ball on the grain consists of compact septate mycelia. The lateral or rarely

Fig. 5.31: False smut of paddy (symptom on grain)

terminal hyphae produce chlamydospores on short sterigma. The young chlamydospores are almost round and smooth, but the mature one becomes rough, olive green colour with granular coating, measuring 3-5 µm x 4-6 µm (Fig. 5.32B). On germination, they develop pear-shaped secondary conidia on branched and septate germ tubes (Fig. 5.32C).

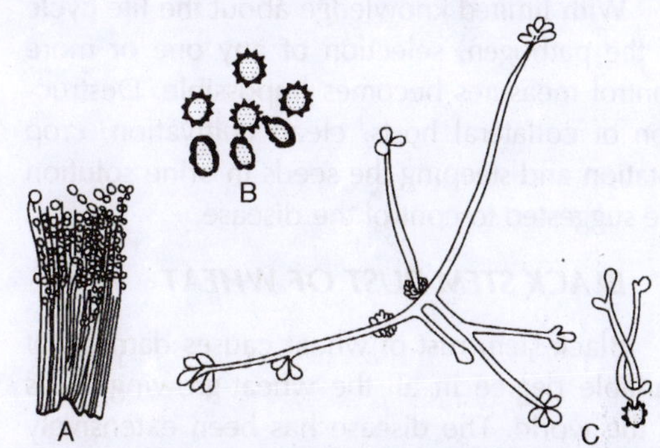

Fig. 5.32: False smut of paddy: A. Section of outer part of sclerotium, B. Chlamydospores and C. Germinating chlamydospores showing development of secondary conidia on germ tubes

Some of the green balls develop into four sclerotia in the centre. The sclerotia overwinter in field and, with the onset of favourable condition i.e., in the next season, they develop stalked stroma (Fig. 5.32A). The stalk of the stromata swells at the tip and becomes more or less globose structure bearing perithecia. Each perithecium is flask-shaped and contains about 300 asci. The asci are cylindrical having a hemispherical apical appendage measuring 180-220 µm x 4 µm. Each ascus contains 8 ascospores. Ascospores are filiform, unicellular, hyaline, measuring 120-180 µm x 0.5-1 µm.

Disease cycle

The incidence of this fungus is reported in wild rice and also in many grasses. Thus, it is presumed that the primary inoculum comes from the collateral host by wind dissemination. Ascospores produced from the sclerotia serves as primary inoculum and causes primary infection. Chlamydospores are air borne, play an important role in the secondary infection and their availability is highly significant at the time of flowering of rice plant. Infection during early flowering causes destruction of the ovary, but if it is late, grain filling takes place. The mycelium invades the endosperm and develops a mass of spores. The disease is not seed borne, thus no success has been achieved to develop infection by artificial floral inoculation.

Disease management

With limited knowledge about the life cycle of the pathogen, selection of any one or more control measures becomes impossible. Destruction of collateral hosts, clean cultivation, crop rotation and steeping the seeds in brine solution are suggested to control the disease.

4. *BLACK STEM RUST OF WHEAT*

Black stem rust of wheat causes damage of variable degree in all the wheat-growing areas of the world. The disease has been extensively studied in Europe, USA and Australia, where it often causes severe losses in yield. In India, it is available in almost all the wheat-growing areas.

The damage from this disease is much more severe in South India than North India. In southern and peninsular India, the disease may appear in the fourth week of November, hence loss is much more here than northern region, where disease may appear in early March at the time the crop reaches the 'dough' stage (plant with developing grain, having milky white liquid endosperm) and thus cause less damage. It may cause about 90% reduction in yield during severe infection showing premature ripening and shrinking of the grains.

Host. *Triticum aestivum* L.

Symptoms

On Wheat. The symptoms appear on the stem, leaf sheaths, leaves and also on ears. The damage of stem is much more severe than the other regions.

In the early part of the season, the symptoms are very common on culms, leaf sheaths and also on leaves as elongated brown pustules, the uredosori containing huge number of uredospores. The uredosori may be quarter of an inch (6 mm) or more in length. After maturation the uredosori burst and expose the brown powder like uredospores. Teleutospores develop towards the end of the season in new sorus, the teleutosorus or in the same sorus along with uredospores — the mixed sorus. The teleutosori are black in colour and expose the mass of black spores — the teleutospores. The teleutosori are more frequent on the stem and other parts, but least on the leaves.

During severe infection plants fail to develop normal ears and the grains become shrivelled with light in weight (For figure see Chapter 4, Fig. 4.100 and 4.101).

On Barberry. The symptoms first appear on the upper surface of the leaflet as small circular yellowish spots of variable numbers. Later on, the spots increase in size up to 5 mm or more. The mature spots are honey-coloured in the centre, margined with red to purple colour, the pycnium. Almost towards the opposite side (i.e., on lower side) of the leaflet, yellowish to orange coloured spots, the aecia or cluster cup appear (For figure see Chapter 4, Fig. 4.103).

Causal organism

Puccinia graminis tritici Erikss. and Henn., is a **macrocyclic** (i.e., large life history cycle),

polymorphic (poly = many, morphs = forms; i.e., various forms in their life cycle) and **heteroecious** (i.e., require two different and unrelated hosts to complete their life cycle) rust.

The different hosts are wheat and barberry.

On Wheat. Uredosori and teleutosori develop on wheat plant and basidiospores on infected plant debris of wheat.

The uredosori contain uredospores which are one-celled, dikaryotic (n + n), brown in colour, oval in shape, having thick echinulate wall and four equatorially placed germ pores. They measure 25–30 μm × 17–20 μm.

The teleutosori contain teleutospores which are two-celled (each cell is dikaryotic i.e., n + n), black, having thick and smooth wall with two germ pores, measuring 40-50 μm × 15-20 μm. The germ pore of the apical cell is situated at the rounded apex, while in the lower cell it is at one side just below the septum. The teleutospores remain with the plant debris and undergo a long period of rest. Before germination they have to be exposed to freezing temperature.

Teleutospores, on germination, produce, a long 4-celled promycelium (the septate basidium) and each promycelial cell develops one sterigma bearing one haploid basidiospore. This is the saprophytic stage of the pathogen. (For figures see Chapter 4, Fig. 4.102).

On Barberry. Pycnia and aecia are produced on barberry plant.

Pycnia (spermatangia) are flask-shaped with an opening — the ostiole. Inner surface of pycnium is lined with elongated pycnophores, develops many small, uninucleate conidia or spermatia. Towards the pore mouth many sterile hyphae, the periphyses, are present, those remain intermingled with much longer and generally forked receptive hyphae.

Aecia are inverted cup-shaped or bell-shaped in structure enclosed by a peridium. The aeciospores are one-celled, dikaryotic (n + n), echinulate having six germ pores. They develop in chain but arrange alternately with disjunctor cell from the basal end of the aecial cup. Aeciospores can infect wheat or any other grass host. (For figure see Chapter 4, Fig. 4.104).

Collateral hosts. Many grass hosts, like *Brachypodium sylvaticum, Bromus patulus, B. coloratus, B. carinatus, B. japonicus, B. mollis, Lolium perenne, Hordeum distichon, H. murinum, Aegilops squarrosa, A. trineclis, A. ventricosa* are infected by rust pathogen. Like wheat, they grow in the winter season and, hence, the above collateral hosts have no importance in the annual recurrence of the black stem rust in the plains.

Disease cycle

The pathogen (*Puccinia graminis tritici*) can survive on straw of wheat plant, (the primary host), for several months as **teleutospores** (also commonly called **teliospores**), which then transfer the disease to the barberry (the alternate host). Teleutospore, on germination, produce the basidiospores of two opposite strains i.e., 2 of (+) type and other 2 of (–) type. **Basidiospores** are dispersed by wind and fall on the barberry plant and infect mainly the leaves. Pycnia (+ and – type) are produced on the upper epidermis of the leaflet by the (+) and (–) basi-diospores, respectively. After spermatization between **pycnospore** and receptive hypha, and with the help of raindrop or insect, dikaryotic condition is established resulting into the formation of aecia and aeciospores on the lower surface of the leaflet. The **aeciospores** developed in different aecium are of the same type. After release from the barberry leaflet with the help of wind, it falls on the surface of wheat plant and after infection it produces uredosorus containing uredospores. Then, **uredospores** infect different region of the same plant or the different plants and develop next crop of uredospores in the same season. This is the secondary cycle which may repeat several times in the growing season. At the end of the growing season i.e., during February to March, infection of uredospores takes place, but instead of developing uredosorus it develops teleutosorus containing teleutospores — and thus complete the disease cycle.

In India, the above cycle does not prevail due to different environmental conditions and the absence of suitable alternate host (For figure see Chapter 4, Fig. 4.105).

Annual recurrence of Black Stem Rust of Wheat in India

Puccinia graminis tritici, the pathogen of Black stem rust of wheat, complete its life cycle with the help of two different and unrelated hosts — the wheat (primary host) and barberry (secondary host).

The disease appears year after year by completion of the life cycle and causes severe damage with great economic loss.

In India, the disease also appears every year and cause great economic loss, but the scenario is different. The possible ways of annual recurrence are discussed here based on the work of different scientists of our country.

Role of Barberry. The role of barberry (*Berberis* spp.) is discussed next:

1. *Berberis* does not occur in the plains of India.
2. *Berberis vulgaris,* the highly susceptible species, is not available in the hills.
3. Only the moderately susceptible species of *Berberis* are available.
4. According to Prasada (1947), no *Berberis* infection observed in nature has been connected with black stem rust.

The above facts rule out the possible role of *Berberis* in the annual recurrence of Black stem rust in India.

Role of environment

Effect on teliospore. Teliospores are formed at the end of March to May and their subsequent exposure to intense summer heat rules out the possibility of their survival. During the appearance of new leaf of *Berberis* in summer, the temperature and humidity are unfavourable for the germination of teliospores, this rules out the possibility of natural infection of *Berberis* by basidiospores.

Effect on uredospores. In Indian plains, the wheat is sown in October to November and harvested in March to May. During the early part of the growing season the conditions are favourable for infection, still the appearance of the disease is delayed by two to three months.

The uredospores developed in the plains will not be able to survive like teleutospores due to intense summer after the end of the season. The above fact rules out the possibility of uredospore development in plains and their role in the annual recurrence of Black stem rust in India.

Now, the uredospores developed in the hills came in focus. Wheat that grows both in the northern and the southern hills develops uredospores. Mehta (1933, 1940) observed in details regarding the survival of the uredospores out of season during summer in the hills. He observed that the disease in the new crop appears earlier in the foothills than the plains. With the comparative studies, he also observed that there is heavier infection in the foothills than the neighbouring plains and the period of rust appearance in the plains and the foothills in different parts of the country is also different. Mehta (1933, 1940, 1952) concluded that the inoculum of the black rust has been introduced in the plains from the hills (1,200-2,400m/4,000-8,000 ft) where the uredospores are able to oversummer. Due to optimum humidity and milder summer, the uredospores oversummers on disease stubbles, volunteer wheat and also on the regular wheat crops of both summer and winter in the Pulney and Nilgiri hills in the south and the submountainous region of the Himalayas in the north. Later observation also includes the Panchgani–Mahabaleshwar range of Western Ghats as another focus of infection. He concluded that the inoculum for the infection in plains comes from north as well as from south.

Later, Joshi *et al.*, (1974) observed that South India gets black stem rust infection earlier than the north. Although the pathogen can survive in the north hills throughout the summer months, a number of factors seem to indicate that the inoculum of the northern hills does not play any significant role as the primary inoculum for the black stem rust infection of the northern plains. The main reason is the temperature of the northern hills. The average minimum temperature in north India from December to February ranges from 7-13°C. The incubation period of the uredospore in this temperature is 40-50 day in January and February, thereby very little time is left for infection as the crop reaches maturity. The southern inoculum develops earlier and its dissemination towards north is the possible principal source of inoculum for the northern wheat belts through Karnataka, M.P., and Maharastra, where the inoculum multiplies much more.

Disease management

The disease can be controlled or reduced by the following methods:

A. Cultural methods

1. Considering the annual recurrence of black rust in India, Mehta (1929) suggested the suspension of summer crop in the Nilgiri and Pulney hills and replacement of wheat and barley cultivation by oats in the hills.
2. Potassium increase the resistance of wheat to rust infection, but nitrogen has the opposite effect. Reduction of nitrogen in NPK ratio helps in reducing the rust incidence in susceptible varieties.
3. Mixed cropping of wheat with non-cereal crops (mustard) reduces the amount of inoculum for secondary infection and further spread of the disease.

B. Chemical control

1. Forsyth and Peturson (1958) reported that effective control of stem rust can be obtained with 4-5 times application of $ZnSO_4$ and Nabam.
2. Experiments done in U.P., by sulphur dusting at the rate of 13.6 kg/acre at an interval of 4 days during the appearance of the disease give good control. To avoid the wastage of sulphur which needs foreign exchange to import, an efficient forecasting system is required.
3. Wallen (1958) proposed to control the rust through the application of an anti-fungal antibiotic, the cycloheximide, at the rate of 50 ppm (182 lt/acre) as a systemic fungicide.
4. Grewal and Dharm Vir (1959) observed that good control of stem rust is possible by spraying of Parazate liquid + $ZnSO_4$, 4 times at fortnight intervals.
5. Edgington and Reinberg (1966) reported that Plantavax and Vitavax are very effective as seed, soil and foliar spray treatment to control stem rust.
6. Brahma and Asir (1988) observed that Propiconazole (Tilt) at 0.1% is effective to control the stem rust by inhibiting completely the germination of uredospores.

C. Disease tolerant varieties. Some varieties are very useful to prevent most of the races of stem rust pathogen. These are S 227, S 307, S 308, S 331, Sonara 63, Sonara 64, Lerma Rojo, HD 2278, WL 614 and HW 741.

5. *LOOSE SMUT OF WHEAT*

Loose smut of wheat is a destructive disease. The disease was recognized and confirmed very early than the other wheat diseases. Maddox (1895) was the first to show the nature of the disease. Due to the characteristic black dusty appearance of diseased inflorescence, it is also called as **smut, black smut, black head, blasted head** and **snuffy ear**.

The disease is available throughout the world and is more abundant in humid and semi-humid regions, cause great loss in yield. In India the loss is much more in cooler and moist northern region than the south.

Mishra and Singh (1969) reported 3-6% loss in some newly developed high yielding varieties of wheat. Later, Gothwal *et al.*, (1972) observed that smutted plant produce 18.7% reduction in height of plants, 22.9% less tillers, 27.5% shorter peduncles and 15.7% reduction in yield. Earlier, Luthra (1953) estimated an annual loss of fifty million rupees only in Punjab.

Host. *Triticum aestivum* L.

Symptoms

The symptom is visible only when the plants develop ears. The ears of the affected plants emerge out of the flag leaf earlier than the

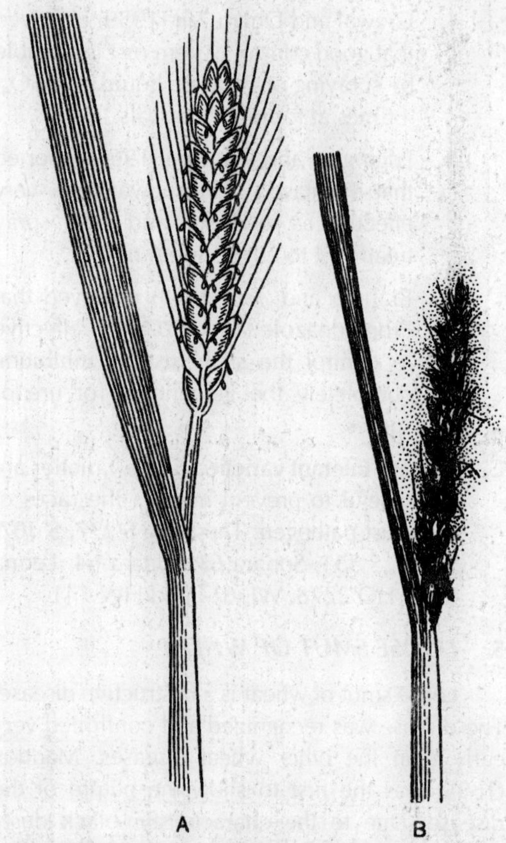

Fig. 5.33: Loose smut of wheat: A. Healthy inflorescence, and B. Diseased inflorescence

healthy ones. In infected plants, generally all the ears and all the spikelets and kernels of each ear are smutted. The spikelets of the affected ear get deformed and filled with dry, black, powdery mass of spores, thus no grains are formed (Fig. 5.33B). Initially the spores are covered by a delicate greyish membrane which soon burst and the spores get free. The spores are blown off by wind leaving only the naked rachis. Rarely some of the tillers developed from infected seed may escape from the disease. The number and length of the tiller is reduced in infected plant.

The growth and general appearance of the plant is not affected. The seeds become infected without showing any change in the external morphology.

Causal organism

The pathogen is *Ustilago segetum* (Pers.) Roussel var. *tritici* [earlier was known as *Ustilago tritici* (Pers.) Rostr.].

The black powdery mass of spores, the chlamydospores, are olive-green, spherical to oval in shape, minutely echinulate walls and 5–9 μm in diameter (Fig. 5.34A). The chlamydospore germinates to develop a promycelium of four uninucleate cells (basidium) and each cell later develops into germ tube or infection thread (Fig. 5.34B-D).

No sporidia are developed from uninucleate cells of promycelium. The spores remain viable for 5–6 months. The fungus is unable to survive saprophytically in the soil.

Disease cycle

The mass of chlamydospores from the infected ear are disseminated by wind or any other agencies to the healthy ear of adjacent plants at the flowering stage (Fig. 5.35). The spores lodge between the glumes and reach the hairy stigmas. The spores germinate on the stigmatic fluid and develop promycelia. The promycelia becomes septate and form four uninucleate cells. The dikaryophase is established by fusion of germ

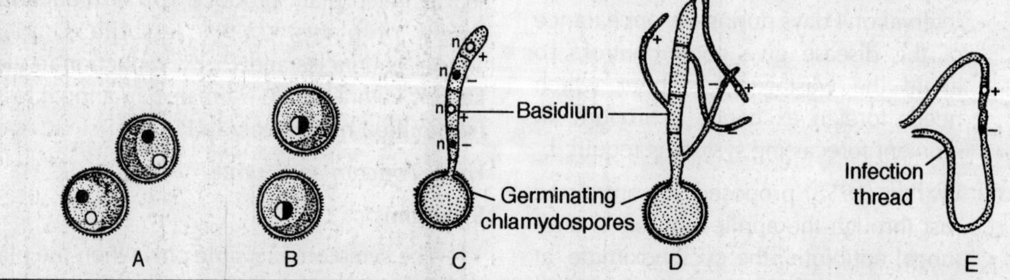

Fig. 5.34: Loose smut of wheat: A. Chlamydospores, B-D. Different stages of Chlamydospore germination, and E. Fusion between infection threads

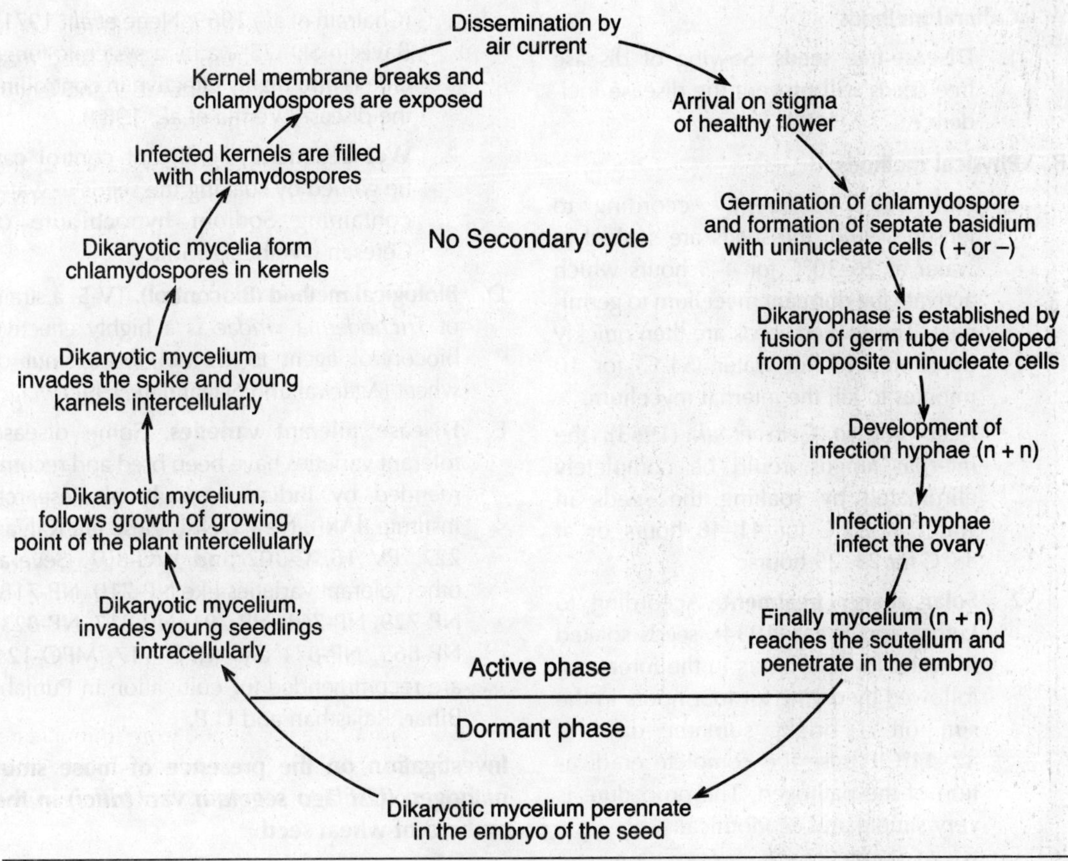

Fig. 5.35: Disease cycle of Loose smut of wheat

tubes (infection threads) derived from the individual uninucleate cells of the promycelium and ultimately infection hyphae are developed (Fig. 5.34E). Earlier it was thought that infection hyphae penetrate the stigma of healthy flowers, but later it has been shown that normally the entry takes place through the young tissue at the base of the ovary. After that, the infection hyphae grow intercellularly and ultimately reach the ovule. The infection hyphae then pass into the space between nucellus and the endosperm and reach the bottom of the endosperm to reach the scutellum and the embryo. The growth of infection hyphae is intercellular and haustoria are not formed. The host cells are not affected even at a minimum level.

Along with the above process of fungal growth; pollination, fertilization and embryo formation normally take place in the ears. The fungal hyphae continue to grow along with the development of seeds. The fungal hyphae become thick walled and remain dormant in the seed until the next growing season. Profuse mycelium is present in the scutellum of the mature infected seed. **Healthy and infected grains cannot be morphologically distinguished.** After sowing, the infected seeds will develop seedlings and simultaneously dormant mycelium becomes active and keeps place behind the growing apex.

There is a great accumulation of hyphae at each ear which ultimately replace the spikelets with masses of black spores. During dry weather, almost all the spores are blown off, leaving the rachis as a bare stalk. After dispersal, the spores again infect the healthy host and continue the cycle.

Disease management

The disease can be controlled or reduced by the following methods:

A. **Cultural methods**

1. **Disease-free seeds**. Sowing of disease free seeds will prevent the disease incidence.

B. **Physical methods**

1. **Hot water treatment**. According to Jensen (1908), the seeds are soaked in water at 26–30°C for 4–5 hours which activate the dormant mycelium to germinate. The soaked seeds are then quickly transferred to hot water (54°C) for 10 minutes to kill the internal mycelium.

   According to Gera *et al*. (1963), the internal fungus could be completely eliminated by soaking the seeds in water at 25°C for 41–48 hours or at 33°C for 23–29 hours.

2. **Solar energy treatment.** According to Luthra and Sattar (1934), seeds soaked in water for four hours in the forenoon, followed by drying for four hours in the sun on a bright summer day (at 42–44°C) cause the complete eradication of the pathogen. The procedure is very simple and economical.

3. **Water soaking treatment**. According to Tyner (1953), loose smut of spring wheat can be effectively controlled by soaking seeds in water at room temperature for 56–64 hours. The death of the mycelium takes place due to accumulation of two quinine compounds, 6-hydroxy quinine and 2, 6-dihydroxy quinine inside the soaked seeds due to anaerobic respiration (Tyner, 1957).

4. **Anaerobic treatment**. Control at significant level can be done by water soaking the seeds for 2-4 hours at 15-21°C (60-70°F), followed by keeping in airtight container for 65-70 hours and then storing after drying them.

C. **Chemical methods**

1. **Dry dressing of seeds**. Loose smut can be controlled at highly significant level by seed dressing with systemic fungicides, like Benomyl and Carboxin at 0.2-0.25% (Chatrath *et al*., 1969; Nene *et al*., 1971). Bavistin SD (25% a.i.), a systemic fungicide is also highly effective in controlling the disease (Verma *et al*., 1984).

2. **Wet treatment.** Significant control can be gained by soaking the seeds in water containing Sodium hypochlorite or Ceresan (Weibel, 1958).

D. **Biological method (Biocontrol)**. TV-5, a strain of *Trichoderma viridae* is a highly effective biocontrol agent to control loose smut of wheat (Agarwal and Nagarajan, 1992).

E. **Disease tolerant varieties**. Some disease tolerant varieties have been bred and recommended by Indian Agricultural Research Institute (IARI), New Delhi; these are Kalyan 227, PV 18, C-302 and WG-307. Several other tolerant varieties like NP-710. NP-718, NP-729, NP-761, NP-791, NP-817, NP-823, NP-865, NP-871 and MPO-117, MPO-125 are recommended for cultivation in Punjab, Bihar, Rajasthan and U.P.

Investigation on the presence of loose smut pathogen (*Ustilago segetum* var. *tritici*) in the embryo of wheat seed

Presence of fungus inside the embryo can be studied by the procedure mentioned below:

Take 100 wheat seeds and dip in 10% KOH solution overnight. The embryo will be separated and the rest becomes a cottony mass. Take out the embryos from the solution and wash in distilled water. Embryo pieces are taken one by one on clean glass slide containing a drop of lactophenol and then covered with cover glass. Slightly tap the cover glass. The cells of the embryonic tissue get separated and thick brown-coloured mycelium becomes visible under compound microscope, if present.

6. *COVERED SMUT OF BARLEY*

The covered smut of barley is widely distributed throughout the world. It is found in all those countries where the barley is grown. It is a common disease of barley in northern India. It causes considerable damage in the Punjab, Uttar Pradesh, Madhya Pradesh and certain parts of

Fig. 5.36: Covered smut of barley: symptom on inflorescence

Bihar. In severe cases of infection, the yield may be lowered by 50 per cent of the crop.

Host. *Hordeum vulgare* L.

Symptoms

The first symptoms may be easily being seen as the ears of the infected plants emerge out. The smutted or blackened ears come out of the boot leaves, which may easily be recognized from a distance. The smutted ears emerge from the boots at the same time as the healthy ones. They usually remain shorter and are retained within the boot leaves for a longer time.

All the ears of diseased plant become infected and all the developing grains turn into smut sorus remain covered by a white, shining, silvery membrane, which is partially developed from host tissue and partially from fungus. Due to this compact covering, the smut spores are not blown off, and thus the disease in named "covered smut" (Fig. 5.36).

The smut spores do not serve as means of propagation of the disease during the growing season. They are liberated by the rupture of the surrounding membrane as the ears get crushed under heavy pressure during threshing. The spores thus released adhere to the rough skinned healthy grains.

Causal organism

The disease is caused by the fungus *Ustilago hordei* (Pers.) Lagerheim. The fungal is an obligate parasite and its body is mycelial, dikaryotic, and grows internally. The hyphae are septate and branched, grow intercellularly. The mycelium grows in the stem as well as in the leaves but generally confined in the stem.

During flowering time, hyphae reach the ear, enter the ovaries, where it grows more vigorously. There they branch repeatedly and become additionally septate. The short, binucleate segments thus formed become round off to form the binucleate smut spores i.e. **teliospores**. The spores remain compacted together into a ball-like mass in each ovary. The teliospores are smooth-walled, brown in colour but lighter on one side and globose to ellipsoidal in shape (Fig. 5.37) and measures 3.6-7.2 microns (more often 4.5

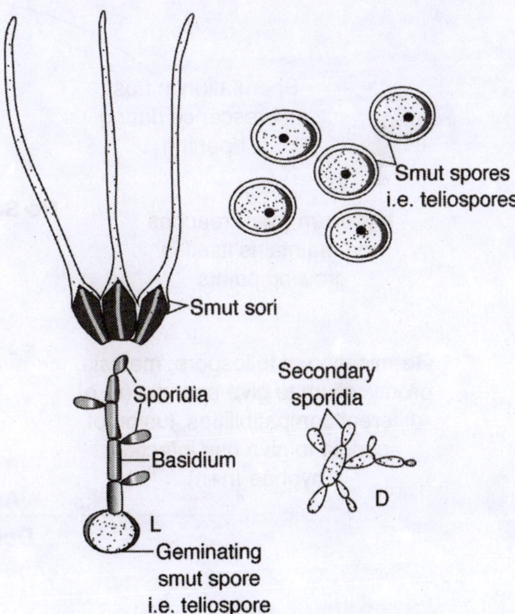

Fig. 5.37: *Ustilago hordei*. A. Smut sori, B. Smut spores, C. Germinating smut spore and D. Secondary sporidia

microns in diameter). The teliospore germinates to form a four-celled basidium bearing four ovate to oblong basidiospores (Fig. 5.37).

In a mass, the teliospores look black. They are liberated at the threshing time by the breaking the walls of the ovary. The spores thus set free remain attached to the surface of healthy grains.

Disease cycle

The disease is externally seed borne and the smut spores i.e. teliospores adhere to the healthy grains during threshing-time (Fig. 5.38).

A. Germination of Teliospores

The binucleate smut spores, or teliospores present in the soil or sticking to the grains, germinate in the presence of warmth and moisture when the contaminated barley grains are sown in the field.

Prior to germination, the two nuclei of dikaryon in the spore fuse to form a single fusion nucleus or the synkaryon. The diploid teliospore germinates to produce a germ tube, the promycelium i.e. the epibasidium.

In the epibasidium, the synkaryon undergoes meiosis to form a row of four haploid nuclei of which two are plus (+) and two are of minus (-) type. Septa are laid down between the nuclei and thus four haploid cells are formed. Each segment of the epibasidium produces one uninucleate sporidium or basidiospore. Two of these basidiospores are of plus and two of minus strain.

Basidiospores are ovate to oblong in form. They are capable of multiplying by budding like yeasts in presence of nutrition. The secondary spores thus produced by budding germinate like the basidiospores and are capable of infecting fresh barley plants.

B. Infection

The basidiospores or secondary spores germinate on the surface of the grain and produces a slender haploid germ tube. The haploid germ tubes by themselves are incapable of infecting the barley seedlings.

The germ tubes from two different basidiospores of opposite strains growing in the vicinity, fuse and one of them becomes dikaryotic. The

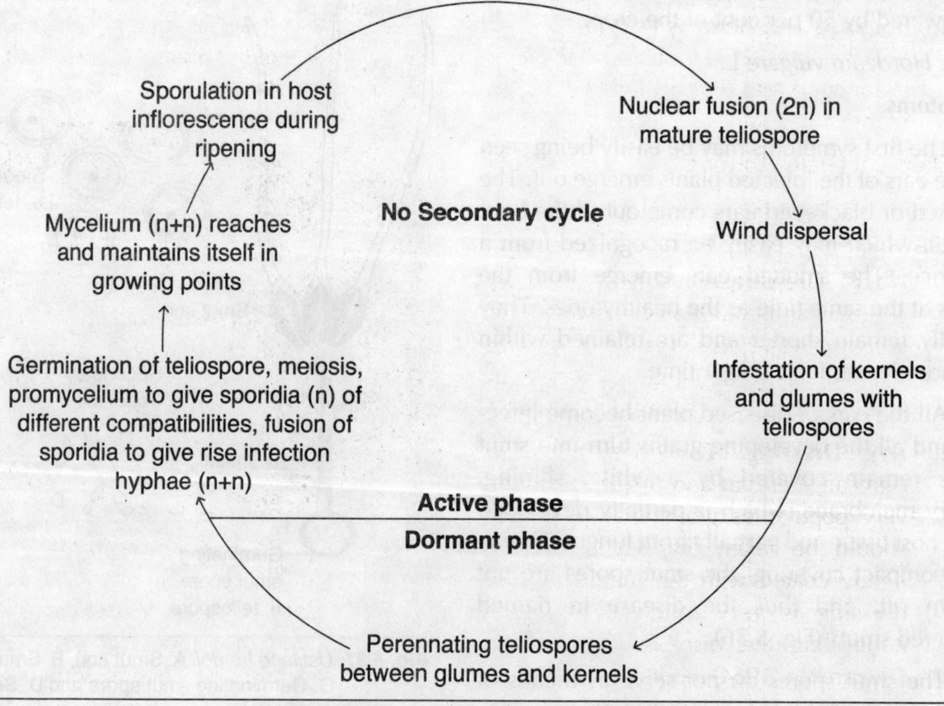

Fig. 5.38: Disease cycle of Covered smut of barley

dikaryotic germ tube or the infection thread infects the barley seedling at a very early stage.

The dikaryotic hypha penetrates the young coleoptile and enters the growing tip of the embryo. There it grows upwards keeping pace with the growing point.

The mycelium grows upwards and grows more vigorously in the stem than in the leaves. During flowering time, the mycelial hyphae reach the inflorescence region and enter the ovaries to produce the smut spores.

Disease Management

Since the disease is surface borne, external application of fungicides to the grain is quite effective. The disease can be controlled or reduced by the following methods:

A. Cultural method

1. **Disease-free seeds**. Use of disease-free seeds will prevent the disease incidence.
2. **Rogueing**. Rogueing of plants with smutted ears is another preventive measure. The infected plants should be uprooted and burnt with great care so that the smut spores may not fall on to the ground and thus reduce the source of infection

B. Chemical Grain Dressing

It may be applied in the form of liquid dressing or in the form of dust.

1. **Liquid Dressing**. Dressing the barley grains with a solution of 3 percent formalin in water gives very good results. The practice of liquid grain dressing is, however, not so popular these days.
2. **Dry Dressing**. Dry dressing with Agrosan, Ceresan and Sulphur dust are very popular as grain dressings. Care should be taken that each grain is lightly coated with the fungicide. Dry dressing is very much effective in controlling the disease. Grewal and Dharam Vir (1964) reported that Ceresan and Agrosan 5 W completely control the disease. Sulphur dusting is also effective in controlling the disease.

C. Disease tolerant varieties

The most effective method is to grow disease resistant varieties. Mathur, *et al*. (1960) reported that K12, K18, K19, C50, C84, and CN249 are moderately resistant.

7. *EARLY BLIGHT OF POTATO*

The disease caused blight symptom which appears towards the early to middle of the season, that's why the name is given. It is a common and serious foliage disease of potato are found everywhere the potatoes are grown. The disease is also known as **target spot**, rarely affects young plants in the vigorously growing season. It is found to appear first in the older leaves. The disease favoured by warm temperature and high humidity.

Mathur *et al*. (1971) reported that there have been outbreaks of the disease during the last 5-6 years in the plains of Uttar Pradesh.

Host. *Solanum tuberosum* L.

Symptoms

Symptoms of the disease appear both on leaf and in tuber potato.

Fig. 5.39: Early blight of potato: Symptoms on leaf (portion).

1. **On foliage**. On the leaflets at 3-4 week stage, symptoms appear as small, isolated, scattered pale brown to dark spots (Fig. 5.39). Spots are oval or angular in shape and

measures 3-4 mm in diameter. Spots are usually delimited by a narrow chlorotic margin.

Chlorotic zone on leaf increases with the increase in size of the spot. The lower i.e. older leaves are affected first and the disease gradually progress upwardly. The spots on leaf often show a series of concentric ridges, a characteristic symptom of the disease.

(a) In favourable condition, the number and size of the spot increases may cover almost the entire leaf surface.

(b) In humid weather, the spots coalesce and big patches of rotting appear on the leaf surface. In dry weather, the spots become hard and leaflets become curl (Singh, 1968).

(c) In advanced stage of the disease, drooping of the leaves may take place leaving only a few leaves at the top. Sometimes there is complete defoliation. Stems and petioles may also develop brown to dark lesions and in advanced stage, the above-ground parts may damage completely.

2. **Inside the Tuber.** Tubers are affected, with dark, circular to irregular spots. The edges of the spots are often raised and purple to dark gray in colour. Internally, the flesh under the spots is usually brown, dry, and leathery or corky in texture. As the disease advances, the potato flesh often becomes water soaked and yellow to greenish yellow.

Causal organism

The causal organism i.e., pathogen is *Alternaria solani* (Ell. and Martin) Jones and Grout of Mitosporic (anamorphic) fungi (earlier form class Deuteromycetes).

The fungus consists of light brown, slender, septate and much branched hyphae, becomes dark coloured with maturity. Initially hyphae grow in the intercellular space but later it penetrates inside the cell (Fig. 5.40).

Later, conidiophores develop from the diseased tissue and emerge through the stomata. Conidiophores are relatively short, measure 50-90 μm x 9 μm and develop conidia from the terminal cell. Conidia are dark coloured, beaked, muriform and multicellular. It measures 120-296 μm x 12-20 μm. Conidia have 5-10 transverse septa and a few longitudinal one. The beak of the conidia is long, septate and rarely branched (Fig. 5.40). Generally, conidia are borne singly but in culture it develops in short chain of two.

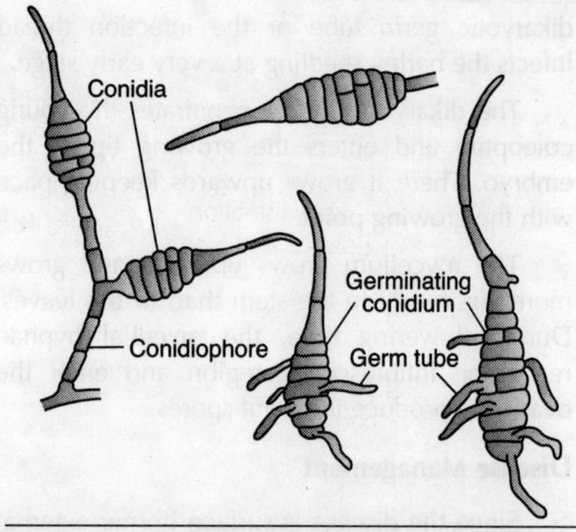

Fig. 5.40: Early blight of potato (C.O. *Alternaria solani*): A. Conidia and B-C. Germinating conidia

Mature conidia are detached and dispersed readily by air current, water and insects. On contact with suitable host, they germinate by putting 5-10 germ tubes and start further infection.

Disease cycle

Both mycelia and conidia of the pathogen remain viable for a considerable time. The mycelia remain viable in the infected plant debris for about one year or more and conidia remain viable for about 17 months in room temperature.

Initially primary inoculum develops from the infected plant debris such as dried leaves and stems, and also from infected tubers (Fig. 5.41). In addition, conidia and mycelia may develop from infected plant debris in the soil. The conidia germinate at the optimum temperature ranges between 28°C-30°C within 35-45 minutes (Walker, 1969). It pushes the germ tube into the lower leaves through stomata or penetrating the leaf and stem directly by piercing through the epidermis. In favourable condition, primary infection and the appearance of the spot takes

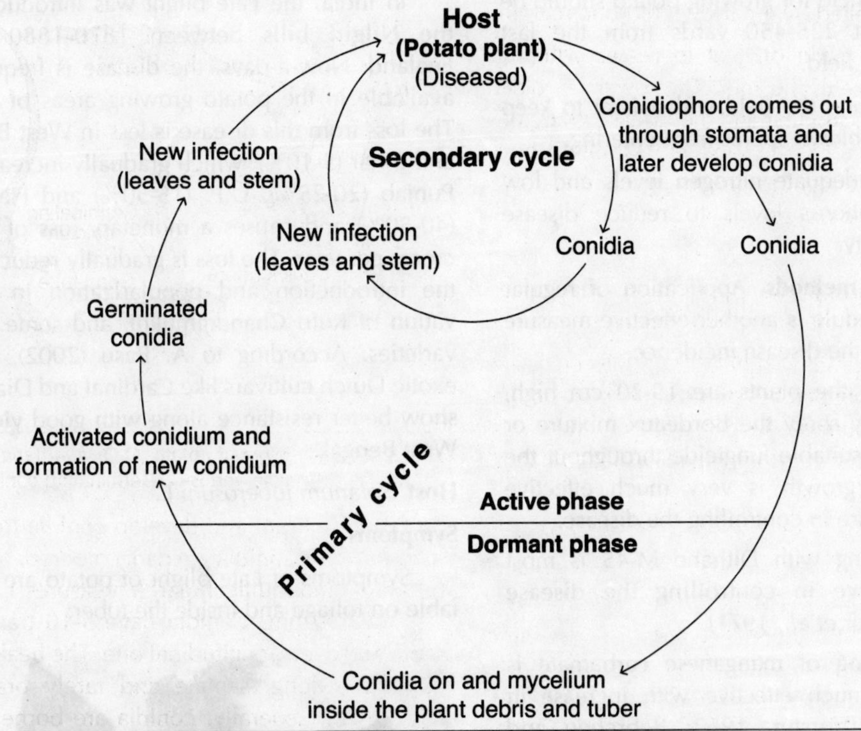

Fig. 5.41: Disease cycle of Early blight disease of potato

place within 2-3 days. Production of secondary inoculum from the primary spot starts within 5-7 days after infection when the spots are about 3 mm in diameter. Heavy dew with frequent rain promotes huge sporulation.

Mature conidia are detached and dispersed through air current, water and insects. Flees and beetles are reported to be helpful in carrying the inoculum as well as helps to penetrate the host through injury made by the insect. The pathogenic effect like chlorosis takes place by the secretion of toxin, alternaric acid; which was isolated from the culture filtrate (Brain *et al.*, 1952).

Predisposing factors/Epidemiology

The climatic factors and soil conditions unfavourable to the host plant, reduces the vigour, and thus tends to promote the disease.

When season starts with abundant moisture followed by high temperature becomes unfavourable to the plant, thus it help the disease to spread and becomes serious.

Weaker plants are more susceptible to the disease than the healthier one.

High soil fertility reduces the severity of the disease. Continued drought also check the spread of the disease.

Disease management

The disease can be controlled or reduced by the following procedures:

A. **Cultural methods**. The methods useful in controlling the early blight of potato are:

1. Seed tubers should not be collected from the diseased field.
2. Tubers should be harvested after the maturity of the tubers (when skin of the tubers comes tough).
3. Previous year's plant debries should be cleaned from the field and it must be burnt outside the field.
4. It is better to follow the crop rotation and not to grow the solanaceous crop, mainly the tomato in the next season.

5. New field for growing potato should be atleast 225-450 yards from the last year's field.
6. Surround the field with wheat to keep wind-blown spores from entering.
7. Use adequate nitrogen levels and low phosphorus levels to reduce disease severity.

B. **Chemical methods**. Application of regular spray schedule is another effective measure to control the disease incidence.

1. When the plants are 15-20 cm high, weekly spray the Bordeaux mixture or other suitable fungicide throughout the plant growth is very much effective measure in controlling the disease.
2. Spraying with Dithane M-45 is most effective in controlling the disease (Mathur *et al.*, 1971).
3. Spraying of manganese carbamate is very much effective with increase in yield (Vorster, 1962; Bebrchen and Brien, 1965).
4. Repeat the application of Zineb is very much effective in controlling the disease (Paharia, 1961; Kadyroa, 1964). Breston 60 is also highly effective in controlling the disease.

C. **Disease tolerant varieties**. Some of diseased tolerant varieties are Kufri Naveen, Kufri Sindhuri and Kufri Jeevan etc.

8. *LATE BLIGHT DISEASE OF POTATO*

The disease causes blight symptom which appears towards middle to late in the season, that's why the name is given. Potato is a native of the North Andes in South America. Initially, late blight of potato was available in that area as an endemic disease. Gradually, by 1840-1847, it spread throughout USA and Europe. The disease became well established in Ireland, Europe and England by 1842. Later, in 1845, it spread as epiphytic disease throughout the Europe and brought **famine in Ireland**, causing the **death of one million (1,000,000) people** and more than one million migrated to other countries.

In India, the Late blight was introduced in the Nilgiri hills between 1870-1880 from England. Now-a-days, the disease is frequently available in the potato growing areas of India. The loss from this disease is less in West Bengal and Bihar (5-10%), which gradually increases in Punjab (20-25%), U.P. (15-50%) and Haryana (40-50%) and causes a monetary loss of Rs. 4 crores per year. The loss is gradually reduced by the introduction and popularization in cultivation of Kufri Chandramukhi, and some other varieties. According to A. Basu (2002), some exotic Dutch cultivars like Cardinal and Diament show better resistance along with good yield in West Bengal.

Host. *Solanum tuberosum* L.

Symptoms

Symptoms of Late blight of potato are available on foliage and inside the tuber:

A

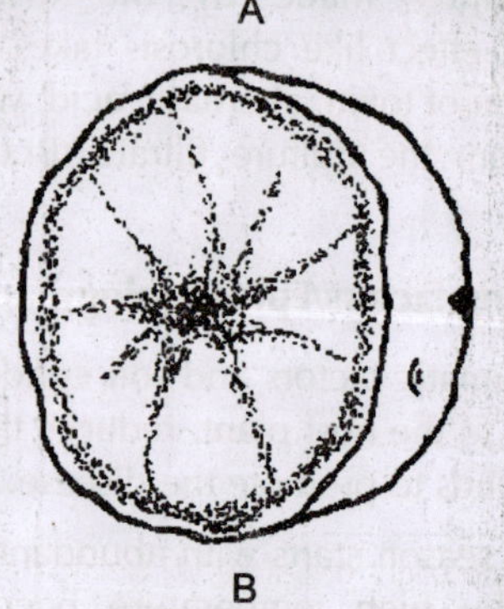
B

Fig. 5.42: Late blight of potato: A. Blighted leaf, B. Diseased tuber (in section)

1. **On foliage**. Symptoms appear as hydrotic areas with indefinite margins at the apex and edges of the leaflets (Fig. 5.42A). The infected regions gradually turn brown to black due to necrosis. In early stage of disease development, chlorotic border develops around the necrotic regions.

During moist weather, the disease progresses very rapidly, causes much more decay of the infected regions and produces a characteristic bad odour. The sporangiophore and sporangia are visible as matty growth on the abaxial surface of the leaflet. The entire plant may be damaged completely within a few days.

During dry weather, the disease progresses very slowly and the affected regions curl and shrivel. The diseased areas become hard and easily break with small disturbance.

2. **Inside the tuber**. The pathogen does not go down through the stem to the tuber. The tuber infection takes place in different ways:

(a) **during growing season**. Hydrotic areas develop on the tuber surface and the regions become necrotic. If condition favours, the entire potato turns brown and becomes damaged completely before harvest.

(b) **during harvest**. During harvesting, the tuber with delicate skin may get ruptured and the tuber may find contact with the infected leaf and cause infection. Later, the symptom becomes visible after cutting the tuber as a wheel marked by small brown dots (Fig. 5.42B).

Causal organism

The causal organism, *Phytophthora infestans,* belongs to the subdivision Mastigomycotina under the Division Eumycota. The plant

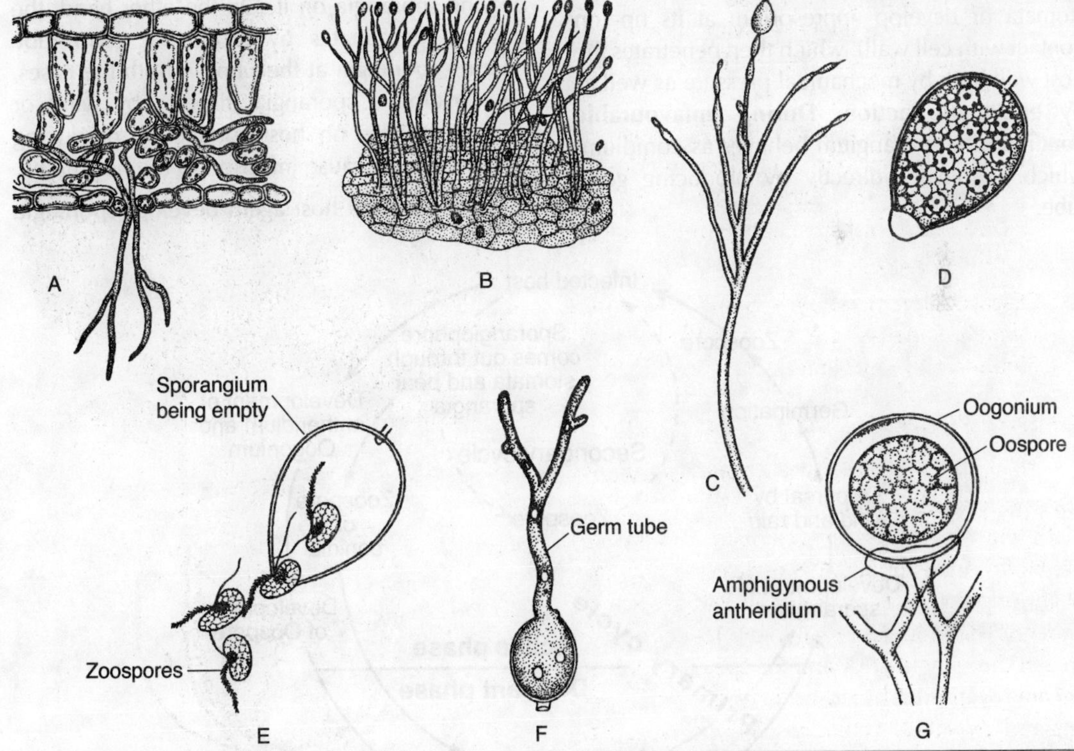

Fig. 5.43: Late blight of potato: A. Section of diseased leaf showing developing sporangiophore coming out through stomata, B. Huge number of sporangiophores coming out through stomata from a diseased tissue, C. Sporangium on branched sporangiophore, D. Single sporangium, E. Germination of sporangium by producing zoospores, F. Direct germination of sporangium, G. Single Oospore

body is a coenocytic mycelium which grows both intra- and intercellularly inside the host tissue. The sporangiophore develops either singly or in groups on the young seedling and also from the lower side of the leaf of adult plant (Fig. 5.43A, B). It is branched and develops sporangia at the apices. Initially, one sporangium develops at the apex (Fig. 5.43C) and, at maturity, the sporangiophore elongates from one side and shifts the sporangium towards the opposite side — another sporangium develops at the apex.

The sporangia are lemon-shaped and colourless with a small stalk below and thin papilla above (Fig. 5.43D). **During favourable condition** sporangium produces a number of bi-flagellate secondary zoospores (Fig. 5.43E). After swimming for some time in dewdrop or thin film of water, the zoospores encyst and then germinate by producing germ tube (Fig. 5.43F). The germ tube penetrates directly through stomata or develop appresorium at its tip (on contact with cell wall), which then penetrates the host wall both by mechanical pressure as well as by enzymatic action. **During unfavourable condition** the sporangium behaves as conidium, which germinates directly by producing germ tube.

Optimum temperature for the growth of mycelium is 16–18°C, for sporangium development is above 10°C, for germination of sporangia by zoospore is 12°C, while at 21°C it germinates directly by germ tube.

Relative humidity of 100% favours the abundant production of sporangium, while below 90% RH inhibits its production.

The fungus is heterothallic. During sexual reproduction oospores are produced inside the aerial parts by the union between antheridium and Oogonium (Fig. 5.43G).

Disease cycle

In the dormant phase, the fungus could perennate in the tuber as mycelium and in plant debris as oospore (Fig. 5.44). During favourable condition, the perennating mycelium becomes active and develops active mycelium. The active mycelium develops branched spo .angiophore bearing sporangia on it. On the other hand, the oospore germinates by producing germ tube bearing sporangia at their tip. In both the cases, the developed sporangia disperse by wind or rain, germinate on host surface by producing zoospores and cause infection to the host tissue.

The infected host again develops sporangia

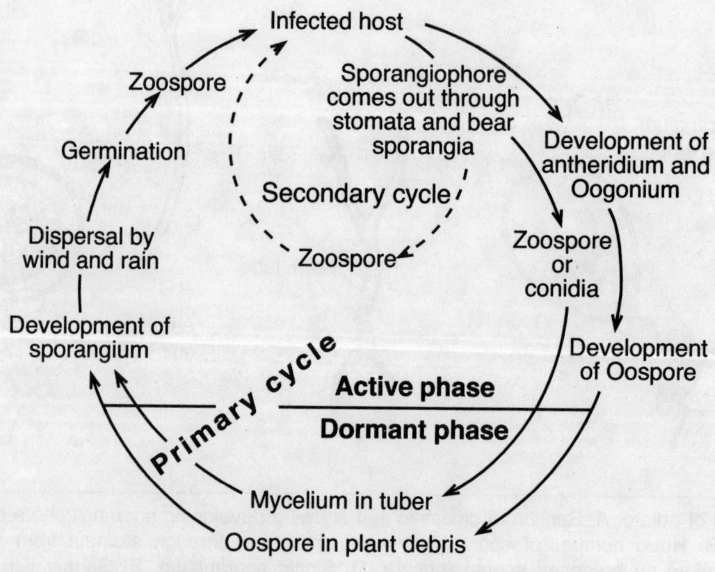

Fig. 5.44: Disease cycle of Late blight disease of potato

on sporangiophore. The sporangia develop zoospore which cause further infection. This process repeats several times in the growing season — this is the secondary cycle.

During favourable condition, the zoospores may come down into the soil and may cause infection to the tubers. The tuber infection may also take place on contact with infected foliage during harvest.

The perennating mycelium inside the tuber remains active, if the infected tubers are kept in the storage for seed potato. In the next season, after sowing the seed, the perennating mycelium becomes active and causes further infection.

Towards the end of the season, inside the aerial parts, both the sex organs — antheridium (male) and oogonium (female) — develop and undergo sexual reproduction of amphigynous type (Fig. 5.43G). The product of sexual reproduction is oospore, which has hard protective covering. The oospore remains dormant inside the host tissue during unfavourable season. During favourable condition in the next season they germinate and produce next crop of zoospores.

Oospore formation is very rare in **USA** and **Canada** and may not play a significant role in the survival of the fungus. But it is frequently available in **Mexico** and indicates the possibility of playing significant role in the annual recurrence. But **in India,** the potato-growing season is followed by hot summer season in the plains, ruling out the possibility of survival of inoculum in the soil and also in the plant debries. The perennating mycelium in the tuber is killed by exposure at 40°C for 4 hours or at 30°C for 65 hours. Thus, in plains, the tubers stored in the house at room temperature becomes free from disease. In the northern hills and also in its surrounding plains, the mycelium may remain active due to high humidity and low temperature and causes disease in more severe form. The disease also appears with more intensity, possibly due to survival of inoculum inside the seed tubers kept in cold storage.

Disease management

The disease can be controlled or reduced by the following procedures:

A. **Cultural methods**. The methods useful in controlling the late blight of potato are:

1. Seed tubers should not be collected from the disease field.
2. Tubers should be harvested after the maturity of the tubers (when skin of the tubers comes tough).
3. In infected field, the aerial part of the plants should be dried completely before harvest of tubers.
4. Infected plant should be cut out and collectively buried deep in the soil.
5. The previous year's plant debris should be cleaned from the field and it must be burnt outside the field.
6. The volunteer potato plants that may harbour the infection mainly in hills should be eradicated during appearance.
7. Planting of alternate row of susceptible and resistant variety will reduce the intensity of the disease.
8. Cultivation of tuber on high ridges and covering the tuber with soil will prevent the fungus to grow out from the infected mother tuber, thus reduce the spread of the diseae.
9. The nitrogen fertilizer should be applied at a reduced rate.
10. During harvesting, the potato plants should be placed on one side and tubers on other side, thereby contact between tuber and infected plant can be avoided.

B. **Physical methods**. Following methods are used to control the disease:

1. The perenating mycelium in the tuber can be killed by exposure to 30°C for 65 hours or to 40°C for 4 hours.
2. The perenating mycelium can be killed by dipping the tuber in water at 45°C for 4-5 hours or at 40°C for 24 hours (Butler, 1903).

C. **Chemical methods**

1. Dutt (1962), while working with 16 different fungicides in hilly region,

found that Bordeaux mixture (5: 5: 50) is the most effective giving better result than Burgundy mixture (Dutt, 1970).

2. Fungicides like **Brestan** (1 kg/ha), **Dithane M-45** (2 kg/ha) and **Difolitan 80WP** (2.5 kg/ha) are used as foliar spray and found highly effective in controlling late blight disease of potato in plains of India.
3. According to Khanna and Sharma (1981), **Metalaxyl** (1 kg/ha), a systematic fungicide, is found highly effective in controlling the disease, thus increasing the yield about 3 times than the control.

D. **Biological methods (Biocontrol)**

1. **Jindal *et al.* (1988)** reported the role of *Epicoccum purpurascens, Trichoderma koningii* and *Stachybotrys atra* in controlling the disease.
2. **Roy *et al.* (1991)** reported the role of *Penicillium aurantiogriseum* and *Myrothecium verrucaria* in late blight disease management.
3. **Arora (1999)** reported the role of *Penicillium viridecatum, Chaetium brasilense* and *Trichoderma viride* in controlling the late blight disease.

E. **Disease tolerant varieties.** It is better to use tolerant varieties. In India, Central Potato Breeding Station (Simla) has developed some tolerant varieties like Kufri Jyoti, Kufri Naveen, Kufri Badshah, Kufri Jeevan, Kufri Neela etc., for commercial cultivation. Cardinal and Diament (Dutch cultivrs) and Kufri Chandramukhi and Kufri Chipsona-1 (Indian cultivars) are the varieties showing good performance in alluvial zone of West Bengal (Basu, 2002).

9. *WART DISEASE OF POTATO*

Wart disease of potato is an important disease of the cultivated potato, also called by various names like black wart, black scab, potato tumor, potato cancer or canker, cauliflower disease, warty disease. It is assumed that *S. endobioticum* spread at the end of the nineteenth century from the center of origin in the Andes first to Europe and North America, and subsequently across the whole potato growing regions of Asia, Africa and Oceania (Smith *et al.* 1997). In India, the disease was first reported from Darjeeling by Ganguly and Paul (1953) and continues to be restricted to that area due to local quarantine.

The disease is so called due to the warty exudations produced on tubers and occasionally on stems, leaves and flowers (Hooker 1981).

Host. *Solanum tuberosum* L.

Symptoms

The symptoms of the disease appear only on underground parts of the plants except roots i.e. on tubers, buds of stems, stem bases and stolons (Fig. 5.45 and 5.46). These are characterised by warty, tuberous and dirty cauliflower like outgrowths on infected parts of plants. The warty outgrowths are of different size from small protuberances to large intricately branched systems.

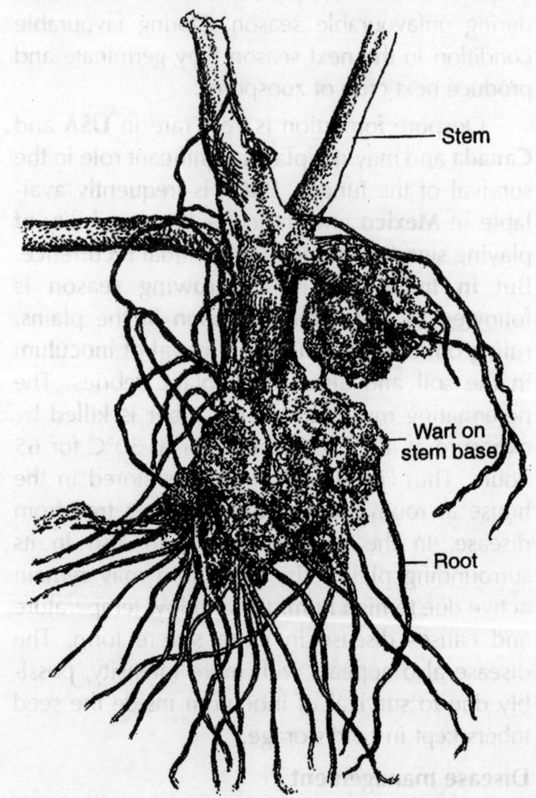

Fig. 5.45: Wart disease of potato. Symptoms on stem base and stolon of potato

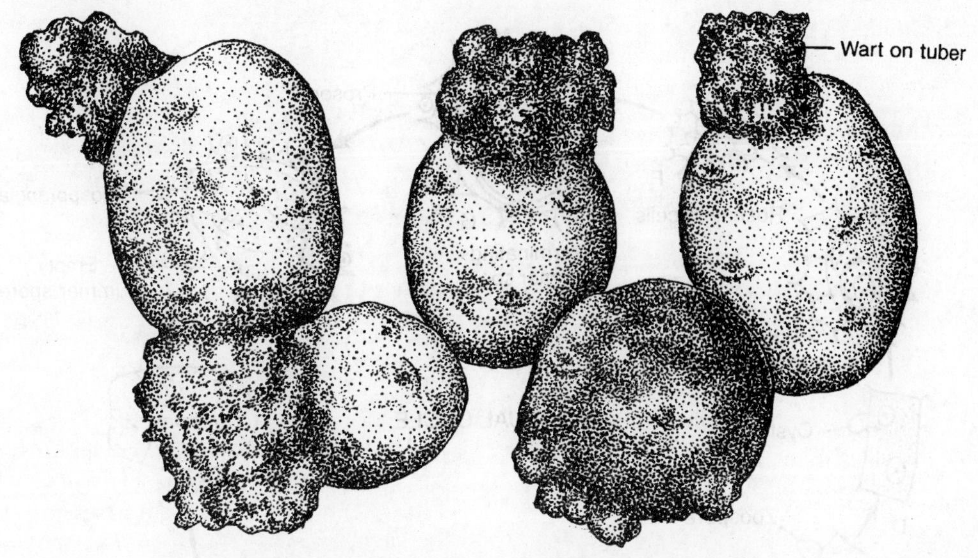

Fig. 5.46: Wart disease of potato – Wart on potato tubers

Warts are cream coloured or black on underground parts.

On the tubers, mostly the warts are larger than the tuber itself covering the whole tuber. In advanced stages, the warts become dark black in colour.

The warts are usually consists of distorted proliferated branched structures grown together into a mass of hypertrophied tissue.

Causal organism

The disease is caused by *Synchytrium endobioticum* (Schilb.) Peres. It is a holocarpic endoparasite and develops no mycelium. The fungus produces a thick walled structure known as a winter sporangium, of 25-75 μm in diameter and contains 200-300 spores. Sporangia are clustered into thin-walled sorus. It produces zoospores with one posterior flagella and measures about 0.5 μm in diameter.

Disease cycle

The primary infection of the young potato takes place by zoospores develops on germination of the resting spores present in the soil (Fig. 5.47).

These zoospores after swimming for a while come in contact on the underground shoot parts, withdraw their flagella and then enter the host cells where they settle down at the bottom of the cells.

Each of the neighbouring cells then starts dividing repeatedly followed by swelling, ultimately results the formation of a wart. These warts are the main symptoms and contain the material for secondary infection. The sporangia contained in the warts produce zoospores upon disintegration of the hosts.

These zoospores are responsible for secondary infection. During favourable condition, the zoospores are formed repeatedly and secondary infections may take place as well causing serious damage to the potato crop.

With the onset of unfavourable condition and as the crop season approaches towards the end, the zoospores function as planogametes, fuse and produce resting sporangia. While the crop is harvested, the resting sporangia remain buried in the soil and perennate till the favourable conditions to return back.

Disease management

It is very difficult to control the disease once it has been introduced in the field. However, some of the disease management procedures are followed to control or reduce the disease incidence:

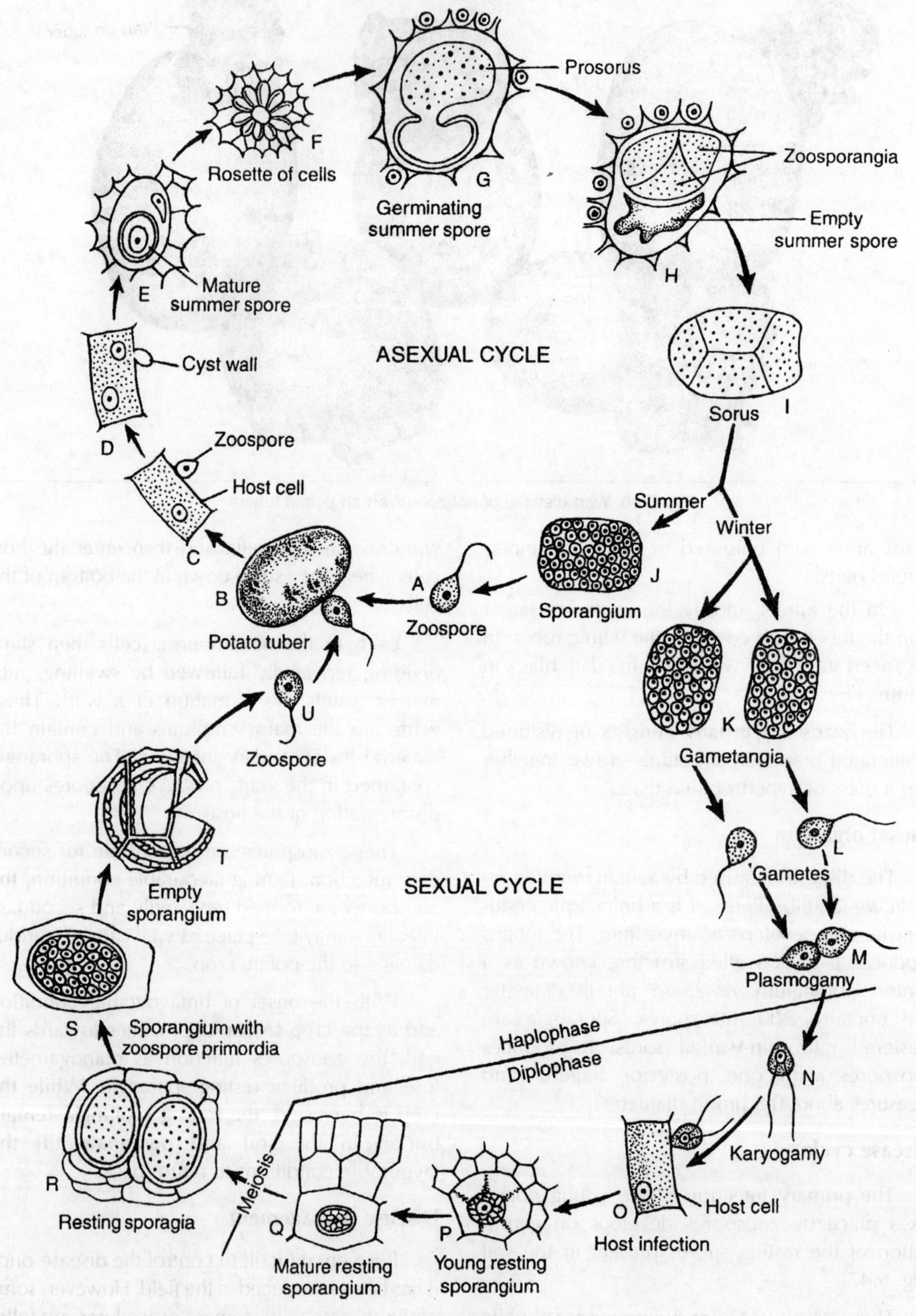

Fig. 5.47: Wart disease of potato. A. Life cycle of *Synchytrium endobioticum*

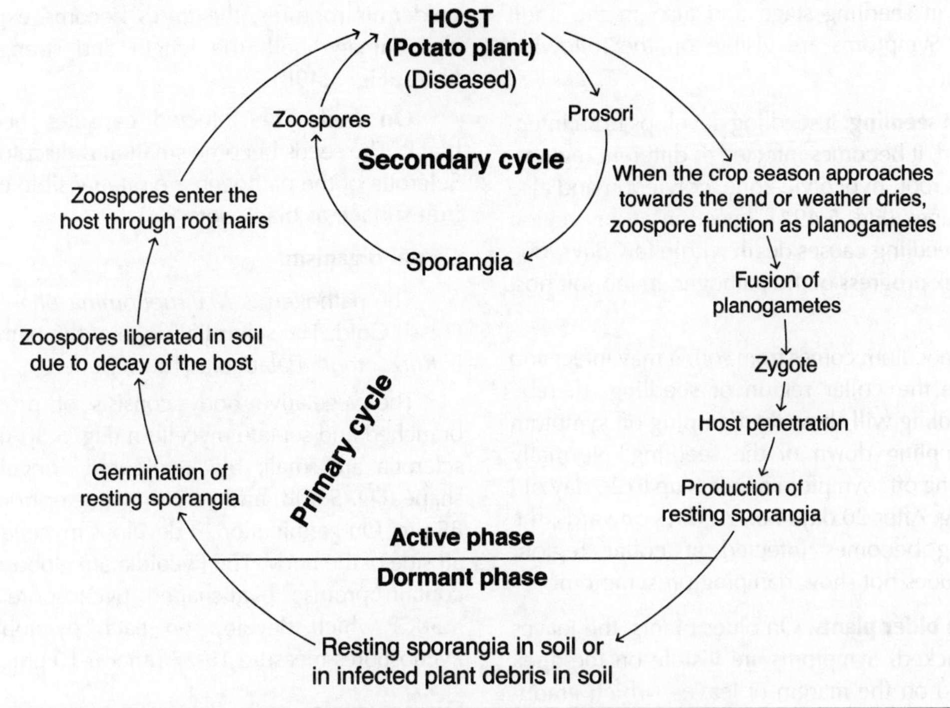

Fig. 5.48: Disease cycle of Wart disease of potato

A. Cultural method

1. **Disease-free seeds.** Use of disease-free seeds will prevent the disease incidence.
2. **Restriction of entry of pathogen.** Entry of diseased material into healthy areas should be prevented.
3. **Precaution for cultivation.** Potatoes and indeed any kind of plants with roots (including bulbs, corms and tubers) for export, should not be grown in fields where *S. endobioticum* has occurred or is still present on account of the fungus longevity factor.
4. **Quarantine measure.** Strict quarantine measure should be followed, to avoid entry of pathogen in new areas. In India, as the disease occurs in Darjeeling, potato tubers should not be exported to other parts of the country.

B. Chemical control

1. **Chemical treatment.** Soil treatment with chemicals may control the disease to a large extent. These include application of mercuric chloride, copper sulphate and 5 percent formaline. But these are very costly. Steam sterilization is also equally effective in controlling the disease.

C. **Disease tolerant varieties.** Cultivation of disease resistant varieties continuously for 8-10 years is the only effective control measure.

10. *STEM ROT OF JUTE*

The disease is commonly observed in Assam, West Bengal, Orissa, Bihar and eastern part of U.P. The disease is also common in the jute growing regions of Bangladesh.

Host. *Corchorus capsularis* L. (Tita jute) and *C. olitorius* L. (Mitha jute).

Symptoms

Plants are attacked mostly on all stages of growth. But the damage from this disease may be severe in seedling stage and also in the adult plants. Symptoms are visible on the following regions:

On seedling. If seedling develops from infected seed, it becomes infected at different regions, such as root, hypcotyle zone, cotyledon and also in the apex (Fig. 5.49A). Infection at any region of the seedling causes death within few days, due to quick progress of the pathogen in the soft host tissue.

If inoculum comes from soil, it may infect and damage the collar region of seedling, thereby the seedling will show the damping off symptom i.e., toppling down of the seedling. Normally "damping off" symptom appears up to 20 day old seedling. After 20 days i.e., 21 days onwards, the seedling becomes infected at collar region, which does not show damping off symptom.

On older plants. On older plants, the leaves are attacked. Symptoms are visible on the apex and also on the margin of leaves, which gradually proceed towards lower side (Fig. 5.49C). The leaves then fall down leaving the jute plants with bare branches. With severe infection, the stems rot and the roots become shredded. On the stem, epidermis ruptures, the fibres become exposed and damage both the length and strength of fibres (Fig. 5.49B).

On fruit. The infected capsules become black. The seeds become small and discoloured. Sclerotia of the pathogen are often visible on the fruit surface as black dots.

Causal organism

The pathogen is *Macrophomina phaseolina* (Tassi) Goid. The sclerotial stage of the pathogen is *Rhizoctonia bataticola* (Taub) Butler.

The vegetative body consists of profusely branched and septate mycelium (Fig. 5.50A). The sclerotia are small, black-coloured, irregular in shape (Fig. 5.50B) and vary in diameter from 40-85 µm. On germination, it develops mycelia from all side of the body. The Pycnidia are globose and contain profuse oval-shaped pycnospore (Fig. 5.50C) which develop on each pycnophore. Pycnospores measure 16-27 µm × 6-10 µm.

Disease cycle

The pathogen can survive in soil and also on host (either living or dead) — so it is called facultative parasite.

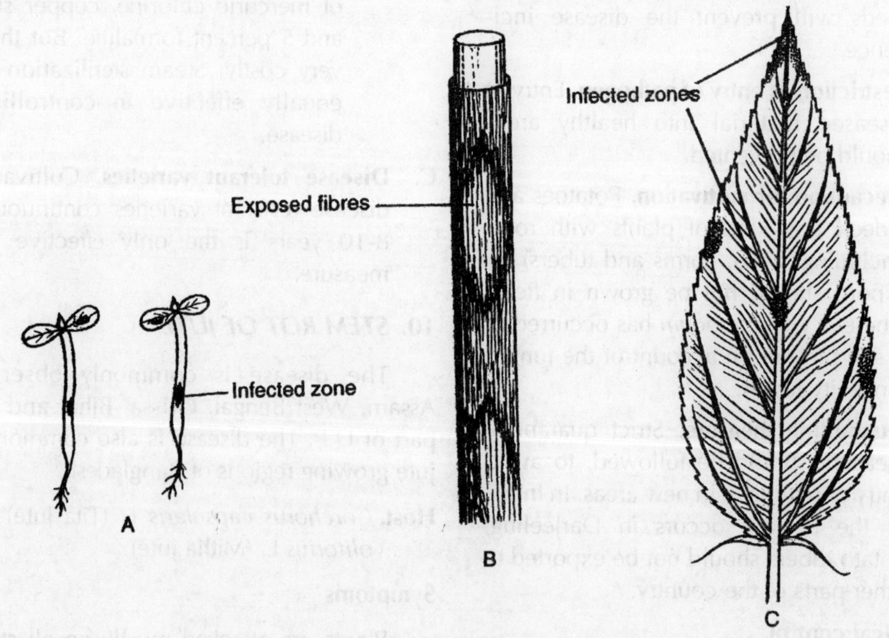

Fig. 5.49: Stem rot of jute : A. Diseased seedlings. B. Diseased stem (portion) showing exposed fibres and C. Infected leaf

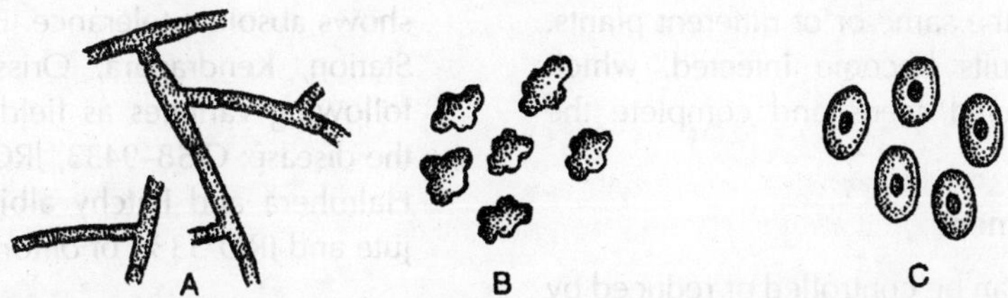

Fig. 5.50: Stem rot of jute: A. Vegetative mycelia. B. Sclerotia and C. Pycnospores

The pathogen may remain as dormant both in infected seed and infected plant debris of the previous year (Fig. 5.51).

If the seeds are infected, depending on the amount of inoculum, they may germinate or remain ungerminated. The ungerminated seeds will increase the amount of inoculum in the soil after sowing. The germinated seeds develop the infected seedlings, which later die and also increase the inoculum in soil. The previous year's infected plant debris will also increase the inoculum in the soil. Inocula from the above three sources may reach the various hosts like cotton, tobacco, potato, sesamum, mulberry etc., along with the jute plant. The inoculum may come to the jute plant from infected host other than jute, in addition to the above three sources.

Pathogen from infected root of jute plant may pass to the root of other healthy jute plant and to the stem and fruit to the same plant by upward movement. Inoculum from the host may reach the jute leaf and its stem gets the infection by passage of inoculum through the petiole. Pycnospores developed on stem may again

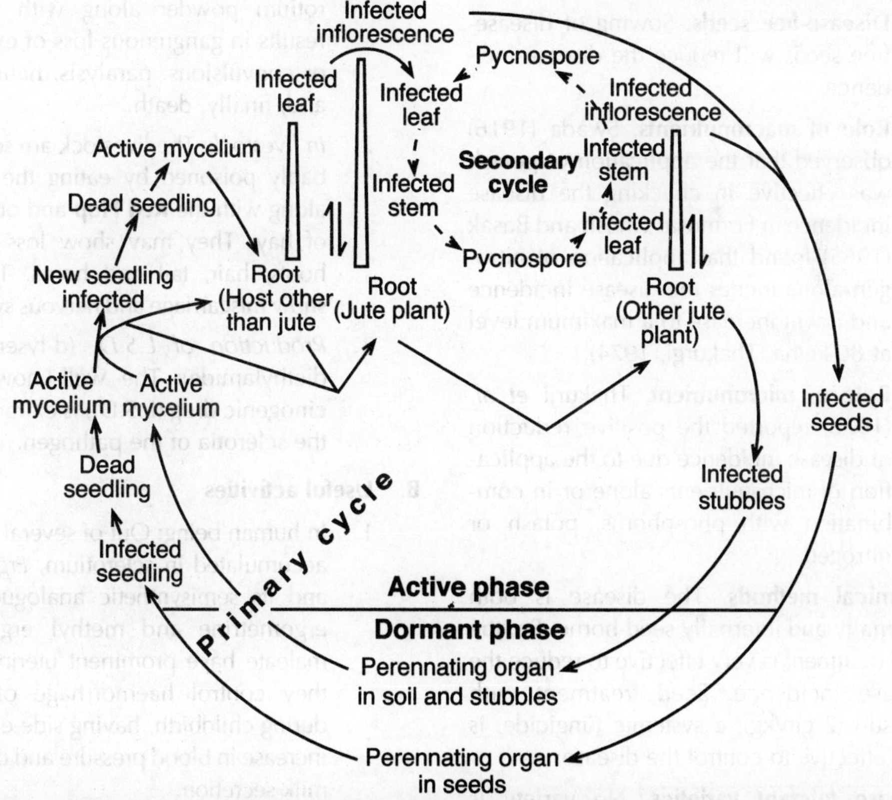

Fig. 5.51: Disease cycle of Stem rot of jute

infect the leaf of the same or of different plants. Gradually the fruits become infected, which develop the infected seeds and complete the cycle.

Disease management

The disease can be controlled or reduced by the following procedures:

A. **Cultural methods**

1. **Sanitation**. Cleaning of the previous year's infected plant debris should be done to reduce the primary source of infection.
2. **Drainage**. Proper drainage of water should be maintained — this reduces the damping off incidence.
3. **Spacing**. Proper spacing between the plants reduces the spread of the disease. *More space may reduce disease, but it induces branching of jute, which will create problem during fibre extraction.*
4. **Disease-free seeds**. Sowing of disease-free seeds will reduce the disease incidence.
5. **Role of macronutrients**. Swada (1916) observed that the application of potash was effective in checking the disease incidence in Formosa. Ghosh and Basak (1965) found that application of nitrogen alone incites the disease incidence and it will increase to a maximum level at 80 kg/ha (Thakurgi, 1974).
6. **Role of micronutrient.** Thakurji *et al.* (1976) reported the positive reduction of disease incidence due to the application of micronutrients alone or in combination with phosphorus, potash or nitrogen.

B. **Chemical methods**. The disease is both externally and internally seed-borne. So, the seed treatment is very effective to reduce the disease incidence. Seed treatment with Bavistin (2 gm/kg), a systemic fungicide, is very effective to control the disease.

C. **Disease tolerant varieties**. No variety is available in the existing germplasms, which shows absolute tolerance. But Jute Research Station, Kendrapara, Orissa, reported the following varieties as field tolerant against the disease: C-58–9433, JRC-9823, Bangkak, Halmhera and Patchy albino of *capsularis* jute and JRO-3331 of *olitorius* jute.

11. *ERGOT DISEASE OF RYE*

The disease occurs throughout the world. In addition to rye the pathogen also grows on bajra, wheat, grasses, barley and oats. About 5-10% of the grains of infected inflorescence become damaged by the pathogen. On infected inflorescence, the ovary becomes replaced by the elongated sclerotium. The sclerotia have great economic importance — both from harmful and useful sides. Though the disease does not reduce much in the yield, it causes various damages to human beings and also to different livestocks.

A. **Harmful activities**

1. *In human being*: Consumption of sclerotium powder along with rye-flour, results in gangrenous loss of extremities or convulsions, paralysis, hallucination and, finally, death.
2. *In livestock*: The livestock are sometimes badly poisoned by eating the sclerotia along with the **Red Hop** and other kinds of hay. They may show loss of teeth, hoofs, hair, tail and horns. They also show miscarriage and nervous symptoms.
3. *Production of L.S.D.* (d-lysergic acid diethylamide): The well-known hallucinogenic drug (L.S.D.) is extracted from the sclerotia of the pathogen.

B. **Useful activities**

1. In human being: Out of several alkaloids accumulated in sclerotium, *Ergometrine* and its semisynthetic analogue methyl ergometrine and methyl ergometrine maleate have prominent uterine action, they control haemorrhage of mother during childbirth, having side-effect with increase in blood pressure and decreased milk secretion.

Host. *Secale cereale* L.

Symptoms

Symptoms of the disease appear in two stages:

1. *Early stage of flowering:* Yellow sticky fluid, secreted from the infected ovarian wall called "**honey dew**" as it is sweet in taste. The fluid contains innumerable number of minute conidia of the pathogen. This conidial stage is also called "sphacelial stage" as the pathogen was initially named as *Sphacelia segetum*.

2. *Later stage*: With the progress of infection the ovarian tissue becomes completely destroyed. Thus the ovary becomes completely replaced by the hard horny structure, made up of fungal mycelia, the sclerotium (Fig. 5.52). The sclerotia are black, elongated and much longer than the size of the normal grain. They protrude out from the glumes. The number of sclerotia varies from 1-many on an inflorescence.

Causal organism

The name of causal organism i.e., pathogen, is *Claviceps purpurea* (Fr.) Tul. In the "sphacelial stage" the pathogen produces innumerable number of conidia. The conidia are minute, oval and uninucleate; those develop singly on the conidiophores present on the outer wall of the infected ovary (Fig. 5.53A). At the later stage, the

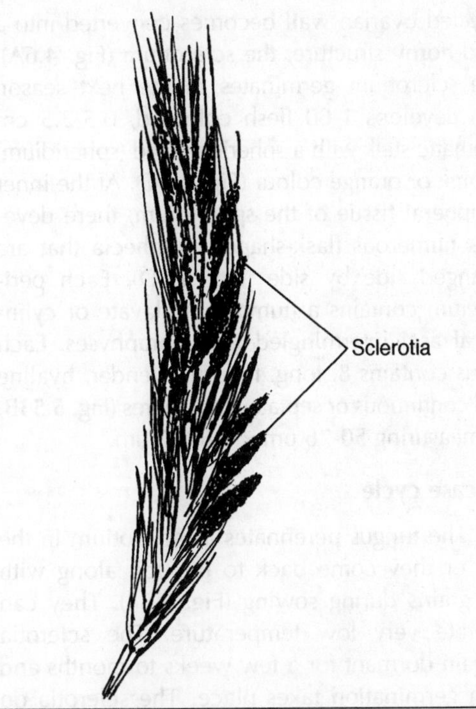

Fig. 5.52: Ergot disease of rye: Infected inflorescence showing many sclerotia

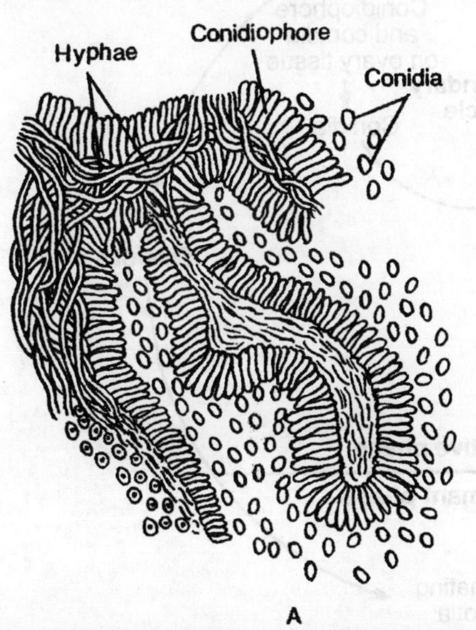

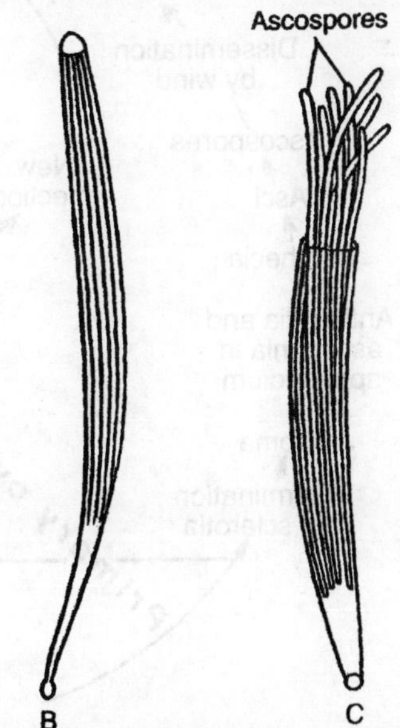

Fig. 5.53: Ergot disease of rye (c.o. *Claviceps purpurea*): A. Sphacelial stage (producing huge conidia on ovarian wall), B. Single ascus and C. Ascus discharging ascospores

infected ovarian wall becomes converted into a hard horny structure, the sclerotium (Fig. 4.6A). The sclerotium germinates in the next season and develops 1-60 flesh coloured, 0.5-2.5 cm stromatic stalk with a spherical head (spheridium) of pink or orange colour (Fig. 4.6C). At the inner peripheral tissue of the spheridium, there develops numerous flask-shaped perithecia that are arranged side by side (Fig. 4.8D). Each perithecium contains a number of clavate or cylindrical asci intermingled with paraphyses. Each ascus contains 8, long, filiform, slender, hyaline and continuous or septate, ascospores (Fig. 5.53B, C) measuring 50-76 μm × 0.6-0.7 μm.

Disease cycle

The fungus perennates as sclerotium in the soil or they come back to the soil along with the grains during sowing (Fig. 5.54). They can tolerate very low temperature. The sclerotia remain dormant for a few weeks to months and then germination takes place. The sclerotia do not remain viable for more than one year. Entire or even the broken sclerotium is able to germinate and can develop stalked spheridium. The number of sclerotium varies from 1-60. Perithecia are developed inside the spheridium. The perithecium contains many asci and each ascus contains filiform ascospores (8), which come out through the ostiole. The ascospores are dispersed by wind or insect and infect the stigma. After germination, the ascospore develops mycelium. The mycelia go inside the ovary and form the sphacelial stage, where huge number of conidia are produced. The conidia are dispersed by wind and cause further infection. This is the secondary cycle. This cycle repeats several times in the growing season.

Disease management

The disease can be controlled or reduced by the following methods:

A. Cultural methods

1. **Sowing of ergot-free seeds.** The seed lot can be made ergot-free by soaking the seeds in 32-37% potassium chloride or in 20-32% sodium chloride solution in a large tub. The seeds will sink at the bottom and the sclerotia will float on

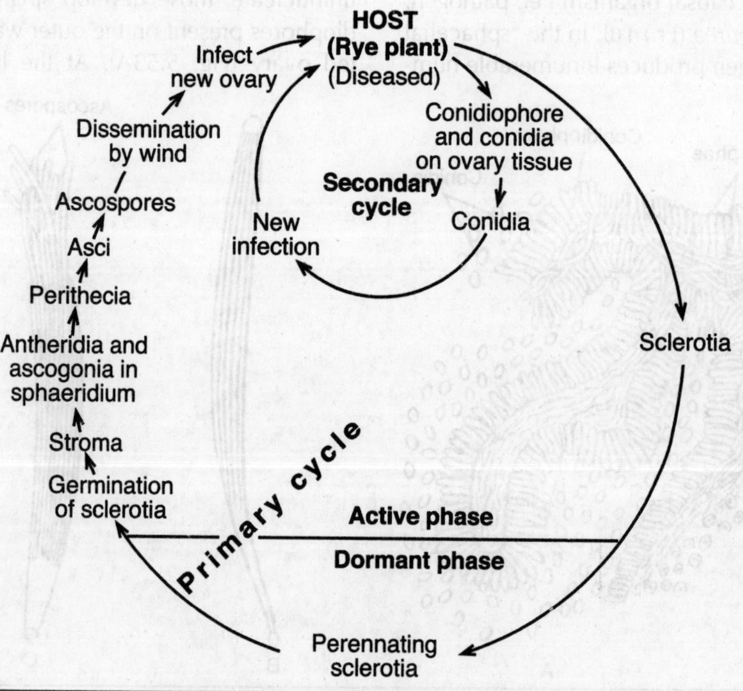

Fig. 5.54: Disease cycle of Ergot disease of rye

the solution and can be separated easily. Immediately the seed lot should be washed in water and then air dried.

2. **Sanitation**. Cleaning the susceptible wild cereal hosts from the surrounding field, reduces the source of secondary inoculum.
3. **Crop rotation**. Rotation of rye with legumes reduces the disease at a very significant level, because the sclerotium can survive in the soil for one year only.
4. **Deep ploughing**. Deep ploughing buried the sclerotia, prevents their germination.

12. *GREY BLIGHT OF TEA*

Like other diseases, Grey blight is an important disease of tea leaves. This disease is available in moderate to severe form in Africa and Asia. It is also available in South America and Australia. In India, the disease is also common in many tea gardens of Darjeeling, Dooars, Tarai, Brahmaputra valley and Nilgiri hills.

Host. *Camelia sinensis* (L) O. Ktz.

Symptoms

Symptoms appear as small brown leaf spots (Fig. 5.55A). The spots gradually enlarge and become the size of 1 cm or more in diameter. The mature spots show greyish centre with light to dark brown margin. The spots are mostly irregular and some of them coalesce to form grey patches of irregular outline, which blight the infected leaf, hence the name grey blight. Concentric rings of dark and light coloured lines are visible on upper surface due to the development of dark coloured acervuli.

Although the leaves are infected at any stage of development, it is much more in older leaves.

Causal organism

Pestalotiopsis theae Swada (formerly known as *Pestalotia theae*). The vegetative body is composed of septate, much-branched hyaline hyphae. The hyphae grow both intra- and intercellularly inside the host tissue. It develops acervulus in the darker zone of concentric rings of infected region and is available on both the

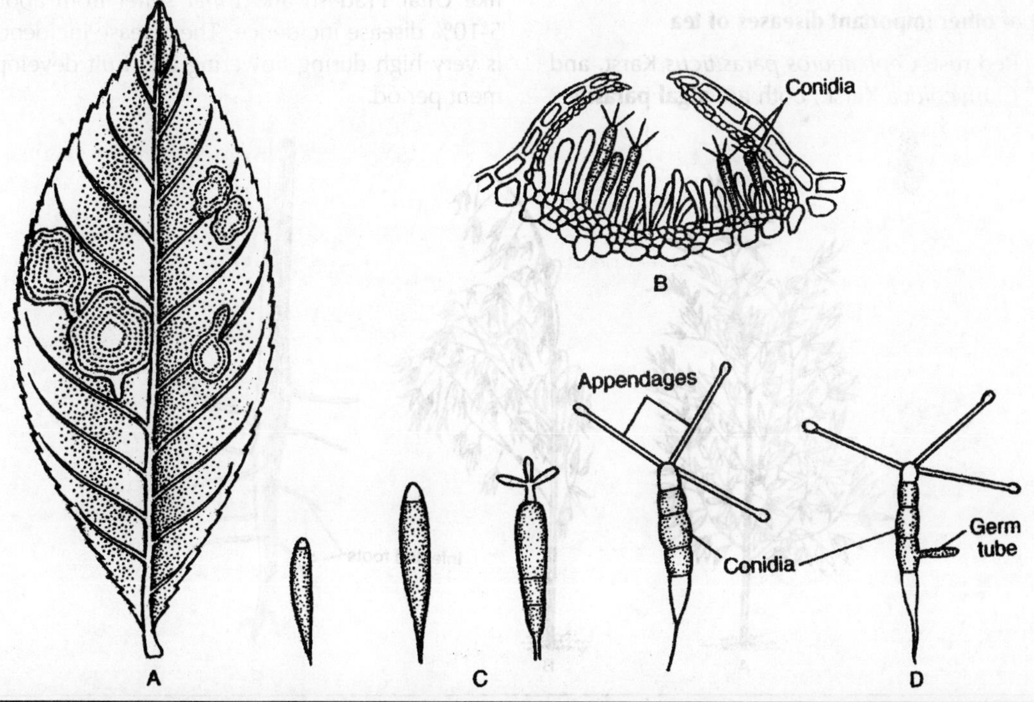

Fig. 5.55: Grey blight of tea: A. Symptoms on leaf. B. Acervulus in vertical section. C. Development of mature conidium, and D. Germinating conidium

surfaces. The acervuli are globose to lenticular in shape, which rupture the epidermis with wide and angular opening.

The acervulus contains huge number of conidia (Fig. 5.55B). The conidia are straight, fusiform in shape and 5-celled. Out of 5 cells, the median 3 cells are darker in colour (Fig. 5.55C). The condia have long hyaline apical appendages with swollen tip. The appendages are usually 3, rarely 2 or 4 in number. The conidia measure 22-23 μm × 6-7 μm and the coloured region being 18-21 μm long.

Disease management

1. During winter, pruning should be done carefully and the infected leaves should be cleaned out from the field and must be burnt in outside. After pruning, one-time spraying of Bordeaux mixture gives good protection from the disease.
2. Two applications of Bordeaux mixture, 1st one in winter (April-Dec.) and 2nd one in April, has been suggested, where pruning has not been done.

List of other important diseases of tea

1. Red rust. *Cephaleuros parasiticus* Karst. and *C. mycoidea* Karst., both are **algal parasite.**
2. Copper blight. *Guignardia camelliae* (Cooke) Butler.
3. Thread blight. *Corticium koleroga.*
4. Brown blight. *Glomerella cingulata* (Stonem) S. and V. S. (Syn. *Colletotrichum camelliae* Mass).
5. Canker. *Nectria cinnabarina* (Tode) fr.
6. Stump rot. *Ustulina deusta.*
7. Root rot. *Rosellinia arcuata* Petch and *Ustulina zonata* (Lev.) Sacc.
8. Brown rot. *Hymenochaete noxia* Berk.
9. Red root. *Sphaerostilbe repens* Berk. and Br.

13. *WILT OF PIGEON PEA*

The disease was first reported by E. J. Butler (1906) in India. It causes severe damage, mostly in the dry regions of India, wherever the crop is grown — Uttar Pradesh, Bihar, Tamil Nadu, Maharashtra etc. According to the survey report of ICRISAT (Andhra Pradesh in India), during 1975-1980, the disease appeared with maximum intensity in Maharashtra (22.6%). Other places like Uttar Pradesh and Bihar suffer from about 5-10% disease incidence. The disease incidence is very high during flowering and fruit development period.

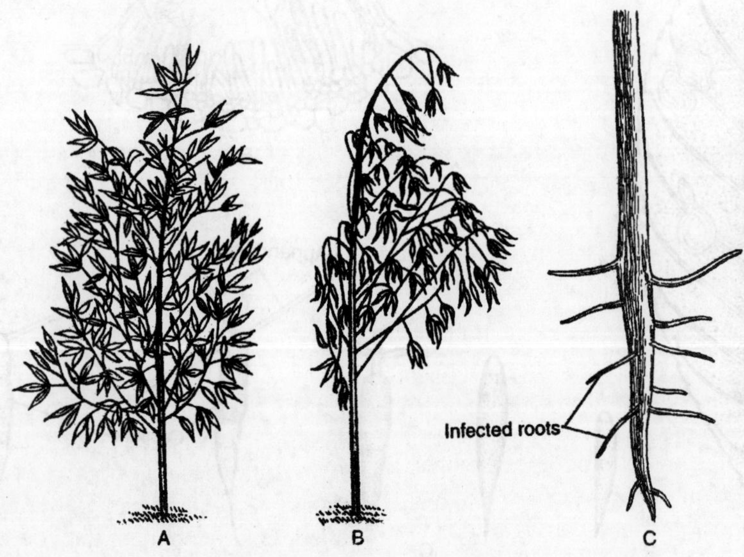

Fig. 5.56: Wilt of pigeon pea (c.o. *Fusarium udum*): A. Healthy plant, B. Infected plant and C. Lower part of plant with bark pealed off to show the infected regions by dark shades

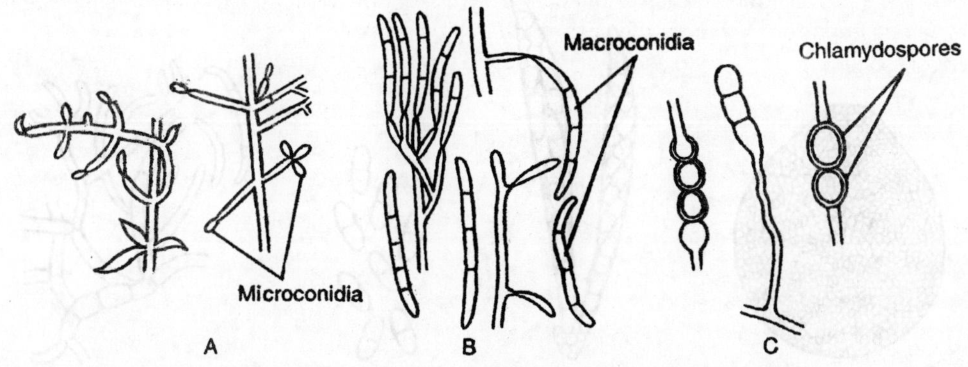

Fig. 5.57: Wilt of pigeon pea (c.o. *Fusarium udum*): A–B. Vegetative body with microconidia (A) and macroconida (B) and C. Chlamydospores

Host. *Cajanus cajan* (L.) Millsp. (Family-Fabaceae i.e., Leguminosae).

Symptoms

The disease appears as wilting in early stages of plant growth (4-6 weeks old), characterised by sudden or commonly by general yellowing and withering of leaves following drying of the entire plant (Fig. 5.56). The disease becomes severe during flowering and pod development.

It is a systemic disease, appears after the lateral roots get infected by the hyphae develop either from conidia, chlamydospores or ascospores. The infection progresses through the xylem vessel where the mycelia collectively plug the vessel and produce conidia and chlamydospores. Later on, the woody tissue becomes brown to black in colour. Partial wilting is also visible when one side of the root system and stem becomes affected by the pathogen.

Causal organism

The disease is caused by the pathogen *Fusarium udum* Butler. The vegetative body is mycelium. The pathogen is a soil-borne facultative parasite. It enters through roots and becomes systemic in plant. The pathogen grows both inter- and intracellularly. The pathogen completely or partially plugs the vessels and interferes the flow of water and causes wilting of the leaves.

The mycelium produces 3 types of spores, both in culture medium and in the host tissue. These are:

1. *Macroconidia*: The macroconidia are linear, thin-walled, curved with pointed ends, notched at the base, 4-5 celled, measuring 15-50 μm × 3-5 μm (Fig. 5.57B).
2. *Microconidia*: The microconidia are small, curved or elliptical, 1-3 celled and measured 5-15 μm × 2-4 μm (Fig. 5.57A).
3. *Chlamydospores*: The chlamydospores are spherical to oval, thick-walled structures; develop either singly or in chain of 2-3 (Fig. 5.57C). They develop either terminally or intercalary. These spores can survive for many years.

The perfect stage of the pathogen, named *Gibberella indica*, was reported by Rai and Upadhyay (1982) found to grow commonly in aggregated form on the exposed roots and collar region of the plant. They grow superficially on the exposed host tissue. The mature perithecia are superficial, globose to subglobose, sessile, smooth-walled, dark violet in colour and measured 350-550 μm in diameter (Fig. 5.58A).

The asci are mostly sub-cylindrical, slightly broader in the middle and contain 8 ascospores (Fig. 5.58B). The ascospores are elliptical to oval, hyaline, 2-celled (Fig. 5.58C), but rarely 3-4 celled. On germination they develop germ tube which produces both macro- and microconidia (Fig. 5.58D).

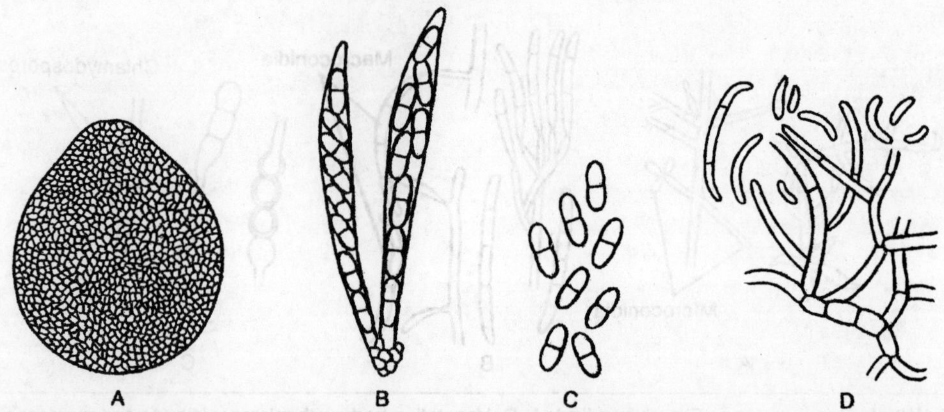

Fig. 5.58: Wilt of pigeon pea (c.o. *Fusarium udum*, perfect state of *Gibberella indica*) : A. Perithecium, B. Asci with ascospores, C. Ascospores and D. Germinating ascospore producing both micro- and macroconidia

Disease cycle

The pathogen can survive in the soil saprophytically in absence of living host for 3-4 years (8-20 years as reported by McRae and Shaw, 1933). The asexual stage is more important and common in nature (Fig. 5.59). The asexual spores like macro- and microconidia can survive in soil for short period, but chlamydospores can survive for long period of time. All the spores develop mycelium on germination and infect a new host. The infection takes place through the fine lateral branches of root and grows both inter- and intracellularly. With further growth, it establishes colonisation in the xylem vessels and completely or partially blocks the water flow, thereby the plant becomes wilted. After the death of the host

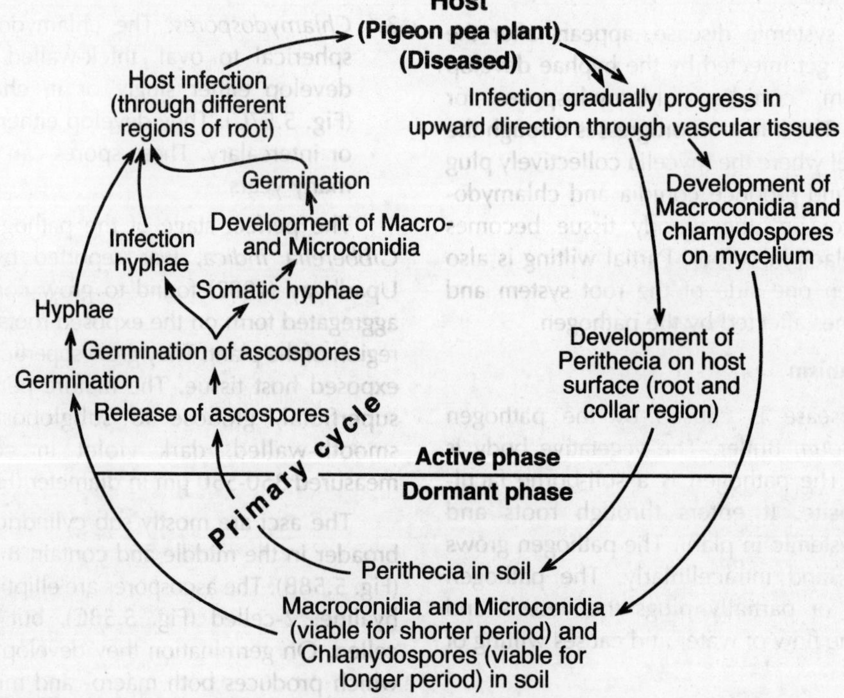

Fig. 5.59: Disease cycle of wilt of pigeon pea

plant, the pathogen passes in the soil and can survive for several years. The duration of survival of ascospores is not yet fully known. The pathogen can survive saprophytically on dead plant debris.

Besides the blockage of xylem vessel by the fungal mycelium, the blockage also takes place due to the production of toxin, fusaric acid. In addition to the above, the pathogen also produces the enzymes like cellulase, pectin methyl esterase and polygalacturonase involved in disease development.

Disease management

The disease can be controlled or reduced by the following procedures:

A. **Cultural methods**. The following cultural methods are used to reduce the disease incidence:

 1. **Crop rotation**. It is the best way to eliminate the soil borne infection. Rotation of 4-5 years is very useful to control the disease. Crop rotation with tobacco is very much effective, possibly due to action of root exudates on pathogen.
 2. **Soil management**. Deep ploughing in the summer months is very useful to eliminate the pathogen.
 3. **Soil solarisation**. Soil solarisation is done by covering the soil with transparent polythene sheet of about 100 μm thick for 6-8 weeks during summer months (April-May), which increases the soil temperature 6-10°C more, causes the reduction of the pathogen (Chauhan *et al*., 1988).
 4. **Mixed cropping**. Mixed cropping with Jowar (*Sorghum*) or *Crotalaria medicaginea* is found to be very much effective in controlling the disease.

B. **Chemical methods**. Application of Bavistin (2000ppm) as soil drench 10 days before inoculation is very much effective to control the disease. Fungicides, like Dithane Z-78, Zinocap and Brassicol are also effective as soil drench.

C. **Biological methods**. Vasudeva *et al*. (1952, 54) demonstrated that *Bacillus subtilis* produces an antibiotic, bulbiformin; which inhibits the growth of *Fusarium udum*. The disease appears more after October, due to the reduction of *B. subtilis* population. The antibiotic production can be enhanced by soil amendment with cotton and groundnut cake.

D. **Disease tolerant varieties**. The varieties recommended to cultivate as disease tolerant are NP (WR)-15, DDN-1, ICP 4769, ICP 7182, ICP 8863, ICP 9168 etc.

14. *WHITE RUST OR WHITE BLISTERS OF CRUCIFERS*

White rust of crucifer is a common disease of several cruciferous crops like mustard, cabbage, radish and turnip; found everywhere in India. The disease is also found in cruciferous weeds like *Capsella bursa-pastoris* and *Nasturtium indicum*. Saharan *et al*. (1984) reported that up to 54.5% loss in yield may take place due to floral infection.

Host. *Brassica campestris* L., *B. alba Hook* and *Thom*, *B. nigra Koch*, *Raphanus sativus* L., etc.

Symptoms

All parts except the root are attacked by the pathogen. Infection is both localized and systemic. Localized infection takes place in adult plant and systemic infection in young one.

Due to localized infection, white or cream yellow coloured pustules of various shapes and sizes commonly appear on the surface of leaves (Fig. 5.60A) and also on stem and inflorescence. Generally spots are developed on the lower surface of leaves. In some plants, the infected leaves become thick, fleshy and inrolled. During severe infection, the growth of leaves and sometimes the entire plant become stunted.

When the young stem and inflorescences are infected, the infection becomes systemic in the tissues and stimulates hypertrophy and hyperplasia resulting into the enlargement and varied distortion of the infected organs. The

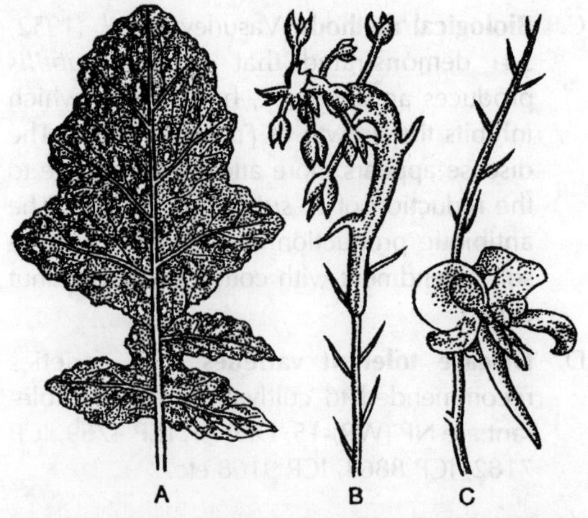

Fig. 5.60: White rust of crucifer: A. Diseased *Brassica* leaf showing white pustules, B–C. Diseased inflorescence and flowers of *Brassica*

inflorescence axis (rachis) becomes distorted, thus suppressed the development of flower or causes malformation and discolouration of floral parts (Fig. 5.60B, C).

Causal organism

White rust is caused by *Albugo candida* (Lev.) Kunze, also known as *Cystopus candidus* Lev. It is an obligate parasite. Mycelial plant body is aseptate, grows intercellularly within the host tissue and develops globular haustoria within the adjacent cell to draw nutrition (Fig. 5.61B, C).

The mycelium grows intercellularly and forms knob-like haustoria in host cells. The hyphae from the internal mycelium collect under the epidermis. From this underlying hyphae numerous short sporangiophores are developed

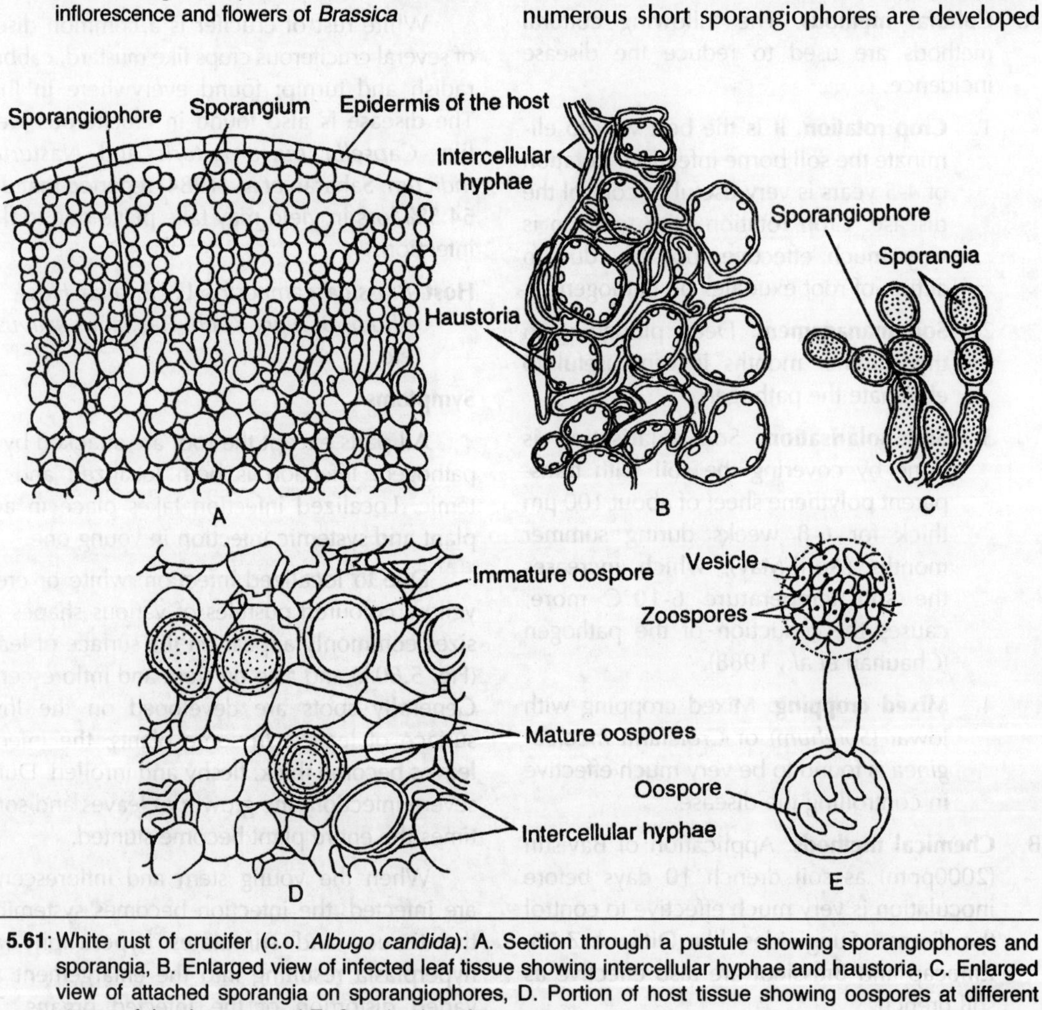

Fig. 5.61: White rust of crucifer (c.o. *Albugo candida*): A. Section through a pustule showing sporangiophores and sporangia, B. Enlarged view of infected leaf tissue showing intercellular hyphae and haustoria, C. Enlarged view of attached sporangia on sporangiophores, D. Portion of host tissue showing oospores at different stages of development and E. Germination of oospore

like a closely compact palisade layer under the epidermis. With further growth blister-like areas developed on the epidermis are called sori. The sporangia developed on sporangiophore are arranged in basipetal succession (Fig. 5.61A). Gelatinous pad developed between the successive sporangia functions as disjunctor. With further growth pressure exerted inside the epidermis, caused the epidermis to rupture and the sporangia are exposed.

Sporangia are hyaline, more or less spherical and measure 14-16 µm × 16-20 µm. They are usually disseminated by wind. The sporangia usually germinate by zoospores or rarely by germ tube. The zoospores are reniform, biflagellate with laterally inserted flagella. After a period of swimming, they are encysted and then germinate to form germ tube. The germ tube, on contact with stomata, penetrates the host tissue and causes infection.

At the end of growing season, oogonia and antheridia are formed from the internal mycelium in the intercellular space and by union they develop oospore (Fig. 5.61D). The oospores are globular and 40-45 µm in diameter. After a long period of rest, they germinate by the formation of zoospores (Fig. 5.61E).

Disease cycle

The pathogen perennates in soil as oospore either in plant debris or in soil, which serves as primary inoculum (Fig. 5.62). With the onset of favourable condition, oospores germinate by zoospores and cause further infection. The sori are developed in the infected region and further develop sporangiophore, sporangium and finally zoospores, those can cause further infection. This cycle repeats several times and spreads the disease more and more. At the end of growing season, oospores are developed inside the host tissue and remain dormant till the next season.

The infection progresses at a faster rate if temperature becomes less than 15°C and humidity becomes more than 65% in months December 15 to February 15 (Lakara and Saharan, 1990).

Disease management

The disease can be controlled or reduced by the following procedures:

A. Cultural methods

1. **Sanitation**. Previous year's plant debris should be cleaned from the field and must be burnt in the field.

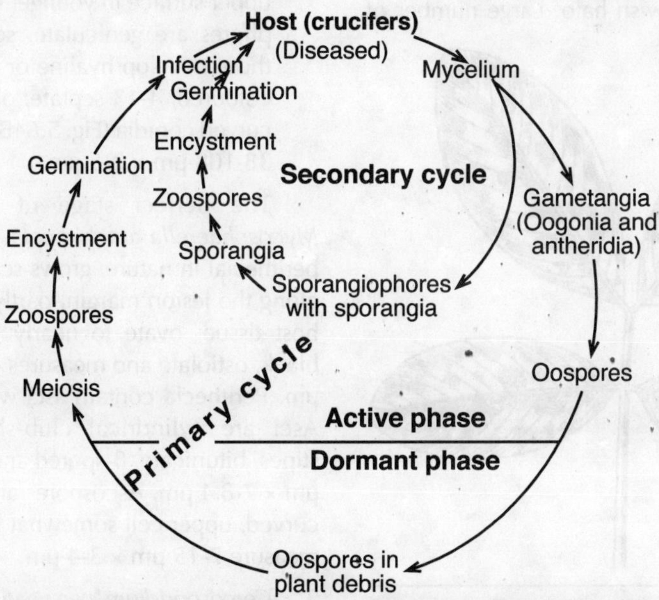

Fig. 5.62: Disease cycle of white rust of crucifers

2. **Destruction of weeds** in and around the field eliminates the source of primary inoculum.
3. **Crop rotation** is a very good way to reduce the disease incidence.

B. Chemical control

1. The fungicides used to control the disease include Bordeaux mixture (0.8%), Dithane M-45 (0.2%), Daconil (0.1%) and Difolatan (0.3%). Kapoor and Sugha (1955) recommended Aliette and Ridonil to control white rust of crucifer.

15. *TIKKA DISEASE OF GROUNDNUT*

The disease, caused by *Cercospora*, is also known as **Cercospora leaf spot**. It is available in all the groundnut growing areas of India and caused about 20-50% reduction in yield. The size as well as the quality of the nuts are generally affected.

Host. *Arachis hypogea* L.

Symptoms

All the aerial parts of the plants attacked by the pathogen show characteristic symptoms. Initially, the lower leaflets get infected showing dark spots and later on, each spot becomes surrounded by yellowish halo. Large number of spots, almost circular in outline, develop on the leaf. With severity and maturation, the spots become dark brown to almost black, particularly on the upper surface of the leaflets (Fig. 5.63).

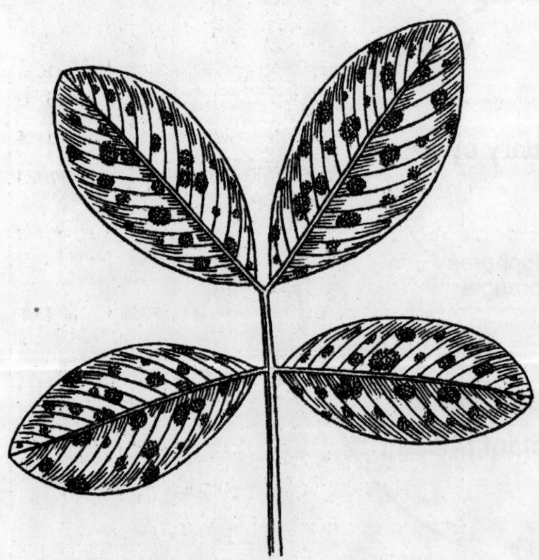

Fig. 5.63: Tikka disease of groundnut (Symptoms on leaflets)

Stems and petioles are also with spots but less in number.

With severity of disease, the spots become coalesced and defoliation of the leaf takes place. Due to defoliation, the size and quality of the nuts become greatly reduced. Infection at younger stage causes defoliation and nuts fail to develop in them. Due to this disease, the ultimate effect is the greater loss in yield.

Causal organisms

Cercospora arachidicola Hori (perfect stage *Mycosphaerella arachidicola* W. A. Jenkins) and *Cercosporidium personatum* (Berk and Curt) Deighton, earlier known as *Cercospora personata* (Berk and Curt) Ell and Eve (perfect stage *Mycosphaerella berkeleyii* W. A. Jenkins).

1. *Cercospora arachidicola* (**early leaf spot**): The mycelial plant body grows both externally and internally, without any haustorium. The mycelia by aggregation form stroma (Fig. 5.64A). The conidiophores are normally amphigenous, present only on the upper surface in younger spots. The conidiophores are geniculate, septate or aseptate, those develop hyaline or slightly olivaceous coloured, 4-13 septate, obclavate and often curved conidia (Fig. 5.64B). Conidia measure 38-108 μm × 2-5 μm.

The perfect stage of the pathogen is *Mycosphaerella arachidicola*. The ascostroma is perithecial in nature, grows scattered, commonly along the lesion margin, partly embedded in the host tissue, ovate to nearly globose in shape, black, ostiolate and measures 47-84 μm × 44 -74 μm. Perithecia contain asci without paraphyses. Asci are cylindrical club-shaped with short stipes, bitunicate, 8-spored and measure 27-37.8 μm × 7-8.4 μm. Ascospores are 2-celled, slightly curved, upper cell somewhat larger, hyaline and measure 7-15 μm × 3-4 μm.

2. *Cercosporidium personatum* (**late leaf spot**): The mycelial plant body of the pathogen grows

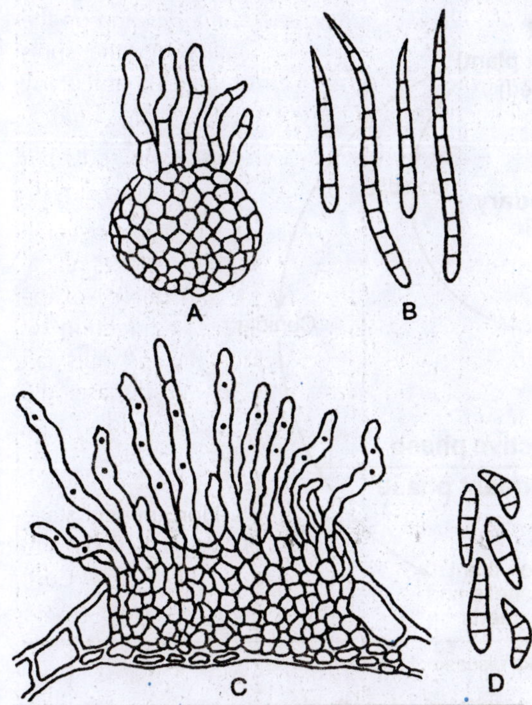

Fig. 5.64: Tikka disease of groundnut: A–B. *Cercospora arachidicola*, A. Stroma bearing conidiophore, B. Conidia, C–D. *Cercospora personata*, C. Section through infected leaf showing conidiophores, D. Conidia

intercellularly and draws nutrition from the neighbouring cells by haustoria. With aggregation, the mycelium forms dense stroma from where long septate to non-septate, geniculate, hypophyllous conidiophores develop (Fig. 5.64C). Conidiophores emerge in tufts by rupturing host epidermis. Conidia develop on conidiophores. Conidia are obclavate or cylindrical, septate, pale-brown and measure 30-50 μm × 5-6 μm (Fig. 5.64D).

The perfect stage of the pathogen is *Mycosphaerella berkeleyii*. The ascostroma is perithecial in nature, grows scattered, commonly along the lesion margin, partly embedded in the host tissue, broadly ovate to globose in shape, black, ostiolate and measured 84-140 μm × 70-112 μm. Perithecia contain asci without paraphyses. Asci are cylindrical club-shaped with short stipes, bitunicate, 8-spored and measured 30-40 μm × 4-6 μm. Ascospores are 2-celled, slightly curved, upper cell somewhat larger, constricted at septum, hyaline and measure 11-19.5 μm × 3-3.8 μm. Differences between *C. personatum* and *C. arachidicola* are mentioned in Table 5.5.

Disease cycle

The pathogen can perennate through conidia on diseased plant debris that remain in the soil, on shell of fruits and also on seeds (Fig. 5.65). Perennating pathogen in all the above three places acts as a source of primary inoculum. It favours high temperature ranges between 26°C to 31°C and high humidity. Continuous low temperature along with dew also favours infection. Infection takes place through stomata or by piercing the host epidermal cell.

Infection commonly takes place through upper epidermis and after entering the host tissue, the pathogen ramifies and by aggregation of mycelium, stroma develops. With further

Table 5.5 : Differences between *Cercosporidium personatum* (earlier known as *Cercospora personata*) and *Cercospora arachidicola*

Cercosporidium personatum	*Cercospora arachidicola*
1. Spots are more circular and small.	1. Spots are not typically circular and larger than *C. personatum*.
2. The yellow halo around the spot develops at maturity.	2. The yellow halo around the spot develops from the beginning.
3. Spots on lower surface are deep black in colour.	3. Spots on lower surface is light brown.
4. Development of disease is slower than *C. arachidicola*.	4. Development of disease is rapid.
5. Spots appear late but dangerous in action.	5. Spots progress fast but action is less than *C. personatum*.

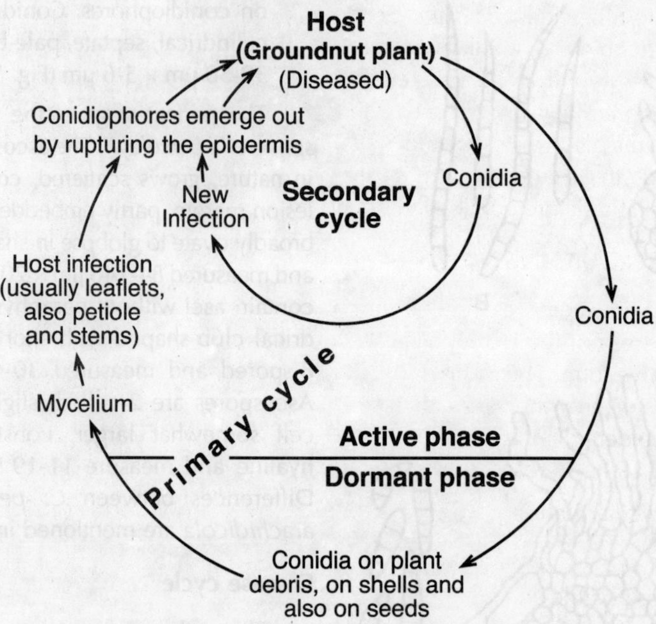

Fig. 5.65: Disease cycle of Tikka disease of groundnut

development, the stroma creates a pressure and the conidiophores come out through ruptured epidermis. Conidia are developed on conidiophores. The conidia after detachment, disseminate through wind to the different regions of the same plant or to the different plants and cause further infection and thus help to spread the disease. This cycle repeats several times in the growing season.

Disease management

Following procedures are adapted to control or reduce the disease incidence:

A. **Cultural methods**

1. **Sanitation.** Cleaning and destruction by burning the previous year's plant debris, reduce the source of primary inoculum, thereby the disease incidence becomes reduced.
2. **Crop rotation**. Crop rotation for 2-4 years is a very good procedure to reduce the rate of infection.
3. **Early sowing.** It is a very useful procedure, where the plants mature early, thus the damage through disease becomes very less.

B. **Chemical methods**

1. **Seed treatment.** As the disease is seed-borne, seed treatment gives good result to control the disease. In this procedure, the seeds are properly separated from the shell and treated with 0.5% $CuSO_4$ solution for 30 minutes before sowing. Dry dressing with Agrosan GN is also very much effective to reduce the source of primary inoculum.
2. **Foliar spray**. To reduce the source of secondary inoculum, foliar spray of chemicals is commonly used. Different chemicals used as foliar spray are:
   (a) Bordeaux mixture (4: 4: 50 or 5: 5: 50). It is used along with linseed oil (as sticker) at an interval of 15 days, which is much effective to reduce the disease to a significant level.
   (b) Other chemicals. Chemicals like Dithane Z-78 (0.2%), Dithane M-45, Cosan and $CuSO_4$ (15 and 25 lb/acre), Fycol 8E etc., are very much effective but requires 5-6 spray.

(c) Systemic fungicides. Bavistin, Benlate, Cercopin etc., are the systemic fungicides found to be very effective in controlling the disease. But, the chemicals are costly and it is advised to calculate the cost-benefit ratio before application.

16. *RED ROT OF SUGARCANE*

The disease is prevalent in most of the sugarcane growing areas of the world in moderate to most destructive form. The disease is also common in sugarcane growing areas of India and appeared as epidemic in U.P. and Bihar during 1939-40 and 1946-47 respectively.

Host. *Saccharum officinarum* L.

Symptoms

The symptoms are visible on and inside stem and also on leaf.

On stem. In early stage of disease development, the affected green stem shows purplish colouration on rind. During the later part of rainy season or still later, the root primordia at the nodal region tends to convert into black dots, the acervuli. The infected plant shows the death of 3rd or 4th leaves and the entire crown becomes dry in severe attack.

Inside stem. The split open diseased stock shows red rot symptom of internal tissues. The infected tissue becomes dull red and interrupted by transverse white patches (Fig. 5.66A-C). The rind becomes shrunken during harvest with longitudinal cavity filled with profuse mycelial growth.

On leaf. The symptom on leaf appears as elongated bright red lesions on the midrib of leaf blade and reddish patches on leaf sheath (Fig. 5.67). The lesions gradually spread throughout the length of midrib. The bright red lesion gradually changed with grayish centre, surrounded by dark reddish-brown margin. The margin becomes occupied by black dots, the acervuli of the fungus.

Causal organism

The causal organism i.e., the pathogen is *Colletotrichum falcatum* went. [Its Perfect stage is *Glomerella tucumanensis* (Speg.) von Arx and Muller., also named as *Physalospora tucumanensis*, an Ascomycetous member].

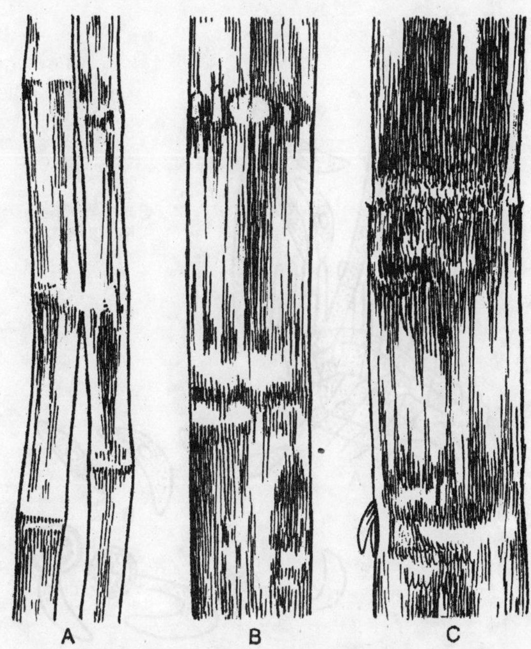

Fig. 5.66: Red rot of sugarcane : A. Dry, shrivelled cane due to pathogenic attack, B and C. Reddening (black in figure) of the pith with white transverse blotches

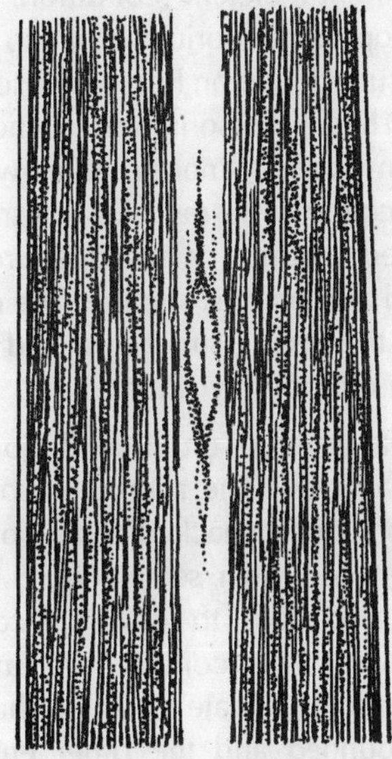

Fig. 5.67: Red rot of sugarcane: Portion of a leaf showing symptom on midrib

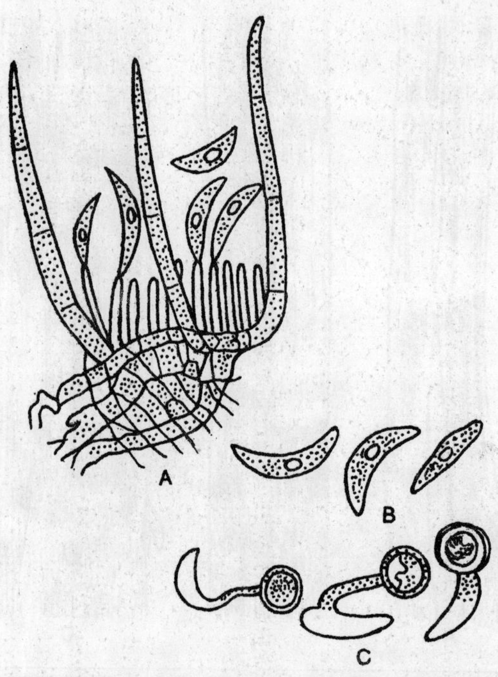

Fig. 5.68: Red rot of sugarcane (c.o. *Colletotrichum falcatum*): A. Acervulus (portion), B. Conidia and C. Germination of conidia

The fungus consists of septate, prostate much branched, thin-walled, hyaline hyphae growing intercellularly.

In nutrient-deficient condition, the germ tube develops appressorium at its tip in contact with hard surface like on host and also in culture medium. The appressoria are round, oval or irregular in shape, smooth, thick-walled and brown in colour. The mycelium also develops thick-walled, round chlamydospores in old culture and also on host surface. The chlamydospores are the resting structures, and on germination they develop mycelium.

Conidia develop on hyphal tip and also inside the acervuli. The acervulus consists of a mass of short, closely packed conidiophores with numerous dark brown setae (Fig. 5.68A). The setae are also able to develop conidia-like conidiophores. The conidia are unicellular, hyaline, generally fulcate or sickle-shaped, with one end rounded and the other end pointed (Fig. 5.68B). The conidia measure about 20-40 μm × 5-7 μm.

The sexual stage of the pathogen i.e., the perithecial stage has been reported from different countries like Australia, Brazil, China, India and Taiwan. This stage was reported in India in 1952 (in culture) and also in 1953 (on leaf attached with plant). The perithecia remain embedded inside the host tissue except the projected ostiole. They are globose and measured between 85-250 μm × 100-260 μm. Perthecia contain numerous hyaline asci that are intermingled with numerous hyaline paraphyses. The asci are hyaline, clavate and measured 70-90 μm x 13-18 μm containing 8 ascospores. Ascospores are unicellular, hyaline, straight or slightly fusoid and measured 18-22 μm × 7-8 μm.

Disease cycle

Pathogen can perpetuate in infected seed sets, in soil, plant debris and collateral hosts. Inoculum is present in the form of conidia, setae, chlamydospores, appressoria, thick walled hyphae and ascospores (Fig. 5.69).

The seed sets of diseased plant are the main source of primary inoculum. The infected seed sets have inoculum inside the host tissue or have dormant mycelium in the bud scales, leaf scars etc. These seed sets invariably developed infected shoots. The acervuli developed on the infected plants produce conidia that are transmitted commonly through irrigation or rain water and rarely by wind and cause further infection.

The inoculum from soil also infects the propagating stock and gradually the young plants. The infected plants develop acervulus on root primordia at nodes and also on leaves. The conidia developed from both the sources are dispersed by rain water or by wind and cause further infection.

The pathogen may survive in soil in the form of chlamydospores, setae, appressoria and also by thick-walled hyphae and causes further infection to the next crop, where there is a gap between the harvesting and next planting. But, the pathogen may remain within the host, where the crops are available in the field round the year.

The perithecia are not common in India, but play a significant role when present.

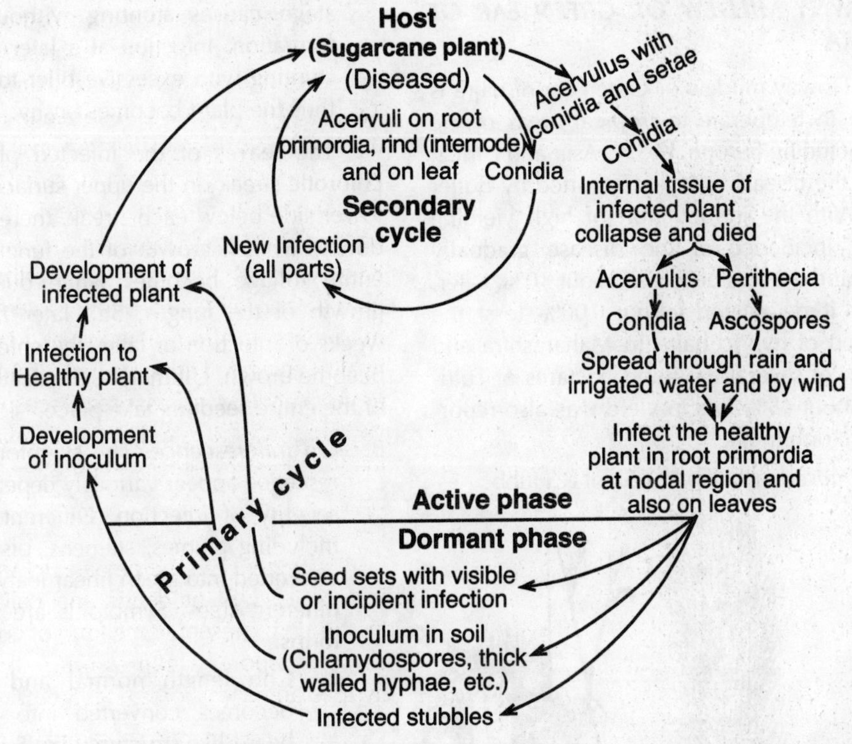

Fig. 5.69: Disease cycle of Red rot of sugarcane

Collateral hosts, like *Sorghum halepense, S. vulgare, Saccharum spontaneum, Leptochloa filiformis* etc., serve a significant role in annual recurrence of disease.

Disease management

The disease can be controlled or reduced by the following methods:

A. **Cultural methods**. These include field sanitation, selection of seed sets, crop rotation and proper drainage in the field.

1. **Field sanitation**. The field should be free from previous year's infected plant debris to reduce the primary source of inoculum.
2. **Selection of seed sets**. The seed sets should be selected from healthy and vigorously growing plants.
3. **Crop rotation**. A long term crop rotation for 2-3 years help in eliminating the pathogen.
4. **Proper drainage**. Water-logging in the field increases disease incidence, so proper drainage is essential to reduce the disease incidence.

B. **Physical methods**. Heat treatment is effective to reduce the disease incidence:

1. Hot water treatment. Seed sets are treated with hot water at 52°C for 18 minutes, to reduce the disease.
2. Hot air treatment. Hot air at 54°C for 8 hours is found very effective to control the disease.

C. **Chemical methods**. Fungicides like Dithane Z-78, Blitox-50 and Carbendazim (0.1%) are used on standing crop and are found very effective to control the disease incidence.

D. **Disease tolerant varieties**. The following disease tolerant varieties are recommended to cultivate — Co449, Co658, Co1158, Co1336, CoC671, CoL9 etc.

17. *DOWNY MILDEW OR GREEN EAR OF BAJRA*

The Downy mildew or Green ear of bajra is available in temperate to tropical areas of the world, including Europe, U.S.A, Asia and Africa. In India, the disease was first reported by Butler (1907). With the introduction of high yielding varieties, incidence of the disease gradually increase and the loss becomes about 30%. Later, in 1975, it devastated (about 100% loss) the kharif crop of hybrid bajra in Maharashtra and Karnataka of several thousand hectares of cultivation. About 45% loss in yield was also reported from North India.

Host. *Pennisetum typhoides* Slapf & Hubb.

Fig. 5.70: Downy mildew or green ear of bajra: Symptoms on diseased ears showing partial (A) and complete transformation (B) into green leafy structures, and C. Inflorescence reduced and converted into beard-like structure without any grain

Symptoms

The symptoms appear in two phases: A. On seedling, and B. On inflorescence.

A. *On seedling*. The infected seedlings turn pale yellow and become stunted due to shortening of internodes. Infection at an early stage causes stunting without any tiller formation. Infection at a later stage causes stunting with excessive tiller formation and thus the plant becomes bushy.

The leaves of the infected plant develop chlorotic streak on the upper surface and on the lower side below each streak, there appears the downy whitish growth of the fungus. Later on, entire foliage becomes white due to downy growth of the fungus. Still later i.e., about 3 weeks of infection or later, the chlorotic streaks become brown. Ultimately, the death and drying of the entire seedling take place.

B. *On inflorescence*. The symptoms on inflorescence appear variously depending on the severity of infection. Different floral parts including glumes, stamens, pistils etc., are converted into green linear leafy structure of different sizes. Symptoms are visible in 3 forms:

1. Cob length normal and lower half becomes converted into green leafy beard-like structure (Fig. 5.70A).
2. Cob length normal and entire cob is converted into green leafy, beard-like structure (Fig. 5.70B), and
3. Entire inflorescence becomes reduced and converted into leafy, beard-like structure without any grain (Fig. 5.70C).

Causal organism

The disease is caused by *Sclerospora graminicola* (Sacc.) Schroet. The fungus is an obligate parasite of Mastigomycotina.

The mycelial plant body is present in roots, stems, leaves and also in inflorescence of infected plants. The hyphae are coenocytic, branched, grow intercellularly and develop small bulbous haustoria to draw nutrition from the neighbouring cells. The hyphae grow freely in the mesophyll of leaves.

The sporangiophores develop in tuft from the tuft of hyphae accumulated below the stomata. The sporangiophores are broad with short hyphae, measuring 100-220 × 15-23 μm. The sporangiophores are initially unbranched, but become branched towards the upper side

(Fig. 5.71A). Each branch ends in a sterigma bearing a single sporangium. The sporangia are hyaline, smooth-walled, broadly elliptical and slightly pointed or papillate at the free end and measured 13-34 μm × 12-23 μm (Fig. 5.71B). Mature sporangia, after detachment, germinate in water and develop reniform, biflagellate zoospores. After a short period of swimming, they become encysted. On germination they develop infection hypha which infects the host.

The sexual stage, i.e., oosporic stage, is very common in India. Oospores are formed by the antheridium and oogonium develop in the infected tissue. Oospores are oval, 3-walled, 31-39 μm and develop in large numbers inside the mesophyll (Fig. 5.71C). Oospores perennate for long period during unfavourable condition and with the onset of favourable condition, they germinate by developing 1-4 germ tubes (Fig. 5.71D).

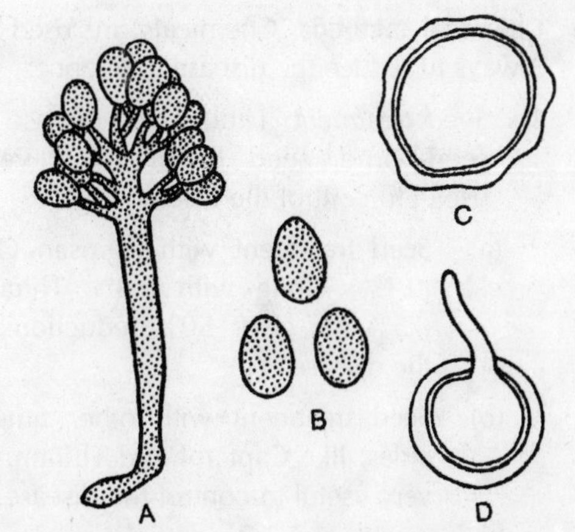

Fig. 5.71: Downy mildew of bajra (c.o. *Sclerospora graminicola*): A. Sporangiophore bearing sporangia, B. Sporangia, C. Oospore and D. Germinating Oospore

Disease cycle

The disease is **systemic**. The pathogen is an obligate parasite and primarily seed-borne. The pathogen survives in nature and primary infection is caused by oospores. **Oospores** are produced in infected floral organs and leaves. The infected parts, after falling on the ground, remain with the plant debris where the oospores perennate during winter. Oospores remain viable for 4 (in the soil) — 8 years (in host tissue). With the onset of favourable condition, the oospores germinate and infect the young seedling (about 9 days) systemically from the underground parts. **Oospores** may be carried with seed during harvesting and thrashing. The oospore also serves as a source of inoculum to infect the young seedling.

The **intercellular mycelium** of the fungus also perennates in the embryo and inside the seed coat. It is presumed that the infection of inflorescence takes place by the intercellular mycelium and it progressively grows towards the apical region.

Secondary inocula, i.e., sporangia are formed on the infected regions favoured by range of temperatures between 15°-25°C. The disease is commonly seen in North India during August-September, when temperature and humidity favour the growth of the pathogen. The sporangia are dispersed by wind, insect and water (both rain and irrigation water). They germinate on the host surface in thin film of water to produce zoospores. The zoospores become encysted and form germ tube which infect the host.

Disease management

The disease can be controlled or reduced by the following methods:

A. **Cultural methods**. Various cultural methods used to reduce the disease incidence are:

1. Cleaning of previous year's infected plant debris,
2. Deep ploughing and sun baking of soil,
3. Early planting,
4. Transplanting of seedling,
5. Removal of infected seedling i.e., roguing,
6. Use of proper dose of nitrogen fertilizer, etc.

B. **Physical methods**. Treatment of seeds in hot water at 55°C for 10 minutes is useful to eliminate the seed-borne pathogen.

C. **Chemical methods**. Chemicals are used in 3 ways to reduce the disease incidence:

1. *Seed treatment*. Though the disease is seed-borne, seed treatment is very useful to control the disease.
   (a) Seed treatment with Agrosan GN (0.1%) along with 0.4% Thiram (75% a.i) gives 50% reduction of the disease.
   (b) Seed treatment with other fungicides, like Captafol and Thiram, is very useful to control the disease.
2. *Foliar spray*. Chemicals of different groups are found to be very effective as foliar spray. These are:
   (a) Mancozeb and Captan are found to be very effective to control the disease.
   (b) Apron, Ridomil etc., are found to be very effective and used as seed treatment and/or foliar spray.
   (c) Seed dressing treatment with Apron SD-35 (8 g/kg seeds), followed by foliar spray with Ridomil MZ-72WP is very useful to control the disease.
3. *Root treatment*. Seedling roots are dipped in water containing 2,000 ppm Difolatan or Aureofungin for 1 hour, just before transplanting gives good protection to adult plants.

D. **Disease tolerant varieties**. WC-C 75, ICTP 8203, ICMH 451, Pusa 23, HB-5, NHB-10 and NHB-14 are the varieties recommended as disease-tolerant cultivars.

18. *DOWNY MILDEW OF SORGHUM (JOWAR)*

The disease is also known as **Leaf shredding disease** is prevalent in many Asian and African countries and in Italy. In India, the disease is commonly found in destructive mode in the states like Tamil Nadu, Karnataka, and Andhra Pradesh and also in Maharashtra. It is also reported from Uttar Pradesh and Madhya Pradesh. The disease is much more severe in fodder varieties than the grain crops. The severity of damage depends on the environmental condition and growing stage of the plant. Severe infection takes place every year in some regions of Maharashtra and in Mysore (Karnataka) showing up to 25% infection. Infected plants rarely develop ears, thereby caused considerable reduction in yield.

Anahosur and Laxman (1991) reported 74% reduction in yield in certain genotype of Sorghum with 100% infection.

Fig. 5.72: Healthy growing plants of Sorghum (Jowar)

Host. *Sorghum bicolor* (L) Moench (Fig. 5.72).

Symptoms

Symptoms of the disease vary with the stage of plant infected. Infection takes place in young seedling and in adult plant.

In Young seedling: When infection is initiated by oospores present in soil or lodged around the young plants, caused **systemic infection**, showing whitish, soft and downy growth mostly on the lower surface of the leaves, as conidia are produced on the growing leaves. When the seedling becomes 5-6 weeks old, white streaks are visible on both the surfaces of the upper leaves. Later on, due to formation of oospores in streaks, it becomes brown coloured due to necrosis. The infected leaves then tear along the veins, caused shredding of leaves the characteristic symptom of the disease, thereby the leaf tissues become separated from the midrib. At this stage the plants become stunted and sterile.

Fig. 5.73: Downy mildew of sorghum (Jowar): Symptom on leaf

In adult plant: In adult plant of 2-3 months old, infection takes place by secondary inoculum i.e. by sporangia develop on systemically infected plants. The apical young leaf and base of the adjacent lower leaves becomes white. At this stage irregular brown or yellow streak appear and the oospores develop abundantly in the infected area (Fig. 5.73). The lower leaves show pale yellow patches, due to sporangial development. Affected plants may not develop ears, but if produced, the ears are small sized bearing undersized grains.

Causal organism

The disease is caused by *Peronosclerospora sorghi* (Ito) Shaw. Earlier it was known as *Sclerospora sorghi*. Mycelium is present systemically throughout the infected plant. The hyphae are coenocytic, much branched and grow intercellularly with small finger-shaped haustoria. Hyphae are present in the parenchymatous tissue of roots, stem and also in mesophyll tissue of the leaves.

Conidiophores develop singly or in cluster through the stomata. Each one has a stout basal region with 1 or 2 transverse septa (Fig. 5.74A, C, D). Conidiophores are 180-300 μm in length, profusely branched and ends in pointed sterigmata bears conidia (Fig. 5.74B, E). The conidiophore along with its branches and conidia are spread out in a hemispherical plane.

Conidia are globose, hyaline and without apical papilla, measures 15-29 μm x 15-27 μm. Conidia germinates directly and never form zoospores (Fig. 5.74F).

The sexual stage, oospores are formed by the antheridium and oogonium in infected tissue. Oospores are oval, 3-walled, 31-39 μm and develop in large number inside the mesophyll. Oospores perennate for long period during unfavourable condition for 3-4 years. With the onset of favourable condition, they germinate by developing 1-4 germ tubes (Fig. 5.74G).

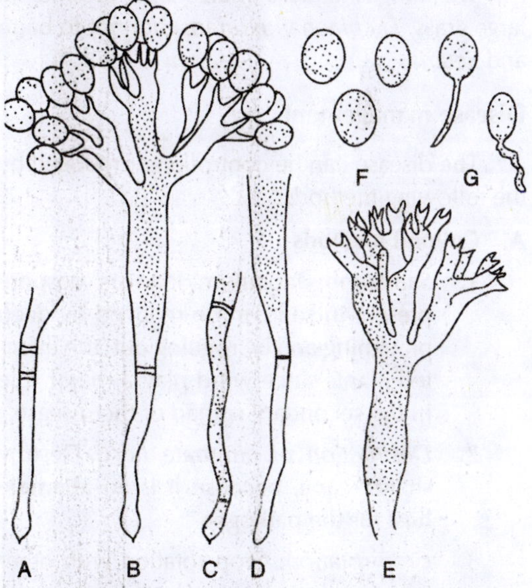

Fig. 5.74: Downy mildew of sorghum (*Sclerospora sorghi*): A, C and D. Septate conidiophore (portion), B. Septate conidiophore with branched apical region bearing profuse conidia, E. Conidiophore after detachment of conidia, F. Conidia and G. Germinating conidia

Disease cycle

Pathogen, *P. sorghi* is both soil and seedborne. Oospore remains in the soil, germinates by producing germ tube during sowing, causes infection to seedlings. Oospores in plant debris may remain viable for about 15 months in exposed field condition. Hay stacks and shady areas in the sorghum field serves as a source of viable inoculum. Some wild species of sorghum may present as weed around the main field become infected and soil-borne inoculum (oospores) can multiply through such wild host. Once the infection is established in the underground parts, the infection spreads upwardly throughout the plant.

The presence of oospores in glumes and pericarp of seeds, established the seed-borne nature of the pathogen. The pathogen becomes established in seed, either directly through the mother plant or by sporangial infection through ovary, style or stigma during spread. Commonly the disease spread in the field by sporangia. Moist part and region of field with more fertilizer suffers much more from the disease.

The fungus attacks maize and toesinte (a large grass, *Euchlaena luxurians*) but not to bajra and *Setaria*.

Disease management

The disease can be controlled or reduced by the following methods:

A. Cultural methods

1. *Sanitation*: Inoculum load as oospore present in soil can be reduced by deep ploughing and by roguing out, the infected plants since wind play a major role in the secondary spread of the disease.
2. *Destruction of alternate host*: Destroy shatter cane, because it is an **alternate host** for the pathogen
3. *Crop rotation*: Crop rotation with other crops viz., pulses and oilseeds is the highly effective procedure in reducing the inoculum intensity in the field.

B. Chemical control

1. *Seed treatment*: Seed treatment with Metalaxyl 25WP (Ridomil) @ 6 g/kg of seed, reduces the source of primary inoculum.
2. *Foliar treatment*: Spraying of Metalaxyl 500 g or Mancozeb 2 kg or Ziram 1 kg or Zineb 1 kg/hectare @ 2 g a.i./litre water at 10 and 40 days interval after seedling emergence reduces the disease significantly.
3. *Soil treatment*: Potassium azide @ 1.12 kg/hectare of soil caused significant reduction of the disease.

C. Disease tolerant varieties

Growing moderately disease tolerant varieties like Co25 and Co26 is very much effective to protect the crop with good yield.

19. *GRAIN SMUT OF SORGHUM (JOWAR)*

The disease is also known as **Covered smut, Kernel smut** and **Short smut** disease is the most common and prevalent in many Asian and African countries and also in Myanmar, Tanganyika, Venezuela, Australia and Northern, Southern and Central region of America. In India the disease is commonly found as destructive mode in states like Tamil Nadu, Karnataka (Mysore), Andhra Pradesh, Uttar Pradesh, Madhya Pradesh and also in Maharashtra.

The disease appears in a destructive mode in certain areas, where about 25% plants have been found to be affected and caused severe losses in yield where the crop is grown without proper control measures.

Host. *Sorghum bicolor* (L) Moench

Symptoms

The pathogen attacks exclusively the ovaries, thereby majority of the grains in an ear are converted into smut sori (Fig. 5.75). The intensity of the damage depends on the susceptibility of the cultivar. In place of normal grain, commonly an oval or cylindrical, dirty grey spore sac (sorus), measuring 3-5 μm x 4-15 μm are found. The spore sacs are surrounded by glumes at the base. The stamens are commonly absent. The sori are covered by a tough membrane and remain intact in early stages of development. Commonly most part of the ovarian wall is rough and rigid, and ruptured only after harvest.

Each sorus has a slender, sometimes curved central column of hard tissue, the columella (Fig. 5.76A, B). The columella are sometimes branched at the apical region. Black powdery mass of teliospores are present surrounding the columella. Surface of the columella having hollow depressions are also full of spores.

Fig. 5.75: Grain smut or Karnel smut of jowar (*Sorghum*): A and B. Symptoms on ears (heads).

Causal organism

The disease is caused by *Sphacelotheca sorghi* (Link) Clint. Teliospores are round to oval, finely echinulate, olivaceous brown in colour, but dark brown in mass and measures 5-9 μm (mostly 6-7 μm) in diameter. The spores remain viable for more than 2 years in storage and as many as 13 years in herbarium specimens. Teliospores on germination develop a septate promycelium, budding off sporidia both laterally and apically (Fig. 5.76D-F). Sporidia are spindle shaped and measures 10-12.5 μm x 2-3 μm. Dikaryotic phase is established by fusion either between sporodia or their germ tubes. Sometimes dikaryotic infection tubes may form by fusion between cells of promycelia.

Disease cycle

The disease is **systemic** and the pathogen borne externally on grains (seeds). Grains become contaminated at the time of threshing. The pathogen, *Sphacelotheca sorghi* perennate as teliospores on contaminated grains.

When contaminated grains are shown in the next cropping season, the teliospores germinate and give rise promycelia and sporidia (Fig. 5.76D). Later on, the dikaryotic infection hyphae are developed by union of + and – sporidia; those infect the young seedling chiefly through the radicle or mesocotyle region. The pathogen grows further in a systemic manner throughout the growing plant and finally reaches the ovaries. The infected ovaries become transformed into smut sori (sing. sorus).

In the next season, during threshing of grains the teliospores are released from the broken sori and contaminate the grains. The teliospores

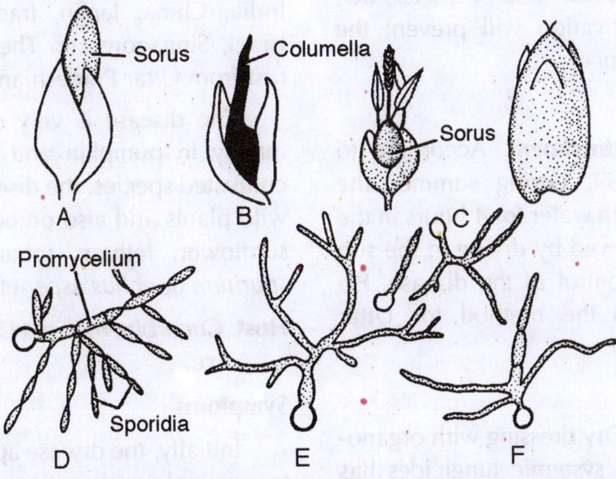

Fig. 5.76: *Sphacelotheca sorghi*: A. Ovary transformed into sorus, B. Columella inside sorus, C. Single flower: Ovary transformed into sorus, without involving stamen, and D-F. Germinated teliospores showing promycelia and budding sporidia

remain dormant along with the grain and remain dormant till the next season (Fig. 5.77).

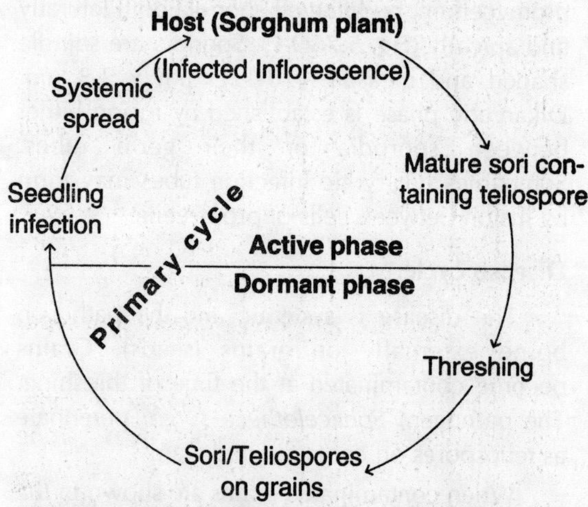

Fig. 5.77: Disease cycle of grain smut of Sorghum

Disease management

The teliospores from the previous crop do not persist in soil (Butler, 1918). As the disease is only grain (seed) borne and infection starts at the seedling stage, therefore treatment of grains is the most effective way to control or reduce the disease incidence.

The disease can be controlled or reduced by the following methods:

A. Cultural method

1. **Disease-free seeds.** Use of disease free seeds for cultivation will prevent the disease incidence.

B. Physical method

1. **Solar energy treatment.** According to Asthana (1947), during summer the grains soaked in water for 4 hours in the morning, followed by drying in the sun gives better control of the disease. He recommended the method for Uttar Pradesh (India).

C. Chemical control

1. **Dry dressing.** Dry dressing with organomercurals and systemic fungicides has been recommended.

   (i) Dry dressing with Agrosan GN @ 2 g/Kg seeds has been found effective and cheaper in controlling the disease.

   (ii) Elemental Sulphur powder used as dust @ 34.4-70.8 g/kg and sulphur containing fungicide like Thiram (trade name Arasan NI) @ 2-4 g/kg is also effective in controlling the disease.

   (iii) *Other fungicides*: Other nonsystemic fungicides like Ceresan, Tillex, Captan and systemic fungicides like Carboxin and Bavistin gives better result in controlling the disease.

2. Wet treatment. Immersion of grains in 3% copper sulphate ($CuSO_4$) solution for 15 minutes or 0.5% solution of formaldehyde for 1-2 hours are the effective methods in controlling the disease.

D. **Disease tolerant varieties.** Some tolerant varieties have also developed and recommended for cultivation are Bilichigan, Nandyal, CSH-5, CSH-7-R, PJ 7K, PJ 23K, SPV-102, SPV-104, SPV-115, SPV-138, SPV-245, SPV-297 and T 29/1.

20. *POWDERY MILDEW OF CUCURBITS*

The disease is observed throughout the world. In South Asia, the disease occurs in India, China, Japan, Iraq, Iran, Saudi Arabia, Israel, Singapore etc. The disease was reported first from Uttar Pradesh and Bihar (Butler, 1918).

The disease is very common in cucurbits, mainly in pumpkin and bottle gourd. Besides cultivated species, the disease also takes place in wild plants and also on other plants like potato, sunflower, lettuce, tobacco seedling, *Sedum spurium, Sonchus asper* etc.

Host. *Cucurbita maxima* Duchesne, *C. pepo* DC. etc.

Symptoms

Initially, the disease appears as minute white to grey coloured spots on leaves as well as on stems. With maturity the spots increase in size and give powdery appearance. Later on, entire

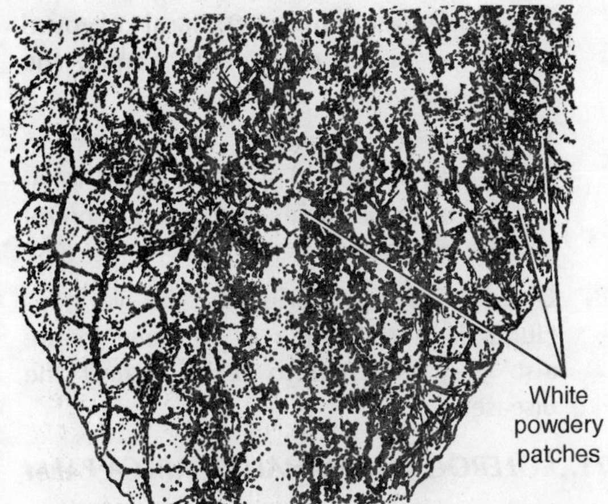

Fig. 5.78: Powdery mildew of cucurbits: Portion of a leaf showing white powdery patches of conidial stage

leaf becomes covered by powdery mass of spores (Fig. 5.78). Towards the end of winter season, minute dark-brown pinhead bodies are unusually found to develop, intermixed with white powdery masses of spores. These pinhead bodies are the ascostroma of cleistothecial type. During severe infection, the leaves fall off and fruits get deformed.

Causal organism

Powdery mildew is a widespread and serious plant disease in India and also in other countries. Both the organisms *Erysiphe cichoracearum* DC and *Sphaerotheca fuliginea* (Sc.) Poll. are responsible for the disease. Regarding the occurrence of pathogens, different observations have been reported. Disease may be caused by **one pathogen** or by **both the pathogens** (but separately).

Disease caused **by one pathogen**:

1. *Erysiphe*: It causes disease in Australia, Brazil, Bolivia, Canada, Egypt, France, Peru, Saudi Arabia, Singapore etc.
2. *Sphaerotheca*: It causes disease in Australia, China, Czeckoslovakia, Greece, Iran, Japan. Malawi, Netherlands, Rumania etc.

Disease caused by **both the pathogens** (but separately):

Both the pathogens cause disease in different countries, dominated either by *Erysiphe* (Bulgaria, U.S.A. and Germany etc.) or by *Sphaerotheca* (Italy, India, Israel, Japan etc.) or by both in equal importance [(Hungary, CIS (former U.S.S.R)].

The pathogens are characterised as:

***Erysiphe cichoracearum* DC**. Mycelial plant body is well-developed, generally evanescent, superficial with well-developed haustoria in host cells. Conidia are ellipsoidal or barrel-shaped and measure 24-44 µm × 14-26 µm, they develop in chain on unbranched and erect conidiophore (Fig. 5.79A). Cleistothecia are globose, scattered or gregarious and 90-130 µm in diameter with 10-20 µm thick wall. Cleistothecia are surrounded by numerous mycelial appendages about 1-4 times larger than its diameter

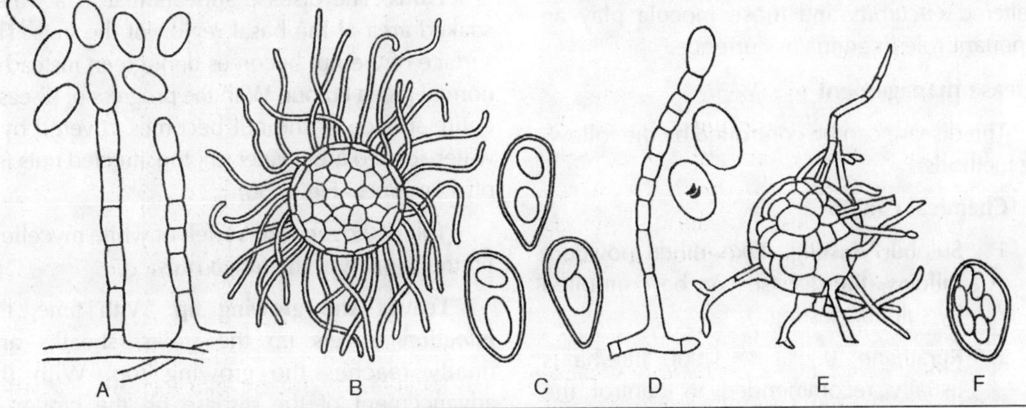

Fig. 5.79: Powdery mildew of cucurbits (causal organisms *Erysiphe cichoracearum* or *Sphaerotheca fuliginea*): A-C. *E. cichoracearum*. A. Conidiophores and conidia, B. Cleistothecium, C. Asci containing ascospores, D-F. *S. fuliginea*. D. Conidiophores and conidia, E. Cleistothecium and F. Ascus containing ascospores

(Fig. 5.79B). Asci are ovate, rarely subglobose, more or less stalked, 12-25 per cleistothecium and measure 60-90 μm × 25-50 μm. Ascospores are 2 in each ascus (rarely 3) and measure 20-30 μm × 12-18 μm (Fig. 5.79C).

***Sphaerotheca fuliginea* (Sc.) Poll**. Mycelial plant body is hyaline, but occasionally becomes brown with maturity, grows intercellularly and develops haustoria to draw nutrition from the neighbouring cell. Conidia are ellipsoid to barrel-shaped, develop in long chain at the apex of simple and short conidiophore (Fig. 5.79D). The oidia measure 24-36 μm × 14-25 μm. Cleistothecia are more or less round and usually 85 μm in diameter with 24μm thick wall (Fig. 5.79E). Asci are elliptical to subglobose, 50-80 μm × 30-60 μm and develop singly inside cleistothecium. **Ascospores** are ellipsoid to nearly spherical, 16-22 μm × 12-19 μm and **8 in number in each ascus** (Fig. 5.79F).

Disease cycle

The appearance of the disease in successive years is not clearly known. It is possible that the cleistothecia are the main source of inoculum. It might be possible that the conidia of fungi that are overgrown in summer in the hills are blown every year to the plains during winter and the reverse situation is observed during summer.

Jhooty (1967) reported that in Punjab (India), the vegetative mycelia and conidia survived on sheltered cucurbits and those inocula play an important role in annual recurrence.

Disease management

The disease can be controlled by the following methods:

A. **Chemical control**

1. Sulphur dusting. Like other powdery mildew, the disease can be controlled by sulphur dusting.
2. Karathane (0.2%) @ 1,000 litre/ha is usually recommended to control the disease (Mandloi and Khare, 1969).
3. Elosal (Wettable powder with 80% elementary sulphur) 0.5% also proved to be highly effective to control the disease.
4. Systemic fungicides, like Bavistin, Benomyl and Triozoles have been found to be highly effective to control the disease.

B. **Disease tolerant varieties**. The varieties Huragola and Diguria of muskmelon are also very effective to resist against the disease.

21. *KOLEROGA OR MAHALI OF ARECA PALM*

Koleroga is an important disease of Areca nut. In Karnataka, it is called as **Koleroga** (kole means rotting, roga means disease), in Kerala it is called **Mahali** (heavy devasting) and also called fruit rot in many places. The disease was first recorded from India in 1906. It occurs in severe form in areas with heavy rainfall.

In **Karnataka**, it is one of the most serious diseases of Areca palm plantations.

There is severe loss due to nut infection but hardly 1% annual loss due to death of plants.

Host. *Areca catechu* L.

Symptoms

The disease appears within 2-3 weeks after the beginning of rains. Symptoms appear on fruits, leaves and growing tip.

Fruits: The disease appears first as a water-soaked area at the basal region of the nuts. The surface of the nut becomes dark-green instead of normal green colour. With the progress of disease, entire surface of the nut becomes covered by a water-soaked area. Later on, the affected nuts fall off from trees (Fig. 5.80).

The pathogen forms a felt of white mycelium on the surface of the fallen nuts.

Leaves and growing tip: With time, the infection travels up the stalks, sheaths and finally reaches the growing top. With the advancement of the disease on the crown of leaves, the tree may dry up, thereby the leaves and bunches withered off.

Fig. 5.80: Koleroga or mahali of Areca palm: A. Infected fruits on bunch of fruit and B. Fruit bunches with less number of fruits and infected fruits

The disease may spread from one tree to the other in the orchard, during favourable condition i.e. with heavy rainfall and moist atmospheric condition.

Causal organism

In most of the areas the disease is caused by *Phytophthora arecae* (Coleman) Pethybridge. Shastry and Hegde (1987) present evidence that support reclassifying *P. arecae* as *P. meadii* and according to them it is one among the most serious diseases of Areca nut in Karnataka.

Beside Areca palm, this fungus, *P. meadii*; also cause diseases in cardamom and cocoa. But the isolate of cardamom is not pathogenic to Areca palm.

The mycelium is coenocytic, grows both inter- and intracellularly, vary in diameter from 8-9 μm and form septa with finger-shaped haustoria.

Sporangiophore develop during asexual reproduction, are about 2.5 μm wide, irregularly branched, forming sporangia at the apices (Fig. 5.81). Sporangia are broadly elliptical or pear-shaped, measure 30-70 μm x 25-40 μm and germinate indirectly to form reniform laterally biflagellate zoospores (Fig. 5.81).

In culture, chlamydospores are formed in variable numbers, sometimes rare or absent, 35-40 μm in diameter.

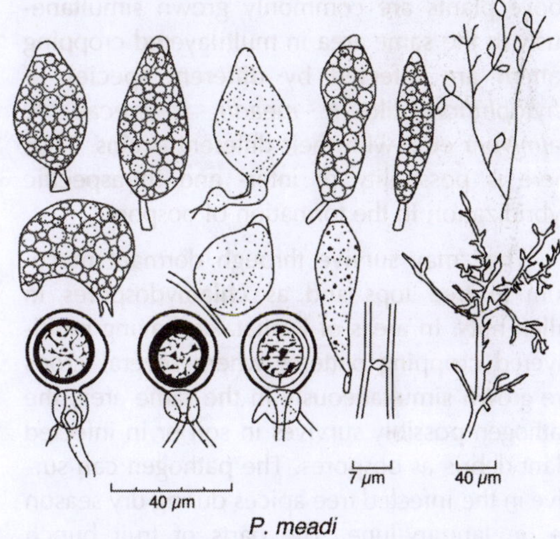

Fig. 5.81: Koleroga or Mahali of Areca palm (C.O. *Phytophthora meadii.*): Somatic and reproductive structures

[Upper row, Obpyriform to ellipsoid sporangia; sporangium with swollen sporangiophore; elongated, hourglass-shaped sporangia; sporangia produced on a simple sympodium (does not conform to size bars).

Centre row, Curved, papillate, caducous sporangium; sporangium sessile on the mycelium; sporangium with an elongated, ellipsoid, abnormal shape.

Lower row, Amphigynous antheridia on globose to obpyriform oogonia (oospores are distinctly aplerotic); old hyphal section with a septum; mycelial pattern. (Courtesy A. Vaziri; Reproduced from Erwin and Ribeiro, 1996)].

Oospores are aplerotic* and yellowish, developed by the union between antheridium and oogonium. It develops rarely in nature but frequently in culture medium. Antheridia are rounded, diclinous, amphigynous and oogonia are spherical, with smooth or wrinkled wall. In culture, oospores formed at 15-20°C, are yellowish, spherical, smooth and thick walled (4 μm) and measures 16-32 μm in diameter. Oospores are also develops on fruit.

Disease cycle

The fungus is endophytic. The species of *Phytophthora* cause diseases in different plants like areca palm, cocoa, black pepper and cardamom are heterothallic in nature.

In Karnataka and other adjoining areas, the above plants are commonly grown simultaneously in the same area in multilayered cropping pattern are infected by different species of *Phytophthora* like *P. meadii, P. arecae, P. palmivora* etc., with their different strains. Thus there is possibility of inter- and intraspecific hybridization in the formation of oospore.

They may survive through dormant mycelium in tree tops and as chlamydospores in fallen nuts. In areas of Karnataka, having multilayered cropping pattern, where several crops are grown simultaneously in the same area, the pathogen possibly survives in soil or in infected plant debris as oospores. The pathogen can survive in the infected tree apices during dry season i.e. in January-June. The parts of fruit bunch attached with the tree and infected nuts also carry pathogen. 2-4 months old nuts are normally infected by the pathogen. The pathogen spread rapidly during intermittent rain with alternating sunshine, but heavy rainfall with high humidity is most favourable for disease development.

Disease management

The disease can be controlled or reduced by the following methods:

* Aplerotic: They have space between oospore wall and oogonium wall.

A. Cultural method

1. **Disease-free seeds**. Use of disease-free seeds will prevent the disease incidence.
2. **Sanitary practice**. Remove and burn all fallen and infected nuts.
3. **Selection of site**. Select sites with deep well drained soil without high water table. Provide adequate irrigation facilities.
4. **Useful cultivation practice**. Selected seed nuts soon after harvest are shown in nursery bed with a spacing of 5-6 cm. Cover the seed nuts with sand and irrigate daily. Transplant 90 day old sprouts having 2-3 leaves to the secondary nursery.

Prepare secondary nursery beds of 150 cm width and of convenient length. Apply cattle manure @ 5 t/ha as basal dose. Transplant sprouts at a spacing of 30 cm x 30 cm. Provide shade by growing banana, *Coccinia indica* etc., or by means of artificial pandal.

5. **Proper irrigation**. Provide irrigation during hot and dry periods and drainage during monsoon.
6. **Weeding and mulching**. Periodical weeding and mulching are necessary for good health of the plants.

B. Chemical control

1. **Wet treatment**

(i) **Contact fungicide**. Spray Bordeaux mixture 1% on all bunches three times in a year, one just before the onset of southwest monsoon and the rest at 40 days intervals. If monsoon season is prolonged give a third spray.

Use rosin soda adhesive to ensure tenacity of the spray deposit on treated substrate.

(ii) **Systemic fungicide**. Areca nut bunches treated with systemic fungicides like metalaxyl mancozeb 72 WP @ 2. gl-1 as spray for twice showed effective reduction of the disease and increase in the yield.

C. **Disease tolerant varieties**. Some tolerant varieties have also developed and recommended for cultivation are Mangala, Sumangala, Sreemangala and Mohitnagar.

22. *RUST OF COFFEE*

Coffee rust has historic as well as economic importance. The historical importance has been reported by Large (1940) and later by Carefoot and Sprott (1967). The fungus is probably of East African origin. During 1870, in Ceylon the coffee cultivation covers an area of about 200 thousand hectares (1ha. = 2.471 acre) producing 50 million kg beans annually but since 1868 the cultivation becomes devastated by *Hemileia vastatrix*, the rust fungus at a faster rate and in 1885 the yield becomes reduced to almost zero, thereby the farmers are forced to replace the coffee cultivation by tea. In India, the disease was reported in 1870 and now it is causing recurring damage to coffee cultivation in Chikladha in M.P. and also in Western Ghats (South India). The disease is also available in other parts of Asia and also in Africa, Brazil, and Afganisthan etc.

In 2012, a coffee leaf rust epidemic struck Central American crops, causing over US $1 billion of damage in just two years (USAID). Even today, producers struggle to see the same yields as they did in 2010-11.

Host. *Coffea arabica* L.

Symptoms

Initially young lesions appear on the lower surface of leaf as chlorotic or pale yellow, small circular spots of 2-5 mm diameter (Fig. 5.82), later it develops as orange to brownish sori but during severe infection the entire lamina may turn brown and then gradually dry up caused premature defoliation. Rarely young shoots are also become affected. Young berries may also become infected, caused reduction in size and fail to ripen.

Causal organism

The disease is caused separately by two species of *Hemileia*, namely *H. vastatrix* and *H. coffeicola*. The former is most common and the latter is found in central and west Africa causing minor damage in Arabica coffee but occasionally become serious with high humidity.

H. vastatrix is a microcyclic and autoecious (as per available information) rust. The fungus produces three types of spores: uredospores, teliospores and basidiospores. *The basidiospores are not kown to infect host and spermogonia and aecia are unknown*. The above information lead to the speculation regarding the existence of an alternate host may have the role in annual recurrence of the disease. Urediniospores are more or less reniform, 26-40 μm × 18-28 μm, with hyaline to pale yellowish, strongly warted (3-7 μm) on the convex side, smooth on the straight or concave side, warts usually longer on spore edges. Telia are pale yellow in colour, teliospores often produced in uredinia; teliospores more or less spherical to limoniform in shape, 26-40 μm × 20-30 μm in diameter, wall hyaline to yellowish, smooth, 1 μm thick, thicker at the apex, pedicel hyaline.

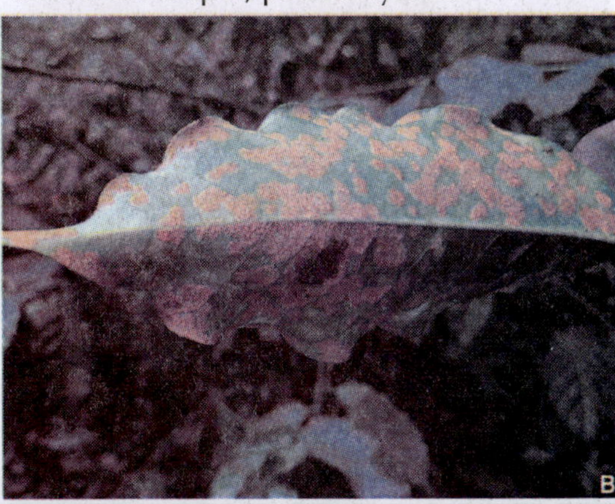

Fig. 5.82: Leaf rust of Coffee: Symptoms on Upper side (A) and Lower side (B) of leaf

Disease cycle

Hemileia life-cycle begins with the germination of uredospores through germ-pores (Fig. 5.83). It mainly attack leaves and is only rarely found on young stems and fruit. Appressoria are produced which in turn develop vesicles from which entry into the substomatal cavity is gained. The infection becomes completed within 24-48 hours. After successful infection, the stomatal cavity is colonized and sporulation starts. One lesion produces 4-6 crops of spores within 3-5 month period and releasing 300-400,000 spores.

Probably it does not complete its life-cycle on Coffee plants, but may live on some stage on some other plants, as yet unknown (i.e. life-cycle may be heteroecious).

An alternative hypothesis is that *H. vestatrix* actually represents early-diverging autoecious rust, in which the teliospores are non-functional and vestigial, and the sexual life cycle is completed by the urediniospores (Fig. 5.83). Hidden meiosis and sexual reproduction (cryptosexuality) has been found within the generally asexual urediniospores. This finding may explain why new physiological races have arisen so often and so quickly in *H. vastatrix*.

Coffee leaf rust is carried in water, rain, or air in the form of thousands of tiny spores. It can survive over long distances, which has helped its spread throughout the world.

Disease management

The disease can be controlled or reduced by the following methods:

1. Bordeaux mixture and other copper fungicides are commonly used to control coffee rust.
2. Systemic fungicides like bavistine etc., are used to control the disease.
3. Chemicals such as triadimefon (Bayleton) and pyracarbolid (Sicarol) are alternated with protectant fungicides like Chlorothalonil etc., to provide a curative treatment.

E. Protozoan diseases

1. *PHLOEM NECROSIS OF COFFEE*

The disease was first reported by Stahel

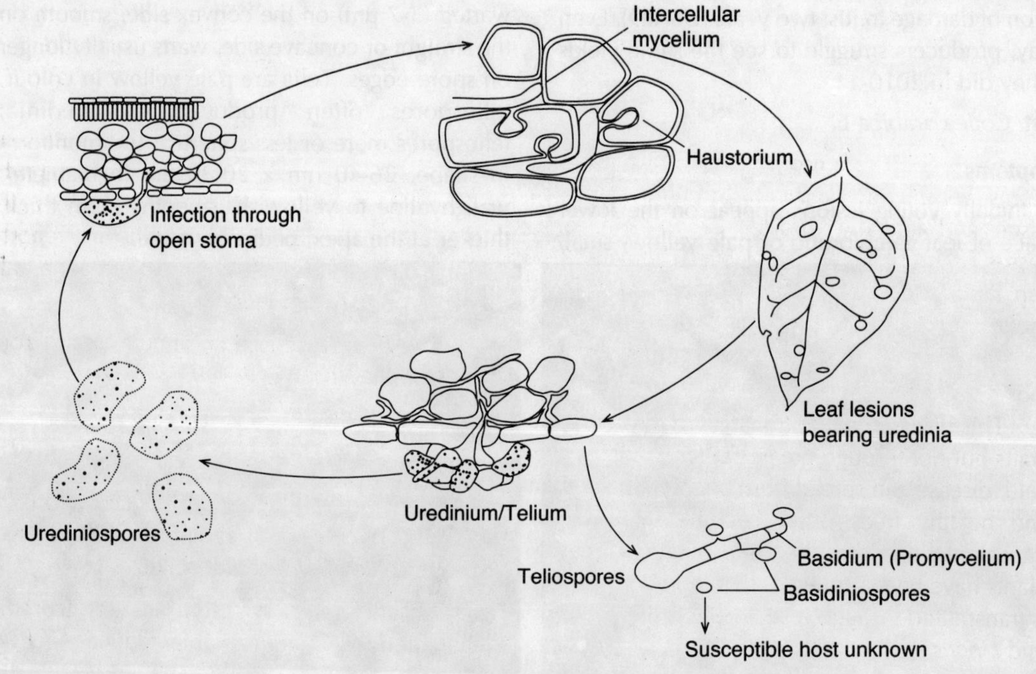

Fig. 5.83: Disease cycle of coffee leaf rust

(1917) in Surinam, USA. The disease has also been reported from British Guyana, and probably Brazil and Columbia. The disease has not been repoted from India.

Host. *Coffea liberica* Hiem, *C. arabica* L.

Symptoms

Infected trees show sparse yellowing and dropping of leaves, and only the young leaves remain on the top otherwise bare branches. The roots begin to die, followed by death of the trees, sometimes within 3-6 weeks.

The flagellates can be traced from the roots upward to trunk, where they seem to migrate vertically in the phloem and laterally through the sieve plates into healthy sieve tubes. The disease is characterized by deposition of callose in the sieve tubes and necrosis.

Causal organism

The disease is caused by *Phytomonas leptovasorum* Stahel. They are unculturable, flagellate, phloem limited, and insect-transmitted plant pathogen. The flagellates have a long, oval or spherical body with a thin, flexible covering membrane or it may be armoured. They have one or more slender flagella, used for locomotion and food capture.

Histopathology

Roots and stems of plants show phloem elements of disorderly structure. Flagellate *Phytoplasmas* is seen from root to stem. The pathogen seems to migrate vertically through phloem tissue and laterally through sieve plates and passes to the healthy sieve tubes. The pathogen can also move downwardly and infect the roots.

Disease cycle

This disease can be transmitted through root grafts but not through green branch grafts. In the field, disease can spread from one tree to another and healthy trees often become infected on transplantation into the field where diseased plants have been removed. In nature, the disease is transmitted through insect vector, the pentatomid *Lincus* sp.

Disease management

The disease can be reduced by the following cultural practices:

1. Immediate removal of diseased plants from the field to reduce the disease progress.
2. Seed-beds preferably prepared on virgin soil at a distance from centres of infection.
3. The field formerly occupied by infected trees should be left fallow for some months.
4. Healthy grafts should be planted far away from the infected area.

F. Mycoplasmal diseases

1. *SANDAL SPIKE DISEASE*

Sandal tree (*Santalum album* L.) is a perennial semi-root-parasitic plant. It is popularly used in India and also enrich the Indian economy through monopoly export the wood and also the products. The heartwood of the tree is valuable as it contains fragrant oil and the wood is used during worship and also in carving industry. The oil is precious and extracted from the heartwood and also from root. It is commonly used as perfumery.

Though the disease was available earlier, but the cause of the disease was confirmed first time by McCarthy (1899) from Coorg in Karnataka. The disease spread to states, Karnataka, Tamil Nadu and Kerala. The disease is most common in Karnataka. It was reported from North Coimbatore in 1903 and from Salem in 1913. In Kerala the disease was first reported from Marayoor in 1980. The disease has also been reported from Indonesia.

The disease caused an annual loss of about Rs. 3 million (Iyenger, 1969).

Host. *Santalum album* L. (Fig. 5.84A).

Symptoms

Plants of all ages get infected from the disease. Two types of symptoms are observed, rosette spike and pendular spike.

Rosette spike symptom is very common characterized by remarkable reduction of leaf size and internodes, caused crowing of leaves on leaf bearing branches. The newly appeared leaves are further reduced in size, become stiff and stand out as spikes (Fig. 5.84B). At an advanced stage i.e. before the death of the tree, the leaves become yellowish or reddish.

Fig. 5.84: Sandal spike disease: A. Healthy plant and B. Diseases plant (Photo by curtsy Dr. Arun A. N. Kumar.)

In infected trees the flowers become leaf-like and it does not bear any fruit. The roots end die out there by the haustorial connection with the host plants get damaged caused the plants become dead.

Pendulous spike symptom appears due to continuous apical growth of individual shoots without proper thickening, caused drooping of shoots. Dormant buds do not grow; thereby no rosette symptoms are developed. During this time, haustorial connection and also the root is not damaged.

Causal organism

Mycoplasma like organisms (MLOs) is the causal organism of the disease demonstrated in the phloem tissue of the infected plant (Dijksta and le, 1969; Hull *et al.*, 1969, and Verma *et al.*, 1969). Size of MLOs, in different host, ranges from 60-750 nm in diameter. They are commonly ovoid to ellipsoidal in shape, measures 180-220 nm x 250-300 nm. Typically, they are devoid of cell walls and the cytoplasm is bounded by 10-12 nm thick unit membrane. Inside the cytoplasm, fibrillar network of DNA and ribosomal bodies are present.

Disease cycle

The MLOs are reported to infect a number of plant species like *Eucalyptus grandis, Zizyphus oenoplea, Vinca rosea* and *Dodonea viscosa.* The disease is transmitted through root contact and insect vectors. The most common insect vectors (leaf hoppers) are *Nephotettix virescens* and *Moonia alhimaculata*. The above report does not give any conclusive evidence. Possibility of transmission by insect (*Redarator bimaculatus*) reported in 1984, yet to confirm.

Natural transmission in the cultivated field usually takes place through haustorial connections. Many other plants with spiked area shows yellow-type symptom may serve as host of sandal tree and transmission takes place from such plants to healthy sandal tree or vice-versa through haustoria. The disease can be transferred from *Vinca rosea* to Sandal (*Santalum album*) and vice-versa through dodder (*Cuscuta* sp.). Sap transmission is also been reported but it is considered as a rare event.

Disease management

No definite method has so far been developed to control the disease. Research is still required to gain the goal.

Nayar (1980) reviewed different methods like the use of heat therapy and chemical treatment by using arsenic, benlate, or tetracyclines to control spike disease. Tetracyclines having the most positive effect on spike disease control. However, the application of tetracyclines in a large cultivated area of a forest is impractical.

Thus, the most effective method to eliminate spike disease from sandalwood tissue are likely to be the use of tissue culture technique to develop disease-free genotype of superior quality (Balasundaran and Muralidharan, 2004), or to employ genetic engineering to introduce sequences that, once overexpressed, will reduce the phytoplasma titre in the forest bed, or the development of plants resistance. It can be controlled by proper management of insect vectors (Weintraub, Beanland 2006; Gurr et al. 2016), removal of infected plants from the field (Gurr *et al*. 2016), and the application of plant growth-promoting microbes these are not only enhance nutrient levels but also provide protection against microbial and insect vectors (Trivedi *et al*. 2016).

2. *LITTLE LEAF OF BRINJAL*

The little leaf of brinjal is a mycoplasmal disease (Anjaneyulu *et al*., 1968; Varma *et al*., 1969). The disease was first reported from Coimbatore (India). It is a graft transmissible disease. The disease is now found in India and Sri Lanka. The disease is now wide spread in India and causes about 90% losses in certain areas.

Host. *Solanum melongena* L.

Symptoms

The characteristic and remarkable symptom of the disease is the reduction of leaf size to small size leaves. Petioles become shortened. Leaf blade becomes soft and pale yellow. Internodes of the top branches are shortened resulting in a bushy appearance of the affected plant. Mostly there is no flowering but if flowers are formed, they remain green. Fruiting is rare in affected plants. However, if any fruit is formed, it becomes hard, tough and fails to mature.

Causal organism

The Mycoplasma-like bodies (MLB) have been detected in phloem cells of roots of brinjal plant affected with little leaf disease.

The size of MLBs varies from 230 to 770 nm. Each MLB contains ribosomes and nuclear material suspended by 16.5nm wide tripple layered unit membrane.

Disease cycle

It is a sap transmissible disease. It occurs in nature on other hosts like *Datura fastuosa, Vinca rosea* etc. The disease is transmitted by the insects (leaf hoppers), *Cestius phycytis* (*Eutettix phycites*) and *Empoosco devostolls*. The pathogen also survives on weed hosts.

Disease management

The disease can be controlled or reduced by the following procedures.

A. Cultural method

1. Disease-free seeds. Use of disease free seeds will prevent the disease incidence.
2. Sowing can be adjusted to avoid main flights of the insects.

B. Physical method

1. Early removal and burning of the infected plants.
2. Use of barriers of trap crops.

C. Chemical control

1. Fort-nightly spray of the insecticides such as Ekalox or Folidol till the fruit set helps to check the spread of disease.
2. The insect vectors can also be controlled by spraying the crop by Dimethoate (Rogor-30 EC) or Oxydemiton methyl (Metasystox 25 EG) or Monocrotophos (monocil) @ 1 ml per litre of water.
3. Disease can be controlled when phorate @ 1.0 kg/ha is applied to the seed beds and seedlings are dipped in acqueous solution of 0.05% tetracyline along with 0.05% monocrotophos.

D. Disease tolerant varieties. Growing disease tolerant varieties, e.g. Brinjal round, Black beauty, Pusa purple cluster etc., are very useful in controlling the disease.

G. Nematode disease

1. *ROOT KNOT OF MULBERRY*

It is a very severe disease of mulberry commonly occurs in China, Japan, Korea, and also in India. Since the pathogen has many number of host plants, the disease is wide spread and cause huge loss. Since the affected part is root system, diagnosis of the disease is very difficult in the early stage and the aerial symptoms will start very late after infection. The different soil factors such as type and texture of soil plays a major role on severity of the disease.

Host. *Morus indica* L.

Symptoms

Symptoms appear as gradual yellowing of leaves and plant becomes stunted. Later, the leaf margins become curls and turns brown and eventually fall off. In severe cases, the affected plants become wilted and die. Many knots can be seen on the roots (Fig. 5.85). The root knots are spherical or nearly spherical in shape. They may be small as a bean or large with irregular surface. The young knots are yellowish white in colour and are smooth and compact. Later the knots gradually turns brown and finally will become black and rotten. In sandy soil, the affected plants will generally die within 2-3 years in severe cases.

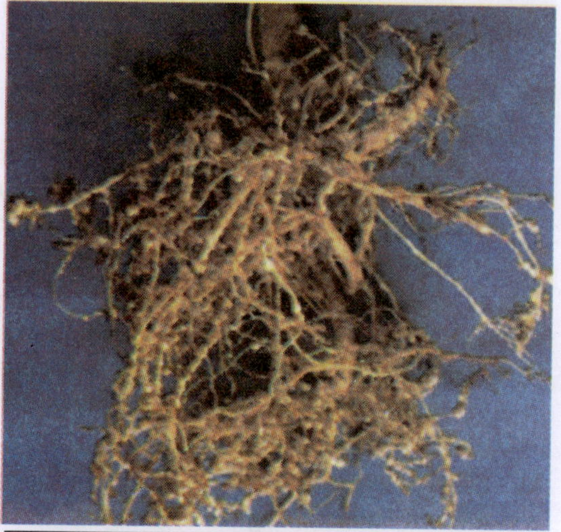

Fig. 5.85: Root-knot disease of mulberry: Knots on root

The affected plants appear to lack water and nutrients. If these knots are dissected, some white and translucent mature female worms can be seen by the naked eye.

Causal organism

The causal organism i.e. pathogen is *Meloidogyne incognita* (Kofoid and White) Chitwood, a nematode belongs to the family Heteroderidae, order Tylenchida under the class Secernentea.

The pathogen *M. incognita* is sexually dimorphic. The female is saccate to globose in structure, 0.4-1.3 mm long, and usually embedded in root tissues which are often become swollen. Its body is soft, pearl white in colour and does not form any cyst. The neck protrudes anteriorly and the excretory pore is anterior to the median bulb and often near the stylet base. Vulva and anus is terminal, flush with or slightly raised from the body contour and the cuticle of the terminal region forms the stunted tail terminus, phasmids. In lateral lines, vulva and anus surrounded by cuticular striae. In female, stylet is shorter, commonly 14-15 μm, and more delicate with small basal knobs. The paired gonads have extensive convoluted ovaries that fill most of the swollen body cavity. Six large unicellular rectal glands are present in the posterior part of the body which produces a gelatinous matrix, and is excreted via the rectum to form an egg sac in which many eggs are deposited.

The male is long, thin, and cylindrical in shape but the lip region has a distinct head cap, which includes a labial disc surrounded by lateral and medial lips. The head skeleton is usually weaker and the stylet is shorter, 18-24 μm long for many species. Infective second stage juveniles, often free in the soil and are usually 0.3-0.5 mm long and are less robust. The stylet is delicate with small basal knobs. The median oesophageal bulb is well developed and the oesophageal glands are extensive. The tail is conoid, often ending in a narrow-rounded terminus, but tail length is variable.

Predisposing factors

The pathogenic nematode overwinters in the affected roots. During the following spring, the second stage larva infests the roots and forms the root knots. The root knot nematode moves slowly in the soil, thus by their active movement the spread of disease is limited. However, the disease is spread to new localities through cultivation, irrigation, contaminated fertilizers and by planting of infected samplings.

The development of disease required temperature ranges from 12- 34°C while optimum is 20-25°C. When the soil is extremely moist or water is in short supply, the nematode is activity curtailed. The requirement of water content is 20-90% with an optimum 70%.

The disease occurs commonly in hilly lands, sandy loam soil, clay loam soil, fertile land or infertile land. The disease occurs quite seriously in sandy soil.

The disease also occurs severely if the preceding crops are jute, hemp, tobacco, soybean, cucumber etc.

Disease cycle

The mature nematode is heterogametic. Males are cylindrical, worm like, colourless, transperant, elastic with stripes across the body. Its caudal part is short and blunt and posterior is finger shaped. The female nematodes have a white saccate spherical and an elongated pyriform body with an elongated neck. The eggs are ellipsoidal and always laid in egg sacs.

The young nematode undergoes four stages. The first stage larva develops and moult within the egg, it breaks the shell and then goes into second stage. The second stage larvae are small, worm like, very active and moves freely in the soil until it invades the root. The time of invasion is referred to as the infection period and the young nematode worm is known as infection stage larva. During the parasitic period, when the nematode attacks the new roots, the worm is referred to as the second stage parasitic larva. Following the second moulting, the worm becomes the third stage larva and resembles a pea pod. The fourth stage larva moults again and becomes a mature worm.

Disease management

The disease can be controlled or reduced by the following methods:

A. Cultural methods

1. *Deep ploughing*. Deep ploughing to a depth of 30-40 cm during summer months helps to kill the nematode eggs and larvae.
2. *Disease free saplings*. Always use disease free saplings (a young tree with slender trunk) for new plantation.
3. Plant marigold (*Tagetus patula*) as intercrop at distance of 30 cm in between mulberry rows.
4. Expose the roots of infected plants by digging to a depth of 15 cm. Cut and remove the bunch of knots on the roots and burn.

B. Physical control

1. Apply neem oil cake @ 2 mt*/ha/year in 4 split doses during fertilizer application/ cultural operation.

C. Chemical control

1. Apply Furadan (Carbofuran 3G) @ 40 kg/ha/year in 4 split doses during fertilizer application (safety period: 40-45 days).

D. Biological control

1. This involves the soil application of 'Bionema' (*Verticillium chlamydosporium*) after mixing with FYM† and neem oil cake in the ratio of 1:200:24. Bionema is a bio-nematicide with no toxic effect on both mulberry and silkworm.

[Use as: Mix one kg Bionema with 200 kg FYM† (for 1000 plants) and 24 kg neem oil cake, and store the mixture under the shade and keep moistened by sprinkling 30-32 liters of water for about one week.]

---

* mt= metric ton.

† FYM= Farm yeard manure

POST-HARVEST DISEASES

The diseases which develop on harvested parts of the plants like seeds, fruits and also on vegetables are the post-harvested diseases. The harvested products may get infected on the way to storage or to market or even before their final consumption. The plant parts may get infected in the field, but expression of symptoms may take place later, at any stage before the final consumption. The plant products may get infected by microorganisms and cause rotting or decaying — partially or totally. The quantity of plant products becomes reduced due to the above infection. The seeds or grains may get damaged by accumulation of toxic substance, the mycotoxin produced by the infected microorganism. The fleshy fruits, vegetables etc., like tomato, banana, citrus, strawberries, rhizome of zinger, bulb of onion, tuber of potato etc., may get damaged. This results in reduction of quantity, quality or both of the affected parts or products as a whole.

The amount or extent of damage depends mainly on the pathogen(s) involved, on the condition of the products and the condition of storage.

The pathogens involved are mainly fungi like *Pythium, Phytophthora, Rhizopus, Aspergillus,* etc., and some bacteria like *Pseudomonas, Erwinia,* etc.

Cristensen and Kaufmann (1965) divided the pathogens into two categories: 1. Field pathogen, and 2. Storage pathogen.

1. **Field pathogen.** The field pathogens are those, which cause infection during development of plants or their products before harvest.
2. **Storage pathogen**. The pathogen which cause infection during storage are the storage pathogen.

Symptoms from infection caused by the 'field pathogens' may be very inconspicuous to be noted at the time of harvest. In fleshy and/or juicy fruits and vegetables, infection by field pathogen continues to develop even after harvest. They may become infected during storage by the same field pathogen(s) or by other pathogen(s). In seeds and grains, the disease caused by field pathogens ceases to develop further soon after harvest. But, they may be infected further by the other pathogens during storage.

Stackmann and Harrar (1957) divided the pathogenic storage diseases into two groups:

1. *Diseases of dry, bulk materials*, such as seeds and grains, and
2. *Diseases of fleshy storage* organs, such as vegetables (tubers, rhizomes, bulbs etc.) and fruits.

Observations of many investigators indicate that the real cause of the spoilage of vegetables and fleshy fruits in transit and also in storage are due to high moisture, high temperature, and injuries caused during marketing. Due to high moisture content and nutrient in harvested vegetables and fruits, they are vulnerable to attack by the pathogenic organisms. Injuries of fruits and vegetables may be caused during harvesting, packing and transposition — they help the pathogen to enter the host and cause damage. But the seeds and grains can be stored for long time due to low moisture content (about 12-14%), where most of the pathogens cannot grow favourably.

I. Diseases of stored seeds and foodgrains

Field fungi, like *Alternaria, Fusarium, Cladosporium, Verticillium, Helminthosporium, Colletotrichum* etc., attack seeds and grains on growing crops, but are unable to grow in storage due to low relative humidity i.e., below 90%. During storage or transit the seeds and grains are damaged by the different species of *Aspergillus* and *Penicillum,* which can grow well at a relative humidity ranges from 70-90%. The commonly available *Aspergillus species are A. repens, A. ruber, A. flavus, A. candidus,* etc.

Aspergillus and a number of other storage fungi invade the embryo of the seeds and grains and they discolour the embryo or seeds as a whole, thereby the germination percentage reduces markedly. In some cases, spoilage of stored grains and seeds results in drastic increase of temperature up to 70°C or more, which encourage the growth of different thermophilic and thermotolerant fungi such as *Aspergillus fumigatus, Absidia* spp., *Mucor pusillus,* etc.

In addition to storage fungi, other micro-organisms may grow in/or on seeds and accelerate the deterioration process. During breeding period of insects, the moisture content and temperature of seeds increase, thereby rapid growth of the pathogen takes place producing enormous amount of spores.

During storage, the fungi produce mycotoxins that cause great damage to both domestic animals and human beings. The important fungi in this respect are *Aspergillus* and *Penicillium,* which produce aflatoxin and other toxins. Most important toxins are:

1. **Yellow rice toxins**. Produced in grains of rice, barley etc., by species of *Penicillium.*
2. **Tremorgenic toxins**. Produced on prepared food during storage in refrigeration or in other places and also on food produced from infected grains and/or seeds.
3. **Penicillic acid.** It is a **carcinogenic substance** produced by the different species of both *Aspergillus* and *Penicillium* in molded cereal grains.

Control. The above-mentioned loss or damage by decay and spoilage of seeds and/or grains by storage fungi can be controlled by the following procedures:

1. **Low moisture**. The moisture content of the rooms for storage should be kept below 70%.
2. **Low temperature**. Temperature in store house should be maintained below 30°C, because most of the storage fungi can grow well at temperatures between 30°C to 55°C.
3. **Ventilation**. Proper ventilation should be maintained during storage and also during holding period before sending to market.
4. **Sanitation**. Proper sanitation should be maintained to keep storage products clean.
5. **Use of insecticide.** Insecticides like methyl bromide and some other fumigants are used to treat the harvested seeds, thereby they regulate the storage fungi and reduce economic loss.
6. **Clean**, uninjured and properly ripened seed should be selected for storage, then only they are able to resist the action of the storage pathogen(s).

II. Diseases of vegetables and fruits

Different members of Ascomycota and Mitosporic fungi (Deuteromycotina) cause the major post-harvest diseases of fruits and vegetables. These are *Alternaria, Botrytis, Fusarium, Penicillium, Sclerotinia* etc.:

1. *Alternaria.* Different species of *Alternaria* cause rot of many fresh fruits and vegetables, e.g., black rot of orange, tuber rot of potato, rot of sweet potato, purple blotch of onion, *Alternaria* rot of onion, *Alternaria* rot of cabbage, etc.
2. *Botrytis.* It causes "**grey mold rots**" of fruits like pear, apple, citrus etc., and vegetables like onion, tomato etc. Every year it causes great economic loss.
3. *Fusarium.* It causes different diseases, commonly called "**pink** or **yellow molds**". Different species of *Fusarium* cause damage to tubers, bulbs, storage roots etc., and frequently on cucurbits, tomato etc. It also cause brown rot of fruits like lemon, orange etc.
4. *Penicillium.* Species of *Penicillium* are commonly called "**blue** or **green molds**", these cause rots of different fruits like onion, sweet potato etc. They also cause spots on different fruits. Under storage, the spotted fruits bear tufts of spores. Though most of the *Penicillium* species prefer relatively high temperature for their growth in storage, they

still remain active near freezing temperature — at a slow rate. A few species produce ethylene which increases respiration of fruits, thereby it reduces the storage life of the fruits. It also produces **patulin** — a mycotoxin — which directly contaminate the sauces and fruit juices prepared from infected partly rotten fruits.

5. *Sclerotinia*. It infects different fruits and vegetables. Most common diseases are cottony rot of lemon, watery soft rot of bean pods, cucurbits etc. Storage diseases like bacterial soft rot of vegetables such as onion, carrot, potato etc., are mainly carried out by different species of *Erwinia*, such as *E. carotovora, E. chrysanthemi* etc.

Control. The diseases can be controlled or reduced following the preventive procedures are:

1. The fruits and vegetables should be harvested and handled carefully to avoid any injury which may facilitate the pathogen to cause infection.
2. The infected region on the vegetables should be cut off to avoid further infection during transportation and storage.
3. Storage container, warehouses etc., should be properly cleaned with $CuSO_4$, formaldehyde etc., to avoid contamination.
4. The crop should be stored or transported at a temperature low enough to slow down the development of disease.
5. Proper ventilation in storage reduces the spread of further development of disease.
6. The crops should be free from insects and other pests, thus creation of new wounds and disease can be avoided.
7. Hot water and hot air treatment help to reduce further spread of the disease.
8. Chemical control. Post-harvest diseases may be controlled by the application of thiabendazole, dichloran, dosa-ash, etc. These chemicals help to prevent infection and suppress the development of pathogen on the host surface. Some other chemicals, such as vapours of acetaldehyde, biphenyle/nitrogen chloride forming chemicals etc., are used as supplementary measures to control the post-harvest diseases during storage and transportation.

BIOPESTICIDE

Pesticide

First of all, one has to know what is pesticide? The substances or the organisms which kill pests are the pesticide. What is pest? Those which caused damage others. In other way you may explain the pest as **"Guest without sweets packet is a pest"**.

Fungi, Bacteria, Virus etc., caused diseases of plants, thus created damage, thereby to control the disease we use fungicide, bactericide, virucide. As the entire organisms caused diseases and consequence damage are all called pests and protective chemicals as fungicide, bactericide, virucide etc. Insects also caused damage to crops and for its control we use insecticide. All are together we called as **pesticide**.

Defination: Substance used for destroying insects or other organisms harmful to the cultivated plants and also to animals are called Pesticide.

Type of Pesticides

The term "-cide" comes from the Latin word "to kill". There are many different types of pesticides. Each one is effective against specific pests. These are:

1. **Algaecides**: These are used for killing and/or slowing the growth of algae.
2. **Antimicrobials**: These pesticides control germs and microbes such as bacteria and viruses.
3. **Biopesticides**: These are made of living organisms come from living organisms, or they are found in nature.
4. **Desiccants**: These are used to dry up living plant tissues.
5. **Defoliants**: This group of pesticide cause plants to drop their leaves.

6. **Disinfectants**: These are used to control germs and microbes such as bacteria and viruses.
7. **Fungicides**: These are used to control fungal problems like molds, mildew, and rust.
8. **Herbicides**: These are used to kill or inhibit the growth of unwanted plants, aka weeds.
9. **Insecticides**: These are chemicals used to control insects.
10. **Insect Growth Regulators**: These are used to disrupt the growth and reproduction of insects.
11. **Miticides**: These are used to control mites that feed on plants and animals. Mites are not insects, exactly.
12. **Molluscicides**: This group of pesticide is designed to control slugs, snails and other molluscs.
13. **Mothballs**: These are insecticides used to kill fabric pests by fumigation in sealed containers.
14. **Natural and Biological Pesticides**: This group of substances control pests using things found in nature, or man-made versions of things found in nature. These are also called as **Biopesticides.**
15. **Ovicides**: These are used to control eggs of insects and mites.
16. **Pheromones**: These are biologically active chemicals used to attract insects or disrupt their mating behavior. The ratio of chemicals in the mixture is often species-specific.
17. **Plant Growth Regulators**: These are used to alter the growth of plants. For example, they may induce or delay flowering.
18. **Repellents**: These are designed to repel unwanted pests, often by taste or smell.
19. **Rodenticides**: These are used to kills rodents like mice, rats, and gophers.
20. **Wood preservatives**: These are used to make wood resistant to insects, fungus and other pests.

Out of the above 20 groups only Biopesticide is discussed below:

Biopesticide

Biopesticide can be defined as "a form of pesticide based on microorganisms or natural products". It is a contraction of 'biological pesticides'; which include different types of pest management process like through predatory, parasitic, or chemical relationships. The term has been associated historically with biological control.

In other way, it can be defined as "The naturally occurring substances that control pests (biochemical pesticides), microorganisms that control pests (microbial pesticides), and pesticide substances produced by plants containing added genetic material i.e. plant-incorporated protectants (PIPs)".

They obtained from organisms including plants, fungi, nematodes, bacteria and other microbes etc. In many cases they act as an important component in the "integrated pest management (IPM) programme", and received much attention as substitutes to synthetic chemical plant protection products (PPPs).

Limited information indicates that usually biopesticides have no function in photosynthesis, growth or other basic aspects of plants physiology. Instead, they are active against biological pests. Many chemical compounds have been identified that are produced by plants to protect them from pests. These materials are biodegradable and renewable alternatives, which have practical use and are economical.

Classification

The Bio-pesticides are classified into three classes. These are as follows:

1. **Microbial pesticides**: This group consists of bacteria, entomopathogenic fungi* or viruses (and sometimes includes the metabolites that bacteria or fungi produce). The entomopathogenic nematodes are often classed as microbial pesticides, even though they are multicellular.

---

* Entomopathogenic fungi are the fungus that can act as a parasite of insects and kills or seriously cripple them.

2. **Bio-derived chemicals**: These are naturally occurring substances that control (or monitor in the case of pheromones) the pests and microbial diseases. Four groups are in commercial use: pyrethrum, rotenone, neem oil, and various essential oils.

3. **Plant-incorporated protectants (PIPs)**: Plant-incorporated protectants (PIPs) have genetic material from other species incorporated into their genetic material (i.e. genetically manipulated or GM crops). They are used as controversial pesticide, used in many developed countries.

Common biopesticides

These are of two types, Microbial agents and naturally occurring materials.

A. Microbial agents

This group includes different microbes which are as follows:

1. *Bacillus thuringiensis*, caused bacterial disease of Lepidoptera, Coleoptera and Diptera, is a well-known insecticide.
2. Entomopathogenic fungi: *Beauveria bassiana, Lecanicillium* and *Metarhizium* spp.,
3. Entomopathogenic viruses: *Cydia pomonella.*
4. Plant disease control agents: *Trichoderma* spp. and *Ampelomyces quisqualis* (a hyperparasite of grape powdery mildew).
5. Beneficial nematodes attacking insect: *Steinernema feltiae.*

B. Naturally occurring materials

This group includes fungal and plant extracts used as bio-pesticides. Some of them are:

1. **Pheromones** and other **semiochemicals** secreted by Insects.
2. **Fermentation products** such as Spinosad (a macro-cyclic lactone) etc..
3. **Chitosan**: A substance whose presence in a plant, naturally induce systemic resistance (ISR), defend itself against pests.
4. **Naturally derived plant products**, include alkaloids, phenolic substances, terpenoides etc. Canola oil (a vegetable oil) is known to have pesticidal properties.

⁂ **Pheromones** are chemical substances released by an organism. On detecting these pheromones the biology and behaviour of another organism, usually a member of the same species is affected. Such as the female Moth (Orgyia antiqua) is flightless and cannot search an opposite sex for mating.

⁂ **Semiochemicals**: chemical substance or mixture that carries a message for purpose of communication.

Application of biopesticide

Biopesticides are biological or biologically-derived substances used in pest management and are ecofriendly. For the effective result with the microbial agents, requires appropriate formulation and proper application.

These are used in different ways like spray, seed treatment and soil amendments.

1. **As spray treatments**: *Bacillus pumilus* is used in Broccoli downy mildew, Pumpkin powdery mildew, Radish downy mildew etc., as spray.
2. **Seed treatment**: Biofungicides are used as seed treatments in controling soil borne fungal pathogens that cause seed rots, damping-off, root rot and seedling blights. They can also be used to control external and internal seed borne fungal pathogens.
3. **Soil amendments**: Microbial pesticides are increasingly used as soil amendments that will target the necessary area of the plant. When microorganisms are added to the soil/plant complex they release families of biochemical molecules in the rhizosphere region. The pesticidal properties of the microbial biochemical excretions then aid the plant in its affected areas.

The most widely used microbial pesticide is the bacteria, *Bacillus thuringiensis* or Bt. Each strain of bacterium produces different proteins that are toxic to certain insects, specifically the larvae of target insect.

Advantages

The advantages of using biopesticides are as follows:

1. Biopesticides are usually inherently less toxic than conventional pesticides.
2. Biopesticides are **biodegradable** i.e. does not produce any harmful residue, thereby these are ecofriendly.
3. It is to some extent cheaper than chemical pesticides.
4. It requires in small quantities and can be more effective than synthetic pesticides in the long-term.
5. Biopesticides generally affect only the target pest and closely related organisms. But the broad spectrum, conventional chemical pesticides that may also affect other different organisms like birds, insects and mammals.
6. As a component of Integrated Pest Management (IPM) programs, biopesticides can greatly reduce the use of conventional pesticides, while crop yields remain high.

⁂ The use of biopesticides is cost effective and safe; however, users need to know a great deal about managing the pests.

Disadvantages

The disadvantages of using bio-pesticides are as follows:

1. **High specificity**: The selection of biopesticide requires an exact identification of the pest and the use of multiple products to be used.
2. **Slow speed in action**: Due to slow speed in action, the biopesticides often unsuitable during severe pest outbreak.
3. **Variable efficiency**: As and when the living organisms are used as biopesticides, their activity may differ based on biotic and abiotic factors, because they need multiplication.
4. **Unwanted consequences**: Broad spectrum biopesticides have lethal and non-lethal risks for non-target native pollinators such as *Melipona quadrifasciata* etc.

Biopesticides in India

In India, during 2005, the overall biopesticide market represents only 2.89% of the overall pesticide market and is expected to increase about 2.3% in the coming years (Thakore, 2006). Only 12 different types have been registered under the Insecticide Act., 1968. Among them only 4 are produced and commonly used. These are Neem based pesticide, *Bacillus thuringiensis, Trichoderma* and NVP (Nuclear Polyhedral Virus). At present the chemical pesticides are more than 190 in number.

The use of biopesticides becomes increased from 219 metric tons (1997) to 683 metric tons (2001). Out of the total products about 85% are produced as neem-based products.

Table 5.10: Annual availability of Biopesticides in India (Kalra and Khanuja, 2007)

Biopesticides/Bioagents	Quantity/year (approx.)
Neem 300 PPM	1000,000 L
Neem 1500 PPM	250,000 L
Bt	50,000 kg
NPV (liquid)	500,000 Le
Beauveria	Meager
Pheromone traps	500,000 nos.
Lures	2 million
Trichogramma	1 million
Chrysoperla and other bio-control insects	Meager
Trichoderma	500 T

Le = Deliverable line item, L = liter, T = metric ton

Some commonly used Biopesticides in India

1. *Bacillus thuringiensis* (Bt)

It is primarily a pathogen of Lepidoptera pests of American bollworm and stem borers of Rice. After it is ingested by larvae of pests, the Bt releases toxins which damage the mid gut of the pest, thereby it is killed. Kathage and Qaim (2012) during their studies between 2002-2008, observed that the Bt caused 24% increase in the yield of cotton due to reduction of pest damage, thereby profit increases about 50%.

2. *Trichoderma* sp.

Different species of *Trichoderma* plays a major role as biocontrol agents, owing to their ability as plant growth promotion along with biopesticidal action. Among the known effective pesticides, *Trichoderma* is exploited much along with great success in grape production. The main problems in grape export are the availability of pesticide residue, SO_2 injury and also decay of berries. To overcome the above problems, "ICAR-National research Centre for Grapes", Pune; isolated a natural strain of *Trichoderma*, the *T. harzianum*, available commercially as "**Grape-guard**". In addition, experiments and pack house studies have revealed that the grapes treated with *Trichoderma*, retain their **freshness** for longer period as compared to fungicide treated grapes.

At present different Govt. Institutions and also private companies of India, regularly producing different biopesticides along with *Trichoderma*. The above data clearly indicate the requirement of biopesticides along with more production of different crops with improvement of quality.

At present 410 biopesticide centres are present in India, out of which 130 leaded by private companies.

3. Neem

Neem (*Azadirachta indica*), is used from long back to cure different diseases but gradually it becomes more attracting the world not only to the scientists but also to the farmers and common people in respect of crop and grain protection. **Azadirachtin**, a tetranortritarpinoid, a major active ingredient isolated from neem. This compound is known to disrupt the metamorphosis of insects. It affects the reproductive and digestive systems of a number of important insect pests.

The following paragraphs deals with some easy methods by which the neem extracts can be prepared and use by the farmers.

A. **Neem seed kernel extract**. For extraction, 50 g kernel is required for 1 liter water. The kernels are pounded gently in such a way that no oil should come out. The outer coat is removed before pounding, used as manure. If it is pounded with seed coat more seeds i.e. about 75 gms are required.

For the preparation of **neem seed kernel extract**, the seed age should be within 3-8 months. Beyond this age, the azadirachtin is quite low and hence the efficiency in controlling pests becomes unsatisfactory.

The kernel powder is collected in a muslin pouch and is soaked overnight in water. The pouch is the squeezed and the extract is filtered gently. After collecting the filtrate, an emulsifier like soap cake powder or soap oil is added. One ml/gram of emulsifier is added with 1 liter water. The emulsifier is added as because it will help the extract to stick well with the leaf surface and/or other plant parts.

B. **Neem leaf extract**. For extraction, 1 kg of green leaf is required for 5 liter water. Initially the leaves are soaked overnight in water. Next morning, the leaves are grounded and the extract is filtered. Emulsifier is added like neem seed kernel extract.

About 80 kg leaves are required to apply the leaf extract in 1 ha of crop. Due to this, it is commonly used in small gardens and nurseries.

The extract is used to control caterpillars, locusts, grubs and grasshoppers.

C. **Neem cake extract**. For extraction, 100 g of neem cake is required for 1 liter water. The cake is put in muslin pouch and soaked in water. It is soaked for overnight before it is used. Emulsifier is added like neem seed kernel extract. It is used on plants by spraying.

D. **Extraction of Neem oil**. Seeds are commonly used for extraction of neem oil, instead of leaf. Preparation of oil includes collection of raw material and selection of extraction methods. Quality of oil depends on the type of extraction. Three processes are commonly used in making neem oil preparations for commercial application. These are as follows:

(a) **1st Process**: Initially ripen seeds are collected from trees. Seeds are then dried in sunlight and then crushed in choppers and semi-powder is produced. The semi-powder is then poured in metal bowl and mixed with water in the ratio 10:1 (crushed seeds: water:: 10:1). After that the mixture is squeezed again and again. Finally, 100% organic oil is produced.

(b) **2nd Process**: After extracting through the 1st process, the neem oil is mixed with other oil.

(c) **3rd Process**: In this process, neem oil extracted with the help of modern machines for commercial production. In this process seeds are used after proper drying in sunlight. The highest concentration of oil available in the market is only 3%.

The neem oil is commonly used by spraying of 0.5-1.0%, which is sufficient enough to use as biopesticide. As and when required the concentration may increase up to 2.0%. The self-life of oil in soil is 3-22 days and about 4 days in water.

Rao *et al.* (2007), in ICRISAT (International Crop Research Institute for the Semi-Arid Tropics), Andhra Pradesh; reported that in field demonstration of bio-intensive pest management, the neem and Helicova NPV are highly effective and these are more effective than synthetic and other biopesticides in reducing the population of larvae and damage of pods in Chickpea (Gram) by Helicova during 2000-2002. The neem based biopesticides are equally or more effective than the synthetic pesticides in reducing the larval population and boll damage in cotton during 2003-2006.

In general, it can be mentioned that in food production system, biopesticides like neem-based products as well as *Trichoderma, Bacillus thuringiensis* and NPV, can safely be used by the food growers during management of pests. In the above context it is assumed that the biopesticides will create new dimensions in the pest management and to produce healthy nontoxic foods without disturbing the global ecosystem.

Selected Readings

Agrios GN. 2005. *Plant Pathology* (5th ed.). Academic Press Inc., California.

Ainsworth GC. 1973. *Introduction and keys to higher taxa. In : The Fungi An Advanced Treatise*, IV B (Ed. GC Ainsworth, FK Sparrow and AS Sussman). Academic Press, New York.

Ainsworth GC, Sparrow FR and Sussman AS. 1973. *(The) Fungi : An Advanced Treatise*. Academic Press, New York.

Alexopoulos CJ. 1962. *Introductory Mycology*. 2nd Asian ed. Toppan Company Ltd. Tokyo, Japan.

Alexopoulos CJ and Mims CW. 1979. *Introductory Mycology (3rd ed.)*. John Wiley and Sons, New York.

Aneja KR. 2001. *Experiments in Microbiology, Plant Pathology, Tissue Culture and Mushroom Production Technology* (3rd ed.) New Age International Publishers, New Delhi.

Banerjee AK and Banerjee N. 2006. *Fundamental of Microbiology and Immunology*. New Central Book Agency (P) Ltd., Kolkata.

Bilgrami KS and Dubey HC. 1976. *A Text book of Modern Plant Pathology*, Vikas Publishing House, New Delhi.

Bold HC and Wynne MJ. 1978. *Introduction to the Algae : Structure and Reproduction*. Prentice-Hall of India, New Delhi.

Bold HC, Alexopoulos CJ and Delevoryas T. 1980. *Morphology of Plants and Fungi*, (4th ed.). Harper and Row, New York.

Burnett JH. 1975. *Mycogenetics : An Introduction to the General Genetics of Fungi*. John Wiley and Sons, New York.

Campbell CL and Madden. LV. 1990. *Introduction to Plant Disease Epidemiology*. Wiley, New York.

Chakraborty P. 2003. *A Text Book of Microbiology* (2nd ed.) New Central Book Agency (P) Ltd., Kolkata.

Chang ST and Hayes WA. 1978. *The Biology of cultivation of edible mushroom*. Academic Press, Inc. New York.

Chapman VJ and Chapman DJ. 1973. *The Algae*. Macmillan, London.

Deacon JW. 1997. *Modern Mycology* (3rd ed.). Backwell Science.

Dubey HC. 1978. *An introduction to Fungi*, Vikas Publishing House, New Delhi.

Dubey RC and Maheswari DK. 1998. *Microbiology*. S. Chand and Co. New Delhi.

Frazier WC and Wethoff DC. 1995. *Food Microbiology*, (2nd ed.) Tata McGraw-Hill Inc., New York.

Fritsch FE. 1935 and 1946. *The structure and reproduction of the Algae*. Vol. I and II, Cambridge, University Press, London.

Gangulee HC and Kar AK. 1989. *College Botany (Vol II)*. New Central Book Agency (P) Ltd., Kolkata.

Goto M. 1992. *Fundamentals of Bacterial Plant Pathology*. Academic Press, London.

Hawksworth DL and Hill DJ. 1984. *The Lichen Forming Fungi*. Blackie and Sons Ltd. Glasgow.

Hawksworth DL, Sutton BC and Ainsworth GC. 1983. *Ainsworth and Bisby's Dictionary of the Fungi* (7th ed.). Commonwealth Mycological Institute, Kew, Surrey.

Ingold CT. 1967. *The Biology of Fungi*. Hutchinson Educational Ltd. London.

Ingold CT. 1971. *Fungal Spores*. Oxford University Press, London.

Kathleen T and Talaro A. 1996. *Foundation in Microbiology*. Wm. C. Brown, London.

Kumar HD and Singh HN. 1982. *A Textbook on Algae*. Affiliated East West Press Pvt. Ltd., New Delhi.

Lee Robert Edward. 1999. *Phycology* (3rd ed.) Cambridge University Press, London

Mehrotra RS and Aggarwall A. 2003. *Plant Pathology* (2nd ed.) Tata McGraw-Hill Publishing company Limited, New Delhi.

Mehrotra RS and Aneja KR. 1990. *An Introduction to Mycology*. Wiley Eastern Limited, New Delhi.

Muller, BM and Litsky W. 1976. *Industrial Microbiology*. McGraw-Hill Book Company, New York.

Pelczar MJ, Chan PCS, and Krieg NR. 1993. *Microbiology*. (5th ed). Tata McGraw-Hill Pub. Com. Ltd., New Delhi.

Prescott GW. 1969. *The Algae: A Review*, Thomas Nelsons and Sons. London.

Prescott M, Harley JP and Donald AK. 2005. *Microbiology* (6th ed.) McGraw-Hill, New York.

Rangaswami G and Mahadevan A. 1999. *Diseases of Crop plants in India* (4th ed). Prentice-Hall of India Private Limited, New Delhi.

Schlegel G 2002. *General Microbiology* (7th ed.) Cambridge University Press, London

Sharma PD. 2001. *Plant Pathology* (1st ed.) Rastogi Publications, India.

Singh DP and Dwivedi SK. 2004. *Environmental Microbiology and Biotechnology*. New Age International Publishers, New Delhi.

Singh RS. 1998. *Plant diseases* (7th ed.) Oxford and IBH Publishing Co. Pvt. Ltd., New Delhi.

Singh US, Mukhopadhyay AN, Kumar J and Chaube HS (Eds.). 1992. *Plant Diseases of International Importance*, Vol. 1-4. Prentice Hall, New Jersey.

Sinha AK. 2001. *Fundamentals of Plant Pathology* (1st ed.). Kalyani Publishers, New Delhi.

Stainer RY, Ingerham JL, Wheelis ML and Painter PR. 1987. *General Microbiology*. Macmillan, London.

Strauss JH and Strauss EG. 2002. *Viruses and Human Disease*. Academic Press, New York.

Webster J. 1970. *Introduction to Mycology*. Cambridge University Press, London.

Index

A

B

C

D

H

I

K

L

Q

R

S

T

U

V

W

Z

λ